ADVANCED

# GENERAL ORGANIC CHEMISTRY

## A Modern Approach

## PART II

For Undergraduate students majoring in Chemistry and for Postgraduate students

# ADVANCED
# GENERAL ORGANIC CHEMISTRY

## A Modern Approach

### PART II

Sachin Kumar Ghosh

Formerly Head of the Department of Chemistry, New Alipore College
Affiliated to University of Calcutta

New Central Book Agency (P) Ltd

LONDON

HYDERABAD ERNAKULAM BHUBANESWAR

NEW DELHI KOLKATA PUNE GUWAHATI

**NCBA**

**REGD OFFICE**
8/1 Chintamoni Das Lane, Kolkata 700 009, India
email: ncbapvtltd@eth.net

**OVERSEAS**
NCBA (UK) Ltd, 149 Park Avenue North, Northampton, NN3 2HY, UK
email: ncbauk@yahoo.com

**EXPORT**
NCBA Exports Pvt. Ltd, 212 Shahpur Jat, New Delhi 110 049
email: ncbaexports@ncbapvtltd.com

**BRANCHES**

2/27 Ansari Road, 1st Floor
Daryaganj, **New Delhi** 110 002
email: ncbadel@ncbapvtltd.com

House No. 3-1-315, 1st Floor
Nimboliadda, Kachiguda, **Hyderabad** 500 027
email: ncbahydb@ncbapvtltd.com

Shop Nos. 3 & 4, Vinayak Towers
681/B Budhwar Peth, Appa Balwant Chowk
**Pune** 411 002
email: ncbapune@ncbapvtltd.com

GSS Shopping Complex, 1st Floor
Opposite College Post Office, Convent Road
**Ernakulam** 682 035
0484-4021054 / 53

Shop No. 15, Bharati Towers, Ground Floor
Forest Park
**Bhubaneswar** 751 001
email: ncbabub@ncbapvtltd.com

Radhamayee Complex [Second Floor]
Opp. Axis Bank, Panbazar, Jaswanta Road
**Guwahati** 781 001
email: ncbaguwahati@ncbapvtltd.com

**ADVANCED GENERAL ORGANIC CHEMISTRY** [Part II] • Sachin Kumar Ghosh



First Published: 1986
Reprinted: 1992, 1994
Second Edition: 1998
Third Edition [thoroughly revised & updated]: 2009
Revised Reprint: 2010
Reprinted: 2011
Revised Reprint: 2013
Reprinted: 2016, 2017, **2019**

PUBLISHER
New Central Book Agency (P) Ltd
8/1 Chintamoni Das Lane, Kolkata 700 009

TYPESETTER
Anin, Kolkata

PRINTER
New Central Book Agency (P) Ltd
Web-Offset Division, Dhulagarh, Sankrail, Howrah

COVER DESIGNER
Soumya Jyoti Ghosh

TECHNICAL EDITOR
Pradip Kr. Chowdhury and Dipan Roy

PROJECT TEAM
Joydeep Ghosh and Prabhat Jas

ISBN: 978 81 7381 441 9

Price of Part I & Part II together: Rs 1500.0

## DEDICATION

To my parents and brother
To my maternal grandfather,
To others, and my wife
Who helped me build my life,
To my sons and daughters
Who bring me joy and cheers,
To my teachers and the taught,
To authors of home and abroad,
To my friends and foes,
To whomsoever the book owes,
To the people who toil.
To make fruitful the soil.
For all times and now
I bow, bow and bow.

THIRD EDITION

To the hands of my grandchildren

Akash, Hindole and Kopai

To the hands of my grandchildren

# CONTENTS



## PART II

*General Chemistry of Hydrocarbons, Heterocycles (with One Heteroatom), Functional Groups, Planning a Synthesis, Amino Acids, Peptides, Proteins, Nucleic Acids and Carbohydrates*

# Foreword

The first edition of the present text, published in 1986, has been quite popular amongst the advanced undergraduate students and teachers of organic chemistry throughout the country, especially in West Bengal. To my knowledge, this has been the only comprehensive text by an Indian author aimed at advanced undergraduate-level, having a wide coverage of almost all major aspects of organic chemistry, viz., modern structural theories, stereoisomerism, reaction mechanism and general functional group chemistry. A large number of problems, solved and unsolved, in most of the chapters have made the relevant topics more intelligible and interesting, arousing scientific curiosity.

The author deserves congratulations for the untiring endeavour he has undertaken to write this text and now to make it updated and more comprehensive in its present second edition. Objectives given at the beginning of each chapter is a characteristic feature of this text, wherein the topics and concepts to be known and understood have been outlined. Many chapters are revised versions of the original ones of the first edition, of which Chapter 4 'Factors Influencing Electron Availability', Chapter 9 'Isomerism', Chapter 11 'Fundamentals of Organic Reaction Mechanism', and Chapter 13 'Addition to C—C Multiple Bond' deserve special mention. The concepts regarding 'The cause of optical activity' and of topicity, prochirality, chirotopicity and stereogenecity introduced in Chapter 9 would give further insight to stereoisomerism. Inclusion of separate chapters on 'Elementary Thermodynamics' (Chapter 5) and 'Kinetics of a Reaction' (Chapter 6) (taking these topics out of Chapter 11 of the first edition) has been appropriate, as only this knowledge can lead to the basic concepts of organic reaction mechanism and the position of equilibrium. Emphasis is usually given on these topics in any good organic chemistry text.

Chapter 21 'Planning a Synthesis' (64 pages) is a noteworthy and very useful addition. This chapter deals with disconnection of different types of simple organic molecules and rational ways of reconnection illustrating chemoselective, regioselective and stereoselective routes. Protection and deprotection of functional groups involved in synthons and synthetic equivalents and common routes to synthesise mono and polyfunctional molecules—aliphatic, aromatic and heterocyclic—have been discussed, incorporating a number of tables in order to enable the readers to assimilate these matters at a glance.

Two more new chapters on 'Chemistry of Amino Acids and Peptides' (Chapter 22) (32 pages) and 'An Introduction to Carbohydrates' (Chapter 23) (55 pages) have been incorporated in the present edition, giving all necessary salient features. These subject-matters are indispensable parts of any modern comprehensive text on organic chemistry. The inclusion of Appendices A, B and C at the end, and the Subject Index and Compound Index, prepared with utmost care, would be useful to the readers for quickly finding out the desired information. Chemistry of a large number of organic compounds of different classes of everyday use (Appendix C), a new addition, will make the subject-matter lively and interesting.

However, the inclusion in an elementary way of some basic mechanistic concepts, as of Curtin-Hammett principle in Chapter 11, of some important topics of stereochemistry like Sharpless epoxidation, Prelog's rule, Octant rule, Axial haloketone rule etc. and also of the principles and applications of $^{1}$H NMR and mass spectra for the structural elucidation of simple organic molecules would have enhanced the standard and utility of this text at the expense of increasing the volume to some extent. I feel that this should be done in a future edition, in which several chapters could be significantly condensed without losing the essence and understandability of the subject-matter. The book is already big enough (~1450 pages) to be split into two volumes for convenience of handling.

The author has made commendable efforts to present the subject-matter in each chapter in lucid language with sufficient numbers of illustrations and problems.

In conclusion, I recommend this textbook to the undergraduate students majoring in chemistry, beginning postgraduate students and also to the teachers of organic chemistry, who, I hope, will find this book very thorough and useful.

July 1998

SUNIL K. TALAPATRA
UGC Professor of Chemistry
Centre of Advanced Studies on Natural Products
University of Calcutta

## Preface to the Third Edition

I am to write the preface to this edition with a deep heart. Meanwhile I have lost three respectable personalities. My elder brother, Late Rathin Ghosh, who taught me and bore the whole financial burden of my education in Kolkata. My wife, Late Jyotsna Ghosh, to whom I could not give my company for long eight years during the preparation of the book. She was a pillar of support to me. And Late Jogendra Nath Sen, the then Director of New Central Book Agency (P) Ltd, who brought me in the limelight. My heart bleeds for them. I have had many predicaments but, inspired by New Central Book Agency (P) Ltd, I have completed the book at last.

I have tried to update the book as far as possible and the following topics have been included in this edition:

1. Homoaromatics [Chapter 4]
2. Phase Transfer Catalyst (PTC) and Crown ethers [Chapter 7]
3. General and Specific acid-base catalysis [Chapter 8]
4. Prelog's Rule [Chapter 9]
5. $^1$HNMR (PMR) Spectroscopy [Chapter 10] and Mass Spectra [Appendix E]
6. Curtain-Hammett Principle [Chapter 11]
7. Designations of addition creaction to $C=C$ [Chapter 13]
8. Sigmatropic [i, j] hydrogen shift [Chapter 14]
9. 1, 3-Dithianes [Chapter 16]
10. Protein and nucleic acids [Chapter 22]
11. Green Chemistry [Appendix D]

Chapters on Organometallics and Active methylenes have not been written separately since organometallics have been discussed in several chapters (2, 16, 18, 20, 21, etc.). Active methylenes have also been discussed in Chapters 9, 16, 20 and 21—as required.

However, the Hammett equation has been dropped from Chapter 17 for personal reasons. Large volume of the book is a problem to handle. Now, the book and its contents have been separated into two parts—I and II. Part II contains all the indexes. The continuity of page numbers is maintained throughout the volumes. Readers are requested to buy both the parts together for fruitful use of the book, since the two parts together complete the whole text. I would request the readers to offer their valuable suggestions. Every suggestion and helpful criticism will be gratefully acknowledged.

In preparing the indexes my elder son, Soumya Jyoti, and his wife, Sayani, have done a lot. Soumya Jyoti has also designed the cover of the book. I am indebted to both of them. I am also indebted to Mr Ranjit Chaudhuri, the father-in-law of my elder son, who presented me a high-class magnifying glass without the help of which the proof reading would not have been possible for me. My younger son, Subhojyoti, has warned me several times not to do so much hard work with ill-health, yet I have completed the book having the full support, physical and mental, from my daughter-in-law and elder son. My grandson, Akash, always keeps me cheerful.

I am thankful to Principal Dilip Kumar Sarkar, formerly of Darjeeling Govt. College for correcting some printing mistakes and language of the revised third edition.

Mr Amitabha Sen, the Director of New Central Book Agency (P) Ltd, Mr Soumen Paul and Mr Prabhat Jas of New Central Book Agency (P) Ltd have taken a keen interest in the publication of the revised third edition. Without their hard labour it was impossible to publish this edition. The second edition is the foundation of the third edition. Mr Arunabha Sen published the second edition. I, most humbly, convey my gratitude to Prof. Sunil Kr. Talapatra for writing the 'Foreward' to the second edition of the book. I am also grateful to him. Dr P. K. Maity and his workers of 'Anin' have done their job neatly. I am really grateful to all of them and convey my heartfelt thanks to them. Good wishes to all.

76B Rishi Arabinda Sarani
Haridevpur, Kolkata 700 082

**AUTHOR**

# Preface to the First Edition

The book is meant for advanced degree students. The syllabi prescribed by the universities for them in general organic chemistry include (i) the modern structural theories of organic chemistry; (ii) the stereoisomerism, (iii) the reaction mechanism; and (iv) the general functional group of chemistry.

Unfortunately, no single book covers all the aspects; naturally, students have to plod through several specific books of eminent authors within a short span of time and, as a consequence, they very often fail to assimilate the subject matter. It is this difficulty of the students that led me to undertake this concise and systematic work on general organic chemistry—in the light of the modern structural theories.

All the four aspects of the syllabi and the theoretical basis of the practical organic chemistry have been dealt with in detail in the book to minimise the labour and save the time of the students. Hundreds of related problems have also been solved to make the subject matter lucid and, at the same time, to develop a mind of scientific inquisitiveness. Thus the book is expected to be a dependable guide to the advanced degree students of organic chemistry.

The field of organic chemistry being inconceivably vast, 'What to Study' and 'How to Study' are the two common questions asked by the students. In chapter one, titled "INTRODUCTION", effort has been made to cope up with such questions. Since organic compounds mostly occur in nature and organic chemistry is an experimental science, isolation and purification of organic compounds, and their analyses—qualitative, quantitative and instrumental—have been discussed in the second, third and tenth chapters, respectively. Analytical portions related to the functional group chemistry have been discussed in their right places in chapter twenty as well as in the Appendix. Chapters four to seventeen deal with the structure, nomenclature, physical properties, electronic condition, acidity-basicity, intermolecular forces, stereo-chemistry and reaction mechanisms of the organic compounds.

Chapters one to seventeen constitute the first part of the book and they deal with the modern basic concepts. The chemistry of a few hydrocarbons, heterocycles with one hetero atom and functional groups, and uses of organic compounds have been included in the second part of the book which have three chapters, namely, chapters eighteen, nineteen and twenty. Notes on a few modern terms and IR spectroscopy and properties of a few organic compounds have been incorporated in the appendix. Thus it is a comprehensive textbook covering all the different aspects of the syllabi. Nevertheless, the students are advised to consult the books on specific aspects written by eminent chemists and established authors.

It is my duty to seek apology of the readers for the printing mistakes in the book which could not be avoided; these are, however, easily detectable and will not create any confusion to the readers. I also hope that all these errors will be rectified in the next edition of the book. In this connection I request the readers to offer their valuable suggestions for making the book more useful and up-to-date. Any such suggestion and helpful criticism will be gratefully acknowledged.

It is a great plesure to thank Dr Ramaprasad Sarkar who excited my feeling to undertake the present work. My thanks are also due to Dr Sankar Chandra Ghosh, Prof. R. N. Sinha Roy, Dr Swapan Kumar Ghosh, Dr Dhananjay Chatterjee, Dr G. P. Chitnis, Dr Niren Chandra Gayen, Dr Gour Gopal Basak and Prof. Bibek Ranjan Sarkar who occasionally read the manuscript and gave valuable suggestions for the betterment of the book.

I also owe a debt of gratitude to my students Arabinda, Manas, Santanu, Falguni, Dhiman, and Manabendra, who supplied me with books and papers on several aspects for preparing the manuscript; and to Sri Jogendranath Sen (Partner, New Central Book Agency Pvt. Ltd), to Sri Biren Das (Artist), to the workers of Books & Allied (P) Ltd and to the staff of New Central Book Agency (P) Ltd who made possible the publication.

I owe the index to Mrs. Suchitra Ganguly (Librarian, New Alipore College), and to my students, Manabendra, Goutam, Sanhita, Nibedita, Arijit, Abhijit, Bhanu, Jyoti and Arup.

I pay my respect to Dr. Atul Chandra Chatterjee who took much pain in editing the language of a few chapters (chapter 9 to chapter 12).

My wife, Smt Jyotsna Ghosh, and my sons, Bubun and Pat, gave me enough time for completing the book, I am indebted to them.

New Alipore College
'L' Block, New Alipore
Calcutta 700 053

AUTHOR

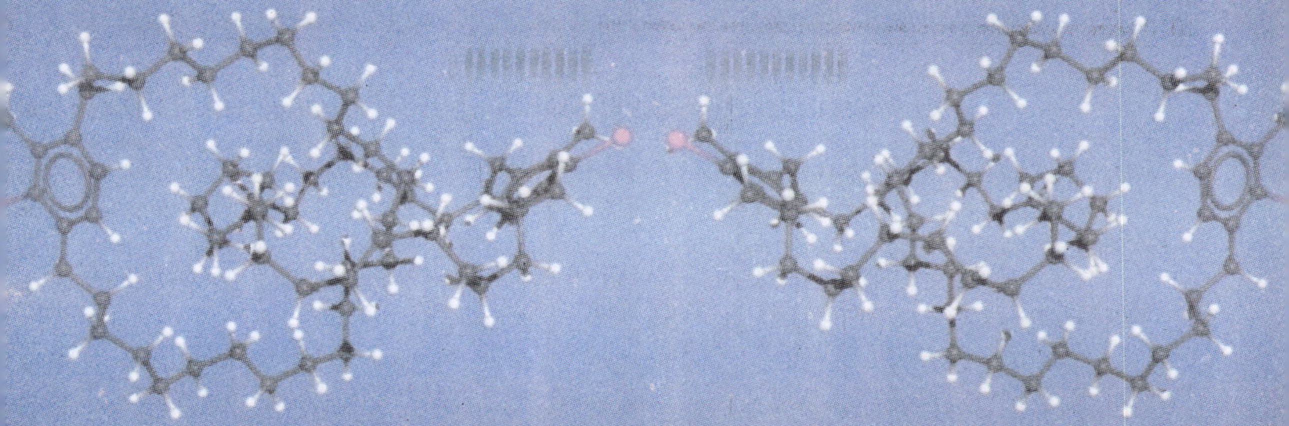

# Part II

## General Chemistry of Hydrocarbons, Heterocycles (with One Heteroatom) Functional Groups, Planning A Synthesis, Amino Acids, Peptides, Proteins, Nucleic Acids and Carbohydrates

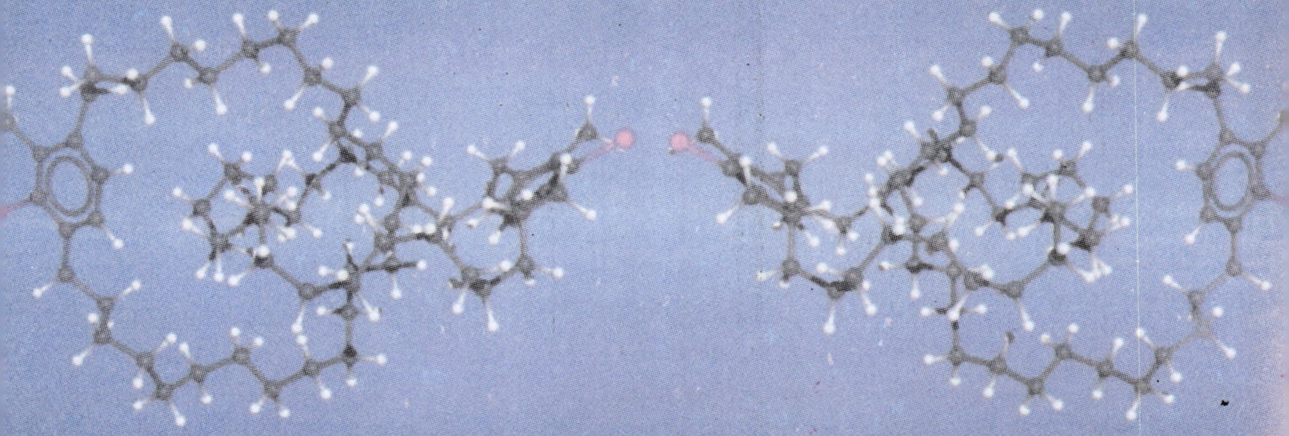

Chapter 18

# Chemistry of Hydrocarbons

## Objectives

- To know the chemistry of :

## 18.1 Alkanes (Paraffins)

Open chain binary compounds of C and H containing C—C and C—H $\sigma$ bonds only are called alkanes. Being inert, they are also called paraffins (meaning little affinity). Since there is no C—C multiple bond, alkanes are saturated hydrocarbons. Usually alkanes are symbolically represented as RH or R—R or R—R′ or R-$CH_3$, etc., where R, R′ are alkyl groups. For example, ethane may be represented as RH (R = $CH_3CH_2$—) or R—R (R = $CH_3$ or R-$CH_3$; R = $CH_3$). Alkanes exhibit chain isomerism (Sec. 9.2); for instance, there are three isomeric pentanes, $C_5H_{12}$ :

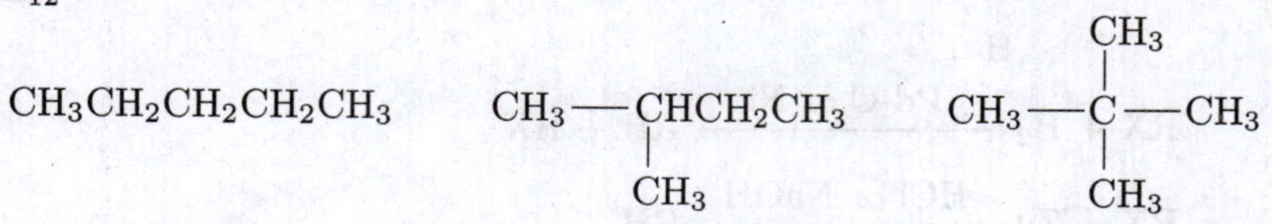

*n*-Pentane *iso*-Pentane *neo*-Pentane

Owing to the rotation about C—C bonds, ethane and higher alkanes have different conformations. Chiral alkanes exhibit enantiomerism.

### 18.1.2 Source

Natural gas and petroleum oil.

### 18.1.3 Formation of alkanes

The methods are given in the following table :

**Table 18.1** Formation of alkanes

| | Reactant | Reagents / conditions | Product |
|---|---|---|---|
| a. | RX | $LiAlH_4$ or $NaBH_4$ or $Ph_3SnH$; Pd-C or Ni and $H_2$ or Zn/$H^{\oplus}$ or Zn/$\bar{O}H$ or Zn—Cu/EtOH, etc. | RH |
| b. | RMgX | $H_2O$ or ROH or $NH_3$ or $RNH_2$ or PhOH, RSH, etc. | RH |
| c. | RLi | $H_2O$ or ROH or $NH_3$ or $RNH_2$ or PhOH, RSH etc. | RH |
| d. | $RCO_2^{-}Na^{\oplus}$ | soda-lime and heat | RH |
| e. | RCH=$CH_2$ | Raney Ni/$H_2$ | $RCH_2CH_3$ |
| f. | RCH=$CH_2$ | 1. $B_2H_6$ 2. HOAc or $AgNO_3$ and NaOH | $RCH_2CH_3$ |
| g. | RX | Wurtz reaction; Na and ether | R—R |
| h. | $RCO_2^{-}K^{\oplus}$ | Kolbe electrolysis | R—R |
| i. | RCOR | Clemmensen reduction; Zn/Hg and HCl | $RCH_2R$ |
| j. | RCOR | Wolff-Kishner reduction; $NH_2NH_2$ and EtONa | $RCH_2R$ |
| k. | RCOR | 1. RSH and dry HCl; 2. Ni/$H_2$ | $RCH_2R$ |
| l. | RX | $R'_2CuLi$/ether | R—R′ |

A L K A N E S

(a) By reduction of aklkyl halides : There are several reagents which reduce alkyl halides to alkanes; e.g., $LiAlH_4$, $NaBH_4$, $Ph_3SnH$, dissolving metals, Pd-C/$H_2$, etc. $LiAlH_4$ reduces 1° and 2° alkyl halides; $NaBH_4$ reduces 2° and 3° alkyl halides, whereas $Ph_3SnH$ reduces 1°, 2° and 3° halides. Pd-C/$H_2$ and Ni/$H_2$ reduce 1°, 2° and 3° chlorides, bromides and iodides. Dissolving metals reduce 1°, 2° and 3° iodides, and 3° chlorides and bromides.

$$\overset{\delta+\delta-}{RX}: + \overset{\oplus}{Li}\overset{-}{Al}H_3(H) \longrightarrow RH + LiAlH_3X$$

$$RX + H_2 \xrightarrow{\text{Pd-C or Ni}} RH + HX$$

$$RX + Zn \xrightarrow{\text{HCl or NaOH}} RH$$

$$Zn \longrightarrow Zn^{++} + 2e;\ \overset{\delta+}{R}—\overset{\delta-}{X} + 2e \longrightarrow R^- + X^-$$

$$R^- + H^{\oplus} \longrightarrow RH \text{ and } X^- + H^{\oplus} \longrightarrow HX$$

Reactivity : 3°RX > 2° > 1° and RI > RBr > RCl.

(b) From the Grignard reagent : Compounds with active H react with the Grignard reagent to give alkanes

$$RMgX + HOH \longrightarrow RH + Mg(OH)X.$$

(c) From alkyllithiums : Alkyllithiums are highly reactive species. They are prepared by treating alkyl chlorides with metallic Li. BuLi on heating with RX gives RLi. However, alkylilithiums react with compounds containing active H to give alkanes :

$$BuCl + Li \longrightarrow BuLi;\ BuLi + RX \longrightarrow RLi + BuX$$

$$RLi + H_2O \longrightarrow RH + LiOH$$

(d) Decarboxylation of salts of carboxylic acids by soda-lime (NaOH + CaO) and heat : Sodium acetate decomposes to methane but other carboxylates give the corresponding alkane, alkanes of lower molecular weight, alkenes and hydrogen. So the reaction has no preparative value.

$$\text{ElcB}: CH_3—\overset{O}{\overset{\|}{C}}—\overset{-}{O}Na^{\oplus} \xrightarrow{NaOH} \overset{-}{C}H_3 + CO_2 \xrightarrow{H_2O} CH_4$$

$$CH_3CH_2CO\overset{-}{O}Na^{\oplus} \xrightarrow{NaOH} CH_3CH_3 + CH_4 + CH_2{=}CH_2 + H_2$$

(e) By hydrogenation of alkenes :

$$CH_2{=}CH_2 \xrightarrow{Ni/H_2} CH_3—CH_3$$

(f) By hydroboronation and hydrolysis :

$$RCH{=}CH_2 + B_2H_6 \longrightarrow (RCH_2CH_2)_3B \xrightarrow{AcOH} 3RCH_2CH_3$$

(g) By the Wurtz reaction : When a solution of an alkyl halide, RX, specially bromides and iodides, in ether is treated with metallic sodium, an alkane, R—R, is obtained and the reaction is known as the Wurtz reaction. The reaction may be carried out with a mixture of two different alkyl halides, RX and R′X, but in this case three different alkanes will be formed; for this reason the mixed Wurtz reaction has no importance.

$$CH_3CH_2Br + 2Na + BrCH_2CH_3 \xrightarrow{\text{ether}} CH_3CH_2.CH_2CH_3$$

$$3RX + 6Na + 3XR' \longrightarrow R—R + R—R' + R'—R' + 6NaX$$

The plausible mechanism of the reaction is :

**Step I.** Halogen exchange between alkyl halide and sodium and formation of alkylsodium having carbanionic character.

$$R—X + 2Na \longrightarrow NaX + \bar{R}Na^{\oplus} \longleftrightarrow R—Na$$

**Step II.** The alkylsodium then displaces the halogen atom of RX by the $S_N2$ pathway.

$$Na\overset{\oplus}{\bar{R}} + R—X \longrightarrow \underset{\text{TS}}{[\overset{\delta-}{R}\cdots R\cdots \overset{\delta-}{X}]}\,Na^{\oplus} \longrightarrow R—R + NaX.$$

The following are the evidences in favour of the mechanism.

**(i)** An alkene forms along with another alkane as side products.

$$Na^{\oplus}\,R—CH_2\bar{C}H_2 + R—\underset{\underset{H}{|}}{CH}—CH_2—X \longrightarrow RCH_2CH_3 + RCH{=}CH_2 + NaX$$

**(ii)** The reaction between benzylsodium and (+)-2-bromobutane occurs with inversion of configuration about $C_2$ atom of the bromobutane and yields an optically active product.

**(h) By the Kolbe electrolysis of carboxylates :** When sodio or potassio salts of carboxylic acids, RCOOH, are electrolysed, alkanes, R—R type, form; alkenes, lower alkanes and esters are obtained as side products and the reaction is known as the Kolbe electrolysis.

$$2RCOONa + 2H_2O \xrightarrow{\text{electrolysis}} R—R + 2CO_2 + 2NaOH + H_2$$

$$CH_3CH_2CH_2COONa + 2H_2O \xrightarrow{\text{electrolysis}} CH_3CH_2CH_2CH_2CH_2CH_3 + CO_2 + H_2 + NaOH$$

Side products :

$$[CH_3CH{=}CH_2 + CH_3CH_2CH_3 + CH_3CH_2CH_2COOCH_2CH_2CH_3]$$

The mechanistic steps are :

$$CH_3CH_2CH_2COONa \rightleftharpoons CH_3CH_2CH_2CO\bar{O} + \overset{\oplus}{Na}$$

| The cathode reactions | The anode reactions |
|---|---|
| $Na^{\oplus} + e \longrightarrow Na$ | $CH_3CH_2CH_2COO^- \longrightarrow CH_3CH_2CH_2COO\cdot + e$ |
| $2Na + H_2O \longrightarrow 2NaOH + H_2$ | $CH_3CH_2CH_2COO\cdot \longrightarrow CH_3CH_2\dot{C}H_2 + CO_2$ |
| | $2CH_3CH_2\dot{C}H_2 \longrightarrow CH_3CH_2CH_2CH_2CH_2CH_3$ |

$$CH_3CH_2\dot{C}H_2 + CH_3CH_2\dot{C}H_2 \xrightarrow{\text{disproportionation}} CH_3CH{=}CH_2 + CH_3CH_2CH_3$$

$$CH_3CH_2\dot{C}H_2 + CH_3CH_2CH_2COO\cdot \xrightarrow{\text{coupling}} CH_3CH_2CH_2COOCH_2CH_2CH_3$$

**(i) By the Clemmensen reduction of carbonyt compounds :** When aldehydes and ketones are reduced by Zn-Hg and conc. HCl, the >C=O group gets reduced to —$CH_2$— and the process is called the Clemmensen reduction. Its mechanism is not understood.

$$RCOR \xrightarrow[\text{and HCl}]{\text{Zn-Hg}} RCH_2R;\ CH_3CHO \xrightarrow[\text{and HCl}]{\text{Zn-Hg}} CH_3CH_3.$$

**(j) By the Wolff-Kishner reduction :**

$$CH_3CH{=}O + NH_2NH_2 \longrightarrow CH_3CH{=}NNH_2 \xrightarrow[\Delta]{\text{NaOEt}} CH_3CH_3$$

**(k) By the catalytic reduction of thioacetals :**

$$R{-}CHO + R'SH \longrightarrow R{-}CH(SR')_2 \xrightarrow{Ni/H_2} RCH_3$$

**(l) By the Corey-House Synthesis :** Lithium dialkylcuprates are the reagents and methyl and 1° alkyl halides are the substrates for the Corey-House synthesis. When methyl or a 1° alkyl halide is treated with a lithium dialkylcuprate, an alkane of the type R—$CH_3$ or R—R′ (R = R′ or R≠R′) is formed and the process is called Corey-House process. The reagent is prepared by treating an alkyl halide (1° or 2° or 3°) in ether with metallic lithium and then adding cuprous iodide.

$$2R{-}X \xrightarrow[\text{ether}]{Li} \underset{2LiX}{\overset{}{2RLi + }} \xrightarrow{CuI} R_2CuLi + LiI$$

Another alkyl halide, methyl or 1°, is then allowed to react with the lithium dialkylcuprate to form an alkane.

$$R_2CuLi + R'{-}X \longrightarrow R{-}R' + RCu + LiX.$$
$$(R' = \text{methyl or } 1°)$$

For example,

$$Me_2CuLi + CH_3(CH_2)_3\,CH_2{-}I \xrightarrow[25\,°C,\ 3\text{-}5\ \text{hrs}]{Et_2O} \underset{(98\%)}{CH_3(CH_2)_3CH_2CH_3}$$

**Note :** However, alkenes cycloalkenes and arenes may also be prepared by the process :

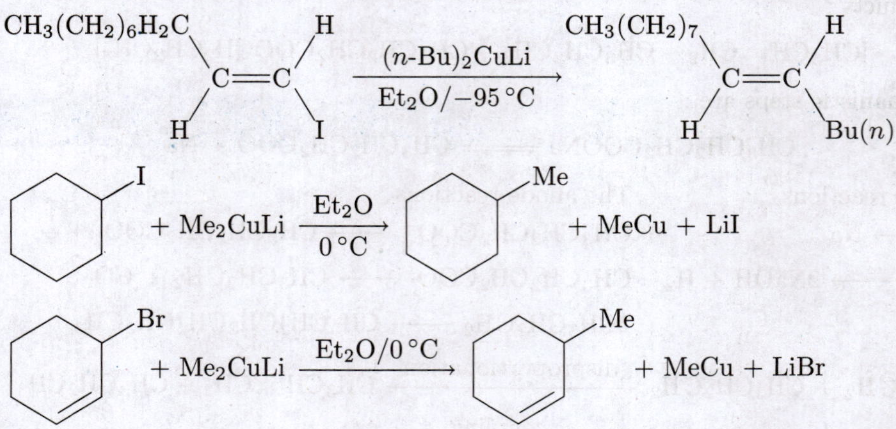

$$\text{Ph—I} + (n\text{-Bu})_2\text{CuLi} \longrightarrow \text{Ph—CH}_2\text{CH}_2\text{CH}_2\text{CH}_3 + n\text{-BuCu} + \text{LiI}.$$

The mechanism for the reaction has not yet been establised. It has been argued that $R^-$ from $R_2CuLi$ displaces $X^-$ from $R'X$ by the $S_N2$ pathway.

## 18.1.4 Intermolecular forces and physical properties

The C—C $\sigma$ bonds in an alkane are non-polar, whereas C—H $\sigma$ bonds are slightly polar because a C atom is more electronegative than an H atom. All the C atoms being tetrahedral, the C—H bond dipoles get cancelled by each other and thus the alkane molecules are of zero dipole moment. Naturally alkanes possess only the London force (Sec. 7.4) as intermolecular force in their molecular crystal lattice. It is a short ranged weak force of the order 4.2 kJ $mol^{-1}$. This makes the crystals very weak and m.p.s and b.p.s of alkanes are very low. $C_1$—$C_4$ alkanes are gases, $C_5$—$C_{16}$ alkanes are liquids and higher members are solids under room temperature and pressure. X-ray diagrams of $n$-alkanes show that the carbon chains are zig-zag. Compounds with an even number of C atoms have end C atoms on the opposite sides of the chain, whereas the chains of odd number of Cs have end C atoms on the same side of the chain (Fig. 18.1). The chains of even number of Cs are packed more closely than those of odd number of Cs in the lattice. This develops greater London force in compounds with even number of Cs than those with odd number of Cs and hence the former compounds have higher m.p.s than the latter compounds. For example m.p. of $n$-hexane is $-24.3\,°C$ and that of $n$-pentane is $-131.5\,°C$.

**Fig. 18.1. Carbon chains in alkanes**

So far as b.p.s are concerned, a normal alkane boils at a higher temperature than its isomeric branched alkane. London force is a surface phenomenon; since branched chain alkanes are more spherical than $n$-alkanes, the former possesses less surface area than the latter. For this reason the amount of London force in an $n$-alkane is greater than that in its branched isomer. Consequently a branched alkane boils at a lower temperature than its isomeric $n$-alkane. For example, the b.p. of $n$-pentane is $36.2\,°C$, whereas the b.p.s of iso- and neo-pentanes are $28\,°C$ and $9.5\,°C$ respectively. However, b.p.s and m.p.s increase with the increasing molecular weight since London force increases as the molecular weight increases.

Being non-polar, alkanes are soluble in non-polar solvents and insoluble in polar solvents. The solvents for alkanes are ether, benzene, etc.; but solubility decreases with increasing molecular weight. Since there is no chromophore which can absorb light in the visible region in alkanes, they are colourless. All the alkanes are lighter than water. The highest density is 0.79 g/cc.

## 18.1.5 Structure and reactivity

All the C—C bonds (bond energy $\sim$ 350 kJ $mol^{-1}$) and C—H bonds (bond energy $\sim$ 410 kJ $mol^{-1}$) being $\sigma$-type, they require high energy to break and thus alkanes are inert to chemical reactions. Usually under drastic conditions, pyrolytically and photochemically, they react via carbon radicals. The reactivity order of the alkyl radicals being $3° > 2° > 1°$, the order of reactivity amongst alkanes is $3° > 2° > 1°$. 3°Hs being the most reactive, they sometime undergo ionic reactions. Since an alkane contains the highest number of 1°H and the lowest number of 3°H, the probability of taking part in the reaction is $1°H > 2°H > 3°H$.

## 18.1.6 Chemical reactions

Chemical reactions of alkanes are given in Table 18.2.

**Table 18.2** Chemical reactions of alkanes

| $C_nH_{2n+2}$ Alkane → reaction | Product | |
|---|---|---|
| combustion | $CO_2 + H_2O$ + heat | a |
| halogenation | mixture of mono, di and polyhaloalkanes | b |
| vapour-phase nitration at 400 °C by $HNO_3$ | mixture of nitroalkanes | c |
| cracking or pyrolysis | mixture of alkanes, alkenes and $H_2$ | d |
| insertion of methylenes or carbenes | $C_{n+1}H_{2n+4}$ | e |
| isomerisation; $AlCl_3$, RCl | isomeric alkane | f |
| addition of alkane to alkene | higher alkane | g |
| sulphonation | $RSO_3H$ | h |
| chlorosulphonation $SO_2 + Cl_2 + h\nu$ | $RSOCl_2$ | i |

(a) Combustion of alkanes : When alkanes are burnt carbon dioxide and water are formed with evolution of heat. So alkanes are used as fuels.

$$2C_nH_{2n+2} + (3n+1)O_2 \longrightarrow 2nCO_2 + 2(n+1)H_2O + \text{heat}$$

$$2C_2H_6 + (3\times 2+1)O_2 \longrightarrow 2\times 2CO_2 + 2(2+1)H_2O + \text{heat}$$

(b) Halogenation : Fluorine, chlorine and bromine undergo exothermic reactions with alkanes. Fluorine reacts vigorously even in the dark. F—F bond dissociation energy is only 151 kJ mol$^{-1}$ and C—F bond formation liberates 438 kJ mol$^{-1}$. In the gaseous state the kinetic energy of molecules is almost sufficient to break the F—F bond and then reaction proceeds. So, it is difficult to check the reaction at the monofluoroalkane stage. When hydrocarbon vapour is passed over freshly prepared cobaltic fluoride in the atmosphere of nitrogen, however, monofluoroalkane is obtained.

$$2CoF_3 + RH \longrightarrow 2CoF_2 + RF + HF$$

The Cl—Cl bond breaking requires 239 kJ mol$^{-1}$ of energy. The reaction between chlorine and an alkane, therefore, requires extra energy from outside and diffused sunlight supplies the energy. Here also the reaction gives rise to several monochloro as well as polychloroalkanes. Methane reacts with chlorine in diffused sunlight to give $CH_3Cl$, $CH_2Cl_2$, $CHCl_3$ and $CCl_4$ (Sec. 11.4.1). However, when a large excess of an alkane is used, an appreciable amount of monochloro product is obtained. For example, when a mixture of $n$-pentane and chlorine in the ratio of 15 : 1 is heated at 300 °C with the rate of flow of the gases is 60 miles per hour (which is faster than the rate of the reaction), 24% 1-chloropentane, 49% 2-chloropentane and 27% 3-chloropentane are obtained. Though a mixture of products is obtained nevertheless chlorination has some industrial value.

$$\underset{15\text{ vol}}{CH_3CH_2CH_2CH_2CH_3} + \underset{1\text{ vol}}{Cl_2} \longrightarrow \underset{24\%}{CH_3(CH_2)_3CH_2Cl} + \underset{49\%}{CH_3\underset{|}{\underset{Cl}{CH}}CH_2CH_2CH_3} + \underset{27\%}{CH_3CH_2\underset{|}{\underset{Cl}{CH}}.CH_2CH_3}$$

Bromine reacts with alkanes like chlorine but the reaction has no practical importance. Iodine reacts reversibly with alkanes and the reaction is endothermic by about 59 kJ mol$^{-1}$; and alkyl iodides can be converted to alkanes by the reaction with HI. Of course, RI forms in the presence of $HIO_3$ or $HNO_3$ or HgO, by the action of $I_2$ on RH.

$$RI + HI \rightleftharpoons RH + I_2$$

RBr and RCl give RI when they are treated with NaI in ethanol.

However, the reaction follows the free radical mechanism. It is needless to mention that the reactivity order and the probability of participation of H atoms are as given in Section 18.1.5.

$SO_2Cl_2$ reacts with RH by free radical mechanism to give alkyl chloride.

$$SO_2Cl_2 \longrightarrow \cdot SO_2Cl + Cl\cdot$$

$$RH + Cl\cdot \longrightarrow R\cdot + HCl$$

$$R\cdot + SO_2Cl_2 \longrightarrow RCl + \cdot SO_2Cl$$

$$R\cdot + \cdot SO_2Cl \longrightarrow RSO_2Cl$$

(c) Nitration of alkanes : The reaction between an alkane and nitric acid may be carried out either in the gas-phase at 400 °C or in the liquid phase. It has no practical importance since a mixture of mono, di, and polynitroalkanes, and lower nitroalkanes are obtained

$$CH_3CH_2CH_3 + HNO_3 \xrightarrow{400\,°C} CH_3CH_2CH_2NO_2 + CH_3\underset{|}{\underset{NO_2}{CH}}\,CH_3 + CH_3CH_2NO_2 + CH_3NO_2$$

The reaction follows the free radical pathway as shown below; here only the main steps are shown.

**Step I.** Initiation : $HONO_2 \longrightarrow \cdot OH + \cdot NO_2$

**Step II.** Propagation : $RH + \cdot OH \longrightarrow R\cdot + H_2O$

$$R\cdot + HONO_2 \longrightarrow RNO_2 + \cdot OH$$

**Step III.** Termination : $R\cdot + O_2NOH \longrightarrow ROH + \cdot NO_2$ (nitrogen dioxide has one odd $e$)

(d) Cracking or pyrolysis of alkanes : Pyrolysis (G; Pyr—fire and lysis—loosing) or cracking is a process in which bonds break at a sufficiently high temperature. For example, at 1200 °C methane decomposes to carbon and hydrogen. Higher alkanes break when they are heated in a tube at a temperature 500-800 °C and yield mixtures of alkenes, lower alkanes and hydrogen. For instance, ethane on cracking generates methane, ethylene and hydrogen. However, the reactions are endothermic and mechanism of cracking involves free radicals.

Since C—C bonds are weaker than C—H bonds, initiation of free radicals occur by homolytic fission of a C—C bond. Homolysis of a C—H bond may also occur in the initiation step.

***Initiation :*** $H_3C—CH_3 \xrightarrow{\text{heat}} H_3C\cdot + \cdot CH_3$

$$CH_3—CH_2—H \xrightarrow{\text{heat}} CH_3\dot{C}H_2 + H\cdot$$

***Propagation :*** $CH_3CH_2—H + \cdot CH_3 \rightleftharpoons CH_4 + CH_3\dot{C}H_2$

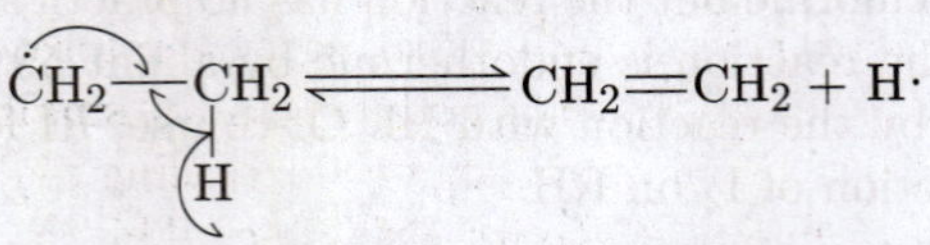

$$CH_3CH_2—H + H\cdot \rightleftharpoons CH_3\dot{C}H_2 + H_2$$

***Termination :*** $\cdot CH_3 + \cdot CH_3 \longrightarrow CH_3CH_3$

$$\cdot CH_3 + \cdot H \longrightarrow CH_4$$

$$\cdot H + \cdot H \longrightarrow H_2$$

The cracking process is of prime importance in refining petroleum since it helps to convert higher alkanes of lesser commercial value to gasoline, the usable fuel.

(e) **Insertion of carbenes :** Singlet carbenes undergo direct insertion to C—H bonds in the liquid phase. Since freshly generated 'hot' carbenes insert themselves to C—H bonds, the reaction is not selective.

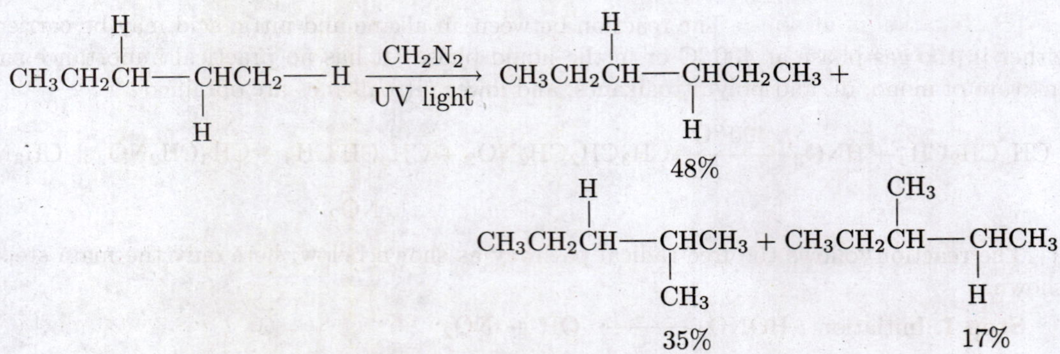

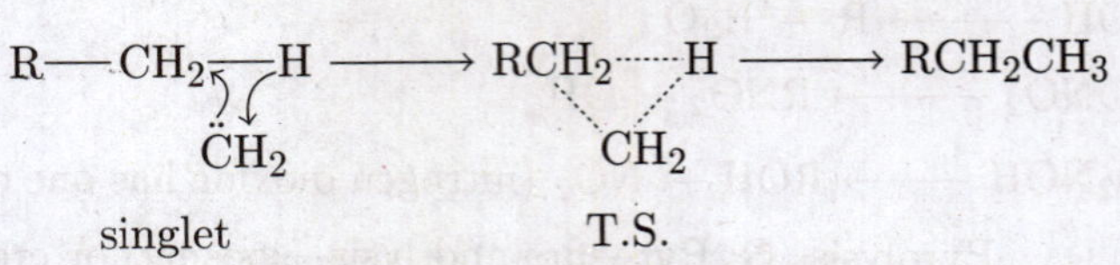

In the gas-phase, triplet carbenes insert themselves by the abstraction-combination process which is selective. The presence of inert gas favours the process, while oxygen inhibits the reaction.

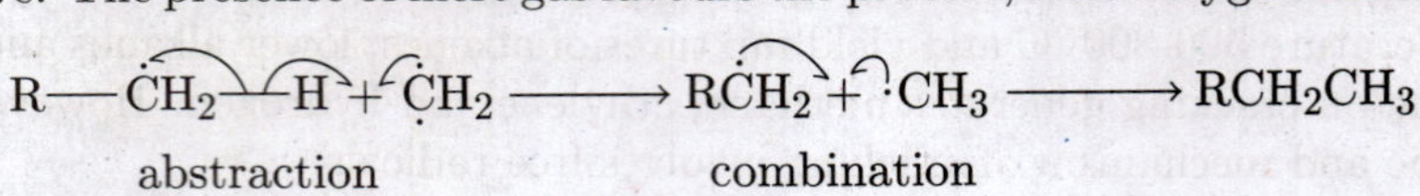

Since the reactivity order of Hs is 3° > 2° > 1°, the selectivity of triplet carbene insertion is :

$$3°C—H > 2°C—H > 1°C—H.$$

**(f) Isomerisation of alkanes :** The conversion of a compound into an isomeric compound is called isomerisation. *n*-Alkanes isomerise to branched chain alkanes and the energy of activation of such a reaction is sufficiently low. Isomerisation of alkanes occur in the presence of a catalytic amount of alkyl chloride and aluminium chloride.

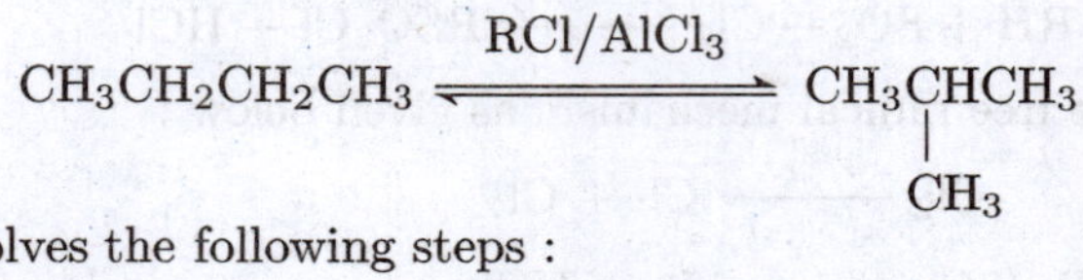

The mechanism involves the following steps :

**Initial steps :** $R—\ddot{Cl}: \curvearrowright AlCl_3 \longrightarrow \overset{\oplus}{R}\overset{-}{A}lCl_4$

**Step I.** Abstraction of a 2°H with its bonded *es* by the $R^{\oplus}$ from the *n*-alkane occurs since a 2°H is more reactive than an 1°H. This develops a carbenium ion as an intermediate.

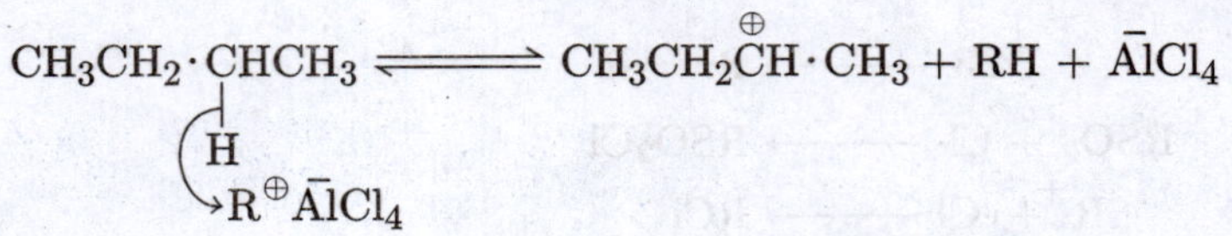

**Step II.** Rearrangement of the carbenium ion by the 1, 2-shift (carbanionic).

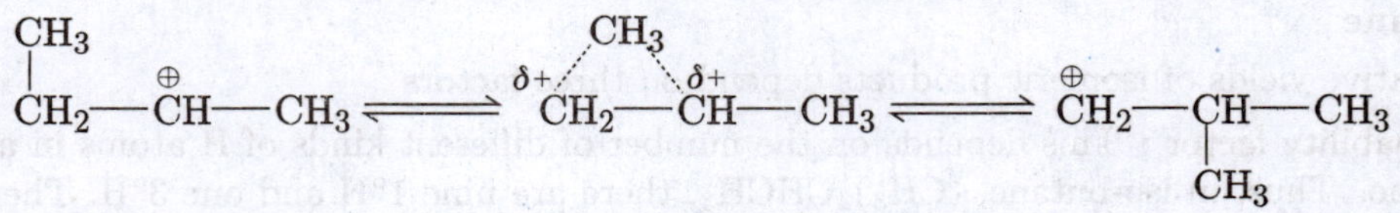

**Step III.** Abstraction of an hydride ion by the rearranged carbenium ion from a 2°C of the *n*-alkane

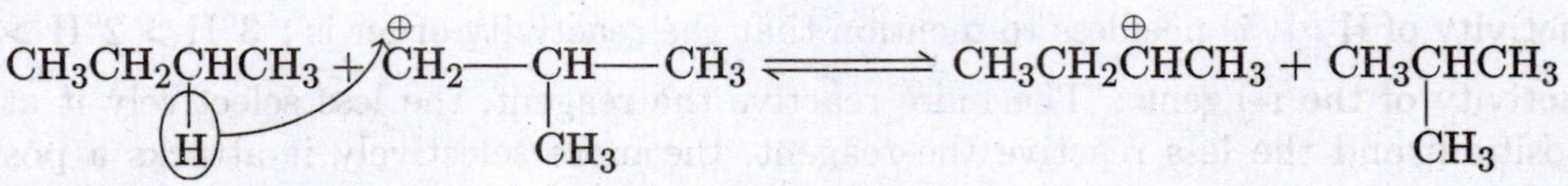

Thus, a secondary carbenium ion regenerated by the process can react further. So, the isomerisation reaction is a self-propagating process.

However, the *n*-alkane and the branched chain alkane remain in equilibrium.

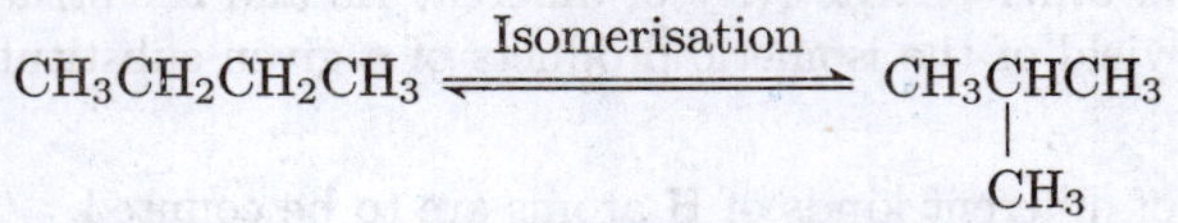

The isomerisation of alkanes are important in the production of gasoline of high octane number.

**(g) Addition of alkanes to alkenes :** See problem 13.9.

**(h) Sulphonation :** Replacement of an H atom by a sulphonic acid group is called sulphonation. 3° butane, 3° pentane and from hexanes onward all alkanes undergo sulphonation reaction when they are treated with oleum ($H_2S_2O_7 \rightarrow H_2SO_4 + SO_3$). The ease of reaction is 3°H > 2° > 1°.

Since optically active alkanes racemise, the mechanism, is supposed to occur via a carbenium ion intermediate :

$$(CH_3)_3CH + (H_2SO_4 + SO_3) \longrightarrow (CH_3)_3C.SO_3H + H_2SO_4$$

(i) Chlorosulphonation : When one of the H atoms of an alkane is replaced by an $—SO_2Cl$ group, the process is called chlorosulphonation. This may be carried out by treating an alkane either with $SO_2Cl_2$ and heat or with $SO_2$ and $Cl_2$ photochemically.

$$RH + SO_2 + Cl_2 \longrightarrow RSO_2Cl + HCl$$

The reaction follows the free radical mechanism as given below :

***Initiation :***

$$Cl_2 \longrightarrow Cl\cdot + Cl\cdot$$

$$RH + Cl\cdot \longrightarrow R\cdot + HCl$$

***Propagation :***

$$R\cdot + SO_2 \longrightarrow R\dot{S}O_2$$

$$R\dot{S}O_2 + Cl_2 \longrightarrow RSO_2Cl + Cl\cdot$$

***Termination :***

$$2R\cdot \longrightarrow R—R$$

$$R\dot{S}O_2 + Cl\cdot \longrightarrow RSO_2Cl$$

$$R\cdot + \cdot Cl \longrightarrow RCl$$

**Note : Calculation of % yield of isomeric products in a given substitution reaction on an alkane**

The relative yields of isomeric products depend on three factors :

(i) Probability factor : This depends on the number of different kinds of H atoms in a given alkane. Thus, in iso-butane, $(CH_3)_2CHCH_3$, there are nine 1°H and one 3°H. Therefore, the probability of reaction involving 1°H is 9/10 since the total number of H atoms to be displaced is 10; the probability of reaction involving 3°H is 1/10.

(ii) Reactivity of H : It is needless to mention that the reactivity order is : 3°H > 2°H > 1°H.

(iii) Reactivity of the reagent : The more reactive the reagent, the less selectively it attacks a position; and the less reactive the reagent, the more selectively it attacks a position. This means a reactive reagent attacks both reactive and unreactive sites of a molecule and, therefore, it is more influenced by the probability factor; while a less reactive reagent attacks selectively and, therefore, the reaction is less influenced by the probability factor.

However, if the relative order of reactivity of different Hs and the structural formula of the alkane are known, the % yield of the isomeric products of a given substitution reaction may be calculated as given below :

(i) First the numbers of different kinds of H atoms are to be counted.

(ii) Then the weighted reactivity of each kind of H atoms is to be calculated by multiplying the number of H atoms of the kind by the relative rate. For example, if the relative rates of reactions of 1°H and 3°H in iso-butane are $x$ and $y$ respectively, the weighted reactivity of 1°H is $9 \times x$ and that of 3°H is $1 \times y$.

(iii) The weighted reactivities are then to be summed up; the summation which is called the total weighted reactivity. Here the total weighted reactivity is $9x + y$.

(iv) The relative proportion of each kind of the isomeric products are then to be calculated by dividing individual weighted reactivity by the total weighted reactivity.

(v) The % yields are then obtained by multiplying the relative proportions by hundred.

(vi) If the relative rates are not given, we shall have to depend on the probability factor and the relative proportions of yield are to be.calculated by dividing the number of each kind of H atoms by total number of H atoms and the % yields will be obtained by multiplying the relative proportions by 100 as usual.

Let us take two examples to understand the process.

(1) Methylene or carbene is a highly reactive species; therefore, it reacts less selectively. On the basis of the Reactivity Selectivity Principle let us calculate the % yield of isomeric products of the insertion reaction between $n$-butane and carbene in which the insertion occurs in C—H bonds.

$CH_3CH_2CH_2CH_3$ contains six 1°H atoms and four 2°H atoms.

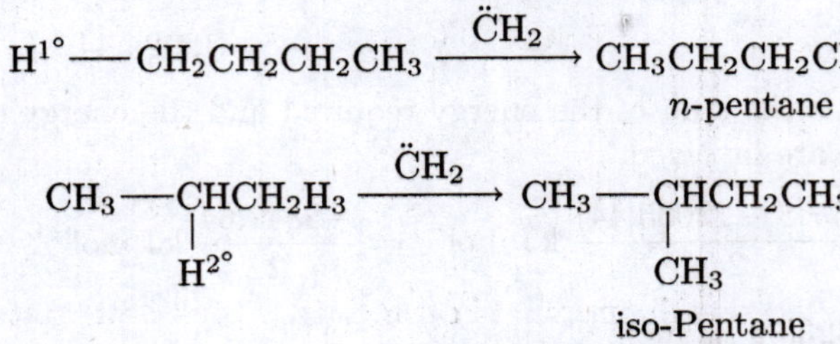

| Product | Kind of H | No. of H | Relative proportion | % yield |
|---|---|---|---|---|
| *n*-Pentane | 1° | 6 | 6/10 | 6/10 × 100 = 60 |
| *iso*-Pentane | 2° | 4 | 4/10 | 4/10 × 100 = 40 |
| | | 10 | | |

(2) Let us now calculate the % yield of monochloroproducts when isopentane is chlorinated. The relative rates of the reaction is : 3°H: 5.0, 2°H: 3.8 and 1°H: 1.0.

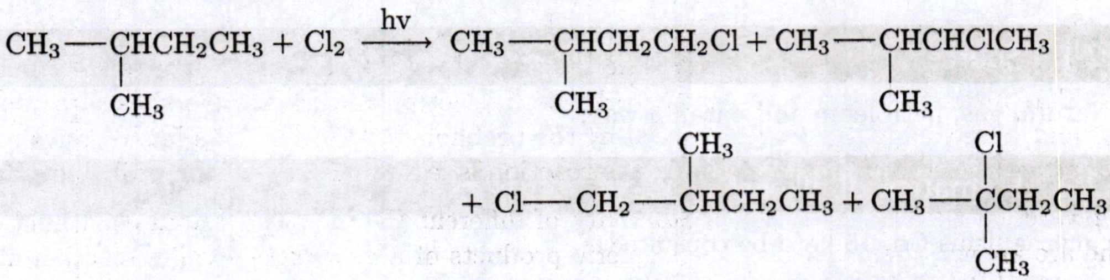

| Product | Kind of H | No. of H | Relative rate | Weighted reactivity | Relative proportion | % yield |
|---|---|---|---|---|---|---|
| $ClCH_2CH(CH_3)CH_2CH_3$ | 1° | 6 | 1.0 | 6.0 | 6/21.6 | 27.77 |
| $(CH_3)_2CHCH_2CH_2Cl$ | 1° | 3 | 1.0 | 3.0 | 3/21.6 | 13.88 |
| $(CH_3)_2CHCHClCH_3$ | 2° | 2 | 3.8 | 7.6 | 7.6/21.6 | 35.18 |
| $(CH_3)_2CClCH_2CH_3$ | 3° | 1 | 5.0 | 5.0 | 5/21.6 | 24.14 |
| | | | | 21.6 | | |

**Problem 18.1.** Calculate the heat of combustion of ethane at 25 °C. The bond energies in kJ mol$^{-1}$ are : C—H : 415.8, C—C : 344.8, O=O : 499.8, C=O : 806.4 and O—H: 464.52.

**Solution :** $2CH_3—CH_3 + 7O_2 = 4CO_2 + 6H_2O$

The energies required to break the bonds.

2×C—C bonds : 2 × 344.8 = 689.6 kJ [∵ 2 moles of $CH_3—CH_3$ are involved]

2 × 6× C—H bonds : 12 × 415.8 = 4989.6 kJ

7× O=O bonds : 7 × 499.8 = 3498.6 kJ

Total energy required = 9177.8 kJ

The energies obtained by the formation of bonds :

4C + 8O ⟶ 4O=C=O; 4 × 2 C=O : 8 × 806.4 = 6451.2 kJ

12H + 6O ⟶ 6H—O—H; 12× O—H : 12 × 464.52 = 5574.24 kJ

Total energy obtained = 12025.44 kJ

The ΔH for the reaction is the sum of the energy required and the energy obtained divided by 2, since 2 moles of ethane are involved.

$$\Delta H = \frac{(9177.8 - 12035.44)}{2} \text{ kJ mol}^{-1} = \frac{-2847.64}{2} \text{ kJ mol}^{-1}$$

$$= -1423.32 \text{ kJ mol}^{-1}$$

Since the enthalpy for the reaction is the heat of combustion, the required heat of combustion is 1423.32 kJ mol$^{-1}$.

## 18.2 Alkenes

*General formula :* $C_nH_{2n}$; where $n = 1, 2, 3$, etc; isomeric with cycloalkanes.
*Functional group :* >C=C<

### 18.2.1 Source

Natural gas, petroleum and paraffin wax.

### 18.2.2 Formation of alkenes

(For mechanisms Ch. 15 is to be consulted)

(a) By the dehydrohalogenation of alkyl halides :

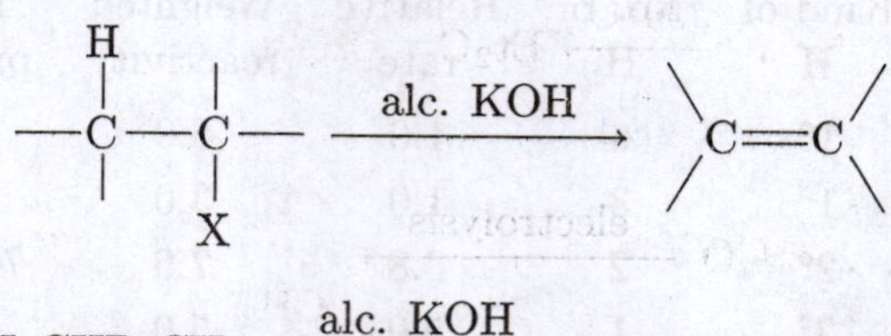

$$CH_3CHBrCH_3 \xrightarrow[\Delta]{\text{alc. KOH}} CH_3—CH{=}CH_2$$

(b) By the dehydration of alcohols :

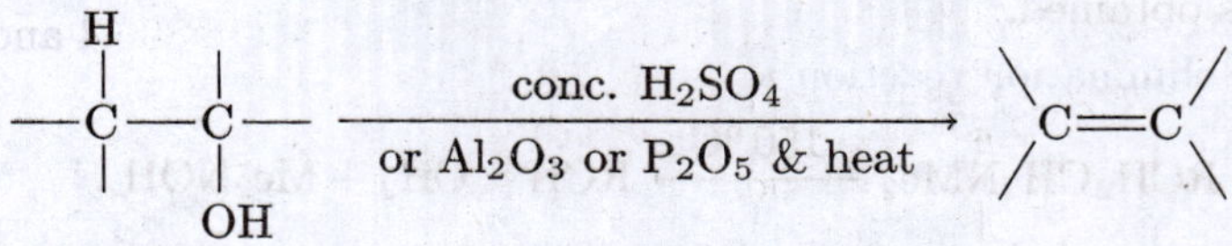

$$CH_3CH_2OH \xrightarrow[170\,°C]{\text{conc. } H_2SO_4} CH_2{=}CH_2$$

(c) By the dehalogenation of vic-dihalides :

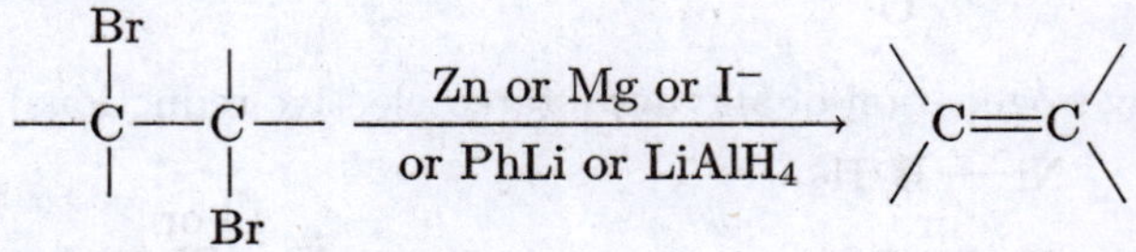

$$CH_3CHBrCH_2Br \longrightarrow CH_3CH{=}CH_2$$

(d) By the Wittig reaction on carbonyl compounds (Sec. 16.8) :

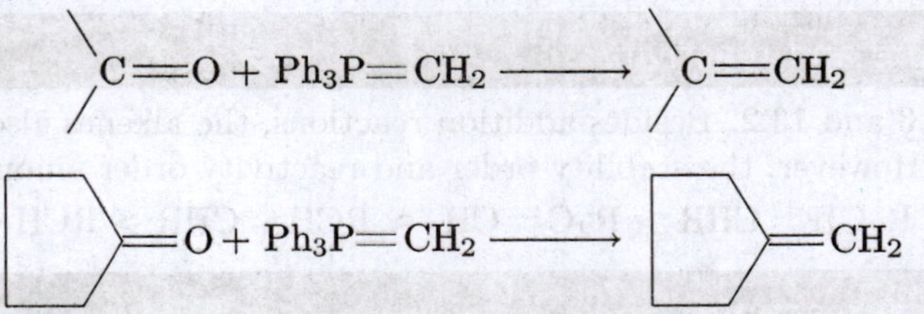

(e) By the pyrolysis of quaternary ammonium hydroxides :

$$Et\overset{\oplus}{N}Me_3OH^- \xrightarrow{\text{heat}} CH_2{=}CH_2 + Me_3N + H_2O$$

(f) By the pyrolysis of acetate esters :

$$R_2CHCH_2OCOMe \xrightarrow{500\,°C} R_2C{=}CH_2 + MeCOOH$$

$$Me_2CHCH_2OCOMe \xrightarrow{500\,°C} Me_2C{=}CH_2 + MeCOOH$$

(g) By the pyrolysis of xanthates :

$$R_2CH.CH_2O\overset{\overset{S}{\|}}{C}SMe \xrightarrow{200\,°C} R_2C{=}CH_2 + MeSH + COS$$

$$Et_2CHCH_2O\overset{\overset{S}{\|}}{C}SMe \xrightarrow{200°C} Et_2C{=}CH_2 + MeSH + COS$$

(h) By the Kolbe electrolysis (Sec. 18.1.3h) :

$$\begin{array}{l} CH_2COONa \\ | \\ CH_2COONa \end{array} + H_2O \xrightarrow{\text{electrolysis}} \begin{array}{c} CH_2 \\ \| \\ CH_2 \end{array} + Na_2CO_3 + CO_2 + H_2$$

When an aqueous and concentrated solution of sodio or potassio salt of succinic acid is electrolysed, ethylene is obtained.

(i) By the Cope elimination reaction :

$$RCH_2CH_2\overset{\oplus}{N}Me_2(-\overset{\ominus}{O}) \xrightarrow{150\,°C} RCH{=}CH_2 + Me_2NOH$$

$$MeCH_2CH_2\overset{\oplus}{N}Me_2(-\overset{\ominus}{O}) \xrightarrow{150\,°C} MeCH{=}CH_2 + Me_2NOH$$

(j) By the partial hydrogenation of alkynes (stereoselective reductions) :

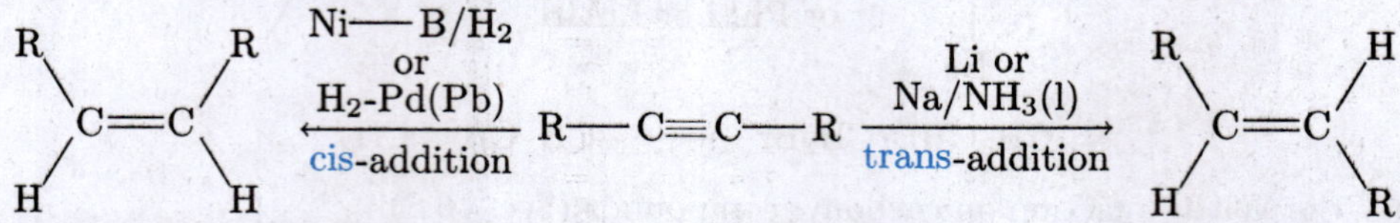


## 18.2.3 Structure and reactivity

See Secs. 2.16.3 and 13.2. Besides addition reactions, the alkenes also undergo radical substitution reactions. However, the stability order and reactivity order among alkenes are :

$$R_2C{=}CR_2 > R_2CH{=}CHR > R_2C{=}CH_2 \approx RCH{=}CHR > RCH{=}CH_2 > CH_2{=}CH_2$$

## 18.2.4 Chemical reactions

(For mechanisms of the following reactions Ch. 13 is to be consulted)

(a) Electrophilic addition of halogens : 1,2-dihaloalkanes are formed in the reaction. The reaction is stereoselective.

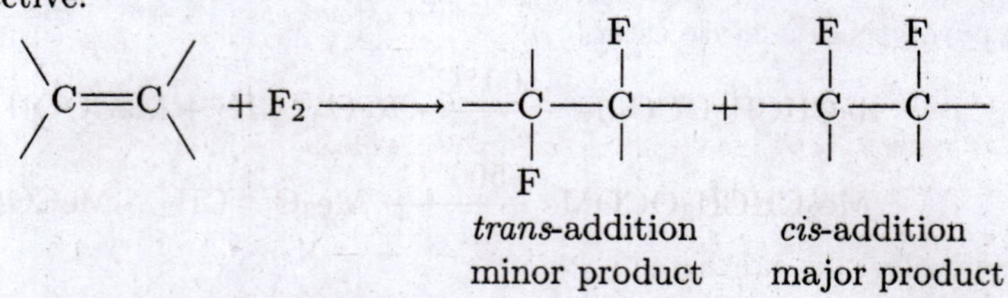


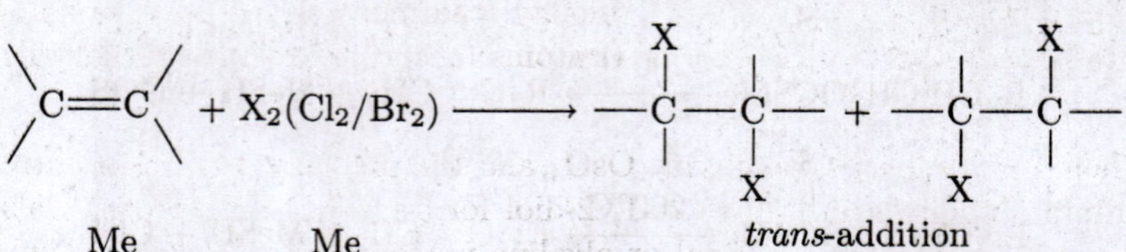


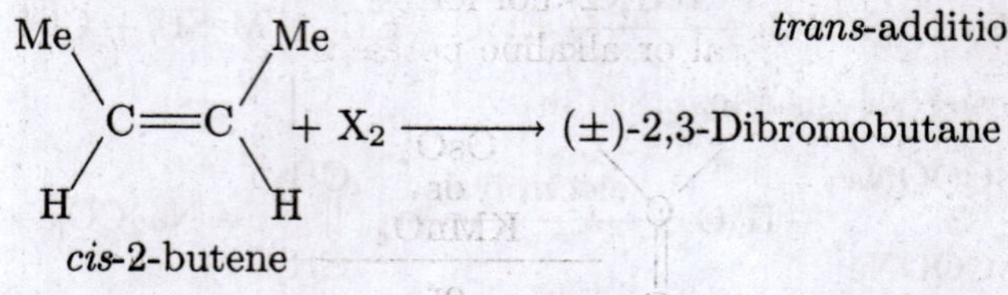


*trans*-2 butene + $X_2$ ⟶ meso-2,3-Dibromobutane

(b) Electrophilic addition of hydrogen halides : 2° and 3° alkyl halides form in this reaction.

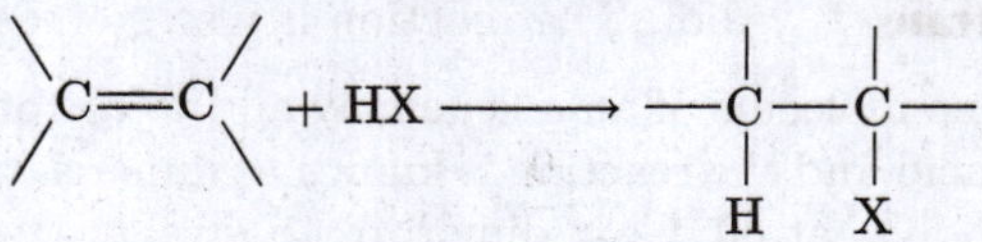

The addition usually follows the Markownikoff's rule. It is less stereoselective than the addition of halogens.

The reactivity order is : HI > HBr > HCl > HF.

$$\mathrm{MeCH{=}CHPh + HBr \longrightarrow MeCH_2CHBrPh}$$

(c) Hydration of alkenes :

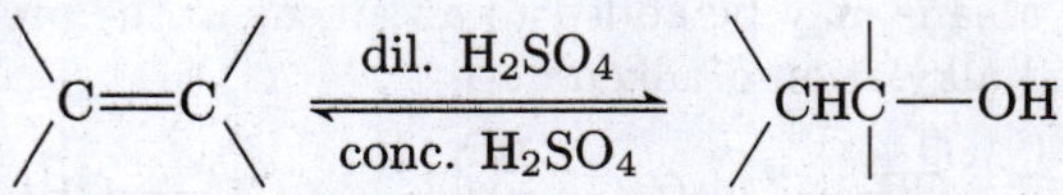

$$\mathrm{CH_3CH{=}CH_2 \underset{conc.\ H_2SO_4}{\overset{dil.\ H_2SO_4}{\rightleftharpoons}} CH_3CHOHCH_3}$$

By this reaction alcohols are formed in industry. This also follows the Markownikoff's rule.

(d) Addition of alcohols : Ethers form by the reaction. Usually the reaction follows the Markownikoff's rule. This is an acid-catalysed reaction.

$$\mathrm{CH_3CH{=}CH_2 + ROH \overset{H^{\oplus}}{\rightleftharpoons} Me_2CHOR}$$

(e) Addition of organic acids : This gives organic esters. The addition follows the Markownikoff's rule. The reaction is catalysed by acids.

$$\mathrm{CH_3CH{=}CH_2 + RCOOH \overset{H^{\oplus}}{\rightleftharpoons} RCOOCHMe_2}$$

(f) Addition of hypohalous acid :

$$\mathrm{CH_3CH{=}CH_2 + HOCl \longrightarrow CH_3CHOHCH_2Cl}$$

In this reaction halonium ions make the electrophilic attack and 2° or 3° halohydrins form. The reaction occurs following the Markownikoff's rule.

(g) Addition of halogen monohalides :

$$\mathrm{CH_3CH{=}CH_2 + ICl \longrightarrow CH_3CHClCH_2I}$$

ICl, IBr, BrCl can add to a C=C moiety; the more electropositive halogen goes to the C atom containing the greater number of H atoms (according to the Markownikoff's rule).

(h) Hydroxylation of alkenes :

(i) When an alkene is treated with $OsO_4$ and the product is refluxed with aqueous ethanolic sodium sulphite solution, a cis-1, 2-diol forms. The same product also forms when the alkene is treated with neutral or alkaline potassium permanganate or chromate solution. The reaction is stereospecific.

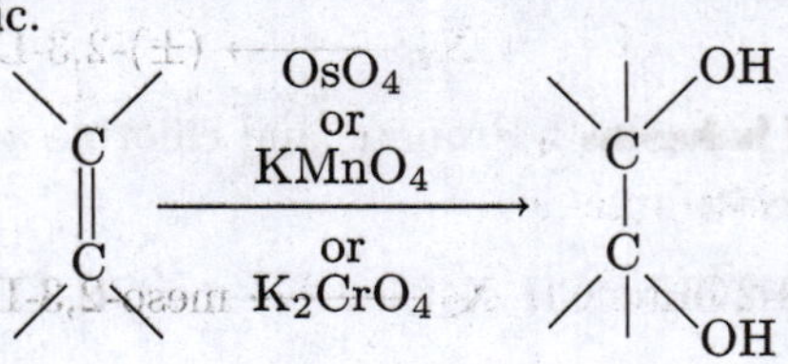

**(ii)** When an alkene is treated with a peracid, an epoxide forms which on acid- or base-catalysed hydrolysis gives **trans**-1, 2-diol. The reaction is also stereospecific.

**(i) Dimerisation :** One molecule of an alkene may add to the other molecule in the presence of concentrated sulphuric acid and the reaction is known as dimerisation reaction. When isobutylene is treated with conc. $H_2SO_4$ at 80 °C, it dimerises to give 2, 4, 4-trimethyl-1-pentene and 2, 4, 4-trimethyl-2-pentene.

$$CH_3-\underset{\displaystyle CH_3}{\underset{|}{C}}=CH_2 \xrightarrow[80\,°C]{H^{\oplus}} CH_3-\overset{\displaystyle CH_3}{\overset{|}{\underset{\displaystyle CH_3}{\underset{|}{C}}}}-CH_2-\overset{\displaystyle CH_3}{\overset{|}{C}}=CH_2 + CH_3-\overset{\displaystyle CH_3}{\overset{|}{\underset{\displaystyle CH_3}{\underset{|}{C}}}}-CH=\underset{\displaystyle CH_3}{\underset{|}{C}}-CH_3$$

**(j) Alkylation :** An alkane may be added to an alkene in the presence of conc. $H_2SO_4$ or HF and the process is called alkylation of alkene.

$$\underset{\text{iso-butylene}}{Me_2C=CH_2} + \underset{\text{iso-butane}}{Me_3CH} \xrightarrow{H^{\oplus}} \underset{\text{iso-octane}}{Me_2CH-CH_2-CMe_3}$$

**(k) Hydroboronation and oxymercuration-deoxymercuration :**

**(i)** The addition of borane across a C—C double bond by the action of diborane on an olefinic compound to form alkylborane is known as hydroboronation. This is a cis-addition reaction and follows the Markownikoff's rule.

$$3\ CH_3CH=CH_2 \xrightarrow{B_2H_6} (CH_3CH_2CH_2)_3B$$

**(ii)** Oxymercuration and deoxymercuration (p. 654; Sec. 13.11).

**(l) Catalytic hydrogenation :** Alkenes add hydrogen in the presence of a catalyst to form alkanes. This is an exothermic process.

$$CH_2=CH_2 + H_2 \xrightarrow{Ni/H_2} CH_3-CH_3 + \text{heat}$$

More or less stereoselective **cis**-addition occurs in this case.

**(m) Addition of methylene :** Methylenes add to alkenes to form cyclopropane derivatives. The addition of singlet methylene is a stereospecific **cis**-addition but the addition of a triplet carbene is non-stereospecific.

**cis**-2-Butane + methylene(singlet) → **cis**-1, 2-Dimethylcyclopropane.

**trans**-2-Butene + methylene(singlet) → **trans**-1, 2-Dimethylcyclopropane

**cis**- or **trans**-2-Butene + methylene(triplet)

→ mixture of **cis**- and **trans**-1, 2-Dimethylcyclopropane.

**(n) Ozonolysis :** Alkenes on ozonolysis form carbonyl compounds :

$$CH_3-\underset{\displaystyle CH_3}{\underset{|}{C}}=CH-CH_3 \xrightarrow{\text{ozonolysis}} CH_3-\underset{\displaystyle CH_3}{\underset{|}{C}}=O + CH_3-CHO$$

**(o) Radical addition of halogens :** Bromine and chlorine add to C—C double bonds as free radicals to form 1, 2-dihalides.

$$CH_3CH=CH_2 + Cl_2 \xrightarrow{h\nu} CH_3CHClCH_2Cl$$

(p) Radical addition of HBr : The addition of HBr to a C—C double bond occurs anti-Markownikoffically in the presence of organic peroxides.

$$CH_3CH{=}CH_2 + HBr \xrightarrow[\text{peroxide}]{\text{organic}} CH_3CH_2CH_2Br$$

(q) Polymerisation reaction : The process by which many molecules of one or more kinds join together to form a bigger molecule with or without giving up any small unit is called polymerisation. The big molecule so obtained is called polymer ('poly' means many and 'mer' means unit; polymer means many-unit and polymerisation means many-unit formation). The polymer may or may not have the same empirical formula as the starting units. Usually the process is capable of proceeding indefinitely. However, the polymerisation reactions are of two types : (i) polyaddition, and (ii) polycondensation. In a polyaddition reaction all the units are added up and in that case the product and the starting units have the same empirical formula; while in a polycondensation reaction small molecules like water, ammonia, etc. are lost during condensation and so the polymer has a different empirical formula from the reactants. Ethylene polymerises to give polythene. This is an example of polyaddition.

$$nCH_2{=}CH_2 \longrightarrow -CH_2-(CH_2-CH_2)_{n-1}-CH_2-$$

Polythene (poly-ethene)

A polymerisation reaction may follow either the free radical pathway or the ionic pathway. The ionic polymerisation may be catalysed by cations or anions (Sec 18.4.4iv). Polymerisation reactions are very important. Nylon, synthetic rubber, plastic, polyester, resins are prepared by this reaction.

(r) Epoxide formation : When alkenes are treated with peracids, $CH_3COOOH$, PhCOOOH etc., epoxides are formed.

Ethylene oxide is industrially prepared by passing a mixture of ethylene and air over a silver catalyst at 200 – 400 °C under pressure.

$$CH_2{=}CH_2 + O_2 \xrightarrow[\substack{200-400^\circ C \\ \text{under pressure}}]{\text{Ag cat.}} H_2C\overset{O}{\frown}CH_2$$

(s) Radical substitution : When an alkene is treated with chlorine at a temperature of 500°C–600°C, an allylic H gets replaced by a Cl atom. This is a radical substitution reaction.

$$Cl-Cl \xrightarrow{h\nu} Cl\cdot + Cl\cdot$$

$$CH_2{=}CH-CH_3 + Cl\cdot \xrightarrow{-HCl} CH_2{=}CH-\dot{C}H_2 \longleftrightarrow \dot{C}H_2-CH{=}CH_2$$

$$CH_2{=}CH-\dot{C}H_2 + Cl_2 \longrightarrow CH_2{=}CH-CH_2Cl + Cl\cdot$$

(t) Vigorous oxidation of alkenes by acid permanganate or dichromate : When an alkene is vigorously oxidised with $KMnO_4$ or $K_2Cr_2O_7$ and conc. sulphuric acid, degradation of the alkene takes place. In such a reaction :

(i) a double-bonded 1°C gives carbon dioxide,

(ii) a double-bonded 2°C gives a carboxylic acid and

(iii) a double-bonded 3°C gives a ketone.

$$CH_3CH{=}CH_2 \xrightarrow[H_2SO_4]{KMnO_4} CH_3COOH + CO_2$$

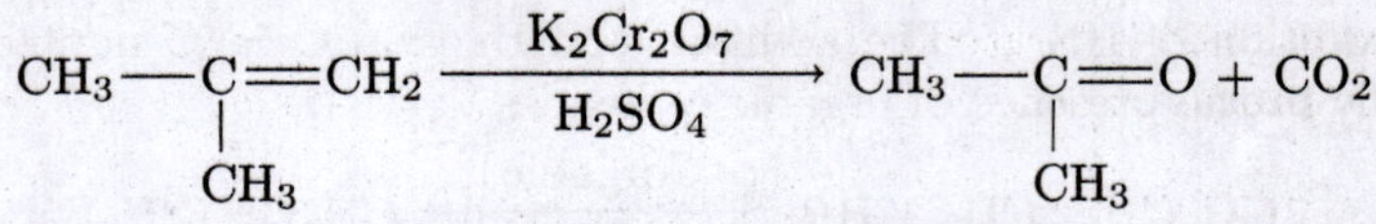

Instead of the acid one may use alkali, but then carbonates, carboxylate salts and ketones will be obtained from 1°, 2° and 3° double-bonded Cs respectively.

$$R_2C{=}CHCH_2CH{=}CH_2 \xrightarrow{\text{alk. } KMnO_4} R_2C{=}O + KOOCCH_2COOK + K_2CO_3$$

## 18.2.5 Analysis of alkenes

**(i) Qualitative test :** Addition of bromine (Sec. 13.3.2) to an alkene and its hydroxylation by neutral $KMnO_4$ (Baeyer's test, Sec. 13.7.2) are the two qualitative tests for C—C double bonds.

**(ii) Determination of position of a double bond :** Ozonolysis experiment is a good method for determining the position of the double bond in an alkene.

**Problem 18.2.** Suggest a structure for compound A,$C_6H_{12}$, which yielded the same product on ozonolysis as on oxidation with acid dichromate.

**Solution :** Since the compound A gives only one product on degradation reactions, it is a symmetrical alkene. Again, since ozonolysis and dichromate oxidation give the same product, the double-bonded Cs. are 3°Cs; this is because in both the reactions only 3°Cs give ketones. So the C-skeleton of the compound A is :

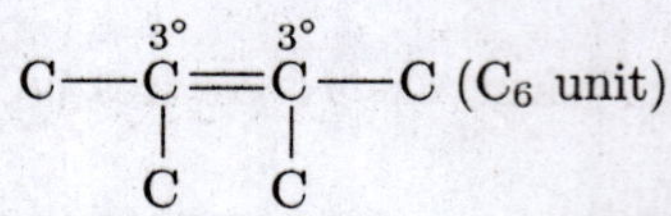

If we put H atoms on the skeleton to satisfy four valences of each C atom, the structure of compound A will be :

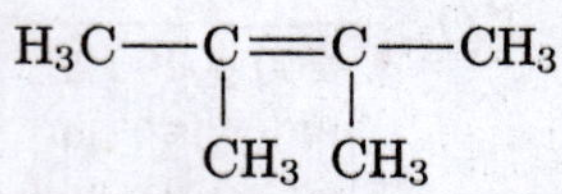

2,3-Dimethylbut-2-ene

## 18.3 Alkynes

*General formula :* $C_nH_{2n-2}$ (where $n = 1, 2, 3,$ etc.); isomeric with alkadienes and cycloalkenes.

*Functional Group :* —C≡C—,

## 18.3.1 Source

Coal gas contains very small amount of acetylene. This is usually prepared from (i) calcium carbide and (ii) methane. Higher members are prepared from acetylene.

## 18.3.2 Formation of alkynes

**(a) From calcium carbide :**

$$CaC_2 + 2H_2O{=}C_2H_2 + Ca(OH)_2.$$

When calcium carbide is treated with water at room temperature, acetylene is obtained. This is also a synthetic process.

(b) From methane :

$$3CH_4 + 3O_2 \longrightarrow CO + C_2H_2 + 5H_2O$$

When a mixture of methane and oxygen (1 : 1 by vol.) is heated at 1500 °C, acetylene is obtained.

(c) From carbon and hydrogen :

$$2C + H_2 \xrightarrow[\text{electric arc}]{1100\,°C} C_2H_2$$

When hydrogen gas is passed over graphite electrodes which are conducting electricity through external circuit, acetylene is formed.

(d) From tetrahaloalkanes :

$$Br_2CHCHBr_2 \xrightarrow{\text{heat/Zn}} HC{\equiv}CH + 2ZnBr_2$$

$$RCX_2CHX_2 \xrightarrow{\text{heat/Zn}} R{-}C{\equiv}CH$$

$$RCX_2CX_2R \xrightarrow{\text{heat/Zn}} RC{\equiv}CR$$

When $\alpha$, $\alpha$, $\beta$, $\beta$-tetrahaloalkanes are heated with zinc, acetylene and substituted acetylenes are obtained.

(e) From vic- or gem-dihalides :

$$CH_2Br{-}CH_2Br \xrightarrow[\text{or NaNH}_2]{\text{alc. KOH}} BrCH{=}CH_2 \longrightarrow CH{\equiv}CH$$

$$CH_3CHCl_2 \xrightarrow[\text{or NaNH}_2]{\text{alc. KOH}} CH_2{=}CHCl \longrightarrow CH{\equiv}CH$$

$$RCHXCH_2X \xrightarrow[\text{or alc. KOH}]{KNH_2} R{-}C{\equiv}CH$$

$$RCH_2CHX_2 \xrightarrow[\text{or alc. KOH}]{KNH_2} R{-}C{\equiv}CH$$

$$RCHXCHXR' \xrightarrow[\text{or alc. KOH}]{KNH_2} R{-}C{\equiv}C{-}R'$$

Acetylenes can be prepared from vic- or gem-dihalides by heating them with alcoholic potassium hydroxide or sodium or potassium amide.

(f) From acetylene :

$$HC{\equiv}C{-}H \xrightarrow{NaNH_2} HC{\equiv}\bar{C}Na^{\oplus} \xrightarrow{RX} HC{\equiv}C{-}R \xrightarrow{NaNH_2} \overset{\oplus}{Na}\bar{C}{\equiv}C{-}R \xrightarrow{R^1X} R^1{-}C{\equiv}C{-}R$$

$$HC{\equiv}C{-}H \xrightarrow{RMgX} HC{\equiv}C{-}MgX \xrightarrow{R^1X} HC{\equiv}C{-}R^1 \xrightarrow{RMgX} MgX{-}C{\equiv}C{-}R^1 \xrightarrow{RX} R{-}C{\equiv}C{-}R^1$$

(g) From haloforms :

$$2\ddot{C}HX_3 \xrightarrow[\text{and heat}]{\text{Ag powder}} CH{\equiv}CH + 6AgX$$

When haloforms are heated with silver powder, acetylene is formed.

(h) From maleic and fumaric acids : When an aqueous solution of sodio or potassio salt of maleic or fumaric acid is electrolysed, acetylene is obtained; this is called the Kolbe electrolysis. Thus benzene can be converted to acetylene through the Kolbe electrolysis.

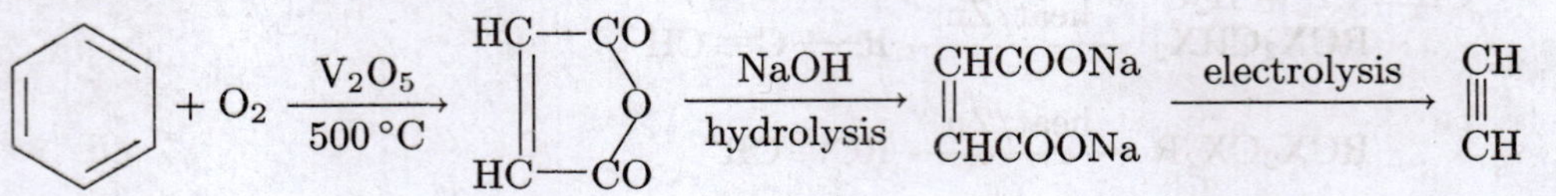

## 18.3.4 Structure and reactivity

See Sec. 2.16.4 and Problem 13.3 p. 628.

## 18.3.5 Chemical reactions

(a) Hydrogenation to alkanes :

$$RC{\equiv}CR \xrightarrow{Ni/H_2} RCH_2CH_2R + \text{heat.}$$

(b) Hydrogenation to alkenes (stereoselective reductions) :

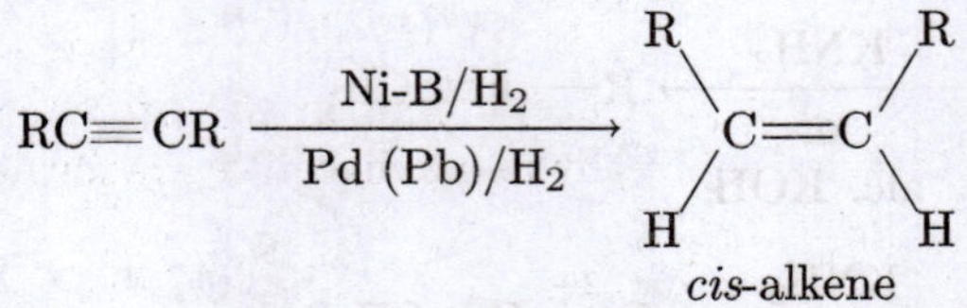

*cis*-alkene

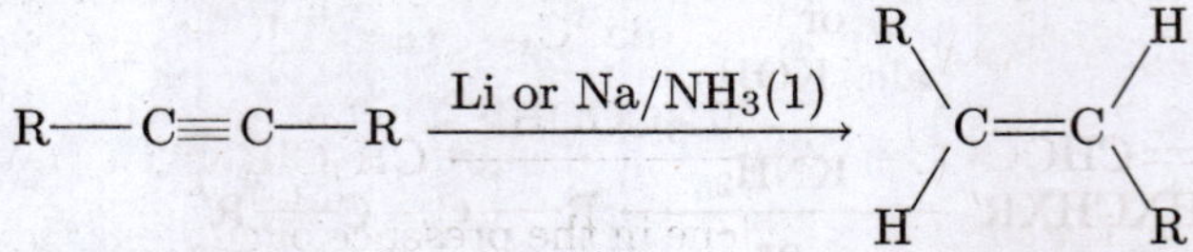

*trans*-alkene

(c) Halogen addition (Sec. 13.3.4) :

$$R{-}C{\equiv}C{-}R + Br_2 \text{ or } Cl_2(g) \longrightarrow RCBr_2{-}CBr_2R \text{ or } RCCl_2CCl_2R$$

$$H—C{\equiv}C—H + Br_2(H_2O) \longrightarrow H—CBr{=}CBr—H$$

$$H—C{\equiv}CH(\text{in EtOH}) + I_2 \longrightarrow CHI{=}CHI$$

Halogens add to acetylene more or less stereoselectively.

(d) Hydrogen halide addition : Halogen acids can add to acetylenes in the dark; but the reaction can be catalysed by light or metallic halides. HF of course adds under pressure. The addition of haloacids follows the Markownikoff's rule.

$$R—C{\equiv}CH + HX \longrightarrow RCX{=}CH_2 \xrightarrow{HX} RCX_2CH_3.$$

$$HC{\equiv}CH + HCl \longrightarrow CHCl{=}CH_2 \xrightarrow{HCl} CHCl_2CH_3.$$

The reactivity order of halides is HI > HBr > HCl > HF.

(e) Hydration (Sec. 13.6) : When an alkyne is treated with dilute sulphuric acid in the presence of catalytic amount of mercuric sulphate at 60 °C, hydration of alkyne occurs and a carbonyl compound forms.

$$CH{\equiv}CH + H_2O \xrightarrow[HgSO_4]{\text{dil. } H_2SO_4} CH_3CHO$$

$$RC{\equiv}CH + H_2O \xrightarrow[HgSO_4]{\text{dil. } H_2SO_4} RCOCH_3$$

$$R^1C{\equiv}CR^2 + H_2O \xrightarrow[HgSO_4]{\text{dil. } H_2SO_4} R^1COCH_2R^2 + R^1CH_2COR^2$$

(f) Addition of acyl chlorides : Acyl halides add to acetylene in the presence of anhydrous $AlCl_3$.

$$RCOCl + CH{\equiv}CH \xrightarrow{AlCl_3} RCOCH{=}CH—Cl.$$

(g) Formation of vinyl derivatives :

(i) At 65 °C dil. HCl adds to acetylene in the presence of mercuric ion as catalyst and forms vinyl chloride which is used for the manufacture of polyvinyl chloride (PVC) plastics.

$$HC{\equiv}CH + HCl \xrightarrow[65\,°C]{Hg^{2+}} CH_2{=}CH—Cl$$

(ii) When acetylene is passed into warm acetic acid in the presence of catalytic amount of $Hg^{2+}$, at first vinyl acetate and then ethylidene diacetate are formed. Vinyl acetate is used in the plastic industry.

$$HC{\equiv}CH + CH_3COOH \xrightarrow[65\,°C]{Hg^{2+}} CH_2{=}CHOOCCH_3$$

$$CH_2{=}CHOOCCH_3 \xrightarrow{CH_3COOH} CH_3CH(OOCCH_3)_2$$

(iii) Hydrogen cyanide adds to acetylene in the presence of cuprous chloride in hydrochloric acid as catalyst to form vinyl cyanide which is used in the manufacture of Buna N, a synthetic rubber (a copolymer of vinyl cyanide and butadiene).

$$HC{\equiv}CH + HCN \xrightarrow{Cu_2Cl_2/HCl} CH_2{=}CH—CN$$

**(iv)** Methyl alcohol adds to acetylene in the presence of 1-2% potassium methoxide at a temperature of 160-200 °C to give methyl vinyl ether which is used for making polyvinyl ether plastics.

$$CH{\equiv}CH + CH_3\bar{O} \longrightarrow \ddot{C}H{=}CHOCH_3 \xrightarrow{CH_3OH} CH_2{=}CH{-}OCH_3$$

**(h) Ethinylation :** Addition of acetylene to unsaturated links or elimination of water by its reaction with certain hydroxy-compounds is called ethinylation, e.g.,

$$HC{\equiv}CH + H_2C{=}O \xrightarrow{Me\bar{O}} HOCH_2C{\equiv}CH \xrightarrow[MeO^-]{CH_2O} \underset{\text{Butynediol}}{HOCH_2C{\equiv}CCH_2OH}$$

$$Me_2NCH_2OH + CH{\equiv}CH \longrightarrow Me_2NCH_2C{\equiv}CH + H_2O$$

$$Me_2C{=}O + HC{\equiv}CH \xrightarrow{NaNH_2} Me_2C(\bar{O}Na^{\oplus}){-}C{\equiv}CH$$

**(i) Addition of hypochlorous acid :** Dichloroacetaldehyde forms when acetylene is passed into hypochlorous acid solution.

$$CH{\equiv}CH + HOCl \longrightarrow CHCl{=}CHOH \xrightarrow{HOCl} Cl_2CH{-}CH(OH)_2$$

$$Cl_2CH{-}CH(OH)_2 \longrightarrow Cl_2CHCHO + H_2O$$

**(j) Addition of borane :** Borane adds to acetylenes to form vinyl boranes among other products.

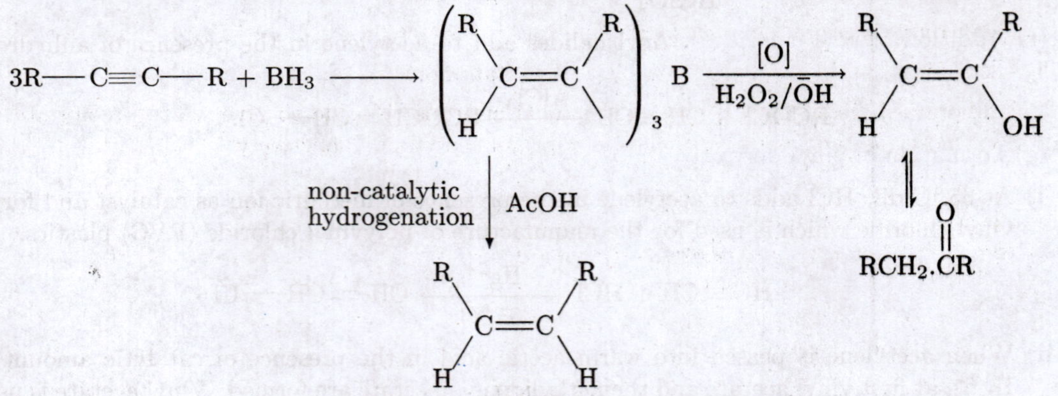

cis-product

$$R{-}C{\equiv}CH \xrightarrow{BH_3} \xrightarrow{H_2O_2/\bar{O}H} RCH_2CHO$$

**(k) Polymerisation :**

**(i)** Benzene is formed in small amount when acetylene is passed through a heated tube. Under suitable conditions acetylene is converted to cyclooctatetraene by heating.

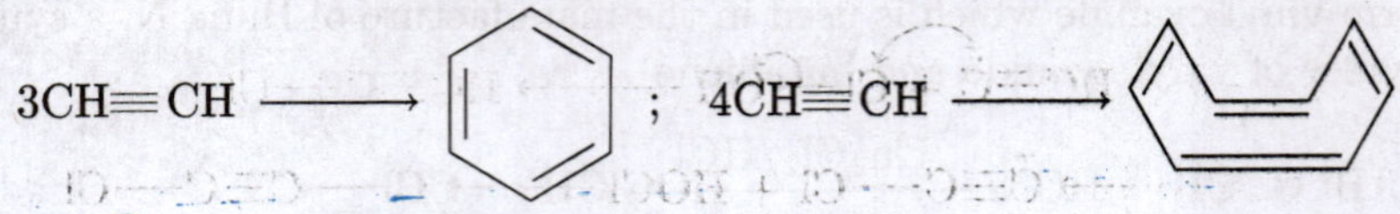

Homologues of acetylene also form benzene derivatives on heating.

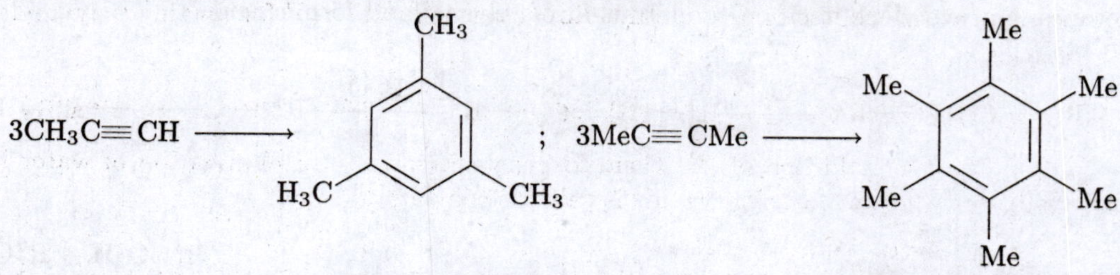

1,3,5-Trimethylbenzene Hexamethylbenzene

**(ii)** When acetylene is passed through ammoniacal cuprous chloride solution, vinyl- and divinylacetylenes are formed.

$$2HC{\equiv}CH \xrightarrow{Cu(NH_3)_2Cl} CH_2{=}CH{-}C{\equiv}CH \xrightarrow{CH\equiv CH} CH_2{=}CH{-}C{\equiv}C{-}CH{=}CH_2$$

When vinylacetylene is passed through concentrated hydrochloric acid in the presence of ammonium chloride and cuprous chloride, chloroprene, 2-chloro-1,3-butadiene, is formed. This polymerises readily to neoprene, a rubber-like polymer.

$$CH_2{=}CH{-}C{\equiv}CH + HCl \longrightarrow CH_2Cl{-}CH{=}C{=}CH_2 \longrightarrow CH_2{=}CCl{-}CH{=}CH_2$$

1. Properties owing to $\equiv$C—H : An H atom on a triple-bonded C atom being acidic, it reacts with :

**(i)** Hot Li or Na to form salts and evolves hydrogen gas.

**(ii)** The Grignard reagent to form alkanes.

**(iii)** Alkali metal amides to form salts and evolve ammonia.

**(iv)** Ammoniacal silver nitrate and ammoniacal cuprous chloride to give white precipitate of silver acetylide and red precipitate of cuprous acetylide respectively.

$$HC{\equiv}CH \xrightarrow{Na} HC{\equiv}CNa \xrightarrow{Na} NaC{\equiv}CNa \xrightarrow{2RX} R{-}C{\equiv}C{-}R$$

$$HC{\equiv}CNa \xrightarrow{CO_2} HC{\equiv}CCO_2Na$$

$$HC{\equiv}CH \xrightarrow{NaNH_2} HC{\equiv}CNa + NH_3 ;$$

$$RC{\equiv}CH \xrightarrow{[Ag(NH_3)_2]NO_3} RC{\equiv}CAg ; \quad RC{\equiv}CH \xrightarrow{Cu(NH_3)_2Cl} RC{\equiv}CCu$$

$$RC{\equiv}CH + CH_3MgI \longrightarrow RC{\equiv}CMgI + CH_4.$$

(m) Halogenation : When acetylene is treated with sodium hypochlorite solution at 0 °C in the absence of air, dichloroacetylene forms.

$$HC{\equiv}C{-}H + \bar{O}Cl \longrightarrow HC{\equiv}\ddot{\bar{C}} + Cl{-}OH \longrightarrow HC{\equiv}C{-}Cl$$

$$H{-}C{\equiv}C{-}Cl + \bar{O}Cl \longrightarrow \bar{C}{\equiv}C{-}Cl + HOCl \longrightarrow Cl{-}C{\equiv}C{-}Cl$$

(n) Ozonolysis : Derivatives of acetylenes on ozonolysis give mixtures of carboxylic acids via diketones; but acetylene itself on ozonolysis forms glyoxal and formic acid.

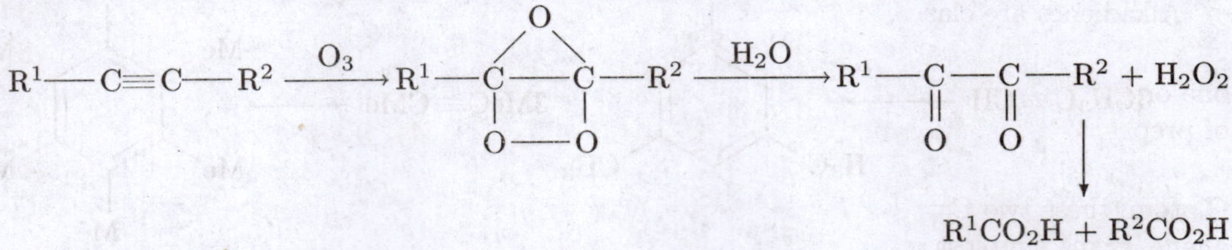

(o) Permanganate oxidation : A 2° triple-bonded C on acid $KMnO_4$ or $K_2Cr_2O_7$ oxidation gets converted to a carboxyl group, whereas 1° triple-bonded Cs form $CO_2$.

$$R{-}C{\equiv}C{-}R \xrightarrow{KMnO_4} R{-}\underset{\underset{O}{\|}}{C}{-}\underset{\underset{O}{\|}}{C}{-}R. \xrightarrow{KMnO_4} RCO_2H + RCO_2H.$$

$$HC{\equiv}CH \xrightarrow{KMnO_4} CO_2 + CO_2.$$

## 18.3.6 Analysis of alkynes

(i) Qualitative tests : a. Alkynes decolourise bromine water. b. 1° Alkynes form precipitates with ammoniacal silver nitrate solution (white ppt.) and ammoniacal cuprous chloride solution (red ppt.); and they evolve hydrogen with Na or Li.

(ii) Position of the triple bond : Joining the acid units obtained by oxidative degradation (by ozonolysis or acid $KMnO_4$), the position of the triple bond can be ascertained as shown below :

$$R{-}C\boxed{OOH + HOO}CR' \longrightarrow RC{\equiv}CR'$$

**Problem 18.3.** Distinguish between each of the following by a single test.

**(a)** 1-Butyne and 2-butyne **(b)** 1-Butyne and 1-butene.

**Solution : (a)** 1-Butyne contains acidic H but 2-butyne does not. So, 1-butyne will give white ppt. with $[Ag(NH_3)_2]NO_3$ but 2-butyne will not.

**(b)** 1-Butene has no acidic H; so it will not give ppt. with ammoniacal silver nitrate but 1-butyne will give.

**Problem 18.4.** Which one will add bromine readily—1-butene or 2-butene? Give reasons for your answer.

**Solution :** Addition of bromine is an electrophilic addition reaction. Electron-donating groups increase the rate of such a reaction. 2-butene contains two electron-donating methyl groups whereas 1-butene contains one electron-donating ethyl group. Therefore, 2-butene will react more readily with bromine than 1-butene.

## 18.4 Alkadienes

*General formula :* $C_nH_{2n-2}$; Alkadienes are isomeric with alkynes and cycloalkenes.

*Functional group :* Two C=C units.

## 18.4.4 Chemical reactions

(a) Reactions of allene :

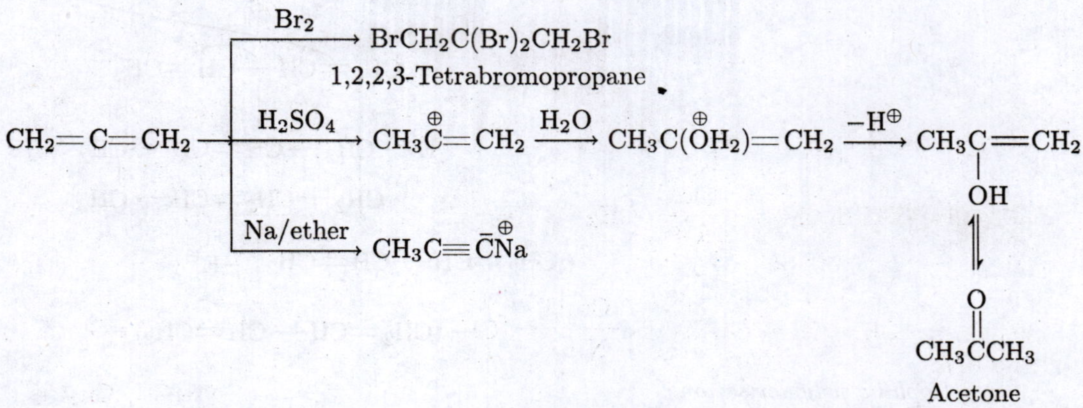


(b) Reactions of conjugated dienes : There are five characteristic properties of conjugated dienes :

(i) 1, 4-Addition and 1, 2-addition :

$$CH_2{=}CH{-}CH{=}CH_2 + Br_2 \longrightarrow \underset{Br}{CH_2}{-}CH{=}CH{-}\underset{Br}{CH_2} + \underset{Br}{CH_2}{-}\underset{Br}{CH}{-}CH{=}CH_2$$

$$CH_2{=}CH{-}CH{=}CH_2 + 2Br_2 \longrightarrow CH_2Br{-}CHBr{-}CHBr{-}CH_2Br$$

(ii) The Diels-Alder reaction (Sec. 13.15.1) :

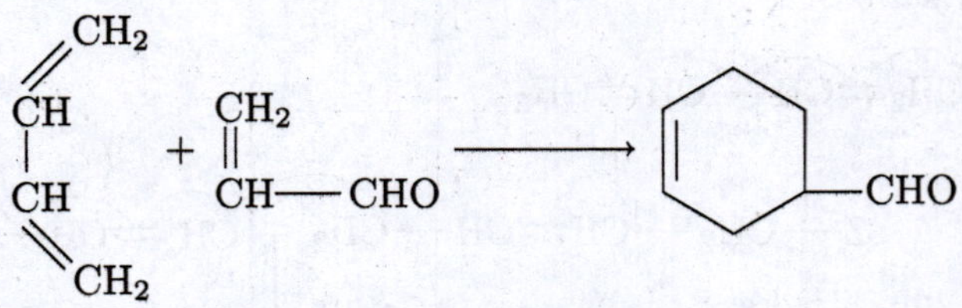


(iii) Cyclic sulphone formation : Conjugated alkadienes combine with sulphur dioxide to form cyclic sulphones

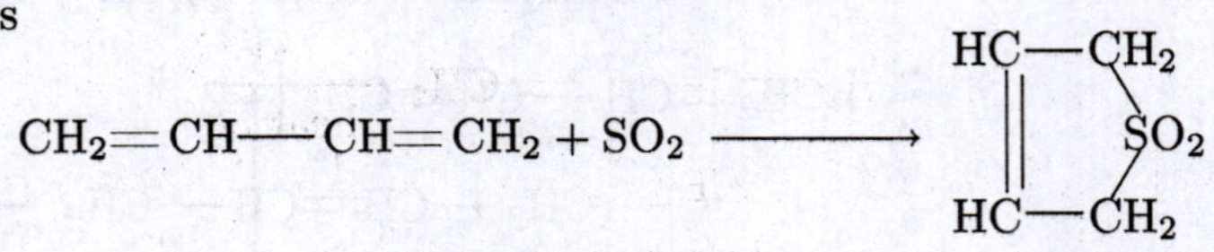


cyclic sulphone

(iv) Polymerisation :

$$CH_2{=}CH{-}CH{=}CH_2 \xrightarrow[\text{or nucleophile or free radical}]{\text{electrophile}} {-}(CH_2CH{=}CH{-}CH_2)_n{-}$$

a unit of polymer

They polymerise readily; ions and radicals are able to polymerise them.

*Electrophilic polymerisation :*

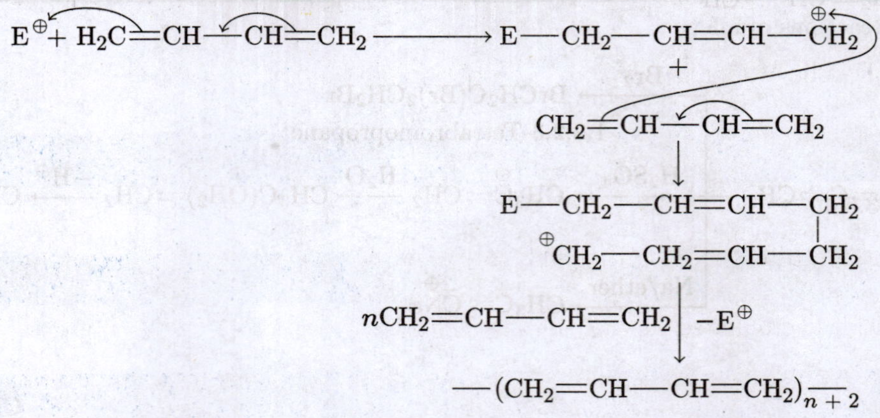

*Nucleophilic polymerisation :*

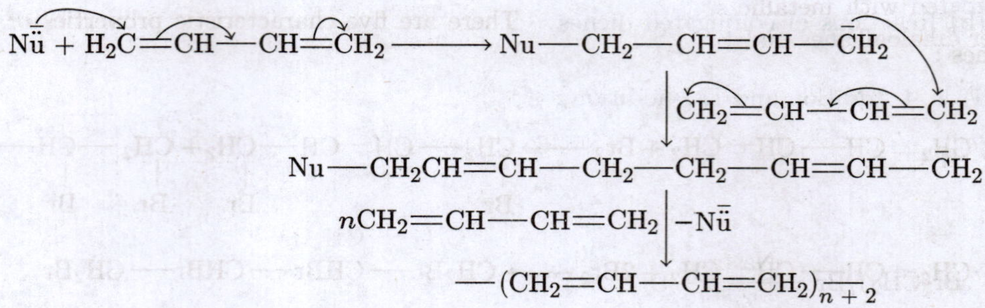

*Radical polymerisation :*

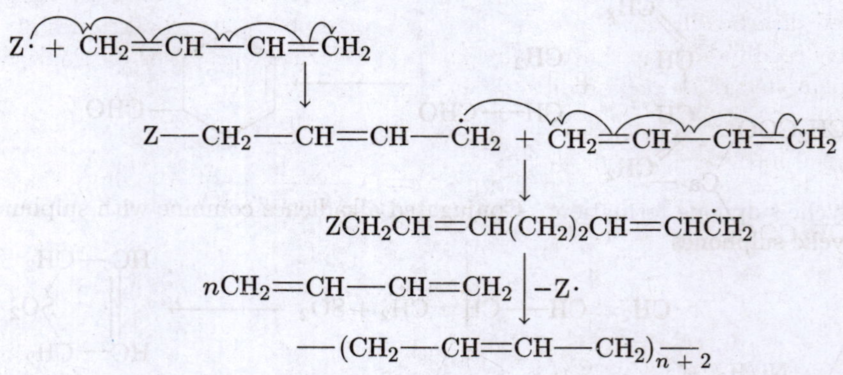

**(v)** Cycloaddition reaction (Sec. 13.15) : The thermal or photochemical addition of alkenes to conjugated alkadienes produce alicyclic products. These reactions do not require any polar solvent or reagent.

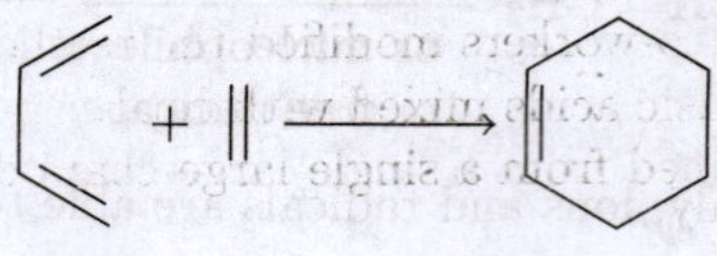

**(vi) Addition of HI to 1, 3-butadiene :**

$$CH_2{=}CH{-}CH{=}CH_2 + HI \longrightarrow ICH_2CH_2CH{=}CH_2 + ICH_2{-}CH{=}CH{-}CH_3$$

## 18.5 Alicyclic Hydrocarbons

### 18.5.1 Introduction and nomenclature

See. Sec. 3.5.4.

### 18.5.2 Stereoisomerism

Conformations and related aspects : See. Sec. 9.20.

### 18.5.3 Formation of alicyclic hydrocarbons

(a) From $\alpha$, $\omega$-dihalogen derivatives of alkanes (Freund) : When 1, 3- to 1, 6-dihaloalkanes are treated with metallic sodium or zinc, the corresponding cycloalkanes are formed; 1, 7-and higher dihaloalkanes undergo the Wurtz reaction under the condition of the experiment.

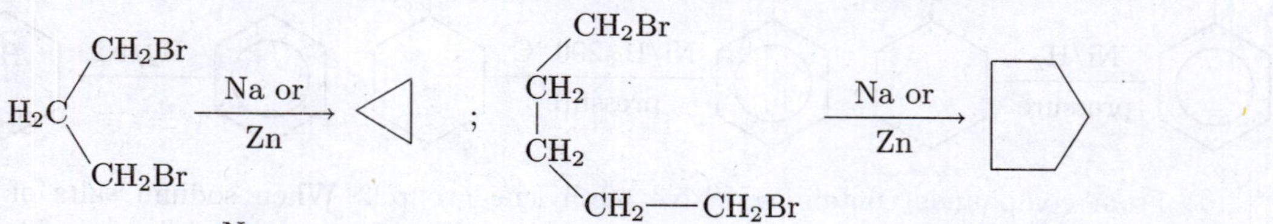


$$Br(CH_2)_7Br \xrightarrow[Zn]{Na\ or} Br(CH_2)_{14}Br$$

(b) From calcium or barium salts of dibasic acids (Wislicenus') : When calcium or barium salts of saturated dicarboxylic acids are distilled, the corresponding cyclic ketones are obtained; the ketones may readily be reduced to cycloalkanes and cycloalkenes; thus, in this way cyclopentane and cyclopentene can be prepared from calcium adipate via cyclopentanone.

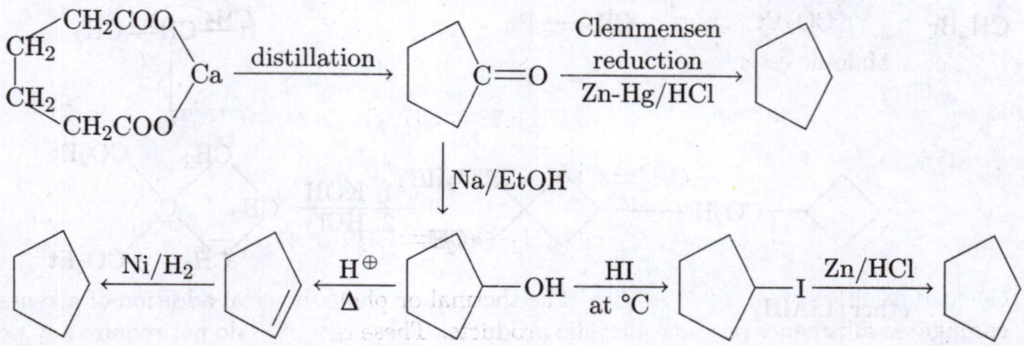


By this method cyclopentanone, cyclohexanone and cycloheptanone can be prepared successfully. However, Ruzika and his co-workers modified the method by heating the thorium, cerium, yttrium salts of large-chain dibasic acids mixed with copper powder at 300 °C in vacuum; usually a number of products are obtained from a single large-chain dibasic acid.

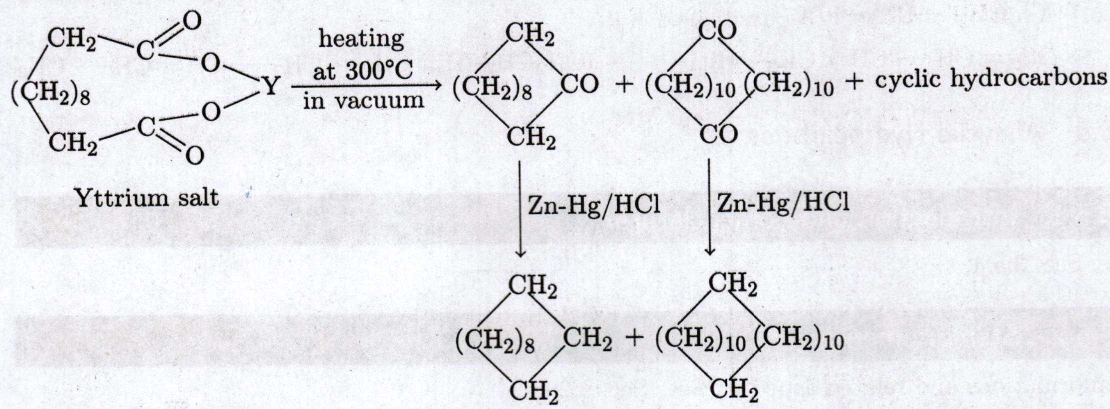

(c) By reduction of aromatic compounds : When benzene, phenol and aniline are catalytically reduced, cyclohexane, cyclohexanol and cyclohexylamine respectively are formed. Catalytic reduction under high pressure using nickel is found to be the best method.

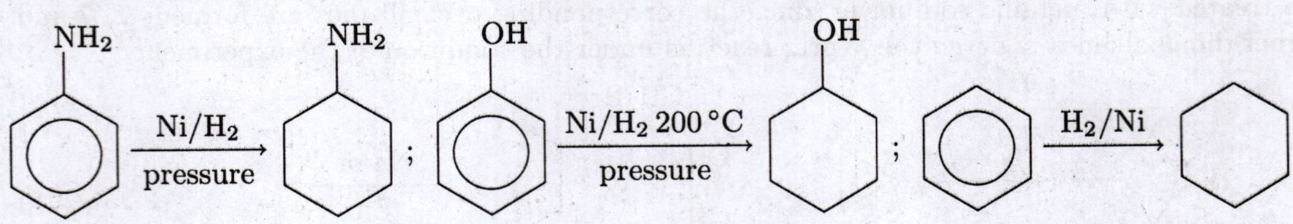

(d) From compounds containing active methylene group : When sodium salts of ethyl acetoacetate, ethyl malonate, acetylacetone, etc. are allowed to react with $\alpha$, $\omega$-dihalides, alicyclic compounds are formed.

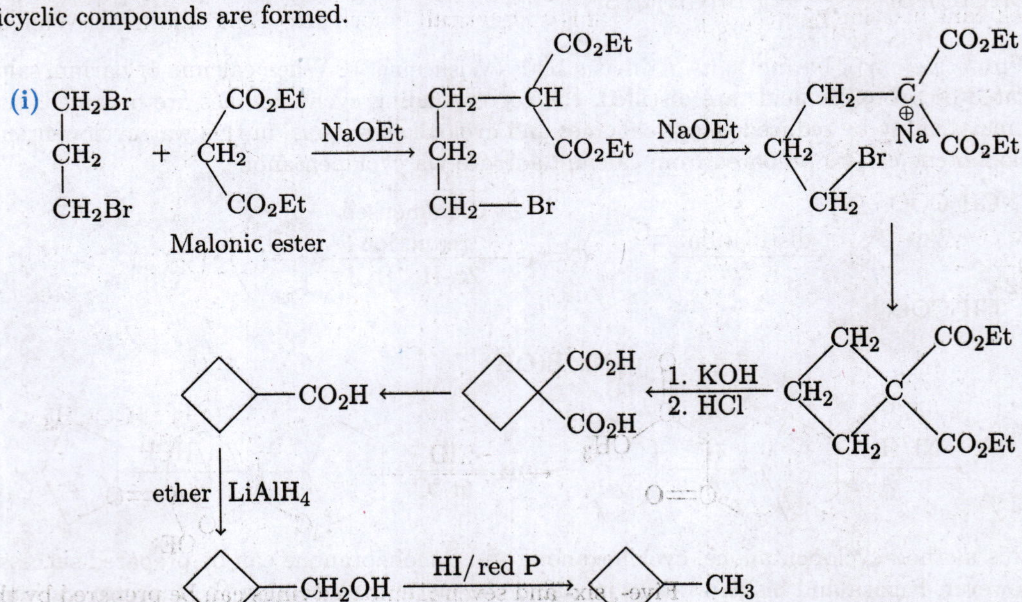

(ii) 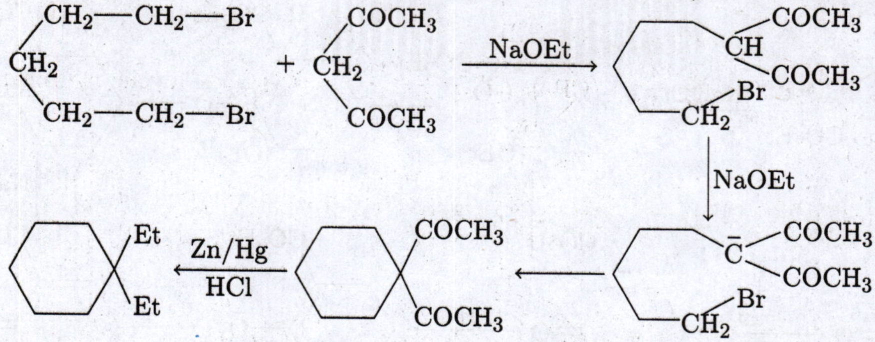

(iii) 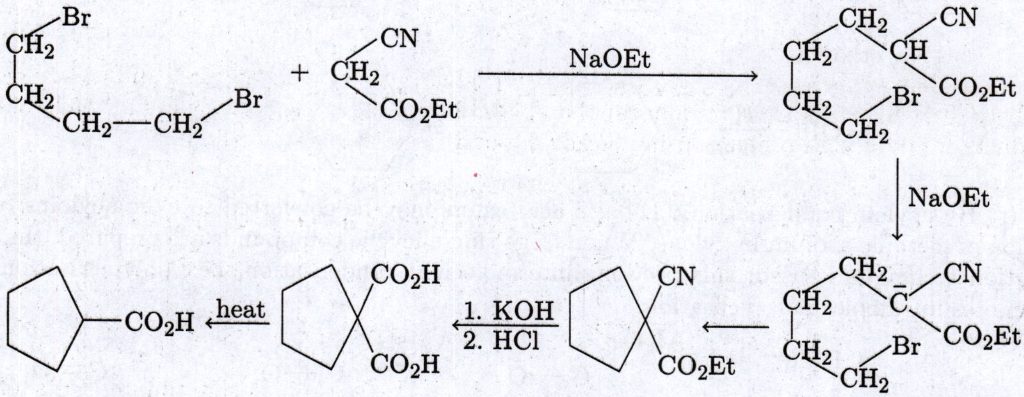

It is here important to note that cyclobutane ring cannot be prepared from ethyl acetoacetate since self condensation of ethyl acetoacetate is energytically more favourable and faster than the reaction.

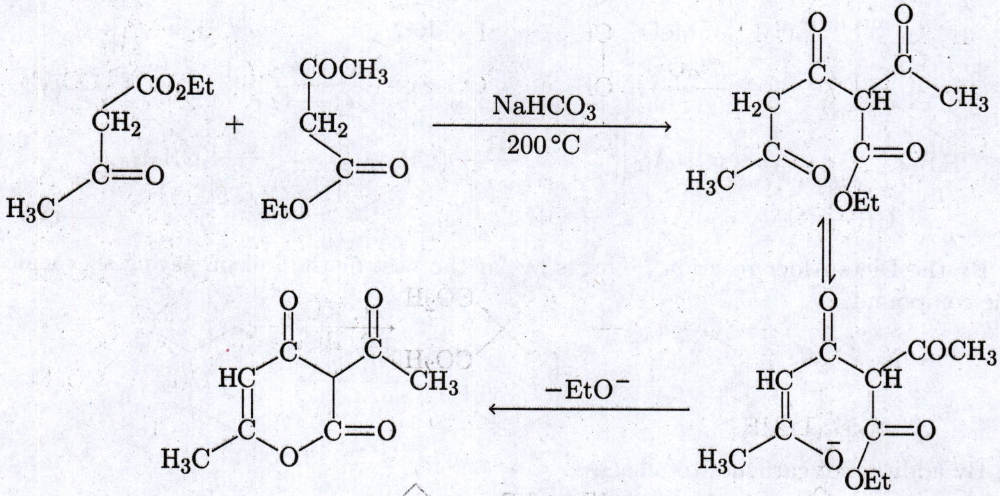

(e) By the Dieckmann reaction : Five-, six- and seven-membered rings can be prepared by the Dieckmann reaction on esters of adipic, pimelic and suberic acids respectively. Usually sodium ethoxide in ethanol is used as the base for the reaction.

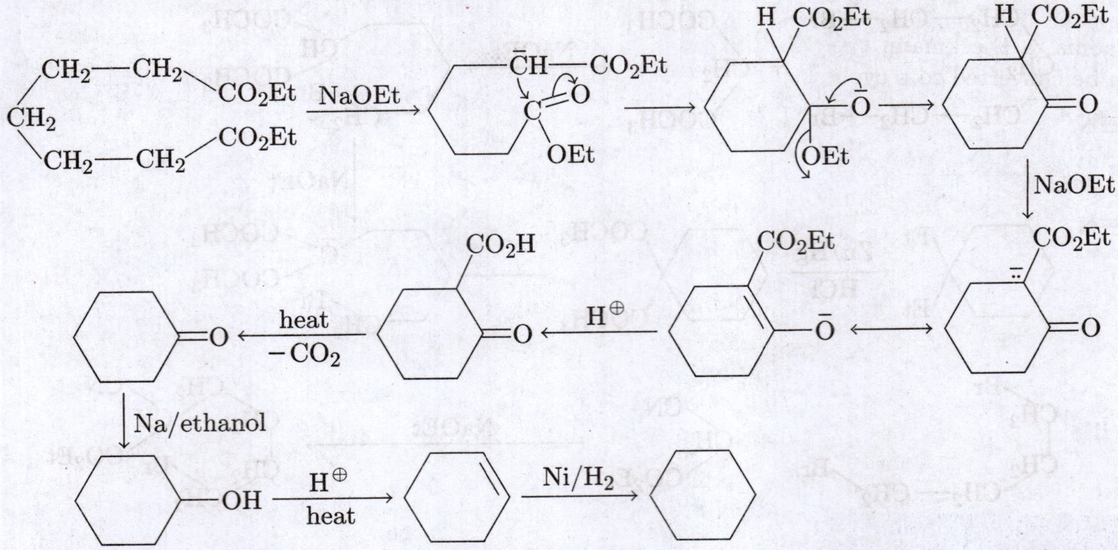

**(f) By acyloin condensation :** Dibasic acid esters may be converted to cylic acyloins by the action of metallic sodium in xylene. When large ring alicyclic compounds are prepared, the high-dilution technique is used; this prevents intermolecular condensation, i.e., polymerisation, and promotes intramolecular cyclisation.

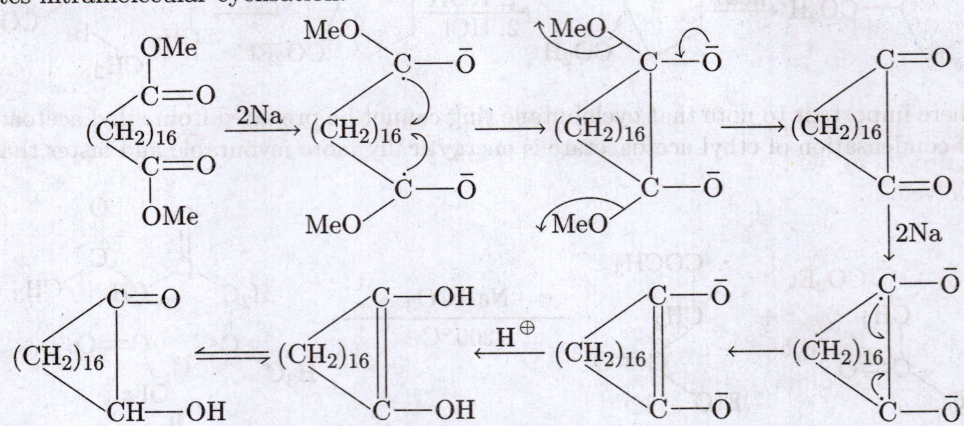

**(g) By the Diets-Alder reaction :** This is by far the best method of preparing six-membered alicyclic compounds.

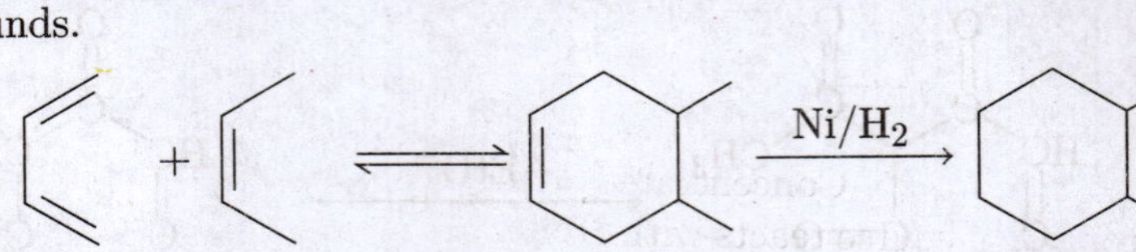

**(h) By addition of carbenes to alkenes :**

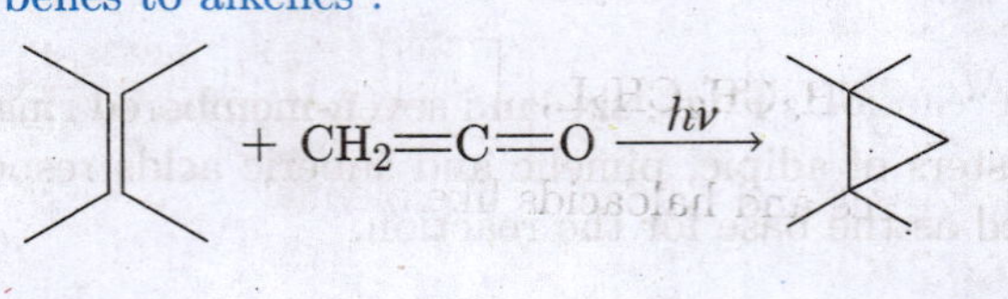

(i) By the Ziegler reaction : When a dinitrile is treated with diethyllithium amide in liquid ammonia, a Dieckmann type of condensation takes place to form an imminonitrile which may then be converted to a cyclic ketone by acid-catalysed hydrolysis and decarboxylation.

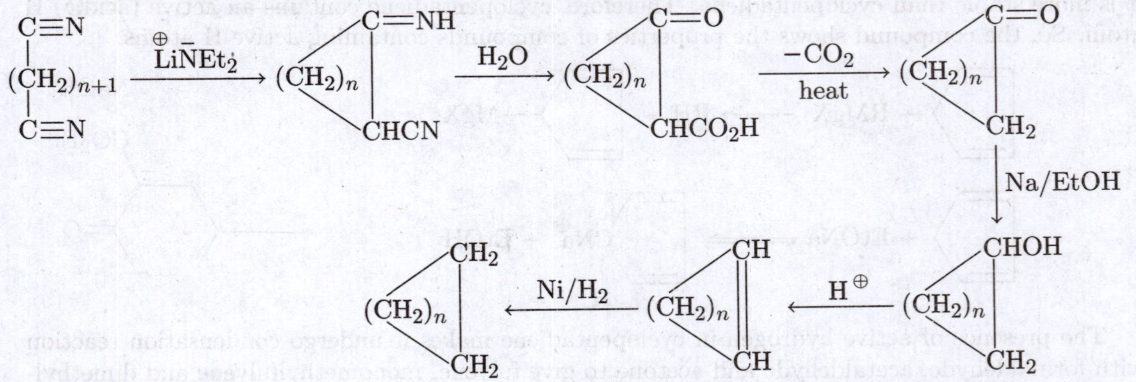


## 18.5.4 Structure and reactivity

Since cyclopropane and cyclobutane bear high angular strain, they are highly reactive and they behave like olefins in some cases; other cycloalkanes and cycloalkenes being stable, they behave like alkanes and alkenes respectively.

## 18.5.5 Chemical reactions

(a) Action of hydrogen : Hydrogen does not react with cycloalkanes except cyclopropane and cyclobutane; the latter two compounds form propane and $n$-butane respectively. Cycloalkenes add on hydrogen in the presence of catalyst to form cycloalkanes; the reaction is exothermic and heat is evolved.

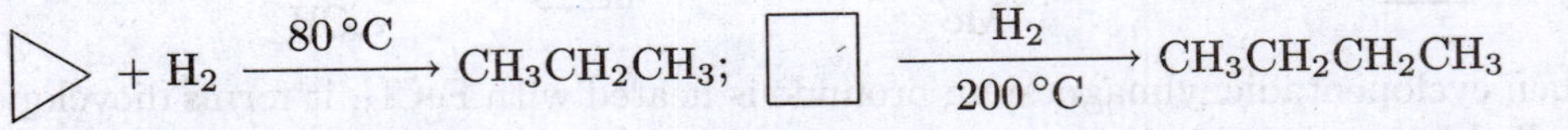


(b) Action of halogens : When chlorine or bromine is treated with cycloalkanes in the presence of catalyst, halogeno-cycloalkanes are obtained. Cyclopropane in this reaction gives a mixture of 1, 1-, 1, 2- and 1, 3-dihalogenopropanes.

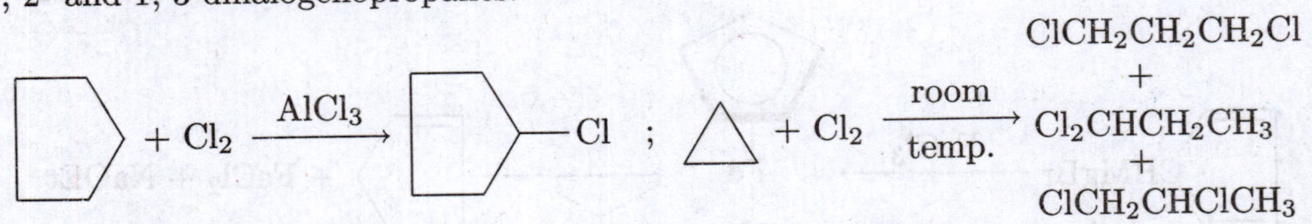


(c) Action of hydroiodic acid : Concentrated hydroiodic acid does not have any action on cycloalkanes. Cyclopropane of course reacts with HI at room temperature to give $n$-propyl iodide; cyclobutane on heating with HI gives $n$-butyl iodide.

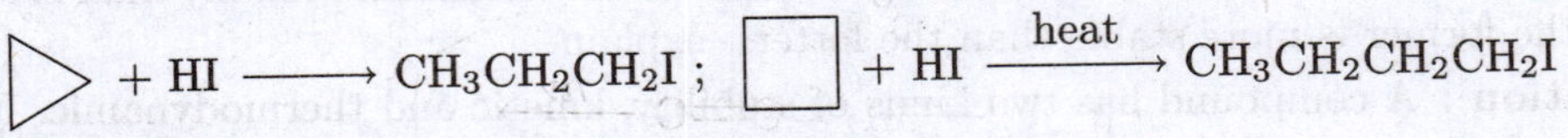


Cycloalkenes add on halogens and haloacids like olefins.

(d) Special properties of cyclopentadiene : Cyclopentadiene is a non-aromatic compound; but it has a tendency to give up a proton from the $sp^3$ hybridised C. This is because when cyclopentadiene donates a proton, cyclopentadienyl anion forms; this being an aromatic species, it is more stable than cyclopentadiene. Therefore, cyclopentadiene contains an active (acidic) H atom. So, the compound shows the properties of compounds containing active H atoms.

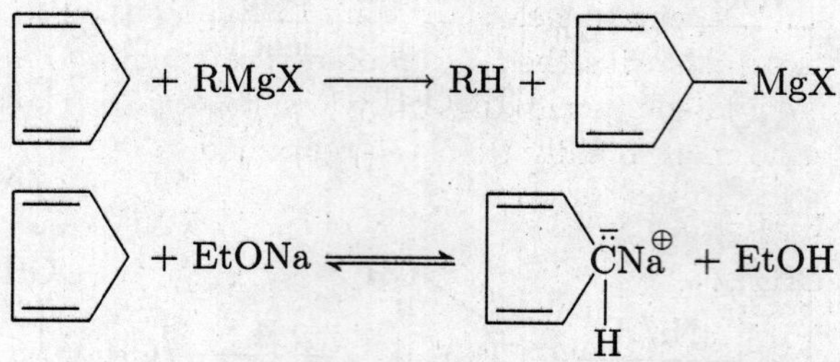


The presence of active hydrogen in cyclopentadiene makes it undergo condensation reaction with formaldehyde, acetaldehyde and acetone to give fulvene, monomethylfulvene and dimethylfulvene respectively. Fulvenes are isomeric with alkylbenzenes but they are unstable and possess colour.

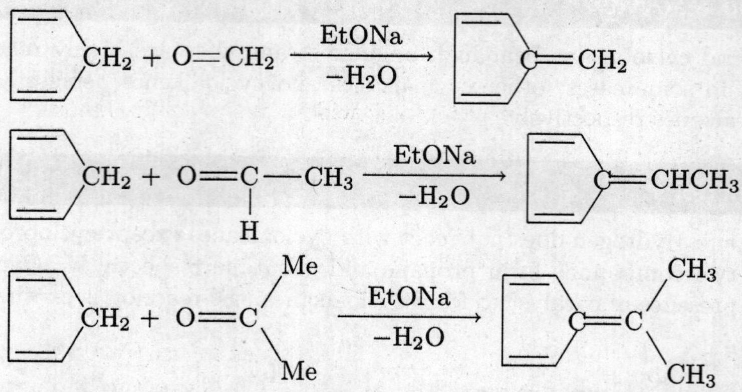


When cyclopentadienylmagnesium bromide is heated with $FeCl_3$, it forms dicyclopentadienyliron, called ferrocene; this is best prepared by the reacion between cyclopentadiene and ferrous chloride in the presence of anhydrous sodium ethoxide. This is an aromatic compound; the name ferrocene was given by Woodward et al. It is an orange solid; its mp is 173 °C.

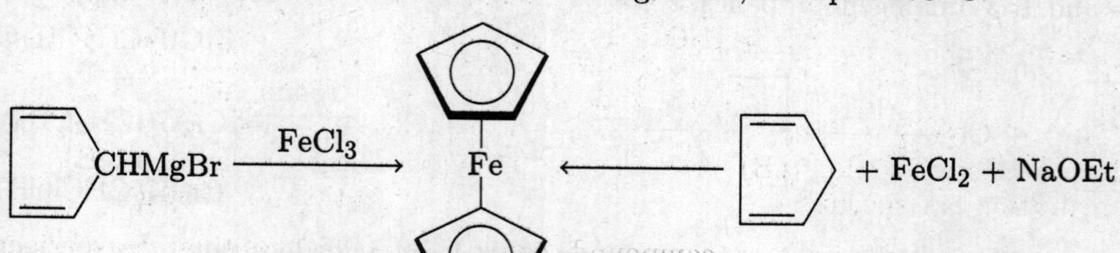


**Problem 18.5.** 1, 3-pentadiene undergoes addition reactions more readily than 1,4-pentadiene though the former is more stable than the latter—explain.

**Solution :** A compound has two kinds of stability, kinetic and thermodynamic. The kinetic stability of a compound depends largely on the energy content of its HOMO. The higher the energy of the HOMO, the higher is the reactivity and the lower is the kinetic stability. The HOMO of 1, 3-pentadiene is of much higher energy than that of 1, 4-pentadiene. So 1, 3-pentadiene is kinetically less stable than 1, 4-pentadiene. On the other hand, the thermodynamic stability of a compound depends on the total energy of the occupied M.O.s. The total energy of the occupied M.O.s of 1,3-pentadiene is less than that of 1,4-pentadiene. Hence 1,3-pentadiene is thermodynamically more stable than 1,4-pentadiene. According to the Frontier M.O. theory HOMO of an alkadiene (or an alkene) interacts with the LUMO of an electrophile to undergo the addition reaction. The HOMO of 1,3-pentadiene being kinetically less stable and being of higher energy, it undergoes the addition reaction more readily than 1,4-pentadiene.

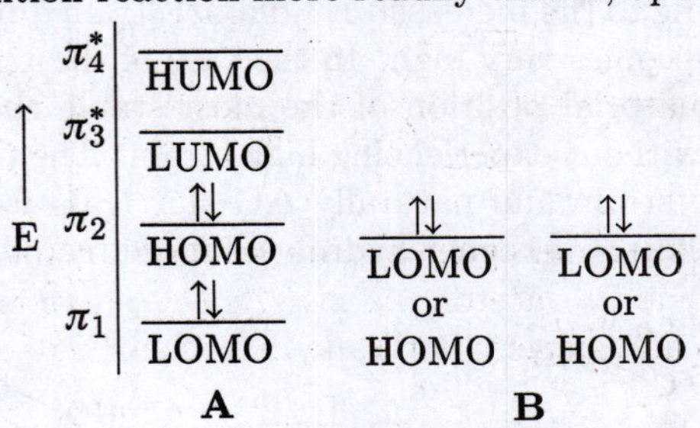


**A** - 1,3-pentadiene
**B** - 1,4-pentadiene

**Problem 18.6.** cis-4-t-Butylcyclohexanecarboxylic acid (1) is less acidic than its trans-isomer in the aqueous medium—explain.

**Solution :**

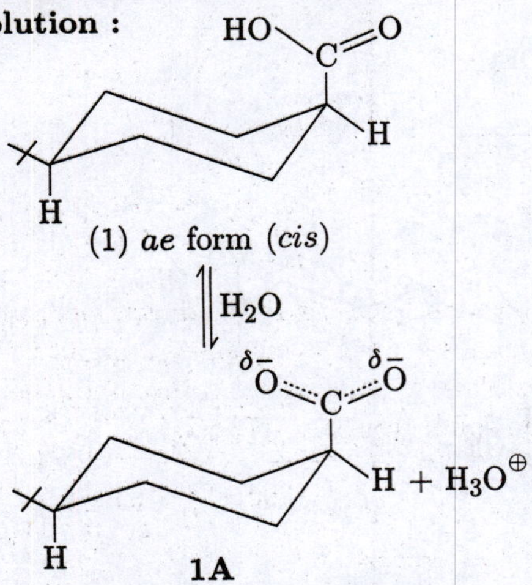


hydrated carboxylate ion (cB)
hydration not shown

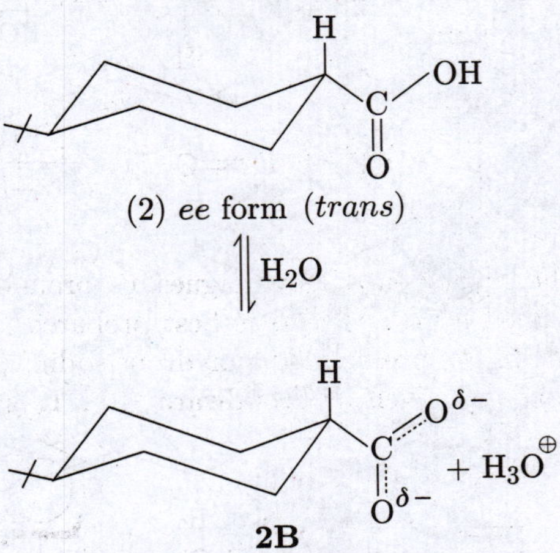


hydrated carboxylate ion (cB)
hydration not shown

The carboxylate group of the cis compound, being in the axial position, it experiences high 1,3-diaxial strain while that of the trans-isomer does not experience any strain. Naturally a $CO_2^-$ group cannot get hydrated since hydrated carboxylate group in the axial position will have to experience much more strain than the base $CO_2^-$ group; whereas $CO_2^-$ group in the equatorial position easily gets hydrated without accepting strain and thus gets stability through hydration. So, the cB of 1 is less stable than the cB of 2 and thus cB of 1 is more basic than the cB of 2. Hence 2 is more acidic than 1.

**Problem 18.7.** Which one of the following compounds hydrolyses more readily in the alkaline medium,—cis-4-t-butylcyclohexanecarbethoxylate or its trans-isomer?

**Solution :** The base-catalysed hydrolysis of carbethoxyl group follows $B_{AC}2$ pathway which involves a tetrahedral T.S. and a tetrahedral intermediate in the R.D. step. The cis-form of the given compound has the bulky carbethoxyl in the axial position of the most stable chair conformation and it experiences strong 1,3-diaxial interactions though the carbonyl C of the carbethoxyl group has a planar trigonal orientation. When $\bar{O}H$ attacks the electron-deficient carbonyl C in the R.D. step, a tetrahedral T.S. develops and the axial position becomes more strained than the substrate owing to the increased 1,3-diaxial strain during transition. Hence $\Delta G^{\ddagger}$ for cis-isomer of the reaction becomes very high. In the case of the trans-isomer the carbethoxyl group being present in the equatorial position of the most stable chair form where very bulky group can accommodate itself without experiencing high strain. The tetrahedral T.S. of the R.D. step does not bear appreciable strain and naturally $\Delta G^{\ddagger}$ for trans-isomer is much less strained than that of the cis-form. So the trans-isomer hydrolyses more readily than the cis-isomer.

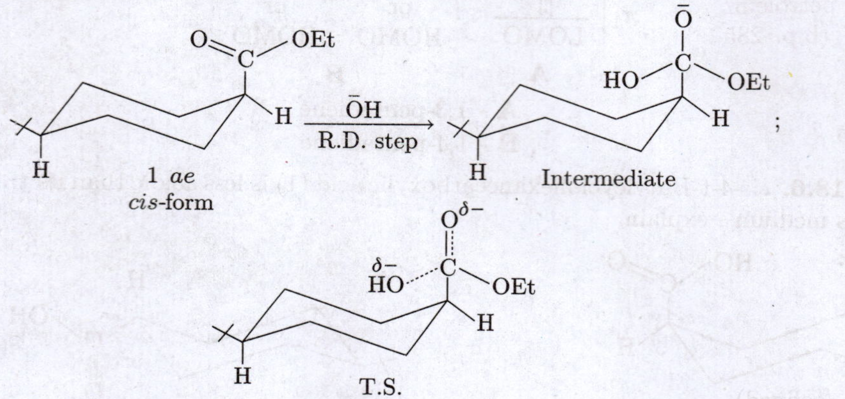


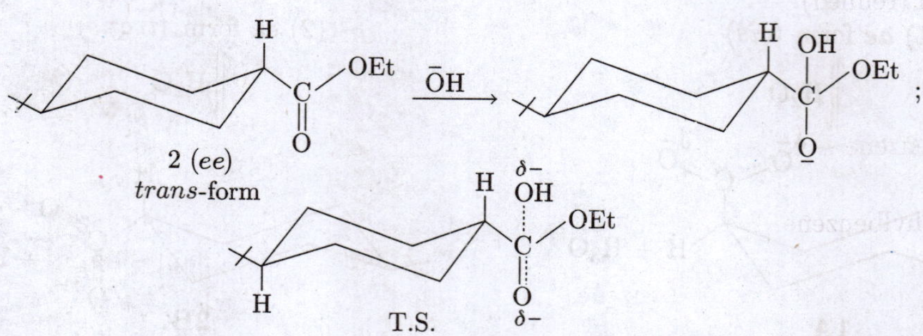


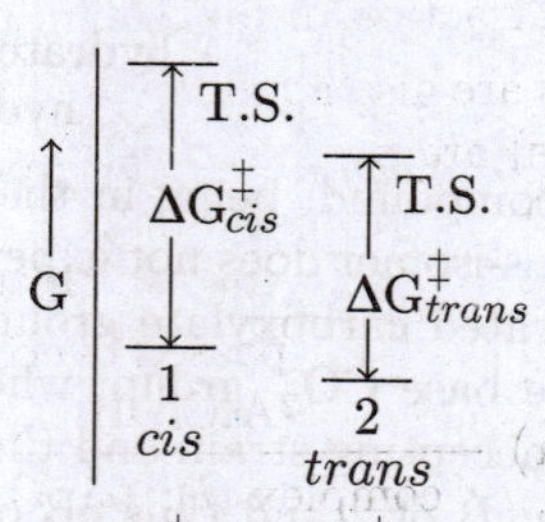


$\Delta G^{\ddagger}_{cis} > \Delta G^{\ddagger}_{trans}$

## 18.6 Benzene

### 18.6.1 Introduction

Arenes are aromatic hydrocarbons. Benzene is the simplest member of the class. Alkylbenzenes, naphthalene and alkylnaphthalenes, anthracene and alkylanthracenes, biphenyls, etc. are some of the arenes (See Table 3.6). Here we shall discuss the chemistry of benzene and alkyl benzenes.

### 18.6.2 Source

Coal gas, coal tar, petroleum oil.

$$\text{Coal gas} \xrightarrow[\text{petroleum fraction (b.p. 285-300 °C)}]{\text{washed with}} \text{oil soln.} \xrightarrow[\text{stripping}]{\text{heat}} \text{vapour} \xrightarrow{\text{condensed}} \text{Benzene + Toluene}$$

Coal tar —high temperature distillation→

- up to 200 °C → light oil ⟶ Benzene, toluene, naphthalene
- 200-250 °C → middle oil ⟶ Naphthalene
- 250-350 °C → heavy oil ⟶ Naphthalene
- 300-350 °C → anthracene oil ⟶ Anthracene, phenanthrene

$$\text{Petroleum (refined)} \xrightarrow[\text{distillation 0-200 °C}]{\text{fractional}} \text{gasolines} \xrightarrow[\text{Pt. cat.}]{\text{reforming}} \text{reformate}$$

reformate —selective extraction with ethylene glycol.→ extract

$$\text{extract} \xrightarrow[\text{distillation}]{\text{fractional}} \text{benzene + toluene + } o\text{-}, m\text{-}, p\text{-xylenes}$$

+ ethylbenzene, 1, 2, 3-trimethylbenzene,
+ 1, 2, 4, 5-tetramethylbenzene.

Mesitylene is obtained from the $C_9$ fraction of petroleum reformates.

However, some of the alkylbenzenes are also commercially produced from benzene; e.g., ethylbenzene and isopropylbenzene (cumene) are produced as shown below :

(a) 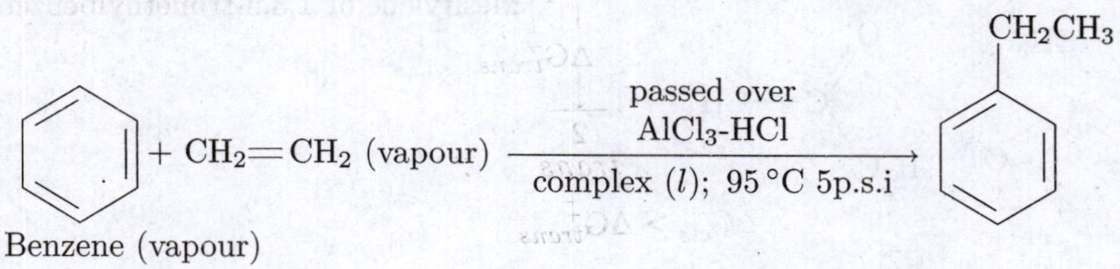

(b) 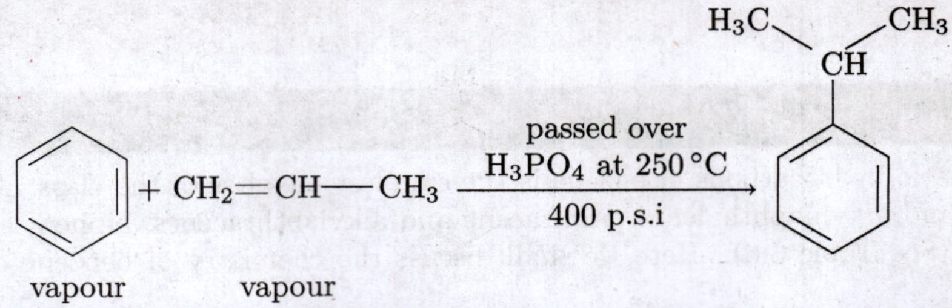


## 18.6.3 Formation of benzene homologues

The methods which will be discussed here may not have preparative value but they will represent relations between benzene homologues and benzene.

(a) **By polymerisation :**

(i) 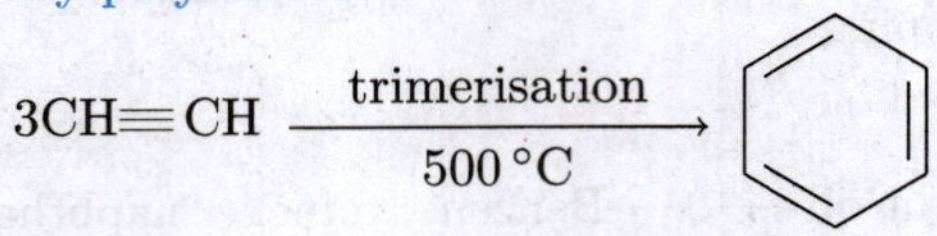


(ii) 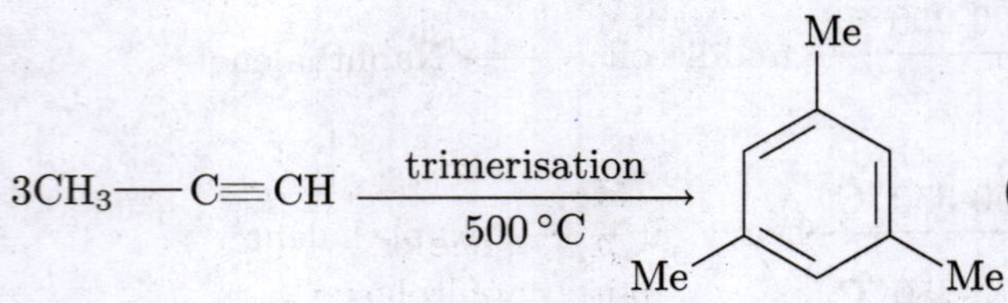


(Mesitylene)

(iii) 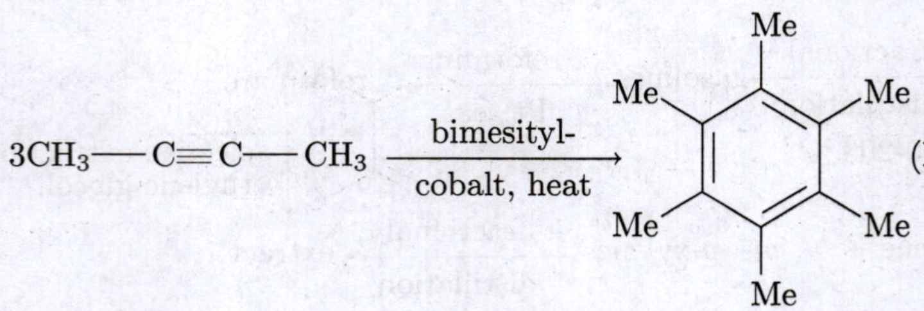


(Hexamethylbenzene)

(iv) 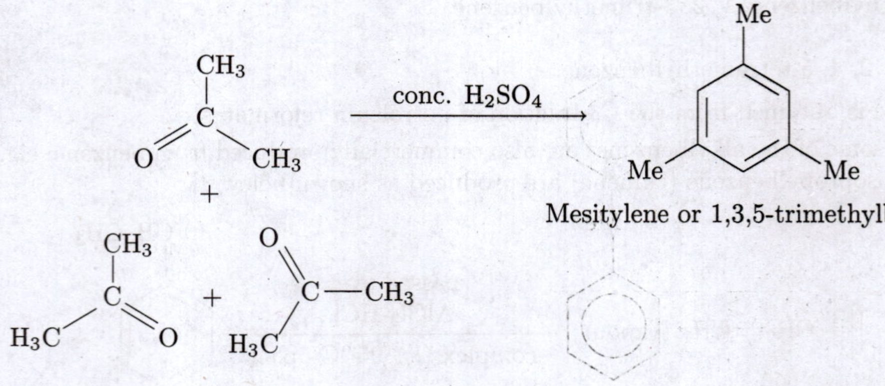


Mesitylene or 1,3,5-trimethylbenzene

(b) By dehydrogenation-cyclisation and aromatisation :

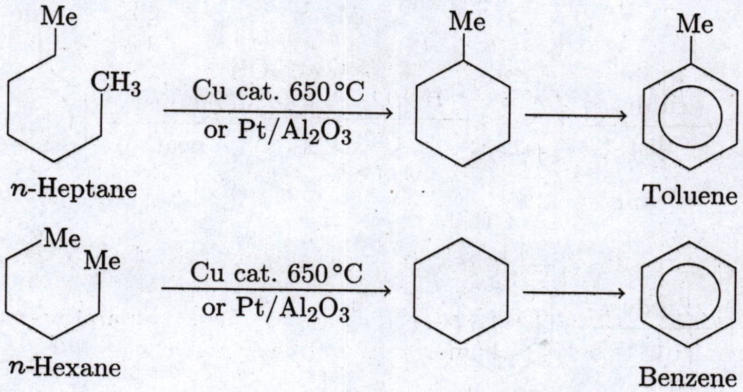


(c) By dehydration :

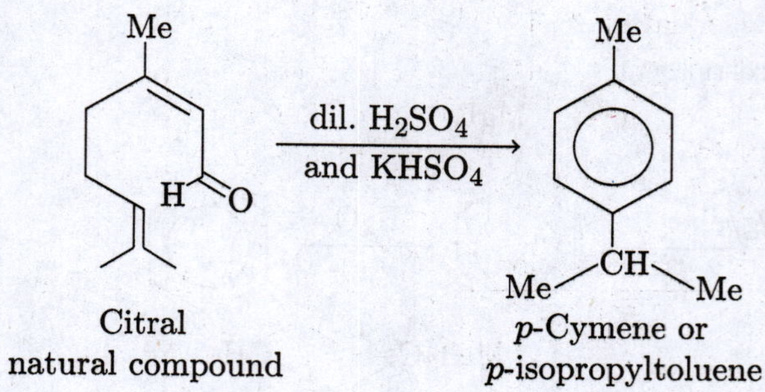


(d) By the Diels-Alder reaction :

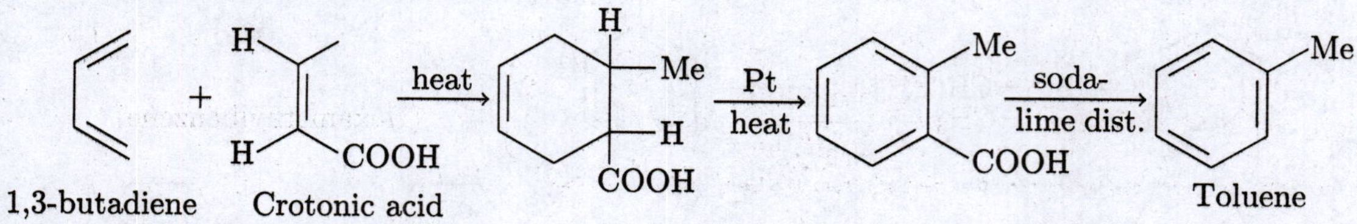


(e) From carboxylic acid salts :

(i) 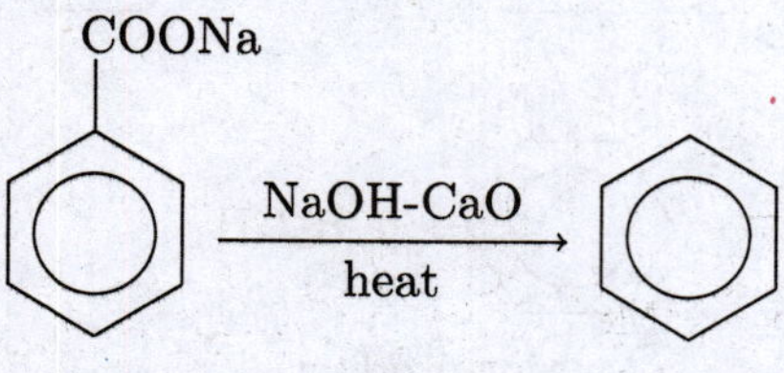


(ii) 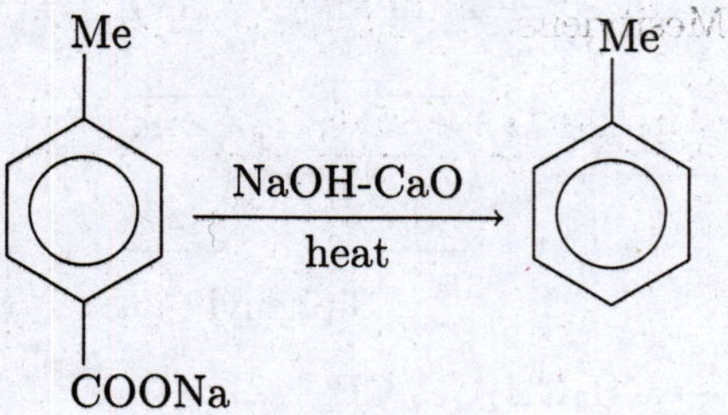

(f) From phenols :

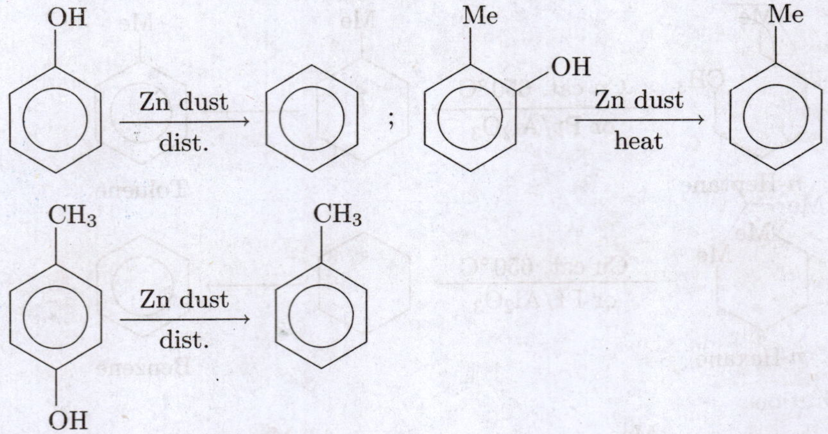


(g) By the Grignard reagent :

(i) 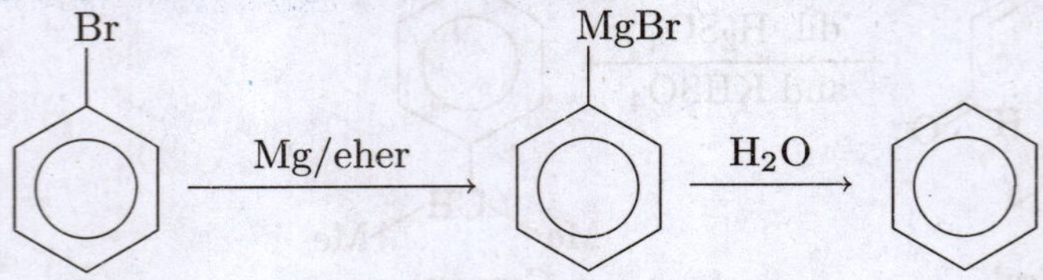


(ii) 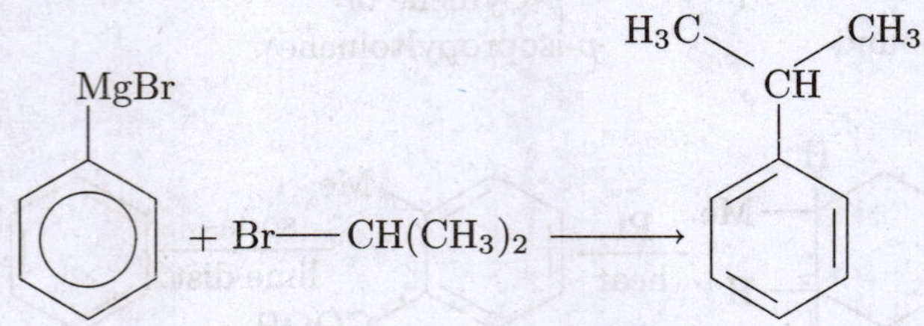


(iii) 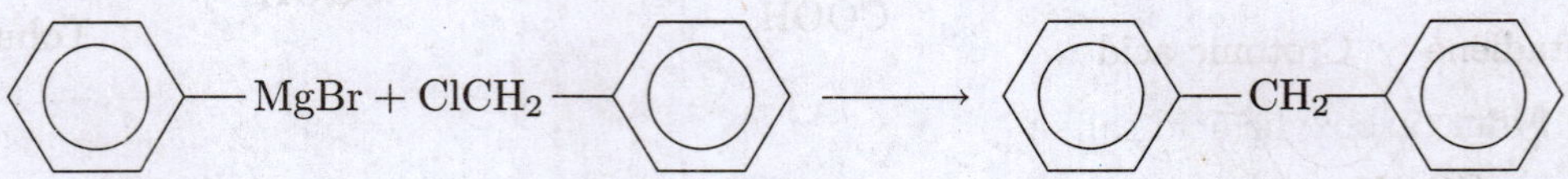


(h) By the Wurtz-Fittig reaction :

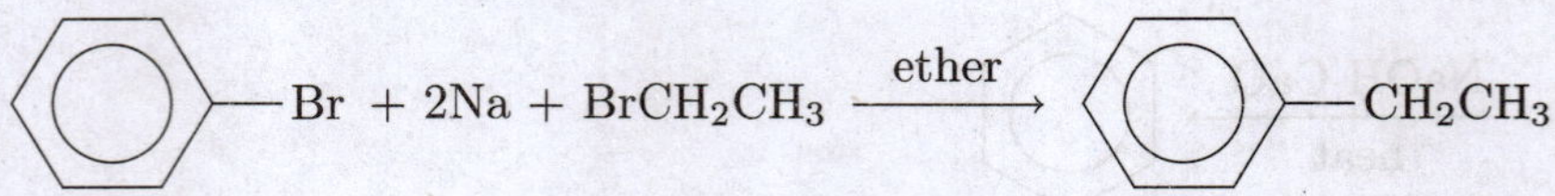


major product

Side products :

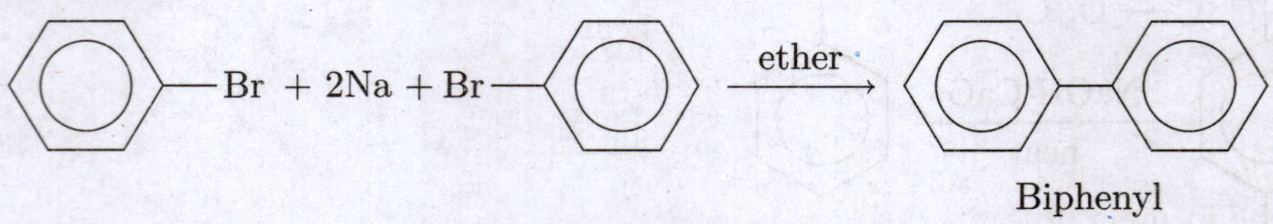


Biphenyl

$$CH_3CH_2Br + 2Na + BrCH_2CH_3 \xrightarrow{\text{ether}} CH_3CH_2CH_2CH_3$$

(i) By the process of desulphonation :

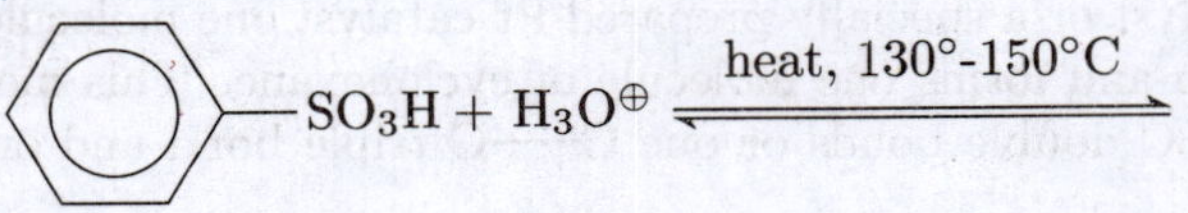


(j) By the reduction of arenediazonium ion :

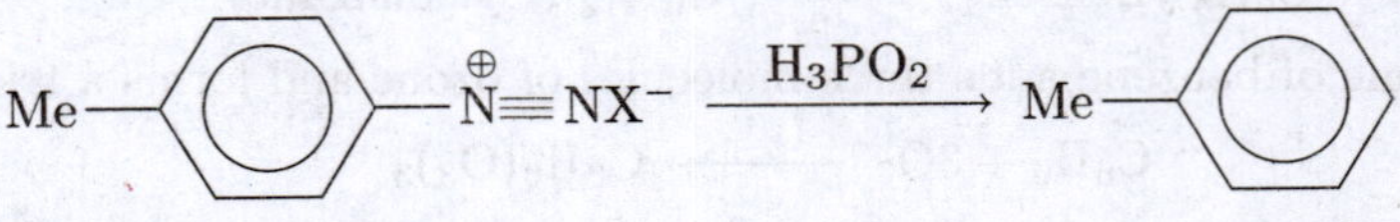


(k) By the reduction of aryl ketones :

(i) 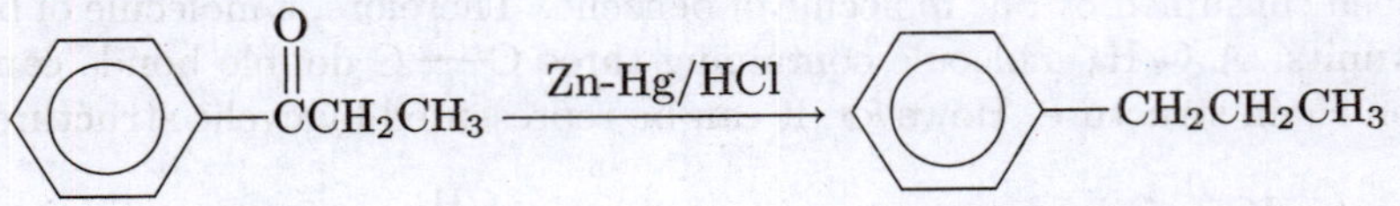


*n*-propylbenzene

(ii) 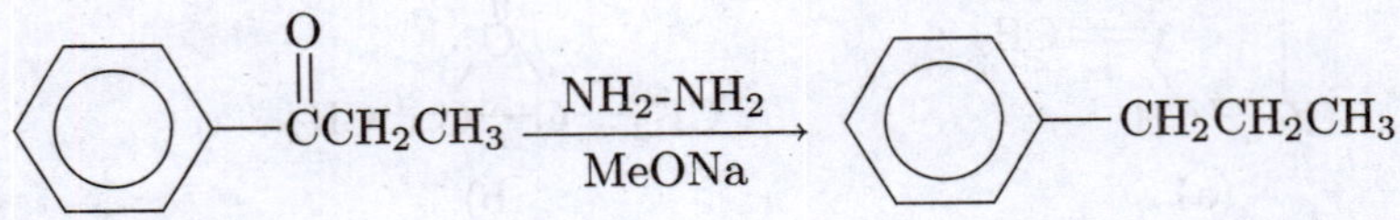


(l) By the Friedel-Crafts reaction :

(i) 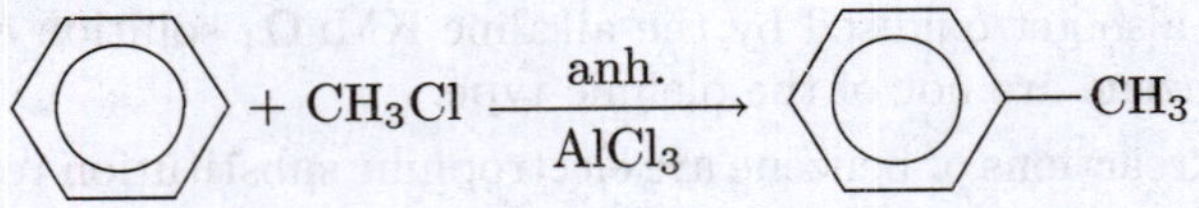


(ii) 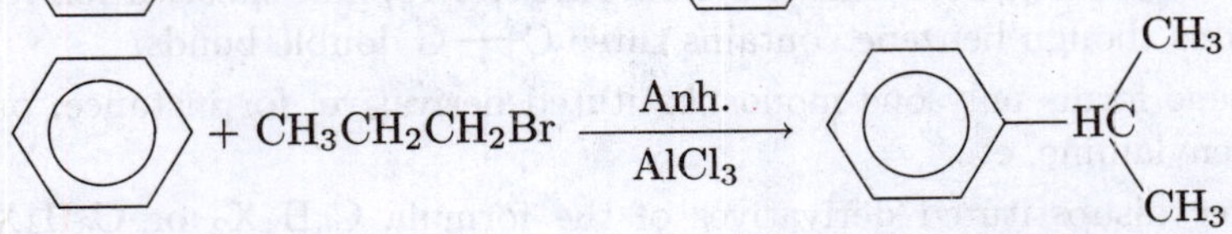


cumene

(m) By the Ullmann reaction :

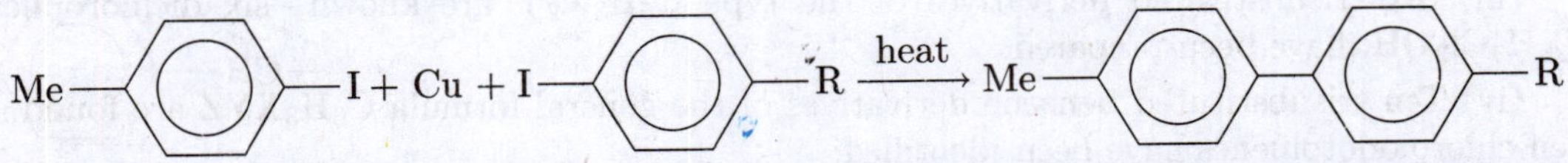


## 18.6.4 Elucidation of structure of benzene

Elemental analysis and vapour density measurements show that the molecular formula of benzene is $C_6H_6$. The molecular formula of a saturated open chain hydrocarbon consisting of six Cs is $C_6H_{14}$. So, benzene is highly unsaturated. The double bond equivalent (D.B.E.), which means number of molecular hydrogen to be taken in by the compound owing to the presence of C—C, C—O, C—N double bonds, C—C, C—N triple bonds and rings. D.B.E. of benzene is found to be four according to the formula as shown below.

$$\text{D.B.E.} = a + 1 - \left(\frac{b-c}{2}\right)$$

where the formula of the compound is $C_aH_bN_cO_d$. If there are halogen atoms, the term '$b$' will be the sum of $b+$ number of halogen atoms.

Benzene does not add hydrogen as readily as simple alkenes. However, in a pressure apparatus in the presence of a Ni catalyst or a specially prepared Pt catalyst one molecule of benzene adds three molecules of hydrogen and forms one molecule of cyclohexane. This means that benzene possesses either three C—C double bonds or one C—C triple bond and one C—C double bond in a cyclic structure.

$$C_6H_6 \text{ (Benzene)} \xrightarrow{3H_2} C_6H_{12} \text{ (Cyclohexane)}$$

Again, one molecule of benzene adds three molecules of ozone and forms a triozonide.

$$C_6H_6 + 3O_3 \longrightarrow C_6H_6(O_3)_3$$

If benzene were a compound containing one C≡C unit and one C=C unit, only two molecules of ozone would have been consumed by one molecule of benzene. Therefore, a molecule of benzene contains three C=C units. A $C_6H_6$ molecule containing three C—C double bonds cannot be represented by an open chain structure. However, it can be represented by cyclic structures, e.g.,

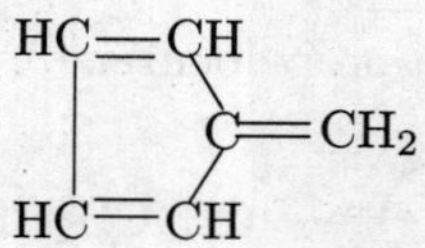

(a)

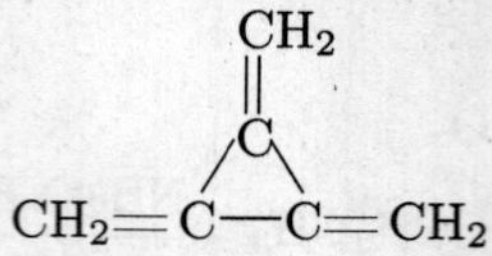

(b)

When benzene is shaken with bromine in carbon tetrachloride, it does not decolourise the brown colour of bromine. It does not also get oxidised by the alkaline $KMnO_4$ solution even on heating. Thus the C=C bonds in benzene are not of the olefinic type.

Over and above these, the common reactions of benzene are electrophilic substitution reactions instead of addition reactions though benzene contains three C—C double bonds.

Furthermore, (i) benzene forms only one monosubstituted derivative; for instance, only one halobenzene, only one phenylamine, etc.

(ii) It forms only three disubstituted derivatives of the formula $C_6H_4X_2$ or $C_6H_4XY$; for example, there are three isomeric dinitrobenzenes and three isomeric nitrotoluenes.

(iii) Six trisubstituted derivatives of the type $C_6H_3X_2Y$ are known—six dichlorophenols, $C_6H_3Cl_2OH$, have been prepared.

(iv) Ten trisubstituted benzene derivatives of the general formula $C_6H_3XYZ$ are found, e.g., ten chloroiodotoluenes have been identified.

All these mean that the isomer numbers of monosubstituted benzene is one, those of disubstituted benzene is three and so on.

The above, written hypothetical cyclic formulae for benzene, *a* and *b*, fail to explain the isomer numbers of substituted benzenes and inertness of benzene toward bromine in carbon tetrachloride, alkaline $KMnO_4$, etc.

However, in 1865 Kekule′ was the first to suggest arbitrarily a ring structure (1) for benzene as given below (1). He had a belief that the structure could explain :

**1**

(i) That benzene contains three double bonds,

(ii) that all the C atoms in benzene are equivalent and consequently there should be only one monosubstituted derivative and three disubstituted derivates of benzene.

Of course, the then scientists raised a few objections against the proposed structure, e.g.,

(i) the proposed structure should add bromine under mild condition,

(ii) it should reduce alkaline $KMnO_4$ and

(iii) the proposed structure should give four disubstituted derivatives instead of three since $C_1$—$C_2$ bond is different form $C_1$—$C_6$ bond; the former has a single bond character, while the latter has a double bond character.

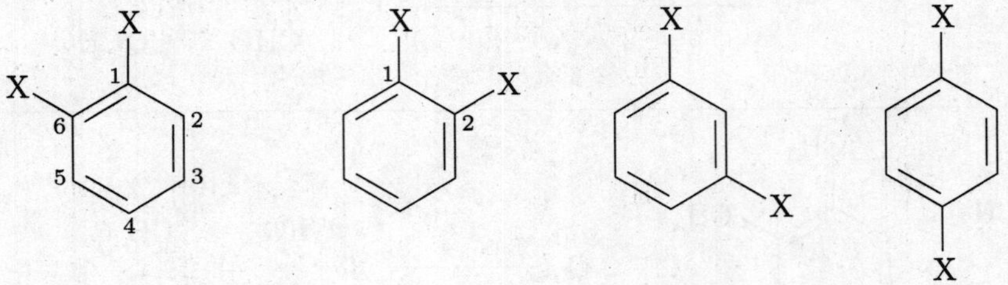


Then Claus (2) Dewar (3), Ladenberg (4) and others (5 & 6) proposed several structures for benzene as given below :

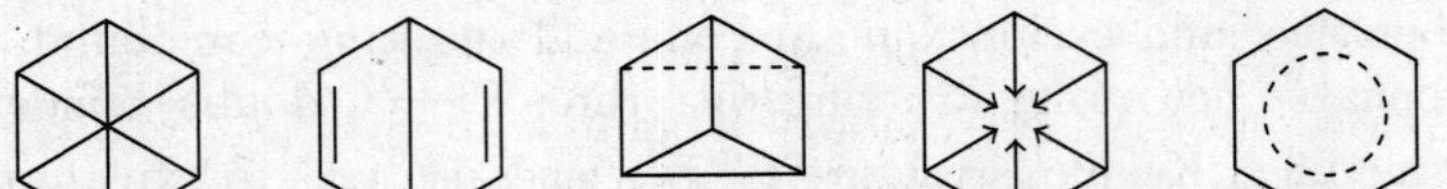

**2** **3** **4** **5** **6**

However, these structures also failed to explain the peculiar properties of benzene and, therefore, the structures were discarded.

In 1872 Kekulé pointed out that the C atoms in benzene are in a state of continuous vibration owing to which each C—C bond sometimes behaves as a C—C single bond and sometimes behaves as a C=C double bond. Thus the two structures (1a and 1b) are in a state of continuous oscillation and the structure of benzene is neither represented by 1a nor by 1b but it is a combination of the two.

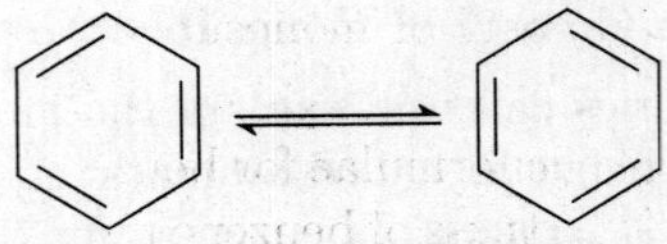

**1a** **1b**

If the modified structure for benzene, as suggested by Kekulé, is considered to be correct, then theoretically one should get three molecules of glyoxal, one molecule of diacetyl and two molecules of pyruvaldehyde form the ozonolysis experiment on two molecules of *o*-xylene. In fact at that time two scientists got these from the ozonolysis of two molecules of *o*-xylene.

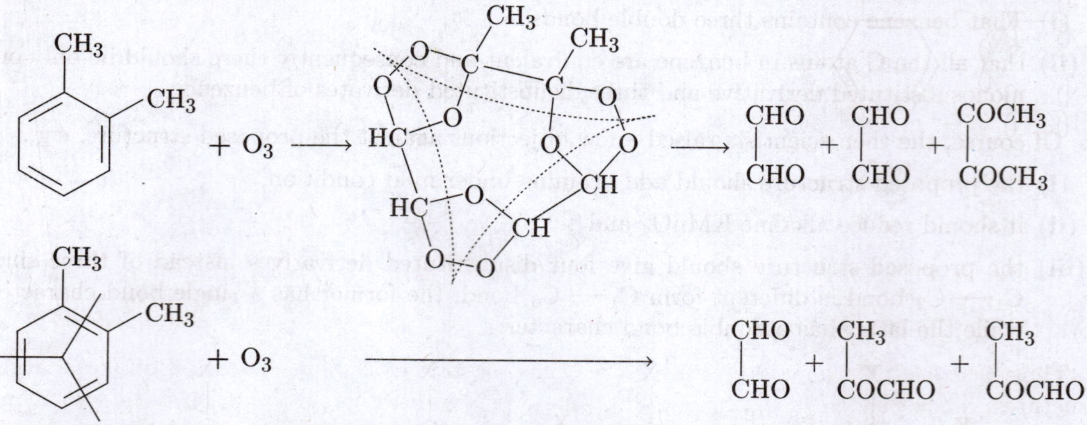

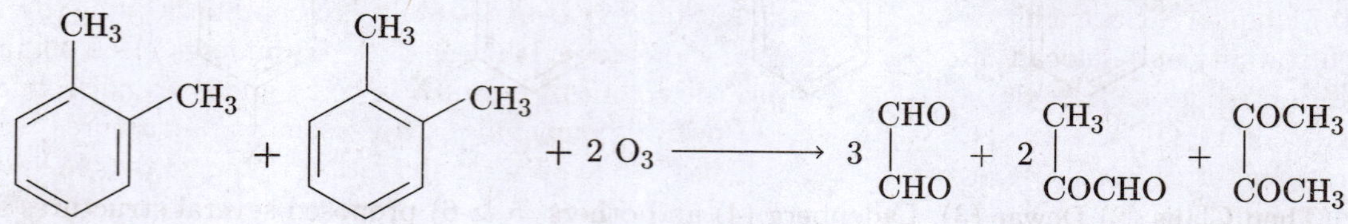

Furthermore, Baeyer proved that the compound obtained from the hydrogenation of benzene, hexahydrobenzene, and cyclohexane are one and the same compound. This experiment established that benzene has a ring structure with three C—C double bonds.

However, the modified Kekule′ structure for benzene also fails to explain the extra stability of benzene, its resistance toward oxidation, etc.

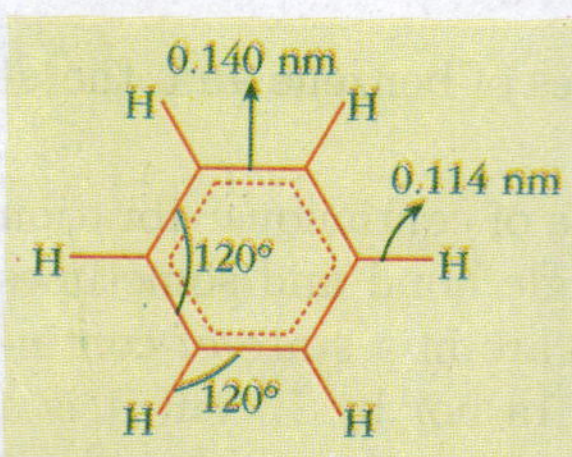

**Fig. 18.2. Geometry of benzene**

Nowadays X-ray diffraction and electron-diffraction studies have led scientists to conclude that benzene has a regular hexagonal planar structure in which the six C atoms are located at the corners of the flat regular hexagon and the six H atoms, one on each C atom, lie on the plane of the ring; the six C—C bond lengths are neither equal to C—C single bond length (0.154 nm) nor they are equal to C—C double bond length (0.134 nm) but they have identical bond lengths of 0.140 nm. The six C—H bond lengths are also same and it is 0.114 nm; the C—C—C and H—C—C bond angles are around 120°.

Both the V.B. and M.O. Theories can well explain the modern structure of benzene as well as the characteristic properties of benzene, called aromatic characters (Sec. 4.7).

To explain the modern structure of benzene the V.B. Theory considers that all the C atoms in benzene are in $sp^2$ hybridised state. Each C atom has three $sp^2$ A.O.s and one $p$ A.O. Each A.O., $p$ or $sp^2$, contains one electron. The $p$ A.O. is perpendicular to the plane of three $sp^2$ A.O.s which are oriented in space in a plane trigonally. Each C atom utilises its three $sp^2$ A.O.s to form two $sp^2$—$sp^2$ $\sigma$ bonds with two adjacent Cs and one $sp^2$—s $\sigma$ bond with an H atom. Thus six C atoms form a planar ring in which C—H bonds are also co-planar. Thus the six $p$ orbitals of the six C atoms are parallel to each other and perpendicular to the plane of the molecule. Any two such $p$ orbitals can form a $\pi$ bond by side-on overlap. In this way three $\pi$ bonds may form in five different ways to give rise to five different structures.

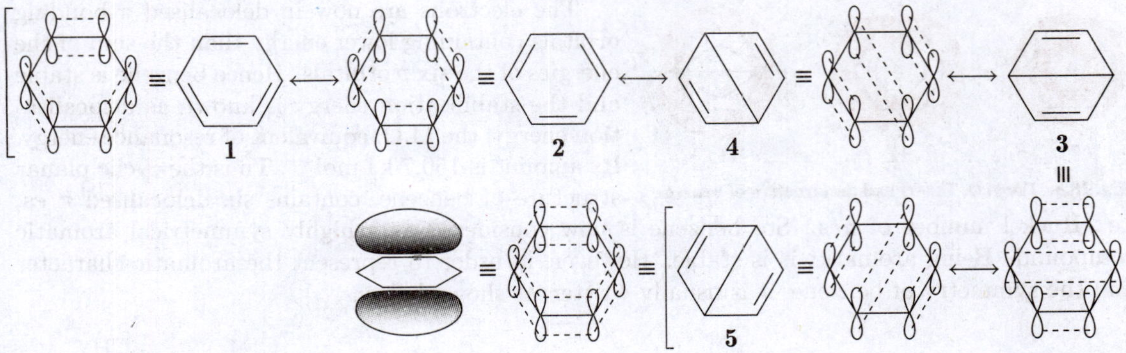

Thus, benzene may be considered as a resonance hybrid of the above-written five resonating structures. Since the delocalisation of $\pi$ *es* occurs over a wide range involving isovalent resonating structures, benzene is expected to be very stable. In fact, the resonance energy of benzene is 150.7 kJ mol$^{-1}$. Such a high amount of resonance energy brings stability to benzene. Owing to the high stability and delocalisation of $\pi$ *es* over a wide range, benzene does not undergo electrophilic addition reactions (i.e., does not decolourise the colour of bromine in $CCl_4$ and does not reduce alkaline $KMnO_4$ solution etc., see later). However, among the five resonating structures, the para-bond structures (3,4,5), the Dewar structures, have less contribution than the two Kekule′ structures (1 and 2) because of the distorted bond length in the former structures. The two Kekule′ structures being equivalent and each C—C single bond in one structure being a C—C double bond in the other structure, all the C—C bonds have identical lengths and it is 0.140 nm. All the C atoms being $sp^2$ hybridised and each C—C and C—H bonds being formed by $sp^2$—$sp^2$ and $sp^2$—$s$ overlap respectively, the C—C—C and H—C—C bond angles are around 120°.

To describe the structure of benzene the M.O. Theory also considers the $sp^2$ hybridisation of the C atoms. The C—C and C—H $\sigma$ bonds are supposed to form in the same way as we have discussed in the V.B. Theory. Therefore, the structure of benzene will be planar hexagonal on which all the C and H atoms will be in one and the same plane. The M.O. Theory further considers that the six *p*. A.O.s overlap to form six delocalised $\pi$ M.O.s—three bonding and three antibonding. The three bonding orbitals will accommodate the six electrons. Fig. 18.3 shows the energy scheme for the $\pi$ electrons in benzene.

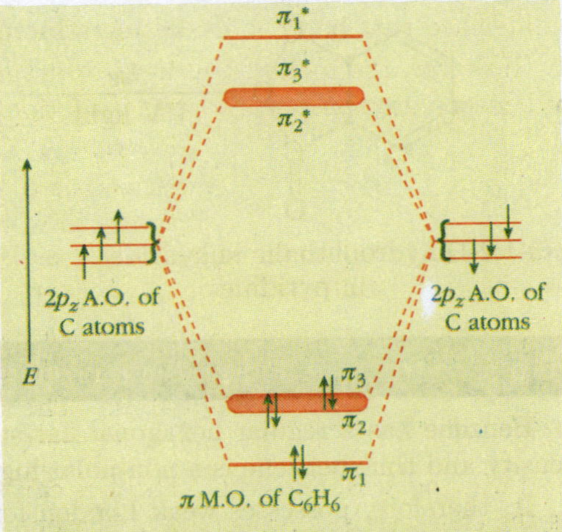


**Fig. 18.3. The M.O. energy diagram of benzene molecule**

The bonding M.O. of the lowest energy ($\pi_1$) is cyclic, it embraces all the six C atoms. $\pi_1$ M.O. has a nodal plane, the plane of carbon ring, and two lobes—one below the nodal plane and the other above it; $\pi_2$ and $\pi_3$ are degenerated M.O.s. They too encompass all the six C atoms. Each of them has another nodal plane which is perpendicular to the plane of the ring. Each of the degenerated M.O.s has four lobes. Altogether the 3 $\pi$ M.O.s has ten lobes. The net result of combination of the ten lobes is an annular electron-cloud; i.e., negative charge sheet, below and above the plane of the ring (Fig. 18.4).

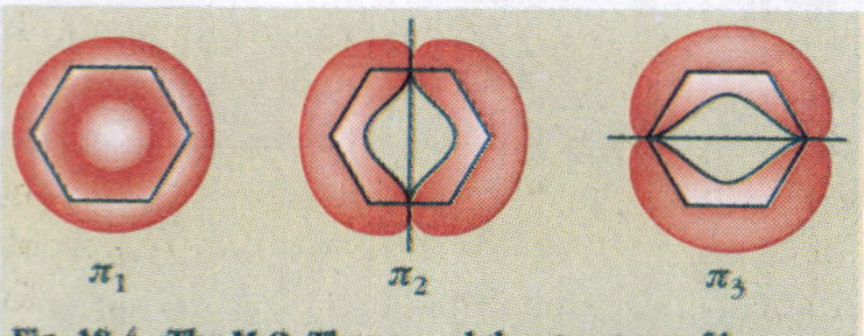

**Fig. 18.4. The M.O. Theory and the structure of benzene**

The electrons are now in delocalised $\pi$-bonding orbitals containing lower energy than the sum of the energies of the six $p$ orbitals. Hence benzene is stable and the stabilisation energy is known as delocalisation energy, the M.O. equivalent of resonance energy. Its amount is 150.7 kJ $mol^{-1}$. Thus the cyclic planar structure of benzene, contains six delocalised $\pi$ es, i.e., Huckel number of $\pi$es. So, benzene is now considered as a highly symmetrical aromatic compound. Being aromatic, it is stable. However, in order to represent the aromatic character and the symmetry of benzene, it is usually written as shown below.

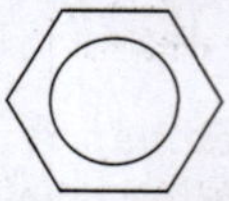

If benzene at all undergoes addition reactions with electrophiles, it will have to utilise two $\pi$es and then there will be only 4 $\pi$ es in its possession. Thus its aromatic character, i.e., its stability will be lost. At the cost of stability it does not usually undergo addition reactions. On the other hand, in a substitution reaction a C—H $\sigma$ bond breaks and a C—E ($E^{\oplus}$ is an electrophile) $\sigma$ bond forms and Huckel numbers of $\pi$ es remain intact i.e., its stability is retained. For this reason the common reactions of benzene are electrophilic substitution reactions.

**Note :** Recently bicyclo [2, 2, 0]hexa-2, 5-diene, the non-planar species corresponding to the Dewar's para bond benzene has been prepared by the following method :

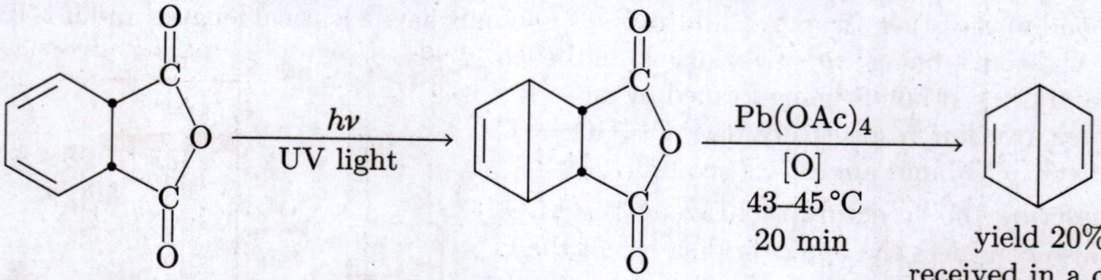


*cis*-1,2-Dihydrophthalic anhydride in pyridine

## 18.6.5 Structure and physical properties

Benzene has a regular hexagonal flat structure in which all the C atoms have equal charge density and thus benzene is a non-polar highly symmetrical molecule.

It, therefore, possesses weak London force. For this reason one should expect its very low m.p.; but its m.p. is 5.5 °C. Monoalkylbenzenes, *o*- and *m*-dialkylbenzenes have dipole moments because of the +I and hyperconjugative effects. *p*-dialkylbenzenes have lower dipole moments than their *o*- and *m*-isomers. The dipole moment of benzene is 0.0D. All these lead us to think that (i) alkylbenzenes should melt at higher temperatures than benzene and (ii) *p*-isomers should melt at lower temperatures than their *o*- and *m*-isomers. Here it is important to note that the more symmetrical a molecule, the more strongly it will fit in a crystal lattice and the better the molecule will fit, the higher will be the m.p.. For this reason the m.p. of benzene is higher than the m.p.s of toluene (−95 °C), *o*-xylene (−25 °C) and *m*-xylene (−48 °C). *p*-xylene being symmetrical and of higher molecular weight than benzene, it has higher m.p. than benzene as well as than *o*-

and *m*-xylenes. In fact, all *p*-dialkyl isomers have higher m.p.s than their *o*- and *m*-isomers. In so far as b.p. is concerned, it increases regularly with the increasing molecular weight. The rise in b.p. per extra —$CH_2$— group is about 20-30 °C. Thus the b.p.s of benzene, toluene and *o*-xylene are 80 °C, 111 °C, 144 °C respectively. Being non-polar or slightly polar, benzene and alkylbenzenes are almost insoluble in water; whereas for the same reason they are highly soluble in organic solvents. Being liquid, they behave as good solvents for organic compounds; e.g., for fats and oils. Almost all alkylbenzenes are less dense than water; they float on it. Aromatic hydrocarbons possess a characteristic smell, known as aromatic smell. These compounds do not have any colour.

## 18.6.6 Structure, reactivity and orientation

[See Secs. 17.3 and 17.4]

## 18.6.7 Chemical reactions

(a) Addition of hydrogen : In the presence of catalysts, Ni , Pt, etc., hydrogen adds to benzene and alkylbenzenes to form alicyclic compounds; the reaction is exothermic and heat evolves.

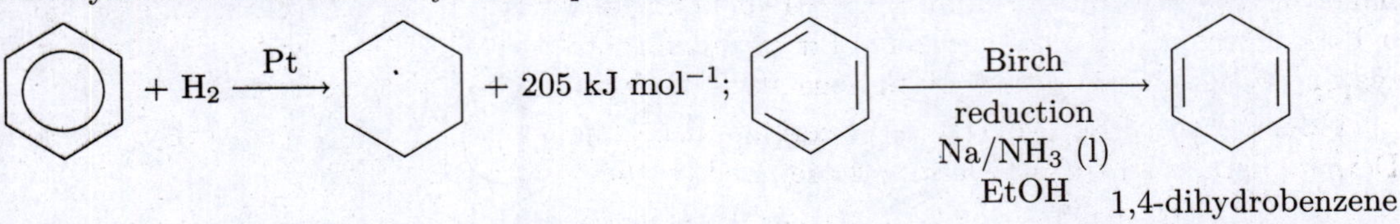


(b) Addition of halogens : Chlorine and bromine add to benzene and alkylbenzenes when a mixture of the halogen and the arene is irradiated with UV light.

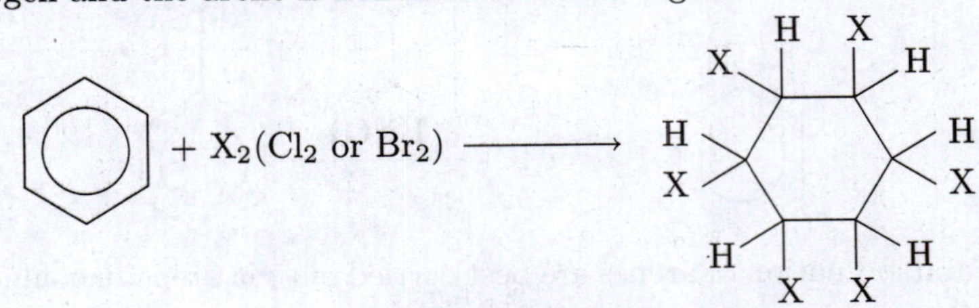


a mixture of isomeric hexahalobenzenes

(c) Addition of ozone : Three molecules of ozone add to one molecule of benzene and alkylbenzenes to form triozonide which on decomposition forms dicarbonyls as usual.

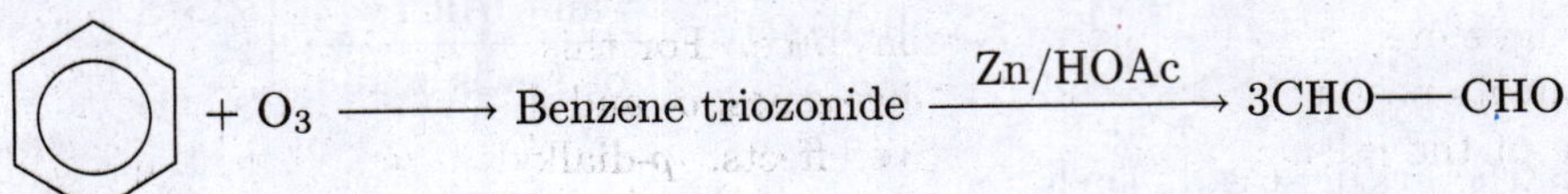


(d) Halogenation :

(i) *Chlorination and bromination :* In the presence of Fe or $I_2$ (catalyst), chlorine and bromine react with benzene and alkylbenzenes to form haloarenes. When 1 mole $Cl_2$ (or $Br_2$) is used per mole of the arene, monosubstitution takes place; in the cases of monoalkyl benzenes, *o*- and *p*-isomers are formed. When two moles of $Cl_2$ (or $Br_2$) are used per mole of the arene, *o*- and *p*-dihaloarenes are formed.

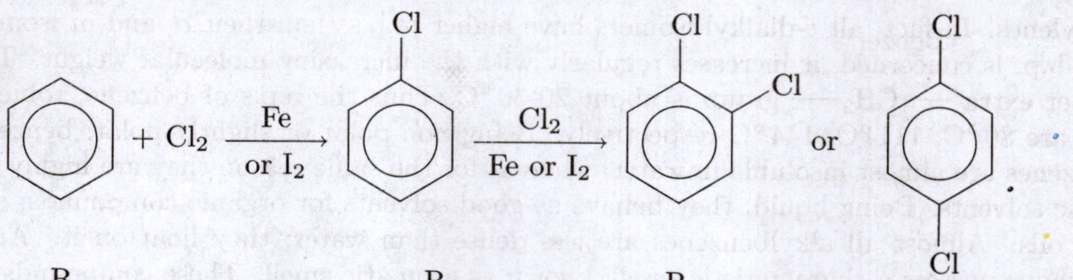

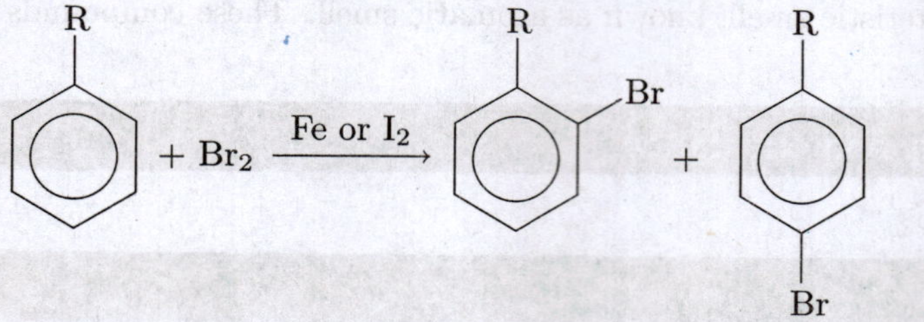

(ii) *Fluorination :* When benzene is treated with xenon difluoride and hydrogen fluoride, monofluorobenzene is obtained.

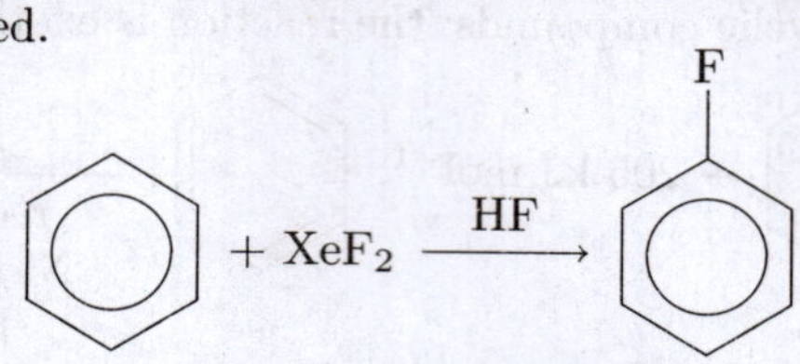

(iii) *Iodination :* Iodination is carried out in the presence of an oxidising agent. e.g., $HNO_3$, HgO, etc.

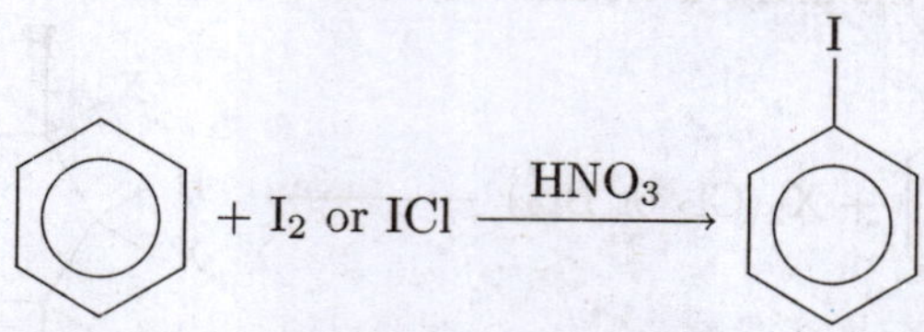

Fluorination and iodination of arenes are best carried out via arenediazonium ion (Sec. 20.18).

(iv) *Alkylation :*

(i) The Friedel-Crafts alkylation :

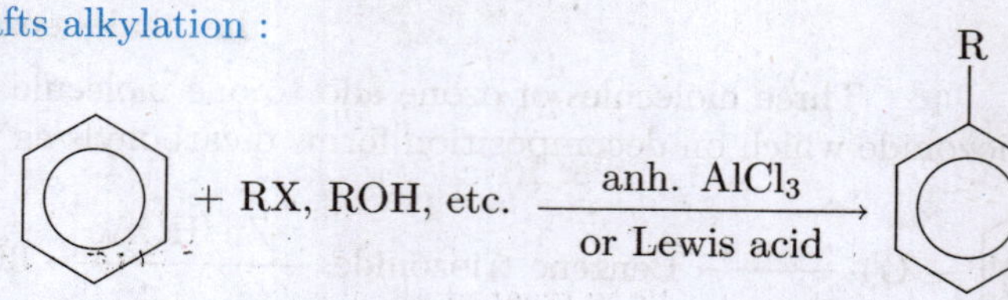

(ii) Via the Friedel-Crafts acylation :

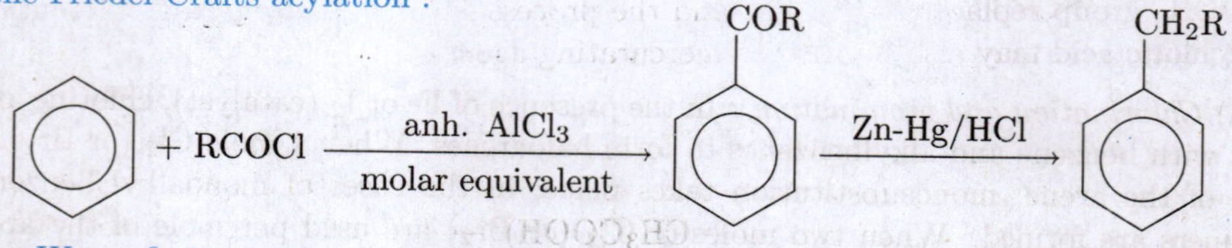

(iii) By the Wurtz-fittig reaction : See Sec. 18.6.3h.

**(e) Acylation :** Benzenes and alkylbenzenes may be acylated by the Friedel-Crafts acylation method.

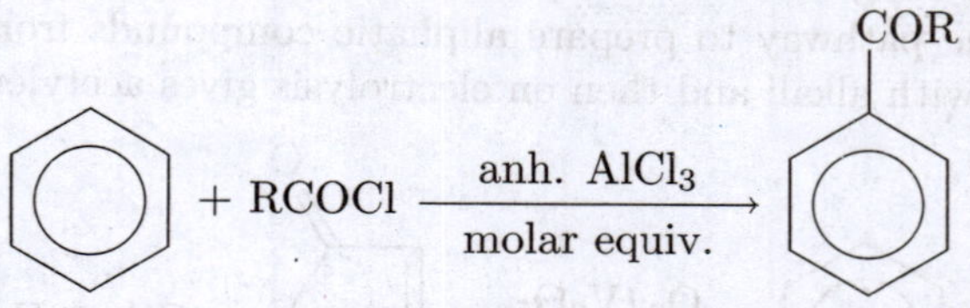


**(f) Nitration :** Several nitrating agents may be used to nitrate arenes.

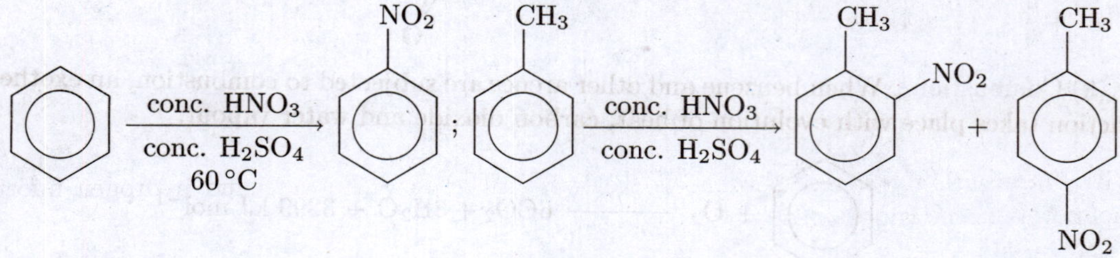


**(g) Sulphonation :** When arenes are treated with conc. sulphuric acid, the corresponding arenesuphonic acids are obtained.

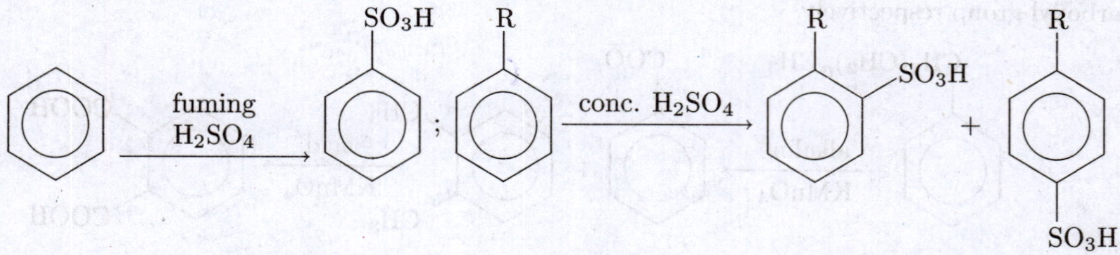


**(h) Metalation :** When alkylmetals are treated with benzene, phenylmetals are formed. This is also an aromatic electrophilic substitution reaction.

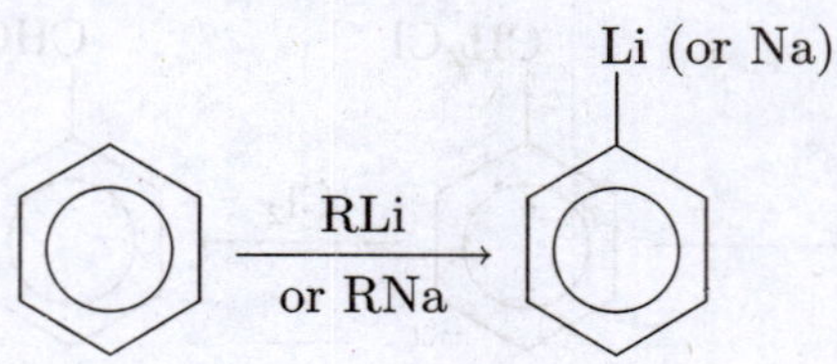


**(i) Direct Mercuration :** When benzene is treated with glacial acetic acid and mercuric oxide, $HgOCOCH_3$ group replaces an H atom and the process is called mercuration. Mercuric acetate and perchloric acid may also be used as mercurating agent.

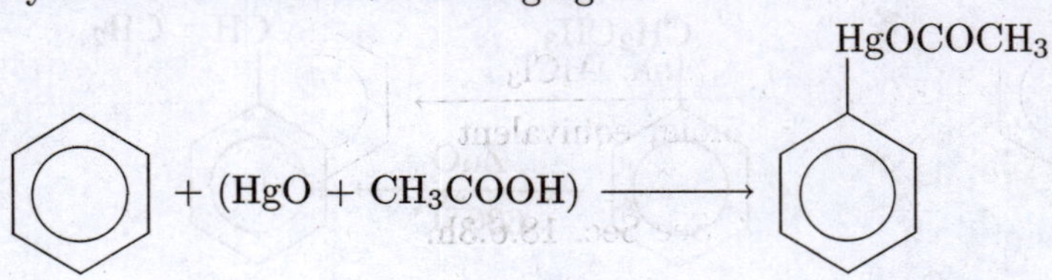

(j) Conversion to maleic anhydride (catalytic oxidation) : When benzene is heated with vanadium pentoxide and oxygen at 500 °C, it is converted to maleic anhydride and $CO_2$ and $H_2O$ evolve out. It is the pathway to prepare aliphatic compounds from benzene since maleic anhydride on hydrolysis with alkali and then on electrolysis gives acetylene.

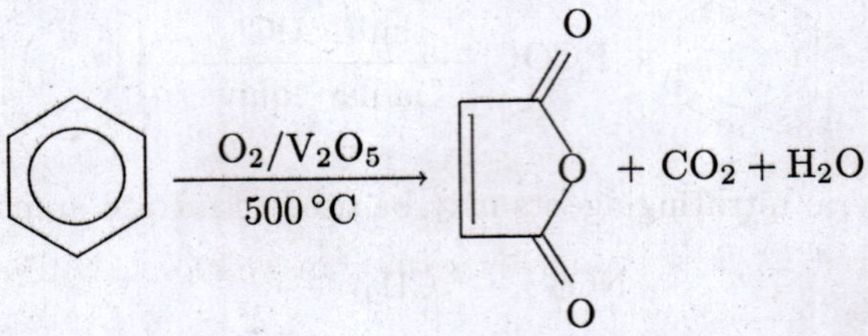


(k) Combustion : When benzene and other arenes are subjected to combustion, an exothermic reaction takes place with evolution of heat, carbon dioxide and water vapour.

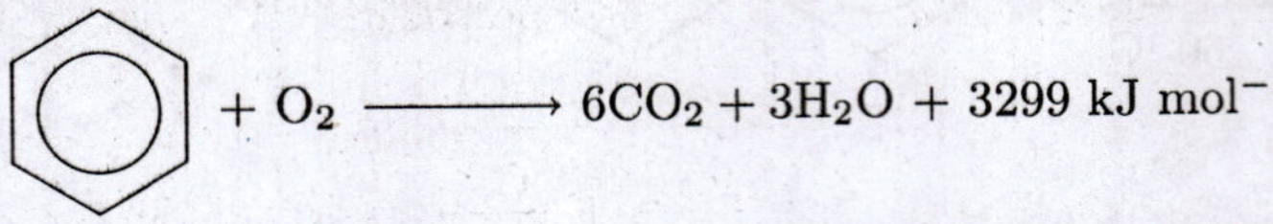


(l) Side-chain oxidation : When alkylbenzenes are refluxed with alkaline or acidic $KMnO_4$ or $K_2Cr_2O_7$ solution, the side chain, whatever may be its length, becomes a carboxylate ion or carboxyl group respectively.

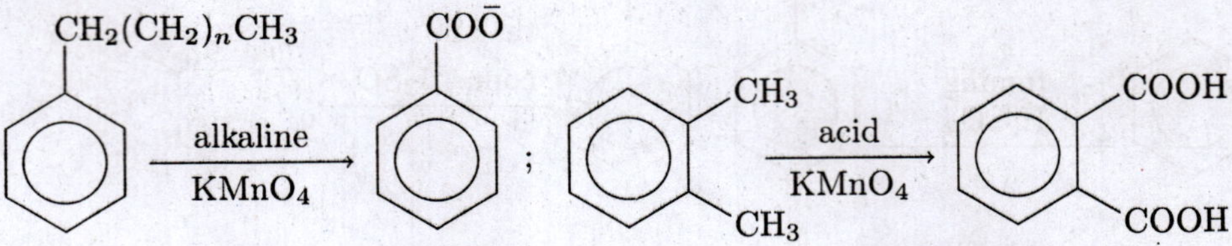


(m) Side-chain cholorination : Boiling toluene reacts with chlorine to form benzyl chloride, benzal chloride and benzotrichloride (a mixture of the three is obtained).

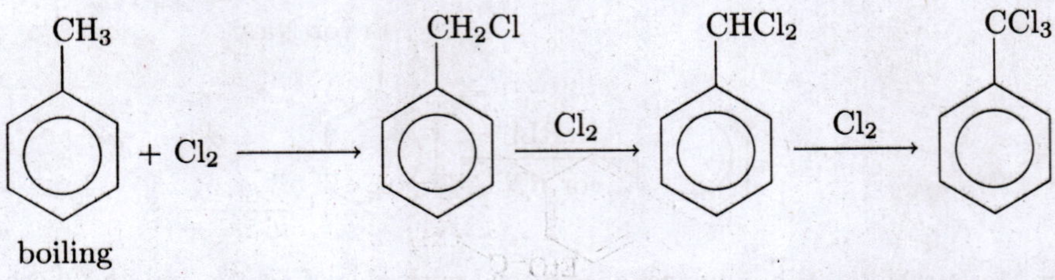


(n) Side-chain dehydrogenation : Ethylbenzene on dehydraogenation with ZnO at 600 °C forms styrene.

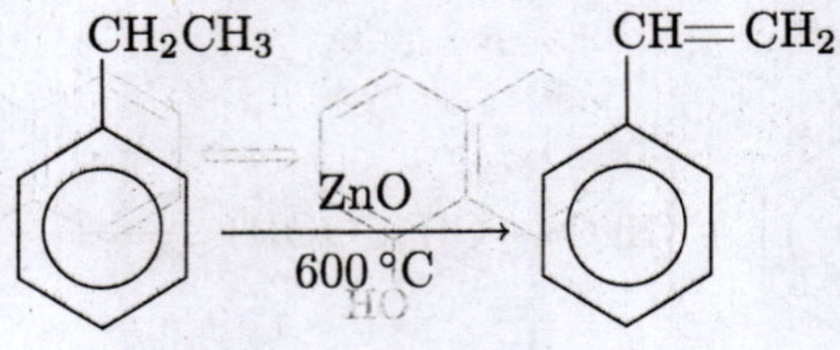

## 18.7 Naphthalene

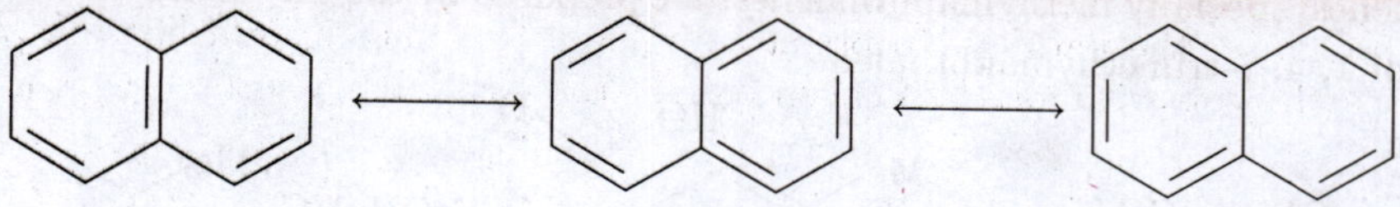

$C_{10}H_8$

### 18.7.1 Introduction

Naphthalene is a bicyclic aromatic hydrocarbon in which two benzene rings are fused together at *o*-positions. Its molecular formula is $C_{10}H_8$.

### 18.7.2 Source and extraction

**(i)** Naphthalene is obtained from coal tar and petroleum oil.

Coal tar —(high temperature distillation)→
- up to 200°C → light oil
- 200–250°C → middle oil
- 250–300°C → heavy oil
- 300–350°C → anthracene oil

middle oil + heavy oil —(cooled)→ napthalene crystals —(oil is pressed out)→ napthalene cake —(melted and treated with $H_2SO_4$)→ napthalene cake (free from basic impurities) —(washed with water and then treated with aqueous NaOH)→ napthalene cake (free from acidic impurities) —(washed with water)→ almost pure napthalene cake —(distilled)→ pure napthalene m.p. 80 °C

**(ii)** Petroleum fractions —(passed over copper catalyst at 680 °C.)→ naphthalene + methylnaphthalene

### 18.7.3 Synthesis of naphthalene and alkylnaphthalenes

(a) From benzaldehyde and ethyl succinate (Fittig and Erdmann synthesis) :

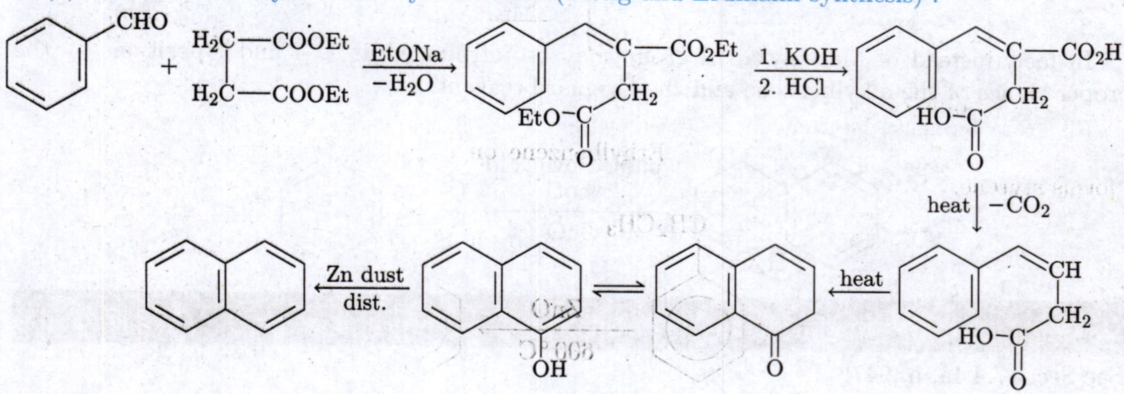

(b) Synthesis from benzene and succinic anhydride—the Haworth synthesis : (Sec. 17.9.2)
Alkylnaphthalenes, specially methylnaphthalenes, are prepared by the Haworth synthesis (p. 983).

Synthesis of 1, 4, 7-trimethylnaphthalene :

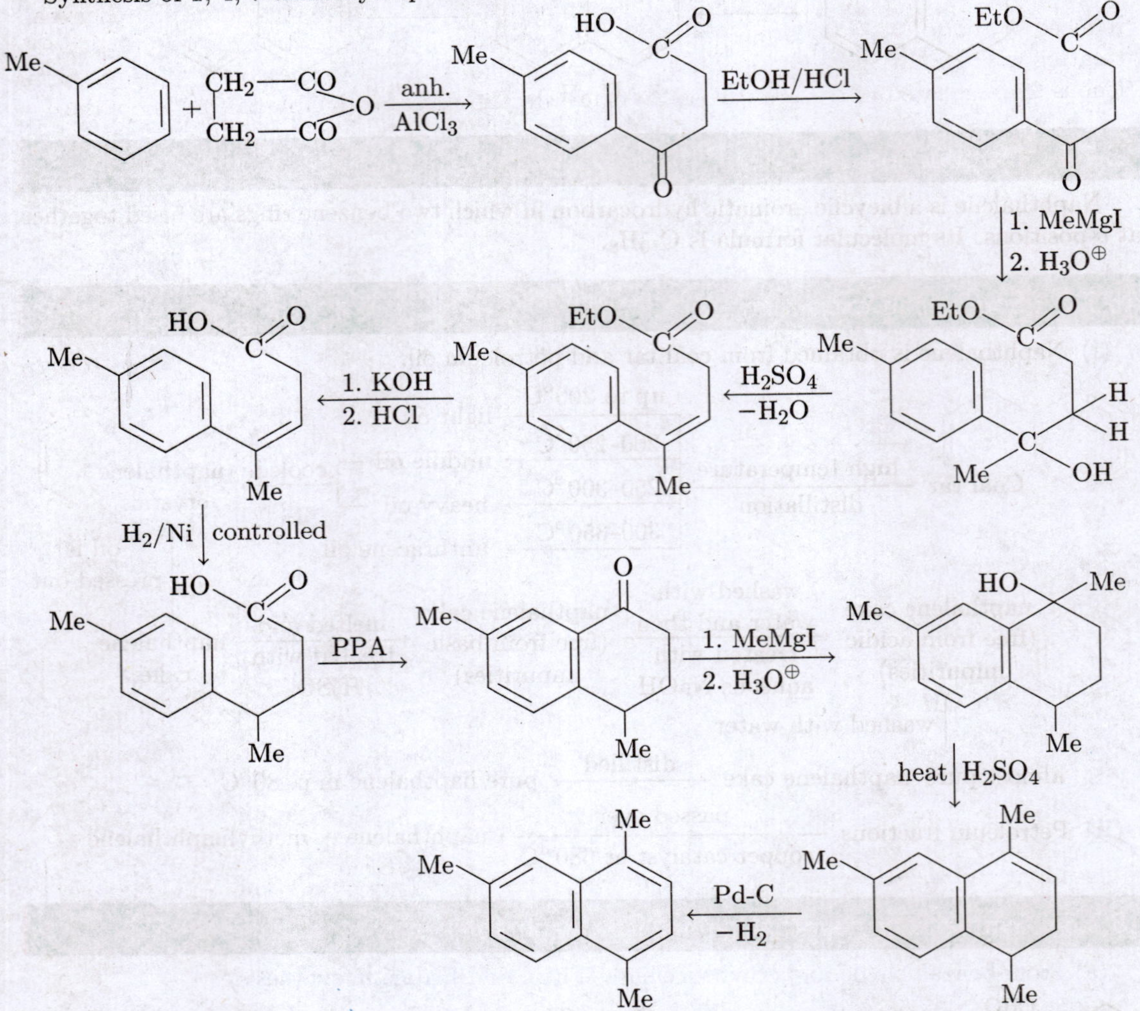


In fact, instead of Me, any alkyl group may be introduced at 1, 4 and 7-positions by the proper choice of the alkylbenzene and the Grignard reagent.

(c) From 4-phenyl-1-butene :

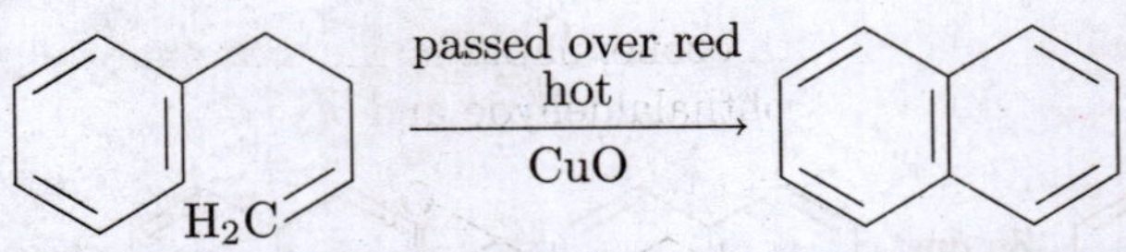


## 18.7.4 Structure, reactivity and orientation

[See Sec. 17.4.4a, p. 947]

## 18.7.5 Chemical reactions

(a) Reduction : Naphthalene being more reactive than benzene, the former is chemically reduced to dialin (1, 4-dihydronaphthalene) and tetralin (1, 2, 3, 4-tetrahydronaphthalene) by treating with Na/EtOH at 78 °C and Na/isoamyl alcohol at 132 °C respectively. When naphthalene is hydrogenated in the presence of Pt or Ni, decalin (decahydronaphthalene) is obtained.

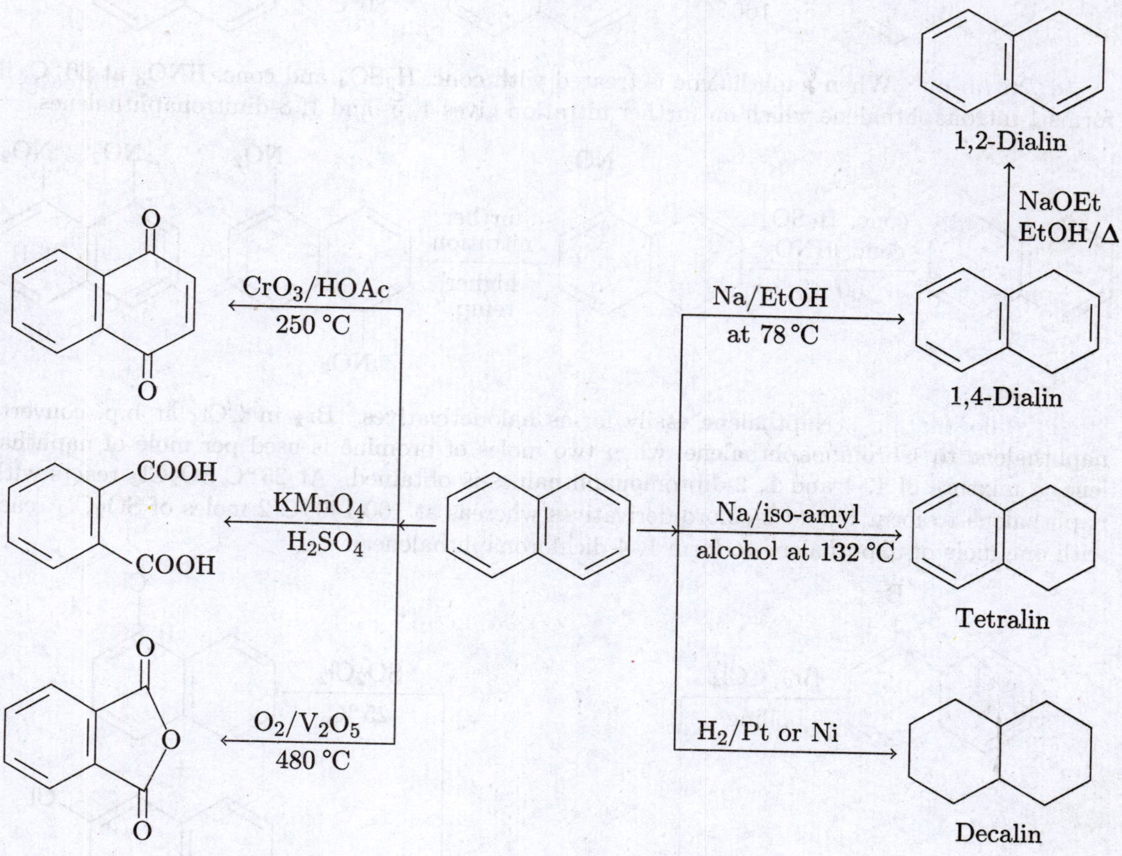


(b) Oxidation :

(i) Naphthalene on oxidation with $CrO_3$/HOAc and $O_2/V_2O_5$ forms 1, 4-naphthaquinone and phthalic anhydride respectively; acid permanganate oxidises naphthalene to phthalic acid.

(ii) When naphthalene is treated with ozone, naphthalene diozonide is formed which on treatment with water and zinc gives phthalaldehyde and glyoxal.

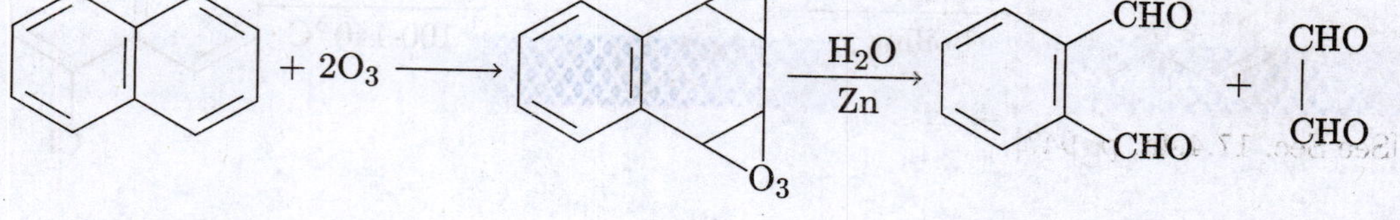

(c) Sulphonation : When naphthalene is sulphonated with conc. $H_2SO_4$, it forms 1-naphthalenesulphonic acid at 80 °C and 2-naphthalenesulphonic acid at 160 °C.

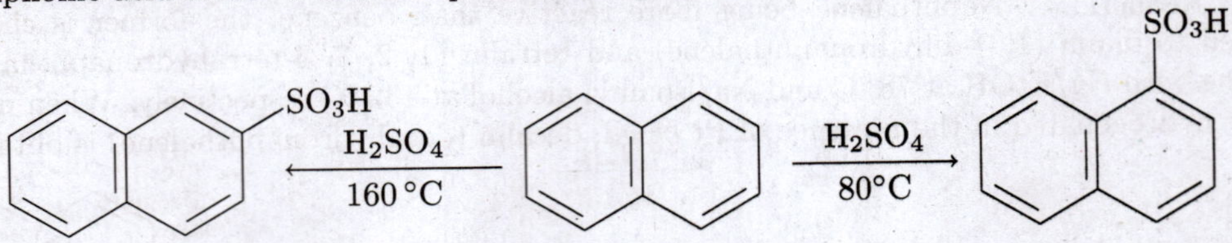


(d) Nitration : When naphthalene is treated with conc. $H_2SO_4$ and conc. $HNO_3$ at 60 °C, it forms 1-nitronaphthalene which on further nitration gives 1, 5- and 1, 8-dinitronaphthalenes.

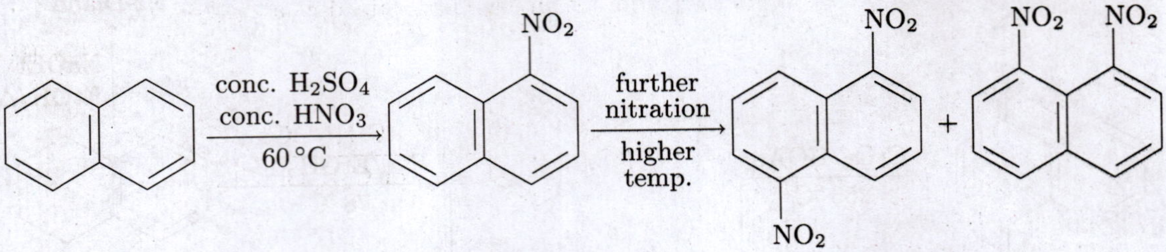


(e) Halogenation : Napthalene easily forms haloderivatives. $Br_2$ in $CCl_4$ at b.p. converts naphthalene to 1-bromonaphthalene; when two moles of bromine is used per mole of naphthalene, a mixture of 1, 4 and 1, 2-dibromonaphthalnes is obtained. At 25 °C $SO_2Cl_2$ reacts with naphthalene to form 1- and 2-chloro derivatives whereas at 100-140 °C 2 moles of $SO_2Cl_2$ react with one mole of naphthalene to form 1, 4-dichloronaphthalene.

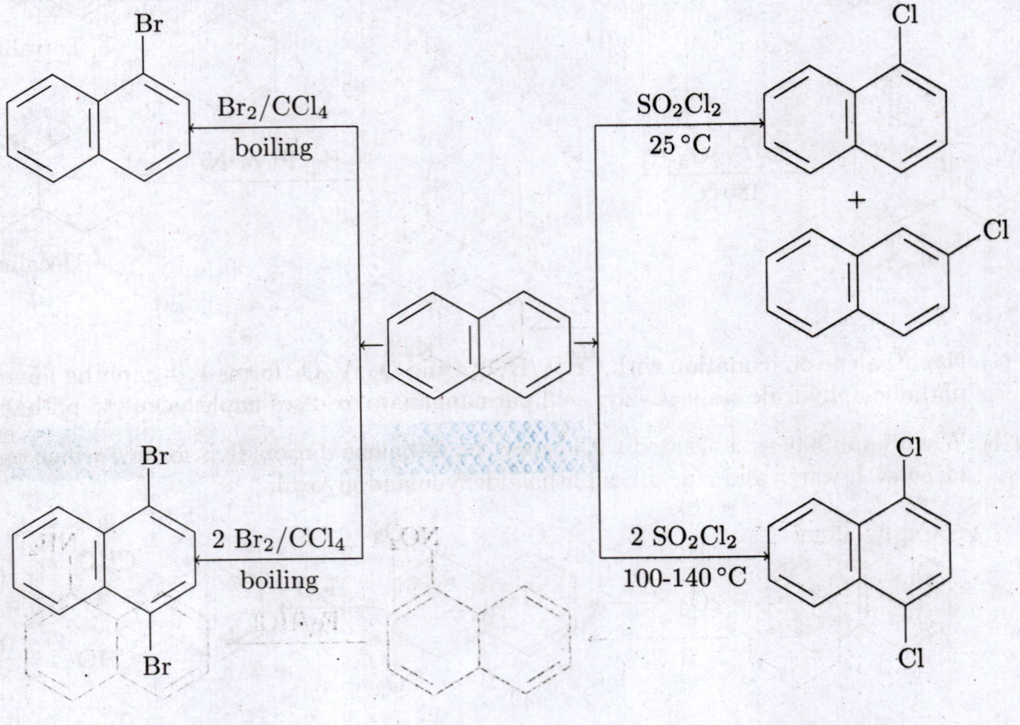

(f) Chloromethylation : When a mixture of paraformaldehyde, hydrochloric acid, glacial acetic acid and phosphoric acid is treated with naphthalene, 1-chloromethylnaphthalene is formed.

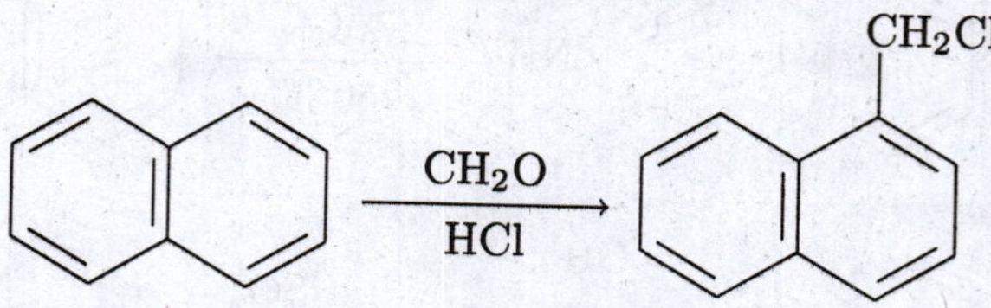


(g) The Friedel-Crafts acylation : The Friedel-Crafts acylation of naphthalene with acetyl chloride in carbon disulphide or tetralchloroethane gives 1-acetylnaphthalene as major product; but when the solvent is nitrobenzene, 2-acetylnaphthalene is obtained as a major product.

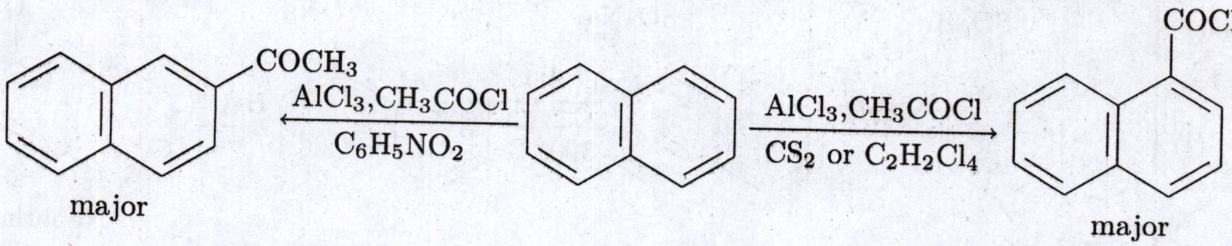


(h) The Friedel Crafts alkylation : Naphthalene being more reactive than benzene, alkylation gives a number of products; but when the reaction is carried out under mild conditions, the Friedel-Crafts alkylation is possible though yields are very poor.

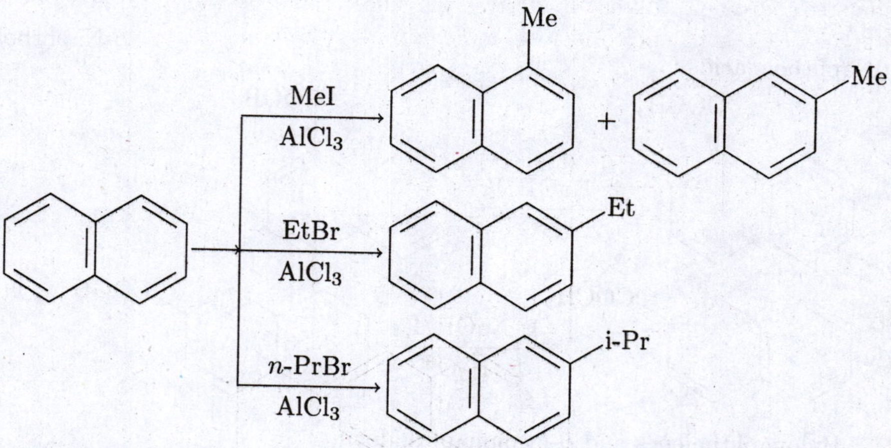


## 18.7.6 Preparation of naphthalene derivatives

(a) Naphthylamines :

(i) $\alpha$-Naphthylamine

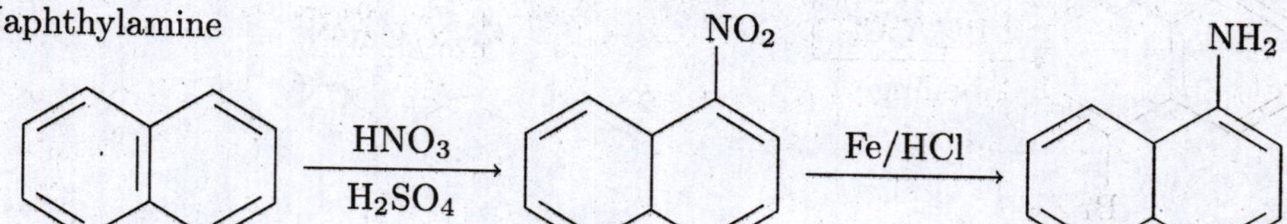

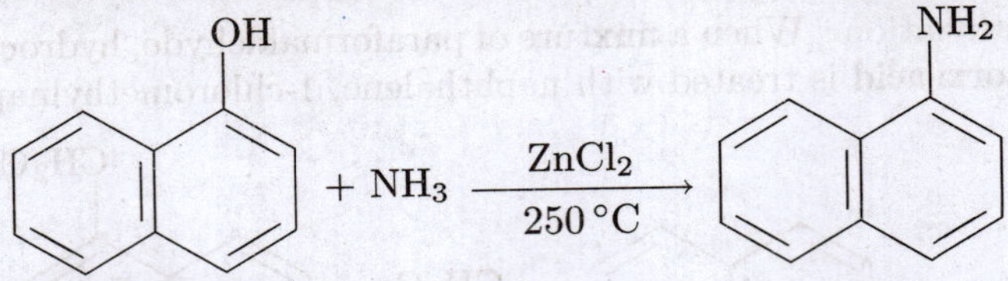


**(ii) $\beta$-Naphthylamine**

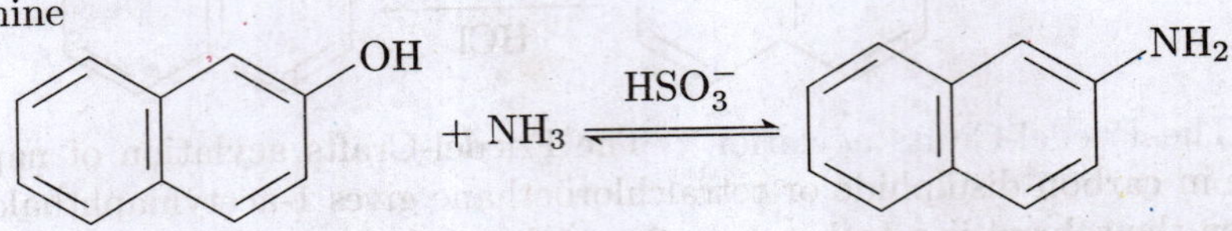


**(b) Naphthols :**

(i)

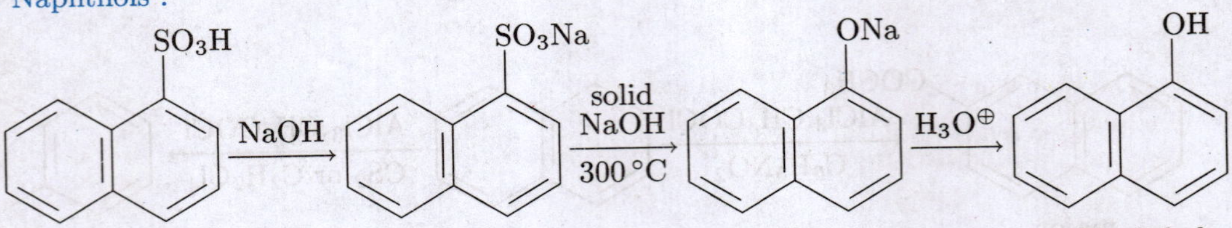


$\alpha$-Naphthol

(ii)

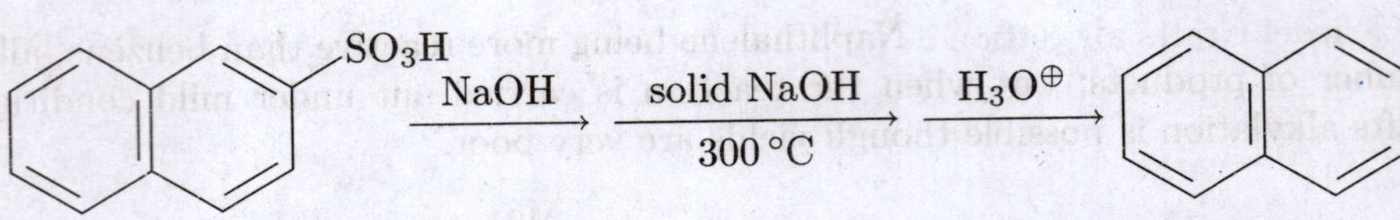


$\beta$-Naphthol

**(c) Naphthoic acids :**

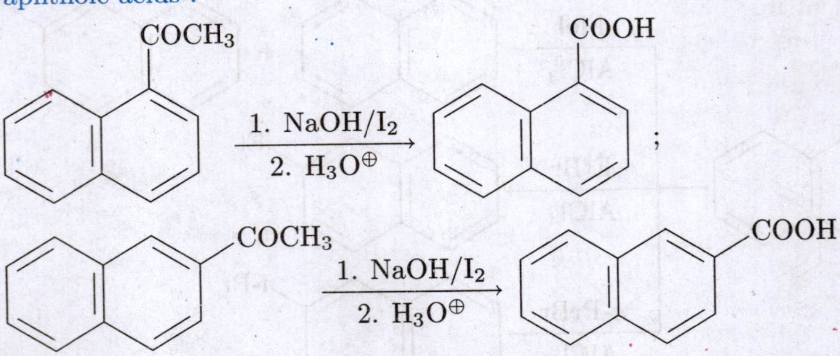


**(d) $\beta$-Halonaphthalenes and $\beta$-cyanonaphthalne :**

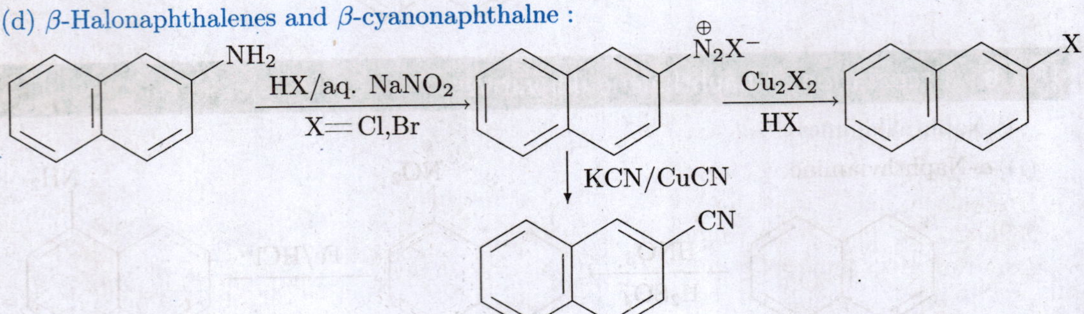

## 18.7.7 Elucidation of the structure of naphthalene

From the analytical data and molecular weight determination the molecular formula of naphthalene is found to be $C_{10}H_8$.

Naphthalene resembles benzene in its chemical reactions; it can be nitrated, sulphonated, halogenated, chloromethylated like benzene. Its hydroxy derivatives also resemble phenols. Furthermore, it is stable and cannot be oxidised by nitric acid and other mild oxidising agents. All these suggest a ring structure for naphthalene as for benzene, and Erlenmeyer in 1866 proposed a fused ring system for naphthalene as shown below (structure 1).

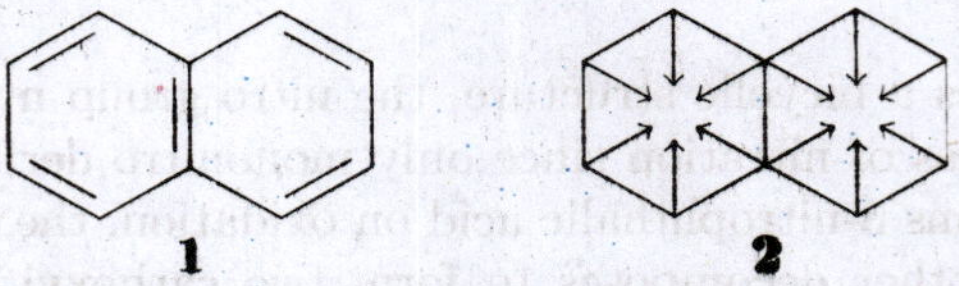

1 2

The then scientists did not accept the structure; they put forward several other structures, one of which is given above (structure 2). Structure 2 cannot explain the stability and isomer numbers of mono and disubstituted naphthalenes and, therefore, cannot be taken under consideration.

However, oxidation of naphthalene yields benzene-1, 2-dicarboxylic acid, phthalic acid. This indicates that naphthalene contains at least a benzene ring with two side chains or a closed chain at *o*-positions.

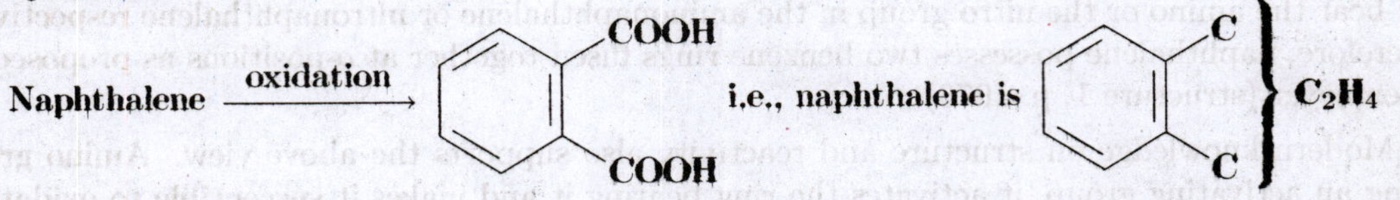


Two independent unsaturated side chains consisting of 4 Cs and 4 Hs can be constructed as shown in the structure below; but this tells us that one mole of naphthalene will add six moles of $H_2$ on hydrogenation whereas one mole of naphthalene actually adds five moles of $H_2$ only. Therefore, this structure does not represent the structure of naphthalene.

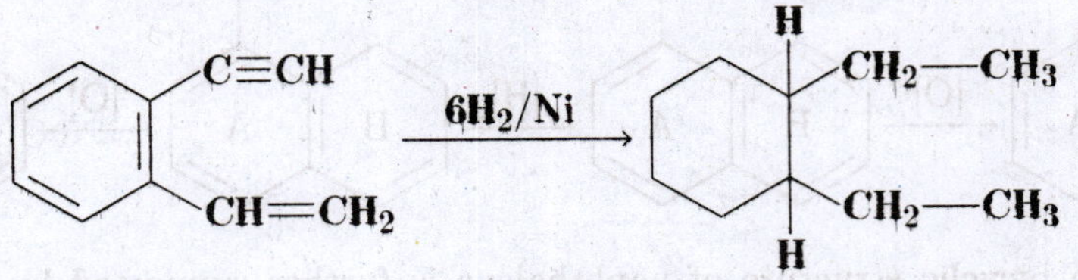


$C_{10}H_8$ $C_{10}H_{20}$

$$\text{Napthalene} \xrightarrow{5H_2/Ni} C_{10}H_{18}$$

Therefore, naphthalene may possess two benzene rings which are fused at *o*-positions as proposed by Erlenmeyer.

Naphthalene on nitration gives mononitronaphthalene which on oxidation forms 3-nitrophthalic acid. On the other hand, nitronaphthalene on reduction to aminonaphthalene and on subsequent oxidation gives only phthalic acid.

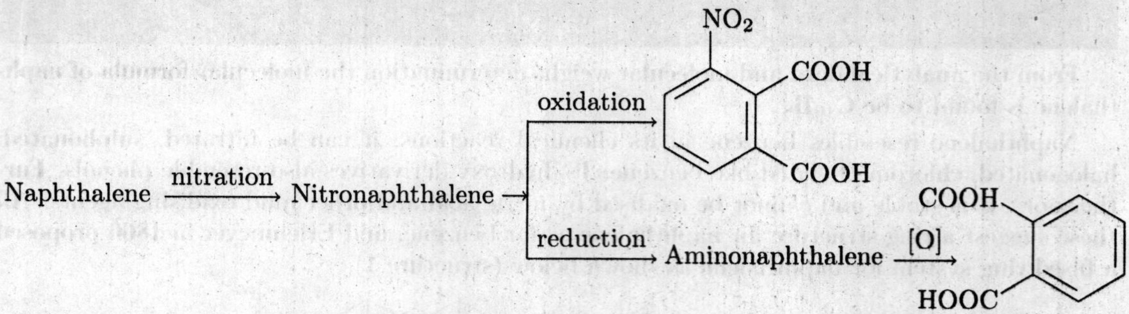


If naphthalene possesses a bicyclic structure, the nitro group must have entered into one of the two rings by the process of nitration since only mononitro derivative forms by the process. Since nitronaphthalene forms 3-nitrophthalic acid on oxidation, the ring holding the nitro group remains intact while the other decomposes to form two carboxyl groups at *o*-positions. The formation of bare phthalic acid on oxidation of aminonaphthalene indicates with certainty that the ring bearing the amino group does not remain intact. Had it been intact, there would have been the formation of 3-aminophthalic acid instead of phthalic acid. This means that in this case the two carboxyl groups of the phthalic acid have come from the ring holding the amino group i.e., the nitro group in the nitronaphthalene. This further tells us that the benzene ring in the phthalic acid is a different one from that in the nitrophthalic acid. The former ring does not bear the amino or the nitro group in the aminonaphthalene or nitronaphthalene respectively. Therefore, naphthalene possesses two benzene rings fused together at *o*-positions as proposed by Erlenmeyer (structure 1, p. 1073).

Modern knowledge on structure and reactivity also supports the above view. Amino group being an activating group, it activates the ring bearing it and makes it susceptible to oxidation; while nitro group being a deactivating group, it deactivates the ring holding it and thereby makes the ring resistant to oxidation. Therefore, the ring containing the nitro group does not open on oxidation whereas the same ring holding the amino group opens on oxidation.

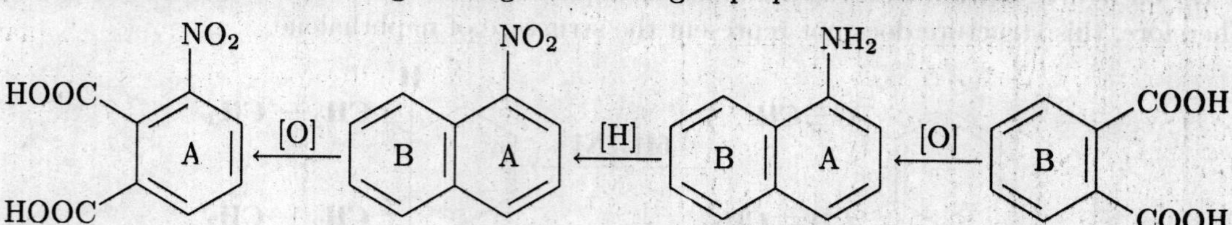


However, the bicyclic structure of naphthalene is further supported by its synthesis from benzene and succinic anhydride by the Haworth's process.

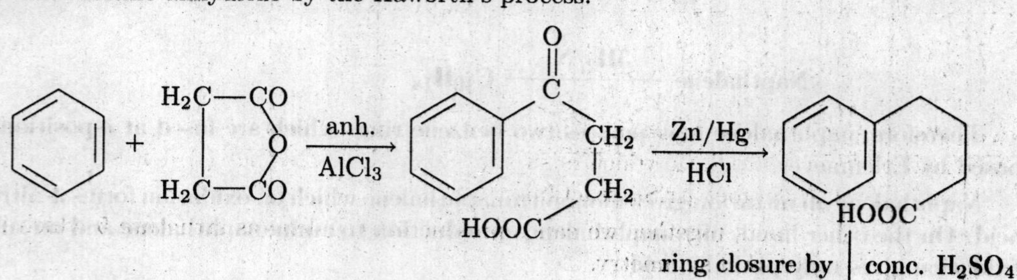

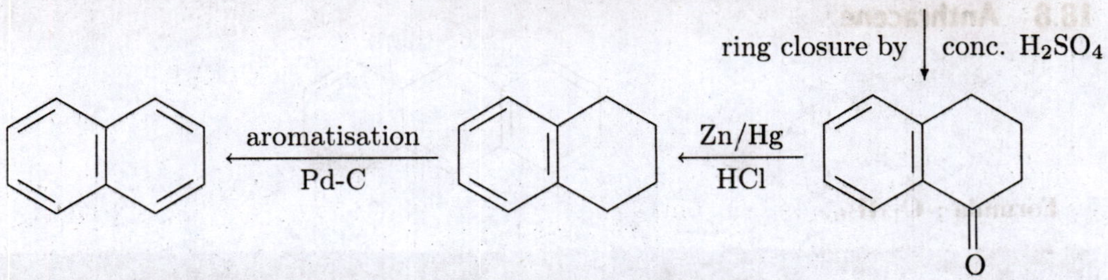


Structure I supports the following observations about naphthalene :

(i) Naphthalene forms two monosubstituted derivatives $\alpha$ and $\beta$.

(ii) It forms ten isomeric compounds of the type $C_{10}H_6X_2$ and fourteen disubstituted isomeric compounds of the formula $C_{10}H_6XY$.

(iii) It behaves as a symmetrical molecule and its dipole moment is zero Debye like benzene.

On the other hand, this structure fails to explain the fact that $C_1—C_2$ and $C_2—C_3$ bond lengths are not equal to C—C double and single bond lengths respectively. X-ray analysis shows that the $C_1—C_2$ bond length is slightly greater than a C—C double bond and $C_2—C_3$ bond length is slightly shorter than a C—C single bond.

According to the V. B. Theory, naphthalene is a resonance hybrid of the three resonating structures and its resonance energy is 255 kJ $mol^{-1}$. This theory, therefore, assumes that the two rings are flat. From the resonating structures it is obvious that the $C_1—C_2$ bond length will be the average of two C—C double bond lengths and one C—C single bond length while that of $C_2—C_3$ will be the average of two C—C single bond lengths and one C—C double bond length. Thus, $C_1—C_2$ bond is shorter than $C_2—C_3$ bond, the former has more double bond character than the latter. This fact supports an early deduction by the then chemists that the bonds in naphthalene appear to have a certain degree of fixation. It is an early finding that 2-naphthol couples with a diazonium salt in alkaline medium at 1-position while 1-substituted 2-naphthol, say 1-methyl-2-naphthol, does not couple with a diazonium salt under the identical conditions. Since the $C_2—C_3$ bond has a less double bond character than the $C_1—C_2$ bond, the 1-position of 2-naphthol has a more enolic character than the 3-position. Since the 3-position is less enolic than the 1-position, 1-methyl- 2-naphthol does not couple. However, structure 1 will contribute maximum to the hybrid since both the rings are benzenoid.

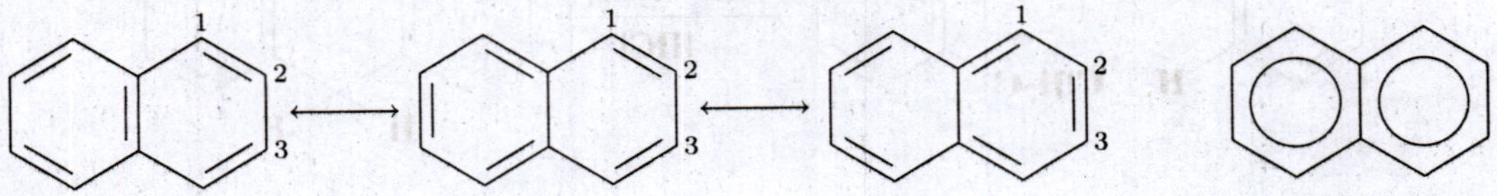


1 3 4 5

The molecular orbital theory tells us that naphthalene possesses 10 delocalised $\pi$es (Huckle no. of $\pi$es) in five bonding M.O.s in a flat bicyclic structure encompassing the two fused rings. Hence naphthalene is aromatic which is supported by the facts that (i) it is stable, (ii) its $\mu$ is 0.00D and (iii) it undergoes Ar.$S_E$ reactions.

Taking all these facts together we can represent naphthalene as structure 5 which indicates its aromatic character and symmetry.

## 18.8 Anthracene

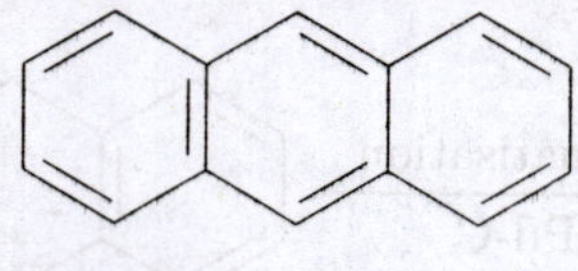

**Formula : $C_{14}H_{10}$.**

### 18.8.1 Introduction

**Anthracene is a tricyclic aromatic hydrocarbon in which three benzene rings are fused together in a linear manner; its molecular formula is $C_{14}H_{10}$.**

### 18.8.2 Source and extraction

**Coal tar.**

**Coal tar ⟶ anthracene oil fraction —(cooled and pressed)⟶ [solid anthracene cake; carbazole + phenanthrene. + anthracene.] ⟶**

**powdered —(treated with solvent naphtha)⟶ [solid anthracene + carbazole] —(washed with pyridine)⟶ anthracene —(sublimed)⟶ pure anthracene m.p. 216 °C.**

### 18.8.3 Synthesis of anthracene

(a) From benzene and phthalic anhydride (phthalylation) : **See Sec. 17.9.2 p. 983.**

(b) From benzyl chloride :

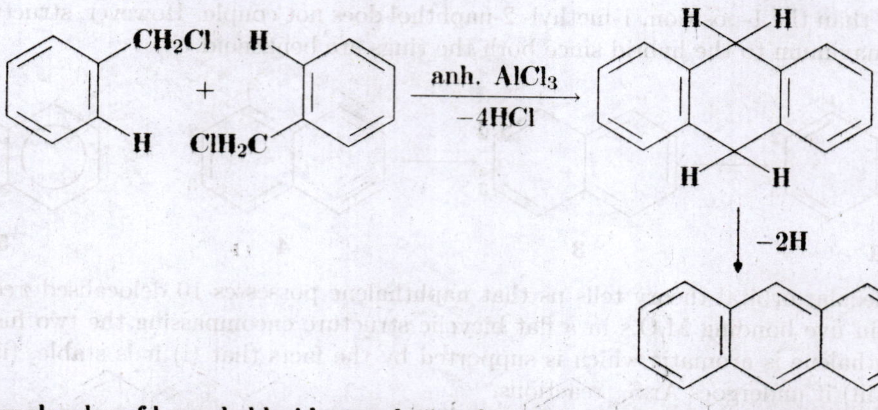


**Two molecules of benzyl chloride may be condensed in the presence of anhydrous aluminium chloride to form 9, 10-dihydroanthracene by the Friedel-Crafts method of alkylation which then readily gives up two H atoms under the condition of the experiment to form anthracene.**

(c) From benzene and methylene dibromide or acetylene tetrabromide : **This also involves the Friedel-Crafts reaction.**

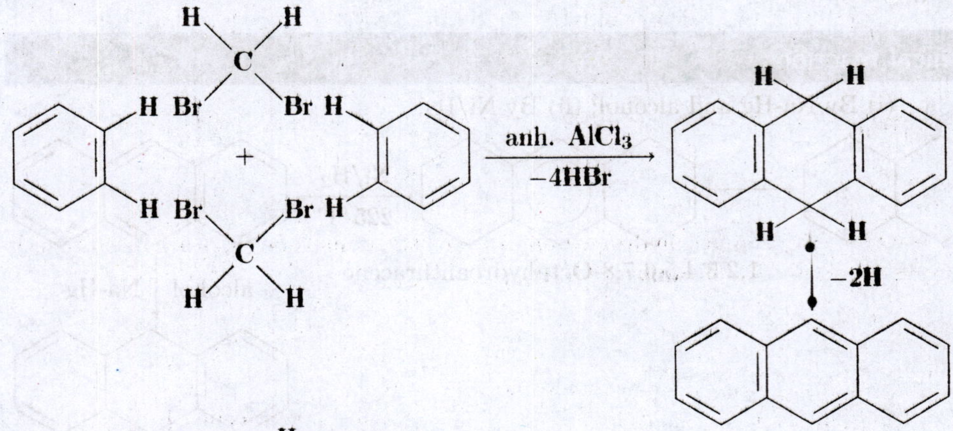

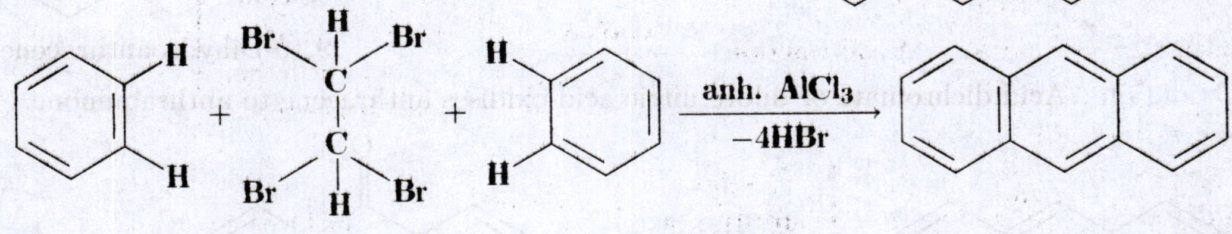

(d) From naphthaquinone and 1, 3-butadiene :

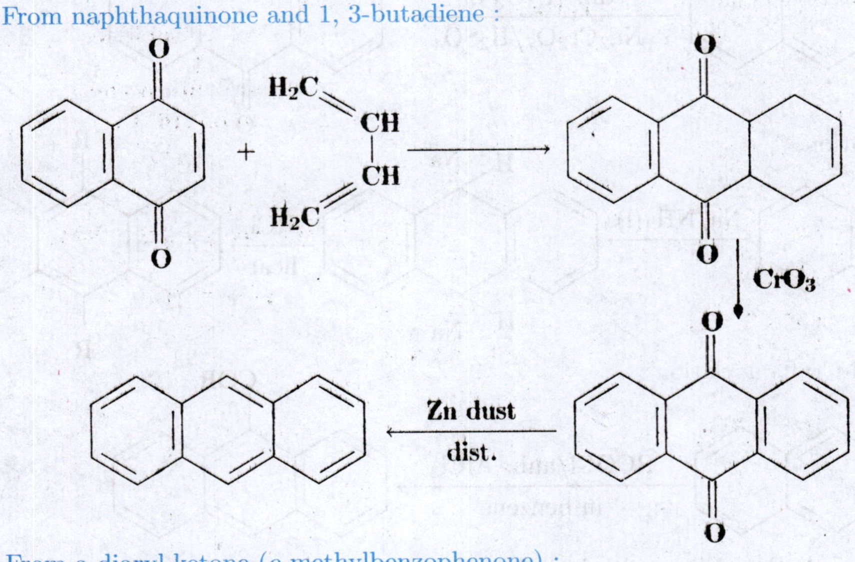

(e) From a diaryl ketone (*o*-methylbenzophenone) :

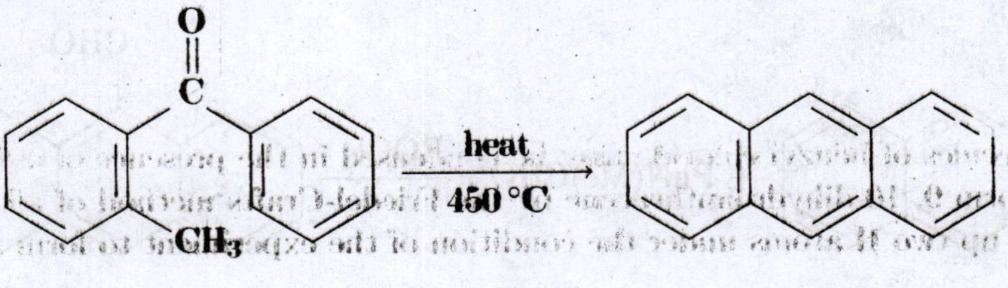

## 18.8.4 Structure, reactivity and orientation

See Sec. 17.4.4b (p. 948).

## 18.8.5 Chemical reactions

(a) Reduction : (i) By Na-Hg and alcohol; (ii) By Ni/$H_2$.

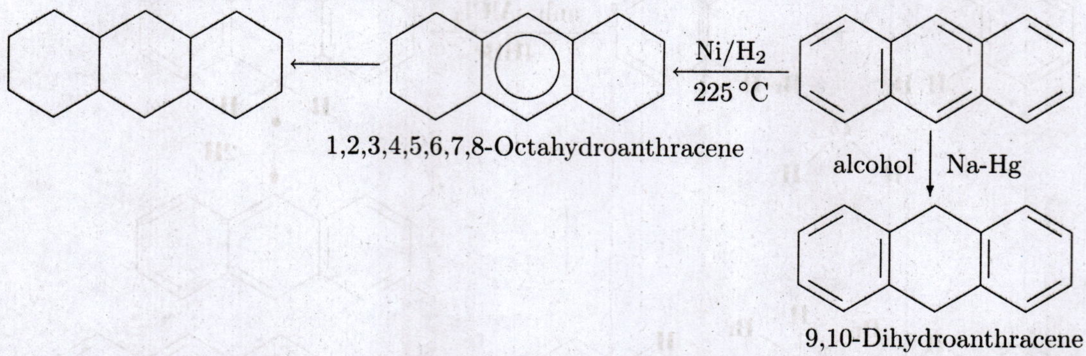


(b) Oxidation : Acid dichromate or dilute nitric acid oxidises anthracene to anthraquinone.

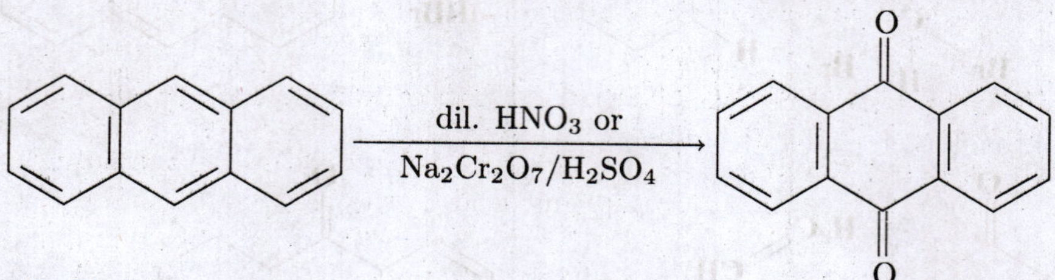


(c) 9, 10-alkylation :

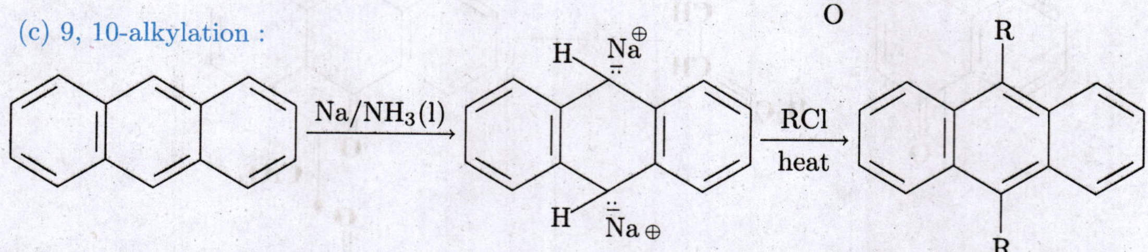


(d) The Friedel-Crafts acylation :

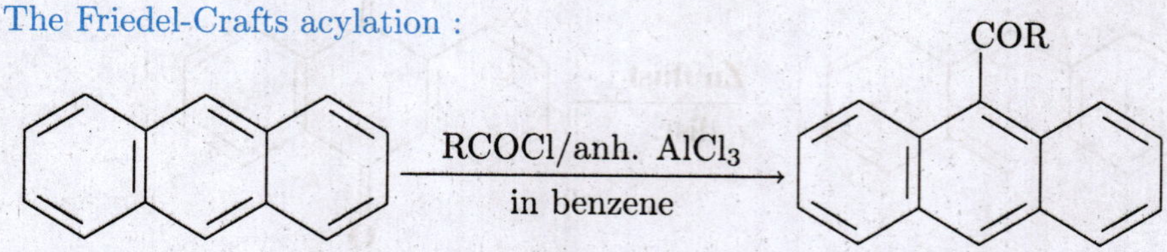


(e) Formylation : Anthracene can be formylated by the Vilsmeier-Haak method :

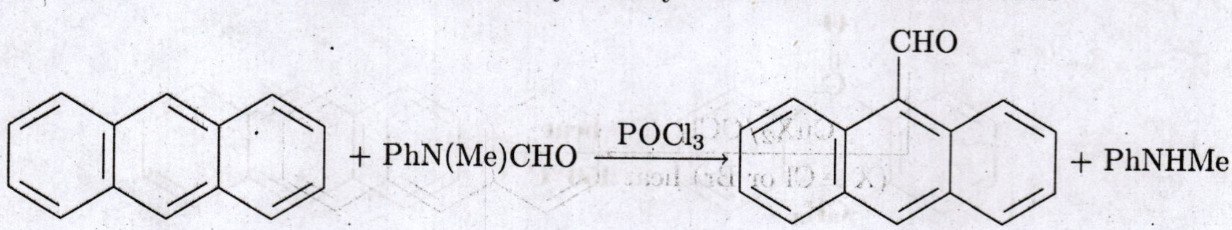

(f) Nitration : **When anthracene is nitrated in acetic anhydride at 15-20 °C, a mixture of 9-nitroanthracene and 9, 10-dinitroanthracene is obtained.**

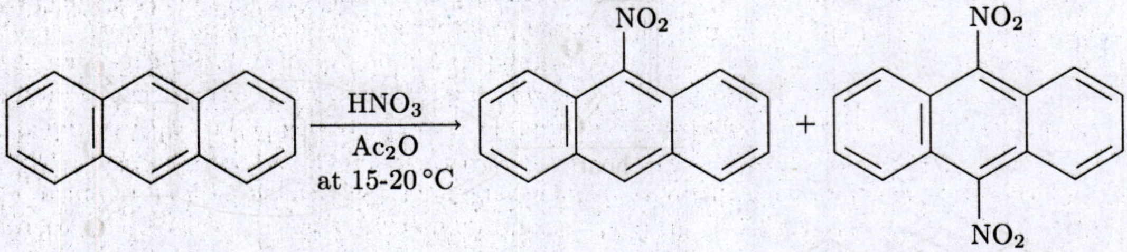


(g) Sulphonation :

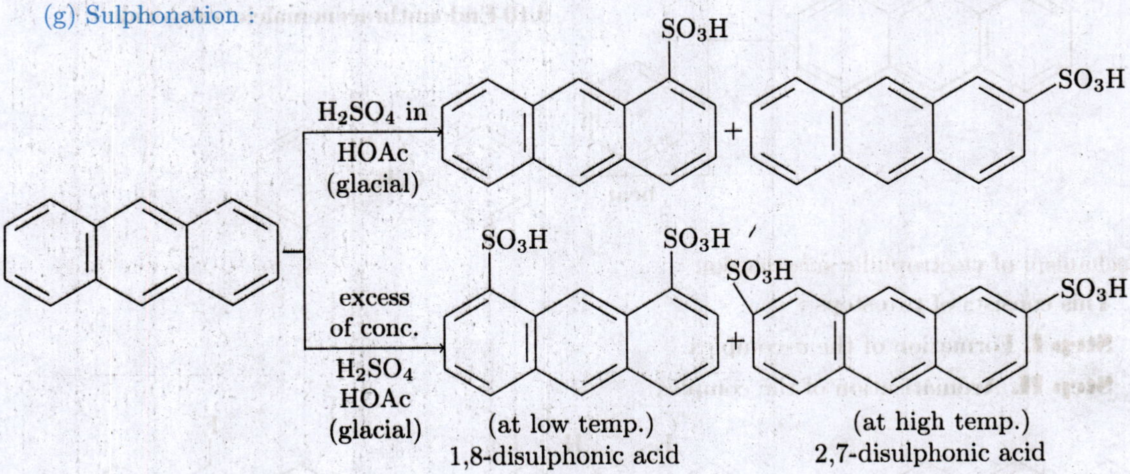


(h) Halogenation :

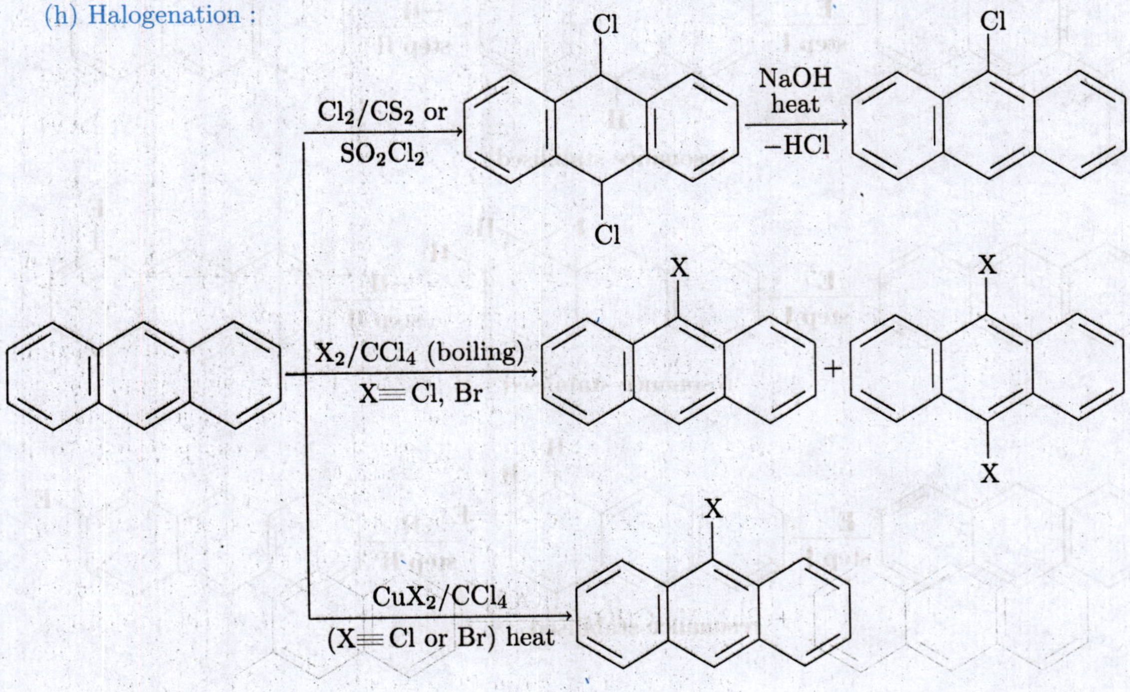

(i) The Diels-Aider reaction :

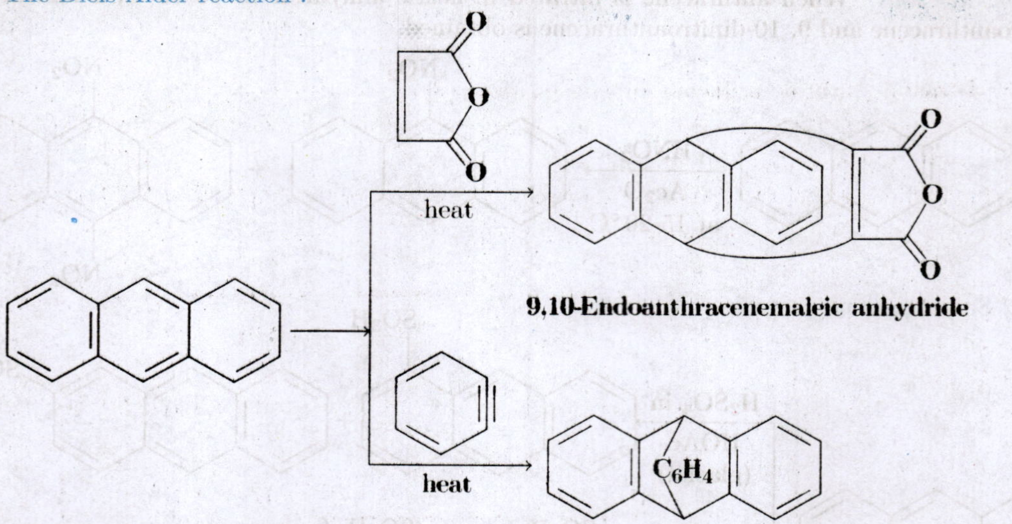


## Mechanism of electrophilic substitution

This consists of two steps :

**Step I.** Formation of the $\sigma$-complex.

**Step II.** Aromatisation of the complex.

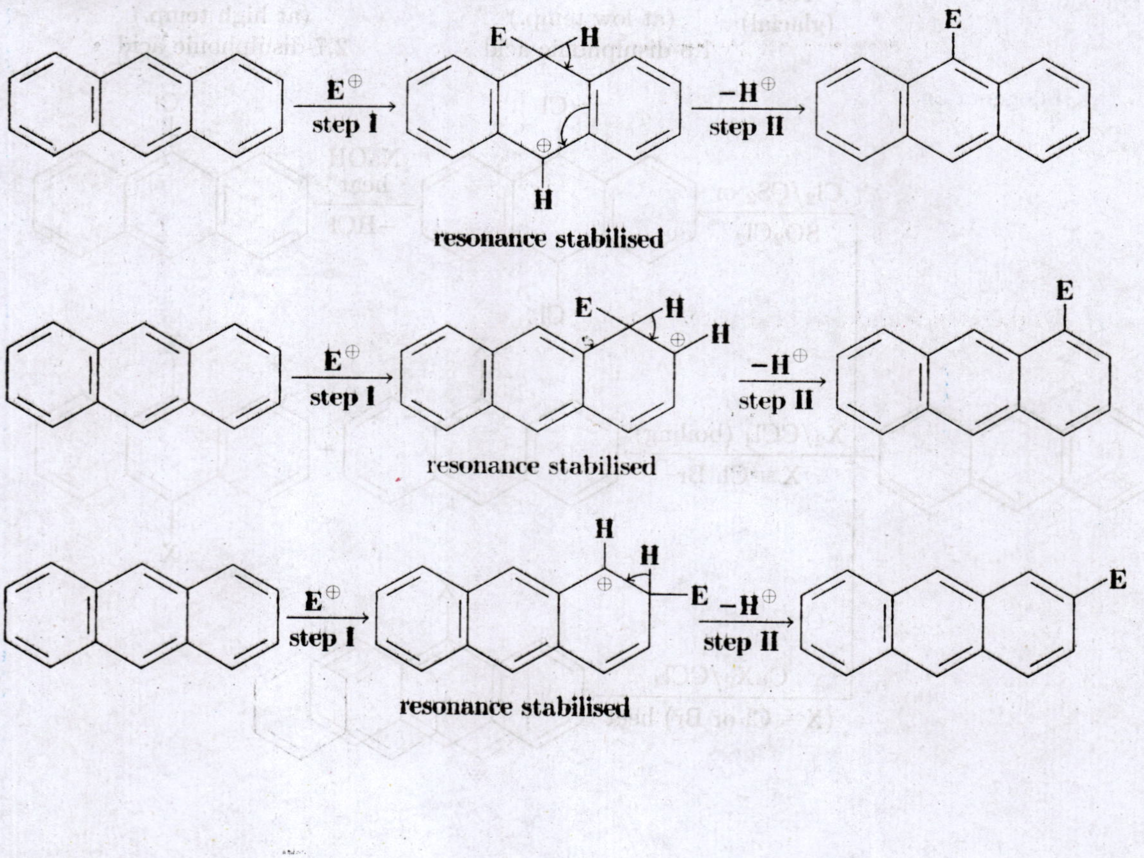

## 18.8.6 Synthesis of anthracene derivatives

(a) **Synthesis** of anthroic acids :

**(i) 1- and 2-Anthroic acids via anthracenenitrile.**

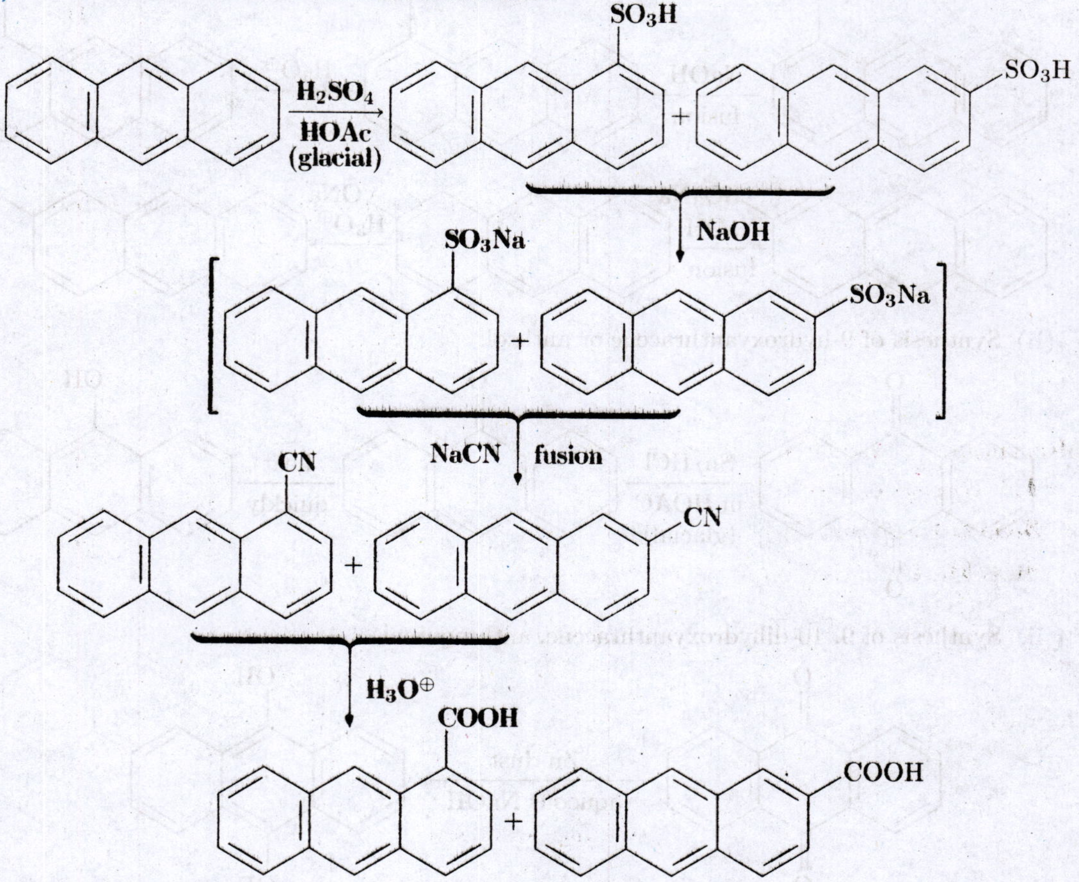


**(ii) Synthesis of 9-anthroic acid via 9-carbonyl chloride.**

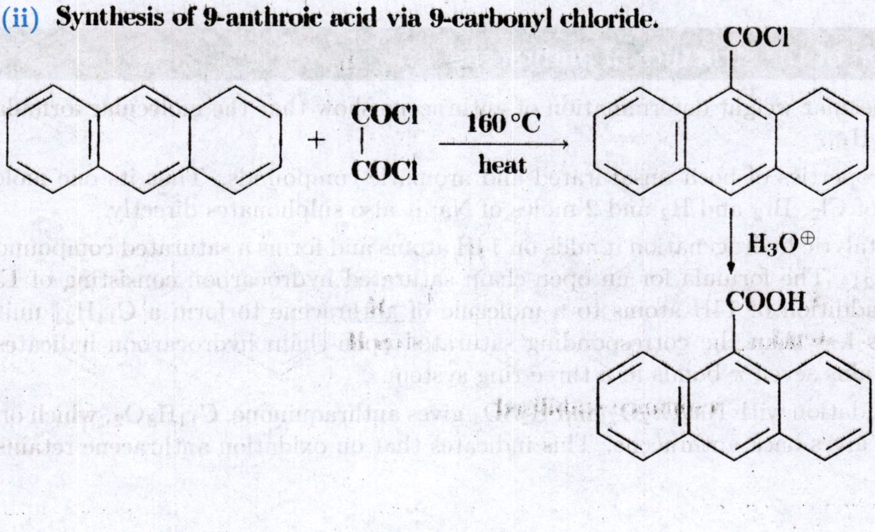

(b) Synthesis of hydroxyanthracenes :

**(i)** Synthesis of 1-, and 2-hydroxyanthracenes (anthrols) :

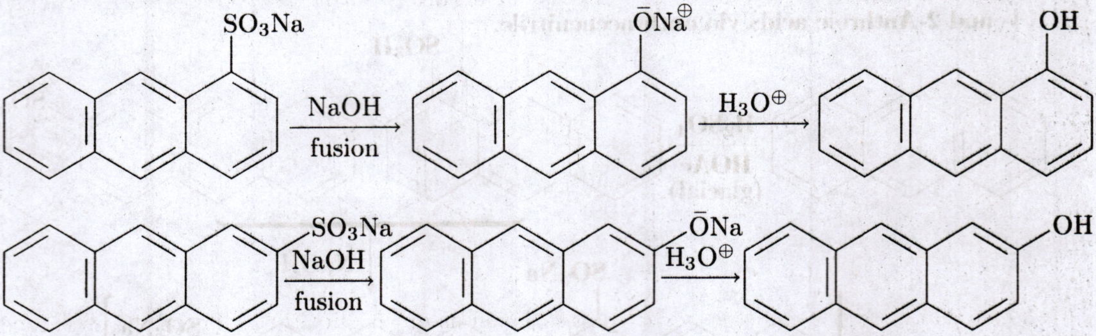


**(ii)** Synthesis of 9-hydroxyanthracene or anthrol.

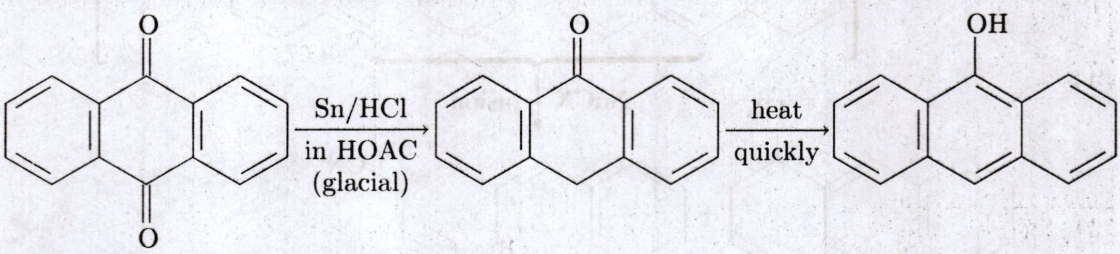


**(iii)** Synthesis of 9, 10-dihydroxyanthracene, anthraquinol.

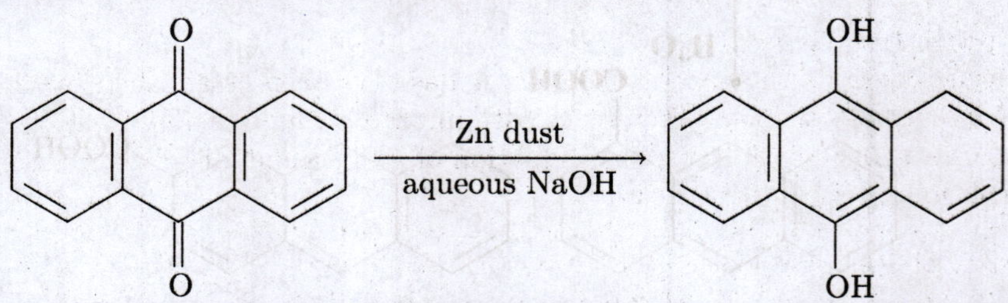


## 18.8.7 Elucidation of the structure of anthracene

Analysis and molecular weight determination of anthracene show that the molecular formula of anthracene is $C_{14}\ H_{10}$.

It exhibits the properties of both unsaturated and aromatic compounds. Thus its one mole readily adds 1 mole of $Cl_2$, $Br_2$ and $H_2$ and 2 moles of Na; it also sulphonates directly.

On exhaustive catalytic hydrogenation it adds on 14H atoms and forms a saturated compound of the formula $C_{14}H_{24}$. The formula for an open-chain saturated hydrocarbon consisting of 14 Cs is $C_{14}H_{30}$. The addition of 14H atoms to a molecule of anthracene to form a $C_{14}H_{24}$ unit containing 6H atoms less than the corresponding saturated open-chain hydrocarbon indicates that anthracene contains seven $\pi$ bonds in a three-ring system.

Anthracene on oxidation with $Na_2Cr_2O_7$ and $H_2SO_4$ gives anthraquinone, $C_{14}H_8O_2$, which on zinc dust distillation gives back anthracene. This indicates that on oxidation anthracene retains its cyclic structure and anthraquinone is a derivative of anthracene. Therefore, if we can establish the structure of anthraquinone, the C-skeleton of anthraquinone will be known and naturally the C-skeleton of anthracene will also be known. The structure of anthraquinone has been established by its synthesis from *o*-benzoylbenzoic acid as shown :

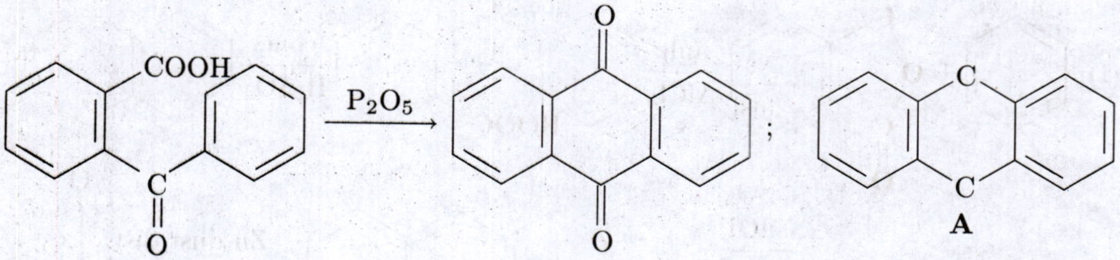


Thus, anthraquinone contains the carbon-skeleton as given above (A). Therefore, anthracene also contains the same carbon-skeleton as in A.

The following degradation reactions also support the presence of the carbon skeleton A in anthracene.

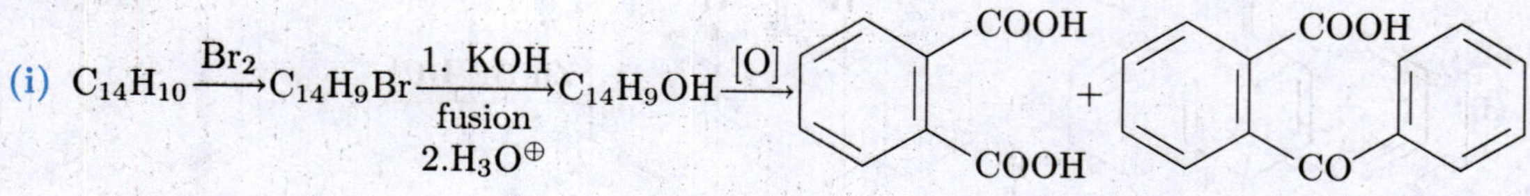


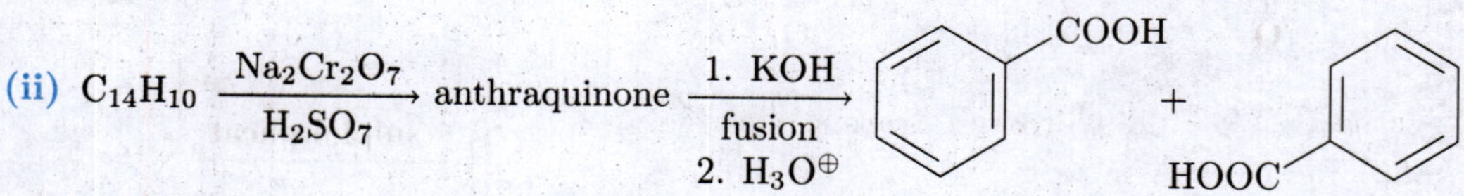


That anthracene contains the carbon-skeleton A has also been established by the fact that when two moles of *o*-bromobenzyl bromide react with metallic sodium, a dihydroanthracene forms which is identical with that obtained by the reduction of anthracene; dihydroanthracene on mild oxidation converts to anthracene.

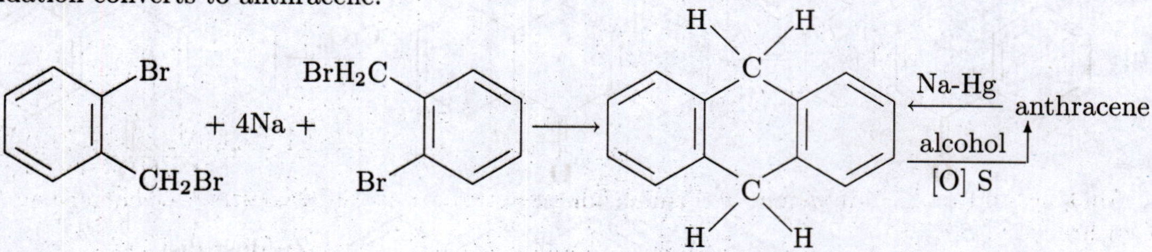


dihydroanthracene
$C_{14}$ $H_{12}$

Taking all these facts under consideration we can represent anthracene as a linear combination of three benzene rings in which the first and the third rings are fused at *o*-positions of the second ring as shown below.

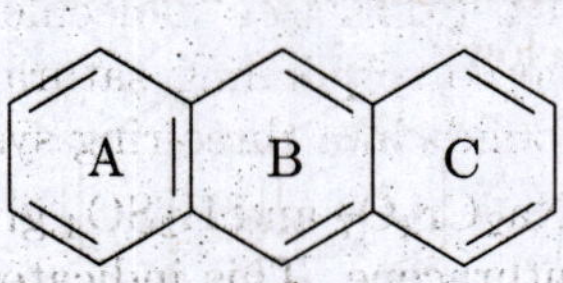


1

**Synthesis of anthracene from (i) benzene and phthalic anhydride, (ii) benzoquinone and butadiene and (iii) naphthaquinone and butadiene supports the structure 1.**

(i) 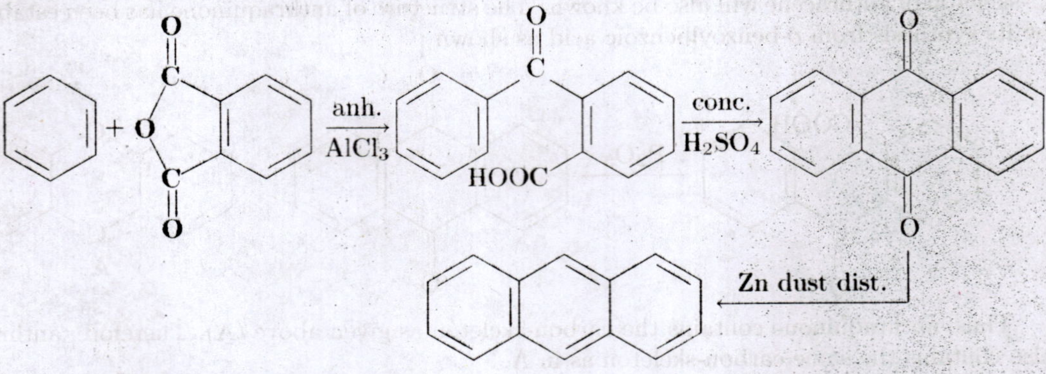


(ii) 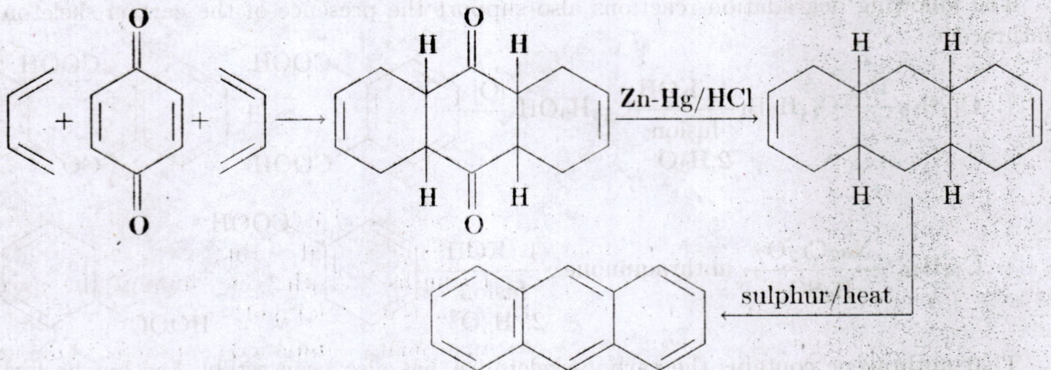


(iii) 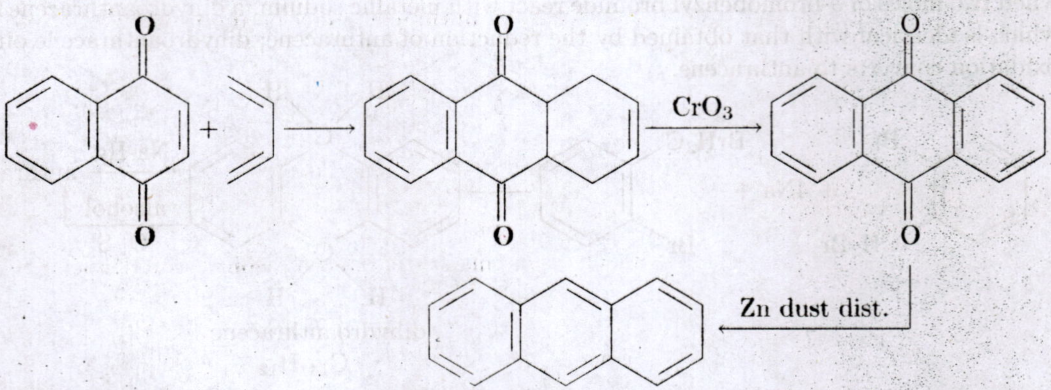


**The higher reactivity of 9- and 10-positions of anthracene and its formation from 2 molecules of benzene and one molecule of tetrabromoacetylene (See 18.8.3.c) lead us to think a para-bond structure of anthracene as given below :**

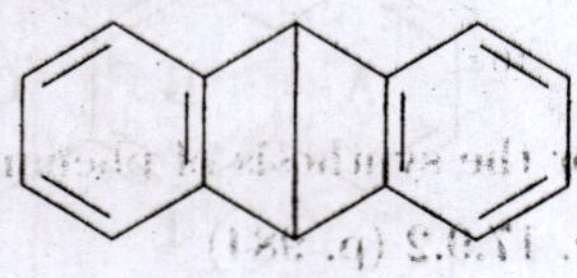

The following facts rule out the possibility of the para-bond structure of anthracene.

(i) It adds maleic anhydride to form 9,10-endoanthracenemaleic anhydride; this suggests the dienecharacter of the middle ring.

(ii) X-ray analysis shows that the distance between $C_9$ and $C_{10}$ atoms is the same as that between $C_1$ and $C_4$ atoms in benzene.

The V.B. Theory tells us that the structure of anthracene is a resonance hybrid of the following four structures.

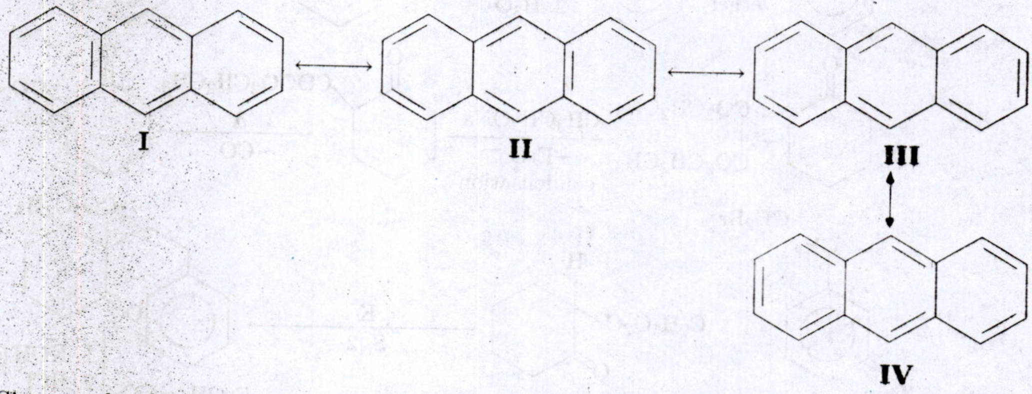


Since each of structures I and II contains two benzonoid rings, they contribute maximum to the hybrid. The isomer number of mono and disubstituted anthracenes, unsaturated and aromatic characters, and different C—C bond lengths in anthracene support the resonance hybrid structure. The resonance energy of anthracene is 351.5 kJ mol$^{-1}$.

From the standpoint of the M.O. Theory it is an aromatic compound containing 14 delocalised $\pi$ es in $7\pi$-bonding orbitals in a flat structure which encompass all the three rings. Thus to represent the aromatic character the structure of anthracene may be represented as :

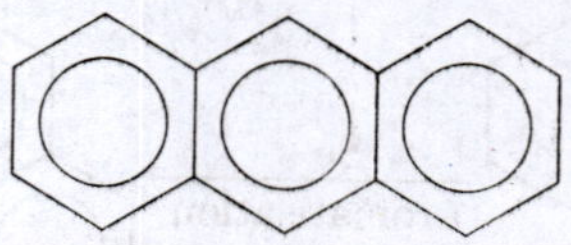

## 18.9 Phenanthrene

Phenanthrene is a tricyclic non-linear fused ring system and an isomer of anthracene.

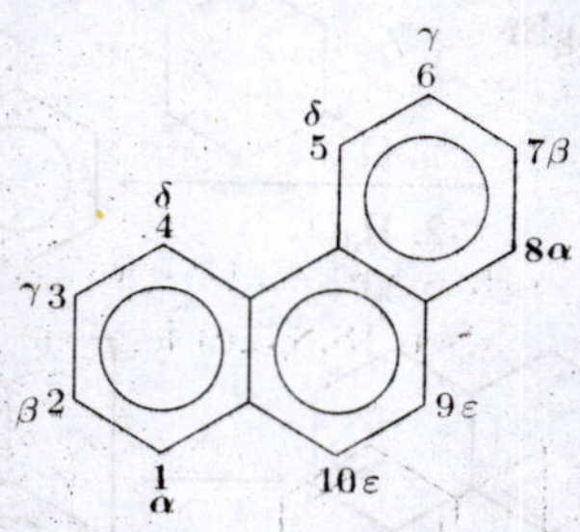


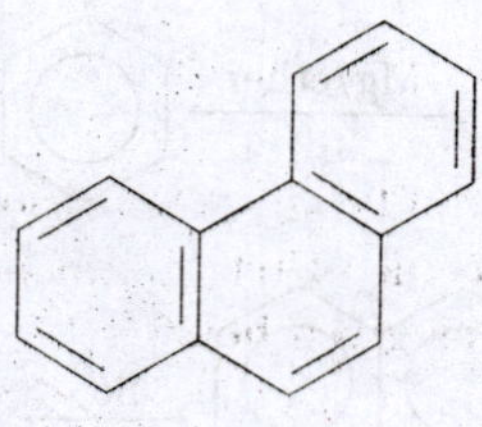

The four important methods for the synthesis of phenanthrene are given below :

(i) Haworth synthesis : See Sec. 17.9.2 (p. 984).

**(ii) Bardhan-Sengupta synthesis :** 1-Bromo-2-phenylethane and ethyl 2-oxo-cyclohexanecarboxylate are the starting materials for the synthesis of phenanthrene in the Bardhan-Sengupta synthesis (1932). The starting materials have been prepared as follows. The two units are then coupled as shown below :

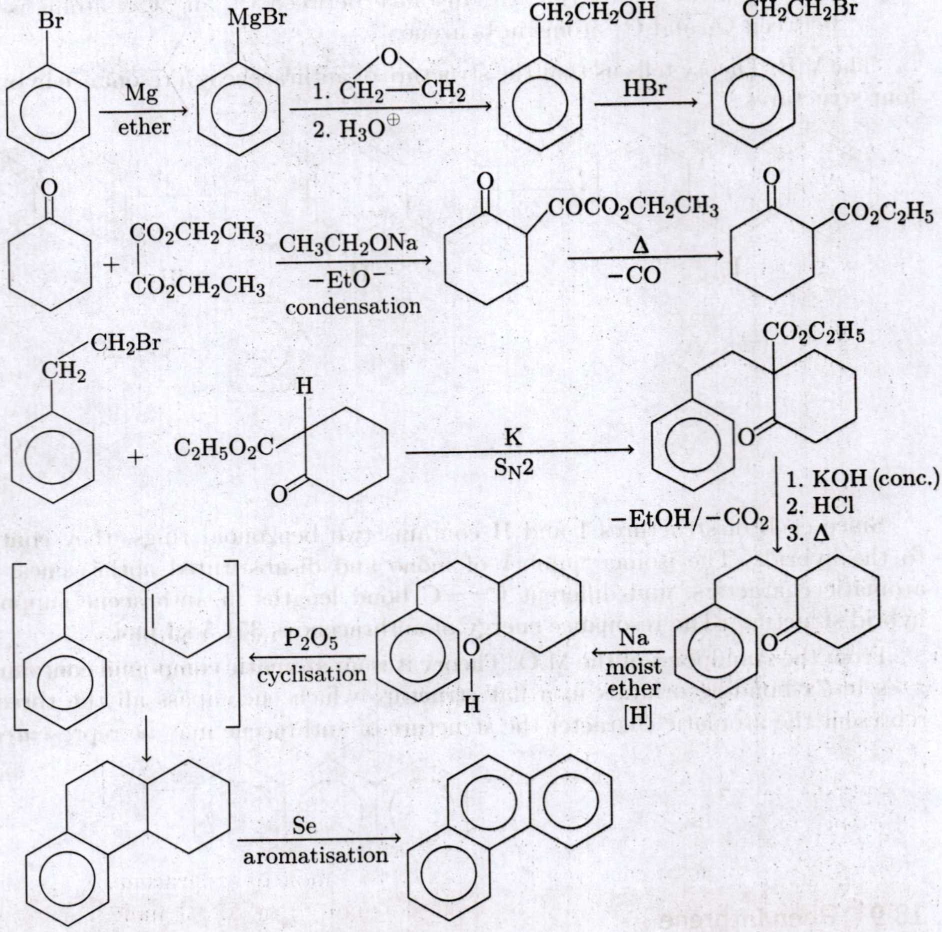


**(iii) Bogert-Cook synthesis :** This is as follows :

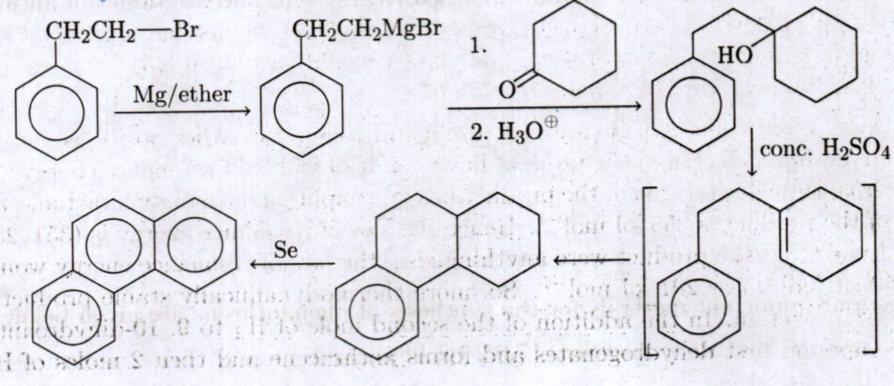

(iv) Pschorr synthesis : In this synthesis *o*-nitrobenzaldehyde and sodium phenylacetate are the starting materials. $Ac_2O$ is used to perform a Perkin-type condensation.

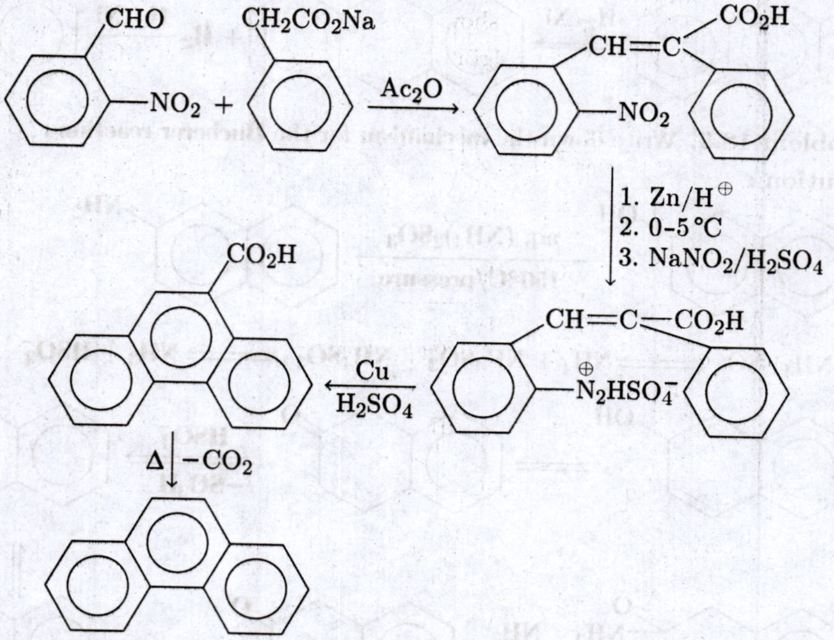


**Problem 18.8.** When 1 mole hydrogen is added to 1 mole of anthracene, 9,10-dihydroanthracene is the product. However, when 2 moles of hydrogen are added to 1 mole of anthracene 1, 2, 3, 4-tetrahydroanthracene is formed—Explain.

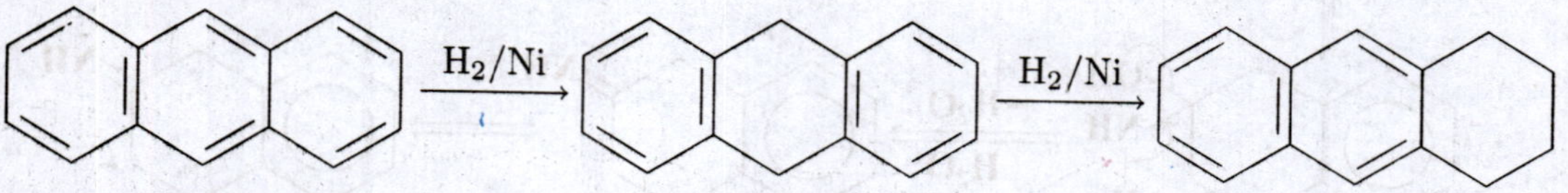


**Solution :** When 1 mole hydrogen is added to 1 mole of anthracene, 9, 10-dihydroanthracene is formed. In this reaction only (351 – 300 kJ $mol^{-1}$) or 51 kJ $mol^{-1}$ of resonance energy is lost since the resonance energy of anthracene is 351 kJ $mol^{-1}$ and that of two benzene rings in 9,10-dihydroanthracene is 300 kJ $mol^{-1}$. If the two H atoms were added to any other positions, only one naphthalene moiety would have been in the product having resonance energy 255 kJ $mol^{-1}$ and in this case the loss of resonance energy would have been [351 – 255 kJ $mol^{-1}$ (for the naphthalene moiety)] or 96 kJ $mol^{-1}$.

However, when 2 moles of $H_2$ are added to 9 and 10 and two other positions, only one benzene ring will remain unaffected in the product having 150 kJ $mol^{-1}$ of resonance energy. Since 1,2,3,4-tetrahydroanthracene is formed, the product has one naphthalene moiety intact and the resonance energy of the product is 255 kJ $mol^{-1}$. Hence the loss of resonance energy is (351–255 kJ $mol^{-1}$) or 96 kJ $mol^{-1}$. If the product were anything else, the loss of resonance energy would have been (351–150 kJ $mol^{-1}$) or 201 kJ $mol^{-1}$. So, more thermodynamically stable products are formed in the given reaction. In the addition of the second mole of $H_2$ to 9, 10-dihydroanthracene, the latter compound first dehydrogenates and forms anthracene and then 2 moles of $H_2$ are added to give 1,2,3,4-tetrahydroanthracene. The dehydrogenation of 9, 10-dihydroanthracene occurs because the catalyst for hydrogenation is also a catalyst for dehydrogenation (p. 847).

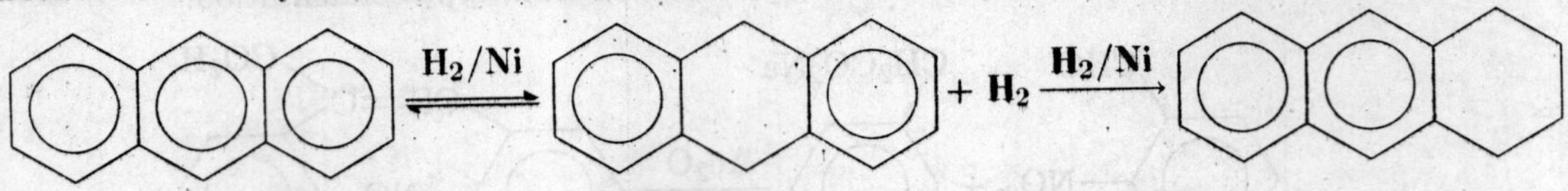


**Problem 18.9.** Write down the mechanism for the Bucherer reaction :

**Solution :**

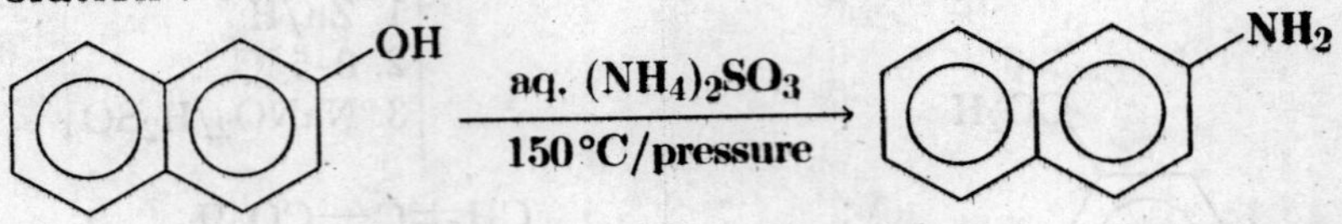


$$(NH_4)_2SO_3 \rightleftharpoons \overset{\oplus}{N}H_4 + NH_4SO_3^- \ ; \ NH_4SO_3^- \rightleftharpoons NH_3 + HSO_3^-$$

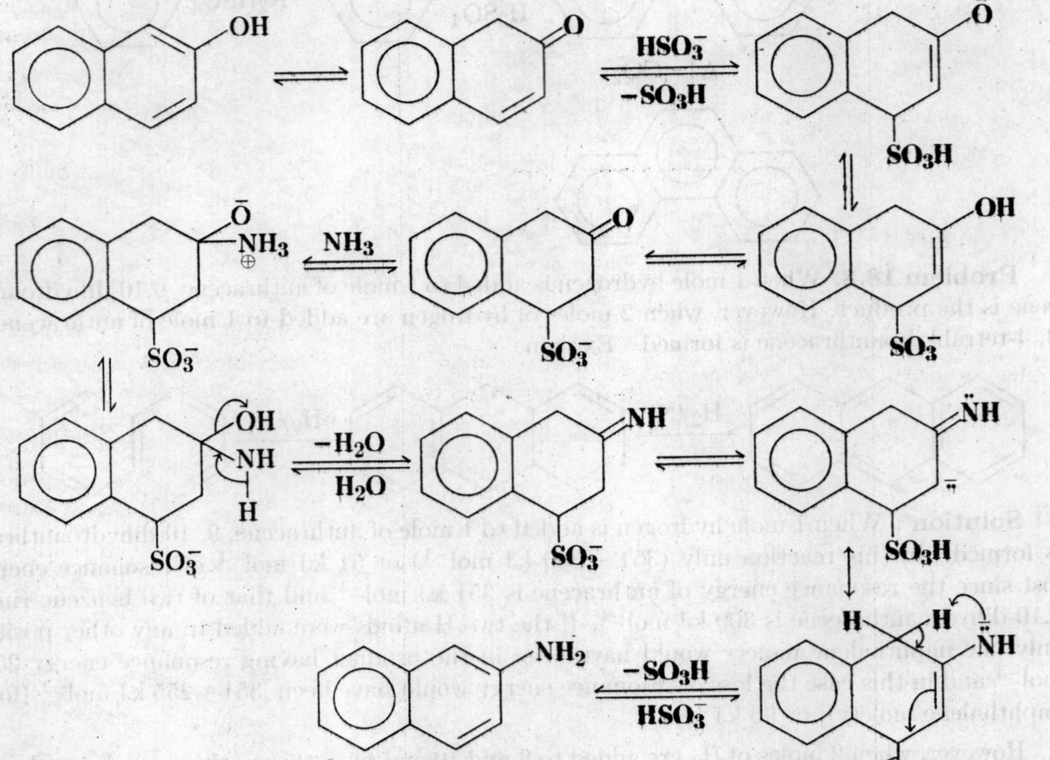


*β*-Naphthylamine

**Problem 18.10.** Name the compounds.

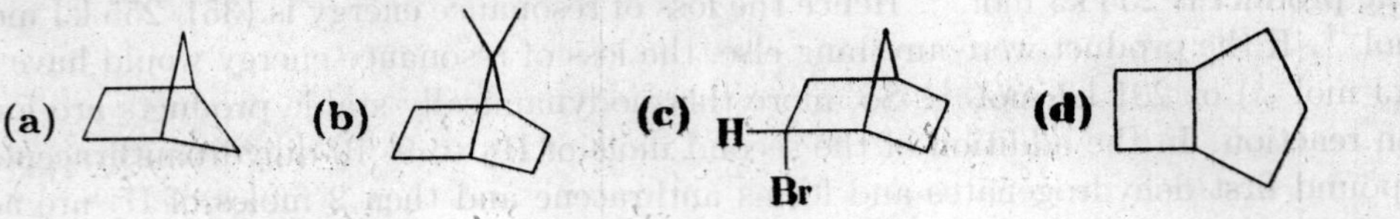

**Solution :** **(i)** The base name for the two fused ring system is the name of the alkane corresponding to the total number of C atoms in the two fused rings. So, the base names are :

**(a)** hexane, **(b)** heptane, **(c)** heptane, **(d)** heptane.

**(ii)** Since the given compounds have two fused rings, bicyclo will be prefixed.

**(iii)** In between 'bicyclo' and the base name the length of bridges are indicated by putting the number of C atoms involved in each bridge within crochets as indicated below. The bridgehead C atoms are not counted to express the chain length. The longer chain is indicated first, then the shorter chain and at last the chain joining the bridgehead—the C atoms common to both the chains are bridgehead carbons. So,

**(a)** Bicyclo [2.1.1] hexane **(b)** Bicyclo [2.2.1] heptane

**(c)** Bicyclo [2.2.1] heptane **(d)** Bicyclo [3.2.0] heptane

**(iv)** Substituents are indicated by prefixing or suffixing their names along with their locants as the cases are. The numbering begins from a bridgehead C through the longest bridge first, then through the second longest chain and back to the first bridgehead.

**(a)** 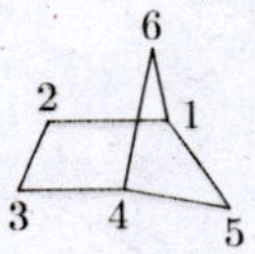


**(b)** 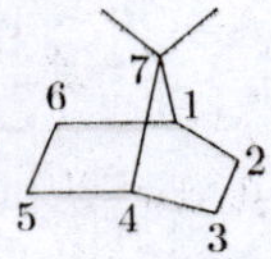


**(c)** 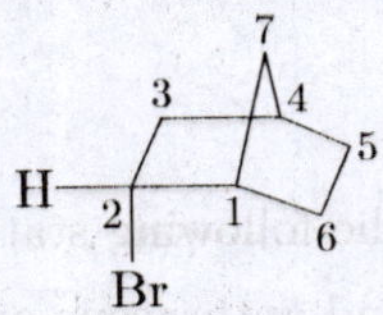


**(d)** 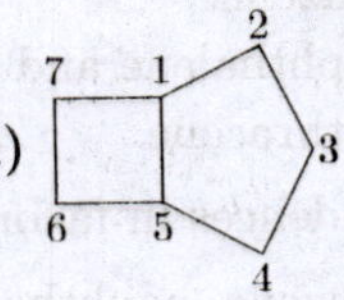


So, the names are

**(a)** Bicyclo [2.2.1] hexane **(b)** 7,7-Dimethylbicyclo [2.2.1] heptane

**(c)** Endo-2-bromobicyclo [2.2.1] heptane **(d)** Bicyclo [3.2.0] heptane.

## Exercise

1. Write notes on
   (a) The Wurtz reaction,
   (b) The Kolbe electrolysis,
   (c) The Clemmensen reduction,
   (d) Isomerisation of alkanes,
   (e) Alkylation of alkenes.
2. Give mechanism of :
   (a) Pyrolysis of alkanes,
   (b) Insertion of carbene to an alkane,
   (c) Isomerisation of alkanes,
   (d) Chlorosulphonation of alkanes,
   (e) Nitration of alkanes,
   (f) Dimerisation of alkenes.
3. Name the reagents for the following conversions :
   (a) Acetone to propane,

(b) *n*-Propyl bromide to propane,
(c) iso-Butane to *t*-butyl alcohol (alkaline $KMnO_4$ is the reagent which oxidises a 3°H to OH),
(d) 2-Butyne to cis-2-Butene and trans-2-butene,
(e) Allyl chloride to 1, 5-hexadiene,
(f) Propene to 1-propanol,
(g) Cyclohexanone to cyclohexane,
(h) Ethylene to cyclopropane, (i) Acetylene to chloroprene.

4. iso-Butylene forms two isomeric alkenes when it is heated with conc. $H_2SO_4$—name the products.
5. How can you extract : benzene, naphthalene and anthracene from coal tar?
6. How could homologues of benzene, naphthalene and anthracene be prepared?
7. Establish the structure of :
   (a) Benzene,
   (b) Naphthalene and
   (c) Anthracene.
8. Give evidences in favour of the following statements :
   (a) Benzene, naphthalene and anthracene are aromatic compounds.
   (b) Naphthalene is a bicyclic compound.
   (c) Anthracene possesses a linearly combined three-ring system.
   (d) C—C bonds in naphthalene are somewhat fixed.
9. Explain :
   (a) Naphthalene is more reactive than benzene.
   (b) 9-position of anthracene is more reactive than any other position.
   (c) Anthracene has a diene character.
   (d) 1, 3-pentadiene is more stable than 1, 4-pentadiene.
   (e) 1, 3-pentadiene is more reactive than 1, 4-pentadiene towards addition reactions.
10. Give the general methods for preparing conjugated alkadienes. What are their characteristic properties?
11. Give the general methods for the preparation of cycloalkanes and cycloalkenes. How could cycloalkane rings be opened?
12. Draw the conformations of *n*-butane and tell which one of them is the most stable. Draw an energy profile for the conformations.
13. Discuss conformations and stereoisomerism of cyclohexane, methylcyclohexane and dimethylcyclohexanes. [See Sec. 9.20].
14. Calculate the heat of combustion of methane : given that the bond energies (in kJ $mol^{-1}$) of C—H: 415.8, C=O: 806.4; O=O: 499.8 and O—H: 464.2. [See Problem 18.1.]
15. Convert :
   (a) Acetylene to benzene and vice versa.
   (b) Naphthalane to benzene.

(c) Adipic acid to cyclopentane.
(d) Ethyl acetate to Ethyl acetoacetate.
(e) Ethyl malonate to 1, 4-cyclohexanedicarboxylic acid.
(f) Cyclobutanecarboxylic acid form ethyl malonate.
(g) Cyclopentanone from succinic acid.

16. What type of isomerism is expected from the compounds of constitutional formula, R—CH=CH—R and RCH=C=CHR. The isomers of the former differ much in their properties but the isomers of the latter do not—explain.

17. Write down the conditions, reagents, products and mechanism of nitration, sulphonation, alkylation, chloromethylation, halogenation and formylation of benzene, naphthalene and anthracene.

18. Write down the products of ozonolysis of the following compounds :

Cyclopentene, 1, 3-butadiene, benzene, *o*-xylene and naphthalene.

19. Which products do you expect when trans- and cis-2-chlorocyclohexanol are separately treated with NaOEt in EtOH.

[Hint :

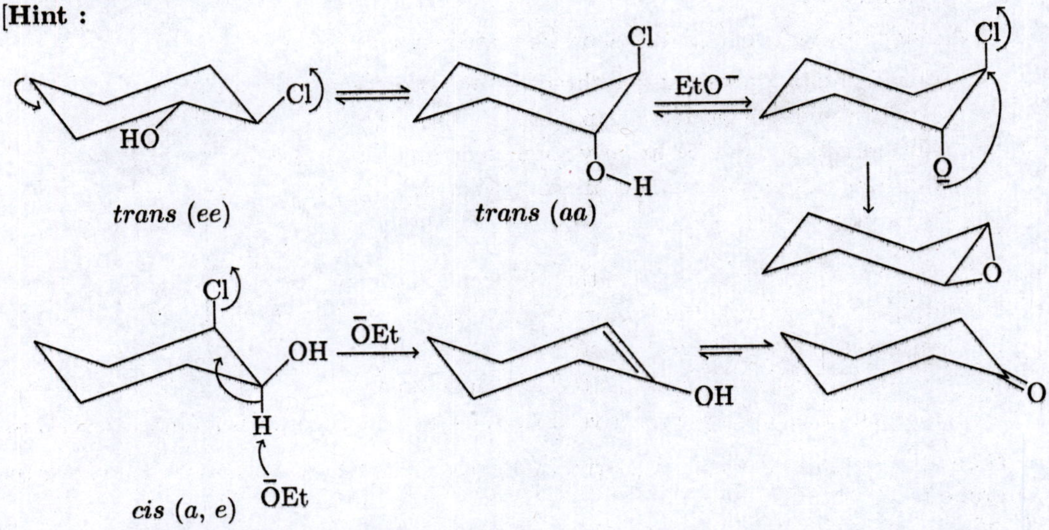


20. Synthesise the anthraquinoid dye, alizarin.

(c) Adipic acid to cyclopentane.

(d) Ethyl acetate to Ethyl acetoacetate.

(e) Ethyl malonate to 1, 4-cyclohexanedicarboxylic acid.

(f) Cyclopentanecarboxylic acid form ethyl malonate.

(g) Cyclopentanone from succinic acid.

16. What type of isomerism is expected from the compounds of constitutional formula, R — CH = CH — R and ROH = C = CHR. The isomers of the former differ much in their properties but the isomers of the latter do not—explain.

17. Write down the conditions, reagents, products and mechanism of nitration, sulphonation, alkylation, chloromethylation, halogenation and formylation of benzene, naphthalene and anthracene.

18. Write down the products of ozonolysis of the following compounds :

Cyclopentene, 1, 3-butadiene, benzene, o-xylene and naphthalene.

19. Which products do you expect when *trans*- and *cis*-2-chlorocyclohexanol are separately treated with NaOEt in EtOH.

**Hint :**

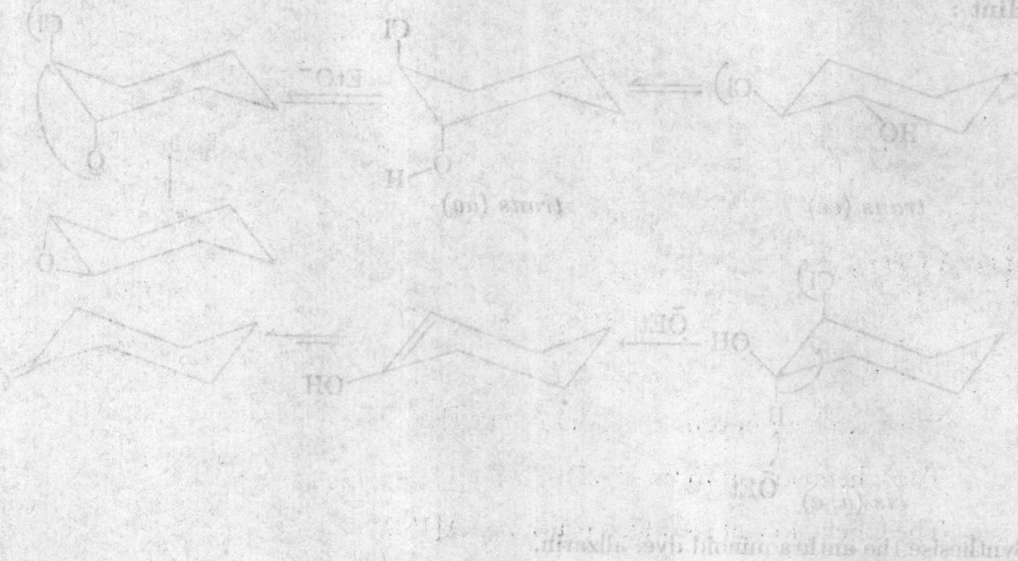


20. Synthesise the anthraquinoid dye alizarin.

# Chapter 19

# A Few Heterocyclic Compounds

## Objectives

- To know :

## 19.1 Introduction

Any cyclic organic compound in which there is at least one atom of an element other than carbon, called heteroatom, among others which constitute the ring is called a heterocyclic compound. According to the above definition of heterocyclic compounds ethylene oxide, succinimide, phthalic anhydride, etc. may be classed under the name of heterocyclic compounds; but their most of the properties are so similar to either aliphatic or aromatic compounds that they are not generally treated as heterocycles. However, heterocyclic compounds are numerous. We shall discuss here some of the heterocyclic compounds containing only one hetero atom and possessing Hückel number of $\pi$es, i.e., $(4n+2)$ delocalised $\pi$es in flat cyclic systems which encompass all the atoms constituting the ring or rings (in the cases of fused ring systems); three of them have five membered ring structures, one of them has a six membered ring structure and three others have fused ring systems. The names and their structures are given below :

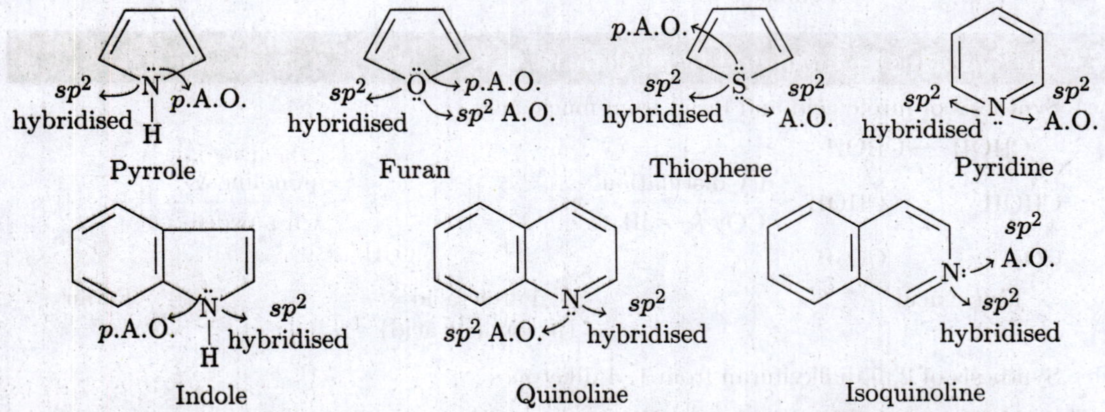


2-Formylfuran, called furfural, is a starting material in the synthesis of nylon; it is also a good insecticide. Pyrrole is a constituent of haemoglobin, an iron complex; it is also a constituent of chlorophyll, the key substance for the synthesis of food for the plants. $\alpha$- and $\beta$-Biotins (Vitamin H) which belong to the class of vitamins contain thiophene moiety. Tryptophan, an essential aminoacid, contains indole moiety. When proteins putrefy, the bad smell is given by skatole which also contains indole ring. Pyridine and quinoline derivatives have immense medicinal use. 8-hydroxyquinolines have antimalerial and antiamebic actions. Pyridine is a very good solvent for organic compounds. Alkaloids contain isoquinoline skeleton. All these information tell us that the above mentioned heterocycles are very important for our own interest and hence they are delt with here in brief.

## 19.2 Furan

R.E. $= 98$ kJ mol$^{-1}$

b.p. $= 32\,°C$

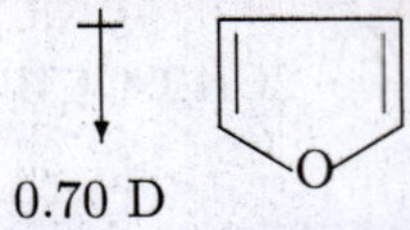


$C_4H_4O$

### 19.2.1 Source

It occurs with its methyl homologues in wood tar, specially in pine-wood tar. It is manufactured from furfural, which, in its turn, is made from oat and rice hulls and corncobs by the treatment of hot hydrochloric acid. Oat hulls, rice hulls and corncobs contain pentosan which on treatment with hot hydrochloric acid gets converted into furfural. When furfural is decomposed in steam at 400 °C in the presence of oxide catalyst, furan is obtained.

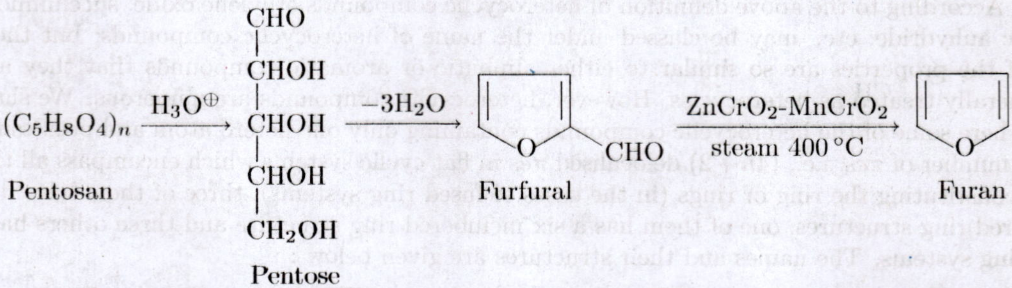

## 19.2.2 Synthesis of furan and furan derivatives

(a) Synthesis of furoic acid and furan from mucic acid :

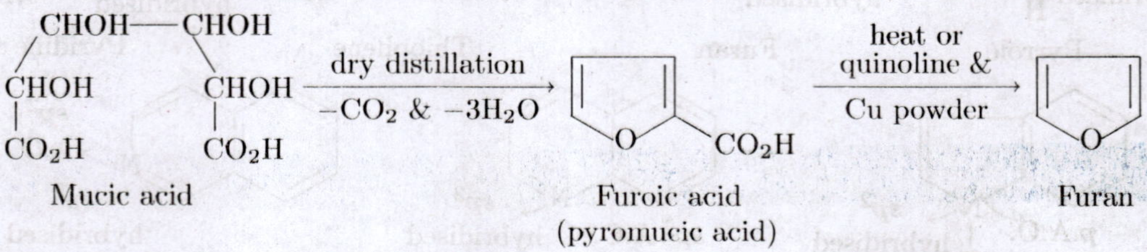

(b) Synthesis of 2, 5-dialkylfuran from 1, 4-diketones :

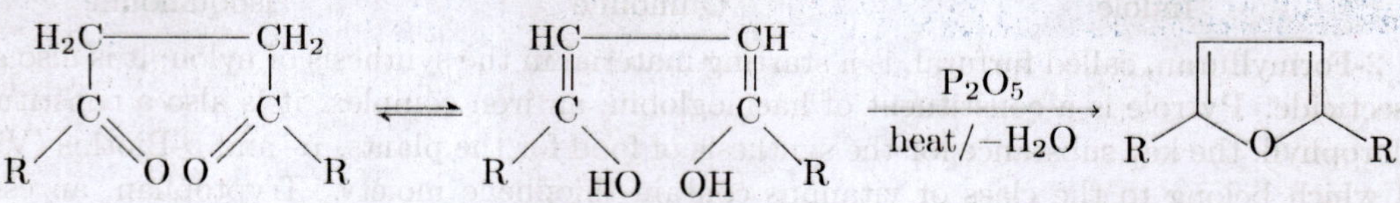

(c) Synthesis of tetrahydrofuran from 1, 4-butanediol :

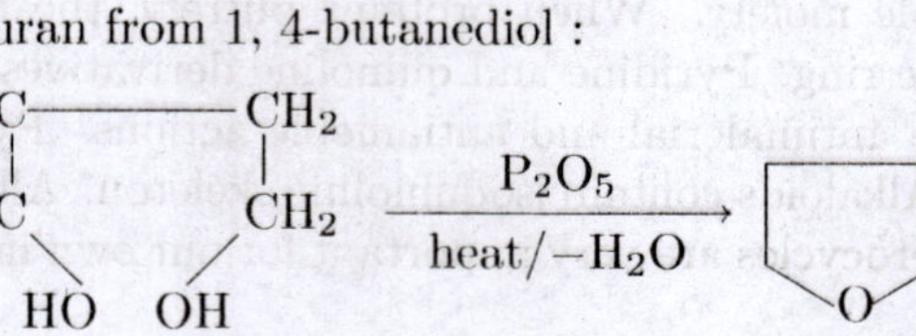

(d) Synthesis of furan derivatives from ethyl acetoacetate :

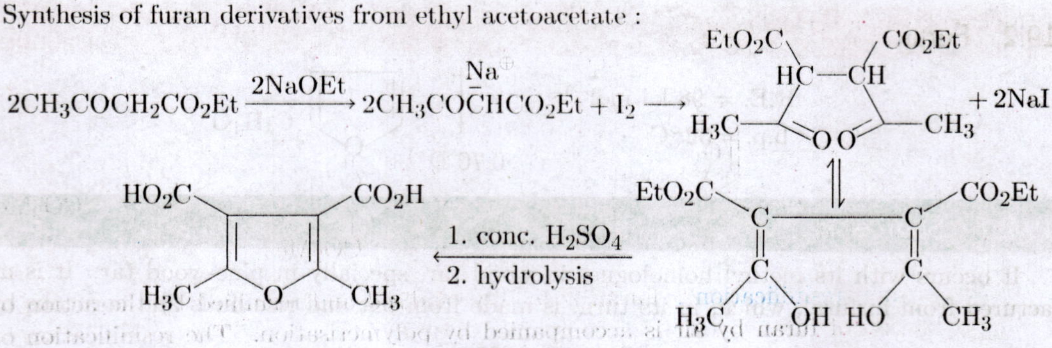

**(e)** Synthesis of furan derivatives from $\alpha$-chloroketones (**Feist-Benary synthesis**) : When an $\alpha$-chloroketone is condensed with a $\beta$-ketoester in the presence of pyridine, a derivative of furan is obtained.

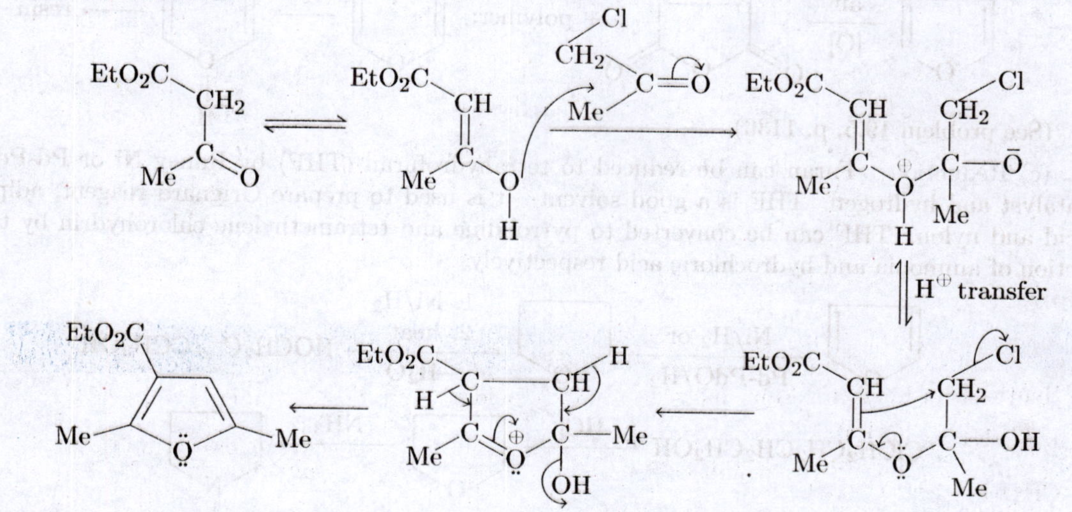

## 19.2.3 Structure, reactivity and orientation

See Sec. 4.7 (p. 182) and Sec. 17.4.4.c (p. 948–949).

## 19.2.4 Chemical properties

Though furan is an aromatic compound yet it shows a few properties of conjugated dienes. It is also less aromatic than thiophene and pyrrole; perhaps the strong electronegative character of the O atom makes it draw the $\pi$-cloud much towards itself and thus makes furan less aromatic than the other two compounds. This is also supported by their resonance energy order as given below :

$$\text{Furan} < \text{pyrrole} \approx \text{thiophene}$$

Of course, the resonance energy of furan is much higher than that of a conjugated diene. However, the following reactions will show its aromatic as well as diene characters. Furan is very weakly basic; it forms oxonium salts with acids; because of this oxonium ion it polymerises in the acid medium.

(a) The Diels-Alder reaction : Among furan, pyrrole and thiophene, only furan undergoes the Diels-Alder reaction, a typical character shown by the conjugated dienes.

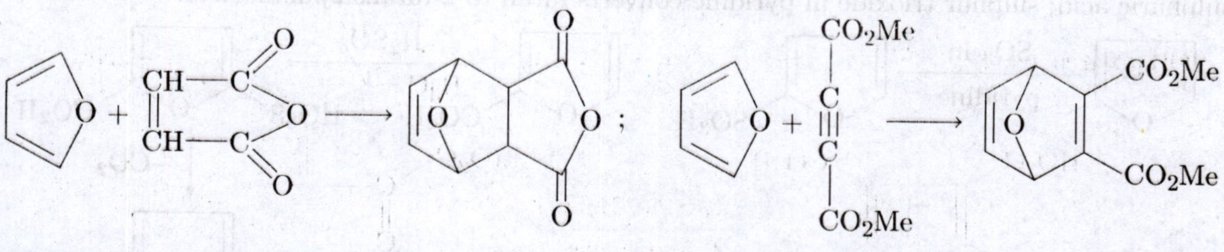

(b) Oxidation and resinification : Furan is readily oxidised and resinified by the action of acids. The oxidation of furan by air is accompanied by polymerisation. The resinification of furan occurs when any mineral acid is added to it. It is needless to mention that polymerisation is also a typical character of 1, 3-dienes.

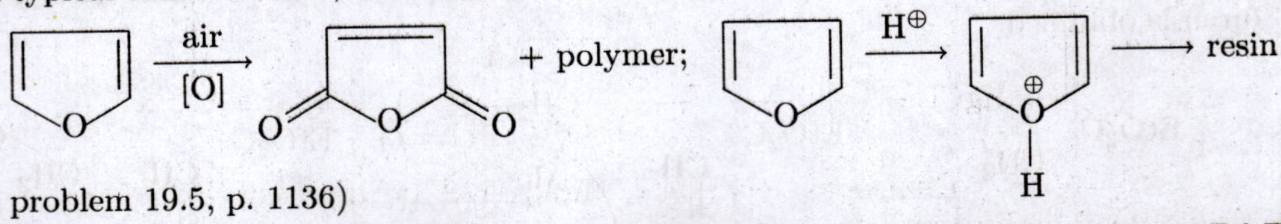


(See problem 19.5, p. 1136)

(c) Reduction : Furan can be reduced to tetrahydrofuran (THF) by Raney Ni or Pd-PdO catalyst and hydrogen. THF is a good solvent—it is used to prepare Grignard reagent, adipic acid and nylon. THF can be converted to pyrrolidine and tetramethylene chlorohydrin by the action of ammonia and hydrochloric acid respectively.

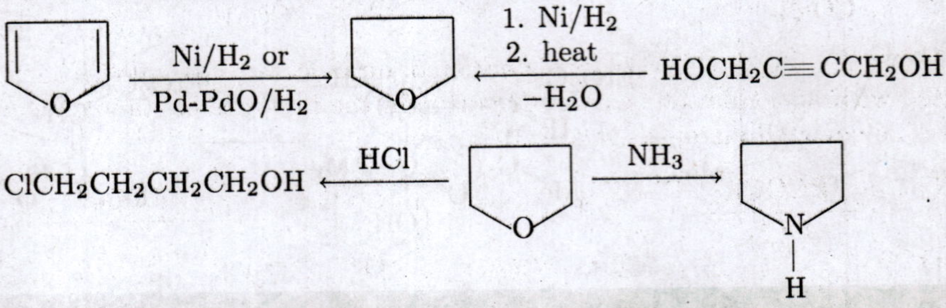


(d) Nitration : Since furan polymerises in the presence of mineral acids, it cannot be nitrated by the mixed acid; 2-nitrofuran may be prepared by nitrating furan with acetylnitrate. On the other hand, 2-nitrofuran may also be prepared by nitrating furoic acid to 5-nitrofuroic acid and then decarboxylating it; the electron-withdrawing effect of the carboxyl group makes furoic acid less susceptible to oxidation than furan.

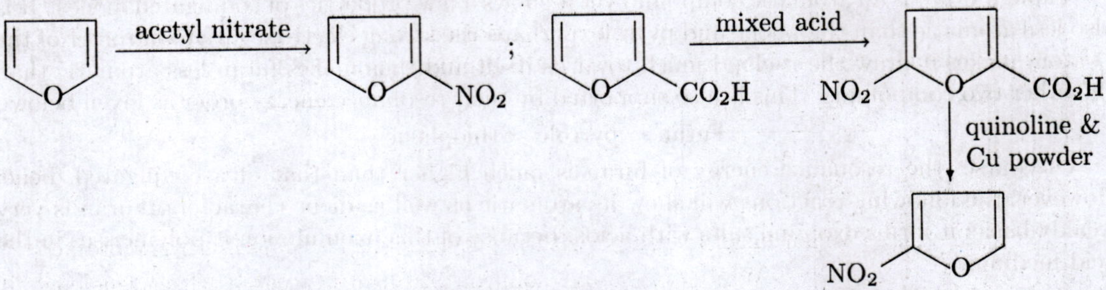


(e) Sulphonation : For the same reason as given for nitration it cannot be sulphonated by sulphuric acid; sulphur trioxide in pyridine converts furan to 2-furansulphonic acid.

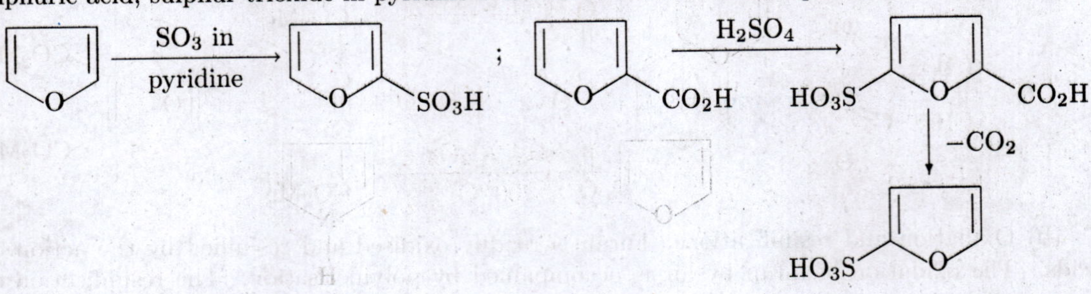

(f) Mercuration : When furan is heated with mercuric chloride in the aqueous sodium acetate medium, 2-chloromercurofuran is obtained.

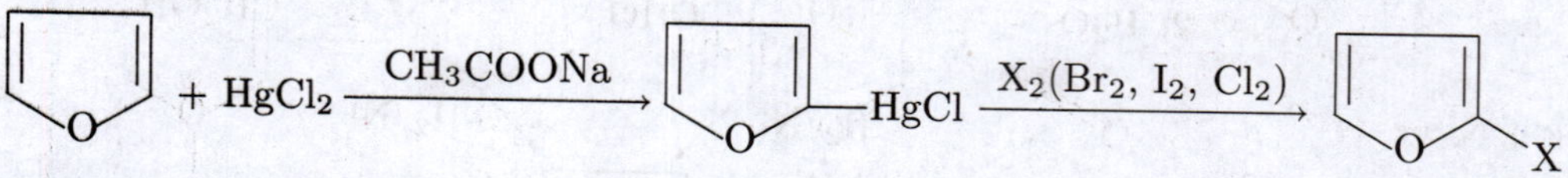

(g) Halogenation : Direct halogenation of furan gives polymer; since the mercuri group is readily replaced by Br and I, 2-bromo or 2-iodofuran can be prepared via 2-chloromercurofuran as shown above. However, 2-chloro- and 2-bromo-derivatives can also be prepared from furoic acid.

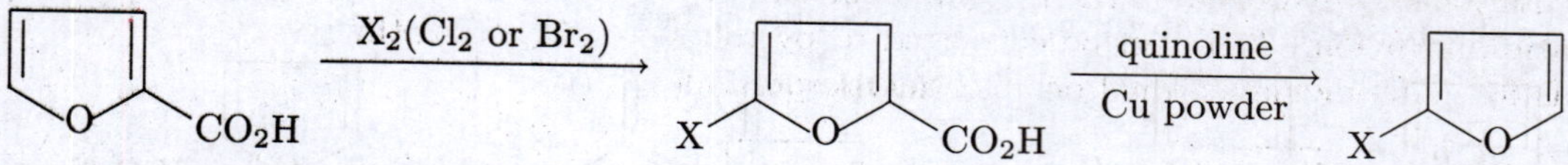

(h) Acylation : The Friedel-Crafts alkylation of furan leads to polymerisation. However, it can be acylated with acyl chloride or acid anhydride in the presence of stannic chloride as catalyst. It can also be acylated via mercuration.

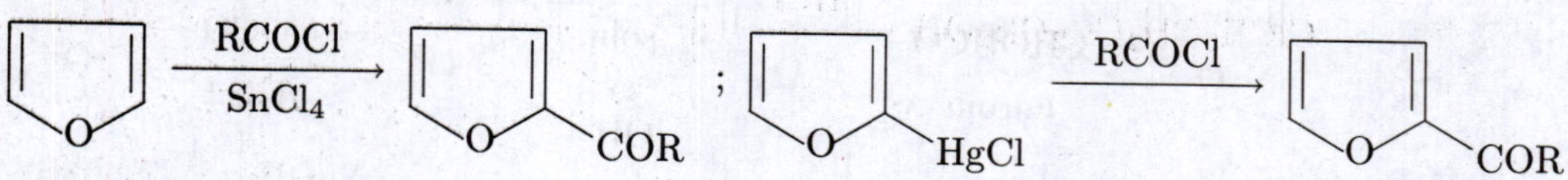

(i) The Gattermann reaction—formylation :

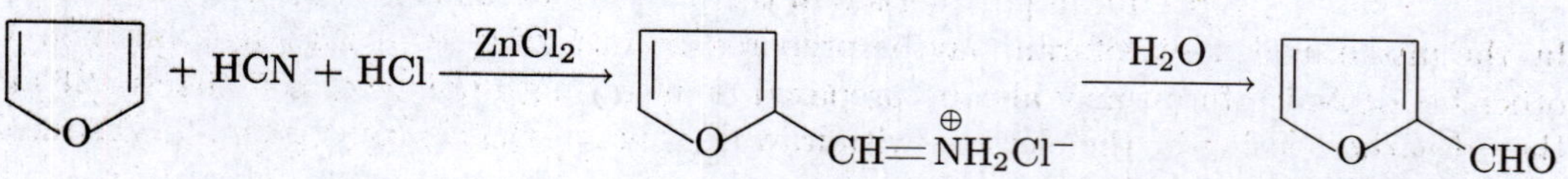

Furfural

(j) Furoic acid from furan : When furfural is oxidised with $Cu_2O$-$Ag_2O$ in aqueous NaOH, it forms furoic acid; this may also be prepared via 2-lithiumfuran.

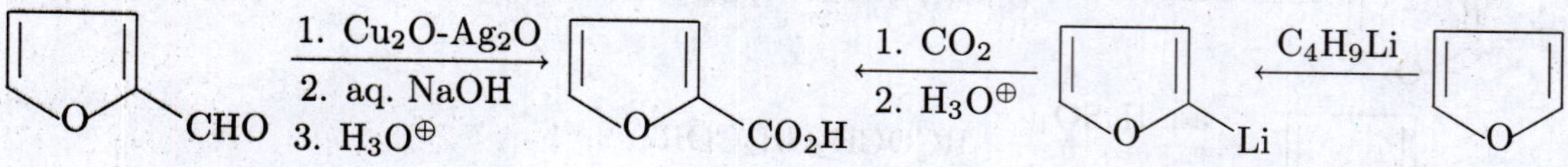

(k) Metalation : When furan is treated with *n*-butyllithium, 2-lithiumfuran is obtained.

(l) Gomberg reaction : When an alkaline solution of furan is treated with arenediazonium salt, 2-arylfuran is obtained.

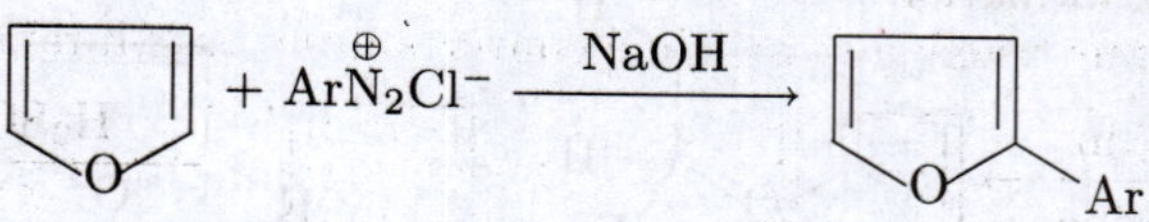

(m) Conversion of furan to pyrrole and pyridine :

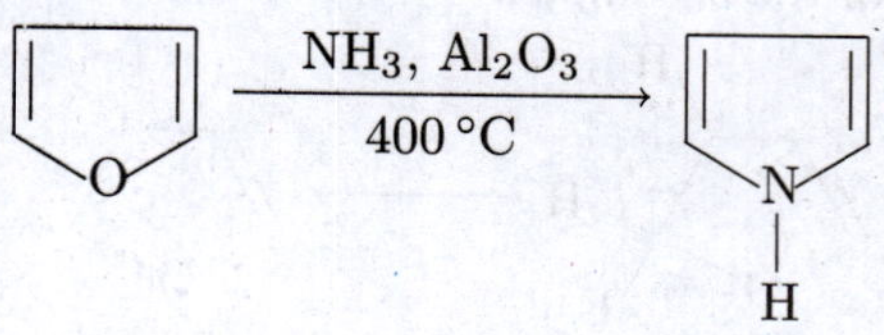

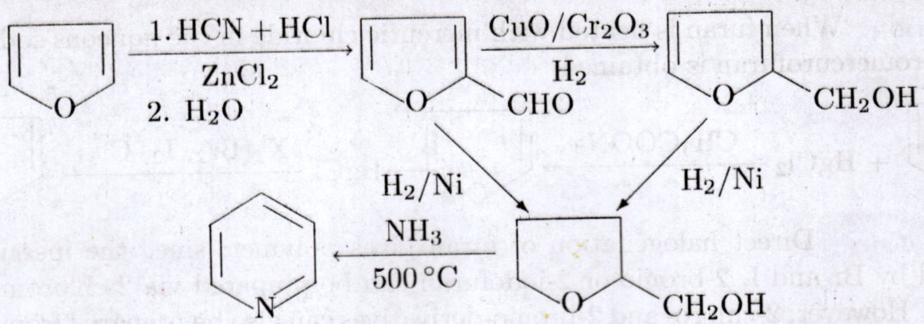


(n) Conversion of furan to furoin, furil and furilic acid, etc :

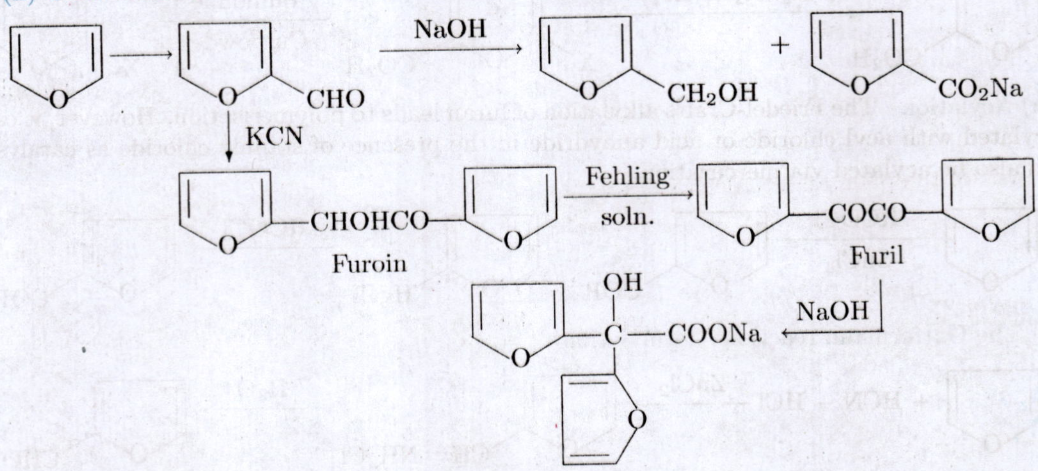


Furilic acid

(o) Ring-opening reactions :

(i)

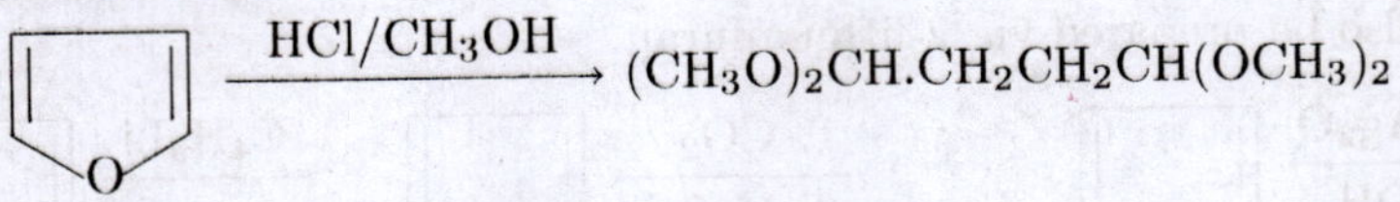


(ii)

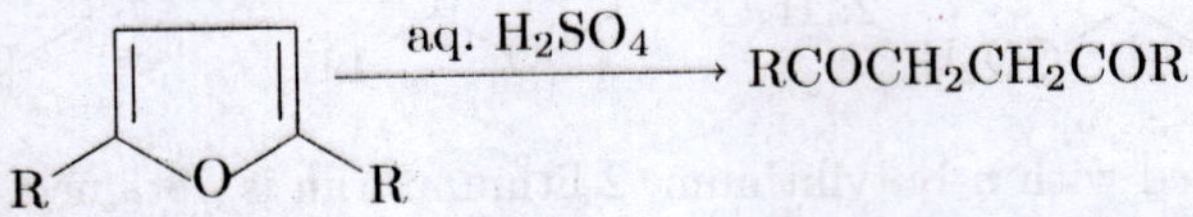


**Mechanism of Ar.S$_E$2 reaction :** It consists of two steps :

**Step I.** $\sigma$-complex formation

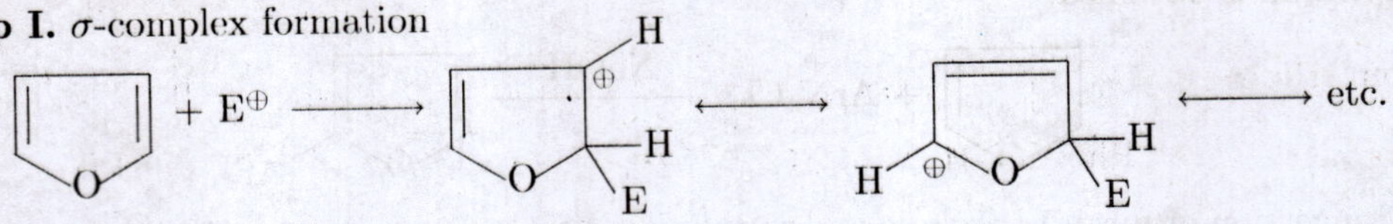


**Step II.** Aromatisation of the $\sigma$-complex

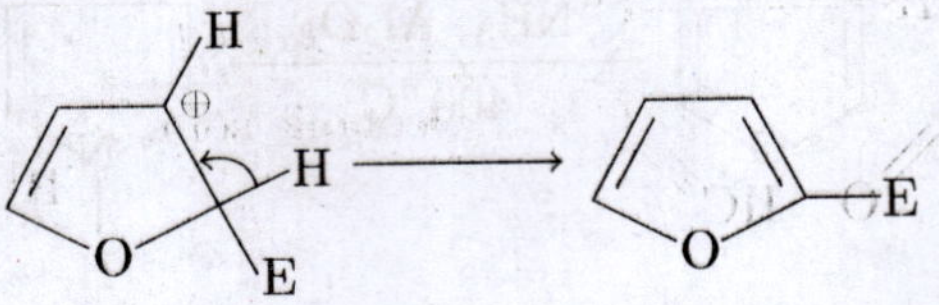

## 19.3 Thiophen (Thiophene)

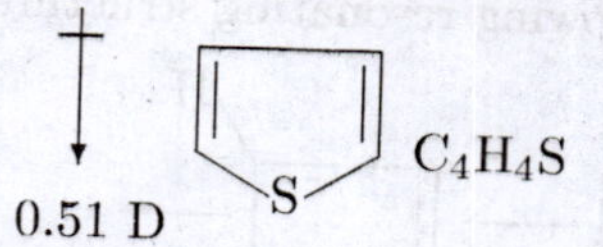

$C_4H_4S$

R.E. = 125.6 kJ $mol^{-1}$
m.p. = −38 °C
b.p. = 84.12 °C

### 19.3.1 Source

Coal tar and Shale oils

Crude benzene from coal tar
Crude benzene contains thiophene

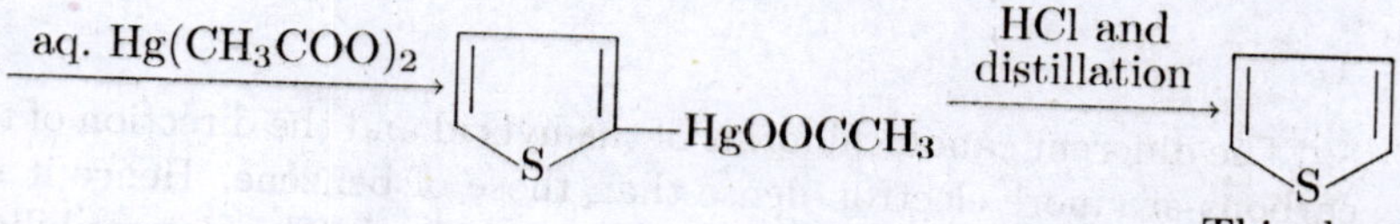

Thiophen

### 19.3.2

Synthesis of thiophen and its derivatives.

**(a)** Commercial processes :

**(i)** From acetylene

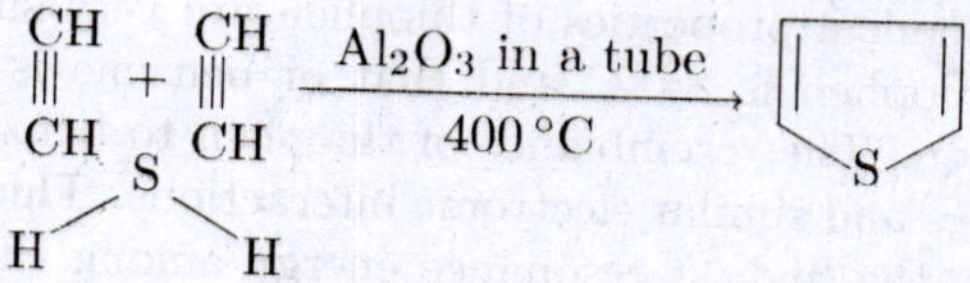

**(ii)** From *n*-butane

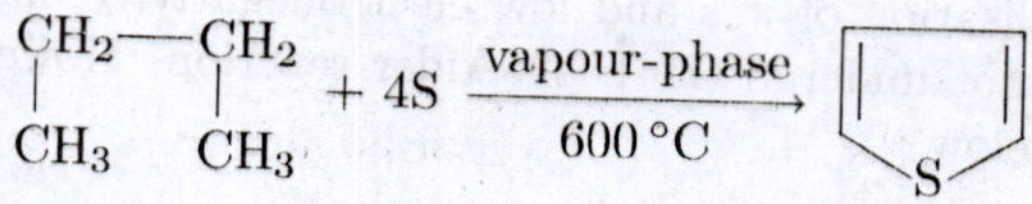

**(b)** From, 1, 4-diketones :

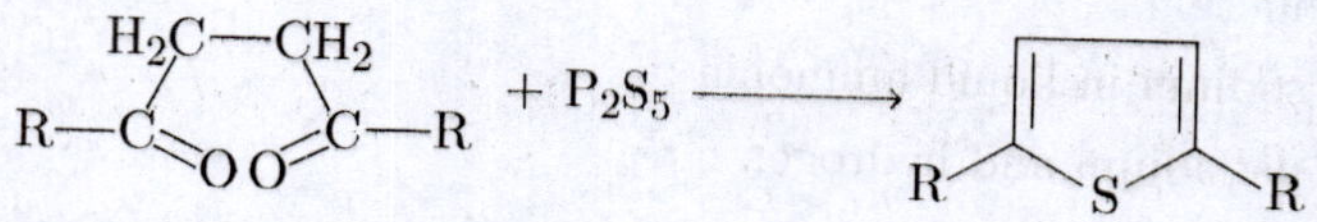

**(c)** From succinate :

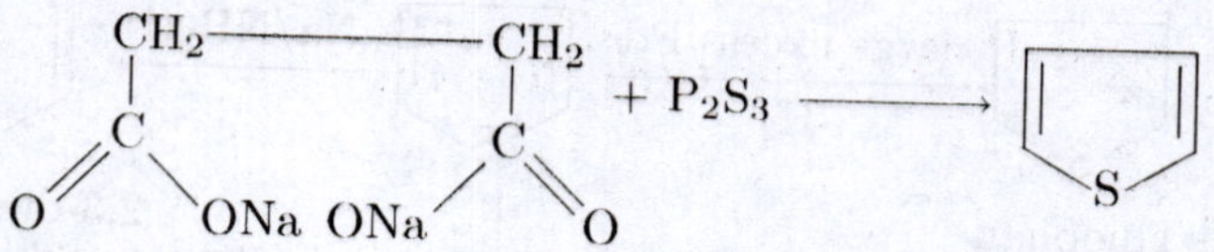

**(d)** From stilbene :

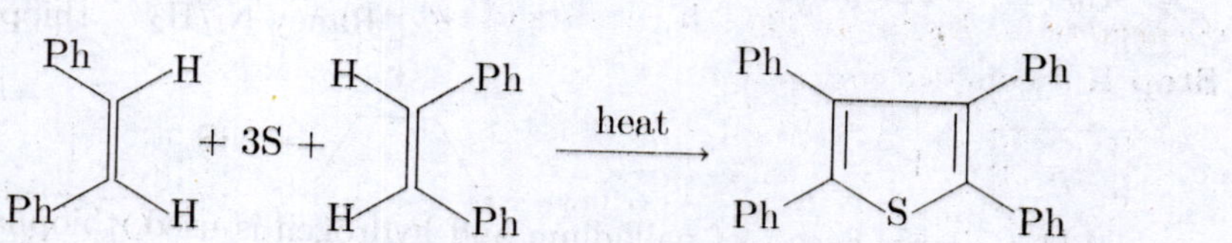

**(e)** From ynones :

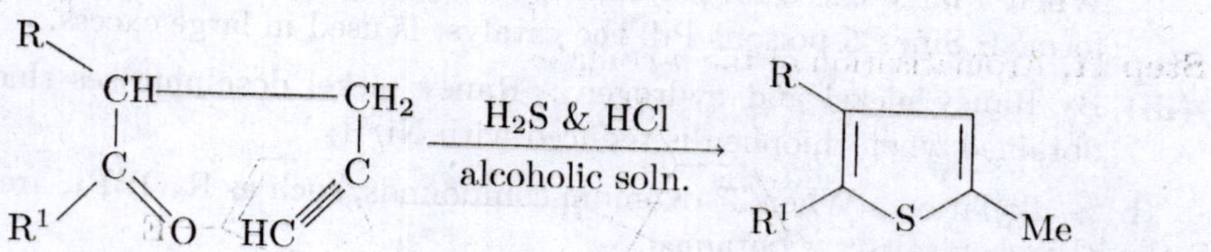

## 19.3.3 Structure, reactivity and orientation

Thiophen is a resonance hybrid of the following resonating structures :

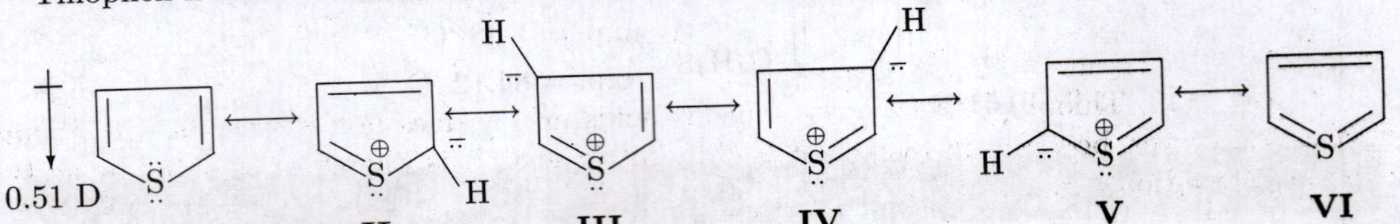

0.51 D I II III IV V VI involving *d* orbital

The different canonical forms of the hybrid and the direction of the moment show that the ring carbons are more electron-dense than those of benzene. Hence it is more reactive than benzene in so far as electrophilic substitution is concerned. The electrophilic substitution occurs at 2- and 5-positions (Sec. 17.4.4c, p. 949). The presence of any group at 2-position directs the incoming electrophile at 5-position.

## 19.3.4 Physical and chemical properties

The physical and chemical properties of thiophen are very similar to those of benzene. For example, the b.p. of thiophen is 84 °C and that of benzene is 80 °C, both are highly stable and resistant to oxidation. The resemblance of thiophen to benzene may be ascribed to similar molecular weights, shapes and similar electronic interactions. Thiophen is more stable than furan and pyrrole since it has the highest resonance energy among the three. The involvement of *d* orbital in the delocalisation of $\pi$es and low electronegativity of S perhaps bring extra stability to thiophen; it does not undergo the Diels-Alder reaction. However, the important reactions of thiophen are given below :

(a) **Reduction :**

**(i)** By sodium in liquid ammonia :

**(ii)** By palladium and hydrogen :

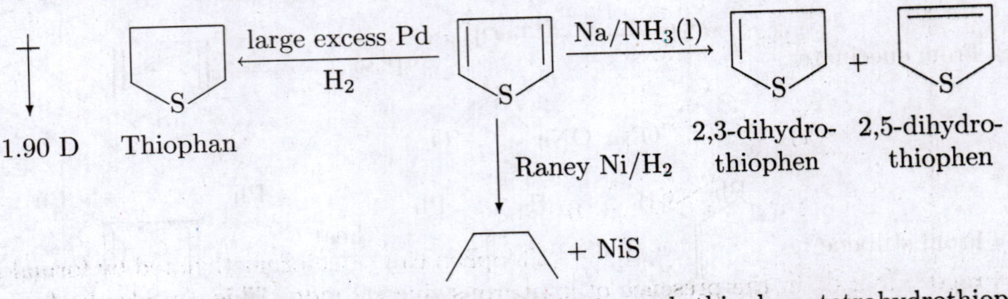



1.90 D Thiophan 2,3-dihydro-thiophen 2,5-dihydro-thiophen

When a large excess of palladium and hydrogen is used, thiophan, tetrahydrothiophen is formed. Since S poisons Pd, the catalyst is used in large excess.

**(iii)** By Raney nickel and hydrogen : Raney nickel desulphurises thiophen and *n*-butane is obtained when thiophen is reduced with $Ni/H_2$.

(b) **S-alkylation :** When 3° oxonium compounds, such as $R_3\overset{\oplus}{O}\overset{-}{B}F_4$, are treated with thiophen, S-alkylthiophen salts are obtained.

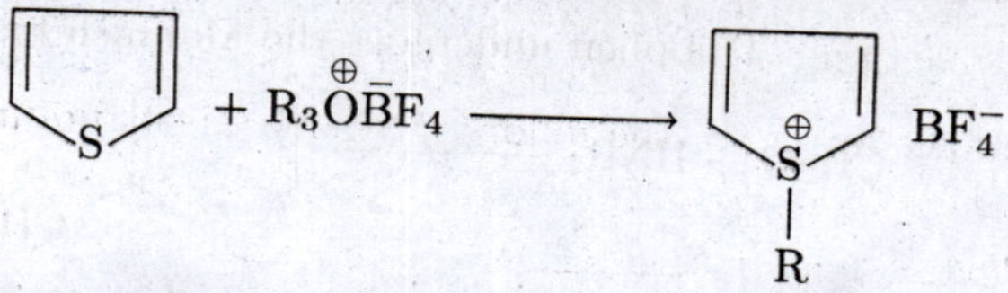

(c) Nitration : Thiophen is stable in aqueous acids and it does not form salts with them; but it can, however, be oxidatively decomposed by nitric acid. It is, therefore, nitrated by acetyl nitrate at 2-position.

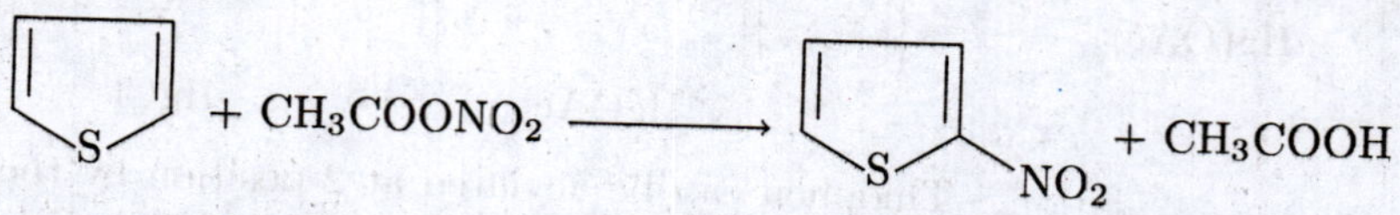

(d) Sulphonation : In contrast to benzene, it is readily sulphonated by cold concentrated sulphuric acid.

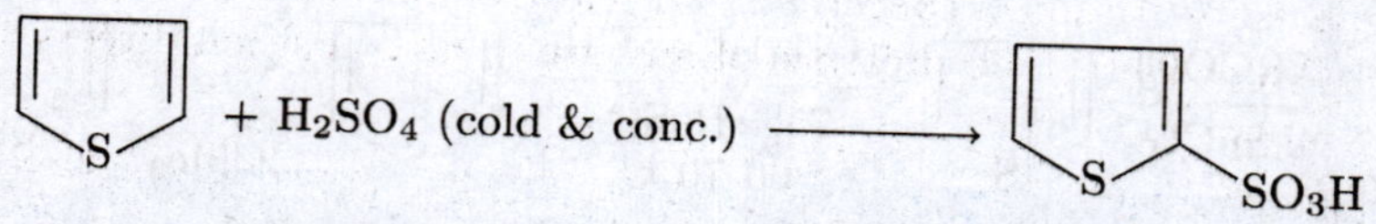

(e) Halogenation : When thiophen is treated with chlorine at −30 °C, 2-chloro- and 2, 5-dichlorothiophens are formed as the main products. However, thiophen can be brominated at 2-position by N-bromosuccinimide (NBS); whereas 2-iodo-derivative may be prepared by treating it with iodine in the presence of yellow mercuric oxide.

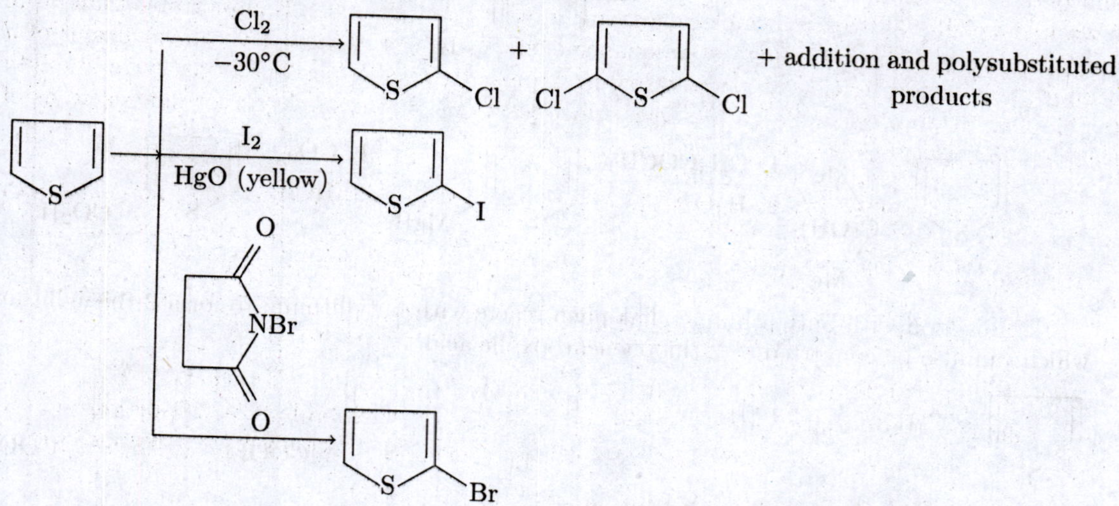

(f) Chloromethylation and formylation : Thiophen can be chloromethylated by formaldehyde and hydrogen chloride in the presence of anhydrous zinc chloride. This can also be formylated by the Vilsmeier-Haak method.

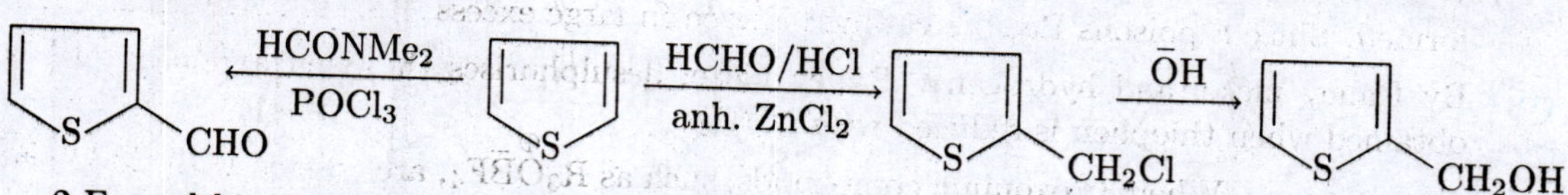

2-Formylthiophen undergoes the Cannizzaro reaction and the benzoin condensation.

(g) The Mannich reaction : Thiophen undergoes the Mannich reaction.

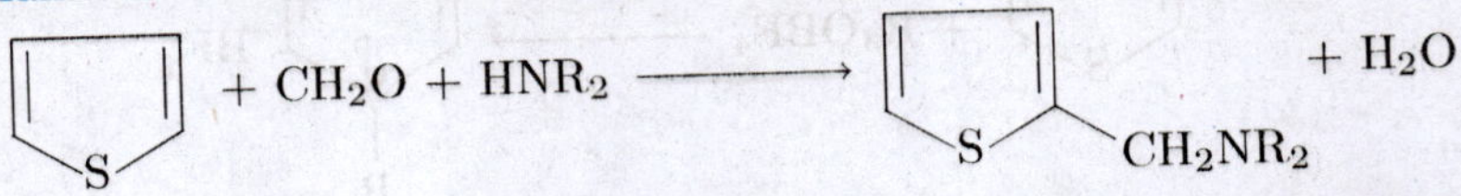

(h) Mercuration : Thiophen, like furan, undergoes the mercuration reaction with murcuric acetate even in the cold but benzene does not.

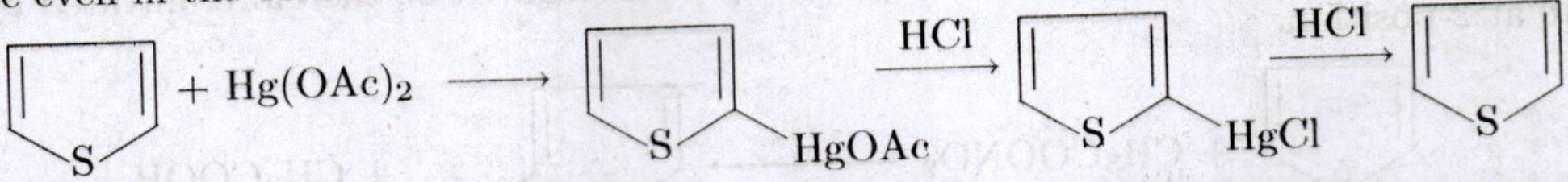

(i) Acylation and alkylation : Thiophen can be acylated at 2-position by the Friedel-Crafts method using stannic chloride as catalyst; but when thiophen is alkylated with isobutylene, a mixture of 2- and 3-t-butyl-thiophens is obtained.

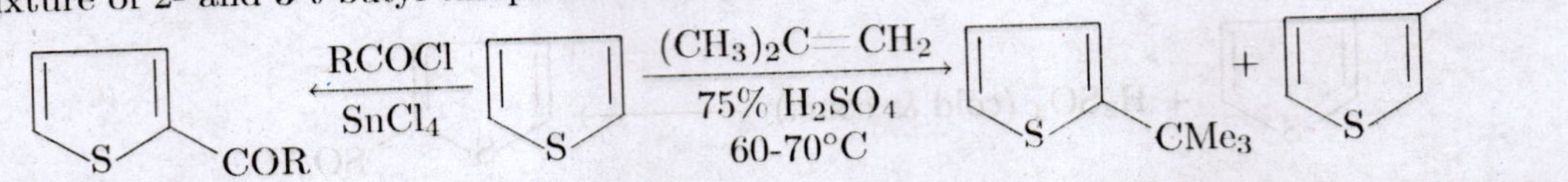

(j) Formation of 2-thienylmagnesium bromide: When 2-bromothiophen is treated with magnesium, 2-thienylmagnesium bromide is formed which can be converted to 2-thiophencarboxylic acid and other thiophen derivatives.

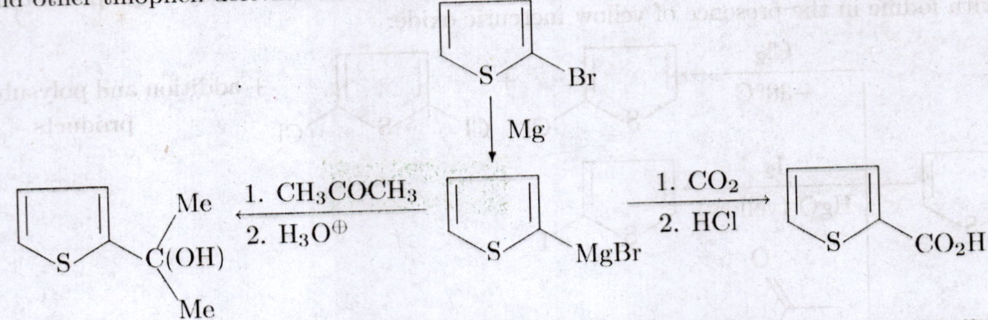

(k) Reaction with butyllithium : Thiophen reacts with butyllithium to form 2-thienyllithium which can also be converted to 2-thiophencarboxylic acid.

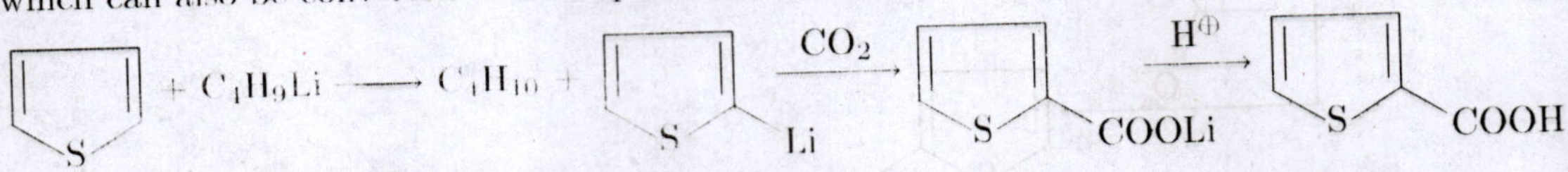

**Mechanism of the electrophilic substitution**

This is a two-step process. The first step involves the formation of the σ-complex and in the second step deprotonation of the σ-complex occurs.

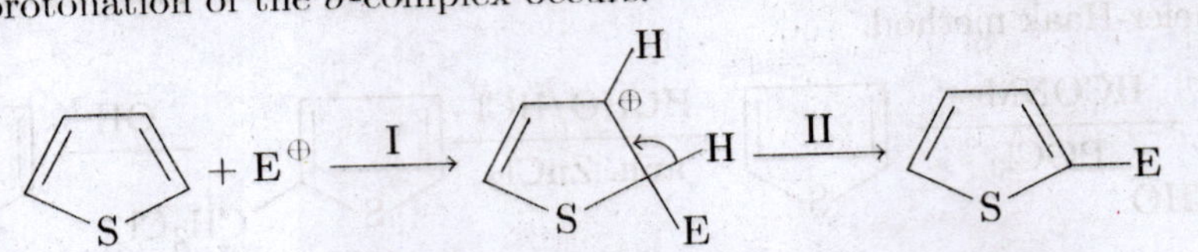

resonance-stabilised σ-complex

**Problem 19.1.** Compare thiophen with furan and benzene.

**Hints :** The following points are to be discussed :

**(i)** Stability; **(ii)** Charge distribution; **(iii)** Diene character; **(iv)** Electrophilic substitutions : **(a)** nitration, **(b)** sulphonation, **(c)** The Friedel-Crafts reaction, **(d)** mercuration, **(e)** halogenation etc., **(v)** orientation in the $S_E$ reaction, **(vi)** base character.

**Problem 19.2.** Furan is less reactive than thiophen so far as electrophilic substitution is concerned—explain with example.

**Solution :** Since oxygen is more electronegative than sulphur, the O atom of furan draws the charge cloud of the ring more towards itself than the S atom of thiophen draws the charge cloud of the thiophen ring. Thus furan ring is slightly deactivated than the thiophen ring; so, the Ar.$S_E$ reactions take place more readily on thiophen than on furan; for example thiophen undergoes the Mannich reaction but furan itself does not—2-methylfuran undergoes the reaction.

## 19.4 Pyrrole

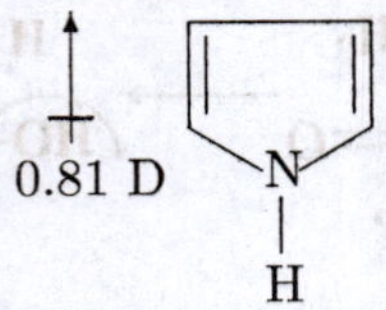

$C_4H_5N$ R.E. = 125.6 kJ mol$^{-1}$

b.p. = 131 °C

### 19.4.1 Source

Coal tar and bone oil.

Bone oil $\xrightarrow[\text{dil. alkali}]{\text{washed with}}$ bone oil free from acidic substances

↓ washed with dilute acid

fraction between 100 °C—150 °C $\xleftarrow[\text{distillation}]{\text{fractional}}$ bone oil free from bases

↓ KOH fusion

potassiopyrrole $\xrightarrow[\text{distillation}]{\text{steam}}$ pure pyrrole b.p. 131 °C.

### 19.4.2 Synthesis of pyrrole and its derivatives

**(a)** Synthesis from succinimide :

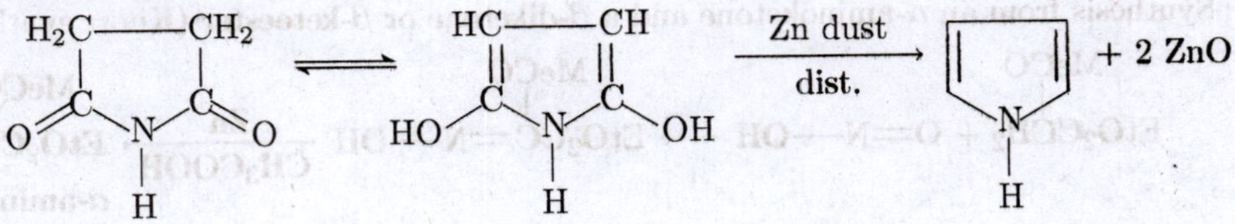

(b) **Synthesis from furan :**

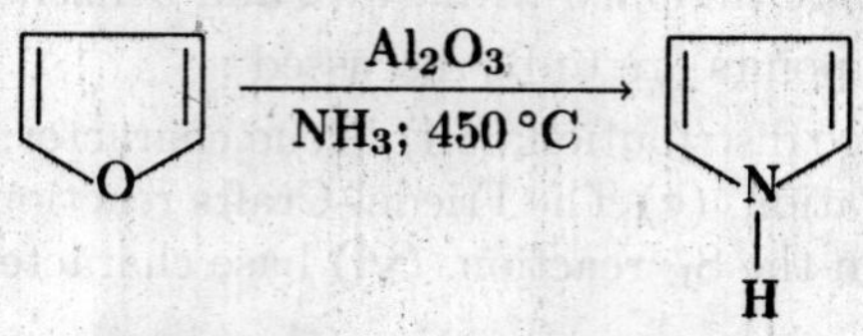

(c) **Synthesis from 1, 4-diketones (Paal-Knorr synthesis) :**

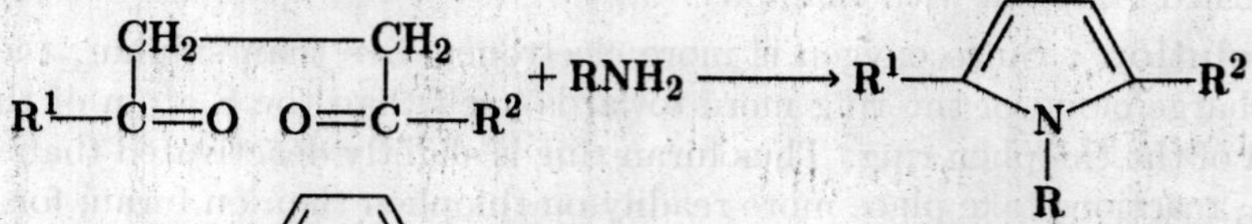

when R = H, we get

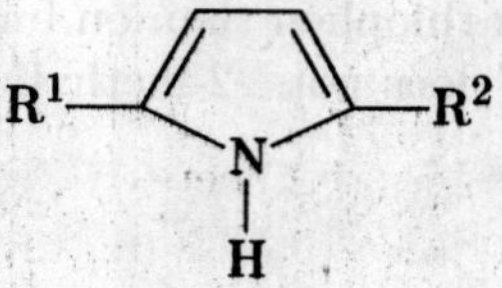

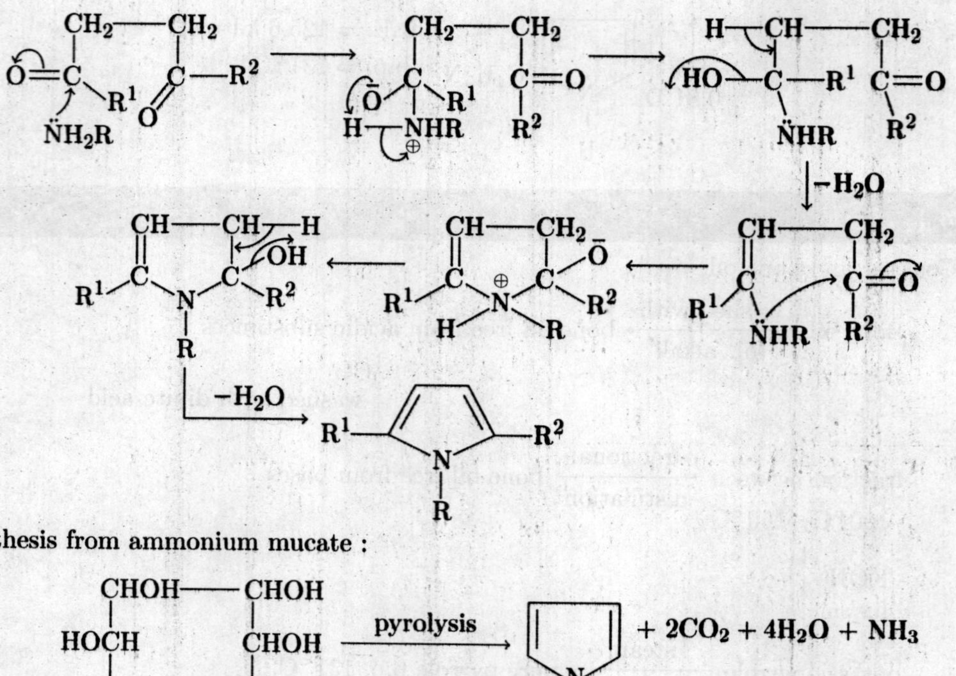

(d) **Synthesis from ammonium mucate :**

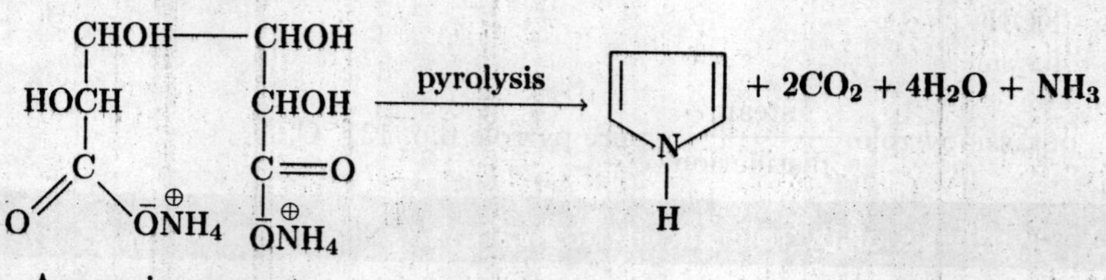

Ammonium mucate

(e) **Synthesis from an α-aminoketone and a β-diketone or β-ketoester (Knorr synthesis) :**

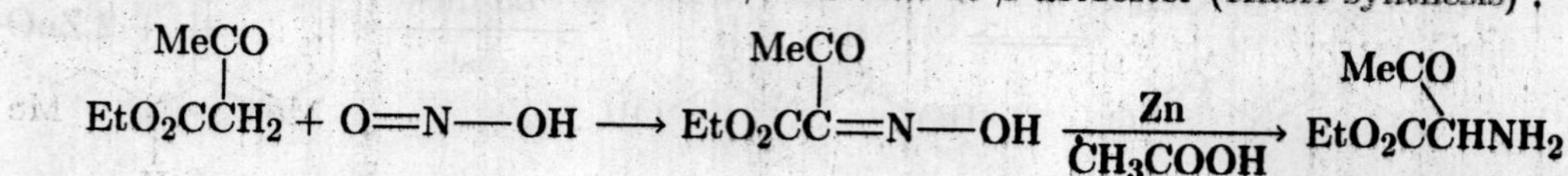

α-aminoketone

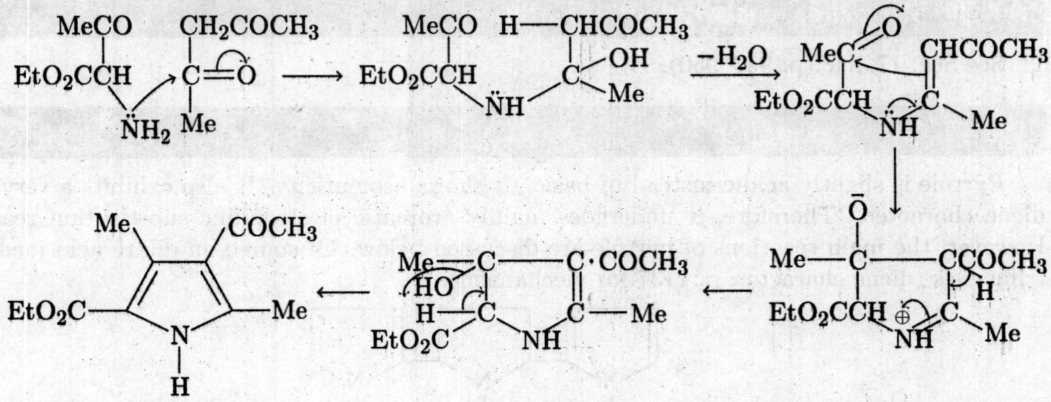

**Instead of $\beta$-diketone if $\beta$-ketoester (say, ethyl acetoacatate) is taken, the product will be 2, 4-dicarbethoxy-3, 5-dimethylpyrrole.**

(f) **Synthesis from acetylene and ammonia :**

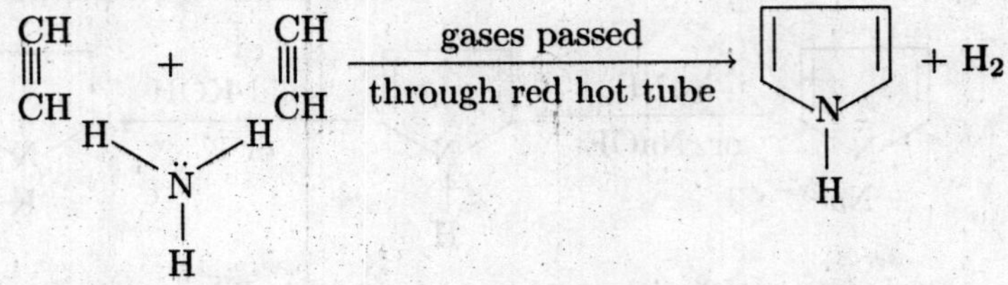

(g) **Synthesis from an $\alpha$-chloroketone (Hantzsch synthesis) :**

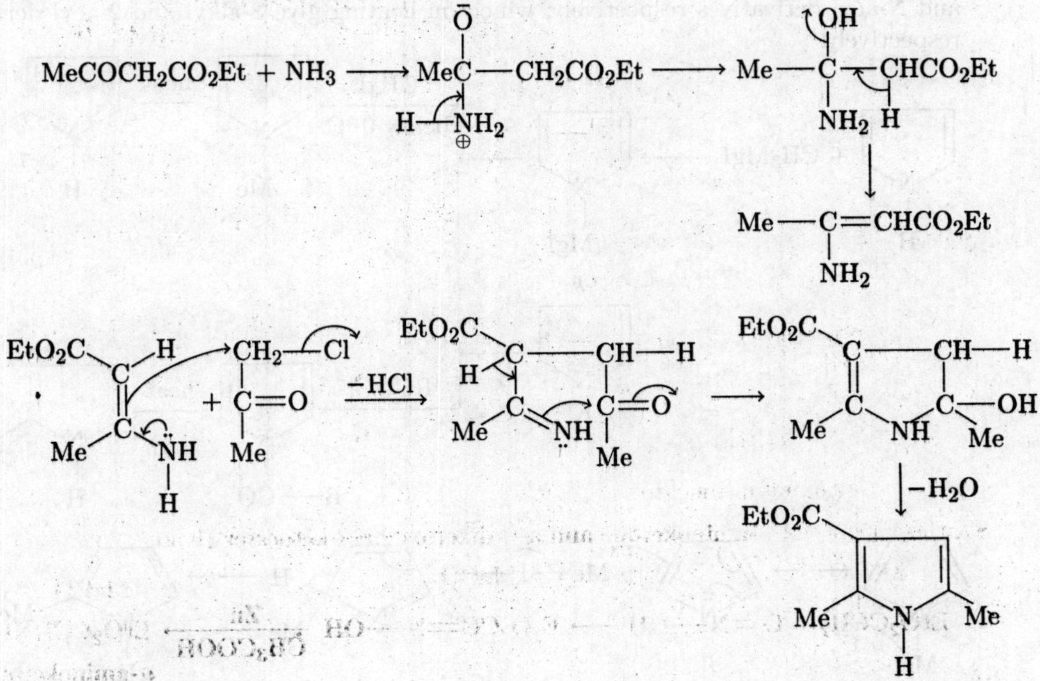

## 19.4.3 Structure, reactivity and orientation

See Sec. 17.4.4c (p. 949–950).

## 19.4.4 Chemical properties

Pyrrole is slightly acidic instead of basic. It shows aromaticity. It also exhibits a very feeble diene character. Therefore, it undergoes mainly aromatic electrophilic substitution reactions. However, the main reactions of pyrrole are discussed below. Of course, in dilute acid medium it trimerises (diene character; p. 1137 for mechanism).

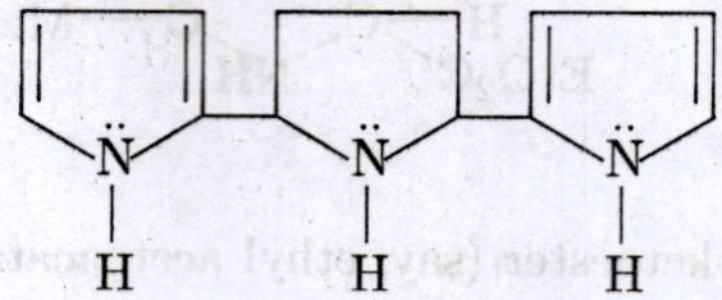

(a) Acid character of pyrrole :

(i) When pyrrole is heated with sodamide or solid KOH or NaOEt, etc., it forms salts.

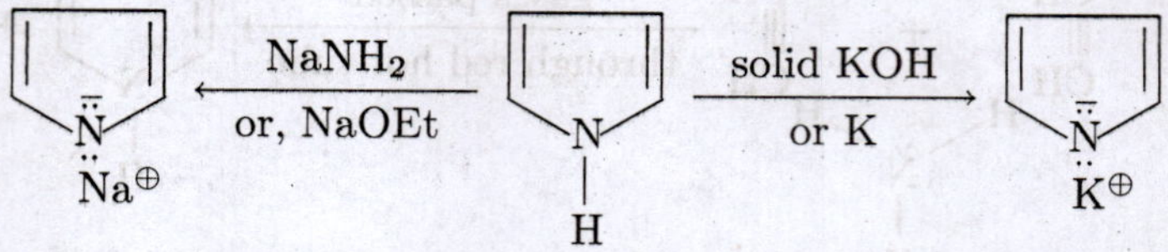


(ii) When pyrrole is treated with methylmagnesium iodide, methane gas evolves and a salt of magnesium iodide forms. This on treatment with alkyl halide and acyl halide forms N-alkyl and N-acyl derivatives respectively, which on heating give 2-alkyl and 2-acyl derivatives respectively.

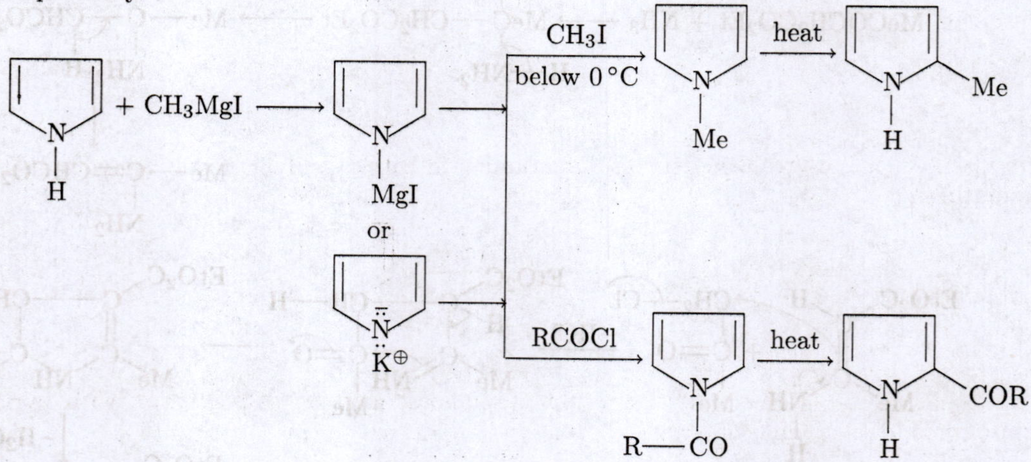


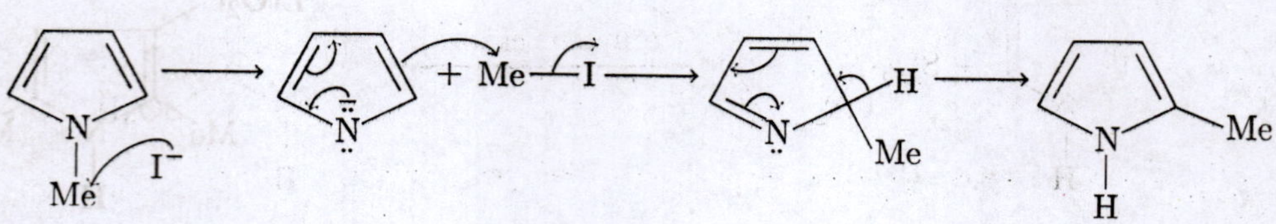

(b) Reduction :

(i) Reduction by nascent hydrogen (Zn/HOAc) :

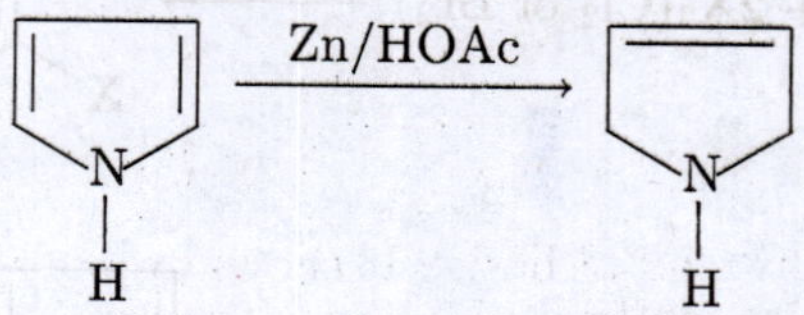

2,5-Dihydropyrrole

(ii) Catalytic reduction :

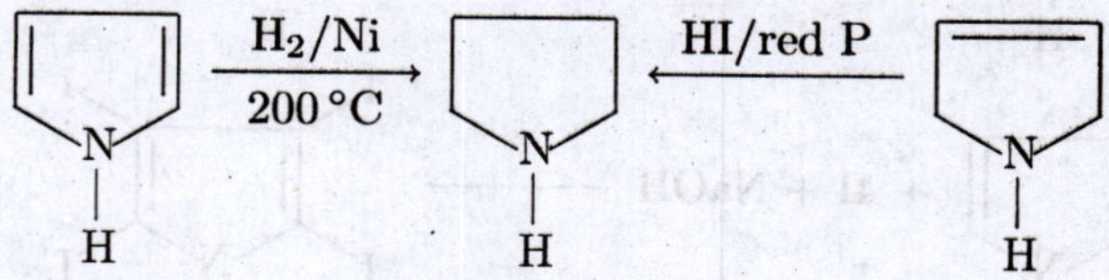

Pyrrolidine

(c) Reaction with triphenylmethyl radical :

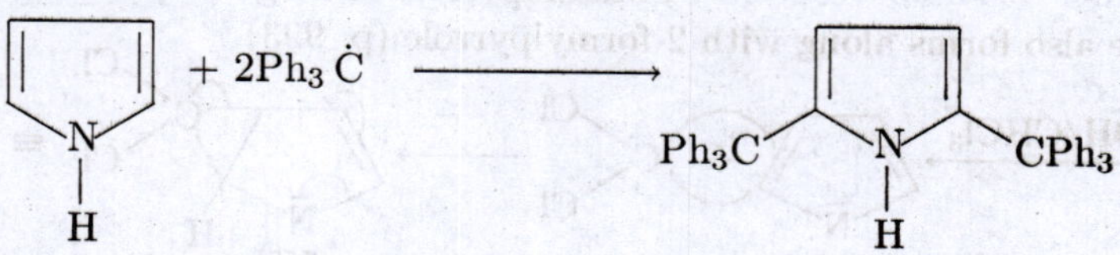

(d) Oxidation :

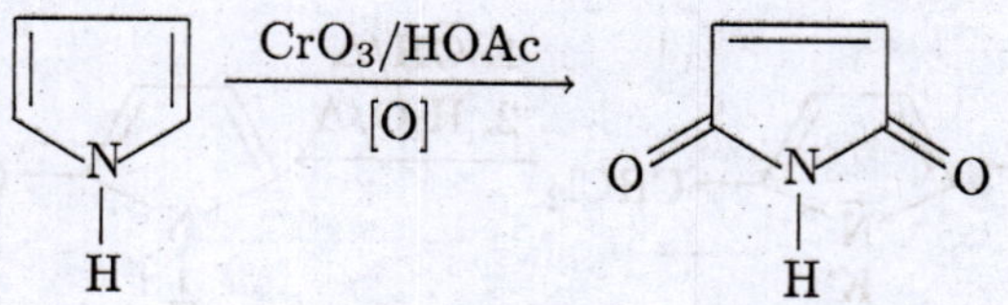

Maleinimide

(e) Nitration : Pyrrole undergoes nitration with nitric acid in acetic anhydride at low temperature.

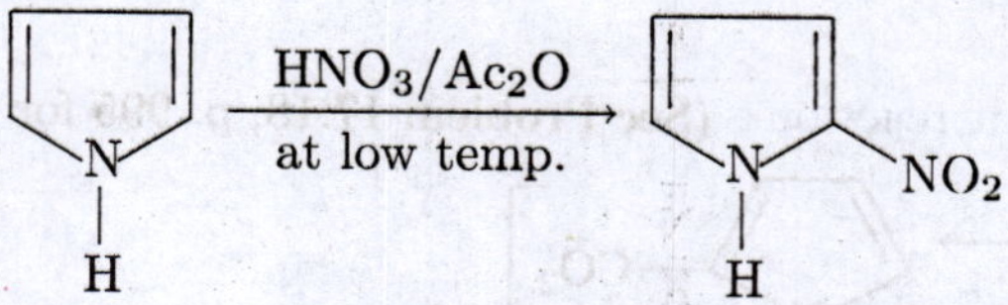

(f) Sulphonation : Pyrrole forms 2-pyrrolesulphonic acid on sulphonation with pyridine-sulphur trioxide in ethylene chloride.

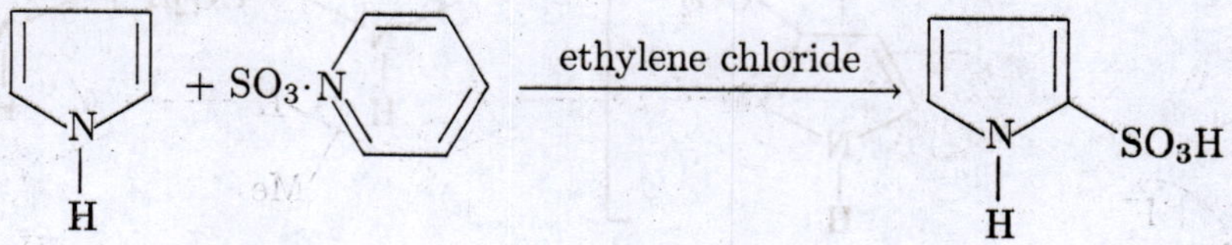

(g) Halogenation :

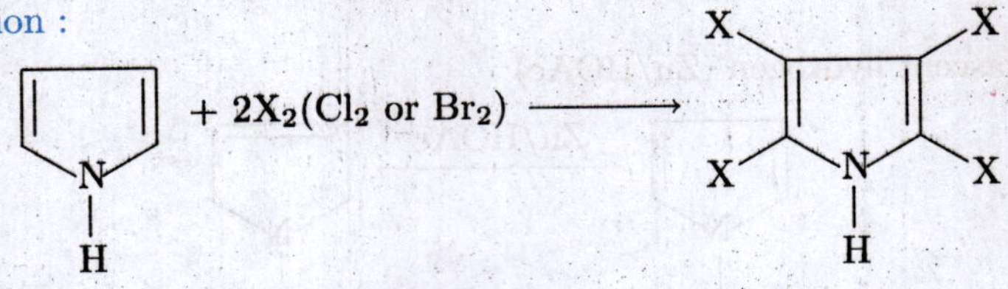

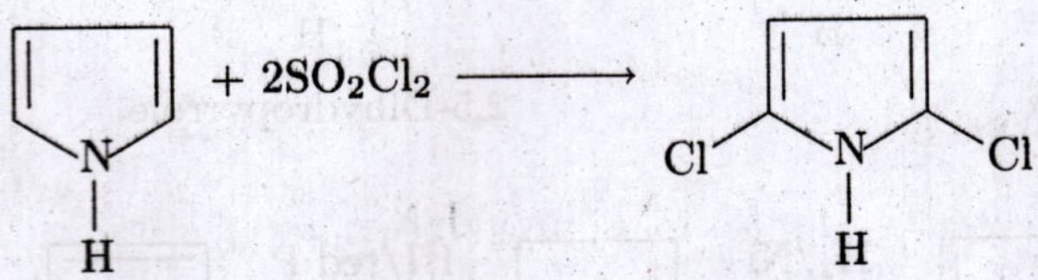

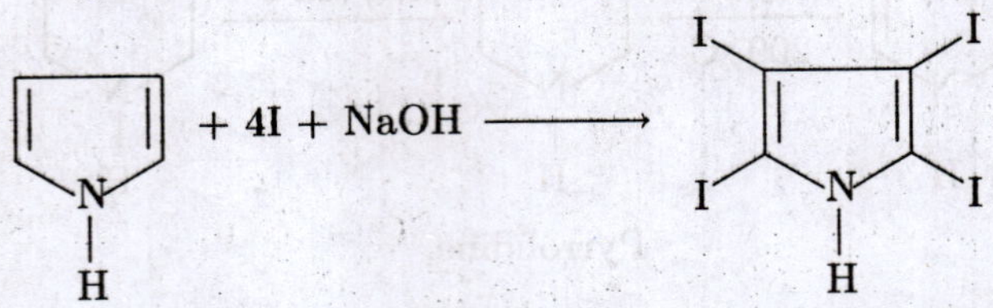

(h) The Reimer-Tiemann reaction : Potassiopyrrole undergoes the Reimer-Tiemann reaction, 3-chloropyridine also forms along with 2-formylpyrrole (p. 993).

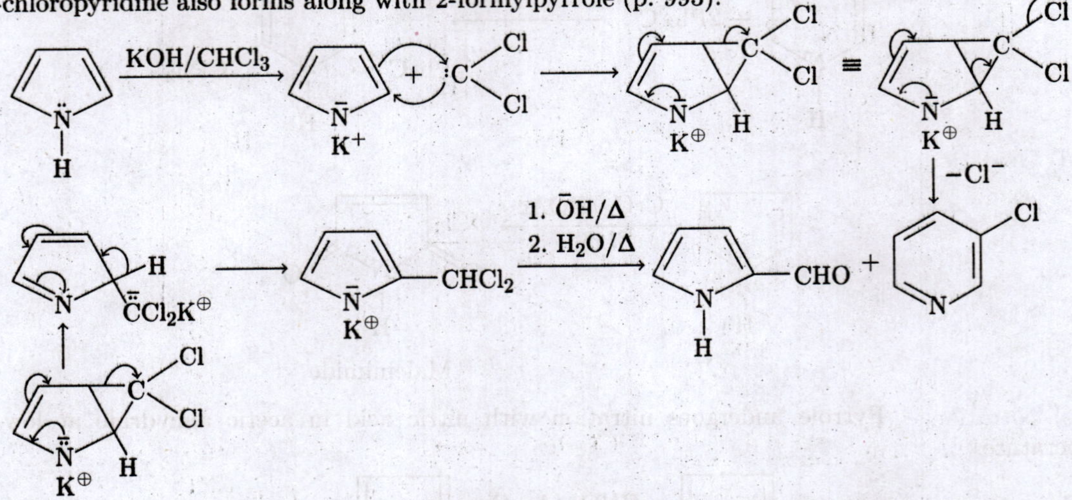

(i) The Kolbe-Schmidt reaction : (Sec Problem 17.18; p. 995 for mechanism)

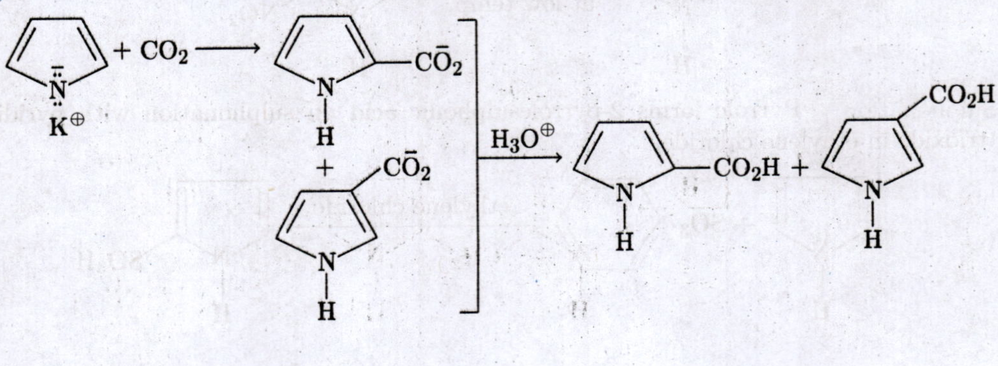

(j) The Houben-Hoesch synthesis :

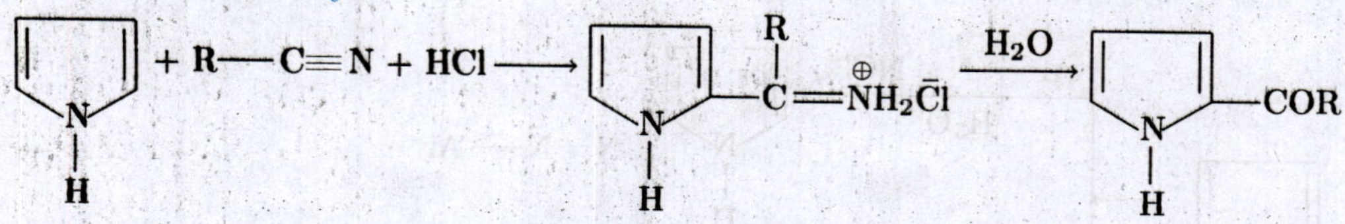

(k) The Mannich reaction and subsequent reaction products :

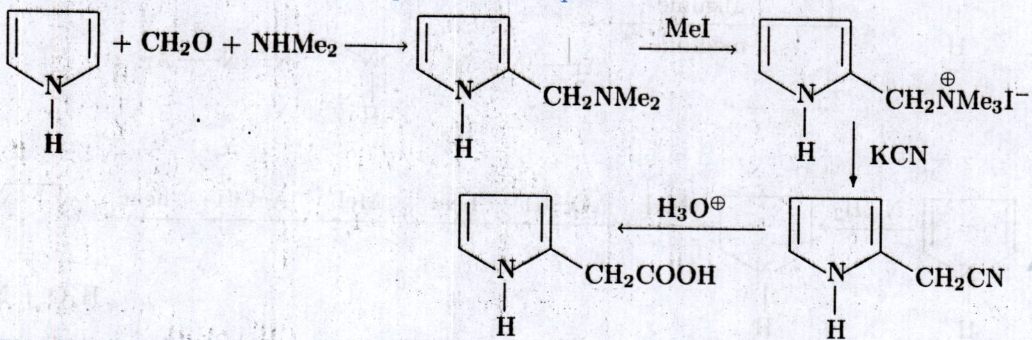

(l) The Gattermann aldehyde synthesis :

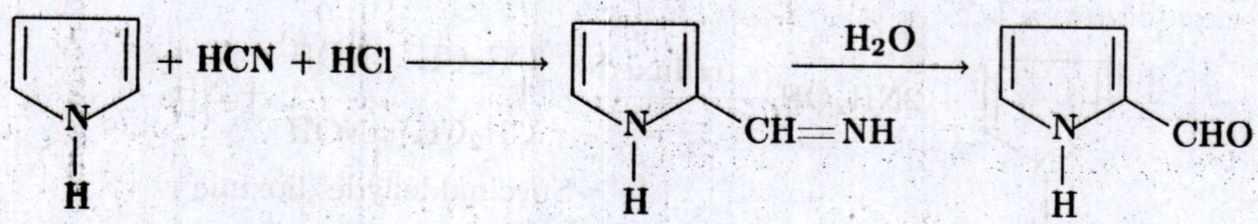

(m) The Vilsmeier-Haak reaction :

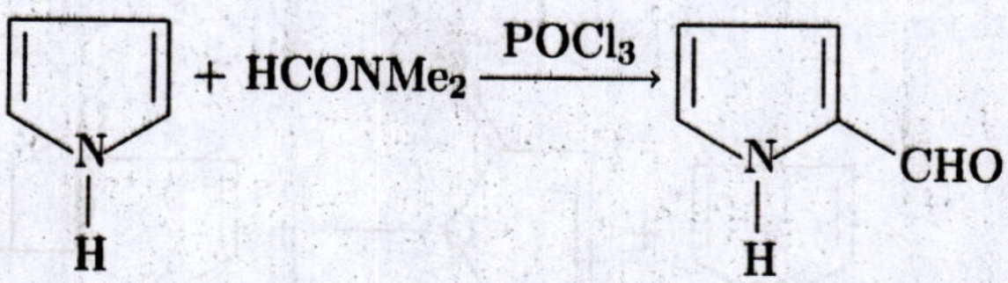

(n) Formation of ester of 2-pyrrolecarboxylic acid :

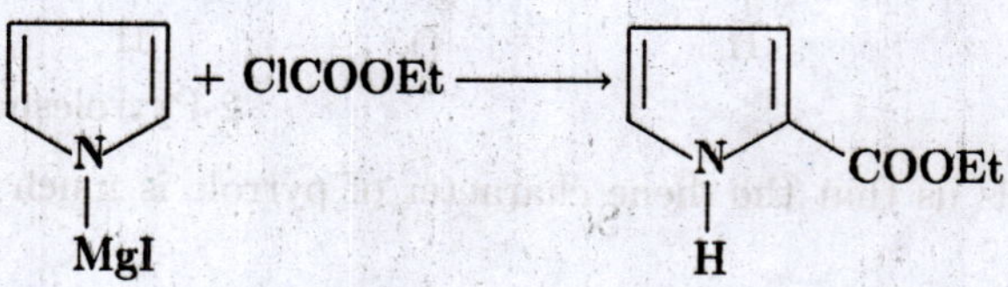

(o) Condensation reaction with formaldehyde :

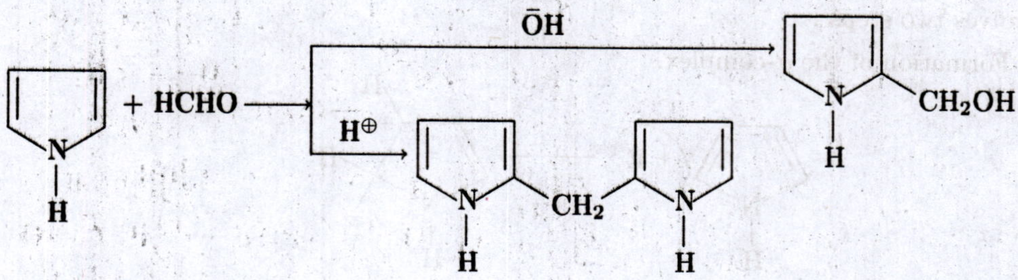

(p) The Diazonium coupling reaction :

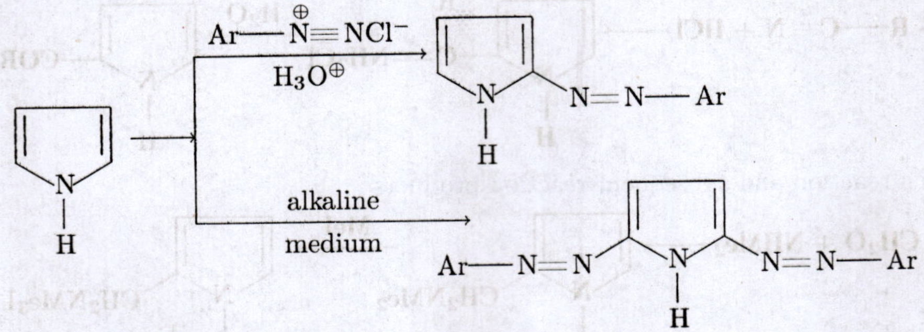

(q) Conversion of pyrrole to butadiene :

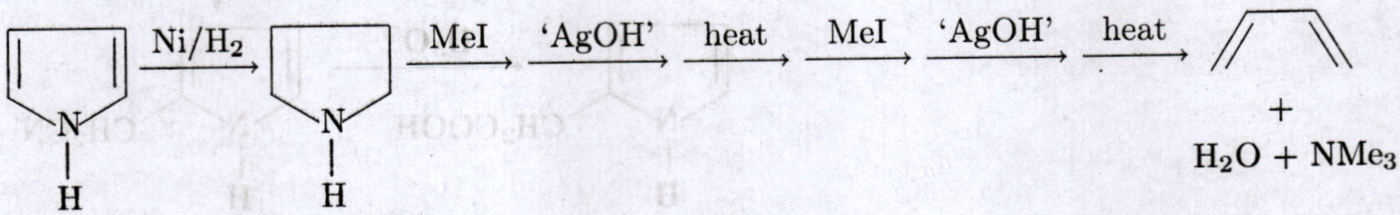

(r) Ring-opening reaction :

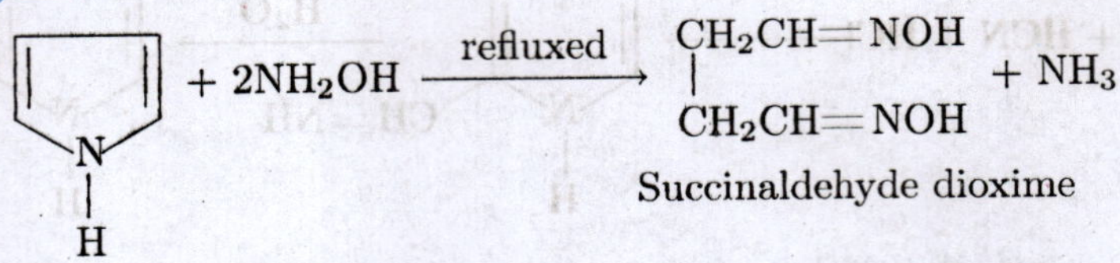

(s) Reaction with maleic anhydride :

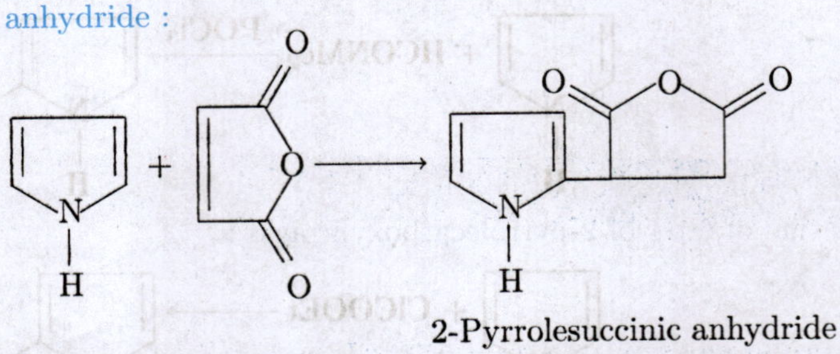

This reaction tells us that the diene character of pyrrole is much less pronounced than that of furan.

**Mechanism of the electrophilic substitution in pyrrole**

This involves two steps :

**Step I.** Formation of the $\sigma$-complex.

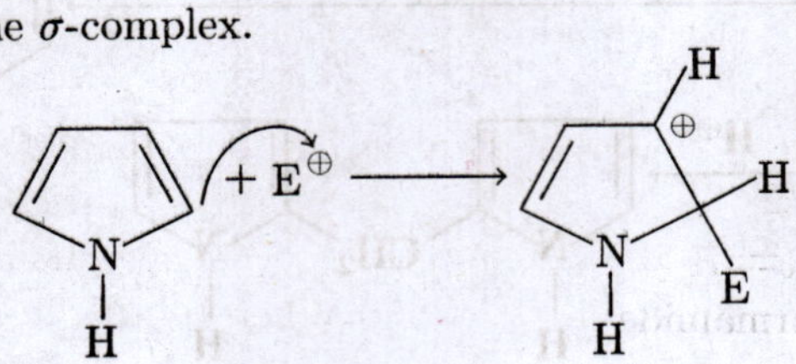

**Step II.** The aromatisation of the σ-complex.

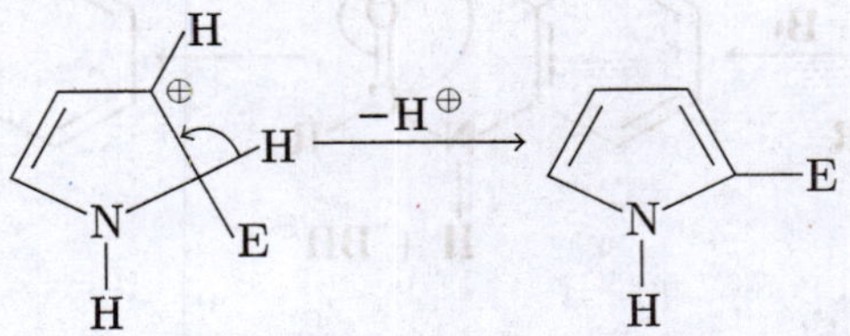

**Problem 19.3.** Pyrrole shows resmemblances to phenol and aniline—explain with examples.

**Hint :** The resemblances of pyrrole to phenol are :

**(a)** Both phenol and pyrrole are acidic; they form salts with bases.

**(b) (i)** Salts of both the compounds are strong nucleophiles and undergo reactions even with weak electrophiles. **(ii)** Both direct the incoming electrophile at the α-position.

**(c)** Both undergo : **(i)** The Reimer-Tiemann reaction, **(ii)** The Kolbe-Schmidt reaction, **(iii)** The Mannich reaction, **(iv)** The diazonium coupling reaction.

The resemblances of pyrrole to aniline are : **(i)** Both direct the incoming nucleophile at the α-position, **(ii)** Both undergo the diazonium coupling reaction, **(iii)** N-Alkyl derivatives of both the compounds undergo the Hofmann-Martius rearrangement. Give example against each point as noted above.

## 19.5 Indole

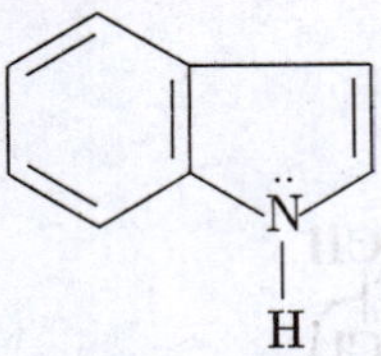

Benzopyrrole

$C_8H_7N$

m.p. = 52 °C

b.p. = 254 °C

### 19.5.1 Source

Coal tar, plant Isatis tinctoria in the form of glucoside of indoxyl (indican), jasmine flowers and orange blossoms.

### 19.5.2 Synthesis of indole and its derivatives

(a) The Madelung synthesis : This involves the intramolecular condensation of N-*o*-tolyl derivatives of acid amides by means of a strong base such as potoassium tert-butoxide, sodamide, etc.

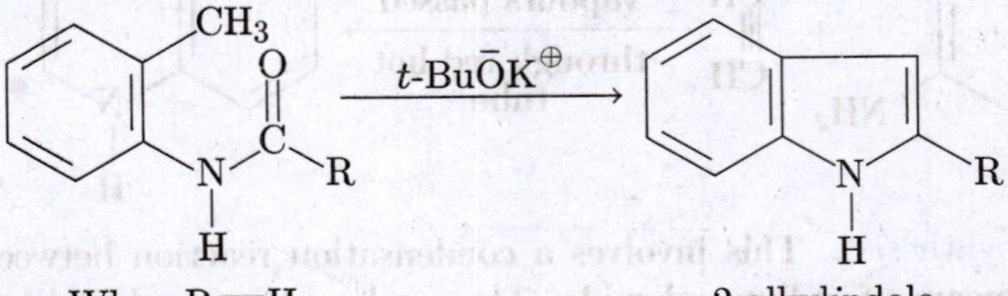

When R═H
N-*o*-tolylformamide

2-alkylindole
When R═H, it is indole.

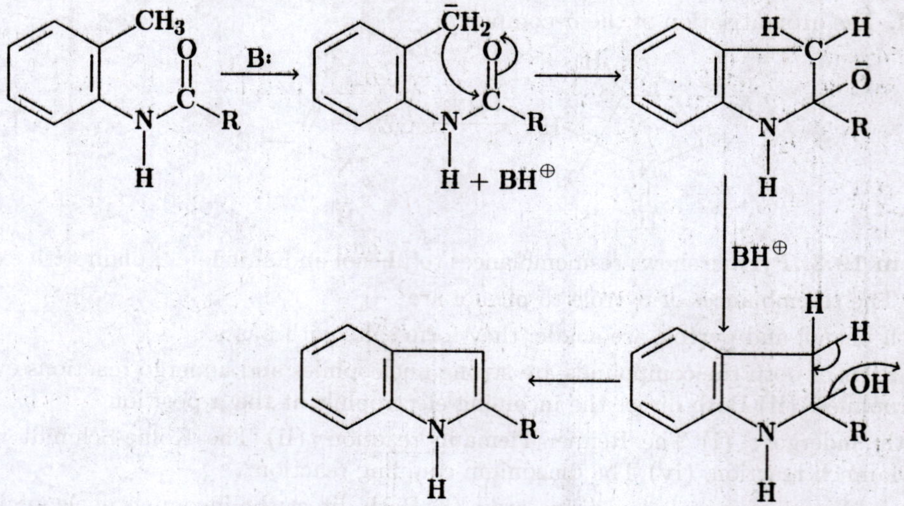

(b) Synthesis from o-amino-ω-chlorostyrene : **This reaction indicates the structure of iodole.**

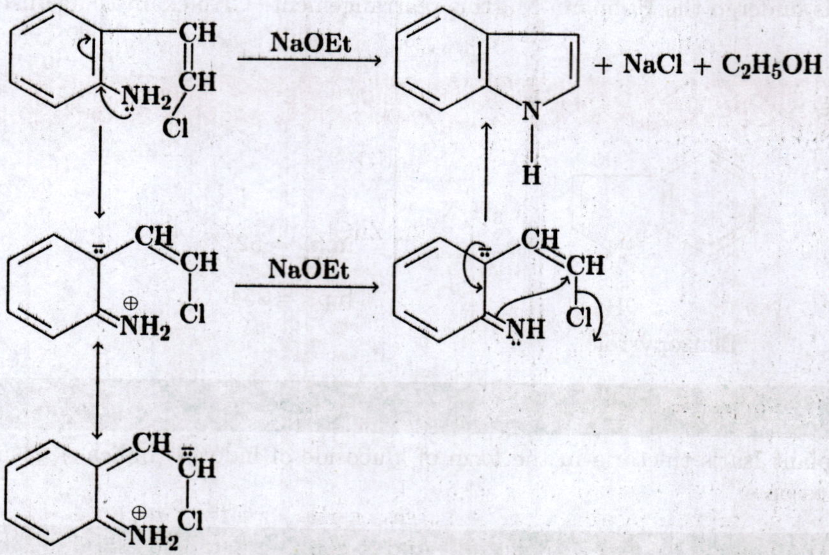

(c) Synthesis from aniline and acetylene :

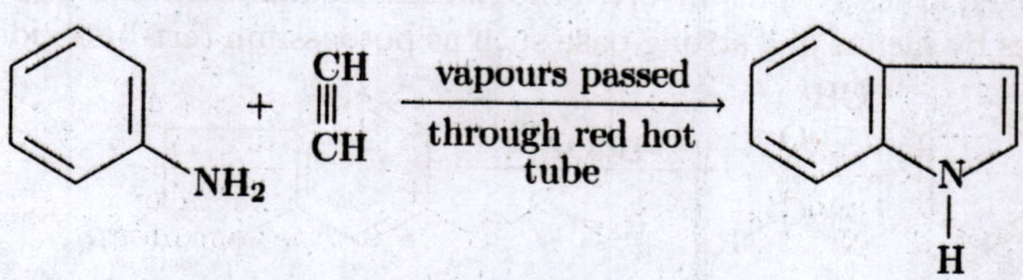

(d) The Reissert synthesis : **This involves a condensation reaction between *o*-nitrotoluene and ethyl oxalate by means of sodium ethoxide. The condensation product is then reduced and cyclised.**

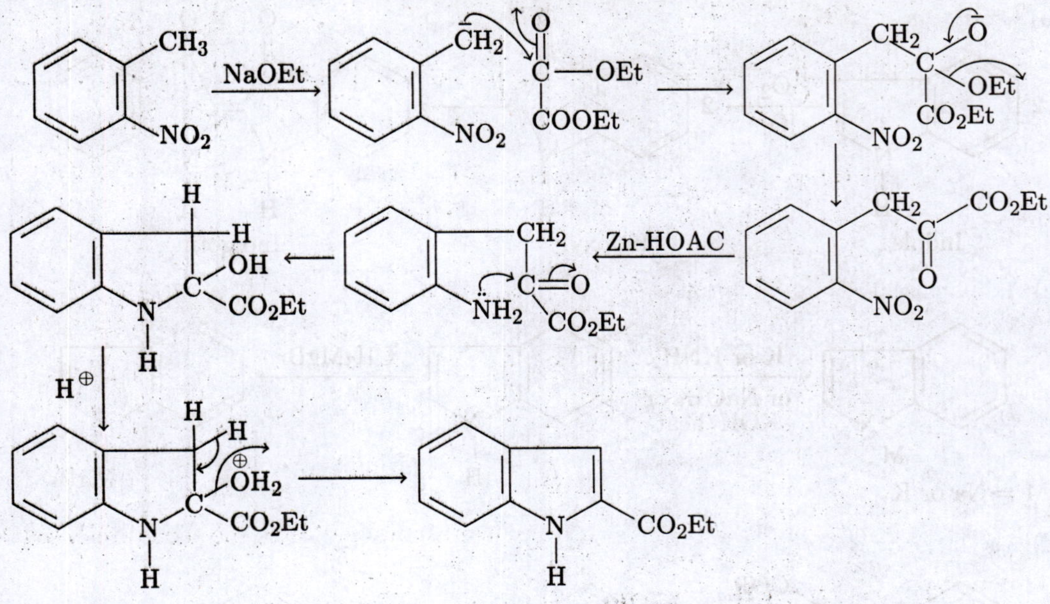


(e) The Fischer indole synthesis (See Sec. 17.18 p. 1004) :

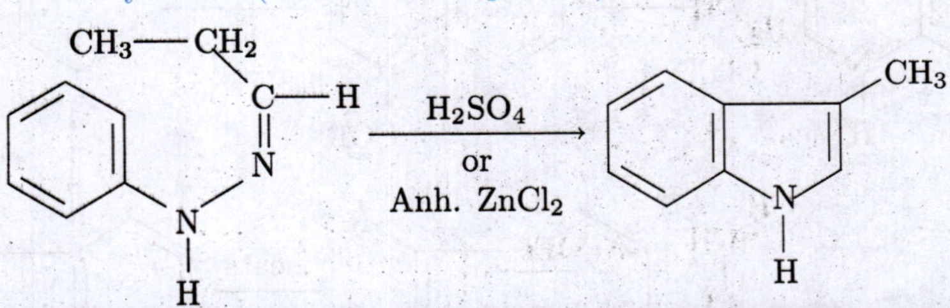


## 19.5.3 Structure, reactivity and orientation

[See Sec. 17.4.4e p. 951]

## 19.5.4 Chemical properties

**Since indole contains a pyrrole ring, it possesses many of the properties of pyrrole. It is an aromatic compound. Its two rings are flat and it contains ten delocalised $\pi$es encompassing both the rings. Therefore, it exhibits aromatic character. It undergoes electrophilic substitution reactions usually at 3-position.**

(a) Reduction :

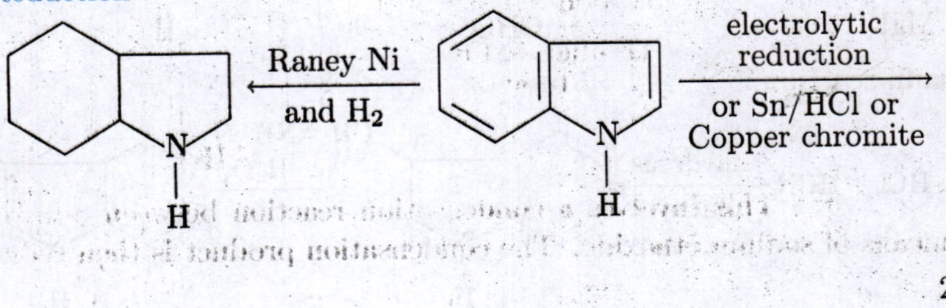


Indoline or
2,3-dihydroindole

(b) Oxidation :

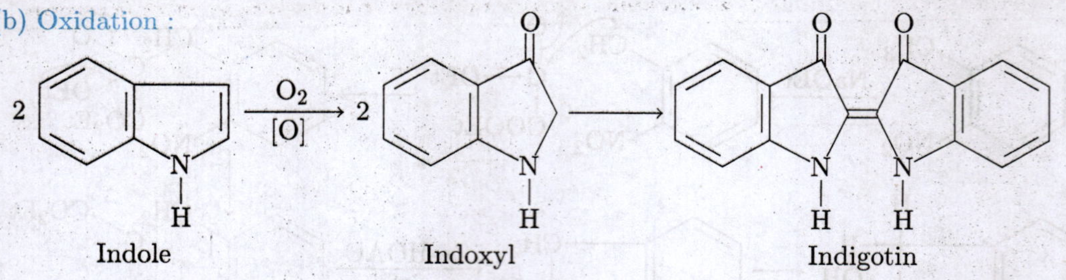


(c) Acid character :

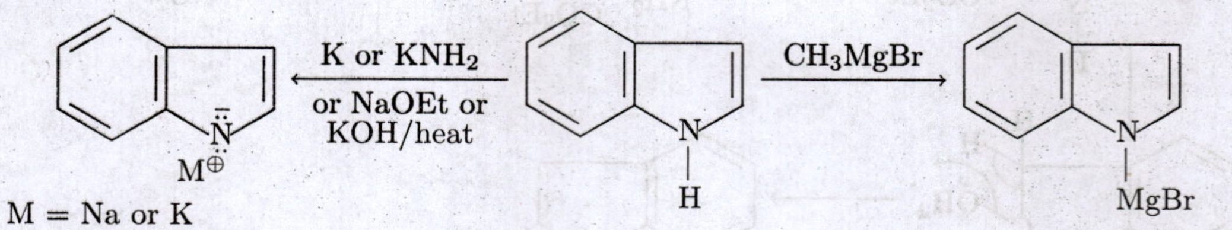


(d) Alkylation and acylation :

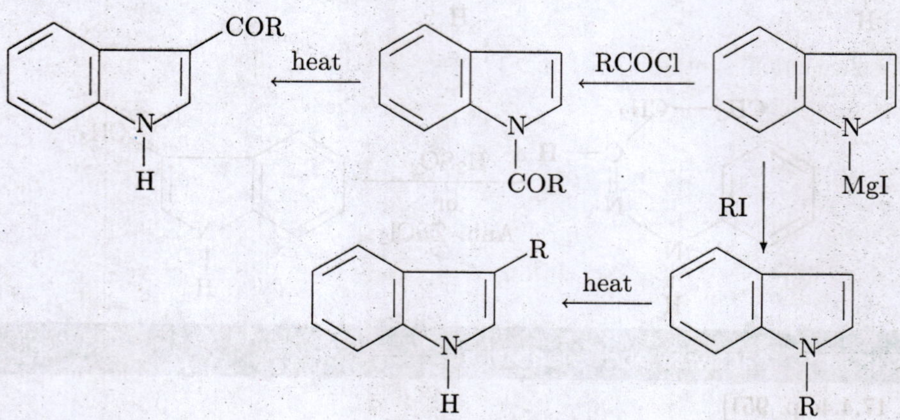


(e) Formylation :

(i) By ethyl formate :

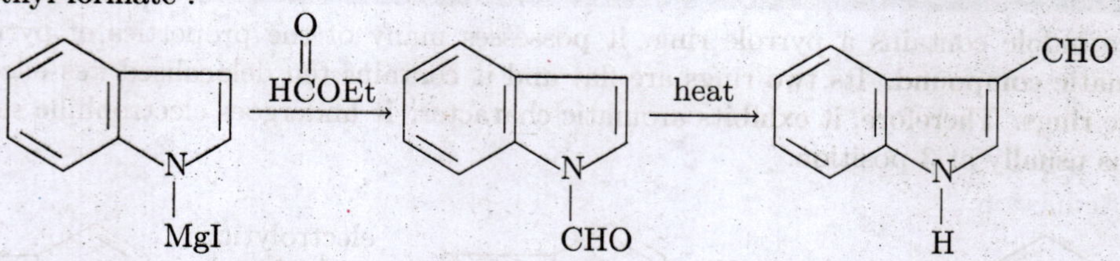


(ii) By the Gattermann reaction :

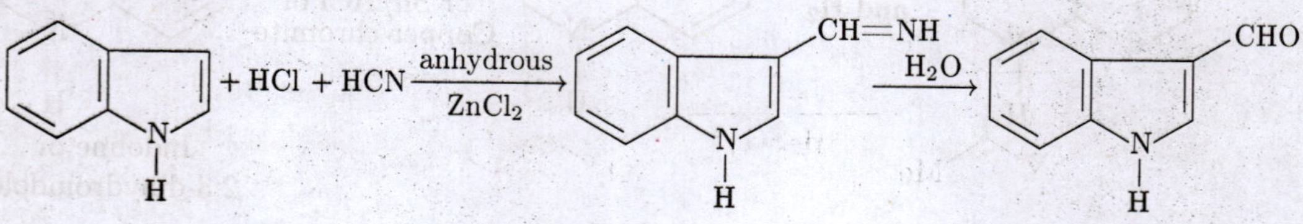

**(iii)** By the Vilsmeier-Haak reaction :

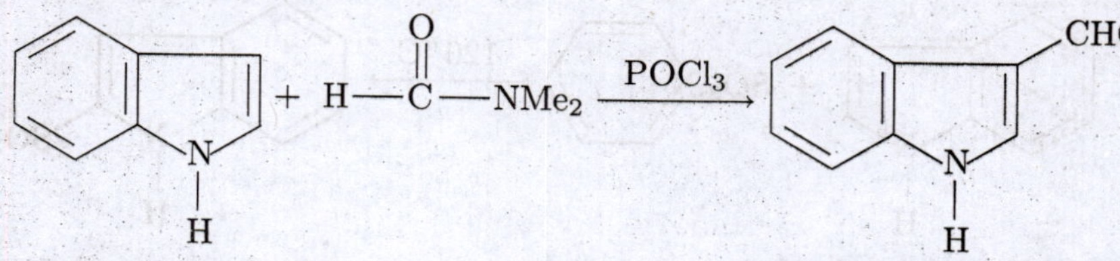

**(iv)** By the Reimer-Tiemann reaction : In this case 3-chloroquinoline also forms. The mechanism of the reaction is as shown in the case of pyrrole (19.4.4.h).

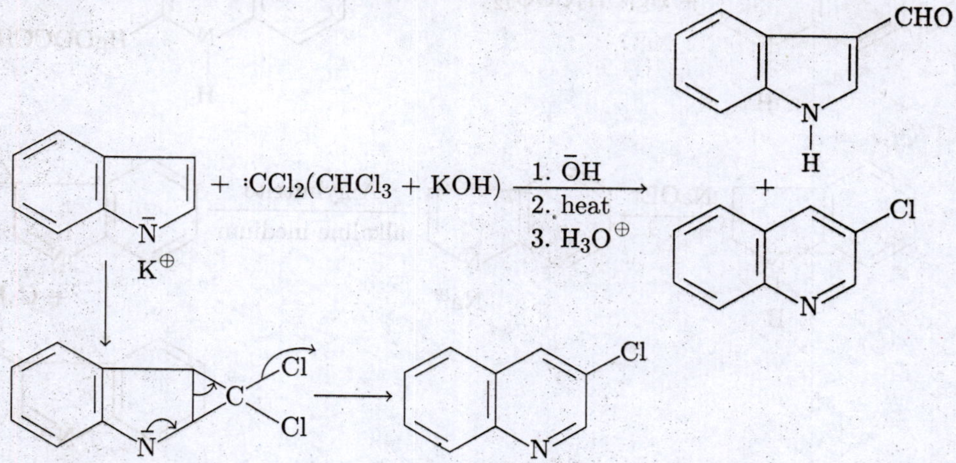

**(f) Halogenation :**

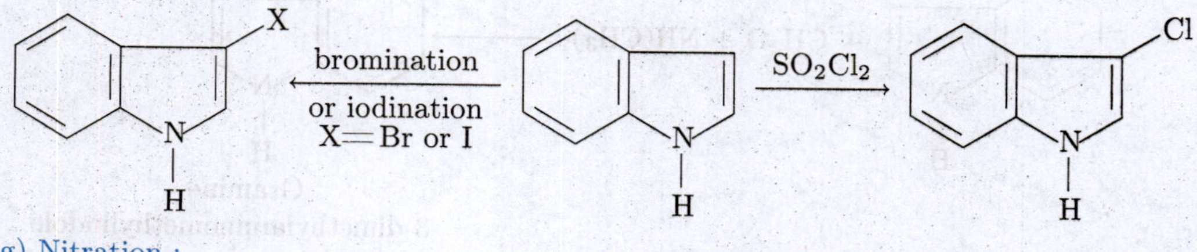

**(g) Nitration :**

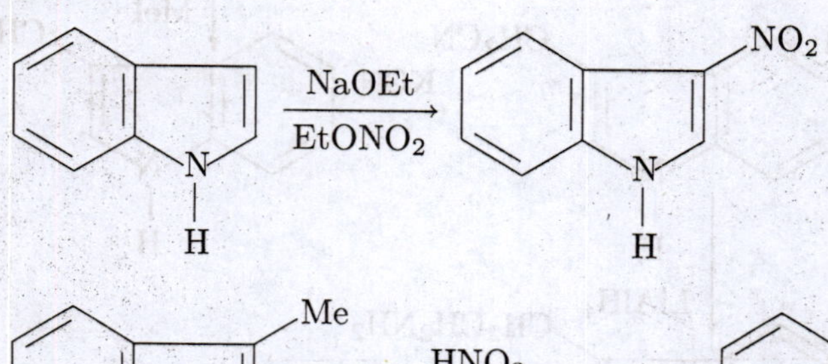

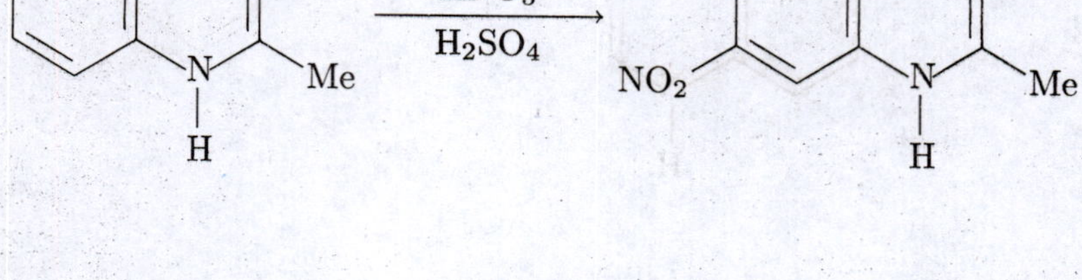

(h) Sulphonation :

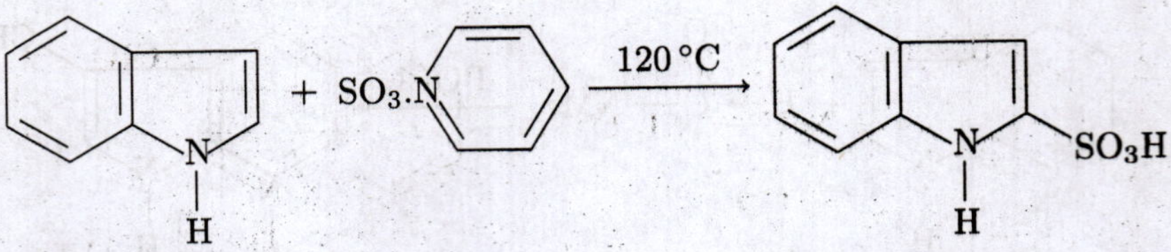

(i) Mercuration :

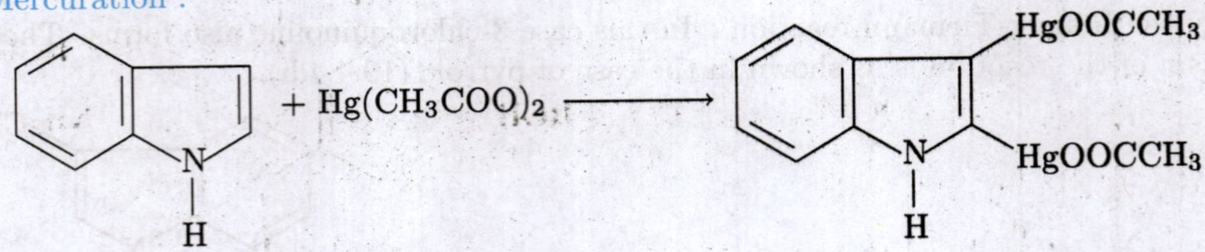

(j) Nitrosation :

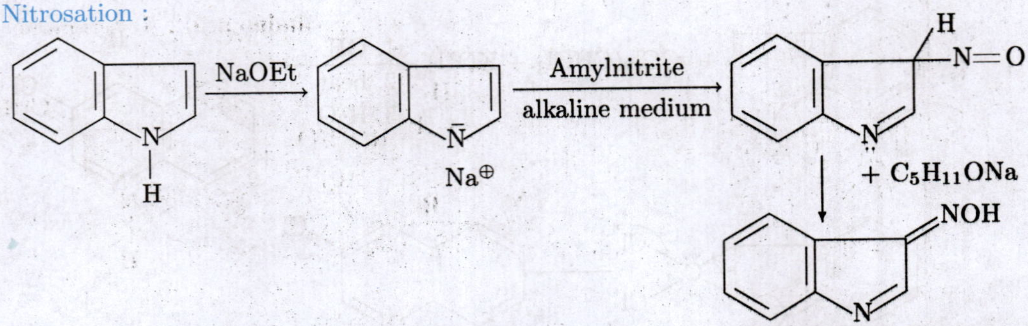

(k) The Mannich reaction and subsequent reaction products :

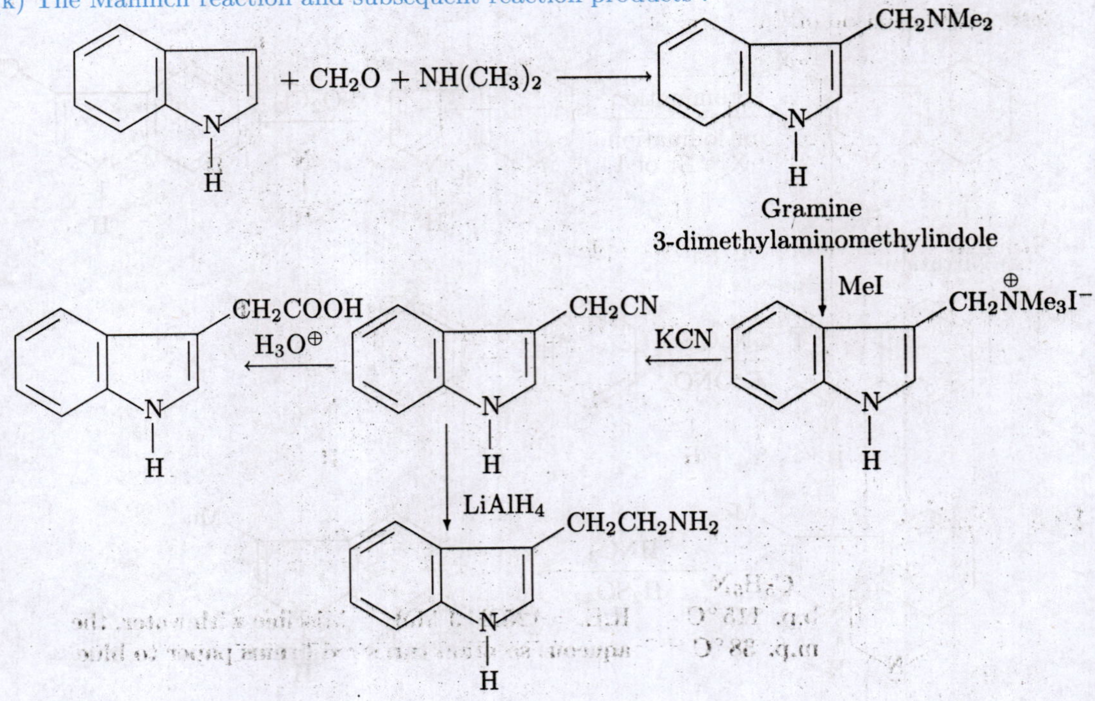

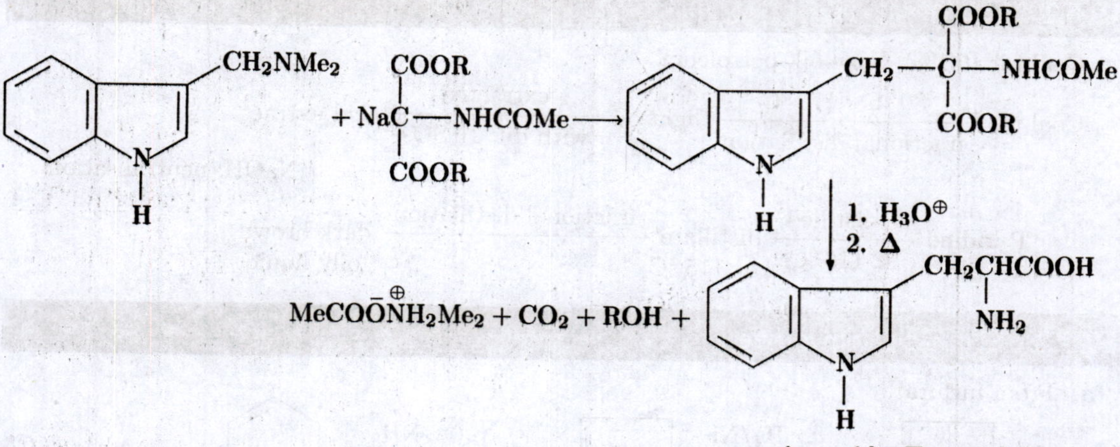


**amino acid : Tryptophan**

(1) Condensation between indole and an aldehyde :

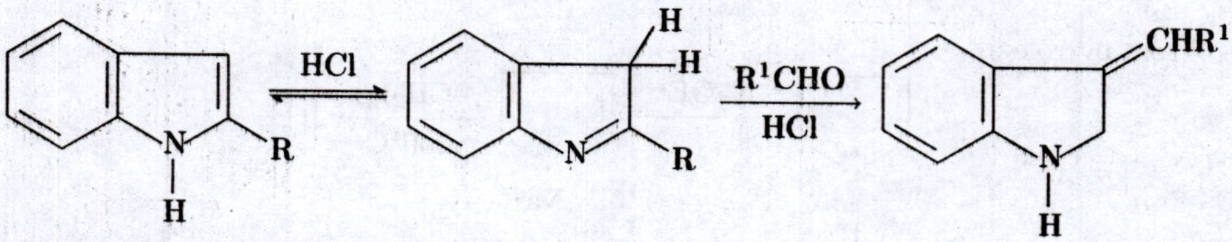


Mechanism of the electrophilic substitution

**It is a two-stepped process :**

**Step I.** **Formation of the $\sigma$-complex :**

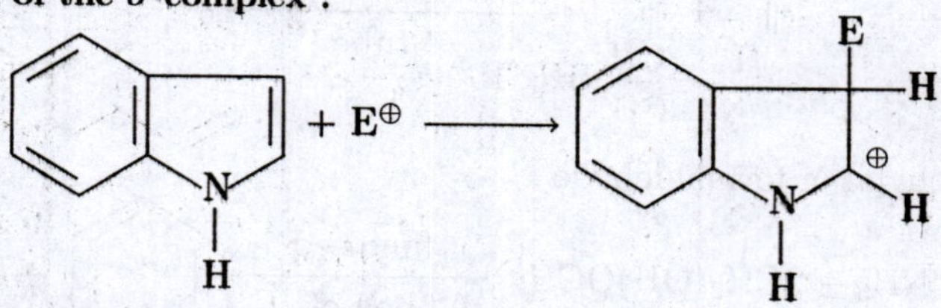


**Step II.** **Aromatisation of the $\sigma$-complex :**

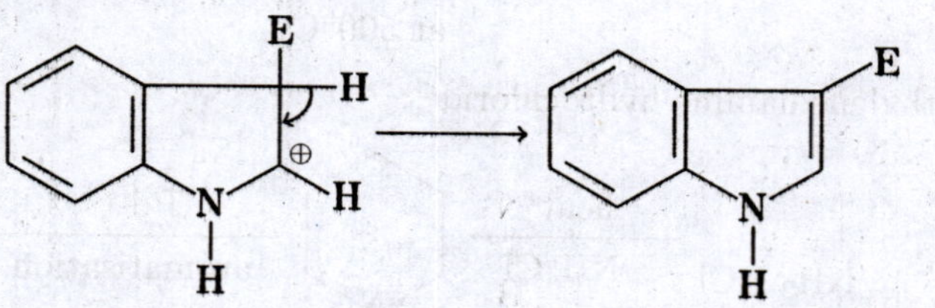


## 19.6 Pyridine

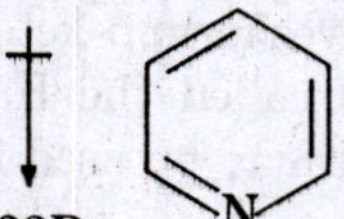

**2.23D**

$C_5H_5N$
b.p. 115 °C
m.p. 38 °C

**R.E. = 125.6 kJ Mol$^{-1}$. Miscible with water, the aqueous solution turns red litmus paper to blue**

## 19.6.1 Source

Coal tar (0.1%), bone oil, petroleum.

Coal tar $\xrightarrow[\text{fractional distillation}]{80\,°C—170\,°C}$ light oil $\xrightarrow[\text{with dil. } H_2SO_4]{\text{extracted}}$ acid extract $\xrightarrow{\text{NaOH; neutralisation}}$ dark brown oily liquid $\xrightarrow{\text{fractional distillation}}$ distillate at 115 °C $\xrightarrow[\text{at 115 °C}]{\text{redistilled}}$ Pyridine

## 19.6.2 Synthesis of pyridine

**(a)** From furfural :

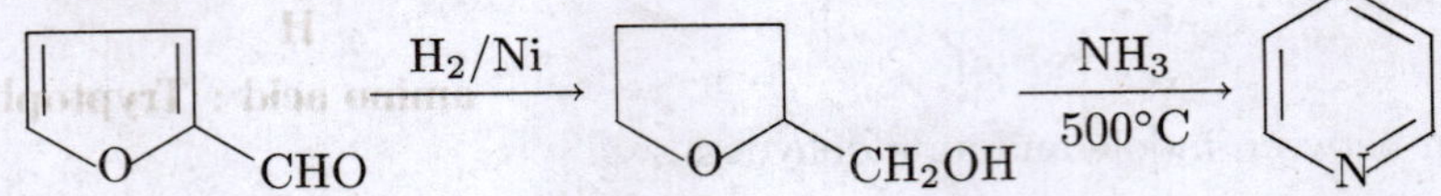


**(b)** From pyrrole :

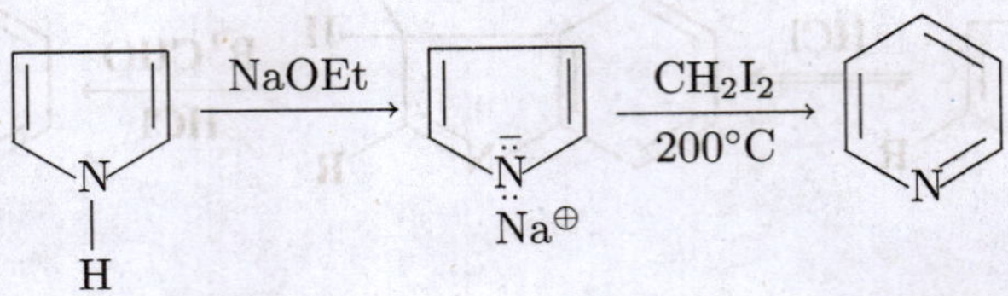


**(c)** From acetylene and hydrogen cyanide :

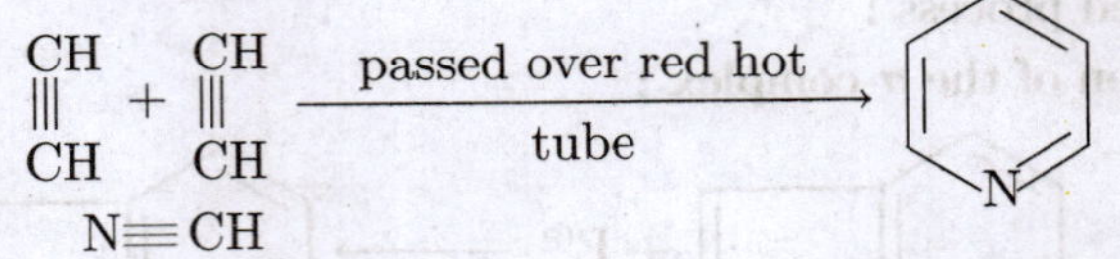


**(d)** From hemimethylal of formaldehyde :

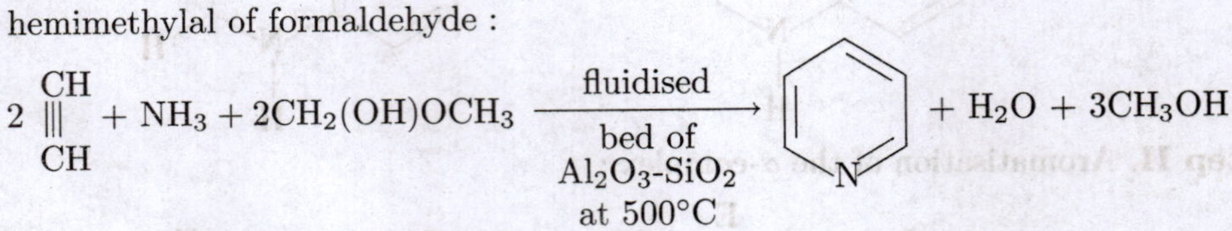


**(e)** From pentamethylenediamine hydrochloride :

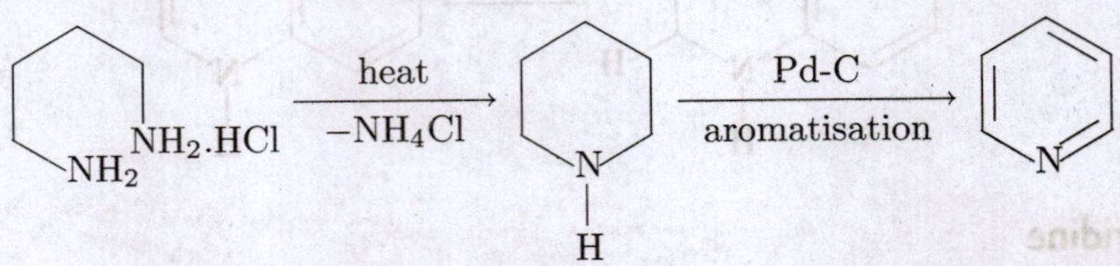


**(f)** By the **Hantzsch synthesis** : The formation of dihydropyridine derivatives form $\beta$-ketoesters, aldehydes and ammonia and subsequent conversion to pyridines is called the Hantzsch synthesis. Nitrous acid here acts as an aromatising agent; nitric acid may be used instead of nitrous acid.

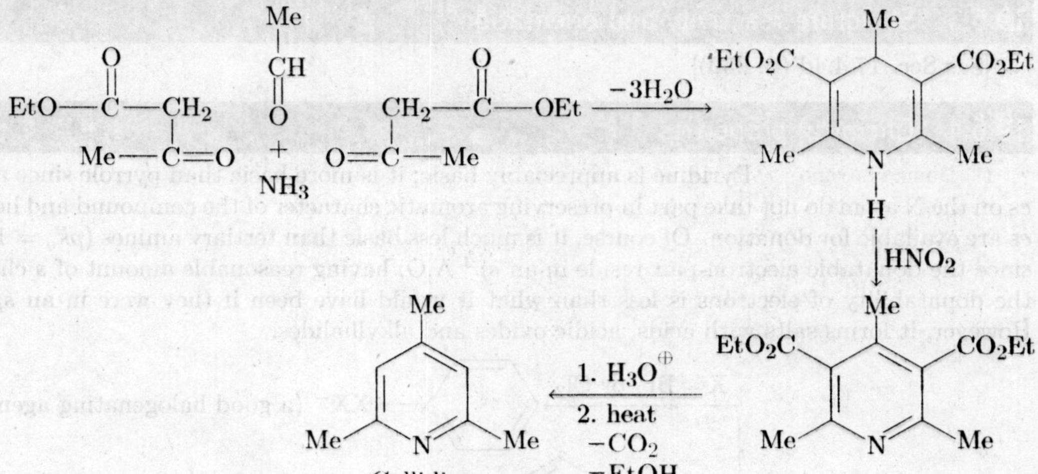

Collidine

***General mechanism :***

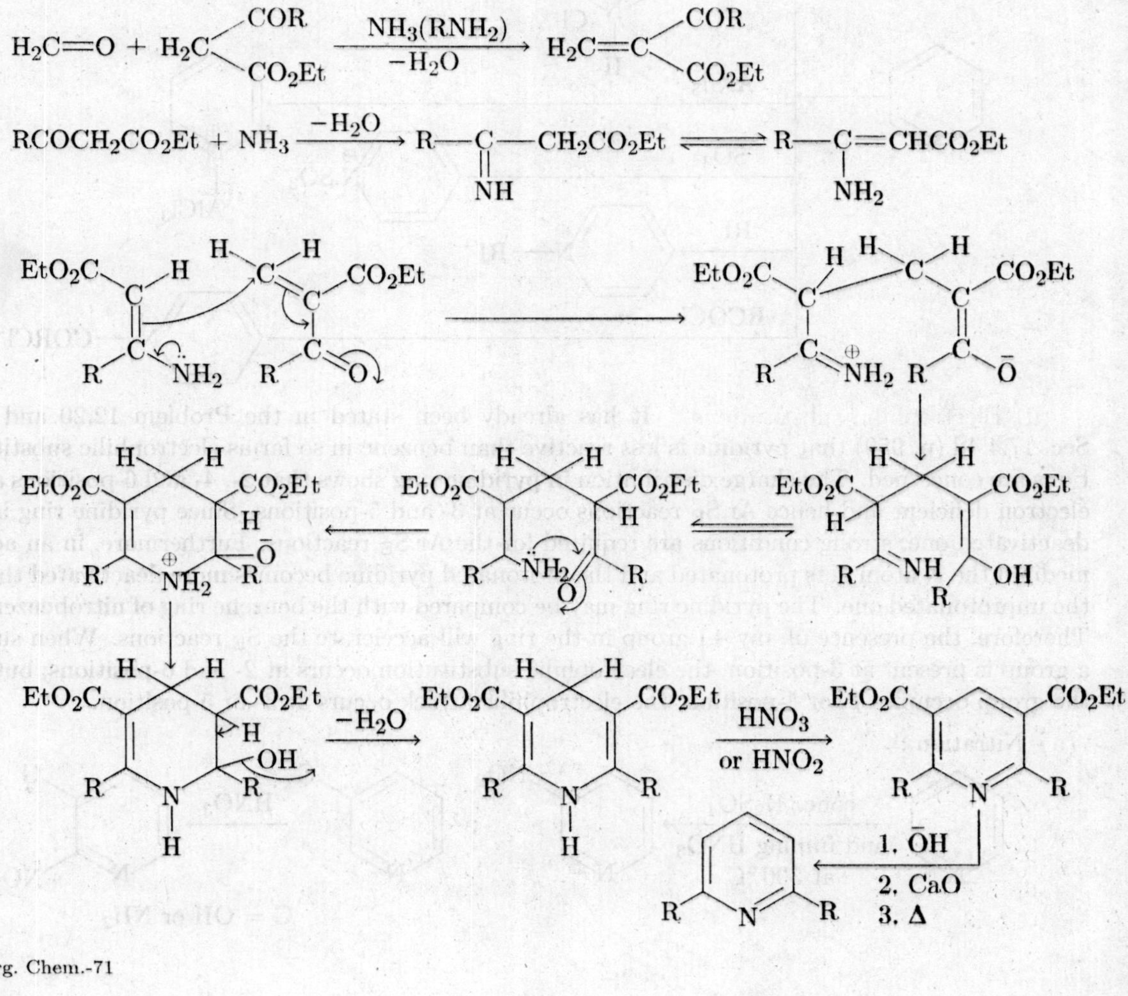

## 19.6.3 Structure reactivity and orientation

[See Sec. 17.4.4d (p. 950)]

## 19.6.4 Chemical properties

(i) Basic character : Pyridine is appreciably basic; it is more basic than pyrrole since a pair of *es* on the N atom do not take part in preserving aromatic character of the compound and hence the *es* are available for donation. Of course, it is much less basic than tertiary amines ($pk_a = 10-11$); since the donatable electron-pair reside in an $sp^2$ A.O. having reasonable amount of *s* character, the donatability of electrons is less than what it would have been if they were in an $sp^3$ A.O. However, it forms salts with acids, acidic oxides and alkylhalides.

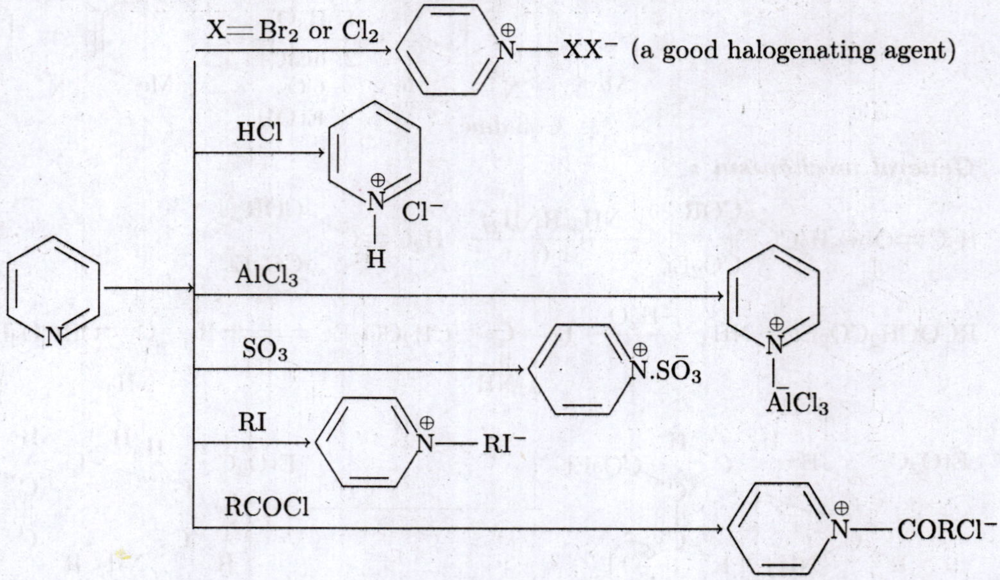


(ii) Electrophilic substitutions : It has already been stated in the Problem 12.20 and in Sec. 17.4.4d (p. 950) that pyridine is less reactive than benzene in so far as electrophilic substitutions are concerned. The charge distribution in pyridine ring shows that 2-, 4- and 6-positions are electron deficient and hence Ar.$S_E$ reactions occur at 3- and 5-positions. Since pyridine ring is a deactivated one, strong conditions are required for the Ar.$S_E$ reactions. Furthermore, in an acid medium the N atom gets protonated and the protonated pyridine becomes more deactivated than the unprotonated one. The pyridine ring may be compared with the benzene ring of nitrobenzene. Therefore, the presence of any +I group in the ring will accelerate the $S_E$ reactions. When such a group is present at 3-position, the electrophilic substitution occurs at 2- and 6-positions; but if the group occupies 1- or 4-position, the electrophilic attack occurs at 3- or 5-position.

(a) Nitration :

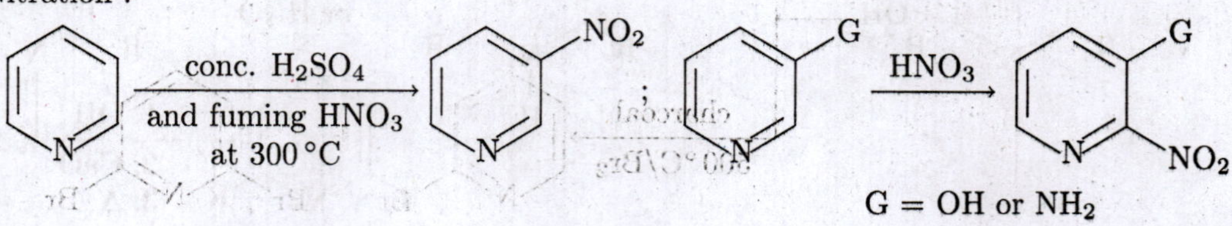

Pyridine is indirectly nitrated by oxidising 2- and 4-aminopyridines with hydrogen peroxide and sulphuric acid.

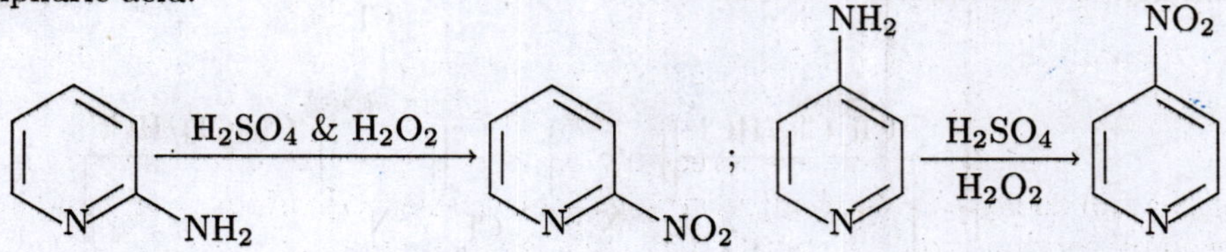


Nitration at 4-position may be done via pyridine-1-oxide.

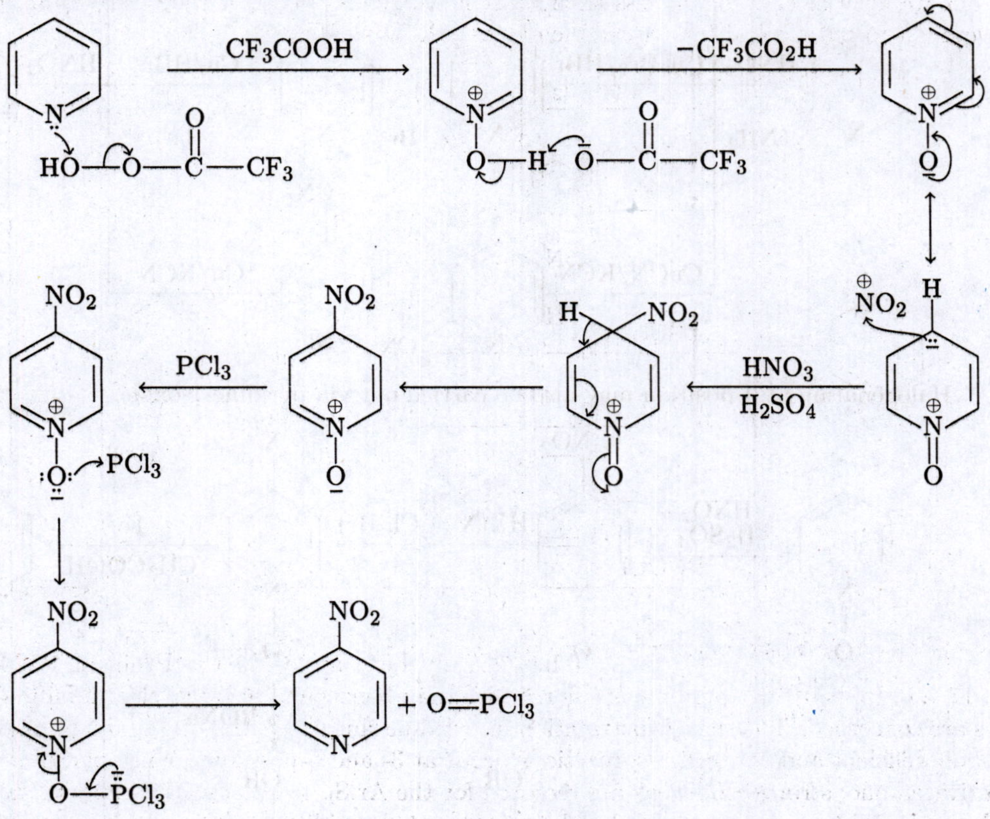


**(b)** Halogenation :

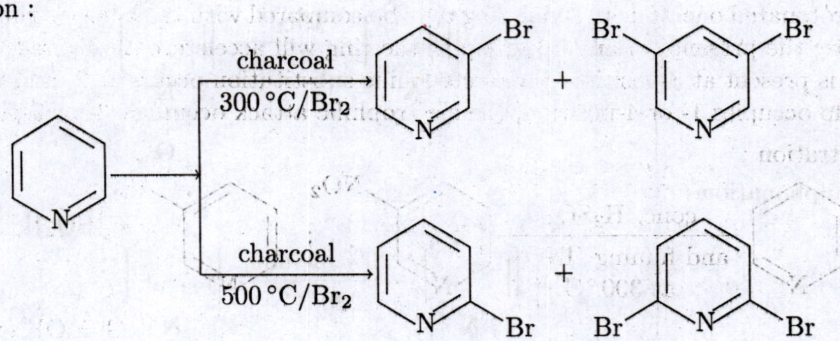

2- and 4-halopyridines are prepared by indirect methods from 2- and 4-amino-pyridines through the Sandmeyer reaction.

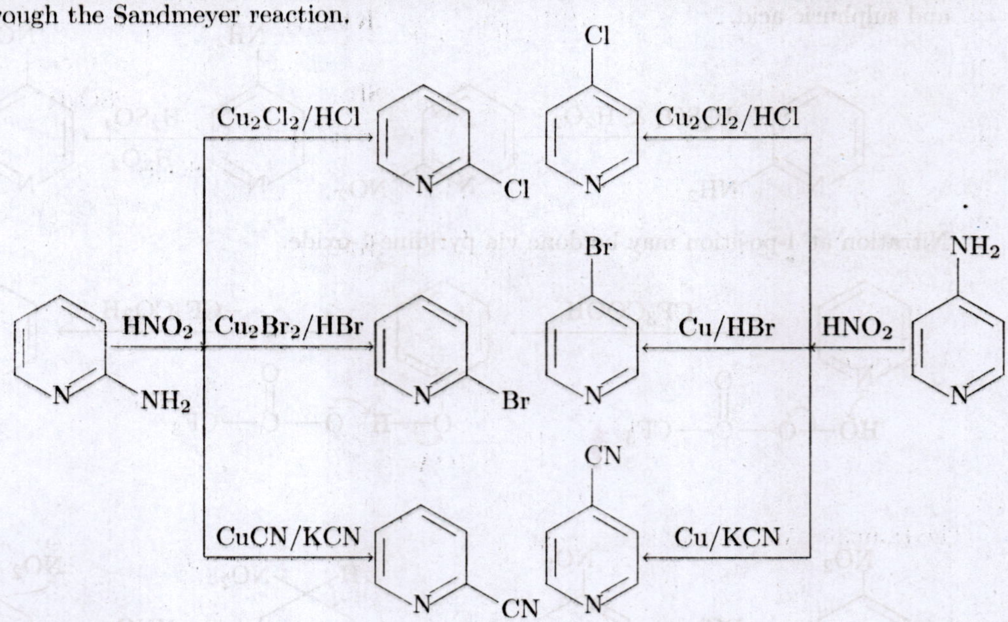


Halogenation at 4-position may also be carried out via pyridine-1-oxide.

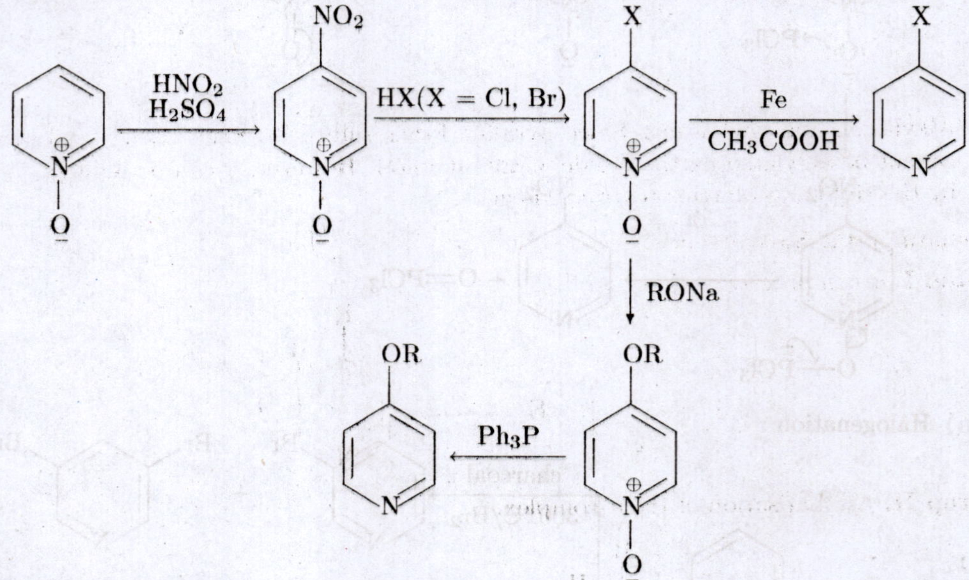


**(c)** Sulphonation :

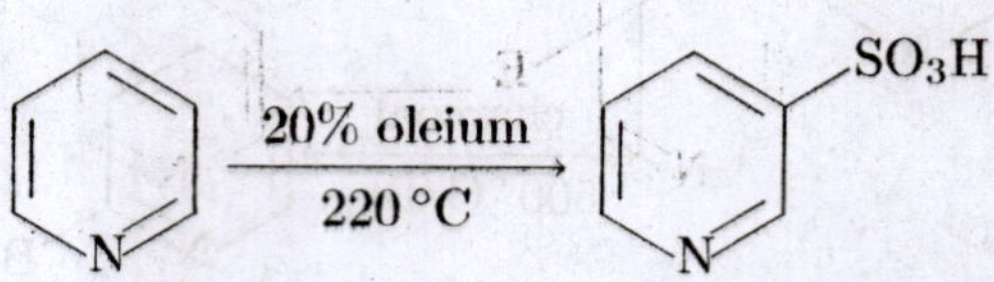

2- and 4-sulphonic acids may indirectly be prepared as shown below.

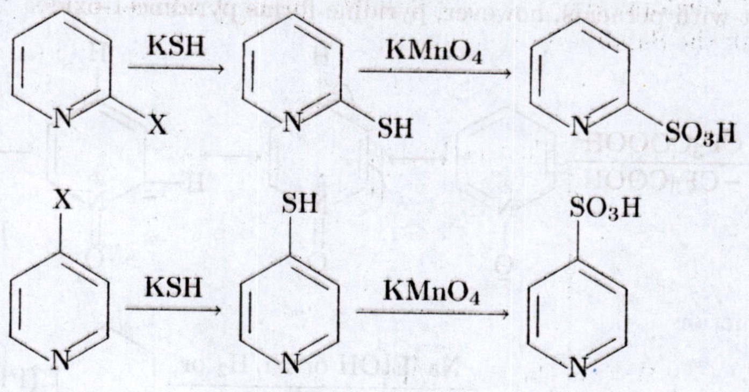

**(d)** Mercuration :

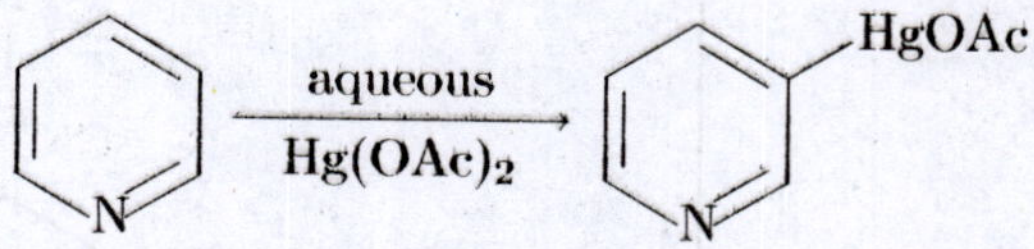

**(e)** The Hofmann-Martius rearrangement :

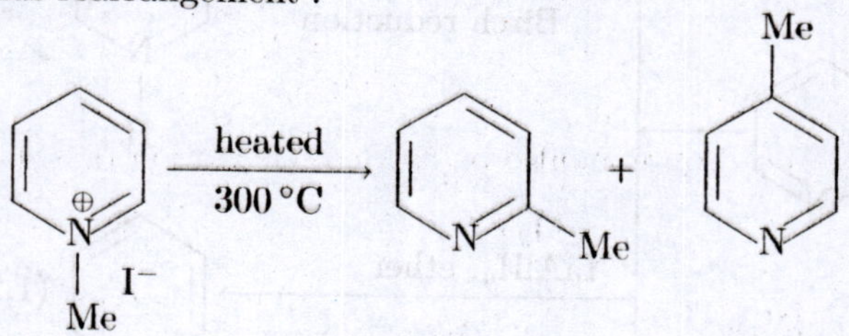

**(f)** Alkylation and acylation : Since pyridine forms salts with acids, Lewis and protonic, it cannot be alkylated by the Friedel-Crafts method. However, it can be indirectly alkylated by the Hofmann-Martius rearrangement.

**Mechanism of the $S_E$ reactions**

**Step I.** $\sigma$-complex formation.

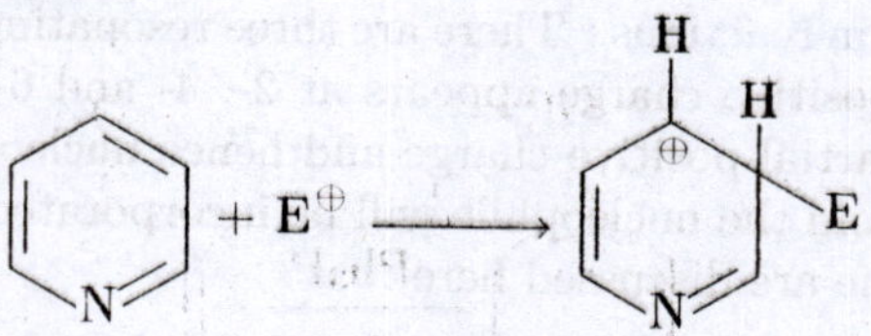

**Step II.** Aromatisation of the $\sigma$-complex.

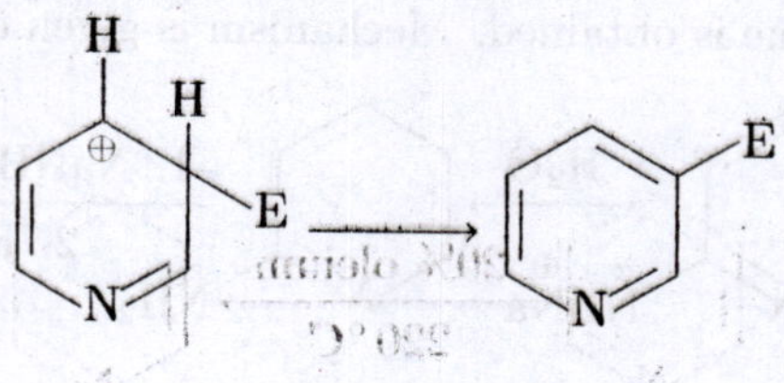

**(iii) Oxidation :** The pyridine ring being deactivated, it is very resistant towards oxidation; on treatment with peracids, however, pyridine forms pyridine-1-oxide.

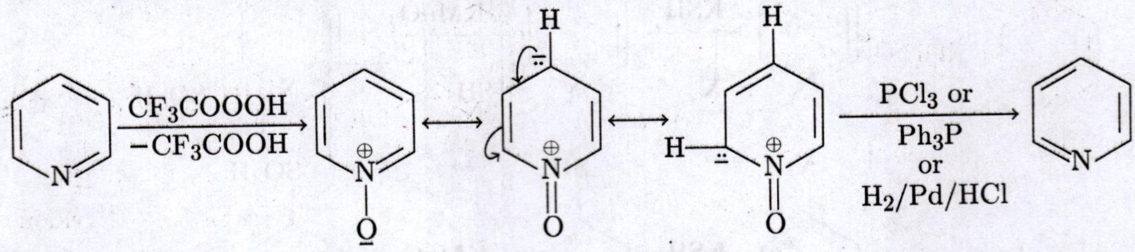

**(iv) Reduction :**

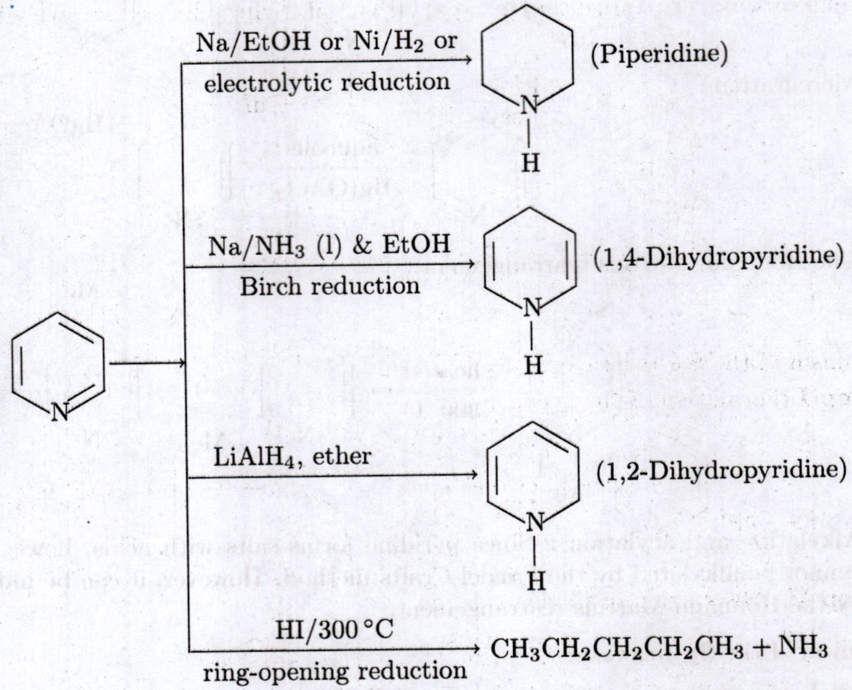

**(v) Nucleophilic substitution reactions :** There are three resonating structures in the resonance hybrid of pyridine in which positive charge appears at 2-, 4- and 6-positions. This means that the ring as a whole bears a partial positive charge and hence nucleophilic substitution reactions are possible on pyridine ring and the nucleophile will be incorporated at 2- or 4- or 6-position. A few $ArS_N$ reactions on pyridine are discussed here :

**(a) Tischitischibabin reaction :** When pyridine is heated with sodamide in toluene sodio salt of 2-aminopyridine is formed. This on hydrolysis gives the amine. When excess of sodamide is used 2,6-diaminopyridine is obtained. Mechanism is given on p. 597.

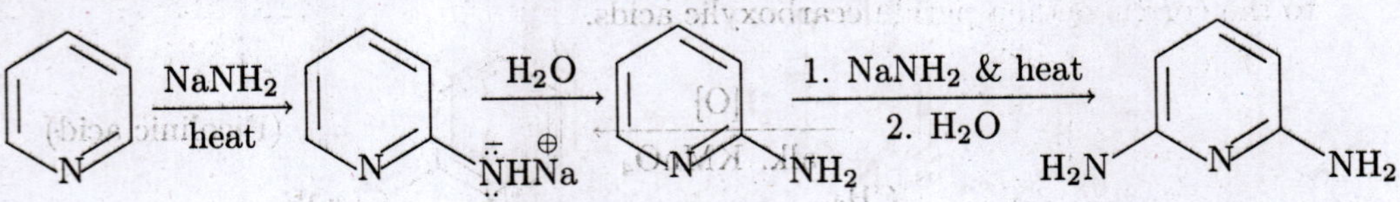

(b) Hydroxylation : When pyridine is heated with solid potassium hydroxide at 400 °C, 2-hydroxypyridine is formed.

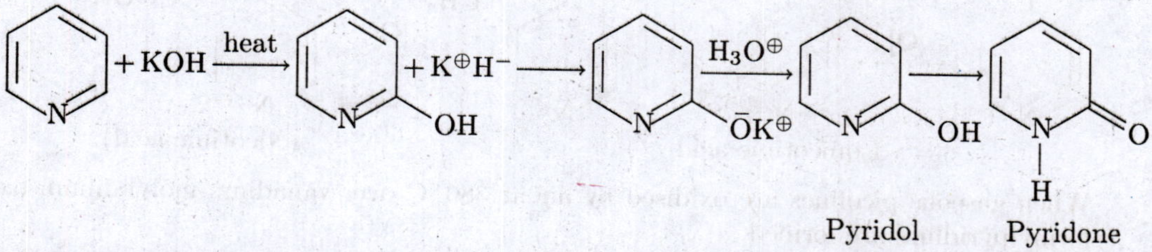

(c) Alkylation and arylation by alkyl- and aryllithiums : When pyridine is heated with *n*-butyllithium or phenyllithium, 2-butyl- or 2-phenylpyridine respectively is formed.

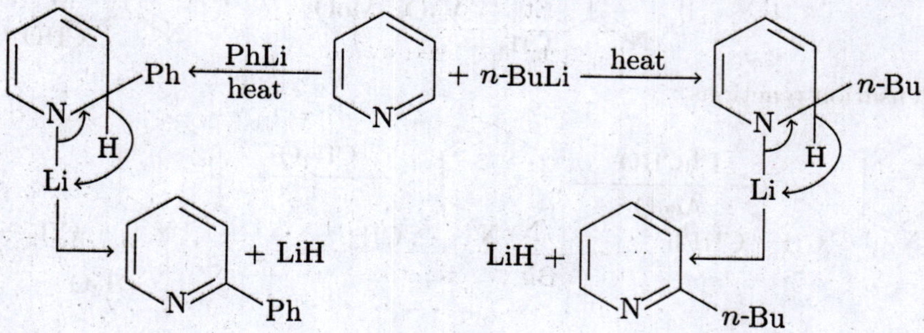

**Mechanism of the $S_N$ reactions :**

**Step I.** Formation of the cyclic anion :

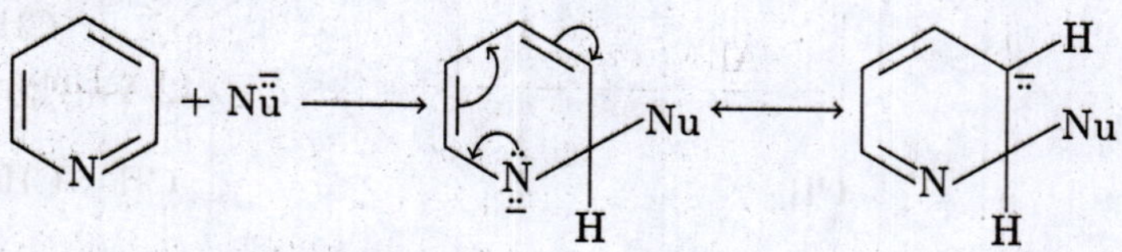

**Step II.** Abstraction of the hydride ion :

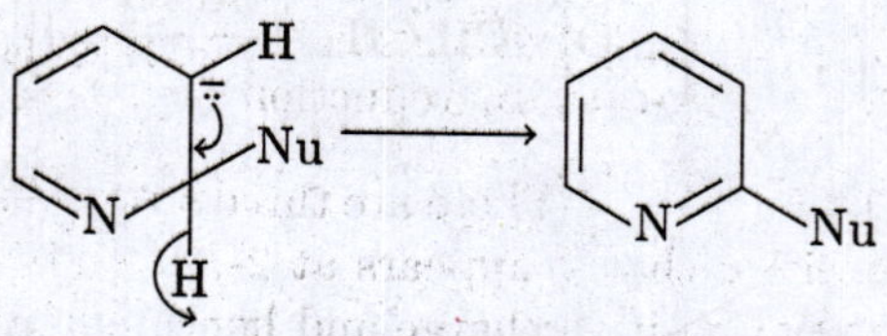

(vi) Reactions on side chains of pyridine homologues : 2-, 3- and 4-methylpyridines are called picolines; they occur in coal tar and in bone oil; 2- and 4-methylpyridines are very reactive and behave as *o*- and *p*-nitrotoluenes respectively. However, the following are their main reactions involving the side chain.

(a) Oxidation : Chromic acid or acid permanganate or alkaline permanganate oxidises picolines to the corresponding pyridinecarboxylic acids.

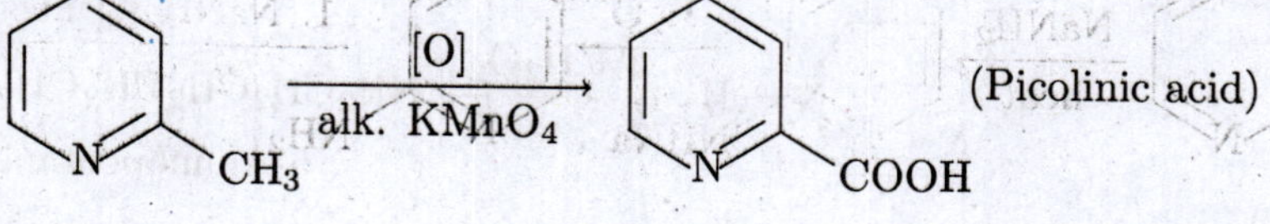

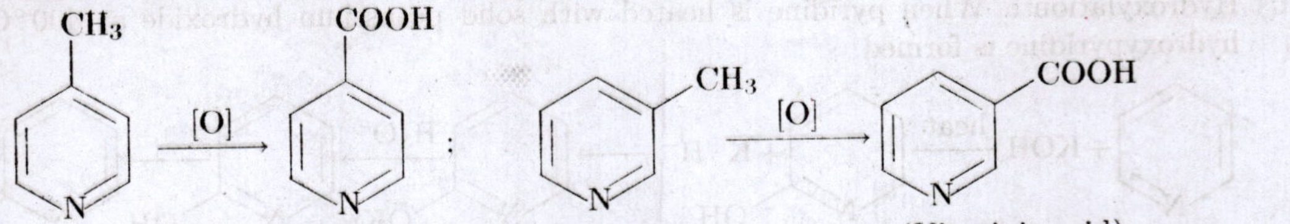

Isonicotinic acid (Nicotinic acid)

When gaseous picolines are oxidised by air at 380 °C over vanadium-molybdinum oxides, formylpyridines are formed.

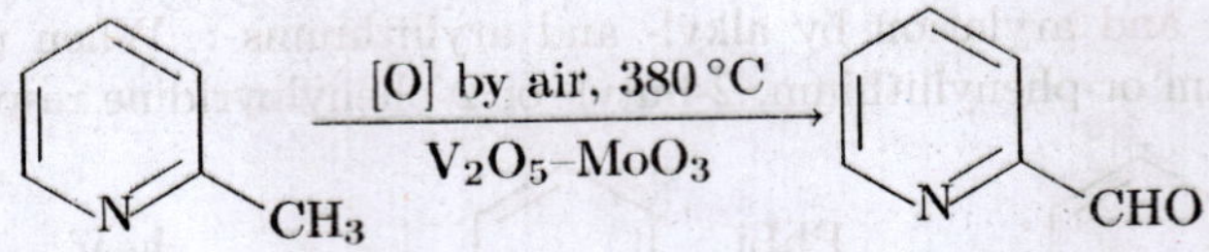

(b) Condensation reactions :

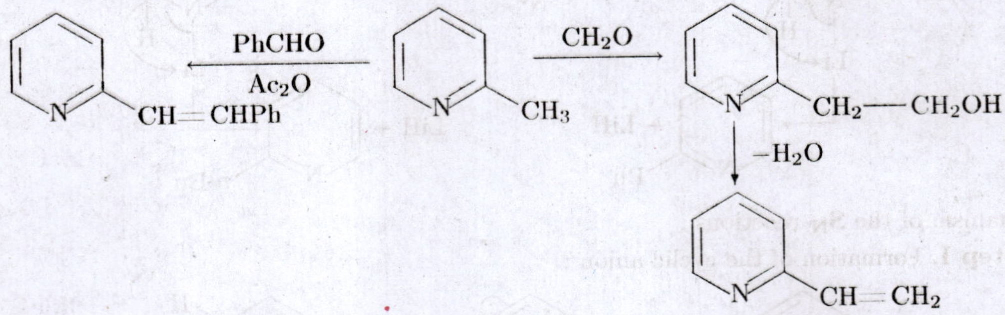

2-Vinylpyridine

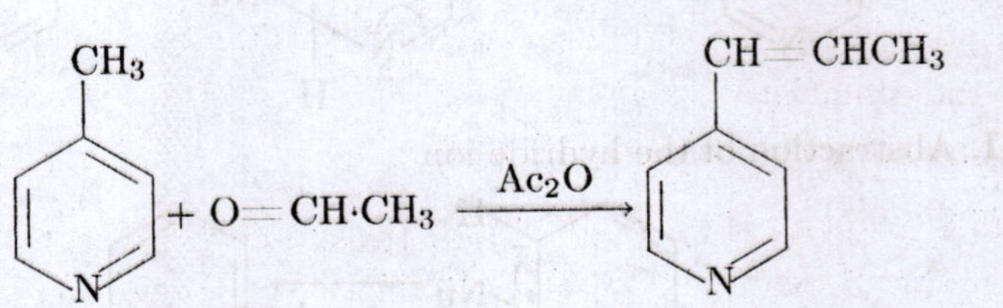

(c) Side-chain carboxylation :

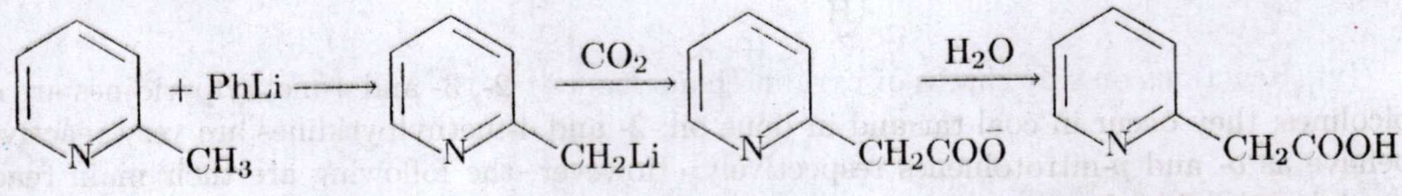

(vii) Ring-fission reactions : Usually the aromatic ring is fast reduced to hydrogenated ring before fission. Then any of the following methods may be used as the ring-opening reaction.

(a) Hydrogen peroxide method :

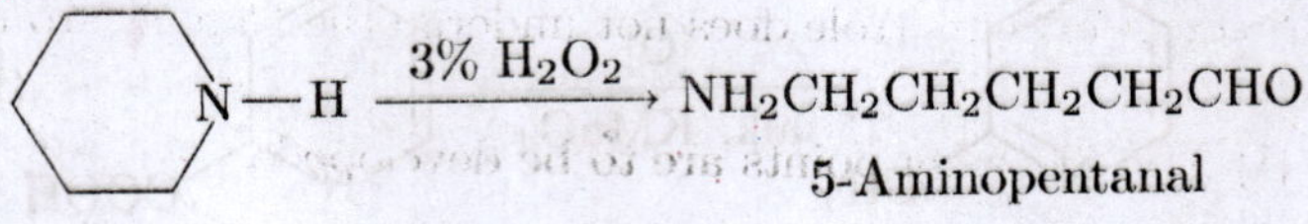

5-Aminopentanal

**(b)** Von Brown's method :

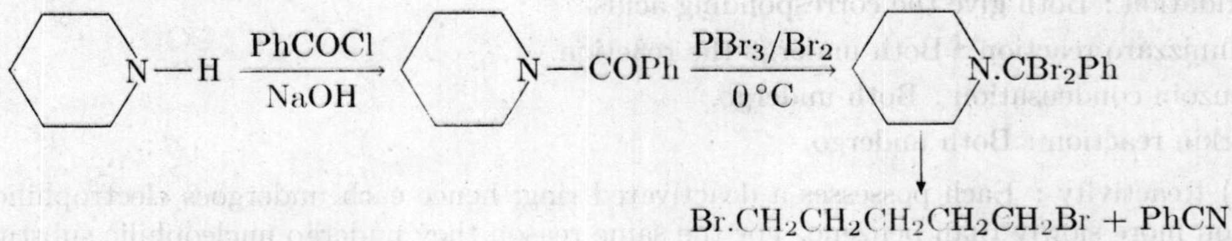


**(c)** Hydroiodic acid method :

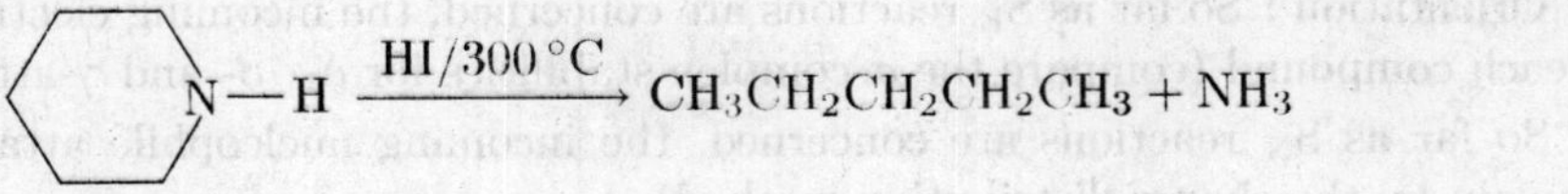


**(d)** The Hofmann exhaustive methylation method :

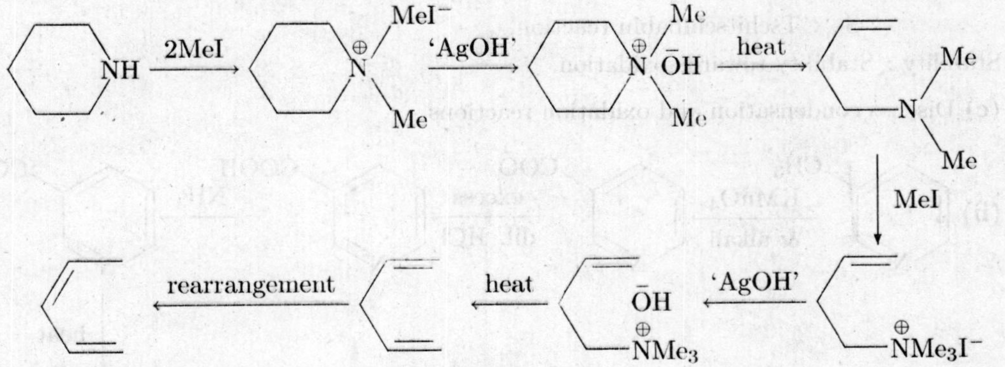


1,3-Pentadiene 1,4-Pentadiene

1, 3-pentadiene being more stable than 1, 4-pentadiene, the latter rearranges to the former under the conditions of the experiments.

**(e)** Von Braun cynogen bromide method

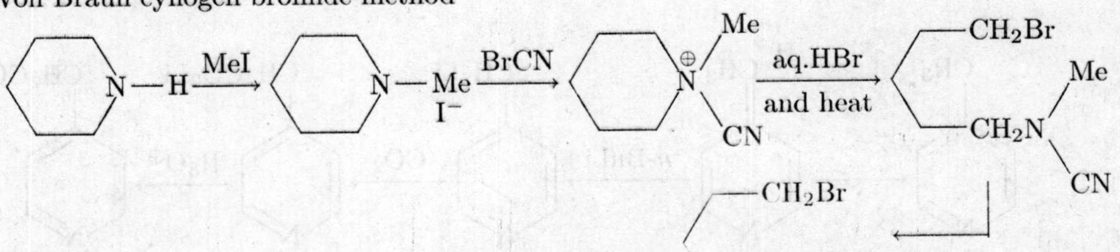


**Problem 19.4. (i)** Compare **(a)** furfural with benzaldehyde and **(b)** pyridine with nitrobenzene and **(c)** $\alpha$-picoline with *o*-nitrotoluene.

**(ii)** Synthesise : **(a)** 3-aminopyridine from $\beta$-picoline, **(b)** 4-pyridylacetic acid from $\gamma$-picoline.

**(iii)** Explain why 2-formylpyrrole does not undergo the Cannizzaro reaction and the benzoin condensation reaction.

**Solution : (i)** The following points are to be developed.

**(a)** Reduction : Both give the corresponding alcohols.

Oxidation : Both give the corresponding acids.

Cannizzaro reaction : Both undergo the reaction.

Benzoin condensation : Both undergo.

Perkin reaction : Both undergo.

**(b)** Reactivity : Each possesses a deactivated ring; hence each undergoes electrophilic substitution more slowly than benzene. For the same reason they undergo nucleophilic substitution reactions (draw resonating structures).

Orientation : So far as $S_E$ reactions are concerned, the incoming electrophile attacks the $\beta$-C of each compound (compare the $\sigma$-complex stabilities for $\alpha$-, $\beta$- and $\gamma$-attacks in each case).

So far as $S_N$ reactions are concerned, the incoming nucleophile attacks the $\alpha$- or the $\gamma$-C (explain by the charge distribution method).

Reactions : $S_E$ : Nitration, sulphonation, halogenation reactions.

$S_N$ : Tschitschibabin reaction.

Stability : Stability towards oxidation.

**(c)** Discuss condensation and oxidation reactions.

**(ii)**

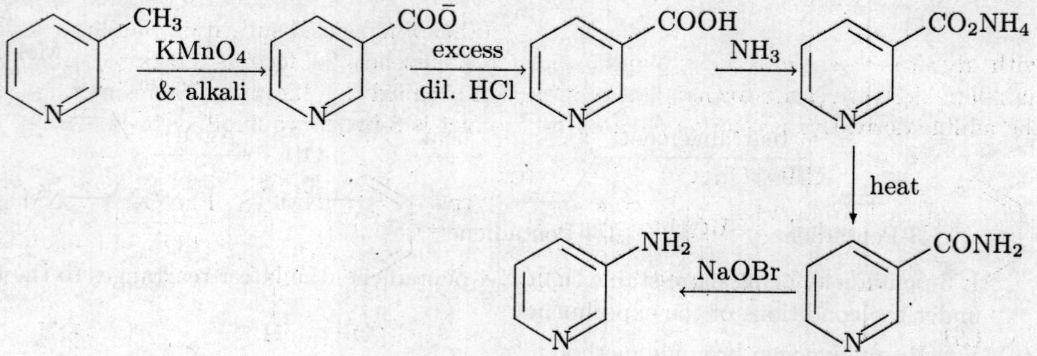


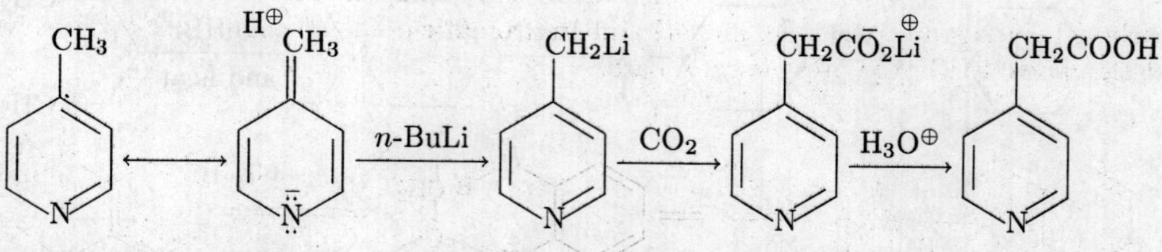


**(iii)** Perhaps the carbonyl C is not much electrophilic for the attack of $\bar{O}H$ or $\bar{C}N$ because of the contribution by the following reasonating structures in the resonance hybrid of 2-formylpyrrole.

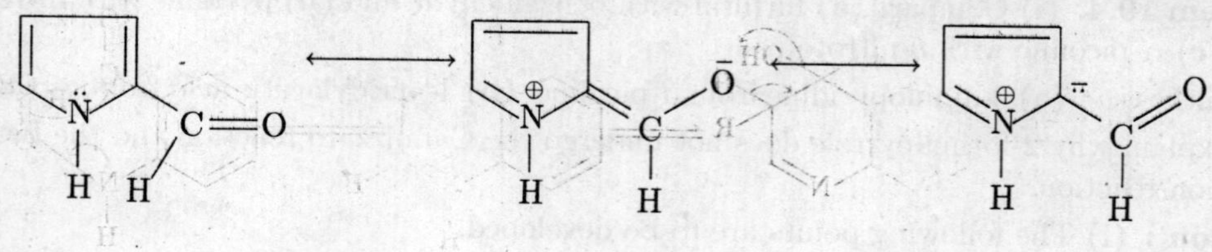

## 19.7 Quinoline

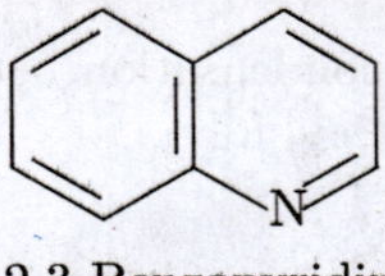

R.E. = 264.9 kJ $mol^{-1}$
insoluble in water

2,3-Benzopyridine
m.p. = 19°C; b.p. = 238°C

### 19.7.1 Source

Coal tar, bone oil and quinine (an alkaloid).

Coal tar ⟶ middle oil fraction (200°-250°C) ⟶ distillate $\xrightarrow[\text{with acid}]{\text{treated}}$ acid extract $\xrightarrow{\text{alkali distillation at 240 °C}}$ Quinoline

### 19.7.2 Synthesis of quinoline

(a) The Skraup synthesis (See Sec. 17.18) : When aniline and substituted anilines are treated with glycerol, concentrated sulphuric acid, nitrobenzene and ferrous sulphate, quinoline and quinoline derivatives are formed and the reaction is called the Skraup quinoline synthesis. When the aniline derivative is *o*-aminophenol, the product is 8-hydroxyquinoline (p. 1006).

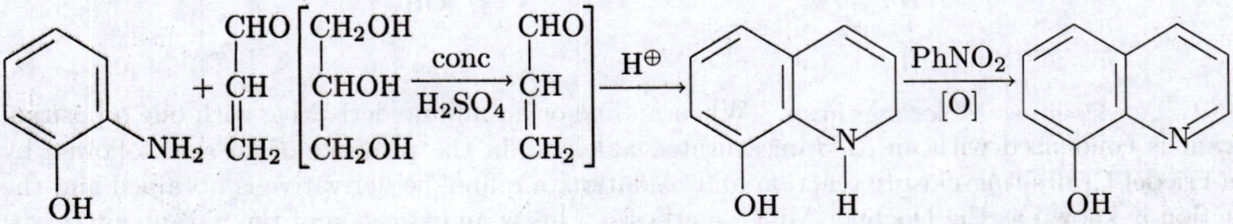

(b) The Friedlander synthesis : This involves a condensation reaction between an aliphatic carbonyl compound containing an $\alpha$-H and an aromatic carbonyl compound with an $NH_2$ group at the *o*-position in the presence of a base.

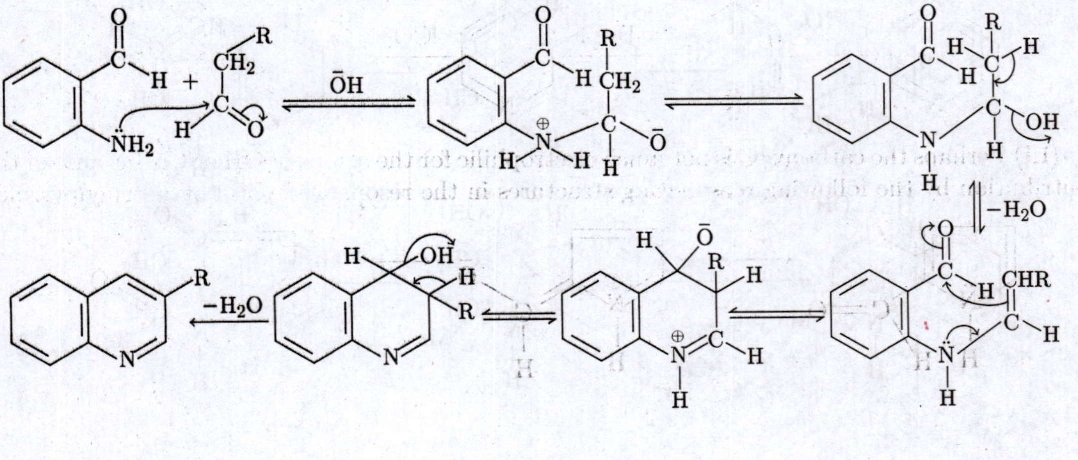

However, *o*-aminobenzaldehydes being very unstable, this method has no practical value.

When R = H, quinoline is obtained.

(c) The Knorr quinoline synthesis : This involves condensation between an aromatic amine and a $\beta$-ketoester followed by a ring-closure reaction.

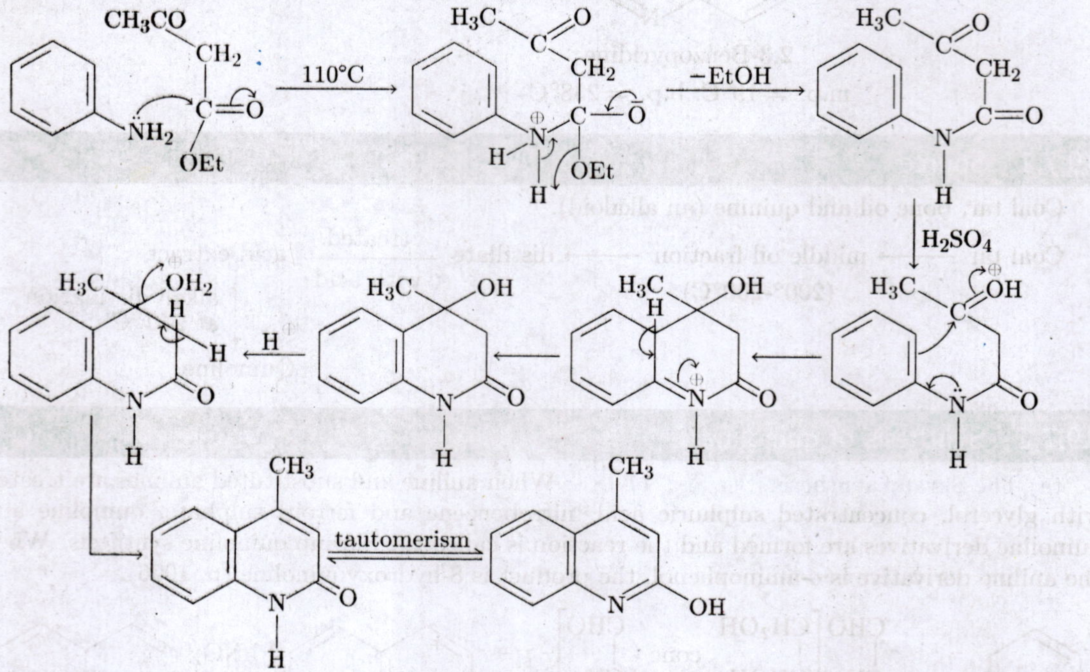

(d) The Doebner-Miller synthesis : When aniline or an aniline derivative with one *o*-position vacant is condensed with an $\alpha$, $\beta$-unsaturated aldehyde in the presence of an acid followed by the Friedel-Crafts ring-closure reaction and oxidation, a quinoline derivative is obtained and the reaction is known as the Doebner-Miller synthesis. This is an extension of the Skraup synthesis; here the oxidising agent used is a Schiff's base instead of nitrobenzene.

$$2\ CH_3CHO \xrightarrow{H_3O^{\oplus}} CH_3CH{=}CHCHO$$

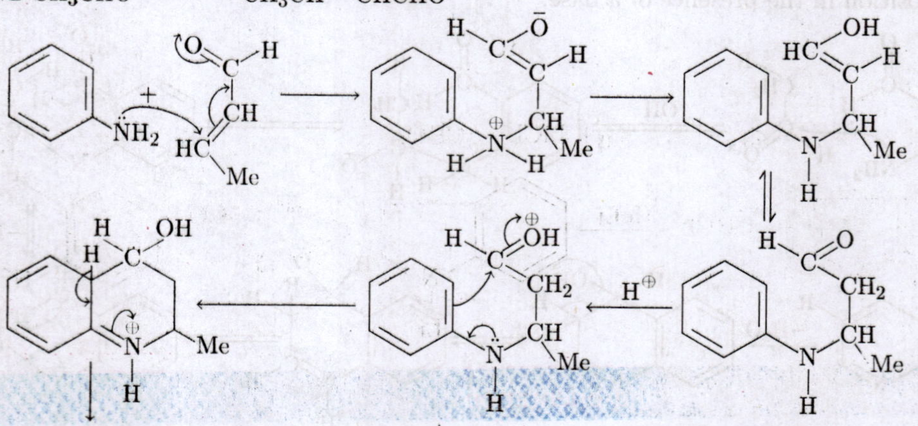

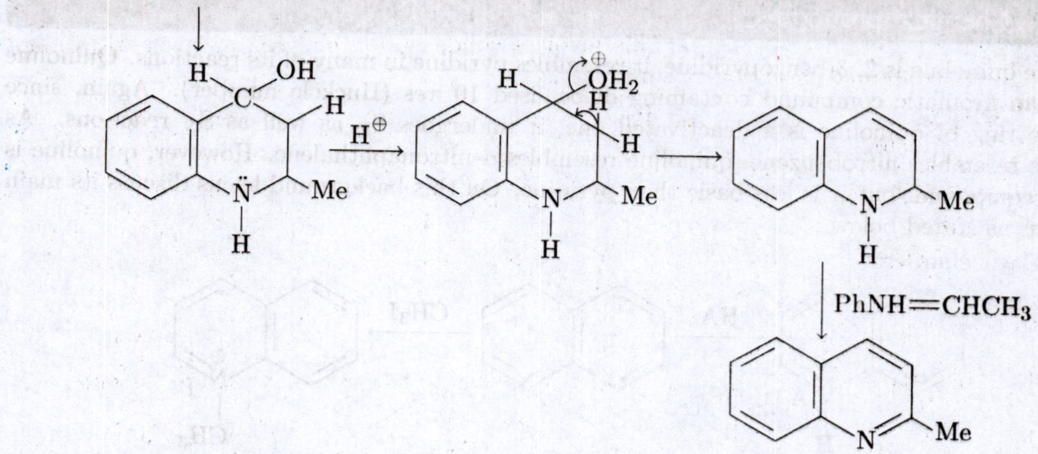

(e) The Conrad-Limpach synthesis : This involves the condensation between aniline or its derivative and a $\beta$-ketoester at room temperature and then ring closure at an elivated temperature.

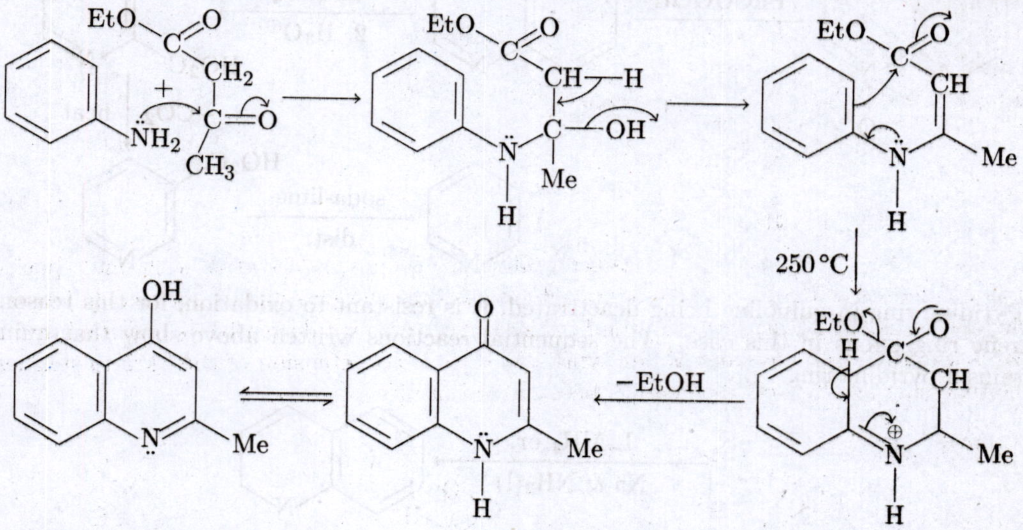

(f) From indole and dichloromethylene in the presence of methyllithium :

$$CH_2Cl_2 + CH_3Li \longrightarrow CH_4 + LiCl + :CHCl$$

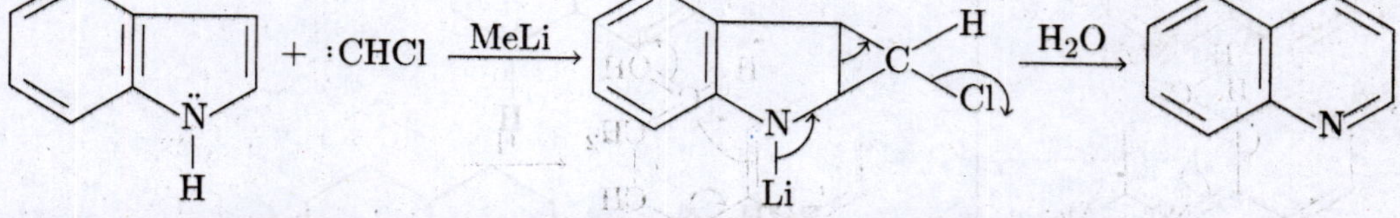

## 19.7.3 Structure, reactivity and orientation

[See Sec. 17.4.4.f (p. 951)]

## 19.7.4 Chemical properties

Since quinoline is 2, 3-benzopyridine, it resembles pyridine in many of its reactions. Quinoline is also an aromatic compound containing delocalised 10 $\pi$es (Huckcle number). Again, since pyridine ring of quinoline is a deactivated one, it undergoes $S_E$ as well as $S_N$ reactions. As pyridine resembles nitrobenzene, quinoline resembles $\alpha$-nitronaphthalene. However, quinoline is a basic compound, but it is less basic than pyridine. On this background let us discuss its main reactions as stated below :

(a) Basic character :

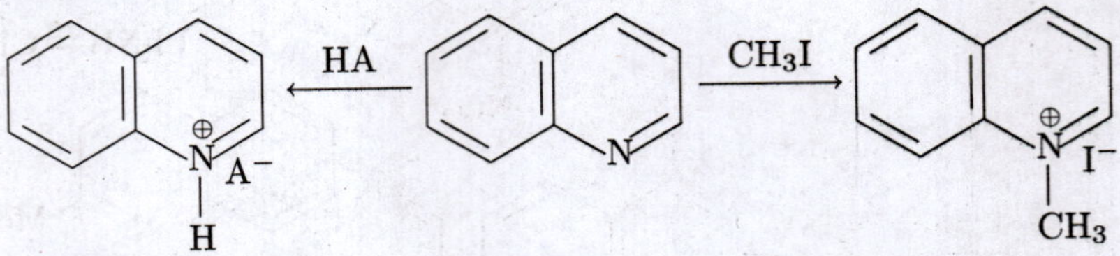


(b) Oxidation :

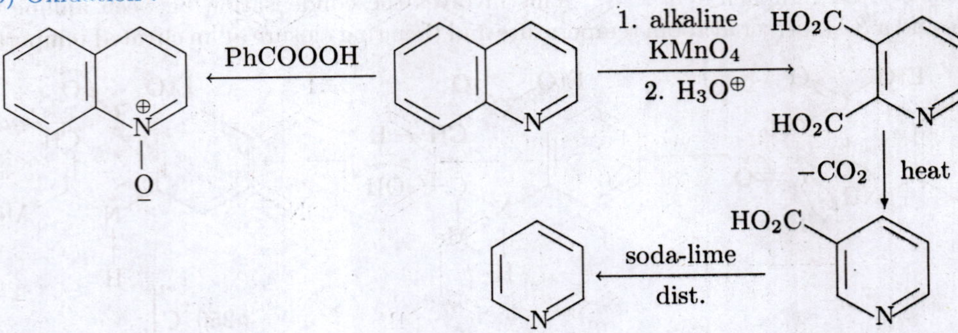


Pyridine ring in quinoline being deactivated, it is resistant to oxidation; for this reason, the benzene ring opens in this case. The sequential reactions written above show that quinoline contains a pyridine ring.

(c) Reduction :

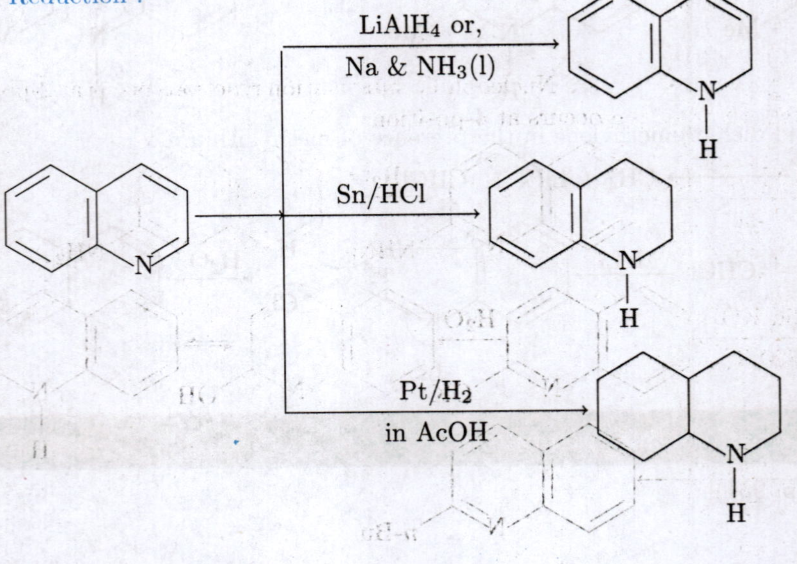

(d) Electrophilic substitution reactions :

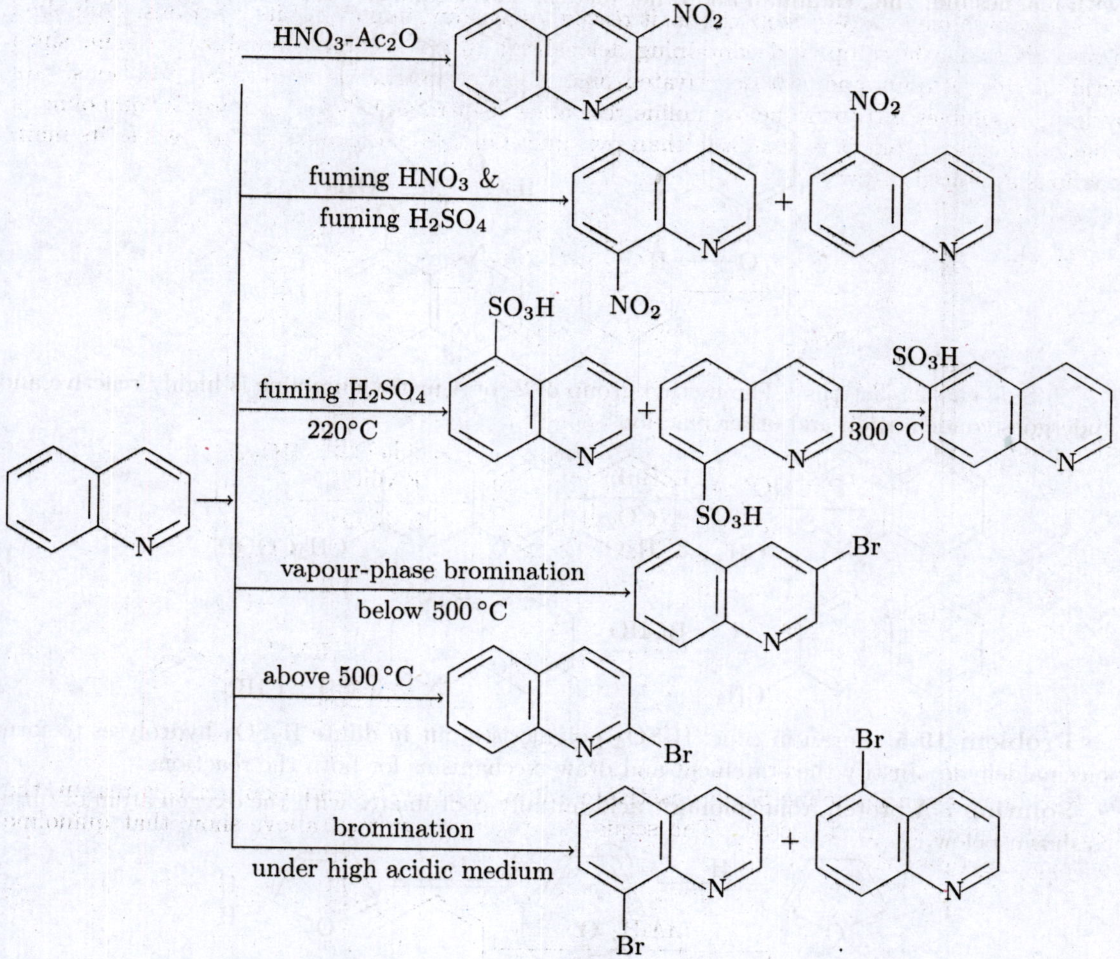


(e) Nucleophilic substitution reactions : Nucleophilic substitution reactions occur at 2-position if this position is blocked, the reaction occurs at 4-position.

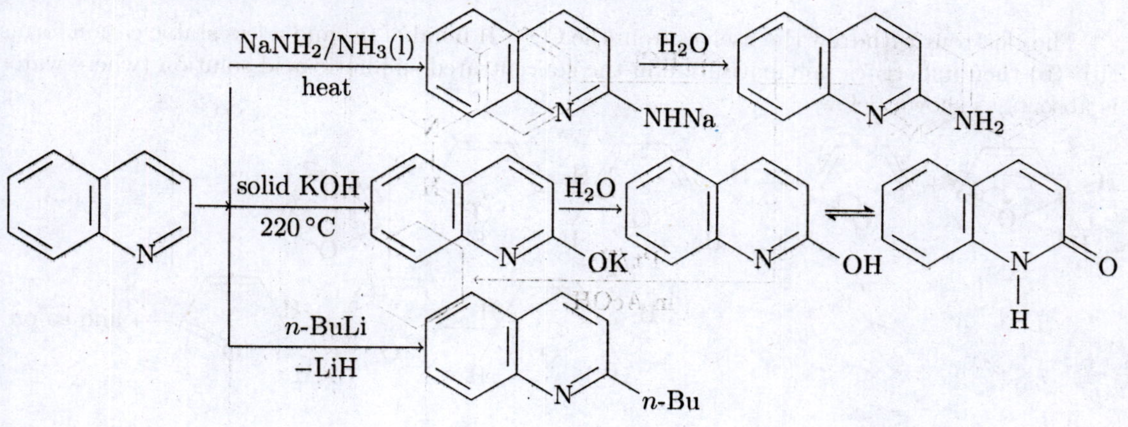

**Note :** 8-hydroxyquinoline is a chelated compound as shown below; it forms complex salts with magnesium, zinc, cadmium and other metals. These salts are used for gravimetric analysis.

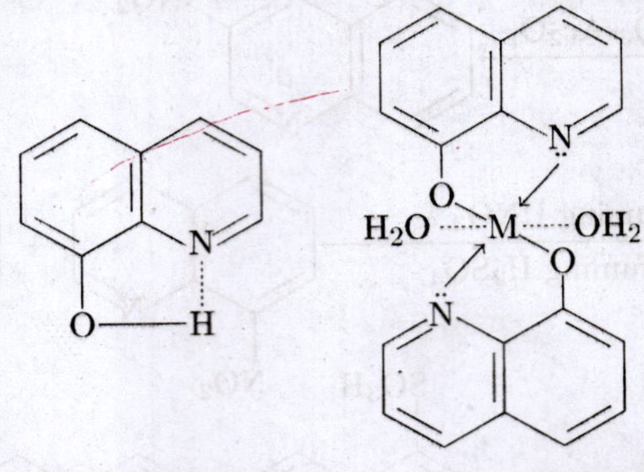

(f) Side-chain reactions : The methyl group of 2- or 4-methylquinoline is highly reactive and undergoes condensation and other reactions.

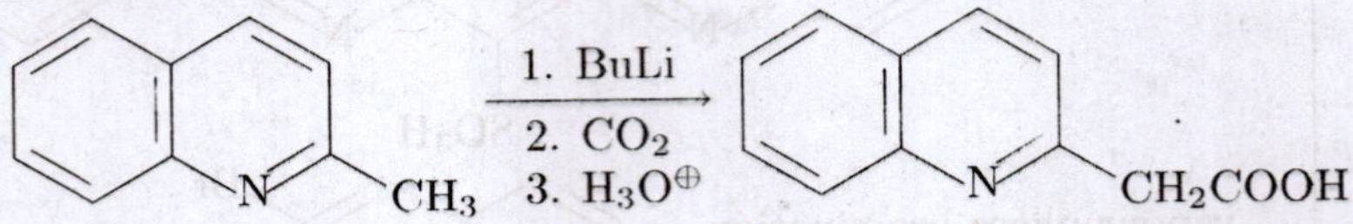

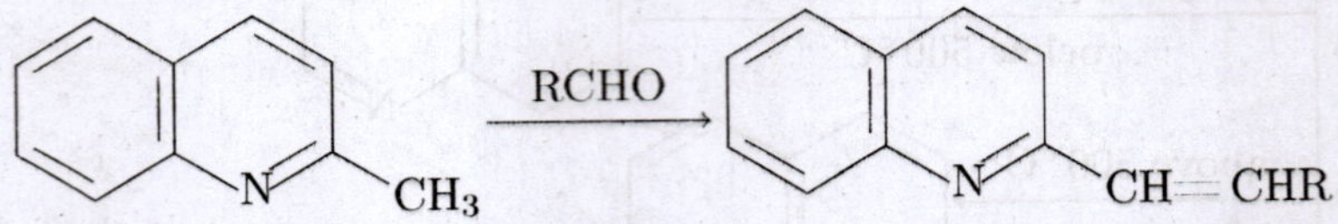

**Problem 19.5.** Furan in conc. $H_2SO_4$ polymerises but in dilute $H_2SO_4$ hydrolyses to form succinaldehyde. Justify the statement and draw mechanisms for both the reactions.

**Solution :** A proton from sulphuric acid initially coordinates with the oxygen atom of furan as shown below :

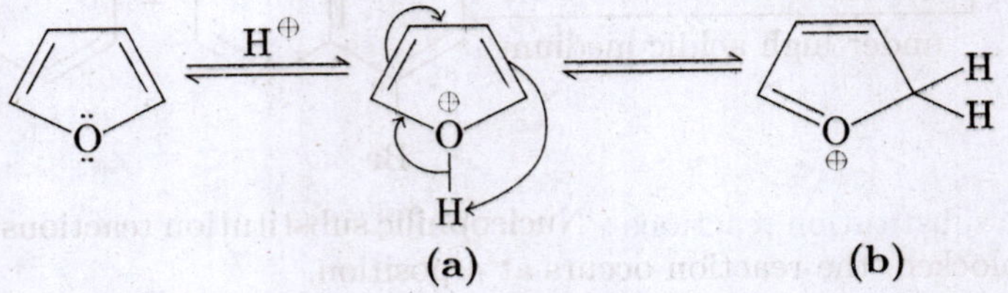

The electrons withdraw themselves from the $\overset{\oplus}{O}$—H bond of (a) and a less stable cation forms (b). (b) then undergoes polymarisation in the concentrated sulphuric acid solution (where water is absent) as shown below :

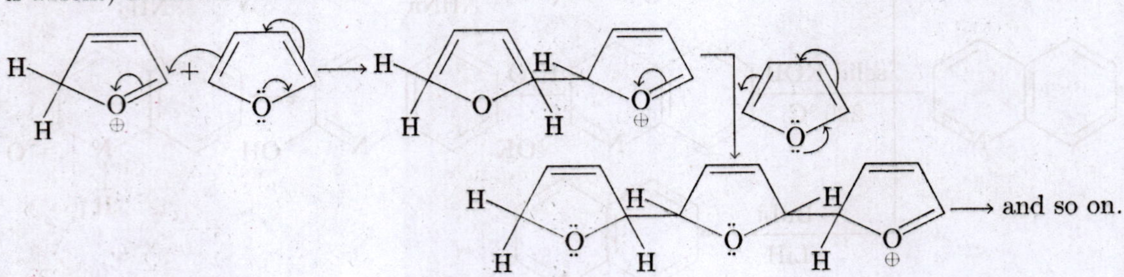

In the dilute sulphuric acid solution the unstable cation (b) is attacked by water at the methylene C and succinaldehyde is formed through several steps.

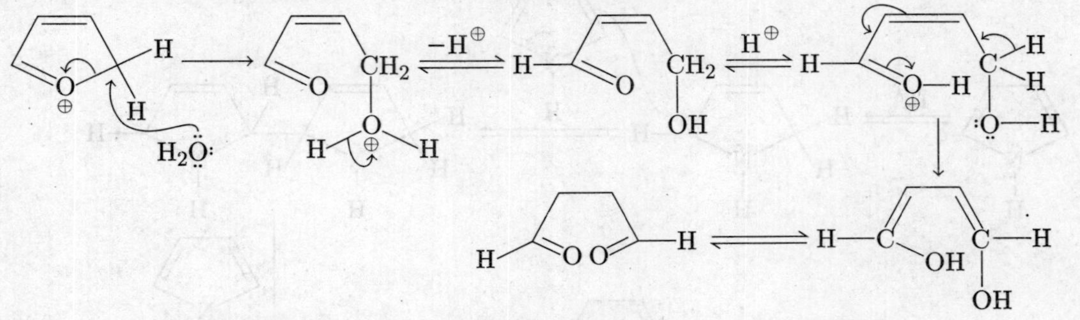

This is the pathway of hydrolysis of enol-ethers. Furan is an enol ether.

However, succinaldehyde may also form from the cation (a) as follows :

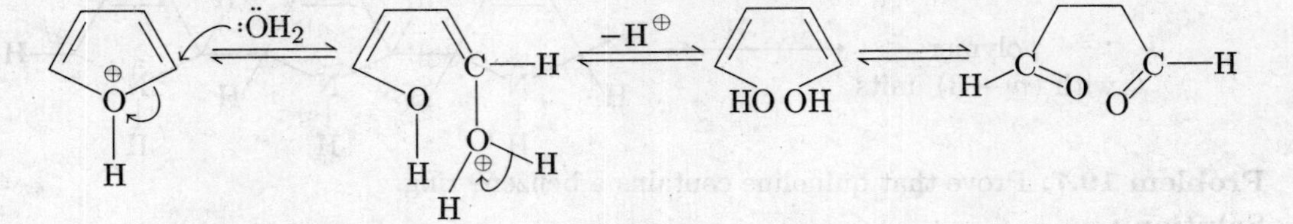

**Problem 19.6.** In the mineral acid medium pyrrole trimerises and polymerises draw the mechanistic steps for the reactions.

**(a) Mechanism for the trimerisation :**

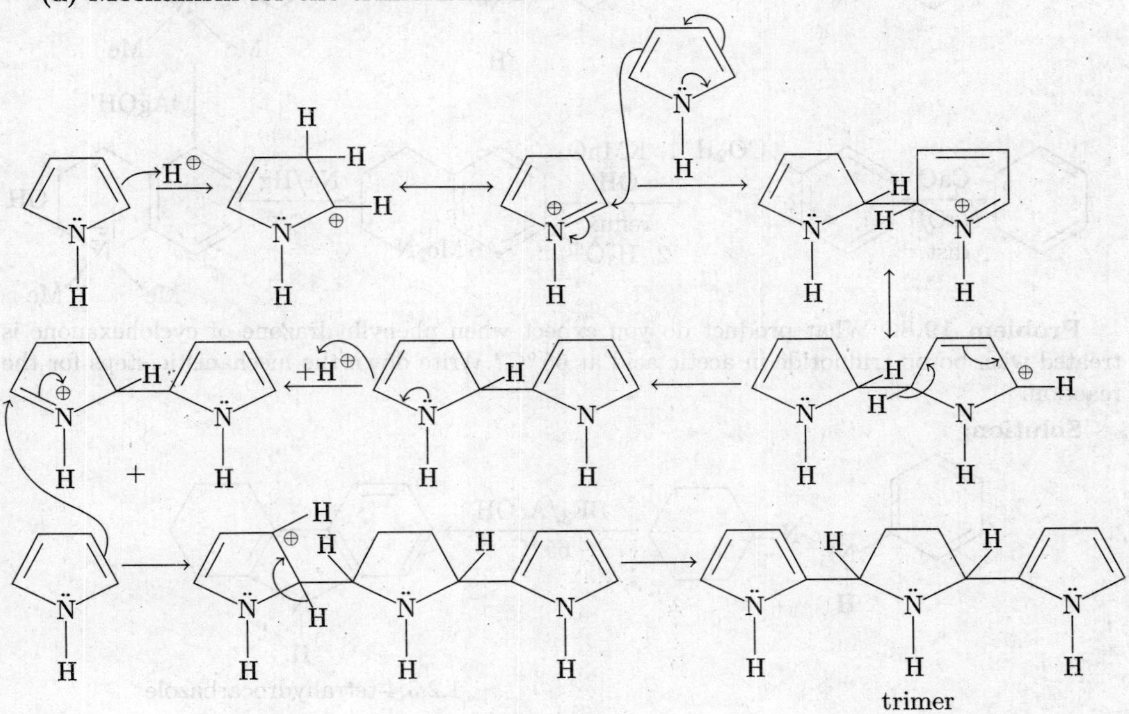

**(b) Mechanism for the polymerisation :**

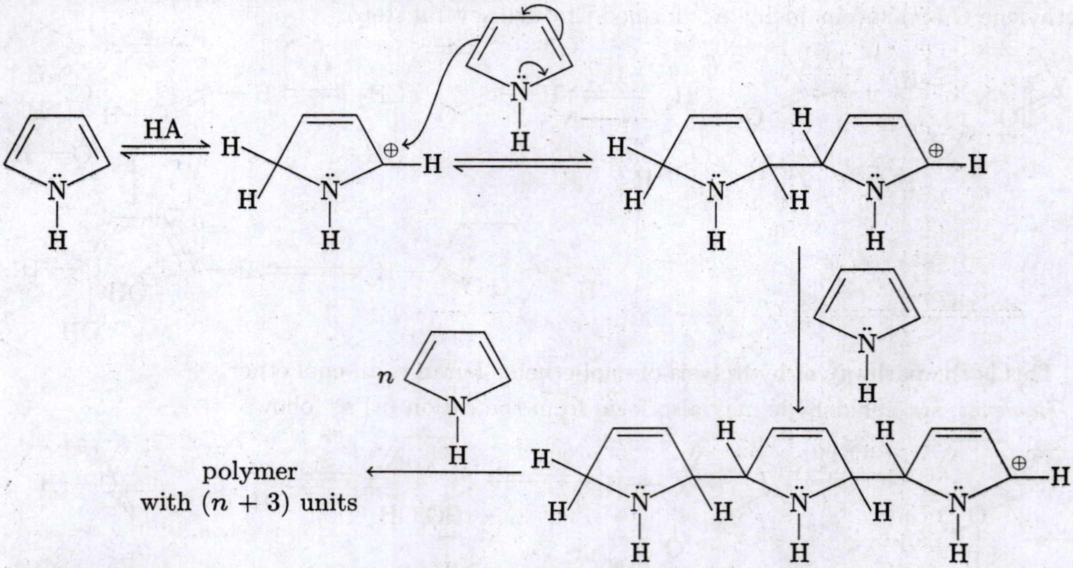


**Problem 19.7.** Prove that quinoline contains a benzene ring.

**Solution :**

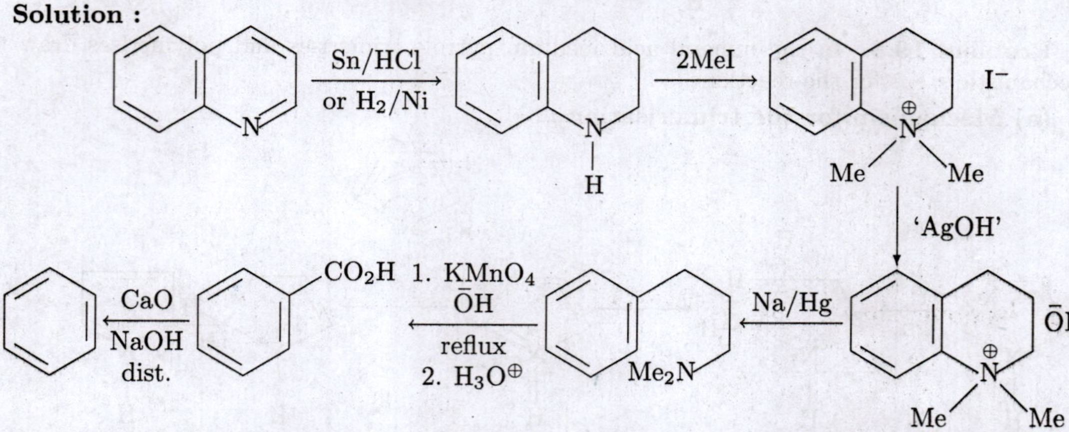


**Problem 19.8.** What product do you expect when phenylhydrazone of cyclohexanone is treated with boron trifluoride in acetic acid at 65 °C? Write down the mechanistic steps for the reaction.

**Solution:**

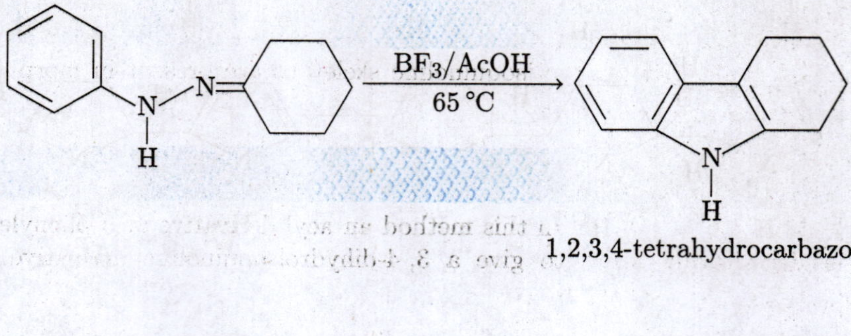


1,2,3,4-tetrahydrocarbazole

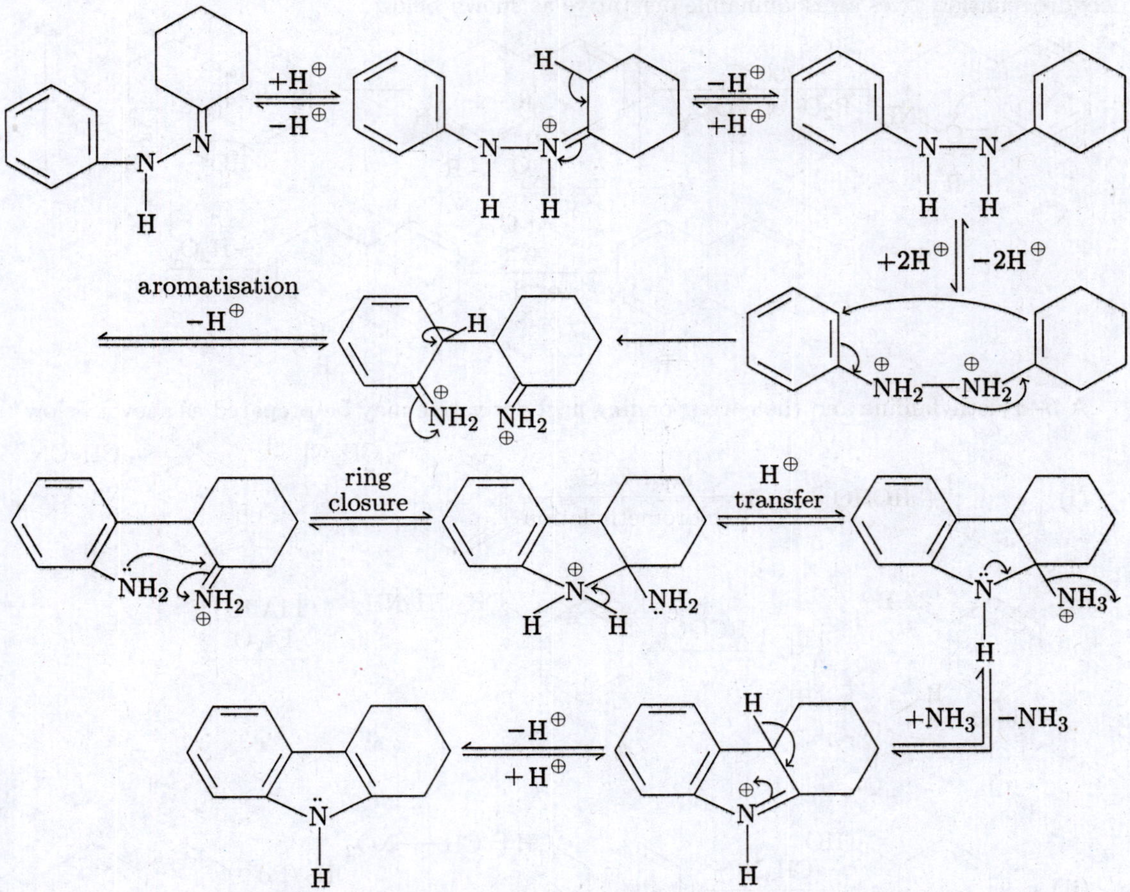

## 19.8 Isoquinoline

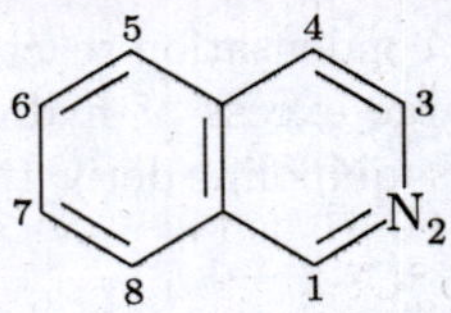

m.p. = 26 °C
b.p. = 240 °C

3, 4-Benzopyridine (considering N as no. 1 atom)

### 19.8.1 Source

Coal tar, bone oil and alkaloids (the isoquinoline skeletons are present in morphine, papaverine, etc.)

### 19.8.2 Synthesis

(a) Bischler-Napieralski synthesis : In this method an acyl derivative of $\beta$-phenylethylamine is heated with a dehydrating agent to give a 3, 4-dihydroisoquinoline derivative which on dehydrogenation gives an isoquinaline derivative as shown below :

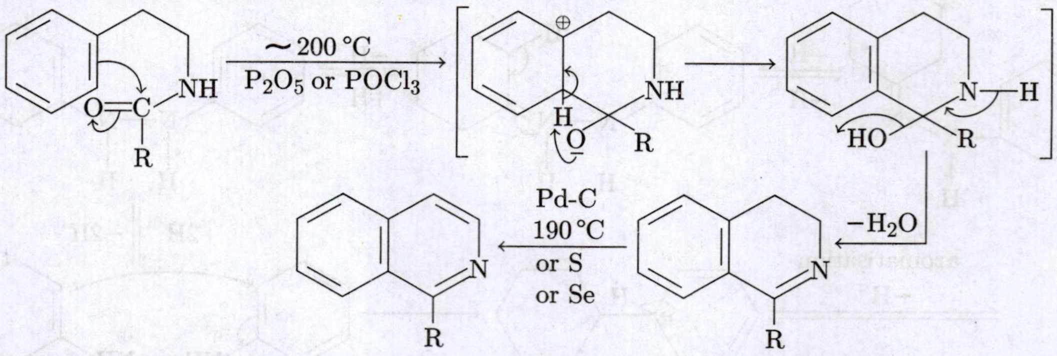

A $\beta$-arylethylamine and the corresponding acyl derivative may be prepared as shown below :

(i)

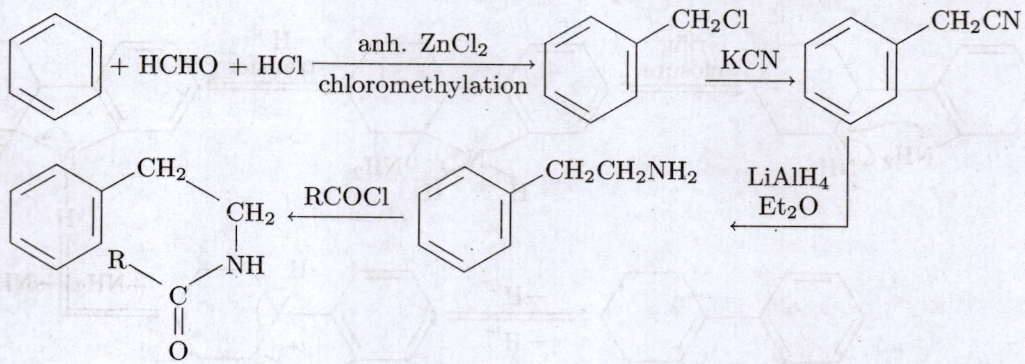

(ii)

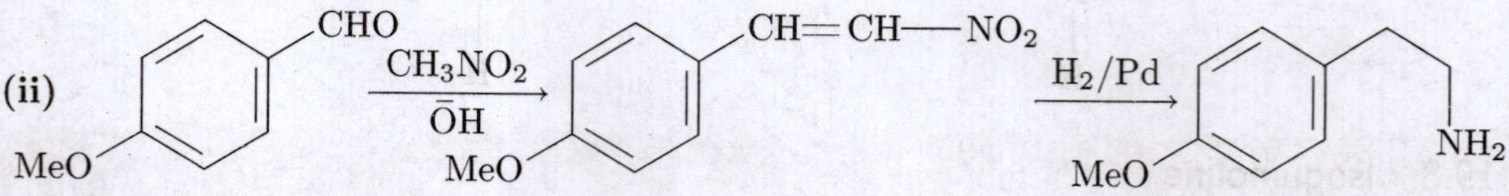

(b) Piclet Spengler reaction : This involves a condensation reaction between a $\beta$-phenylethylamine and an aldehyde in the presence of a large excess of hydrochloric acid at 100 °C. The condensation reaction gives 1, 2, 3, 4-tetrahydroisoquinoline derivative which on dehydrogenation forms isoquinoline derivative.

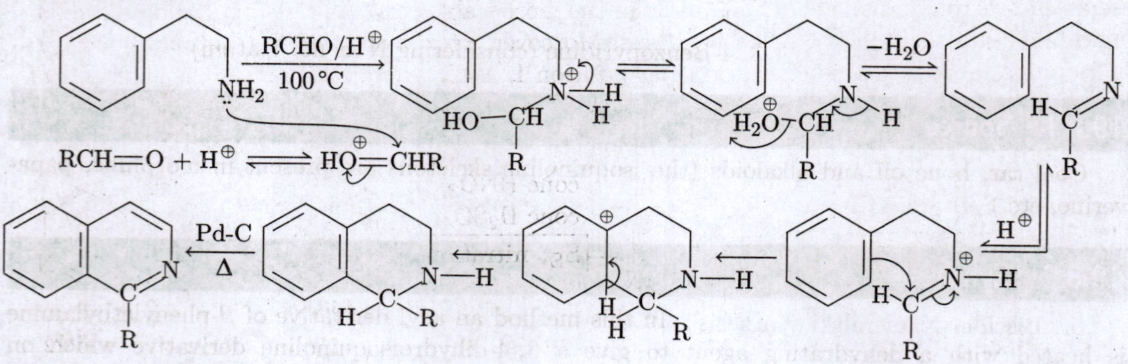

(c) From the oxime of cinnamaldehyde :

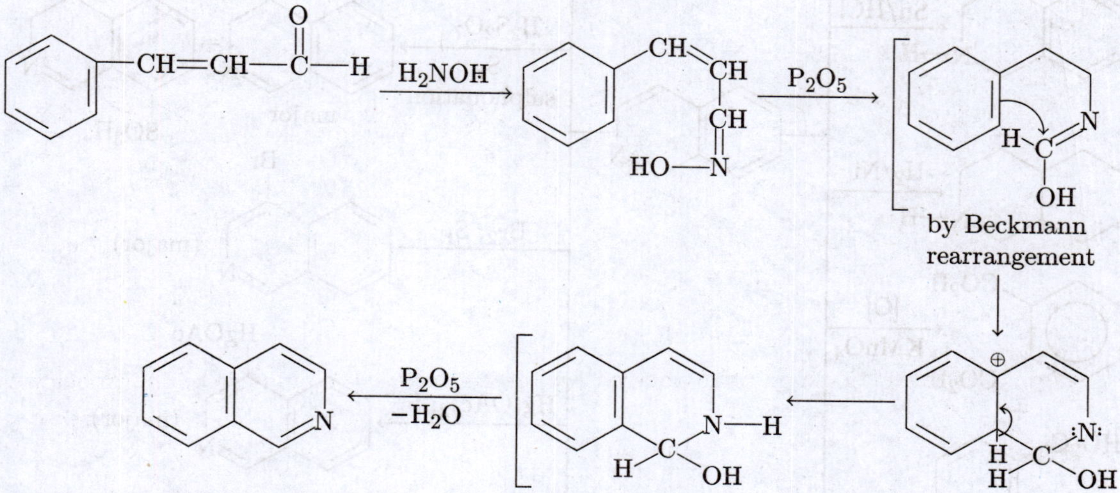


## 19.8.3 Structure, reactivity and orientation

Isoquinoline on potassium permanganate oxidation produces phthalic acid and cinchomeronic acid (3, 4-pyridine dicarboxylic acid). The formation of two side chains both in the benzene ring and pyridine ring prove that isoquinoline is a fused ring system involving benzene and pyridine; the products further tells us about the relative positions of the two rings.

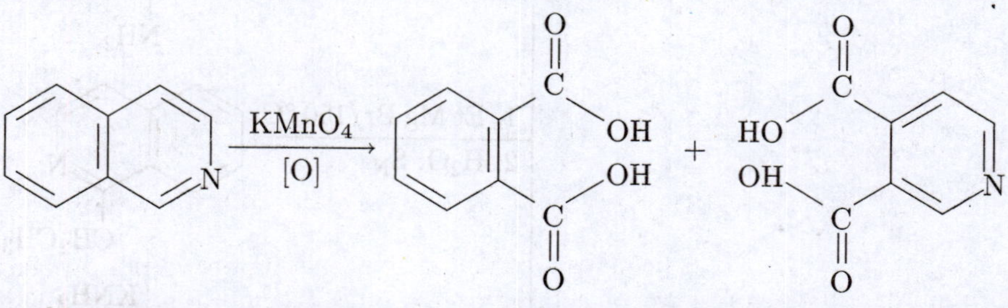


Isoquinoline is more reactive than pyridine insofar as Ar $S_E2$ reactions are concerned. These reactions mainly occur at position 5 and to a small extent at 8. In this regard isoquinoline resembles 2-nitronaphthalene. This is because the more stable pyridine ring is being preserved on electrophilic substitution at 5 and 8 positions. Moreover, a few reactions also occur at 4-position. Nucleophilic substitution reactions occur at position 1. 1-alkylated isoquinoline has acidic Hs on the α-C of the alkyl group.

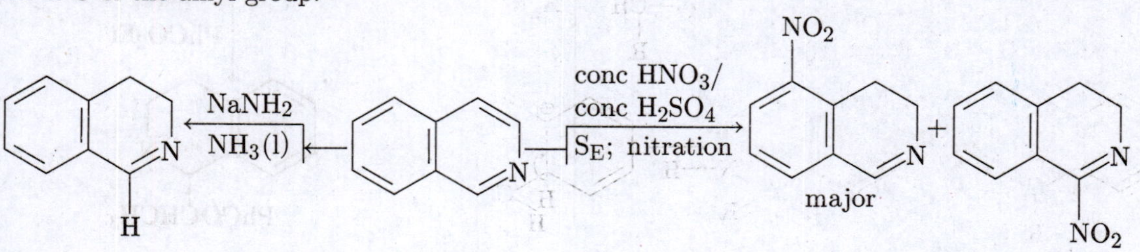

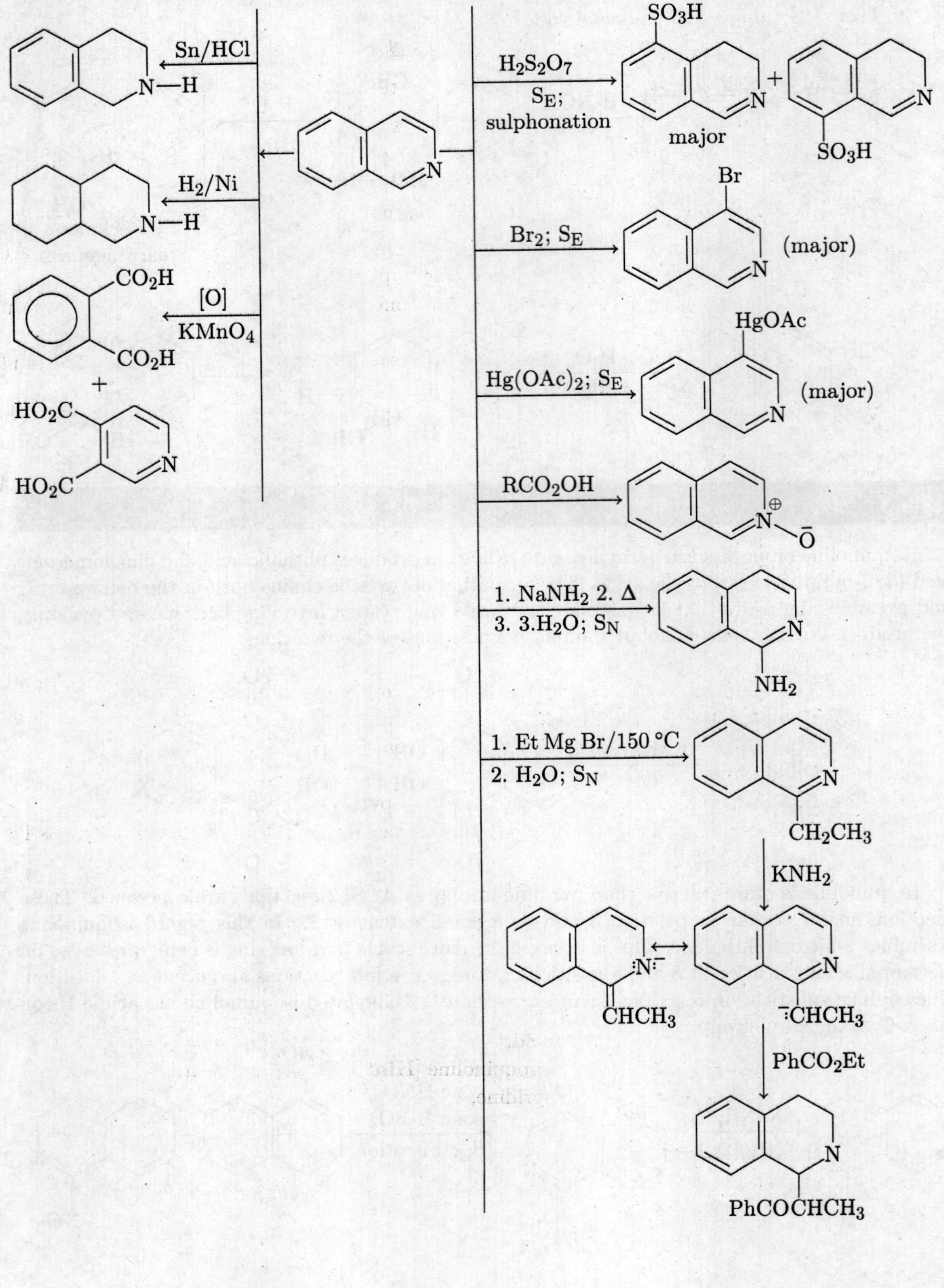

Sn/HCl
N—H
H2/Ni
N—H
CO2H
[O]
KMnO4
CO2H
+
HO2C
HO2C
N
H2S2O7
SE;
sulphonation
SO3H
N
+
N
major
SO3H
Br
Br2; SE
N
(major)
HgOAc
Hg(OAc)2; SE
N
(major)
RCO2OH
N
O
1. NaNH2 2. Δ
3. 3.H2O; SN
N
NH2
1. Et Mg Br/150 °C
2. H2O; SN
N
CH2CH3
KNH2
N
CHCH3
N
CHCH3
PhCO2Et
N
PhCOCHCH3

## Exercise

1. Explain the following considering the stability of the corresponding T.S.
   - (i) Furan, thiophen and pyrrole undergo Ar.$S_E$ reactions mainly at 2-position.
   - (ii) Pyridine undergoes Ar.$S_E$ reactions mainly at 3-position.
   - (iii) Pyridine undergoes the nucleophilic substitution at 2-, 4- and 6-positions.
2. From the consideration of charge distribution discuss the following statements :
   - (i) So far as electrophilic substitutions are concerned the reactivity order among the following is : pyrrole > furan > benzene > pyridine.
   - (ii) So far as nucleophilic substitutions are concerned the order of reactivity between pyridine and benzene is pyridine > benzene.
   - (iii) Pyridine-1-oxide and quinoline-l-oxide direct the incoming nucleophile at 2- and 4-positions.
3. Write notes on the following synthetic processes with plausible mechanism :
   - (i) The Fischer indole synthesis,
   - (ii) The Skraup quinoline synthesis,
   - (iii) The Friedlander's synthesis,
   - (iv) The Knorr quinoline synthesis,
   - (v) The Doebner-Miller synthesis,
   - (vi) Synthesis of pyridine by the Hantzsch process,
   - (vii) The Reissert indole synthesis.
4. Give mechanistic steps for :
   - (i) The Mannich reaction on pyrrole, indole thiophen and 2-methylfuran.
   - (ii) The Reimer-Tiemann reaction on pyrrole and indole,
   - (iii) The Vilsmeier-Haak formylation on thiophen,
   - (iv) The Houben-Hoesch reaction on pyrrole,
   - (v) The Hofmann-Martius rearrangement on N-pyridylalkyl chloride,
   - (vi) Conversion of pyridine to 1, 3-pentadiene through the Hofmann's exhaustive methylation,
   - (vii) The Gattermann reaction on furan.
5. Convert :
   - (i) Furan to pyrrole,
   - (ii) Furan to pyridine,
   - (iii) Quinoline to pyridine,
   - (iv) Pyridine to 4-pyridylacetic acid,
   - (v) 2-Hydroxyquinoline to 2-chloroquinoline [**Hint** : $POCl_3$ and heat.]
   - (vi) β-Picolinic acid to 3-aminopyridine,
   - (vii) Furan to furilic acid,
   - (viii) Indole to quinoline,
   - (ix) Pyrrole to pyridine,
   - (x) Furan to thiophen.

6. Explain :
   (i) 2-Formylfuran undergoes many of the reactions of benzaldehyde but 2-formylpyrrole does not.
   (ii) Quinoline resembles 1-nitronaphalene [See Problems 19.4].
   (iii) So far as electrophilic substitution reactions are concerned pyrrole is more reactive than furan.
   (iv) Furan possesses diene character.
   (v) Pyridine is a stronger base than pyrrole.
   (vi) Pyridine ring does not open when quinoline is subjected to oxidative degradation.
7. Synthesise :
   (a) Pyrrole, furan and thiophen,
   (b) 2, 4-Dimethylpyridine,
   (c) $\beta$-Cyanopyridine $\left[\beta\text{-picoline} \longrightarrow \beta\text{-picolinic acid} \longrightarrow \text{ammonium salt} \xrightarrow[\text{heat}]{P_2O_5}\right]$
   (d) 5-Aminoquinoline from quinoline,
   (e) 8-Hydroxyquinoline,
   (f) 3-Indoleacetic acid,
   (g) 2-Carbethoxyindole.
8. Into what position will a substituent enter in the following substitution reactions?
   (a) Furoic acid is treated with bromine,
   (b) Pyridine is heated with sodamide and the product is treated with water,
   (c) Pyrrole is treated with benzenediazonium ion,
   (d) Quinoline-1-oxide is treated with the mixed acid,
   (e) Pyrrole is treated with NaOI.
9. Write down the outlines of the ring-fission reactions given in this chapter.
10. Explain why indole and quinoline are said to be aromatic compounds. Give examples to show their aromatic character.
11. How indole itself can be synthesised from a phenylhydrazine derivative by Fischer indole synthesis.

**Ans. :**

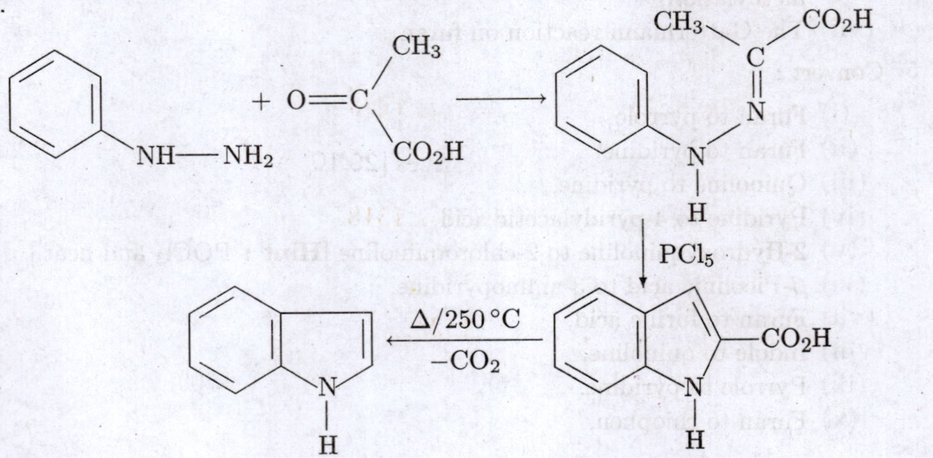

# Chapter 20

# Chemistry of Functional Groups

## Objectives

- To know the Chemistry of :

## 20.1 Organic Halo Compounds

### 20.1.1 Introduction

In this section we shall discuss the introduction and reactions of halo groups in acyclic, alicyclic and aromatic compounds.

### 20.1.2 Formation of organic halo compounds

Chart 20.1 represents the methods :

(a) Large excess of alkane $\xrightarrow[\text{direct chlorination}]{Cl_2}$ mixture of monochloroalkanes

(b) $ROH \xrightarrow{HX} RX + H_2O$

(c) $ROH \xrightarrow{PX_3} RX + H_3PO_3$

(d) $ROH \xrightarrow{SOCl_2} RCl + HCl + SO_2$

(e) $RH \xrightarrow{SO_2Cl_2} RCl + HCl + SO_2$

(f) 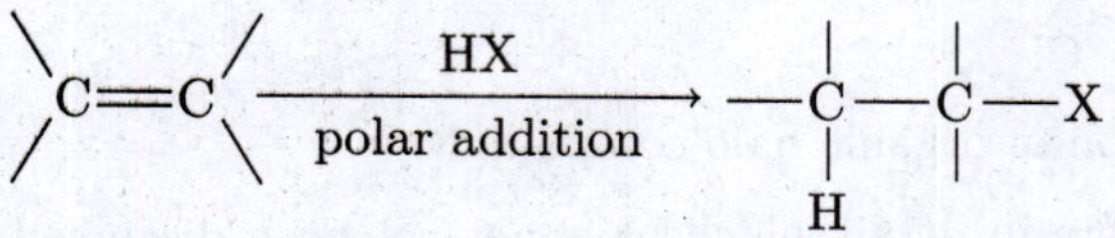

(g) 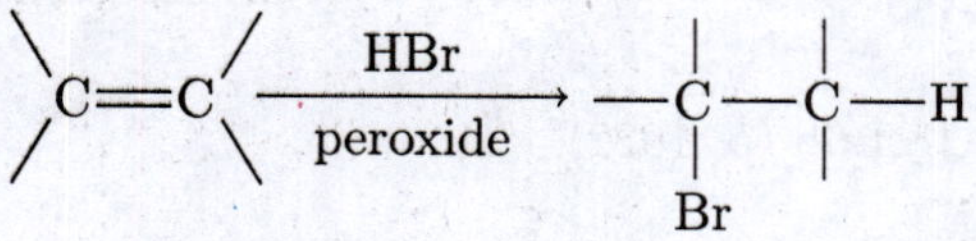

(h) $-CH{=}CHCH_2- \xrightarrow[600\,°C]{Cl_2} -CH{=}CH-CHCl-$

(i) $-CH{=}CHCH_2- \xrightarrow{NBS} -CH{=}CH-CHBr-$

(j) $-CH_2CO- \xrightarrow{X_2/H^{\oplus}} -CHX-CO-$

(k) $-CH_2-COOH \xrightarrow{X_2/PX_3} -CHX-COOH$

(l) (i) $R-CO_2Ag + X_2 \longrightarrow RX + AgX + CO_2$

(ii) $(R-CO_2)_4\,Pb + X_2 \longrightarrow 2RX + RCO_2R$

(m) Arene $\xrightarrow[X_2/Fe \text{ or } I_2]{\text{halogenation}} Ar-X$

(n) $Ar\overset{\oplus}{N}_2 \xrightarrow[\text{or } Cu/HX]{CuX/HX(X{=}Cl,\ Br)} Ar-X$

(o) $Ar\overset{\oplus}{N}_2 \xrightarrow{KI} Ar-I$

(p) $Ar\overset{\oplus}{N}_2 \xrightarrow[\text{or } HPF_6]{HBF_4/\text{gentle heat}} Ar—F$

(q) $Ph—CH_3 \xrightarrow{Cl_2/\text{sunlight}} PhCH_2Cl + PhCHCl_2 + PhCCl_3$
(boiling)

(r) 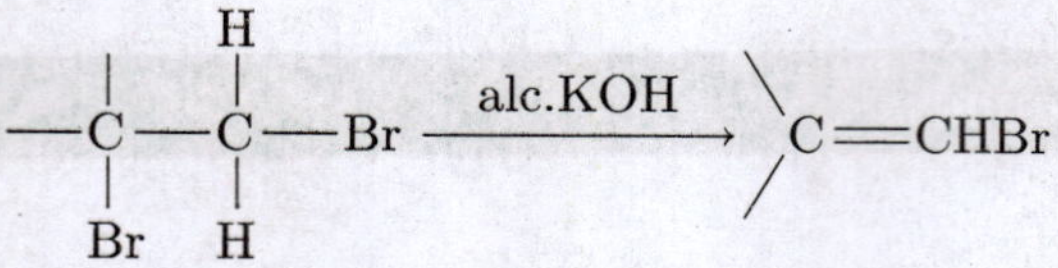

(s) $ArH \xrightarrow[HCHO \;\&\; HCl/ZnCl_2(\text{anh.})]{\text{chloromethylation}} Ar—CH_2Cl$

(t) 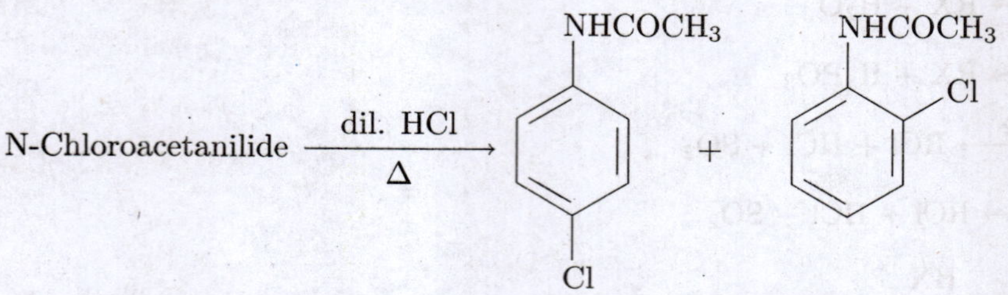

**Chart 20.1** Formation of organic halo compounds.

Only those methods will be discussed here in detail which have not yet been discussed anywhere else.

**(a) By direct chlorination of alkanes :** See sec. 18.1.6b (p. 1022).

$$CH_4 + Cl_2 \text{ (large excess)} \xrightarrow{h\nu} CH_3Cl + CH_2Cl_2 + CHCl_3 + CCl_4$$

**(b) From alcohols and hydrohaloacids :**

$$ROH + HX \longrightarrow RX + H_2O$$

This is a nucleophilic substitution reaction. If the R is 3°, $S_N1$ reaction occurs; when the R is 1°, the reaction is $S_N2$ and in the case of 2°R the reaction may take both the pathways. When the hydroxyl group is an allylic one, the reaction follows $S_N1'$ and $S_N2'$ pathways along with $S_N1$ and $S_N2$ pathways. The reactivity order among alcohols is : allylic ≈ benzylic > 3° > 2° > 1°; among haloacids the order is HI > HBr > HCl; HF does not convert alcohols to R—F. Usually HX is generated in situ from the sodio salt and $H_2SO_4$ or $H_3PO_4$.

$$CH_2{=}CH_2—CH_2OH \xrightarrow{HCl} CH_2{=}CH—CH_2Cl + ClCH_2—CH{=}CH_2 + H_2O$$

$$PhCH_2OH \xrightarrow{HBr} PhCH_2Br + H_2O.$$

**(c) From alcohols and phosphorus halides :** Alcohols may be converted to alkyl halides by $PCl_5$ or $PX_3$ (where, X= Cl, Br. & I). Straight chain 1° alcohols give unrearranged products while many 2° alcohols give rearranged products.

**(i)** $ROH + PCl_5 \longrightarrow RCl + HCl + POCl_3$

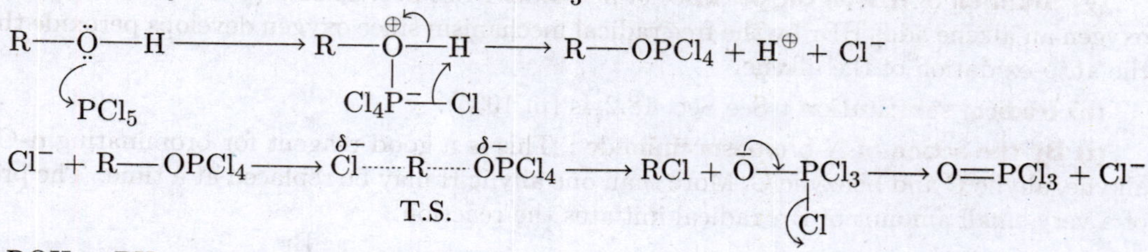

**(ii)** $ROH + PX_3 \longrightarrow RX + H_3PO_3$

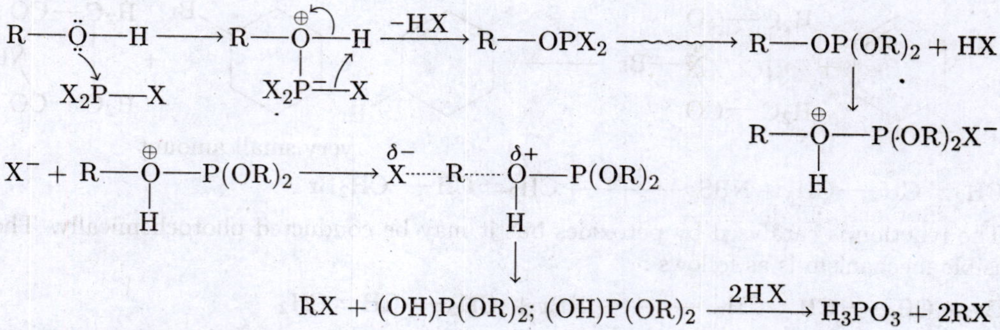

**(d) From alcohols and thionyl chloride :**

$ROH + SOCl_2 \xrightarrow{\text{reflux}} RCl + SO_2 + HCl$ ($S_Ni$; See Sec. 12.5. p. 585)

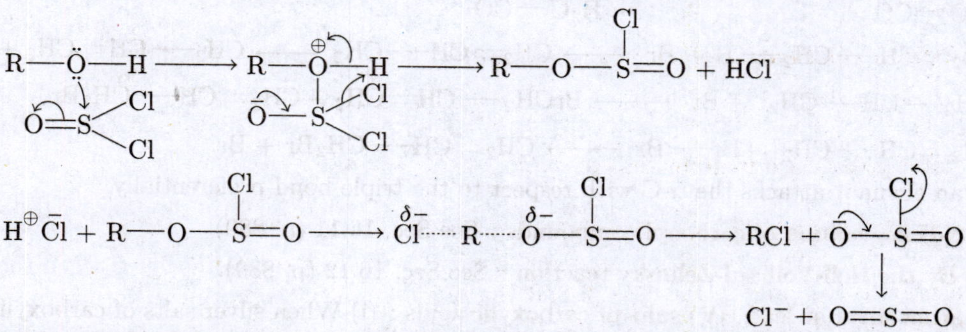

If the reaction is carried out in the presence of a 3° amine or pyridine, the reaction will follow the $S_N2$ pathway as shown above.

**(e) From alkanes and sulphuryl chloride :** See Sec. 18.1.6b (p. 1025).

$$C_2H_5H + SO_2Cl_2 \longrightarrow C_2H_5Cl + SO_2 + HCl$$

The reaction may be carried out photocatalytically or in the presence of peroxides. In both the cases the reaction follows a free radical pathway.

$(R^1CO_2)_2 \longrightarrow 2\,R^1\dot{C}O_2;\ R^1\dot{C}O_2 \longrightarrow \dot{R}^1 + CO_2$

$\dot{R}^1 + SO_2Cl_2 \longrightarrow R^1Cl + \cdot SO_2Cl;\ \cdot SO_2Cl \longrightarrow SO_2 + Cl\cdot$

$R{-}H + \cdot Cl \longrightarrow R\cdot + HCl;\ R\cdot + Cl{-}SO_2Cl \longrightarrow RCl + \cdot SO_2Cl$

**(f) Addition of hydrogen halides :** See Sec. 13.4 (p. 625).

**(g) Addition of HBr in the presence of peroxides :** See Sec. 13.12.2 (p. 659). In the presence of oxygen an alkene adds HBr by the free radical mechanism since oxygen develops peroxide through the auto-oxidation of the alkene.

**(h) Radical substitution :** See Sec. 18.2.4s (p. 1035).

**(i) By the action of N-bromosuccinimide :** This is a good reagent for brominating $\alpha$-C of an alkyne, allylic C and benzylic C. More than one allylic H may be replaced at a time. The presence of a very small amount of Br· radical initiates the reaction.

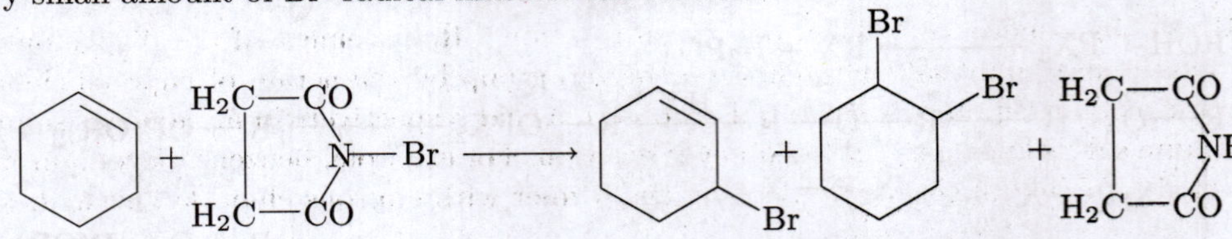

very small amount

$$CH_2{=}CH{-}CH_3 + NBS \longrightarrow CH_2{=}CH{-}CH_2Br$$

The reaction is catalysed by peroxides but it may be conducted photochemically. The most plausible mechanism is as follows :

$$Br\cdot + CH_3{-}CH{=}CH_2 \longrightarrow HBr + \cdot CH_2{-}CH{=}CH_2$$

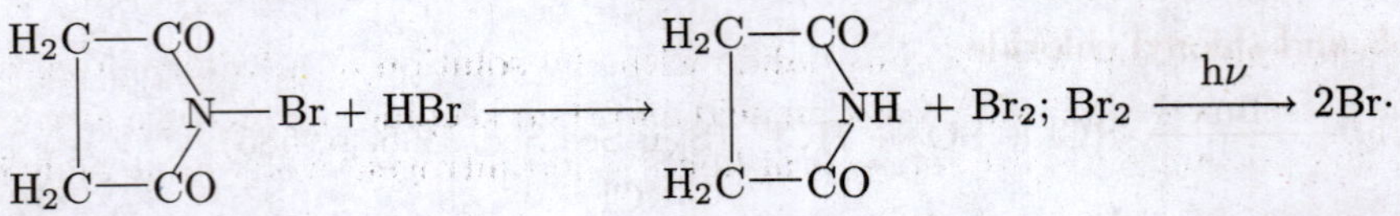

$$CH_2{=}CH{-}CH_2{-}H + Br\cdot \longrightarrow CH_2{=}CH{-}\dot{C}H_2 \longleftrightarrow \dot{C}H_2{-}CH{=}CH_2 + HBr$$

$$[CH_2{\cdots}CH{\cdots}CH_2]^{\cdot} + Br\cdot \longrightarrow BrCH_2{-}CH{=}CH_2 + CH_2{=}CH{-}CH_2Br$$

$$CH_2{=}CH{-}CH_2\cdot + Br{-}Br \longrightarrow CH_2{=}CH{-}CH_2Br + Br.$$

Of an enyne it attacks the $\alpha$-C with respect to the triple bond preferentially.

**(j) $\alpha$-Halogenation of carbonyl compounds :** See Sec. 16.11 (p. 880).

**(k) By the Hell-Volhard-Zelinsky reaction :** See Sec. 16.12 (p. 889).

**(l) From silver or lead (IV) salts of carboxylic acids :** (i) When silver salts of carboxylic acids in $CCl_4$ are decomposed by chlorine or bromine, the corresponding halides form (Hunsdiecker reaction). The 2° and 3° acids rearrange and so the yields of 2° and 3° RX are less than that of 1° RX. The yield order is : 1° > 2° > 3°. However, bromine in the presence of HgO (red) gives better yield. Iodine does not form RI with $RCO_2Ag$; it forms an ester and silver iodide.

$$ArCOOAg + Br_2 \longrightarrow Ar{-}Br + CO_2 + AgBr$$

$$RCOOAg + Br_2(Cl_2) \longrightarrow RBr\ (RCl) + CO_2 + AgBr(AgCl)$$

$$2RCOOAg + I_2 \longrightarrow RCO_2R + CO_2 + 2AgI$$

The mechanism of the reaction is uncertain. However, it may be as follows :

$$RCOOAg + Br_2 \longrightarrow RCOOBr + AgBr$$

$$RCOOBr \longrightarrow RCOO\cdot + Br\cdot\ ;\quad RCOO\cdot \longrightarrow R\cdot + CO_2$$

$$R\cdot + RCOOBr \longrightarrow RBr + RCOO\cdot\ ;\quad RCOO\cdot \longrightarrow R\cdot + CO_2$$

(ii) When carboxylic acids in benzene solution are treated with lead tetraacetate and lithium chloride (or $X_2$), alkyl halides are formed. 1°, 2° and 3° acids give 1°, 2° and 3° alkyl halides respectively; in this case no rearrangement occurs.

$$\underset{\text{in benzene}}{RCOOH + Pb(OAc)_4} \xrightarrow{LiCl} RCl; \quad RCOOH + Pb(OAc)_4 \xrightarrow{X_2} RX$$

(m) Aromatic halogenation : See Sec, 17.8 (p. 968).

(n) By the Sandmeyer and the Gattermann reaction : Replacement of the diazonium group of an aromatic diazonium salt by a chloro or bromo group by the action of cuprous chloride in hydrochloric acid or cuprous bromide in hydrobromic acid respectively in an aqueous solution of the diazonium salt is known as the Sandmeyer reaction, For example, benzene diazonium salt can be converted into chlorobenzene by treating the former with cuprous chloride and hydrochloric acid.

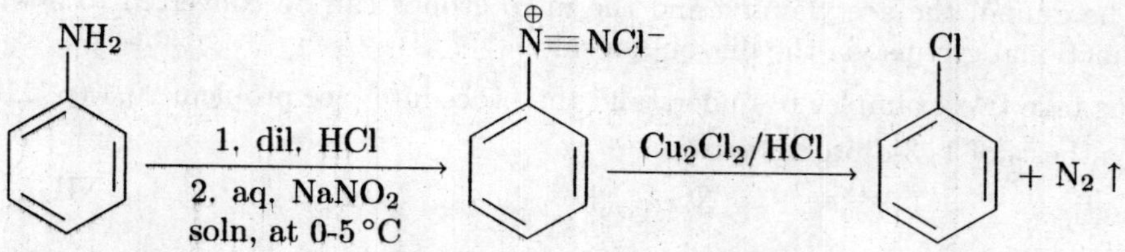


The reaction is carried out by mixing cuprous halide with the solution of a freshly prepared aromatic diazonium salt in the corresponding halogen acid and then keeping the mixture at room temperature or at an elevated temperature for several hours when nitrogen evolves out and the aryl halide forms.

A special case of the Sandmeyer reaction is the conversion of an aromatic diazonium salt to the corresponding aryl cyanide by treating the salt solution with cuprous cyanide dissolved in aqueous potassium cyanide.

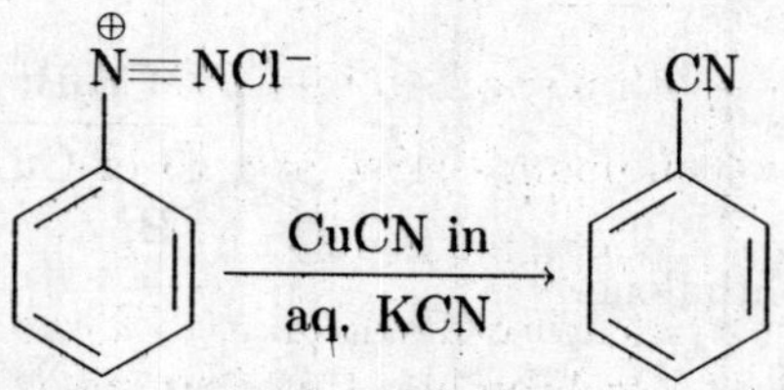


The Gatterman reaction is the modification of the Sandmeyer reaction; in this reaction aromatic diazonium salts are treated with copper powder and halo acids instead of cuprous halide for the replacement of the diazonium group by the halo group. In order to replace the diazonium group by the cyano group an aqueous solution of potassium cyanide and copper powder is used.

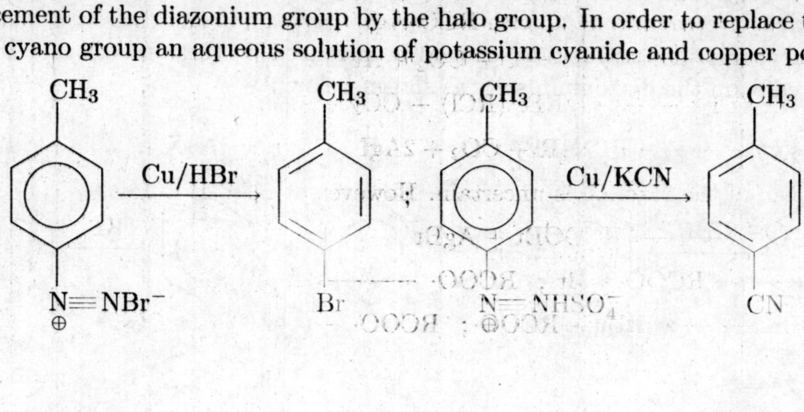

The Sandmeyer and the Gattermann reactions are good means for introducing a wide variety of functional groups into the aromatic ring. The system involves a few sequential steps as given below :

(a) Nitration of the aromatic compound.

(b) Reduction of the nitro compound to the corresponding amine.

(c) Diazotisation of the aromatic amine.

(d) The Sandmeyer or the Gattermann reaction on the diazonium ion.

In order to carry out synthesis involving the Sandmeyer reaction the following points are to be remembered :

(i) The amino and the acetylamino ($CH_3CONH$—) groups are *o*-, *p*-orienting.

(ii) The nitro group is a good *m*-orienting group.

(iii) The amino, the acetylamino and the nitro groups can be converted to a wide variety of functional groups via the diazonium salt.

Let us take two examples to understand the procedure [See problem 20.4 (p. 1167)].

1. Synthesis of 1, 3-dibromobenzene.

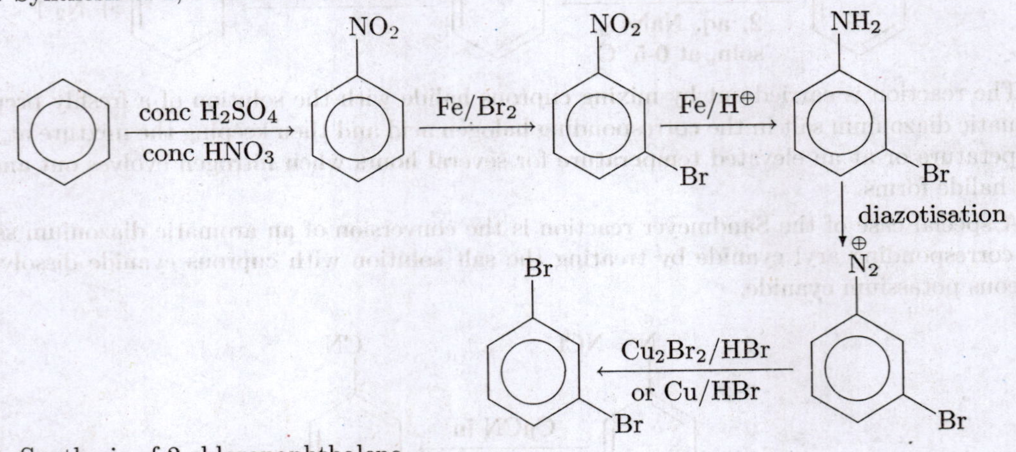


2. Synthesis of 2-chloronaphthalene.

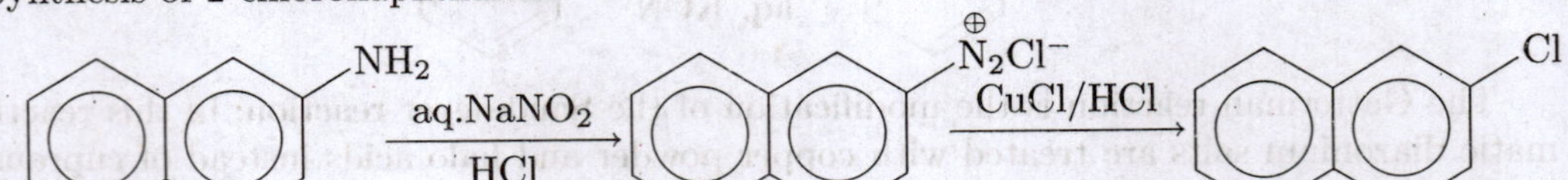


However, aryl fluorides and iodides cannot be prepared by these reactions. Of course, these can be prepared form the diazonium salts as shown below :

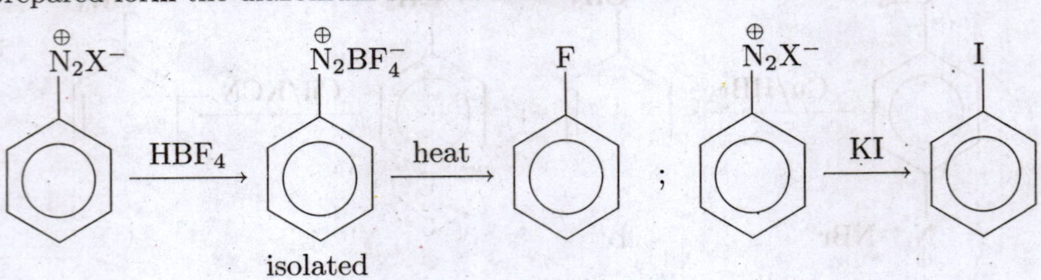

The reaction involves the following mechanistic steps :

**Step I :** Complexation between the diazonium ion and the cuprous halide and the transfer of one electron from $CuCl_2^-$, to the diazonium ion through a chloride bridge T.S.

$$Ar\overset{\oplus}{N}_2Cl^- + CuCl \longrightarrow Ar\overset{\oplus}{N}_2[CuCl_2]^- \longrightarrow Ar{-}\overset{\oplus}{N}{\equiv}\ddot{N}\cdots\overset{\delta-}{Cl}{-}Cu{-}\overset{\delta-}{Cl}$$

'chloride bridge' T.S.

$$Ar{-}\overset{\oplus}{N}{\equiv}\ddot{N}\cdots\overset{\delta-}{Cl}{-}Cu{-}\overset{\delta-}{Cl} \longrightarrow Ar\cdot + N_2 + CuCl_2$$

This is a slow step and hence this is the rate-determining step.

**Step II :** Radical abstraction of chlorine atom of the cupric chloride by the aryl radical. This is a fast step.

$$Ar\cdot + Cl{-}Cu{-}Cl \longrightarrow ArCl + CuCl$$

$$\text{The rate} = k[Ar\overset{\oplus}{N}_2]\,[CuCl_3^{2-}]$$

Thus the reaction is a free-radial decomposition reaction involving one-electron transfer in which the oxidation state of copper changes from +I to +II and then back to +I.

**(o) Aryl fluoride from diazonium salt :** See Sec. 20.1.2n (p. 1152) (limitation of the Sandmeyer reaction).

**(p) Aryl iodide from diazonium salt :** See Sec. 20.1.2n (p. 1152) (limitation of the Sandmeyer reaction).

**(q) Side-chain halogenation :** When chlorine gas is passed through boiling toluene, benzyl chloride, benzal chloride and benzotrichloride are obtained.

$$PhCH_3 \xrightarrow{Cl_2} PhCH_2Cl;\quad PhCH_2Cl \xrightarrow{Cl_2} PhCH_2Cl;\quad PhCHCl_2 \xrightarrow{Cl_2} PhCCl_3$$

**(r) Formation of substituted vinyl halides from 1, 2-dihalides :** When 1,2-dihalides are treated with ethanolic KOH, substituted vinyl halides, chlorides and bromides, are formed.

$$CH_3CHBrCH_2Br \xrightarrow{\text{ethanolic KOH; }-HBr} CH_3CH{=}CHBr$$

**(s) Chloromethylation of arenes :** See Sec. 17.10 (p. 986).

**(t) Formation of *o*-, *p*-chloroacetanilides from N-chloroacetanilide :** See See. 14.13 (p. 744).

## 20.1.3 Structure, reactivity and orientation

In alkyl halides, cycloalkyl halides and allyl halides, the C—Cl bond is appreciably polar because of the greater electronegativity of the halogen atom than that of the carbon. Naturally the dipole moment of those compounds has a reasonable value. In the C—Cl bond dipole C is the positive pole and Cl is the negative pole. This makes the C atom susceptible to nucleophilic attack and $S_N$ reactions occur there; a 3° RX undergoes the $S_N1$ reactions and a 1° RX the $S_N2$ reactions, while 2°RX undergoes $S_N$ reactions by both the pathways. Since $S_N$ reactions compete with elimination and rearrangement reactions, they also undergo elimination and rearrangement reactions. Allyl halides also undergo $S_N$ reactions by $S_N1$, $S_N1'$, $S_N2$ and $S_N2'$ pathways. On the other hand, in aryl and vinyl halides the halogen atom is bonded to an $sp^2$ hybridised C and the *p* electrons of X are involved in *p*-$\pi$ conjugation which gives the C—X bond a double bond character. This also reduces the dipole moment of the C—X bond. So in these compounds the C atom of the C—X bond is not appreciably positively charged and naturally $S_N$ reactions do not occur readily; this requires drastic conditions. These compounds do not form precipitate with silver nitrate because the C—X bond of these compounds does not ionise. However, vinyl halides may be dehydrohalogenated to acetylenes but aryl halides on dehydrohalogenation form unstable reaction intermediates, the benzynes; benzynes immediately undergo addition reactions to form aromatic compounds. Since the halogen atom of an aryl halide exerts −I and +R effects, it deactivates the ring in an electrophilic substitution reaction and its +R effect directs the attacking electrophile at the *o*- and *p*-positions.

Owing to the +R effect of the halogen atom in vinyl halides, the C atom bearing the halogen atom takes up another halo group when HX is added to it; while the −I effect of the halo group makes allyl halide to add unsymmetrical reagents anti-Markownikoffically. On the basis of these findings the following reactions will be discussed here as given in Chart 20.2.

| | Reagent/conditions | Product | |
|---|---|---|---|
| R—X | *n*-BuLi | RLi + BuX | (i) |
| | Na/Pb | $R_4Pb$ | (ii) |
| | metal alloy | organometallics | (iii) |
| | $H_2$/Pd-C or dissolving metal | RH | (iv) |
| | $LiAlH_4$ or $R_3SnH$ | RH | (v) |
| | Mg/ether | RMgX | (vi) |
| | Na/ether; Wurtz reaction | R—R | (vii) |
| | alcoholic KOH elimination | $>C=C<$ | (viii) |
| | $AlCl_3$(cat) 300 °C | isomerization | (ix) |
| | $Me_2S=O$ | aldehyde | (x) |
| | $NH_3$ or an amine 1 mole | $R\overset{\oplus}{N}H_3X^-$ | (xi) |
| | $NH_3$ or an amine RX in large excess | mixture of amines quaternary salt | (xii) |
| | $\bar{O}H/S_N$ reaction | ROH | (xiii) |
| | $\bar{S}H/S_N$ reaction | RSH | (xiv) |
| | $NaCN/S_N$ reaction | RCN + RNC (trace) | (xv) |
| | $AgCN/S_N$ reaction | RNC + RCN (trace) | (xvi) |
| | $\bar{O}R^1/S_N$ reaction | $ROR^1$ | (xvii) |
| | $\bar{S}R/S_N$ reaction | RSR | (xviii) |

*Contd.*

$$R—X \xrightarrow{Na_2SO_3/S_N \text{ reaction}} RSO_3Na \quad \text{(xix)}$$

$$R—X \xrightarrow{AgONO/S_N \text{ reaction}} R—NO_2 + RONO \quad \text{(xx)}$$

$$R—X \xrightarrow{R^1CO_2Ag/\text{ethanol}} R^1CO_2R \quad \text{(xxi)}$$

$$—CH{=}CH—X \xrightarrow{N\bar{u}/S_N \text{ reaction}} —CH{=}CH—Nu \quad \text{(xxii)}$$

$$—CH{=}CH—CH_2—X \xrightarrow{N\bar{u}/S_N \text{ reaction}} —CHNu—CH{=}CH_2 + —CH{=}CH—CH_2—Nu \quad \text{(xxiii)}$$

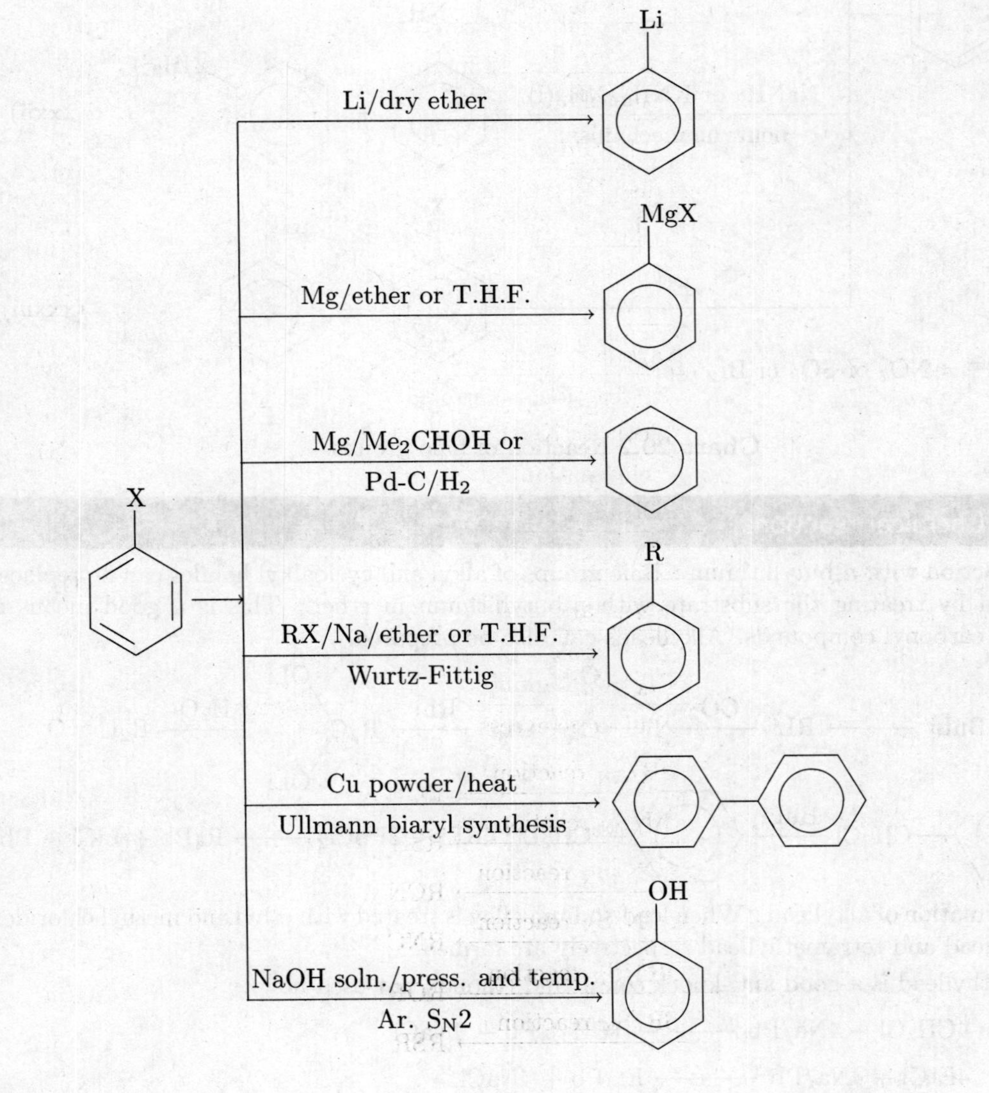


(xxiv)

(xxv)

(xxvi)

(xxvii)

(xxviii)

(xxix)

*Contd.*

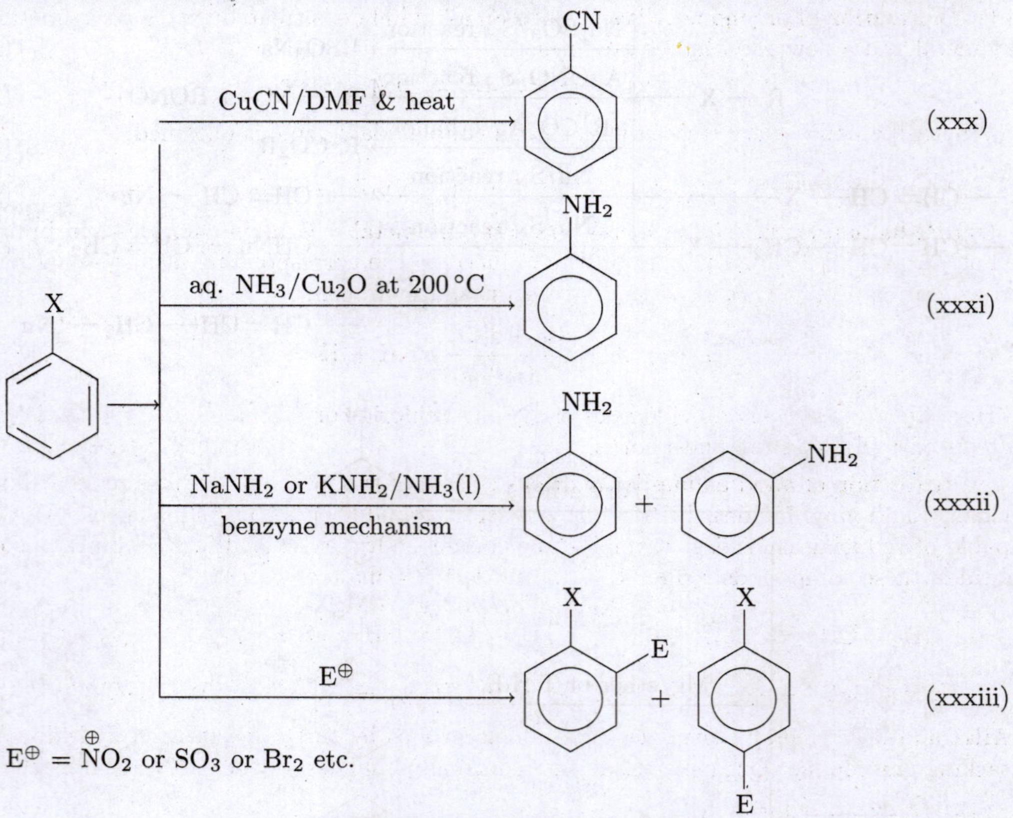


**Chart 20.2** Reaction of halo groups.

## 20.1.4 Chemical reactions

**(i) Reaction with *n*-butyllithium :** Halo groups of alkyl and cycloalkyl halides can be replaced by lithium by treating the substrate with *n*-butyllithium in ether. This is a good means of preparing carbonyl compounds. Alkylleads can also be prepared.

$$RX + BuLi \longrightarrow RLi \xrightarrow{CO_2} R{-}\overset{O}{\overset{\|}{C}}{-}OLi \xrightarrow{RLi} R_2C(OLi)_2 \xrightarrow{H_2O} R_2C{=}O$$

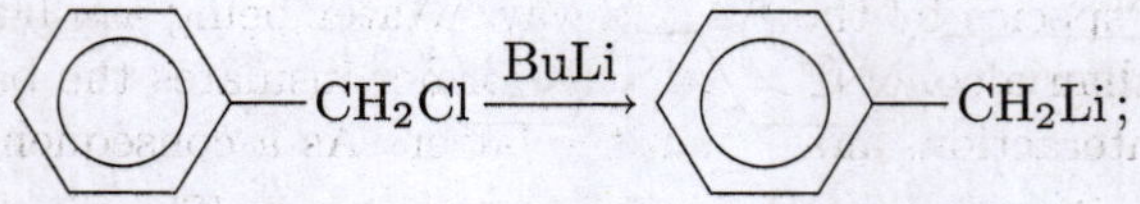


$; 4RLi + 2PbCl_2 \longrightarrow R_4Pb + LiCl + Pb$

**(ii) Formation of alkyllead :** When lead-sodium alloy is treated with ethyl and methyl chlorides, tetraethyllead and tetramethyllead respectively are formed.

Tetraethyllead is a good anti-knock compound (though not used now).

$$4CH_3Cl + 4Na/Pb \longrightarrow (CH_3)_4Pb + 4NaCl$$

$$4EtCl + 4Na/Pb \longrightarrow Et_4Pb + 4NaCl$$

**(iii) Formation of organometallics :** When an alkyl halide is treated with an alkylmetal, a new alkylmetal and a new alkyl halide are formed.

$$RX + R^1M \longrightarrow RM + R^1X$$

When alkyl iodides are treated with Zn-Cu couple, dialkylzinc is obtained.

$$2C_2H_5I + 2Zn \xrightarrow{Cu} Et_2Zn + ZnI_2$$

**(iv) Reduction by dissolving metal and catalytic metal :** 3° alkyl chlorides and bromides as well as 1°, 2° and 3° alkyl iodides can be reduced to the corresponding alkanes by either metal and acid or catalytically using Pd-C as catalyst and hydrogen.

$$R{-}X \xrightarrow[M/H^{\oplus}]{H_2/Pd\text{-}C} RH + H{-}X$$

Here, RX = 1°, 2°, 3°-alkyl iodides or, 3° alkyl chloride or bromide; an alkyl halide may also be reduced by HI and red phosphorus.

**(v) Reduction of alkyl halides by hydrides :** Most of the metallic hydrides reduce alkyl, allyl, cycloalkyl and vinyl halides. $LiBH_4$-LiH is a selective reducing agent for halogens. All these are capable of reducing chlorides, bromides and iodides. Organotin hydrides ($R_3SnH$) also reduce fluorides; these compounds reduce 1, 1-di- and 1, 1, 1-trihalides partially.

$$CH_2{=}CH{-}CH_2Cl \xrightarrow{LiBH_4\text{-}LiH} CH{=}CH{-}CH_3$$

$$CH_3CH_2F \xrightarrow{R_3SnH} CH_3CH_3;\quad CH_3CHCl_2 \xrightarrow{R_3SnH} CH_3CH_2Cl$$

$LiAlH_4$ and other metallic hydrides reduce halo groups by the $S_N2$ pathway; hydride ion is the attacking nucleophile. Organotin hydrides reduce alkyl halides by the free radical mechanism.

$$LiAlH_4 \longrightarrow Li^{\oplus} + \bar{A}lH_4$$

$$H_3\bar{A}l{-}H + RX \longrightarrow H_3\overset{\delta-}{Al}\cdots\overset{\delta-}{H}\cdots R\cdots\overset{\delta-}{X} \longrightarrow \bar{A}lH_3X + HR$$

The reaction involves inversion of configuration and the rate order is 1° > 2° > 3° RX; these facts support the $S_N2$ pathway.

**(vi) Formation of the Grignard reagent :** See See. 16.4.5 (p. 839).

**(vii) Formation of alkanes by the Wurtz reaction :** See Sec. 18.1.3g (p. 1020).

**(viii) Formation of alkenes and alkynes by alc. KOH :** See p. 786, Sec. 15.2.2. Here it is interesting to note that aqueous KOH is used in the $S_N$ reactions whereas alcoholic KOH is used in the elimination reactions. There are several opinions in this respect. According to some authors in alcoholic KOH, $EtO^-$ forms; this is a stronger base than aqueous $\bar{O}H$ and actually as a base $EtO^-$ abstracts a proton and $X^-$ leaves the species by the E2 pathway. Water being a solvent of much higher dielectric constant ($\varepsilon = 80$) than alcohol ($\varepsilon = 24$), the former insulates the base and the acidic H, which involve hard-hard interaction, more than the latter. As a consequence, in water $\bar{O}H$ cannot abstract the proton readily while in alcohol $EtO^-$ does so. Thus in the alcoholic KOH, $EtO^-$ acts as a base as well as a neucleophile.

$$KOH + EtOH \rightleftharpoons Et\bar{O}K^{\oplus} + H_2O;$$

$$CH_3CHBrCH_3 + EtO^-K^{\oplus} \longrightarrow CH_3CH{=}CH_2 + EtOH + KBr$$

In the aqueous KOH, $\bar{O}H$ acts as a nucleophile only.

**(ix) Isomerisation :** 1° and 2° alkyl halides may isomerise to 3° alkyl halides by anhydrous $AlCl_3$ at 300 °C.

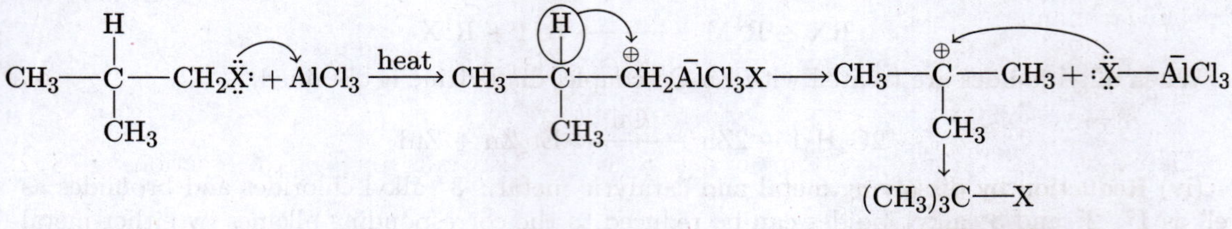

isomerised product

**(x) Oxidation of 1° and 2° halides to aldehydes by DMSO :** 1° and 2° alkyl and cycloalkyl chlorides, bromides and iodides can be oxidised to the corresponding aldehydes and ketones by DMSO.

$$CH_3CH_2Cl + Me_2SO \longrightarrow CH_3CHO + Me_2S + HCl$$

$$CH_3CHBrCH_3 + Me_2SO \longrightarrow CH_3COCH_3 + Me_2S + HBr$$

The reaction follows the $S_N2$ pathway.

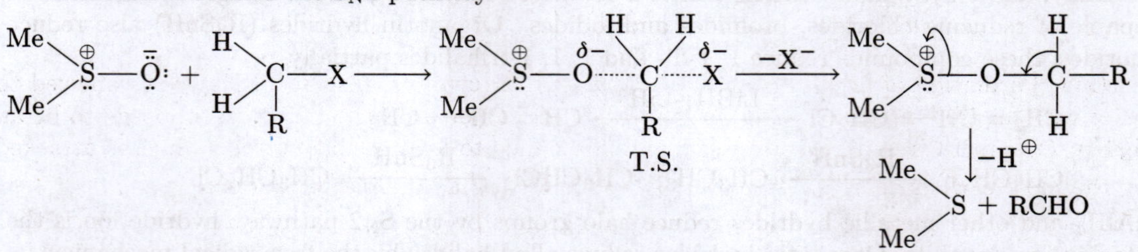

**(xi) Formation of alkylammonium salts :** When alkyl halides are treated with 1 mole ammonia or an amine, an alkylammonium salt is formed.

$$RX + NH_3 \longrightarrow R\overset{\oplus}{N}H_3X^-; \; RX + R^1NH_2 \longrightarrow R^1\overset{\oplus}{N}H_2RX^-$$

This is also an $S_N2$ reaction (See Sec. 15.4, p. 691).

**(xii) Ammonolysis of alkyl halides :** When a large excess of alkyl halide is treated with ammonia, a mixture of 1°, 2°, and 3° amines and quaternary ammonium halide is obtained. This is an industrial method of preparing amines; amines are then separated by fractional distillation. This is also an $S_N2$ reaction (See Sec. 15.4) p. 810.

$$R{-}X + NH_3 \longrightarrow RNH_2.HX; \; RNH_2 + RX \longrightarrow R_2NH.HX$$

$$R_2NH + RX \longrightarrow R_3N.HX; \; R_3N + RX \longrightarrow R_4\overset{\oplus}{N}X^-$$

**(xiii) Formation of alcohols :** When alkyl halides are treated with aqueous alkali or moist silver oxide, alcohols are formed. 3° alkyl halides undergo the reaction even by the solvent (water) and the process is said to be solvolysis. However, the reaction has no preparative value, since RX is prepared form ROH; the former is costlier than the latter.

$$Me_3CCl + H_2O \longrightarrow Me_3COH \text{ (solvolysis)}; \quad CH_3CH_2Cl + HO^- \longrightarrow CH_3CH_2OH$$

When R is 3°, the reaction follows the $S_N1$ pathway; when it is 1°, the reaction is $S_N2$ and 2° RX follows both the pathways. Allyl halides usually follow $S_N1$ and $S_N1'$ pathways. The detail of the mechanisms have been discussed in Sec. 12.3.3. p. 659. Problem 12.23 p. 600.

**(xiv) Formation of thiols :** 1° and 2° alkyl halides react with sodium and potassium bisulphides to form thiols; 3° alkyl halides fail to react. Sodium monoalkylsulphates undergo the reaction in the similar manner.

$$R—X + NaSH \longrightarrow RSH + NaX \text{ (R is 1° or 2°)}$$

$$R—OSO_2ONa + KSH \longrightarrow RSH \text{ (R is 1° or 2°)}$$

Here again the reactions are $S_N$ reactions. Since as 3° RX fails to react, the reactions are $S_N2$.

$$NaSH \longrightarrow Na^{\oplus}HS^{-}$$

$$HS^{-} + R—X \longrightarrow \underset{\text{T.S.}}{\overset{\delta-}{HS}\cdots R\cdots \overset{\delta-}{X}} \longrightarrow HS—R + X^{-}$$

**(xv) Formation of alkyl cyanides :** When alkyl halides are treated with sodium cyanide, alkyl cyanides along with very small amount of alkyl isocyanides form. 1° Alkyl halides give better yield and a 2° RX reacts but gives poor yield; a 3° RX fails to yield cyanide, instead it gives an alkene by the E2 pathway. The reaction follows the $S_N2$ pathway.

$$R—X + NaCN \longrightarrow RCN + \underset{\text{trace}}{RNC} + NaX \text{ [R = 1° or 2°]}$$

**(xvi) Formation of alkyl isocyanides :** When silver cyanide, instead of NaCN, is allowed to react with an alkyl halide, an alkyl isocyanide is the major product. It is now supposed to be an $S_N1$ reaction; the presence of silver salt aids the halide to form carbenium ion which is attacked by the more electronegative N atom of $\bar{C}N$ to form RNC as the major product.

$$RX + AgCN \longrightarrow RNC + \underset{\text{trace}}{RCN} + NaX \text{ [R = 1° or 2°]}$$

$$AgCN \rightleftharpoons \underset{\text{in soln.}}{AgCN} \rightleftharpoons Ag^{\oplus} + \bar{C}N$$

$$R—X \xrightarrow{Ag^{\oplus}} R^{\oplus} + AgX;\ R^{\oplus} + :\bar{C}N: \longrightarrow \underset{\text{major}}{R—NC} + RCN$$

**(xvii) The Williamson synthesis :** This is the best method of preparing unsymmetrical and symmetrical dialkyl and aryl alkyl ethers. This involves the treatment of an alkyl halide with sodium alkoxide or phenoxide respectively.

$$RX + Na^{\oplus}\bar{O}R^1 \longrightarrow R—O—R^1;\ RX + Na^{\oplus}\bar{O}R \longrightarrow R—O—R$$

$$RX + Na^{\oplus}\bar{O}Ar \longrightarrow R—O—Ar$$

Since in aryl halides C—X bond has a double-bond character, they do not react with alkoxides or phenoxides. It is better to choose 1° alkyl halide for the reaction since 2° and 3° alkyl halides undergo E2 reactions in the presence of strong alkoxide and phenoxide bases. The alkoxide ion may be 2° and 3° but 3° alkoxides react very slowly. It is here important to note that $Ar_3C$—X reacts even with alcohol, a weak nucleophile, to give ether; this is because $Ar_3C$—X gives carbenium ion readily and, therefore, in this case the reaction is $S_N1$. In the case of a 1° RX the reaction is $S_N2$.

$$R^1\bar{O} + R—X \longrightarrow \underset{\text{T.S.}}{R^1\overset{\delta-}{O}\cdots R\cdots \overset{\delta-}{X}} \longrightarrow R^1—O—R$$

Another reason for non-participation of a 3° RX is that the T.S. will have to bear high steric strain because of the bulk of the 3° R group; this means that the T.S. will be of high energy and, therefore, it will be difficult to form the T.S.

**(xviii) Formation of thioethers :** When potassium sulphide is heated with an alkyl halide, thioether is formed.

$$K_2S + 2RX \xrightarrow{\text{heat}} R—S—R$$

$$K_2S \longrightarrow 2K^{\oplus} + S^{2-}$$

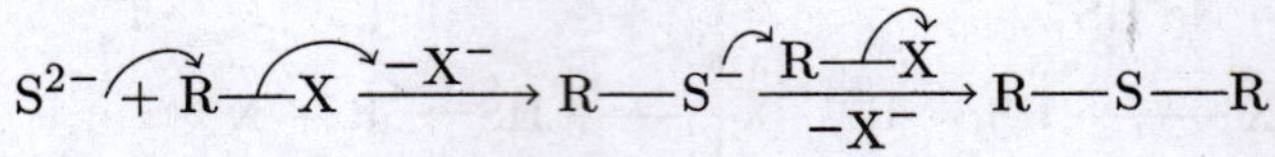

Simple thioethers are obtained in good yields.

**(xix) Formation of alkyl sulphonates :** Sodium sulphite reacts with alkyl halides to form sodium salt of alkanesulphonic acids.

It is expected that 1° and 2° alkyl halides will undergo the reaction.

$$R—X + Na_2SO_3 \longrightarrow RSO_3Na + NaX$$

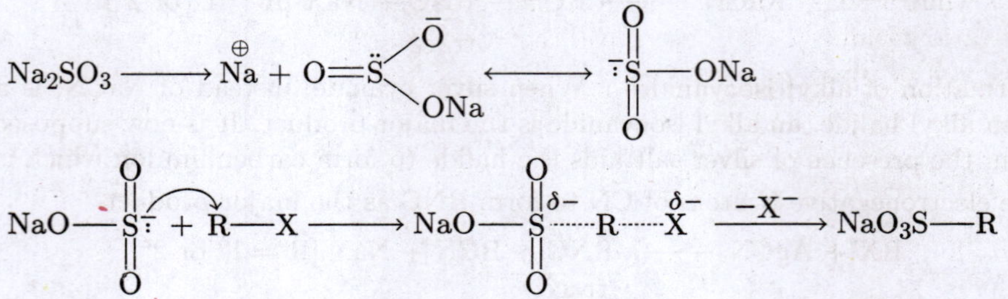

T.S.

**(xx) Formation of alkyl nitrites and nitroalkanes :** When alkyl iodides are treated with $NaNO_2$ and $AgNO_2$ solutions separately, a mixture or alkylnitrites and nitroalkanes are obtained in both cases; the former product predominates in the case of $AgNO_2$ while the latter product is the major product when $NaNO_2$ is the reagent.

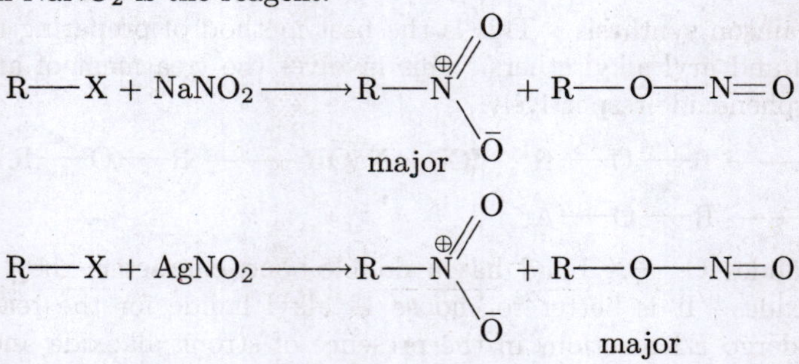

There are evidences that nitrite ion reacts with RX by the $S_N2$ pathway when the reagent is $NaNO_2$, while the same ion attacks the alkyl halide by the $S_N1$ pathway when the reagent is $AgNO_2$. In the $S_N2$ reaction $NO_2^-$ ion attacks the C atom of RX through the N atom but in the $S_N1$ reaction the $NO_2^-$ ion attacks the C atom through the O atom. The action of cyanides and nitrites on alkyl halides have already been discussed under ambident nucleophiles (Chapter 11.5.1d; p. 535) as hard-hard and soft-soft interactions.

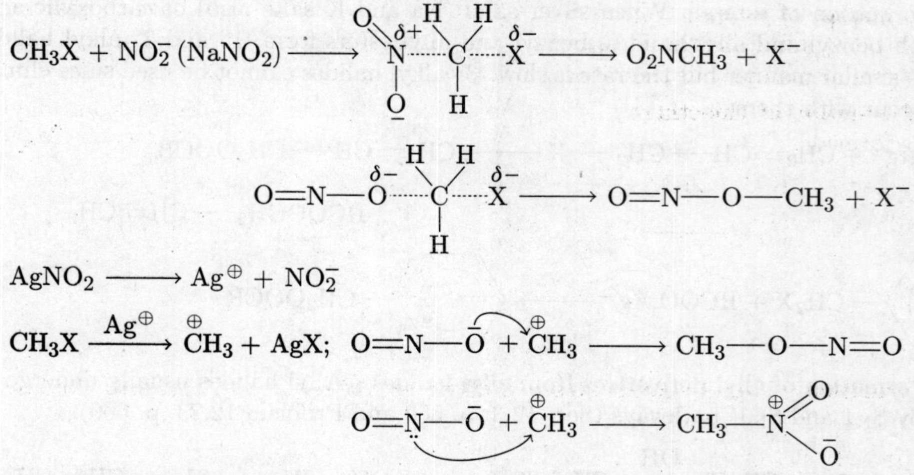

When *p*-methoxybenzyl bromide is treated with $AgNO_2$ solution, the nitrite-nitro ratio is 2 : 1; whereas when a benzyl bromide with a nitro group in the *p*-position is treated with $AgNO_2$ solution, the ratio becomes 1 : 2. It is argued that the +R effect of the MeO group aids to develop a benzyl cation which undergoes the reaction by the $S_N1$ pathway and so the nitrite-nitro ratio is 2 : 1; the −R effect of the nitro group increases the positive charge on the benzylic C atom but fails to develop carbenium ion and makes it undergo the $S_N2$ reaction, and consequently the nitrite-nitro ratio becomes 1 : 2.

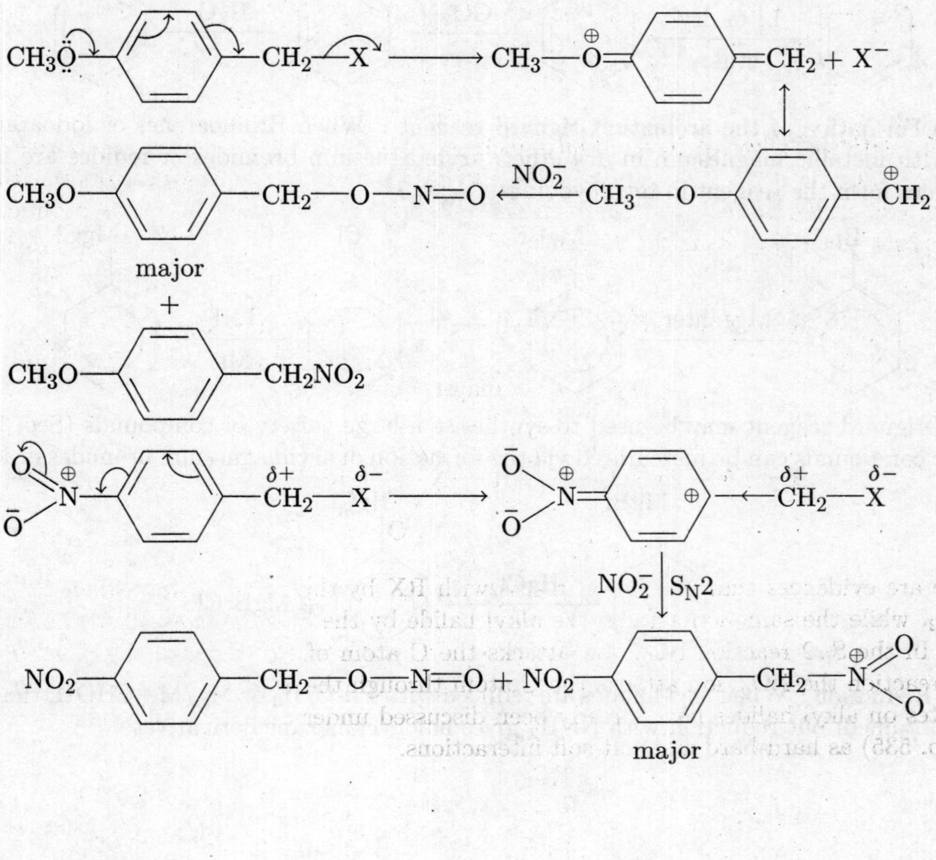

**(xxi) Formation of esters :** When silver salts (Na and K salts also) of carboxylic acids are treated with benzyl and allyl halides, benzyl and allyl esters form; 1° and 2° alkyl halides also react in the similar manner but the rate is slow. 3° alkyl halides cannot be used since elimination reactions occur with them.

$$RCO\bar{O}Ag^{\oplus} + CH_2{=}CH{-}CH_2{-}X \longrightarrow CH_2{=}CH{-}CH_2OOCR + RCOOCH_2{-}CH{=}CH_2$$

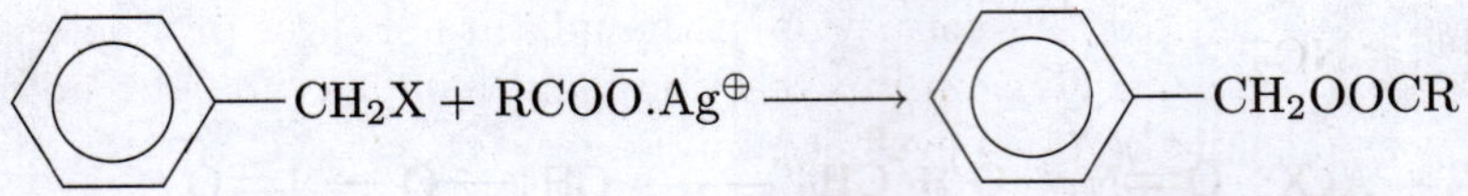

**(xxii) Formation of allyl derivatives from allyl halides :** Allyl halides usually undergo the $S_N$ reactions by $S_N1$ and $S_N1'$ pathways (Sec. 12.3, p. 569 and Problem 12.23, p. 600).

$$CH_2{=}CH{-}CH_2Cl \xrightarrow{\bar{O}H} CH_2{=}CH{-}CH_2OH + HO{-}CH_2{-}CH{=}CH_2$$

**(xxiii) Formation of phenyllithium :** When chlorobenzene is treated with lithium in dry ether, phenyllithium is formed. This is a good method for replacing chloro group by carboxyl group via phenyllithium; PhLi on treatment with carbon dioxide followed by hydrolysis gives benzoic acid.

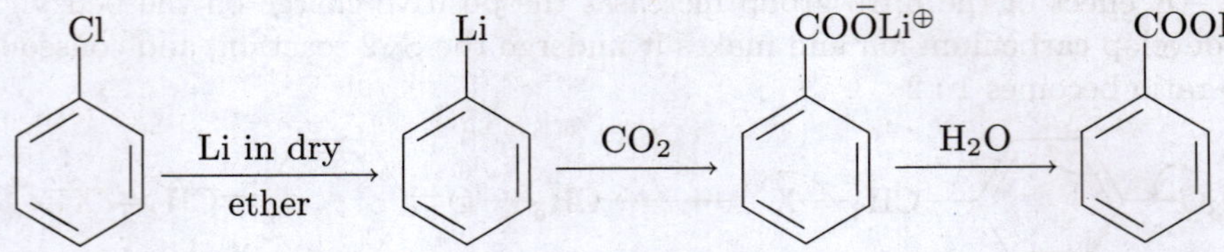

**(xxiv) Formation of the aromatic Grignard reagent :** When Bromoarenes or iodoarenes are treated with metallic magnesium in dry ether, arylmagnesium bromides or iodides are formed; chloroarenes form the reagent in tetrahydofuran.

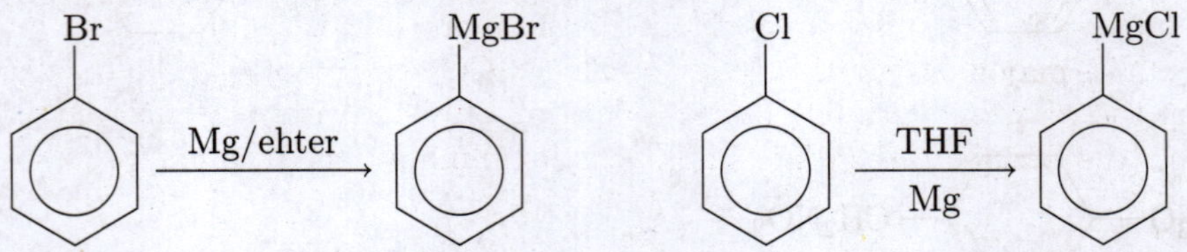

The Grignard reagent may be used to synthesise a large variety of compounds (Sec. 16.4.5). Aromatic compounds can be mercurated via the formation of arylmagnesium bromides or iodides.

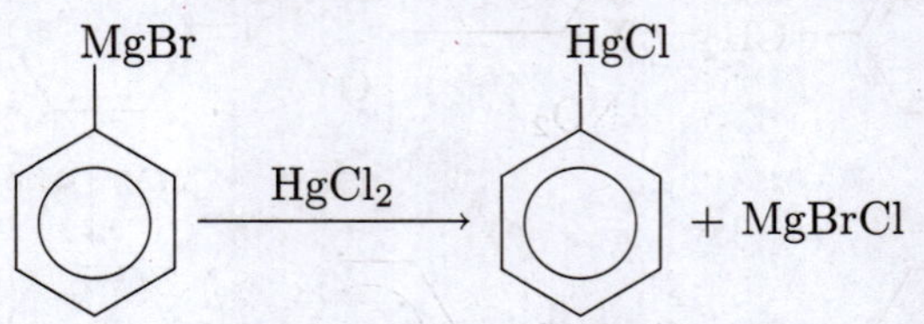

**(xxv) Reduction :** When aryl halides are reduced with Pd-C/$H_2$ or Mg/$Me_2CHOH$, the parent arene is obtained; but reduction with Ni/$H_2$ gives halocyclohexane derivatives.

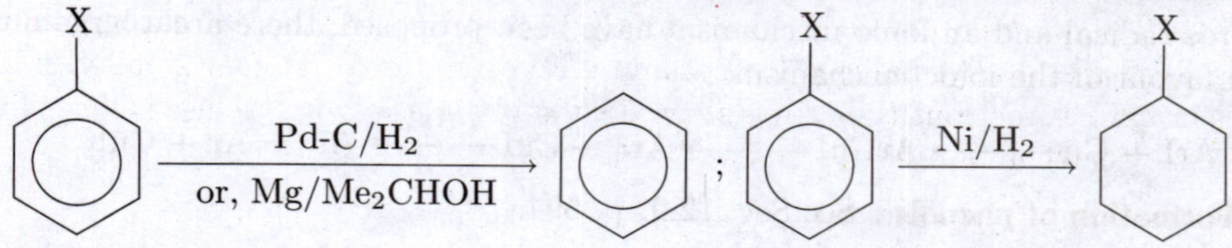

**(xxvi) The Wurtz-Fittig reaction :** When a mixture of aryl and alkyl halides and sodium is warmed in an ethereal solution, alkylarenes are formed along with alkanes and biaryls and the reaction is called the Wurtz-Fittig reaction. The formation of the mixture of three products is a demerit of the reaction. On the other hand, since in this reaction long chain alkyl halides do not undergo rearrangement, large side chains can be developed by the reaction.

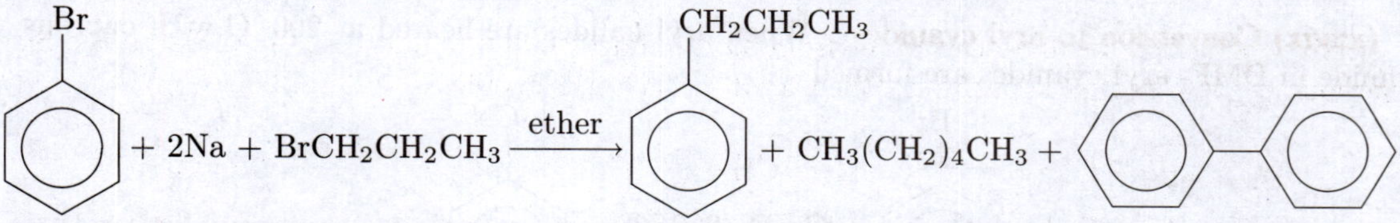

Both radical and ionic mechanisms have been proposed for the reaction. The formation of biaryl supports the radical mechanism. However, the ionic mechanism is more plausible which is shown below :

$$PhX + 2Na \longrightarrow Ph^{-}\,Na^{\oplus} + NaX;\; Ph^{-}\,Na^{\oplus} + R—X \longrightarrow PhR + NaX$$

$$R—X + 2Na \longrightarrow R^{-}\,Na^{\oplus} + NaX;\; R^{-}\,Na^{\oplus} + Ph—X \longrightarrow PhR + NaX$$

$$Ph^{-}\,Na^{\oplus} + Ph—X \longrightarrow Ph—Ph;\; R^{-}\,Na^{\oplus} + R—X \longrightarrow R—R$$

The reaction is an extension of the Wurtz reaction.

**(xxvii) The Ullmann biaryl synthesis :** The Ullmann reaction involves the coupling of aryl halides in the presence of copper powder; DMF is found to be the best solvent for the reaction. Aryl iodides react readily; but chlorides and bromides may also be used. In these cases the presence of nitro, alkyl, alkoxy, etc. groups in the *o*-position is helpful. It is interesting to note that —$NO_2$ group activates the ring while —COOH, —$SO_2NH_2$, etc. deactivate the ring and stop the reaction; —$NO_2$ in the *o*-position only activates the ring. OH, $NH_2$, NHR, NHCOR, etc. (the deactivating groups in $S_N$ reactions) inhibit the reaction.

$$Ar\,I + Cu + I\,Ar \longrightarrow Ar—Ar$$

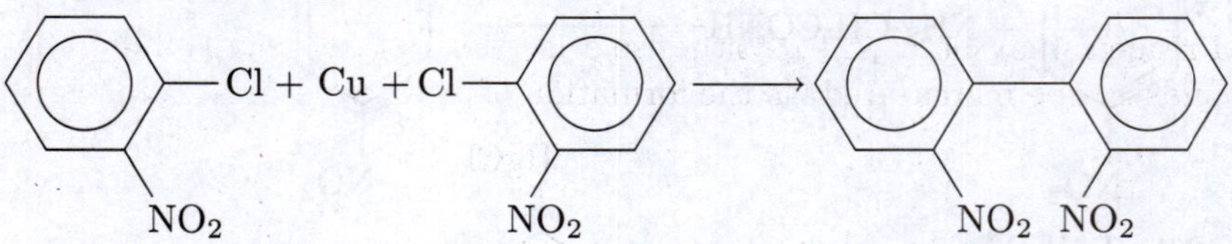

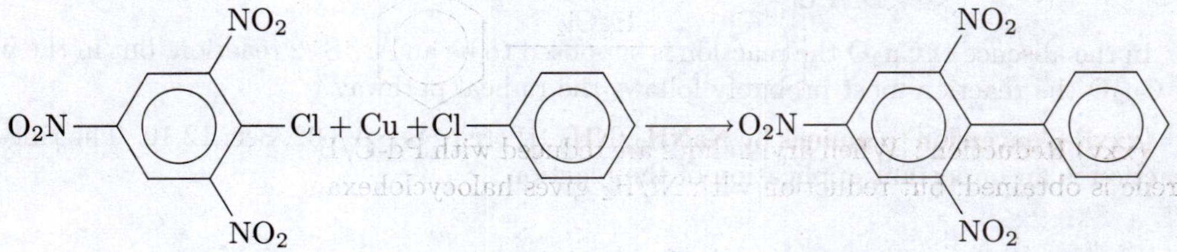

Both a free radical and an ionic mechanism have been proposed; there are a large number of evidences in favour of the ionic mechanism.

$$ArI + Cu \longrightarrow ArCuI \longleftrightarrow Ar^- + \overset{\oplus}{Cu}I \xrightarrow{ArI} Ar{-}Ar + CuI_2$$

**(xxviii) Formation of phenols :** See Sec. 12.9. p. 594.

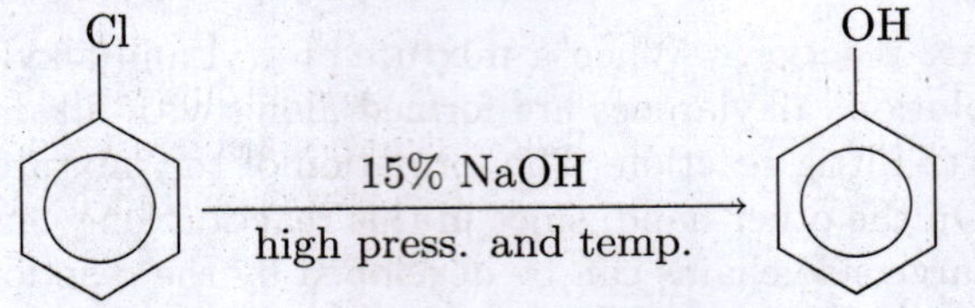

**(xxvix) Conversion to aryl cyanides :** When aryl halides are heated at 200 °C with cuprous cyanide in DMF, aryl cyanides are formed.

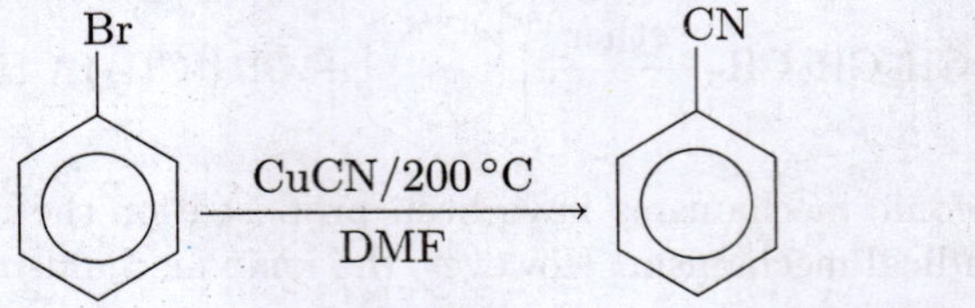

The mechanism of the reaction is uncertain and probably it does not follow $S_N$ pathways.

**(xxx) Conversion to arylamines :** Ammonia, 1° and 2° amines react with haloarenes to form arylamines; 1° and 2° amines usually give better results than ammonia.

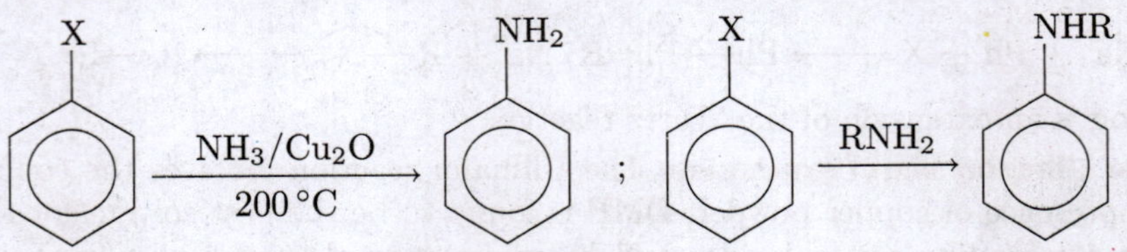

The important application of the reaction is :

Tagging of the amino end of a peptide or protein chain.

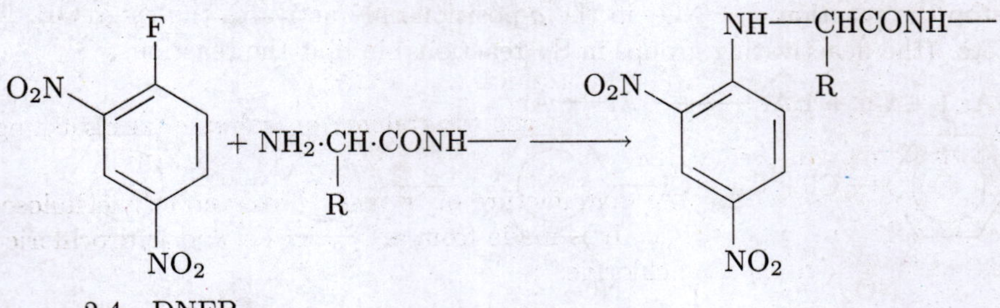

2,4 - DNFB

In the absence of $Cu_2O$ the reaction is supposed to be an Ar. $S_N2$ reaction, but in the presence of $Cu_2O$ the reaction most probably follows the radical pathway.

**(xxxi) Conversion to amines by $NaNH_2/NH_3$ (l) etc. :** See p. 602, Sec. 12.10. The ring-closure reaction is an important application of the reaction.

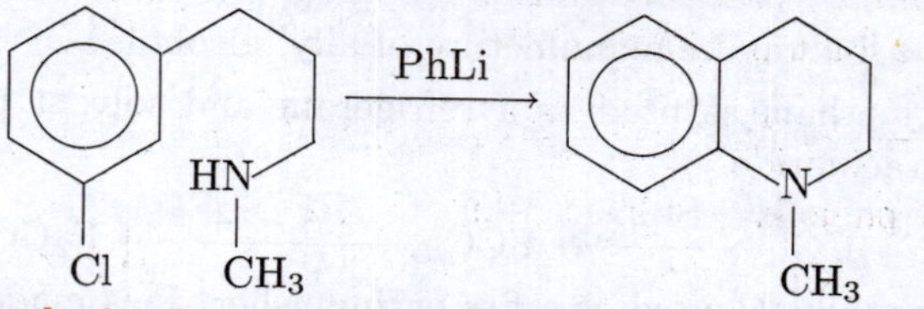


**(xxxii) Electrophilic substitution reactions :** A halo group is a deactivating group; it deactivates the ring. So, nitration, halogenation, sulphonation are successfully carried out on haloarenes, but they fail to undergo the Friedel-Crafts reaction; but if there are activating groups in the ring, they can undergo the reaction. Halo groups direct the incoming electrophile at *o*- and *p*-positions.

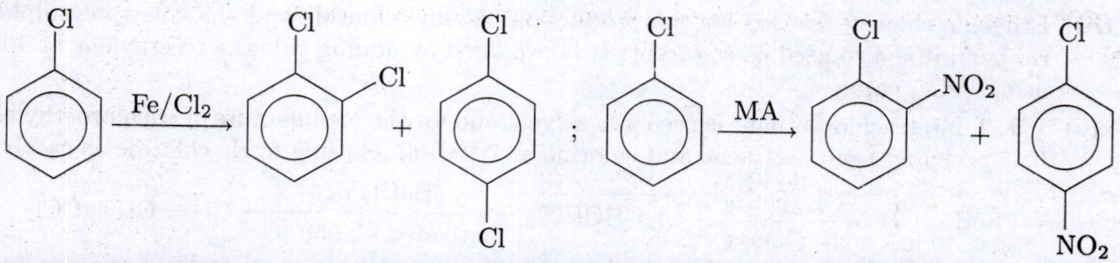


## 20.1.5 Tests

**(i)** All halides (aliphatic and aromatic) except fluorides, impart a green flame when they are thrust into a flame by means of a copper wire.

**(ii)** Except aryl halides and fluorides all other halo derivatives give precipitate with alcoholic silver nitrate solution. The reactivity order is; 3°, allyl > 2° > 1° > vinyl and RI > RBr > RCl.

## 20.1.6 Uses of halo compounds

**(i)** Methyl chloride is used for synthesis of silicones, tetramethyllead and methylcelluloses; it is also used as a solvent for butyl rubber. It is prepared by heating methanol with HCl.

$$[CH_3Cl + Si \xrightarrow[200\,°C]{Cu} \underset{\text{byproducts}}{\underset{+}{(CH_3)_2\,SiCl_2}}\ (70\text{-}90\%) \xrightarrow{N_2O} \text{Silicone polymers}]$$

**(ii)** Methyl bromide is used as a poison gas for rodent control and extinguishing fire. It is made from methanol and HBr.

**(iii)** Ethyl chloride is used for the manufacture of tetraethyllead and ethylcelluloses. It is also used as a general anaesthetic. It is made from ethyl alcohol and hydrochloric acid in the presence of anhydrous zinc chloride.

**(iv)** Amyl chlorides are used to prepare pentasols, amylamines, amyl mercaptans and amyl sulphides. Amyl chlorides are prepared by the chlorination of *n*-pentane and *i*-pentane.

**(v)** Chlorobenzene is used as an intermediate in the manufacture of phenol, aniline, DDT and dyes.

**(vi)** *p*-Dichlorobenzene is used as a larvicide.

**(vii)** *o*-Dichlorobenzene is used as a solvent for the preparation of 2, 6-tolylene diisocyanate which is used to manufacture urethan plastic.

**(viii)** Allyl chloride is used for the manufacture of allyl alcohol.

**(ix)** Dichlorodifluoromethane is used as a refrigerant and solvent propellant. It is prepared from carbon tetrachloride.

$$HF + CCl_4 \xrightarrow{SbF_5} FCCl_3 \xrightarrow{HF/SbF_5} F_2CCl_2 \text{ (Freon-12)}$$

**(x)** Bromotrifluoromethane is used as a fire extinguisher. It is made from fluoroform.

**(xi)** Methylene chloride is used as a paint and varnish remover and as a solvent for cellulose triacetate in the manufacture of film and fibre. It is made by chlorinating methyl chloride.

**(xii)** Choroform is mainly used as an intermediate in the synthesis of chlorodifluoromethane (Freon-22) which is used as a refrigerant.

**(xiii)** Ethylene chloride is used for the production of vinyl chloride and thiokols (polysulphide rubber); it is also used as a solvent. It is prepared by adding chlorine to ethylene at 40°C using $FeCl_3$ catalyst.

**(xiv)** 1, 1, 2, 2-tetrachloroethane is used as a solvent and for the manufacture of trichloroethylene. It is prepared from acetylene and chlorine at 70 °C-95 °C using ferric chloride as catalyst.

$$CH \equiv CH \xrightarrow[70\,°C\text{-}90\,°C]{Cl_2/FeCl_3} Cl_2CHCHCl_2 \xrightarrow[\text{activated C, } 250\,°C]{BaCl_2 \text{ or}} Cl{-}CH{=}CCl_2$$

**(xv)** Tetrachloroethylene is used as a solvent for dry-cleaning plants and coin-operated machine. It is prepared from carbon tetrachloride by pyrolysis.

$$2\,CCl_4 \xrightarrow{800\,°C\text{-}900\,°C} Cl_2C{=}CCl_2 + 2Cl_2$$

**(xvi)** Vinyl chloride is used for the manufacture of resins and plastic which are used in floor tiles, shoe soles, rain-coats, pipe, tubing, wire insulators, etc. It is made from ethylene chloride.

$$ClCH_2CH_2Cl \xrightarrow{400\,°C} CH_2{=}CH{-}Cl$$

**(xvii)** Vinylidene chloride is used for the manufacture of sarans, copolymers with vinyl chloride acrylonitrile, etc.

$$HC \equiv CH \xrightarrow[135\,°C]{Cl_2} H_2C{=}CCl_2 \text{ (Vinylidene chloride)}$$

**Problem 20.1.** Compare ethyl chloride with chlorobenzene.

**Solution :**

| Points | EtCl | PhCl |
|---|---|---|
| 1. Structure | C—Cl bond is highly polar. | C—Cl bond has some double bond character and much less polar than that in an aliphatic compound. |
| 2. $S_N$ reaction | Undergoes $S_N2$ reaction readily. | Undergoes Ar.$S_N2$ reaction under drastic condition. |
| 3. Alc. $AgNO_3$ | AgCl precipitates. | AgCl does not precipitate. |
| 4. Elimination reaction | Undergoes elimination reaction with alc. KOH to form ethylene. | In the presence of $NaNH_2/NH_3(l)$ it forms very unstable benzyne which ultimately forms aniline. |
| 5. $S_E$ reactions | Does not undergo $S_E$ reaction. | Undergoes $S_E$ reactions. The attacking electrophile takes *o*- and *p*-positions. |

**Problem 20.2.** Give the mechanism of the conversion of allylsulphonyl chloride to allyl chloride.

**Solution :** This is an example of $S_Ni'$ reaction.

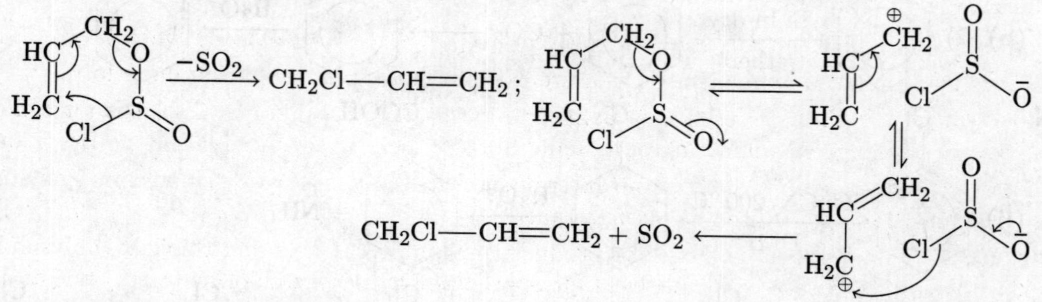


**Problem 20.3.** What happens when ethyl chloride is treated with **(i)** sodium azide, **(ii)** $CH\equiv C—Na$, **(iii)** $CH_3COCH_2CO_2Et$ in the presence of a base.

Do these reagents react with chlorobenzene? Give reasons for your answer.

**Solution : (i)** $CH_3CH_2Cl + NaN_3 \longrightarrow CH_3CH_2N_3$

**(ii)** $CH_3CH_2Cl + NaC\equiv CH \longrightarrow CH_3CH_2—C\equiv CH$

**(iii)** $CH_3COCH_2CO_2Et \xrightarrow{NaOEt} CH_2CO\overset{..}{\bar{C}}HCO_2Et$

**(iv)** $CH_3—CO\bar{C}HCO_2Et + CH_3—CH_2—Cl \longrightarrow CH_3COCH(CH_2CH_3)CO_2Et$

Since the C—Cl bond in chlorobenzene is not ionic and since it has a double bond character, the given reagents do not react with chlorobenzene.

**Problem 20.4.** How can you synthesise the following compounds from chlorobenzene **(a)** *p*-chloroiodobenzene. **(b)** benzoic acid. **(c)** *p*-chlorobenzoic acid.

**Solution :**

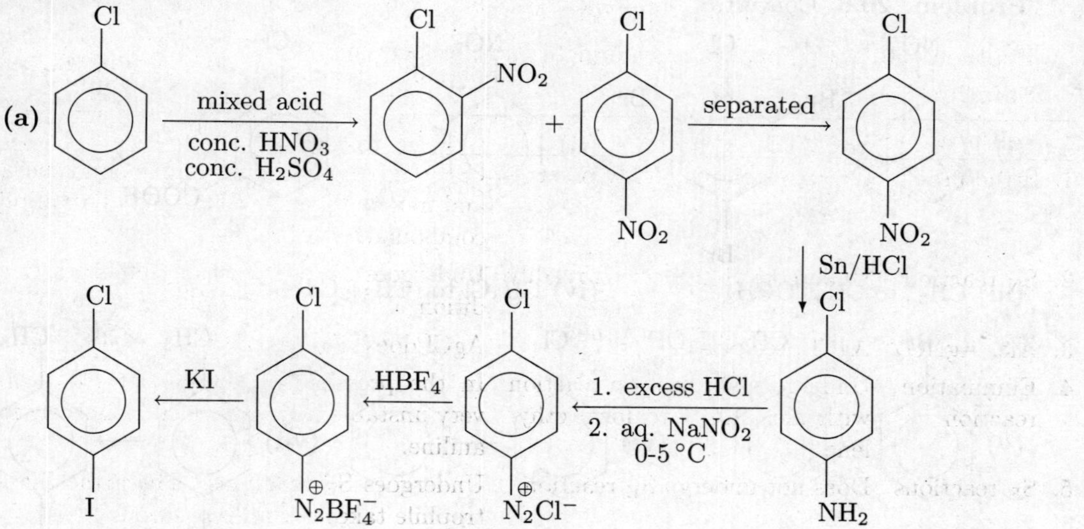

(b) (i) 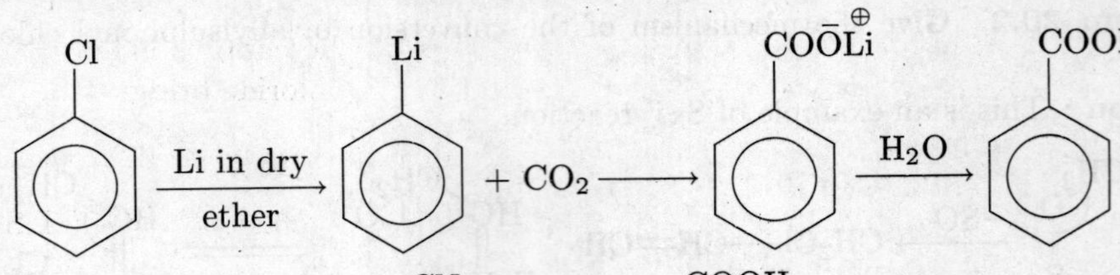

(ii) 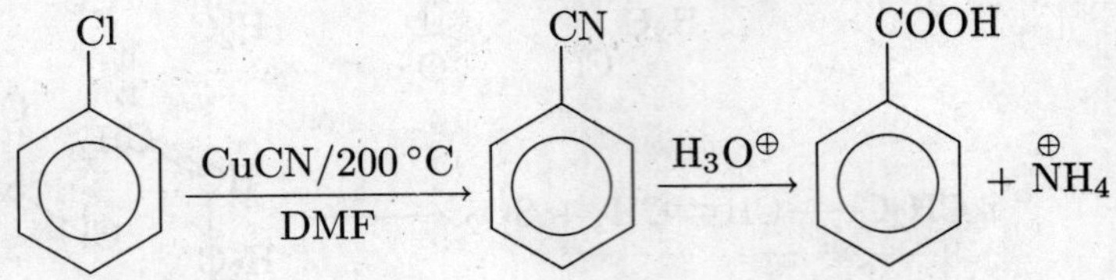

(c) 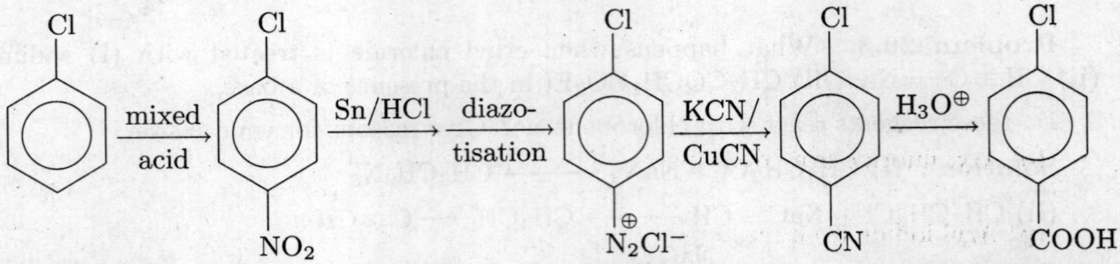

**Problem 20.5.** When an alkyl halide is treated with silver nitrite solution both a nitroalkane and an alkyl nitrite are formed. Give the mechanism of the reaction and tell which one will be formed predominantly.

**Solution :** Silver nitrite will produce $Ag^{\oplus}$ and $NO_2^-$ ions in the solution. In the presence of silver ion, alkyl cation will form from the alkyl halide. Then $S_N1$ reaction will take place to form both nitrite and nitroalkane. Since in an $S_N1$ reaction an ambident ion attacks through the more electronegative atom, here the ambident nitrite ion will attack the carbenium ion through the anionic O atom and alkyl nitrite will be the major product; attack through N atom will produce nitroalkane, (See ambident nucleophile. Sec. 11.5. 1d)

**Problem. 20.6.** Convert.

(i) 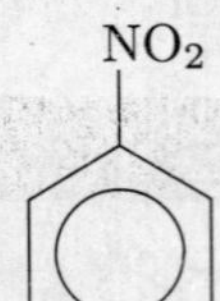 to 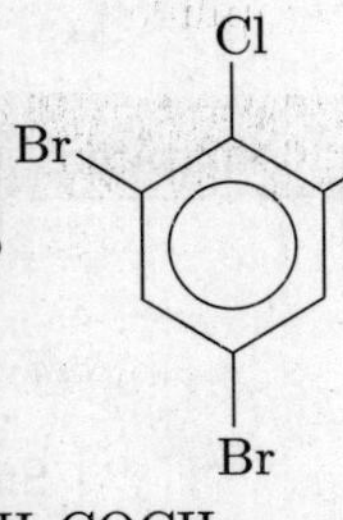 (ii) 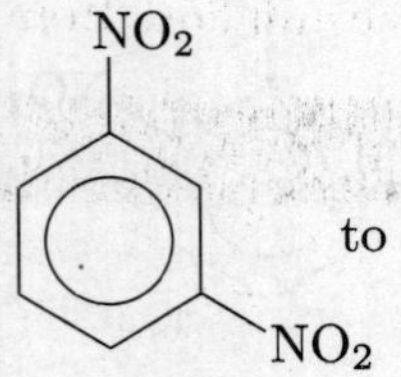 to 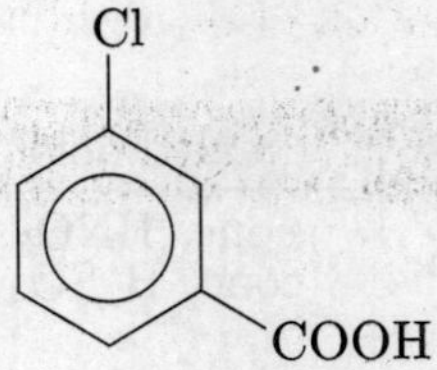

(iii) $CH_3I$ to $CH_3COCH_3$ (iv) $CH_3Cl$ to $(CH_3)_2Cd$

(v)  to 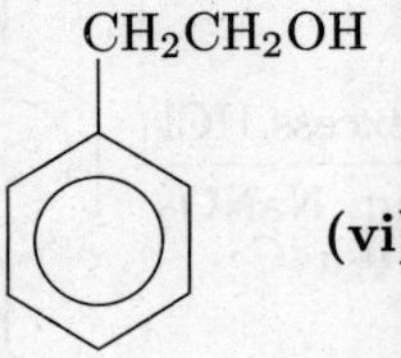 (vi) 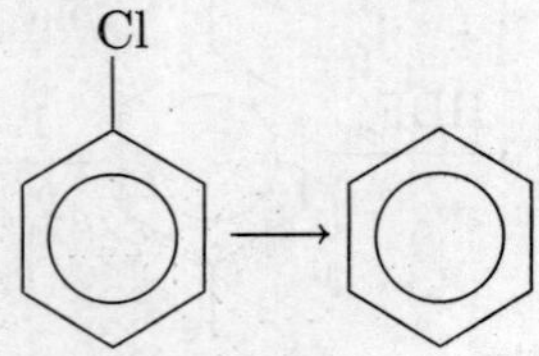 (vii) 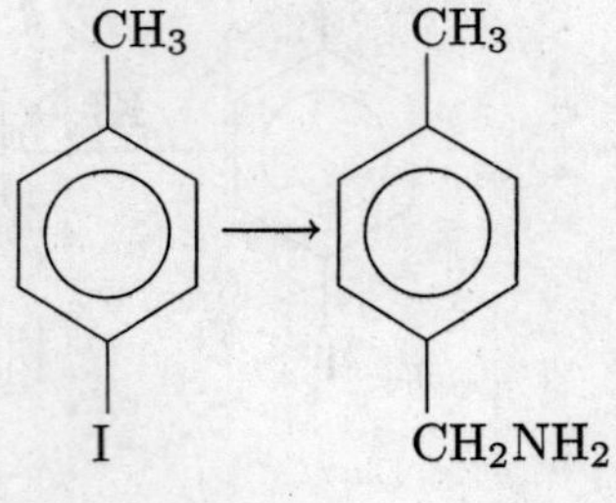

**Solution :**

(i) 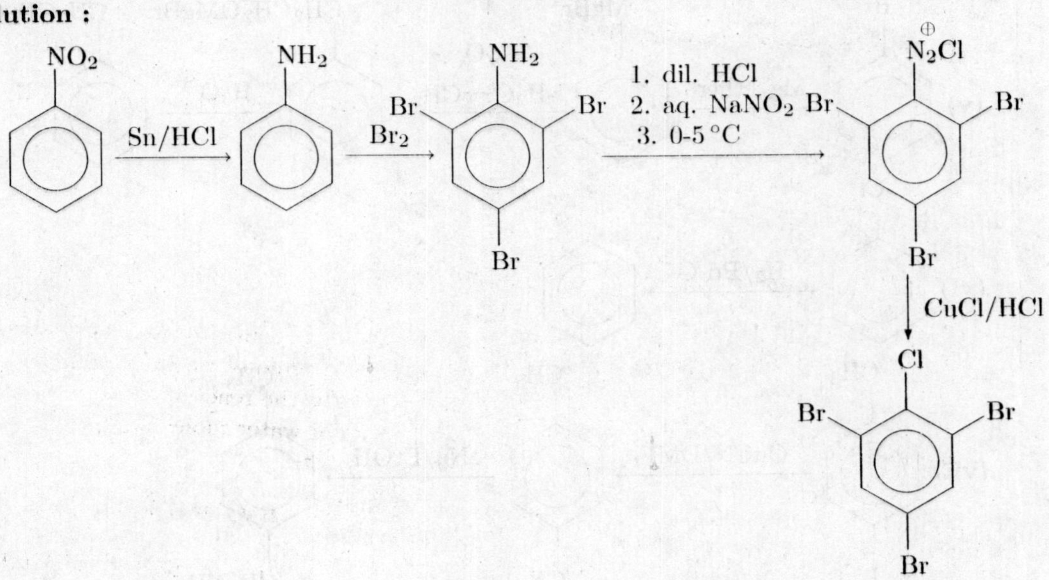


(ii) 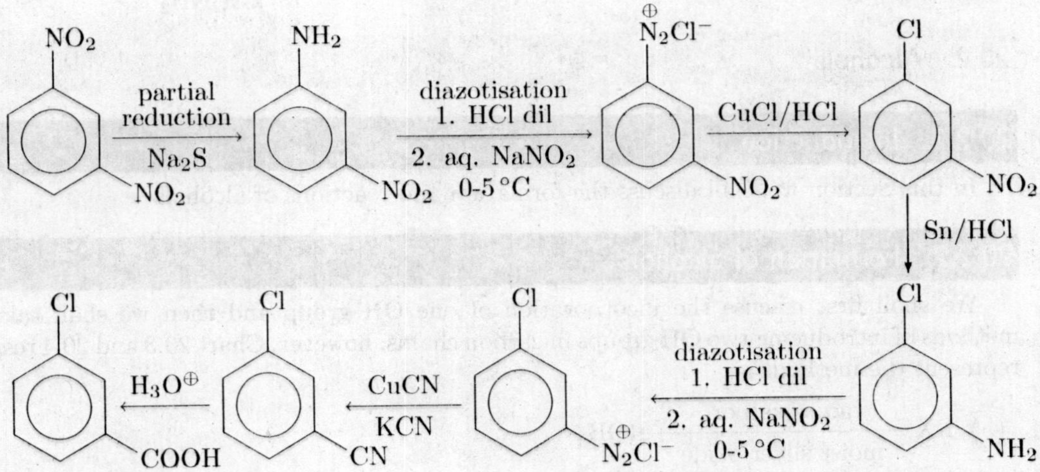


(iii) 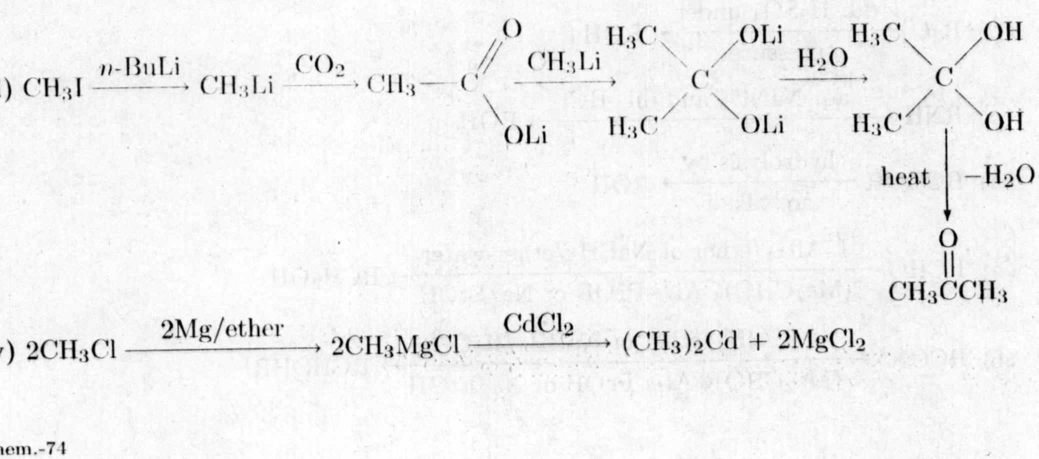


(iv) $2CH_3Cl \xrightarrow{2Mg/ether} 2CH_3MgCl \xrightarrow{CdCl_2} (CH_3)_2Cd + 2MgCl_2$

(v) 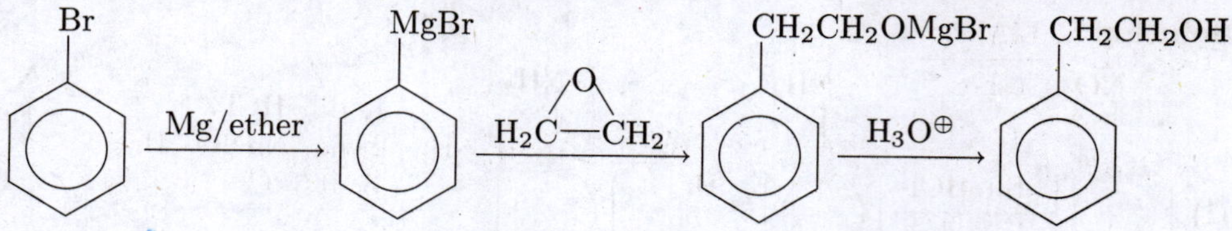

(vi) 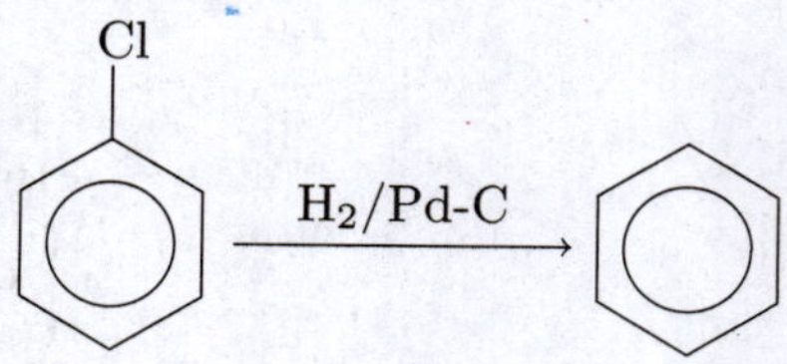

(vii) 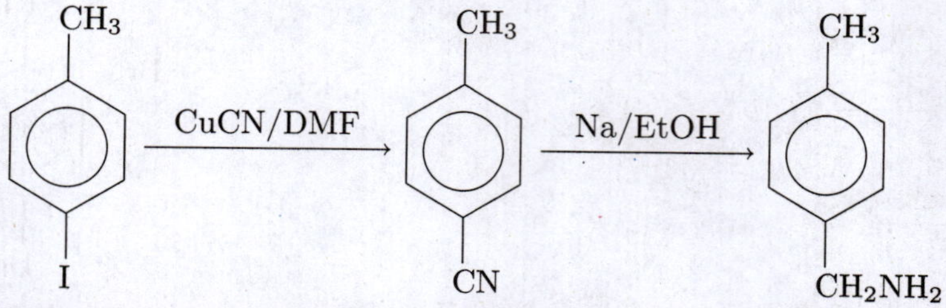

## 20.2 Alcohols

### 20.2.1 Introduction

In this Section we shall discuss the formation and reactions of alcohols.

### 20.2.2 Formation of alcohols

We shall first discuss the incorporation of one OH group and then we shall take up the methods of introducing two OH groups in carbon chains; however, Chart 20.3 and 20.4 respectively represent the methods :

(a) $RX \xrightarrow[\text{moist silver oxide}]{\text{aq. alkali or}} ROH$

(b) $R_2O \xrightarrow[\text{pressure}]{\text{dil. } H_2SO_4 \text{ under}} ROH$

(c) $RNH_2 \xrightarrow{\text{aq. } NaNO_2 \text{ and dil. HCl}} ROH$

(d) $RCOOR \xrightarrow[\text{aq. alkali}]{\text{hydrolysis by}} ROH$

(e) $RCHO \xrightarrow[(Me_2CHO)_3\,Al/i\text{-PrOH or Na/EtOH}]{LiAlH_4/\text{ether or } NaBH_4/\text{ether-water}} RCH_2OH$

(f) $RCOR^1 \xrightarrow[(Me_2CHO)_3\,Al/i\text{-PrOH or Na/EtOH}]{LiAlH_4/\text{ether or } NaBH_4/H_2O} RCHOHR^1$

**(g)** $RCOOH \xrightarrow[\text{ether}]{LiAlH_4} RCH_2OH$

**(h)** $RCOOR^1 \xrightarrow{LiAlH_4;\ \text{ether or } Na/EtOH} RCH_2OH + R^1OH$

**(i)** $RMgX \xrightarrow{1.\ HCHO\ 2.\ H_3O^{\oplus}} RCH_2OH$

**(j)** $RMgX \xrightarrow{1.\ R^1CHO\ 2.\ H_3O^{\oplus}} R{-}CHOH{-}R^1$

**(k)** $RMgX \xrightarrow{1.\ R^1COR^2\ 2.\ H_3O^{\oplus}} R{-}C(OH)(R^2){-}R^1$

**(l)** $RMgX \xrightarrow[2.\ H_3O^{\oplus}]{H_2C{-}CH_2\ (\text{epoxide, }{-}O{-})} RCH_2CH_2OH$

**(m)** $>C{=}C< \xrightarrow[H_3O^{\oplus}]{\text{Hydration}} >CH{-}\overset{|}{\underset{|}{C}}OH$

**(n)** $>C{=}C< \xrightarrow[1.\ B_2H_6\ 2.\ H_2O_2/\bar{O}H]{\text{hydroboronation oxidation}} >CH{-}\overset{|}{\underset{|}{C}}OH$

**(o)** $>C{=}C< \xrightarrow{\text{ozonide formation; reduction}} {-}\overset{|}{\underset{|}{C}}{-}OH + HO{-}\overset{|}{\underset{|}{C}}{-}$

**(p)** $CO + H_2 \xrightarrow[\text{Zinc chromite + alkali metal}]{\text{heat/pressure}} \text{mixture of alcohols}$

**(q)** $\text{Alkene} + CO + H_2 \xrightarrow[CoH(CO)_4]{\text{oxoprocess}} \text{aldehydes} \xrightarrow[Zn\text{-}Cu/H_2]{\text{reduction}} \text{alcohols}$

**(r)** $RCOCH{=}\overset{\oplus}{N}{=}\bar{N}{:} \xrightarrow{H_3O^{\oplus}} RCOCH_2OH$

**Chart 20.3** Formation of monohydric alcohols.

**(a) Conversion of halo group to —OH group :** See Sec. 20.1.4. xiii (p. 1158).

**(b) Conversion of ether linkage to —OH group by dil. $H_2SO_4$ :** When an ether is warmed with dilute sulphuric acid under pressure, an alcohol is obtained.

$$R_2O \xrightarrow[\text{pressure}]{\text{dil. } H_2SO_4} 2ROH;\ CH_3CH_2OCH_2CH_3 \xrightarrow[\text{pressure}]{\text{dil. } H_2SO_4} 2CH_3CH_2OH$$

$$R{-}O{-}R \xrightarrow{H^{\oplus}} R{-}\underset{\underset{H}{|}}{\overset{\oplus}{O}}{-}R \xrightarrow{H_2O} \underset{\text{T.S.}}{\overset{\delta+}{H_2O}\cdots R\cdots \underset{\underset{H}{|}}{\overset{\delta+}{O}}{-}R} \longrightarrow R{-}\underset{\underset{H}{|}}{\overset{\oplus}{O}}H + ROH \longrightarrow 2ROH$$

**(c) Conversion of —$NH_2$ group into —OH group by $HNO_2$ :** When 1° aliphatic amines are treated with nitrous acid, the —$NH_2$ group gets converted into —OH group and alcohols are formed. Nitrous acid is very unstable and usually it is prepared in situ by treating aqueous and dilute sodium nitrite solution with dilute acids, HCl, $H_2SO_4$, acetic acid, etc.

$$RNH_2 + HNO_2 \longrightarrow ROH + N_2\uparrow + H_2O; \; CH_3CH_2NH_2 + HNO_2 \longrightarrow CH_3CH_2OH + N_2 + H_2O$$

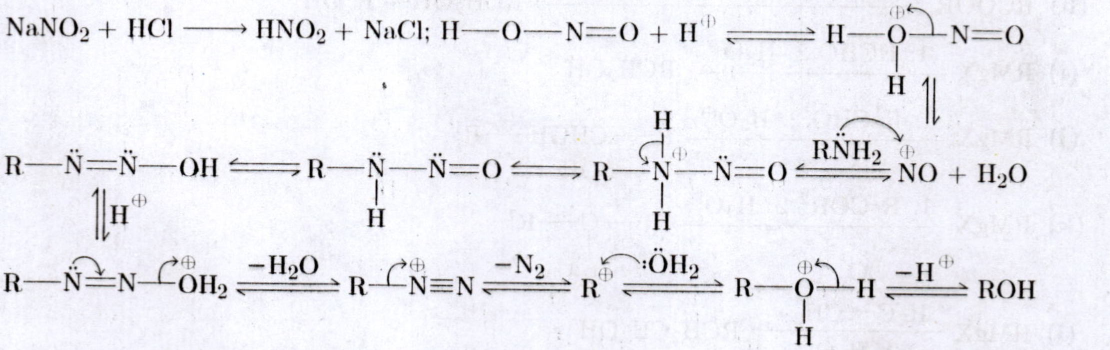

However, carbenium ions being the intermediates of the reaction, several side-products and rearranged products are obtained as shown below (See Sec. 14.4 the Demyanov reaction, p. 712).

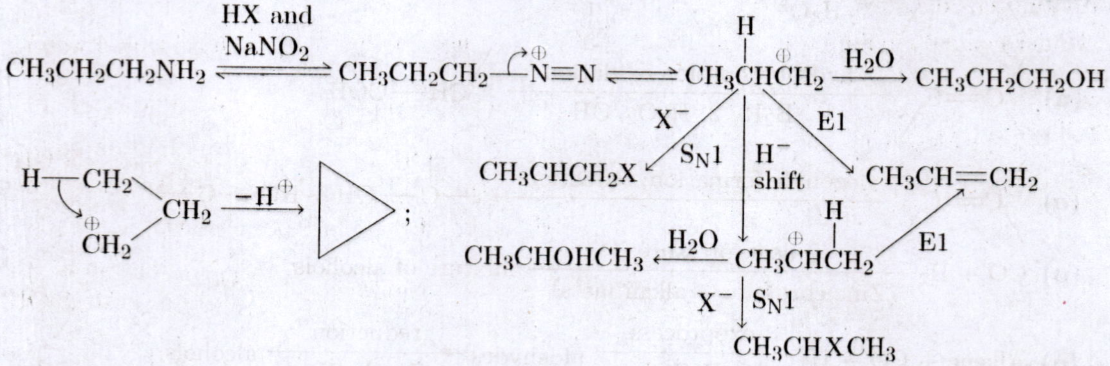

Thus, in this reaction isomeric alcohols are obtained through the nucleophilic attack by water on the carbenium ions. They may undergo the nucleophilic attack by the $X^-$ to form inorganic esters. Isomeric alkenes also form through the El reaction of the carbenium ions; alkenes and esters are the side products of the reaction. Besides these, a small amount of cycloalkane is also obtained. Since molecular nitrogen evolves out in the reaction, this is used as a test to detect 1° amines.

**(d) Conversion of ester group to —OH group by hydrolysis :** When organic esters are hydrolysed by dilute sulphuric acid or alkali, alcohols are obtained. Natural fats and oils are esters of glycerol and higher fatty acids; glycerol and soap are manufactured from fats and oils by hydrolysing them with alkali.

Since through alkaline hydrolysis soaps are formed, the process is called saponification. The reaction of esters with dilute sulphuric acid is a reversible reaction. The forward reaction is called hydrolysis and the backward reaction, which takes place in the presence of hot concentrated sulphuric acid, is called esterification.

$$R^1COOR \xrightarrow{\bar{O}H} R^1CO\bar{O} + ROH; \; RCOOR^1 \underset{H^{\oplus}}{\overset{H_3O^{\oplus}}{\rightleftharpoons}} R^1COOH + ROH$$

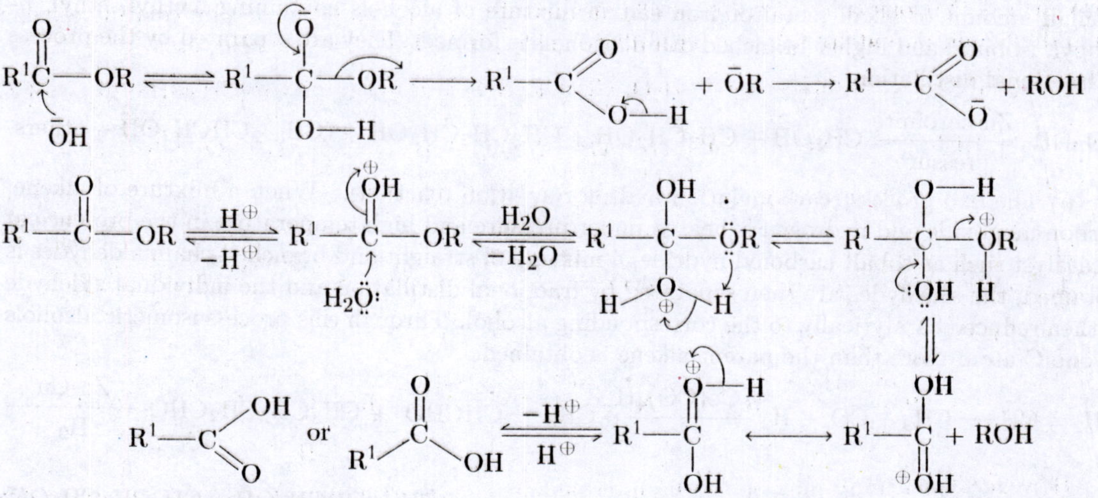

Tracer study is a strong evidence in favour of the above mechanism (See Sec. 11.8.2, p. 418). For detail see 20.9.2.

Under neutral medium esters can also be hydrolysed using trimethylsilyl chloride and water :

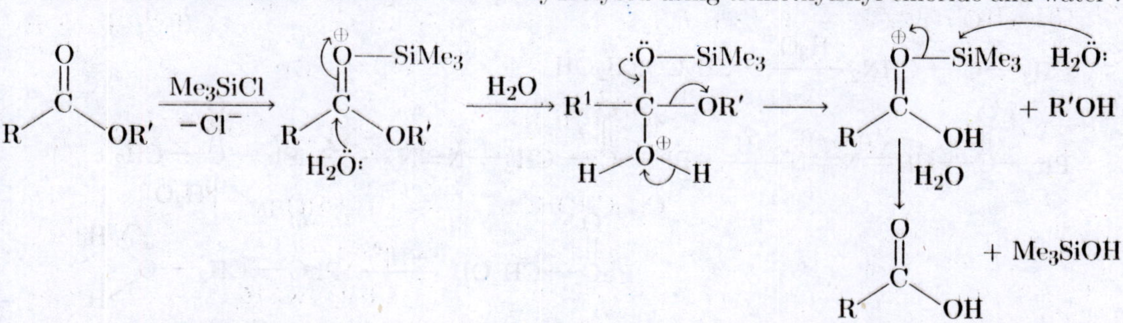

**(e), (f), (g) and (h) Conversion of —CHO, —CO—, —COOH and —COOR groups to —OH group by reduction :** The —CHO, —CO— and —COOR groups may be reduced to —COH group by metallic hydrides in ethers and/or water, aluminium isopropoxide in isopropyl alcohol and by metallic sodium and alcohol; —COOH group may be reduced to —$CH_2OH$ group by $LiAlH_4$ (See Sec. 16.6.1, p. 854, 16.6.2, p. 857 and 16.9.1, p. 871).

**(i), (j), (k), (l) Formation of —OH group by the reaction of the Grignard reagent :** Carbonyl groups and epoxy linkage may be converted to —OH group by the reaction of the Grignard reagents on carbonyl compounds and epoxides (See Sec. 16.4.5, p. 838).

**(m) Introduction of alcoholic OH group to C—C multiple bonds by hydration :** See Sec. 13.5, p. 628. This is a commercial method of preparing alcohols from natural gas.

**(n) Introduction of alcoholic OH group to a C—C multiple bond by hydroboronation oxidation :** See Sec. 13.10, p. 650.

**(o) Introduction of alcoholic OH group to a C—C unit by ozonolysis and reduction :** See Sec. 13.8, p. 642.

**(p) Synthesis of alcohols from carbon monoxide and hydrogen :** When a mixture of carbon monoxide and hydrogen is heated under pressure in the presence of zinc-chromite catalyst and a small amount of alkali metal or iron salt, a mixture of alcohols containing methyl, ethyl, *n*-propyl, isobutyl and higher branched chain alcohols is formed. They are separated by the process of fractional distillation.

$$CO + H_2 \xrightarrow[\text{pressure}]{\text{Zn-chromite}} CH_3OH + CH_3CH_2OH + CH_3CH_2CH_2OH + (CH_3)_2CHCH_2OH + \text{others.}$$

(q) **The oxo process (carbonylation hydroformylation reaction) :** When a mixture of alkene, carbon monoxide and hydrogen is heated under pressure and high temperature in the presence of a catalyst such as cobalt carbonyl hydride, a mixture of straight and branched-chain aldehydes is obtained; the aldedydes are then separated by fractional distillation and the individual aldehyde is then reduced catalytically to the corresponding alcohol. Through this process isomeric alcohols of one C atom more than the parent alkene is obtained.

$$CH_2{=}CH{-}CH_3 + CO + H_2 \xrightarrow{Co(CO)_4H} CH_3{-}\underset{\large CH_3}{\underset{|}{C}}HCHO + CH_3CH_2CH_2CHO \xrightarrow[H_2]{\text{Zn-Cu}}$$

$$(CH_3)_2CHCH_2OH + CH_3CH_2CH_2OH.$$

(r) **Conversion of diazo group to OH group :** When diazoketones are hydrolysed with dilute acids, 1° alcohols with a carbonyl group at $\beta$-position are formed.

$$CH_3{-}\overset{\overset{\large O}{\|}}{C}{-}CHN_2 \xrightarrow{H_3O^{\oplus}} CH_3COCH_2OH$$

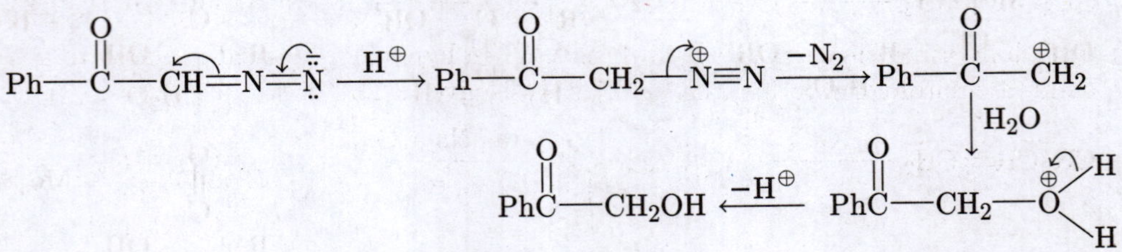

## 20.2.3 Formation of 1, 2-glycols

The methods are given in Chart 20.4.

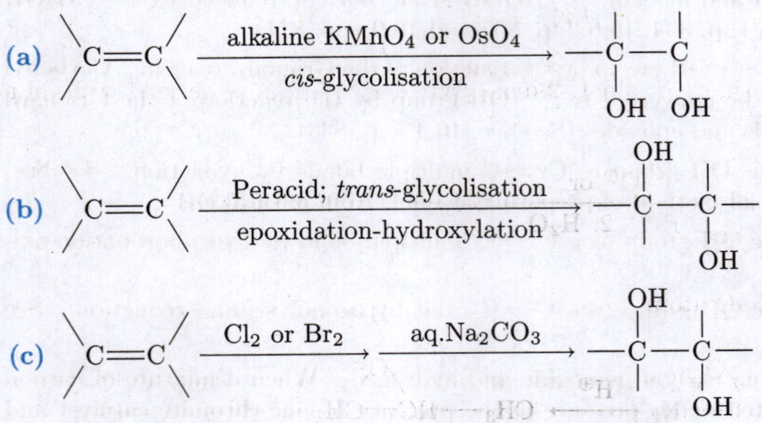

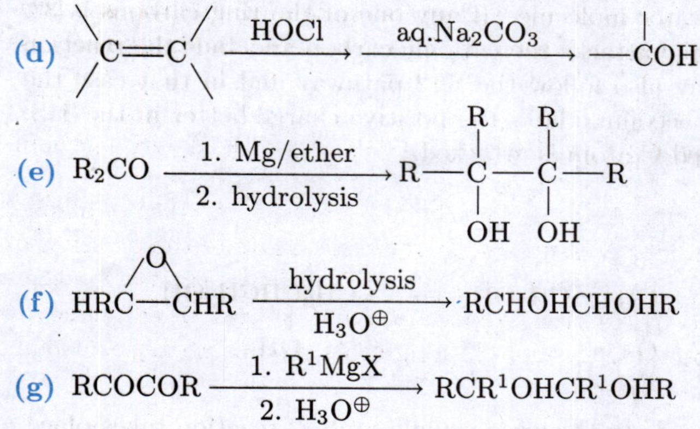

**Chart 20.4** Formation of 1, 2-glycols.

Let us now discuss the methods of introducing two OH groups at 1, 2-positions.

**(a) Cis-glycolisation by $OsO_4$ and $KMnO_4$ :** See Sec. 13.7.1, p. 633, and Sec. 13.7.2, p. 635.

**(b) Trans-glycolisation by peracids :** See Sec. 13.7.3, p. 637.

**(c) By halogen addition and hydrolysis :** When chlorine or bromine is added to a C—C double bond by polar mechanism, trans-dihaloalkane is formed; this on hydrolysis with aqueous sodium carbonate solution gives a 1, 2-diol.

$$CH_3CH{=}CH_2 \xrightarrow{Cl_2} CH_3CHCl{-}CH_2Cl \xrightarrow{aq.Na_2CO_3} CH_3CHOHCH_2OH$$

**(d) By halohydrin formation and hydrolysis :** When hypochlorous acid is added to a C—C double bond, a chlorohydrin forms; this on hydrolysis gives a 1, 2-diol.

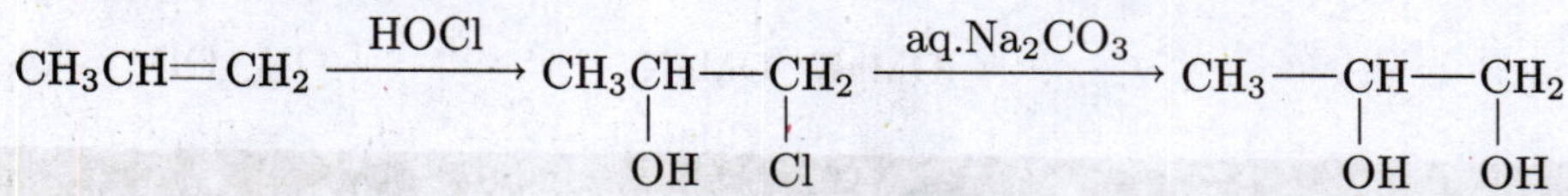

**(e) By bimolecular reduction of carbonyl compounds and hydrolysis :** See p. 872, Sec. 16.9.2.

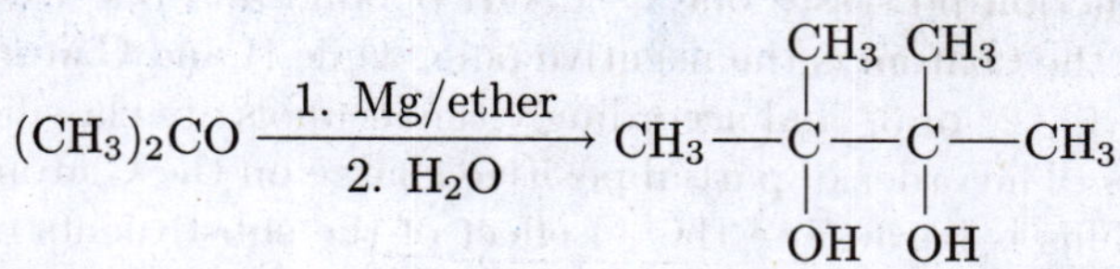

**(f) Conversion of an epoxy linkage to a 1, 2-diol :** When an epoxide is warmed with dilute acid solution, a 1, 2-diol forms.

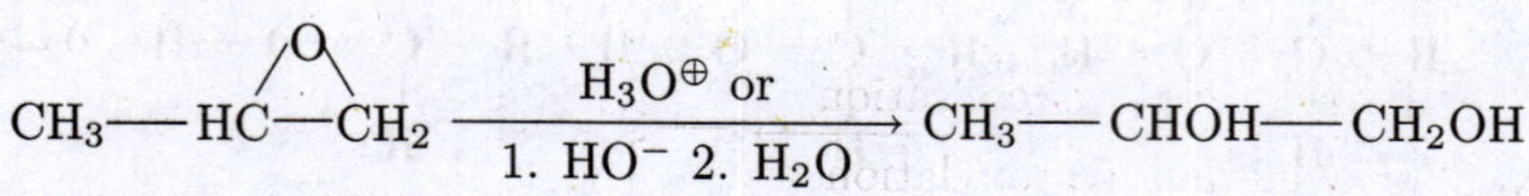

The mechanistic steps are :

**Step I :** Protonation of the O atom

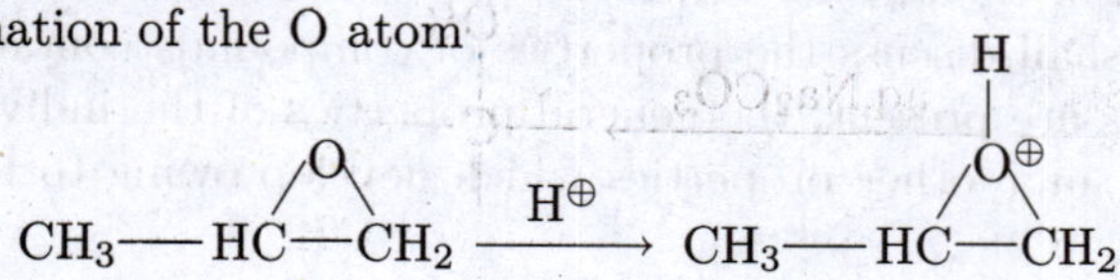

**Step II :** Nucleophilic attack by the water molecule. If any one of the ring carbons is 3°, a 3° carbenium ion forms and the nucleophile attacks the cationic carbon and thus the reaction follows the $S_N1$ pathway. The reaction may also follow the $S_N2$ pathway and in that case the nucleophile will attack that C atom which accommodates the positive charge better in the T.S. Thus in both the cases, the more substituted C atom is attacked.

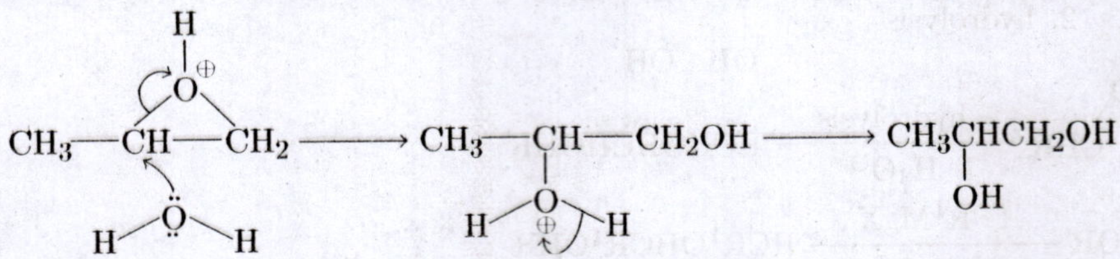

The hydrolysis may also be base catalysed. In this case usually an $S_N2$ reaction takes place. Since a 1 °C undergoes an $S_N2$ attack more readily than a 2 °C or 3 °C, usually the less substituted C atom of the epoxy ring is attacked by the $HO^-$ ion.

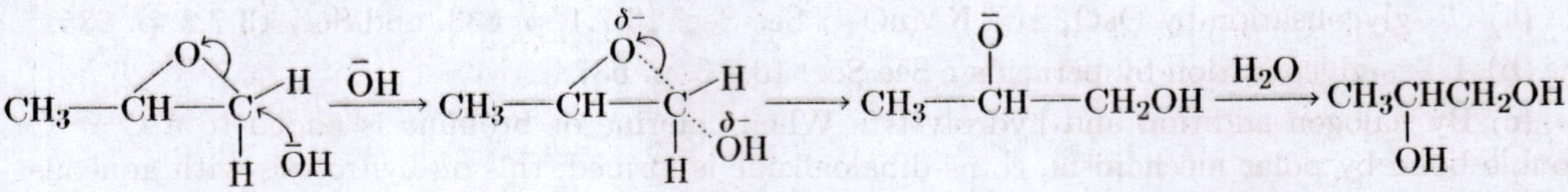

(g) **Conversion of 1, 2-dicarbonyls to 1, 2-diols by the Grignard reagent :** 1, 2-dicarbonyls on treatment with the Grignard reagent followed by hydrolysis give 1, 2-diols.

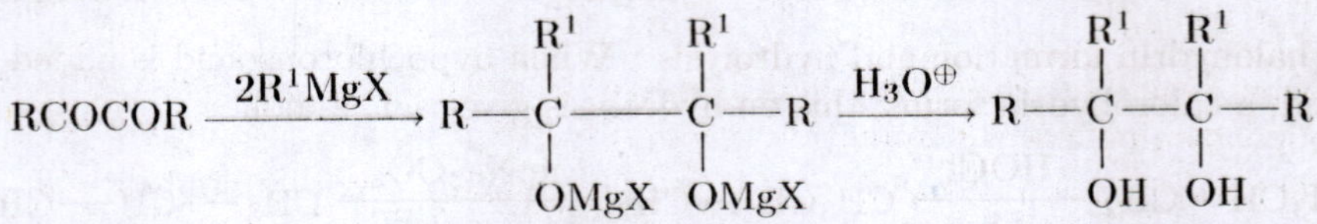

## 20.2.4 Structure, reactivity and physical properties of monohydroxy alcohols

Alcoholic OH function possesses one C—OH $\sigma$ bond and one CO—H $\sigma$ bond; both the bonds are polar and the O atom is the negative pole, while H and C atoms are the positive poles. The C atom may be 1°, 2° or 3° and accordingly the alcohols are classified as primary, secondary and tertiary alcohols. The order of partial positive charge on the C atoms bearing the OH group is : 1° > 2° > 3°. This is because of the +I-effect of the substituents on the involved C atoms; the +I-effect also increases the negative charge density on the O atom as shown below :

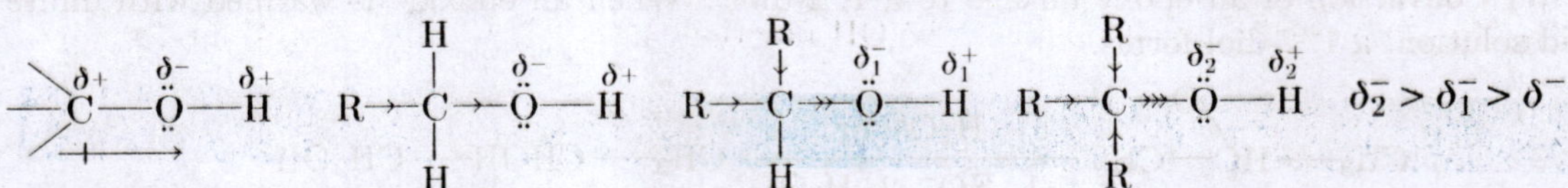

$\mu = 1.65\text{–}1.70\ \mu$

Furthermore, the O atom of the alcohol bears two pairs of non-bonded *es*. On the basis of these structural features we shall discuss the properties of compounds containing alcoholic OH group. When two OH groups are present, the general properties of the individual OH function will be exhibited along with some other properties which develop owing to the presence of the two OH groups.

Since the OH group has a polar bond, alcohol molecules form intermolecular H-bonding (See. Sec. 7.5) and thereby get associated among themselves. Owing to the association, compounds containing OH function have high m.p. and b.p. (See Sec. 7.7.1) and low volatility. Since there is H-bonding among water molecules and since 'like dissolves like', hydroxy compounds of low molecular weight dissolve in water (See sec. 7.7.2). As the molecular weight increases, the hydrocarbon character and the London force increase and as a consequence the m.p. and b.p. increase but solubility in water decreases. In so far as the structure of C—O—H moiety is concerned, the C—OH as well as CO—H bond can undergo cleavage in a reaction. Since among 1°, 2° and 3° alcohols, the negative charge density order on O atom is 3° > 2° > 1°. The ease of the C—OH bond fission order is 3° > 2° > 1° and the ease of the CO—H bond fission order is 1° > 2° > 3°. This means acidity order is 1° ROH > 2° ROH > 3° ROH and the basicity order is the reverse one; since the O atom can donate its *n es*, an alcohol is also basic. However, if the C—OH bond breaks heterolytically, $\bar{O}H$ ion and a carbenium ion will be formed; but C—O bond dissociation energy is high and $\bar{O}H$ is a very poor leaving group because of its high basicity. The reactions involving C—OH bond cleavage are usually acid catalysed. The O atom coordinates with the proton of the acid through its lone pair of *es*; $-C-\overset{\oplus}{O}H_2$ moiety now holds a very good leaving group $H_2O$ and the $S_N$ reactions can occur readily. The H atom of the O—H group being acidic, it is active. An active H in a compound undergoes the following reactions in general :

(i) Alkali metals react with it to form salts.

(ii) The Grignard reagents react to form hydrocarbons and salts.

(iii) Acid halides react to form acyl and benzoyl derivatives.

Besides these reactions, the β-H is slightly active because of the −I-effect of the O group; for this reason alcohols undergo β-elimination reaction to form alkenes.

Again a C atom in a compound may have the maximum oxidation number +IV (as in $CO_2$) and the minimum oxidation number −IV (as in $CH_4$). The oxidation number of carbinol C atom in 1°, 2° and 3° alcohols are −I, 0 and +I respectively.

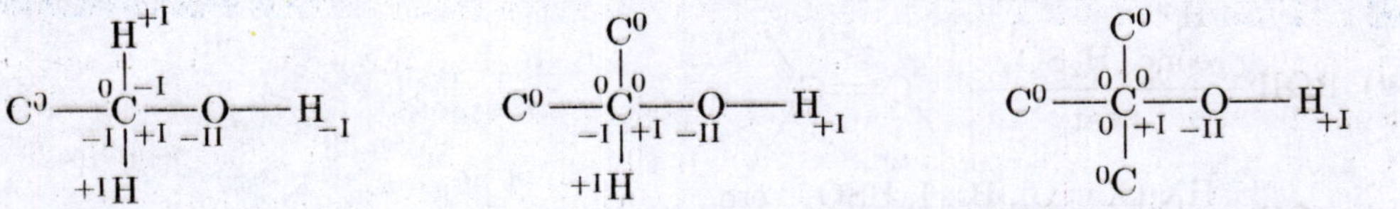

O.N = 0 + (−1) + (−1) + (+1) = −1    O.N = 0 + 0 + 1 − 1 = 0    O.N = 0 + 0 + 0 + 1 = +1

Therefore, alcohols undergo both oxidation and reduction reactions. However, the following are the important reactions of the alcoholic OH function (Chart 20.5).

## 20.2.5 Chemical reactions

(i) $ROH \xrightarrow[\text{Zn-Cu or Cu}/\frac{1}{2}O_2]{\text{Dehydrogenation}} -CHO$ (from 1°ROH), $>C{=}O$ (from 2°ROH)

$>C{=}C<$ (from 3°ROH)

(ii) $ROH \xrightarrow[K_2Cr_2O_7/H_2SO_4]{[O]} -CHO$ (from 1°ROH), $>C=O$ (from 2°ROH)

$>C=C<$ (from 3°ROH)

(iii) $ROH \xrightarrow{Na} RONa + H_2$

(iv) $ROH \xrightarrow{R^1Li} R^1H + ROLi$

(v) $ROH \xrightarrow{R^1MgX} R^1H + ROMgX$

(vi) $ROH \xrightarrow[(R^1CO)_2O]{R^1COCl \text{ or}} ROCOR^1$

(vii) $ROH \xrightarrow{PhCOCl} ROCOPh$

(viii) $ROH \xrightarrow[\text{conc. } H_2SO_4]{R^1COOH} R^1COOR$

(ix) $ROH \xrightarrow[H^{\oplus}]{R^1COOR^2} R^1COOR + R^2OH$

(x) $ROH \xrightarrow{D_2O} ROD + HOD$

(xi) $ROH \xrightarrow[\text{heat}]{\text{conc. } H_2SO_4} ROR + H_2O$

(xii) $ROH \xrightarrow[Al(OEt_3)]{CH_2N_2} ROCH_3 + N_2\uparrow$

(xiii) $ROH \xrightarrow[H^{\oplus}]{\text{ethylene oxide}} HOCH_2CH_2OR$

(xiv) $ROH \xrightarrow[\text{heat}]{\text{conc. } H_2SO_4} >C=C<$

(xv) $ROH \xrightarrow{HX\ (X = Cl,\ Br,\ I,\ HSO_4^-,\ \text{etc.})} RX$

(xvi) $ROH \xrightarrow{PCl_5} RCl + POCl_3 + HCl$

(xvii) $ROH \xrightarrow{PX_3} RX + H_3PO_3$

(xviii) $ROH \xrightarrow{SOCl_2} RCl + SO_2 + HCl$

(xix) $ROH \xrightarrow{NH_3} RNH_2$ (ROH in excess 1°, 2° & 3° amines)

(xx) $ROH \xrightarrow[H^{\oplus}]{>C=C<} -\overset{|}{C}H-\underset{|}{C}OR$

Several mechanisms have been proposed for the reaction; the most plausible one is given below :

$$K_2Cr_2O_7 + H_2SO_4 \longrightarrow H_2Cr_2O_7 + K_2SO_4$$

$$H_2Cr_2O_7 \longrightarrow H_2CrO_4 + CrO_3$$

$$H_2CrO_4 \downarrow \; HCrO_4^- + H^{\oplus}$$

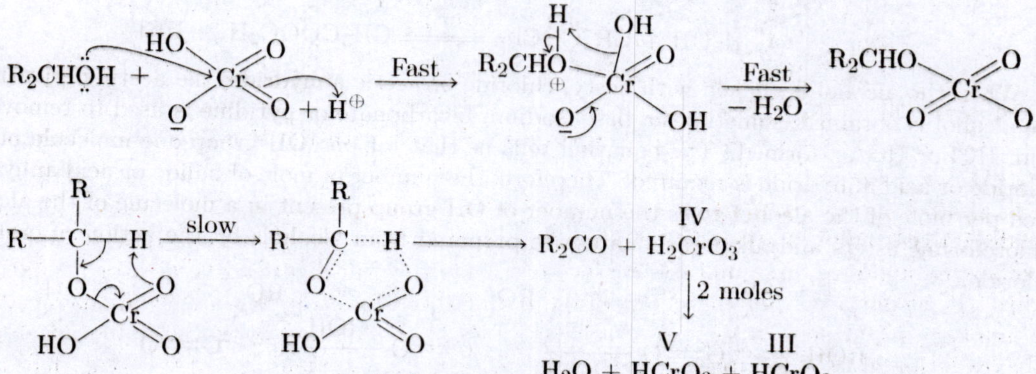

cyclic T.S.

MeCDOHMe reacts 7 times slower than MeCHOHMe; this primary kinetic isotope effect tells us that the $\alpha$-H is removed in the rate-determining step.

**(iii) Reaction with active metals :** Na, Ca, Al, etc. react with alcohol to form alkoxide and evolve hydrogen gas; water also reacts similarly with these metals. The reaction shows the acid character of alcohols.

$$2ROH + 2Na \longrightarrow 2\,R\bar{O}\overset{\oplus}{Na} + H_2;\; R\bar{O}\overset{\oplus}{Na} + H_2O \rightleftharpoons ROH + Na^{\oplus}\bar{O}H$$

Alcohols being extremely weak acids, the alkoxides hydrolyse in water to alcohols and alkalis. The reactions of an alcohol with calcium and aluminium are much slower than the reaction with sodium. Aluminium alkoxides have some synthetic use and are prepared by the action of aluminium amalgam on anhydrous alcohols. The acidity order of the three types of alcohols being 1° > 2° > 3°, a 1° alcohol reacts with sodium faster than the corresponding 2° alcohol and the latter, in its turn, reacts more readily than the corresponding 3° alcohol.

Sodium alkoxides are used in the Williamson synthesis and aluminium alkoxides are used in the Meerwein-Pondorf-Verley reduction and Oppenauer oxidation.

**(iv) Reaction with alkyllithium :** Alkyllithiums react with alcohols and form alkanes and lithium alkoxides.

$$ROH + R^1Li \longrightarrow R^1H + ROLi$$

This reaction also indicates the acidic character of alcohols.

**(v) Reaction with the Grignard reagent :** The Grignard reagents react with alcohols to form alkanes. When methylmagnesium iodide is used, methane is formed; from the volume of methane collected and the amount of alcohol used, the number of hydroxyl groups present in the alcohol molecule can be calculated (See Zerewitinoff's method of active hydrogen determination Sec. 16.4.5, p. 840).

**(vi) Acylation of the OH group :** The alcoholic OH group can be acylated by the action of acyl halides or acid anhydrides. Alcohols form esters on acylation and in fact, this is by far the best method for the preparation of esters in general. When acetyl chloride or acetic anhydride is used, the reaction is called acetylation.

$$ROH + R^1COCl \rightleftharpoons R^1COOR + HCl$$

$$ROH + (R^1CO)_2O \rightleftharpoons R^1COOR + R^1COOH$$

$$C_2H_5OH + CH_3COCl \rightleftharpoons CH_3COOC_2H_5 + HCl$$

When the alcohol is shaken with acetyl chloride or acetic anhydride, the acetyl derivative of the alcohol is obtained; sometimes a base, sodium bicarbonate or pyridine is used to remove the acid, HCl or HOAc, formed. The equation tells us that for one OH group one molecule of acyl chloride or acid anhydride is required. Therefore, the number of mole of halide or acid anhydride used per mole of the alcohol gives the number of OH group present in a molecule of the alcohol. Chloroformic esters and alkyl carbonates are prepared from alcohols by the action of carbonyl chloride.

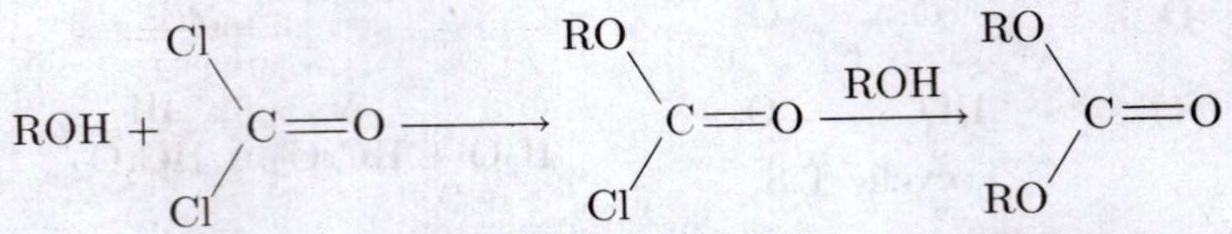

Chloroformic ester Alkyl carbonate

Carbobenzoxy chloride, an important reagent for protecting the amino function of an amino acid for the synthesis of peptides, is synthesised by this reaction.

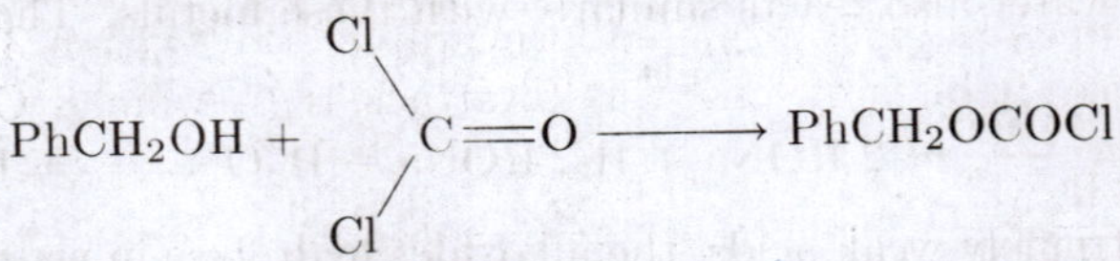

The order of reactivity of alcohols in this reaction is 1° > 2° > 3°. Acetic anhydride reacts with 3° alcohols very slowly; cyclic 3° alcoholic OH groups are resistant towards acetylation.

**(vii) Benzoylation :** When alcohols are shaken with benzoyl chloride, benzoate esters of the alcohols form; pyridine or dil. NaOH solution may be used to remove the acid formed.

$$ROH + PhCOCl \rightleftharpoons RO.COPh + HCl.$$

Both acylation and benzoylation of alcohols follow the tetrahedral $S_N$ mechanism (Sec. 12.6) as shown below :

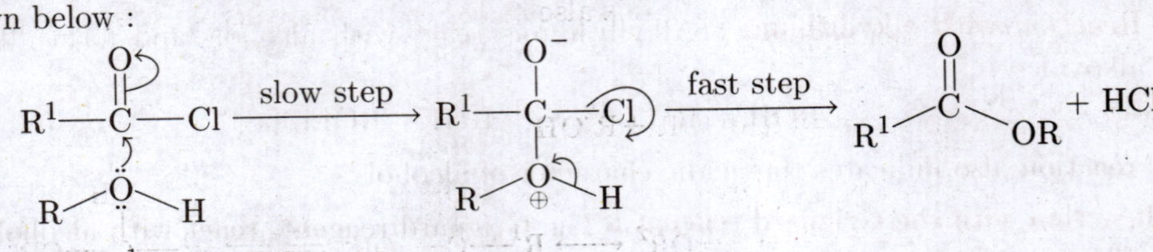

here $R^1 = R$ or Ph or Me

**(viii) Esterification :** When a carboxylic acid is boiled with an alcohol in the presence of concentrated sulphuric acid or dry HCl gas, an ester forms. It is a reversible reaction; the reverse reaction is called the hydrolysis of ester. Concentrated sulphuric acid catalyses the reaction as well as absorbs the water formed in the reaction.

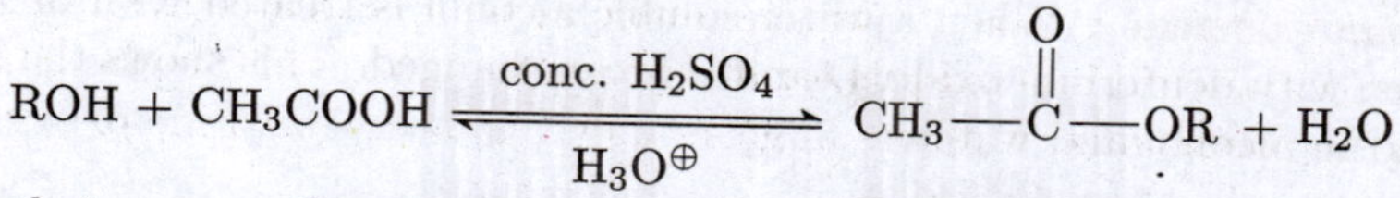

The reaction being reversible, it never goes to completion and as a consequence the reaction is not always economical. However, either rapid distillation of the ester (when its b.p. is low) or removal of the water by forming azeotropic mixture with benzene, toluene or xylene gives a better yield.

Steric effect plays an important role in the esterification reaction. Carboxylic acids with $\alpha$- or $\beta$-branching react more slowly than those without such branching; the reactivity order among alcohols is $1° > 2° > 3°$ which is their acidity order. When acetic acid is esterified with 1°, 2° and 3° alcohols, the yields are approximately 47%, 22% and 1.5 to 3% respectively.

The mechanism of the reaction is the reverse of the hydrolysis of ester reaction. Let us show the mechanism with an alcohol labelled with O-18 atom.

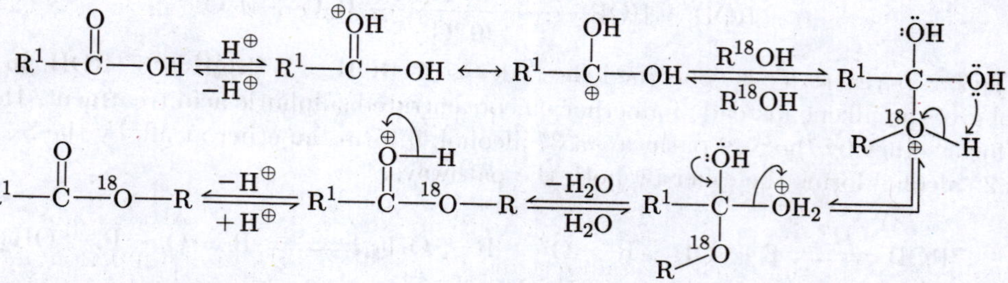

The labelling experiment shows that in an esterification reaction the alcohol supplies the H atom and the acid supplies the OH group to form water. However, a 3° or any other alcohol, which generates a stable carbenium ion, usually attack the carbonyl O of the acid as shown below :

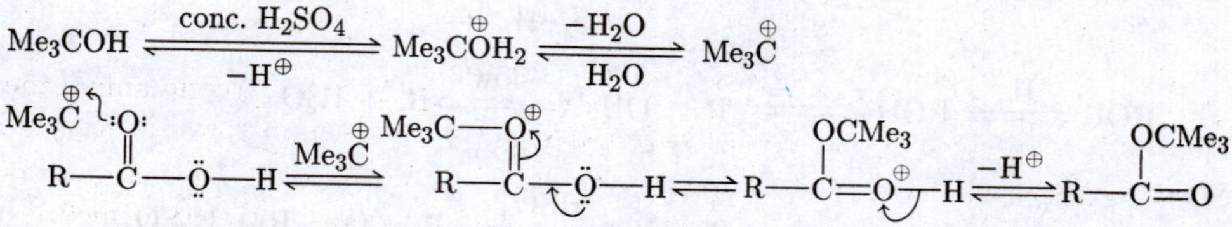

**(ix) Transesterification reaction :** When in an ester the alkoxy part is replaced by another alkoxy moiety, the reaction is called transesterification. When an ester is treated with an alcohol in the presence of a catalytic amount of mineral acid or NaOEt or t-BuOK, the transesterification takes place; this is a reversible reaction and it is also called the alcoholysis of ester. If the R group is 3°, the reaction frequently fails.

$$R^1COOR^2 + ROH \overset{H^\oplus}{\rightleftharpoons} R^1COOR + R^2OH$$

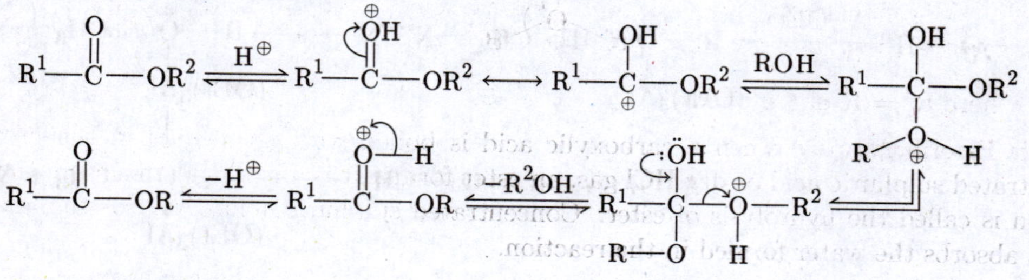

**(x) Deuterium exchange :** When a water-soluble alcohol is treated with or a water-insoluble alcohol is shaken with deuterium oxide, D and H are exchanged. This shows the identical polarity of O—H bond in alcohol and water.

$$ROH + D_2O \rightleftharpoons ROD + HOD$$

$$ROH + D{-}O{-}D \rightleftharpoons R{-}\overset{\oplus}{O}(D){-}H + \bar{O}D \rightleftharpoons ROD + HOD$$

**(xi) Etherification :** When an alcohol is heated with concentrated sulphuric acid at an elevated temperature, an ether is formed by the loss of one molecule of water from two molecules of the alcohol. In this case concentrated acid not only catalyse the reaction but also absorbs the water. Since an ether is formed from an alcohol by the loss of water molecule, the former is called the anhydride of the latter.

$$ROH + HOR \xrightarrow[140\,°C]{\text{conc. } H_2SO_4} R_2O + H_2O$$

In so far as reactivity is concerned the order is 1° ROH > 2° ROH > 3° ROH, In fact, a 3° alcohol gives an alkene instead of an ether on concentrated sulphuric acid treatment. However, 1° ROH forms ether by the $S_N2$ pathway; a 3° alcohol, if forms an ether at all, by the $S_N1$ pathway and a 2° alcohol forms the ether by both the pathways.

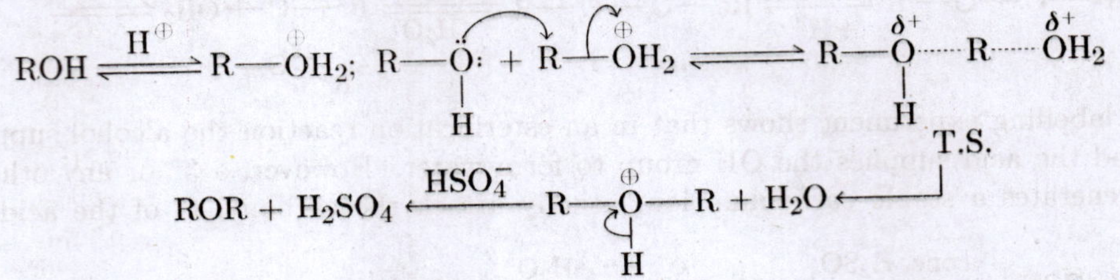

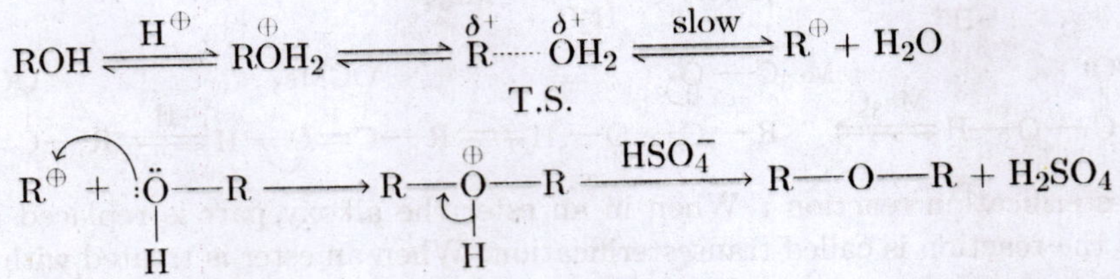

**(xii) Reaction with diazomethane :** In the presence of a Lewis acid diazomethane reacts with an alcohol to form methyl ether. H atom of the alcoholic OH being extremely weakly acidic, diazomethene cannot react directly; the Lewis acid coordinates with the O atom of the alcohol and thereby increases the acidity of the H atom.

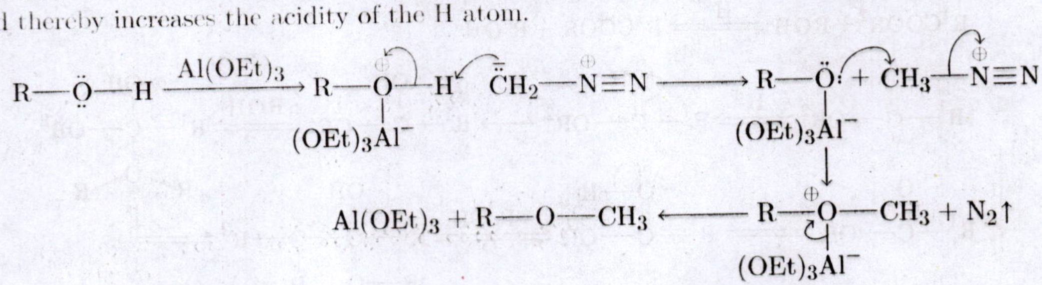

**(xiii) Reaction with epoxides :** The acid- or base-.catalysed reaction between an alcohol and an epoxide, which leads to the formation of $\beta$-hydroxy ether, may be called alcoholysis of epoxide. Important solvents like diethyleneglycol, cellosolve, etc. have been prepared by the process. Conversion of aziridines to $\beta$-aminoethers is a similar reaction. The mechanism is very similar to that of the hydrolysis of epoxide to 1, 2-diol by dilute sulphuric acid (See. Sec. 20.2.3f).

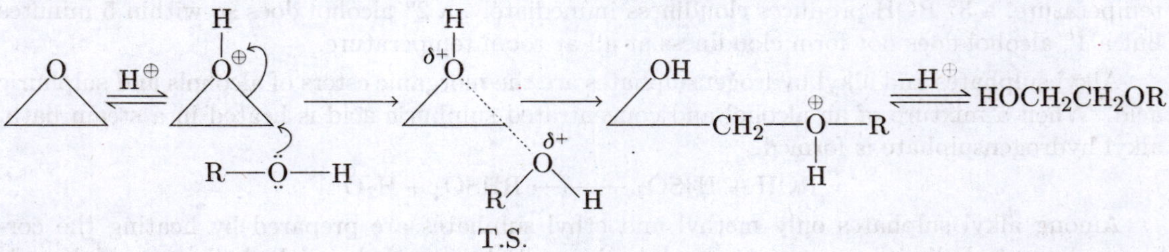

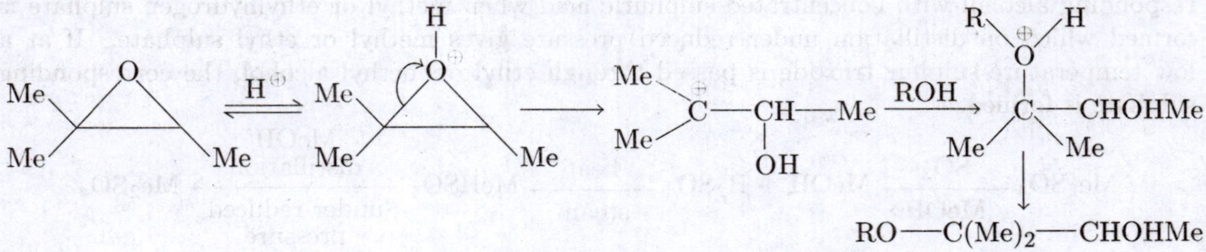

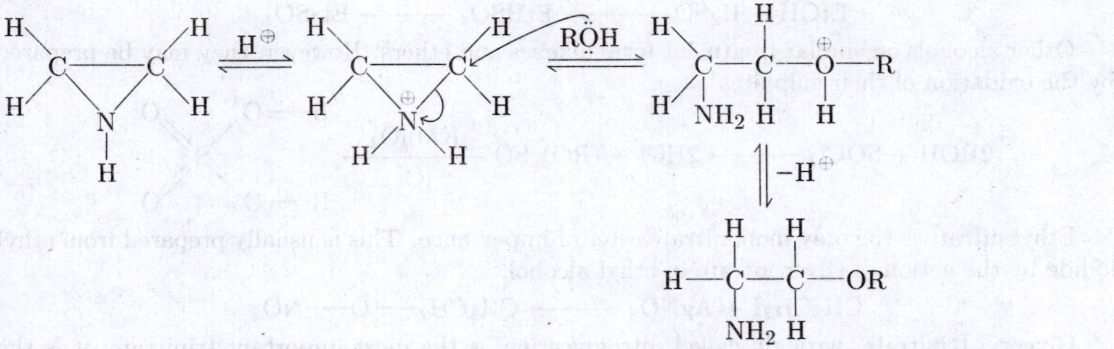

**(xiv) $\beta$-elimination reaction :** In the presence of concentrated acid or $Al_2O_3$ alcohols are dehydrated at a high temperature by a $\beta$-elimination reaction (El) and alkenes form (Sec. 15.3, p. 807) :

$$CH_3\text{—}CHOHCH_3 \underset{}{\overset{H^{\oplus}}{\rightleftharpoons}} CH_3\underset{\overset{\oplus}{O}H_2}{CH}CH_3 \overset{-H_2O}{\rightleftharpoons} CH_3\text{—}\overset{\oplus}{C}H\text{—}CH_2(H) \overset{-H^{\oplus}}{\rightleftharpoons} CH_3\text{—}CH{=}CH_2$$

**(xv) Formation of inorganic esters :** At a very low temperature dry HCl or HBr forms alkoxonium salts with alcohols.

$$CH_3CH_2OH \xrightarrow[-28\,°C]{HBr} CH_3CH_2OH.HBr \text{ (m.p. } -65\,°C) \text{ or } CH_3CH_2\overset{\oplus}{O}H_2Br^-$$

Alcohols form inorganic esters with dry halo acids and concentrated sulphuric acid.

$$ROH + HCl \longrightarrow RCl + H_2O;\ CH_3CH_2OH + H_2SO_4 \xrightarrow{100\,°C} CH_3CH_2HSO_4 + H_2O$$

In this case the inorganic acid supplies the proton and the alcohol supplies the OH group to form water. The removal of water increases the yield by stopping the hydrolysis of the ester. Since ROH supplies the OH group by breaking C—OH bond, the reactivity order is 3° ROH > 2° ROH > 1° ROH. The reaction constitutes the Lucas test which tells us whether an alcohol is 1° or 2° or 3°. The given alcohol is mixed with anhydrous $ZnCl_2$ in concentrated HCl at room temperature; a 3° ROH produces cloudiness immediately, a 2° alcohol does so within 5 minutes but a 1° alcohol does not form cloudiness at all at room temperature.

Alkyl sulphates and alkyl hydrogensulphates are the inorganic esters of alcohols and sulphuric acid. When a mixture of an alcohol and concentrated sulphuric acid is heated in a steam bath, alkyl hydrogensulphate is formed.

$$ROH + H_2SO_4 \longrightarrow RHSO_4 + H_2O$$

Among alkyl sulphates only methyl and ethyl sulphates are prepared by heating the corresponding alcohol with concentrated sulphuric acid when methyl or ethylhydrogen sulphate is formed which on distillation under reduced pressure gives methyl or ethyl sulphate. If at a low temperature sulphur trioxide is passed through ethyl or methyl alcohol, the corresponding sulphate is formed.

$$Me_2SO_4 \xleftarrow[\text{MeOH}]{SO_3} MeOH + H_2SO_4 \xrightarrow[\text{steam}]{\text{heat}} MeHSO_4 \xrightarrow[\text{under reduced pressure}]{\text{MeOH distillation}} Me_2SO_4$$

$$EtOH + H_2SO_4 \longrightarrow EtHSO_4 \longrightarrow Et_2SO_4$$

Other alcohols on similar treatment form alkenes and ethers. However, they may be prepared by the oxidation of their sulphites.

$$2ROH + SOCl_2 \longrightarrow 2HCl + (RO)_2SO \xrightarrow[\text{[O]}]{KMnO_4}$$ 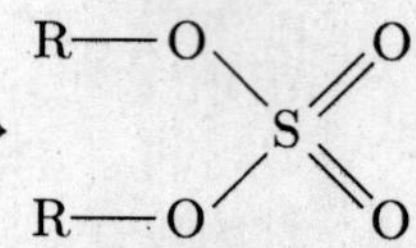

Ethyl nitrate is the only mononitrate ester of importance. This is usually prepared from ethyl iodide by the action of silver nitrate in ethyl alcohol.

$$CH_3CH_2I + AgNO_3 \longrightarrow CH_3CH_2—O—NO_2$$

Glyceryl trinitrate, wrongly called nitroglycerine, is the most important trinitrate; it is the compound that explodes in dynamite (glycerol trinitrate in kieselguhr or in wood pulp and solid ammonium nitrate). This is prepared by adding glycerol in a thin stream to a cold mixture of concentrated nitric acid and concentrated sulphuric acid.

$$CH_2OHCH\ OHCH_2OH \xrightarrow[\text{conc. } HNO_3 \text{ cold}]{\text{conc. } H_2SO_4}$$ 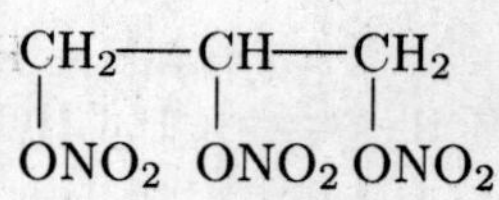

Among alkyl nitrites, ethyl and iso-amyl nitrites deserve mention. These are prepared by adding concentrated hydrochloric or sulphuric acid to an aqueous solution of sodium nitrite and the corresponding alcohol in cold.

$$NaNO_2 + H_2SO_4 \longrightarrow HNO_2 + Na_2SO_4; \quad H_2SO_4 \rightleftharpoons H^{\oplus} + HSO_4^-$$

$$H—\ddot{O}—N=O \underset{}{\overset{H^{\oplus}}{\rightleftharpoons}} H—\overset{\oplus}{O}(H)—N=O \rightleftharpoons \overset{\oplus}{N}O + H_2O$$

$$R-\ddot{O}-H + \overset{\oplus}{N}O \rightleftharpoons R-\overset{\oplus}{\underset{H}{O}}-N{=}O \rightleftharpoons R-O-N{=}O + H^{\oplus}$$

Alkyl sulphates are used as alkylating agents for compounds containing active H atoms, such as phenols, alcohols, thiols, amines, carboxylic acids, etc. The reaction is carried out in alkaline medium.

Ethyl nitrite and isoamyl nitrite are used to prepare nitrous acid in anhydrous media.

$$Et-O-N{=}O \xrightarrow[HCl]{dry} EtCl + HNO_2$$

**(xvi) Conversion to chloro group by phosphorus pentachloride :** For detail see Sec. 20.1.2.c.

$$EtOH + PCl_5 \longrightarrow EtCl + POCl_3 + HCl$$

**(xvii) Conversion to halo group by phosphorus trihalide :** See Sec. 20.1.2.c.

$$3CH_3OH + PI_3 \longrightarrow 3CH_3I + H_3PO_3$$

**(xviii) Conversion to chloro group by thionyl chloride :** See Sec. 12.5 and Sec. 20.1.2.d.

**(xix) Conversion to an amino group by ammonolysis :** When a mixture of an alcohol and ammonia in equimolecular amounts is heated under pressure in the presence of a catalyst like alumina or copper chromite, a primary amine is formed.

$$ROH + NH_3 \xrightarrow[\text{Cu-chromite under pressure}]{Al_2O_3 \text{ or}} RNH_2 + H_2O$$

When a large excess of alcohol is used, a mixture of 1°, 2° and 3° amines is obtained.

$$ROH + NH_3 \xrightarrow[cat.]{-H_2O} RNH_2 \xrightarrow[\substack{cat.\\ -H_2O}]{-H_2O} R_2NH \xrightarrow[\substack{cat.\\ -H_2O}]{-H_2O} R_3N$$

**(xx) Addition of an alcohol to an alkene :** See Sec. 13.5, p. 630 (synthesis of ether).

**(xxi) Acetal formation :** 16.4.4, p. 835.

$$CH_3CHO + CH_3OH \xrightarrow{dry\ HCl} CH_3CH(OCH_3)OH \xrightarrow[dry\ HCl]{CH_3OH} CH_3CH(OCH_3)_2$$

**(xxii) Reaction with isocyanates :** Alcohols added to cyanic acid and alkyl isocyanates to form a class of compounds is called urethans or carbamates. These are often used as derivatives of alcohols.

$$CH_3OH + H-N{=}C{=}O \longrightarrow H-NH-\underset{OCH_3}{\underset{|}{C}}{=}O$$

$$ROH + R^1-N{=}C{=}O \longrightarrow R^1-NH-\underset{OR}{\underset{|}{C}}{=}O$$

The mechanism is almost unknown but this may be represented as shown below :

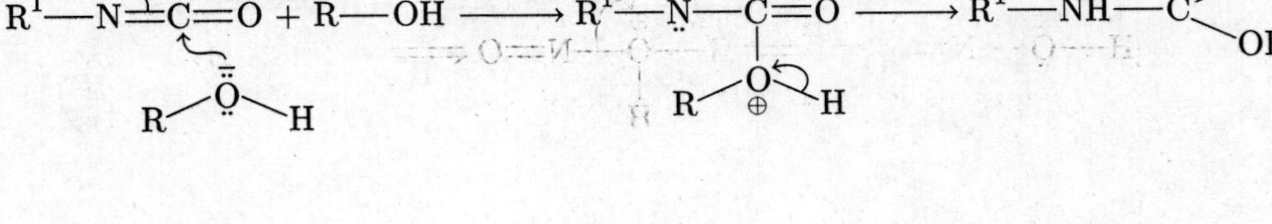

**(xxiii) Reaction with cyanides :** Alcohols react with cyanides, aliphatic, aromatic and heterocyclic, in the presence of dry HCl to give imino ester salts. If the alcohol is taken in large excess, ortho esters are formed. On the other hand, when aqueous HCl is used, carboxylic esters form.

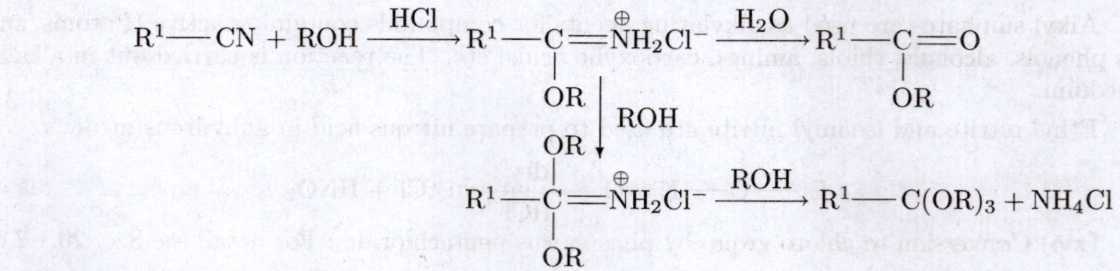

The mechanistic steps are :

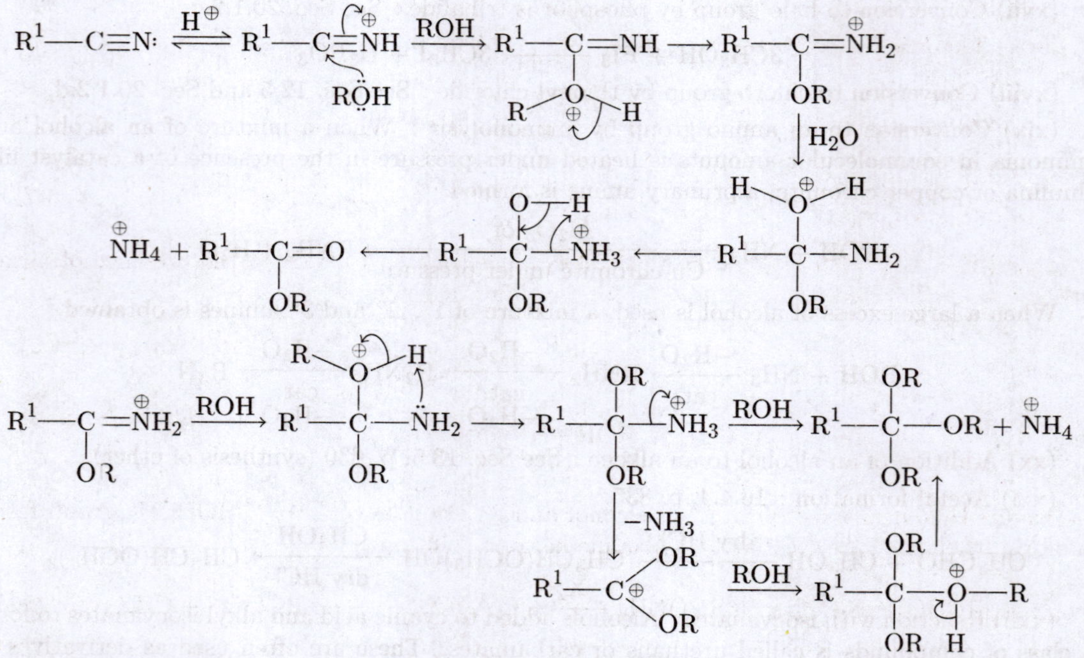

Alcohols may also be added to cyanides in the following way to form an N-substituted amide.

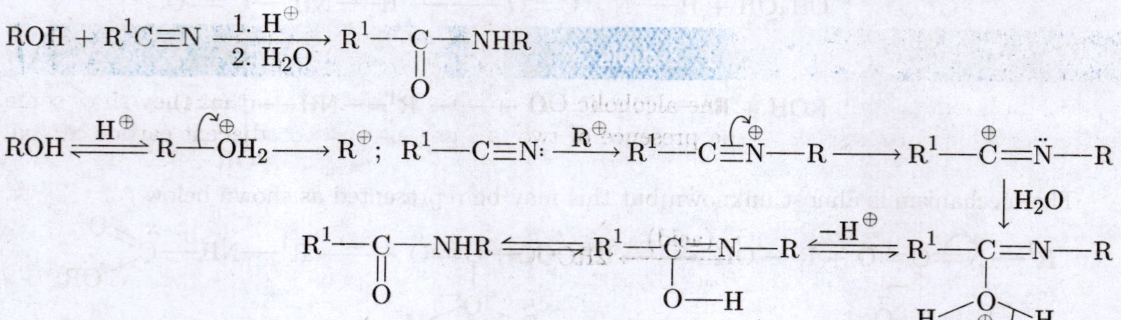

**(xxiv) Reaction with carbon disulphide :** In the presence of a base, alcohols add to carbon disulphide and form xanthate salts.

$$ROH + CS_2 \xrightarrow{\bar{O}H} R-O-\underset{\underset{S}{\|}}{C}-\bar{S}$$

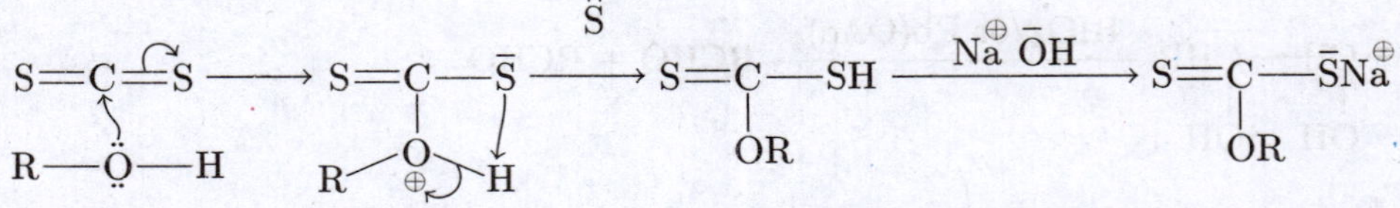

**(xxv) Carbonylation-esterification :** When a palladium chloride complex of an olefin is treated with carbon monoxide in alcohol, an ester is formed

$$CH_2=CH-CH_3 \xrightarrow[CO\,+\,ROH]{Pd/HCl} Me_2CH-\underset{\underset{OR}{|}}{C}=O$$

**(xxvi) The Friedel-Crafts reaction :** See Sec. 17.9.1, p. 975.

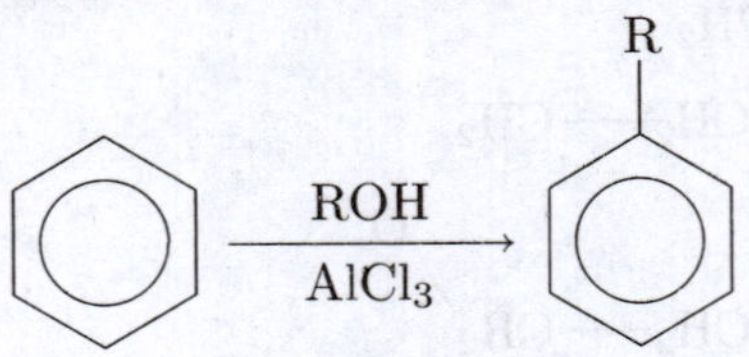

**(xxvii) Reaction with aryl iodide :** Alcohols react with aryl iodide in the presence of nickel carbonyl at an elevated temperature and form carboxylic esters.

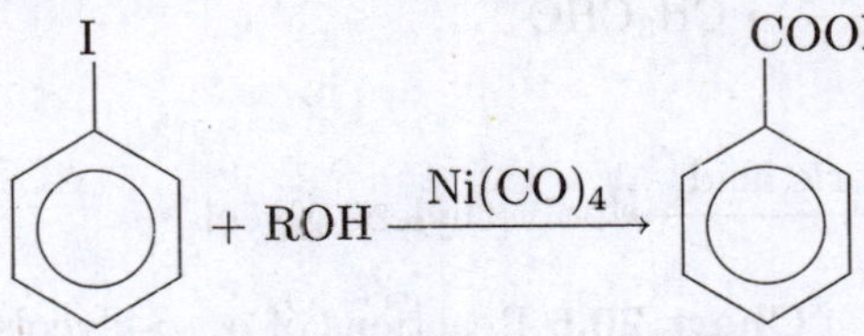

**(xxviii) Haloform reaction :** Ethyl alcohol and 2° alcohols with $-CHOHCH_3$ group form haloform when they are treated with alkaline halogen solution :

$$RCHOHCH_3 \xrightarrow[{[O]}]{X_2} RCOCH_3 \xrightarrow{X_2} RCOCX_3 \xrightarrow{\bar{O}H} RCOO^- + CHX_3$$

**(xxix) Conversion of $-CHOH-$ group to $>C=O$ group :** See the Oppenauer oxidation, Sec. 16.6.2, p. 859.

## 20.2.6 Chemical reactions of 1, 2-diols

1, 2-diols exhibit properties of the alcoholic OH groups. In addition to that they show some properties which arise owing to the presence of two OH groups on two adjacent carbon atoms. These are listed in Chart 20.6.

$$\text{(a)}\quad R-\underset{\underset{OH}{|}}{\overset{\overset{H}{|}}{C}}-\underset{\underset{OH}{|}}{\overset{\overset{H}{|}}{C}}-R \xrightarrow{KMnO_4\ (acid)} 2RCOOH$$

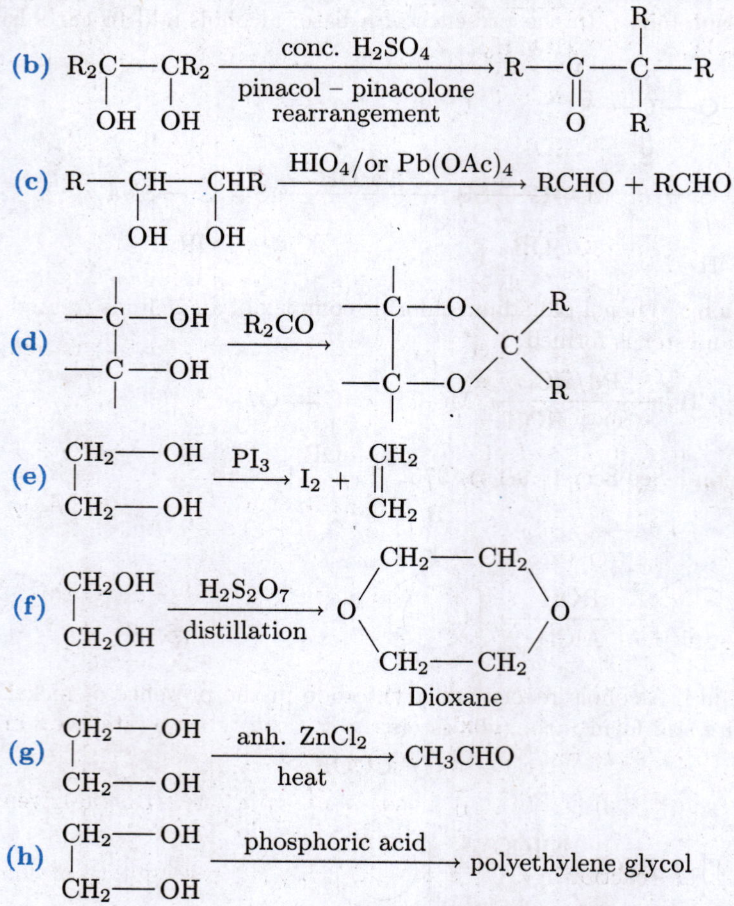

**Chart 20.6** Reactions of $\alpha$, $\beta$-glycols.

(a) **Oxidation by potassium permanganate :** In the presence of concentrated sulphuric acid potassium permanganate or dichromate oxidises a 1, 2-diol to a mixture of acids but the reaction has no synthetic value.

(b) **Pinacol-pinacolone rearrangement :** See Sec. 14.2, p. 701.

(c) **Oxidation by periodic acid or lead tetraacetate :** Periodic acid or lead tetraacetate breaks C(OH) —— C(OH) bond oxidatively and forms 2 molecules of carbonyl compounds. In this reaction, a 1° carbinol C gives formaldehyde, a 2° carbinol C forms aldehyde other than formaldehyde and a 3° carbinol C forms a ketone. If there are three OH groups on three adjacent C atoms, the middle C atom gets converted to formic acid. However, the reaction is widely used in oxidative degradations. The size of the ring of a sugar can be determined through the oxidative degradation reaction involving periodic acid.

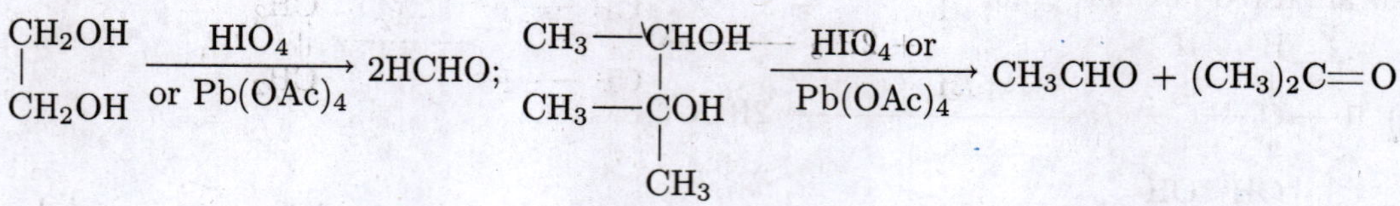

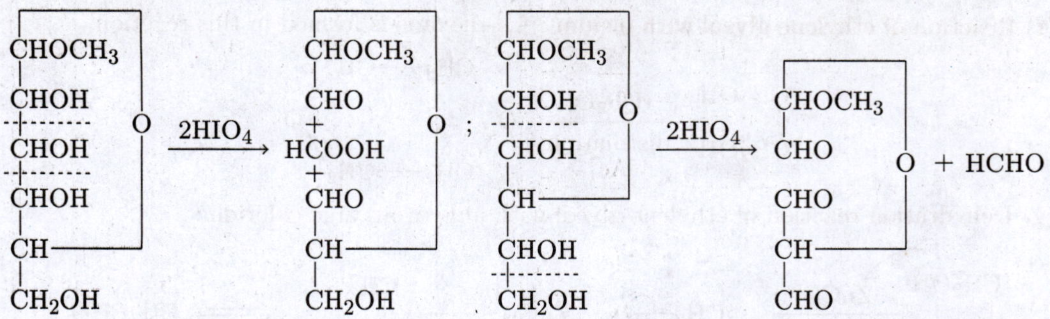

In this reaction $n$ molecules of $HIO_4$ are required per $(n+1)$ OH groups on $(n+1)$ adjacent carbon atoms and $(n-1)$ molecules of formic acid form.

The mechanism of lead tetraacetate oxidation is :

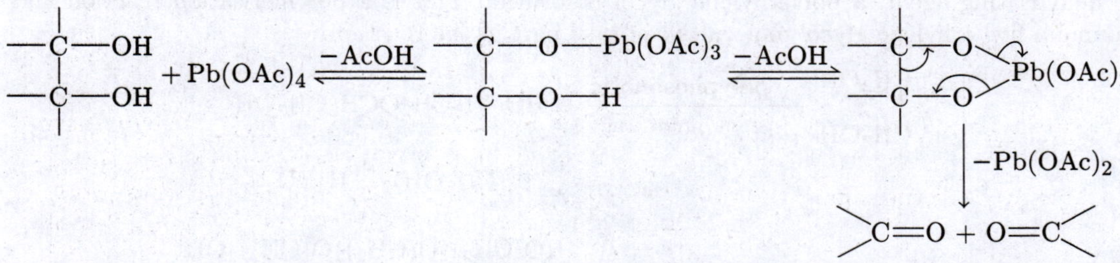

Since lead possesses an inert pair of es, it is a good oxidising agent; lead tetraacetate takes in two electrons and comes down to Pb(II) state from Pb(IV) state. However,

(i) The rate of the reaction$=k$ [substrate) $[Pb(OAc)_4]$ which is in conformity with the given mechanism.

(ii) When acetic acid is added, the reaction retards which supports the reversibility of the reaction.

(iii) cis-Diols react much faster than the trans-diols.

This is a strong evidence in favour of the cyclic intermediate.

The mechanism of perodic acid oxidation is similar to that of the lead tetraacetate :

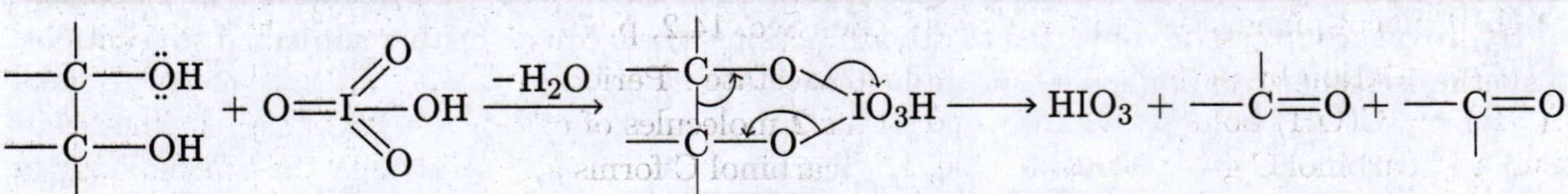

**(d) Reaction with a carbonyl compound :** See Sec. 16.4.4, p. 833.

**(e) Reaction with phosphorus triiodide :** Ethyleneglycol reacts with $PI_3$ and forms ethylene.

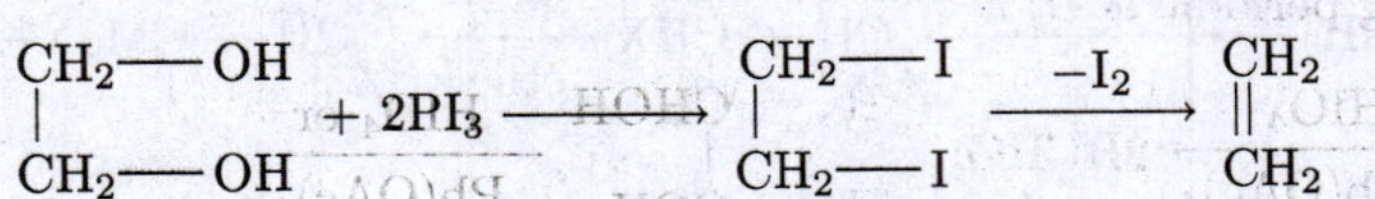

**(f) Reaction of ethylene glycol with oleium :** 1,4-dioxane is formed in this reaction.

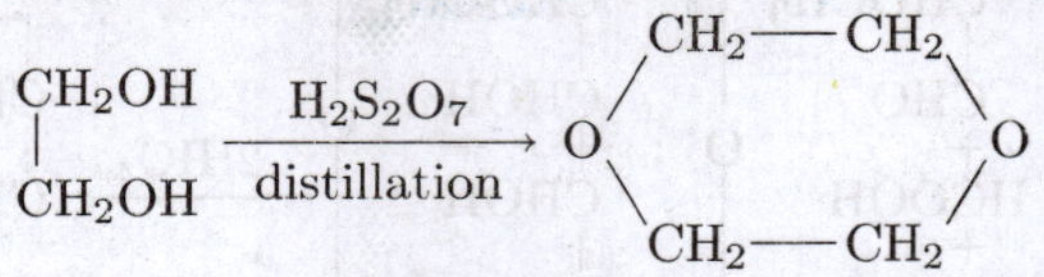

**(g) Dehydration reaction of ethylene glycol with anhydrous zinc chloride :**

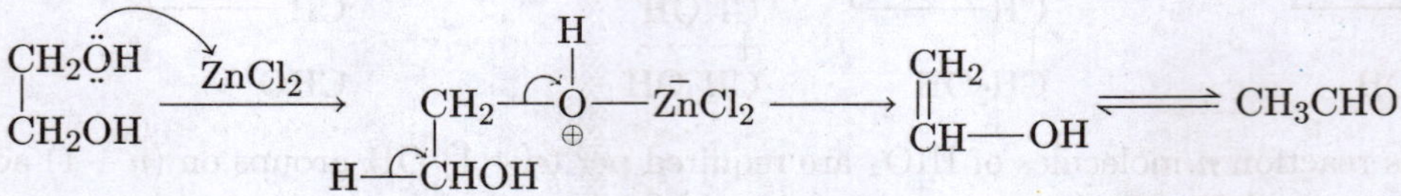

**(h) Dehydration reaction of ethylene glycol with phosphoric acid :** When phosphoric acid is the dehydrating agent, a polyethylene glycol is formed. This is a polymerisation reaction and maximum five ethylene glycol molecules can take part in the reaction.

$$2\,(CH_2OH)_2 \xrightarrow[\text{acid}]{\text{polyphosphoric}} HOCH_2CH_2OCH_2CH_2OH$$

$$HOCH_2CH_2OCH_2CH_2OH \xrightarrow{(CH_2OH)_2\ \ H_3PO_4} HO(CH_2)_2O(CH_2)_2O(CH_2)_2OH$$

## 20.2.7 Test

The following reactions may be used as quantitative tests for alcohols :

(i) Reaction with sodium, (ii) Lucas test, (iii) reaction with acetyl chloride, (iv) reaction of dichromate and conc. sulphuric acid (the solution turns blue-green opaque within a few seconds, 3° ROH does not give the test). (v) reaction with $CH_3MgI$—methane evolves, (vi) Victormeyer's test (See problem 20.7, p. 1194).

## 20.2.8 Alcoholic hydroxyl group and synthesis of organic compounds

Hydroxyl group can be converted to other functions which can be further subjected to reactions to synthesise other compounds as shown below :

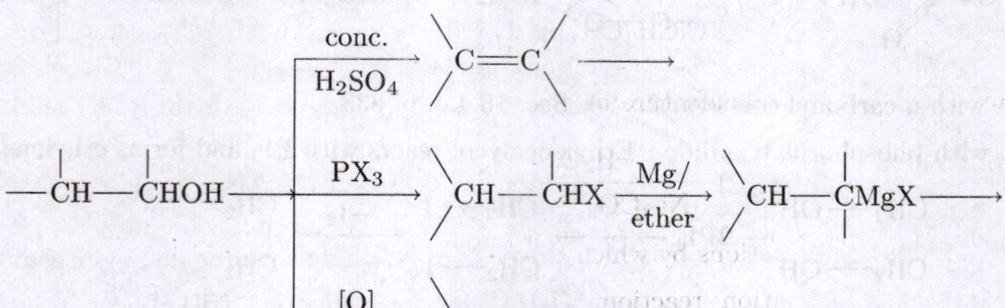

### 20.2.9 Uses of alcohols

(i) **Methyl alcohol** is mainly used for the manufacture of formaldehyde and other chemicals. It is also used as a jet fuel radiator, an antifreeze and as a solvent. It is manufactured from carbon monoxide and hydragen.

$$CO + 2H_2 \xrightarrow[\text{catalyst}]{ZnO\text{-}Cr_2O_3} CH_3OH$$

(ii) **Ethyl alcohol** is mainly used as a beverage and as a solvent. It is manufactured from ethylene by the process of hydration and from starchy materials and molasses by the process of fermentation.

(iii) iso-Propyl alcohol is chiefly used for the manufacture of acetone and hydrogen peroxide. Isopropyl alcohol is manufactured from propylene by the process of hydration.

$$CH_3CHOHCH_3 + O_2 \xrightarrow[\substack{30\text{–}80\text{ sec}\\ \text{contact time}}]{420\,°C\text{-}460\,°C} CH_3COCH_3 + H_2O_2$$

(iv) ***n*-Butyl alcohol** is chiefly used for the manufacture of *n*-butyraldehyde and *n*-butyric acid. It is also used as a solvent for the manufacture of esters.

(v) **Benzyl alcohol** is used to prepare perfumes of jasmine of gardenia type (benzyl acetate).

(vi) **Phenethyl alcohol** is a component of rose oil. It is synthesised from benzene and ethylene oxide by the Friedel-Crafts reaction.

(vii) **Cinnamyl alcohol** is also used in perfumery which has hyacinth-like odour. It is made from cinnamaldehyde by the Meerwein-Ponndorf-Verley reduction.

$$PhCH{=}CH.CHO \xrightarrow{Al(O\text{-}i\text{-}Pr)_3} PhCH{=}CHCH_2OH$$

(viii) **Ethylene glycol** is used as a non-volatile antifreeze for radiators, and as a cooling agent for airplane motors. It is also used as a solvent and for the manufacture of polyester fibre and glycol nitrate.

(ix) **Glycerol** is used to prepare glyceryl trinitrate (so-called nitro-glycerine) used in dynamite. It is also used as medicine and in cosmetics. It is used for the manufacture of alkyd resins and for the softening of cellophane. This may be manufactured as follows :

(a) $CH_2{=}CHCH_2OH + H_2O_2 \xrightarrow[65\,°C]{WO_3} HOCH_2CHOHCH_2OH$

(b) $CH_2{=}CHCH_2OH \xrightarrow{Cl_2,H_2O} ClCH_2CHOHCH_2OH \xrightarrow[NaOH]{Na_2CO_3} HOCH_2CHOHCH_2OH$

(x) **Allyl alcohol** is used in the manufacture of glycerol. It is prepared by hydrolysing allyl chloride.

$$CH_2{=}CHCH_2Cl \xrightarrow[Na_2CO_3,\ NaOH]{H_2O} CH_2{=}CH.CH_2OH$$

**Problem : 20.7.** Name the reactions by which three types of alcohols can be distinguished.

**Solution :** (i) Dehydrogenation reaction, **(ii)** Oxidation reaction, **(iii)** Lucas test, **(iv)** Organic ester formation reaction. **(v)** reaction with metallic sodium and **(vi)** Victormeyer's test.

**Victormeyer's test :** This, test involves the following sequential reactions :

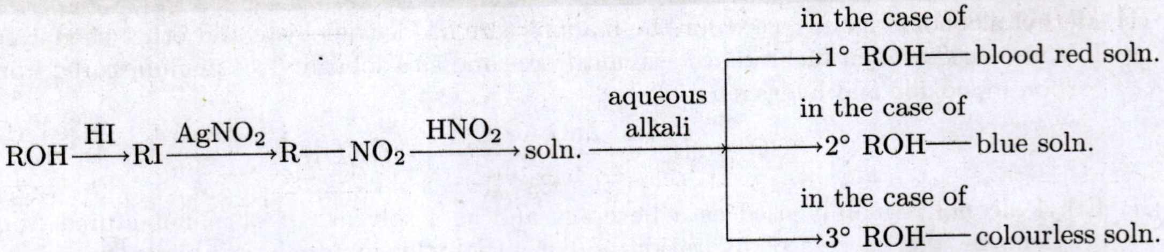


Since through these reactions 1°, 2° and 3° ROH gives red, blue and colourless solutions respectively, this test is also called red-blue-white test. However, the involved chemical reactions are shown below :

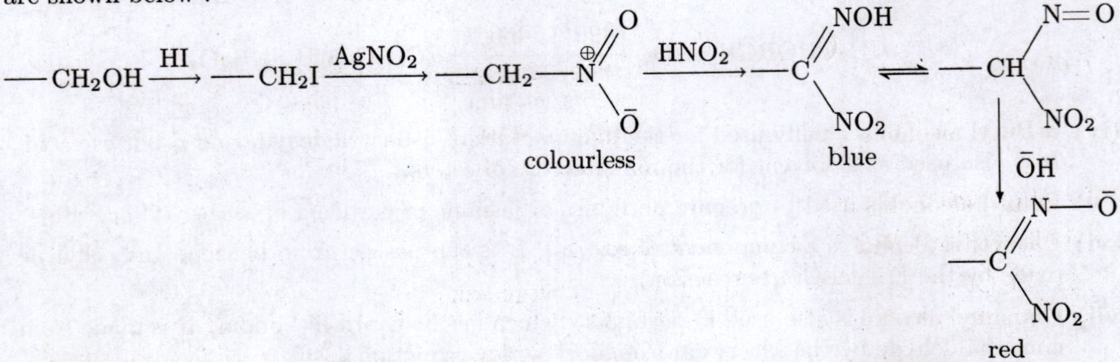


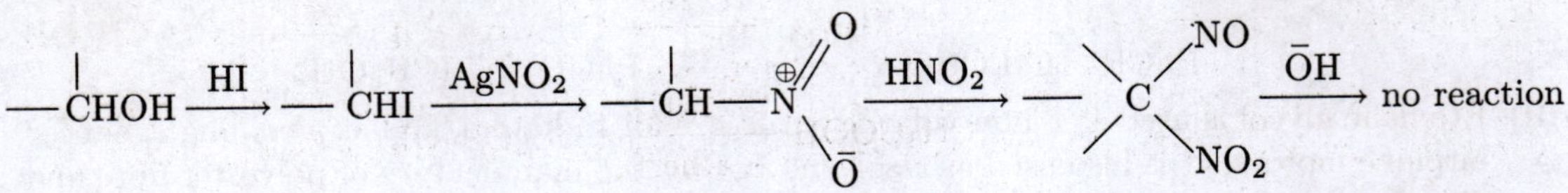


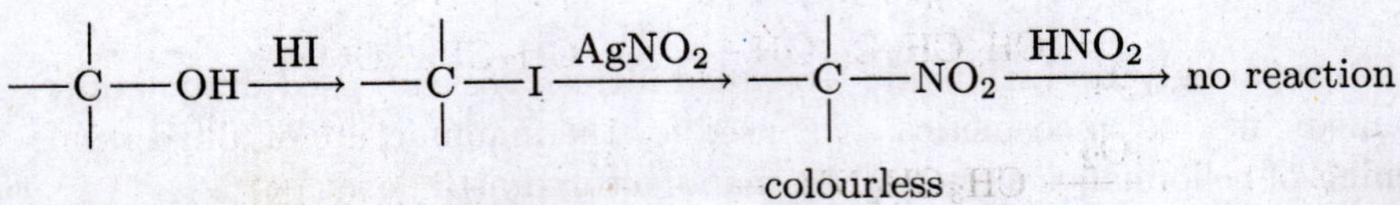


**Problem 20.8.** Convert :

**(i)** $CH_3CHOH\ CH_2CH_3$ to $CH_3CH_2CH_2CH_2OH$

**(ii)** $CH_3CH_2CH_2CH_2OH$ to $CH_3CH_2CHOHCH_3$

**(iii)** 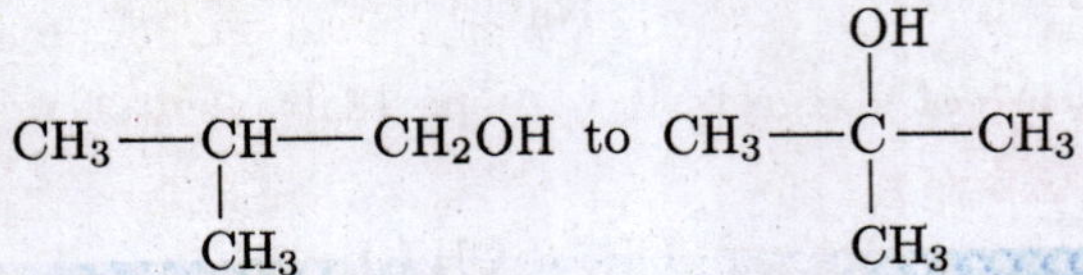


**(iv)** 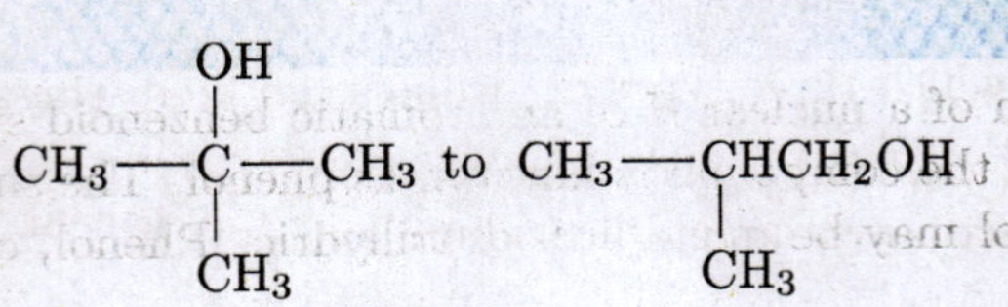

**Solution :**

(i) $CH_3CHOHCH_2CH_3 \xrightarrow[\text{heat}]{\text{conc. } H_2SO_4} CH_3CH{=}CHCH_3 \xrightarrow[\text{and heat}]{B_2H_6} (CH_3CH_2CH_2CH_2)_3B \xrightarrow{H_2O_2/\bar{O}H} CH_3CH_2CH_2CH_2OH$

(ii) $CH_3CH_2CH_2CH_2OH \xrightarrow[\text{heat}]{\text{conc. } H_2SO_4} CH_3CH_2CH{=}CH_2 \xrightarrow{H^{\oplus}/H_2O} CH_3CH_2CHOHCH_3$

(iii) $CH_3{-}CH(CH_3)CH_2OH \xrightarrow[\text{heat}]{\text{conc. } H_2SO_4} CH_3{-}C(CH_3){=}CH_2 \xrightarrow[\text{2. 'AgOH'}]{\text{1. HBr}} CH{-}C(OH)(CH_3){-}CH_3$

(iv) $CH_3{-}C(OH)(CH_3){-}CH_3 \xrightarrow[\text{heat}]{\text{conc. } H_2SO_4} CH_3{-}C(CH_3){=}CH_2 \xrightarrow[\text{2. heat}]{\text{1. } B_2H_6} (Me_2CHCH_2)_3B \xrightarrow{H_2O_2/\bar{O}H} Me_2CHCH_2OH$

**Problem 20.9.** Convert :

(i) $CH_3OH$ to $CH_3CH_2OH$ (step-up reaction)

(ii) $CH_3CH_2CH_2OH$ to $CH_3CH_2OH$ (step-down reaction)

**Solution :**

(i) $CH_3OH \xrightarrow{\text{red } P/I_2} CH_3I \xrightarrow[\text{ether}]{Mg} CH_3MgI \xrightarrow{HCHO} CH_3CH_2OMgI \xrightarrow{H_3O^{\oplus}} CH_3CH_2OH$

or $CH_3I \xrightarrow{KCN} CH_3CN \xrightarrow{H_3O^{\oplus}} CH_3COOH \xrightarrow[\text{ether}]{LiAlH_4} CH_3CH_2OH$

(ii) $CH_3CH_2CH_2OH \xrightarrow[\text{conc. } H_2SO_4]{K_2Cr_2O_7} CH_3CH_2COOH \xrightarrow{NH_3} CH_3CH_2COONH_4 \xrightarrow{\text{heat}} CH_3CH_2CONH_2$

$CH_3CH_2CONH_2 \xrightarrow[\bar{O}H]{NaOBr} CH_3CH_2NH_2 \xrightarrow{HNO_2} CH_3CH_2OH$

or $CH_3CH_2CH_2OH \xrightarrow[H_2SO_4]{\text{conc.}} CH_3CH{=}CH_2 \xrightarrow[\text{2. Zn/HOAc}]{\text{1. } O_3} CH_3CHO + CH_2O$

$CH_3CHO \xrightarrow[\text{ether}]{LiAlH_4} CH_3CH_2OH$

## 20.3 Phenols

### 20.3.1 Introduction

When a hydroxyl group takes the position of a nuclear H of an aromatic benzenoid system, it is said to be a phenolic hydroxyl group and the compound is knwown as phenol. The simplest member is PhOH, known as phenol. A phenol may be mono-, di- or trihydric. Phenol, cresols, xylenols, naphthols occur in coal tar.

## 20.3.2 Formation of phenols

The general methods are given in Chart 20.7.

(i) $ArSO_3H \xrightarrow[\text{solution}]{NaOH} ArSO_3^-Na^{\oplus} \xrightarrow[\text{fusion}]{NaOH(s)} Ar\bar{O}Na^{\oplus} \xrightarrow[H_2O]{SO_2} ArOH$

(ii) $ArNH_2 \xrightarrow[\substack{2.\ \text{aq. } NaNO_2 \\ 0\text{-}5\,°C}]{1.\ \text{dil. } H_2SO_4} Ar\overset{\oplus}{N}_2HSO_4^- \xrightarrow[50\text{-}60°C]{H_2O} ArOH$

(iii) $ArH \xrightarrow{TI(OOCCF_3)_3} ArTI(OOCCF_3)_2 \xrightarrow[2.\ Ph_3P]{1.\ Pb(OAc)_4} ArOOCCF_3 \xrightarrow[3.\ H_3O^{\oplus}]{1.\ H_2O\ 2.\ \bar{O}H} ArOH$

(iv) Nitroaryl halides $\xrightarrow{\bar{O}H}$ nitrophenols

(v) $ArMgBr \xrightarrow{O_2} ArO_2MgBr \xrightarrow{ArMgBr} 2\ ArOMgBr \xrightarrow{H_3O^{\oplus}} 2\ ArOH$

(vi) Hydroxybenzoic acids $\xrightarrow[\text{distillation}]{\text{sodalime}}$ phenoxides $\xrightarrow{H_3O^{\oplus}}$ phenols

(vii) $C_6H_6 \xrightarrow[\text{cat./heat}]{HCl,\ O_2} C_6H_5Cl \xrightarrow[360\,°C/320\ atm]{\bar{O}H} PhOH$

(viii) $C_6H_6 + C_3H_6 \xrightarrow{AlCl_3} Ph\text{—}CHMe_2 \xrightarrow{O_2} Ph\text{—}C(Me)_2\text{—}O\text{—}O\text{—}H \xrightarrow{H_3O^{\oplus}} PhOH + O{=}CMe_2$

(ix) $PhCH_3 \xrightarrow[Mn\ (II)\text{-}Cu(II)]{\text{air } (O_2)} PhOH + CO_2$

(x) $PhOCOCH_3 \xrightarrow[\text{heat}]{AlCl_3}$ phenolic ketones (xi) $PhOCH_2\text{—}CH{=}CH_2 \xrightarrow{\text{heat}}$ allylphenols

**Chart 20.7** Formation of phenols.

**(i) From benzenesulphonates :** When sodium benzenesulphonate (s) is fused with solid sodium hydroxide, sodium phenate and sodium sulphite are formed. The latter is used to convert fresh batch of benzenesulphonic acid to the sodio salt whereby sulphur dioxide evolves which on passing through sodium phenate gives phenol.

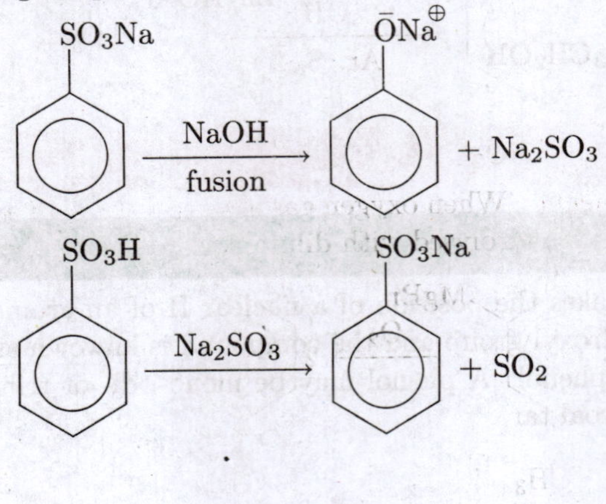

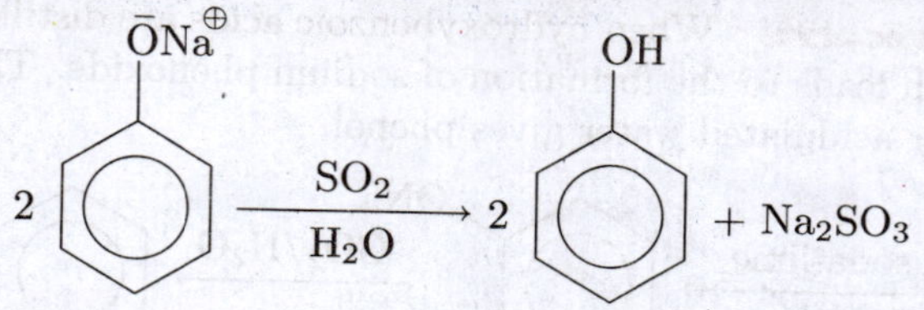

**(ii) From aromatic 1° amines :** This involves diazotisation of the amine with dil. $H_2SO_4$ and aqueous sodium nitrite solution in cold (0-5 °C) and then hydrolysis of the diazonium ion by warming the solution at 50-60 °C.

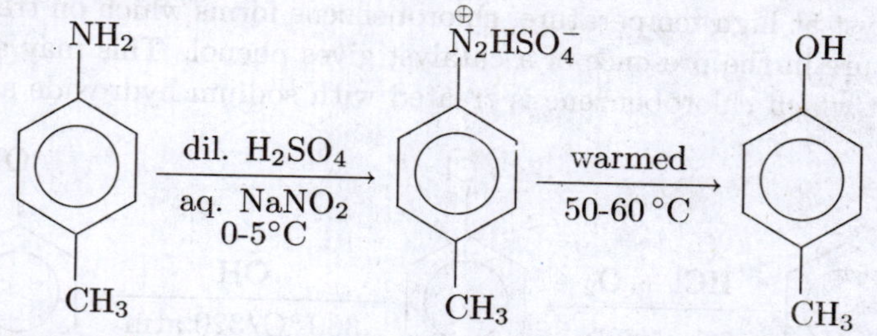

**(iii) Via thallation :** When aromatic hydrocarbons are treated with thallium salt of trifluoroacetic acid, ring-thallated intermediates form. These on controlled oxidation with lead tetra-acetate give aryl trifluoroacetates which on hydrolysis with dilute aqueous alkali and subsequent acidification yield phenols.

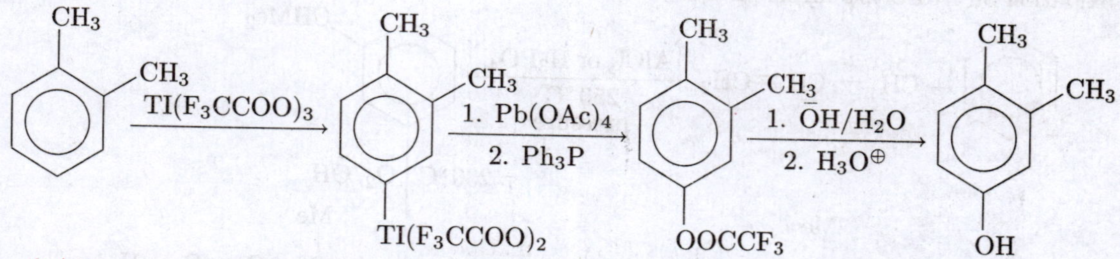

**(iv) By aromatic $S_N2$ reactions :** Nitroaryl halides on hydrolysis with alkali get converted into nitrophenols.

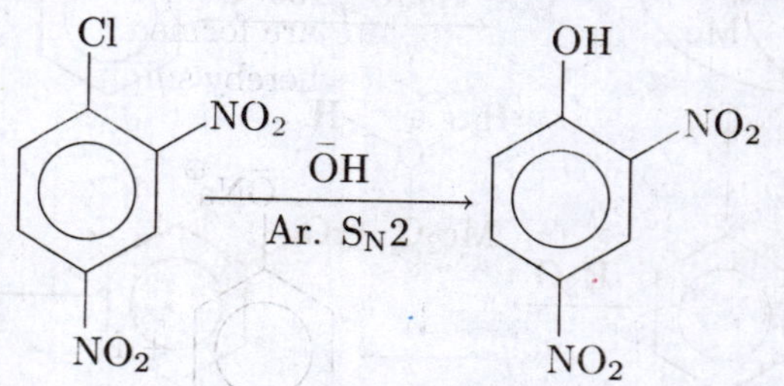

**(v) From the Grignard reagent :** When oxygen gas is passed through arylmagnesium bromides and the addition complexes are hydrolysed with dilute acid solution, phenols are obtained.

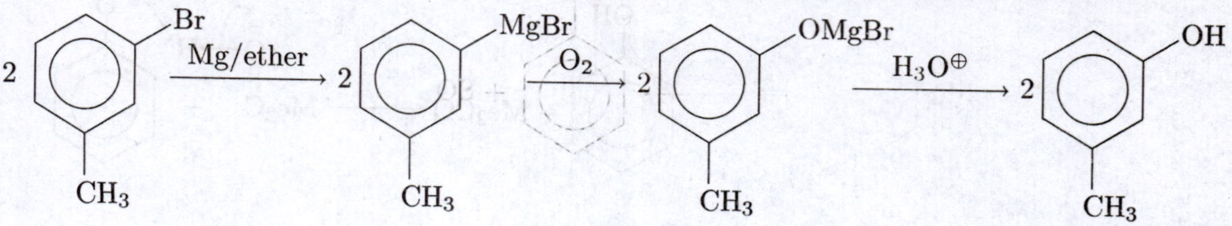

**(vi) From hydroxybenzoic acids :** When hydroxybenzoic acids are distilled with soda-lime, decarboxylation occurs which leads to the formation of sodium phenoxide. The latter on treatment with carbon dioxide gas or acidulated water gives phenol.

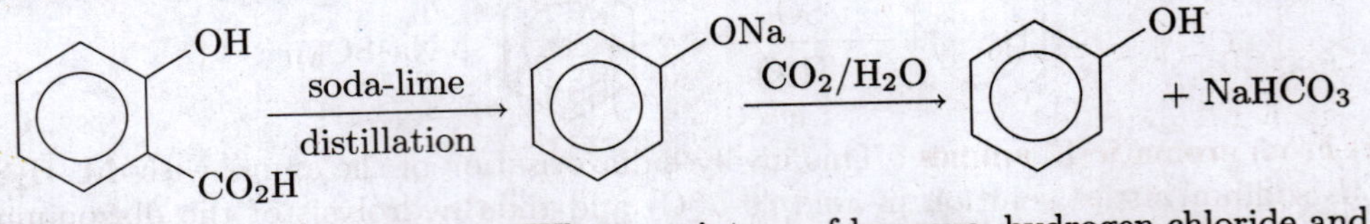


**(vii) By the Raschig-Dow process :** When a mixture of benzene, hydrogen chloride and air is passed over a catalyst at high temperature, chlorobenzene forms which on treatment with steam at a high temperature in the presence of a catalyst gives phenol. This may also be obtained by the Dow process in which chlorobenzene is treated with sodium hydroxide at 300 °C under 320 atm pressure.

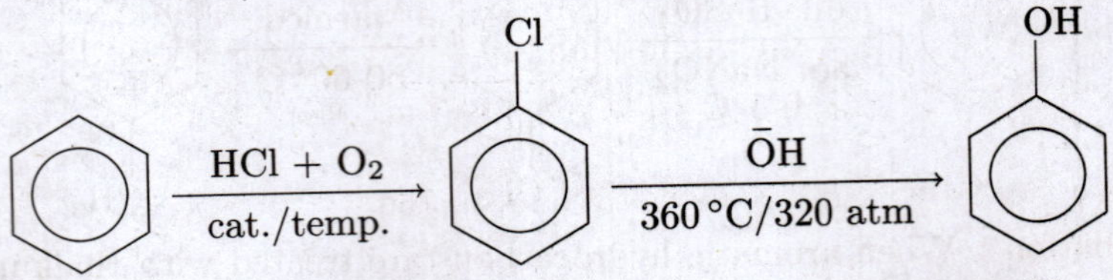


**(viii) From cumene :** Cumene is first converted to cumene hydroperoxide which on heating with conc. $H_2SO_4$ at 100 °C undergoes the Baeyer-Villiger (Dakin) oxidation and the product of oxidation on hydrolysis forms phenol.

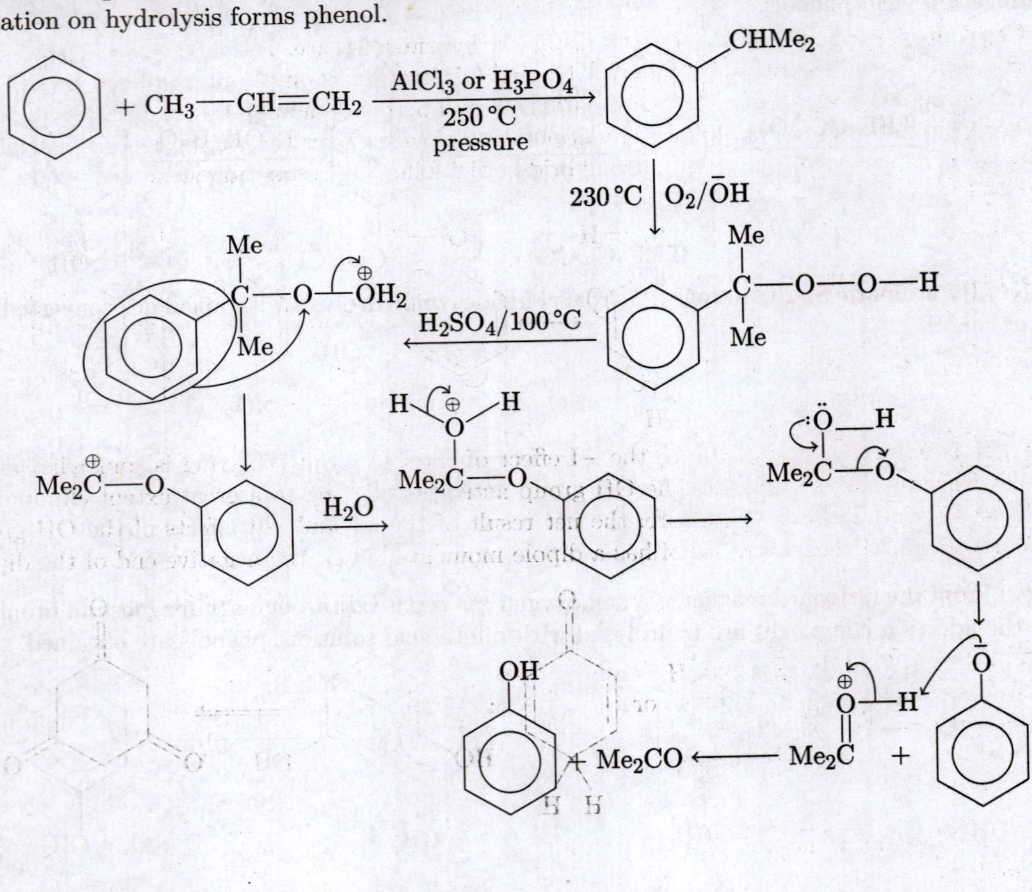

The rearrangement step is shown here as a concerted one; according to some authors this occurs in two steps :

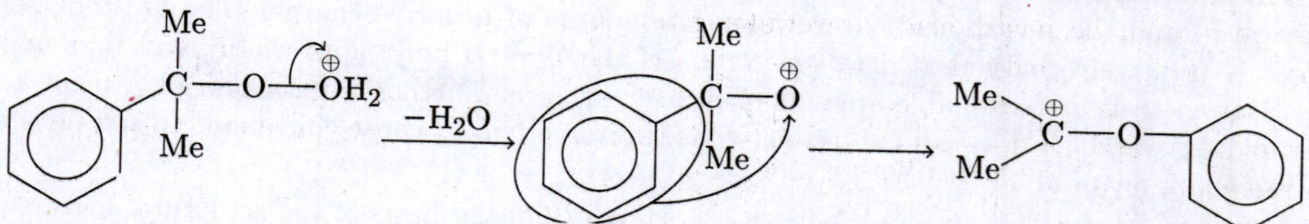

**(ix) From toluene :** When toluene is oxidised by air at an elevated temperature in the presence of Mn (II) - Cu (II) salts, phenol is obtained.

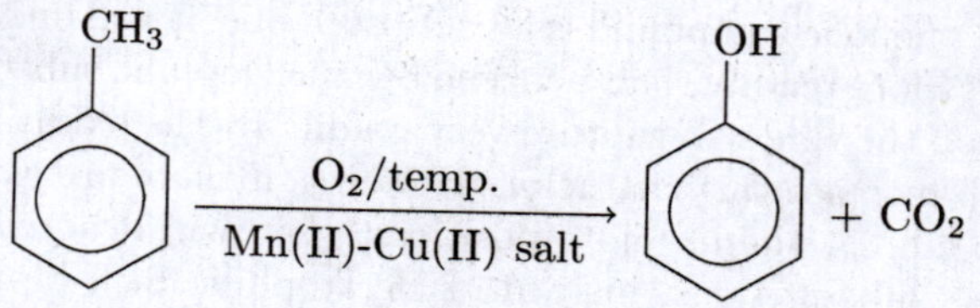

**(x) By the Fries rearrangement :** See Sec. 14.19, p. 769.

**(xi) By the Claisen rearrangement :** See Sec. 14.17.1, p. 753.

## 20.3.3 Structure, reactivity, orientation, and physical properties

The O atom of the phenolic OH group is in $sp^2$ hybridised state. It holds two lone pairs of *es*, one in a *p* A.O. and the other in an $sp^2$ A.O.; the OH group lies on the plane of the ring. The *p* A.O. is parallel to the *p* A.O.s of the aromatic ring and perpendicular to the ring. The *p es* of the OH group, therefore, undergo *p*-$\pi$ conjugation and delocalise over the ring. Thus, the structure of phenol is supposed to be a resonance hybrid of the following resonating structure :

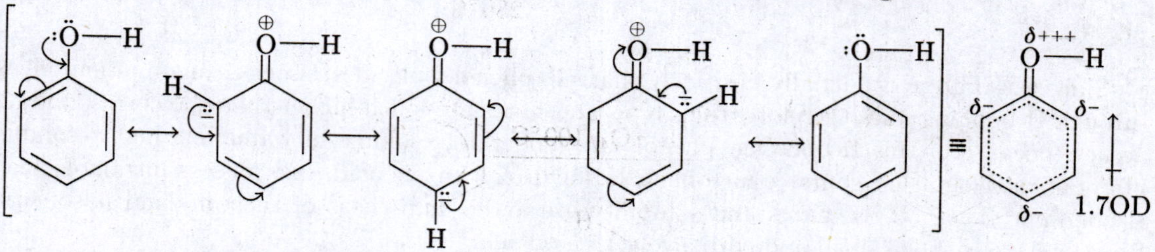

There is no doubt that owing to the −I effect of the OH group, the ring is somewhat deactivated; whereas the +R-effect of the OH group activates the ring to a great extent eliminating the deactivating −I effect. Therefore, the net result of the −I and +R effects of the OH group is the activation of the ring. Phenol has a dipole moment 1.70 D, the negative end of the dipole

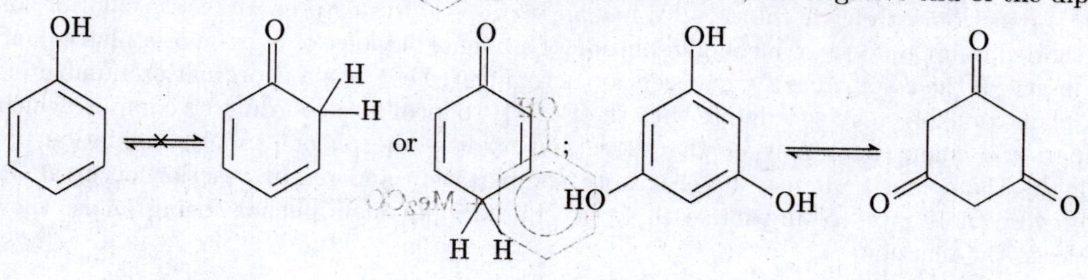

points away from the ring owing is the presence of $n$ es on O atom. Thus the O—C bond has some double bond character; the O atom also bears a very small amount of positive charge while the two *o*- and the *p*-carbons bear very small amounts of negative charge. The O atom being partially positively charged, it draws electrons of the O—H bond appreciably and thus makes the H atom to bear reasonable amount of positive charge. Furthermore, a phenol (except 1, 3, 5-trihydroxybenzene) does not exhibit keto-enol tautomerism; because conversion to ketone is not energetically favourable.

So; a phenol does not show properties of a ketone. On the basis of the structural features as stated above we shall discuss the reactivity, orientation and physical and chemical properties of phenol.

Since the ring of a phenolic compound is an activated one, it will undergo aromatic electrophilic substitution reactions more readily than benzene but nucleophilic substitution reactions less readily; for the same reason the ring will rupture very readily by the oxidising agents. The presence of activating groups will increase this character. However, if there are electron-withdrawing groups e.g., nitro, carboxyl, etc. at the *o*- and *p*-positions, they will deactivate the ring and as a consequence, nucleophilic substitutions, instead of electrophilic, will be favoured and the ring will be resistant towards oxidation. Since the *o*- and *p*-positions are partially negatively charged, the electrophilic substitution in phenol will occur at *o*- and *p*-positions.

Since the O and H atoms of the OH group bear partial positive charge, the H atom has a tendency to leave the species as a proton and, therefore, a phenol exhibits the properties of compounds containing active hydrogen. For the same reason and for the +R effect it does not show the properties of a base.

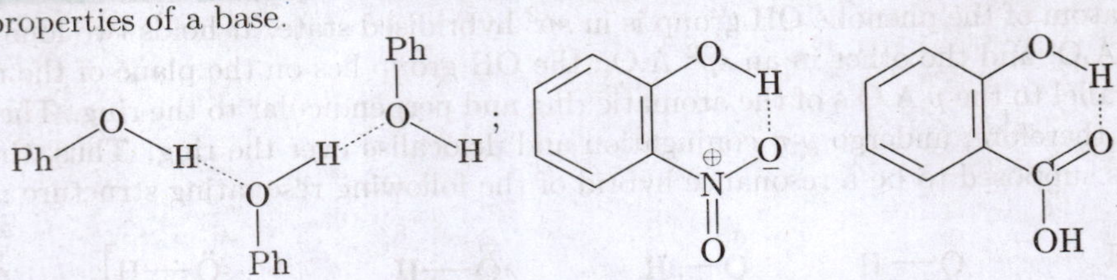

Since the H atom is partially positively charged, phenol molecules associate among themselves through H-bonding and, therefore, their b.p.s and m.p.s are much higher than the corresponding arenes; because of this H-bonding, phenols are moderately soluble in water but highly soluble in organic solvents. Of course *o*-nitrophenol, *o*-hydroxybenzoic acid, etc. possess intramolecular H-bonding. Their, b.p.s, m.p.s and solubility are lower than those of their *m*- and *p*-isomers. Phenols have carbolic smell and corrosive action on skin.

### 20.3.4 Chemical reactions

The reactions are given in Chart 20.8.

**(i) Reaction with ferric chloride :** When a few drops of freshly prepared ferric chloride solution are added to an aqueous or alcoholic solution of a phenol, a violet or a green or a blue colouration is observed; the colour may be transient or permanent. $Fe^{3+}$ has a coordination number six and six phenoxide ions may coordinate with the Fe (III) to form hexacoordinated complex which may impart the colour or it may be that ferric ion oxidises the phenol and the oxidation product exhibits the colour or it may be that both complexation and red-ox reaction occur at a time. However, enols give colouration with ferric chloride. In fact, phenols being enols, they also undergo the reaction.

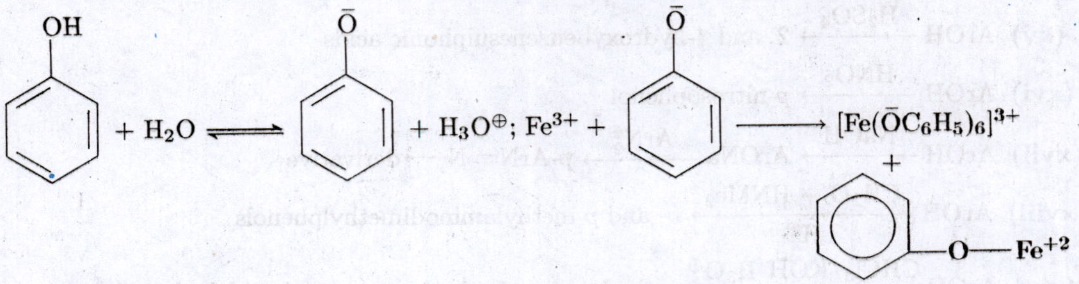

**(ii) Reaction with bases :** PhOH is weakly acidic, its $pKa$ value is 9.95. Any base whose conjugate acid is weaker than phenol will react with the latter to form salt. Thus NaOH, KOH etc. react with phenol to form sodium or potassium phenoxide and water; water is the conjugate acid of the base NaOH or KOH which is a weaker acid than phenol.

$$ArOH + NaOH \rightleftharpoons Ar\bar{O}\ Na^{\oplus} + H_2O$$

(i) $ArOH \xrightarrow{\text{aq. } FeCl_3}$ violet or green or etc. colouration

(ii) $ArOH \xrightarrow{\text{aq. } NaOH} ArONa + H_2O$

(iii) $ArOH \xrightarrow[Ac_2O]{CH_3COCl \text{ or}} ArOCOCH_3$

(iv) $ArOH \xrightarrow{RMgX} ArOMgX + RH$

(v) $ArOH \xrightarrow[(CH_3)_2SO_4]{CH_2N_2 \text{ or}} ArOCH_3$

(vi) $ArOH \xrightarrow{NaOH} \xrightarrow{RI} ArOR$

(vii) $ArOH \xrightarrow[PCl_5]{POCl_3 \text{ or}} (ArO)_3PO + HCl$

(viii) $ArOH \xrightarrow{P_2S_5} ArSH$

(ix) $ArOH \xrightarrow[AlCl_3]{RCl}$ *o*- and *p*-alkyl derivatives

(x) $ArOH \xrightarrow{RCOCl} ArOCOR \xrightarrow[AlCl_3]{\text{heat}}$ *o*- and *p*-acyl derivatives

(xi) Nitrophenol $\xrightarrow[ZnCl_2]{HCHO/HCl}$ chloromethyl derivatives

(xii) $ArOH \xrightarrow{Hg(OAc)_2}$ *o*-acetoxymercuriphenol

(xiii) $ArOH \xrightarrow{\text{dil. } HNO_3}$ *o*- and *p*-nitro derivatives

(xiv) $ArOH \xrightarrow[H_2O]{Br_2}$ 2, 4, 6-tribromo derivative

(xv) $ArOH \xrightarrow{H_2SO_4}$ 2, and 4-hydroxybenzenesulphonic acids

(xvi) $ArOH \xrightarrow{HNO_2}$ *p*-nitrosophenol

(xvii) $ArOH \xrightarrow{NaOH} ArONa \xrightarrow{ArN_2^{\oplus}}$ *p*-ArN=N— derivative

(xviii) $ArOH \xrightarrow[H^{\oplus}]{CH_2O\ +\ HNMe_2}$ *o*- and *p*-methylaminodimethylphenols

(xix) $ArOH \xrightarrow{CHCl_3/KOH} \xrightarrow{H_3O^{\oplus}}$ *o*-hydroxy and *p*-hydroxyaromatic aldehydes

(xx) $ArOH \xrightarrow{\bar{O}H/CO_2} \xrightarrow{H_3O^{\oplus}}$ *o*-hydroxy and *p*-hydroxyaromatic acids

(xxi) Polyhydroxyphenols $\xrightarrow[ZnCl_2]{RCN/HCl} \xrightarrow{H_3O^{\oplus}}$ acylphenols

(xxii) $ArOH \xrightarrow[ZnCl_2]{HCN/HCl} \xrightarrow{H_3O}$ hydroxyaldehyde

(xxiii) $ArOH \xrightarrow{Ni/H_2}$ derivative of cyclohexanol

(xxiv) $ArOH \xrightarrow[dist.]{Zn\ dust} Ar—H$

(xxv) $ArOH \xrightarrow{K_2S_2O_8} \xrightarrow{H_3O^{\oplus}}$ dihydric phenol

(xxvi) $ArOH \xrightarrow[ZnCl_2]{NH_3} ArNH_2$

(xxvii) $ArOH \xrightarrow[R\ may\ be\ H;\ R;\ Ar]{RCHO}$ condensation product

(xxviii) ArOH 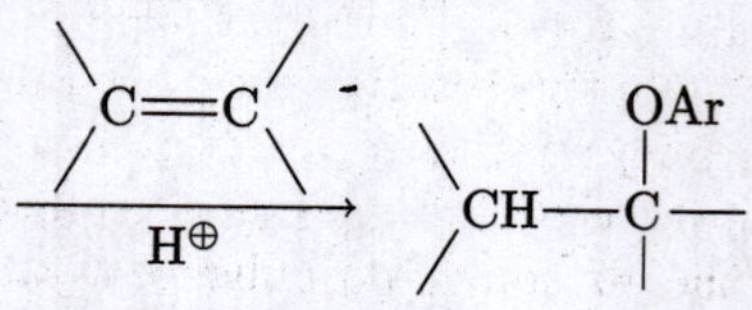

(xxix) $ArOH \xrightarrow{\beta\text{-keto ester}} \xrightarrow{AlCl_3}$ coumarins

(xxx) $ArOH \xrightarrow{phthalic\ anhydride}$ phenolphthalein

**Chart 20.8** Reactions of phenols.

Since carbonic acid ($pK_a = 6$) is a stronger acid than phenol, sodium bicarbonate does not form salt and evolve $CO_2$ with phenols (except di- and trinitrophenols). This fact has been utilised to differentiate carboxylic acids from phenols. Carboxylic acids ($pKa = 4.5$) being stronger than carbonic acid, the former compounds react with sodium bicarbonate to form sodio salt and evolve carbon dioxide. If a saturated aqueous solution of sodium bicarbonate evolves $CO_2$ on treatment with an organic sample, it is expected that the organic sample is either a carboxylic acid or a sulphonic acid. However, alcohols are weaker acids than phenols ; the conjugate bases of alcohols are $R\bar{O}$, the alkoxide ions; the negative charge on the O atom has no scope to delocalise and so they are strong bases and naturally alcohols are very weak acids. On the other hand, the negative

$$PhOH + NaHCO_3 \rightleftharpoons\!\!\!\!\!\times\ PhONa + H_2CO_3 \text{(carbonic acid)}$$

charge on the O atom of the phenoxide ion, the conjugate base of the phenol, delocalises through $p$-$\pi$ conjugation which brings stability to the ion and as a consequence phenol is acidic. However, the acidity of phenol increases as the presence of electron-withdrawing groups at $o$- and $p$-positions increases, while the acidity decreases as the presence of electron-releasing groups increases in the ring. Furthermore, if there is steric inhibition of resonance involving either the +R effect of the phenolic OH group or the electron-withdrawing −R group, the acidity of the phenol decreases. The $pKa$ values of the following phenols support the explanation as stated above.

| | $pKa$ | Reason |
|---|---|---|
| 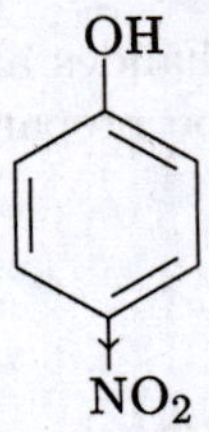 | 7.15 | −I and −R effects of the $NO_2$ group pull in the $p$ $e$s of the OH. |
| 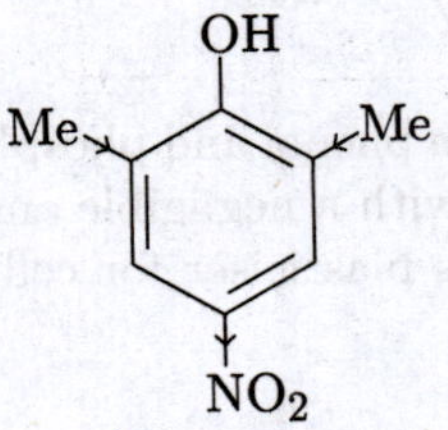 | 7.22 | −I and −R effects of the $NO_2$ group pull in the $p$ $e$s of the OH but +I effect of the two Me groups push in electrons. |
| 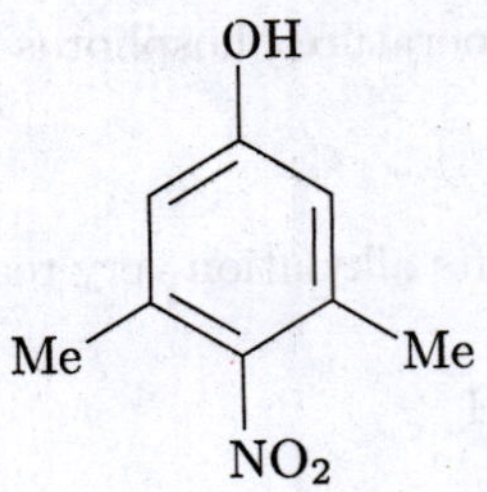 | 8.25 | Since the two Me groups and the $NO_2$ group are very close to each other, they cannot lie in the same plane owing to their bulk. The $NO_2$ group rotates round the C—N bond to avoid steric strain. Owing to this rotation the $p$ orbital of the N atom does no longer remain parallel to the $p$ A.O.s of the ring carbons. Thus, in the 3rd compound the $NO_2$ group cannot involve itself properly in the $p$-$\pi$ conjugation; hence it cannot withdraw the lone pair of $e$s of the hydroxyl group appreciably and as a consequence 3, 5-dimethyl-4-nitrophenol is less acidic than our |

expectation. The bulk of the OH group being smaller than that of the $NO_2$ group, steric inhibition of resonance does not play its role properly in 2, 6-dimethyl-4-nitrophenol which is a stronger acid than 3, 5-dimethyl-4-nitrophenol.

(iii) **Ester formation :** When phenols are warmed with acyl halides or acid anhydrides in the presence of a base, they form ester; but the reaction is much slower than the esterification of alcohols by the same reagents. The reaction follows the tetrahedral mechanism in which phenols and alcohols are nucleophiles. Since the lone pair of $e$s of the O atom in phenol gets delocalised through $p$-$\pi$ conjugation, they are not readily available for nucleophilic attack and consequently the reaction takes place slowly.

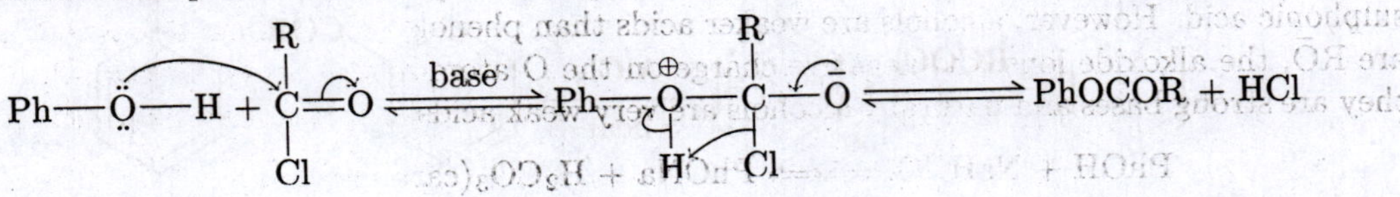

**(iv) Reaction with the Grignard reagent :** Since phenols are acidic, they will react with the Grignard reagent to form hydrocarbons.

$$ArOH + RMgX \longrightarrow ArOMgX + RH$$

**(v) Reaction with diazomethane :** A methyl aryl ether is formed when diazomethane or dimethyl sulphate is allowed to react with phenols. No Lewis acid is required for the reaction (alcohols require). In the case of dimethyl sylphate aqueous NaOH is required.

$$ArOH + \ddot{C}H_2{-}N{\equiv}\overset{\oplus}{N} \longrightarrow Ar\overset{-}{O} + CH_3{-}\overset{\oplus}{N}{\equiv}N \longrightarrow ArOCH_3 + N_2$$

$$ArOH + (CH_3)_2SO_4 \xrightarrow{NaOH} ArOCH_3 + CH_3HSO_4$$

**(vi) The Williamson synthesis and the Ullmann reaction :** When alkyl halides are treated with sodio salt of phenols, alkyl aryl ethers are formed. Diaryl ethers may be prepared by the Ullmann reaction.

$$ArOH \xrightarrow{NaOH} ArONa \xrightarrow{RI} ArOR + NaI$$

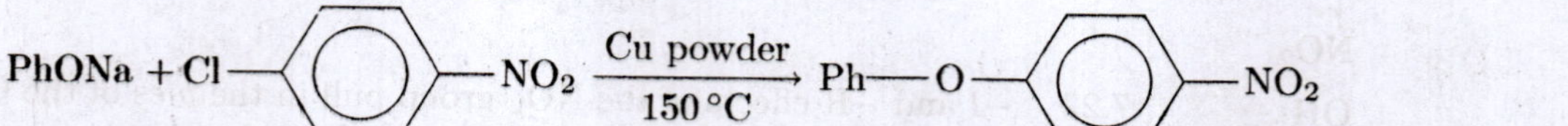

**(vii) Reaction with phosphorus pentachloride :** The reaction between a phenol and phosphorus pentachloride or phosphorus oxychloride gives rise to triaryl phosphate with a negligible amount of chloroarene. Triphenyl and tricresyl phosphates together are used as plasticiser for celluloid and similar materials.

$$ArOH + \xrightarrow{POCl_3} (ArO)_3\ PO + 3HCl$$

**(viii) Reaction with phosphorus pentasulphide :** At an elevated temperature phosphorus pentasulphide converts phenols to thiophenols.

$$ArOH + P_2S_5 \longrightarrow ArSH + P_2O_5$$

**(ix) The Friedel-Crafts alkylation :** Phenols undergo the Friedel-Crafts alkylation very readily and form *o*- and *p*-alkylphenols.

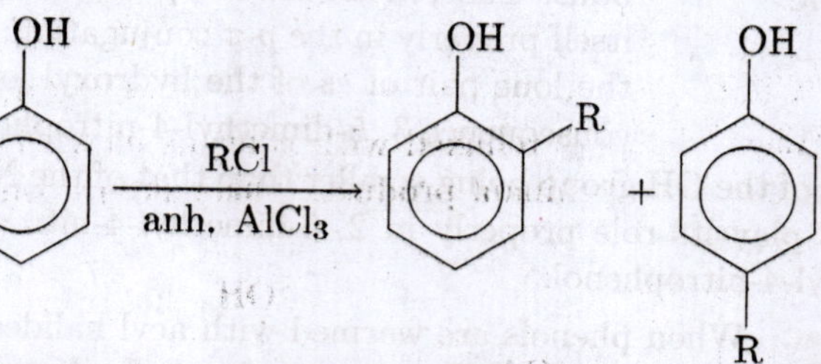

**(x) The Fries rearrangement :** When phenols are heated with acyl halides in the presence of anhydrous aluminium chloride, esters are formed which then undergo the Fries rearrangement to form *o*- and *p*-acylphenols.

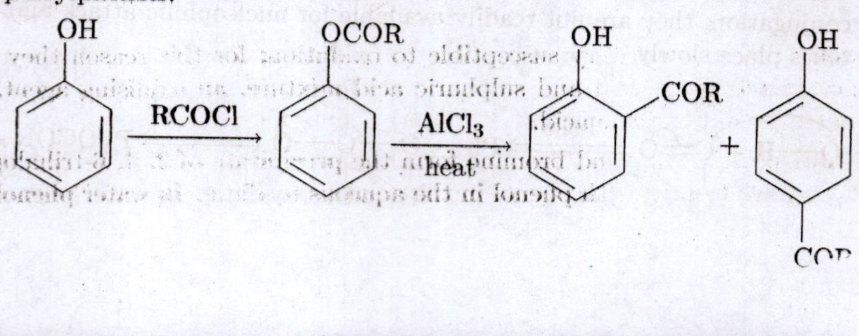

**(xi) Chloromethylation :** Phenols and their derivatives with electron-releasing substituents on chloromethylation form chloromethyl derivatives which polymerise immediately; on the other hand, derivatives of phenols with electron-withdrawing substituents, e.g., —$NO_2$, undergo chloromethylation at *o*- and *p*-positions with respect to the phenolic OH group. However, phenols

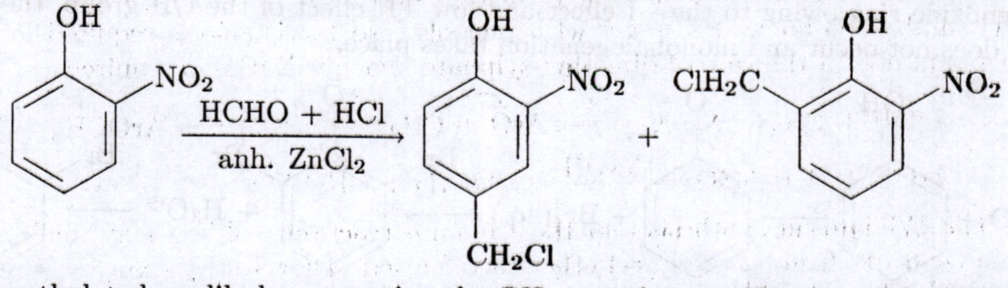


can be chloromethylated readily by protecting the OH group by esterification with chloroformic ester and then carrying out the chloromethylation.

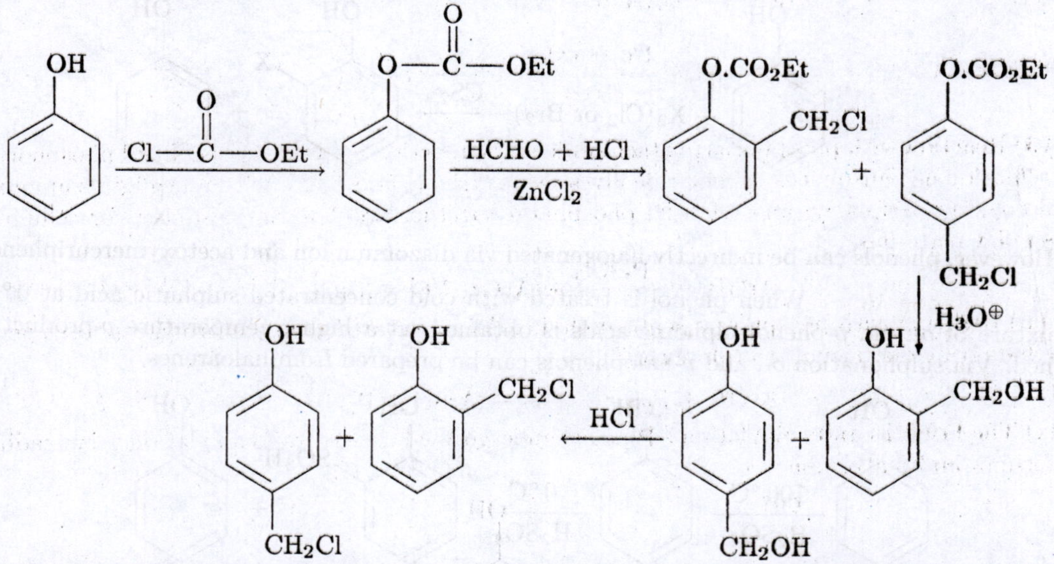


**(xii) Mercuration :** When phenol is refluxed with aqueous mercuric acetate, usually *o*-actoxymercuriphenol is formed as a predominant product. This mercuri compound can be converted to iodophenol.

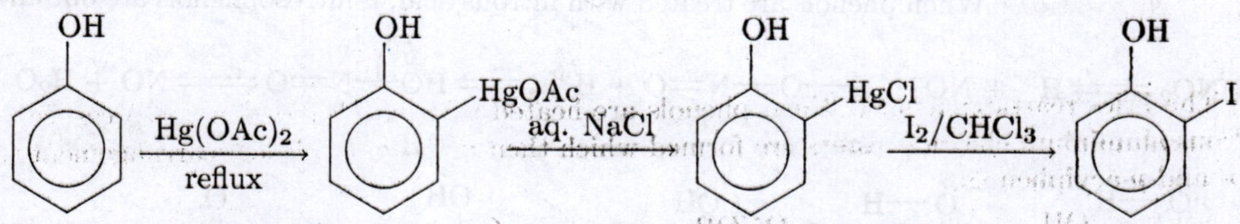


**(xiii) Nitration :** Phenols are very susceptible to oxidation; for this reason they cannot be nitrated by concentrated nitric acid and sulphuric acid mixture, an oxidising agent. However they can be nitrated by dilute nitric acid.

**(xiv) Halogenation :** Chlorine and bromine form the precipitate of 2, 4, 6-trihalophenol immediately when they are treated with phenol in the aqueous medium. In water phenol ionises to phenoxide ion. The $Ph\bar{O}$ ion has no $-I$ effect but it has a strong $+R$ effect. Owing to the strong $+R$ effect, the ring of the phenoxide ion is highly activated and can undergo electrophilic substitution reaction readily (much readily than phenol itself). When chlorination or bromination is carried out in non-aqueous solvents, e.g., in carbon disulphide, the ring being less activated than the phenoxide ring owing to the $-I$ effect and low $+R$ effect of the OH group, the ionisation of phenol does not occur and monohalogenation takes place.

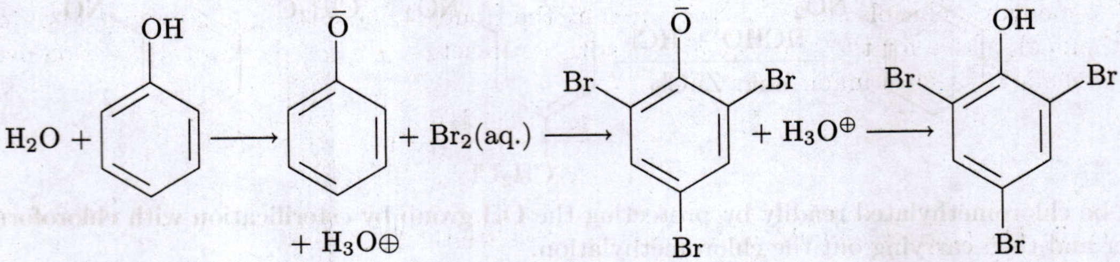

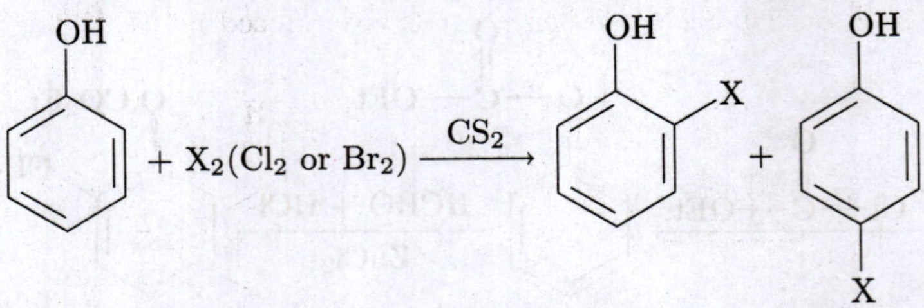

However, phenols can be indirectly halogenated via diazonium ion and acetoxymercuriphenol.

**(xv) Sulphonation :** When phenol is treated with cold concentrated sulphuric acid at 0 °C, a mixture of *o*- and *p*-phenolsulphonic acids is obtained; at a higher temperature *p*-product is formed. Via sulphonation *o*-, and *p*-halophenols can be prepared from haloarenes.

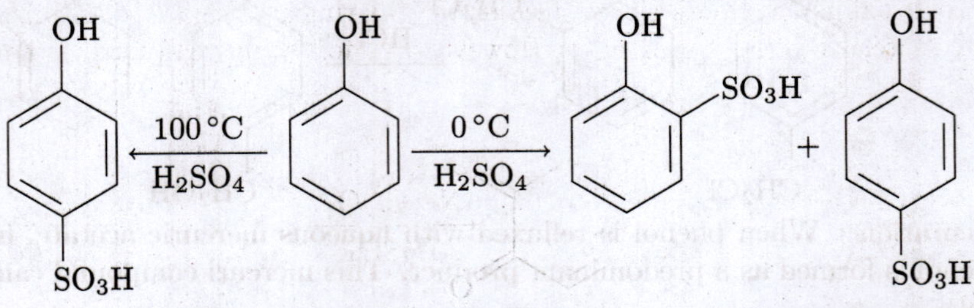

**(xvi) Nitrosation :** When phenols are treated with nitrous acid, *p*-nitrosophenols are obtained.

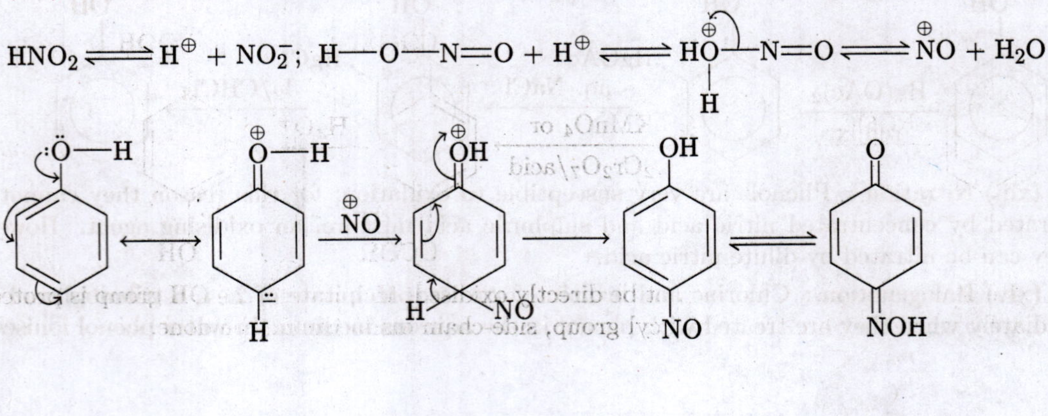

**(xvii) Diazonium coupling reaction :** See Sec. 17.16, p. 997.

**(xviii) The Mannich reaction :** See Sec. 16.13.8, p. 912.

**(xix) The Reimer-Tiemann reaction :** See Sec. 17.13, p. 990.

**(xx) The Kolbe-Schmidt reaction :** See Sec. 17.14, p. 994.

**(xxi) The Houben-Hoesch reaction :** See Sec. 17.9.2, p. 986.

Monohydric phenols do not form ketones by the Houben-Hoesch reaction; perhaps the ring is not enough active for the electrophilic substitution reaction. In these cases the reaction occurs at phenolic O to form imidoesters.

$$\text{PhOH} + \text{RCN} + \text{HCl} \xrightarrow{\text{anh. ZnCl}_2} \text{PhO}-\underset{\substack{|\\ \text{R}}}{\text{C}}=\overset{\oplus}{\text{N}}\text{H}_2\text{Cl}^-$$

**(xxii) The Gattermann aldehyde synthesis :** See Sec. 17.12, p. 989.

**(xxiii) Catalytic reduction :** When phenols are reduced by $Ni/H_2$, derivatives of cyclohexanol are obtained.

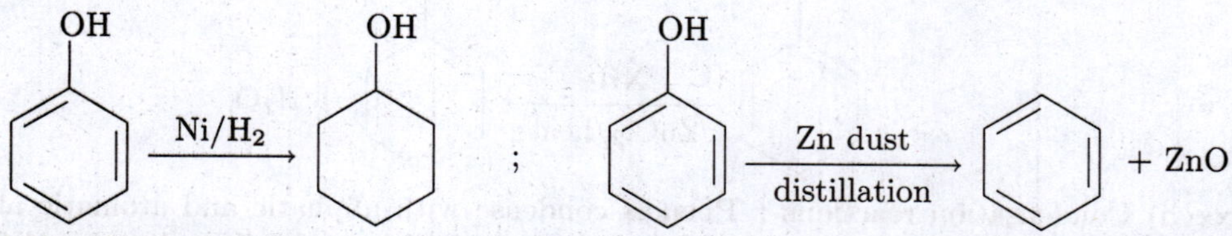

**(xxiv) Zinc dust distillation :** When phenols are distilled with Zinc dust, the OH group is replaced by an H atom.

**(xxv) Oxidation :** It has already been said that phenols are very susceptible to oxidation. Hydrogen peroxide converts phenols to catechol, hydroquinol, pyrogallol and quinone in small amounts. Strong oxidising agents either decompose the ring or bring about the union of several rings to tarry materials of no importance. However, the following reactions take place :

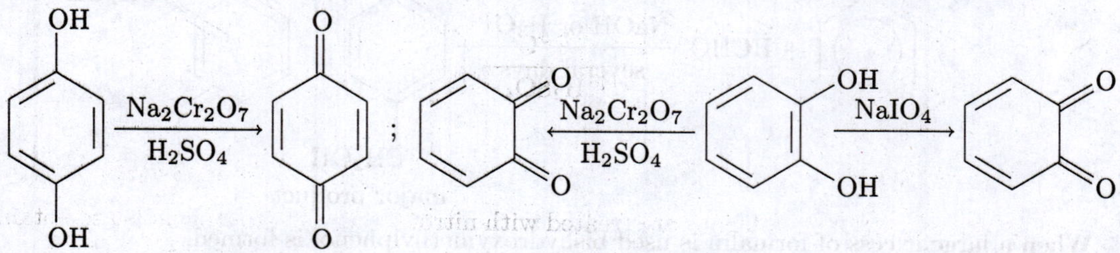

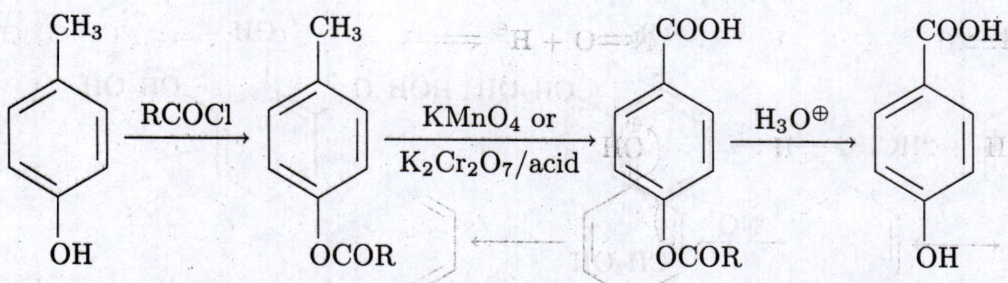

The side chain of a phenol cannot be directly oxidised. If, however, the OH group is protected either by benzensulphonate or by acyl group, side-chain oxidation can be done.

Monohydric phenols may be oxidised to dihydric phenols by alkaline potassium perdisulphate. Usually hydroxylation occurs at the *p*-position. When the *p*-position is blocked, *o*-hydroxylation also occurs. This oxidation process is known as the Elbs persulphate oxidation and this may be mechanistically represented as :

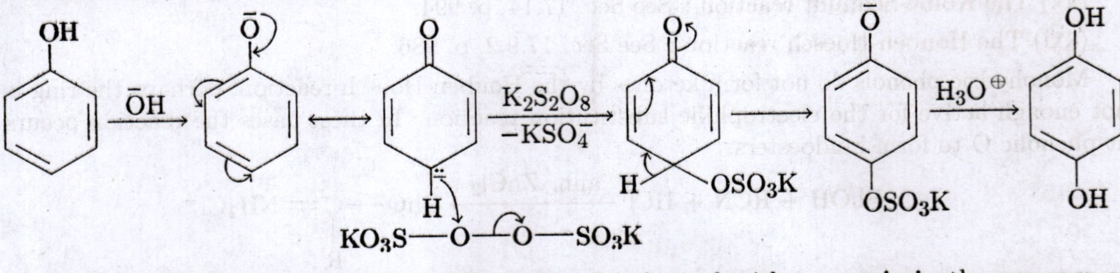

**(xxvi) Reaction with ammonia :** When phenol is heated with ammonia in the presence of anhydrous zince chloride, aniline is formed; the reaction may be called ammonolysis of phenol.

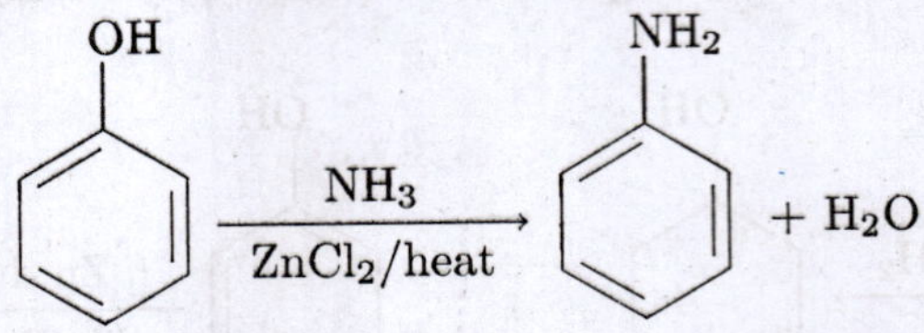

**(xxvii) Condensation reactions :** Phenols condense with aliphatic and aromatic aldehydes. Phenol-formaldehyde resins and plastics are manufactured by condensing phenol and formaldehyde in a basic medium. However, formalin (40% aqueous solution of formaldehyde) condenses with phenol at a low temperature in the presence of a dilute acid or alkali and the reaction is called Lederer-Manasse reaction.

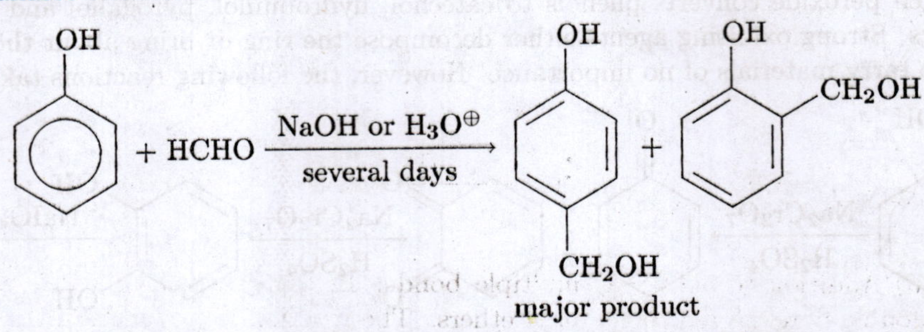

When a large excess of formalin is used bishydroxymethylphenol is formed.

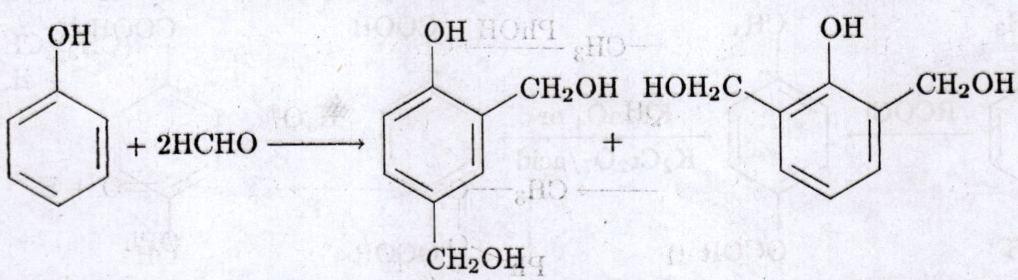

Aldehydes, formaldehyde and others, usually condense with phenol to give *pp*′-dihydroxydiphenylmethane derivatives.

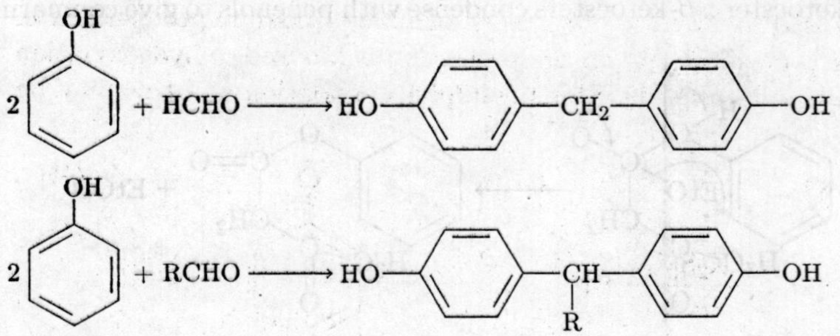

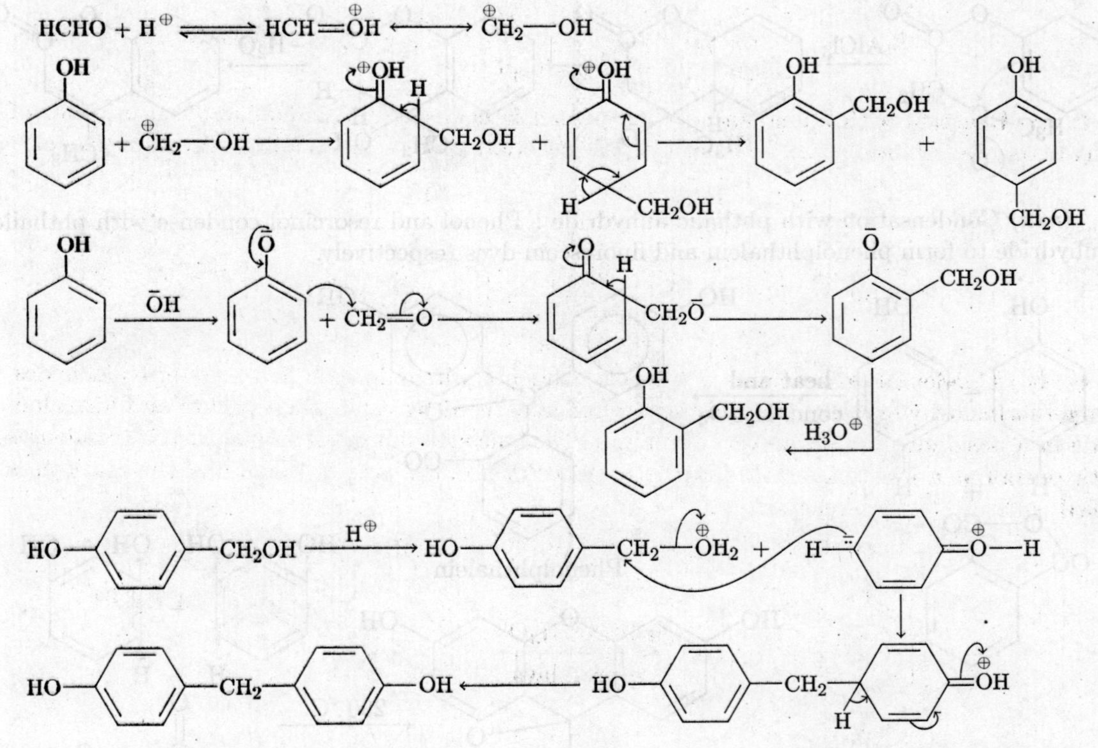

**(xxviii) Addition to a C—C multiple bond :** In the presence of acids, phenols add to a C—C double bond to form aryl alkyl ethers. The reaction is slower than the similar reaction of the alcohols. It adds to ketenes to form esters.

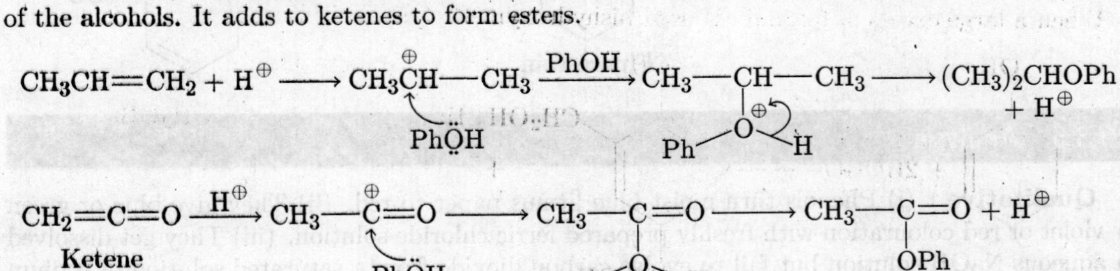

**Alcohols also add to ketenes.**

**(xxix) Condensatiojm with β-ketoester :** β-ketoesters condense with pehenols to give coumarins.

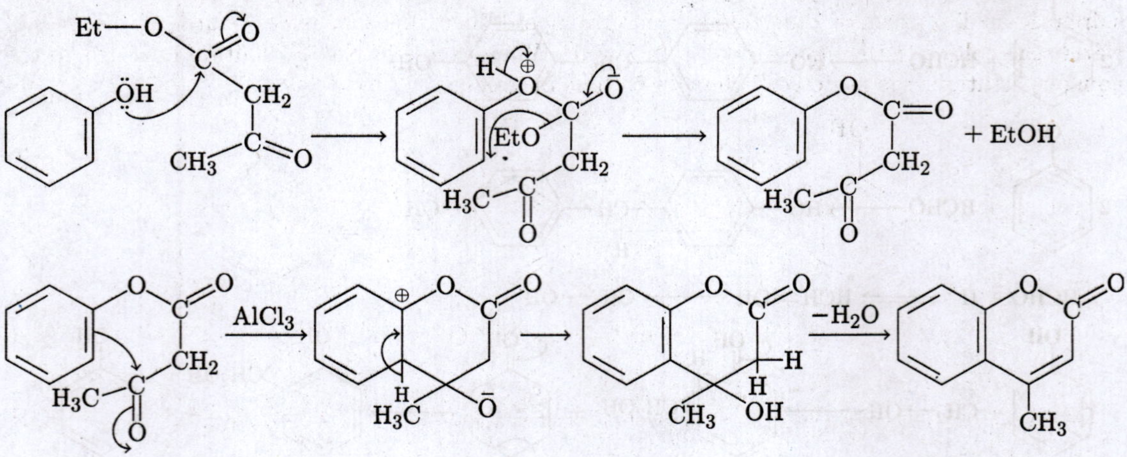

**(xxx) Condensation with phthalic anhydride :** Phenol and resorcinol condense with phthalic anhydride to form phenolphthalein and fluorescein dyes respectively.

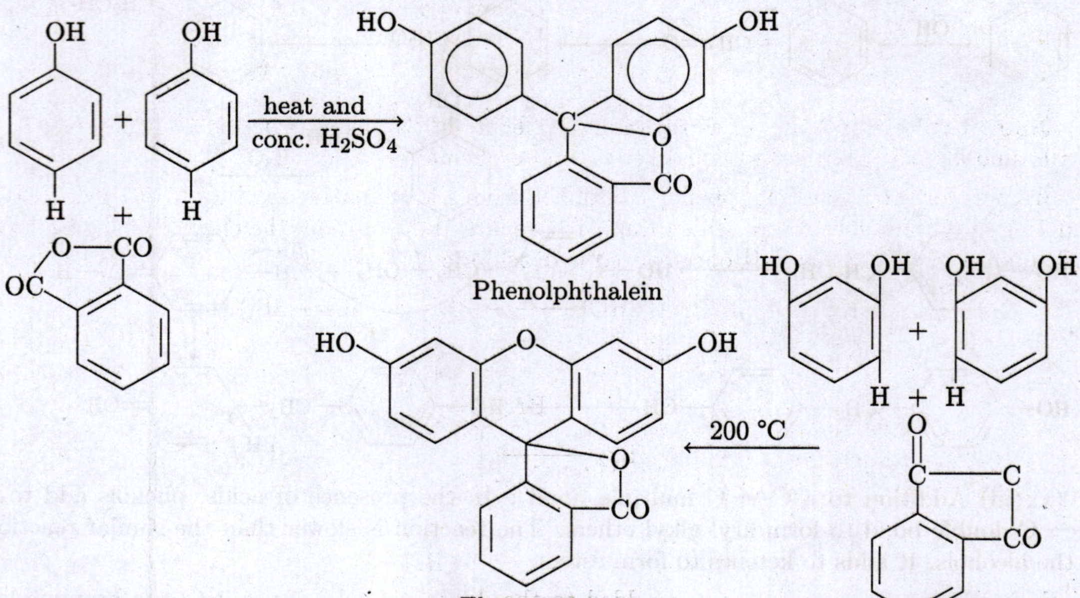

Phenolphthalein

Fluorescein

## 20.3.5 Analysis

**Qualitative :** (i) Phenols turn moist blue litmus paper to red. (ii) They give blue or green or violet or red colouration with freshly prepared ferric chloride solution. (iii) They get dissolved in aqueous NaOH solution but fail to evolve carbon dioxide from a saturated solution of sodium bicarbonate. (iv) They give immediate precipitate of bromoderivatives with aqueous bromine solution and (v) They give the **Libermann's test** which is as follows :

When a few drops of phenol is heated with a few crystals of sodium nitrite and 1 ml conc. sulphuric acid, a green or blue colouration develops. When this solution is poured into excess of water, the colour of the aqueous solution is red. When excess of NaOH solution is added to the aqueous solution, the red colour changes to blue or green.

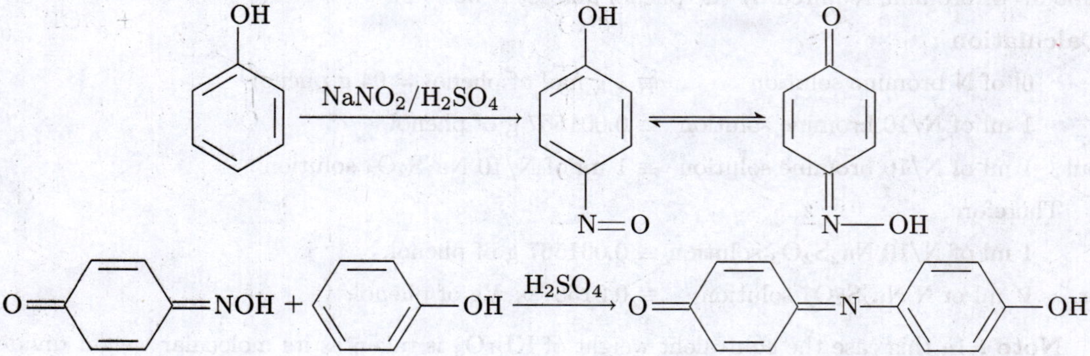


Indophenol; brown solid in strong acid looks blue or green

$\downarrow H_2O$

alkaline solution, blue or green $\xleftarrow[\text{in excess}]{\text{NaOH}}$ aqueous solution, red

**Quantitative analysis :** The reaction of phenol with aqueous bromine solution gives 2, 4, 6-tribromophenol; this reaction is used to estimate phenol.

Bromine for bromination of phenol is obtained from a brominating mixture (5.567 g of $KBrO_3$ and 75 g of KBr dissolved in distilled water to prepare 1l of solution; the chemicals should be of A.R. quality) by the action of concentrated hydrochloric acid.

$$KBrO_3 + 5KBr + 6HCl \longrightarrow 3Br_2 + 6KCl + 3H_2O$$

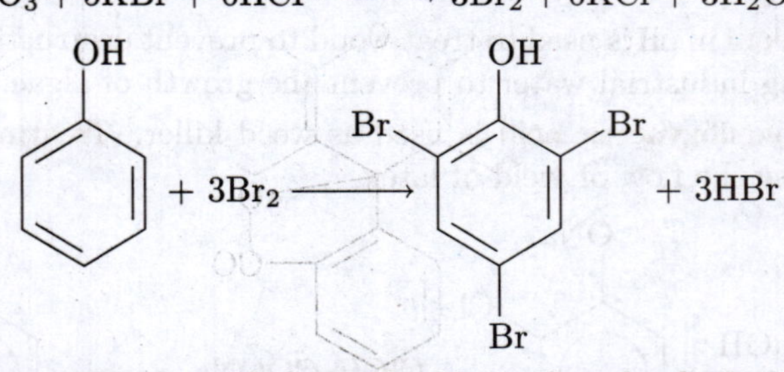


Excess of brominating mixture is added to the dilute phenol solution and the bromine left unreacted is made to react with KI and the liberated iodine is titrated with standard $Na_2S_2O_3$ solution using starch as indicator.

$$Br_2 + 2KI \longrightarrow I_2 + 2KBr; \quad 2Na_2S_2O_3 + I_2 \longrightarrow 2NaI + Na_2S_4O_6$$

Given phenol is diluted to 100 ml in a volumetric flask with addition of distilled water. 10 ml of this solution is taken in a 250 ml glass stoppered flask and to that 20 ml of N/10 brominating mixture and 5 ml of concentrated HCl are added, the stopper is inserted immediately and swirled occassionally for half-an-hour and then the flask is allowed to stand for 15 minutes. Then 10 ml 10% KI solution is added to it and shaken thoroughly and titrated with a standard N/10

$Na_2S_2O_3$ solution until a faint yellow colour remains. A few ml of freshly prepared cold starch solution is added and the titration is completed with vigorous shaking. The end point is the sharp disappearance of the blue colour. The same procedure is repeated with 10 ml distilled water instead of phenol solution. The difference between the two titre values represents the amount of bromine required by the phenol and let it be $V$ ml

**Calculation :**

6l of N bromine solution $\equiv$ 1 g mol of phenol = 94 g phenol

1 ml of N/10 bromine solution $\equiv$ 0.001567 g of phenol

but, 1 ml of N/10 bromine solution $\equiv$ 1 ml of N/10 $Na_2S_2O_3$ solution

Therefore,

1 ml of N/10 $Na_2S_2O_3$ solution $\equiv$ 0.001567 g of phenol.

or, $V$ ml of N $Na_2S_2O_3$ solution $\equiv 0.01567 \times$ Vg of phenol.

**Note :** In this case the equivalent weight of $KBrO_3$ is taken as its molecular weight divided by six. This is because in this reaction the total change of oxidation number of bromine is six (from +5 in $KBrO_3$ to −1 in $Br^-$ since molecular bromine is formed via bromide ion).

## 20.3.6 Uses of phenols

**(i) Phenol** is chiefly used for the manufacture of phenol-formaldehyde resins and plastics, called phenolics. It is also used to manufacture bisphenol-A and other compounds used to manufacture resins and plastics.

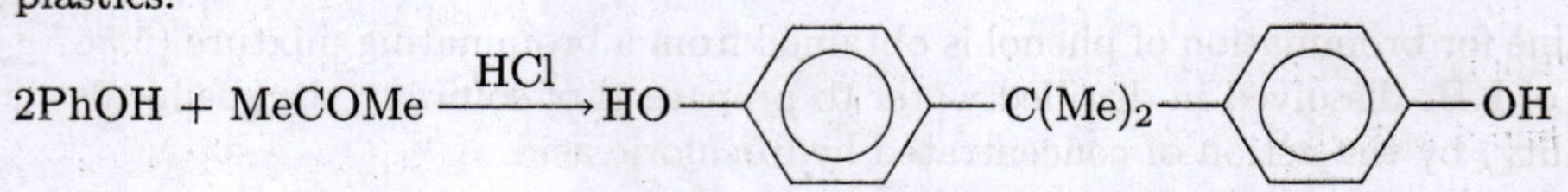


Bisphenol-A

**(ii) Pentachlorophenol** in oil is used to treat wood to prevent destruction by fungi. The sodium salt is used for treating industrial water to prevent the growth of algae.

**(iii) 2, 4-Dichlorophenoxyacetic acid** is used as weed killer. Treatment of rubber trees with this compound increase the flow of yield of latex.

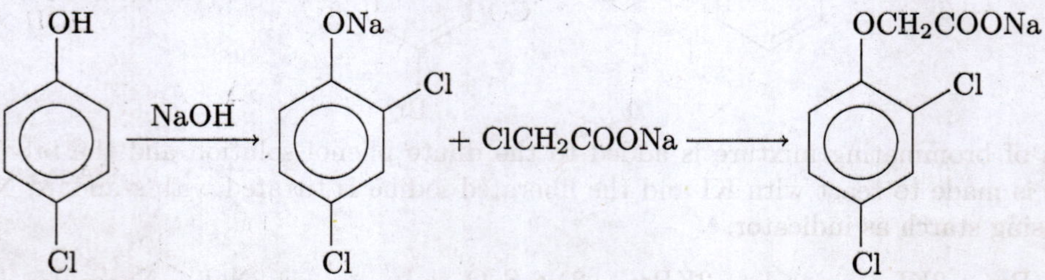


**(iv) *p*-Nitrophenol** is used to synthesise parathion, an insecticide.

**(v) Trinitrophenol (picric acid)** is used to identify and isolate hydrocarbons. It is used as a yellow dye for silk and as an explosive.

**(vi) Phenyl phosphate** is used in photographic film base and other arylphosphates are used as plasticisers and flame retardants.

**(vii) Cresols** are present in creosote oil which is used to preserve wood and to treat lung congestion.

**(viii) Thymol** has antiseptic property and pleasant odour and so, it is used in antiseptic creams. It is also used to synthesise menthol.

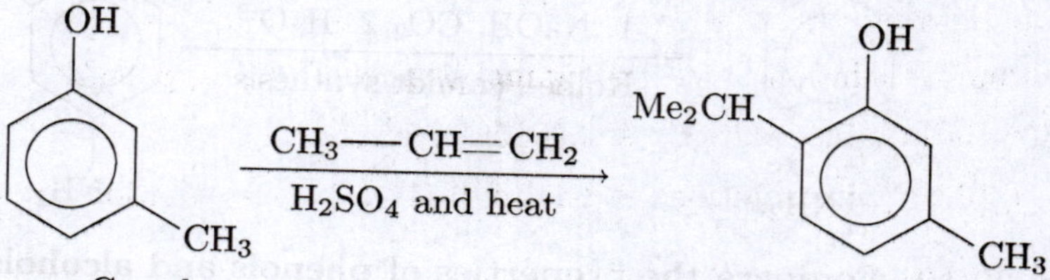

*m*-Cresol  Menthol

**(ix) Catechol** is used as a photographic developer.

**(x) Styphnic acid** is used for identifying certain compounds.

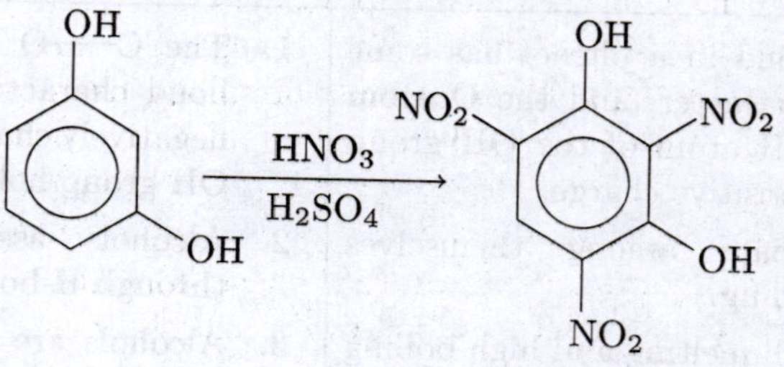

Styphnic acid

**(xi) Quinol** is used as photographic developer.

**(xii) Phenyl salicylate** (Salol) is used as a medicinal coating. It is also used as a sun-screening agent and stabiliser of plastics. It is prepared by heating salicylic acid with phenol and phosphoryl chloride at 200 °C.

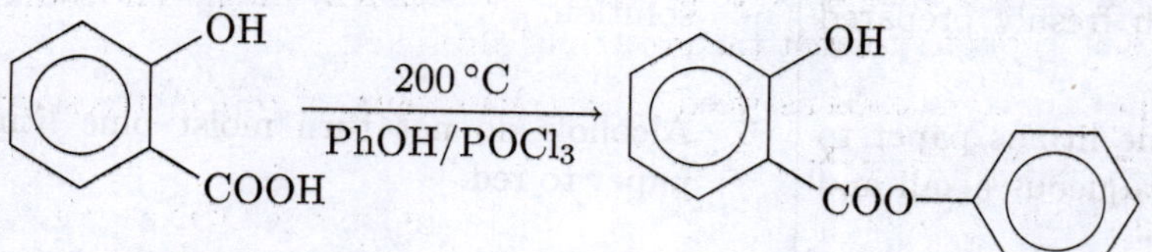

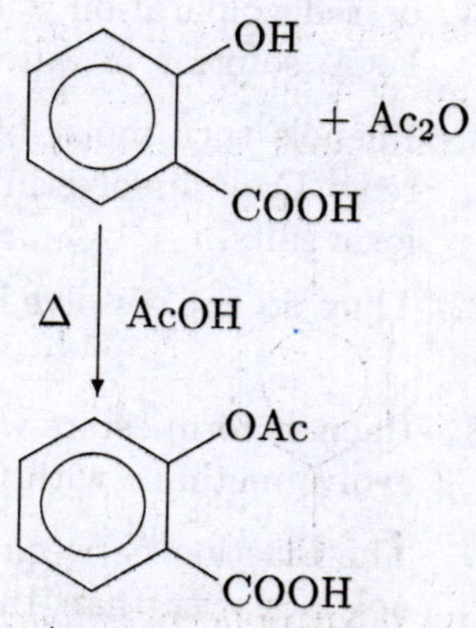

Aspirin

**(xiii) Acetyl salicylate** (Aspirin), an analgesic and antipyretic, is prepared by heating salicylic acid, acetic anhydride and glacial acetic acid. Aspirin is also anti-coagulant of blood and it is used as a medicine for thrombosis.

(xiv) *p*-Aminosalicylic acid (PASA) is used in the treatment of tuberculosis along with streptomycin and isoniazide.

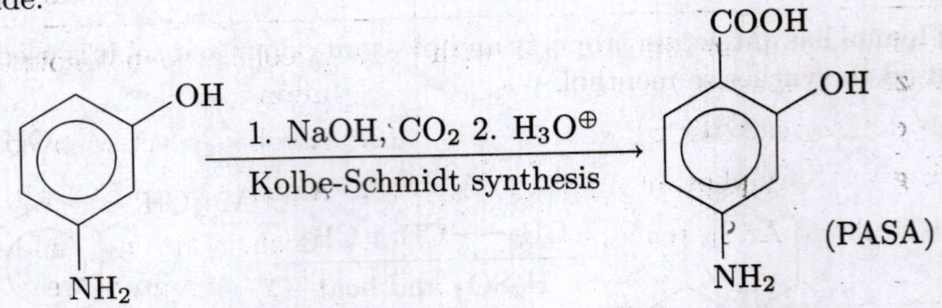

**Problem 20.10. Compare the properties of phenols and alcohols :**

**Solution :**

| Phenols | Alcohols |
|---|---|
| 1. The C—O bond in a phenol has some double bond character, and the O atom as well as the H atom of the OH group bear a partial positive charge. | 1. The C—O bond in alcohol has single bond character. The O atom is slightly negatively charged while the H atom of the OH group holds a partial positive charge. |
| 2. Phenols associate among themselves through H-bonding. | 2. Alcohols associate among themselves through H-bonding. |
| 3. Phenols are high melting and high boiling compounds because of the intermolecular H-bonding. | 3. Alcohols are also high melting and high boiling compounds because of the H-bonding. |
| 4. Lower members of the phenols are moderately soluble in water. | 4. Lower members of the alcohols are highly soluble in water. |
| 5. Phenols usually give blue or green or violet or red colouration with freshly prepared $FeCl_3$ solution in water. | 5. Alcohols do not give colour with $FeCl_3$ solution. |
| 6. Phenols turn moist blue litmus paper to red. They dissolve in aqueous alkali and form salts. | 6. Alcohols do not turn moist blue litmus paper to red. |
| 7. They do not dissolve in $NaHCO_3$. | 7. Alcohols do not form salts with dil. aqueous alkali or with $NaHCO_3$. |
| 8. Phenols form esters with acyl halides and evolve methane with $CH_3MgI$. | 8. Alcohols form esters with RCOCl and evolve methane with $CH_3MgI$. |
| 9. The reaction between RCOCl and a phenol is slower than that between RCOCl and an alcohol. | 9. The reaction is faster than the reaction between phenol and RCOCl. |
| 10. Phenols on oxidation give complex coloured products of unknown composition. | 10. Alcohols smoothly oxidise to aldehydes and ketones. |

*Contd.*

| Phenols | Alcohols |
|---|---|
| 11. Phenols in alkaline medium couple with arenediazonium ions to form azo dyes. | 11. Alcohols do not couple with arenediazonium ions. |
| 12. Phenols do not react with HX. | 12. Alcohols react with HX and form RX. |
| 13. Phenols give triarylphosphate with $PCl_5$. | 13. Alcohols form RCl with $PCl_5$. |
| 14. Phenols undergo Ar.$S_E$ reactions. | 14. Alcohols do not undergo $S_E$ reactions. They undergo $S_N$ reactions. |

**Problem 20.11. Synthesise.**

**(a)** Resorcinol, **(b)** Catechol, **(c)** *m*-Cresols, **(d)** *m*-Nitrophenol, **(e)** 2-Phenoxyacetic acid.

**Solution :**

(a)

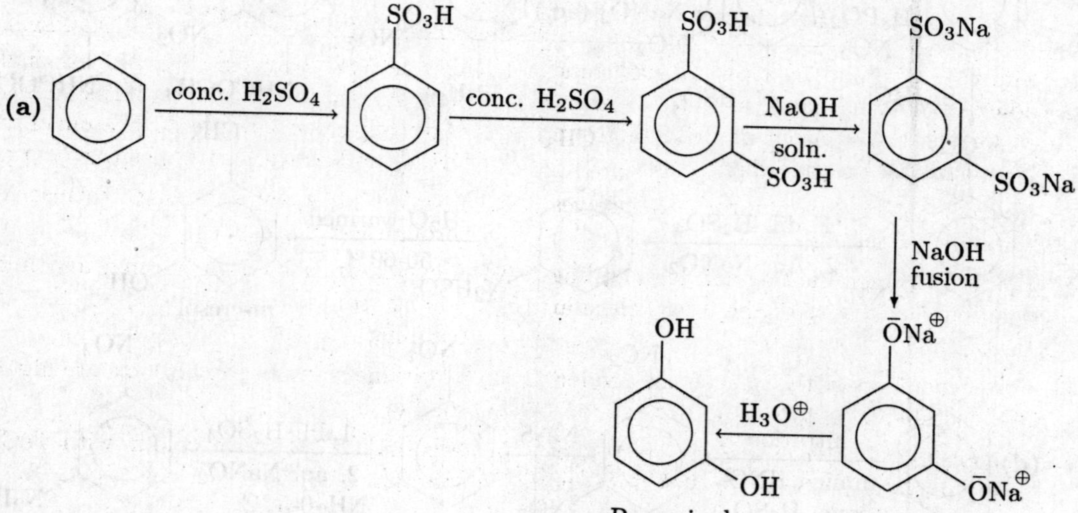


Resorcinol

(b)

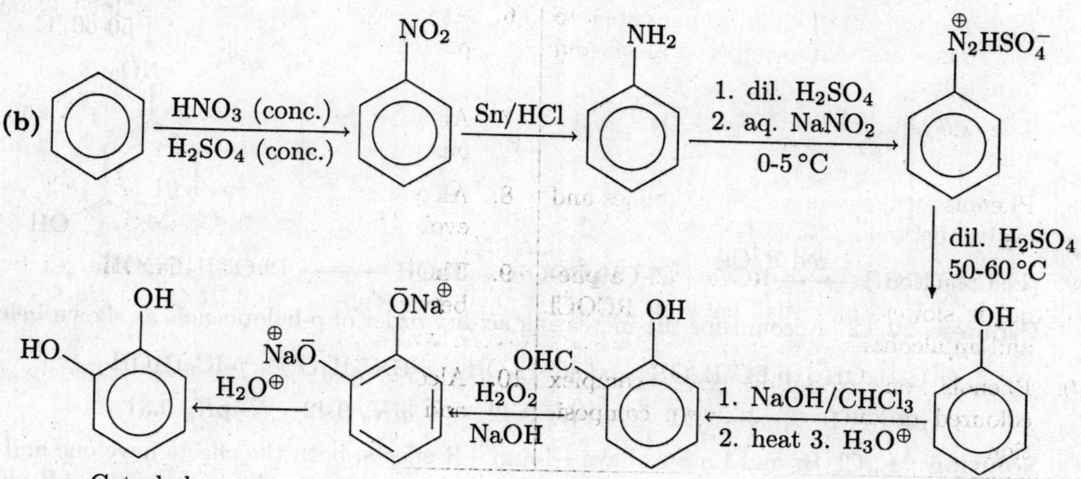


Catechol

(c) 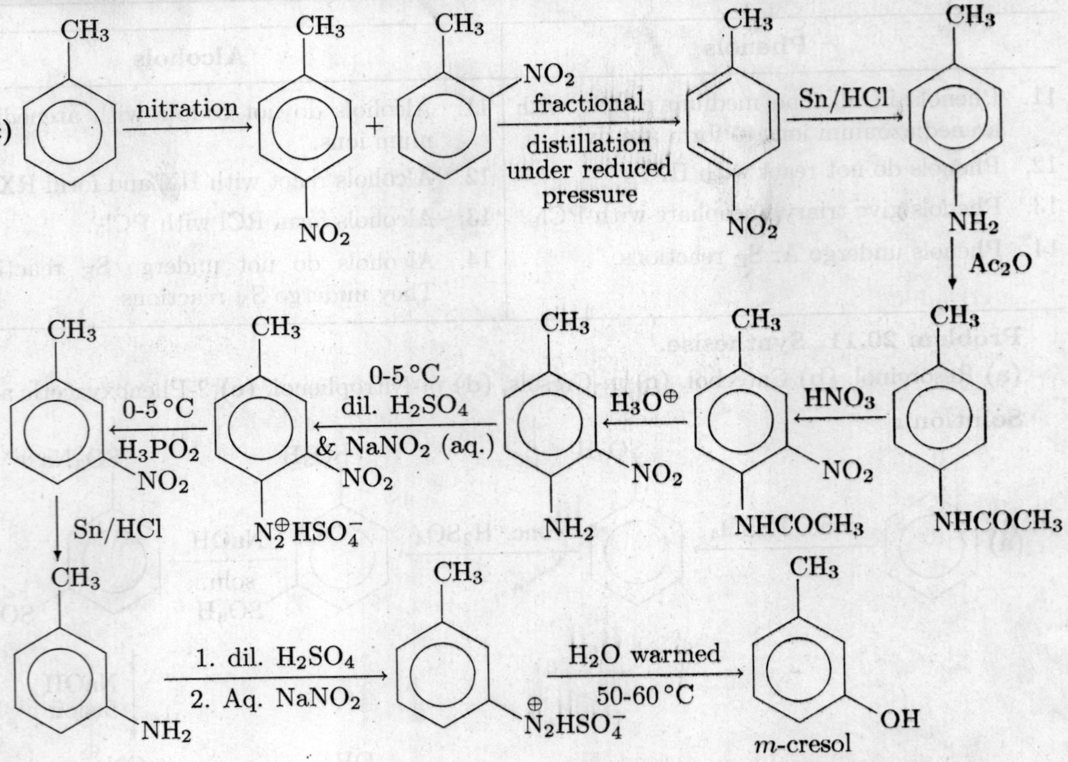

(d) 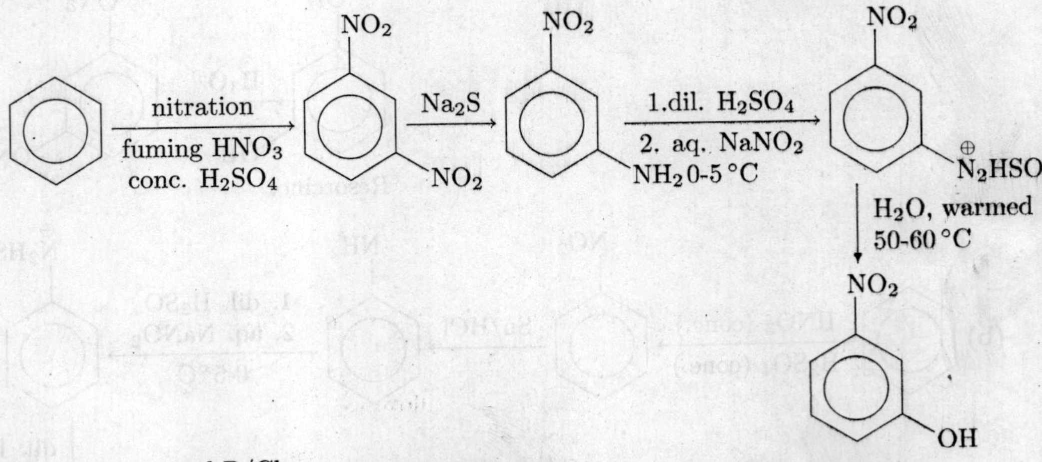

(e) $$CH_3COOH \xrightarrow{\text{red P/Cl}_2} Cl\,CH_2COOH + PhOH \longrightarrow PhOCH_2COOH$$

**Problem 20.12.** Account for the increasing acidity order of *p*-halophenols as shown below :

| $C_6H_5OH$ | *p*-$FC_6H_4OH$ | *p*-$ClC_6H_4OH$ | *p*-$BrC_6H_4OH$ | *p*-$IC_6H_4OH$ |
|---|---|---|---|---|
| $pK_a$ 10.0 | $pK_a$ 9.91 | $pK_a$ 9.42 | $pK_a$ 9.42 | $pK_a$ 9.31 |

**Solution :** F, Cl, Br and I atoms have –I and +R effects; both the effects have one and the same order which is : F>Cl>Br>I. The –I effect, however, predominates over the +R effect;

for this reason the *p*-halophenoxide ions, specially *p*-fluorophenoxide, are more stable than the phenoxide ion and hence they are less basic than the latter. So the *p*-halophenols are more acidic than phenol. However, in the cases of *p*-chloro-, *p*-bromo- and *p*-iodophenoxide ions Cl, Br and I atoms act as $d$-$\pi$ acceptor ($-$R effect) and thereby the negative charge on the O atom gets delocalised to a large extent and this effect being predominant, the stability of the phenoxides increases. Since $d$-$\pi$ accepting ability order is I>Br>Cl, *p*-iodophenoxide is the most stable base among the three. The greater $-$I effect of Cl than that of Br and the greater $d$-$\pi$ accepting ability of Br than that of Cl make their phenoxides equally stable. Naturally *p*-iodophenol is the strongest acid having the lowest $pK_a$ among the *p*-halophenols. The stability of *p*-chlorophenoxide and *p*-bromophenoxide being equal, the corresponding phenols have identical $pK_a$ values. F atom being unable to act as a $d$-$\pi$ acceptor because of the absence of $d$ orbital, the *p*-fluorophenoxide ion is the least stable and most basic among *p*-halophenoxides. So, *p*-fluorophenol is the least acidic among *p*-halophenols and possesses the highest $pK_a$ value.

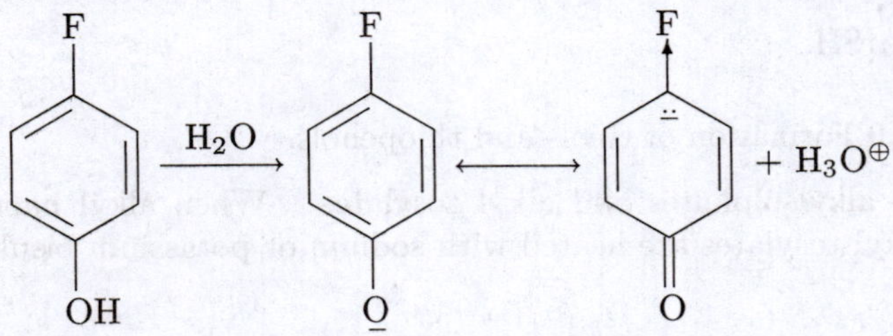


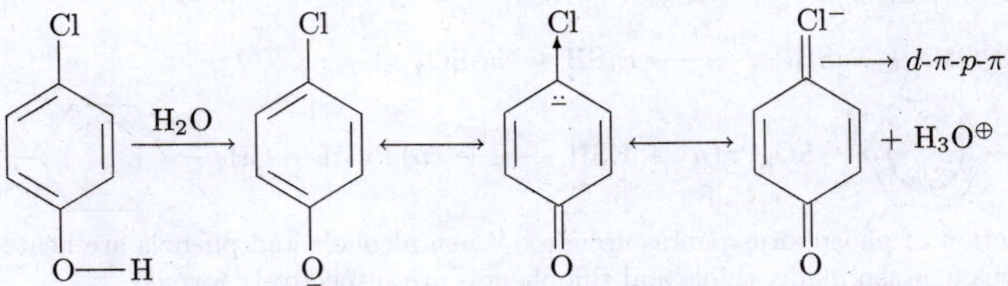


## 20.4 Thiols and Thiophenols

### 20.4.1 Introduction

Thiols and thiophenols are the sulphur analogues of alcohols and phenols respectively. They may be represented as RSH and ArSH. Petrolium contains thiols.

### 20.4.2 Formation of thiols and thiophenols

The methods by which these compounds can be formed are given in Chart 20.9.

(i) NaSH or KSH $\xrightarrow{\text{RX or ROSO}_2\text{ONa}}$ RSH (ii) $P_2S_5$ $\xrightarrow{\text{ROH or ArOH}}$ RSH or ArSH

or

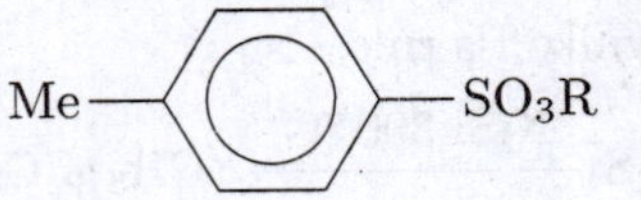

(iii) $H_2S \xrightarrow[ThO_2;\ 400\,°C]{ROH\ (vapour)} RSH$

(iv) 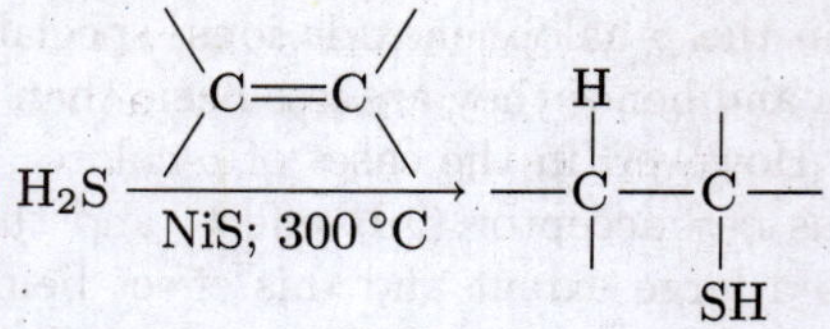

(v) $RBr \xrightarrow[2.\ NaOH]{1.\ Thiourea} RSH$

(vi) $RMgX \xrightarrow[2.\ H_3O^{\oplus}]{1.\ S} R{-}SH$

(vii) $ArBr \xrightarrow{NaSH} ArSH$

(viii) $ArN_2^{\oplus} \xrightarrow{NaSH} ArSH$

(ix) $Xanthates \xrightarrow[100\text{-}250\,°C]{heat} RSH$

(x) $RSSR \xrightarrow{Zn/dil\ acid} 2RSH$

(xi) $ArSO_2Cl \xrightarrow[or\ LiAlH_4,\ ether]{Zn/H_2SO_4} ArSH.$

**Chart 20.9** Formation of thiols and thiopenols.

(i) **From alkyl halides, sodium alkylsulphates and alkyl tosylates :** When alkyl halides or sodiumalkyl sulphates or alkyl tosylates are heated with sodium or potassium bisulphide, thiols are formed.

$$CH_3Br + NaSH \xrightarrow{heat} CH_3SH + NaBr$$

$$EtOSO_2ONa + NaSH \xrightarrow{heat} EtSH + Na_2SO_4$$

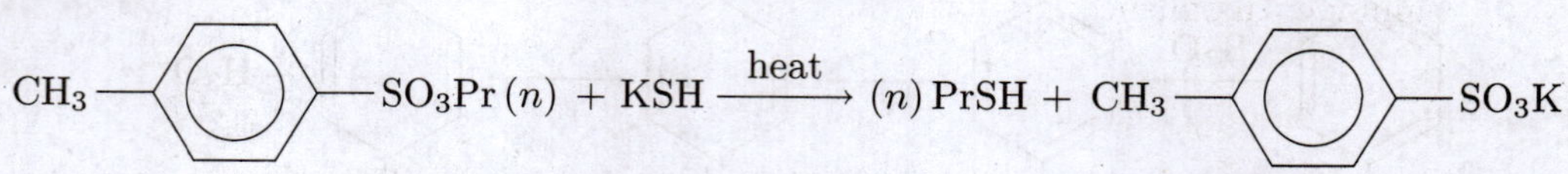

(ii) **By the action of phosphorus pentasulphide :** When alcohols and phenols are heated with phosphorus pentasulphide, thiols and thiophenols are respectively formed.

$$ROH + P_2S_5 \xrightarrow{heat} RSH + P_2O_5;$$

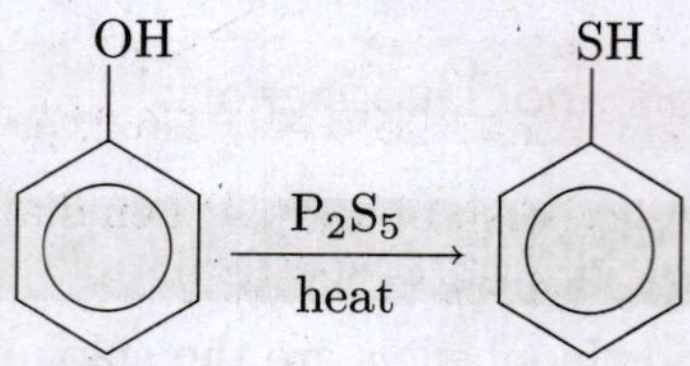

(iii) **By the action of hydrogen sulphide on alcohols :** When a mixture of alcohol vapour and gaseous hydrogen sulphide is passed over thoria as catalyst at a temperature 400 °C, a thiol is formed.

$$ROH + H_2S \xrightarrow[400\,°C]{ThO_2} RSH + H_2O$$

(iv) **By the addition of hydrogen sulphide to C—C double bonds :** When $H_2S$ is passed through an alkene in the presence of nickel (II) sulphide at 300 °C, a thiol forms; the addition occurs according to the Markownikoff's rule.

$$CH_3{-}CH{=}CH_2 + H_2S \xrightarrow{NiS;\ 300\,°C} (CH_3)_2\,CH\,SH$$

**(v) From thiourea :** When thiourea is treated with alkyl halides, S-alkylisothiouronium salts are formed; these on treatment with aqueous sodium hydroxide solution give thiols. This method gives pure thiols.

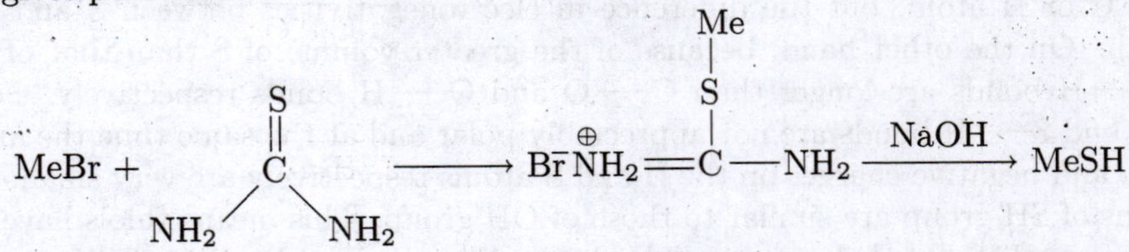

**(vi) From the Grignard reagent :** When the Grignard reagents are treated with sulphur, thiocomplexes are formed which on treatment with acidulated water give thiols.

$$\text{EtMgBr} + \text{S} \longrightarrow \text{Et—S—MgBr} \xrightarrow{\text{H}_3\text{O}^{\oplus}} \text{EtSH} + \text{Mg(OH) Br}$$

**(vii) Thiopheonols from aryl halides :** When active aryl halides are heated with sodium or potassium hydrogen sulphide thiophenols are obtained.

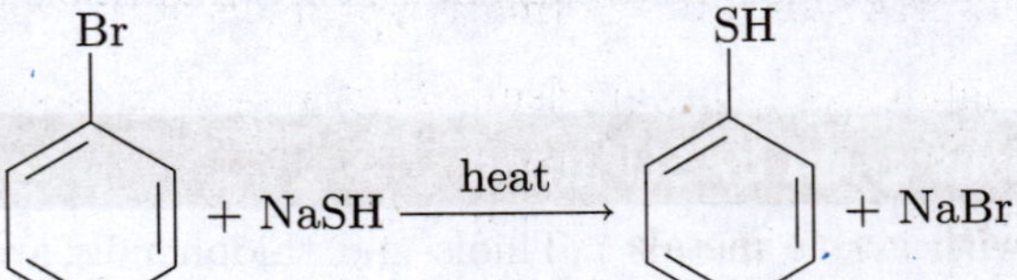

**(viii) From arenediazonium ion :** When arenediazonium ions are treated with sodium hydrogen-sulphide, thiophenols are formed.

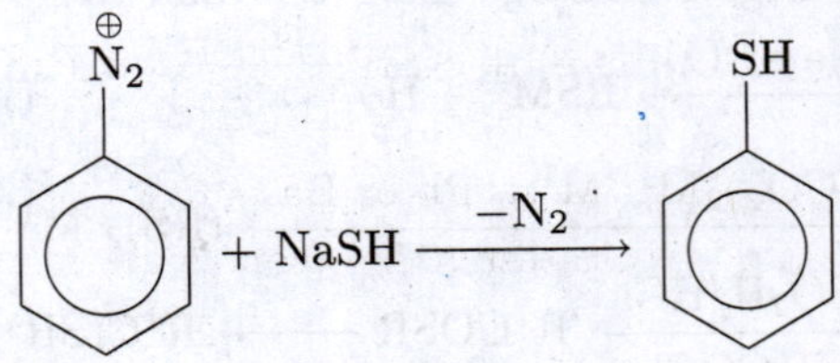

**(ix) From xanthates :** See Sec. 15.6.2, p. 814.

**(x) From dialkyl disulphides :** When dialkyl disulphides are reduced with zinc and dilute acids, thiols are formed.

$$\text{R—S—S—R} \xrightarrow{\text{Zn/acid}} 2\text{RSH}$$

**(xi) From sulphonyl chlorides :** Usually thiophenols are prepared from arenesulphonyl chlorides.

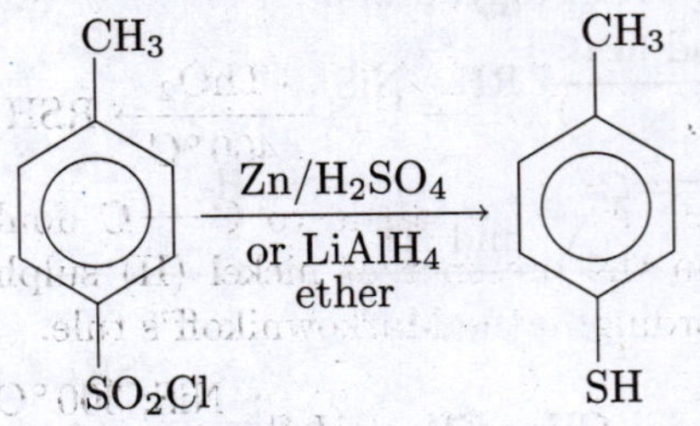

## 20.4.3 Structure, reactivity, orientation and physical properties

Thiols contain C—S—H linkage. The electronegativity of a sulphur atom is greater than that of C or H atom; but the difference in electronegativities between S and C or S and H is not high. On the other hand, because of the greater volume of S than that of O atom, C—S and S—H bonds are longer than C—O and O—H bonds respectively. For these reasons, C—S and S—H bonds are not appreciably polar and at the same time the amounts of partial positive and negative charges on the H and S atoms respectively are very small. Nevertheless the reactions of SH group are similar to those of OH group. This means thiols have similarities with alcohols and thiophenols have resemblances to phenols. The H atom of SH group is an active H and naturally the SH group exhibits the properties of active H atoms. However, the reactions of the SH group are given in Chart 20.10.

S—H bond being not enough polar, thiols and thiophenols cannot form H-bonding and do not remain associated among themselves and, therefore, their m.p.s and b.p.s are much lower than those of alcohols and phenols, the oxygen analogues of comparable molecular weights. For the same reason they are not soluble in water since they cannot form H-bond with water. Alcohols have sweet smell, and phenols have carbolic smell, while thiols and thiophenols have nausiating smell.

## 20.4.4 Reactions given by SH group

**(i) Reaction with active metals :** Thiols and thiophenols are more acidic than their oxygen analogues. The S—H bond being longer than the O—H bond, the former is weaker than the latter. So an S—H bond breaks more readily than an O—H bond. Furthermore, the volume of S atom being greater than the volume of O atom, $RS^-$ & $ArS^-$ can better accommodate the negative charge than $RO^-$ and $ArO^-$ respectively.

(i) $RSH \xrightarrow{M(Na\text{ or }Li)} R\bar{S}M^{\oplus} + H_2$

(ii) $RSH \xrightarrow{HgO} (RS)_2Hg + H_2O$

(iii) $RSH \xrightarrow{(R^1CO_2)_2M^1;\ M^1=Pb\text{ or }Ba} (RS)_2M^1$

(iv) $RSH \xrightarrow[\text{or } R^1COCl]{R^1CO_2H/HCl} R^1COSR \longrightarrow R^1C(SR)_3$

(v) $RSH \xrightarrow[\text{or } Na_2PbO_2 + S]{H_2O_2\text{ or }CuCl_2\text{ or }NaOCl} R{-}S{-}S{-}R$

(vi) $RSH \xrightarrow{HNO_3} RSO_3H$

(vii) $RSH \xrightarrow[R^2 = H\text{ or alkyl group}]{R^1COR^2\ \&\ \text{dry HCl}} R^1R^2\,C(SR)_2$

(viii) $RSH \xrightarrow{\text{Ni and heat}} RH + NiS$

(ix) $RSH \xrightarrow{>C{=}C<\ \text{and } H^{\oplus}} -\overset{H}{\underset{|}{\overset{|}{C}}}-\underset{SR}{\underset{|}{\overset{|}{C}}}-$

(x) $RSH \xrightarrow{ArOH/H^{\oplus}} ArSR$

**Chart 20.10** Reaction of SH group.

Therefore, $RS^-$ and $ArS^-$ are weaker bases than $R\bar{O}$ and $Ar\bar{O}$ respectively; hence RSH is a stronger acid than ROH and same is the relation between ArSH and ArOH. However, their acid characters are shown by their reaction with active metals.

$$2RSH + 2Na \longrightarrow 2R\bar{S}\overset{\oplus}{Na} + H_2\uparrow;\ 2ArSH + 2Na \longrightarrow 2Ar\bar{S}\overset{\oplus}{Na} + H_2\uparrow$$

Active metals form salts with thiols and thiophenols. Thiol salts hydrolyse in water. Thiophenols like phenols dissolve in alkali :

$$R\bar{S}\overset{\oplus}{Na} + H_2O \longrightarrow RSH + NaOH;\ ArSH + NaOH \longrightarrow Ar\bar{S}\overset{\oplus}{Na} + H_2O$$

**(ii) Reaction with mercuric oxide :** —SH group reacts with mercuric oxide and forms mercaptides. Since an —SH group captures mercury, it is named as mercapto group and the compound bearing the SH group is called mercaptan.

$$2RSH + HgO \longrightarrow (RS)_2Hg + H_2O$$

**(iii) Reaction with heavy metal salts :** When heavy metal salts are treated with thiols, heavy metal mercaptides precipitate.

$$(CH_3COO)_2\ Pb + 2\ MeSH \longrightarrow (MeS)_2\ Pb + 2CH_3COOH$$

$$(CH_3COO)_2\ Ba + 2\ EtSH \longrightarrow (EtS)_2\ Ba + 2CH_3COOH$$

**(iv) Ester formation :** Thiols react with acyl halides and form S-alkyl esters which undergo further reaction to produce ortho esters.

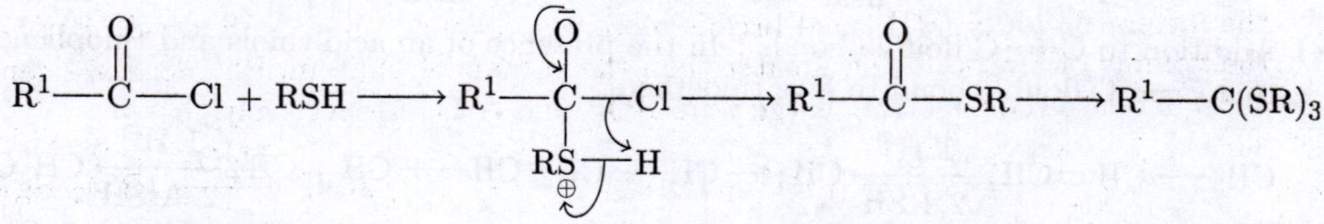

**(v) Action of mild oxidising agents :** Mild oxidising agents like $H_2O_2$, $CuCl_2$, NaOCl, $Na_2PbO_2/S$ (**Doctor's solution**) convert thiols to dialkyl disulphides.

$$2RSH \xrightarrow[\text{mild}]{[O]} R{-}S{-}S{-}R;\ 2RSH + Na \longrightarrow 2R\bar{S}\overset{\oplus}{Na} \xrightarrow{I_2} RS{-}SR$$

Sodio and potassio salts of thiols on treatment with iodine also give dialkyl disulphides.

**(vi) Reaction with nitric acid :** Nitric acid oxidises the SH group to $SO_3H$ group. Thiophenols are oxidised to arenesulphonic acids by potassium permanganate.

$$RSH \xrightarrow{HNO_3} RSO_3H;\ ArSH \xrightarrow{KMnO_4} ArSO_3H$$

**(vii) Reaction with carbonyl compounds :** Carbonyl compounds form thioacetals and thioketals in the presence of dry hydrogen chloride.

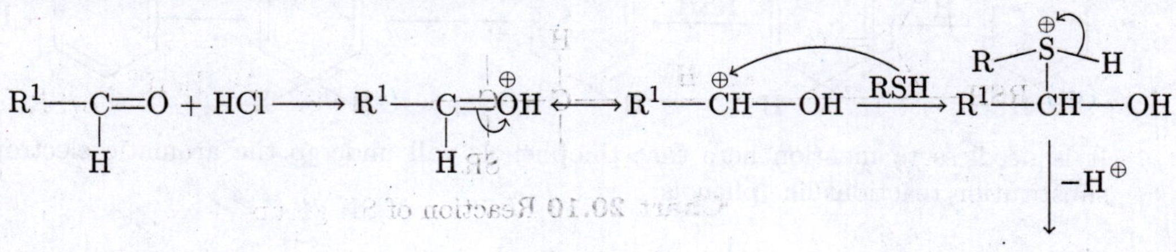

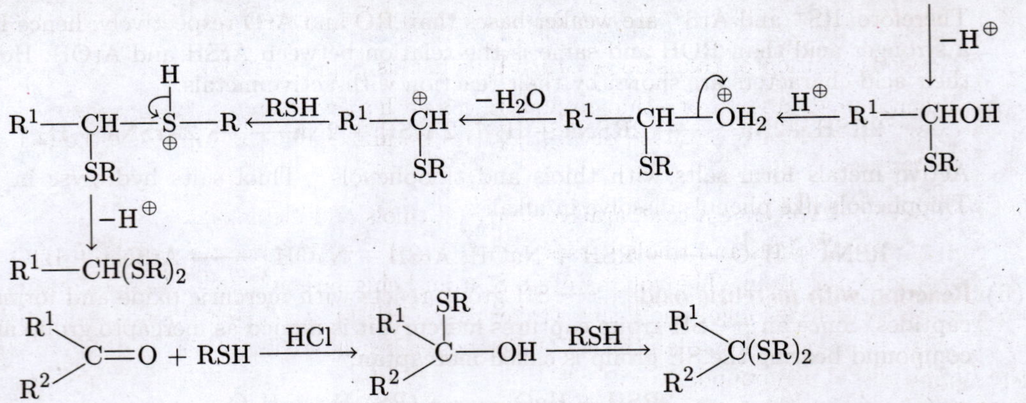

When thioacetals and ketals are refluxed with EtOH/Raney Ni, —SR groups are replaced by H atoms. This is, therefore, a method of converting >C=O function to $>CH_2$ group.

$$Me_2C(SR)_2 \xrightarrow{Ni/EtOH} MeCH_2Me$$

**(viii) Desulphurisation** : When thiols and thiophenols are heated with nickel, desulphurization occurs.

$$RSH \xrightarrow[heat]{Ni} RH + NiS \; ; \; ArSH \xrightarrow[heat]{Ni} ArH + NiS$$

**(ix) Addition to C—C double bonds** : In the presence of an acid thiols and thiophenols add to a C—C double bond to give thioethers.

$$CH_3—CH{=}CH_2 \xrightarrow[2.\ RSH]{1.\ H^{\oplus}} CH_3—\underset{\displaystyle SR}{\underset{|}{CH}}—CH_3;\; CH_3—CH{=}CH_2 \xrightarrow[2.\ ArSH]{1.\ H^{\oplus}} CH_3\underset{\displaystyle SAr}{\underset{|}{C}}HCH_3$$

Epoxides also add thiols and thioethers in the acid medium.

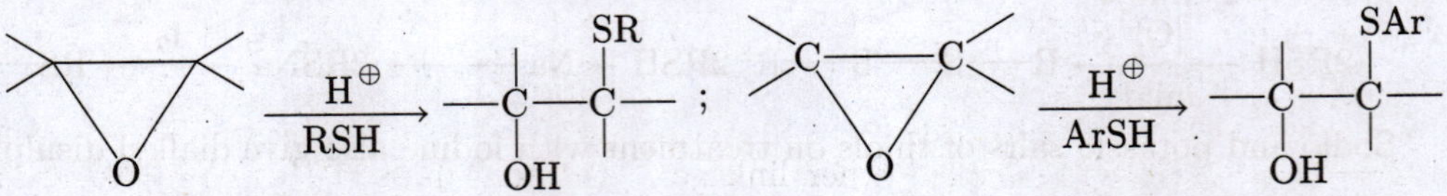

**(x) Aromatic nucleophilic substitution** : When phenols are treated with thiols, alkyl aryl thioethers are formed.

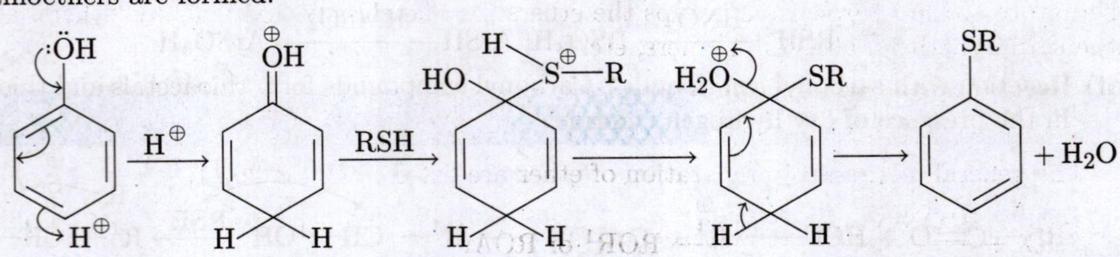

It is needless to mention here that thiophenols will undergo the aromatic electrophilic substitution reactions like phenols.

### 20.4.1 Test

(a) Aqueous solution of thiols do not turn blue litmus paper to red.

(b) When a small amount of a thiol is shaken well with a few drops of aqueous mercuric chloride solution, a white precipitate of $RS.HgCl_2$ is obtained and the solution becomes acidic to litmus.

**Problem 20.13.** Give a comparative study on thiols and alcohols.

**Solution :** Alcohols and thiols possess —OH and —SH functions respectively; both form a $\sigma$ bond with a C atom. The volume of an S atom being larger than that of a O atom, C—S and S—H bonds are longer and hence weaker than C—O and O—H bonds respectively. The electronegativity of S is less than that of O. So there are similarities as well as differences between the two classes of compounds.

| Alcohols | | Thiols | |
|---|---|---|---|
| 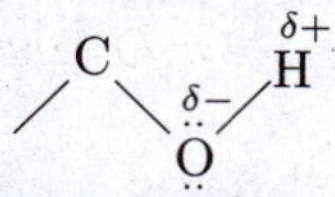 | | 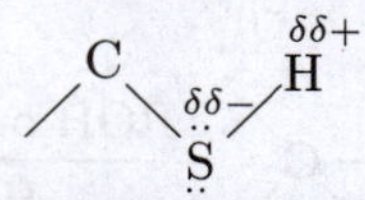 | |
| 1. | Form intermolecular H-bond. | 1. | Do not form intermolecular H-bond. |
| 2. | O—H bond is more polar. | 2. | S—H bond is less polar. |
| 3. | Less acidic | 3. | More acidic |
| 4. | Pleasant smell | 4. | Unpleasant smell |
| 5. | Higher m.p. and b.p. | 5. | Lower m.p. and b.p. |
| 6. | More soluble in water. | 6. | Less soluble in water. |
| 7. | Form acetals. | 7. | Form thioacetals. |
| 8. | Get oxidised to aldehydes and ketones. | 8. | Get oxidised to sulphonic acids. |
| 9. | Evolve $H_2$ with Na, K, Li, etc. | 9. | Evolve $H_2$ with Na, K, Li etc. |
| 10. | Do not react with $HgCl_2$ and HgO. | 10. | React with $HgCl_2$ and HgO. |

## 20.5 Ethers and Thioethers

### 20.5.1 Introduction

C—O—C linkage is called ether linkage : An atom flanked by two hydrocarbon groups or their derivatives constitute the ether linkage. Compounds containing ether linkage are called ethers. The hydrocarbon groups may be alkyl, aryl, cycloalkyl, allyl, benzyl, etc. Depending on the nature of the hydrocarbon groups the ethers may be classified as : dialkyl ethers, alkyl aryl ethers, diaryl ethers, alkyl allyl ethers, etc.

### 20.5.2 Formation of ethers

The general methods of preparation of ether are listed in Chart 20.11.

(a) $$RX \xrightarrow[\text{Williamson synthesis}]{R^1\bar{O}\overset{\oplus}{Na} \text{ or } Ar—\bar{O}\overset{\oplus}{Na}} ROR^1 \text{ or } ROAr$$

(b) $$ArBr \xrightarrow[\text{Ullmann synthesis}]{Ar\bar{O}K^{\oplus}/Cu} Ar—O—Ar$$

(c) 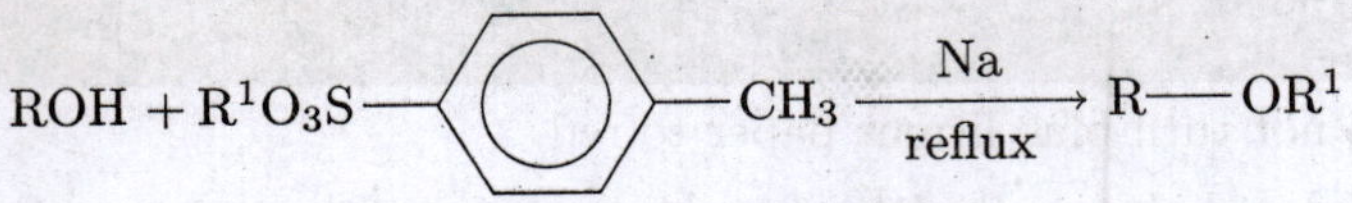

(d) $$ROH + CH_2N_2 \xrightarrow[\text{Lewis acid}]{BF_3 \text{ or } Al(OEt)_3} ROCH_3$$

(e) $$ArOH + CH_2N_2 \longrightarrow ArOCH_3$$

(f) $$ArOH + Me_2SO_4 \longrightarrow ArOMe$$

(g) $$ROH \xrightarrow[\text{heat}]{\text{conc. } H_2SO_4} ROR$$

(h) 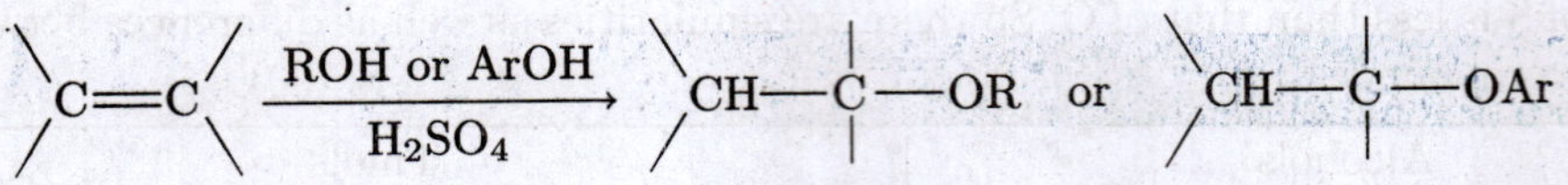

(i) 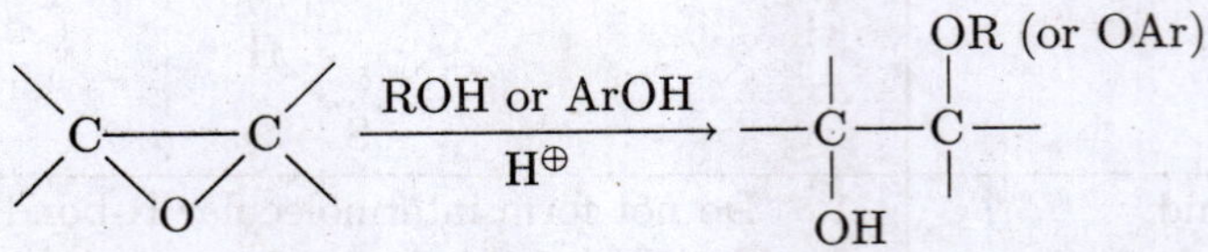

**Chart 20.11** Formation of ethers.

**(a) The Williamson synthesis** : See Sec. 20.1.4 xvii, p. 1159.

**(b) The Ullmann diaryl ether synthesis** : When aryl bromides or iodides are heated with phenoxides in the presence of copper powder, diaryl ethers are formed and the reaction is called the Ullmann diaryl ether synthesis.

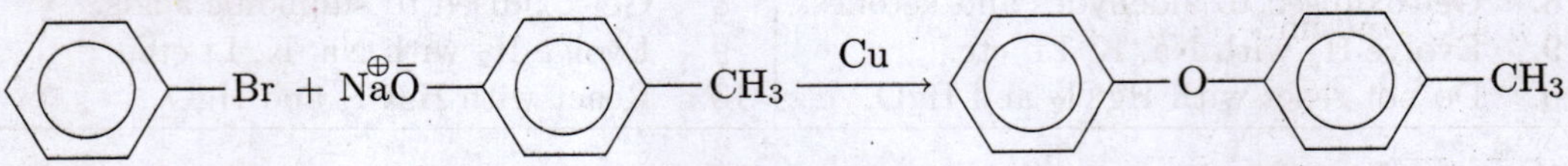

**(c) From alcohol and alkyl *p*-toluenesulphonate** : When an alcohol is refluxed with metallic sodium and an alkyl *p*-toluenesulphonate, a dialkyl ether is formed.

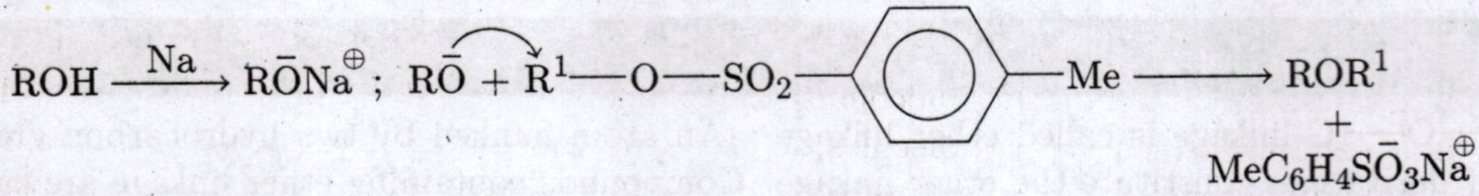

**(d) Conversion of an alcoholic OH group to a methoxyl group** : When alcohols are treated with diazomethane in the presence of a Lewis acid, e.g. an $Al(OEt)_3$ or $BF_3$, a methyl ether forms (See Sec. 20.2.5. xii p. 1184).

$$ROH + CH_2N_2 \xrightarrow{Al(OEt)_3} ROCH_3 + N_2$$

**(e) Conversion of a phenolic OH group to a methoxyl group by diazomethane** : Phenols react with diazomethane in the absence of any Lewis acid to form aryl methyl ether [See Sec. 20.3.4.(v) p. 1204].

$$Ar—O—H + CH_2N_2 \longrightarrow ArOCH_3 + N_2$$

**(f) Conversion of a phenolic OH group to methoxyl group by dimethyl sulphate** : See Sec. 20.3.4 v. p. 1204.

**(g) Etherification of alcohols :** See sec. 20.2.5xi, p. 1184.

**(h) Conversion of a C—C double bond into an ether linkage :** See Sec. 13.5, p. 630 and Sec. 20.3.5 xxviii, p. 1038.

**(i) Conversion of an epoxide to a β-hydroxy ether :** See Sec. 20.2.5. xiii, p. 1185. Alcohols and phenols can be converted to β-hydroxy ethers by their reaction with epoxides.

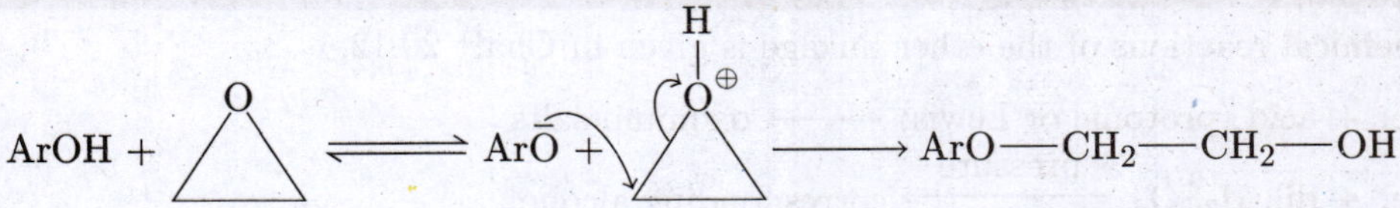

## 20.5.3 Structure and reactivity and physical properties

Ether linkages have angular shape; the C—O—C bond angles are usually 110° to 118°. The O atom holds 2 lone pairs of electrons. Owing to the greater electronegativity of O atom than that of C atom, C—O bonds are polar. Since the shape is angular and the bonds are polar, ether molecules have dipole moments of the order 1.12—1.18D.

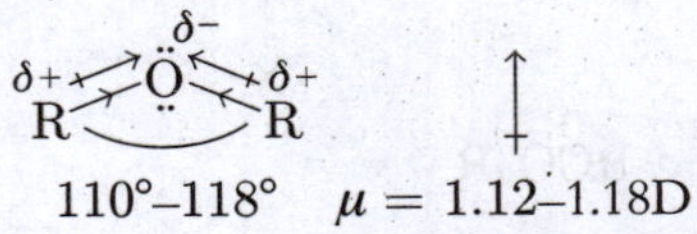

The O atom may be in $sp^2$ of $sp^3$ hybridised state. In an alkyl ether it is in $sp^3$ hybridised state, the two lone pairs of *es* occupy two $sp^3$ A.O.s; while in an aryl or aryl alkyl ether the O atom may assume $sp^2$ hybridised state and as a consequence one of the two lone pairs of *es* occupy a *p* A.O. and undergo *p*-π conjugation with π bonds of the aryl part.

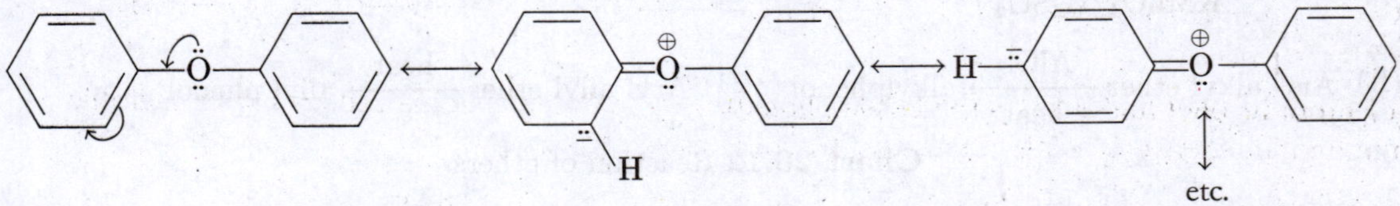

Since there is no O—H bond (cf. alcohols and phenols) in an ether, it cannot form H-bond and as a result of which ether molecules do not associate amongst themselves; but they experience dipole-dipole interaction (because of their dipolar nature) which is weaker than the H-bonding force. Hence ethers are low melting, low boiling and highly volatile. their m.p.s and b.p.s are much lower than those of alcohols of comparable molecular weights, e.g., methyl ether (b.p. −24°C); propane (b.p. −42°C) and ethanol (b.p. 78°C). The m.p. and b.p. increase with increasing molecular weight; b.p. of ethylether is 35°C.

However, they can form H-bond with water molecules and for this reason lower members are slightly soluble in water; the solubility decreases with the increasing molecular weight.

The structure of an ether molecule tells us that it will be weakly basic and it will form salt with acids. This also tells us that an ether can undergo reactions involving C—O bond cleavage and the cleavage requires a protonic or Lewis acid since OR is a bad leaving group. As the O atom draws electrons from the alkyl group, the α-H atoms are expected to be slightly acidic and hence they can be substituted by electrophiles. The acidity order of the α-H atoms is 1° > 2° > 3°.

Since in aryl ethers $p$-$\pi$ conjugation takes place, an aryl ether undergoes electrophilic substitution reactions besides C—O bond breaking reactions. Since the ether linkage has a +R effect, it is an *o*-, *p*-directing group. We shall discuss the general reactions of the ether linkage on the basis said above.

## 20.5.4 Chemical reactions

The chemical reactions of the ether linkage is given in Chart 20.12.

(a) Ether + acid (protonic or Lewis) $\longrightarrow$ oxonium salts

(b) Ether + dil. $H_2SO_4 \xrightarrow{\text{pressure}}$ corresponding alcohol.

(c) (i) $ROR \xrightarrow{\text{HX in cold (X=Br, I)}} RX + ROH$

(ii) $ROR \xrightarrow{\text{HX in hot condition}} 2RX + H_2O$

(iii) $ArOR \xrightarrow{HX} ArOH + RX$

(d) $RCH_2—O—CH_2R \xrightarrow{X_2\ \text{(X=Cl or Br.)}} RCH(X)—O—CHR(X)$

(e) $R_2O + CO \xrightarrow[BF_3]{150\,°C;\ 500\ \text{atm}} RCO_2R$

(f) $RCH_2OCH_2R \xrightarrow[\text{hydroperoxydation}]{O_2} RCH(OOH)—OCH_2R$

(g) Ether $\xrightarrow[KMnO_4/H_2SO_4]{K_2Cr_2O_7/H_2SO_4}$ oxidation products of the corresponding alcohols.

(h) Aryl alkyl ether $\xrightarrow[\text{heat}]{AlCl_3}$ alkylphenol (i) Aryl allyl ether $\xrightarrow{\text{heat}}$ allyl phenol

**Chart 20.12** Reaction of ethers.

**(a) Formation of oxonium salts :** Ethers are basic ($pKa$ of $R_2\overset{\oplus}{O}H$ is approximately —35). They form salts with protonic and Lewis acids at a low temperature. If a solution of hydrogen chloride in dry diethyl ether is cooled to a low temperature, a crystalline oxonium salt is obtained.

$$CH_3—O—CH_3 + HCl \longrightarrow CH_3—\overset{H}{\overset{|}{\underset{\oplus}{O}}}—CH_3Cl^- \ (\text{m.p.} -97\,°C)$$

$$Et—O—Et + HCl \longrightarrow Et—\overset{H}{\overset{|}{\underset{\oplus}{O}}}—EtCl^- \ (\text{m.p.} -92\,°C)$$

$$Et—O—Et + 2HCl \longrightarrow [Et—\overset{H}{\overset{|}{\underset{H}{\underset{|}{O}}}}—Et]^{\oplus\oplus}2Cl^- \ (\text{m.p.} -88\,°C)$$

$$R{-}O{-}R + BF_3 \rightleftharpoons R{-}\overset{\oplus}{O}(R){-}\overset{-}{B}F_3$$

$$R{-}O{-}R + H_2SO_4 \rightleftharpoons R_2\overset{\oplus}{O}{-}HHSO_4^-$$

The coordinating ability of the O atom of ether is the cause for dissolution of the Grignard reagents in some ethers (See Sec. 16.4.5, p. 839). However, trialkyl oxonium salts may be prepared by treating a dialkyl ether with an alkyl bromide in the presence of silver fluoroborate.

$$Et_2O + EtBr + AgBF_4 \longrightarrow Et_3\overset{\oplus}{O}\overset{-}{B}F_4 + AgBr$$

It is because of the salt formation, the ethers are soluble in concentrated aqueous solutions of strong acids. When the acid solutions are diluted with water, ethers are thrown out from the solution. When water is added to an oxonium salt solution, the anions get hydrated by the water molecules and the oxonium ion gives up the proton to the water molecules since water solvates the proton better than the ether, the better solvating effect is attributed to smaller volume of hydronium ion and higher charge density on its O atom. $H_3O^{\oplus}$ forms $H_9O_4^{\oplus}$.

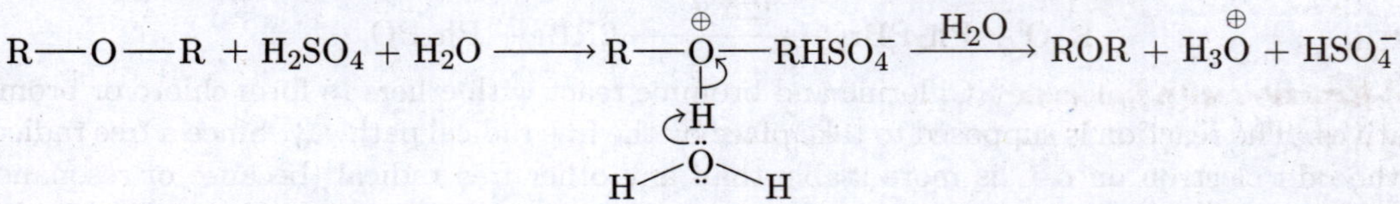

**(b) Reaction with dilute sulphuric acid :** When ethers are heated with dilute sulphuric acid under pressure, alcohols are formed.

$$R_2O + H_2O \xrightarrow[\text{pressure}]{\text{dil. } H_2SO_4} 2ROH$$

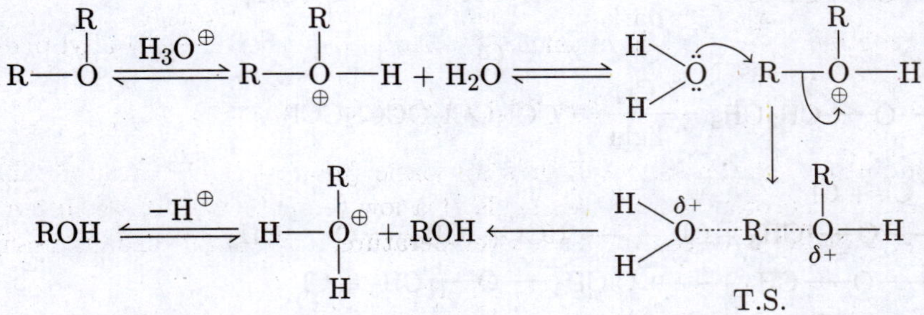

**(c) Reaction of hydrohaloacids :** When ethers are treated with hydroiodic acid or hydrobromic acid in cold, an alcohol and an alkyl halide form.

$$R{-}OR^1 + HI\ (HBr) \longrightarrow R{-}I\ (RBr) + R^1OH$$

$$Ar{-}OR^1 + HI \longrightarrow ArOH + R^1I$$

Usually methyl iodide forms when one of the hydrocarbon groups is methyl. However, if there is a 3° alkyl group, a 3° alkyl iodide and methyl alcohol are formed. In this condition it is important to note that HI is more reactive than HBr. The reaction may follow the $S_N1$ or the $S_N2$ pathway.

$$R_2O + HI \longrightarrow R_2\overset{\oplus}{O}HI^-;\ R_2\overset{\oplus}{O}H \xrightleftharpoons{\text{slow}} ROH + R^{\oplus};\ R^{\oplus} + I^- \longrightarrow RI$$

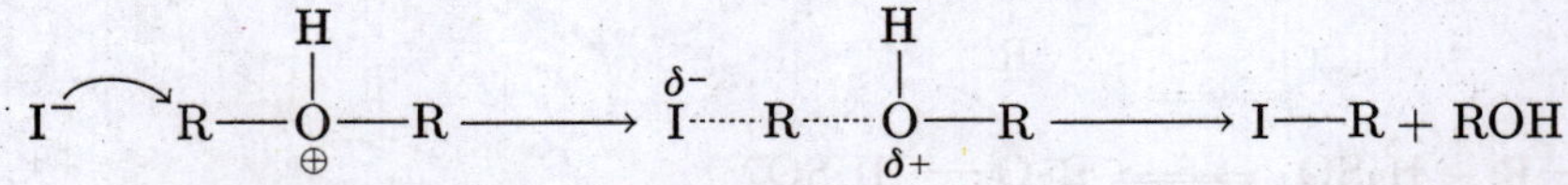

When ethers are heated with HI (HBr) two moles of alkyl iodides (bromides) are obtained; at first an alcohol and an alkyl iodide form and then the alcohol changes to the alkyl iodide again either by the $S_N1$ or by the $S_N2$ pathway.

$$R—OH + HI \longrightarrow RI + H_2O$$

Ethers are also readily cleaved by Lewis acids.

$$R_2O + AlCl_3 \longrightarrow R_2\overset{\oplus}{O}—\bar{A}lCl_3 \xrightarrow{\Delta} RCl + R—O—AlCl_2 \xrightarrow{R_2O} RCl + R—O—Al(Cl)—OR \xrightarrow{R_2O} RCl + (RO)_3Al$$

Triphenyl phosphine dibromide also splits ethers.

$$R_2O + Ph_3PBr_2 \xrightarrow{125\,°C} 2RBr + Ph_3PO$$

**(b) Reaction with halogens :** Chlorine and bromine react with ethers to form chloro or bromo derivatives. The reaction is supposed to take place by the free radical pathway. Since a free radical with the odd electron on $\alpha$-C is more stable than any other free radical (because of resonance stabilization), $\alpha$-Hs are most active in this reaction. However, the extent of halogenation in this process depends on the condition of the reaction. For example, when ethyl ether is treated with chlorine in the dark, $\alpha$, $\alpha'$-dichloro derivative is obtained; on the other hand, in the presence of light, perchloro or decachlorodiethyl ether forms.

$$CH_3CH_2—O—CH_2CH_3 \xrightarrow[\text{dark}]{Cl_2} CH_3CH(Cl)—O—CH(Cl)CH_3$$

$$CH_3CH_2—O—CH_2CH_3 \xrightarrow[\text{light}]{Cl_2} CCl_3CCl_2OCCl_2CCl_3$$

$$Cl_2 \xrightarrow{h\nu} Cl\cdot + Cl\cdot$$

$$H—CH_2—O—CH_3 + Cl\cdot \longrightarrow HCl + \cdot CH_2—\ddot{O}—CH_3$$

$$Cl_2 + \dot{C}H_2—\ddot{O}—CH_3 \longrightarrow ClCH_2—O—CH_3 + Cl\cdot$$

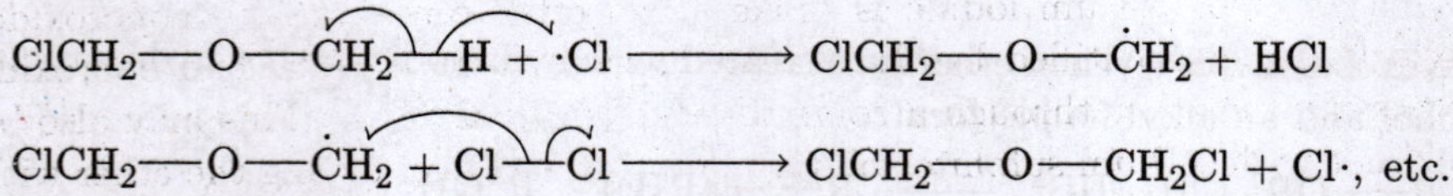

**(c) Ester formation :** In the presence of boron trifluoride and water, ethers add to carbon monoxide to form esters.

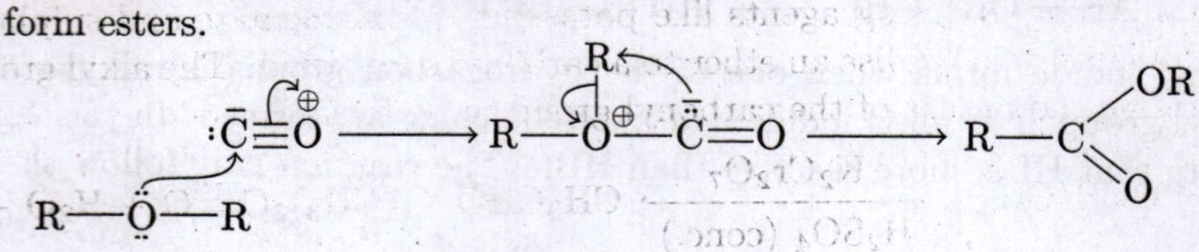

When acyl halides are heated with an ether and a Lewis acid, e.g., anhydrous $ZnCl_2$ and $AlCl_3$, esters and alkyl halides are formed.

$$R^1COCl + AlCl_3 \rightleftharpoons R^1\overset{\oplus}{C}O\bar{A}lCl_4;\ R^1\overset{\oplus}{C}O + R_2O \longrightarrow R^1-\underset{\overset{|}{\underset{R\ \ \oplus\ \ R}{O}}}{C}=O$$

$$\longrightarrow R^1-\underset{\underset{OR}{|}}{C}=O + R^{\oplus};\ R^{\oplus} + AlCl_4^- \longrightarrow RCl + AlCl_3$$

**(f) Auto-oxidation :** When an ether is left exposed to air, oxygen of the air reacts with it and forms hydroperoxide. Thus *i*-propyl ether form di-*i*-propyl hydroperoxide. The $\alpha$-H atoms are

$$(CH_3)_2CH-O-CH(CH_3)_2 + O_2 \longrightarrow (CH_3)_2CH-O-\underset{\underset{CH_3}{|}}{\overset{\overset{CH_3}{|}}{C}}-O-O-H$$

involved in the auto-oxidation, i.e., in the hydroperoxide formation. Molecular oxygen itself is a poorly reactive radical. However, if a radical is present in any system, it reacts with oxygen molecule to form peroxy radical which then abstracts an H radical to form hydroperoxide.

$$CH_3CH_2OCH_2CH_3 \xrightarrow{h\nu} CH_3\dot{C}HOCH_2CH_3 + H\cdot$$

$$CH_3\dot{C}HOC_2CH_3 + O_2 \longrightarrow CH_3\underset{\underset{O-O\cdot}{|}}{C}HOCH_2CH_3$$

$$CH_3\underset{\underset{O-O\cdot}{|}}{C}H-O-CH_2CH_3 + CH_3CH_2-O-CH_2CH_3 \longrightarrow CH_3\underset{\underset{O-O-H}{|}}{C}H-O-CH_2CH_3 + CH_3\dot{C}HOCH_2-CH_3, \text{ etc.}$$

Hydroperoxides are less volatile than ethers. Therefore, the % of hydroperoxide increases when an ether is used repeatedly. The compound is also explosive. So an ether contaminated with hydroperoxide may undergo explosion. To avoid this, an ether must be tested before use. When an acidified solution of potassium iodide is shaken with ether containing hydroperoxide, iodine is liberated which can be identified by the starch solution as usual. The hydroperoxide may be removed by passing the ether through a column packed with alumina. This may also be done by treating the ether with sodium sulphite solution in water and then drying the ether with fused calcium chloride.

**(g) Oxidation :** Strong oxidising agents like potassium permanganate and sulphuric acid or dichromate and sulphuric acid oxidise an ether to a carbonyl compound. The alkyl groups present in the ether determine the nature of the carbonyl group to be formed.

$$CH_3CH_2-OCH(CH_3)_2 \xrightarrow[H_2SO_4\ (conc.)]{K_2Cr_2O_7} CH_3CHO + (CH_3)_2C=O + H_2O + H^{\oplus}$$

**(h) Rearrangement of aryl alkyl ethers :** Certain aryl alkyl ethers on heating with a Lewis acid rearrange to *o*- and *p*-alkylphenols (See Fries rearrangement; p. 769, Sec. 14.19, a similar reaction).

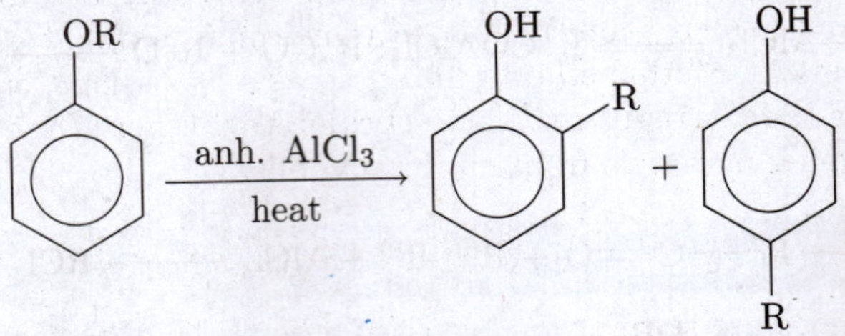

**(i) Rearrangement of aryl allyl ethers :** See Sec. 14.17.1, p. 753 (The Claisen Rearrangement).

## 20.5.5 Synthesis of a few Important ethers

**(a) Synthesis of vinyl ether :** This is usually synthesised from $\alpha$, $\alpha'$-dichlorodiethyl ether by the action of alcoholic KOH.

$$CH_2(H)—CH(Cl)—O—CH(Cl)—CH_2(H) \xrightarrow{\text{alc. KOH}} CH_2{=}CH—O—CH{=}CH_2 + 2H_2O + 2KCl$$

This is used as an anaesthetic.

**(b) Synthesis of allyl ether :** This may be synthesised by the Williamson synthesis; the starting materials are allyl iodide and sodium derivative of allyl alcohol.

$$CH_2{=}CH—CH_2\bar{O}\overset{\oplus}{Na} + I—CH_2—CH{=}CH_2 \longrightarrow (CH_2{=}CH—CH_2)_2O + NaI$$

**(c) Synthesis of dichloroethyl ether :** This is prepared from ethylene via ethylene chlorohydrin.

$$2CH_2{=}CH_2 + 2HOCl \longrightarrow 2HOCH_2CH_2Cl \xrightarrow{HCl} ClCH_2CH_2—O—CH_2CH_2Cl$$

**(d) Synthesis of diglime and tetrahydrofuran :** See p. 1239 and p. 1096 & 1098 respectively.

## 20.5.6 Analysis

**(a) Qualitative test :** The ether linkage is a very unreactive functional group. Ethers have similarities with hydrocarbons. Ethers do not react with metallic sodium and phosphorus halides; whereas isomeric alcohols (ethers are constitutional isomers of alcohols) react with them. On the other hand, ethers are soluble in concentrated sulphuric acid at a low temperature whereas hydrocarbons are insoluble in concentrated sulphuric acid.

**(b) Estimation of methoxyl and ethoxyl groups (The Zeisel method) :** Methyl or ethyl ether is refluxed with HI and the methyl iodide or ethyl iodide so formed is distilled out to a vessel containing alcoholic solution of silver nitrate whereby silver iodide precipitates. The precipitate of AgI is filtered, dried and weighed. The vessel containing the RI and silver nitrate is weighed before the experiment and after it the weight of the distillate is obtained by subtracting the two weights. When the weight of the iodine is subtracted from the weight of the distillate, the weight of the alkyl group is obtained.

## 20.5.7 Uses of ethers

Ethers are mostly used as solvents and extracting agents for organic compounds. Methyl ether, ethyl ether, *i*-propyl ether, *i*-amyl ether, tetrahydrofuran, dioxane and diglime deserve to be mentioned in this respect. Methyl ether is also used as a propellant for aerosol sprays. Others are used as solvents for the Grignard reagents. Dichlorethyl ether is used as a selective solvent for removing sludgeforming materials from lubricating oils.

## 20.5.8 Formation of thioethers

Thioethers or sulphides are sulphur analogues of ethers. Their formation are listed in Chart 20.13.

(a) $2R{-}X \xrightarrow[\Delta]{K_2S} R_2S + 2KX$

(b) $5R_2O + P_2S_5 \xrightarrow{\Delta} 5R_2S + P_2O_5$

(c) $RX + R^1\bar{S}Na + R{-}S{-}R^1 + NaX$

(d) $2RSH \text{ (vapour)} \xrightarrow[Al_2O_3\text{–}ZnS \text{ at } 300\,°C]{\text{passed over}} R_2S + H_2S.$

(e) $R^1CH{=}CH_2 \xrightarrow[\text{organic peroxide}]{RSH} R^1CH_2CH_2SR$

(f) $ArH \xrightarrow{SCl_2/AlCl_3 \text{ cat.}} Ar_2S$

(g) $2RMgX + S \longrightarrow R_2S$ (h) $ArBr + R\bar{S}Na^{\oplus} \longrightarrow ArSR$

(i) $2ArN_2^{\oplus}X^- + K_2S \longrightarrow Ar_2S + N_2 + 2KX$

(j) 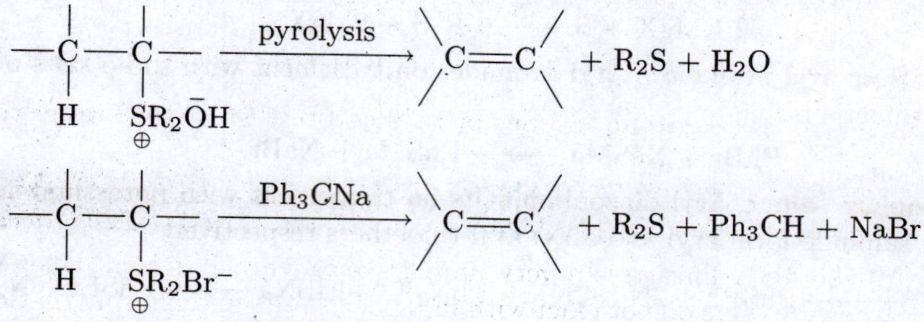

(k) 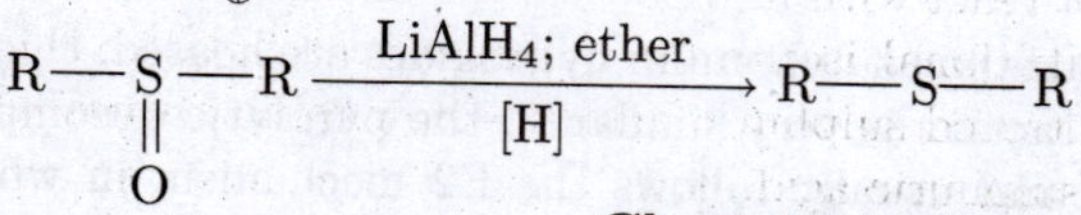

**Chart 20.13** Formation of Thioethers.

(a) **From alkyl halides :** Alkyl halides on heating with alkali metal sulphides form thioethers via thiol salts.

$$K_2S + R{-}X \longrightarrow KX + R{-}\bar{S}K^{\oplus}$$

$$R{-}\bar{S}K^{\oplus} + R{-}X \longrightarrow R{-}S{-}R + KX$$

Depending on the nature of the alkyl group, the reaction may follow the $S_N2$ or $S_N1$ pathway.

**(b) From ethers :** When ethers are heated with phosphorus pentasulphide, thioethers form. The yield is poor.

$$Et_2O + P_2S_5 \longrightarrow Et_2S + P_2O_5$$

**(c) By the Williamson synthesis :** When alkyl halides are heated with sodium or potassium salts of thiols, thioethers are formed. The yield is very good.

$$R^1X + R\bar{S}\overset{\oplus}{Na} \longrightarrow R^1—S—R$$

**(d) From thiols :** When vapour of thiols are passed over alumina and zinc sulphide catalyst at 300 °C, thioethers are formed.

$$EtSH \xrightarrow[300\,°C]{Al_2O_3\text{-}ZnS} Et_2S + H_2S$$

**(e) By the free radical addition of thiols to alkenes :** In the presence of organic peroxides thiols add to alkenes anti-Markowinkoffically to form thioethers.

$$R^1CH{=}CH_2 + RSH \xrightarrow{(R^2CO_2)_2} R^1CH_2CH_2SR$$

$$(R^2CO_2)_2 \xrightarrow{\Delta} 2R^2\cdot + 2CO_2; \quad R^2\cdot + RSH \longrightarrow RS\cdot + R^2H$$

$$R^1CH{=}CH_2 + RS\cdot \longrightarrow R^1\dot{C}H—CH_2SR$$

$$R^1\dot{C}H—CH_2SR + RSH \longrightarrow R^1CH_2CH_2SR + RS\cdot, \text{ etc.}$$

$$RS\cdot + \cdot SR \longrightarrow RS—SR, \text{ etc.}$$

**(f) By sulpharisation :** When arenes are heated with sulphur dichloride in the presence of a Lewis acid, e.g., $AlCl_3$, aryl thioethers form.

$$2ArH + SCl_2 \xrightarrow[heat]{2AlCl_3} Ar_2S + 2H^{\oplus}\ \bar{A}lCl_4$$

**(g) From the Grignard reagents :** When excess of the Grignard reagent is treated with sulphur, thioethers form.

$$2\ RMgX + S \longrightarrow R_2S + 2\overset{\oplus}{Mg}X$$

**(h) Aryl thioethers from aryl bromide :** Aryl bromides on treatment with sodio salts of thiols give alkyl aryl thioethers.

$$PhBr + \overset{\oplus}{Na}\bar{S}Me \longrightarrow PhSMe + NaBr$$

**(i) From aryl diazonium salts :** Aryl diazonium salts on treatement with potassium sulphide or sodiumalkyl sulphide form aryl and alkyl aryl thioethers respectively.

$$2\ Ph\overset{\oplus}{N}_2X^- + K_2S \longrightarrow PhSPh + 2N_2 + 2KX; \quad Ph\overset{\oplus}{N}_2X^- + Et\bar{S}\overset{\oplus}{Na} \longrightarrow PhSEt + N_2 + NaX$$

**(j) From sulphonium compound :** When trialkyl sulphonium hydroxides are heated, thioethers, alkenes and water are formed. The reaction is very similar to the pyrolytic decomposition of quaternary ammonium hydroxide and usually follows the E2 mechanism in which the least sterically hindered and the most acidic H gets eliminated from an anti-periplanar arrangement.

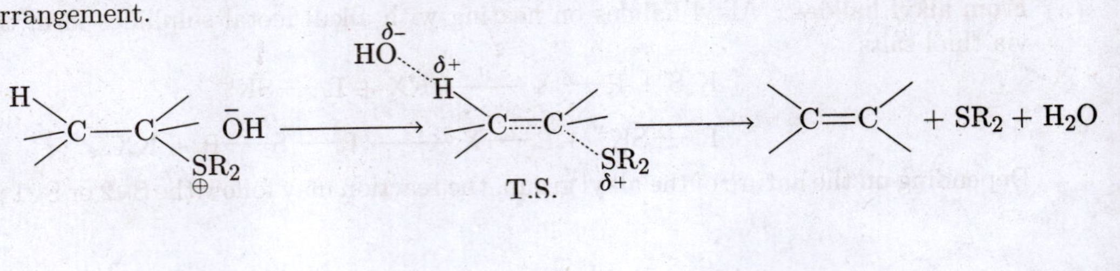

If dimethyl *t*-alkyl sulphonium hydroxide is a substrate, the reaction is supposed to take place via a carbenium ion and thus follows the E1 pathway.

$$Me_3C—\overset{\oplus}{S}Me_2\bar{O}H \longrightarrow Me_2S + Me_2\overset{\oplus}{C}—\overset{H}{\overset{|}{C}}H_2 \xrightarrow{\bar{O}H} Me_2C{=}CH_2 + H_2O$$

Sulphonium bromides on heating with sodiumtriphenyl methide, form alkyl ethers. This also follows the E2 pathway. However, a dimethyl *t*-alkyl sulphonium ion undergoes the E1 reaction.

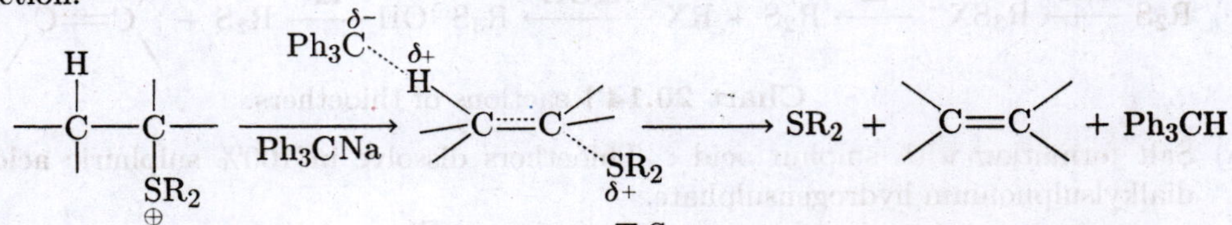

T.S.

**(k) From dialkyl sulphoxide :** When dialkyl sulphoxides are reduced with lithium aluminium hydride, alkyl thioethers are formed.

$$Me—\overset{O}{\overset{||}{S}}—Me \xrightarrow[\text{ether}]{LiAlH_4} Me_2S$$

## 20.5.9 Structure, reactivity and physical properties

Thioethers are sulphur analogues of ethers; so thioethers and ethers have similar structures, reactivity and physical properties. C—S bonds are polar and $R_2S$ has angular shape. So molecules of alkyl thioethers have dipole moments. The S atom bears two pairs of non-bonded *e*s. In alkyl ethers S is in the $sp^3$ hybridised state; while in aryl or alkyl aryl ethers, it remains in $sp^2$ hybridised state and one of the two pairs of *e*s occupies a *p*. A.O. which is parallel to the *p*. A.O.s of the aromatic ring and thus the *e*s get delocalsied. Owing to delocalisation, the *o*- and *p*-positions become slightly negatively charged and, therefore, RS group (like —OR group) is an *o*-/*p*-directing group. Thioethers can donate their non-bonded *e*s and hence they are basic; since the volume of S atom is greater than that of O atom, the electronic charge density on the S atom of a thioether is less than that on the O atom in an ether and so thioethers are less basic than ethers. For the greater volume of S than O the C—S bond in a thioether is weaker than the C—O bond in an ether. Therefore, the C—S bond breaks more easily than the C—O bond.

Thioether molecules do not contain any active H atom but the molecules are polar. So they cannot associate amongst themselves but they experience dipole-dipole interactions. Thus thioethers have low m.p.s and b.p.s : $Me_2S$ (b.p. 38 °C), MeSEt (b.p. 67 °C) and $Et_2S$ (b.p. 92 °C). Thioethers are unpleasant-smelling oils. They are insoluble in water but soluble in 100 per cent sulphuric acid and in organic solvents.

## 20.5.10 Chemical reactions

The reactions of thioethers are listed in Chart 20.14.

**(a)** $R_2S \xrightarrow{100\%\ H_2SO_4} R_2SH^{\oplus}HSO_4^-$ **(b)** $R_2\ddot{S} \xrightarrow{HgCl_2} R_2\overset{\oplus}{S}—Hg\bar{C}l_2$

(c) $R_2S \xrightarrow[HIO_4 \text{ at } 0\,°C]{H_2O_2 \text{ or}} R_2S{=}O \xrightarrow[\text{or } KMnO_4]{H_2O_2}$ 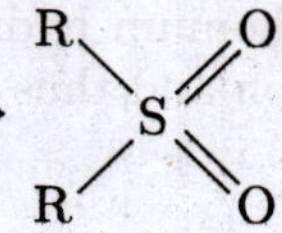

(d) $RSR^1 \xrightarrow{H_2/Ni} RH + R^1H + NiS$

(e) $R_2S \xrightarrow{Br_2} R_2SBr_2$

(f) $R_2S \xrightarrow{RX} R_3\overset{\oplus}{S}X^- \xrightarrow{\Delta} R_2S + RX \xrightarrow{\text{'AgOH'}} R_3S^{\oplus}\bar{O}H \xrightarrow{\Delta} R_2S + \;>C{=}C<\; + H_2O$

**Chart 20.14** Reactions of thioethers.

(a) **Salt formation with sulphur acid :** Thioethers dissolve in 100% sulphuric acid to form dialkylsulphonium hydrogensulphate.

$$Et_2S + H_2SO_4 \longrightarrow Et_2\overset{\oplus}{S}HHSO_4^-$$

(b) **Formation of metal complexes :** Thioether form metal complexes with metal salts.

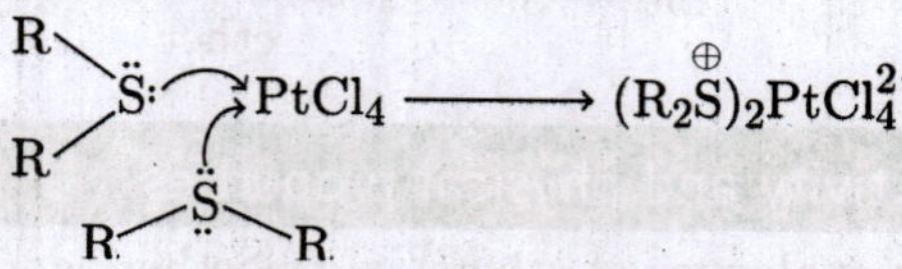

(c) **Oxidation of thioethers :** Hydrogen peroxide can oxidize thioethers to sulphones via sulphoxides. Periodic acid at 0 °C oxidises thioethers to sulphoxides; which can be converted by the same reagent to sulphone at a high temperature. Peracids oxidise a thioether to the corresponding sulphoxide.

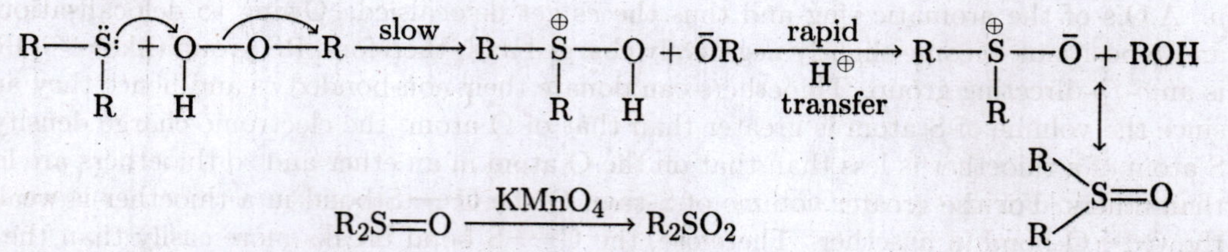

$$R_2S{=}O \xrightarrow{KMnO_4} R_2SO_2$$

(d) **Desulphurisation :** When a mixture of hydrogen and thioether vapour is passed over Raney Ni, alkanes are formed.

$$MeSEt \xrightarrow{Ni/H_2} CH_4 + C_2H_6 + NiS$$

(e) **Reaction with bromine :** Bromine adds to thioethers and forms dialkyl dibromides.

$Me\,SEt + Br_2 \longrightarrow$ 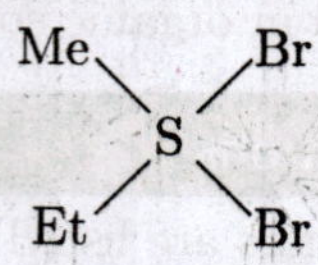

(f) **Formation of sulphonium salt :** Alkyl halides add to thioethers and form sulphonium salts which on heating give thioethers.

## 20.6 Oxiranes or Epoxides

### 20.6.1 Introduction

Oxiranes contain three membered rings in which two are C atoms and one is a O atom and hence oxiranes may be called derivatives of ethylene oxide. They may also be called epoxides or alkene oxides.

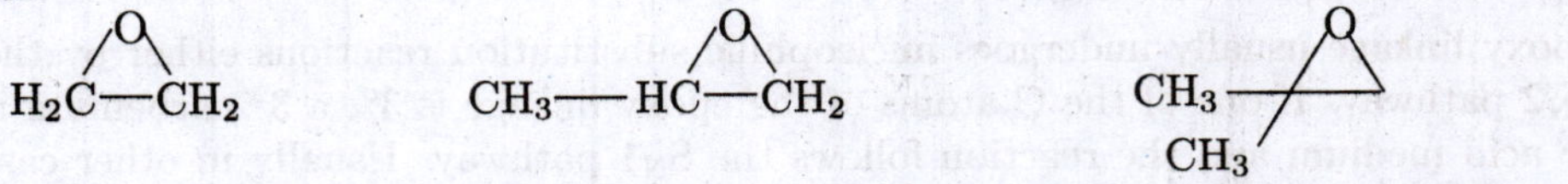

Oxacyclopropane — Ethylene oxide; 2-Methyloxacyclopropane — Propylene oxide; 2, 2-Dimethyloxacyclopropane — is-butylene oxide

### 20.6.2 Formation of epoxy-linkage

The methods are given below :

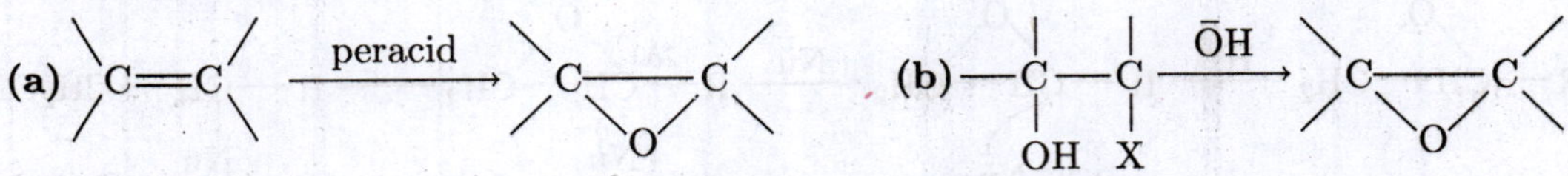

**(a) By the action of peracids on alkenes :** See Sec. 13.7.3, p. 637.

**(b) From $\alpha$-halohydrins :** When $\alpha$-halohydrins are treated with bases, $\alpha$-haloalkoxides are formed. These then undergo internal Williamson synthesis and form epoxides.

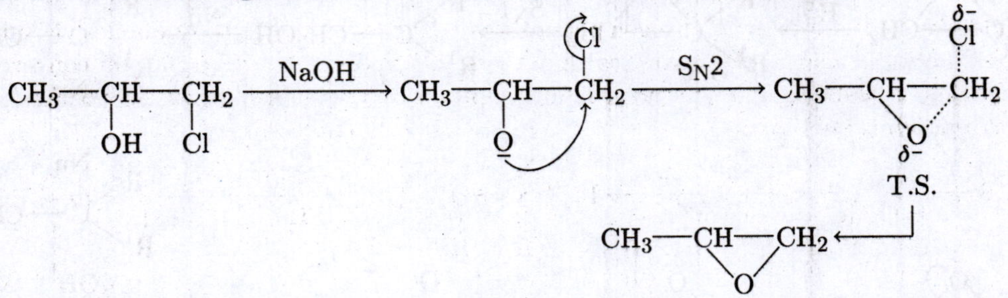

Since halohydrins are prepared from alkenes by the action of halogen-water, the epoxides are prepared from alkenes via halohydrins.

### 20.6.3 Structure, reactivity and orientation

Oxiranes possess triangular planar ring structure. The C—O—C and O—C—C bond angles are approximately 67° and 57° respectively. The C and O atoms are in $sp^3$ hybridised states. Therefore, the expected bond angels are 109°28′. Thus the molecules have high I-strain (small angle-strain).

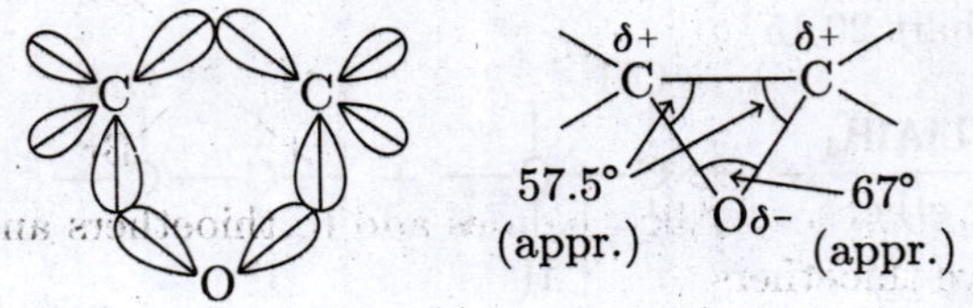

The C—C and C—O $\sigma$ bonds are formed by the overlap of $sp^3$ orbitals in a manner which is neither end-on nor side-on but an intermediate between the two; this form of overlapping makes the C—O and C—C bonds very weak. Furthermore; because of the greater electronegativity of the O atom than that of the C atom, the C—O bonds are polar. The I-strain, the weak $\sigma$ bond character and the polarity of the C—O bonds make the oxiranes very unstable and highly reactive. They undergo acid-catalysed as well as base-catalysed reactions, whereas ethers only undergo acid-catalysed reactions.

Epoxy linkage usually undergoes nucleophilic substitution reactions either by the $S_N1$ or by the $S_N2$ pathway. If one of the C atoms of the epoxy linkage is 3°, a 3° carbenium ion develops in the acid medium and the reaction follows the $S_N1$ pathway. Usually in other cases reactions are $S_N2$. In the acid-catalysed $S_N2$ reactions, the nucleophile attacks that C atom which better accommodates the positive charge in the T.S.; whereas in the base-catalysed $S_N2$ reactions the order of the nucleophilic attack is as usual, i.e., 1° > 2° > 3°. Thus usually less substituted epoxy-C is attacked by the nucleophile in the case of a base-catalysed reaction.

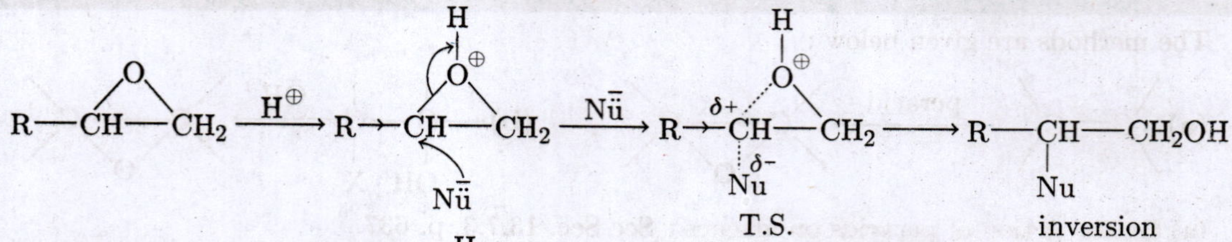

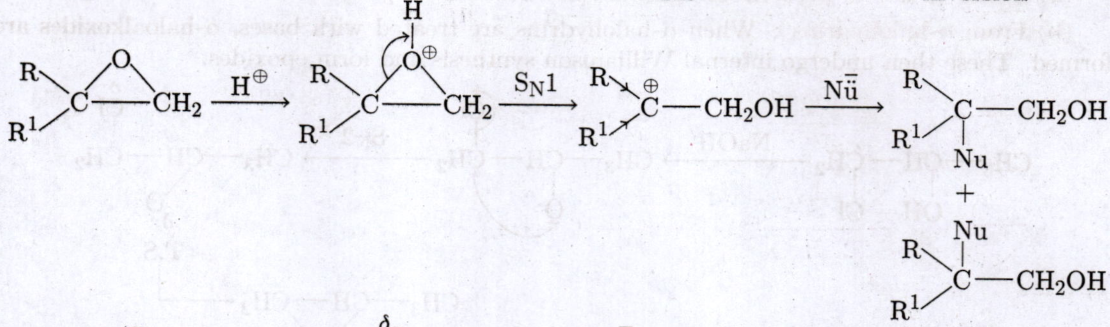

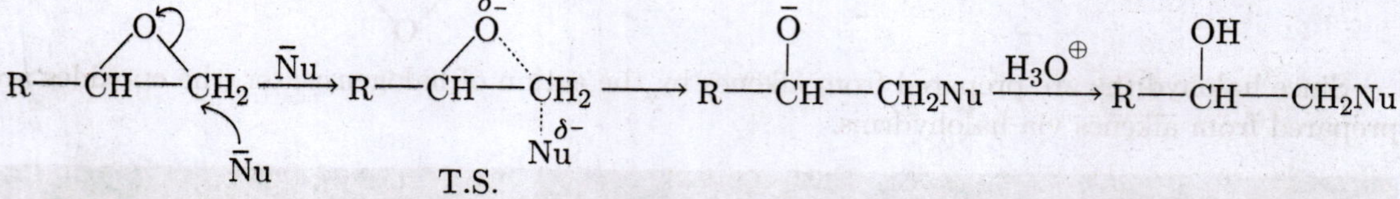

Therefore, the stereochemistry of the $S_N$ reactions of epoxy linkage is the stereocheniistry of $S_N1$ and $S_N2$ reactions.

## 20.6.4 Reactions of the expoxy-linkage

These are listed in Chart 20.15.

(a) 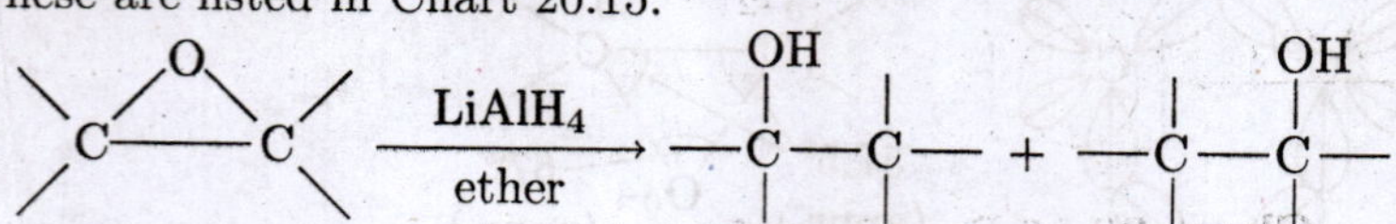

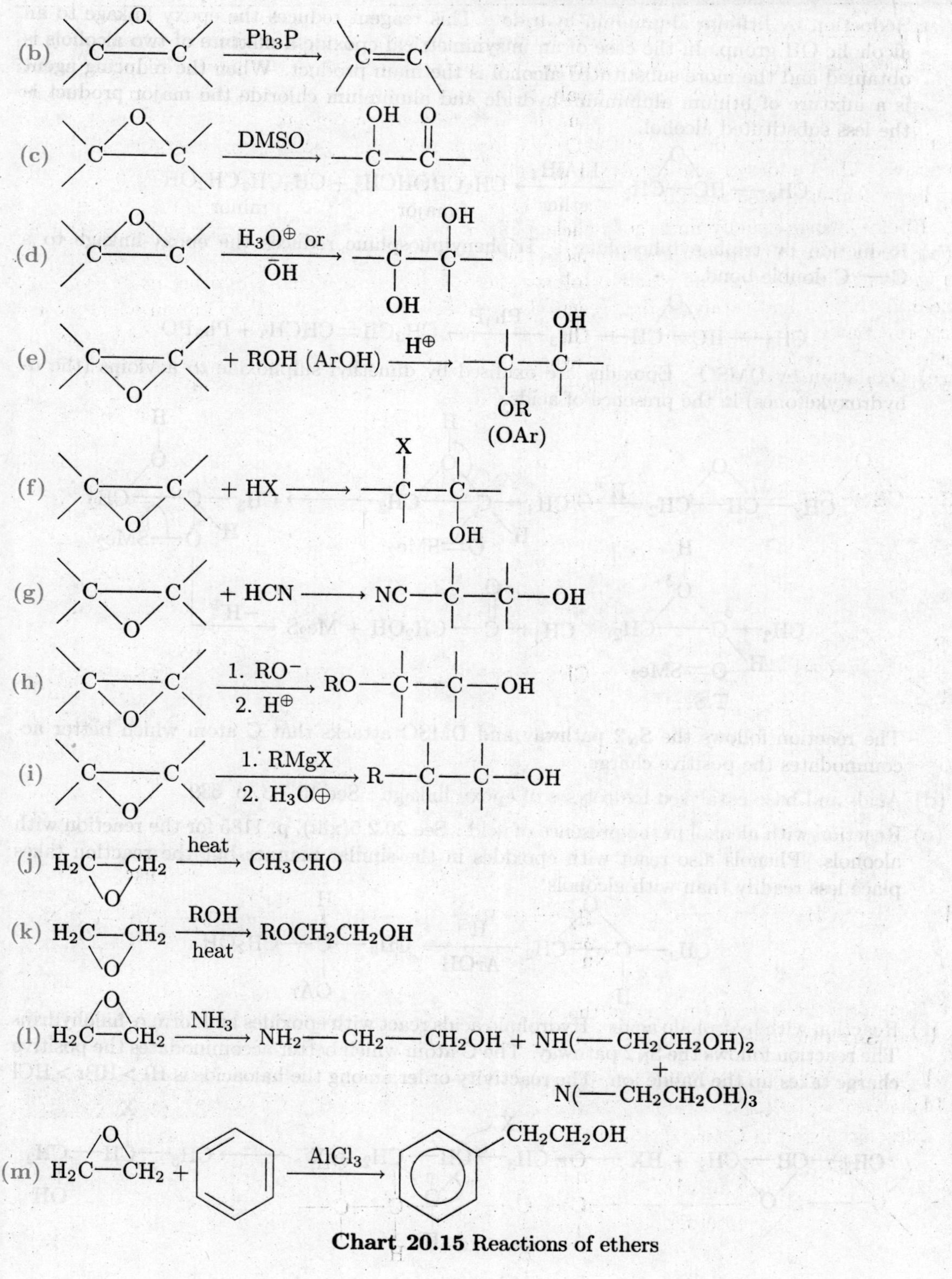

**Chart 20.15** Reactions of ethers

(a) **Reduction by lithium aluminium hydride :** This reagent reduces the epoxy linkage to an alcoholic OH group. In the case of an unsymmetrical epoxide a mixture of two alcohols is obtained and the more substituted alcohol is the main product. When the reducing agent is a mixture of lithium aluminium hydride and aluminium chloride the major product is the less substituted alcohol.

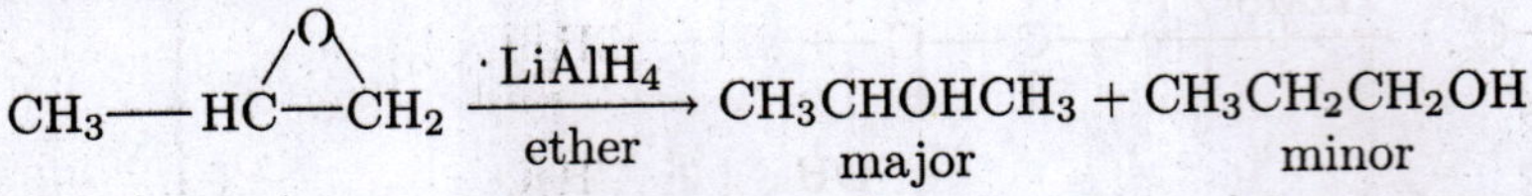

(b) **Reduction by triphenylphosphine :** Triphenylphosphine reduces the epoxy-linkage to a C—C double bond.

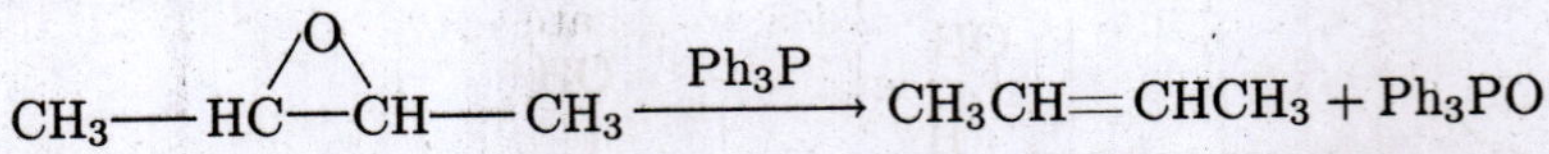

(c) **Oxidation by DMSO :** Epoxides are oxidised by dimethyl sulphoxide to acyloins (the α-hydroxyketones) in the presence of acids.

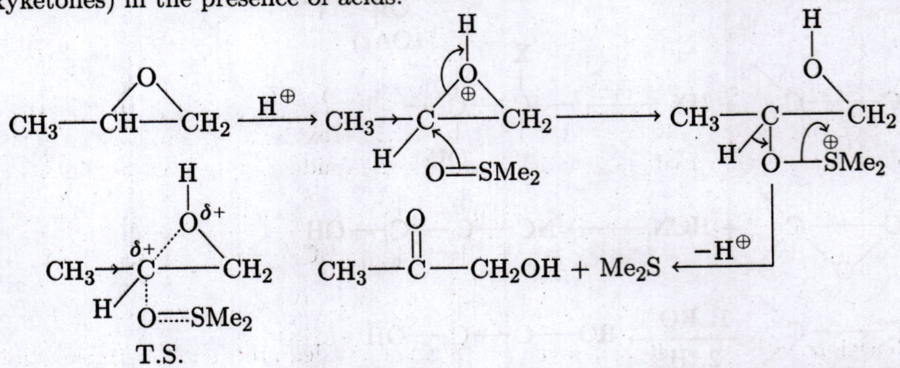

The reaction follows the $S_N2$ pathway and DMSO attacks that C atom which better accommodates the positive charge.

(d) **Acid- and base-catalysed hydrolyses of epoxy linkage :** See 13.7.3, p. 639.

(e) **Reaction with alcohol in the presence of acid :** See 20.2.5(xiii), p. 1185 for the reaction with alcohols. Phenols also react with epoxides in the similar manner but the reaction takes place less readily than with alcohols.

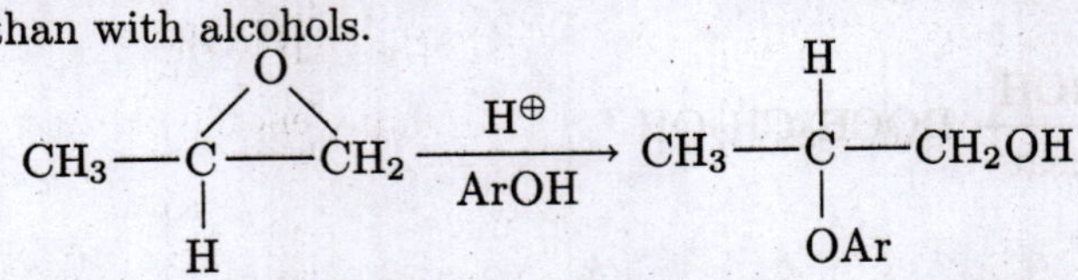

(f) **Reaction with hydrohalo acids :** Hydrohalo acids react with epoxides and form α-halohydrins. The reaction follows the $S_N2$ pathway. The C atom which better accommodates the positive charge takes up the halide ion. The reactivity order among the haloacids is HI > HBr > HCl

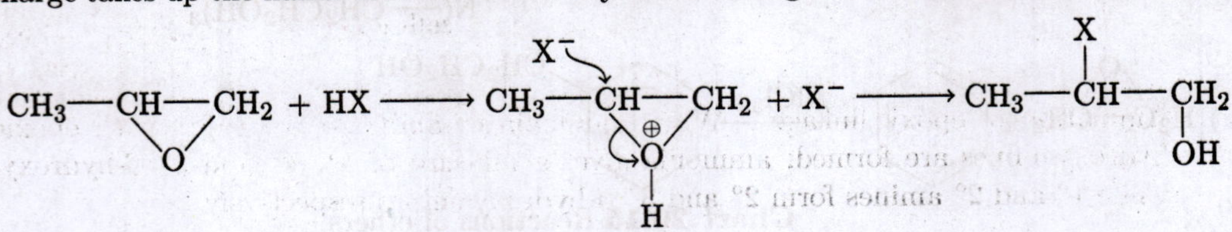

(g) **Reaction with hydrogen cyanide :** In the presence of a base HCN adds to epoxides and forms α-cyanohydrins.

$$HCN + \bar{O}H \rightleftharpoons \bar{C}N + H_2O$$

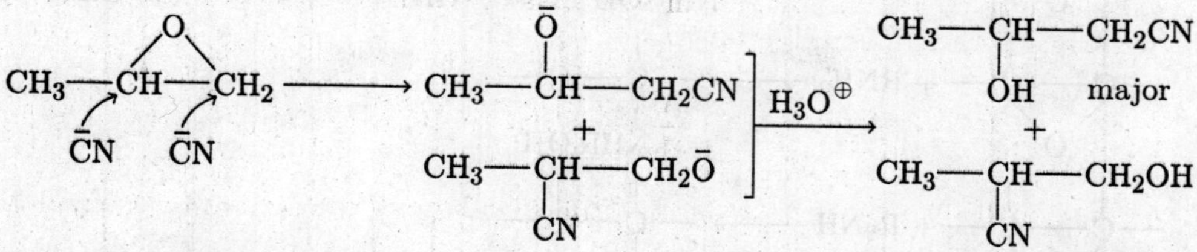

(h) **Reaction with alkoxides :** When epoxides are treated with alkoxides in the corresponding alcohols, β-hydroxyethers are obtained.

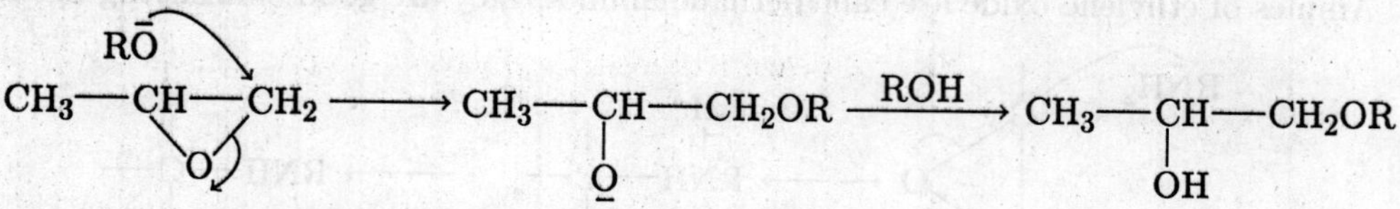

(i) **Reaction with the Grignard reagent :** See 16.4.5. p. 845.

(j) **Action of heat on ethylene oxide :** When ethylene oxide is heated, a 1, 2-hydride shift takes place and acetaldehyde is obtained through the anionatropic rearrangement.

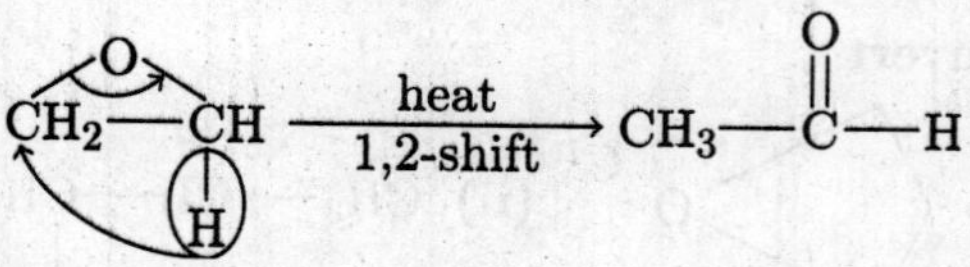

(k) **Reaction with alcohols and glycols :** When epoxides are heated with alcohols or glycols, hydroxyethers are obtained. For example, ethylene oxide reacts with methyl alcohol, ethyl alcohol and ethylene glycol and form methyl cellosolve, ethyl cellosolve and diethylene glycol respectively.

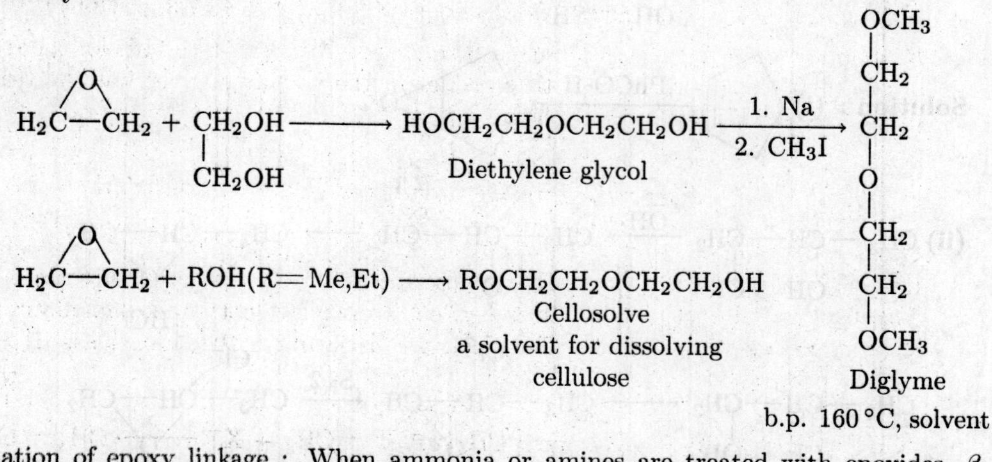

(l) **Amination of epoxy linkage :** When ammonia or amines are treated with epoxides, β-hydroxyamines are formed; ammonia gives a mixture of 1°, 2° and 3° β-hydroxyamines while 1° and 2° amines form 2° and 3° β-hydroxyamines respectively.

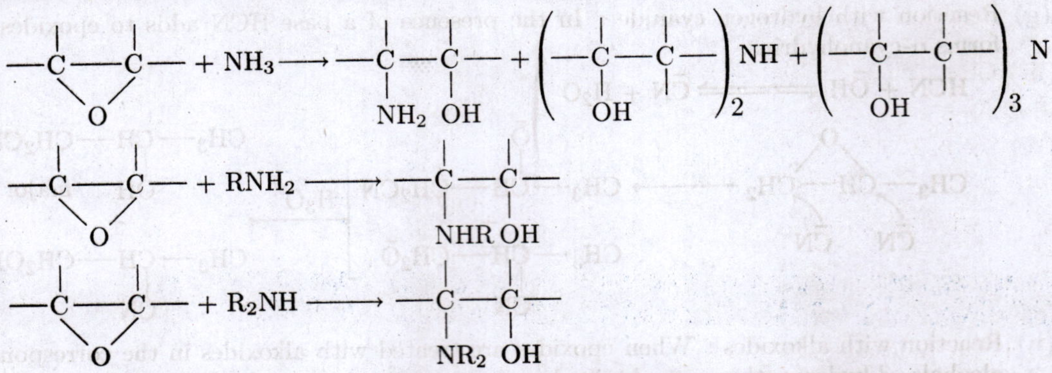

Amines of ethylene oxide are called ethanolamines; they are good emulsifying agents.

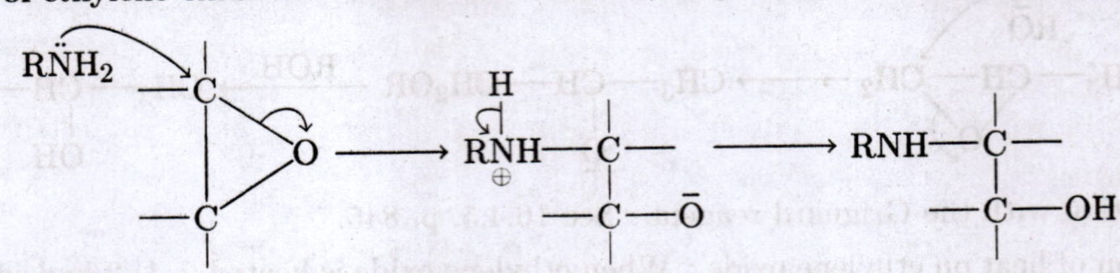

**(m) The Friedel-Crafts reaction :** See Prob. 17.14.iv, p. 981.

**Problem 20.14. Convert**

**(i)** 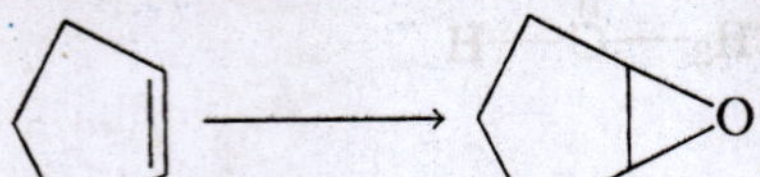 **(ii)** 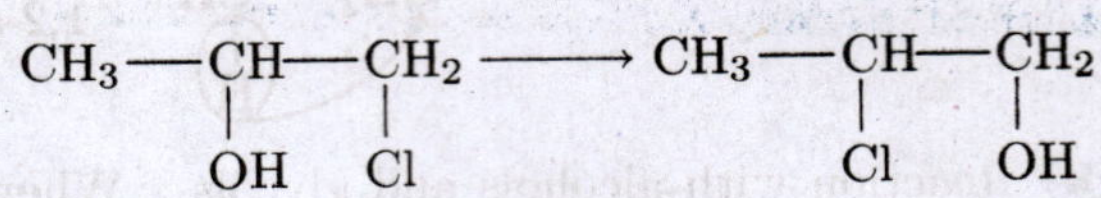

**(iii)** 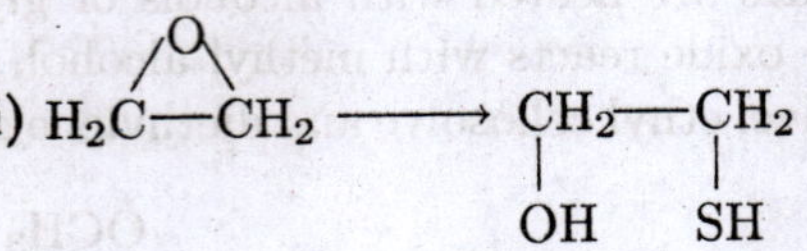 **(iv)** Ethylene oxide to diglyme

**Solution : (i)** 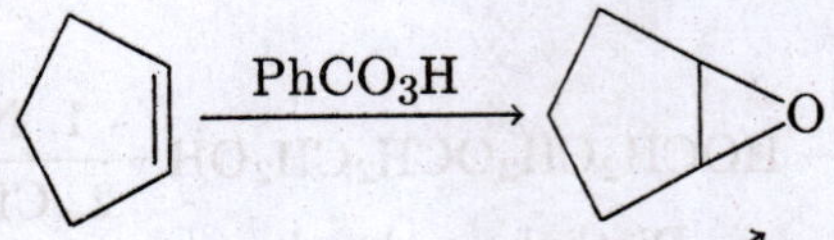

**(ii)** 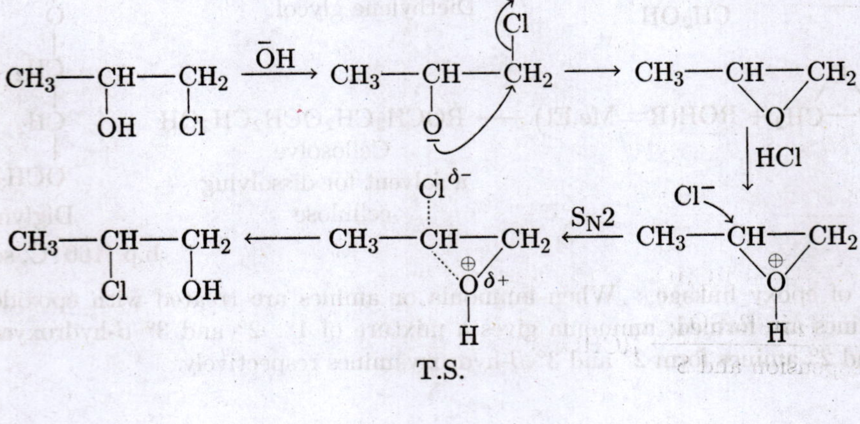

**(iii)** 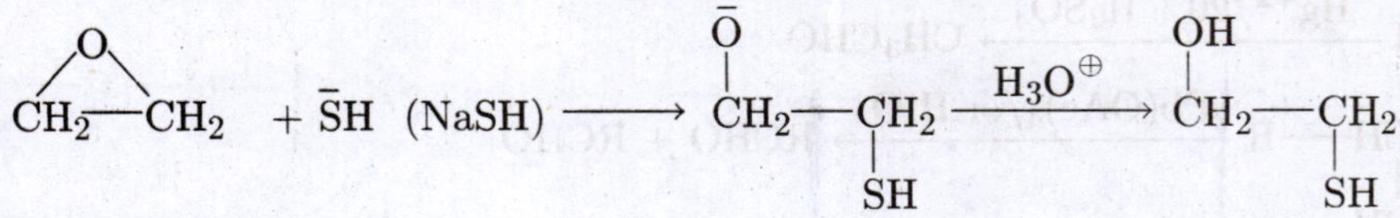

**(iv)** See p. 1239.

## 20.7 Carbonyl Compounds

### 20.7.1 Introduction

Carbonyl functions are of two types, formyl (—CHO) and keto (>C=O). Compounds containing formyl group are called aldehydes whereas keto group containing compounds are called ketones. The simplest member of the aldehydes is formaldehyde. Its properties differ in many ways from the properties of the other aldehydes; so formaldehyde is not a typical aldehyde and does not represent the class. Acetaldehyde represents the class. Ketones are of two types—simple ketones and mixed ketones. A mixed ketone contains a ketonic carbonyl group flanked by two different hydrocarbon groups whereas the ketonic carbonyl group of a simple ketone has two identical hydrocarbon groups. However, acetone is the simplest typical ketone.

| HCHO | $CH_3CHO$ | $CH_3COCH_3$ | PhCHO | $PhCOCH_3$ |
|---|---|---|---|---|
| Formaldehyde | Acetaldehyde | Acetone | Benzaldehyde | Acetophenone |

### 20.7.2 Formation of carbonyl compounds

We shall discuss this under two heads—the formation of aldehydes and the formation of ketones. The methods are given in Chart 20.16 and in Chart 20.17 respectively.

**(a)** $RCH_2OH \xrightarrow[\text{dehydrogenation}]{\text{Cu or Ag/heat}} RCHO$

**(b)** $RCH_2OH \xrightarrow[\text{oxidation}]{K_2Cr_2O_7/\text{conc. } H_2SO_4} RCHO$

**(c)** $RCH{=}CHR \xrightarrow{1.\ O_3\ 2.\ Zn/HOAc} RCHO + RCHO$

**(d)** $RCHX_2 \xrightarrow{1.\ \bar{O}H\ 2.\ \Delta} RCHO$

**(e)** $(RCO_2)_2Ca + (HCO_2)_2Ca \xrightarrow{\text{distillation}} RCHO$

**(f)** $RCO_2H + HCO_2H \xrightarrow[300\,°C]{ThO_2/MnO} RCHO$

**(g)** $RCN \xrightarrow{1.\ SnCl_2/HCl\ 2.\ H_3O^{\oplus}} RCHO$

**(h)** $HCO_2Et \xrightarrow{RMgX} RCHO$

**(i)** $RCOCl \xrightarrow[\text{suspension and S}]{Pd/H_2 \text{ in } BaSO_4} RCHO$

(j) $HC \equiv CH \xrightarrow{Hg^{+2}/20\%\ H_2SO_4} CH_3CHO$

(k) $\underset{OH}{RCH}-\underset{OH}{CH}-R \xrightarrow{Pb(OAc)_4/\text{or } HIO_4} RCHO + RCHO$

(l) $PhCH_2Cl \xrightarrow{\text{aq. } Cu(NO_3)_2 \text{ or } Pb(NO_3)_2} PhCHO$

(m) $PhCH_3 \xrightarrow[\text{Etard reaction}]{CrO_2Cl_2} PhCHO$

(n) $PhCH_3 \xrightarrow[500\,°C]{\text{metal oxides } CrO_3/Ac_2O} PhCHO$

(o) $PhH \xrightarrow{CO + HCl/AlCl_3(\text{cat.})} PhCHO$

(p) $PhH \xrightarrow{HCl + HCN/ZnCl_2} PhCHO$

(q) $Ph\overset{\oplus}{N}_2Cl^- \xrightarrow{1.\ CH_2{=}NOH\ 2.\ H_3O^{\oplus}} PhCHO$

(r) $PhCH_2Cl \xrightarrow{1.\ (CH_2)_6N_4\ 2.\ H_3O^{\oplus}} PhCHO$

(s) $PhH \xrightarrow{PhN(Me)CHO/POCl_3} PhCHO$

(t) $PhOH \xrightarrow{1.\ CHCl_3\ \&\ \bar{O}H\ 2.\ H_3O^{\oplus}} HO-C_6H_4CHO$

*o-* and *p*-products

**Chart 20.16** Formation of aldehydes.

(a) By the dehydrogenation of 1° alcohols : See Sec. 20.2.5(i), p. 1179.

(b) By the oxidation of 1° alcohols : See Sec. 20.2.5(ii), p. 1180.

(c) By the ozonolysis of symmetrical alkenes : See Sec. 13.8, p. 641.

(d) From terminal gem-dihalides : When terminal gem-dihalides are hydrolyzed with alkali and the product of hydrolysis are heated, aldehydes are formed. Simplified mechanism is

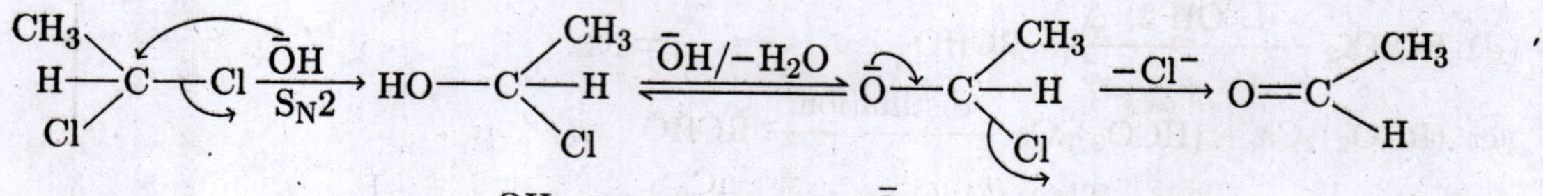


$$PhCHCl_2 \xrightarrow{\bar{O}H} Ph\overset{OH}{C}HCl \underset{}{\overset{\bar{O}H/H_2O}{\rightleftharpoons}} Ph\overset{\bar{O}}{C}H-Cl \xrightarrow{-Cl^-} PhCHO$$

(e) From calcium salts of carboxylic acids : When calcium salts of fatty acids are distilled with calcium formate, aldehydes are formed.

$$(RCOO)_2Ca + Ca(HCOO)_2 \longrightarrow 2RCHO + 2CaCO_3$$

$$(CH_3COO)_2Ca + (HCOO)_2Ca \longrightarrow 2CH_3CHO + 2CaCO_3$$

**(f) From carboxylic acids :** When a mixture of a fatty acid and formic acid vapour is passed over thoria or manganous oxide catalyst at 300 °C, an aldehyde is formed.

$$\text{HCOOH} + \text{RCOOH} \xrightarrow[300\,°\text{C}]{\text{ThO}_2 \text{ or MnO}} \text{RCHO}$$

$$\text{HCOOH} + \text{HCOOH} \xrightarrow[300\,°\text{C}]{\text{ThO}_2 \text{ or MnO}} \text{HCHO}$$

**(g) By the Stephen's reaction :** When alkyl or aryl cyanides are reduced with stannous chloride and hydrochloric acid in ethereal solution, and then the products are hydrolysed with water, the corresponding aldehydes are obtained and the reaction is known as the Stephen's reaction. Sterically hindered cyanides do not undergo the reaction; for example, *o*-tolyl cyanide and other *o*-substituted benzonitriles give negligible yield of aldehydes. Ethyl formate or ethyl acetate may also be used as a solvent instead of ether.

$$\text{CH}_3\text{CN} \xrightarrow[\text{2. H}_2\text{O}]{\text{1. SnCl}_2/\text{HCl}} \text{CH}_3\text{CHO}$$

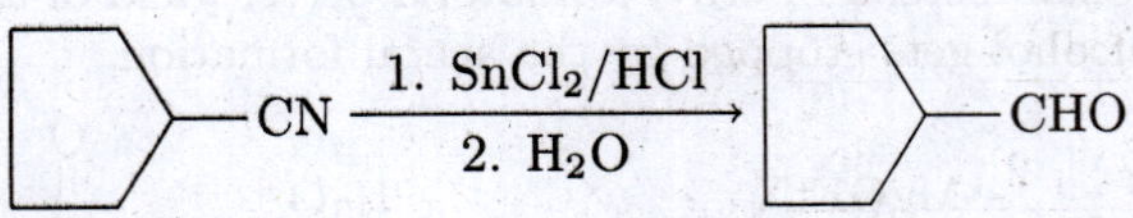


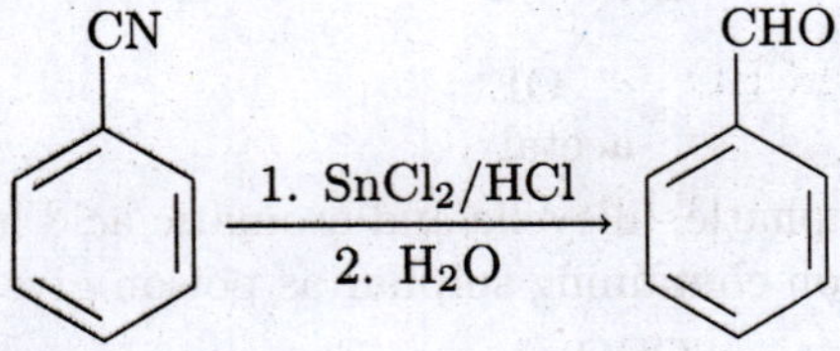


The mechanism of the reaction is as follows :

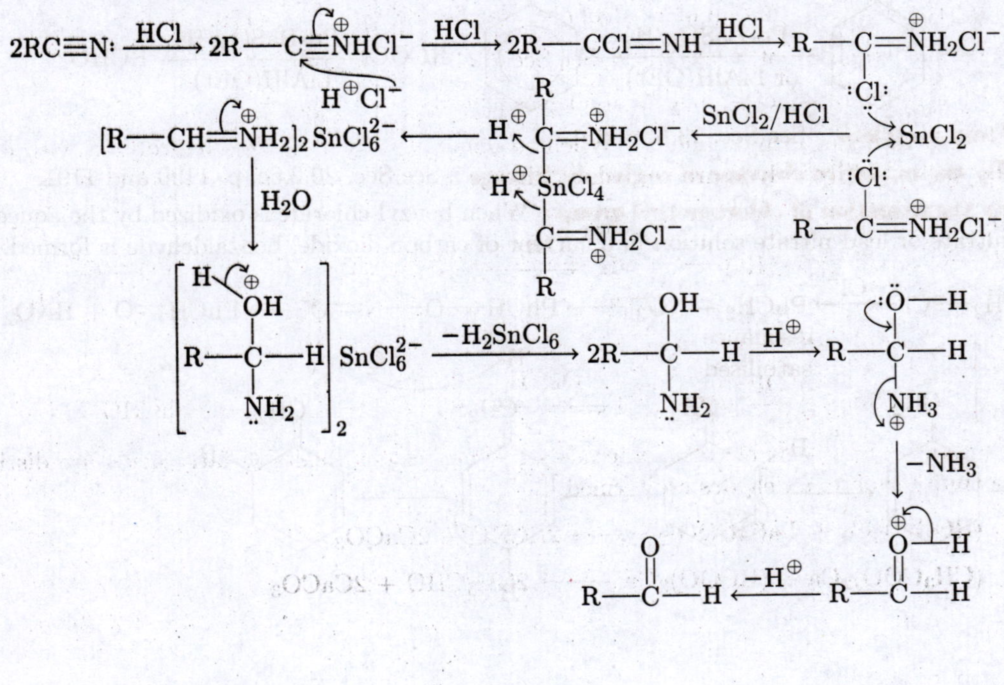

Cyanides may also be reduced by other reagents to aldehydes. For exmple, metal hydrides and sodium hypophosphite in the presence of Raney nickel reduce cyanides to aldehydes.

$$\mathrm{RC{\equiv}N \xrightarrow[\text{ether}]{LiAlH(OEt)_3} R{-}\underset{\displaystyle H}{\underset{|}{C}}{=}\overset{-}{\ddot{N}}{:}\ \overset{\oplus}{L}iAl(OEt)_3 \xrightarrow{5H_2O} RCHO + NH_3 + LiOH + Al(OH)_3 + 3EtOH}$$

$$\mathrm{R{-}C{\equiv}N \xrightarrow[\text{H}_2\text{O}\quad 2.\ \text{Ni}]{1.\ NaH_2PO_2} RCHO + NaHP(ONH_4)O_2}$$

**(h) By the action of Grignard reagent on ethyl formate :** A Grignard reagent reacts with ethyl formate to form an aldehyde. In this case a 2° alcohol is also obtained.

$$\mathrm{RMgX + H{-}\overset{\displaystyle O}{\overset{\|}{C}}{-}OEt \longrightarrow H{-}\overset{\displaystyle OMgX}{\overset{|}{\underset{\displaystyle R}{\underset{|}{C}}}}{-}OEt \xrightarrow{H_3O^{\oplus}} H{-}\overset{\displaystyle O}{\overset{\|}{C}}{-}R}$$

If ethyl orthoformate is taken instead of ethyl formate, a better yield of aldehyde is obtained since the formation of a 2° alcohol gets stopped by the acetal formation.

$$\mathrm{H{-}C(OEt)_2{-}OEt + RMgX \xrightarrow{-MgOEtX} \underset{\text{acetal}}{HC(R)(OEt){-}OEt} \xrightarrow{H_3O^{\oplus}} H{-}\overset{\displaystyle O}{\overset{\|}{C}}{-}R + 2EtOH}$$

**(i) By the Rosenmund reduction :** Aliphatic, alicyclic and aromatic acid halides on reduction with $Pd/H_2$ in barium sulphate suspension containing sulphur as poison give aldehydes.

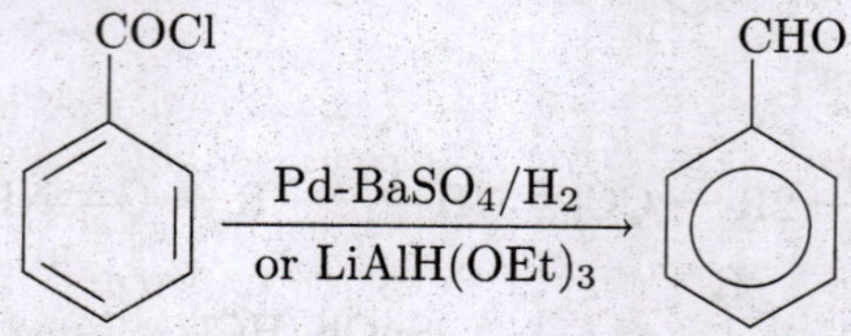


$$\mathrm{;\ RCOCl \xrightarrow[LiAlH(OEt)_3]{Pd\text{-}BaSO_4/H_2} RCHO}$$

**(j) From acetylene :** See Sec. 13.6, p. 631.

**(k) By the oxidative cleavage of α-glycolic linkage :** See Sec. 20.2.6c, p. 1190 and 1191.

**(l) By the oxidation of chloromethyl group :** When benzyl chloride is oxidized by the aqueous cupric nitrate or lead nitrate solution in a current of carbon dioxide, benzaldehyde is formed.

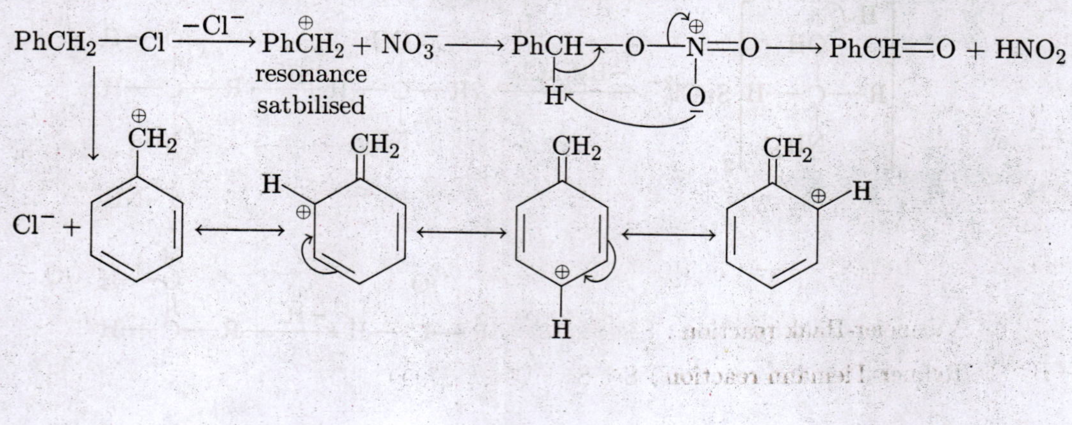

**(m) By the Etard reaction :** When toluene is treated with chromyl chloride in carbon tetrachloride, a complex is formed. This on treatment with water gives benzaldehyde and the reaction is called the Etard reaction.

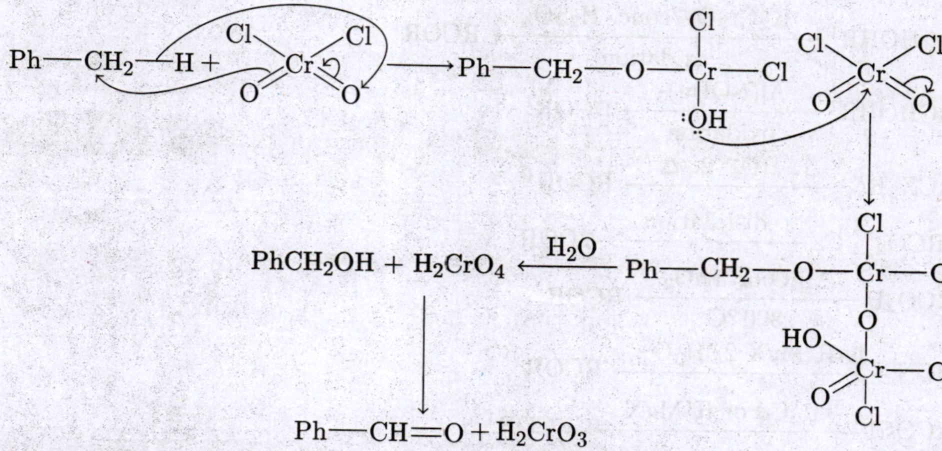

**(n) By the oxidation of side chain of arenes :** The side chain of an arene may also be oxidised to the aldehyde stage by chromium trioxide in acetic anhydride.

$$\text{Ar—CH}_3 \xrightarrow[\text{Ac}_2\text{O}]{\text{CrO}_3} \text{Ar—CH(OAc)}_2 \xrightarrow{\text{H}_2\text{O}} \text{Ar—CHO}$$

Commercially benzaldehyde is prepared from toluene vapour by oxidising the latter in air diluted with nitrogen at 500 °C. The catalysts used for the purpose are oxides of Mn, Zr, Mo, etc.

$$\text{Ph—CH}_3 + \text{O}_2 \xrightarrow[500\,^\circ\text{C}]{\text{MnO}} \text{PhCHO} + \text{H}_2\text{O}$$

**(o) By the Gattermann and Koch aldehyde synthesis :** See Sec. 17.11, p. 988.

**(p) By the Gattermann aldehyde synthesis :** See Sec. 17.13. p. 989.

**(q) From arenediazonium salts :** When arenediazonium salts are treated with formaldoxime and the products of the reaction are hydrolysed with water, aromatic aldehydes are formed.

$$\text{PhN}_2^{\oplus}\text{Cl}^- \xrightarrow{\text{HCH=NOH}} \text{PhCH=NOH} \xrightarrow{\text{H}_2\text{O}} \text{PhCHO}$$

**(r) By the Sommelet reaction :** Both aliphatic and aromatic aldehydes can be prepared by this reaction. The Sommelet reaction involves treatment of aliphatic amine hydrochlorides or derivatives of benzyl chloride with hexamethylenetetramine in aqueous ethanolic solution followed by acidification and steam distillation.

$$\text{PhCH}_2\text{Cl} + (\text{CH}_2)_6\text{N}_4 \longrightarrow \text{PhCHO};\ \text{Me(CH}_2)_5\overset{\oplus}{\text{N}}\text{H}_3\text{Cl}^- \xrightarrow[\text{2. AcOH}]{\text{1. (CH}_2)_6\text{N}_4} \text{Me(CH}_2)_4\text{CHO}$$

**(s) By the Vilsmeier-Haak reaction :** See Sec. 17.15, p. 996.

**(t) By the Reimer-Tiemann reaction :** See Sec. 17.13, p. 990.

For ketones :

**(i)** $RCHOHR^1 \xrightarrow[\text{dehydrogenation}]{\text{Cu or Ag/heat}} RCOR^1$

**(ii)** $RCHOHR^1 \xrightarrow[\text{oxidation}]{K_2Cr_2O_7/\text{conc. } H_2SO_4} RCOR^1$

**(iii)** $RCHOHR^1 \xrightarrow[\text{oxidation}]{Al(t\text{-}OBu)_3} RCOR^1$

**(iv)** $RCX_2R^1 \xrightarrow{1.\ HO^-\ 2.\ \Delta} RCOR^1$

**(v)** $(RCO_2)_2\,Ca \xrightarrow{\text{distillation}} RCOR$

**(vi)** $RCO_2H \xrightarrow[300\,°C]{MnO \text{ or } ThO_2} RCOR$

**(vii)** $RCN \xrightarrow{1.\ R^1MgX\ 2.\ H_3O^{\oplus}} RCOR^1$

**(viii)** $RCOCl \xrightarrow{R^1_2Cd \text{ or } R^1MgX} RCOR^1$

**(ix)** $\underset{(Ar)}{R}—C\equiv CH \xrightarrow{Hg^{2+}/20\%\ H_2SO_4} R—COCH_3\ (ArCOCH_3)$

**(x)** $R—C(R^1)(OH)—C(R^1)(OH)—R \xrightarrow[\text{or } HIO_4]{Pb(OAc)_4} 2RCOR^1$

**(xi)** $R—C(R)(OH)—C(R)(OH)—R \xrightarrow{H^{\oplus}} R_3C—\overset{O}{\overset{\|}{C}}—R$

**(xii)** $ArH \xrightarrow[\text{anh. } AlCl_3]{RCOCl \text{ or } (RCO)_2O} ArCOR$

**(xiii)**

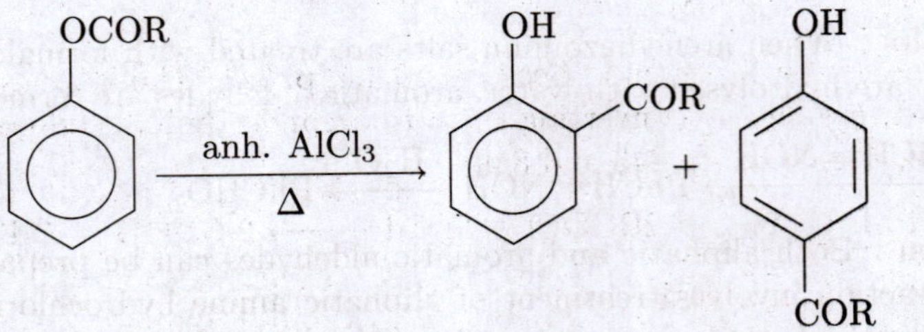


**(xiv)**

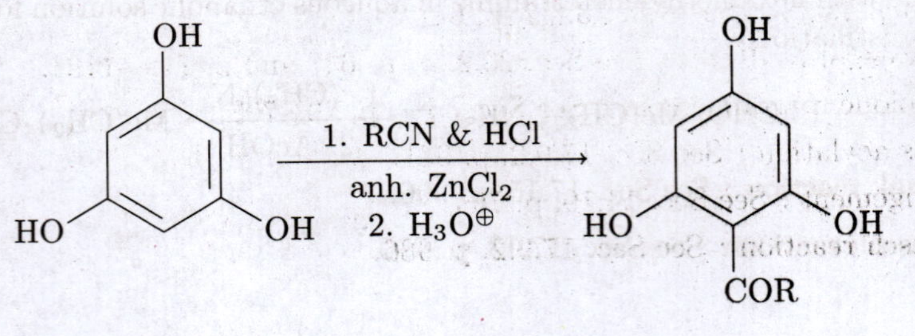

**(xv)** $\beta$-keto acid $\xrightarrow{\text{heat}}$ ketone

**(xvi)** $2RLi + CO_2 \longrightarrow R_2C(OLi)_2 \xrightarrow{H_2O} R_2CO$

**Chart 20.17** Formation of ketones.

(i) By the dehydrogenation of 2° alcohols : See Sec. 20.2.5(i), p. 1179.

(ii) By the oxidation of 2° alcohols : See Sec. 20.2.5.(ii), p. 1180.

(iii) By the Oppenauer oxidation : See Sec. 16.6.2, p. 859.

(iv) From gem-dihalides : When gem-dihalides of the type $RCX_2R'$ ($R{=}R'$ or $R{\neq}R'$) are hydrolysed with dilute alkali and then heated, ketones form.

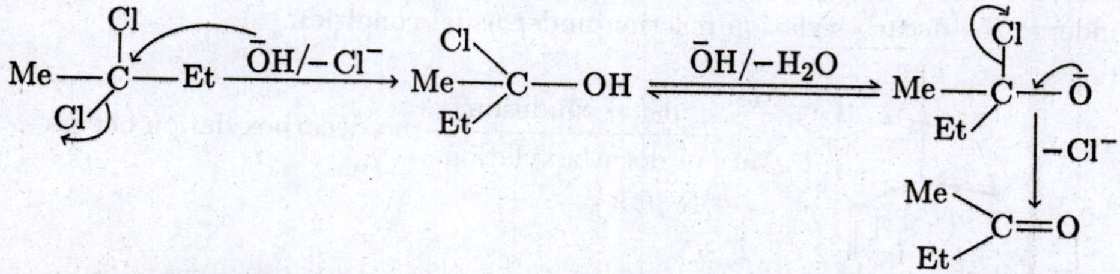

(v) From calcium salts of carboxylic acids : When calcium salts of carboxylic acids are distilled, ketones are formed.

$$(CH_3COO)_2Ca \xrightarrow{\text{distillation}} CH_3COCH_3$$

$$(PhCO_2)_2Ca + (AcO)_2Ca \xrightarrow{\text{distillation}} PhCOCH_3$$

(vi) From fatty acids : When fatty acid vapours are passed over $ThO_2$ or MnO at 300 °C, ketones are formed. If a mixture of two acid vapours are passed, a mixed ketone is obtained.

$$CH_3COOH \xrightarrow[300\,°C]{MnO \text{ or } ThO_2} CH_3COCH_3$$

$$CH_3COOH + PhCOOH \xrightarrow[300\,°C]{MnO \text{ or } ThO_2} PhCOCH_3$$

(vii) From cyanides : See Sec. 16.4.5.

(viii) From acid halides and the Grignard reagents : See Sec. 16.4.5, p. 841. The ketones may also be prepared from the Grignard reagents by converting them to organocadmium compounds and then treating the latter with acid halides (p. 843 and 849).

$$2R^1MgX + CdCl_2 \longrightarrow R^1_2Cd + MgXCl; \quad 2RCOCl + R^1_2Cd \longrightarrow 2\,RCOR^1 + CdCl_2$$

$$2\,ArCOCl + R^1_2Cd \longrightarrow 2\,ArCOR^1 + CdCl_2$$

(ix) By the hydration of acetylenes : See Sec. 13.6.2, p. 631.

(x) By the oxidative cleavage of $\alpha$-glycols : See Sec. 20.2.6c, p. 637 and p. 1190-1191.

(xi) By the pinacol-pinacolone rearrangement : See Sec. 14.2, p. 701-703.

(xii) By the Friedel-Crafts acylation : See Sec. 17.9.2, p. 981.

(xiii) By the Fries rearrangement : See Sec. 14.19, p. 769.

(xiv) By the Houben-Hoesch reaction : See Sec. 17.9.2, p. 986.

**(xv) From $\beta$-keto acids :** When $\beta$-keto acids are heated, ketones are formed. On the basis of this fact $\beta$-keto esters are converted to ketones via $\beta$-keto acids.

$$CH_3\overset{\beta}{C}O\overset{\alpha}{C}H_2\ CO_2H \xrightarrow{\text{heat}} CO_2 + CH_3\ COCH_3$$

$$RCOCR^1R^2CO_2Et \xrightarrow[\text{2. dil. } H_2SO_4]{\text{1. dil. aq. or alc. KOH}/\Delta} RCOCR^1R^2CO_2H \xrightarrow[-CO_2]{\text{heat}} RCOCHR^1R^2$$

$$CH_3COCMe_2CO_2Et \xrightarrow{\text{1. dil. aq. or alc. KOH}/\Delta;\ \text{2. dil. } H_2SO_4;\ \text{3. } \Delta} CH_3COCHMe_2$$

However, bicyclic $\beta$-keto acids in which the carboxyl group is bonded to a 'bridge-head' carbon do not undergo the decarboxylation reaction under usual condition.

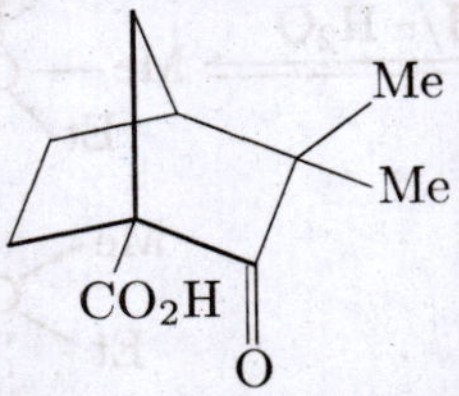


$\xrightarrow[\text{of decarboxylation}]{\text{usual condition}}$ no decarboxylation occurs

The decarboxylation reaction of acetoacetic acid and $\alpha$, $\alpha$-dimethylacetoacetic acid have been studied in detail and it has been found that they follow a rate law as given below :

$$\text{the rate} = k\ [\text{keto acid}] + k'\ [\text{keto acid anion}]$$

The above mentioned rate equation tells us that $\beta$-keto acids may undergo the decarboxylation in two distinct ways, and neither of the ways involve the enol form (or the enolic anion) of the $\beta$-keto acid in the first stage since $\alpha$, $\alpha$-dimethylacetoacetic acid does not have an $\alpha$-H but it undergoes the reaction. The rate equation further suggests that the decarboxylation of a $\beta$-keto acid or a $\beta$-keto acid anion is a unimolecular process. However, the decarboxylation process forms an enol (in the case of a $\beta$-keto acid) or an enolate ion (in the case of a $\beta$-keto acid anion) in the slow rate-determining step which then gets converted to a ketone through a fast step.

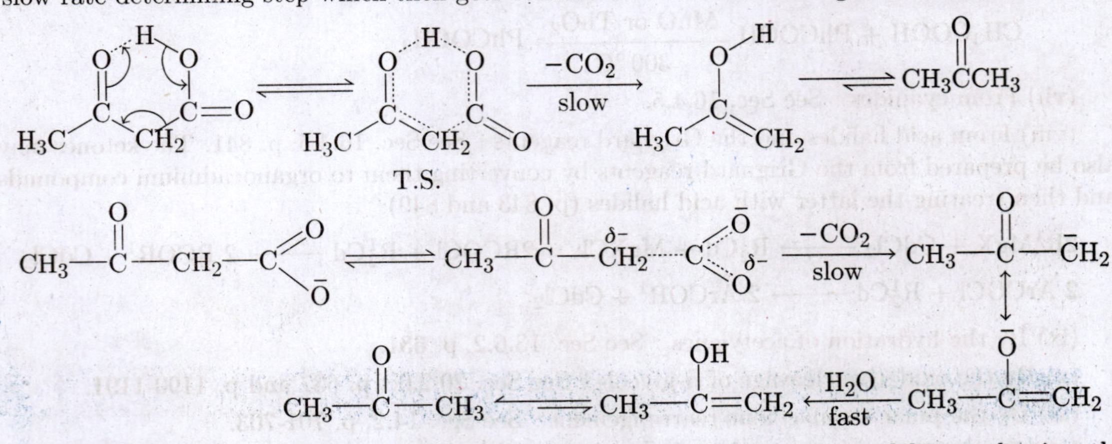


[That the enol or the enolate ion is the intermediate has been proved by conducting the decarboxylation reaction in the presence of bromine or iodine and then measuring the rate of the disappearance of the $\beta$-keto acid which is found to remain unchanged. The product of the reaction is found to be monobromo (or monoiodo) ketone. The bromination (or iodination) may take place either before the decarboxylation or after the formation of the ketone or after the decarboxylation and before the formation of the ketone. Since the rate of the decarboxylation remains unchanged, the bromination cannot occur before the decarboxylation process. Again since ketones cannot be brominated under the condition of decarboxylation, the bromination must have taken place before the ketone formation. This is possible if an enol or an enolate ion forms as an intermediate.]

Since an enol or an enolate ion is an intermediate and since bicyclic $\beta$-keto acids in which the carboxyl group is bonded to a 'bridge-head' carbon cannot develop enols because of their high strain, they do not undergo decarboxylation reaction under usual conditions.

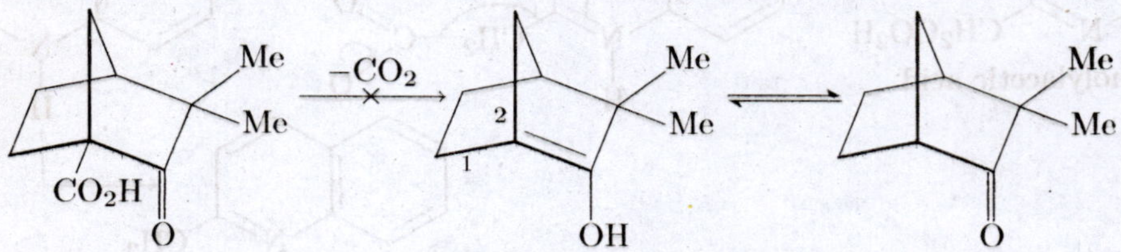

(a) (b) (c)

If the compound (a) undergoes the decarboxylation reaction, the intermediate enol will be (b) and the product will be (c), a ketone. In the compound (b) the double-bonded carbon atoms are $sp^2$ hybridised and naturally the C—C bonds, which are marked by 1 and 2 must be coplanar and the expected bond angle between them should be 120° or nearly 120°; but a 'bridged carbon' displays 109°28′ bond angle. Therefore, if the compound (b) at all forms, it will be highly strained and its energy content will be very high; and as a consequence the T.S. of $\beta$-keto carboxylic acid (a) which will lead to the enol (b) will be of very high energy. Thus the energy of the T.S. being very high, bicyclic $\beta$-keto acids of the type (a) undergo the decarboxylation with extreme difficulty through rearrangement.

**(xvi) By the reaction between alkyllithium and carbon dioxide :** When an alkyllithium is treated with carbon dioxide in an atmosphere of nitrogen and the product of reaction is hydrolysed with water and heated, a ketone is formed :

$$RLi + O{=}C{=}O \longrightarrow R{-}\overset{\overset{\displaystyle O}{\|}}{C}{-}OLi \xrightarrow{RLi} R_2C(OLi)_2$$

$$R_2C(OLi)_2 \xrightarrow[\text{heat}]{H_2O} R_2C{=}O$$

**Problem 20.15.** Explain

**(i)** $CF_3COCH_2CO_2H$ does not decarboxylate on heating.

**(ii)** 2-Quinolylacetic acid decarboxylates more readily than quinoline-2-carboxylic acid.

**Solution : (i)** The decarboxylation of $\beta$-keto acids involve a T.S. in which the H atom of the carboxyl group forms H-bond with the oxygen atom of the carbonyl group. If the carbonyl O atom is not reasonably anionic, the H-bond cannot form and as a consequence the transfer of the H atom from the carboxyl group to the carbonyl group cannot take place to form the enol. In the intermediate of the decarboxylation reaction of $CF_3COCH_2CO_2H$, the $-I$ effects of the three F atoms make the F-bonded C atom strongly positively charged which in its turn draws electrons from the carbonyl moiety; thus the O atom of the carbonyl group becomes less electron dense than required for the H transfer, and H does not actually get transferred. So, the compound does not decarboxylate.

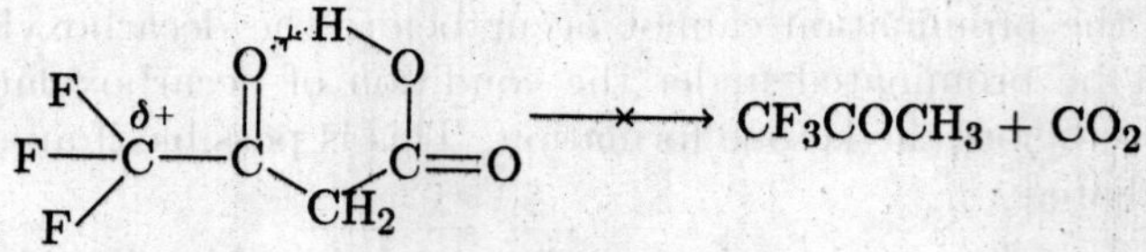

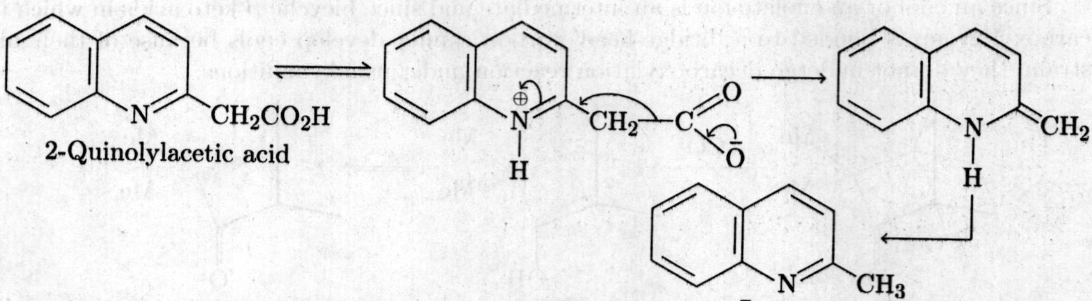

2-Quinolylacetic acid

I

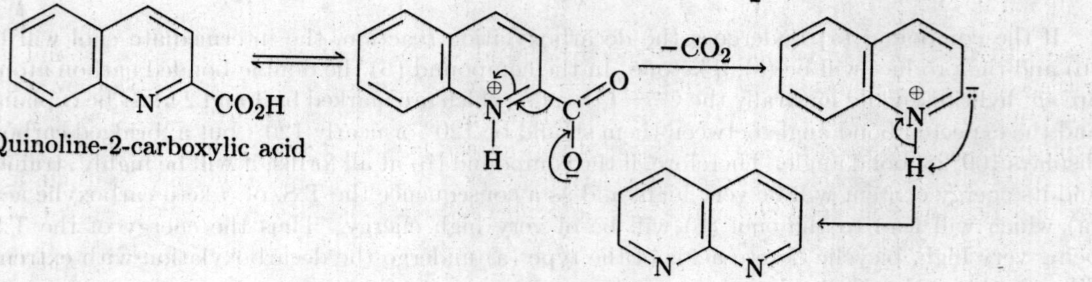

Quinoline-2-carboxylic acid

II

In the decarboxylation of the former compound the intermediate is uncharged while that in the latter case is charged. So the former intermediate is more stable than the latter intermediate and hence the former decarboxylates more readily than the latter.

**Problem 20.16.** How can you synthesise

$$CH_3COCH_2COCH_3 \text{ and } CH_3COCH(CH_3)—CH_3 \text{ from ethyl acetoacetate?}$$

**Solution :**

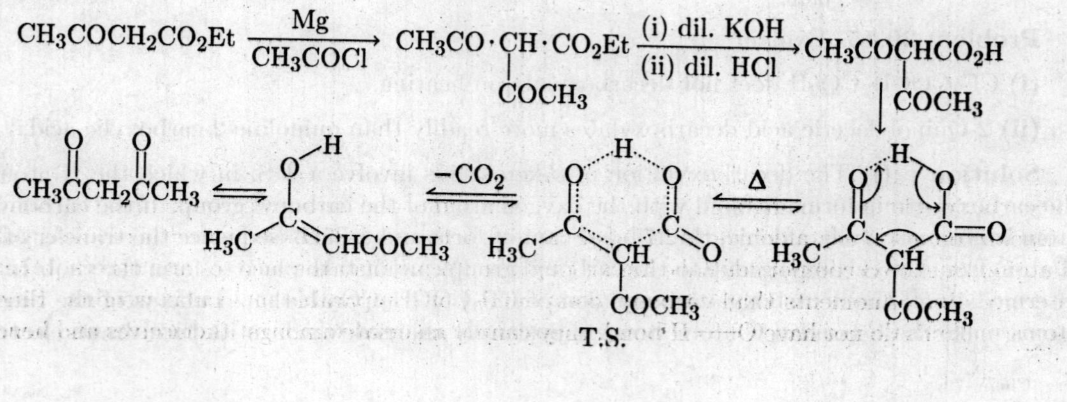

T.S.

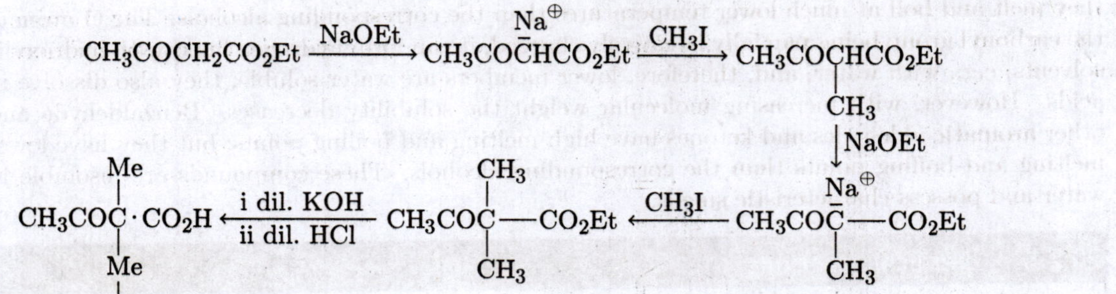

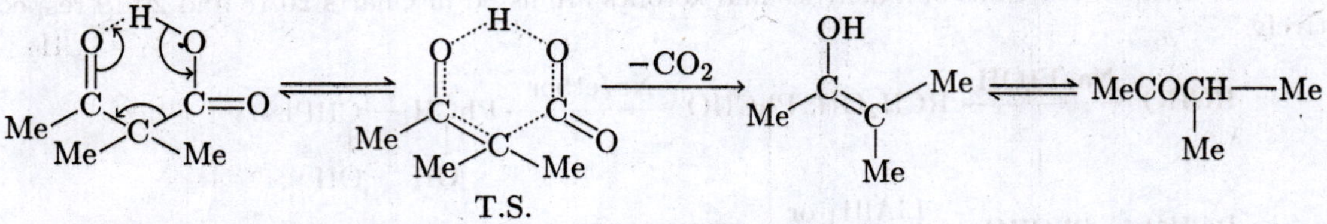

## 20.7.3 Structure, reactivity and orientation

See p. 825-826, Sec. 16.1. Aromatic aldehydes and ketones undergo aromatic electrophilic substitution reactions. Since —CHO and —CO groups have –I and –R effects, they are *m*-directing groups. Since these groups have deactivating effects, substrates containing carbonyl groups undergo the aromatic electrophilic substitution reactions more slowly than benzene. Aromatic rings being stable to oxidation, aromatic aldehydes and ketones on oxidation give aromatic acids in which the side chain containing the carbonyl group gets oxidised to the carboxyl group only.

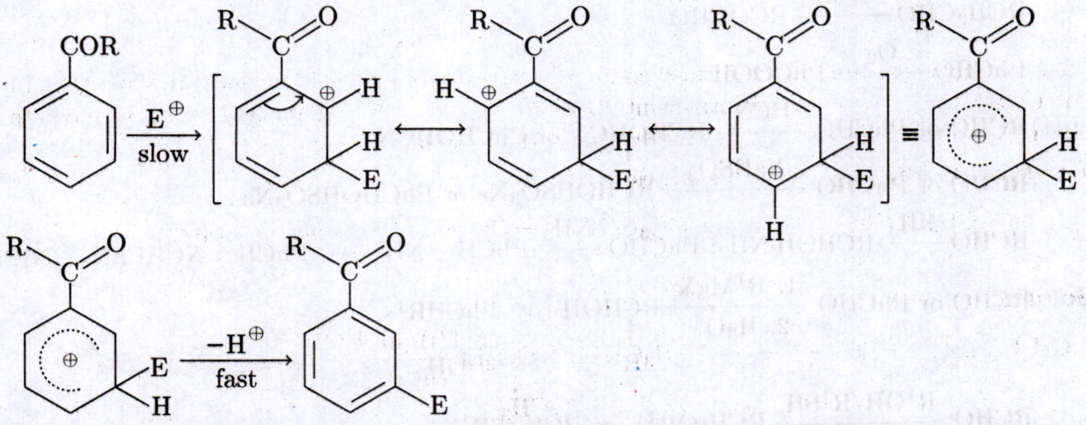

## 20.7.4 Physical properties

C—O double bond being highly polarisable, aldehydes and ketones possess dipole moment; the dipole moment of acetaldehyde is 2.7D and that of acetone is 2.85D. Owing to the dipole-dipole interaction, carbonyl compounds have higher m.p.s and b.p.s than the alkanes and ethers (ethers have lower dipole moments than carbonyl compounds) of comparable molecular weights. Since these compounds do not have O—H bond, they cannot associate amongst themselves and hence they melt and boil at much lower temperatures than the corresponding alcohols. The O atom of the carbonyl group being partially negatively charged, it can form hydrogen bond with hydroxylic solvents, e.g., with water, and, therefore, lower members are water soluble; they also dissolve in acids. However, with increasing molecular weight the solubility decreases. Benzaldehyde and other aromatic aldehydes and ketones have high melting and boiling points; but they have lower melting and boiling points than the corresponding alcohols. These compounds are insoluble in water and possess characteristic smells.

## 20.7.5 Chemical reactions

The chemical reactions of aldehydes and ketones are listed in Charts 20.18 and 20.19 respectively.

(i) $RCHO \xrightarrow{Na/EtOH} RCH_2OH$; $PhCHO \xrightarrow{Na/ether} PhCH(OH){-}CH(OH)Ph$

(ii) $RCHO$ or $PhCHO \xrightarrow[(Me_2CHO)_3Al]{LiAlH_4\text{ or }} RCH_2OH$ or $PhCH_2OH$

(iii) $RCHO \xrightarrow[\text{or } NaBH_4/H_2O\text{-}Et_2O]{\text{catalytic reduction}} RCH_2OH$; $PhCHO \xrightarrow[\text{reduction}]{\text{catalytic}}$ cyclohexylcarbinol

(iv) $RCHO$ or $PhCHO \xrightarrow[NH_2\text{-}NH_2/\text{base}]{Zn\text{-}Hg/HCl\text{ or }} RCH_3$ or $PhCH_3$

(v) $RCHO$ or $PhCHO \xrightarrow[\text{conc. } H_2SO_4]{K_2Cr_2O_7\ \&} RCOOH$ or $PhCOOH$

(vi) $RCH_2CHO \xrightarrow{SeO_2} RCOCHO$

(vii) $PhCHO \xrightarrow{O_2} PhCOOH$

(viii) $RCHO$ or $PhCHO \xrightarrow{HCN} RCHOHCN$ or $PhCHOHCN$

(ix) $RCHO$ or $PhCHO \xrightarrow{NaHSO_3} RCHOHSO_3Na$ or $PhCHOHSO_3Na$

(x) $RCHO \xrightarrow{NH_3} RCHOH{\cdot}NH_2$; $PhCHO \xrightarrow{NH_3} PhCH{=}NH \longrightarrow PhCH{=}NCHPhN{=}CHPh$

(xi) $RCHO$ or $PhCHO \xrightarrow[2.\ H_3O^{\oplus}]{1.\ R^1MgX} RCH(R^1)OH$ or $PhCHR^1(OH)$

(xii) $RCHO \xrightarrow[H^{\oplus}]{R^1OH/R^1SH} RCH(OR^1)_2$ or $RCH(SR^1)_2$
(PhCHO) $\quad PhCH(OR^1)_2$ or $PhCH(SR^1)_2$

(xiii) $RCHO$ or $PhCHO \xrightarrow{H_2NOH} RCH{=}NOH$ or $PhCH{=}NOH$

(xiv) $RCHO$ or $PhCHO \xrightarrow{NH_2NH_2} RCH{=}NNH_2$ or $PhCH{=}N{-}N{=}CHPh$

(xv) $RCHO$ or $PhCHO \xrightarrow{\text{Schiff's reagent}}$ rose colouration

(xvi) $RCHO \xrightarrow[\text{solution}]{\text{Fehling}} RCO\bar{O} + Cu_2O\text{(red ppt.)}; PhCHO \xrightarrow[\text{solution}]{\text{Fehling}} \text{no reaction}$

(xvii) $RCH_2CHO \xrightarrow{\text{dil. }\bar{O}H \text{ or dil. acid}} RCH_2CHOH{\cdot}\underset{|}{\underset{R}{C}}HCHO \xrightarrow{-H_2O} RCH_2CH{=}\underset{R}{\underset{|}{C}}CHO$

(xviii) $PhCHO \xrightarrow{\text{dil. }\bar{O}H\text{; warm}} PhCH_2OH + PhCOO^-$

$RCH_2CHO \xrightarrow[\text{warm}]{\text{conc. }\bar{O}H} \text{resin; } HCHO \xrightarrow{\bar{O}H} HCH_2OH + HCOO^-$

(xix) $RCHO \text{ or } PhCHO \xrightarrow{R^1NH_2} RCH{=}NR^1 \text{ or } PhCH{=}NR^1$

(xx) $RCHO \text{ or } PhCHO \xrightarrow{PCl_5} RCHCl_2 \text{ or } PhCHCl_2$

(xxi) $RCHO \text{ or } PhCHO \xrightarrow[\text{ammoniacal}]{AgNO_3} Ag + RCOO^- \text{ or } PhCOO^-$

(xxii) $RCH_2CHO \xrightarrow{X_2} RCX_2CHO; PhCHO \xrightarrow{X_2} PhCOX$

(xxiii) $RCHO \xrightarrow[\text{conc. } H_2SO_4]{HN_3 \text{ \&}} RCN + RNHCHO$

(xxiv) $\underset{(RCHO)}{PhCHO} \xrightarrow[\text{2. } H_3O^{\oplus}]{\text{1. } (RCH_2CO)_2O; RCH_2COONa} \underset{\text{(no reaction)}}{PhCH{=}CH{\cdot}COOH}$

(xxv) $PHCHO \xrightarrow{KCN} PhCHOHCOPh; RCHO \xrightarrow{KCN} \text{no benzoin-like condensation}$

(xxvi) Other condensation reactions

(xxvii) $PhCHO \xrightarrow{E^{\oplus}} m\text{-}EC_6H_4CHO$

**Chart 20.18** Reactions of aldehydes.

**(i) Reduction by the Bouvault-Blanc method :** Sodium and ethyl alcohol reduce aldehydes to 1° alcohols. When benzaldehyde is treated with sodium in an inert solvent and the product is treated with water, hydrobenzoin forms (See pinacol synthesis; Sec. 16.9.2. p. 872).

$$RCHO \xrightarrow{Na/OEtOH} RCH_2OH; 2PhCHO \xrightarrow[\text{2. } H_2O]{\text{1. Na/ether}} PhCH(OH){-}CH(OH){-}Ph$$

**(ii) Reduction by metal hydrides & aluminium isopropoxide :** See. Sec. 16.6.1 & Sec. 16.6.2

**(iii) Catalytic reduction :** Aliphatic aldehydes on catalytic reduction form alcohols while aromatic aldehydes give derivatives of cyclohexylcarbinols.

$RCHO \xrightarrow{Ni/H_2} RCH_2OH;$

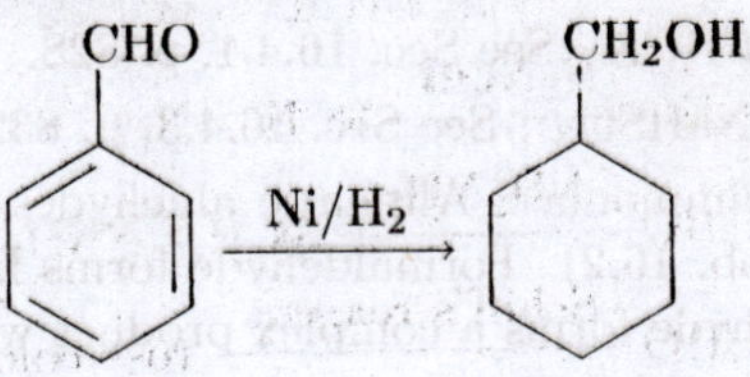

(iv) (a) **The Clemmensen reduction** : See Sec. 18.1.3.(i), p. 1022.

(b) **The Wolff-Kishner reduction** : See Sec. 16.7, p. 865.

(v) **Chromic acid oxidation** : On chromic acid oxidation both aliphatic and aromatic aldehydes form the corresponding carboxylic acids. The oxidation may also be carried out by potassium permanganate and concentrated sulphuric acid.

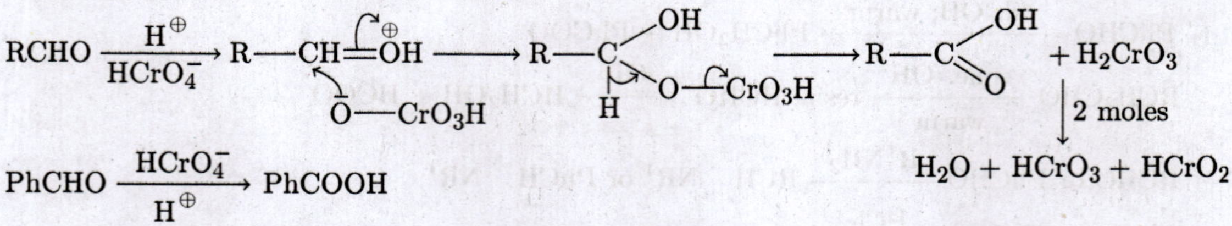

(vi) **Oxidation by selenium oxide** : Aldehydes or ketones having a methyl or methylene group adjacent to the carbonyl C atom are readily oxidised by selenium dioxide at room temperature to dicarbonyl compounds.

$$RCH_2CHO + SeO_2 \longrightarrow RCOCHO$$

$$CH_3CHO + SeO_2 \longrightarrow CHOCHO$$

$$PhCH_2CHO + SeO_2 \longrightarrow PHCOCHO$$

The reaction is carried out by treating the aldehyde or ketone in acetic acid solution with selenous acid. The mechanism has not yet been setteled. However, this may be as shown below :

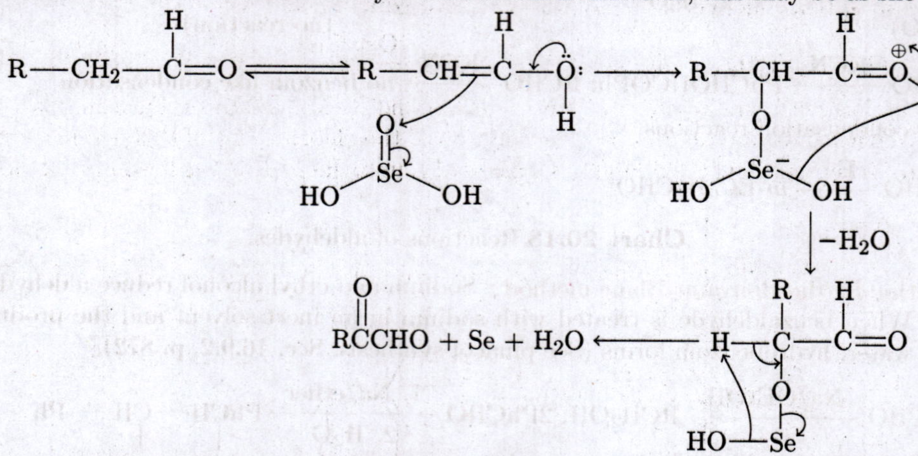

(vii) **Autooxidation of aromatic aldehydes** : In the presence of air aromatic aldehydes get oxidised to the corresponding carboxylic acid.

$$PhCHO \xrightarrow[\text{autooxidation}]{O_2} PhCOOH$$

(viii) **Addition of HCN** : See Sec. 16.4.1, p. 828.

(ix) **Addition of $NaHSO_3$** : See Sec. 16.4.3, p. 832.

(x) **Addition of ammonia** : Aliphatic aldehydes add up ammonia to form aldehyde ammonia (See. p. 830, prob. 16.2). Formaldehyde forms hexamethylenetetramine instead of aldehyde ammonia. Benzaldehyde forms a complex product with ammonia, called hydrobenzamide.

$$RCHO + NH_3 \longrightarrow RCH(OH)NH_2$$

$$6HCHO + 4NH_3 \longrightarrow H_2O +$$ 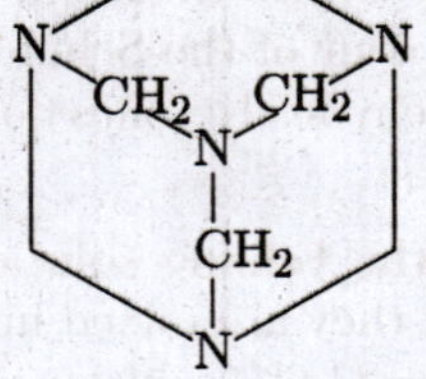

$$PhCHO + NH_3 \longrightarrow Ph{-}CH(OH){-}NH_2 \xrightarrow{-H_2O} PhCH{=}NH$$

$$3PhCH{=}NH \longrightarrow PhCH{=}N{-}CH(Ph){-}N{=}CHPh + NH_3$$

Aldehydes may be regenerated from their ammonia derivatives by treatment with a dilute acid.

(xi) **Reaction with the Grignard reagent** : See Sec. 16.4.5, p. 841.

(xii) **Reaction with alcohols and thiols** : See Sec. 16.4.4, p. 833. Aromatic aldehydes usually do not undergo the reaction. However, with 1, 3-diol they react.

(xiii) **Reaction with hydroxylamine** : See Sec. 16.5, p. 851.

(xiv) **Reaction with hydrazine** : See Sec. 16.5, p. 851.

In the case of benzaldehyde two molecules of hydrazine reacts.

$$Ph{-}CH{=}O + H_2NNH_2 \longrightarrow Ph{-}CH{=}NNH_2 \xrightarrow{O{=}CHPh} PhCH{=}N{-}N{=}CHPh$$

(xv) **Reaction with the Schiff's reagent** : Rosaniline or Mangenta or Fuchsine is a triarylmethane dye with three amino groups. The solution of the chloride salt of this dye has a rose red colouration. When sulphur dioxide gas is passed through the chloride salt solution, the colour

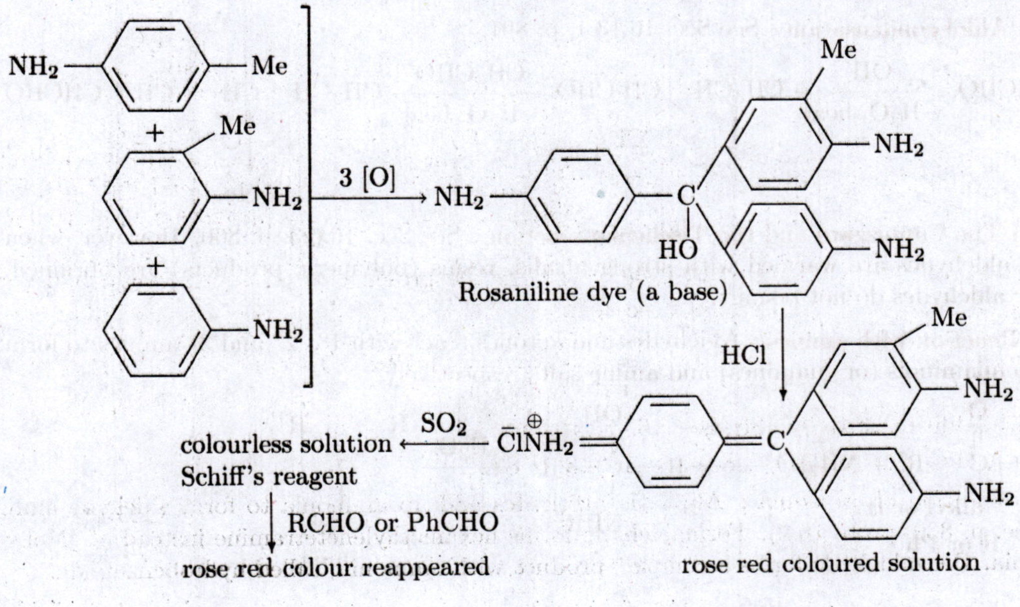

disappears and the colourless solution is called the Schiff's reagent. When an aliphatic or an aromatic aldehyde is treated with the Schiff's reagent, the rose red colour of the dye salt reappears. The appearance of the colour of the Schiff s reagent by an organic compound indicates that the compound may be an aldehyde. Ketones do not give this test and hence they can be distinguished from aldehydes by this test.

**(xvi) Reaction with the Fehling solution :** Fehling solution is a mixture of two solutions, Fehling A and Fehling B; they are mixed in equal volumes. Fehling A is a 7% solution of $CuSO_4$, $5H_2O$ and Fehling B is a 34.6% solution of sodium potassium tartrate (Rochelle salt) in 24% caustic potash solution. The mixture of the two is a blue coloured solution. When an aliphatic aldehyde is added to the Fehling solution and warmed in a water bath, a yellow or brown or red precipitate of cuprous oxide is obtained. A ketone or an aromatic aldehyde does not give this test; so an aliphatic aldehyde can be distinguished from a ketone by this test. When Fehling A and Fehling B are mixed together in equal volumes, a complex salt, known as sodium cuppritartrate is formed which may be considered as CuO or $Cu^{\oplus\oplus}$.

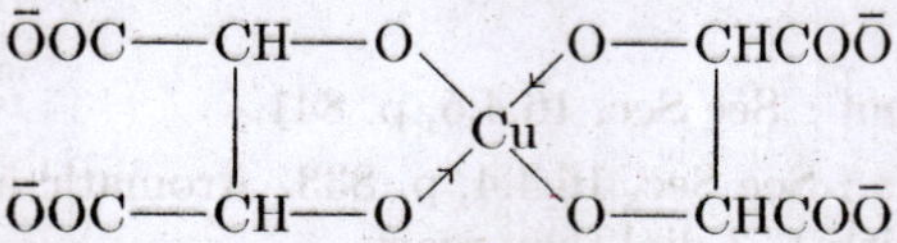

$\equiv CuO$ or $Cu^{\oplus\oplus}$

blue solution

When an aldehyde reacts with sodium cuppritartrate, the former gets oxidised to a carboxylate ion while the latter reduces to cuprous salt which precipitates as yellow or brown or red particles.

$$CH_3-\overset{H}{\overset{|}{C}}=O + 2\,CuO \text{ (from the Fehling solution)} \longrightarrow CH_3CO\bar{O} + Cu_2O \text{ red ppt.}$$

$$CH_3-\overset{H}{\overset{|}{C}}=O + 2\,Cu^{\oplus\oplus} + 5\,OH^- \longrightarrow CH_3CO\bar{O} + Cu_2O + 3\,H_2O$$

**(xvii) Aldol condensation :** See Sec. 16.13.1, p. 891.

$$2CH_3CHO \xrightarrow[-H_2O,\ heat]{\bar{O}H} CH_3CH{=}CH{\cdot}CHO \xrightarrow[-H_2O,\ heat]{CH_3CHO} CH_3CH{=}CH{-}CH{=}CHCHO \longrightarrow \text{etc.}$$

**(xviii) The Cannizzaro and the Tischenko reaction :** See Sec. 16.6.3. p. 860. However, when aliphatic aldehydes are warmed with strong alkalis, resins (polymeric products) are obtained. Aromatic aldehydes do not resinify.

**(xix) Reaction with amines :** Aldehydes and ketones react with 1°, 2° and 3° amines to form imines, hemiaminols (or enamines) and amine salts respectively.

$$R-\overset{O}{\overset{\|}{C}}-R^1 + NH_2R^2 \longrightarrow R-\underset{NHR^2}{\underset{|}{\overset{OH}{\overset{|}{C}}}}-R^1 \xrightarrow{-H_2O} \underset{\text{imine}}{\underset{NR^2}{\underset{\|}{\overset{R\quad R^1}{C}}}}$$

$R^1 =$ alkyl or H

$R = R$ or Ph

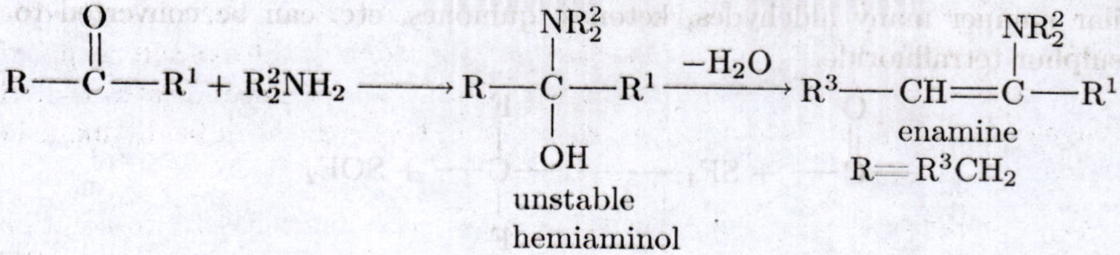

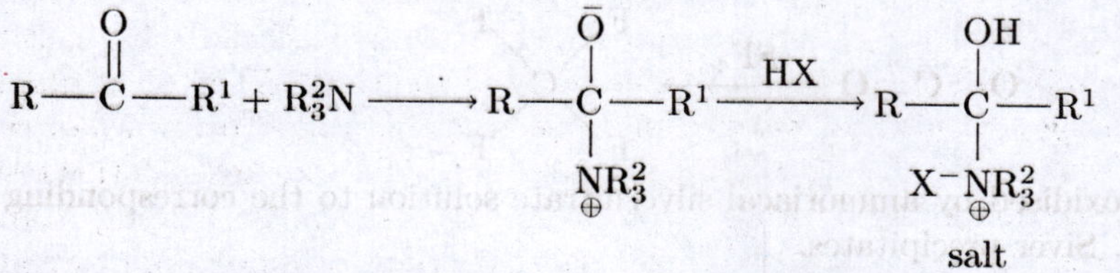

Imines, the reaction products of 1° amines and carbonyl compounds, are unstable excepting those which have an aryl group; aryl imines are called the Schiff's bases. The imine formation reaction is used as a ring-closure reaction.

Secondary amines react with carbonyl compounds to give unstable hemiaminols. If the latter contains an H atom on the $\alpha$-C (w.r.t. the carbonyl C), loss of water molecule occurs and enamine forms. The water is removed by forming constant boiling mixture. However, 3° amines form salts only.

**(xx) Reaction with phosphorus pentachloride** : Both aliphatic and aromatic carbonyl compounds, aldehydes and ketones, react with phosphorus pentachloride to form gem-dihalides.

$$RCHO + PCl_5 \longrightarrow RCHCl_2 + POCl_3$$

$$RCOR^1 + PCl_5 \longrightarrow RCCl_2R^1 + POCl_3$$

$$PhCHO + PCl_5 \longrightarrow PhCHCl_2 + POCl_3$$

$$RCH_2CHO \xrightarrow{PCl_5} RCH_2CHCl_2 \xrightarrow{-HCl} RCH{=}CHCl$$

If the gem-dihalide contains an $\alpha$-H with respect to the C atom bearing two Cl atoms, one molecule of HCl may get eliminated to form a derivative of vinyl chloride.

The halogenation occurs step by step; each step follows a two-step $S_N1$ path way as shown below :

$$2PCl_5 \rightleftharpoons \overset{\oplus}{P}Cl_4\bar{P}Cl_6 \ ; \ \bar{P}Cl_6 \rightleftharpoons PCl_5 + Cl^-$$

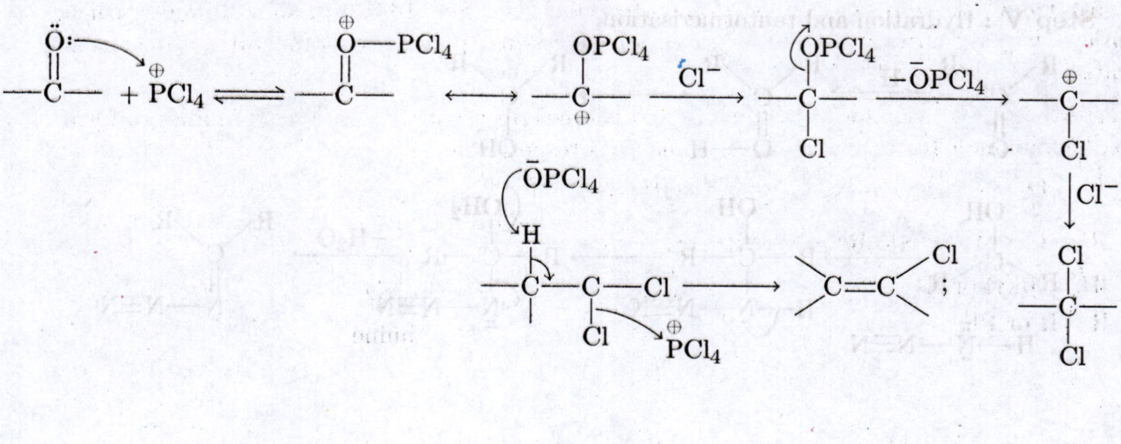

In the similar manner many aldehydes, ketones, quinones, etc. can be converted to gem-difluorides by sulphur tetrafluoride.

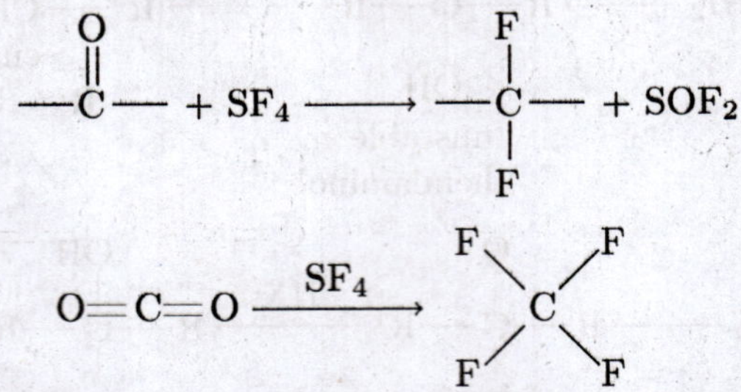

**(xxi)** Aldehydes get oxidised by ammoriacal silvernitrate solution to the corresponding acid salts but ketones do not. Siver precipitates.

$$\underset{(PhCHO)}{RCHO} \xrightarrow[\text{ammoniacal}]{AgNO_3} \underset{(PhCO\bar{O})}{RCO\bar{O}} + Ag\downarrow;\ CH_2{=}CH{-}CHO \xrightarrow[\text{ammoniacal}]{AgNO_3} CH_2{=}CH{-}CO\bar{O} + Ag\downarrow$$

**(xxii) *α*-halogenation :** Aldehydes and ketones react with halogen to give mono, di or tri *α*-halogeno products (See Sec. 16.11. p. 880) However, aromatic aldehydes give aromatic acid chlorides on reaction with chlorine in the absence of a halogen carrier.

$$PhCHO + Cl_2 \longrightarrow PhCOCl + HCl$$

**(xxiii) The Schmidt reaction :** When carbonyl compounds are treated with hydrazoic acid in the presence of concentrated sulphuric acid, aldehydes form a mixture of a nitrile and a formyl derivative of 1° amines, while ketones form amides.

$$RCHO + HN_3 \xrightarrow{\text{conc. } H_2SO_4} RCN + RNHCHO + N_2$$

$$RCOR + HN_3 \xrightarrow{\text{conc. } H_2SO_4} RCONHR + N_2$$

The probable mechanism involves the following steps :

**Step I :** Protonation of the carbonyl O atom.

**Step II :** Nucleophilic attack on the carbonyl C by the N atom of the hydrazoic acid.

**Step III :** Elimination of water molecule.

**Step IV :** Elimination of nitrogen molecule and migration of the anti-alkyl group with bonded es form the C atom to the N atom.

**Step V :** Hydration and tautormerisation.

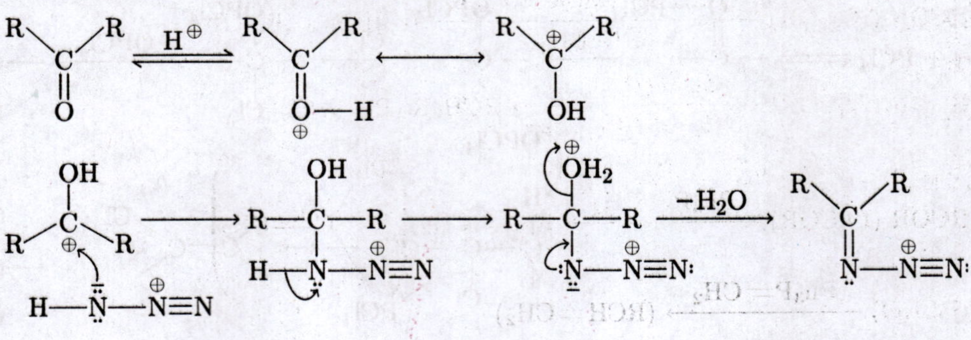

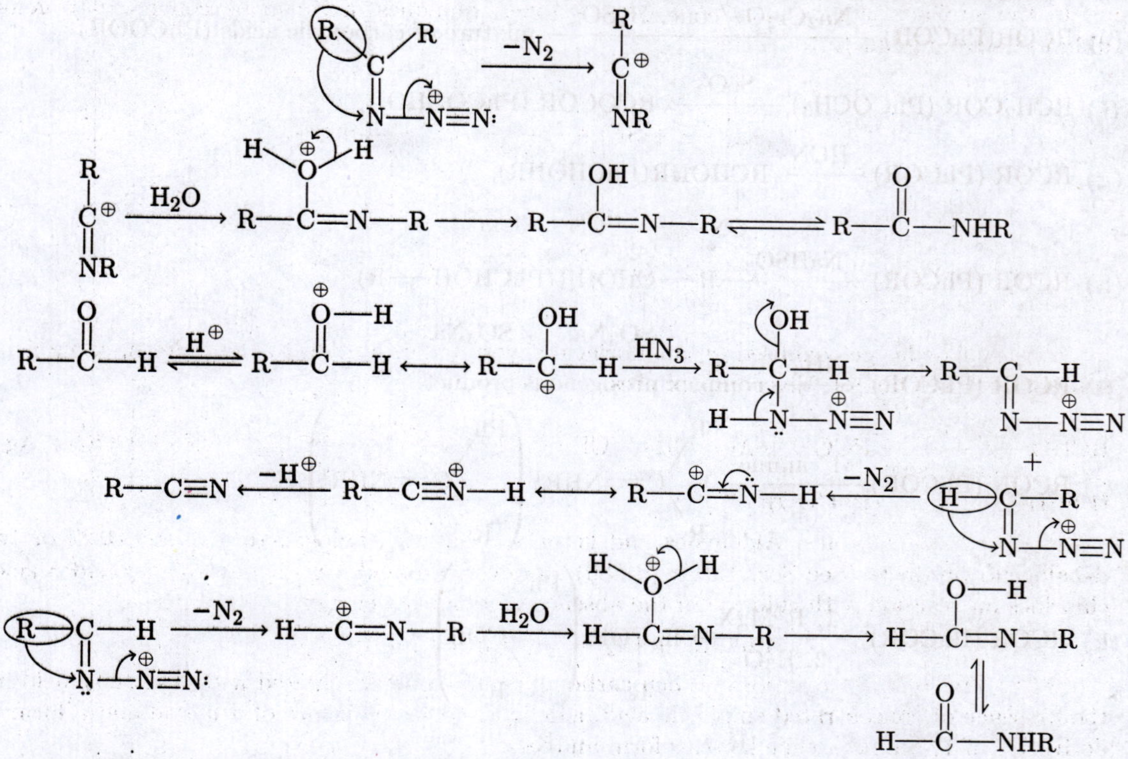

**(xxiv) The Perkin reaction :** See Sec. 16.13.6, p. 903.

**(xxv) The Benzoin condensation :** See Sec. 16.13.5, p. 900.

**(xxvi) Other condensation reactions :** These include (a) the Claisen condensation, (b) the Glycidic ester condensation, (c) the Dieckman condensation, (d) the Michael condensation (p. 664, Sec. 13.13.2), (e) the Knovenagel condensation. For these See p. 890-911, Sec. 16.13.

**(xxvii)** Formyl group is a *m*-directing group in an Ar.$S_E2$ reaction

$$PhCHO \xrightarrow{E^{\oplus}} m\text{-E-}C_6H_4CHO$$

(a) $RCOR\ (PhCOR) \xrightarrow{Na/EtOH} RCHOHR\ (PhCHOHR)$

(b) $RCOR\ (PhCOR) \xrightarrow[\text{or } NaBH_4]{LiAlH_4 \text{ or } (Me_2CHO)_3Al} RCHOHR\ (PhCHOHR)$

(c) $RCOR\ (PhCOR) \xrightarrow[NH_2-NH_2/\text{base}]{\text{Zn-Hg/HCl or}} RCH_2R\ (PhCH_2R)$

(d) $RCOR\ (PhCOR) \xrightarrow{Ph_3P{=}CH_2} R_2C{=}CH_2$ $\left(\begin{matrix}Ph\\ \quad\diagdown\\ \quad\quad C{=}CH_2\\ \quad\diagup\\ R\end{matrix}\right)$

$(RCHO) \xrightarrow{Ph_3P{=}CH_2} (RCH{=}CH_2)$

(e) $RCOR(PhCOR) \xrightarrow{Na_2Cr_2O_7/\text{conc. } H_2SO_4}$ mixture of carboxylic acids (PhCOOH)

(f) $RCH_2COR\ (PhCOCH_3) \xrightarrow{SeO_2} RCOCOR\ (PhCOCHO)$

(g) $RCOR\ (PhCOR) \xrightarrow{HCN} R\underset{\overset{|}{CN}}{C}HOHR(Ph\underset{\overset{|}{CN}}{C}HOHR)$

(h) $RCOR\ (PhCOR) \xrightarrow{NaHSO_3} R{-}\underset{\overset{|}{SO_3Na}}{C}HOHR(Ph\underset{\overset{|}{SO_3Na}}{C}HOH{-}R)$

(i) $RCOR\ (PhCOR) \xrightarrow{NH_3}$ complex nitrogenous product

(j) $RCOR\ (PhCOR) \xrightarrow{1^\circ \text{ amine}} R_2C{=}NHR^1\ \left(\begin{matrix}Ph\\R\end{matrix}\!\!>C{=}NHR^1\right)$

(k) $RCOR\ (PhCOR) \xrightarrow[2.\ H_3O^{\oplus}]{1.\ R^1MgX} R_2\underset{\overset{|}{R^1}}{C}OH\ \left(\begin{matrix}Ph\\R\end{matrix}\!\!>\underset{\overset{|}{R^1}}{C}OH\right)$

(l) $RCOR \xrightarrow[HCl]{R^1OH \text{ or } R^1SH} \begin{matrix}R\\R\end{matrix}\!\!>C(OR^1)_2$ or $\begin{matrix}R\\R\end{matrix}\!\!>C(SR^1)_2$

(m) $RCOR\ (PhCOR) \xrightarrow{H_2NOH} \begin{matrix}R\\R\end{matrix}\!\!>C{=}NOH\ \left(\begin{matrix}Ph\\R\end{matrix}\!\!>C{=}NOH\right)$

(n) $RCOR\ (PhCOR) \xrightarrow{NH_2NH_2} \begin{matrix}R\\R\end{matrix}\!\!>C{=}NNH_2\ \left(\begin{matrix}Ph\\R\end{matrix}\!\!>C{=}NNH_2\right)$

(o) $RCOR\ (PhCOR) \xrightarrow[\text{dil. acid}]{\text{dil. } \bar{O}H}$ aldol

(p) $RCOR\ (PhCOR) \xrightarrow{PCl_5} RCCl_2R\ (PhCCl_2R)$

(q) $RCOCH_3 \xrightarrow{X_2/\bar{O}H} RCOO^- + CHX_3$

(r) $RCOR\ (PhCOR) \xrightarrow{HN_3/\text{conc. } H_2SO_4} R\overset{\overset{O}{\|}}{C}NHR(RCONHPh\ \&\ PhCONHR)$

(s) $RCOR\ (PhCH_2COR) \xrightarrow{\text{peracid}} R{-}COOR\ \left(\begin{matrix}Ph \\ Me\end{matrix}\!\!>CHOCOR\right)$

(t) $R_2CO \xrightarrow{\text{Mg/ether}} R_2C(OH){-}C(OH)R_2$

$(Ph_2CO) \qquad Ph_2C(OH){\cdot}C(OH)Ph_2$

(u) $RCOCH_3 \xrightarrow{HNO_2\ \&\ EtONO/HCl} RCOCH{=}NOH + H_2O \xrightarrow{H^{\oplus}} RCOCHO$

(v) $RCOR \xrightarrow{CHCl_3/KOH} R_2C(OH)(CCl_3)$

(w) $R{-}\overset{\overset{O}{\|}}{C}{-}CH_2R \xrightarrow{NaNH_2} RCH{=}\overset{\overset{\bar{O}Na^{\oplus}}{|}}{C}{-}R$

(x) $PhCOCH_3 \xrightarrow{(NH_4)_2S_x} PhCH_2CONH_2 + PhCH_2COONH_4$

(y) $PhCOCH_3 \xrightarrow{HCHO\ +\ (CH_3)_2\overset{\oplus}{N}H_2Cl^-} PhCOCH_2CH_2\overset{\oplus}{N}HMe_2Cl^-$

(z) $PhCOPh \xrightarrow{\text{Na/ether}} Ph_2\dot{C}{-}\bar{O}\overset{\oplus}{Na}$

Other condensation reactions.

**Chart 20.19** Reactions of ketones.

**(a) The Bouvault-Blanc reduction :** See. Sec. 16.9.1. p. 871

**(b) Reduction by metal hydrides :** See. Secs. 16.6.1. and 16.6.2. p. 854 and p. 857.

When acetophenone and benzophenone are reduced with $LiAlH_4$-$AlCl_3$ mixture in ether, ethylbenzene and diphenylmethane are respectively obtained. Zinc dust distillation also gives the same products.

$$PhCOCH_3 \xrightarrow{LiAlH_4/AlCl_3;\ \text{ether}} PhCH_2CH_3;\ PhCOPh \xrightarrow[\text{or, Zn dist.}]{LiAlH_4/AlCl_3;\ \text{ether}} PhCH_2Ph$$

**(c) The Clemmensen reduction and the Wolff-Kishner reduction :** See Sec. 18.1.3i (p. 1020) and sec. 16.7. p. 865.

**(d) The Wittig reaction :** See Sec. 16.8. p. 867.

**(e) Chromic acid oxidation :** Ketones on chromic acid oxidation give a mixture of carboxylic acids each of which contains a fewer number of C atoms than that of the parent ketone. In this reaction C—CO and CO—C bonds break.

$$RCH_2COCH_2CH_2R^1 \xrightarrow[\text{conc. } H_2SO_4]{Na_2Cr_2O_7} RCOOH + R^1CH_2CH_2COOH + RCH_2CO_2H + R^1CH_2CO_2H$$

**(f) Oxidation by Selenium dioxide :** See Sec. 20.7.5. vi, p. 1254.

**(g) Hydrogen cyanide addition :** See Sec. 16.4.1 (p. 828). Benzophenone does not form cyanohydrin.

**(h) Sodium bisulphite addition :** See Sec. 16.4.3 (p. 832). Acetophenone and benzophenone do not form bisulphites.

**(i) Reaction with ammonia :** Ketones form complex nitrogenous organic compounds with ammonia. Acetophenone forms acetophenoneammonia with ammonia.

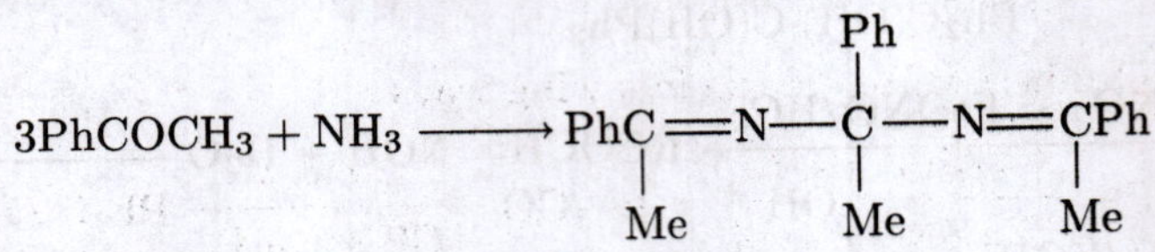


**(j) Reaction with amines :** See Sec. 20.7.5(xix), p. 1256.

**(k) Reaction with the Grignard reagent :** See Sec. 16.4.5, p. 841. However, acetophenone and benzophenone with both *o*-positions occupied do not undergo the reaction owing to steric hindrance.

**(l) Reaction with alcohols and thiols :** See Sec. 16.4.4 (p. 833). Ketones seldom react with alcohol; but they form thioketals (p. 836) readily.

**(m) Reaction with hydroxylamine :** See Sec. 16.5. However, acetophenone and benzophenone with both *o*-positions occupied do not form oximes owing to steric hindrance.

**(n) Reaction with hydrazine and hydrazine derivatives :** See Sec. 16.5, p. 851.

**(o) Aldol condensation :** See Sec. 16.13.1, p. 891.

In the presence of aluminium *t*-butoxide, acetophenone undergoes self-condensation to form dypnone; whereas in the presence of hydrochloric acid, 1, 3, 5-triphenylbenzene is formed.

When benzophenone is fused with solid potassium hydroxide, benzene and potassium benzoate are formed.

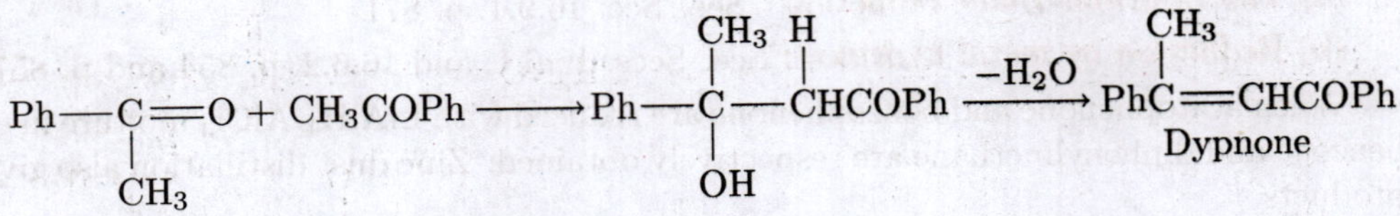


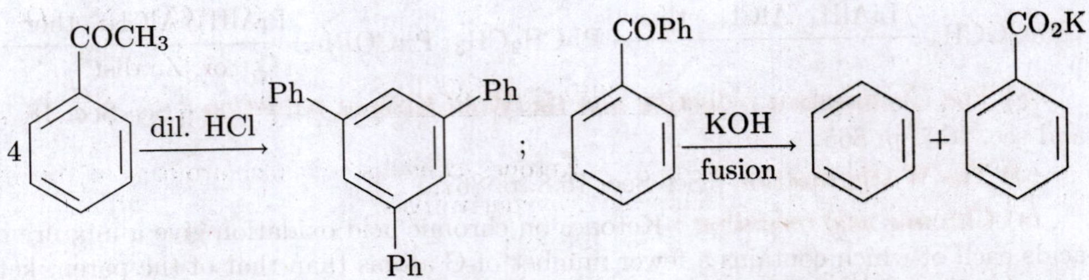


**(p) Reaction with phosphorus pentachloride :** See Sec. 20.7.5(xx), p. 1257.

**(q) Reaction with halogen :** See Sec. 20.7.5(xxii), p. 1258.

**(r) The Schmidt reaction :** See Sec. 20.7.5(xxiii), p. 1258.

**(s) Baeyer-Villigar reaction :** See Sec. 14.20, p. 773.

**(t) Bimolecular reduction of carbonyl compounds :** See Sec. 16.9.2 (p. 872). When benzophenone is treated with a mixture of magnesium and magnesium iodide, benzopinacol is obtained.

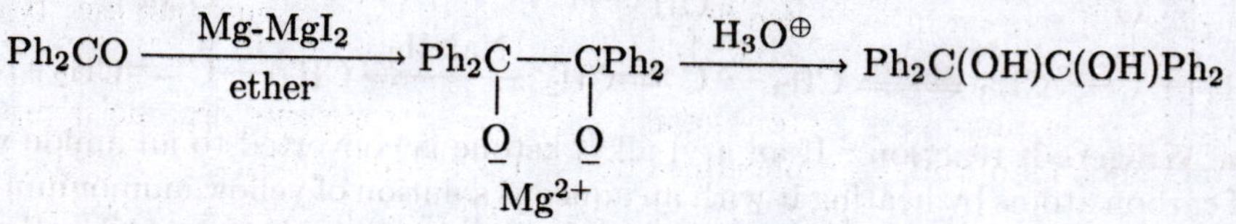

Benzopinacol may also be formed by the following ways :

$$Ph_2C(OH)C(OH)Ph_2 \xleftarrow[\text{a few drop of AcOH, sunlight}]{Me_2CHOH} Ph_2CO \xrightarrow[CH_3CO_2H]{Zn} Ph_2C(OH)C(OH)Ph_2$$

However, when the acetic acid solution of benzopinacol is heated in the presence of iodine catalyst, triphenylacetophenone is formed.

$$Ph_2C(OH)C(OH)Ph_2 \xrightarrow[I_2\text{ cat.}]{\text{in acetic acid}} Ph_3CCOPh + H_2O$$

**(u) Reaction with nitrous acid and ethyl nitrite :** Carbonyl compounds with —$CH_2CO$— moiety react with nitrous acid in the presence of ethyl nitrite to form oximino derivatives and the reaction is used to detect the presence of a ketomethylene group. The oximino derivatives on hydrolysis form $\alpha$, $\beta$-dicarbonyls.

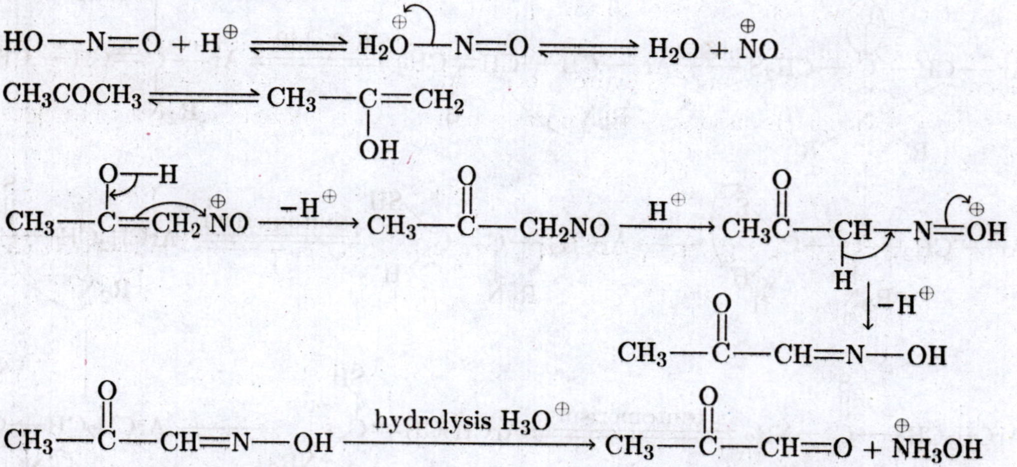

**(v) Condensation with chloroform :** Ketones condense with chloroform in the presence of potassium hydroxide to form trichlorohydroxyderivatives.

$$CH_3COCH_3 + CHCl_3 \xrightarrow{KOH} Me_2C(OH)CCl_3$$

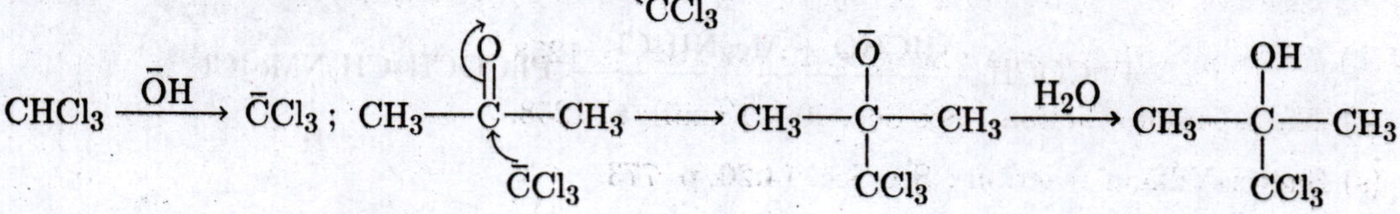

**(w) Action of sodamide on ketones :** Ketones containing $\alpha$-H give enolates on reaction with sodamide.

$$CH_3-\overset{O}{\overset{\|}{C}}-CH_3 \rightleftharpoons CH_3-\overset{OH}{\overset{|}{C}}=CH_2 \xrightarrow{NaNH_2} CH_3-\overset{\overset{\ominus}{O}Na}{\overset{|}{C}}=CH_2 + NH_3$$

**(x) The Willgerodt reaction :** If an aryl alkyl ketone is converted to an amide with the same number of carbon atoms by heating it with an aqueous solution of yellow ammonium polysulphide, the reaction is called the **Willgerodt reaction**; for example, acetophenone forms phenylacetamide along with a small amount of ammonium phenylacetate.

$$PhCOCH_3 \xrightarrow{(NH_4)_2S_x} PhCH_2CONH_2 + PhCH_2COONH_4$$

The reaction is very useful for preparing aryl derivatives of aliphatic acids. The reaction is modified (**Kindler**) by treating the aryl alkyl ketone with a 2° amine e.g., morphine and sulphur in equimolecular amounts. In this case a thioamide forms, which on hydrolysis with alkali or acid gives the required acid. However, the mechanism of the reaction may be as follows :

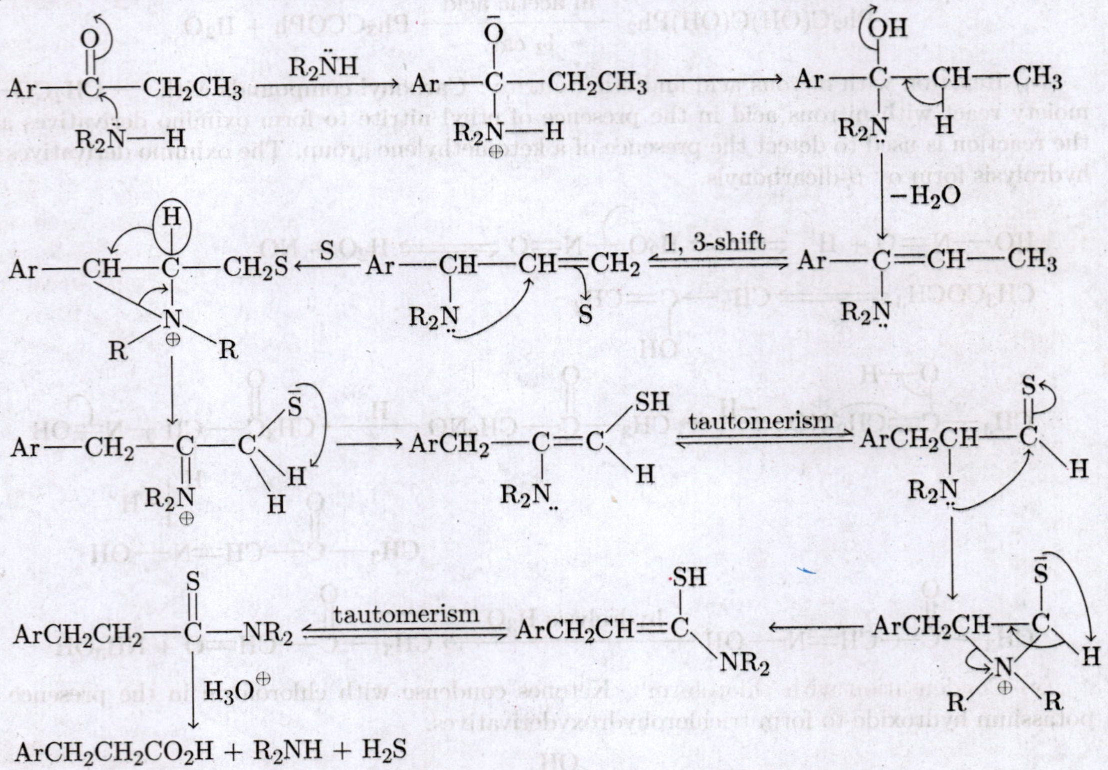

**(y) The Mannich reaction :** See Sec. 16.13.8.. p. 911.

$$PhCOCH_3 \xrightarrow{HCHO + Me_2\overset{\oplus}{N}H_2Cl^-} PhCOCH_2CH_2\overset{\oplus}{\underset{H}{\underset{|}{N}}}Me_2Cl^-$$

**(z) Ketyl formation :** Sodium gets dissolved in the ethereal solution of benzophenone without evolving hydrogen and forms a blue solution containing ketyl, an anionic radical.

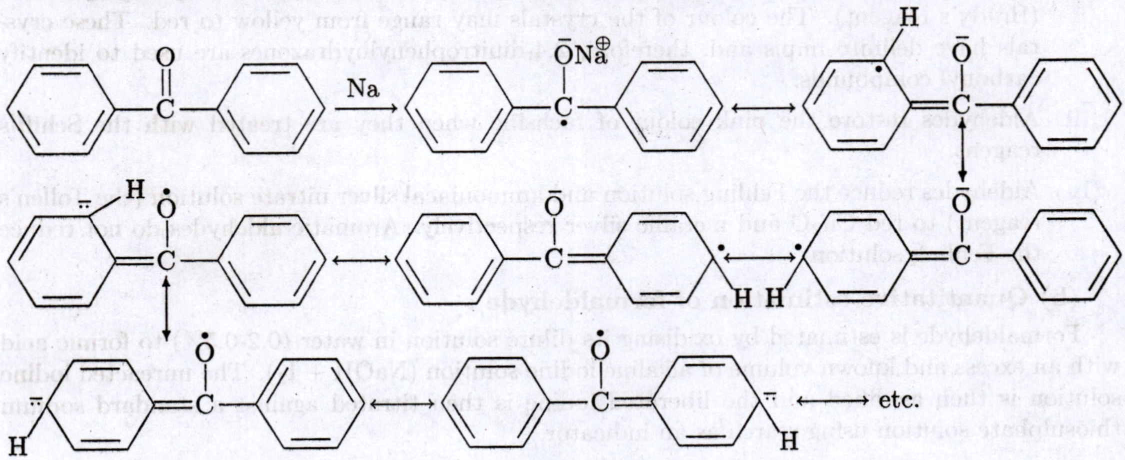

**Other condensation reactions :** Acetophenone and other aryl alkyl ketones having α-H condense with acetic anhydride in the presence of boron trifluoride to form benzoylacetones.

$$\mathrm{PhCOCH_2R + (MeCO)_2O \xrightarrow{BF_3} PhCOCH(R){-}COCH_3 + CH_3CO_2H}$$

Acetophenone undergoes the Claisen condensation reaction with ethyl benzoate and ethyl acetate in the presence of sodium ethoxide to form β-dicarbonyls.

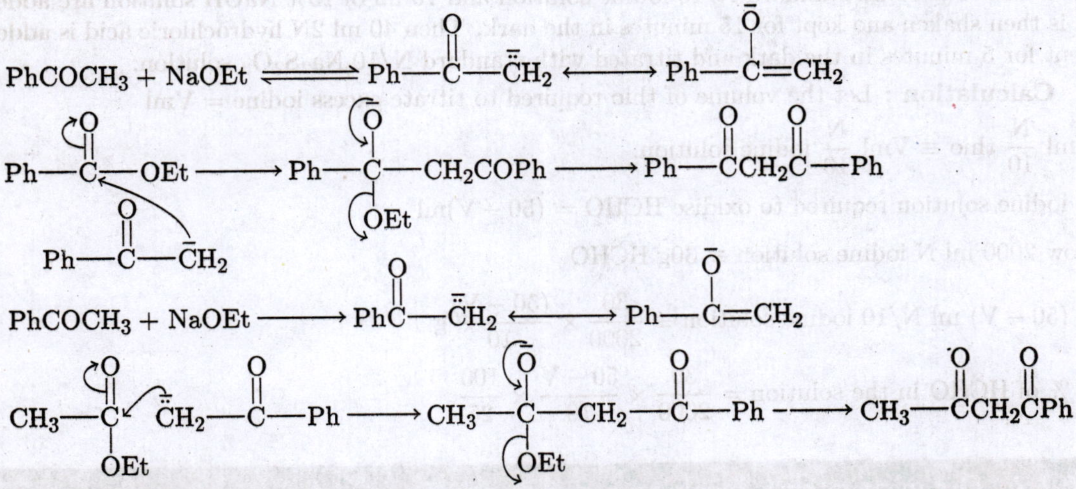

## 20.7.6 Analysis

**(a) Qualitative tests for aldehydes and ketones :**

(i) Aldehydes and sterically unhindered methyl ketones form white precipitate with aqueous saturated solution of sodium bisulphite.

(ii) An alcoholic solution of a carbonyl compound gives white precipitate with phenylhydrazine or semicarbazide. They form heavy crystalline precipitate with 2, 4-dinitrophenylhydrazine (Brady's reagent). The colour of the crystals may range from yellow to red. These crystals have definite m.p.s and, therefore, 2,4-dinitrophenylhydrazones are used to identify carbonyl compounds.

(iii) Aldehydes restore the pink colour of fuchsine when they are treated with the Schiff's reagent.

(iv) Aldehydes reduce the Fehling solution and ammoniacal silver nitrate solution (the Tollen's reagent) to red $Cu_2O$ and metallic silver respectively. Aromatic aldehydes do not reduce the Fehling solution.

**(b) Quantitative estimation of formaldehyde :**

Formaldehyde is estimated by oxidising its dilute solution in water (0.2-0.5%) to formic acid with an excess and known volume of alkaline iodine solution ($NaOH + I_2$). The unreacted iodine solution is then acidified and the liberated iodine is then titrated against a standard sodium thiosulphate solution using starch as an indicator.

$$I_2 + NaOH \longrightarrow NaI + NaOI + H_2O; \; HCHO + NaOI \xrightarrow{NaOH} HCOO^- + NaI + H_2O$$

$$NaOI + NaI + 2HCl \longrightarrow 2NaCl + H_2O + I_2$$

$$I_2 + 2Na_2S_2O_3 \longrightarrow Na_2S_4O_6 + 2\,NaI$$

**Reagents :** Standard N/10 $Na_2S_2O_3$ solution, standard N/10 iodine solution, concentrated 2N hydrochloric acid and starch solution.

**Procedure :** 25 ml of 0.2-0.5% aqueous formaldehyde solution is taken in a 250 ml conical flask and to it 50 ml standard N/10 iodine solution and 10 ml of 10% NaOH solution are added. It is then shaken and kept for 15 minutes in the dark. Then 40 ml 2N hydrochloric acid is added, kept for 5 minutes in the dark and titrated with standard N/10 $Na_2S_2O_3$ solution.

**Calculation :** Let the volume of thio required to titrate excess iodine = Vml

$$\text{Vml } \frac{N}{10} \text{ thio} \equiv \text{Vml } \frac{N}{10} \text{ iodine solution.}$$

$\therefore$ iodine solution required to oxidise HCHO = (50 – V)ml

Now 2000 ml N iodine solution $\equiv$ 30g HCHO.

$$\therefore (50 - V) \text{ ml N/10 iodine solution} \equiv \frac{30}{2000} \times \frac{(50-V)}{10} \text{g}$$

$$\therefore \% \text{ of HCHO in the solution} = \frac{30}{2000} \times \frac{50-V}{10} \times \frac{100}{25}.$$

## 20.7.7 Organic synthesis involving carbonyl groups

Carbonyl groups can be converted to a large number of other functional groups; keeping the carbonyl group intact, other functional groups may also be introduced in a carbonyl compound. Carbon chains of carbonyl compounds can be lengthened through the condensation reactions. Thus carbonyl compounds are key materials from which several other compounds can be synthesised. A few synthetic chains are given below :

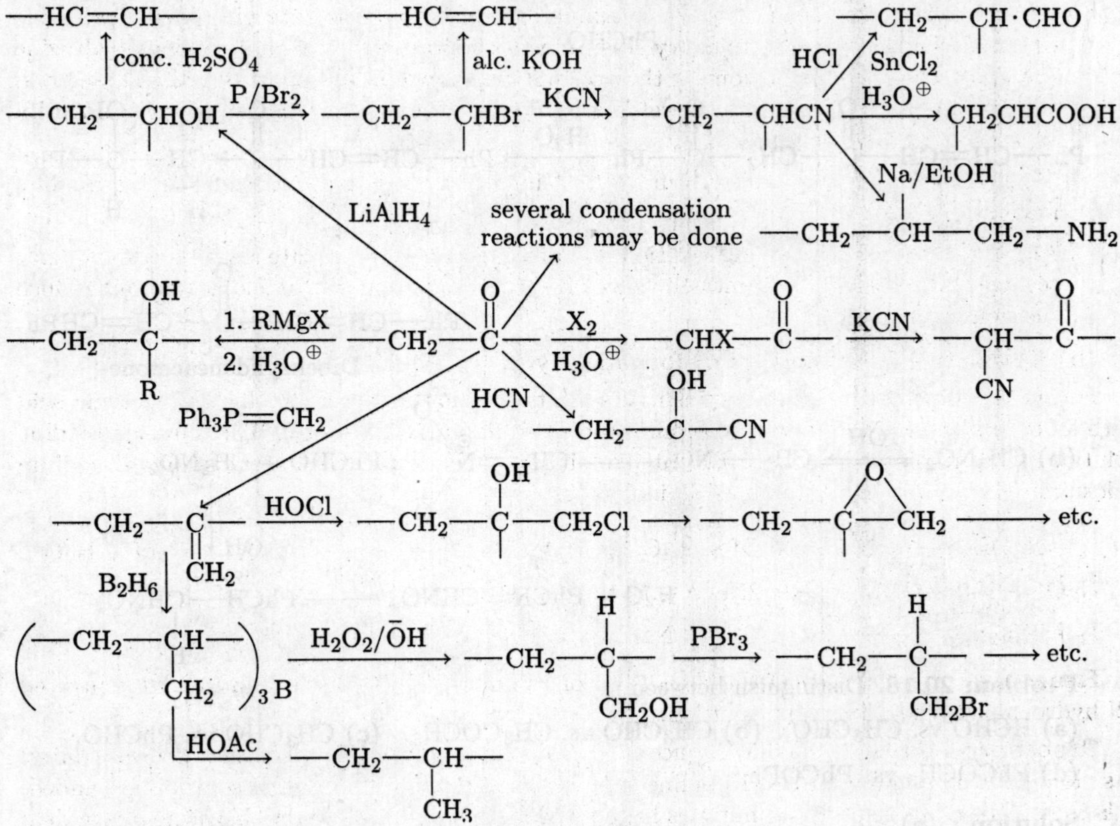

**Problem 20.17.** What happens when

**(a)** Benzaldehyde is condensed with acetone in aqueous ethanolic sodium hydroxide.

**(b)** Benzaldehyde is condensed with nitromethane.

**Solution : (a)** Benzylidene acetone is first formed which then gets converted to dibezylidene acetone.

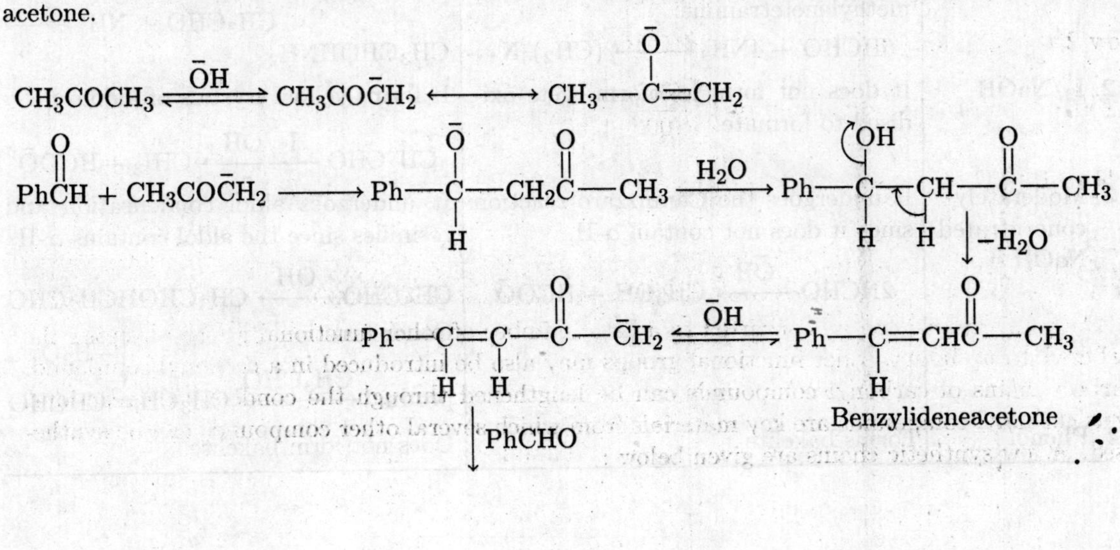

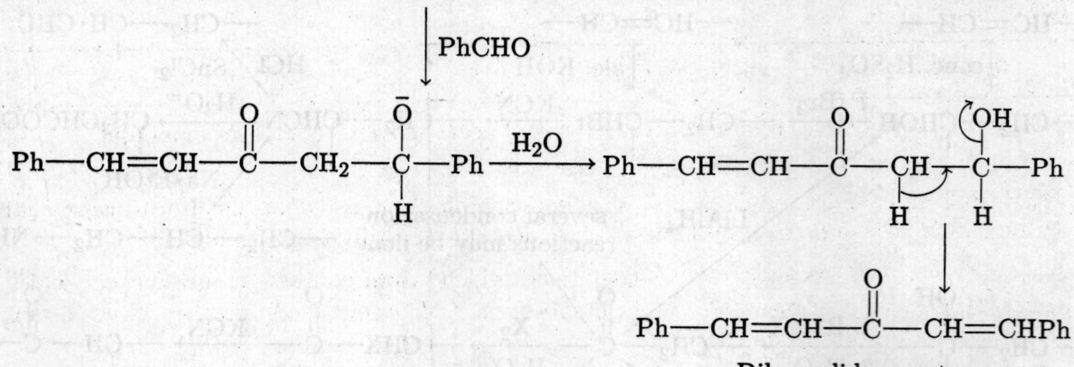

Dibenzylideneacetone

(b)

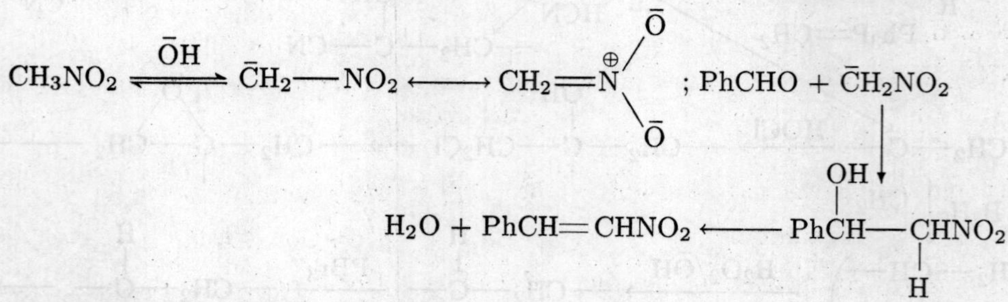

**Problem 20.18.** Distinguish between :

**(a)** HCHO vs. $CH_3CHO$, **(b)** $CH_3CHO$ vs. $CH_3COCH_3$, **(c)** $CH_3CHO$ vs. PhCHO,
**(d)** $PhCOCH_3$ vs. PhCOPh.

**Solution : (a)**

| Reagent | HCHO | $CH_3CHO$ |
|---|---|---|
| 1. Ammonia | It reacts with ammonia to form a solid crystalline substance, called hexamethylenetetramine. $6HCHO + 4NH_3 \longrightarrow (CH_2)_6N_4$ | It gives addition product called acetaldehyde ammonia. $CH_3CHO + NH_3 \longrightarrow CH_3CHOHNH_2$ |
| 2. $I_2$/NaOH | It does not form iodoform; gets oxidised to formate. | It forms yellow crystals of iodoform. $CH_3CHO \xrightarrow{I_2/\bar{O}H} CHI_3 + HCO\bar{O}$ |
| 3. Moderately concentrated NaOH | It undergoes the Cannizzaro reaction since it does not contain $\alpha$-H. $2HCHO \xrightarrow{\bar{O}H} CH_3OH + HCO\bar{O}$ | It undergoes aldol condensation and resinifies since the aldol contains $\alpha$-H. $CH_3CHO \xrightarrow{\bar{O}H} CH_3CHOHCH_2CHO$ Condensation $\xrightarrow{-H_2O} CH_3CH{=}CHCHO \xrightarrow{CH_3CHO}$ product |
| 4. Phenol | Forms bakelite | Does not form bakelite. |

**(b)**

| Reagent | $CH_3CHO$ | $CH_3COCH_3$ |
|---|---|---|
| 1. $NH_3$ | Gives addition product, called acetaldehyde ammonia. | Gives a complex compound. |
| 2. ROH/HCl and RSH/$H_2SO_4$ | Forms full acetal via hemiacetal, and thioacetal via hemithioacetal. | Usually does not form ketal but readily forms thioketal. |
| 3. $Na_2Cr_2O_7$/ conc. $H_2SO_4$ | Gets oxidised to acetic acid which contains the same number of C atoms as $CH_3CHO$. | Gets oxidised to acetic acid which contains less number of C atoms than acetone. |
| 4. $LiAlH_4$/ether | Forms $CH_3CH_2OH$, a 1° alcohol. | Forms $CH_3CHOHCH_3$, a 2° alcohol. |
| 5. RMgX & $H_3O^{\oplus}$ | Forms 2° alcohol. | Forms 3° alcohol. |
| 6. $[Ag(NH_3)_2]NO_3$ | Reduces ammoniacal silver nitrate to metallic silver which forms black precipitate or mirror. | No reaction. |
| 7. Fehling solution | Forms red precipitate. | No reaction. |
| 8. Alkali | Resinifies. | Does not resinify. |
| 9. Schiff's reagent | Restores the pink colour of the fuchsine. | Does not restore the pink colour. |
| 10. Polymerisation | Readily polymerises. | Does not polymerise. |

**(c)**

| Reagent | $CH_3CHO$ | PhCHO |
|---|---|---|
| 1. $Na_2Cr_2O_7$ conc. $H_2SO_4$ | Forms acetic acid. | Forms benzoic acid. |
| 2. $LiAlH_4$/ether | Forms ethyl alcohol. | Forms benzyl alcohol. |
| 3. Fehling solution | Red precipitate of $Cu_2O$ is formed. | No reaction. |
| 4. $I_2$/NaOH | Forms yellow crystals of iodoform | Does not form iodoform but forms benzoate. |
| 5. $NH_3$ | Forms aldehyde ammonia. | Form hydrobenzamide. |
| 6. Moderately concentrated NaOH | Aldol condensation and resinification take place. | Undergoes the Cannizzaro reaction. |
| 7. Perkin reaction : $(MeCO)_2O$ and AcONa | No reaction. | Forms cinnamic acid. |
| 8. Benzoin condensation : aq. KCN | No reaction. | Undergoes benzoin condensation and forms benzoin. |

(d)

| Reagent | $PhCOCH_3$ | $PhCOPh$ |
|---|---|---|
| 1. $I_2/NaOH$ | Forms yellow crystals of iodoform. | Does not form iodoform. |
| 2. HCN | Forms cyanohydrin. | Does not form cyanodydrin. |
| 3. $NaHSO_3$ | Forms bisulphite. | Does not form bisulphite. |
| 4. $(NH_4)_2S_x$ | Undergoes the Willgerodt reaction and forms $PhCH_2CONH_2$. | Does not undergo the reaction. |
| 5. $HCHO + NHMe_2$ + HCI | Undergoes the Mannich reaction and forms $PhCOCH_2CH_2\overset{\oplus}{N}HMe_2Cl^-$ | Does not undergo the reaction. |
| 6. $Mg/MgI_2$/ether | Does not form pinacol. | Forms pinacol $Ph_2C(OH)\ C(OH)\ Ph_2$. |

## 20.7.8 Uses of carbonyl compounds

(i) **Formaldehyde** : Formalin (40% formaldehyde) is used as a disinfectant and for the preservation of biological specimens. Its chief use is for the manufacture of synthetic resins through condensation with phenol, urea, melamine. Paraformaldehyde [$HO(CH_2)_xH$, $x = 30$; this is prepared by concentrating formalin at reduced pressure] and hexamethylenetetramine; sometimes used to prepare resins. Formaldehyde is used to prepare pentaerythritol nitrate which is used as an explosive. Pentaerythritol is itself used to dry oil. Sodium formaldehyde sulphoxylate (prepared by the reaction of zinc dithionite on formaldehyde and then treating the reaction product with sodium carbonate) is a strong reducing agent which is used to destroy azo dyes in the discharge printing of cloth.

(ii) **Acetaldehyde** is used as an intermediate to synthesise acetic acid, acetic anhydride (by air-oxidation), ethyl acetate (by the Tischenko reaction) and *n*-butyl alcohol (by aldol condensation followed by catalytic reduction), acrylonitrile, lactic acid and acrylic esters. Acetaldehyde, the paraldehyde (trimer) and the metaldehyde (tetramer) are used for the manufacture of rubber accelerators. Acetaldehyde is also used to manufacture DDT. Chloral hydrate, a sleep-producer, is formed by chlorinating acetaldehyde.

(iii) **Acetone** is mainly used as a solvent for lacquers, resin, cellulose acetate and acetylene. It is also used for the manufacture of methyl iso-butyl ketone and methyl methacrylate.

(iv) **Methyl ethyl ketone (MEK)** is used chiefly as a solvent for dewaxing lubricating oils.

(v) **Benzaldehyde** and **veratral** (3,4-dimethoxybenzaldehyde) are chiefly used in organic synthesis.

(vi) **Vanillin** (*p*-**hydroxy**-*m*-**methoxybenzaldehyde**) and **etharan** (*p*-**hydroxy**-*m*-**ethoxybenzaldehyde**) are used as flavouring agents. Vanillin is obtained from the liquors from the extraction of lignin. It may also be synthesised from 2-methoxyphenol.

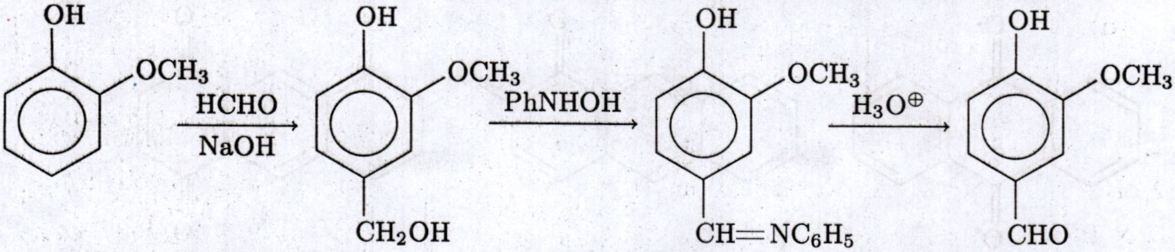

(vii) Acetophenone and benzophenone are used as intermediates in organic synthesis.

(viii) Chloroacetophenone is used as tear gas for dispersing mob.

(ix) Michler ketone is used in dye industry.

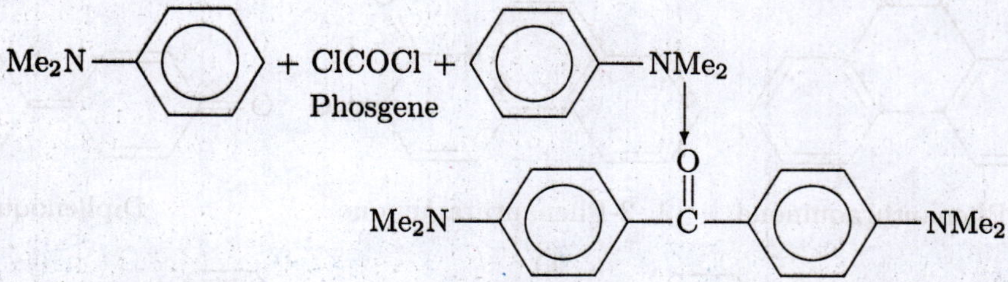

## 20.8 Quinones

### 20.8.1 Introduction

Quinones are cyclic diketones in which ketonic carbonyl groups are conjugated with C—C double bonds and hence they are not aromatic compounds but they can be prepared from and converted to aromatic compounds. The most familiar quinones are *p*-benzoquinone, 1, 4-naphthaquinone, 9, 10-anthraquinone. However, the following quinones are known :

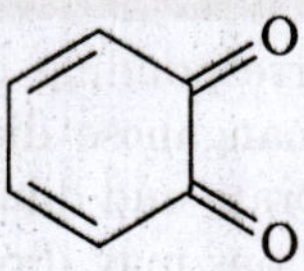

*o*-Benzoquinone or, 1, 2-Benzoquione

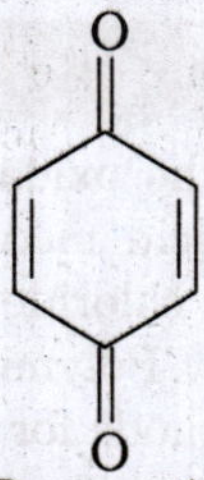

*p*-Benzoquinone or, 1, 4-Benzoquinone

*m*-Benzoquinone cannot exist.

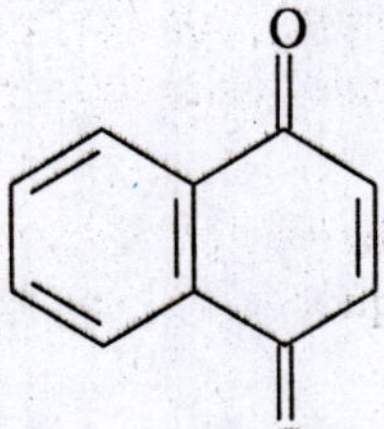

1, 4-Naphthaquinone

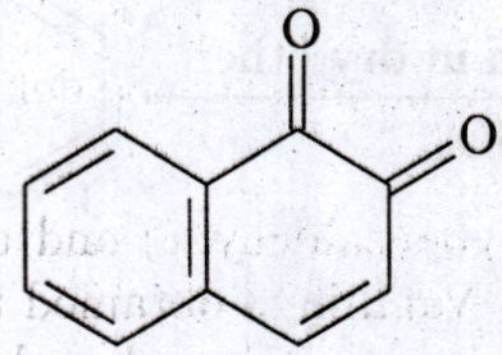

1, 2-Napthaquinone

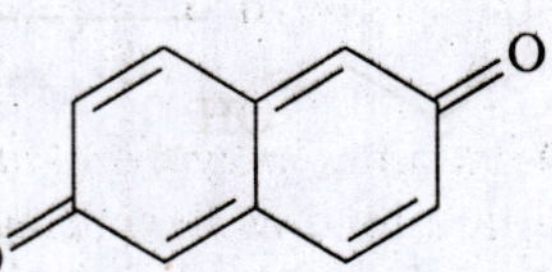

2, 6-Napthaquinone

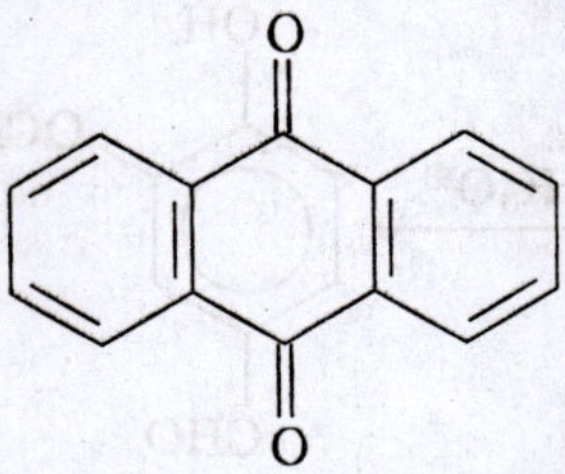

9, 10-Anthraquinone

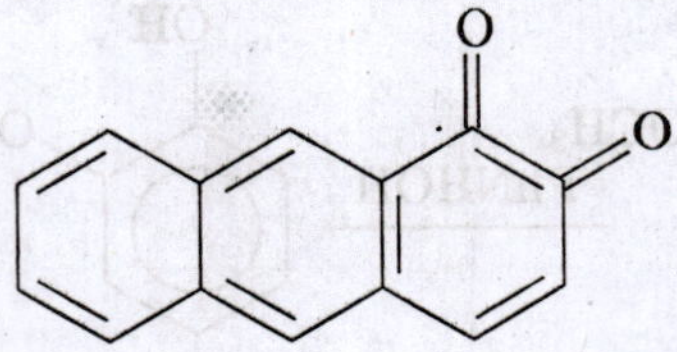

1, 2-Anthraquinone

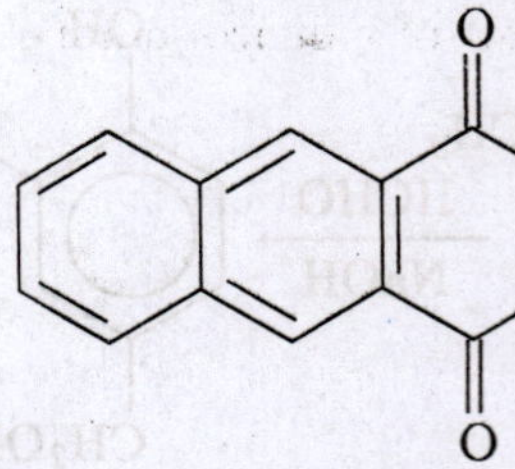

1, 4-Anthraquinone

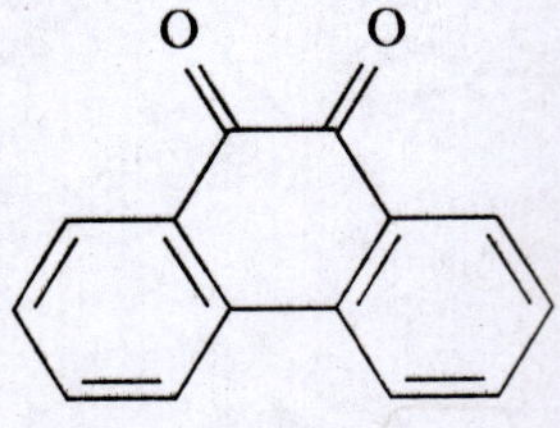

9, 10-Phenanthraquinone

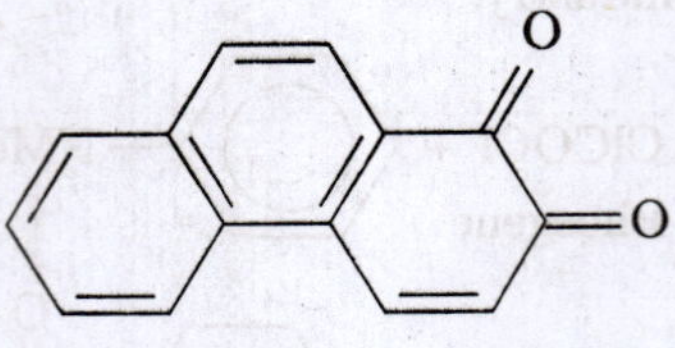

1, 2-Phenanthraquinone

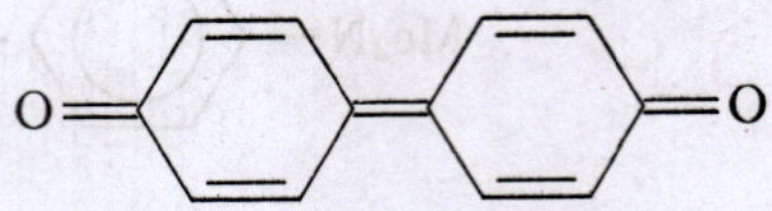

Diphenoquinone

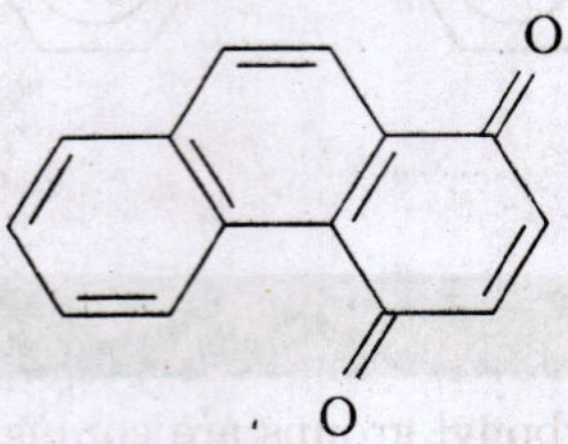

1, 4-Phenanthraquinone

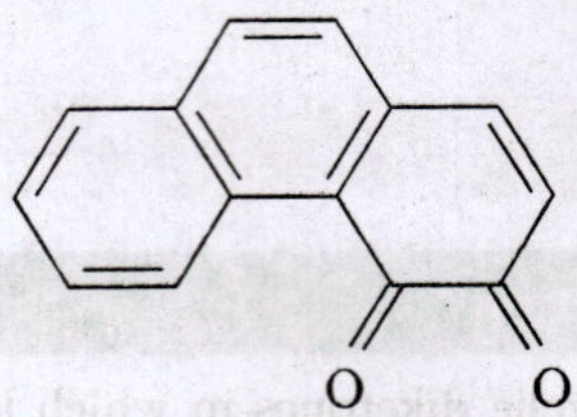

3, 4-Phenanthraquinone

From the above list it is clear that quinones are named after the parent hydrocarbons.

## 20.8.2 Preparation of quinones

Quinones are usually prepared by the oxidation of the corresponding dihydroxy, diamino and aminohydroxy compounds in the acid medium with (i) manganese dioxide or potassium dichromate and sulphuric acid, (ii) ferric chloride or sodium chlorate and dilute sulphuric acid in the presence of vanadium pentoxide, etc. Polynuclear hydrocarbons may directly be oxidised to quinones. However, the preparatory methods for individual members are given below :

(i) Synthesis of *o*-benzoquinone

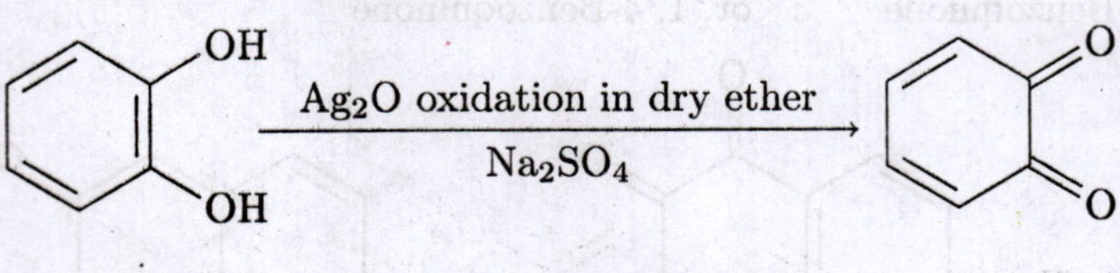

**(ii) Synthesis of *p*-benzoquinone**

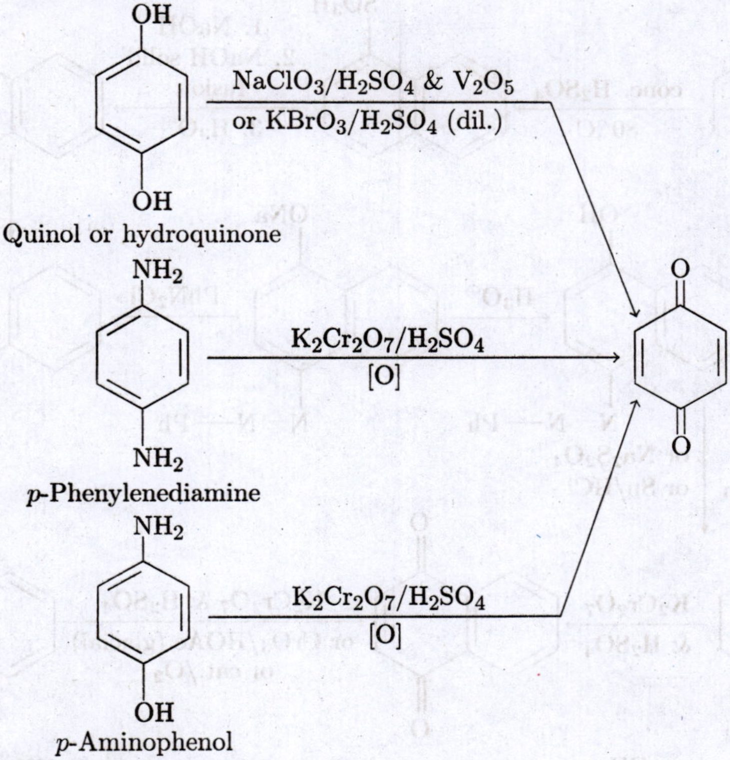


**(ii) Synthesis of 1, 2-naphthaquinone**

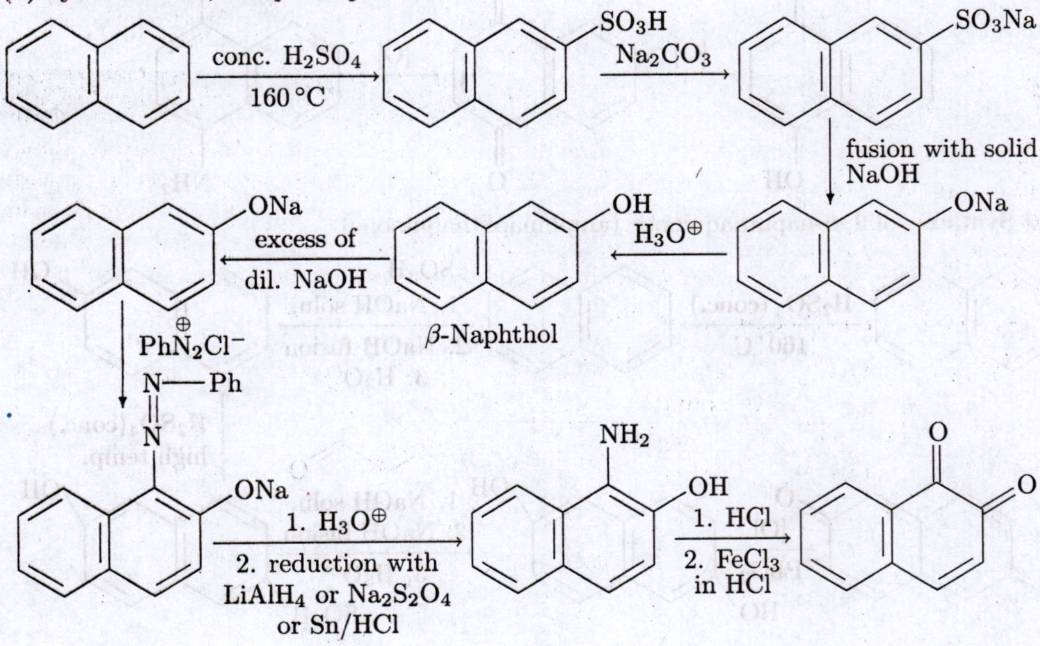

**(iv) Synthesis-of 1, 4-naphthaquinone**

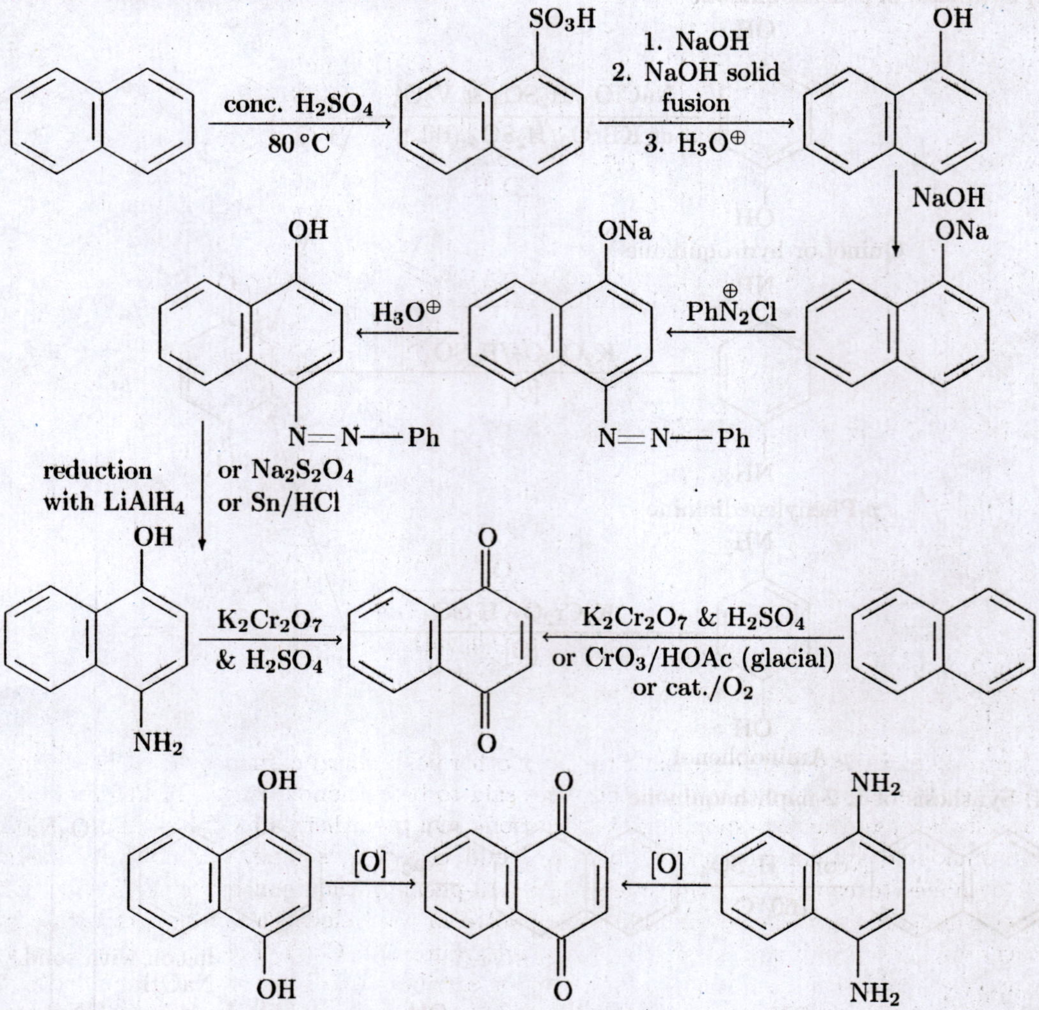


**(v) Synthesis of 2, 6-naphthaquinone (amphinaphthaquinone)**

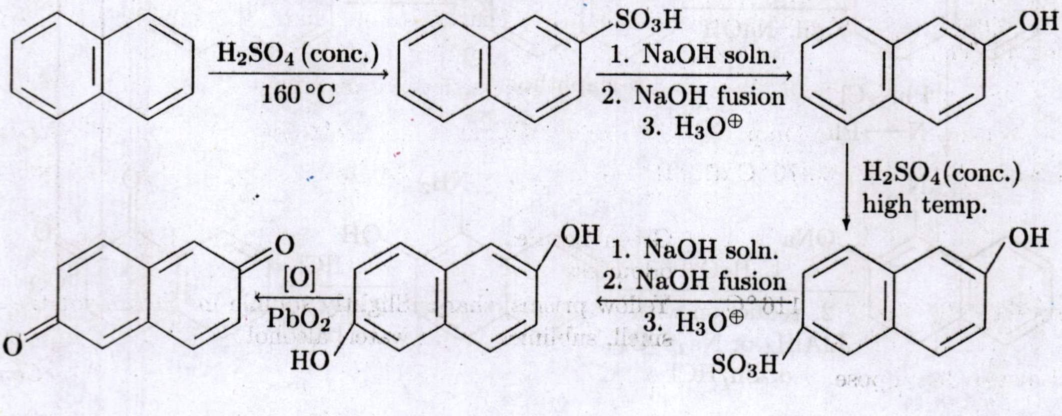

**(vi) Synthesis of 9, 10-anthraquinone**

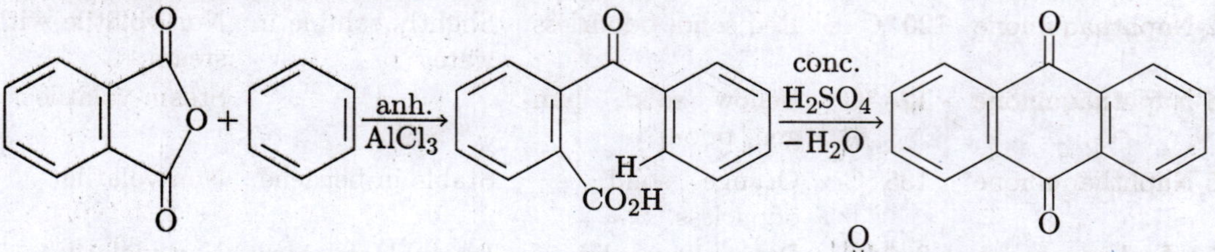

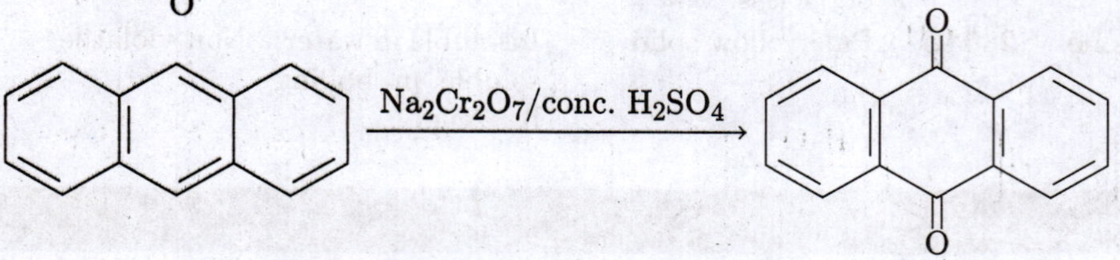

**(vii) Synthesis of 9, 10-phenanthraquinone**

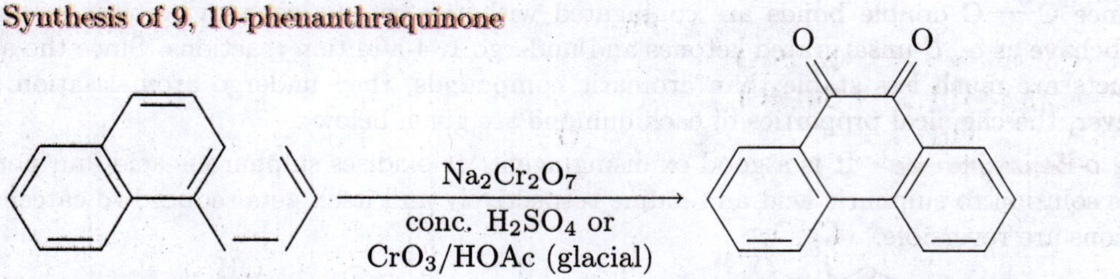

## 20.8.3 Structure and physical properties

Benzoquinones do not have aromatic rings. In other polynuclear quinones the rings bearing the carbonyl groups are not aromatic and these are said to be quinonoid rings. Therefore, quinones are less stable than the corresponding hydrocarbons. On the other hand, the $\pi$ orbitals involved in the quinonoid system embrace $\pi$ bonded C and O atoms, so they are not very unstable; in all quinones (except in 1, 2-quinones and 9, 10-phenanthraquinone) one of the two C=C is cross conjugated and hence this does not involve in the delocalisation of electronic charge. However, owing to conjugation between carbonyl groups and C=C bonds, quinones are good chromophores and they bear colour; the colour is attributed to $n \rightarrow \pi^*$ transition. They can form weak H-bonding with water and alcohol. This makes them slightly soluble in water but readily soluble in alcohol. Some of the polynuclear quinones are found to be stable in benzene only. Usually 1, 4-quinones are steam volatile. A chart of colour, m.p., and solubility is given in the following table :

Table 20.1 Physical properties of quinones

| Name | m.p. | Colour & Odour | Solubility | Steam-volatility |
|---|---|---|---|---|
| 1, 2-Benzoquinone. | 60-70 °C d*. | Red palate | Alcohol | Hot steam-volatile Stable |
| | | Green needles, odourless | | Unstable |
| 1, 4-Benzoquinone | 116 °C | Yellow prisms, sharp smell, sublimes | Slightly soluble in water, alcohol | Steam-volatile |

* d means decompose

*Contd.*

| Name | m.p. | Colour & Odour | Solubility | Steam-volatility |
|---|---|---|---|---|
| 1, 2-Naphthaquinone | 120 °C d. | Red solid, odourless | Slightly soluble in water | Non-volatile with steam. |
| 1, 4-Naphthaquinone | 125 °C | Yellow solid, pungent odour | | Steam-volatile |
| 2, 6-Naphthaquinone | 135 °C | Orange solid, odourless. | Stable in benzene | Non-volatile |
| 9, 10-Anthraquinone | 286 °C | Pale yellow solid | Insoluble in water, soluble in boiling benzene. | Non-vlolatile |

## 20.8.4 Chemical reactions

Since C—C double bonds are conjugated with the ketonic carbonyl groups in quinones, they behave as $\alpha$, $\beta$-unsaturated ketones and undergo 1, 4-addition reactions. Since the addition products are much less stable than aromatic compounds, they undergo aromatisation readily. However, the chemical properties of each quinone are given below :

**(i) *o*-Benzoquinone :** It is a good oxidising agent. It oxidises sulphurous acid and potassium iodide solution to sulphuric acid and iodine respectively and itself gets reduced to catechol—the reactions are reversible.

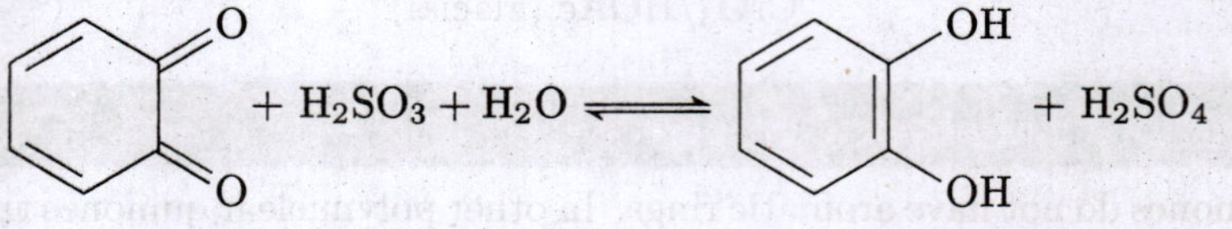


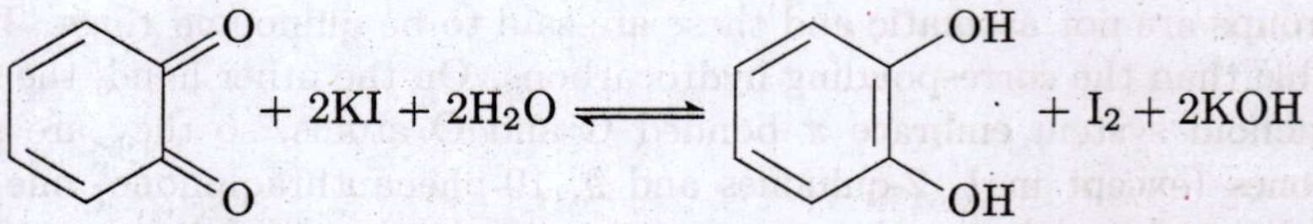


1, 2-Benzoquinone is very unstable in water. It polymerises readily in water and forms several types of coloured products.

**(ii) *p*-Benzoquinone :**

(a) Oxidation of *p*-benzoquinone :

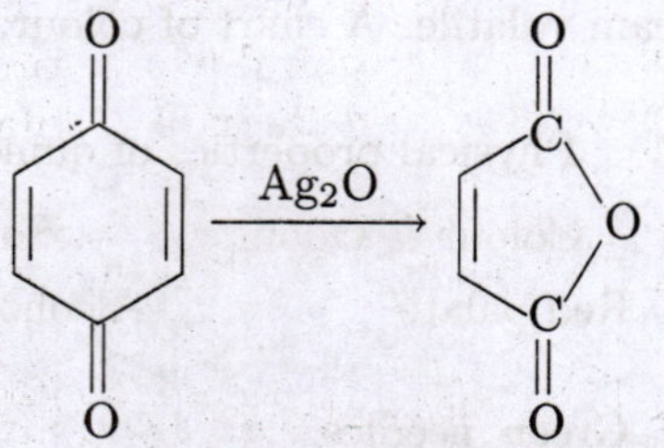


Maleic anhydride

(b) Reduction : All quinones get readily reduced and hence they are often used as oxidising agents. *p*-Benzoquinone and its tetrachloro derivative, chloranil, (See p. 1280) deserve special mention in this regard.

All quinones, except 1, 4-naphthaquinone and 9, 10-anthraquinone, are reduced by (i) $H_2SO_3$, (ii) excess $PhNHNH_2$, (iii) aqueous hydroxylamine or zinc dust and acetic acid. *p*-Benzoquinone liberates iodine from acidified potassium iodide solution.

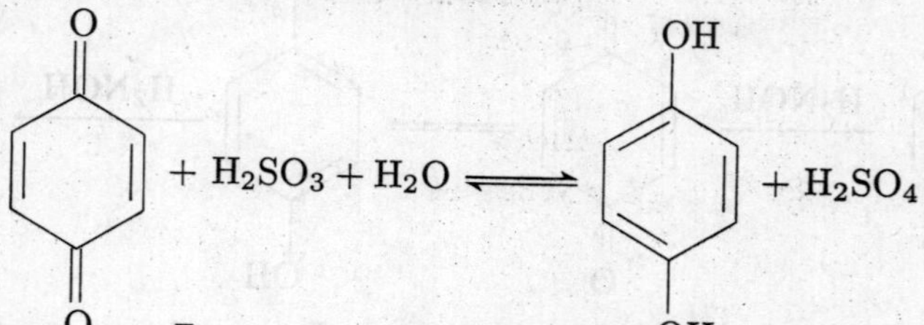

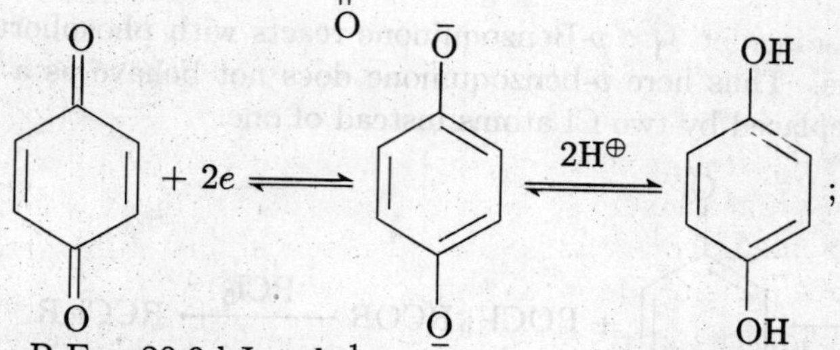

; $\Delta$ R.E. = (20.9 − 125.5) kJ $mol^{-1}$

R.E ≈ 20.9 kJ $mol^{-1}$ R.E ≈ 125.5 kJ $mol^{-1}$

$2I^- \longrightarrow 2I + 2e;\ 2I \longrightarrow I_2$

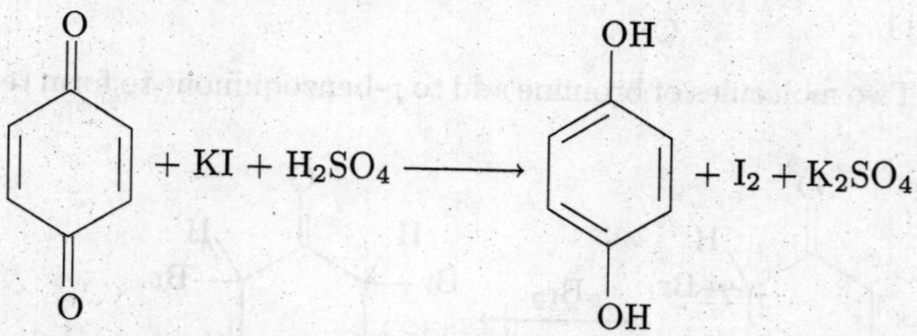

The oxidising power of the quinones is attributed to the fact that by accepting 2*e*s a quinone gets aromatised with a large resonance energy.

The red-ox potential of quinol/*p*-benzoquinone system is appreciably high and it is 0.70 volt; of course, that of catechol/*o*-benzoquinone system is 0.79 volt. It has been found that −I substituents and +I substituents in a quinone respectively increase and decrease oxidation-reduction potential of the system. In chloranil 4Cl atoms are present and they exert strong −I effect. This raises the oxidation-reduction potential of the chloranil-tetrachloroquinol system and hence chloranil (See p. 1280) is used as a strong oxidising agent. In fact, chloranil is used as an aromatising agent in the synthesis of polynuclear hydrocarbons.

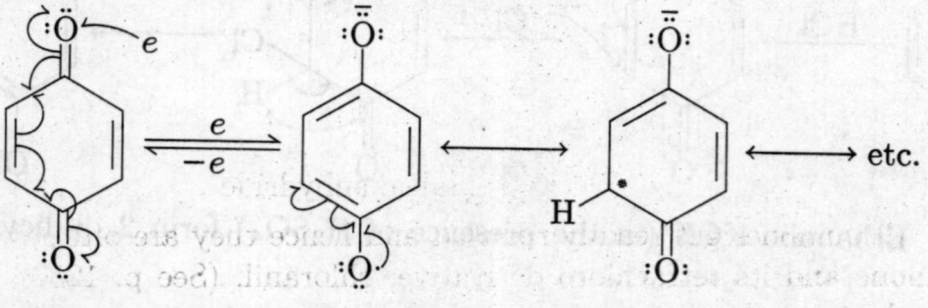

However, on electrolytic reduction, *p*-benzoquinone forms a resonance-stabilised anionic free radical as shown above.

(c) Oxime formation : *p*-Benzoquinone reacts with hydroxylamine to form mono- and di-oxime. The mono-oxime is tautomeric with *p*-nitrosophenol.

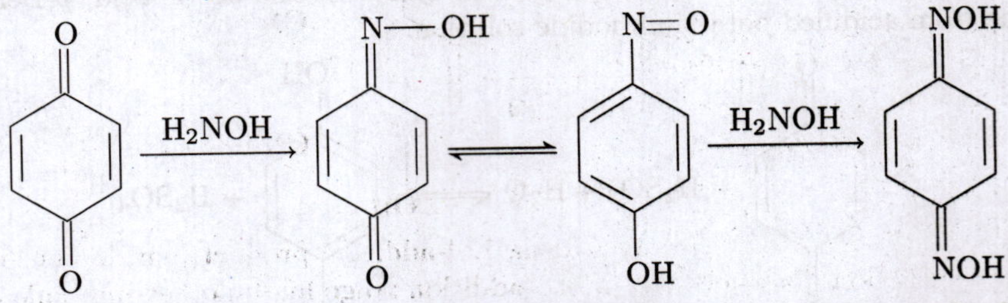

(d) Reaction with phosphorus pentachloride : *p*-Benzoquinone reacts with phosphorus pentachloride to form *p*-dichlorobenzene. Thus here *p*-benzoquinone does not behave as a typical ketone in which each O atom gets replaced by two Cl atoms instead of one.

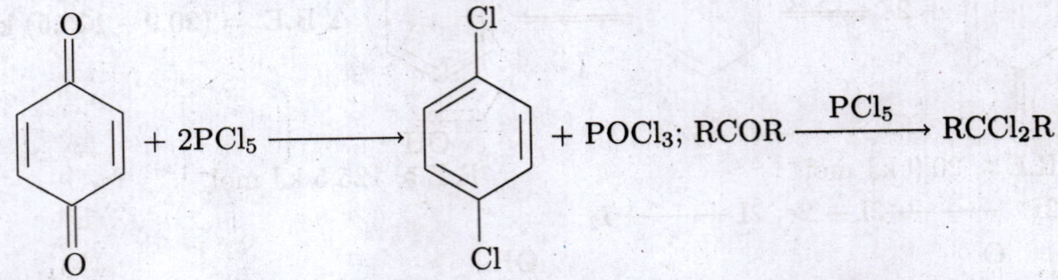

(e) Addition of bromine : Two molecules of bromine add to *p*-benzoquinone to form tetrabromo derivative.

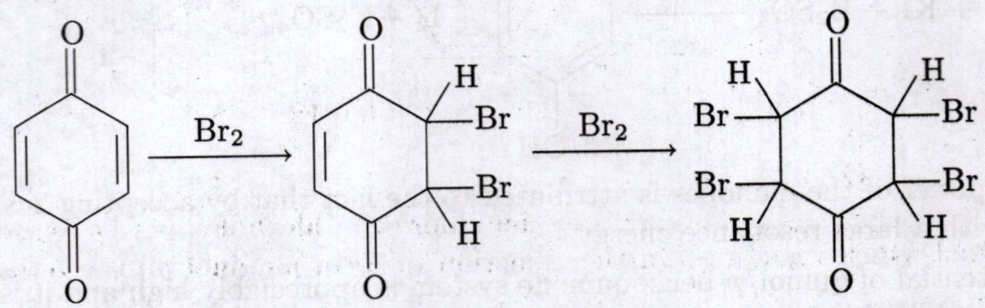

(f) 1, 4-Addition reactions : With HCl : HCl adds to *p*-benzoquinone to form chloroquinol (m.p. 106 °C).

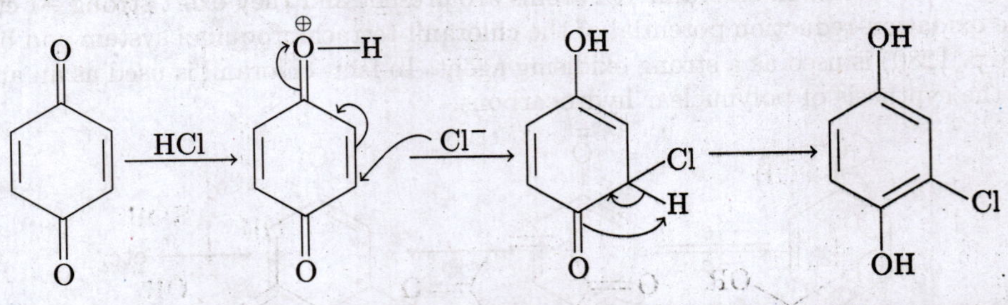

With KCN : Ethanolic KCN (in the presence of $H_2SO_4$) form 2, 3-dicyanoquinol with *p*-benzoquinone.

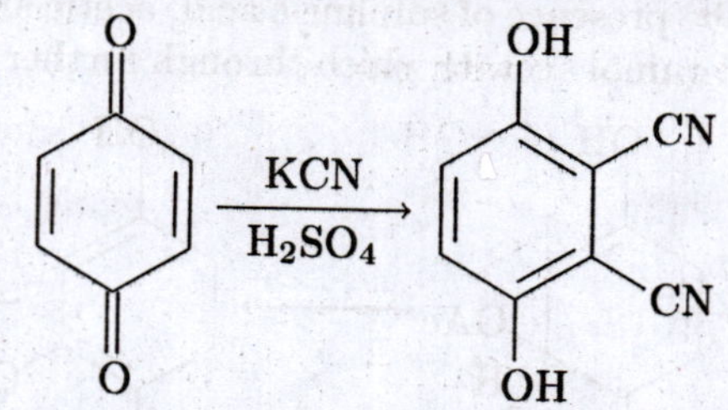

With 1° and 2° amines : 1° and 2° amines form, 1, 4-addition products with *p*-benzoquinone; but in this case the reaction does not stop at the addition stage and another molecule of amine and *p*-benzoquinone take part in the reaction to form a substitution product, 2, 5-diamino-*p*-benzoquinone. For example, when an ethanolic solution of aniline is treated with *p*-benzoquinone 2, 5-dianilino-*p*-benzoquinone is obtained via monoanilinoquinol as shown below :

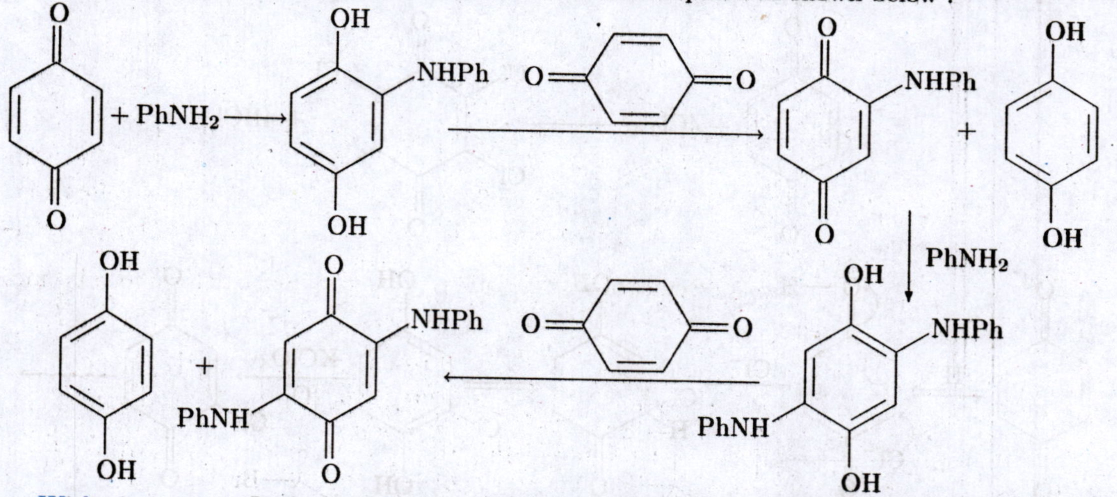

With 1° alcohols : In the presence of anhydrous zinc chloride 1° alcohols add to *p*-benzoquinone to form *o*-alkoxyquinol which undergoes further reaction to form *p*-quinol and 2, 5-dialkoxy-*p*-benzoquinone.

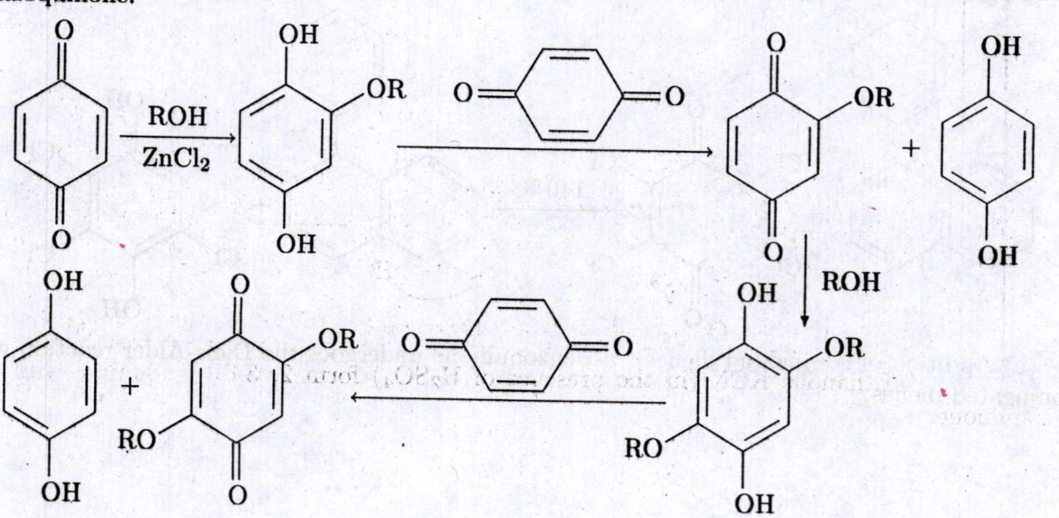

Thiele acetylation : In the presence of sulphuric acid, acetic anhydride adds to *p*-benzoquinone to form nucleus-substituted quinol acetate which through further reaction forms quinol-triacetate.

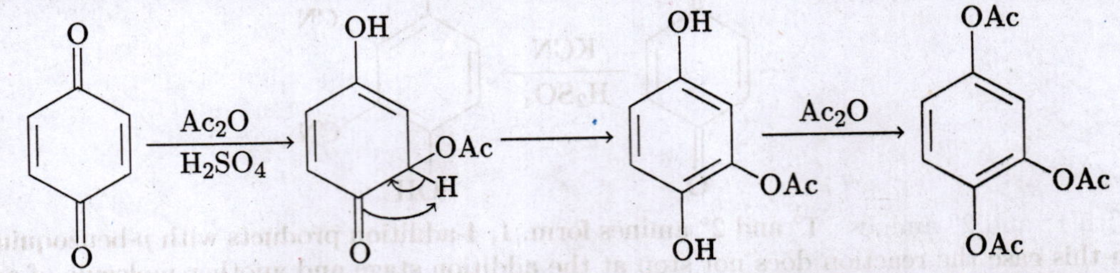

(g) Reaction with chlorine : Chlorine undergoes substitution reaction with *p*-benzoquinone and forms chloranil, tetrachloro-*p*-benzoquinone. Chloranil is obtained, however, by heating benzoquinone with potassium perchlorate and hydrochloric acid as shown below :

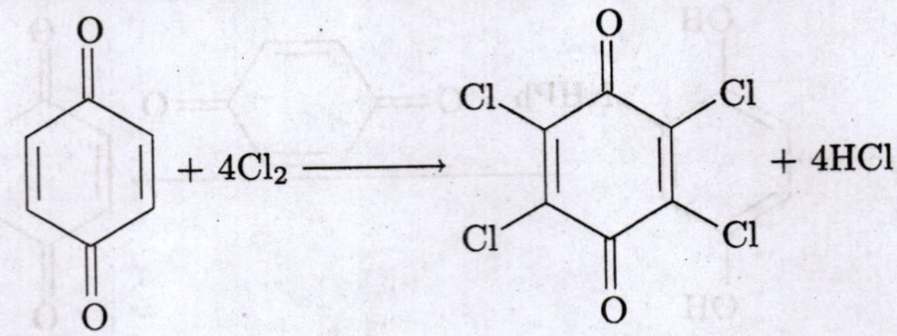

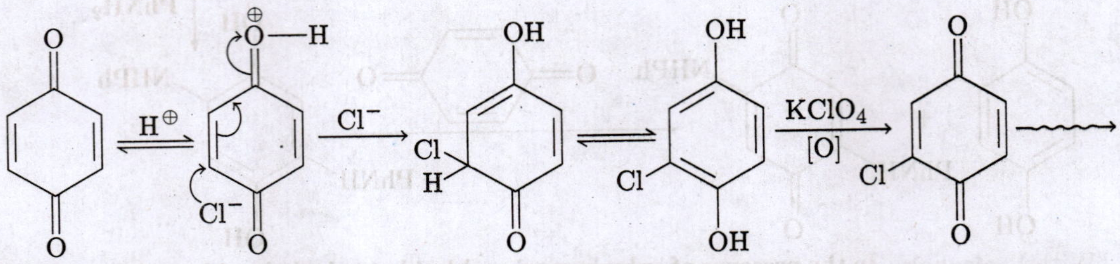

Chloranil is a good fungicide, it is also a good oxidising agent and so, it is used as an aromatising agent.

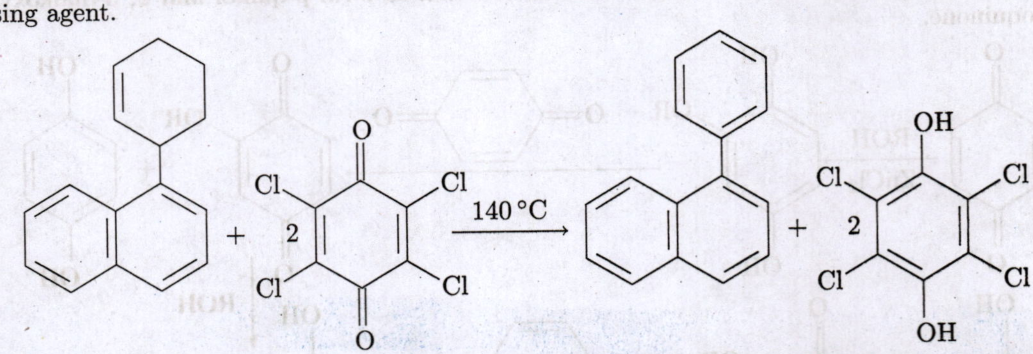

(h) Addition of conjugated dienes : *p*-Benzoquinone undergoes the Diels-Alder reaction with conjugated dienes.

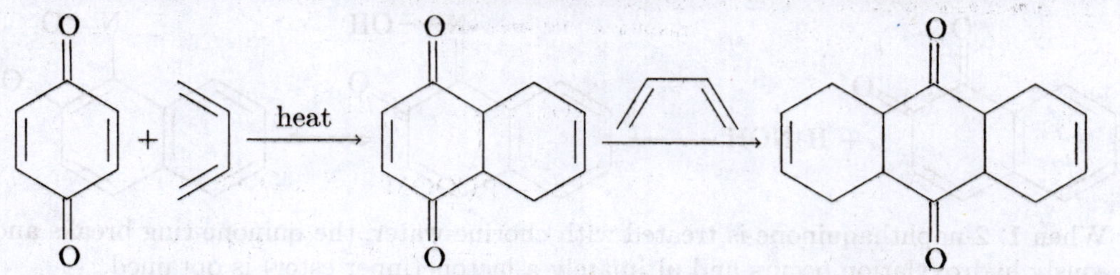

(i) Formation of quinhydrone : When a concentrated alcoholic solution of quinone is added to a concentrated alcoholic solution of hydroquinone (quinol), crystals of quinhydrone are obtained. Quinhydrone is a dark green solid with metallic lusture and its m.p. is 121 °C.

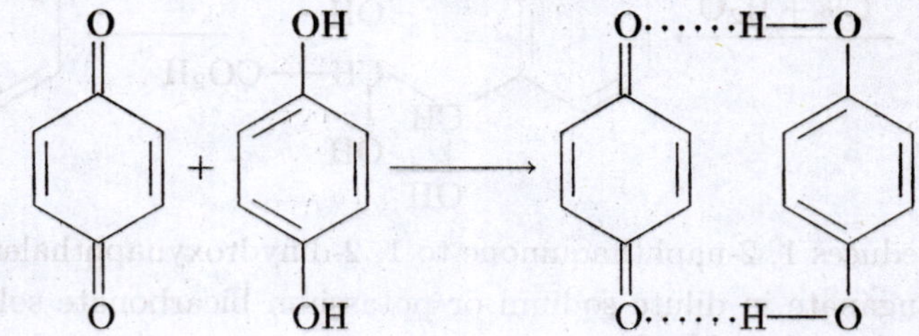

Quinhydrone

In fact, quinhydrones are a class of charge-transfer molecular complexes having colour and formed from quinones and phenols; for example, *p*-benzoquinone and phenol form phenoquinone.

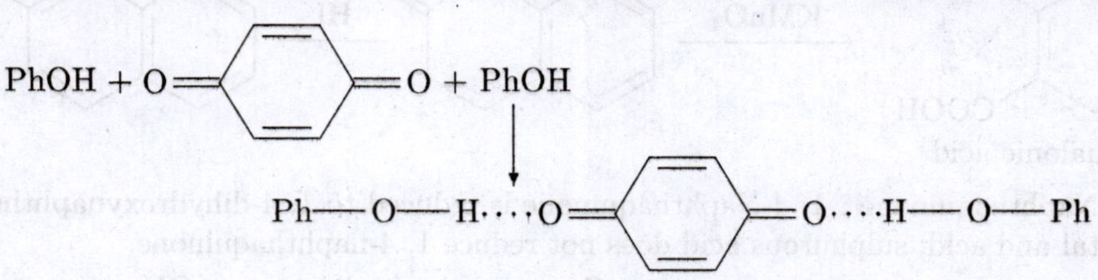

Phenoquinone

Quinhydrone in solution provides, through its partial dissociation, equimolecular concentrations of quinone and hydroquinone. A platinum electrode dipped into a quinhydrone solution having a given $H^{\oplus}$ ion concentration forms a half-cell. The potential of this half-cell can be measured by connecting it with a calomel half-cell and its value may be calculated by the equation,

$$E = E_0 + 0.05912 \log [H^{\oplus}] + \frac{0.05912}{2} \log \frac{[\text{quinone}]}{[\text{hydroquinone}]}$$

If $[H^{\oplus}] = 1g/l$, $0.05912 \log [H^{\oplus}] = 0$ and

$$E = E_0 + \frac{0,05912}{2} \log \frac{[\text{quinone}]}{[\text{hydroquinone}]}$$

Now if $E_0$ is known, the electrode potential becomes a measure of $H^{\oplus}$ ion concentration in any solution under study.

**(iii) 1, 2-Naphthaquinone :** 1, 2-Naphthaquinone forms mono-oxime with hydroxyl amine, the mono-oxime is tautomeric with 1-nitroso-2-naphthol.

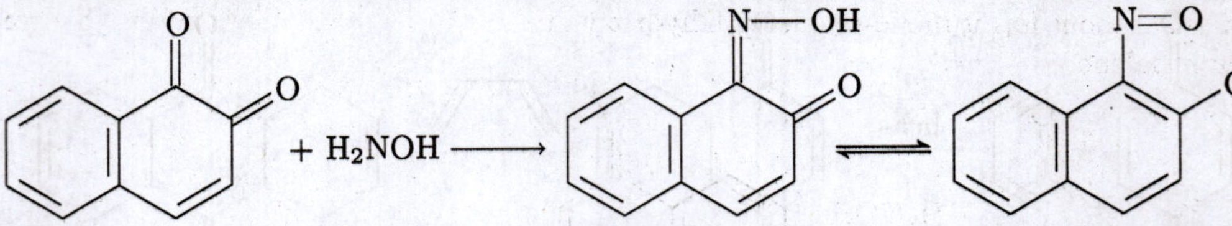

When 1, 2-naphthaquinone is treated with chorine-water, the quinone ring breaks and simultaneously hydroxylation occurs and ultimately a lactone(inner ester) is obtained.

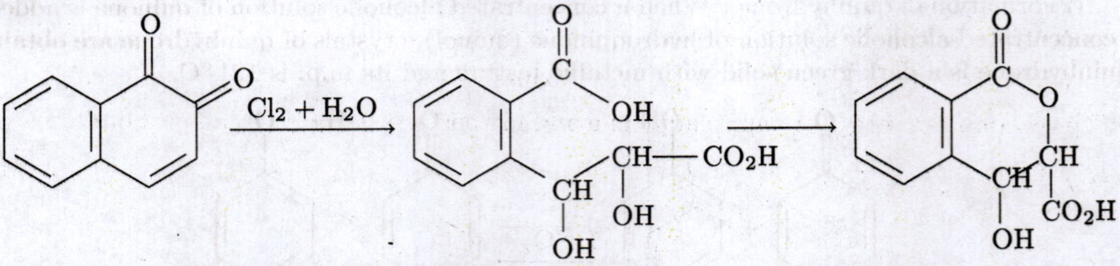

Hydroiodic acid reduces 1, 2-naphthaquinone to 1, 2-dihydroxynaphthalene

Potassium permanganate in dilute sodium or potassium bicarbonate solution oxidises 1, 2-naphthaquinone to phthalonic acid.

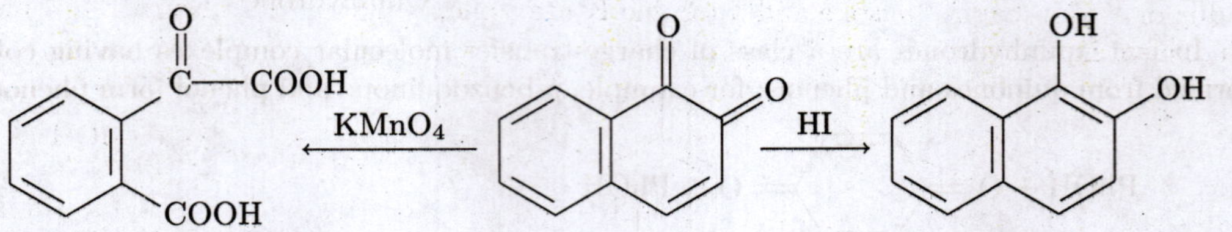

Phthalonic acid

(iv) 1, 4-Naphthaquinone : 1, 4-Naphthaquinone is reduced to 1, 4-dihydroxynaphthalene by HI or by metal and acid; sulphurous acid does not reduce 1, 4-naphthaquinone.

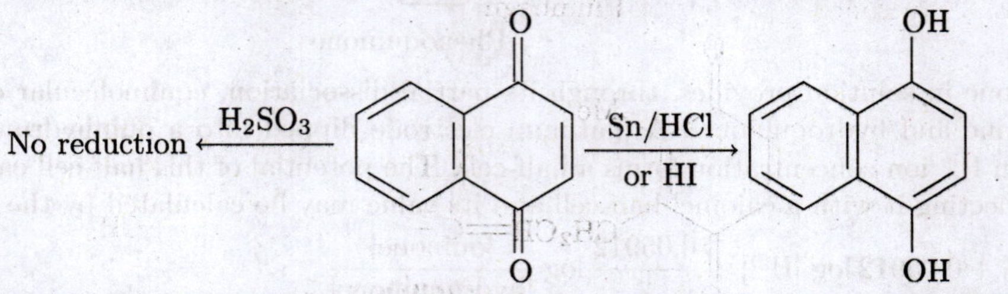

On oxidation with nitric acid or acidified potassium permanganate, 1, 4-naphthaquinone forms phthalic acid.

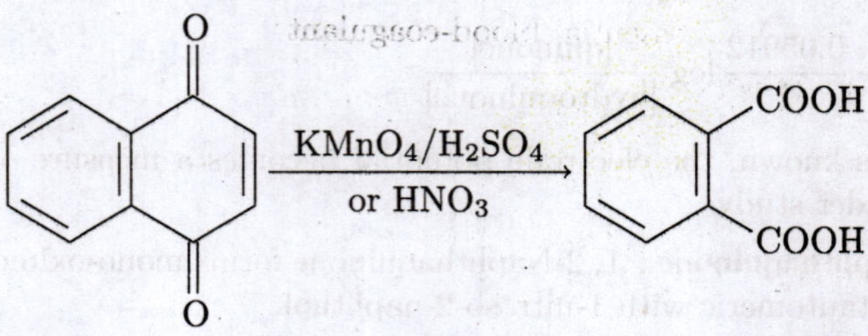

1,4-naphthaquinone forms mono-oxime with hydroxylamine which is tautomeric with 1-hydroxy-4-nitorsonaphthalene.

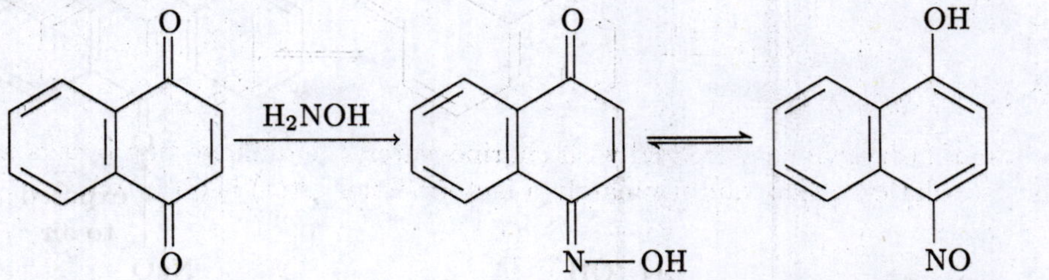


oxime

4-Nitroso-1-naphthol

The reaction between 1, 4-naphthaquinone and nitrous acid forms 1, 3-indanedione.

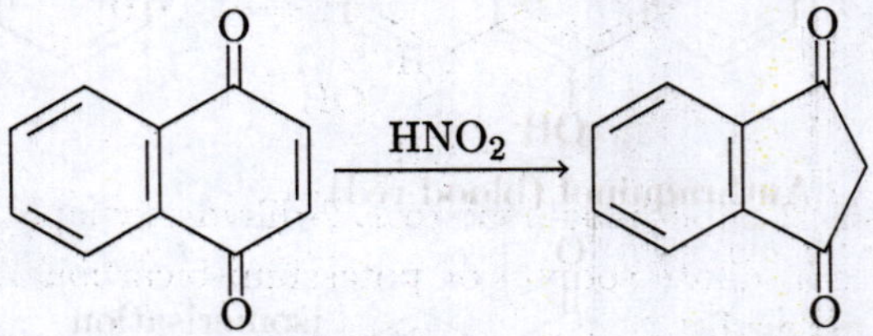


Juglone, plumbagin, lausone and vitamine K are some of the natural occuring 1, 4-naphthaquinones of importance.

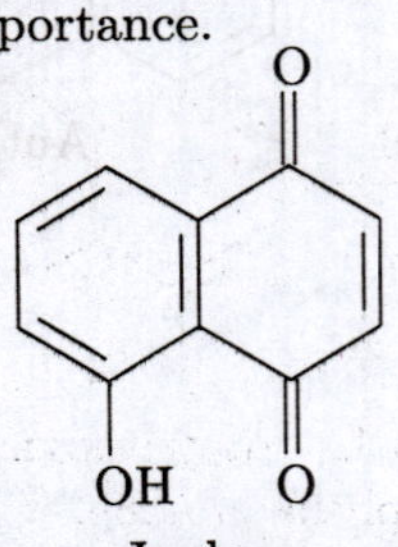

Juglone

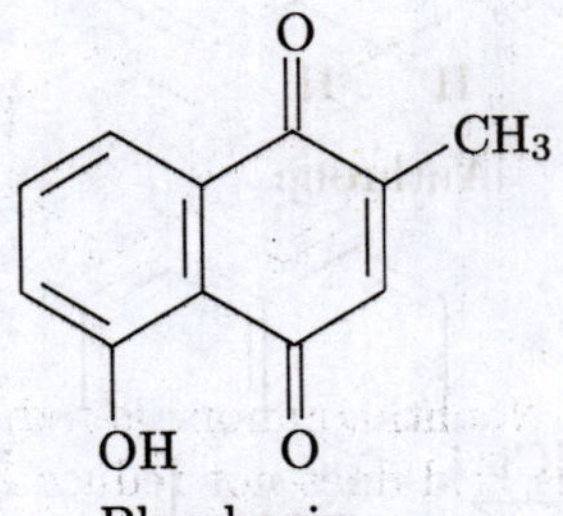

Plumbagin

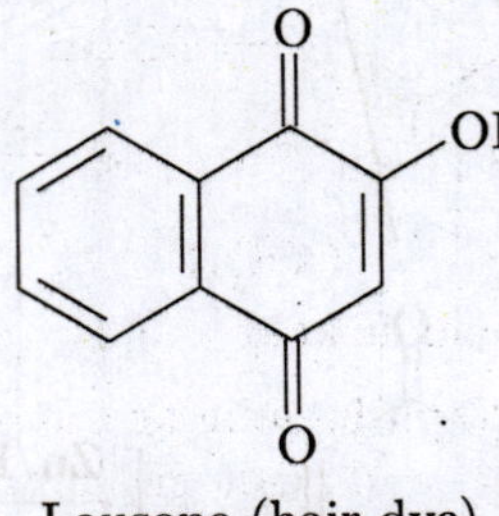

Lausone (hair dye)

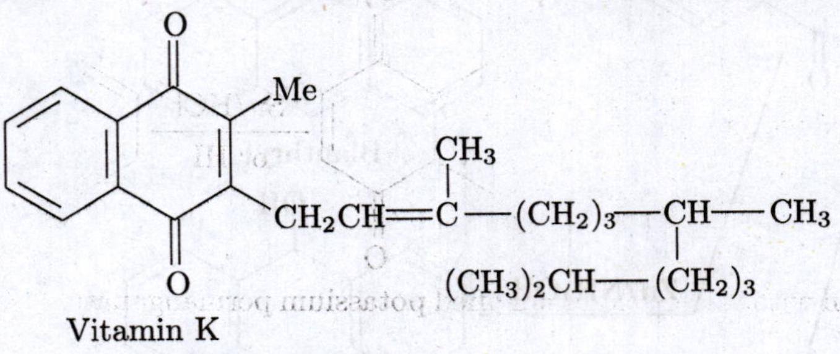

Vitamin K

essential blood-coagulant

(v) 9, 10-Anthraquinone

(a) Reduction :

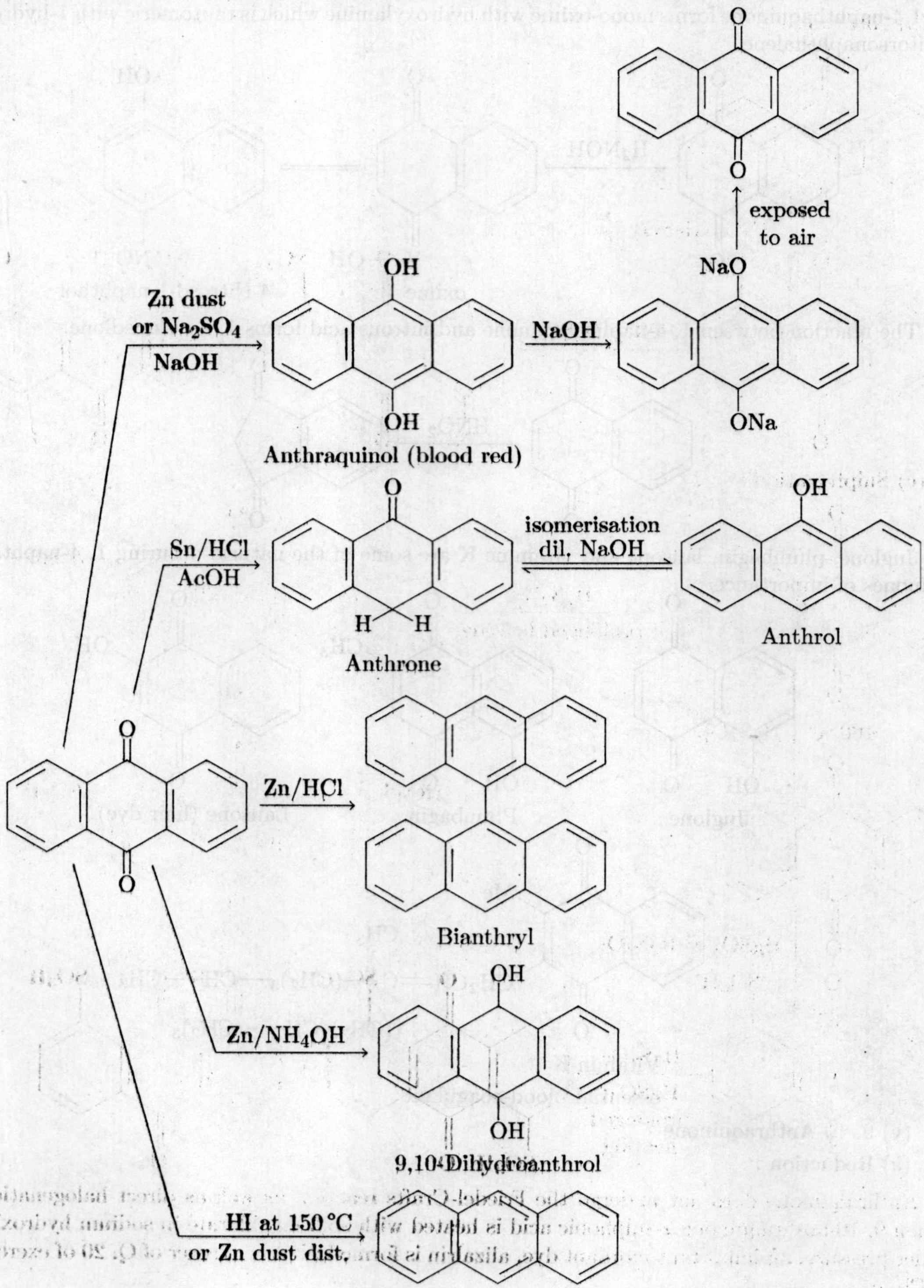

exposed
to air
NaO
ONa
NaOH
OH
Zn dust
or $Na_2SO_4$
NaOH
OH
Anthraquinol (blood red)
O
OH
isomerisation
dil. NaOH
Sn/HCl
AcOH
H
H
Anthrone
Anthrol
O
Zn/HCl
O
Bianthryl
OH
Zn/$NH_4OH$
OH
9,10-Dihydroanthrol
HI at 150 °C
or Zn dust dist.

**(b) Nitration :**

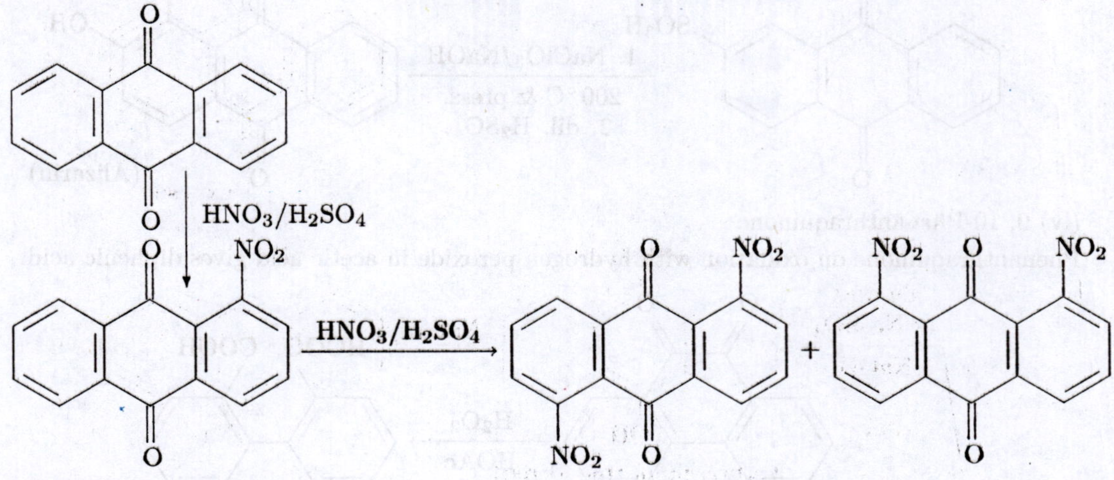

**(c) Sulphonation :**

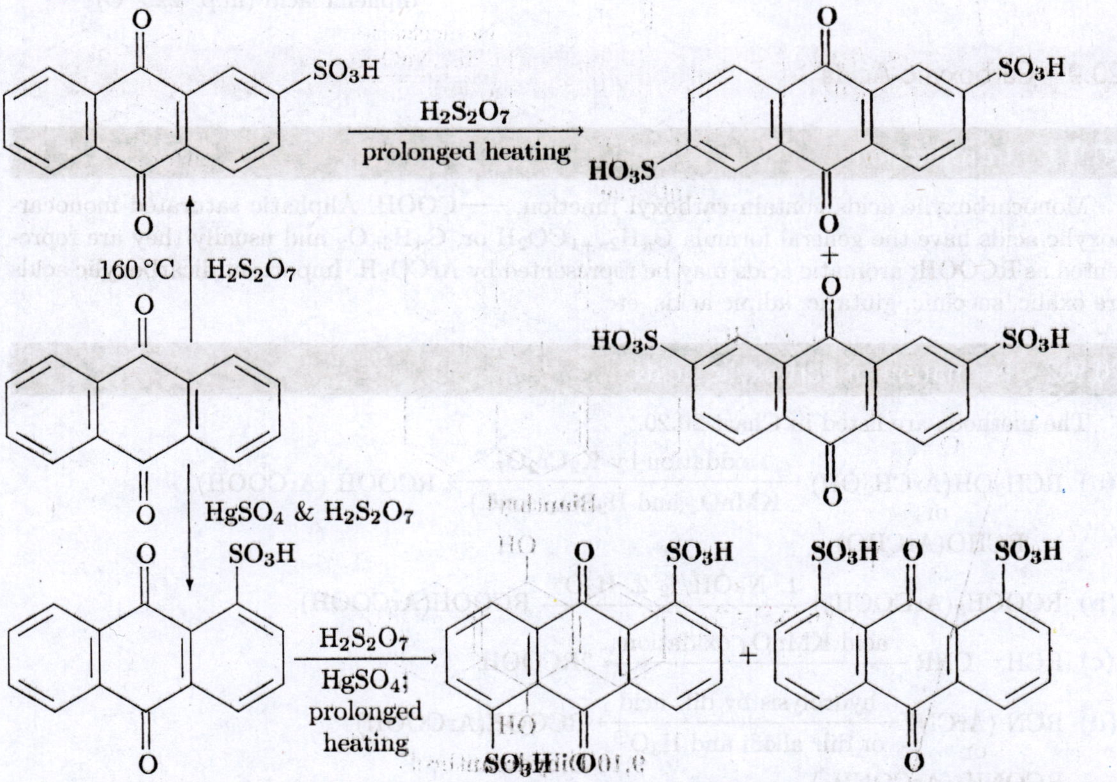

Anthraquinone does not undergo the Friedel-Crafts reaction as well as direct halogenation. When 9, 10-anthraquinone-2-sulphonic acid is heated with sodium chlorate in sodium hydroxide under pressure, an important mordant dye, alizalrin is formed (for the answer of Q. 20 of exercise of Ch. 18).

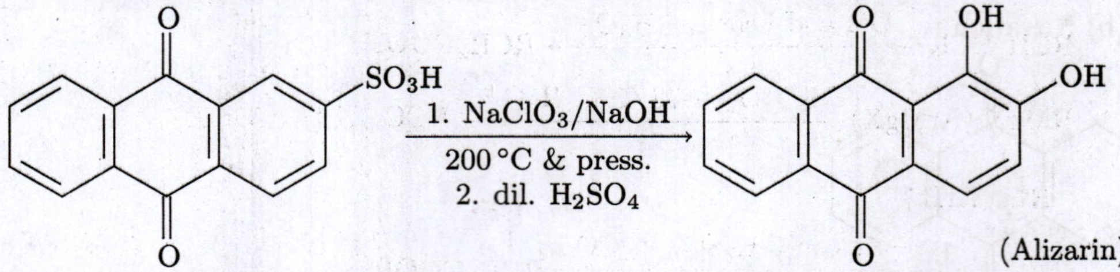


(iv) 9, 10-Phenanthraquinone

Phenanthraquinone on oxidation with hydrogen peroxide in acetic acid gives diphenic acid.

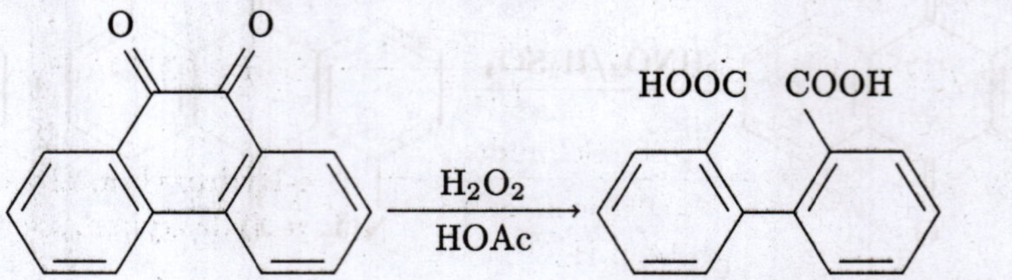


diphenic acid (m.p. 229 °C)

## 20.9 Carboxylic Acids

### 20.9.1 Introduction

Monocarboxylic acids contain carboxyl function, —COOH. Aliphatic saturated monocarboxylic acids have the general formula $C_nH_{2n+1}CO_2H$ or, $C_nH_{2n}O_2$ and usually they are represented as RCOOH; aromatic acids may be represented by $ArCO_2H$. Important dicarboxylic acids are oxalic, succinic, glutaric, adipic acids, etc.

### 20.9.2 Formation of carboxylic acids

The methods are listed in Chart 20.20.

(a) $RCH_2OH(ArCH_2OH)$ or $RCHO(ArCHO)$ $\xrightarrow[\text{KMnO}_4\text{ and H}_2\text{SO}_4\text{(conc.)}]{\text{oxidation by K}_2\text{Cr}_2\text{O}_7}$ RCOOH (ArCOOH)

(b) $RCOCH_3(ArCOCH_3)$ $\xrightarrow{\text{1. NaOH/I}_2\text{ 2. H}_3\text{O}^\oplus}$ RCOOH(ArCOOH)

(c) RCH$=$CHR $\xrightarrow{\text{acid KMnO}_4\text{ oxidation}}$ 2RCOOH

(d) RCN (ArCN) or $RCONH_2(ArCONH_2)$ $\xrightarrow[\text{or dil. alkali and H}_3\text{O}^\oplus]{\text{hydrolysis by dil. acid}}$ RCOOH(ArCOOH)

(e) RCOCl(ArCOCl) $\xrightarrow{\text{H}_2\text{O}}$ RCOOH (ArCOOH)

(f) $RCOOR^1$ or $ArCOOR^1$ $\xrightarrow{\bar{\text{O}}\text{H/or H}_3\text{O}^\oplus\text{ reflux}}$ RCOOH or ArCOOH

(g) $RCH_2CCl_3 \xrightarrow{\bar{O}H \text{ and heat}} \xrightarrow{H_3O^{\oplus}} RCH_2COOH$

(h) $RMgX\ (ArMgX) \text{ or } RLi\ (ArLi) \xrightarrow{1.\ CO_2,\ -78\,^{\circ}C\ 2.\ H_3O^{\oplus}} RCOOH\ (ArCOOH)$

(i) $RCH{=}CH_2 \xrightarrow[3.\ H_2O]{1.\ \text{conc. } H_2SO_4\ 2.\ CO} RCH(CH_3)COOH$

(j) $ArH + COCl_2 \xrightarrow{\text{anh. } AlCl_3} \xrightarrow{H_3O^{\oplus}} ArCOOH$

(k) $ArCHO + (CH_3CO)_2O \xrightarrow[2.\ H_3O^{\oplus}]{1.\ CH_3COONa} ArCH{=}CHCO_2H$

(l) $PhOH + CO_2 \xrightarrow[\text{i. NaOH or ii. KOH, then } H_3O^{\oplus}]{\text{Kolbe-Schmidt}}$ i. *o*-Hydroxybenzoic acid (major) ii. *p*-Hydroxybenzoic acid (major)

(m) $\text{Cyclohexanone} \xrightarrow{HNO_3\ [O]} HOOCCH_2CH_2CH_2CH_2COOH$

(n) $Br(CH_2)_nBr \xrightarrow{KCN} CN(CH_2)_nCN \xrightarrow{H_3O^{\oplus}} HOOC(CH_2)_nCOOH$

(o) $CH_3COCR_2CO_2Et \xrightarrow[\text{boiled } 2.\ H_3O^{\oplus}]{1.\ \text{conc. ethanolic KOH}} CH_3COOH + R_2CHCO_2H + EtOH$

(p) $R_2C(CO_2Et)_2 \xrightarrow{1.\ KOH\ 2.\ HCl\ 3.\ \text{heat}} R_2CHCO_2H$

(q) $CH_2(CO_2Et)_2 \xrightarrow[2.\ KOH\ 3.\ HCl\ 4.\ \text{heat}]{1.\ Br\ CH_2\ CH_2\ Br} HOOCCH_2CH_2CH_2CH_2CO_2H$

(r) $NCCH_2CO_2Et \xrightarrow[2.\ KOH\ 3.\ HCl\ 4.\ \text{heat}]{1.\ Br\ CH_2\ CH_2\ Br} HOOCCH_2{-}CH_2{-}CH_2{-}CH_2COOH$

**Chart 20.20** Formation of carboxylic acids.

**(a) By the oxidation of 1° alcohols and aldehydes :** See Sec. 20.2.5, p. 1254 and Sec. 20.7.5, p. 1180–1181.

**(b) By the haloform reaction of methyl ketones :** See Sec. 16.11. p. 883–884.

**(c) By the oxidation of symmetrical alkenes :** See Sec. 18.2.4t. p. 1033

**(d) By the hydrolysis of alkyl and aryl cyanides and amides :** Alkyl and aryl cyanides are hydrolysed by either moderately dilute acids or alkalis; the corresponding carboxylic acids or carboxylate ions are respectively formed.

Mechanistic steps for the acid-catalysed hydrolysis of cyanides are given below :

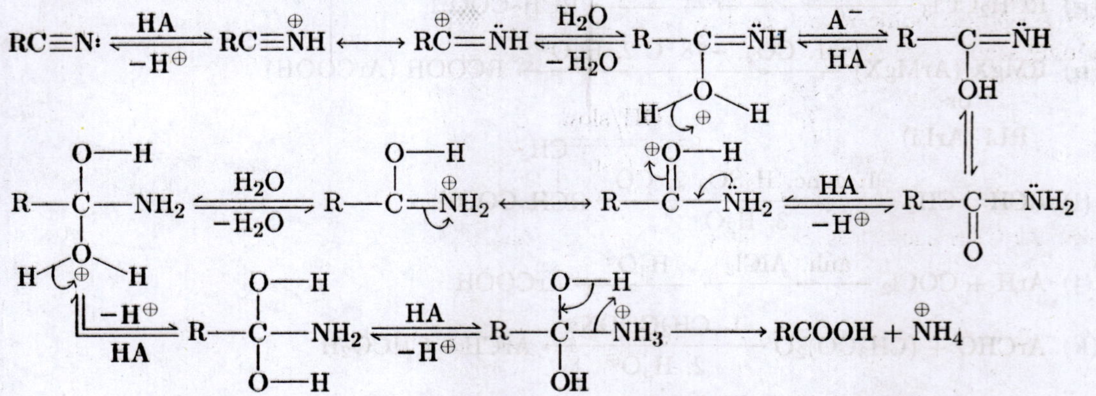

The base-catalysed hydrolysis of cyanides involves the following mechanistic steps :

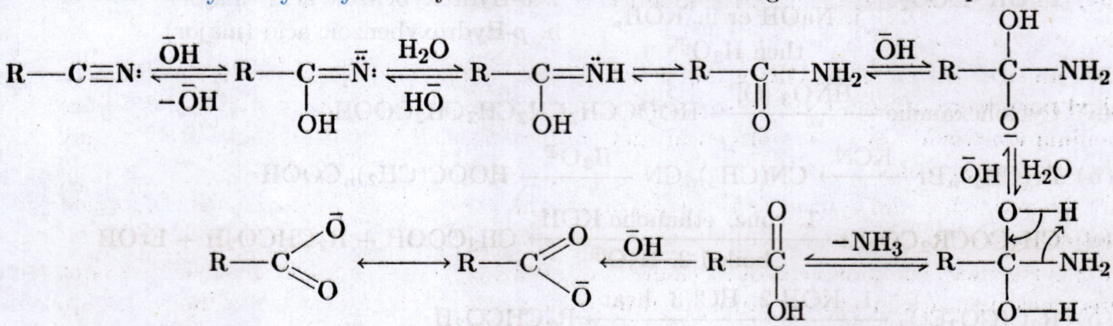

**(e) By the hydrolysis of acid halides :** Acid halides (aliphatic) undergo hydrolysis with water even in the absence of acid and base catalysts to give the corresponding carboxylic acids as shown below.

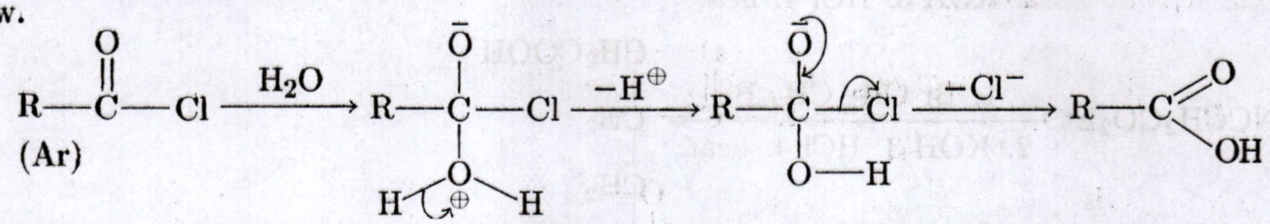

**(f) By the hydrolysis of esters :** Hydrolysis of esters may be carried out by refluxing them either with dilute acid or with dilute alkali. Natural esters of higher fatty acids form soap and glycerol through the base-catalysed process; soaps are sodium or potassium salts of higher fatty acids.

$$RCOOR^1 \underset{-H_2O}{\overset{H_3O^{\oplus}}{\rightleftharpoons}} RCOOH + R^1OH; \; RCOOR^1 \xrightarrow{\bar{O}H} RCO\bar{O} + R^1OH$$

There are several possible mechanisms for the hydrolysis of esters. These are : (i) $B_{AC}1$, (ii) $B_{AC}2$, (iii) $A_{AC}1$, (iv) $A_{AC}2$, (v) $B_{AL}2$, (vi) $B_{AL}1$ (vii) $A_{AL}1$, (viii) $A_{AL}2$, (ix) in neutral medium (p. 1173).

Here, B stands for base-catalysed hydrolysis while A stands for acid-catalysed hydrolysis; AC stands for acyl—O (RCO—OR′) bond fission and AL indicates O—alkyl (RCOO—R′) bond fission whereas 1 and 2 indicate respectively unimolecularity and bimolecularity of the reactions. Some of the above said mechanisms will be discussed here in brief.

**BAc2 Mechanism** (Base-catalysed Acyl-oxygen bond cleavage bimolecular) :

**Step I : Nucleophilic attack by the base on the carboxylate C atom. This is a slow rate-determining step.**

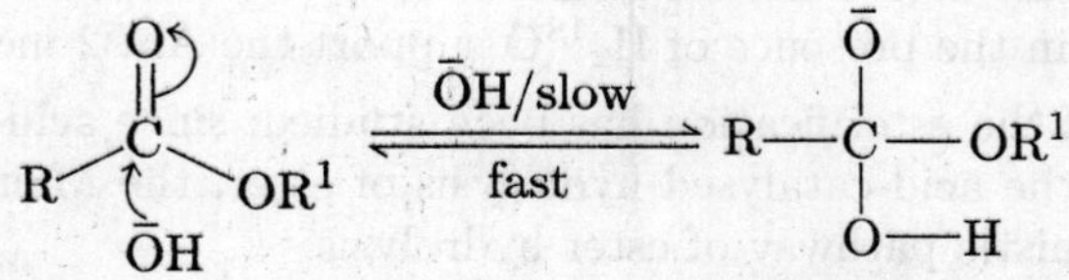

**Step II :** Cleavage of the Acyl—$OR^1$ bond; this is a fast step.

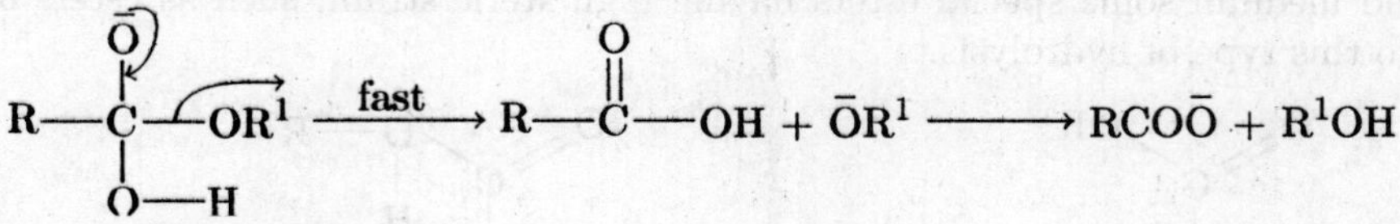

The reaction follows the second order kinetics.

The rate = $k$ [ester] [$\bar{O}H$]

When an ester with a chiral $OR^1$ group is hydrolysed, an optically active alcohol is obtained. If O-alkyl bond were broken, a racemised alcohol would have been a product of hydrolysis since a carbenium ion would have been formed in that case; an optically active alcohol can form only when acyl-O bond breaks.

Furthermore, when the hydrolysis is carried out in the presence of $H_2{}^{18}O$, the alcohol formed does not contain $^{18}O$.

BAc2 is the most common type of mechanism of ester hydrolysis. If the R group of the ester contains an electron-withdrawing group, the rate of the reaction increases; while if the $R^1$ group, is bulky, the rate of hydrolysis decreases owing to the steric effect.

**AAc2 Mechanism** (Acid-catalysed Acyl-oxygen bond cleavage bimolecular). This is also the most common type of mechanism.

**Step I :** 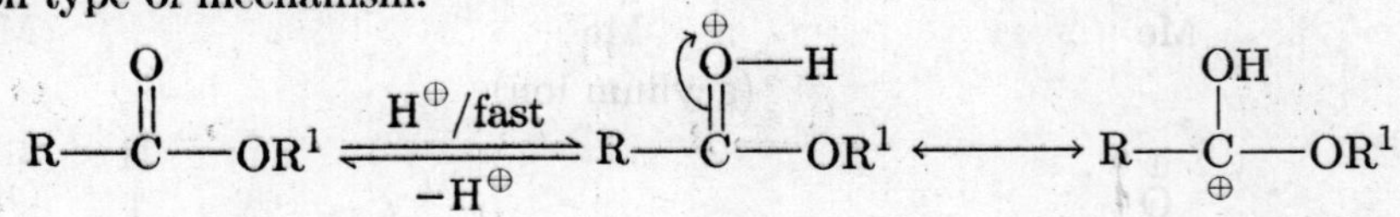

**Step II :** 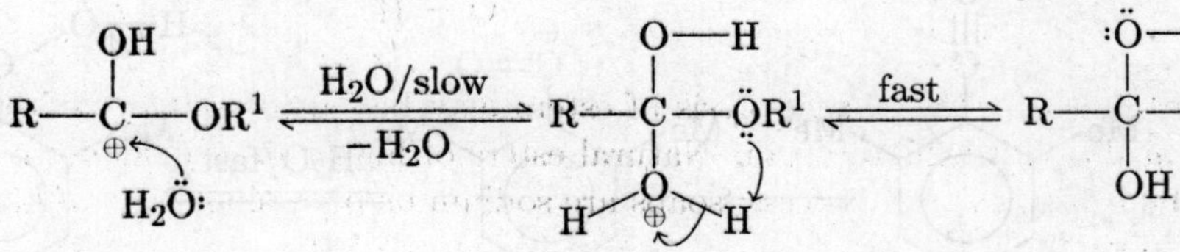

**Step III :** 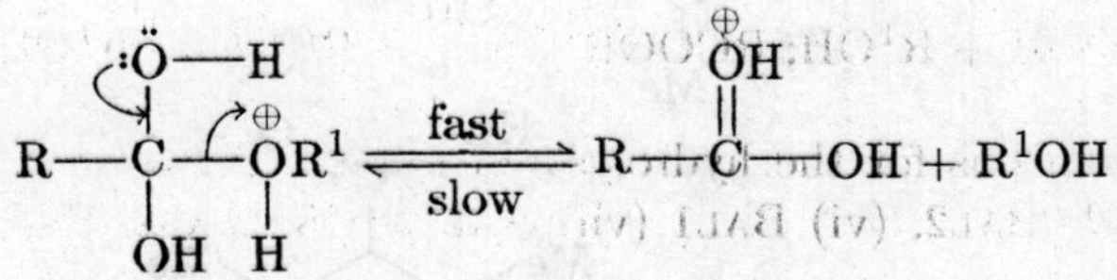

**Step IV :** 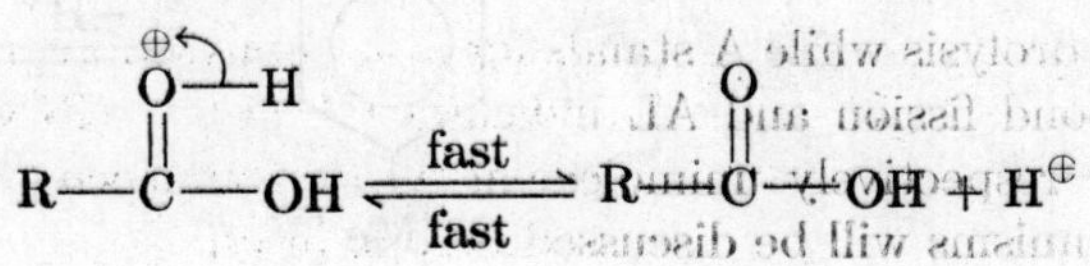

The rate = $k$ [ester] [$H_2O$] [$H^{\oplus}$], when the solvent is not water. If the solvent is water, the rate is, $k$ [ester] [$H^{\oplus}$].

The formation of optically active alcohol from an ester with chiral $R^1$ group and formation of $R^1OH$ (and not $R^{1\,18}OH$) in the presence of $H_2{}^{18}O$ support the AAC2 mechanism.

Actually mechanism of the esterification has been studied; since acid-catalysed esterification is the reverse reaction of the acid-catalysed hydrolysis of ester, the Microscopic Reversibility is applied to give the mechanistic pathway of ester hydrolysis.

**AAC1 mechanism** (Acid-catalysed Acyl-oxygen bond cleavage, unimolecular) : In the concentrated acid medium some special esters having high steric strain, such as esters of mesitoic acid, undergo this type of hydrolysis.

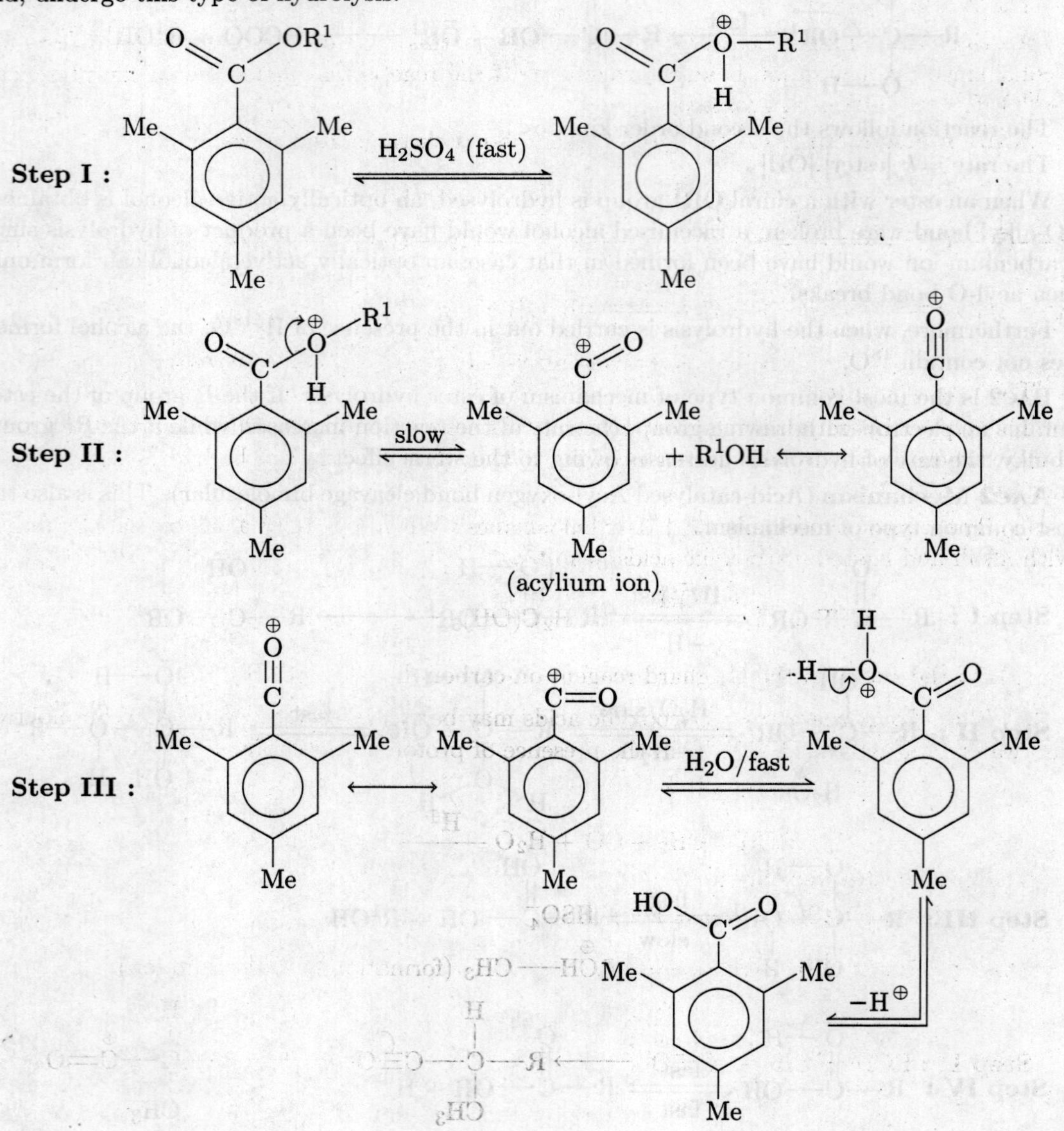

**BAL1 mechanism** (Base-catalysed O-Alkyl bond cleavage, unimolecular) : When O—$R^1$ group of the substrate ester is capable of forming very stable carbenium ion, e.g., t-butyl cation, then one may find this type of hydrolysis in a neutral or faintly alkaline solution of strongly ionising solvent.

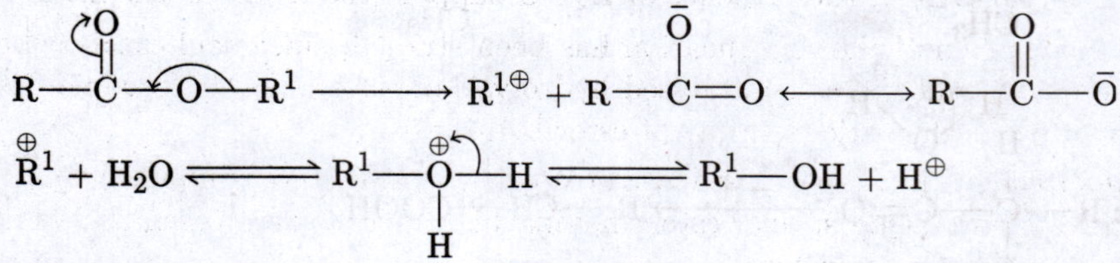

The rate = $k$ [ester].

An ester containing a chiral $R^1$ group undergoes this type of hydrolysis to form racemised alcohol since carbenium ion is an intermediate. If the reaction is carried out in the presence of $H_2{}^{18}O$, $R^1$—$^{18}OH$ is also formed.

**AAL1 mechanism** (Acid-catalysed O-Alkyl bond cleavage, unimolecular).

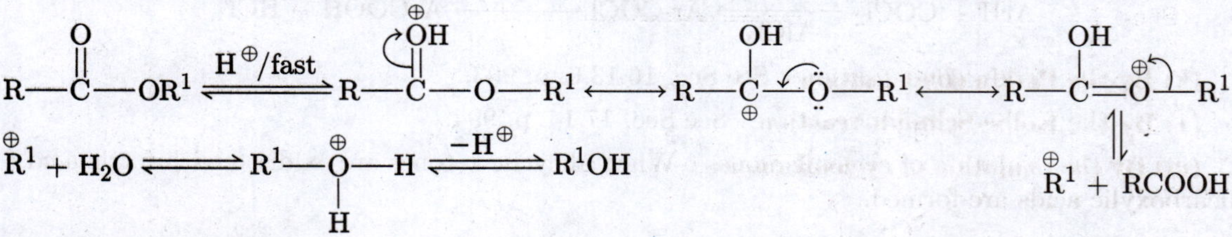

The rate = $k$ [ester].

This type of hydrolysis is relatively common. Esters with $R^1$ group which is capable of forming stable carbenium ion usually undergo the AAL1 hydrolysis. Esters can be hydrolysed in neutral medium (p. 1173).

**(g) By the basic hydrolysis of 1, 1, 1-trihaloalkanes :** When 1, 1, 1-trihaloalkanes are hydrolysed with alkali and heated, carboxylic acids form.

$$RCH_2CCl_3 \xrightarrow{\bar{O}H} RCH_2C(OH)_3 \xrightarrow[-H_2O]{\Delta} RCH_2COOH$$

**(h) By the action of the Grignard reagent on carbon dioxide :** See Sec. 16.4.5.

**(i) The Koch reaction :** Carboxylic acids may be synthesised by adding carbon monoxide and water to an alkene at −25 °C in the presence of protonic acid catalyst.

$$RCH{=}CH_2 + CO + H_2O \xrightarrow{H^{\oplus}} R{-}\underset{}{\overset{CH_3}{\overset{|}{C}}}HCO_2H$$

The mechanistic steps are :

**Initial Step :** $H_2SO_4 \rightleftharpoons H^{\oplus} + HSO_4^-$

**Step I :** $RCH{=}CH_2 + H^{\oplus} \rightleftharpoons R\overset{\oplus}{C}H{-}CH_3$ (formation of carbenium ion)

**Step II :**

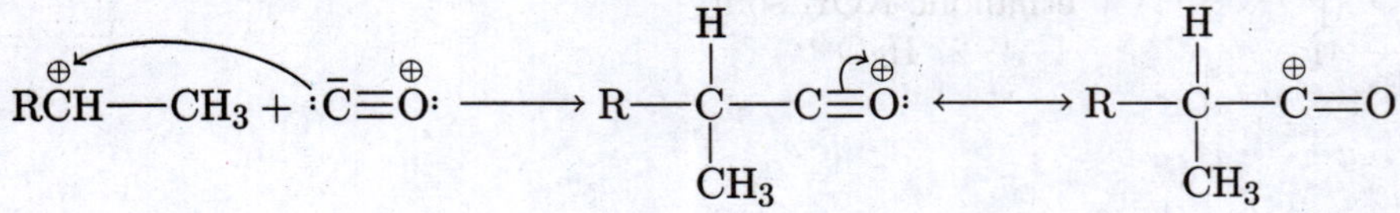

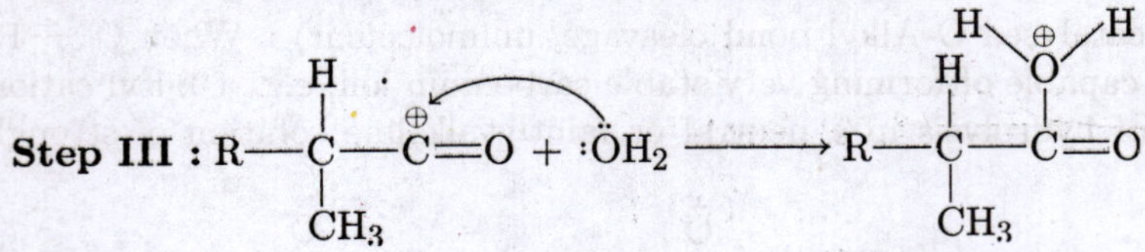

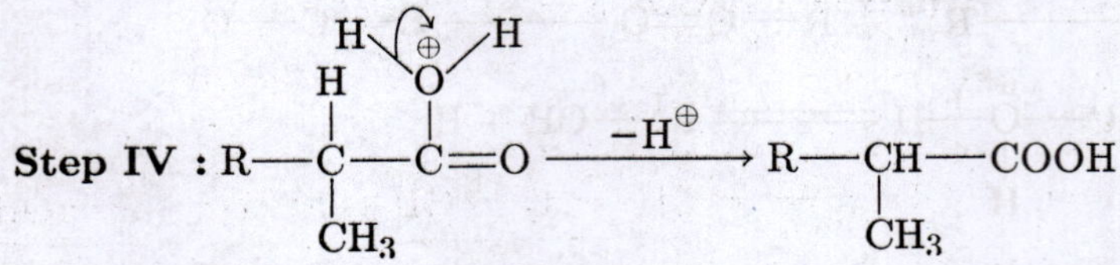

**(j) By the Friedel-Crafts reaction and hydrolysis :** When an aromatic compound is subjected to the Friedel-Crafts reaction with carbonyl chloride and then the product of reaction is hydrolysed with water, an aromatic carboxylic acid forms.

$$\mathrm{ArH + COCl_2 \xrightarrow[\text{AlCl}_3]{\text{anh.}} Ar{\cdot}COCl \xrightarrow{H_2O} ArCOOH + HCl}$$

**(k) By the Perkin condensation :** See Sec. 16.13.6. p. 903.

**(1) By the Kolbe-Schmidt reaction :** See Sec. 17.14. p. 994.

**(m) By the oxidation of cycloalkanones :** When alicyclic ketones are oxidised with nitric acid, dicarboxylic acids are formed.

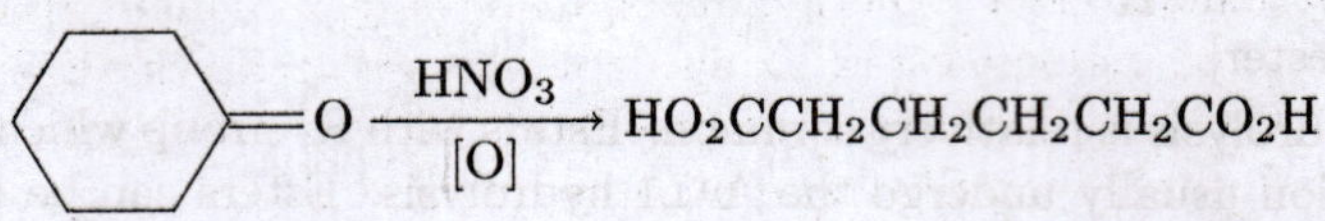

**(n) From α, ω-dihaloalkanes :** Saturated dicarboxylic acids can be prepared by treating α, ω-dihaloalkanes with potassium cyanide and then hydrolysing the dicyano derivative to saturated dicarboxylic acids by a moderately dilute mineral acid solution.

$$\mathrm{Br(CH_2)_2Br \xrightarrow{KCN} CN(CH_2)_2CN \xrightarrow{H_3O^{\oplus}} HO_2C(CH_2)_2CO_2H}$$

**(o) From acetoacetic ester :** Acetoacetic ester contains an active methylene group. The H atoms of the active methylene group being acidic, they can be replaced by alkyl groups on treating acetoacetic ester with a strong base and then treating the salt with alkyl halide. The alkyl derivative of the acetoacetic ester thus formed gives an acid on hydrolysis with concentrated ethanolic caustic potash solution under boiling condition.

$$\mathrm{CH_3COCH_2CO_2Et \xrightarrow{EtOH/EtO^-} CH_3CO\overset{Na^{\oplus}}{\ddot{C}}HCO_2Et \xrightarrow{RI} CH_3CO\underset{\displaystyle R}{\underset{|}{C}}HCO_2Et}$$

$$\mathrm{CH_3CO\underset{\displaystyle R}{\underset{|}{C}}HCO_2Et \xrightarrow[\text{ethanolic KOH soln.}\quad 2.\ H_3O^{\oplus}]{\text{1. boiled with conc.}} RCH_2CO_2H + CH_3COOH}$$

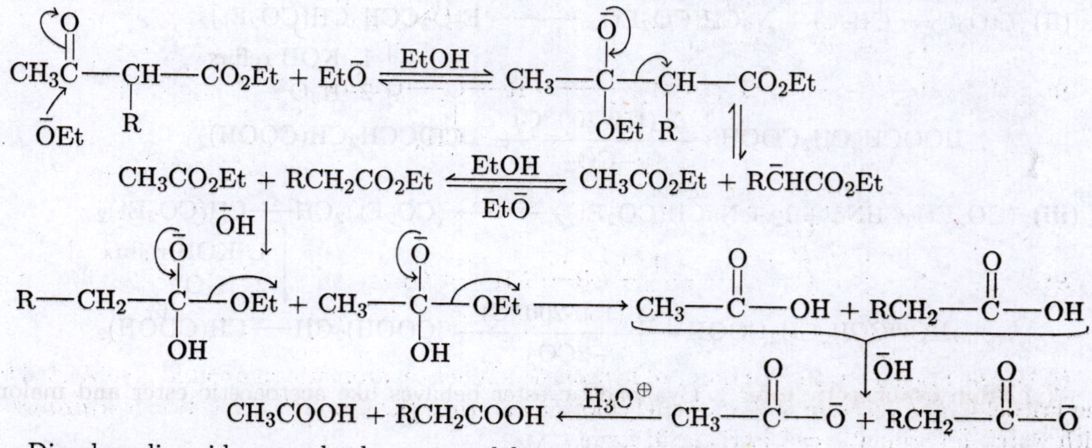

Dicarboxylic acids may also be prepared from acetoacetic ester :

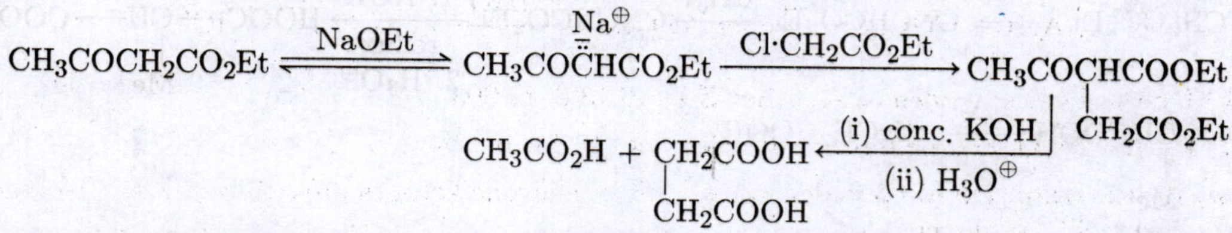

**(p) & (q) From malonic ester :** Diethyl malonate or ethyl malonate, the ethyl ester of malonic acid, is popularly called malonic ester. It also contains an active methylene group and the two acidic H atoms of the group are easily replaced by alkyl and acyl groups. The derivatives of malonic ester on hydrolysis with alkali and on subsequent acidification give derivatives of malonic acid which on heating readily docarboxylate and form monocarboxylic acids.

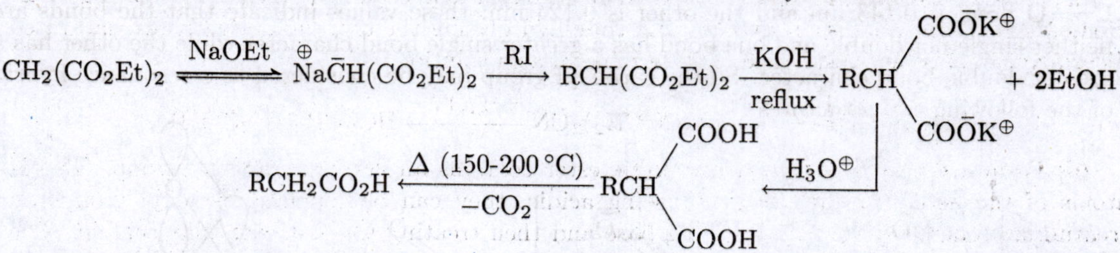

This process may also be used to prepare dicarboxylic acids :

(i)

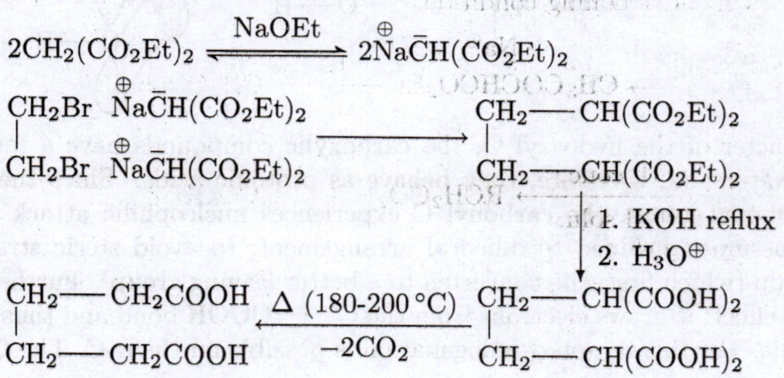

**(ii)** $EtO_2C—CH_2Cl + \overset{\oplus}{Na}\overset{-}{C}H(CO_2Et)_2 \longrightarrow EtO_2CCH_2CH(CO_2Et)_2$

$\xrightarrow{\text{1. KOH reflux; 2. } H_3O^{\oplus}} HOOCCH_2CH(COOH)_2 \xrightarrow[-CO_2]{\Delta\ (150\text{-}200\ ^\circ C)} HOOCH_2CH_2COOH$

**(iii)** $(CO_2Et)_2\overset{-}{C}H\overset{\oplus}{Na} + I_2 + \overset{\oplus}{Na}\overset{-}{C}H(CO_2Et)_2 \longrightarrow (CO_2Et)_2CH—CH(CO_2Et)_2$

$\xrightarrow{\text{1. KOH reflux; 2. } H_3O^{\oplus}} (COOH)_2CH—CH(COOH)_2 \xrightarrow[-2CO_2]{\Delta\ (150\text{-}200\ ^\circ C)} HOOCCH_2CH_2COOH$

**(r) From cyanoacetic ester :** Cyanoacetic ester behaves like acetoacetic ester and malonic ester.

$$CNCH_2CO_2Et \xrightleftharpoons{NaH} CN\overset{-}{C}HCO_2Et \xrightarrow{CH_3I} CN\underset{}{\overset{Me}{\overset{|}{C}}}HCO_2Et \xrightarrow[\text{2. } H_3O^{\oplus}]{\text{1. KOH reflux}} HOOC—\underset{\underset{Me}{|}}{CH}—COOH$$

$$HOOC\underset{\underset{Me}{|}}{C}HCOOH \xrightarrow{\Delta} MeCH_2COOH$$

## 20.9.3 Structure, reactivity and orientation

Carboxylic acids contain carboxyl group which consists of a carbonyl moiety and a hydroxyl group. The carboxyl C atom is $sp^2$ hybridised and it forms a $\sigma$ and a $\pi$ bonds with one O atom and a $\sigma$ bond with the other O atom. The C—O bond length measurements show that one C—O bond is 0.143 nm and the other is 0.124 nm; these values indicate that the bonds are neither single nor double and one bond has a greater single bond character while the other has a greater double bond character. So the carboxyl group may be considered as a resonance hybrid of the following two structures :

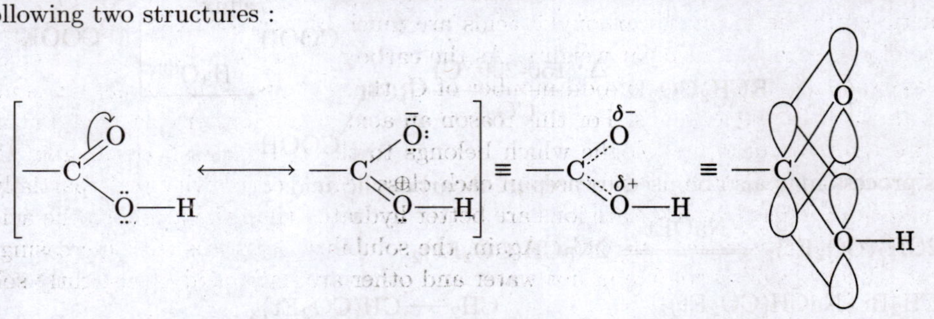

Owing to the $\delta+$ character of the hydroxyl O, the carboxylic compounds have a tendency to give up its proton to water and, therefore, they behave as protonic acids. Since there is a carbonyl group in the carboxyl group, the carbonyl C experiences nucleophilic attack by the nucleophiles and usually assumes strained tetrahedral arrangement; to avoid steric strain the latter gives up the OH group (which first gets converted to a better leaving group). Furthermore, the carbonyl group has $-I$ effect; it draws electrons from the $C^{\alpha}$—COOH bond and thus makes the $\alpha$-H atom slightly acidic. For this reason $\alpha$-halogenation is possible on the $\alpha$-C. The C atom of the carboxyl group has an oxidation number + III. Its oxidation number may increase to + IV and decrease to – IV. Therefore, the carboxyl group can be oxidised and reduced. The carboxyl group of the aromatic carboxylic acids exerts –I and –R effects and thus deactivates the aromatic ring. Owing to deactivation, the aromatic carboxylic acids undergo the electrophilic substitution reactions more slowly than benzene. For the same reason, carboxyl group is a *m*-directing group in so far as aromatic electrophilic substitutions are concerned.

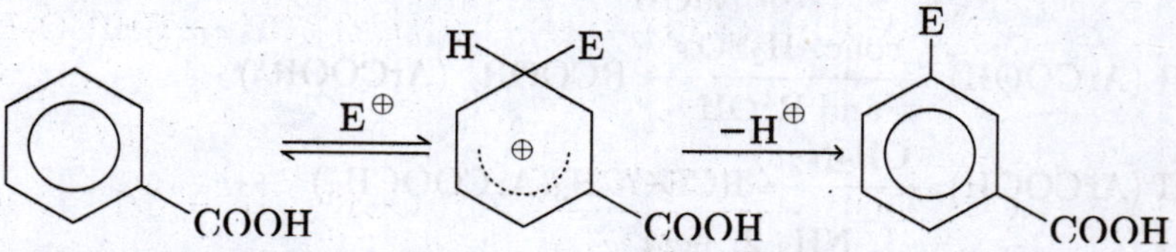

## 20.9.4 Physical properties

In all the three states—solid, liquid and vapour, and in solution in non-polar solvents carboxylic acids usually remain in dimeric state through H-bonding between the hydroxyl H and the carbonyl O atoms as shown below :

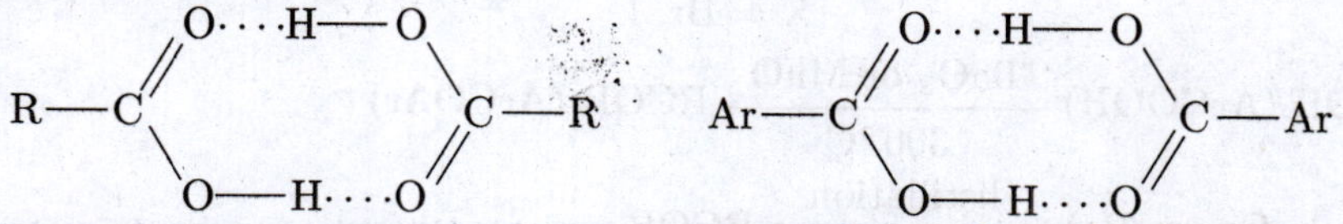

The dimeric state of a carboxylic acid has been determined by the molecular weight determination of acetic and benzoic acids by vapour density method in which the respective molecular weights are found to be 120 and 244 (instead of 60 and 122). Owing to this dimerisation through H-bonding, the vapour pressure of carboxylic acids gets reduced and, therefore, they boil at higher temperatures than the corresponding alcohols, aldehydes and alkanes of comparable molecular weights, e.g., b.p. of acetic acid = 118 °C, b.p. of acetaldehyde = 20 °C; b.p. of *n*-propyl alcohol = 98 °C, b.p. of *n*-butane = –55 °C. However, the b.p. increases with the increasing molecular weight since the London force increases; the increase of b.p. is about 18 °C per C atom. For the same dimeric state the m.p.s of carboxylic acids are much higher than the alcohols, aldehydes and alkane of comparable molecular weights. As the carboxylic acids of even number of C atoms are more symmetrical than those of odd number of C, the crystals of the former acids are more rigid than those of the latter acids. For this reason an acid of the former class melts at a higher temperature than the next homologue which belongs to the latter class. Here again the m.p. increases as the molecular weight increases in each class. Since carboxylic acids partially ionise in water and since ionised carboxylate ions are better hydrated than the alcohols, the acids have higher solubilities in water than alcohols. Again, the solubility decreases with increasing molecular weight. Benzoic acid is soluble in hot water and other aromatic acids are slightly soluble in water.

## 20.9.5 Chemical reactions

The eighth chapter of this book deals with acidity and basicity of the organic acids and bases. Here we shall discuss the general properties of carboxylic acids. These are listed in Chart 20.21.

(a) $\text{RCOONa (ArCOONa)} \xrightarrow{\text{soda-lime \& heat}} \text{RH(ArH)}$

(b) $RCOOH\ (ArCOOH) \xrightarrow[\text{ether}]{LiAlH_4} RCH_2OH \text{ or } ArCH_2OH$

(c) $RCOOH\ (ArCOOH) \xrightarrow{M = Na, Li, K} RCO\bar{O}M^{\oplus}\ (ArCO\bar{O}M^{\oplus})$

(d) $RCOOH\ (ArCOOH) \xrightarrow[\text{or } NaOH]{Na_2CO_3 \text{ or } NaHCO_3} RCO\bar{O}Na^{\oplus}\ (ArCO\bar{O}Na^{\oplus})$

(e) $RCOOH\ (ArCOOH) \xrightarrow[\text{and } R^1OH]{\text{conc. } H_2SO_4} RCOOR^1\ (ArCOOR^1)$

(f) $RCOOH\ (ArCOOH) \xrightarrow{CH_2N_2} RCOOCH_3(ArCOOCH_3)$

(g) $RCOOH\ (ArCOOH) \xrightarrow[3.\ P_2O_5\ \&\ \text{heat}]{1.\ NH_3\ 2.\ \text{heat}} 1.\ RCOONH_4(ArCOONH_4)$
$2.\ RCONH_2(ArCONH_2)\ 3.\ RCN(ArCN)$

(h) $RCOOH \text{ or } ArCOOH \xrightarrow{PCl_5 \text{ or } SOCl_2} RCOCl\ (ArCOCl)$

(i) $RCH_2COOH\ (ArCH_2COOH) \xrightarrow[X = Br, I]{\text{red P and } X_2} R\ CX_2COOH\ (ArCX_2COOH)$

(j) $RCOOH\ (ArCOOH) \xrightarrow[300\,°C]{ThO_2 \text{ or } MnO} RCOR\ (ArCOAr)$

(k) $(RCO_2)_2\ Ca\ (\text{or } Ba) \xrightarrow{\text{distillation}} RCOR$

(l) $(CH_2)_n \begin{matrix} COO \\ COO \end{matrix} Ba(Ca) \xrightarrow[n \geq 4 \leq 19]{\text{heat}} (H_2C)_nC{=}O$

(m) $RCOOH\ (ArCOOH) \xrightarrow{\text{heat}/Ac_2O} (RCO)_2O\ [(ArCO)_2O]$

(n) $RCOOH\ (ArCOOH) \xrightarrow{\text{conc. } H_2O_2} RCOOOH\ (ArCOOOH)$
in $CH_3SO_3H$

(o) $RCOOK \xrightarrow{\text{electrolysis}} R{-}R$

(p) $RCOOH \xrightarrow[\text{conc. } H_2SO_4]{HN_3 \text{ and}} RNH_2$

(q) $RCOOAg \xrightarrow{Br_2} RBr + AgBr + CO_2.$

**Chart 20.21** Reaction of carboxylic acids.

**(a) Decarboxylation reaction :** See Sec. 18.1.3.d.

**(b) Reduction by lithium aluminium hydride :** See Prob. 16.9.ii, p. 732.

**(c) Reaction with metals :** Active metals, i.e., metals appearing at the top of the electrochemical series (Li, Na, Ka, etc.) react with carboxylic acids and evolve hydrogen gas and form carboxylate salts.

$$2\ RCOOH + 2M\ (Na, Li, K) = 2\ RCO\overset{\oplus}{\bar{O}}M + H_2$$

These salts being ionic, they are soluble in water. However salts of the high molecular weight carboxylic acids, such as stearic, are only slightly soluble in water.

**(d) Reaction with carbonates and bicarbonates :** Organic acids in general are stronger acids than carbonic acid. When aqueous solutions or suspensions of metal carbonates and bicarbonates are treated with organic acids, carbon dioxide gas evolves.

$$\text{RCOOH (ArCOOH)} + \text{Na}_2\text{CO}_3 \longrightarrow \text{RCO}\bar{\text{O}}\overset{\oplus}{\text{Na}}(\text{ArCO}\bar{\text{O}}\overset{\oplus}{\text{Na}}) + \text{CO}_2$$

$$\text{RCOOH (ArCOOH)} + \text{NaHCO}_3 \longrightarrow \text{RCO}\bar{\text{O}}\overset{\oplus}{\text{Na}}(\text{ArCO}\bar{\text{O}}\overset{\oplus}{\text{Na}}) + \text{CO}_2$$

However, stearic and other high molecular weight carboxylic acids react very slowly.

**(e) Esterification :** Esterification is the reverse reaction of the ester hydrolysis. When a mixture of an organic acid and an alcohol is heated with a few drops of concentrated sulphuric acid, an ester forms (for detail See Sec. 20.9.3(f), p. 1288.

$$\text{RCOOH (ArCOOH)} + \text{R}^1\text{OH} \underset{\text{H}_3\text{O}^{\oplus}\text{ or }\bar{\text{O}}\text{H}}{\overset{\text{conc. H}_2\text{SO}_4}{\rightleftharpoons}} \text{RCOOR}^1(\text{ArCOOR}^1)$$

**(f) Reaction with diazomethane :** Diazomethane reacts with carboxylic acids to form methyl ester :

$$\text{RCOOH(ArCOOH)} + \text{CH}_2\text{N}_2 \longrightarrow \text{RCOOCH}_3(\text{ArCOOCH}_3) + \text{N}_2\uparrow$$

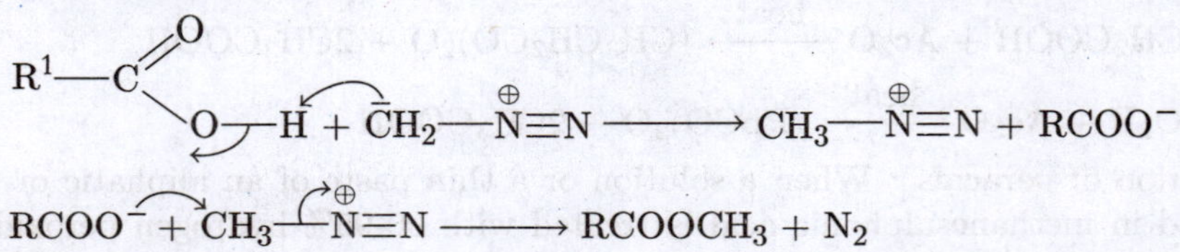

**(g) Reaction with ammonia :** When carboxylic acids are treated with ammonia, the corresponding ammonium salts are formed. The ammonium salts on heating give amides. The amides on further heating with phosphorus pentoxide give nitriles.

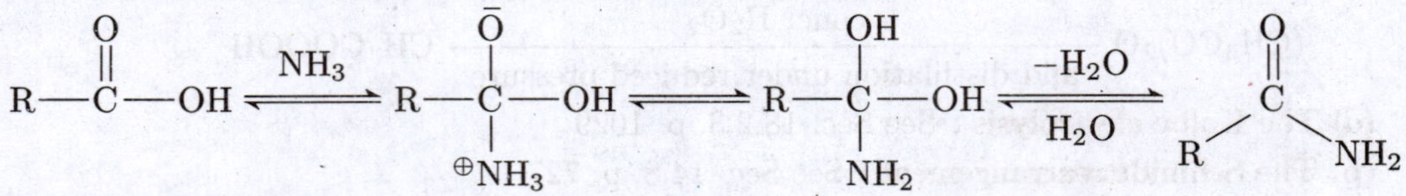

$$\underset{(\text{ArCOOH})}{\text{RCOOH}} + \text{NH}_3 \longrightarrow \text{RCOONH}_4\,(\text{ArCOONH}_4)$$

$$\text{RCOONH}_4 \xrightarrow[-\text{H}_2\text{O}]{\text{heat}} \text{RCONH}_2;\ \text{ArCOONH}_4 \xrightarrow[-\text{H}_2\text{O}]{\text{heat}} \text{ArCONH}_2$$

$$\text{RCONH}_2 \xrightarrow[-\text{H}_2\text{O}]{\text{heat/P}_2\text{O}_5} \text{RC}\equiv\text{N};\ \text{ArCONH}_2 \xrightarrow[-\text{H}_2\text{O}]{\text{heat/P}_2\text{O}_5} \text{ArCN}$$

**(h) Reaction with phosphorus chlorides and thionyl chloride :** On reaction with phosphorus chloride or thionyl chloride carboxylic acids form acid halides; in the cases of $PCl_5$ and $SOCl_2$ HCl evolves out.

$$\text{RCOOH (ArCOOH)} \xrightarrow{\text{PCl}_3} \text{RCOCl (ArCOCl)} + \text{H}_3\text{PO}_3$$

$$\text{RCOOH (ArCOOH)} \xrightarrow{\text{PCl}_5} \text{RCOCl (ArCOl)} + \text{POCl}_3$$

$$\text{RCOOH (ArCOOH)} \xrightarrow{\text{SOCl}_2} \text{RCOCl (ArCOCl)} + \text{SO}_2 + \text{HCl}$$

**(i) The Hell-Volhard-Zelinsky reaction :** See Sec. 16.12. p. 889

**(j) & (k) Formation of ketons :** See Sec. 20.7.2. v & vi.

**(l) Formation of cycloalkanones :** When calcium, barium or thorium salts of dicarboxylic acids containing 6 to 19 C atoms are heated, cycloalkanones having one C atom less than the parent acid are formed.

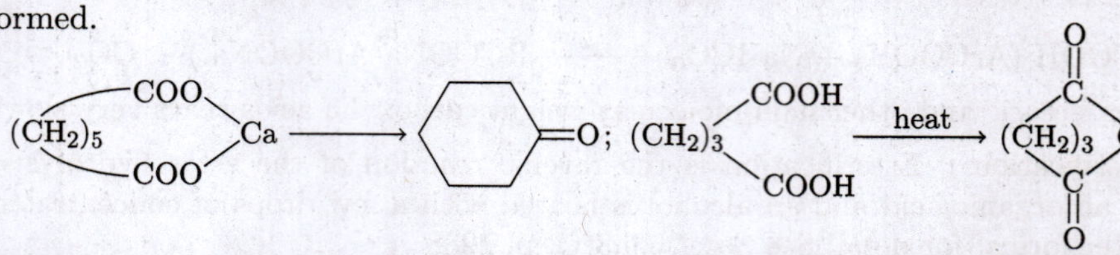

When dicarboxylic acids containing 6 or 7 C atoms are heated alone or with acetic anhydride at 300 °C, cyclopentanones or cyclohexanones are respectively formed. On the other hand, dicarboxylic acids of 4 or 5 C atoms give cyclic anhydride on the same treatment, whereas those with 8 or more C atoms do not react on heating (this is known as the Blanc's rule).

**(m) Formation of acid anhydrides :** When carboxylic acids are heated with acetic anhydride, the carboxylic acid anhydrides are formed.

$$2CH_3CH_2COOH + Ac_2O \xrightarrow{\text{heat}} (CH_3CH_2CO)_2O + 2CH_3COOH$$

$$2PhCO_2H + Ac_2O \xrightarrow{\text{heat}} (PhCO)_2O + 2CH_3COOH$$

**(n) Formation of peracids :** When a solution or a thin paste of an aliphatic or an aromatic carboxylic acid in methanesulphonic acid is treated with 90-95% hydrogen peroxide in cold, a peracid is formed. Peracetic acid may, however, be prepared by treating acetic anhydride with concentrated hydrogen peroxide solution and then distilling the mixture under reduced pressure.

$$RCOOH\ (ArCOOH) \xrightarrow{90\text{-}95\%\ H_2O_2} RCOOOH\ (ArCOOOH)$$

$$(CH_3CO)_2O \xrightarrow[\text{and distillation under reduced pressure}]{\text{conc. } H_2O_2} CH_3COOOH$$

**(o) The Kolbe electrolysis :** See Sec. 18.2.3. p. 1029.

**(p) The Schmidt rearrangement :** See Sec. 14.8. p. 727.

**(q) Formation of alkyl halides from silver salts :** See Sec. 20.1.2.l. p. 1150.

## 20.9.6 Analysis

**(i) Qualitative tests :**

(a) Aliphatic and aromatic acids turn moist blue litmus paper to red.

(b) They evolve carbon dioxide gas when a saturated solution of sodium bicarbonate is treated with a carboxylic acid.

(c) They liberate iodine from an aqueous mixture of potassium iodide and iodate. The liberation of iodine is detected by adding a few drops of freshly prepared starch solution which gets turned to blue.

(d) The confirmatory test for a carboxyl group is the esterification reaction. When 1g of a carboxylic acid is heated with 1 to 2 ml of ethyl alcohol and a few drops of concentrated sulphuric acid and then the solution is poured into a beaker containing distilled water and a pinch of sodium bicarbonate, an oil with a fruity smell of ester separates.

**(ii) Quantitative analysis :** Equivalent weight of a carboxylic acid may be determined by titrating a solution of the acid with standard caustic soda solution. If the basicity of the acid is known, the molecular weight of the acid will be the product of the basicity and the equivalent weight. However, gravimetrically the molecular weight of a carboxylic acid may be determined by converting the acid into the silver salt.

**Problem 20.19. Distinguish between**

(a) HCOOH and $CH_3COOH$. (b) PhCOOH and PhOH.

**Solution : (a)**

| HCOOH | $CH_3COOH$ |
|---|---|
| 1. 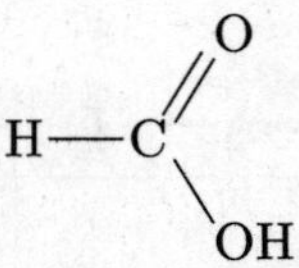<br>Formic acid contains both a formyl and a carboxyl group. | 1. 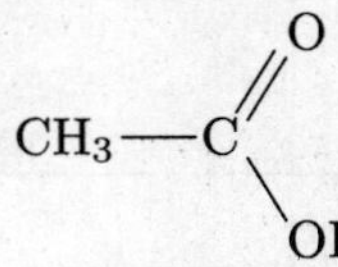<br>Acetic acid contains a carboxyl group; it does not contain a formyl group. |
| 2. When formic acid is heated with conc. $H_2SO_4$, carbon monoxide gas evolves out; the gas burns with a blue flame $HCOOH + H_2SO_4 \longrightarrow CO + [H_2O, H_2SO_4]$; $CO + O \longrightarrow CO_2$. | 2. When acetic acid is heated with conc. $H_2SO_4$, acetic anhydride forms.<br>$2CH_3COOH + H_2SO_4 \longrightarrow (CH_3CO)_2O + H_2O, H_2SO_4$ |
| 3. When sodium formate is heated, sodium oxalate is formed.<br>$HCOONa \xrightarrow{\text{heat}} \begin{matrix} COONa \\ \vert \\ COONa \end{matrix} + H_2$ | 3. When sodium acetate is heated, no characteristic reaction takes place. |
| 4. Formic acid reduces the Fehling solution, the Tollen's reagent and mercuric chloride solution to form red precipitate, silver mirror and black precipitate respectively.<br>$HCOO^- + [2Cu^{++} + 5OH^-] \rightarrow CO_3'' + Cu_2O$<br>$HCOO^- + [2\ Ag^+ + 3OH^-] \rightarrow CO_3'' + 2Ag + 2H_2O$<br>$HCOOH + 2HgCl_2 \rightarrow Hg_2Cl_2 \rightarrow 2Hg$<br>It also decolourises the pink colour of the acidified potassium permanganate solution<br>$5HCOOH + 2MnO_4' + 6H^{\oplus} \longrightarrow 2Mn^{++} + 5CO_2 + 8H_2O$ | 4. Acetic acid has no reducing property since it does not contain a formyl group. |

*Contd.*

| HCOOH | $CH_3COOH$ |
|---|---|
| 5. Formic acid does not form formyl chloride when it is treated with phosphorus pentachloride.<br>$HCOOH \xrightarrow{PCl_5} \times$ | 5. Acetic acid forms acetyl chloride when it is treated with phosphorus pentachloride.<br>$CH_3COOH \xrightarrow{PCl_5} CH_3COCl$ |
| 6. No obnoxious odour is obtained on heating sodium formate with arsenious oxide. | 6. A characteristic odour of cacodyl oxide is obtained when sodium salt of acetic acid is heated with arsenious oxide.<br>$CH_3COONa + As_2O_3 \longrightarrow Me_2As—O—AsMe_2$ |

**(b)**

| PhCOOH | PhOH |
|---|---|
| 1. 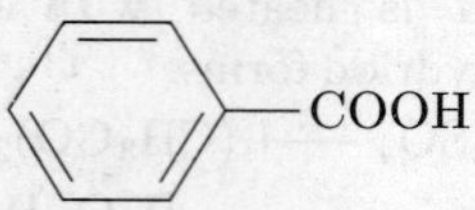<br>Benzoic acid is a much stronger acid than phenol; it evolves $CO_2$ from a saturated solution of sodium bicarbonate. | 1. 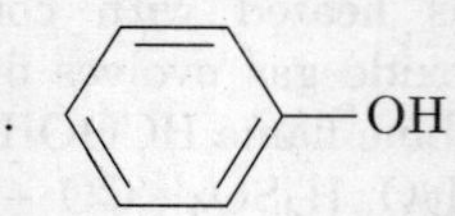<br>It does not evolve $CO_2$ from a saturated solution of sodium bicarbonate. |
| 2. An incoming electrophile takes the *m*-position w.r.t. the carboxyl group; electrophilic substitution takes place in PhCOOH much slower than that in benzene since COOH is a deactivating group. | 2. An incoming electrophile takes the *o*- and *p*-positions w.r.t. the OH group. The $S_E$ reaction takes place in PhOH much faster than that in benzene since OH group is an activating group. |
| 3. On soda-lime distillation it forms benzene. | 3. It forms phenoxide. |
| 4. When a few drops of freshly prepared ferric chloride solution is added to a hot aqueous solution of benzoic acid, a buff-coloured precipitate is obtained. | 4. When a few drops of freshly prepared ferric chloride solution is added to an aqueous solution of phenol, a violet coloured solution is obtained. |
| 5. An alkaline solution of benzoic acid does not undergo coupling reaction with an acidic solution of benzenediazonium salt. | 5. An alkaline solution of phenol undergoes coupling reaction with an acidic solution of benzenediazonium ion to form a red azo dye. |

**Problem 20.20.** Write down the acidity order among benzoic acid, *p*-hydroxybenzoic acid and *p*-nitrobenzoic acid.

**Solution :**

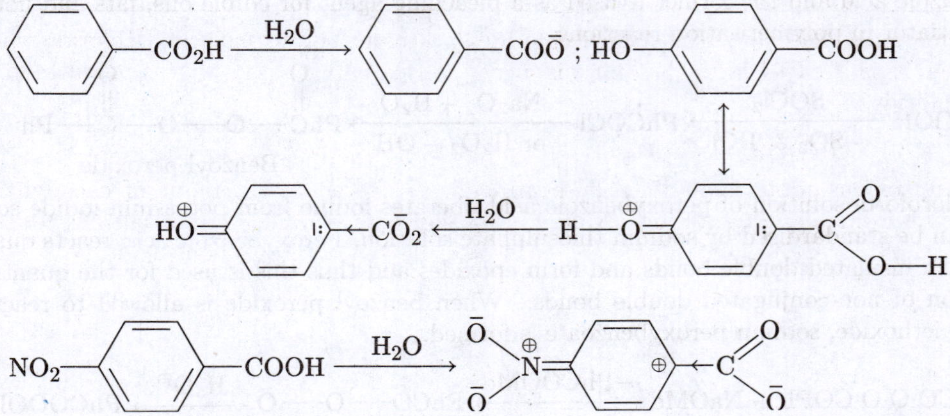

From the above structures it is clear that −R effect of the nitro group makes the C atom (which bears $CO_2^-$ group) positively charged while +R effect of the OH group makes that C atom negatively charged; these in their turn make $CO_2^-$ group less and more electron dense respectively than the $CO_2^-$ group of benzoic acid. The greater the electron density on $CO_2^-$ group, the more is its base character and naturally the less is the acid character of the original acid. Thus the acidity order is :

*p*-nitrobenzoic acid > benzoic acid > *p*-hydroxybenzoic acid.

## 20.9.7 Uses of carboxylic acids and their derivatives

**(i) Formic acid** is used where an acid stronger than acetic acid but weaker than a mineral acid is needed. Sodium formate is used to manufacture formic and oxalic acids. It is also used as a reducing agent. Sodium formate is manufactured from carbon monoxide and caustic soda.

$$CO + NaOH \xrightarrow[200\,°C]{6\text{-}10\text{ atm}} HCOONa \xrightarrow{400\,°C} \begin{matrix} COONa \\ | \\ COONa \end{matrix} \xrightarrow{H_3O^{\oplus}} \begin{matrix} COOH \\ | \\ COOH \end{matrix}$$

**(ii) Acetic acid** is the chief component of vinegar which is almost exclusively used as preservative and condiment. It is produced by the fermentation of fruit juice or malt using acetobactor and air as oxidising agent.

$$CH_3CH_2OH \xrightarrow[O_2\text{ (air)}]{\text{acetobactor}} CH_3COOH$$

The most important use of acetic acid is the preparation of acetic anhydride. It is also used in the preparation of metallic salts, esters and specially cellulose acetate. To precipitate casein from milk, and rubber from their emulsions, acetic acid is used. Sodium acetate is used to reduce the acidity of a reaction medium. Lead acetate is used to prepare other lead salts, and aluminium acetate is used to impregnate cotton fibres and cloths with aluminium hydroxide before dyeing and the process is known as mordanting.

**(iii) Propionic and butyric acids** are used in the manufacture of cellulose acetate-propionate and cellulose acetate-butyrate. Calcium propionate is used in bread to prevent moulding.

(iv) **Benzoic acid** is used as sodium benzoate to preserve food. Benzoyl peroxide is prepared from benzoic acid and the former is used as a bleaching agent for edible oils, fats and flour and as an initiator in polymerisation reactions.

$$\text{PhCOOH} \xrightarrow[-SO_2\ \&\ HCl]{SOCl_2} \text{PhCOCl} \xrightarrow[\text{or } H_2O_2 + \bar{O}H]{Na_2O_2 + H_2O} \underset{\text{Benzoyl peroxide}}{\text{Ph}\overset{O}{\overset{\|}{C}}\text{—O—O—}\overset{O}{\overset{\|}{C}}\text{—Ph}}$$

A chloroform solution of peroxybenzoic acid liberates iodine from potassium iodide solution which can be standardised by sodium thiosulphate solution. Peroxybenzoic acid reacts quantitatively with insulated double bonds and form epoxides and thus this is used for the quantitative estimation of non-conjugated double bonds. When benzoyl peroxide is allowed to react with sodium methoxide, sodium peroxybenzoate is formed.

$$\text{PhCO·O·O·COPh} + \text{NaOMe} \xrightarrow{-\text{PhCOOMe}} \text{PhCO—O—}\bar{O} \xrightarrow{H_3O^{\oplus}} \text{PhCOOOH}$$

(v) **N, N-Dimethyltoluamide** is an important insect repellent.

(vi) **3-Ethanamido-2, 4, 6-triiodobenzoic acid and 3, 5-diethanamido-2, 4, 6-triiodobenzoic acid** are used for X-ray examination of urinary tract.

(vii) **Anthranilic** acid is used as an intermediate for the synthesis of thioindigo and several other dyes. Methyl anthranilate is used to produce grape flavours.

(viii) ***p*-Aminobenzoic acid** (PABA) is used in vitamin B complex; it is a part of the folic acid. This is prepared as follows :

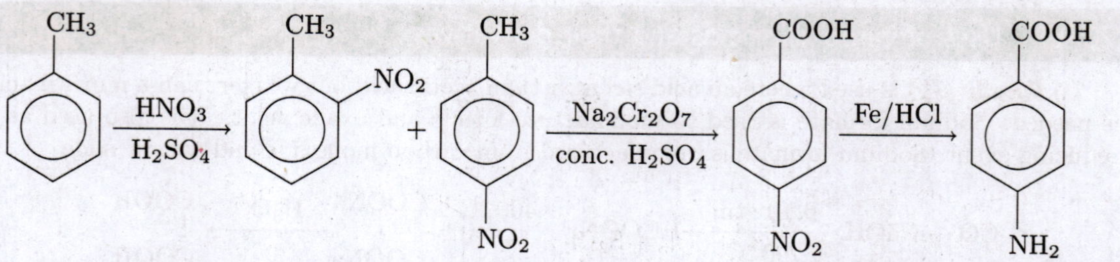


Some esters of *p*-aminobenzoic acid have a local anaesthetic action. Anaesthesin (ethyl *p*-aminobenzoate) and *n*-butyl *p*-aminobenzoate (butesin) are used as pain reliever of burns and open wounds. Novocain blocks the transmission of pain by the nerve trunks.

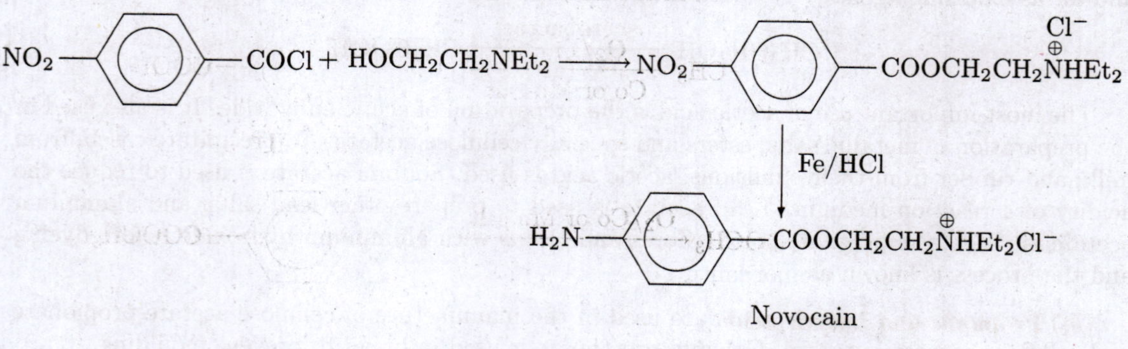


Novocain

**Pontocaine** is used for the surface anaesthesia of the eye and for the anaesthesia of the spine.

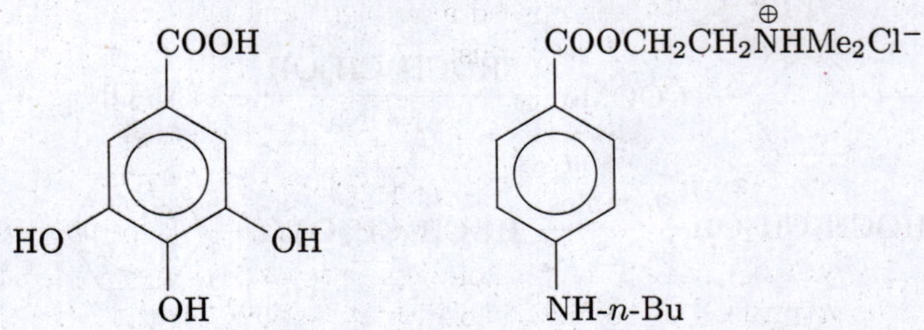


Gallic acid Pontocaine

**(ix) Gallic acid and tannic acid** are used in the manufacture of permanent writing ink. They form colourless and water-soluble salts with ferrous ion which get oxidised by air to a black water-insoluble ferric salt, a constituent of the writing ink.

**(x) Methyl, ethyl and higher esters of phthalic acid** are used as plasticisers of synthetic polymers such as polyvinyl chloride (PVC).

**(xi) Phaltan and Captan** are powerful agricultural fungicides.

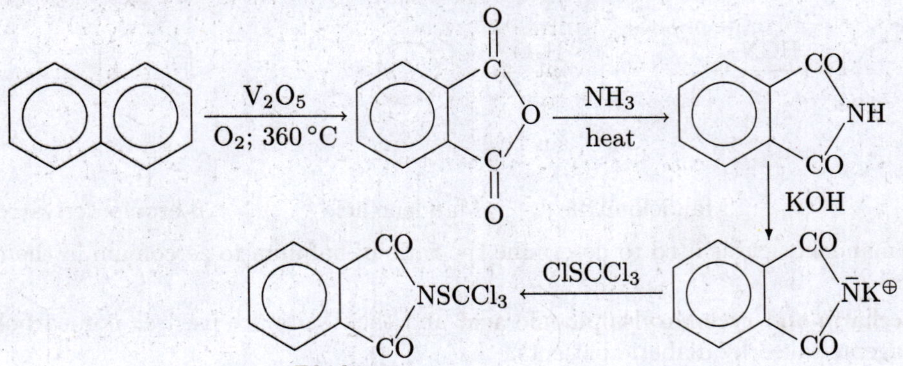


$$2CS_2 + 5Cl_2 \longrightarrow 2ClSCCl_3 + S_2Cl_2;$$

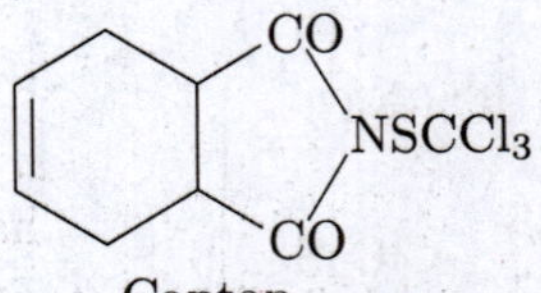


Captan

**(xii) Methyl terephthalate** is used in the manufacture of Terylene or Dacron.

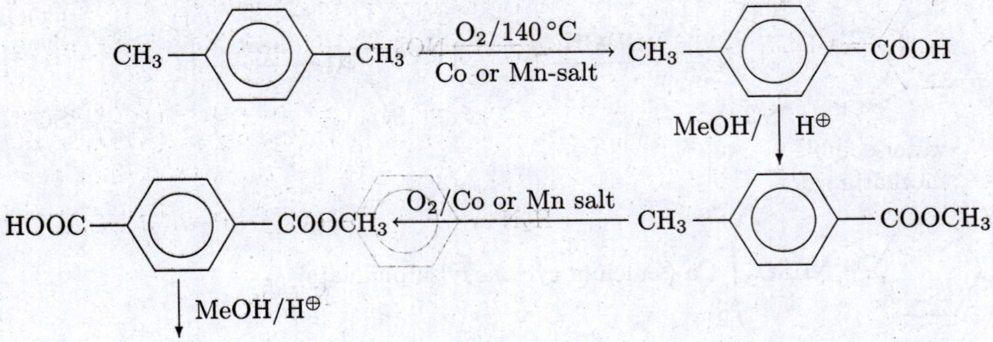

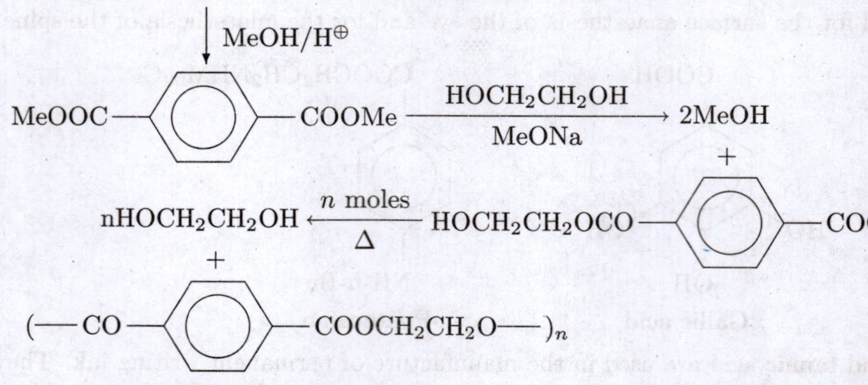


Terylene or Dacron

**(xiii) Mandelic acid** is used in the treatment of urinary infections. It is prepared from benzaldehyde.

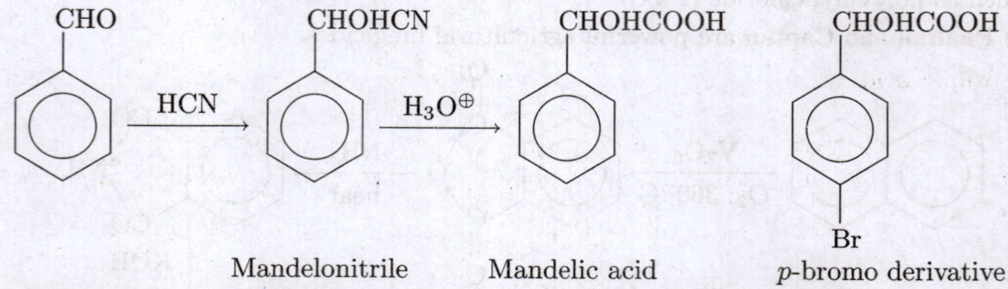


*p*-Bromomandelic acid is used to determine the ratio of hafnium to zerconium in the mixture of their salts.

**(xiv) Saccharin and cyclohexylsulphamic acid** and their salts are used as non-carbohydrate sweetening agents taken by diabetic patients.

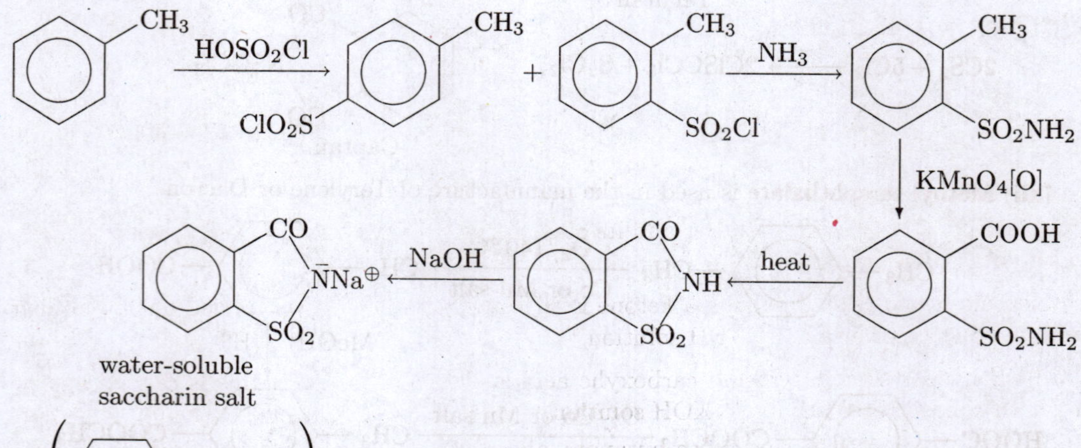


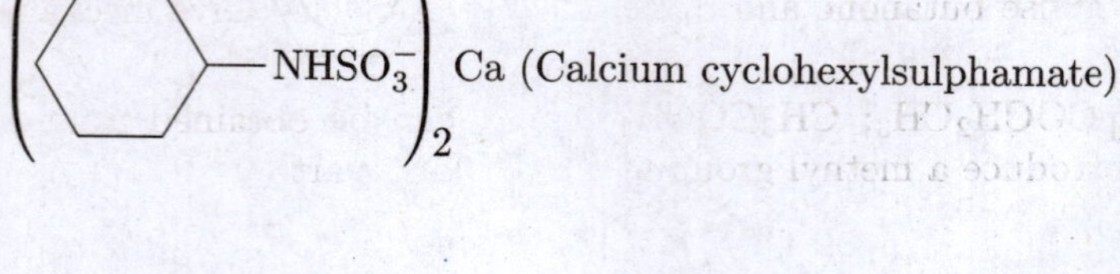

**(xv) Coumarin** is used in perfumery.

**(xvi) N-Bromosuccinimide** is a good brominating agent at the $\alpha$-position to a C=C double bond or an aromatic ring. This is prepared as follows :

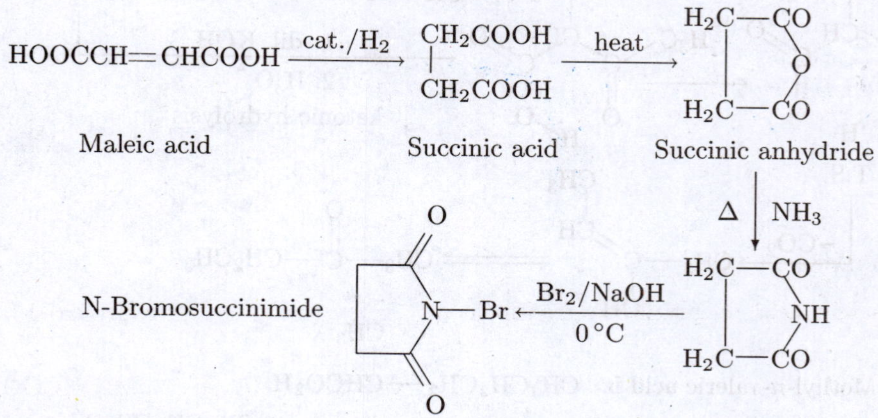


**(xvii) Adipic acid** is used as an intermediate in the synthesis of Nylon 6-6; the latter is a polyamide. It is formed when the hexamethylenediamine salt of adipic acid is heated.

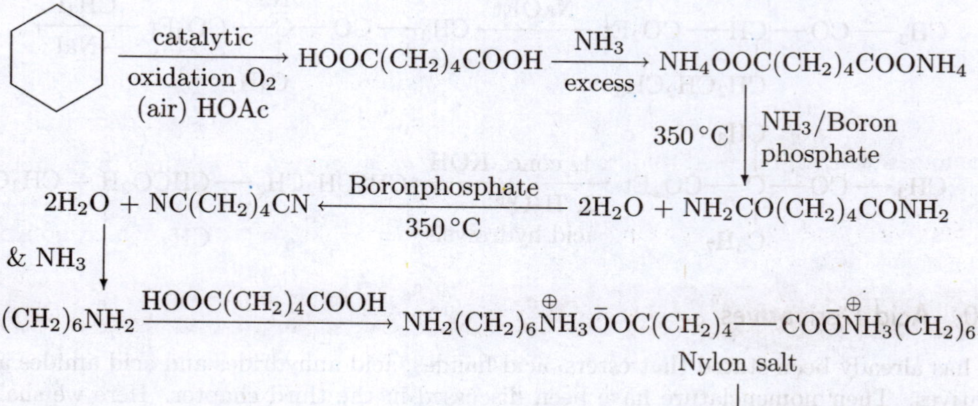


**Problem 20.21.** Acetoacetic ester and its alkyl derivatives can undergo two types of hydrolysis with potassium hydroxide:

**(i)** Ketonic hydrolysis : in which a ketone is the main product this is carried out by boiling with dilute aqueous or ethanolic KOH solution.

**(ii)** Acid hydrolysis : in which a carboxylic acid is the main product; this is carried out by boiling with concentrated ethanolic KOH solution.

Utilise these processes to synthesise butanone and $\alpha$-methyl-$n$-valeric acid. Give mechanism in each case.

**Solution :** Butanone is $CH_3COCH_2CH_3$; $CH_3COCH_2$ moiety can be obtained from acetoacetic ester. So we require to introduce a methyl group in $CH_3COCH_2$ part.

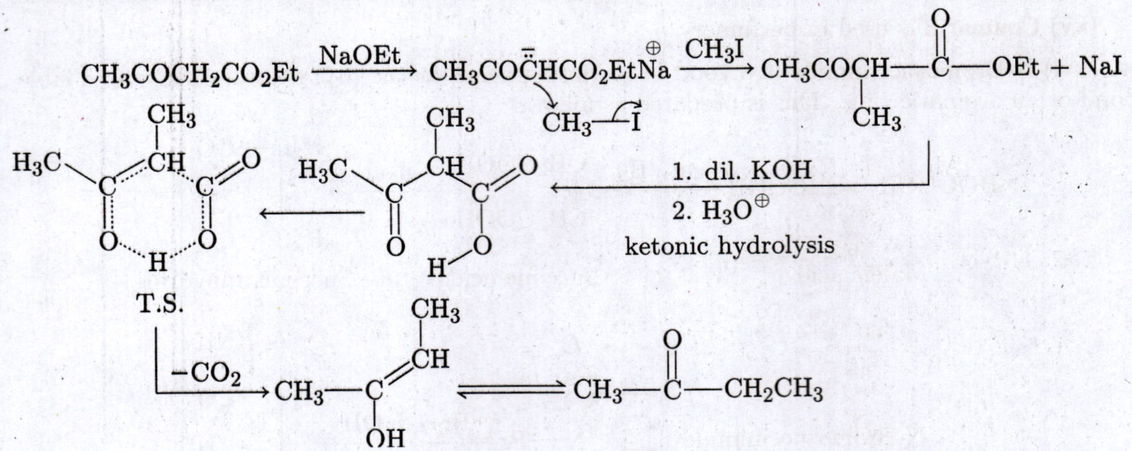

α-Methyl-*n*-valeric acid is $CH_3CH_2CH_2{-}CH(CH_3)CO_2H$

$$CH_3COCH_2CO_2Et \xrightarrow{NaOEt} CH_3CO\bar{C}HCO_2Et\,Na^{\oplus} \xrightarrow[-NaBr]{CH_3CH_2CH_2Br}$$

$$CH_3{-}CO{-}CH(CH_2CH_2CH_3){-}CO_2Et \xrightarrow{NaOEt} CH_3{-}CO{-}\bar{C}(Na^{\oplus})(C_3H_7){-}CO_2Et \xrightarrow[-NaI]{CH_3I}$$

$$CH_3{-}CO{-}C(CH_3)(C_3H_7){-}CO_2Et \xrightarrow[\text{2. } H_3O^{\oplus}\text{ acid hydrolysis}]{\text{1. conc. KOH}} CH_3CH_2CH_2{-}CH(CH_3)CO_2H + CH_3CO_2H$$

## 20.10 Acid Derivatives

It has already been stated that esters, acid halides, acid anhydrides and acid amides are acid derivatives. Their nomenclature have been discussed in the third chapter. Here we shall make lists of their general methods of formation and chemical reactions only since the processes and the methods have already been discussed in other places.

### 20.10.1 Formation of esters

(a) $RCOOH(ArCOOH) + R^1OH \xrightarrow[-H_2O]{\text{conc. } H_2SO_4} RCOOR^1(ArCOOR^1)$

(b) $RCOCl(ArCOCl) + R^1OH \xrightarrow{-HCl} RCOOR^1(ArCOOR^1)$

(c) $(RCO)_2O + R^1OH \longrightarrow RCOOR^1 + RCOOH$

(d) $R^1CO_2Ag + R^2Br \xrightarrow{-AgBr} R^1CO_2R^2(ArCO_2R^2)$
$(ArCO_2Ag)$

(e) $RCO_2H(ArCO_2H) \xrightarrow{CH_2N_2} RCO_2CH_3(ArCO_2CH_3) + N_2$

(f) $R_2O + CO \xrightarrow[125\text{–}180\,°C;\ 500\ atm]{BF_3 + H_2O} RCO_2R$

(g) $RCHO \xrightarrow[\text{reaction}]{Al(EtO)_3\ \text{Tischenko}} RCOOCH_2R$

(h) $RCOR \xrightarrow{R^1COOOH} RCOOR$

(i) $RCH_2CO_2Et \xrightarrow[EtOH]{NaOEt} RCH_2\text{—}C(=O)\cdot CH(R)\cdot CO_2Et$

(j) $CH_2{=}C{=}O \xrightarrow{ROH} MeCOOR$

**Chart 20.22** Formation of esters.

## 20.10.2 Structure and physical properties of esters

Ester group has an $sp^2$ hybridised C linked to (i) a double-bonded O atom, (ii) an O atom of alkoxy or aryloxy group and (iii) another C atom. Since esters do not possess any active H, they do not undergo molecular association through H-bonding. For this reason they are low boiling than the corresponding acids; they boil approximately at the same temperatures as the alkanes of comparable molecular weights boil. Although there are 2 moles of O atom per mole of ester, yet they have less solubility in water than the corresponding acids. Esters of low molecular weight are fairly soluble in water through H-bonding but the solubility decreases with increasing molecular weight. However, esters are either liquids or solids and they have pleasant smell.

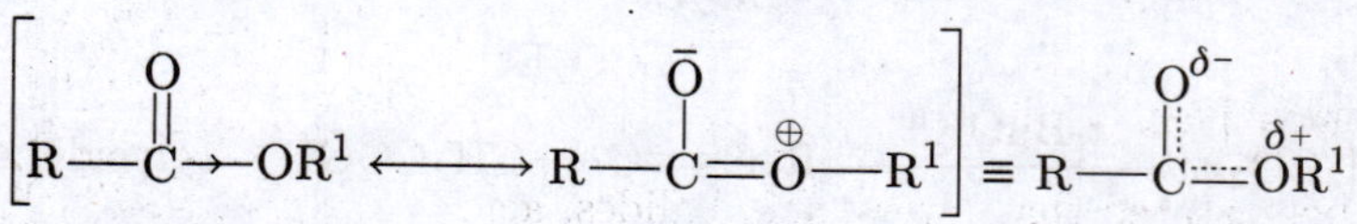

## 20.10.3 Chemical reactions of esters

Owing to the greater electronegativity of the O atom than the C atom, the latter in an ester group is a site for nucleophilic attack. After the attack the tetrahedral intermediate suffers from steric congestion and gives up the $OR^1$ group to form a derivative of carboxylic acid through $S_N$ reaction. Owing to the –I effect of the —COOR group, the $\alpha$-H atoms have acidic character and, therefore, they are removable by a strong base to generate carbanions which can undergo condensation reactions. The aromatic esters undergo $S_E$ reactions. The $CO_2R$ group being a deactivating group, aromatic esters undergo the $S_E$ reactions more slowly than benzene and the incoming electrophiles take the *m*-positions. Besides this, the aromatic esters undergo the Fries rearrangement. However, the chemical reactions are listed in Chart 20.23.

(a) $\underset{(Ar)}{RCOOR^1} \xrightarrow[2.\ H_3O^{\oplus}]{1.\ LiAlH_4} \underset{(Ar)}{RCH_2OH} + R^1OH$

(b) $\underset{(Ar)}{RCOOR^1} \xrightarrow{Na/ether} \text{Acyloin}$

(c) $\underset{(Ar)}{RCOOR^1} \xrightarrow[\bar{O}H/H_3O^{\oplus}]{H_3O^{\oplus}} \underset{(Ar)}{RCOOH} + R^1OH$

(d) $\underset{(Ar)}{RCOOR^1} \xrightarrow{PCl_5/SOCl_2} \underset{(Ar)}{RCOCl} + R^1Cl$

(e) $\underset{(Ar)}{RCOOR^1} \xrightarrow{NH_3} \underset{(Ar)}{RCONH_2} + R^1OH$

(f) $\underset{(Ar)}{RCOOR^1} \xrightarrow{H_2NNH_2} \underset{(Ar)}{RCONHNH_2} + R^1OH$ (acyl hydrazine)

(g) $\underset{(Ar)}{RCOOR^1} \xrightarrow{R^2OH} \underset{(Ar)}{RCOOR^2} + R^1OH$ (trans esterification)

(h) $RCOOR^1 \xrightarrow{R^2COOH} R^2COOR^1 + RCOOH$ (trans esterification)

(i) $\underset{(Ar)}{RCOOR^1} \xrightarrow{R^2MgX} \underset{(Ar)}{RCOR^2}$

(j) $RCOOCH_2CH_2R^1 \xrightarrow{heat} RCOOH + CH_2{=}CHR^1$

(k) $RCH_2CO_2Et \xrightarrow[EtOH]{NaOEt} R{-}CH_2\overset{O}{\overset{\|}{C}}{-}\underset{R}{\underset{|}{CH}}{-}CO_2Et$

(l) 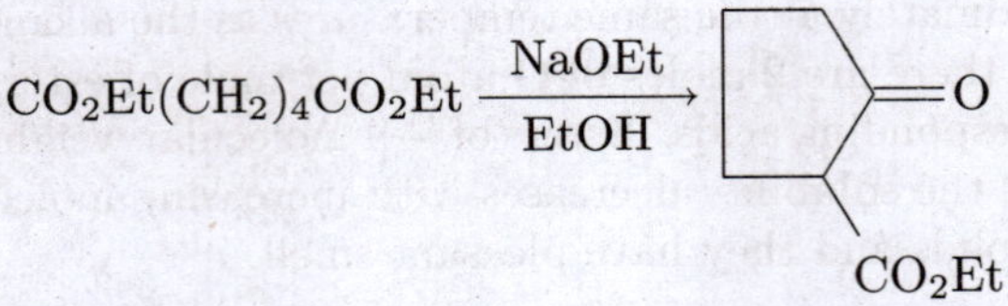

(m) $R_2CO + \underset{CH_2CO_2Et}{\underset{|}{CH_2CO_2Et}} \xrightarrow{t\text{-}Bu\bar{O}K^{\oplus}} R_2C{=}\overset{CO_2Et}{\overset{|}{C}}{-}CH_2CO\bar{O}K^{\oplus}$ (glycidic ester salt)

(n) 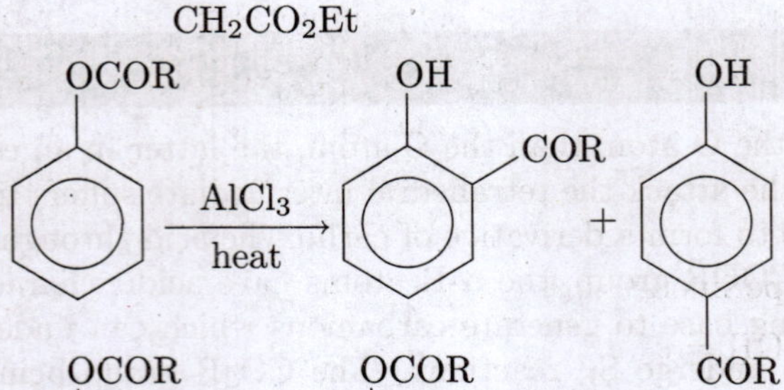

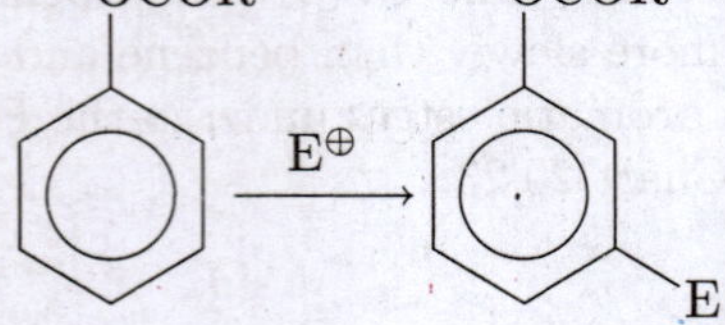

**Chart 20.23** Reactions of esters.

### 20.10.4 Uses

Esters of aliphatic acids are mostly used as solvents especially for cellulose nitrate in the formulations of lacquers. High boiling esters are used as softening agent for resins, plastics which are used as synthetic fibres such as, dacron, alkyd and polyester resins.

Some esters have specific fruity odour and they are used as synthetic flavouring agents such as isoamyl acetate (odour of banana), isoamyl valerate (odour of apple), butyl butyrate (odour of pine apple).

### 20.10.5 Analysis of esters

**(i) Test :** To 0.2 g of an ester 0.2 g of solid $NH_2OH.HCl$ and 5 ml of 10% NaOH solution are added. The mixture is then boiled for two minutes. It is then cooled and acidified with dilute HCl. To the cold acidified solution 2-3 drops of freshly prepared ferric chloride solution is added $\longrightarrow$ formation of violet or deep reddish brown colour $\longrightarrow$ indicates the presence of ester group.

$$R^1COOR + NH_2OH{\cdot}HCl \longrightarrow R^1CONHOH{\cdot}HCl + ROH$$

$3R^1CONHOH{\cdot}HCl + FeCl_3 \longrightarrow$ 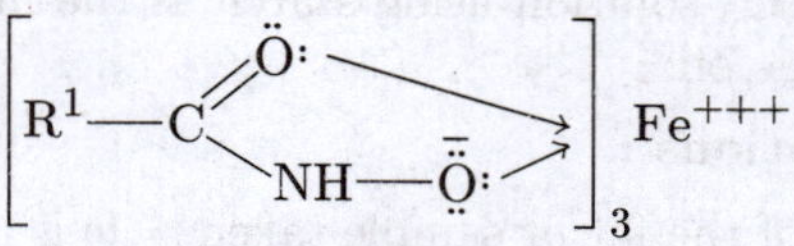

coloured complex

**(ii) Quantitative analysis :**

**(a) Determination of saponification value :** By the saponification value of an ester we mean the number of milligrams of KOH required to hydrolyse (saponify) completely one gram of ester. Usually saponification values of fats and oils (which are glycerides of higher fatty acids) are determined as given below :

About 0.5 g of the ester is accurately weighed and taken in a 250 ml conical flask fitted with an aircondenser. The ester is then dissolved in 10 ml of absolute alcohol. To that 10 ml of 2.5N KOH solution is added. The mixture is then refluxed on a sand bath for about two hours. The solution is then cooled, a few drops of phenolphthalein indicator is then added and then unreacted potassium hydroxide is titrated with standard N/2 oxalic acid solution untill the pink colour of the solution first disappears. The experiment is repeated and also done without the ester (blank experiment).

**Calculations :**

Weight of the ester taken $= w$ g

Vol. of N/2 oxalic acid used for blank experiment $= v_1$ ml

Vol. of N/2 oxalic acid used for excess KOH $= v_2$ ml

Vol. of N/2 oxalic acid equivalent to used KOH $= (v_1 - v_2)$ ml

$$\equiv (v_1 - v_2) \text{ ml N/2 KOH.}$$

1000 ml of N KOH contains 56 g of KOH (g-eq of KOH = 56).

$(v_1 - v_2)$ ml of $\frac{N}{2}$ KOH contains $\frac{56 \times (v_1 - v_2)}{1000 \times 2}$ g of KOH

$$\text{Hence the saponification value of the ester} = \frac{56(v_1 - v_2) \times 1000}{1000 \times 2 \times w} \text{ (mg)}$$
$$= \frac{56(v_1 - v_2)}{2w} \text{ mg.}$$

**(b) Iodine value of fat or oil :** By the iodine value of fat or oil we mean the number of grams of iodine taken up by 100 g of the fat or oil. The iodine value is determined by reacting a known amount of oil or fat with a known volume of excess iodine monochloride in acetic acid (known as the Wijs' solution) and then titrating back the unreacted iodine with standard sodium thiosulphate.

About 1 g of the accurately weighted fat or oil is dissolved in 25 ml of $CHCl_3$ or $CCl_4$ in a 500 ml conical flask. To this 25–50 ml of the Wijs' solution (prepared by dissolving 13.0 g of pure iodine in 1*l* of glacial acetic acid (A.R.) and then passing dry chlorine gas through the solution until on titration of this solution with N/10 $Na_2S_2O_3$ shows the titre value double of that obtained by titrating the iodine solution before passing chlorine). The conical flask is covered, shaken well and kept in the dark for about one hour with occasional shakings. Then about 20–40 ml of KI solution in 100 ml water is added. The liberated iodine is then titrated with standard N/10 $Na_2S_2O_3$ solution using starch as the indicator. A blank test is performed without taking the fat or the oil.

**Calculations :**

Weight of the fat or sample taken $= w$ g

Normality of the $Na_2S_2O_3$ solution $= \frac{N}{x}$

Vol. of $Na_2S_2O_3$ solution used for the blank exp. $= v_1$ ml

Vol. of $Na_2S_2O_3$ solution used for the sample $= v_2$ ml

$$\text{Iodine value} = \frac{(v_1 - v_2) \times N/x \times 127 \times 100}{w \times 1000}.$$

## 20.10.6 General formation of acid chlorides

Acid chlorides are usually prepared by treating acids with $PCl_5$, $PCl_3$ or $SOCl_2$ and the reactions are given in Section 20.9.5.h, p. 1296.

## 20.10.7 Structure and physical properties of acid chlorides

Structure of an acid halide is very alike to that of an ester. In this case the carbonyl C is more electrophilic than that in an ester because of the electron-withdrawing effect of the Cl atom. Since there is no active H atom, they do not form intermolecular H-bond and hence they boil at lower temperatures than the corresponding acids. The members are colourless liquids or solids of irritating odour.

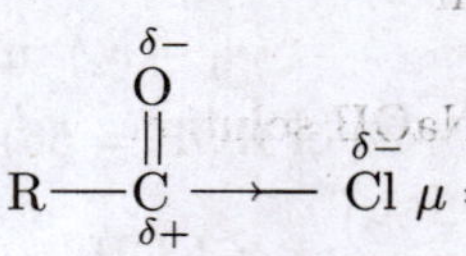

$\mu = 2.7$D (approx.)

## 20.10.8 Chemical reactions

Strong electrophilic character of the carbonyl C makes it undergo $S_N$ reaction through tetrahedral intermediate; strong $-I$ effect makes the $\alpha$-H to undergo $S_E$ reactions. The reactions are given in Chart 20.24.

(a) $RCOCl \text{ or } ArCOCl \xrightarrow[\text{suspension}]{H_2/Pd;\ BaSO_4} RCHO \text{ or } ArCHO$

(b) $RCOCl\ (ArCOCl) \xrightarrow[3.\ R^1OH\ 4.\ R^1NH_2]{1.\ H_2O\ 2.\ NH_3} 1.\ RCOOH\ 2.\ RCONH_2\ 3.\ RCOOR^1\ 4.\ RCONHR^1$

(c) $RCH_2COCl \xrightarrow{Cl_2(Br_2),\ \text{red P}} RCHClCOCl\ (ArCHClCOCl)$
$(ArCH_2COCl)$

(d) $RCOCl \xrightarrow[R^1Li]{R^1MgX \text{ or}} RCOR^1\ (ArCOR^1)$
$(ArCOCl)$

(e) $RCOCl \text{ or } ArCOCl \xrightarrow[\text{ether}]{LiAlH_4} RCH_2OH(ArCH_2OH)$

(f) $RCOCl + R^1COONa \longrightarrow RCO{\cdot}O{\cdot}COR^1$

(g) $PhH + RCOCl \xrightarrow[AlCl_3]{\text{anh}} PhCOR$

**Chart 20.24** Reactions of acid halides

## 20.10.9 Analysis of acid chlorides

**(i) Tests :** Acid chlorides on treatment with alcoholic silver nitrate give white precipitate of silver chloride immediately.

**(ii) Quantitative analysis :** Determination of molecular weight: volumetric method: when 1 mole of monocarboxylic acid chloride is hydrolysed, 1 mole of carboxylic acid and 1 mole of hydrogen chloride form which are equivalent to 2000 ml of (N) NaOH solution.

$$RCOCl + H_2O \longrightarrow RCOOH + HCl \xrightarrow{NaOH} RCOONa + 2H_2O + NaCl$$

About 1g of accurately weighed acid chloride is hydrolysed completely and the solution is then cooled and then titrated with standard (N) NaOH solution using phenolphthalein as indicator.

Let the weight of the sample taken $= w$ g

Volume of NaOH solution required $= v$ ml

1 g mol wt. of the acid halide $\equiv$ 2000 ml (N) NaOH solution.

$w$ g of the sample $= v$ ml N NaOH solution.

1 g of the sample $= \dfrac{v}{w}$ ml N NaOH solution

1 g mol wt. of the sample $\equiv \dfrac{M \times v}{w}$ ml N NaOH solution

$$\therefore \quad \frac{M \times v}{w} = 2000 \quad \text{or,} \quad M = \frac{2000 \times w}{v}.$$

## 20.10.10 Formation of acid anhydrides

(a) $RCOCl + RCO\bar{O}\overset{\oplus}{Na} \longrightarrow RCO{\cdot}O{\cdot}COR$

(b) $CH_3CO_2H + C_2H_2 \xrightarrow{Hg^{2+}} CH_3CH(OCOCH_3)_2 \xrightarrow{heat} (CH_3CO)_2O$

(c) $CH_2{=}C{=}O + CH_3CO_2H \longrightarrow (CH_3CO)_2O$

(d) $RCO_2H + (CH_3CO)_2O \longrightarrow (RCO)_2O + AcOH$

(e) $RCOCl + RCO_2H \xrightarrow[\text{pyridine}]{-HCl} (RCO)_2O$

**Chart 20.25** Formation of acid anhydrides

## 20.10.11 Structure and physical properties of acid anhydrides

The structure of an acid anhydride may be considered as a resonance hybrid of the following resonating structures :

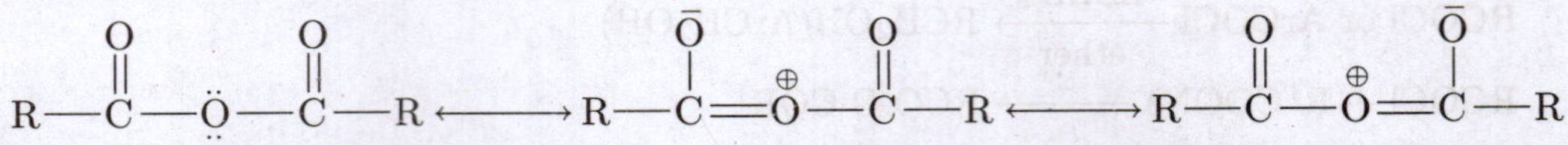

Since there is no H-bonding among acid anhydride molecules, they boil at low temperatures compared to their molecular weights. Since an anhydride possesses 125 kJ mol$^{-1}$ of resonance energy, and since negative charge is not localised, it cannot form H-bond with water and, therefore, anhydrides are insoluble in water.

## 20.10.12 Chemical reactions of acid anhydrides

Acid anhydrides undergo $S_N$ reactions through tetrahedral intermediates. The reactions are listed in Chart 20.26.

(a) $(RCO)_2O \xrightarrow[\bar{O}H/H_3O^{\oplus}]{H_2O \text{ or}} 2RCO_2H$ (b) $(RCO)_2O \xrightarrow{R^1OH} RCOOR^1 + RCOOH$

(c) $(RCO)_2O \xrightarrow{NH_3} RCONH_2 + RCOOH$ (d) $(RCO)_2O \xrightarrow{HCl} RCOCl + RCO_2H$

(e) $(CH_3CO)_2O \xrightarrow{CH_3CHO} CH_3CH(OCOCH_3)_2$

(f) $(CH_3CO)_2O \xrightarrow{N_2O_5} CH_3COONO_2$ (Acetyl nitrate)

**Chart 20.26** Reactions of acid anhydrides.

In fact, acetic anhydride is the only anhydride which deserves special mention. The above reactions are given by acetic anhydride. It reacts more slowly than acetyl chloride.

## 20.10.13 Uses of acetic anhydride

Acetic anhydride is used as an acetylating agent. It is used in the production of cellulose acetate, acetanilide, aspirin, acetyl nitrate, etc. Acetyl nitrate is used as a good nitrating agent for nitrating some aromatic and heterocyclic compounds.

(a) $RCOONH_4 \xrightarrow{\text{heat}} RCONH_2$ (Ar) (Air) (b) $RCOOR^1 \xrightarrow{NH_3} RCONH_2 + R^1OH$ (Ar) (Ar)

(c) $RCOCl(ArCOCl) \xrightarrow{NH_3} RCONH_2(ArCONH_2) + NH_4Cl$

(d) $(RCO)_2O[(ArCO)_2O] \xrightarrow{NH_3} RCONH_2(ArCONH_2)$

(e) $RCN \xrightarrow{\bar{O}H/H_2O_2} RCONH_2$ (p. 1325 for mechanism)

(f) $RCOOH + CO(NH_2)_2 \longrightarrow RCONH_2(ArCONH_2)$

(g) $R_2C{=}CH_2 \xrightarrow[3.\ H_2O]{1.\ H_2SO_4\ 2.\ R^1CN} R_2C(CH_3){-}NHCOR^1$

(h) $RCH{=}CH_2 \xrightarrow[2.\ CO\ 3.\ R^1NH_2(NH_3)]{1.\ H_2SO_4} RCH(CH_3){-}C(NHR^1){=}O\ (RCH(CH_3){-}C(NH_2){=}O)$

**Chart 20.27** Formation of acid amides

The structure may be represented as a resonance hybrid of two resonating structures as shown below :

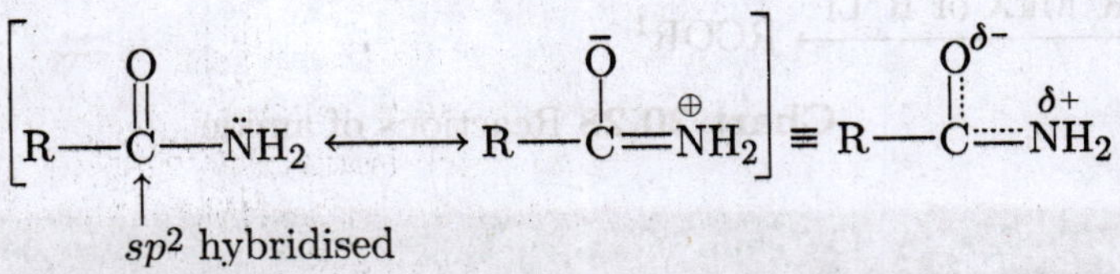

Amides remain in tautomeric forms; the system is called amido-imidol system.

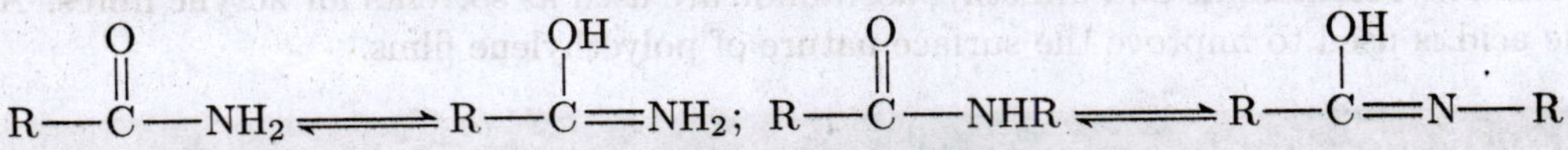

Unsubstituted and N-monosubstituted amides have active H. So, these amides can associate among themselves and as a consequence their melting and boiling points are high. N, N-disubstituted amides have high dipole moments and hence they boil at high temperatures, e.g., N, N-dimethyl formamide has a $\mu$ value 3.34D and its boiling point is 155 °C. However, N, N-disubstituted amides boil at lower temperatures than N-monosubstituted and unsubstituted amides.

$HCONH_2$(m.p. 25 °C); b.p. 193 °C, HCONHMe : b.p. 180 °C, $HCONMe_2$ : b.p. 155 °C

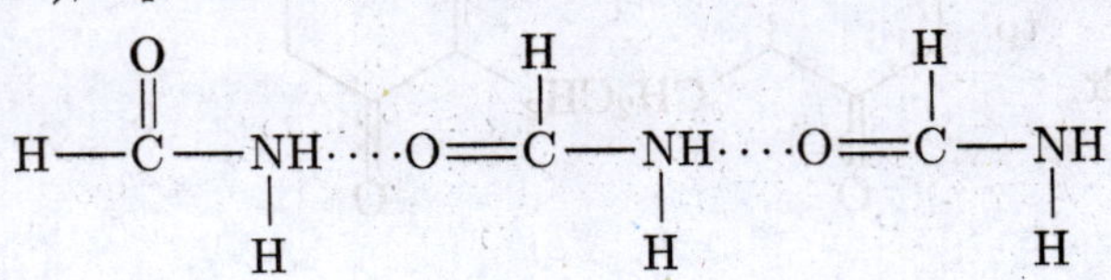

Since unsubstituted and monosubstituted amides can form H-bond with hydroxylic solvents, they are soluble in those solvents. These amides up to 5 Cs are soluble in water. Since amides have high dielectric constants (e.g., for N, N-dimethyl formamide $\varepsilon = 37.6$), they are good solvents for covalent as well as ionic compounds. For the same reason, $S_N$ reactions take place more rapidly in disubstituted amides than in water or alcohol.

## 20.10.16 Chemical reactions

These are given in Chart 20.28.

(a) $RCONH_2 \xrightarrow{H_3O^{\oplus}/\bar{O}H} RCOOH + \overset{\oplus}{N}H_4(RCO\bar{O} + NH_3)$

(b) $RCONH_2 \xrightarrow[\text{feeble base character}]{H^{\oplus}} R{-}\overset{\overset{\oplus}{O}-H}{\overset{\|}{C}}{-}\ddot{N}H_2 \longleftrightarrow R{-}\overset{OH}{\overset{|}{C}}{=}\overset{\oplus}{N}H_2$ or $R{-}\overset{O}{\overset{\|}{C}}{-}\overset{\oplus}{N}H_3$

(c) $R{-}\overset{O}{\overset{\|}{C}}NH_2 \xrightarrow[\text{feeble base character}]{NaNH_2} R{-}\overset{O}{\overset{\|}{C}}{-}\ddot{\bar{N}}H \longleftrightarrow R{-}\overset{\bar{O}}{\overset{|}{C}}{=}NH$

(d) $RCONH_2 \xrightarrow{Na/EtOH} RCH_2NH_2$

(e) $RCONH_2 \xrightarrow{LiAlH_4;\ ether} RCH_2NH_2$

(f) $RCONH_2 \xrightarrow{HNO_2} RCOOH + N_2 + H_2O$

(g) $RCONH_2 \xrightarrow[\text{heat}]{P_2O_5 \text{ or } PCl_5} R{-}C{\equiv}N$

(h) $RCONH_2 \xrightarrow{Br_2/NaOH} RNH_2$

(i) $HCONMe_2 \xrightarrow{R^1MgX \text{ or } R^1Li} R^1CHO$

(j) $RCONMe_2 \xrightarrow{R^1MgX \text{ or } R^1Li} RCOR^1$

**Chart 20.28** Reactions of amide

## 20.10.17 Uses of amides

Dimethyl formamide and dimethyl acetamide are used as solvents for acrylic fibres. Amide of oleic acid is used to improve the surface nature of polyethylene films.

## 20.10.18 Analysis

**(i) Test :** When unsubstituted amides are heated with 50% NaOH solution, ammonia gas evolves which turns white mercurous nitrate paper to black.

**Problem 20.22.** Convert :

**(a)** Propene to $CH_3COCH_2COCO_2Et$

**(b)** $\begin{array}{l} CH_2CH_2CO_2Et \\ | \\ CH_2CH_2CO_2Et \end{array}$ to

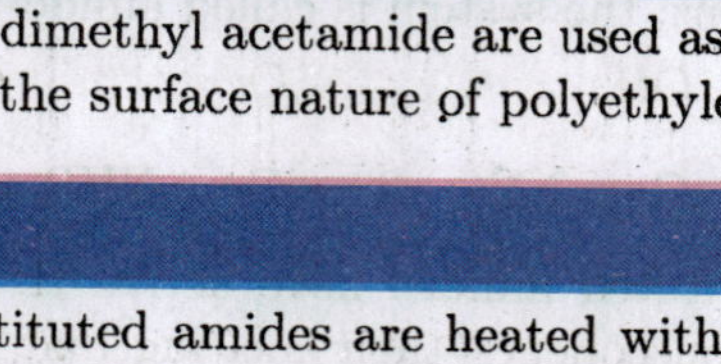

**(c)**

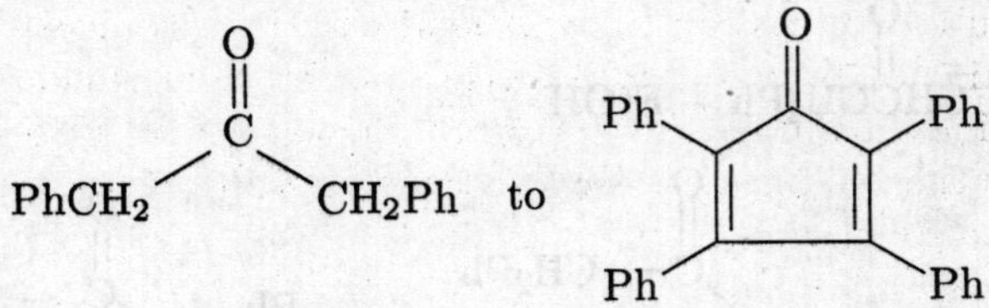

**(d)** $CH_2(COEt)_2$ to

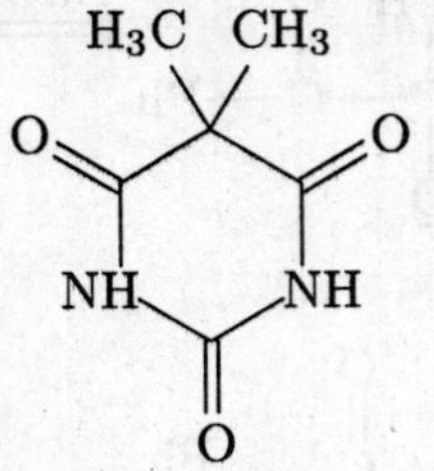

**Solution :**

**(a)**

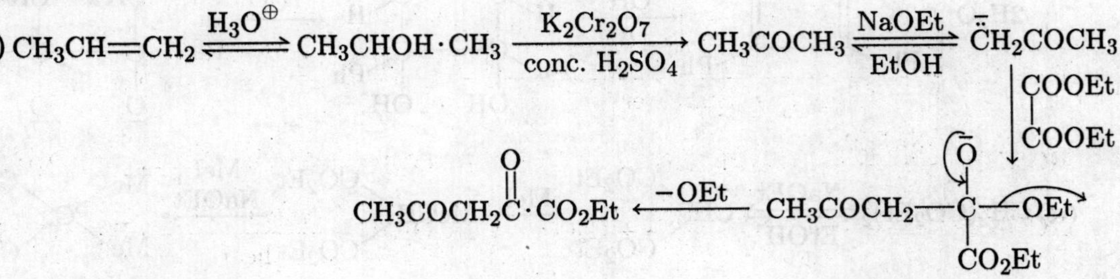

**(b)**

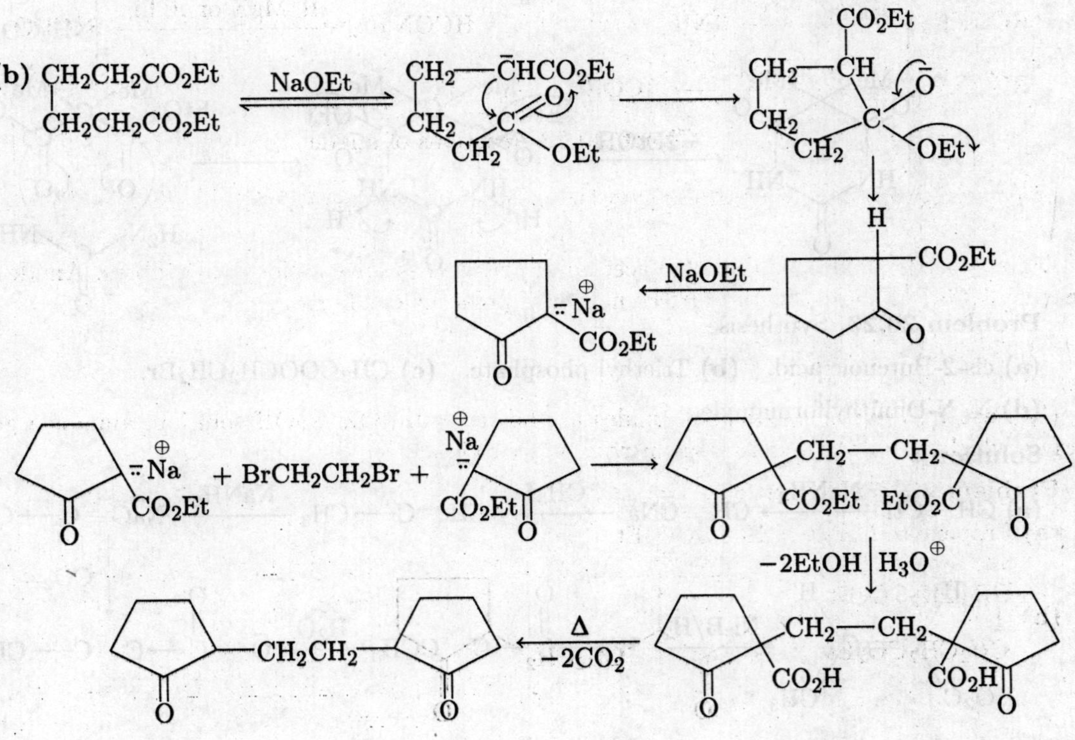

(c) 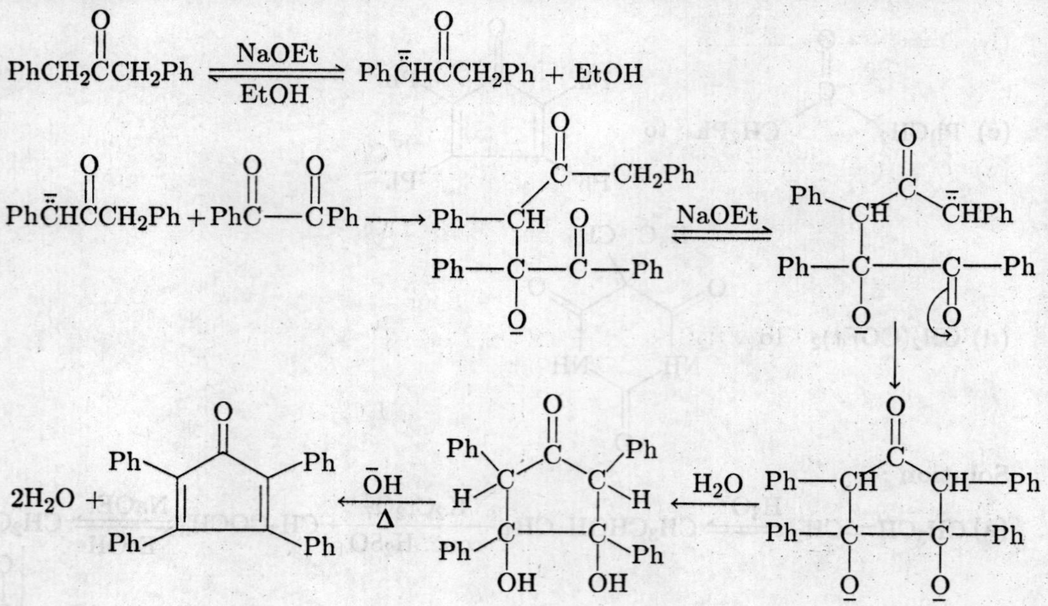

(d) 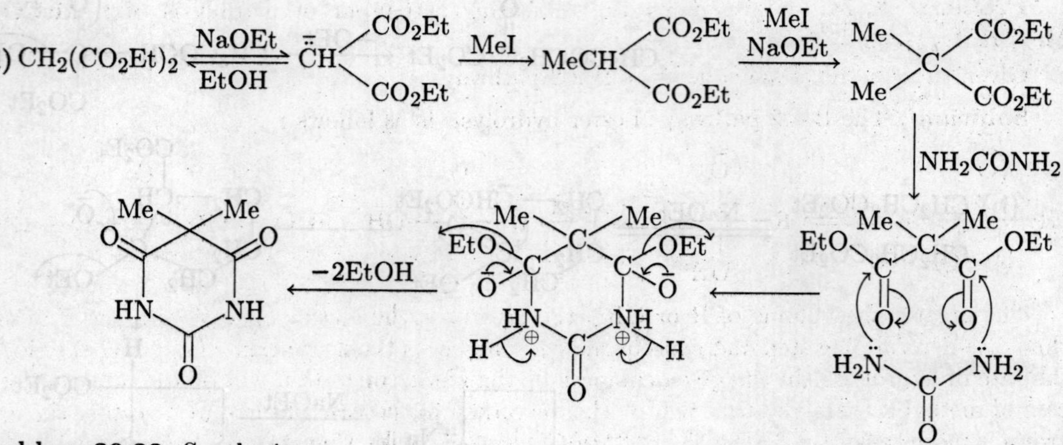

**Problem 20.23.** Synthesise

**(a)** cis-2-Butenoic acid. **(b)** Triethyl phosphate. **(c)** $CH_3COOCH_2CH_2Br$.

**(d)** N, N-Dimethylformamide.

**Solution :**

(a) 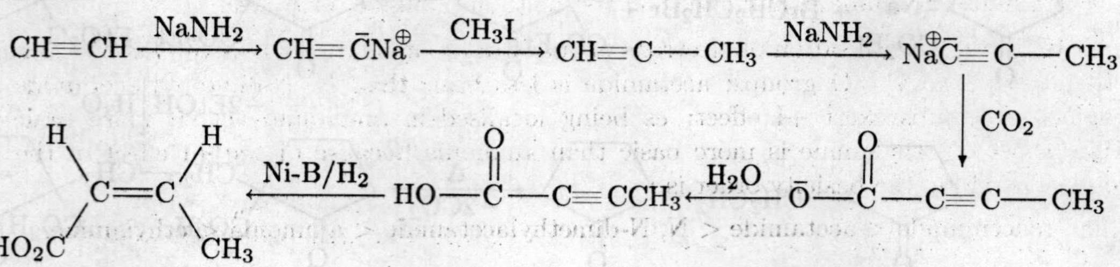

(b) 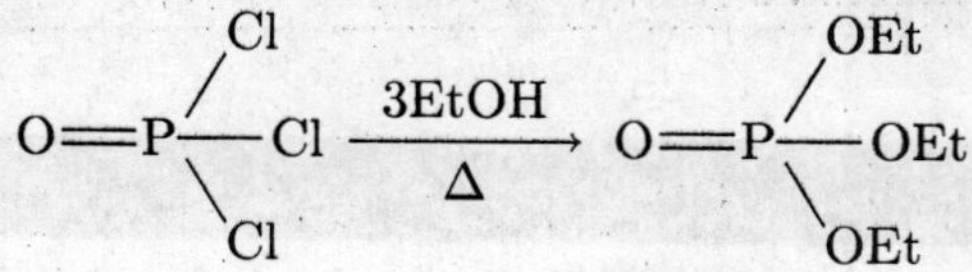

(c) 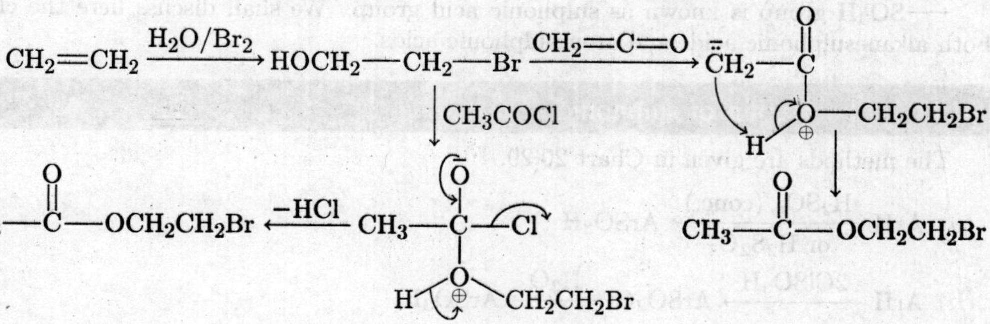

(d) 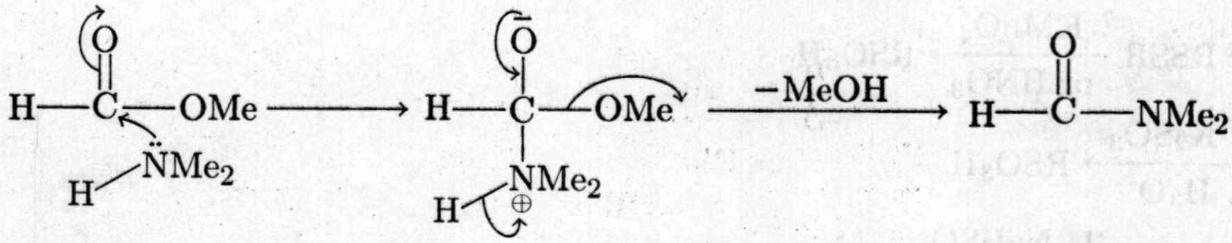

**Problem 20.24.** Write down the increasing rate-order of hydrolysis of : $MeCO_2Me$, $Me_2CHCO_2Me$ and $MeCO_2Bu$ (t).

Given that the reactions follow the BAC2 pathway.

**Solution :** The BAC2 pathway of ester hydrolysis is as follows :

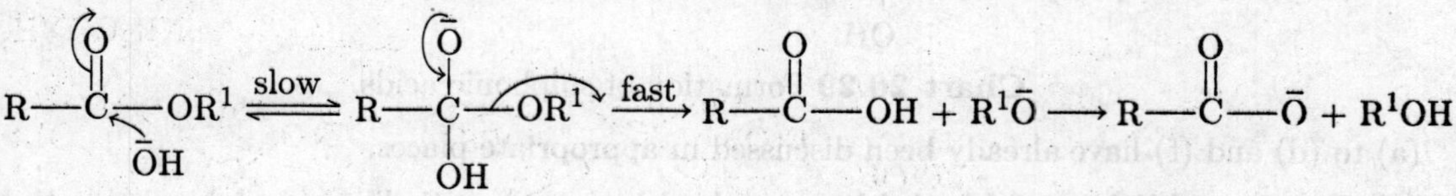

The greater the volume of R or $OR^1$, the greater is the steric congestion in the T.S. of the slow rate-determining step and consequently the higher is the activation energy and the slower is the rate of hydrolysis. In the given compounds the steric congestion will be the minimum in the case of methyl acetate and that will be the maximum in the case of tert-butyl acetate since t-Bu group is more bulky than $Me_2CH$, and $Me_2CH$ is more bulky than Me. Therefore, the increasing rate-order of hydrolysis is :

$$MeCO_2Bu(t) < Me_2CHCO_2Me < MeCO_2Me.$$

**Problem 20.25.** Write down the basicity order of the following bases.

Acetamide, succinimide, ammonia, N, N-dimethylacetamide, ethylamine.

**Solution :** Delocalisation of electrons of the N atom of succinimide is the maximum since it is flanked by two C=O groups; acetamide is less basic than N, N-dimethyl acetamide since two methyl groups exert +I effect; *e*s being localised in ammonia, this is more basic than $CH_3CONMe_2$. Ethylamine is more basic than ammonia because of the +I effect of the ethyl group. Therefore, the basicity order is :

succinimide < acetamide < N, N-dimethylacetamide < ammonia < ethylamine.

## 20.11 Sulphonic Acids

### 20.11.1 Introduction

—$SO_3H$ group is known as sulphonic acid group. We shall discuss here the chemistry of both alkanesulphonic acids and arenesulphonic acids.

### 20.11.2 Formation of sulphonic acids

The methods are given in Chart 20.29.

(a) $$\text{ArH} \xrightarrow[\text{or } H_2S_2O_7]{H_2SO_4\ (\text{conc.})} \text{ArSO}_3\text{H}$$

(b) $$\text{ArH} \xrightarrow{2\text{ClSO}_3\text{H}} \text{ArSO}_2\text{Cl} \xrightarrow{H_2O} \text{ArSO}_3\text{H}$$

(c) $$\text{RSH or RSSR} \xrightarrow[\text{or HNO}_3]{\text{KMnO}_4} \text{RSO}_3\text{H}$$

(d) $$\text{RX} \xrightarrow[2.\ H_3O^{\oplus}]{1.\ K_2SO_3} \text{RSO}_3\text{H}$$

(e) $$\text{RCH}{=}\text{CH}_2 \xrightarrow[2.\ H_2O\ 3.\ H^{\oplus}]{1.\ \text{NaHSO}_3} \text{RCH}_2\text{CH}_2\text{SO}_3\text{H}$$

(f) $$\text{R—}\overset{\overset{\displaystyle O}{\|}}{C}\text{—R} \xrightarrow[2.\ H_2O^{\oplus}]{1.\ \text{NaHSO}_3} \text{R}_2\underset{\underset{\displaystyle OH}{|}}{C}\text{—SO}_3\text{H}$$

**Chart 20.29** Formation of sulphonic acids.

(a) to (d) and (f) have already been discussed in appropriate places.

**(e) By the addition of bisulphites to double bonds :** Sodium bisulphite adds to a C—C double bond anti-Markownikoffically by a free radical mechanism and the reaction takes place in the presence of oxygen or oxides of nitrogen.

$$\text{NaHSO}_3 \rightleftharpoons \text{Na}^{\oplus} + \bar{\text{O}}\text{SOOH}$$

$$\bar{O}\text{—}\underset{\underset{\displaystyle O}{\|}}{S}\text{—OH} + \cdot O\text{—}O\cdot(\text{or}\cdot\text{NO}) \longrightarrow \bar{O}\text{—}\underset{\underset{\displaystyle O}{\|}}{S}O\cdot + \text{HOO}\cdot(\text{or HNO}\cdot)$$

$$\text{RCH}{=}\text{CH}_2 + \bar{O}\text{—}\underset{\underset{\displaystyle O}{\|}}{S}O\cdot \longrightarrow \text{R}\dot{\text{C}}\text{H—CH}_2\text{—O—}\underset{\underset{\displaystyle O}{\|}}{S}\text{—}\bar{O}$$

$$\text{R}\dot{\text{C}}\text{H—CH}_2\text{—O—}\underset{\underset{\displaystyle O}{\|}}{S}\text{—}\bar{O} + \bar{O}\underset{\underset{\displaystyle O}{\|}}{S}\text{—OH} \longrightarrow \text{RCH}_2\text{CH}_2\text{O}\underset{\underset{\displaystyle O}{\|}}{S}\bar{O} + \cdot\text{SO}_3$$

$$\text{RCH}_2\text{CH}_2\text{O}\underset{\underset{\displaystyle O}{\|}}{S}\bar{O} + H_3O^{\oplus} \longrightarrow \text{RCH}_2\text{CH}_2\text{O}\underset{\underset{\displaystyle O}{\|}}{S}\text{OH}$$

## 20.11.3 Structure, reactivity and physical properties

In sulphonic acids the S atom is bonded to a C atom.

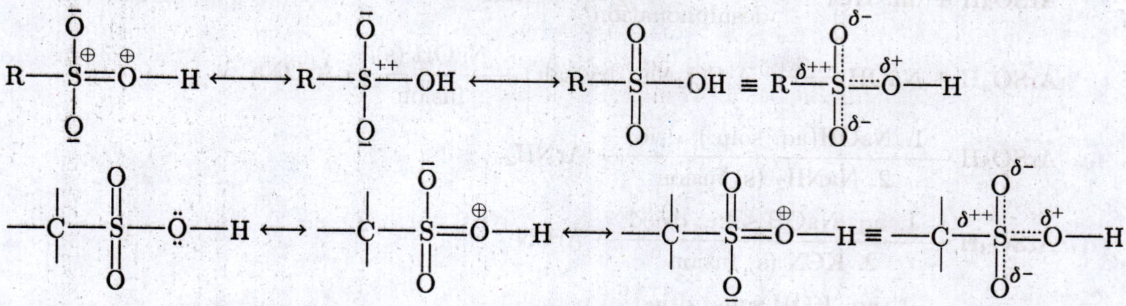

—$SO_3H$ group has −I effect and −R effect. Owing to the $\delta^+$ charge on the O atom of the OH part in the —$SO_3H$ group, $SO_3H$ group has a tendency to give up a proton; S—OH bond may also break in a chemical reaction. In arenesulphonic acids the sulphonic acid group deactivates the rings and directs the incoming nucleophile to the $\gamma$-position ($m$-position in benzene nucleus). The —$SO_3H$ group can also be removed from a compound very easily.

However, sulphonic acids being highly polar, their properties are characteristic of highly polar compounds. Because of this polar character and because of the complete ionisation of the —$SO_3H$ group, they are highly soluble in water. For the same reason they are insoluble in the usual organic solvents but soluble in some other polar solvents. Sulphonic acids are of low volatility and they decompose below their boiling points when they are heated.

$$RSO_3H \xrightarrow{H_2O} RSO_3^- + H_3O^{\oplus}; \; ArSO_3H \xrightarrow{H_2O} ArSO_3^- + H_3O^{\oplus}$$

Alkanesulphonic acids are usually highly deliquescent thick liquids while arenesulphonic acids are deliquescent solids.

## 20.11.4 Chemical reactions

Sulphonic acids are strong acids ($K_a = 6.3 \times 10^{-5}$). They dissociate into sulphonates and hydronium ion in the aqueous medium. They react with active metals and alkalis to form salts.

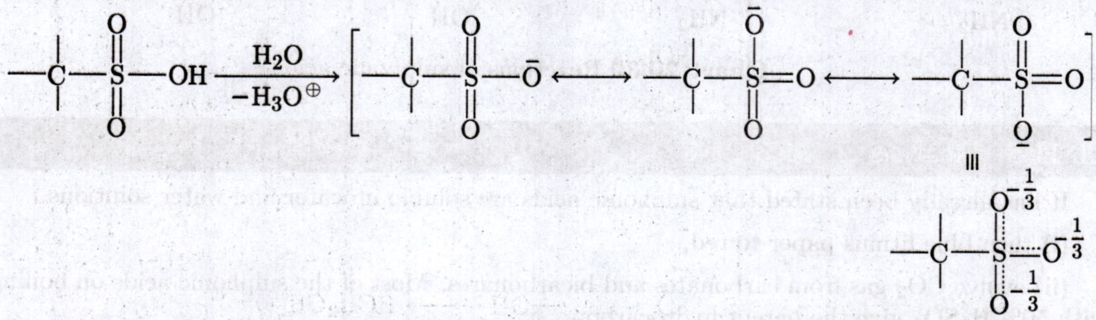

As acids, aqueous solution of sulphonic acids turn blue litmus paper to red and evolve carbon dioxide gas from saturated solutions of carbonates and bicarbonates.

$$R{-}SO_3H + NaHCO_3 \longrightarrow RSO_3Na + CO_2 + H_2O$$

$$Ar{-}SO_3H + NaHCO_3 \longrightarrow ArSO_3Na + CO_2 + H_2O$$

However, other properties of sulphonic acids are listed in Chart 20.30.

(a) $ArSO_3H + \text{dil. } HCl \xrightarrow[\text{desulphonation}]{150\,°C \text{ \& pressure}} ArH$

(b) $ArSO_3H + NaOH \longrightarrow ArSO_3Na \text{ (dry salt)} \xrightarrow[\text{fusion}]{NaOH \text{ (s)}} ArONa$

(c) $ArSO_3H \xrightarrow[\text{2. } NaNH_2 \text{ (s) fusion}]{\text{1. NaOH(aq. soln.); dried}} ArNH_2$

(d) $ArSO_3H \xrightarrow[\text{2. KCN (s) fusion}]{\text{1. aq. NaOH soln.; dried}} ArCN$

(e) $ArSO_3H \xrightarrow[\text{2. KSH (s) fusion}]{\text{1. aq. KOH soln.; dried}} ArSH$

(f) (i) $RSO_3H \xrightarrow{PCl_5} RSO_2Cl \xrightarrow{H_2O} RSO_3H$

(ii) $ArSO_3H \xrightarrow{PCl_5} ArSO_2Cl \xrightarrow{NH_3} ArSO_2NH_2 \xrightarrow{ROH/NaOH} ArSO_3R$

(g) $ArSO_3H + PhH \xrightarrow{\text{heat}} ArSO_2Ph + H_2O$

(h) $PhSO_3H \xrightarrow[200\,°C]{\text{fuming } H_2SO_4}$ 1, 3-benzenedisulphonic acid

(i) $PhSO_3H \xrightarrow[\text{heat}]{P_2O_5} PhSO_2—O—SO_2Ph$, sulphonic acid anhydride

(j) (k)

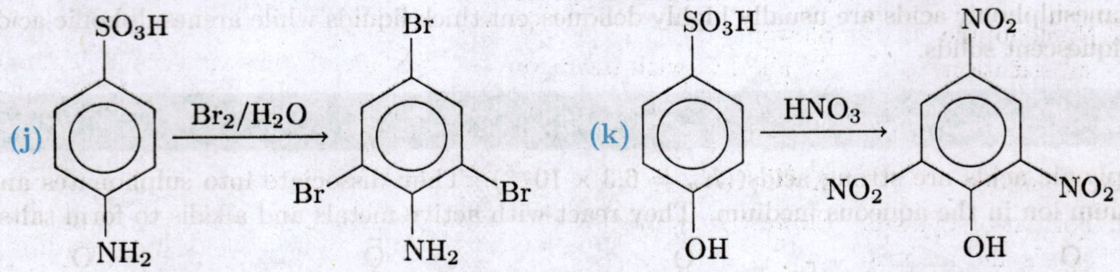


**Chart 20.30** Reactions of sulphonic acids.

## 20.11.5 Analysis

It has already been stated that sulphonic acids are soluble in water and water solutions :

(i) turn blue litmus paper to red,

(ii) evolve $CO_2$ gas from carbonates and bicarbonates. Most of the sulphonic acids on boiling with 50% $H_2SO_4$ give the parent hydrocarbon.

However, individual sulphonic acids may be identified by converting one to the corresponding sulphonamide or sulphanilide and then determining its melting point.

**Problem 20.26.** Draw a comparison between the chemistry of sulphonic acids and that of carboxylic acids.

**Solution :** In sulphonic acids the S atom has four groups joined to it while in carboxylic acids the carboxyl C has three groups. $RSO_2$— or $ArSO_2$— group has stronger –I and –R effects than RCO— or ArCO— group. In the case of nucleophilic substitution the carboxyl C has enough space for nucleophilic attack while the sulphonyl group experiences steric hindrance when nucleophilic attack occurs on the S atom since this involves bipyramidal sulphur while in the case of RCOOH the nucleophilic substitution involves tetrahedral arrangement. In the case of a carboxylic acid ester usually RCO—O (ArCO—O) bond breaks whereas in the case of a sulphonic acid ester O—R′ bond breaks. However, the following points of differences may be noted.

| Compound | Sulphonic acids R- or Ar-$SO_3H$ | Carboxylic acids R- or Ar-$CO_2H$ |
|---|---|---|
| 1. Acidity | Stronger | Weaker |
| 2. Formation of derivative. | Indirect via sulphonyl chloride | Direct and indirect |
| a. Esters | R- or Ar-$SO_3R'$ | R- or Ar-$CO_2R'$ |
| i. Preparation | $ArSO_2Cl \xrightarrow[ArOH]{ROH} ArSO_3R(Ar)$ | $RCOOH \xrightarrow[R'OH]{H_2SO_4} RCOOR'$<br>$RCOCl \xrightarrow{R'OH} RCOOR'$ |
| ii. Hydrolysis | Slower<br>$ArSO_3R + H^{18}OH$<br>↓<br>$ArSO_3H + R^{18}OH$<br>R—O bond cleavage occurs | Faster<br>$RCOOR' + H_2{}^{18}O$<br>↓<br>$RCO^{18}OH + R'OH$<br>RCO—OR′ bond cleavage occurs |
| iii. Nucleophilic substitution | Slower<br>At alkyl C with inversion | Faster<br>At acyl C with retention of configuration or racemisation of the alkyl group. |
| b. Acyl halides | $ArSO_2Cl$ | RCOCl |
| i. Formation of acids, esters and amides | Slower. Requires base | Faster<br>With water; ArCOCl requires base. |
| c. Amides | $ArSO_2NH_2$ | $RCONH_2$ |
| i. Hydrolysis | Slow, acid-catalysed | Rapid, acid- or base-catalysed. |
| ii. Formation of acid halide. | Slower | Faster. |

## 20.12 Nitriles (Cyanides)

### 20.12.1 Introduction

Cyano derivatives of alkanes and arenes are called alkanenitriles (alkyl cyanides) and arenenitriles (aryl cyanides) respectively. They are isomeric with isonitriles. However, nitriles may be designated as RCN or ArCN where R and Ar represent alkyl and aryl groups respectively.

## 20.12.2 General methods of formation

The methods are given in Chart 20.31.

(a) $RCONH_2(ArCONH_2) \xrightarrow[\text{dehydration}]{P_2O_5/\text{heat}} RCN\ (ArCN)$

(b) $RCO_2H\ (ArCO_2H) + NH_3 \xrightarrow[\text{industrial method}]{Al_2O_3/500\,°C} RCN\ (ArCN)$

(c) $RCO_2H\ (ArCO_2H) + NH_3 \xrightarrow[\text{industrial method}]{O_2/\text{cat.}500\,°C} RCN\ (ArCN)$

(d) $RCH{=}NOH \xrightarrow[\text{dehydration}]{P_2O_5 \text{ or } Ac_2O} RCN\ (ArCN)$
(ArCH=NOH)

(e) $RX + KCN \xrightarrow[\text{medium}]{\text{aqueous ethanolic}} RCN + RNC\ (\text{trace})$

(f) $ROSO_2K + KCN \xrightarrow{\text{warmed}} RCN + RNC\ (\text{trace})$

(g) 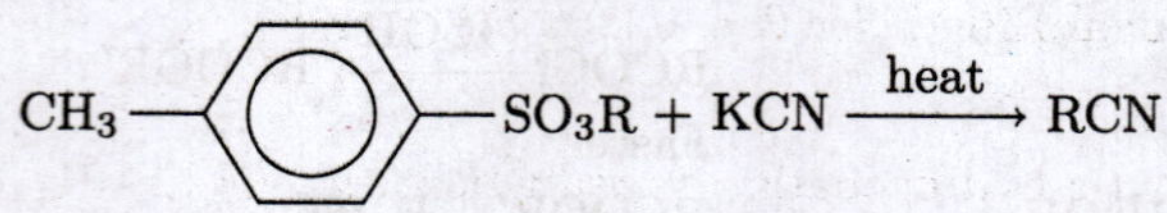

alkyl *p*-tosylate

(h) $RMgCl\ (ArMgBr) + ClCN \longrightarrow RCN\ (ArCN)$

(i) $RNC \xrightarrow{\text{prolonged heating}} RCN$

**Chart 20.31** Formation of nitriles.

## 20.12.3 Structure, physical properties and reactivity

The C atom of the cyano group is bonded to an $sp^3$C or an $sp^2$C according as the nitrile is alkanenitrile ot arenenitrile. The cyano C and N atoms are in *sp* hybridised state and the lone

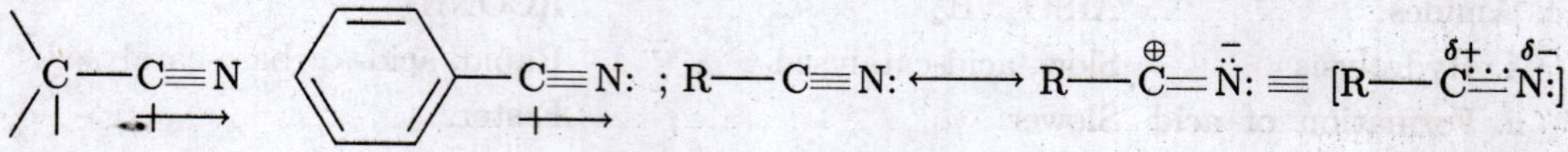

pair of *e*s occupy strongly-bound *sp* A.O. Alkanenitriles and Arenenitriles have high dipole moment and naturally they have a strong polar character, e.g., the $\mu$ of propanenitrile is 4.0D.

Lower members of the nitriles are liquid. They have pleasant odour and are moderately toxic. However, $\alpha$-hydroxy and $\alpha$-aminonitriles are highly toxic; perhaps because of easy loss of hydrogen cyanide make them toxic. Owing to the high dipole moment they melt and boil at higher temperatures than alkanes and ethers of comparable molecular weights; e.g., propane ($\mu = 0.0$D): $-42\,°C$ b.p.; methyl ether ($\mu = 1.3$D): $-24\,°C$ b.p. methyl cyanide ($\mu = 4.0$D): $82\,°C$ b.p. Since nitriles have no scope to form intermolecular H-bonding, they have lower b.p.s and m.p.s than the corresponding carboxylic acids. However, as the molecular weight increases among the nitriles, their b.p.s and m.p.s increase because of the increasing London force. Since the N atom of a nitrile bears a lone pair of *e*s, it is able to form H-bond with water and thus cyanides of low molecular weights are soluble in water. However, they are less soluble than the corresponding amines; since the N atom is *sp* hybridised and since the lone pair of *e*s occupy an *sp* A.O., the donatability of *e*s is very small and hence effective H-bonding with water is not possible. For the same reason the nitriles are very weak bases and as a consequence they do not dissolve appreciably in dilute aq. acids. Nitriles being weak bases, their conjugate acids, $R{-}C{\equiv}\overset{\oplus}{N}H$ or $Ar{-}C{\equiv}\overset{\oplus}{N}H$, are reasonably strong.

Since the cyano C and N atoms have electrophilic and nucleophilic characters respectively, they undergo addition reactions. Strong $-$I effect of the cyano group also makes $\alpha$-H atoms slightly acidic. Increasing +I effect of the alkyl groups, however, decreases the electrophilicity of the C atom while the same effect increases the nucleophilicity of the N atom. Since nitriles are electrophiles as well as nucleophiles, both acids and bases can catalyse addition reactions.

## 20.12.4 Chemical reactions

These are given in Chart 20.32.

**(a) Hydrolysis of alkyl and aryl cyanides :** See Sec. 20.9.2 (d), p. 1288.

**(b) The Stephen's reaction :** See p. 1243.

**(c) Reduction of cyanides :** Catalytic hydrogenation or reduction with (i) $LiAlH_4$/ether, or (ii) Na/EtOH, etc, reduces nitriles to 1° amines.

$$R{-}C{\equiv}N: \xrightarrow[\substack{Na/EtOH \text{ or} \\ H_2 \text{ and cat.(Pd/Pt or Ni)}}]{LiAlH_4 \text{ or}} RCH_2NH_2$$

**(d) Formation of esters :** Nitriles combine with alcohols in the presence of acids to produce the salts of imidoesters which on hydrolysis give esters.

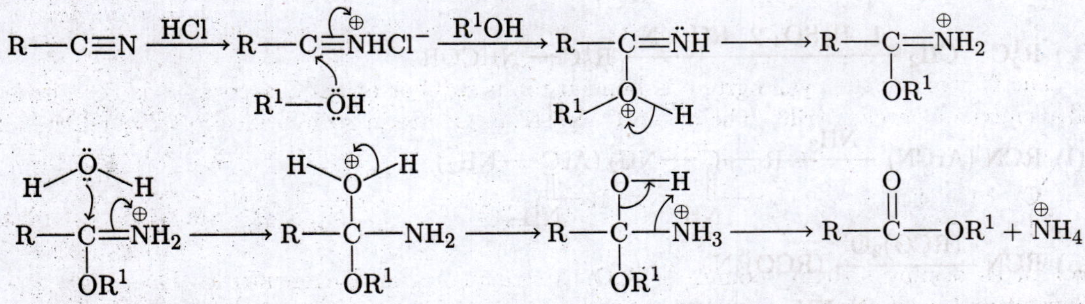

**(e) Formation of ketones :** Grignard reagents add to nitrites and the addition products on hydrolysis give ketones.

$$R{-}C{\equiv}N \xrightarrow[\text{ether}]{R^1MgX} R{-}\underset{R^1}{\underset{|}{C}}{=}NMgX \xrightarrow{H_2O} R{-}\underset{R^1}{\underset{|}{C}}{=}O$$

(a) (i) $R{-}C{\equiv}N\ (ArCN) \xrightarrow[\text{conc. acid}]{\text{moderatety}} RCO_2H\ (ArCO_2H) + \overset{\oplus}{N}H_4$

(ii) $R{-}C{\equiv}N\ (ArCN) \xrightarrow{\bar{O}H} RCO\bar{O}\ (ArCOO^-) + NH_3$

(b) $R{-}C{\equiv}N\ (ArCN) \xrightarrow[{[H]}]{SnCl_2/HCl} RCHO\ (ArCHO)$

(c) (i) $R{-}C{\equiv}N\ (ArCN) \xrightarrow[Na/EtOH]{LiAlH_4 \text{ or}} RCH_2NH_2\ (ArCH_2NH_2)$

(ii) $RCN \xrightarrow{\text{catalytic reduction}} RCH_2NH_2$

(d) $RCN\ (ArCN) + R^1OH \xrightarrow[2.\ H_2O]{1.\ \text{conc. } H_2SO_4 \text{ or } HCl} RCO_2R^1(ArCO_2R^1) + \overset{\oplus}{N}H_4$

(e) $RCN\ (ArCN) + R^1MgX \longrightarrow R{-}\underset{\large R^1}{\underset{|}{C}}{=}NMgX(Ar\underset{\large R^1}{\underset{|}{C}}{=}NMgX) \xrightarrow{H_3O^{\oplus}} R{-}\underset{\large R'}{\underset{|}{C}}{=}O(ArCOR^1)$

(f) $RCH_2CN + RCH_2CN \xrightarrow[\text{Thorpe nitrile condensation}]{Na/\text{ether or } NaOEt} RCH_2{-}\underset{\large NH}{\underset{\|}{C}}{-}\underset{\large R}{\underset{|}{C}HCN}$

(g) $ArH + RCN \xrightarrow[2.\ H_2O]{1.\ HCl} Ar{-}COR$

(h) $CH_2{=}CH{-}CN \xrightarrow[Z\ =\ RO^-,\ RS^-,\ RCOO^-,\ RNH^-,\ \text{etc.}]{HZ \text{ and base}} ZCH_2CH_2CN$

(i) $RCN\ (ArCN) \xrightarrow{H_2O_2/\bar{O}H} RCONH_2\ (ArCONH_2)$

(j) $RCH_2CN \xrightarrow{1.\ NaNH_2\ 2.\ R^1X} \underset{\large R^1}{\underset{|}{RCHCN}}$

(k) $R^1_2C{=}CH_2 \xrightarrow{1.\ H_2SO_4\ 2.\ RC{\equiv}N} R^1_2\underset{\large CH_3}{\underset{|}{C}}{-}NHCOR$

(l) $RCN\ (ArCN) \xrightarrow{NH_3} R{-}\underset{\large NH}{\underset{\|}{C}}{-}NH_2\ (Ar\underset{\large NH}{\underset{\|}{C}}{-}NH_2)$

(m) $RCN \xrightarrow{(RCO)_2O} (RCO)_3N$

(n) $RCH_2CN \xrightarrow[2.\ R^1CO_2Et]{1.\ NaNH_2} R^1CO\underset{\large R}{\underset{|}{C}HCN} + EtOH$

**Chart 20.32** Reactions of nitriles.

**(f) The Thorpe nitrile condensation :** In the presence of a base (NaOEt; EtOH) or metallic sodium and ether two molecules of a nitrile condense to form an iminonitrile in which the $\alpha$-C of one molecule is involved and the reaction is known as the Thorpe nitrile condensation.

The reaction is analogous to the aldol condensation reaction. The iminonitrile on treatment with water gives a $\beta$-ketonitrile.

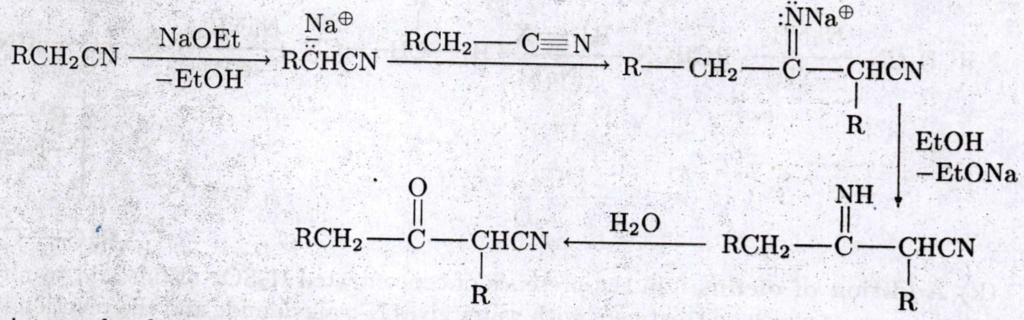

The intramolecular or internal Thorpe condensation is also known and the reaction is said to be the Thorpe-Ziegler reaction. However, the product of this extended reaction is not an iminonitrile, but the tautomeric enaminonitrile.

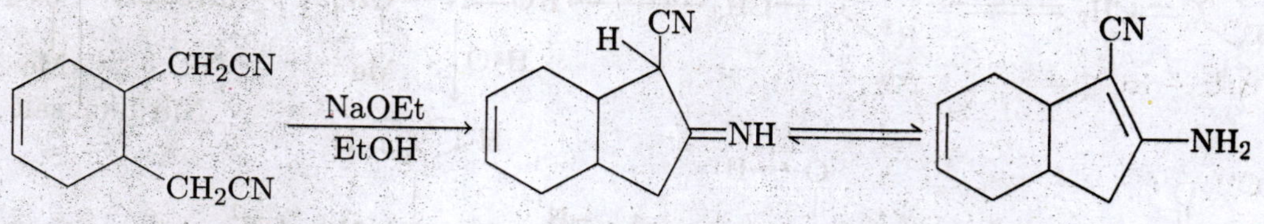

The Thorpe nitrile synthesis gives a good method of preparing $\beta$-ketonitrile

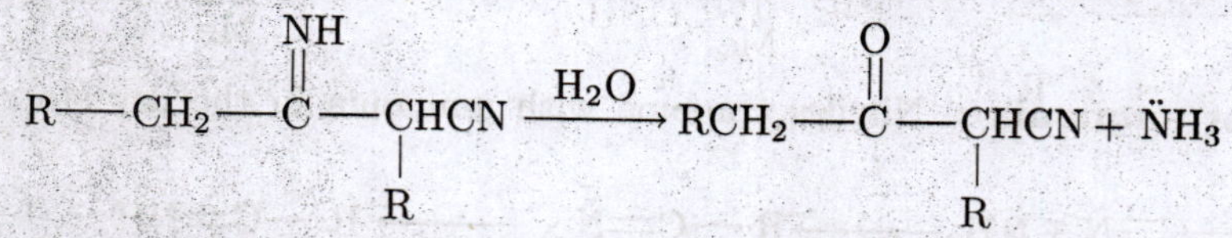

The Thorpe-Ziegler reaction is a useful method for closing large rings. Five to eight-membered rings and rings containing fourteen or more carbons can be closed successfully with high yields provided high dilution techniques are used.

**(g) The Houben-Hoesch reaction :** See Sec. 17.9.2, p. 986.

**(h) The Cyanoethylation :** See Sec. 13.13.1, p. 663.

**(i) Hydrolysis of a nitrile to an amide :** A 1° nitrile can be hydrolysed to the amide, stage very safely with alkaline hydrogen peroxide. This is because the rate of nucleophilic attack on the cyano C by hydroperoxide ion is almost ten thousand times faster than that on the amido C by the hydroxide ion. As a consequence the amide forms much faster than it gets hydrolysed to the corresponding acid ($\alpha$-effect; p. 533).

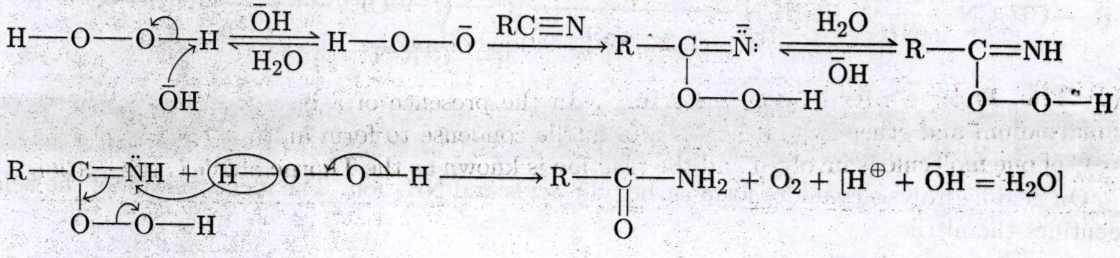

**(j) Alkylation :** When a 1° or 2° nitrile is heated with alkyl halide and sodamide, α-H atoms get replaced by alkyl groups.

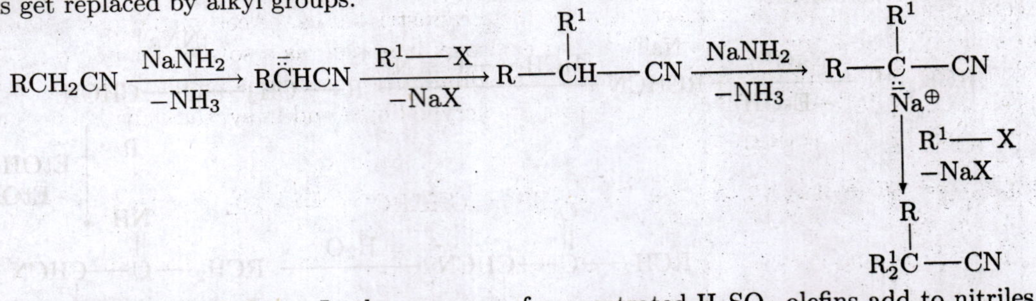

**(k) Addition of olefins :** In the presence of concentrated $H_2SO_4$ olefins add to nitriles to give nitrilium salts which on treatment with water gives N-t-alkylamide and the reaction is called the Ritter reaction.

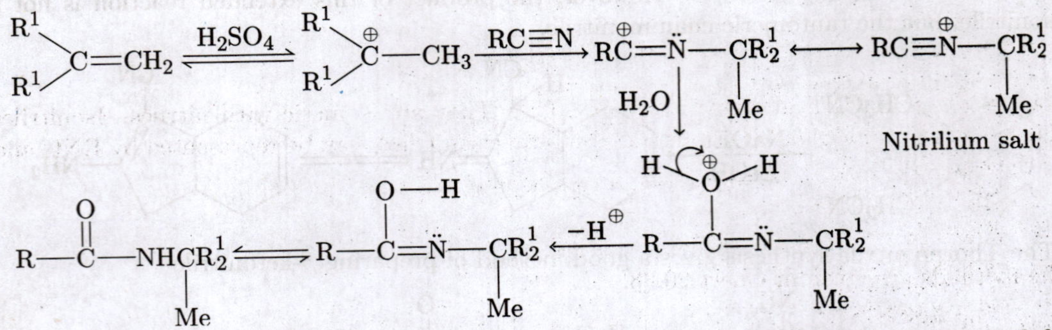

**(l) Addition of ammonia :** Nitriles combine with ammonia in the dry state and form amidines.

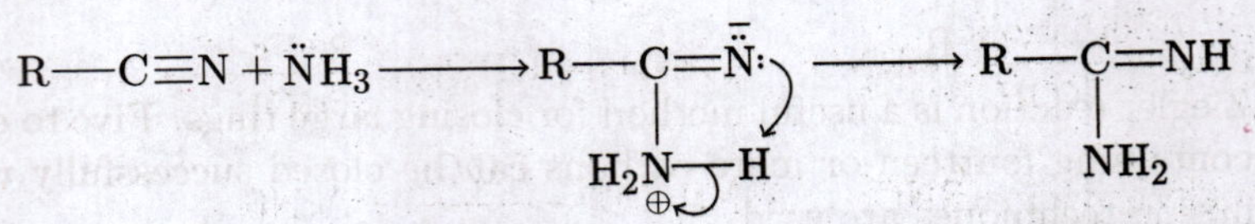

**(m) Reaction with acid anhydrides :** When nitriles are heated with acid anhydrides, tertiary acid amides are formed.

$$R—C\equiv N + (RCO)_2O \longrightarrow (RCO)_3N:$$

**(n) Formation of β-ketonitriles :** 1° or 2° nitriles on treatment with sodamide and an ester give β-ketonitriles. This is also a condensation reaction.

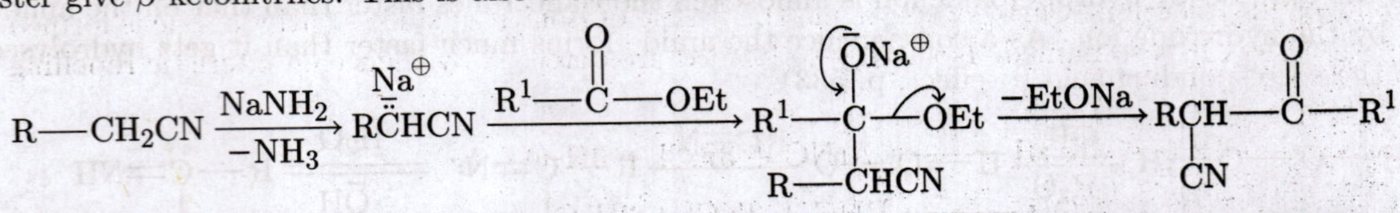

## 20.12.5 Analysis

On acid hydrolysis cyanides form carboxylic acids and $\overset{\oplus}{N}H_4$ ion. The identification of the acid identifies the nitrile.

### 20.12.6 Uses

Acetonitrile, adiponitrile and acrylonitrile have industrial uses. Acetonitrile has extensive use in separating and purifying butadiene and isoprene; it is used as a solvent for polymers and for extraction, crystallisation, and as a reaction medium. Adiponitrile is an intermediate in the synthesis of nylon. Acrylonitrile is used to prepare acrylic fibres and in synthesising rubbers and plastics. This is manufactured as follows :

$$O_2 + CH_2{=}CH{-}CH_3 \xrightarrow[400\,°C]{MnO_3\text{-}Bi_2O_3} CH_2{=}CHCHO \xrightarrow{NH_3} CH_2{=}CHCH{=}NH$$

$$CH_2{=}CHCH{=}NH \xrightarrow{O_2,\ MnO_3\text{-}Bi_2O_3} CH_2{=}CHCN$$

## 20.13 Isonitriles

### 20.13.1 Introduction

Isonitriles contain isocyano group ($-\overset{\oplus}{N}{\equiv}\bar{C}{:}$). They are isomeric with nitriles. Isonitriles are also called isocyanides; alkaneisonitriles and areneisonitriles may be represented by RNC and ArNC respectively.

### 20.13.2 Formation of isonitriles

The methods are given in Chart 20.33.

(a) $RI \xrightarrow[\text{(aq. ethanolic)}]{AgCN} RNC + RCN\ \text{(trace)}$

(b) $RNH_2 + CHCl_3 \xrightarrow[KOH]{\text{ethanolic}} RNC;\quad ArNH_2 + CHCl_3 \xrightarrow[KOH]{\text{ethanolic}} ArNC$

(c) $RNHCHO \xrightarrow[\text{dehydration}]{POCl_3} RNC;\quad ArNHCHO \xrightarrow{POCl_3} ArNC$

**Chart 20.33** Formation of nitriles.

**(a) By the reaction between alkyl iodide and silver cyanide :** See 11.5.1.d, p. 536.

**(b) By the carbylamine reaction :** When a 1° amine, aliphatic or aromatic, is treated with ethanolic solution of caustic potash, an isonitrile forms and the reaction is popularly known as carbylamine reaction. Carbylamine is another name of isonitrile; since a carbylamine is formed in the reaction, it is so named. Isonitriles so formed are detected by their characteristic repelling odour.

$$ArNH_2 + CHCl_3 + 3KOH \longrightarrow ArNC + 3KCl + 3H_2O$$

$$RNH_2 + CHCl_3 + 3KOH \longrightarrow RNC + 3KCl + 3H_2O$$

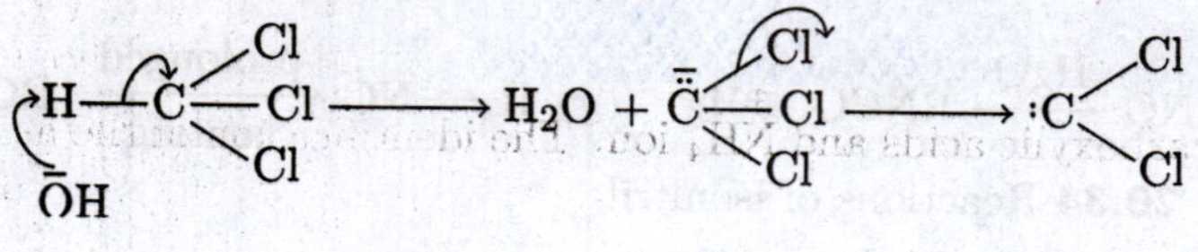

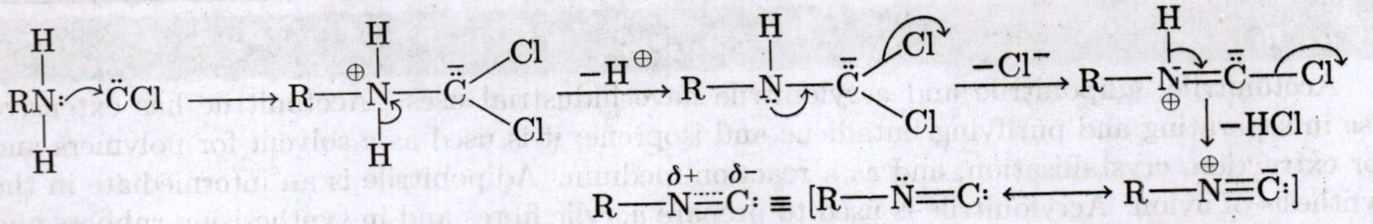

**(c) From N-alkyl formamide :** When N-alkyl formamides are dehydrated by phosphoryl chloride in the presence of pyridine alkyl isocyanides are formed.

$$RNHCHO \xrightarrow{POCl_3} RNC$$

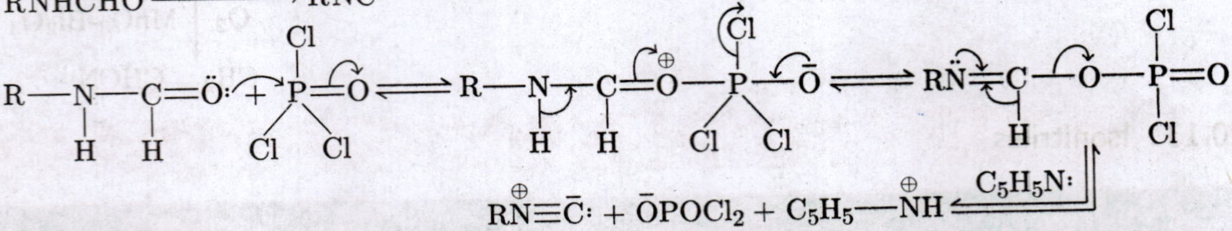

## 20.13.3 Structure, reactivity and physical properties

The structure of an isonitrile is as follows :

Since there are two $\pi$ orbitals, they can undergo addition reactions. The C atom of the isocyano group bears a negative charge while the more electronegative N atom holds a positive charge. For this reason the dipole moments of the isonitriles are less than our expectation and it is of the order ~3D while those of isomeric nitriles are of the order ~4D. However, since the less electronegative element carries the negative charge, the compound is appreciably basic and gets dissolved in dilute acids and ultimately gets hydrolysed. Since there is no active H and since the more electronegative element N is positively charged, isonitriles cannot form H-bond with water and hence the lower members are not much soluble in water. Since isonitriles have lower dipole moments than the isomeric nitriles, the former compounds boil at lower temperatures than the latter compounds; the b.p. of MeCN is 81°C but that of MeNC is 60 °C, and EtCN and EtNC boil at 97°C and 78°C respectively. Alkaneisonitriles are very poisonous and obnoxious smelling liquids; their smell can be detected if they are present even in minute amounts. Areneisonitriles also have a very disagreeable smell.

## 20.13.4 Chemical reactions

The general reactions are given in Chart 20.34.

(a) $R{-}NC \xrightarrow{MCN,\ 100\,°C} RCN + MNC$ (b) $R{-}NC \xrightarrow{H_3O^{\oplus}} R\overset{\oplus}{N}H_3 + HCOOH$

(c) $R{-}NC \xrightarrow[Na/EtOH]{H_2/Pt} RNHCH_3$ (d) $R{-}NC \xrightarrow{X_2} RN{=}CX_2$

(e) $R{-}NC \xrightarrow{S} RNCS$ (f) $RNC \xrightarrow{HgO} RNCO + Hg$ (g) $R{-}NC \xrightarrow[heating]{prolonged} RCN$

**Chart 20.34** Reactions of isonitriles.

**(a) Reaction with metal cyanides :** When isonitriles are heated with metal cyanides at 100 °C the more stable nitriles are formed through nucleophilic displacement.

$$R-\overset{\oplus}{N}\equiv\bar{C} + \overset{\oplus}{M}C\bar{N} \longrightarrow RCN + MNC$$

**(b) Hydrolysis of isonitriles :** Isonitriles get hydrolysed by dilute acids readily to form 1° amines and formic acid.

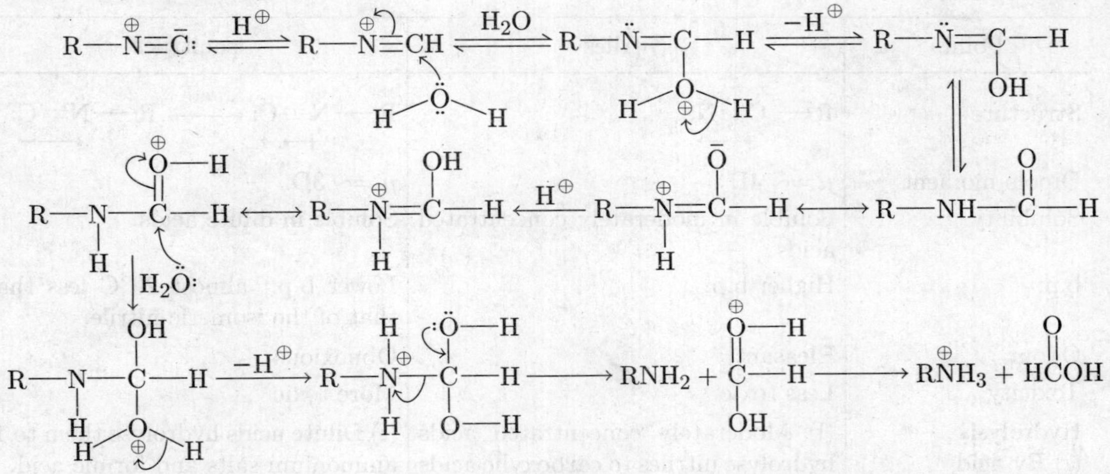

However, isonitriles do not get hydrolysed by alkalis. This is because the C atom bears a negative charge which cannot coordinate with $\bar{O}H$ ion; while the positively charged N atom being tetracovalent, its outer shell is saturated and as a consequence it has no capacity to accommodate two more electrons in that shell.

**(c) Reduction :** Isonitriles are reduced catalytically to 2° methylamines. The reduction may also be carried out by dissolving metals.

$$R-\overset{\oplus}{N}\equiv\bar{C}: \xrightarrow{H_2/Pt} RNHCH_3$$

**(d) Addition of halogen :** Isonitriles add on halogen to form alkyl and aryliminocarbonyl halides.

$$R-\overset{\oplus}{N}\equiv\bar{C}: + X_2 \longrightarrow R-N=CX_2$$

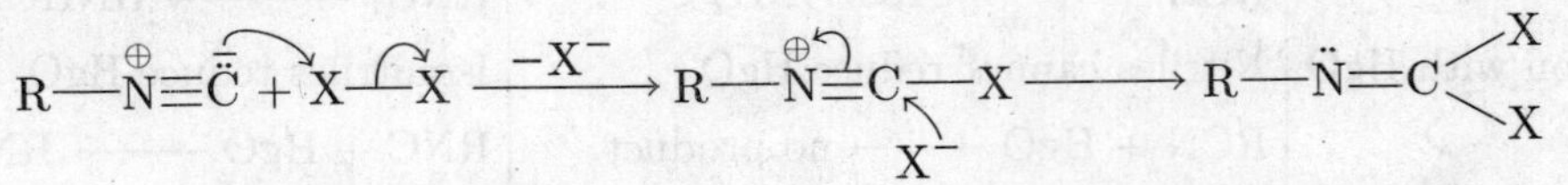

**(e) Addition of sulphur :** Isonitriles react with sulphur to form isothiocyanates.

$$R-\overset{\oplus}{N}\equiv\bar{C}: + S \longrightarrow R-\ddot{N}=C=S$$

**(f) Reaction with mercuric oxide :** When isonitriles are treated with mercuric oxide, isocyanates are formed and mercuric oxide is reduced to metallic mercury.

$$R-NC + HgO \longrightarrow R-N=C=O + Hg$$

**(g) Action of heat :** When isonitriles are heated strongly for a long time, they rearrange to the more stable cyanides.

$$RNC \xrightarrow[\text{heat}]{\text{prolonged}} RCN$$

## 20.13.5 Test

Isonitriles are easily detected by their characteristic disagreeable odour. They liberate metallic mercury when they are heated with mercuric oxide.

**Problem 20.27.** Distinguish between nitriles and isonitriles.

**Solution :**

| Points | Nitriles | isonitriles |
|---|---|---|
| Structure | $R{-}C{\equiv}N{:}$ (dipole arrow $\longleftrightarrow$ toward N) | $R{-}\overset{\oplus}{N}{\equiv}\bar{C}{:} \longleftrightarrow R{-}\ddot{N}{=}C{:}$ |
| Dipole moment | $\mu =\sim 4D$ | $\mu =\sim 3D$. |
| Solubility | Soluble in moderately concentrated acids. | Soluble in dilute acids. |
| b.p. | Higher b.p. | Lower b.p.; almost 20°C less than that of the isomeric nitrile. |
| Odour | Pleasant | Obnoxious |
| Toxicity | Less toxic | More toxic |
| Hydrolysis (i) By acid | (i) Moderately concentrated acids hydrolyse nitriles to carboxylic acids and ammonium ion. $RCN \xrightarrow{H_3O^{\oplus}} RCOOH + \overset{\oplus}{N}H_4$ | (i) Dilute acids hydrolyse them to 1° ammonium salts and formic acid. $RNC \xrightarrow{H_3O^{\oplus}} R\overset{\oplus}{N}H_3 + HCOOH$ |
| (ii) By alkali | Aqueous alkali hydrolyses it to a carboxylate ion and ammonia. $RCN \xrightarrow[H_2O]{\bar{O}H} RCOO^- + NH_3 \uparrow$ | Alkalis fail to hydrolyse isonitriles. $RNC \xrightarrow{\bar{O}H} \times$ |
| Reduction | On catalytic reduction nitriles form 1° amines $RCN \xrightarrow{Pt/H_2} RCH_2NH_2$ | On catalytic reduction isonitriles form methylamine. $RNC \xrightarrow{Pt/H_2} RNHCH_3$ |
| Reaction with HgO | Nitriles cannot reduce HgO $RCN + HgO \longrightarrow$ no product. | Isonitriles reduce HgO $RNC + HgO \longrightarrow RNCO + Hg$. |

**Problem 20.28.** How can you remove (i) a trace of RNC from RCN and (ii) a trace of RCN from RNC.

**Solution : (i)** RNC is soluble and gets hydrolysed in very dilute acids, but RCN hydrolyses in moderately concentrated acids. So when RCN contaminated with a trace of RNC is treated with a dilute acid, RNC hydrolyses but RCN does not and the latter can be distilled out from the solution; RCN can be extracted with ether from the distillate.

**(ii)** RCN gets hydrolysed by a base but RNC does not. So when RNC contaminated with RCN is treated with aqueous caustic soda solution, RCN dissolves out. RNC may be recovered by distillation and ethereal extraction of the distillate.

## 20.14 Nitro Compounds

### 20.14.1 Introduction

We shall discuss here the general chemistry of nitroalkanes and nitroarenes. All of these compounds contain nitro $\left( -\overset{\oplus}{N}(=O)-O^{-} \right)$ group. Nitroalkanes and nitroarenes are represented by the general formulae R—$NO_2$ and Ar—$NO_2$ respectively. Nitro compounds are isomeric with alkyl and aryl nitrites which contain O—N=O group.

### 20.14.2 The general methods of formation

The methods are given in Chart 20.35.

(a) $RH \xrightarrow[\text{or, vapour phase nitration at 150-400 °C.}]{\text{liquid phase nitration at 140 °C}} \text{mixture of nitroalkanes}$

(b) $RX \xrightarrow[AgNO_2/DMF \text{ or, } AgNO_2/DMSO]{\text{aq. ethanolic } AgNO_2 \text{ or}} R-\overset{\oplus}{N}(=O)-O^{-} + R-O-N=O$

(c) $ClCH_2CO_2Na \xrightarrow[\text{solution of } NaNO_2]{\text{boiled in an aqueous}} CH_3NO_2$

(d) $R_3C-NH_2 \xrightarrow{\text{oxidation by } KMnO_4} R_3C-NO_2$ (3° nitroalkane)

(e) $RCH=NOH \xrightarrow{CF_3COOOH} RCH_2NO_2$ ($R_2CH-NO_2$)
($R_2C=NOH$)

(f) $\text{Arene} \xrightarrow[\text{or, fuming } HNO_3]{\text{conc. } HNO_3/\text{conc. } H_2SO_4} \text{nitroarene}$

(g) $Ar-NH_2 \xrightarrow[2.\ NaNO_2/Cu]{1.\ NaNO_2/HBF_4} Ar-NO_2$

**Chart 20.35** Formation of nitro compounds.

(a) **By Nitration of alkanes :** See Sec. 18.1.6.c, p. 1025.

(b) **From alkyl halides :** See Sec. 20.1.4.xx, p. 1160.

(c) **From sodium chloroacetate :** When an aqueous solution of sodium chloroacetate is boiled with an aqueous solution of sodium nitrite, nitromethane forms through $S_N$ reaction, decarboxylation and abstraction of proton from water as shown below :

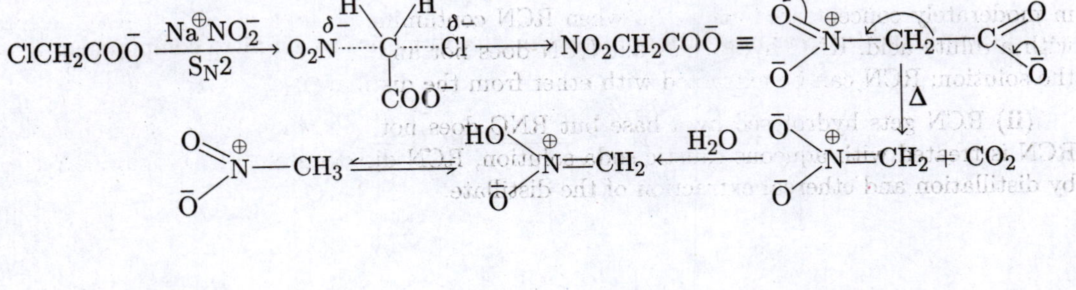

**(d) By the oxidation of 3° alkylamines :** When tert-alkyl amines are oxidised by potassium permanganate, tert-nitroalkanes are formed.

$$Me_3C—NH_2 \xrightarrow{KMnO_4} Me_3C—NO_2$$

**(e) By the oxidation of oximes :** When aldoximes and ketoximes are oxidised by trifluoroperacetic acid, 1° and 2° nitro compounds respectively are formed.

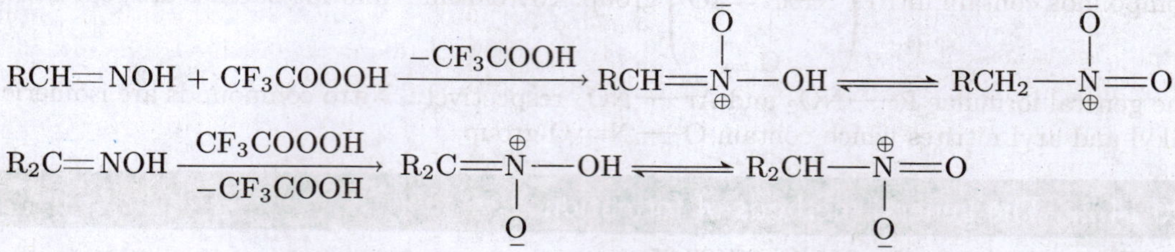

**(f) By nitration of aromatic compounds :** See Sec. 17.6, p. 955.

**(g) From arenediazonium compounds :** When aromatic primary amines are diazotised with sodium nitrite and hydrogen borofluoride and the arenediazonium salts so formed are treated with copper powder and sodium nitrite, nitroarenes are formed.

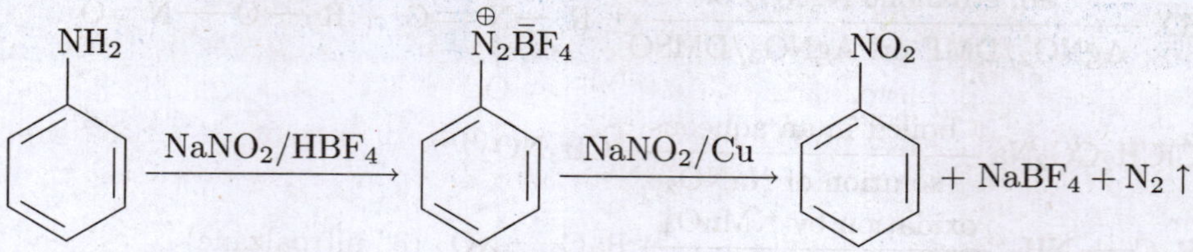

## 20.14.3 Structure, reactivity and physical properties

Nitro compounds bear a C—N bond and the two O atoms are bonded to the N atom; the N atom is supposed to be $sp^2$ hybridised. The nitro group is a resonance hybrid of two equivalent resonating structures as shown below. One of $sp^2$ A.O.s of N atom holds a lone pair of es which coordinates with an O atom. 1-electron $p$ A.O. of N forms a $\pi$ bond with the other O atom.

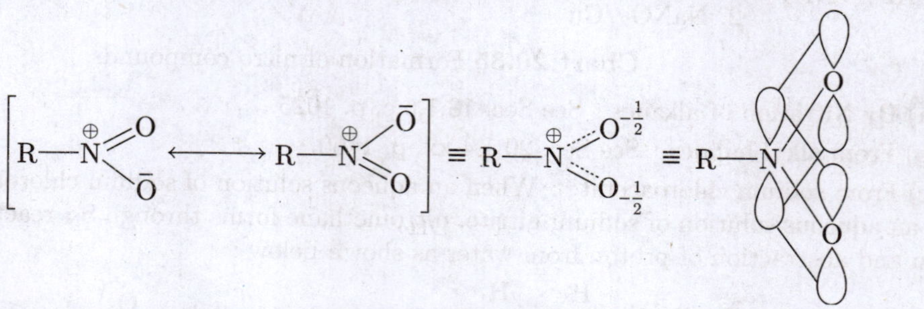

In the liquid state or in solution nitroalkanes tautomerise to aci-nitro compounds.

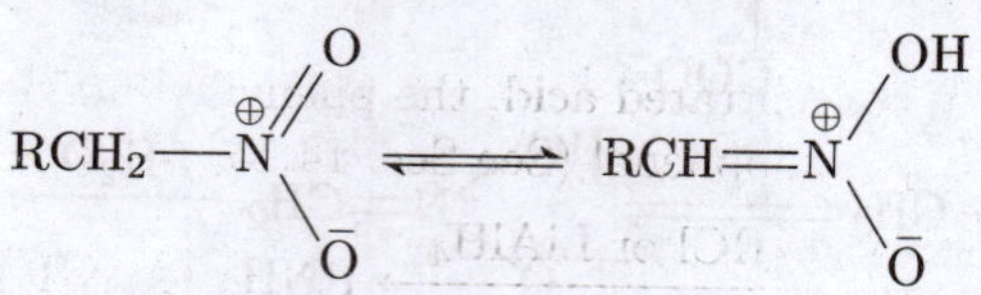

The nitro groups has −I and −R effects. Because of these effects, the $\alpha$-H atoms are acidic and can be abstracted by a base to form a carbanion. The carbanion then can undergo condensation reactions. Nitro group being −I and −R groups, it deactivates an aromatic ring and directs an incoming electrophile to the *m*-position. Since a nitro group deactivates an aromatic ring, aromatic nitro compounds undergo nucleophilic substitution reactions. In some reactions the nitro group gets replaced by other nucleophiles and in some other reactions an *o*- or a *p*-group other than nitro gets displaced in the $S_N$ reaction. The nitro group can be reduced to a number of other functions, e.g., —$NH_2$, —NHOH, etc.; different reducing agents have different effects on the nitro group and so different products of reduction are obtained using different reagents.

The strong −I effect of the O atoms and the semipolar character of the nitro group develop a high dipole moment in nitroalkanes and in nitroarenes ($\mu$ nitroalkane =~ 3.6D, $\mu$ nitrobenzene = 3.98D). For this high dipole moment the nitro compounds have high b.p.s., e.g., b.p. of nitromethane : 101 °C and that of nitroethane : 115 °C. Since two O atoms of the nitro group bear a partial negative charge and since they can form H-bond with water, nitroalkanes are soluble in water. Of course the solubility decreases with increasing molecular weight. However, nitroarenes are almost insoluble in water. Nitroalkanes and nitroarenes have high dielectric constant ($\varepsilon = 25 - 35$); so they are good solvents for polar compounds.

## 20.14.4 Chemical reactions

The reactions are given in Chart 20.36.

**(a) Reduction by metal and acid or lithiumaluminium hydride :** Both aliphatic and aromatic nitro compounds can be reduced to amines by Zn or Sn or Fe and HCl or by $LiAlH_4$ in ether.

$$RCH_2NO_2 \xrightarrow[\text{HCl}]{\text{Zn or Sn or Fe}} RCH_2NH_2$$

The metal and acid reduction of aromatic nitro compounds perhaps follows the free radical pathway as shown below :

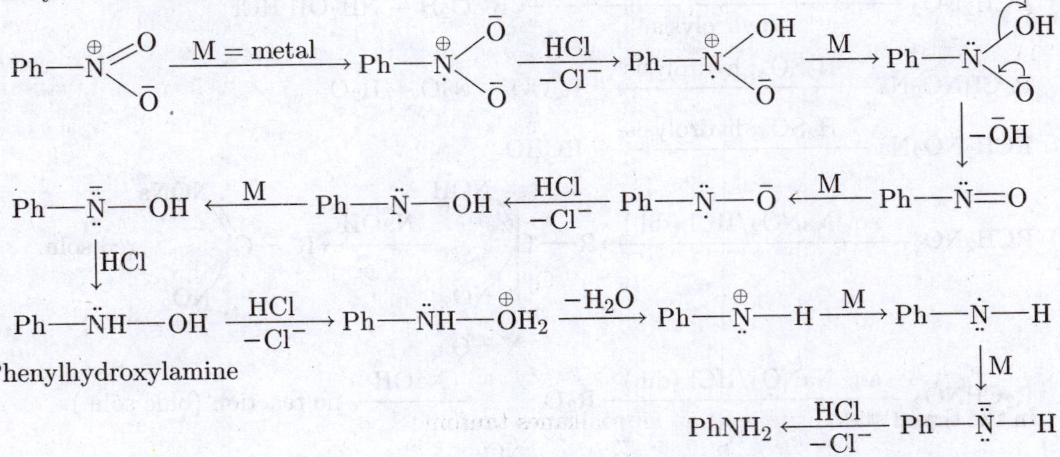

However, in the presence of concentrated acid, the phenylhydroxylamine, which is formed as an intermediate, rearranges to *p*-aminophenol (See Sec. 14.15, p. 747).

(a) $R\text{—}NO_2\ (ArNO_2) \xrightarrow{\text{Sn or Fe/HCl or LiAlH}_4} RNH_2\ (ArNH_2)$

**(b)** $RCH_2NO_2 \xrightarrow{SnCl_2/HCl} RCH_2NHOH + RCH{=}NOH$

$RCH{=}N\bar{O}_2Na^{\oplus} \xrightarrow{SnCl_2/HCl} RCH{=}NOH$

$RR'C{=}N\bar{O}_2Na^{\oplus} \xrightarrow{SnCl_2/HCl} RR'C{=}NOH$

$ArNO_2 \xrightarrow{SnCl_2 \text{ or } TiCl_2/HCl} ArNH_2$ (selective)

**(c)** $R{-}NO_2(ArNO_2) \xrightarrow{Zn/NH_4Cl} RNHOH\ (ArNHOH)$

**(d)** $ArNO_2\ (RNO_2) \xrightarrow{Pd/Pt \text{ or Raney Ni}/H_2} ArNH_2(R{-}NH_2)$

**(e)**

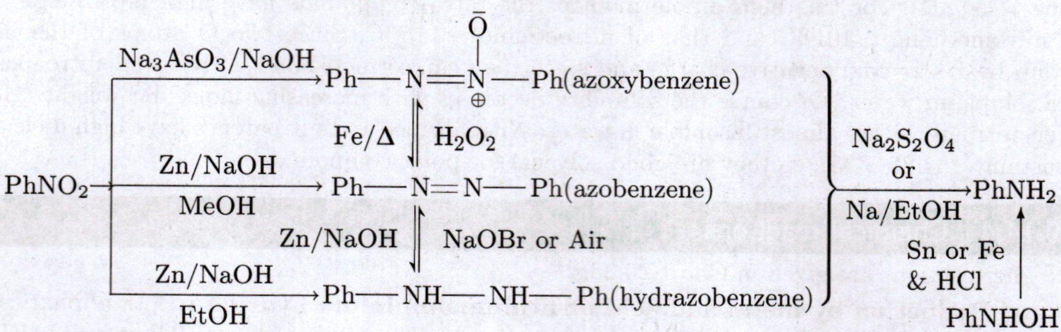

**(f)** $NO_2{-}Ar{-}NO_2 \xrightarrow{(NH_4)_2S \text{ or } Na_2S \text{ or } NaHS, \text{ etc.}} NH_2ArNO_2$

**(g)** $R{-}NO_2(1° \ \&\ 2°) \xrightarrow{NaOH}$ salt

**(h)** $RCH_2NO_2 \xrightarrow[\text{hydrolysis}]{\text{boiling HCl or 85\% } H_2SO_4} RCO_2H + NH_2OH{\cdot}HCl$

$R_2CHN\bar{O}_2Na^{\oplus} \xrightarrow{H_2SO_4;\ \text{hydrolysis}} R_2CO + N_2O + H_2O$

$RCH_2N\bar{O}_2Na^{\oplus} \xrightarrow{H_2SO_4;\ \text{hydrolysis}} RCHO$

**(i)** $RCH_2NO_2 \xrightarrow{\text{aq. } NaNO_2/HCl \text{ (dil.)}} R{-}C({=}NOH)(NO_2) \xrightarrow{NaOH} R{-}C({=}N\bar{O}Na^{\oplus})(NO_2)$ red soln.

$R_2CHNO_2 \xrightarrow{\text{aq. } NaNO_2/HCl \text{ (dil.)}} R_2C(N{=}O)(NO_2) \xrightarrow{NaOH}$ no reaction (blue soln.)

$R_3CNO_2 \xrightarrow{\text{aq. } NaNO_2/HCl \text{ (dil.)}}$ no reaction (colourless).

**(j)** $RCH_2NO_2(R_2CHNO_2) \xrightarrow{NaOH/Br_2} RCBr_2NO_2(R_2CBrNO_2)$

(k) $RCH_2NO_2 + ArCHO \xrightarrow{\bar{O}H} ArCH{=}C(R)NO_2$

$RCH_2NO_2 + R^1CHO \xrightarrow{\bar{O}H} RCH(NO_2){-}CH(OH)R^1$

(l) $RCH_2NO_2 \xrightarrow[\text{Mannich reaction}]{Et_2NH\ \&\ HCHO/H^{\oplus}} RCH(NO_2)CH_2NEt_2$

(m) $RCH_2NO_2 \xrightarrow[\text{2. HCl; Michael cond.}]{\text{1. } \bar{O}Et/R^1_2C{=}CHCOR^2} R{-}CH(NO_2){-}CR^1_2{-}CH_2COR^2$

(n) $ArNO_2 \xrightarrow{E^{\oplus}} m\text{-}E{-}Ar{-}NO_2$

(o) 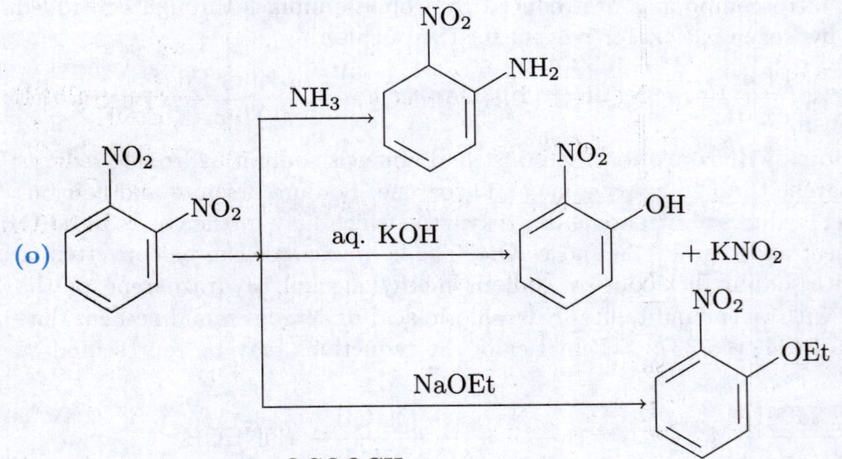


(p) 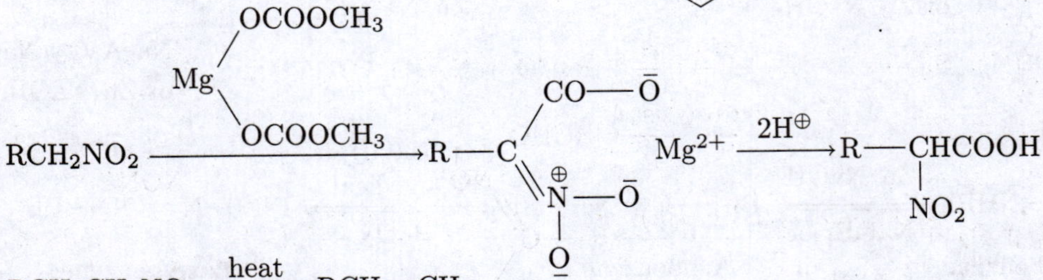


(q) $RCH_2CH_2NO_2 \xrightarrow{\text{heat}} RCH{=}CH_2$

**Chart 20.36** Reactions of nitro compounds.

**(b) Reduction by stannous chloride in hydrochloric acid :** Stannous chloride in hydrochloric acid reduces nitroalkanes to a mixture of hydroxylamine and oxime; whereas sodium salts of 1° and 2° nitroalkanes form only oximes through stannous chloride reduction. On the other hand, the reagent reduces a nitroarene to an aromatic amine. Instead of $SnCl_2$, $TiCl_2$ may be used to carry out the reduction.

$$CH_3CH_2NO_2 \xrightarrow{SnCl_2/HCl} CH_3CH_2NHOH + CH_3CH{=}NOH$$

$$CH_3CH_2NO_2 \xrightarrow{NaOH} CH_3CH{=}\overset{\oplus}{N}(OH)(\overset{\ominus}{O}Na^{\oplus}) \xrightarrow{SnCl_2/HCl} CH_3CH{=}N{-}OH$$

$$PhNO_2 \xrightarrow[\text{and HCl}]{SnCl_2 \text{ or } TiCl_2} PhNH_2$$

**(c) Reduction by neutral reducing agent :** When nitroalkanes and nitroarenes are reduced by metallic zinc and ammonium chloride in aqueous or alcoholic suspension, hydroxylamines are formed which may be separated by the 'salting out' process.

$$CH_3NO_2 \xrightarrow{Zn/NH_4Cl} CH_3NHOH;\ PhNO_2 \xrightarrow{Zn/NH_4Cl} PhNHOH$$

**(d) Catalytic reduction :** Raney nickel or platinum or paladium and hydrogen can reduce the nitro compounds to amines. Unsaturated aliphatic nitro compounds are reduced to saturated amines whereas aromatic nitro compounds are reduced to aromatic amines through controlled reduction. Paladium and hydrogen is a better reagent for the purpose.

$$CH_2{=}CH{-}CH_2NO_2 \xrightarrow[\&\ H_2]{\text{Pt or Ni}} CH_3CH_2CH_2NH_2;\ PhNO_2 \xrightarrow[\&\ H_2]{\text{Pt or Ni or Pd}} PhNH_2$$

**(e) Alkaline reduction of nitroarenes :** Zn dust and aqueous sodium hydroxide reduces a nitroarene to a hydrazoarene through azoxyarene and azoarene. Sodium arsenite and aqueous sodium hydroxide, however, reduces a nitroarene to azoxyarene which may further be reduced to hydrazoarene by Fe and heat and then by Zn and NaOH. The hydrazoarene can be converted to arylamine by reduction with sodium dithionate or sodium in ethyl alcohol. Hydrazoarene, on the other hand, on treatment with warm and dilute hydrochloric acid undergoes semidine-benzidine rearrangements (See Sec. 14.11. p. 733). Schematically the reductions may be represented as shown below.

$$PhNO_2 \xrightarrow[\text{or } Zn/NaOH]{Na_3AsO_3/NaOH} Ph{-}N{=}O \xrightarrow[Zn/NaOH]{Na_3AsO_3/NaOH} PhNHOH$$

$$PhNHOH + Ph{-}N{=}O \xrightarrow[\text{or } Zn/NaOH]{Na_3AsO_3/NaOH} \text{Azoxybenzene}$$

$$\underset{\text{Hydrazobenzene}}{PhNH{-}NHPh} \underset{\text{NaOBr or air}}{\overset{Zn/NaOH}{\rightleftharpoons}} \underset{\text{Azobenzene}}{Ph{-}N{=}N{-}Ph} \underset{H_2O_2}{\overset{\text{Zn/NaOH or Fe and heat}}{\rightleftharpoons}} \underset{\text{Azoxybenzene}}{Ph{-}N{=}\overset{\oplus}{N}(\overset{\ominus}{O}){-}Ph}$$

However, when nitrobenzene is reduced by electrolysis in concentrated sulphuric acid solution, at first phenylhydroxylamine forms which undergoes rearrangement reaction as soon as it is formed; *p*-aminophenol is obtained as the end product.

$$PhNO_2 \xrightarrow[\text{electrolysis}]{\text{conc. } H_2SO_4} PhNHOH \xrightarrow{\text{rearrangment}} p\text{-}HO{-}C_6H_4{-}NH_2$$

**(f) Selective reductions :** When an aromatic nitro compound contains more than one nitro group, one of them may selectively be reduced to an amino group by the reducing action of sodium or ammonium sulphide or bisulphide.

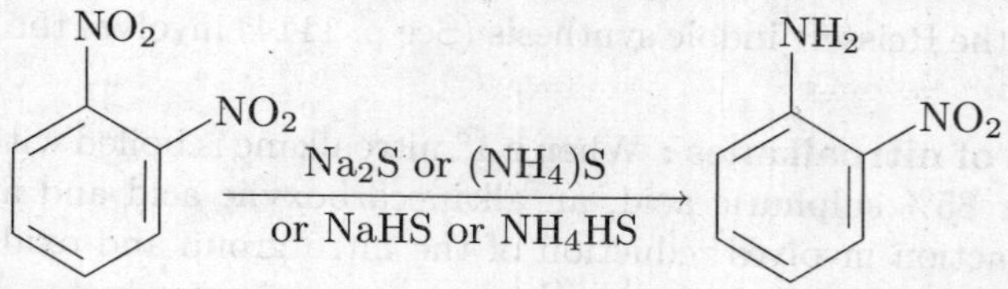

It has been found that these reagents are unable to reduce a nitro group ortho to an alkyl group but $SnCl_2/HCl$ can selectively reduce a nitro group to an amino group in the ortho position to an alkyl group. $SnCl_2/HCl$, however, if cannot reduce a nitro group in the *p*-position.

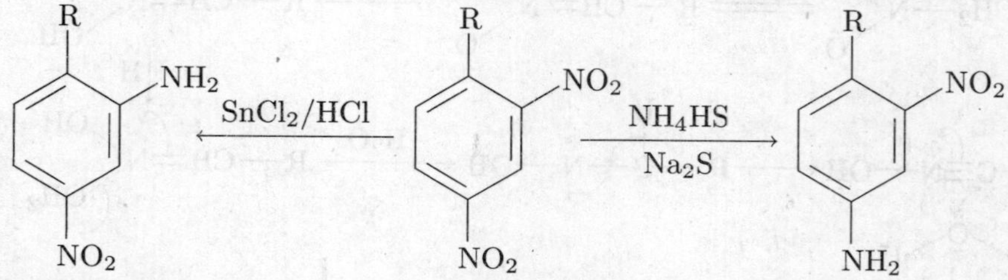

**(g) Salt formation reactions :** 1° and 2° nitroalkanes are acidic ($pKa \sim 9$). Since nitro group has a strong −I effect, the *a*-H atoms of 1° and 2° nitroalkanes are acidic and leave the species as proton when they are treated with bases, the conjugate base of the nitroalkane, a carbanion, gets stability through resonance. The carbanion being resonance stabilised, a nitroalkane is a strong acid.

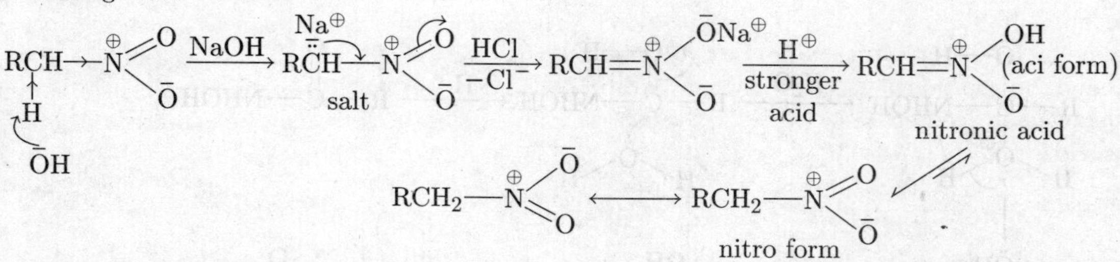

When an acid, which is stronger than nitroalkane, is added to the salt of nitroalkane, at first nitronic acid is formed which then tautomerises to nitroalkane. Thus nitroalkane exhibits tautomerism and the tautomers are called aci form (nitronic acid) and nitro form. The acidity of the aci form is about as great as that of the carboxylic acids ($pKa$ : 4 to 6). However, in the equilibrium between the nitro and the aci forms, the latter remains in small amount and so the aqueous solution of a nitroalkane is slightly acidic, the pH is about 4.3. It is needless to mention that 3° nitro compounds being devoid of *a*-H, they do not form salts with bases.

Owing to the −I and −R effects of the nitro group in *o*-nitrotoluene, methyl group forms a salt with sodium ethoxied as shown below.

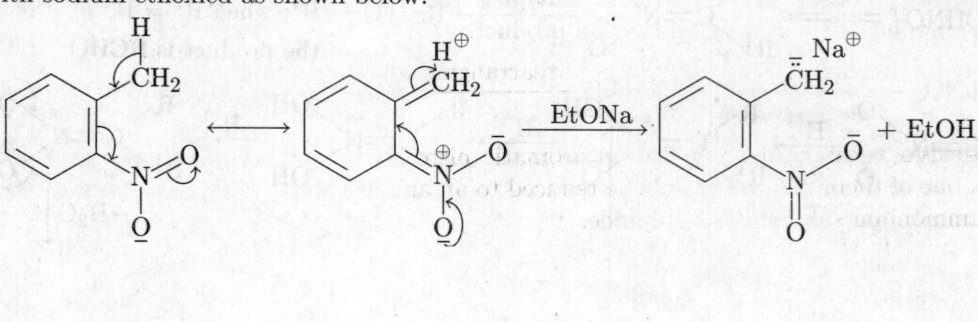

The first step of the Reissert indole synthesis (See p. 1114) involves the salt formation reaction of *o*-nitrotoluene.

**(h) Hydrolysis of nitroalkanes :** When a 1° nitroalkane is boiled with concentrated aqueous hydrochloric acid or 85% sulphuric acid, an alkanecarboxylic acid and a salt of hydroxylamine are formed. The reaction involves reduction of the nitro group and oxidation of the methylene group to form hydroxamic acid which then hydrolyses to the products. This is the most simple and cheapest process for the manufacture of hydroxylamine.

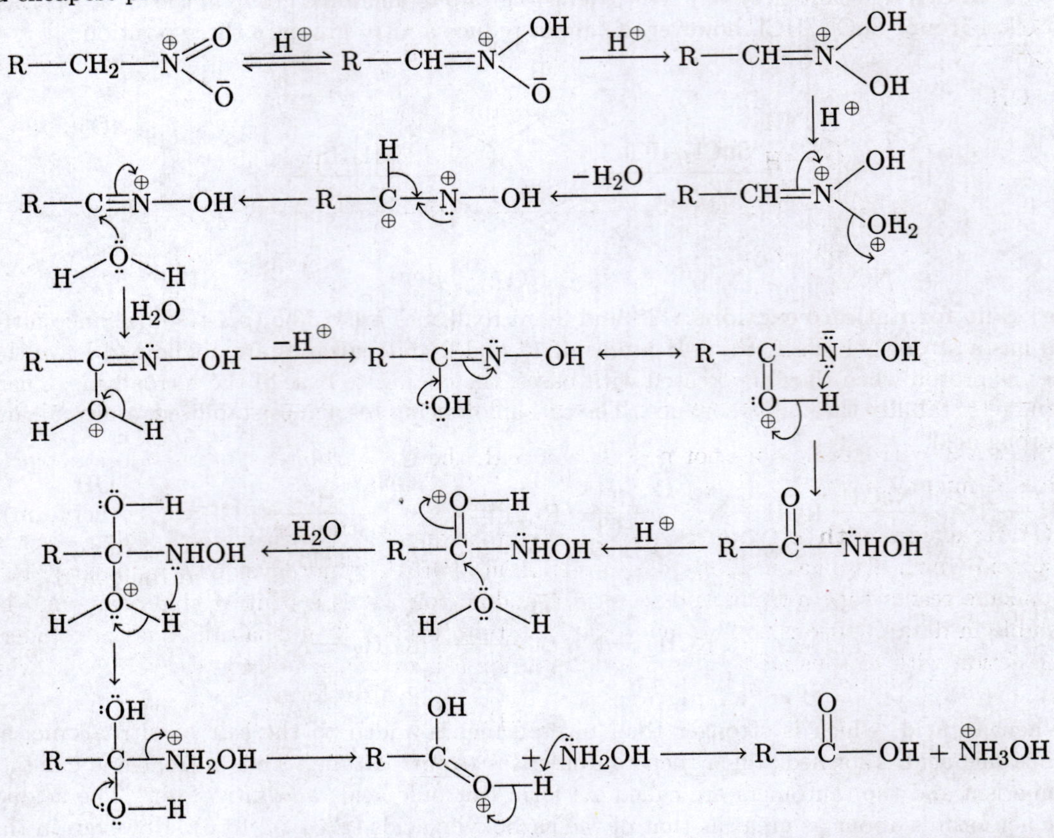

When the salts of 1° and 2° nitroalkanes are treated with sulphuric acid, aldehydes and ketones are respectively formed and the reaction is popularly called the Nef carbonyl synthesis. This is a good method of synthesising aldehydes and ketones.

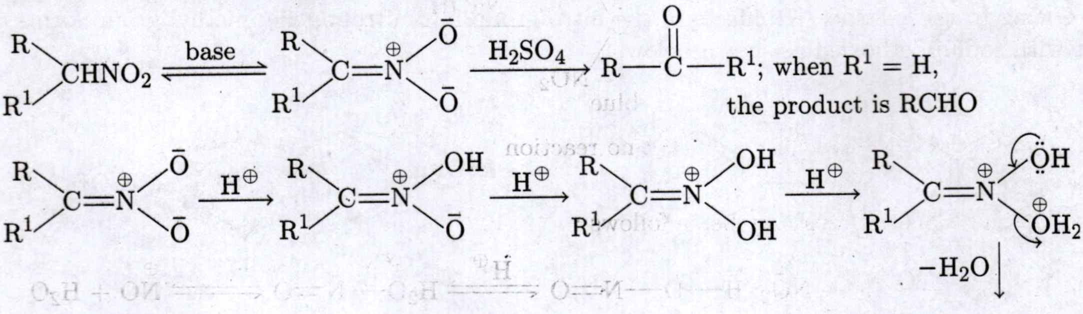

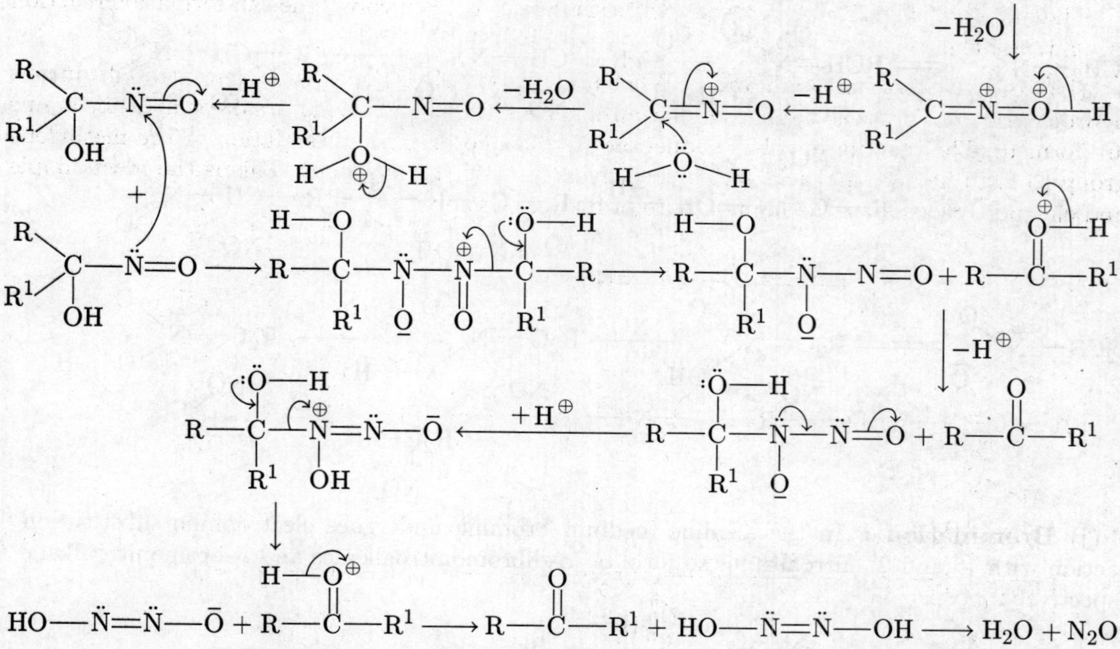

Since a 3° nitroalkane does not possess any $\alpha$-H, the Nef carbonyl synthesis does not occur with a 3° nitroalkane.

**(i) Reaction with nitrous acid :** A 1° nitroalkane reacts with nitrous acid to form nitrolic acid which dissolves in dilute aqueous sodium hydroxide and develops a red colour. A 2° nitroalkane reacts with nitrous acid to form pseudonitrole. It is a neutral blue compound but insoluble in dilute aqueous sodium hydroxide solution, while a 3° nitroalkane does not undergo the reaction with nitrous acid. This reaction is the basis of Victor Meyer's red, blue and white test.

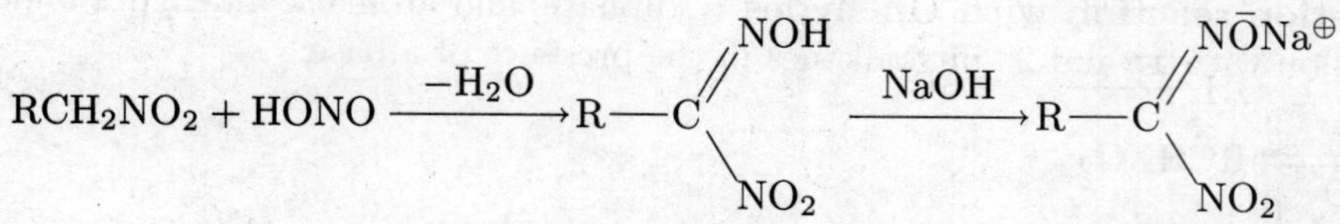

Nitrolic acid — red coloured salt

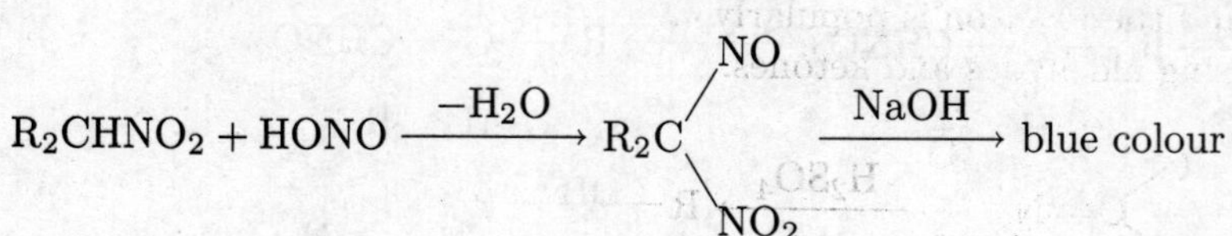

blue

$R_3C—NO_2 + HONO \longrightarrow$ no reaction
colourless

The mechanistic steps may be as follows :

$HNO_2 \rightleftharpoons H^{\oplus} + \bar{N}O_2;\ H—O—N{=}O \overset{H^{\oplus}}{\rightleftharpoons} H_2\overset{\oplus}{O}—\ddot{N}{=}O \rightleftharpoons \overset{\oplus}{N}O + H_2O$

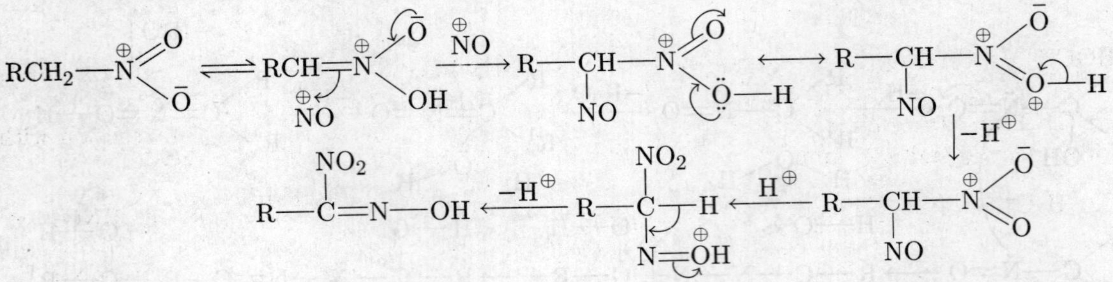

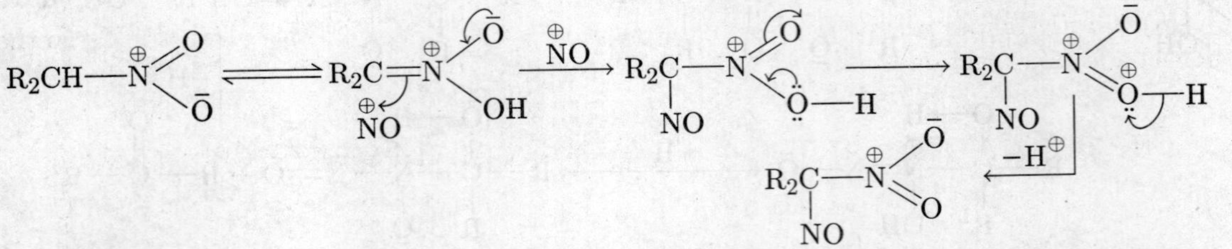

**(j) Bromination :** In the alkaline medium bromine undergoes electrophilic substitution reaction with 1° and 2° nitroalkanes to form $\alpha$, $\alpha$-dibromonitroalkanes and $\alpha$-bromonitroalkane respectively.

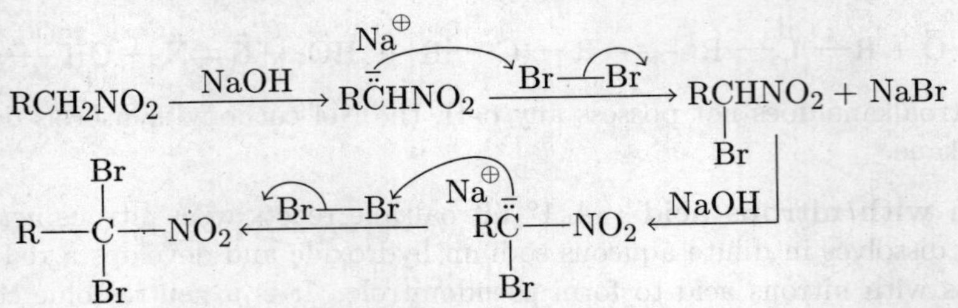

$$R_2CHNO_2 \xrightarrow[2.\ Br_2]{1.\ NaOH} R_2C(Br)NO_2$$

**(k) Condensation reaction with aldehydes :** Aliphatic and aromatic aldehydes undergo condensation reaction with 1° and 2° nitroalkanes in the presence of a base.

$$RCH_2NO_2 \underset{H_2O}{\overset{\bar{O}H}{\rightleftharpoons}} R\ddot{\bar{C}}HNO_2;$$

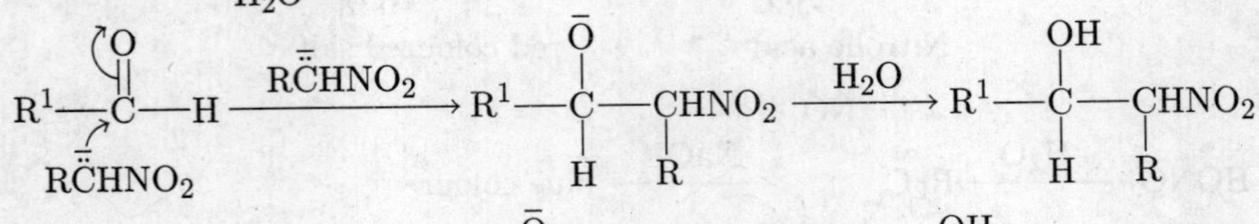

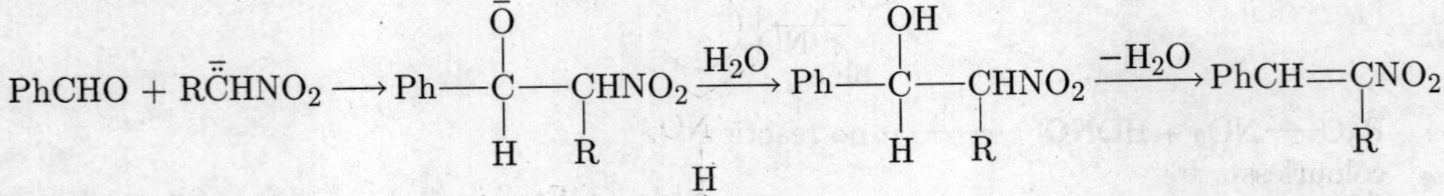

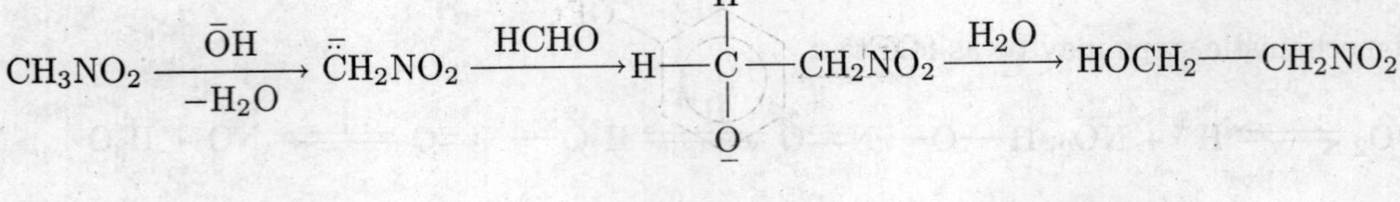

$$HOCH_2—CH_2NO_2 \xrightarrow[\text{2. HCHO; 3. } H_2O]{\text{1. } \bar{O}H} NO_2CH(CH_2OH)_2 \xrightarrow[\text{3. } H_2O]{\text{1. } \bar{O}H;\ \text{2. HCHO}} NO_2—C(CH_2OH)_3$$

**(l) The Mannich reaction :** 1° and 2° nitroalkanes undergo the Mannich reaction with formaldehyde and a 2° amine.

$$H_2C{=}O + R^1_2NH \longrightarrow H_2C(OH)—\ddot{N}R^1_2 \xrightarrow[-H_2O]{H^{\oplus}} H_2C{=}\overset{\oplus}{N}R^1_2 \xrightarrow[\text{NaOEt}]{RCH_2NO_2} RCH(NO_2)—CH_2NR^1_2$$

**(m) The Michael addition :** In the presence of a base 1° and 2° nitroalkanes undergo the Michael addition reaction with $\alpha$, $\beta$-unsaturated carbonyl compounds.

$$CH_3CH_2NO_2 + CH_2{=}CHCOMe \xrightleftharpoons{\text{NaOEt, EtOH}} CH_3CH(NO_2)—CH_2—CH{=}C(OH)Me \rightleftharpoons CH_3—CH(NO_2)—CH_2CH_2\overset{O}{\overset{\|}{C}}Me$$

**(n) Ar.S$_E$2 reactions of aromatic nitro-compounds :** Aromatic nitro compounds undergo electrophilic substitution reactions like halogenation, nitration sulphonation, etc.

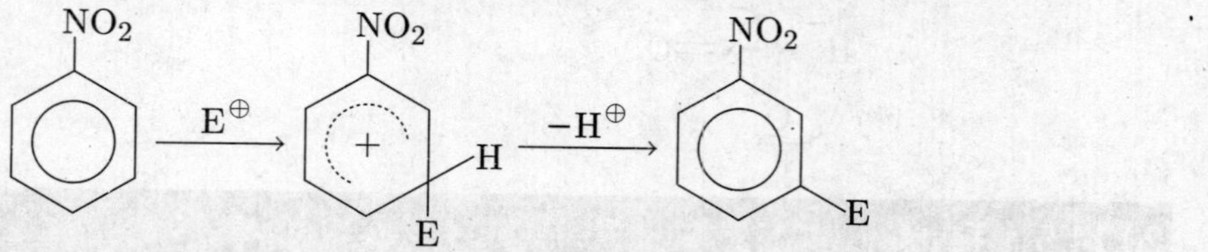


**(o) Ar.S$_N$2 reactions of aromatic nitro compounds :** Aromatic nitro compounds undergo the bimolecular nucleophilic substitution reactions.

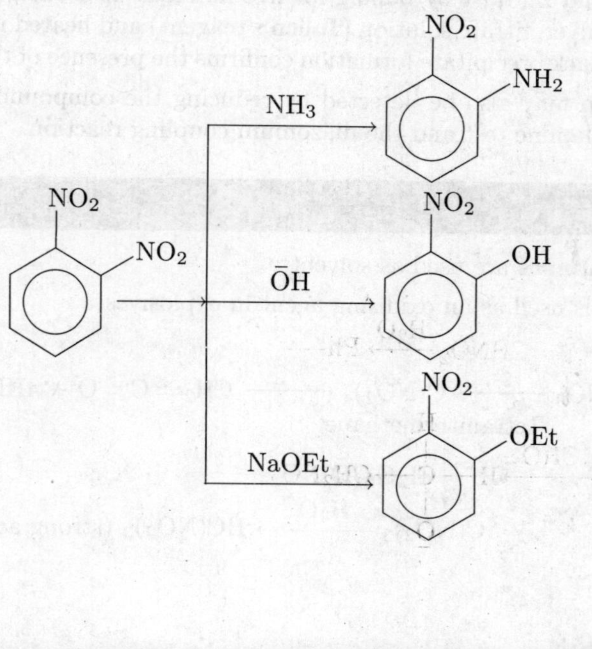

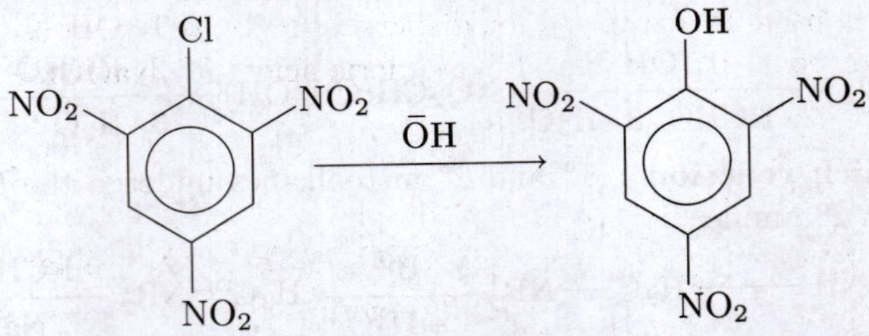


**(p) Carboxylation :** A 1° nitroalkane reacts with magnesium methylcarbonate and the product on acidification gives $\alpha$-nitrocarboxylic acid :

$$RCH_2NO_2 + Mg^{++}(^{-}OCOOMe)_2 \xrightarrow[-CO_2]{-MeOH} R{-}\underset{NO_2}{\overset{H}{C}}{-}COO^- \xrightarrow{H^{\oplus}} R{-}\underset{NO_2}{\overset{H}{C}}{-}COOH$$

**(q) Action of heat on 1° and 2° nitroalkanes :** 1° and 2° nitroalkanes decompose on heating to form alkenes.

$$R{-}CH_2{-}CH_2{-}NO_2 \xrightarrow[\text{above } 300\ ^\circ C]{\text{heat}} RCH{=}CH_2 + HNO_2$$

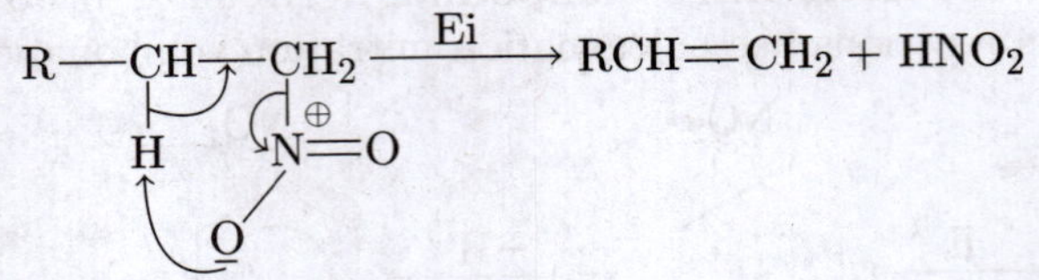


## 20.14.5 Tests

**Mulliken's test :** A little of the sample is dissolved in 2 ml alcohol and it is reduced with Za/$NH_4Cl$ or $CaCl_2$ solution and Zn dust by boiling for five minutes. It is then filtered, cooled and treated with ammoniacal silver nitrate solution (**Tollen's reagent**) and heated in a water bath $\longrightarrow$ white to gray and then black precipitate formation confirms the presence of the nitro group.

The presence of nitro group may also be detected by reducing the compound with Zn/HCl and then conducting the carbylamine test and the diazonium coupling reaction.

## 20.14.6 Uses

**(i) Nitrobenzene and nitroalkanes** are used as solvents.

**(ii) Nitroform**, $CH(NO_2)_3$, is used as an oxidising agent in explosives.

It is prepared as follows :

$$(CH_3CO)_2O + 4HNO_3 \longrightarrow \underset{\text{Tetranitromethane}}{C(NO_2)_4} \longleftarrow CH_2{=}C{=}O + 4HNO_3$$

$$C(NO_2)_4 \xrightarrow{H_2O_2/\bar{O}H} \bar{C}(NO_2)_3 \xrightarrow{H_3O^{\oplus}} HC(NO_2)_3 \text{ (strong acid)}$$

**(iii) Chloropicrin** is used as a tear gas for dispersing mob. It is also used as a disinfectant of soil for agricultural purpose. It is prepared from picric acid with NaOCl.

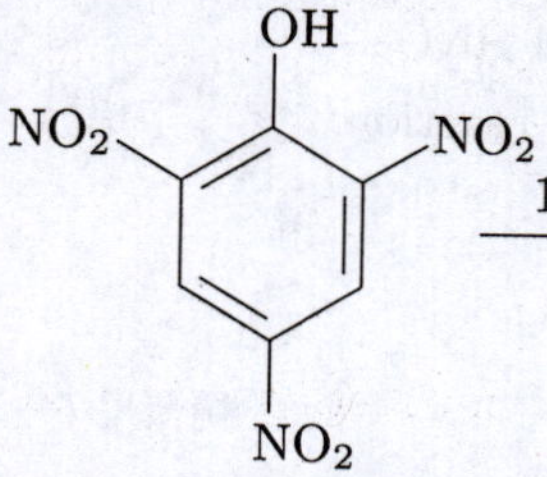

$$\xrightarrow{11\ NaOCl} \underset{\text{Chloropicrin}}{3Cl_3CNO_2} + 3Na_2CO_3 + 3NaOH + 2NaCl$$

**(iv) T.N.B.** (1, 3, 5-trinitrobenzone) is a powerful explosive.

It is prepared from **T.N.T.** (2, 4. 6-trinitrotoluene) which is also used as an explosive.

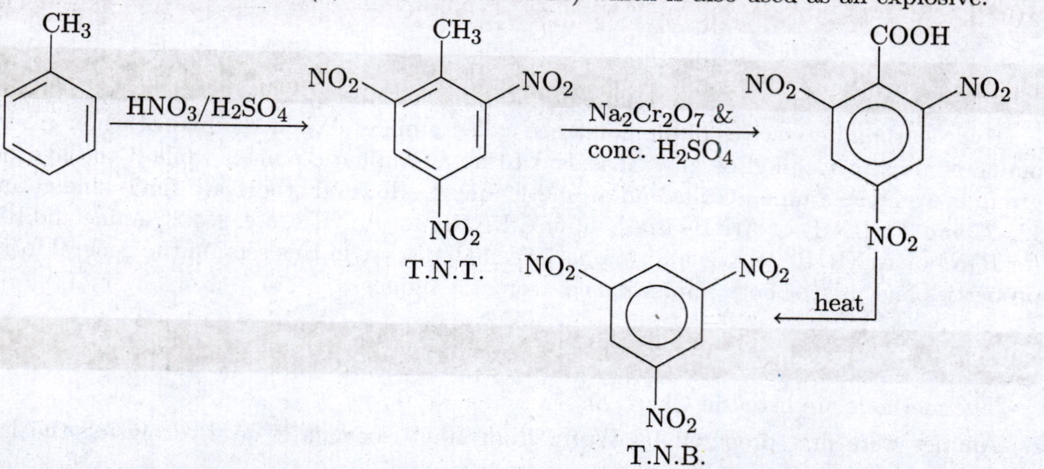


**Problem 20.29.** Distinguish between nitroalkane and alkyl nitrite.

**Solution :** The structure of the two isomeric compounds are :

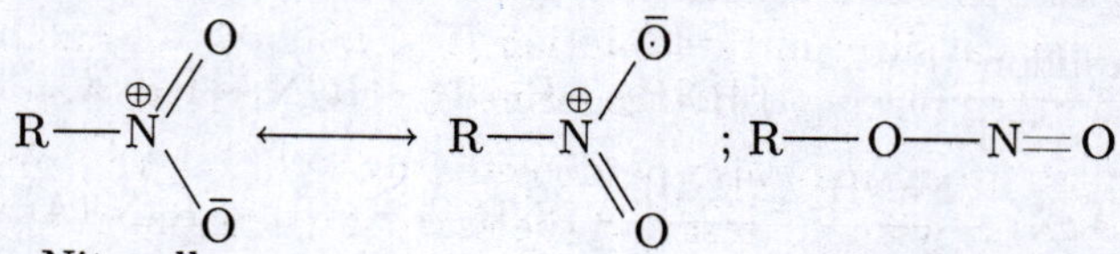


| Points | Nitroalkane | Alkyl nitrite |
|---|---|---|
| 1. Reduction $LiAlH_4$ & ether or Ni-$H_2$, etc. | Forms $RNH_2$. | Does not form $RNH_2$ but gives ROH + $NH_3$ + $H_2O$. |
| 2. Basic hydrolysis | Does not hydrolyse but 1° and 2° nitroalkanes form salts. $RCH{=}\overset{\oplus}{N}(O^-Na^{\oplus}){-}\bar{O}$ & $R_2C{=}\overset{\oplus}{N}(O^-Na^{\oplus}){-}\bar{O}$ | Hydrolyses to form alcohol. $R{-}O{-}N{=}O + NaOH \longrightarrow ROH + NaNO_2$ |

*Contd.*

| Points | | Nitroalkane | Alkyl nitrite |
|---|---|---|---|
| 3. | Acid hydrolysis | A 1° nitroalkane forms RCOOH | Alkyl nitrites form alcohol, ROH and $HNO_2$. |
| 4. | $HNO_2$ | A 1° Nitroalkane forms nitrolic acid which develops red colour with NaOH. A 2° nitroalkane forms pseudonitrole which is blue in colour. A 3° nitroalkane does not react. | No reaction. |

## 20.15 Amines

### 20.15.1 Introduction

Hydrocarbon derivatives of ammonia are called amines. When the hydrocarbon group of an amine is alkyl, cycloalkyl or allyl, it is said to be an aliphatic amine; while if the hydrocarbon group is aryl, the amine is called an aromatic amine. However, there are three kinds of amines, 1°, 2° and 3°. $RNH_2$ or $ArNH_2$ is a 1° amine, $RNHR^1$ or ArNHR, etc. is a 2° amine and $RNR^1R^2$ (or $R_3N$) or $ArNR^1R^2$, etc. is a 3° amine. We shall discuss in this section the general formation, properties and uses of both aliphatic and aromatic amines.

### 20.15.2 Formation of amines

The methods are listed in Chart 20.37.

Amines were first prepared by Wurtz from alkyl isocyanates by hydrolysing the latter in alkaline medium.

$$\text{Et—N=C=O} + 2KOH \longrightarrow EtNH_2 + K_2CO_3$$

However, the general methods for the preparation of amines are discussed below :

(a) (i) $RX \xrightarrow[NH_3]{\text{ammonolysis}} \text{(i) } RNH_2 + R_2NH + R_3N + R_4\overset{\oplus}{N}X^-$

(ii) $ROH\ (ArX) \xrightarrow{NaNH_2/NH_3(l)} RNH_2 + R_2NH + R_3N\ (ArNH_2)$

(b) $RCHO \xrightarrow[NH_3;\ H_2/Ni]{\text{reductive amination}} RCH_2NH_2 + (RCH_2)_2NH$

(c) (i) $RNO_2(ArNO_2) \xrightarrow[H_2/Pt \text{ or } Ni]{Zn/HCl \text{ or}} RNH_2(ArNH_2)$

(ii) $NO_2\text{—}C_6H_4\text{—}NO_2 \xrightarrow[\text{partial reduction}]{Na_2S \text{ or } NaHS} NH_2C_6H_4NO_2$

(d) $RCN(ArCN) \xrightarrow{LiAlH_4 \text{ or } Na/EtOH} RCH_2NH_2(ArCH_2NH_2)$

(e) $RCONH_2(ArCONH_2) \xrightarrow{LiAlH_4} RCH_2NH_2(ArCH_2NH_2)$

(f) $RCH{=}NOH(ArCH{=}NOH) \xrightarrow{Na/EtOH} RCH_2NH_2(ArCH_2NH_2)$

(g) Phthalimide $\xrightarrow{\text{1. }\bar{O}H\text{ 2. RX 3. KOH}} RNH_2(1°;\ \text{pure})$

(h) $ArNH_2 \xrightarrow{\text{1. RX 2. }HNO_2\text{ 3. KOH}} R_2NH\ (2°;\ \text{pure})$

(i) $RMgCl \xrightarrow{ClNH_2} RNH_2$

(j) *p*-Tolylsulphonyl chloride $\xrightarrow[H_3O^{\oplus}]{RNH_2/\bar{O}H;\ R^1X} R^1RNH$

(k) $CaNCN \xrightarrow{NaOH} Na_2NCN \xrightarrow{2\ RX} R_2NCN \xrightarrow{H^{\oplus}\text{ or }\bar{O}H} R_2NH$

(l) $RCON_3 \xrightarrow{\text{heat}} R{-}N{=}C{=}O \xrightarrow{\bar{O}H,\ H_2O} RNH_2$

(m) $RCONHOH \xrightarrow{H^{\oplus}\text{ or }\bar{O}H} R{-}N{=}C{=}O \xrightarrow{\bar{O}H,\ H_2O} RNH_2$

(n) $RCONH_2 \xrightarrow{NaOH/Br_2} R{-}N{=}C{=}O \xrightarrow{\bar{O}H,\ H_2O} RNH_2$

(o) $\begin{matrix}R\\R^1\end{matrix}\!\!>C{=}O \xrightarrow[HCONH_2/\Delta]{HCOONH_4\text{ or}} \begin{matrix}R\\R^1\end{matrix}\!\!>CHNH_2$

(p) $R_3COH \xrightarrow{H^{\oplus}} \xrightarrow{NaCN} \xrightarrow{H_2O} \xrightarrow{\bar{O}H} R_3CNH_2$

**Chart 20.37** Formation of amines.

**(a) (i) By the process of ammonolysis :** When a mixture of an alkyl halide and an ethanolic solution of ammonia is heated in a sealed tube, a mixture of three kinds of amino salts, 1°, 2° and 3°, and quaternary ammonium halide is obtained. The amine salts obtained by the process are converted to amines by distilling with strong alkali.

$$RX + NH_3 \longrightarrow RNH_2 + HX \longrightarrow R\overset{\oplus}{N}H_3X^-$$
$$RNH_2 + RX \longrightarrow R_2NH + HX \longrightarrow R_2\overset{\oplus}{N}H_2X^-$$
$$R_2NH + RX \longrightarrow R_3N + HX \longrightarrow R_3\overset{\oplus}{N}HX^-$$
$$R_3N + RX \longrightarrow R_4\overset{\oplus}{N}X^-$$

If a large excess of ammonia is allowed to react with a small amount of the alkyl halide, a 1° amine forms predominantly. However, a 3° alkyl halide gives an alkene by ammonolysis. The reactivity order of alkyl groups and halides are :

$$\text{Methyl} > 1° > 2° > 3° \text{ and Iodide} > \text{bromide} > \text{chloride.}$$

**(ii)** In a modified process a mixture of vapour of an alcohol and ammonia is passed over a heated catalyst (300 °C) in a reaction tube whereby a mixture of three types of amines is obtained.

$$ROH + NH_3 \xrightarrow[300\,°C]{ThO_2} RNH_2 + H_2O; \quad RNH_2 + ROH \xrightarrow[300\,°C]{ThO_2} R_2NH + H_2O$$
$$R_2NH + ROH \xrightarrow[300\,°C]{ThO_2} R_3N + H_2O$$

When a haloarene (chloro and bromo) is heated with sodium amide or potassium amide in liquid ammonia, a 1° aromatic amine is formed through an aryne.

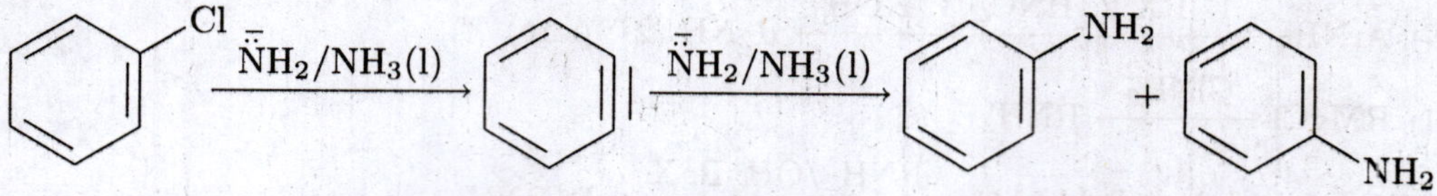

The detail of the reaction and the mechanism have already been discussed in Sec. 12.10, p. 602.

When halobenzenes are heated with ammonia or a 1° or a 2° amine in the presence of cuprous oxide as catalyst, an aromatic amine is obtained.

$$\text{PhCl} \xrightarrow[\text{Cu}_2\text{O}/\Delta]{\text{NH}_3} \text{PhNH}_2 \;;\; \text{PhBr} \xrightarrow[\text{Cu}_2\text{O}/\Delta]{\text{RNH}_2} \text{PhNHR}$$

**(b) Reductive amination of carbonyl compounds :** When a mixture of an aldehyde or a ketone and a large excess of ammonia and hydrogen is passed over Raney Ni at 40-50 °C under a pressure of 20-150 atm, a 1° amine is formed. If ammonia is not taken in excess, a mixture of 1° and 2° amines is obtained. The reaction may be carried out under 3 atm pressure in the presence of excess of ammonium chloride and Adam's platinum-platinum oxide catalyst.

$$\text{R}_2\text{C}{=}\text{O} + \text{NH}_3 + \text{H}_2 \xrightarrow{\text{Ni}} \text{R}_2\text{CHNH}_2 \xrightarrow{\text{R}_2\text{C}=\text{O}} (\text{R}_2\text{CH})_2\text{NH} \xrightarrow{\text{R}_2\text{C}=\text{O}} (\text{R}_2\text{CH})_3\text{N}$$

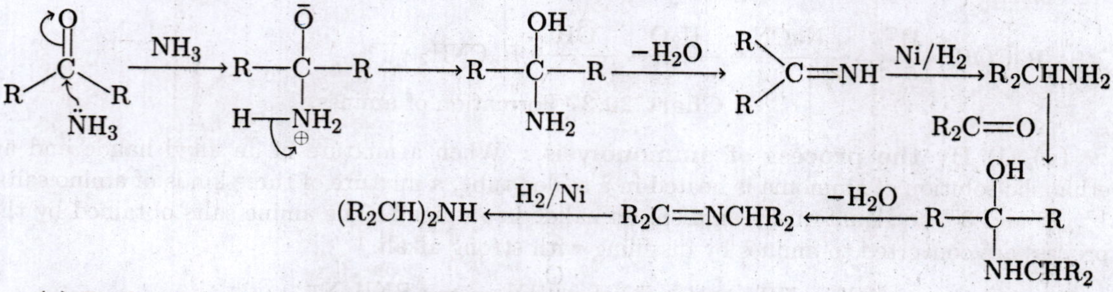

**(c) Reduction of nitro compounds :** See Sec. 20.14.4(a), (d) and (f).

**(d) Reduction of alkyl cyanide :** See Sec. 20.12.4(c). p. 1323.

**(e) Reduction of amides :** Amides on reduction with $LiAlH_4$/ether form 1° amines.

$$\text{CH}_3\text{CONH}_2 \xrightarrow[\text{ether}]{\text{LiAlH}_4} \text{CH}_3\text{CH}_2\text{NH}_2 \;;\; \text{PhCONH}_2 \xrightarrow[\text{Na}]{\text{EtOH}} \text{PhCH}_2\text{NH}_2$$

**(f) Reduction of oximes :** Oximes on reduction with sodium in ethanol form 1° amines.

$$\text{CH}_3\text{CH}{=}\text{NOH} \xrightarrow[\text{or LiAlH}_4/\text{ether or cat./H}_2]{\text{Na/EtOH}} \text{CH}_3\text{CH}_2\text{NH}_2$$

**(g) The Gabriel's phthalimide synthesis :** The method consists of : (i) conversion of phthalimide to its salt by ethanolic potassium hydroxide, (ii) heating of the salt with alkyl halide to form N-alkylphthalimide and (iii) hydrolysis of the N-alkyl derivative to phthalic acid and a 1° amine (corresponding to the alkyl halide); the hydrolysis is carried out either by heating the N-alkyl-phthalimide with 20% hydrochloric acid under pressure, or by refluxing with potassium hydroxide solution. When hydrolysis is not an easy task, the amine is obtained by treating the N-alkyl derivative with hydrazine.

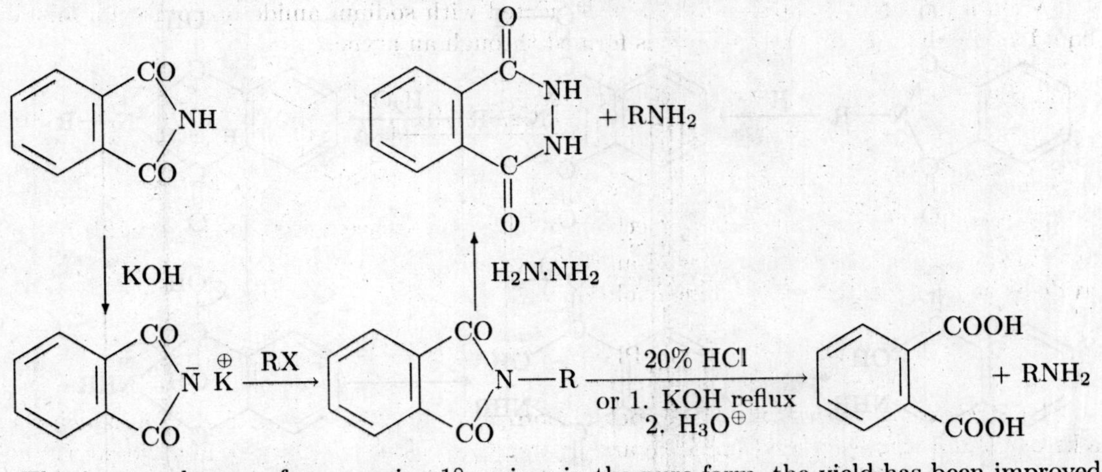

This is a good means for preparing 1° amines in the pure form, the yield has been improved by using DMF as the solvent.

The mechanism of the reaction may be represented as shown below :

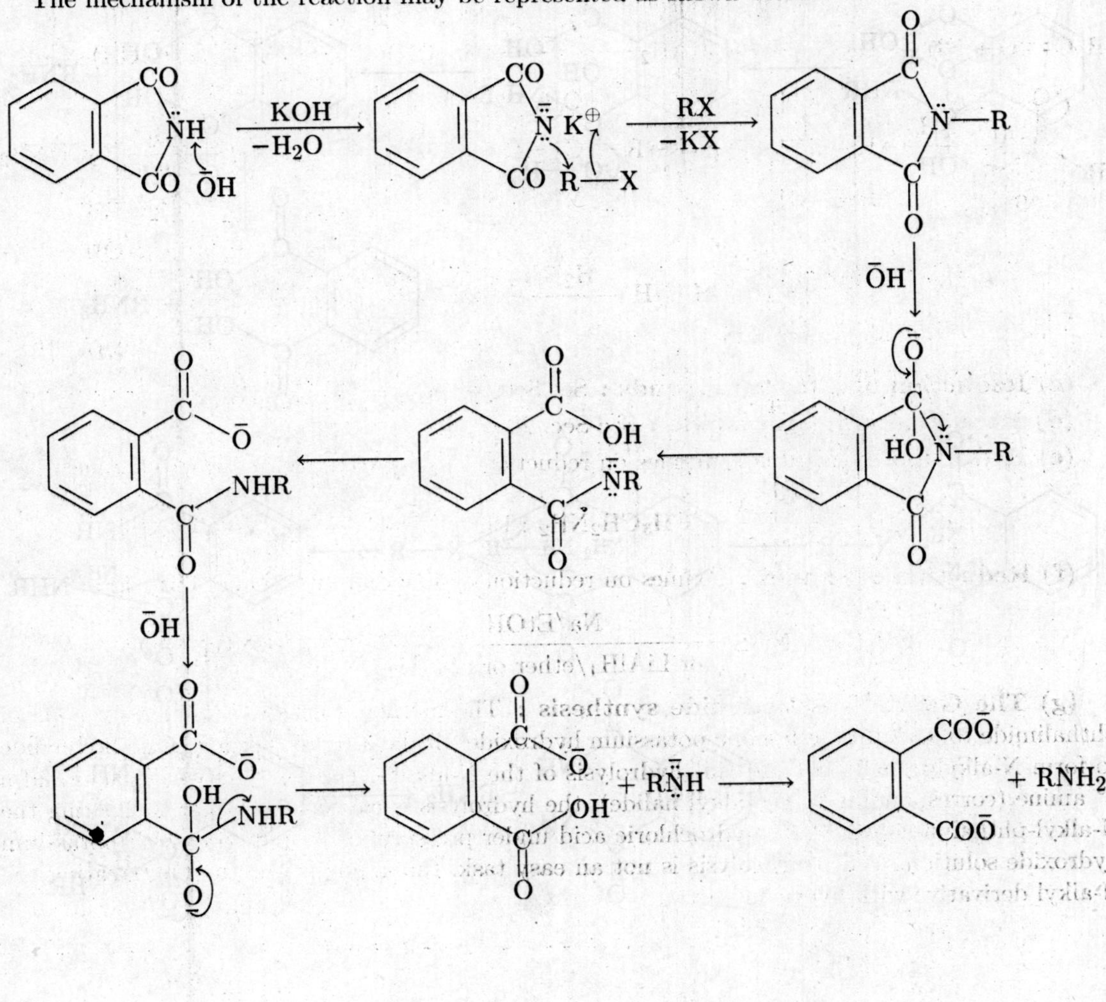

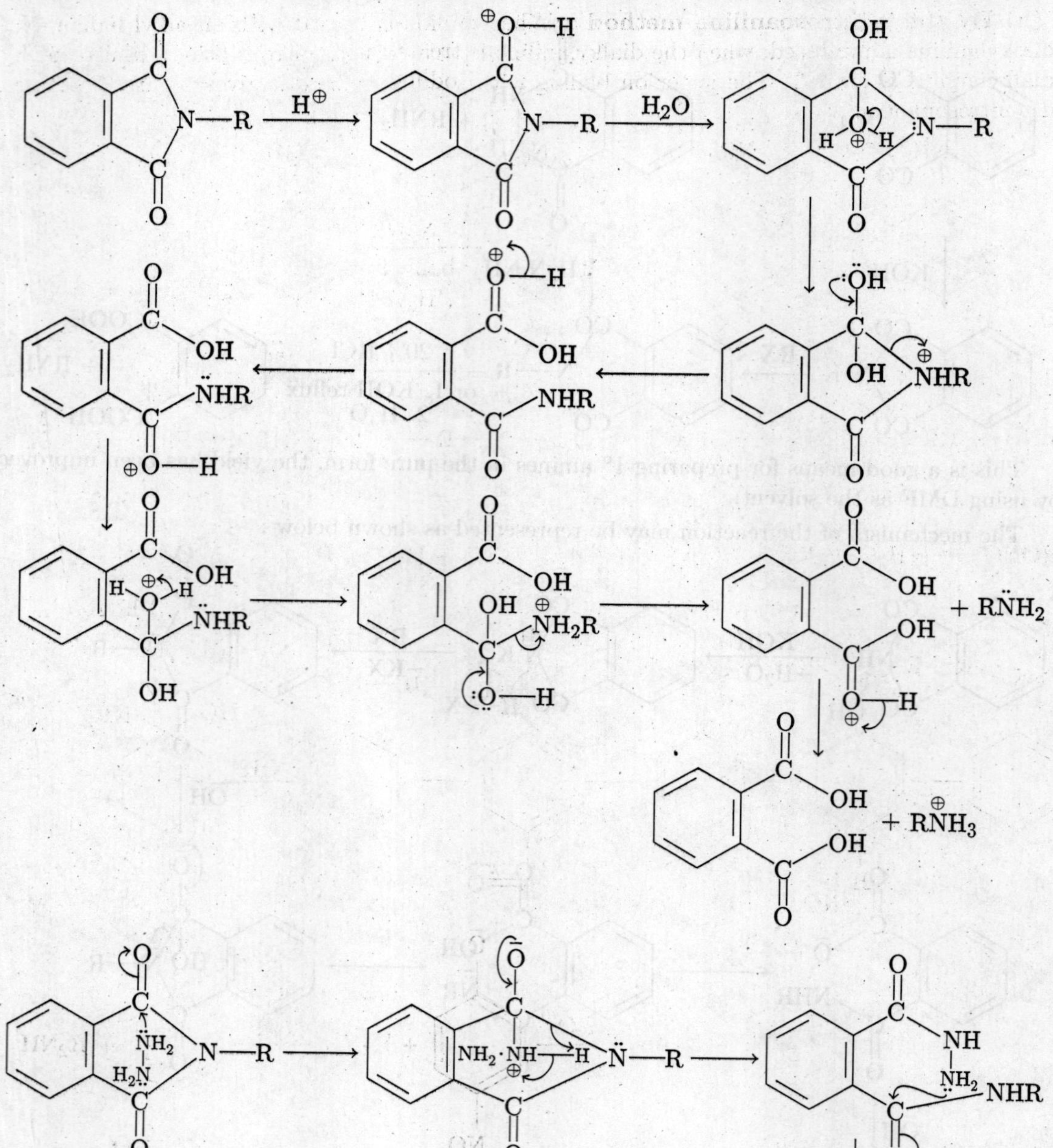

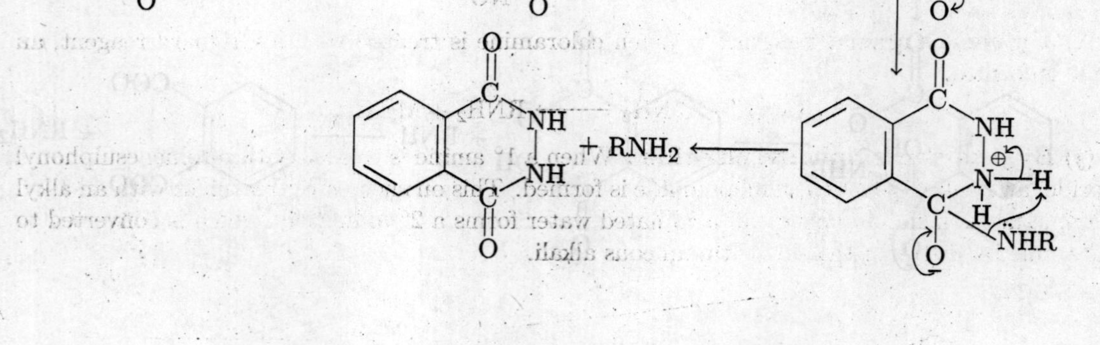

**(h) By the *p*-nitrosoaniline method :** When aniline is heated with an alkyl halide, N, N-dialkylaniline is produced; when the dialkylaniline is treated with nitrous acid, a *p*-nitroso-N, N-dialkylaniline is formed. The latter on boiling with sodium hydroxide gives a pure 2° amine and *p*-nitrosophenol.

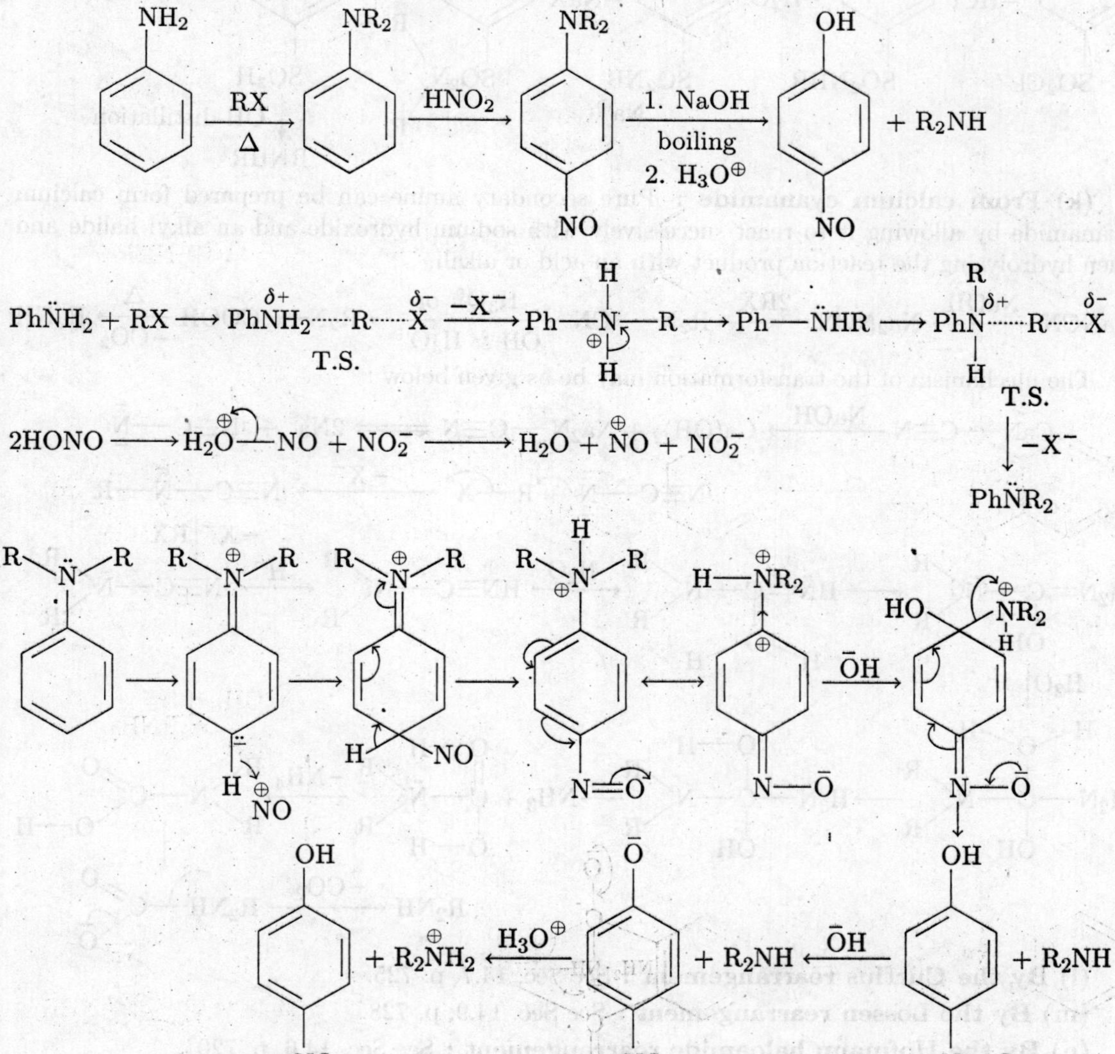

**(i) By the Grignard reagent :** When chloramine is treated with a Grignard reagent, an amine is formed.

$$RMgX + ClNH_2 \longrightarrow RNH_2 + MgXCl$$

**(j) By *p*-toluenesulphonyl chloride :** When a 1° amine is treated with *p*-toluenesulphonyl chloride, an N-alkyl-*p*-toluenesulphonamide is formed. This on successive treatment with an alkyl halide, aqueous alkali and excess of acidulated water forms a 2° amine salt which is converted to a 2° amine by distilling the salt with aqueous alkali.

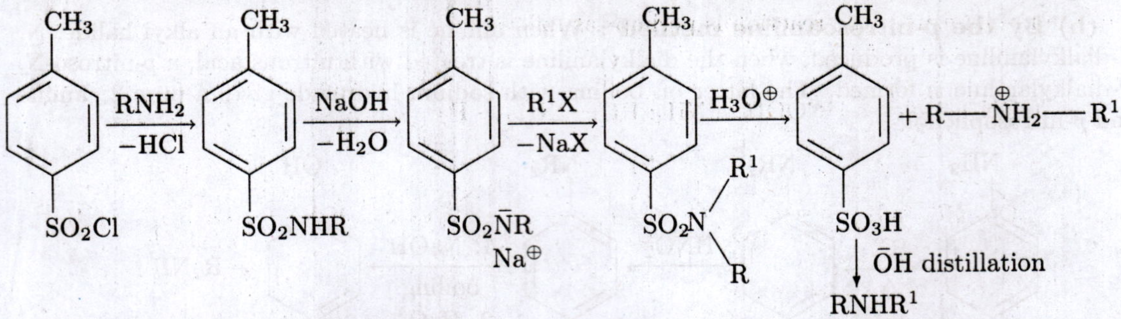

**(k) From calcium cyanamide :** Pure secondary amine can be prepared form calcium cyanamide by allowing it to react successively with sodium hydroxide and an alkyl halide and then hydrolysing the reaction product with an acid or alkali.

$$\text{CaNCN} \xrightarrow{\text{NaOH}} \text{Na}_2\text{NCN} \xrightarrow{2\text{RX}} \text{R}_2\text{N—CN} \xrightarrow[\bar{\text{O}}\text{H \& H}_3\text{O}^{\oplus}]{\text{H}_3\text{O}^{\oplus}\text{ or}} \text{R}_2\text{N—COOH} \xrightarrow[-\text{CO}_2]{\Delta} \text{R}_2\text{NH}$$

The mechanism of the transformation may be as given below :

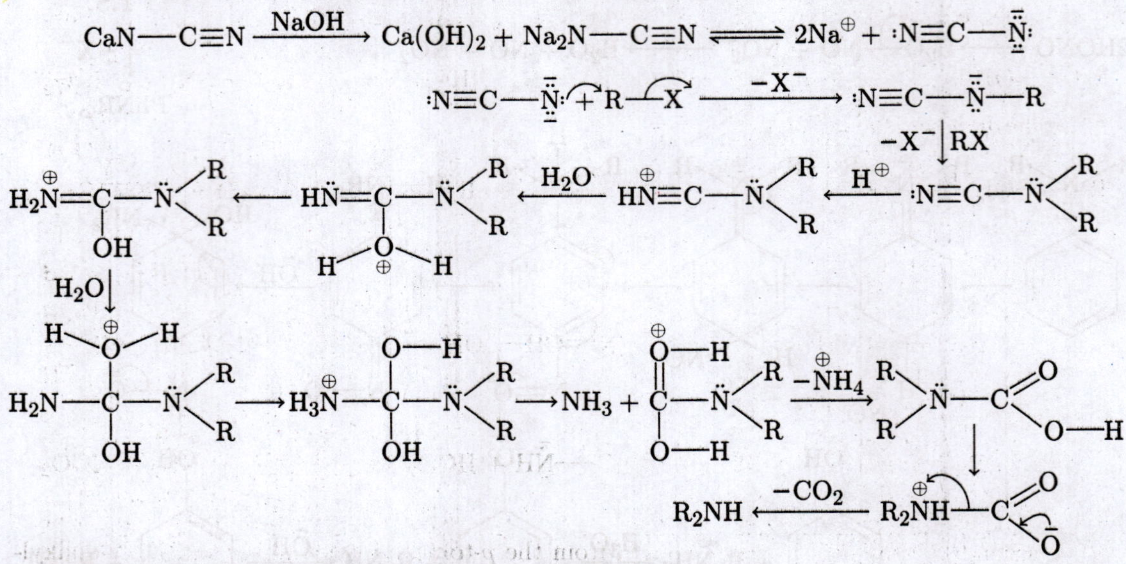

**(l) By the Curtius rearrangement :** See Sec. 14.7, p. 725.

**(m) By the Lossen rearrangement :** See Sec. 14.9, p. 728.

**(n) By the Hofmann haloamide rearrangement :** See Sec. 14.6, p. 720.

**(o) By the Leuckart reaction :** The reductive amination of high boiling ketones and aldehydes by ammonium formate or formamide is called the Leuckart reaction. It is carried out by heating the carbonyl compound with ammonium formate or formamide; α-phenylethylamine (α-methylbenzylamine) is usually prepared in the laboratory from acetophenone by the Leuckart reaction. This is a good means for preparing primary amines (N-formyl derivatives often form).

$$\text{PhCOCH}_3 + \text{HCOO}\overset{\oplus}{\text{N}}\text{H}_4 \xrightarrow{\text{heat}} \underset{\substack{|\\ \text{CH}_3}}{\text{PhCHNH}_2}$$

The mechanism of the reaction is perhaps as follows :

$$HCOONH_4 \rightleftharpoons HCOOH + NH_3;\ H\overset{O}{\overset{\|}{C}}-NH_2 + H_2O \longrightarrow HCOOH + NH_3$$

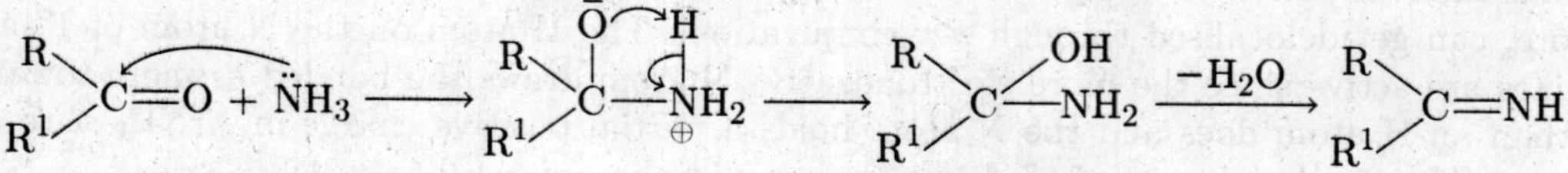

$R^1$ = H or alkyl or aryl group

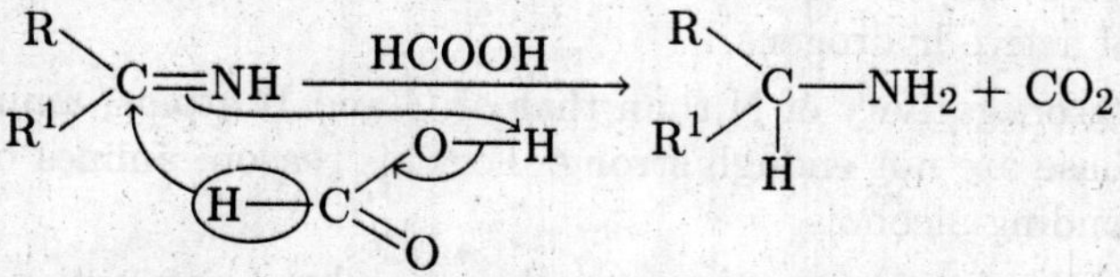

**(p) By the Ritter reaction :** When 3° alcohols or alkenes are treated with alkali metal cyanides (or hydrogen cyanide) in the presence of sulphuric acid, amides form which on hydrolysis give primary amines.

$$R_3C-OH \xrightarrow{H^{\oplus}/\bar{C}N} R_3C-\overset{\oplus}{N}\equiv\bar{C}: \xrightarrow{H_2O} R_3CNHCHO \xrightarrow{\bar{O}H} R_3C-NH_2 + HCO_2^-$$

The mechanism is as follows :

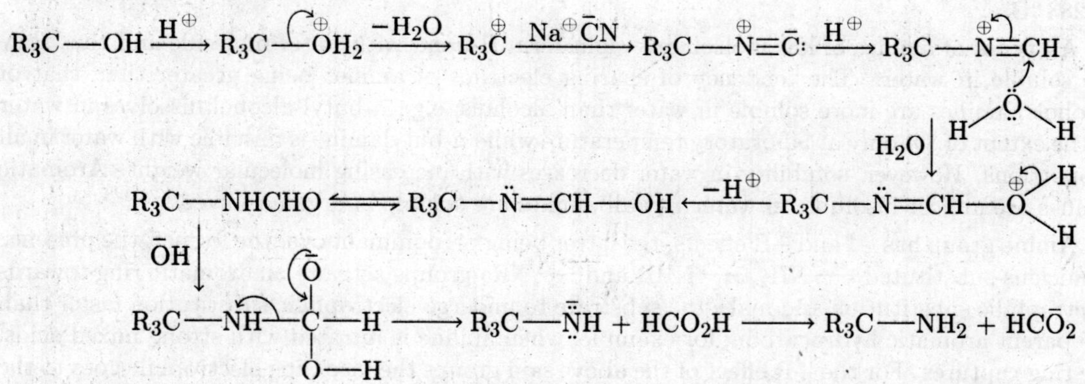

The secondary amines are best prepared from the *p*-toluenesulphonamide and N, N-dialkyl-*p*-nitrosoanilines and the tertiary amines are prepared by the ammonolysis of alkyl halides and reductive amination of carbonyl compounds using excess of alkyl halide and carbonyl compounds respectively.

Quaternary ammonium salts are prepared by heating 3° amines with alkyl halides.

$$R_3N + RX \longrightarrow R_4\overset{\oplus}{N}X^-;\ Me_3N + EtI \longrightarrow Me_3\overset{\oplus}{N}EtI^-$$

Quaternary ammonium hydroxides may be parpared by treating quaternary ammonium salts with aqueous alkali or moist silver oxide.

$$R_4\overset{\oplus}{N}X^- + NaOH \longrightarrow R_4\overset{\oplus}{N}\bar{O}H + NaX$$

$$R_4\overset{\oplus}{N}X^- + \text{'AgOH'} \longrightarrow R_4\overset{\oplus}{N}\bar{O}H + AgX$$

### 20.15.3 Structure, reactivity and physical properties

Alkyl amines possess an $sp^3$ hybridised N atom which bears a lone pair of *e*s in an $sp^3$ A.O. Alkyl amines have pyramidal shape. However, the N atom of an aromatic amine is $sp^2$ hybridised and holds the lone pair of *e*s in a *p* A.O. which is parallel to the *p* A.O.s of the aromatic ring and, therefore, can get delocalised through $p$-$\pi$ conjugation. The H atom on the N atom of 1° and 2° amines are active since the more electronegative N atom draws the bonded *e*s more towards itself than an H atom does and the N atom holds a partial positive charge in $ArNH_2$ through resonance. The dipole moment of aliphatic amines is of the order 1.0 to 1.3D.

So, the properties of amines are those owing to the presence of donatable electrons of the N atom and those owing to the presence of active hydrogens.

However, because of the greater electronegativity of N than that of C and H atoms, amines can form intermolecular H-bonds but those are not enough strong. For this reason, amines boil at lower temperatures than the corresponding alcohols.

Since 1° and 2° amines have active H and since a 3° amine does not have any active H, 1° and 2° amines remain associated among themselves through H-bonding but a 3° amine does not. This is reflected in their b.p.s. Methylamine boils at −7 °C, dimethylamine boils at +7 °C while trimethylamine boils at +4 °C. Though the latter has greater polarisability and molecular weight than dimethylamine; dimethylamine boils at a higher temperature than trimethylamine because of the H-bonding. However, the b.p. of amines increases with increasing molecular weight. Aromatic amines boil at much higher temperatures than its aliphatic analogue, e.g., aniline boils at 283 °C.

All types of amines of low molecular weight form H-bond with water molecule and thus they are soluble in water. The tendency of sharing electrons of amines being greater than that of alcohols, amines are more soluble in water than alcohols; e.g., *n*-butyl alcohol dissolves in water to the extent of 8% only at laboratory temperature while *n*-butylamine is miscible with water in all proportions. However, solubility in water decreases with increasing molecular weight. Aromatic amines are almost insoluble in water but all amines are soluble in organic solvents.

Amino group has −I and +R effects, the latter being predominant over the former, the presence of nucleus-substituted —$NH_2$, —NHR and —$NR_2$ groups activate an aromatic ring towards electrophilic substitution and make the substrate to undergo electrophilic substitution faster than the parent aromatic hydrocarbon; for example, when aniline is nitrated with strong mixed acids, the ring ruptures. For the +R effect of the above said groups the incoming electrophile goes to the $\beta$ and $\delta$-positions (*o*- and *p*- in the case of benzene ring). Furthermore, owing to this activating effect, the aromatic amines have no tendency to undergo nucleophilic substitution reactions.

However, the mechanism of the Ar.$S_E2$ reactions of an aromatic amine may be shown as given below :

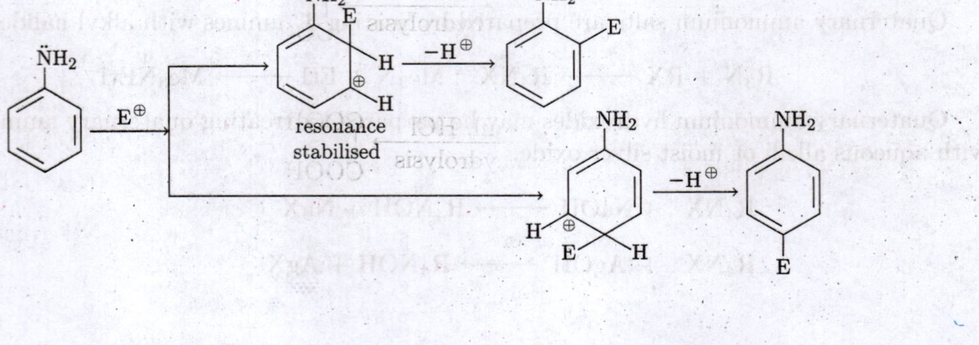

## 20.15.4 Separation of three types of amines from their mixture

We have seen that three types of amines and quaternary ammonium derivatives are formed when alkyl halides and alcohols are ammonolysed, and carbonyl compounds are subjected to the reductive amination reaction (in the cases of alcohols and reductive amination quaternary compounds do not form). So, a separational procedure is needed in order to obtain the three types of amines in the pure state. When their mixture with quaternary ammonium hydroxide is distilled, the three types of amines distil out and the quaternary ammonium hydroxide is left in the distilling flask; in this way the latter is separated from the amines. A mixture of amine salts and quaternary ammonium salts may thus be separated by treating the mixture with an aqueous solution of sodium hydroxide and then distilling out the amines.

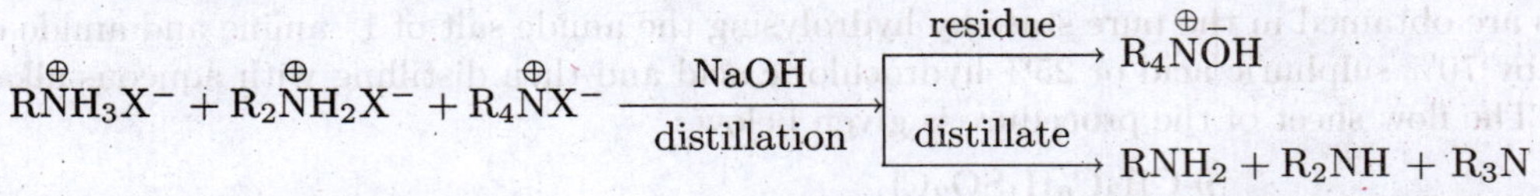

The three types of amines may then be separated from their mixture by any one of the following processes (reductive amination of 3° amines does not form) :

**(a) Fractional Distillation :** The process is used in industry and this cannot be applied in the laboratory.

**(b) Hofmann's method :** The process utilises the fact that a 1° amine contains two active H atoms, a 2° amine one active H atom and a 3° amine none. So when a mixture of three types of amines is treated with ethyl oxalate, a 1° amine forms a solid oxamide, a 2° amine forms a liquid oxamic ester while a 3° amine remains unreacted. So the 3° amine is distilled out after treating the mixture with ethyl oxalate. The oxamide and the oxamic ester being solid and liquid respectively, they are separated by the process of filtration. The 1° and 2° amines are then obtained from the corresponding oxamide and oxamic ester by separately hydrolysing them with hydrochloric acid and then distilling them out with an aqueous solution of sodium hydroxide. This is a laboratory process. However, the process is obsolete nowadays. The flow sheet of the separation is given below :

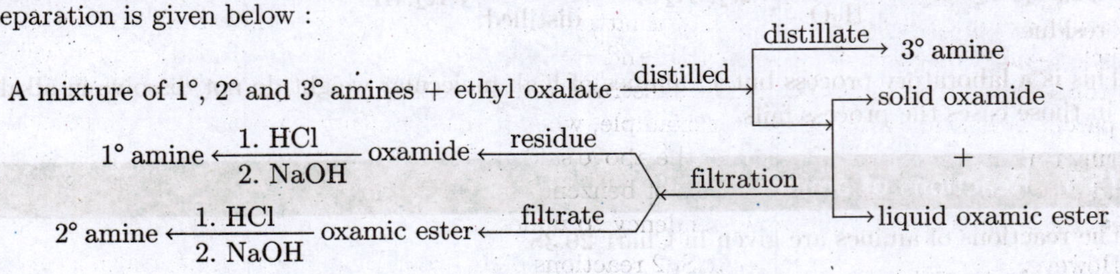

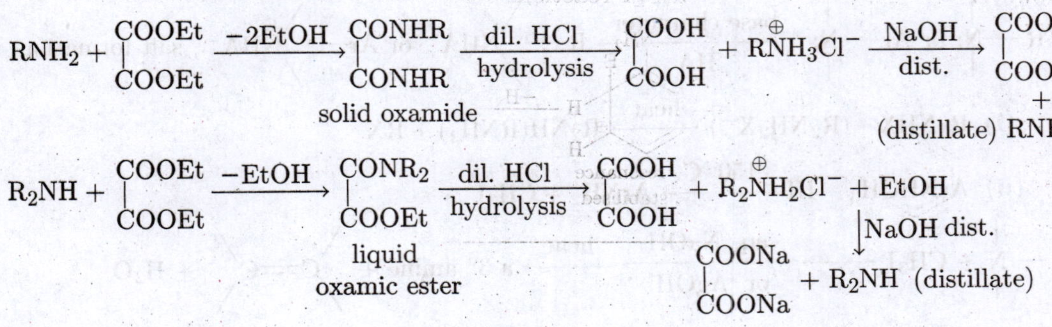

**(c) Hinsberg's method :** This also utilises the presence of active H atoms in 1° and 2° amines and its absence in a 3° amine. In this case the reagent is *p*-toluenesulphonyl chloride. This reacts with a 1° amine to form alkali-soluble N-alkyl-*p*-toluenesulphonamide, *p*-$CH_3C_6H_4SO_2NHR$; since the product contains an H atom bonded to the N atom which, in its turn, is bonded to sulphur containing two double-bonded O atoms, the H atom is acidic and the compound dissolves in alkali to form salt. In the case of a 2° amine N, N-dialkyl-*p*-toluenesulphonamide (*p*-$CH_3C_6H_4SO_2NR_2$) forms by the reaction with *p*-toluenesulphonyl chloride. This amide does not contain any acidic H and hence insoluble in the aqueous alkali. A 3° amine does not at all react with the reagent since it has no active H. The amide of a 1° amine being soluble in aqueous alkali while that of a 2° amine being insoluble, the two is separated by treating their mixture with aqueous alkali and filtering the alkaline solution. The corresponding amines are obtained in the pure state by hydrolysing the amide salt of 1° amine and amide of 2° amine by 70% sulphuric acid or 25% hydrochloric acid and then distilling with aqueous alkali as usual. The flow sheet of the procedure is given below :

$$R_3N + R_2NH + RNH_2 \xrightarrow{p\text{-}CH_3C_6H_4SO_2Cl} \underbrace{R_3N + p\text{-}CH_3C_6H_4SO_2NHR + p\text{-}CH_3C_6H_4SO_2NR_2}_{\text{distilled}}$$

$$p\text{-}CH_3C_6H_4SO_2NR_2 + p\text{-}CH_3C_6H_4SO_2\bar{N}RK^{\oplus}\ (\text{salt}) \xleftarrow[\text{KOH}]{\text{aq.}} \underset{\text{(in the distilling flask)}}{p\text{-}CH_3C_6H_4SO_2NHR + p\text{-}CH_3C_6H_4SO_2NR_2} + R_3N\ (\text{distillate})$$

filtered:

$$\underset{\text{filtrate}}{p\text{-}CH_3C_6H_4SO_2\bar{N}R\,K^{\oplus}} \xrightarrow[H_2O]{25\%\ HCl} R\overset{\oplus}{N}H_3Cl^- \xrightarrow[\text{distilled}]{\text{aq. NaOH}} RNH_2$$

$$\underset{\text{residue}}{p\text{-}CH_3C_6H_4SO_2NR_2} \xrightarrow[H_2O]{25\%\ HCl} R_2\overset{\oplus}{N}H_2Cl^- \xrightarrow[\text{distilled}]{\text{aq. NaOH}} R_2NH$$

This is a laboratory process but 1° amines of high molecular weight do not dissolve in alkali and in those cases the process fails.

## 20.15.5 Reactions of amines

The reactions of amines are given in Chart 20.38.

(a) $R{-}\overset{|}{\underset{|}{N}}{:}$ or $Ar{-}\overset{|}{\underset{|}{N}}{:}$ $\xrightarrow[HA]{\text{base character}}$ $R{-}\overset{\oplus}{N}HA^-$ or $Ar{-}\overset{\oplus}{N}HA^-$ salt formation

(b) (i) $R_3\overset{\oplus}{N}HX^-\ (R_2\overset{\oplus}{N}H_2X^-) \xrightarrow{\text{heat}} R_2NH(RNH_2) + RX$

(ii) $ArNHCH_3 + HI \xrightarrow{150\,°C} ArNH_2 + CH_3I$

(c) $-\overset{|}{\underset{|}{N}} + CH_3I \longrightarrow \xrightarrow[\text{or 'AgOH'}]{\text{aq. NaOH}} \xrightarrow{\text{heat}}$ a 3° amine + $\gt C{=}C\lt$ + $H_2O$

**(d)** Ethyl oxalate $\xrightarrow{RNH_2 \text{ or } R_2NH} \begin{matrix} CONHR \\ | \\ CONHR \end{matrix} \left( \text{or } \begin{matrix} COOEt \\ | \\ CONR_2 \end{matrix} \right) + EtOH$

**(e)** *p*-toluenesulphonyl chloride $\xrightarrow[\text{R may be Ar}]{RNH_2 \text{ or } R_2NH}$ N-alkyl (or N, N-dialkyl)-*p*-toluenesulphonamide

**(f)** **(i)** $RNH_2 + HONO \longrightarrow ROH + N_2 + H_2O$ + many side products

**(ii)** $R_2NH + HONO \longrightarrow R_2N{\cdot}NO + H_2O$ **(iii)** $R_3N + HONO \longrightarrow R_3N{\cdot}HNO_2$

**(iv)** $ArNH_2 + HONO \longrightarrow Ar—\ddot{N}{=}\overset{\oplus}{N}Cl^- \longleftrightarrow Ar—\overset{\oplus}{N}{\equiv}NCl^- \longleftrightarrow$ etc.

**(v)** $PhNR_2 + HONO \longrightarrow$ N, N-dialkyl-*p*-nitrosoaniline

**(g)** **(i)** $RNH_2 \xrightarrow[\text{benzoylation}]{\text{acetylation or}} RNHCOMe \text{ (or } RNHCOPh) \xrightarrow[\text{benzoylation}]{\text{acetylation or}} RN(COMe)_2$ or $RN(COPh)_2$

**(ii)** $R_2NH \xrightarrow[\text{benzoylation}]{\text{acetylation or}} R_2NCOMe \text{ or } R_2NCOPh$

**(h)** **(i)** $RNH_2 + CS_2 \longrightarrow S{=}C\begin{matrix} NHR \\ SH \end{matrix} \xrightarrow{HgCl_2} HgS\text{(black ppt.)} + HCl$

**(ii)** $R_2NH + CS_2 \longrightarrow S{=}C\begin{matrix} NH_2 \\ SH \end{matrix} \xrightarrow{HgCl_2}$ no black ppt.

**(i)** $RNH_2 \text{ or } R_2NH \xrightarrow[\text{active H compound and a base or acid}]{\text{HCHO \& an}}$ Mannich base or salt (Mannich reaction)

**(j)** **(i)** $R_3C—NH_2 \xrightarrow[\text{[O]}]{KMnO_4} R_3C—NO_2$ **(ii)** $R_3C—NH_2 \xrightarrow[\text{[O]}]{H_2SO_5} R_3C{\cdot}NO$

**(iii)** $R_3N \xrightarrow[\text{or } H_2O_2]{H_2SO_5 \text{ or } O_3} R_3\overset{\oplus}{N}—\bar{O} \xrightarrow[\text{reaction}]{\text{Cope}} \rangle C{=}C \langle + R_2NOH$

**(k)** $RNH_2 \text{ or } ArNH_2 \xrightarrow[\text{ethanolic KOH}]{CHCl_3 \text{ and}} RNC \text{ or } ArNC$

**(l)** Aniline $\xrightarrow{Br_2}$ 2, 4, 6-tribromoaniline

**(m)** Aniline $\xrightarrow{H_2SO_4} Ph\overset{\oplus}{N}H_3HSO_4^- \xrightarrow{180\,°C} PhNHSO_3H \xrightarrow[\text{3 hr}]{180\,°C}$ sulphanilic acid

**(n)** N-alkylanilium hydrochloride $\xrightarrow[\text{Hofmann-Martius reaction}]{250°C}$ *p*-alkylanilinium salt

**(o)** $RCHO \text{ or } ArCHO + H_2NR \longrightarrow RCH{=}NR \text{ (or } ArCH{=}NR)$

**Chart 20.38** Reactions of amines.

**(a) Base character of amines and their salt-formation reactions :** Owing to the presence of a lone pair of *es* on the N atom of amines, they are basic. Their relative basicities may be discussed considering the usual effects such as inductive effect, resonance effect, solvation effect, steric effect, etc. and these have already been discussed in the eighth chapter. We shall discuss here the basicity of a few specific compounds only. Let us first take the case of guanidine which is the strongest base among the organic compounds so far known to us. The conjugate acid of guanidine is very stable since it is a resonance hybrid of three equivalent resonating structures. Being very stable, the conjugate acid is a weak acid and consequently the parent base is very strong.

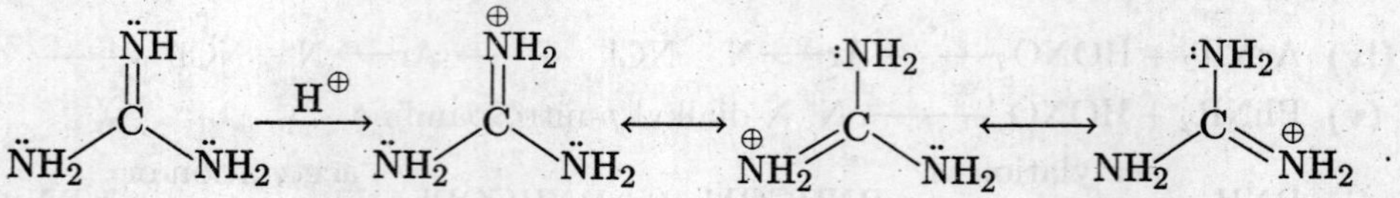

The resonance hybrid of the conjugate acid is :

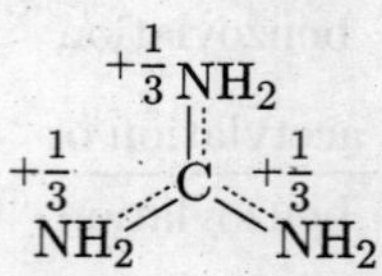

The relative basicities of the *p*-haloanilines deserve special discussion. The *pKa* value of the *p*-haloanilines are given below :

| Compounds | *pKa* value |
|---|---|
| *p*-fluoroaniline | 4.65 |
| *p*-chloroaniline | 3.98 |
| *p*-bromaniline | 3.86 |
| *p*-iodoaniline | 3.78 |

We know that the lower the $pK_a$ value, the higher is the acidity. So, *p*-iodoaniline is the strongest acid and the *p*-fluoroaniline is the weakest among haloanilines. This means that *p*-fluoroaniline is the strongest base among them and the basicity decreases down the series though the halogens exhibit decreasing $-$I effect in the order F > CI > Br > I. This is because iodine atom is the best *d*-$\pi$-acceptor whereas chlorine atom is the worst *d*-$\pi$-acceptor and fluorine has no *d*-$\pi$-acceptor character since it has no *d* orbital. Thus the lone pair of *es* on the N atom of haloanilines delocalises through resonance interaction involving *p*-$\pi$—*d*-$\pi$ bonding as shown below :

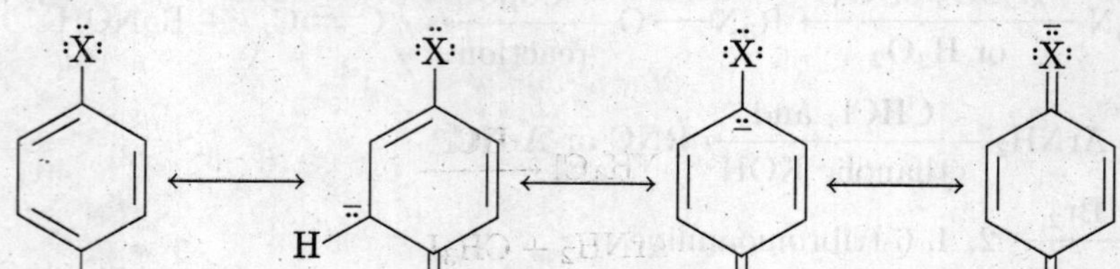

X = I, Br, Cl

Iodine atom being the best *d*-$\pi$-acceptor, the delocalisation of *es* is maximum in the case of *p*-iodoaniline and hence it has the least electron donatability and, therefore, it shows the weakest base character and the strongest acid character.

It has also been found that 2, 6-dimethyl-4-nitroaniline is less basic than 3, 5-dimethyl-4-nitroaniline, though the +I effect of two methyl groups in the *o*-positions (with respect to the amino group) is expected to increase the electron density on the —$NH_2$ group of the former compound in comparison to the +I effect of the methyl groups in *m*-positions in the latter compound. Since the bulk of —$NO_2$ group is much greater than that of —$NH_2$ group, steric inhibition of resonance occurs in 3, 5-dimethyl-4-nitroaniline to a greater degree than in 2, 6-dimethyl-4-nitroaniline. Owing to the steric inhibition of resonance, the lone pair of *e*s on the N atom of the —$NH_2$ group in 3, 5-dimethyl-4-nitroaniline delocalises to a lesser extent than those in 2, 6-dimethyl-4-nitroaniline. This makes the former a stronger base than the latter.

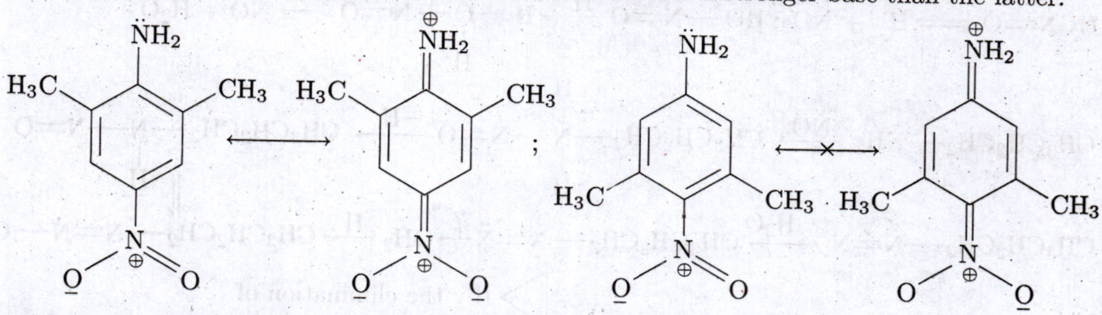

However, amines being basic, they form salts with acids.

$$RNH_2 + HA \longrightarrow R\overset{\oplus}{N}H_3A^-;\; R_2NH + HA \longrightarrow R_2\overset{\oplus}{N}H_2A^-$$

$$R_3N + HA \longrightarrow R_3\overset{\oplus}{N}HA^-;\; ArNH_2 + HA \longrightarrow Ar\overset{\oplus}{N}H_3A^-$$

Amines also form complex salts with chloroauric and chloroplatinic acid; from these salts the equivalent weight and hence the molecular weight of an amine can be determined.

**(b) Pyrolysis of ammonium salts :** When tertiary ammonium salts are heated, 2° amines are formed; whereas a 2° ammonium salt on heating gives a 1° amine. However, arylmethylammonium iodides on heating at 150 °C give methyl iodide and derivatives of aniline. Thus, estimating the amount of methyl iodide formed, the methyl-imino group of a compound can be estimated. This constitutes the basis of the Herzig-Meyer method of methyl-imino group estimation.

$$R_3NH\,X^- \xrightarrow{\Delta} R_2NH + RX\;;\; R_2NH + HCl \longrightarrow R_2\overset{\oplus}{N}H_2Cl^- \xrightarrow{\Delta} RNH_2 + RCl$$

$$RNH_2 \xrightarrow{HCl} R\overset{\oplus}{N}H_3Cl^- \xrightarrow{\Delta} NH_3 + RCl \xrightarrow{HCl} NH_4Cl$$

$$ArNHCH_3 \xrightarrow{HI} Ar\overset{\oplus}{N}H_2CH_3I^- \xrightarrow{\Delta} ArNH_2 + CH_3I$$

**(c) The Hofmann's exhaustive methylation :** See Sec. 15.4, p. 809.

**(d) Reaction with ethyl oxalate :** This reaction helps us to differentiate three types of amines and forms a basis of separating them from their mixture (which is known as the Hofmann's reaction and has already been discussed under the Sec. 20.15.4. A 3° amine does not undergo the reaction because of the absence of active H atoms.

**(e) Reaction with *p*-toluenesulphonyl chloride :** This is known as the Hinsberg reaction and has already been discussed under the Sec. 20.15.4. This reaction also differentiate three types of amines; a 3° amine does not undergo the reaction since it has no active H atoms.

**(f) Reaction with nitrous acid :** This reaction is a good means for differentiating three types of amines. Aliphatic 1° amines on reaction with nitrous acid give alcohols and evolve molecular nitrogen along with several other side products as shown below :

$$NaNO_2 + HCl \longrightarrow NaCl + HNO_2$$

$$CH_3CH_2CH_2—\ddot{N}H_2 + HNO_2 \longrightarrow CH_3CH_2CH_2OH + N_2\uparrow + H_2O$$

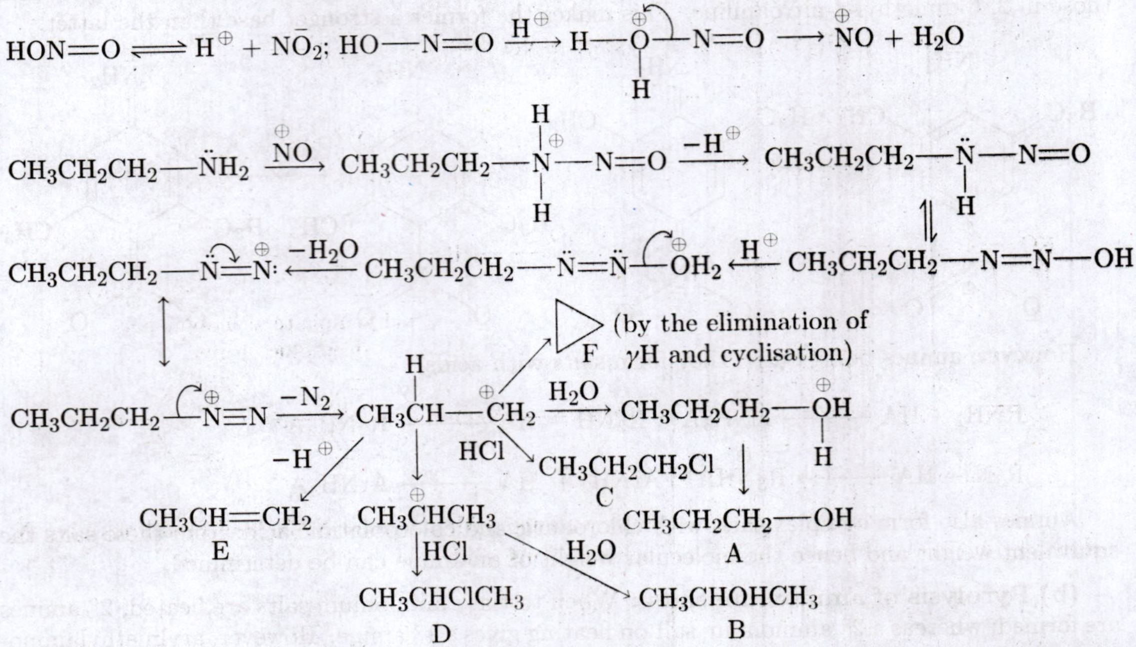

Here it is important to note that nitrous acid being unstable, this cannot be kept in the laboratory and it is always generated in situ by the reaction of mineral acids on nitrites; the mineral acid reacts further with diazonium ion to form alkyl derivatives of the acid.

However, an aromatic 1° amine forms arenediazonium salt on treatment with nitrous acid. This is because :

**(i)** arenediazonium salts at a low temperature (0-5 °C) are much more stable than aliphatic diazonium salts since resonance stabilises the former more than the latter.

**(ii)** Phenyl cation is very unstable; since the vacant $sp^2$ orbital is perpendicular to *p* A. O.s. of the aromatic system, the positive charge developed on a C atom cannot get delocalised.

However, when a 1° aromatic amine in 2.5 to 3 equivalent of dilute HCl or $H_2SO_4$ is treated with dilute aqueous sodium nitrite solution at 0-5 °C, an arenediazonium salt forms.

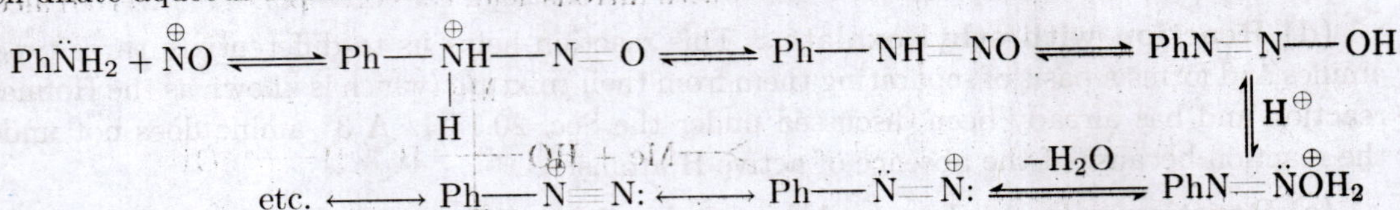

A 2° amine, aliphatic or aromatic or mixed, forms an N-nitrosoamine when it is treated with nitrous acid. These are steam volatile yellow oils and they may be identified by the **Libermann's**

nitrosoamine test. This test involves heating of a mixture of nitrosoamine, a few crystals of phenol and concentrated sulphuric acid when a deep blue or green solution develops which on treatment with aqueous sodium hydroxide turns to a deep blue solution. Therefore, the treatment of a 2° amine with nitrous acid and the subsequent Libermann's nitrosoamine test constitute a good test for a 2° amine.

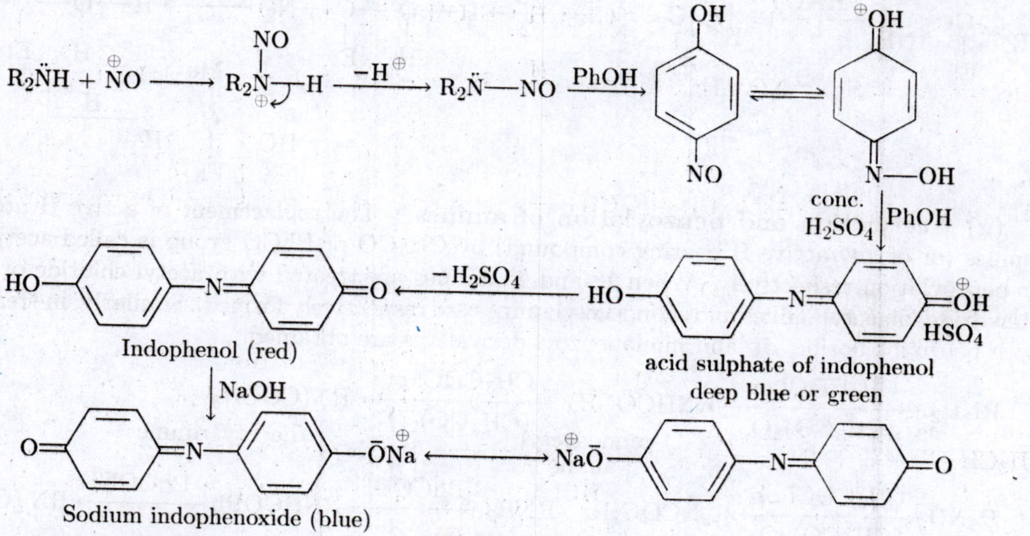


An aliphatic 3° amine forms a salt with nitrous acid; whereas N, N-dialkylanilines give N, N-dialkyl-*p*-nitrosoanilines on treatment with nitrous acid.

$$R_3\ddot{N} + HNO_2 \longrightarrow R_3\overset{\oplus}{N}HN\bar{O}_2 \equiv R_3N \cdot HNO_2$$

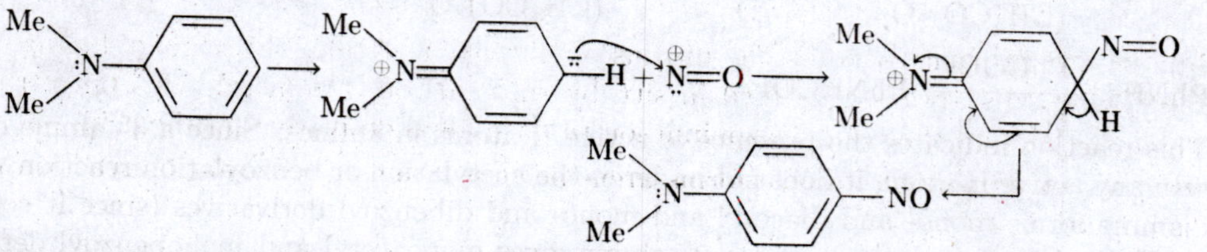

So, the three types of aliphatic amines give different products with nitrous acid :

(i) a 1° amine evolves molecular nitrogen, (ii) a 2° amine forms nitrosoamine, (iii) a 3° amine forms nitrite salt.

The mechanism given on p. 1358 for the reaction between nitrous acid and 1° amine is not supported by the following fact.

When an optically active amine is treated with nitrous acid, the corresponding inverted alcohol is formed in major amount.

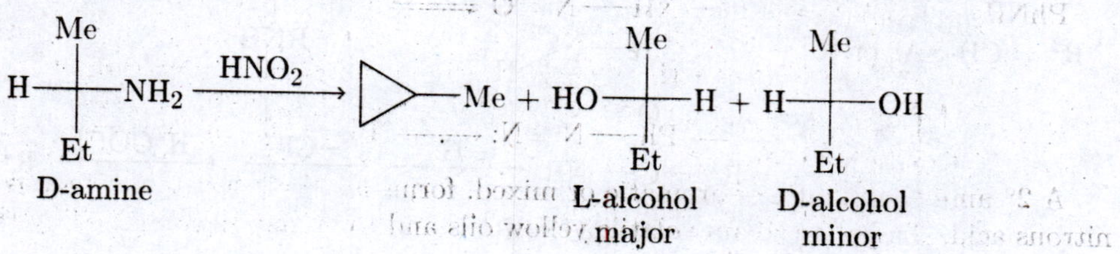

This suggests that a carbenium ion is not an intermediate, but an $S_N2$ reaction takes place on the diazonium ion in which $N_2$ is the leaving group and $H_2O$ is the attacking nucleophile as shown below :

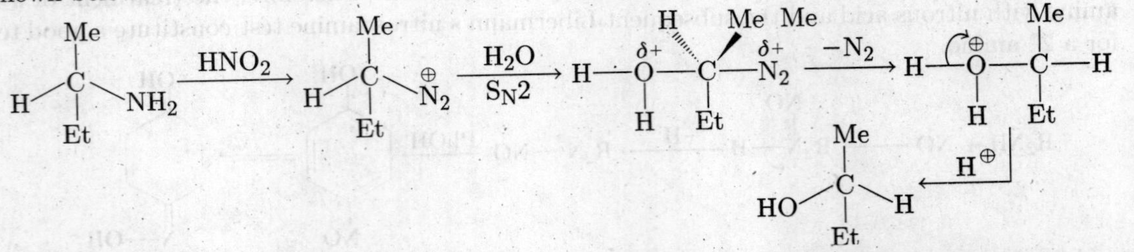

**(g) Acetylation and benzoylation of amines :** The replacement of active H atoms of amines (or of any active H bearing compound) by $CH_3CO$ or PhCO group is called acetylation or benzoylation respectively. When 1° and 2° amines are treated with acetyl chloride or acetic anhydride, diacetylamines and monoacetylamines are respectively formed. Similarly, in treatment with benzoyl chloride, di- and monobenzoyl derivatives are obtained.

$$RNH_2 \xrightarrow[(CH_3CO)_2O]{CH_3COCl \text{ or }} \underset{\text{monoacetyl amine}}{RNHCOCH_3} \xrightarrow[(CH_3CO)_2O]{CH_3COCl \text{ or }} \underset{\text{diacetyl amine}}{RN(COCH_3)_2}$$

$$R_2NH \xrightarrow[(CH_3CO)_2O]{CH_3COCl \text{ or }} R_2NCOCH_3; \quad RNH_2 \xrightarrow{PhCOCl} RNHCOPh \xrightarrow{PhCOCl} RN\,(COPh)_2$$

$$R_2NH \xrightarrow{PhCOCl} R_2NCOPh$$

$$PhNH_2 \xrightarrow[(CH_3CO)_2O]{CH_3COCl \text{ or }} PhNHCOCH_3 \xrightarrow[(CH_3CO)_2O]{CH_3COCl \text{ or }} PhN(COCH_3)_2$$

$$PhNH_2 \xrightarrow{PhCOCl} PhNHCOPh$$

This reaction indicates the presence of active H atoms in amines. Since a 3° amine does not contain any active H atom, it does not undergo the acetylation or benzoylation reaction; whereas a 1° amine forms mono- and diacetyl and mono- and dibenzoyl derivatives (since it contains 2 active H atoms per molecule), and a 2° amine gives monoacetyl and monobenzoyl derivatives. It is needless to mention that diacetyl and dibenzoyl derivatives are a bit difficult to prepare since the N atom becomes sterically hindered after monoacetylation or monobenzoylation stage. However, acetylation is a special case of acylation; so amines can be acylated by acyl chlorides or anhydrides. The reaction follows a tetrahedral mechanism as shown below :

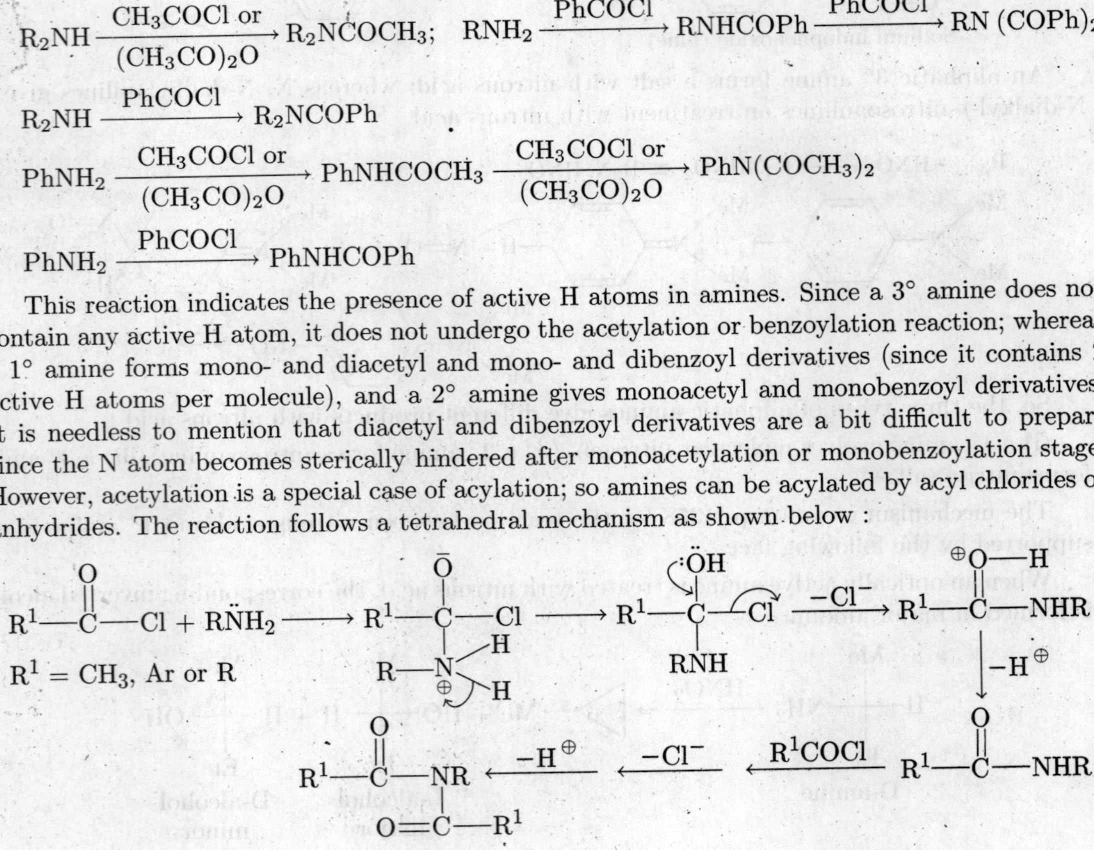

Similarly,

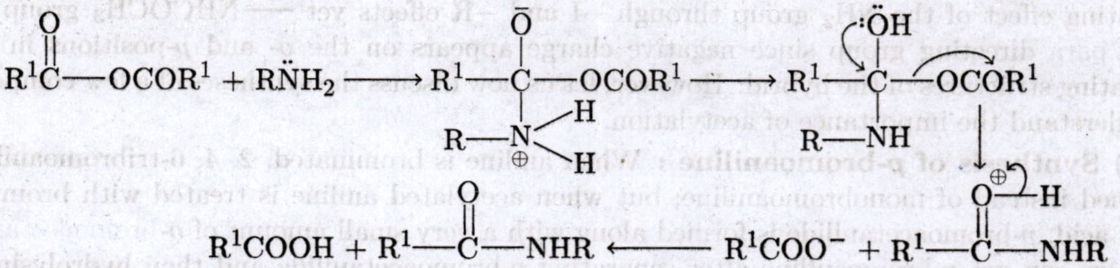

The acylation and benzoylation reactions are reversible and acids (HCl or RCOOH) are formed. So the acyl and benzoyl derivatives get hydrolysed to the corresponding amine under the conditions of the reaction. In order to avoid hydrolysis the acid formed is removed by pyridine as pyridinium salt. However, diacetyl and dibenzoyl derivatives are unstable and they get hydrolysed during crystallisation.

Acetylation and benzoylation processes have analytical as well as synthetic values.

Acetyl and benzoyl derivatives of compounds containing active H are crystalline solids with sharp melting points. So phenols and amines are identified from their acetyl and benzoyl derivatives.

Since aromatic amines and phenols are highly reactive owing to the presence of activating groups ($NH_2$ and OH respectively), they very often form polysubstituted products through aromatic electrophilic substitution reactions and sometimes ring rupturing also occurs. When activating groups are acetylated or benzoylated, the activity of the group decreases owing to the $-I$ and $-R$ effects of the carbonyl group; because of the low activity of the acetylated (or benzoylated) group, monosubstituted products are usually obtained when they are subjected to electrophilic substitution reactions. Acetylated and benzoylated groups have ortho/para directing effect; but such groups being very bulky, para products are obtained predominantly. Thus acetylation and benzoylation are used to protect the activating group in order to introduce substituents in aromatic amines and phenols. After the introduction of the substituent on the acetylated or benzoylated derivative, the compound is hydrolysed to remove the acetyl or benzoyl group.

The following syntheses will explain the importance of the acetylation or benzoylation process. However, the effect of the acetyl group on the activating character of the amino group is first represented below :

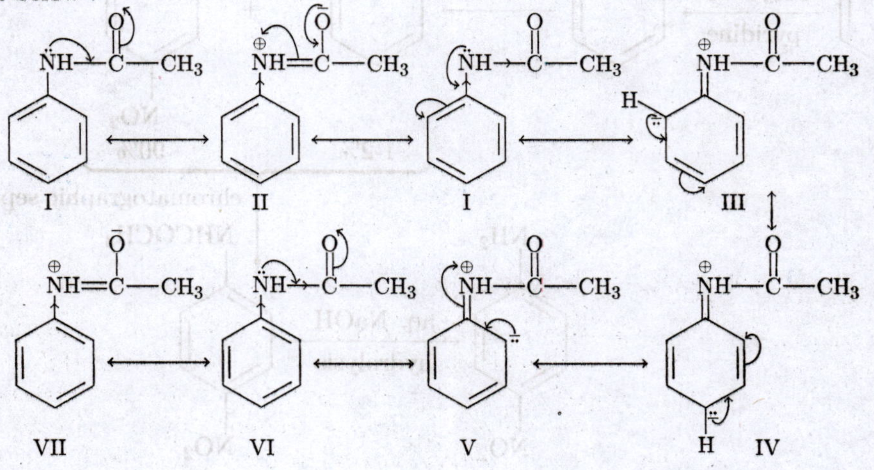

The resonance-hybrid of the acetanilide shows that though the acetyl group decreases the activating effect of the $NH_2$ group through $-I$ and $-R$ effects yet —$NHCOCH_3$ group is an ortho/para directing group since negative charge appears on the *o*- and *p*-positions in three resonating structures of the hybrid. However, let us now discuss the syntheses of a few compounds to understand the importance of acetylation.

**(i) Synthesis of *p*-bromoaniline :** When aniline is brominated, 2, 4, 6-tribromoaniline is obtained instead of monobromoaniline; but when acetylated aniline is treated with bromine in acetic acid, *p*-bromoacetanilide is formed along with a very small amount of *o*-bromoacetanilide. Thus we can get *p*-bromoaniline after separating *p*-bromoacetanilide and then hydrolysing the latter with aqueous alkali.

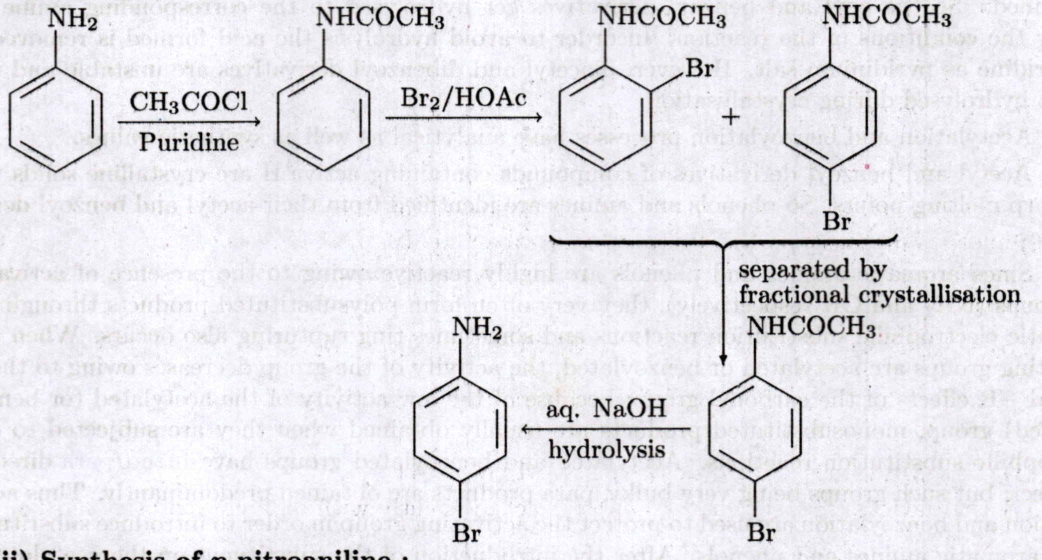


**(ii) Synthesis of *p*-nitroaniline :**

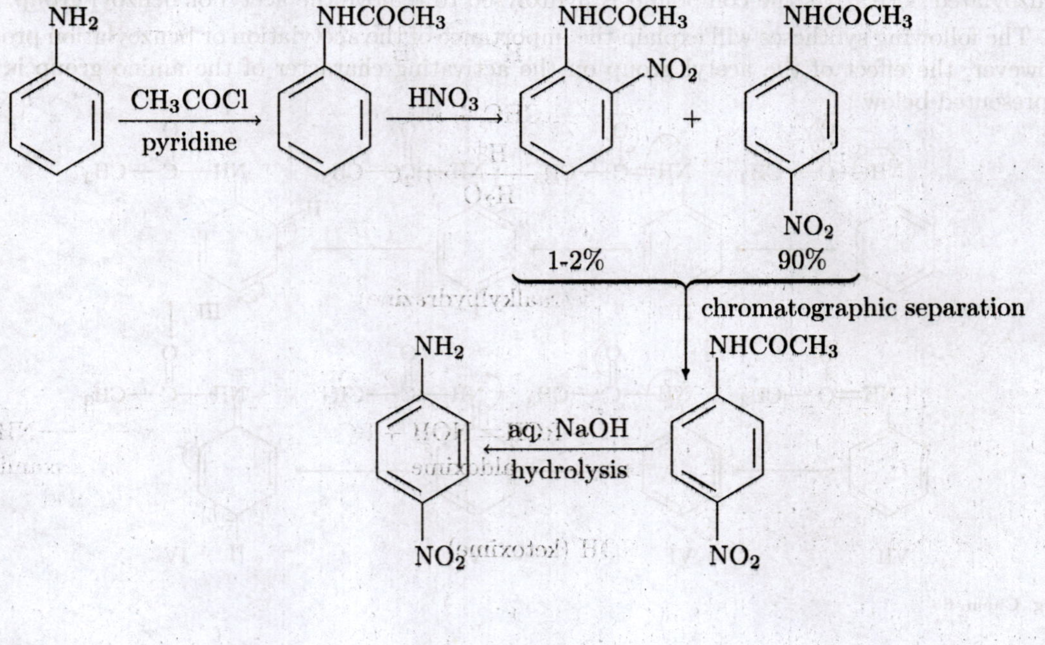

**(h) Reaction with carbon disulphide :** This is also a differential reaction; a 1° amine reacts with carbon disulphide to form N-alkyl-dithiocarbamic acid which gets decomposed to alkyl isothiocyanate and mercuric sulphide in warming with mercuric chloride solution, while a 2° amine forms N, N-dialkyl-dithiocarbamic acid which does not decompose to isothiocyanate on treatment with mercuric chloride solution and a 3° amine does not at all react with carbon disulphide—this reaction is popularly known as the Hofmann's mustard oil reaction.

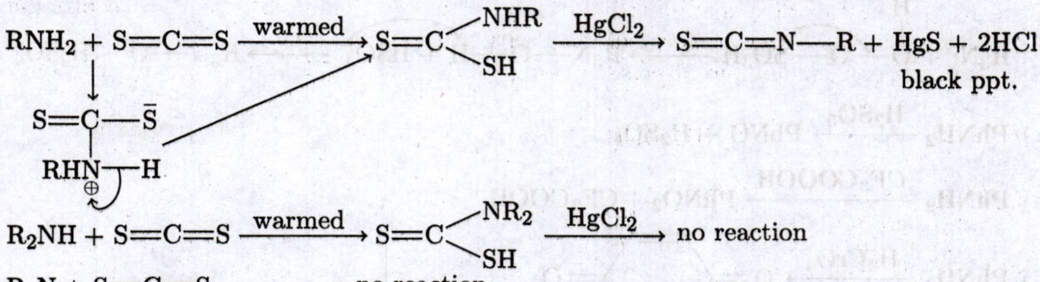

$$R_3N + S{=}C{=}S \longrightarrow \text{no reaction}$$

However, when aniline is refluxed with ethanolic solution of carbon disulphide and solid potassium hydroxide solution, N, N′-diphenyl thiocarbanilide (m.p. 154 °C) is obtained. This is used as a rubber accelerator and forms phenyl isothiocyanate when it is treated with concentrated hydrochloric acid.

$$CS_2 + 2\,PhNH_2 + 2\,KOH \longrightarrow S = C(NHPh)_2 + K_2S + H_2O$$

$$S = C(NHPh)_2 \xrightarrow{\text{HCl (conc.)}} PhNCS$$

**(i) The Mannich reaction :** See Sec. 16.13.8, p. 911.

**(j) Oxidation :** Three types of amines react differently with different oxidising agents.

(i) 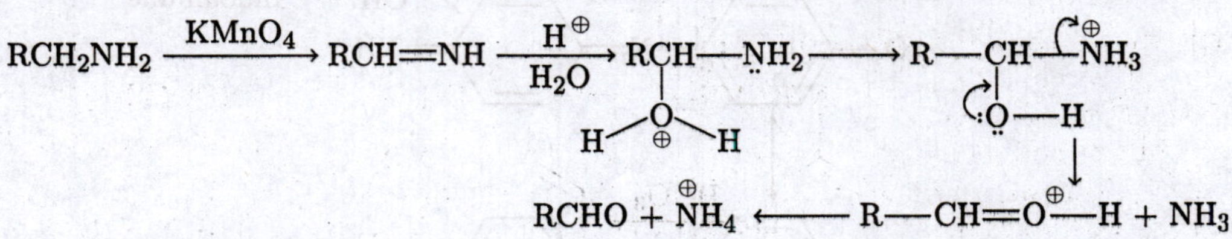

(ii) $R_2CH{-}NH_2 \xrightarrow{KMnO_4} R_2C{=}NH \xrightarrow[H_2O]{H^{\oplus}} R_2C{=}O + \overset{\oplus}{N}H_4$

(iii) $R_3C\,NH_2 \xrightarrow{KMnO_4} R_3CNO_2$

(iv) $R_2NH \xrightarrow{KMnO_4} R_2N{-}NR_2$ (tetraalkylhydrazine)

$R_3N \xrightarrow{KMnO_4}$ no reaction

(v) $RCH_2NH_2 \xrightarrow{H_2SO_5} RCH_2NHOH + RCH{=}NOH + RC(OH){=}NOH \rightleftharpoons R{-}C(=O){-}NHOH$

N-alkylhydroxylamine; aldoxime; hydroxamic acid

(vi) $R_2CHNH_2 \xrightarrow{H_2SO_5} R_2C{=}NOH$ (ketoxime)

(vii) $R_3CNH_2 \xrightarrow{H_2SO_5} R_3CNO$ (3° nitrosoalkane)

(viii) $R_2NH \xrightarrow{H_2SO_5} R_2NOH$ (N, N-dialkylhydroxylamine)

(ix) $R_3N \xrightarrow[\text{or } O_3]{H_2SO_5 \text{ or } H_2O_2} R_3\overset{\oplus}{N}—O^- \xrightarrow{\Delta}$ Cope reaction

$$R_3N\!: + \overset{H}{\underset{|}{O}}—O—SO_3H \longrightarrow R_3\overset{\oplus}{N}—O—H + HSO_4^- \longrightarrow R_3\overset{\oplus}{N}—\bar{O} + H_2SO_4$$

(x) $PhNH_2 \xrightarrow{H_2SO_5} PhNO + H_2SO_4$

(xi) $PhNH_2 \xrightarrow{CF_3COOOH} PhNO_2 + CF_3COOH$

(xii) $PhNH_2 \xrightarrow{H_2CrO_4}$

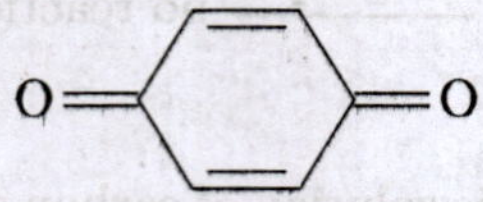


(xiii) $PhNH_2 \xrightarrow{NaOCl}$

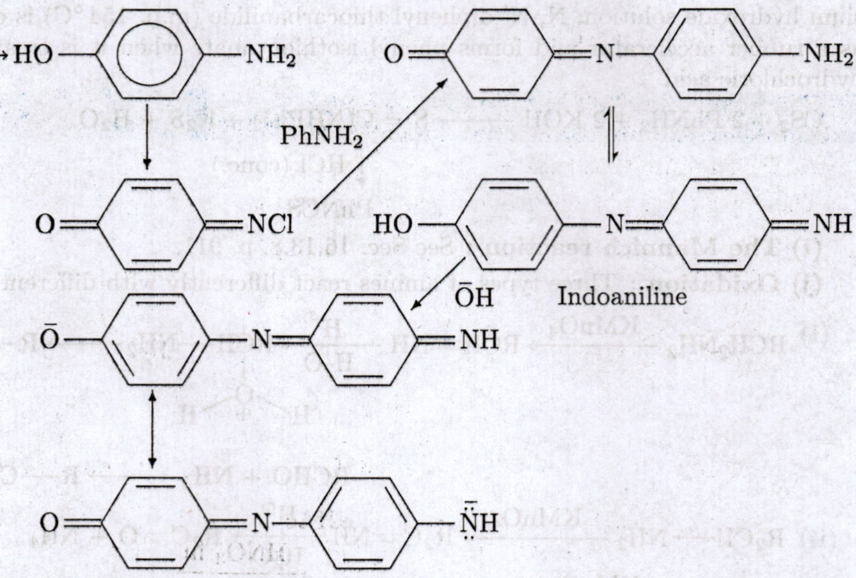


**(k) Formation of isocyanides :** See Sec. 20.13.2.b.

**(l) Reaction of aniline with bromine :**

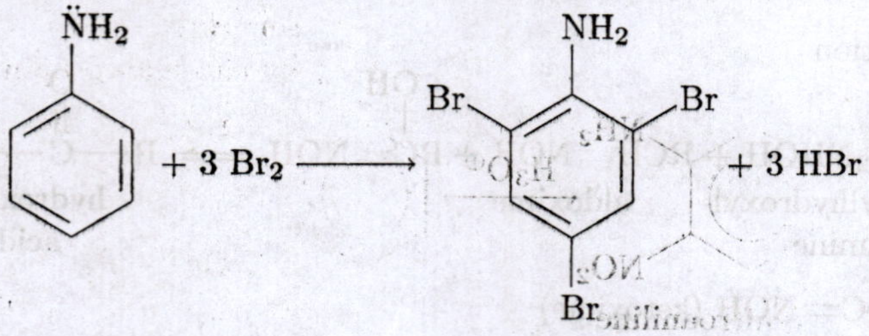

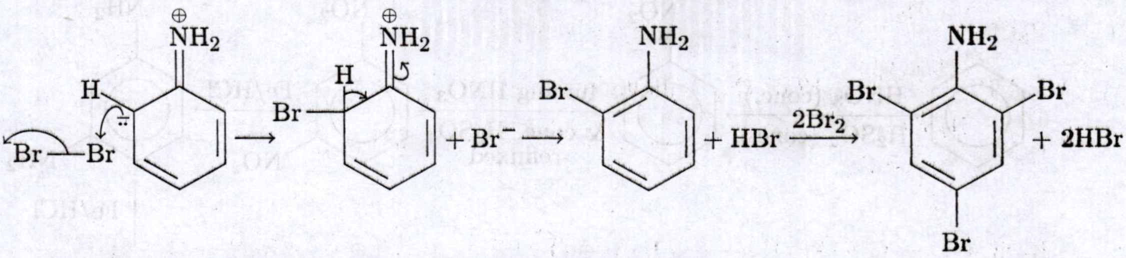


**(m) Sulphonation of aniline :** See Prob. 17.9. p. 967.

**(n) The Hofmann-Martius reaction :** See Sec. 14.12, p. 742.

**(o) N-alkyl- or arylaldimine formation :**

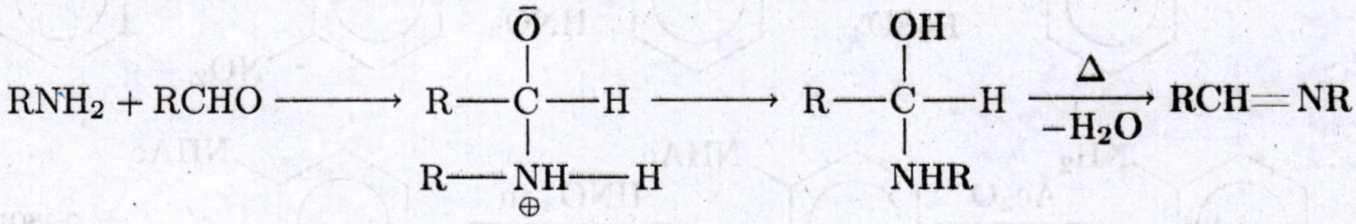


$$ArNH_2 + ArCHO \xrightarrow{-H_2O} ArN{=}CHAr$$

## 20.15.6 A few derivatives of aniline

**A. Phenylenediamines :**

There are three phenylenediamines, *o*-, *m*- and *p*-.

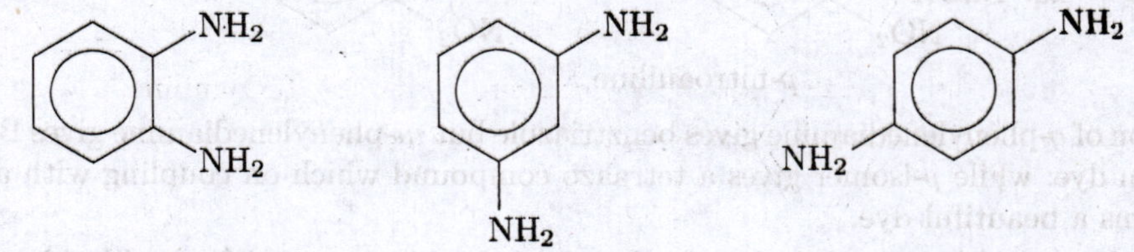


| ortho | meta | para |
|---|---|---|
| m.p. 102 °C | m.p. 63 °C | m.p. 147 °C |

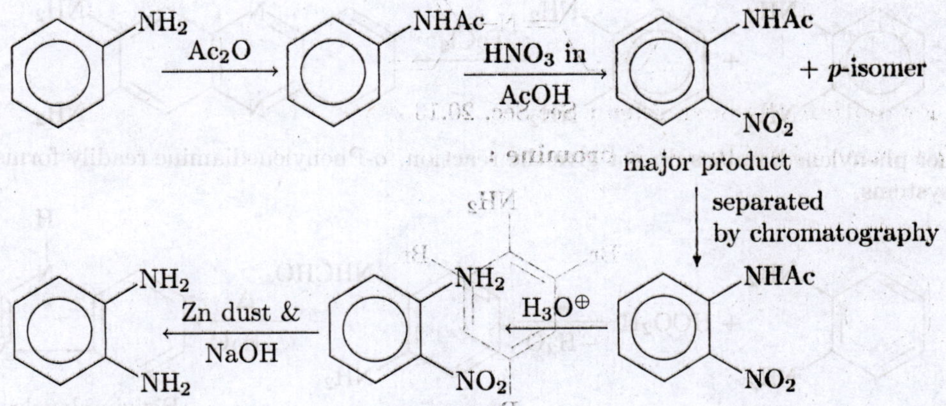

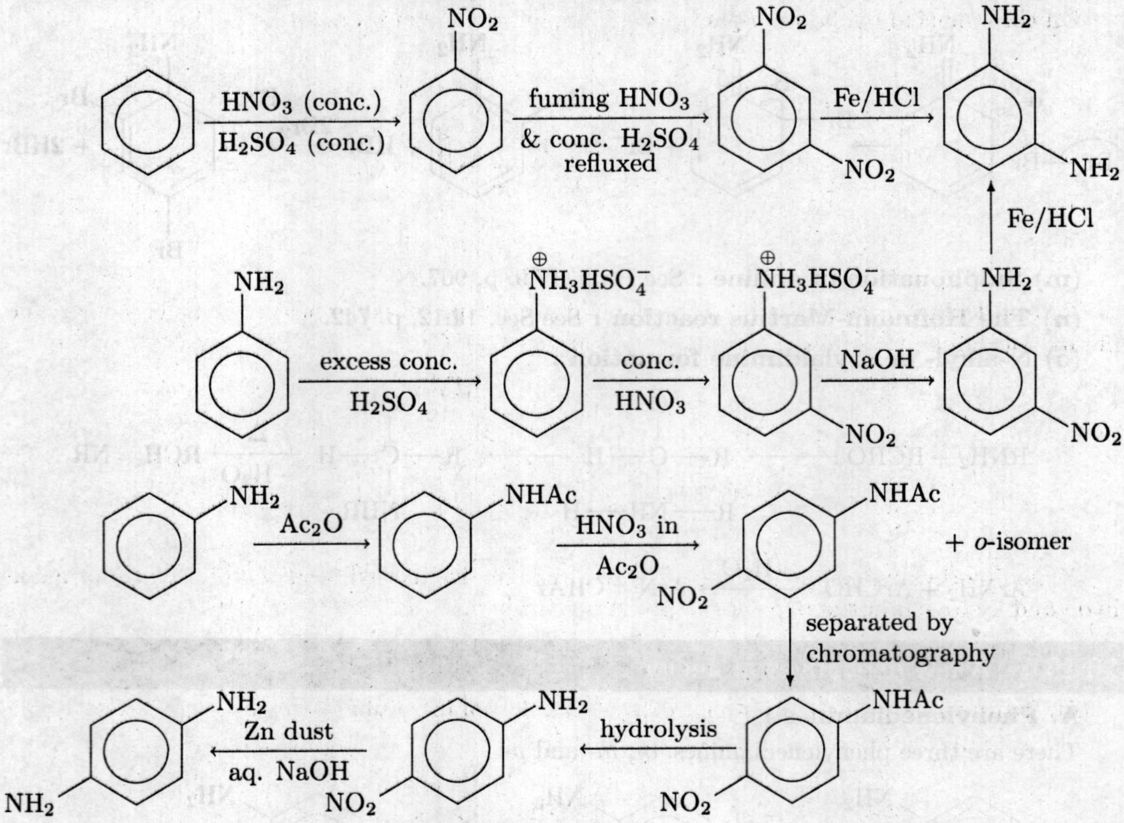

*p*-nitroaniline

Diazotisation of *o*-phenylenediamine gives benztriazole but *m*-phenylenediamine gives Bismark Brown, a brown dye; while *p*-isomer gives a tetraazo compound which on coupling with alkaline $\beta$-naphthol forms a beautiful dye.

*o*-Phenylenediamine developes a dark red colour with freshly prepared ferric chloride solution owing to the formation of 2, 3-diaminophenazine.

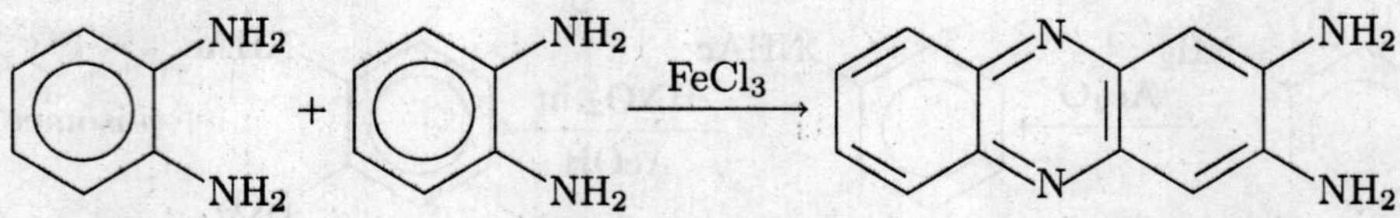

Other phenylenediamines do not give the reaction. *o*-Phenylenediamine readily forms heterocyclic systems.

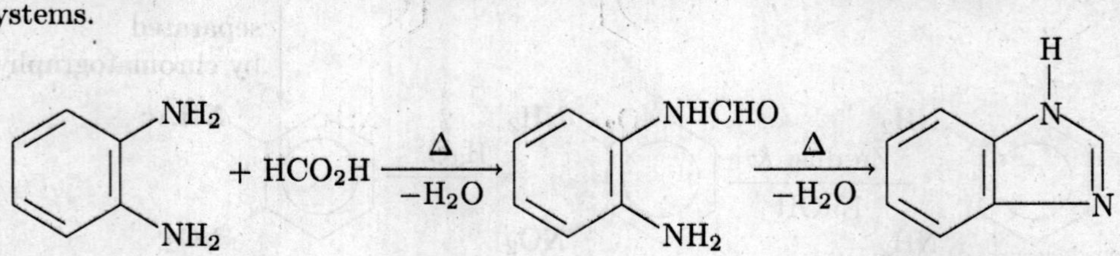

Benzimidazoles

On chromic acid oxidation *p*-phenylenediamine forms *p*-benzoquinone.

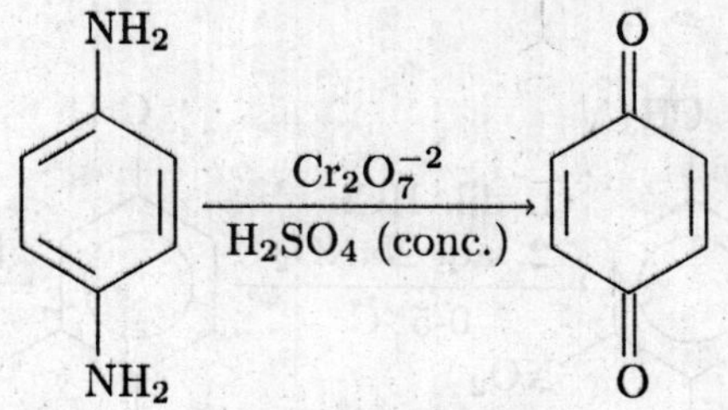


**B. N-alkylanilines :** In industry methylanilines are prepared by heating a mixture of aniline, methyl alcohol and sulphuric acid under pressure at 230 °C.

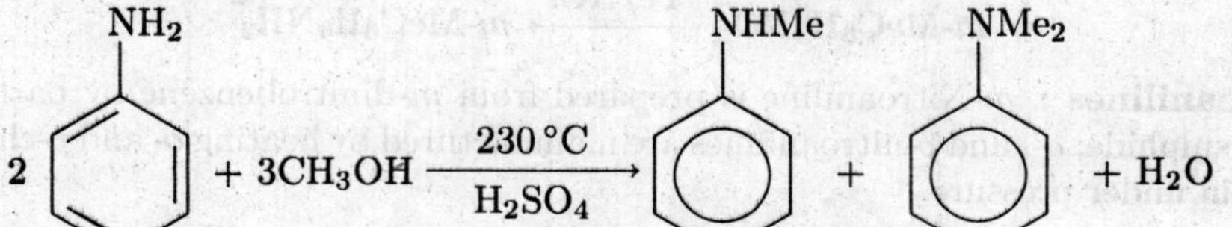


N, N-dimethylaniline may also be prepared by reductive methylation of aniline with formaldehyde and hydrogen in the presence of platinum catalyst.

$$PhNH_2 + HCHO + H_2 \xrightarrow{Pt} PhNMe_2 + H_2$$

**C. Toluidines :** These are prepared by reduction of nitrotoluenes.

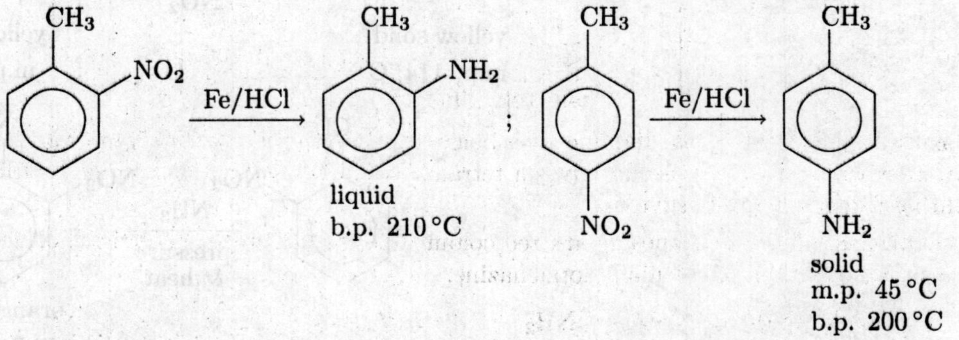


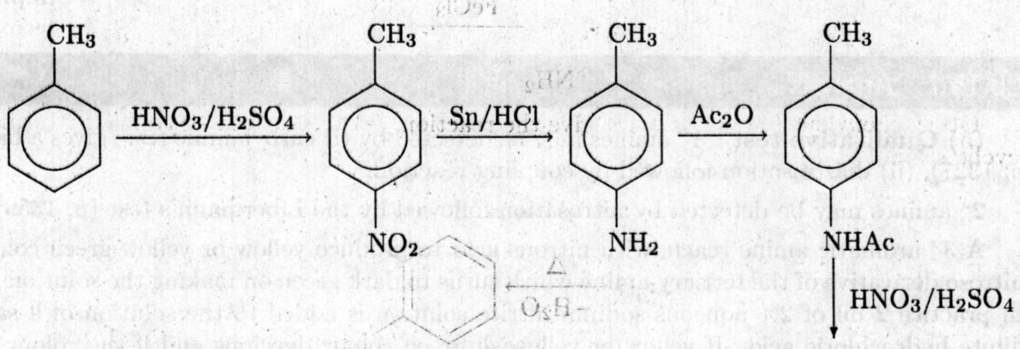

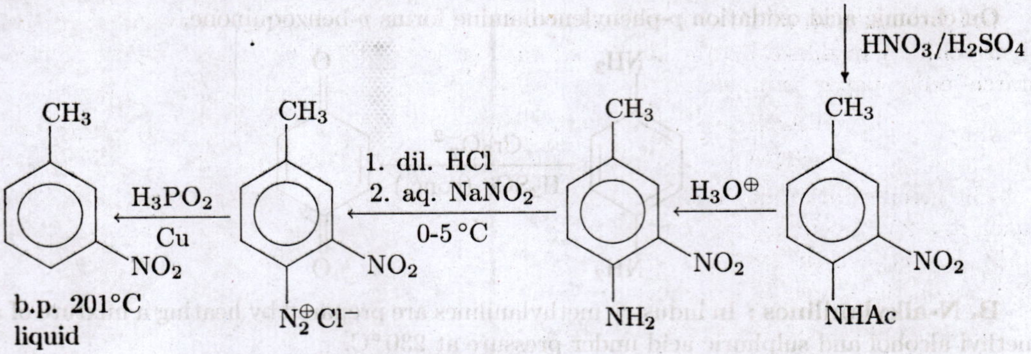


$$m\text{-}MeC_6H_4NO_2 \xrightarrow{Fe/HCl} m\text{-}MeC_6H_4NH_2$$

**D. Nitroanilines :** *m*-Nitroaniline is prepared from *m*-dinitrobenzene by partial reduction with sodium sulphide; *o*-, and *p*-nitroanilines are manufactured by heating *o*- and *p*-chloronitrobenzene with ammonia under pressure.

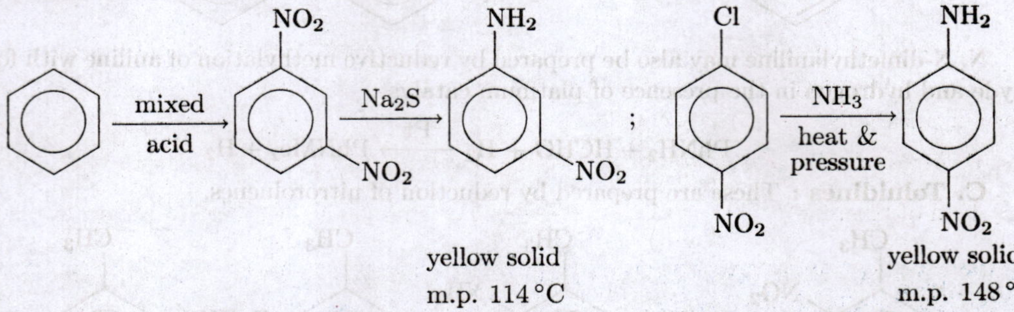


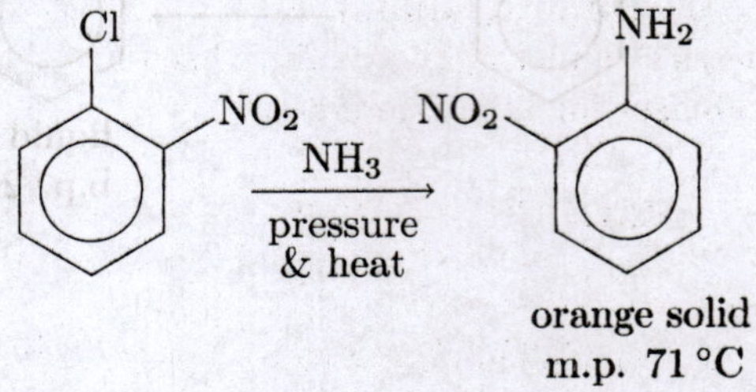


## 20.15.7 Analysis

**(a) Qualitative test :** 1° amines may be detected by (i) carbylamine test (gives a bad smell, p. 1327), (ii) diazotisation followed by coupling reaction.

2° amines may be detected by nitrosation followed by the Libermann's test (p. 1359).

A 3° aromatic amine reacts with nitrous acid to produce yellow or yellow-green colour of *p*-nitroso derivative of the tertiary amine which turns to dark green on making the solution alkaline. In practice 2 ml of 2% aqueous sodium nitrite solution is added to the solution of a sample in dilute hydrochloric acid. If yellow or yellowish green colour develops and if the colour deepens on making the solution alkaline, the sample is expected to be a 3° aromatic amine.

**(b) Estimation of 1° aromatic amines :** These compounds may be estimated by titrating their solution in dilute hydrochloric acid with standard aqueous sodium nitrite solution using starch-iodide paper as indicator.

$$ArNH_2 + NaNO_2 + 2HCl \longrightarrow ArN_2Cl + NaCl + 2H_2O.$$

$$2HNO_2 + 2HI \longrightarrow 2NO + I_2 + 2H_2O.$$

The given solution of the amine is diluted up to 100 ml with water. 10 ml of this solution is taken in a 250 ml conical flask; to this 10 ml conc. HCl and 75 ml of water are added and cooled to 5 °C. It is then titrated with standard N/10 $NaNO_2$ solution in water until distinct blue ring appears on the starch iodide paper five minutes after the last addition of $NaNO_2$.

**Aqueous sodium nitrite solution is standardised by titrating 25 ml of standard $KMnO_4$ solution (mixed with 50 ml of water and 5 ml of concentrated sulphuric acid). The reaction is carried out after warming the $KMnO_4$ solution to 50 °C. At the end point the pink colour disappears.**

Potassium permanganate is standardised by standard oxalic acid solution as usual :

$$2KMnO_4 + 5NaNO_2 + 3H_2SO_4 = K_2SO_4 + 2MnSO_4 + 5NaNO_3 + 3H_2O$$

$$\text{2 ml of (N/10) } KMnO_4 = \text{2 ml of (N/10) } NaNO_2 = \text{1 ml of (M/10) } NaNO_2$$

$$\text{1 ml of (M/10) } NaNO_2 = \frac{\text{mol. wt. of amine}}{10\,000} \text{ g of amine.}$$

However, aniline may be estimated by the potassium bromide-bromate method as done in the case of phenol (See p. 1211; Sec. 20.3.5). In this case

$$\text{60 litres of KBr-KBrO}_3 \text{ solution} = \text{93 g of aniline}$$

$$\text{or} \quad \text{1 ml of bromine} = \frac{93}{60\,000} \text{ g of aniline} = \text{0.00155 g of aniline.}$$

## 20.15.8 Uses of amines

**(i) Dimethylamine** is mainly used to prepare DMF, DMA (dimethylacetamide). Unsymmetrical dimethylhydrazine is used as a rocket missile fuel; dimethyllaurylamine oxide is a foam stabiliser for liquid detergents.

$$HCOOCH_3 + NH(CH_3)_2 \longrightarrow H{-}\overset{\overset{\displaystyle O}{\|}}{C}{-}NMe_2 + MeOH$$

$$CH_3COOCH_3 + NH(CH_3)_2 \longrightarrow CH_3{-}\overset{\overset{\displaystyle O}{\|}}{C}{-}NMe_2 + MeOH$$

$$(CH_3)_2NH + ClNH_2 \longrightarrow (CH_3)_2N{\cdot}NH_2{\cdot}HCl \text{ (UDMH, unsymmetrical dimethylhydrazine)}$$

**(ii) Methylamine** is mostly used for the manufacture of Sevin, an insecticide.

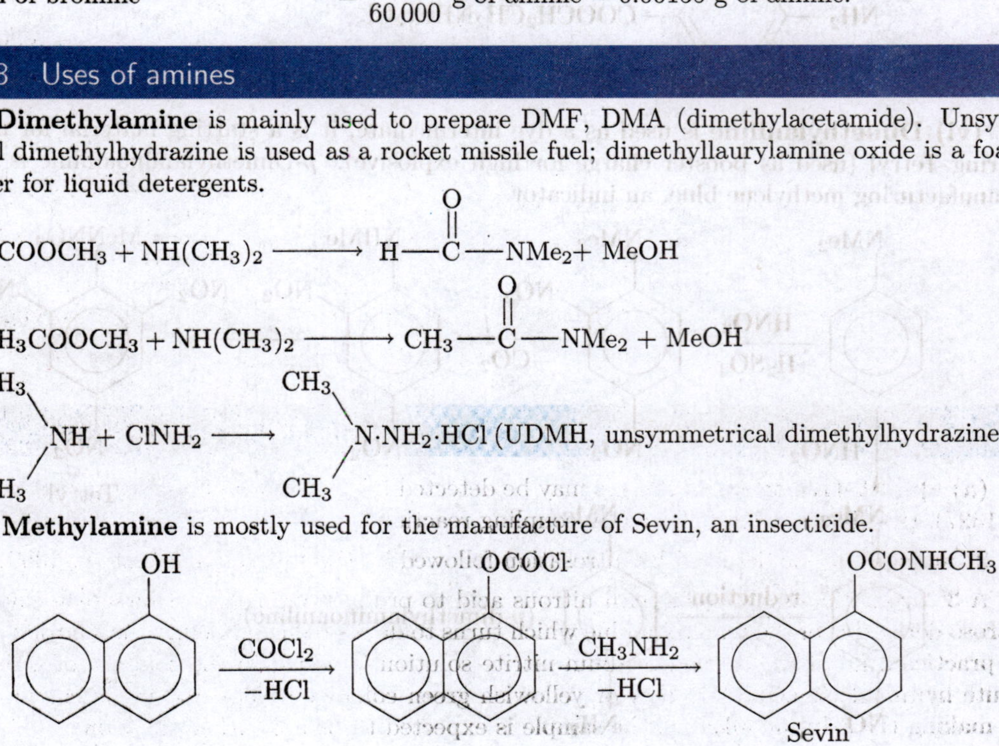

**(iii) Trimethylamine** is chiefly used to prepare choline chloride which is used in animal feed.

$$ClCH_2CH_2Cl + NMe_3 \longrightarrow ClCH_2CH_2—\overset{\oplus}{N}Me_3Cl^-$$

Choline chloride

**(iv) *n*-Butylamines and amylamines** are widely used as anti-oxidants, corrosion inhibitors, starting materials in the manufacture of oil-soluble soaps and as absorbents for acid gases.

**(v) Lidocaine and Novocaine** are used as local anaesthetic.

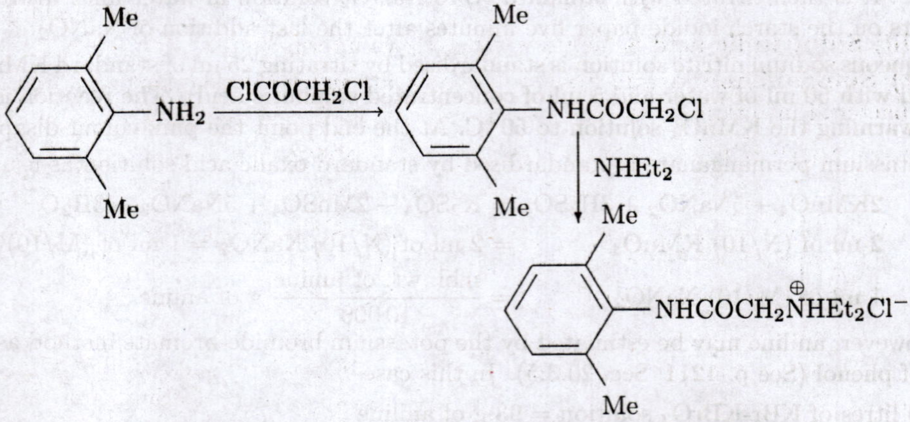

Lidocaine

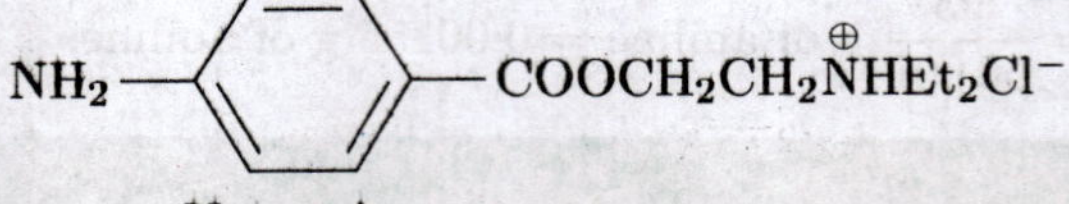

Novocaine

**(vi) Dimethylaniline** is used as a dye intermediate, it is a starting material for manufacturing Tetryl (used as booster charge for high explosive). *p*-Dimethylaminoaniline is used for manufacturing methylene blue, an indicator.

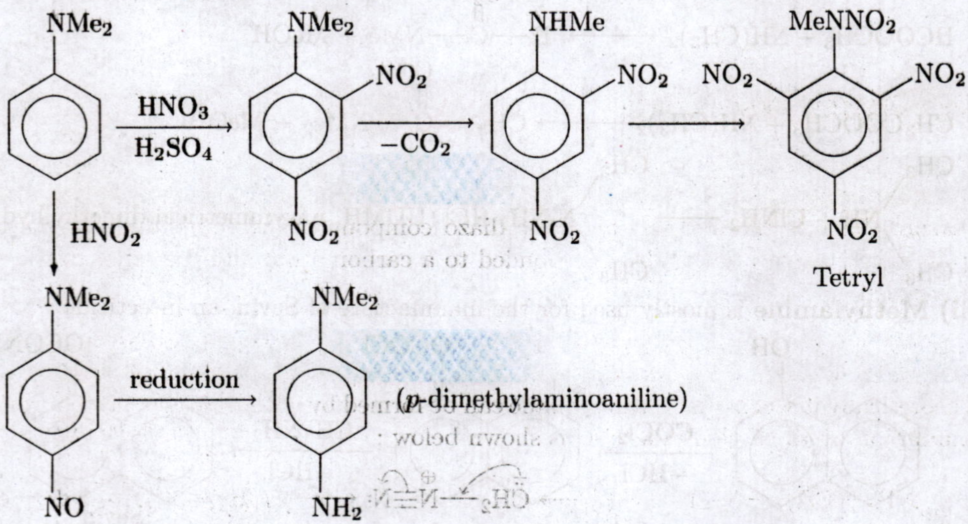

Tetryl

(*p*-dimethylaminoaniline)

**(vii) Diphenylamine** stabilises smokeless powder, the propellant explosive for guns. It is also used to prepare phenothiazine (used as an intestinal disinfectant for animals).

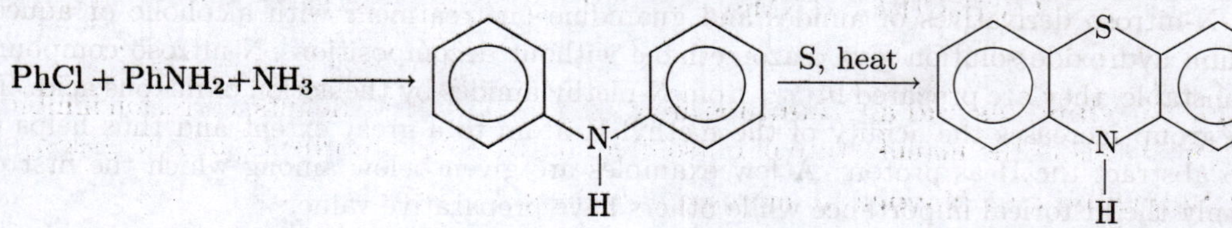

Phenothiazine

**(viii) Sulphanilic acid** is mainly used as a dye intermediate.

**(ix) Sulphanilamide and its derivatives** are used against pneumonia, gonorrhoea and other bacterial infections.

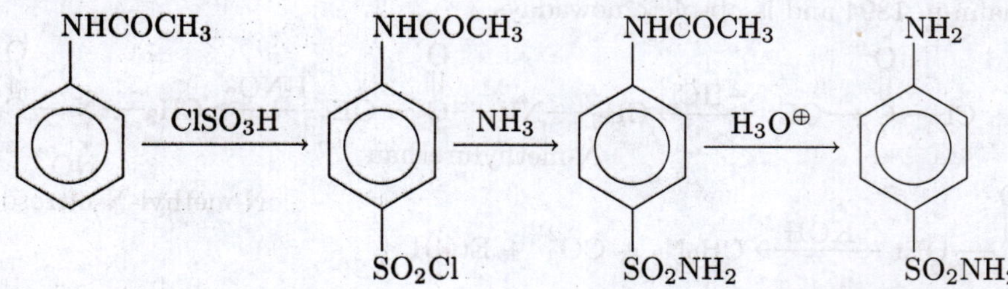

Sulphanilamide

**(x)** ***p, p′*****-diaminodiphenyl sulphone** is used in leprosy treatment.

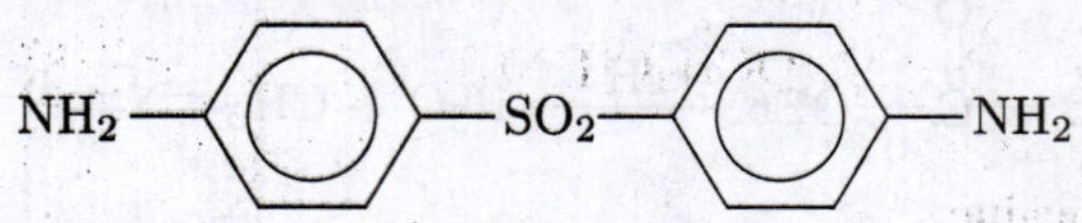

**(xi) N-methylformanilide** (b.p. 253 °C) is used as a formylating agent in the preparation of aldehydes. The compound is prepared by distilling a mixture of methylaniline and formic acid in toluene.

$$PhNHCH_3 + HCO_2H \longrightarrow PhN(Me)—CHO + H_2O$$

## 20.16 Diazomethane ($CH_2N_2$)

### 20.16.1 Introduction

Diazomethane is the simplest aliphatic diazo compound; any compound containing $>CN_2$ group in which one N atom is directly bonded to a carbon atom and the other to the former N atom is called a diazo compound.

### 20.16.2 Formation of diazomethane

Theoretically it seems that diazomethane can be formed by withdrawing a proton from methyldiazonium ion by the action of a base as shown below :

$$B{:} \; H—CH_2—\overset{\oplus}{N}{\equiv}N{:} \xrightarrow[-BH^{\oplus}]{B} \overset{\ominus}{\ddot{C}}H_2—\overset{\oplus}{N}{\equiv}N{:} \longleftrightarrow CH_2{=}\overset{\oplus}{N}{=}\overset{\ominus}{\ddot{N}}{:} \longleftrightarrow \text{etc.}$$

However, no such base is known which converts methyldiazonium chloride to diazomethane but does not decompose diazomethane under the reaction condition. On the other hand, N-methyl-N-nitroso derivatives of amides and guanidine on treatment with alcoholic or aqueous potassium hydroxide solution give diazomethane without decomposition. N-nitroso compounds being unstable, they are prepared in situ from N-methylamides by the action of nitrous acid. The nitroso group increases the acidity of the methyl H atom to a great extent and thus helps the base to abstract the H as proton. A few examples are given below among which the first one bears only the historical importance while others have preparative value.

**1. From N-methylurethan :** When N-methylurethan is treated with nitrous acid in ether medium, N-methyl-N-nitrosourethan forms; the latter on warming with methanolic potassium hydroxide solution gives diazomethane which is collected in cold ether. N-methylurethan is prepared by the reaction between methylamine and ethyl chloroformate. The method is discovered by Von Pechmann in 1894 and is obsolete nowadays.

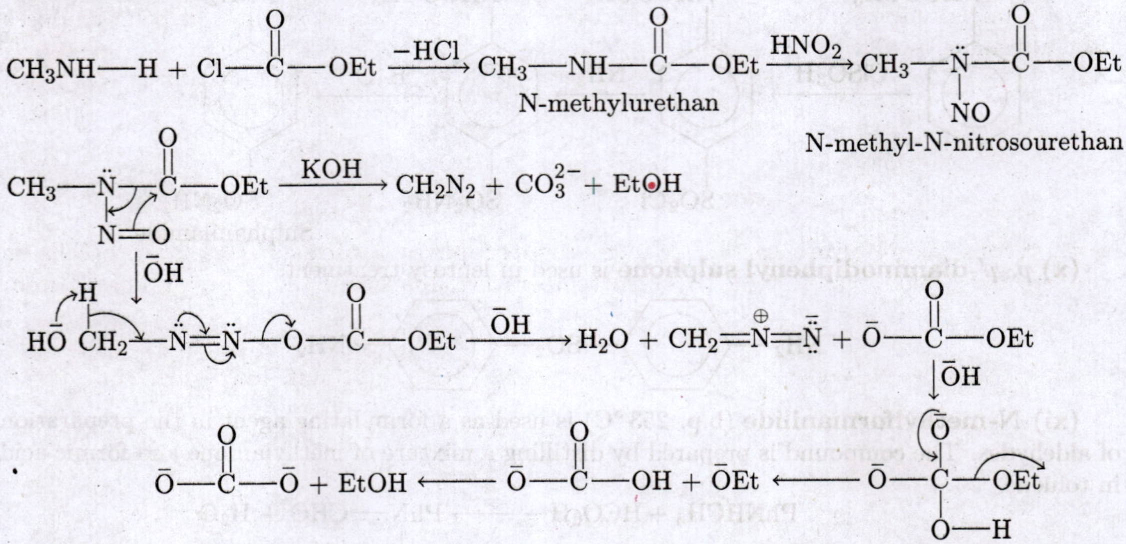

However, diazomethane can be prepared from methylurea by the Von Pechmann method.

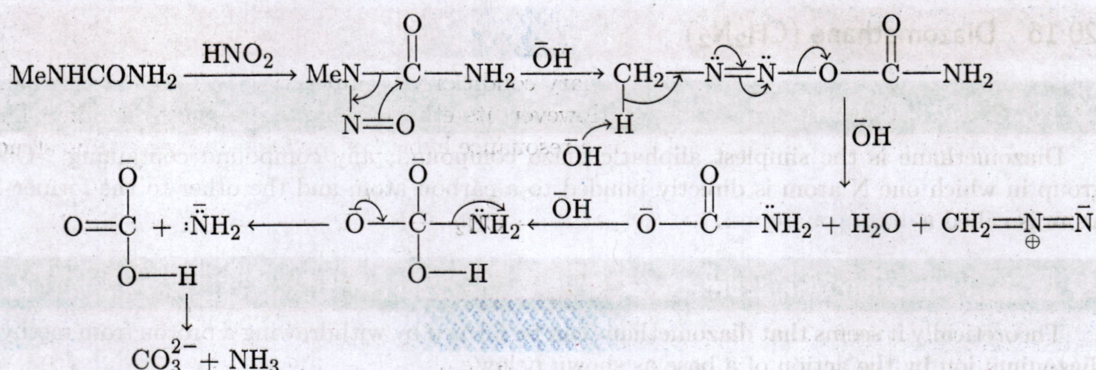

**2. From N-methyl-N-nitroso-N′-nitroguanidine :** Methylammonium chloride on reaction with nitroguanidine in potassium hydroxide solution forms N-methyl-N′-nitroguanidine.

This on treatment with nitrous acid gives N-methyl-N-nitroso-N′-nitroguanidine which on warming with potassium hydroxide solution produces diazomethane.

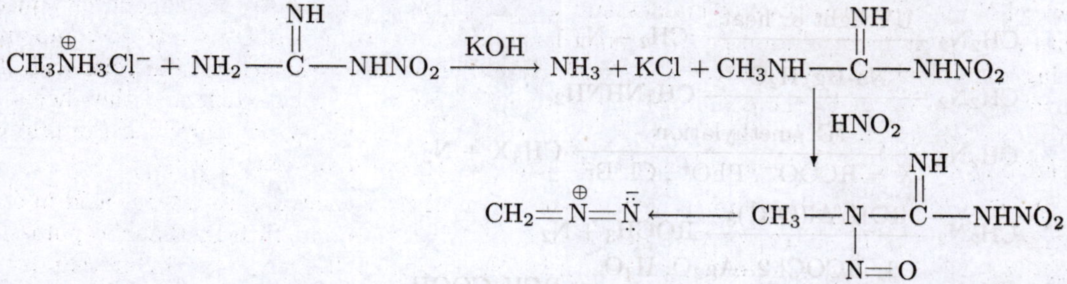

**3. From *p*-toluenesulphonamide :** *p*-toluenesulphonyl chloride on treatment with methylamine gives N-methyl-*p*-toluenesulphonamide. This on nitrous acid treatment forms N-methyl-N-nitroso-*p*-toluenesulphonamide which on warming with ethanolic potassium hydroxide produces diazomethane.

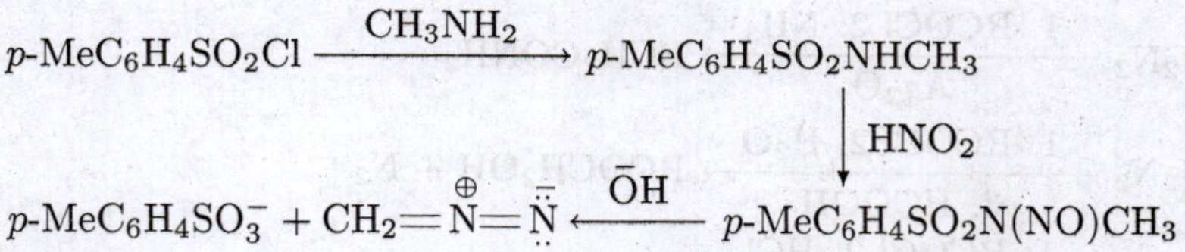

**4. From N, N′-dimethyl-N, N′-dinitrosoterephthalamide (NTA) :** Since NTA is commercially available and stable at room temperature, diazomethane is best prepared from NTA.

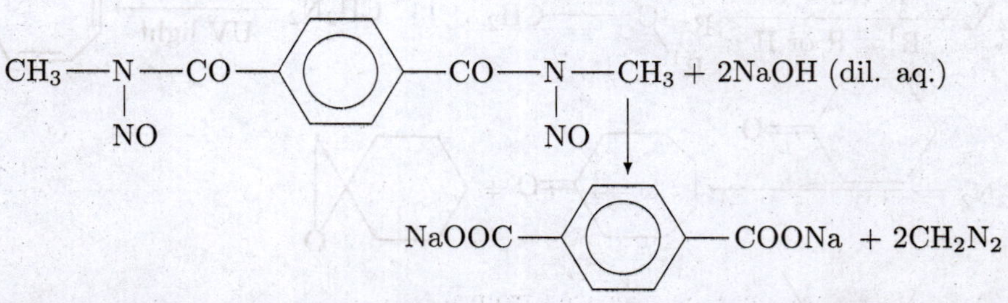

## 20.16.3 Physical properties and structure

Diazomethane is a yellow gas under ordinary condition of temperature and pressure. It boils at −24 °C. It is poisonous and explosive. However, its ethereal solution is safe to handle. The structure of diazomethane is supposed to be a resonance hybrid of the following resonating structures :

$$\underset{\text{I}}{CH_2{=}\overset{\oplus}{N}{=}\bar{\ddot{N}}} \longleftrightarrow \underset{\text{II}}{\bar{C}H_2{-}\overset{\oplus}{N}{\equiv}N:} \longleftrightarrow \underset{\text{III}}{\overset{\oplus}{C}H_2{-}N{=}\bar{N}:} \longleftrightarrow \underset{\text{IV}}{\bar{C}H_2{-}N{=}\overset{\oplus}{N}:}$$

## 20.16.4 Reactions of diazomethane

The resonance hybrid of diazomethane shows that it is a dipolar ion; it can attack as an electrophile through C atom (structure III) as well as a nucleophile through the same atom (structure IV). Since molecular nitrogen is very stable, diazomethane (structure II) explodes to give $N_2$ and carbene; the carbene then can undergo addition and insertion reactions. However, the reactions of diazomethane are listed in Chart 20.39.

1. $CH_2N_2 \xrightarrow{\text{UV light or heat}} :CH_2 + N_2\uparrow$

2. $CH_2N_2 \xrightarrow{\text{Na-Hg/}H_2O} CH_3NHNH_2$

3. $CH_2N_2 \xrightarrow[X\,=\,RCOO^-,\ PhO^-,\ Cl,\ Br]{\text{HX/methylation}} CH_3X + N_2$

4. $CH_2N_2 \xrightarrow{ROH/Al(OR^1)_3} ROCH_3 + N_2$

5. $CH_2N_2 \xrightarrow[\text{Arndt-Eistert synthesis}]{\text{1. RCOCl 2. }Ag_2O,\ H_2O} RCH_2COOH$

6. $CH_2N_2 \xrightarrow[Ag_2O]{\text{1. RCOCl 2. }R^1OH} RCH_2CO_2R^1$

7. $CH_2N_2 \xrightarrow[Ag_2O]{\text{1. RCOCl 2. }NH_3} RCH_2CONH_2$

8. $CH_2N_2 \xrightarrow[\text{3. HCOOH}]{\text{1. RCOCl 2. }H_2O} RCOCH_2OH + N_2$

9. $CH_2N_2 \xrightarrow{\text{1. RCOCl 2. HCl}} RCOCH_2Cl + N_2$

10. $CH_2N_2$ 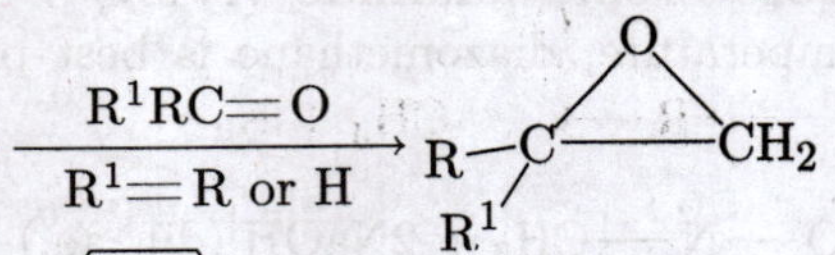

11. $CH_2N_2$ 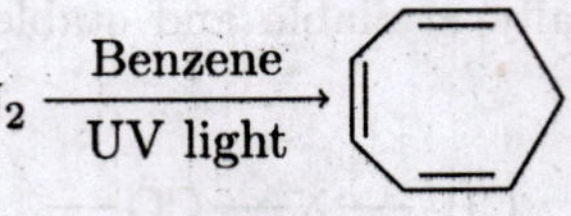

12. $CH_2N_2$ 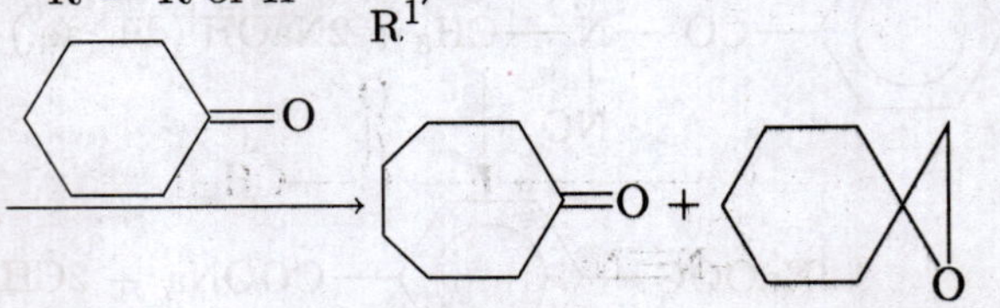

13. $CH_2N_2$ 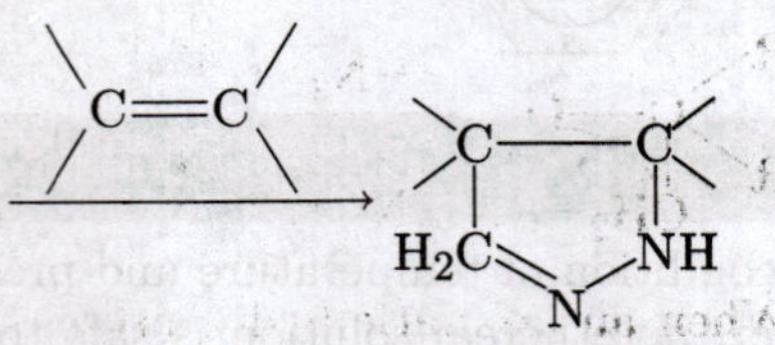

14. $CH_2N_2$ 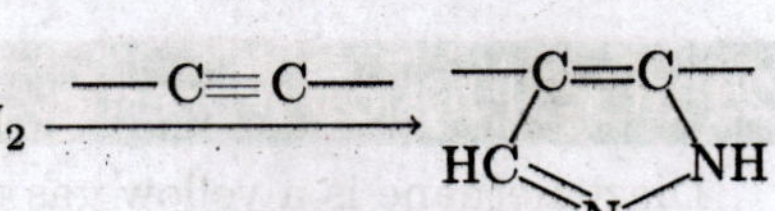

**Chart 20.39** Reactions of diazomethane.

**1. Generation of carbene :** When diazomethane is irradiated with UV light or heated, carbene forms and nitrogen gas evolves out.

$$\bar{\ddot{C}}H_2-\overset{\oplus}{N}\equiv N: \longrightarrow :CH_2 + N_2\uparrow$$

**2. Reduction by sodium amalgam :** Sodium amalgam reduces diazomethane to methylhydrazine.

$$CH_2=\overset{\oplus}{N}=\overset{-}{\ddot{N}}: \xrightarrow[H_2O]{Na/Hg} CH_3NHNH_2$$

**3. Methylation by diazomethane :** Haloacids, carboxylic acids, phenols, etc. are readily methylated by diazomethane to form methyl halides, methyl esters, methyl phenyl ethers respectively. The general equation and mechanism for the reaction are as follows :

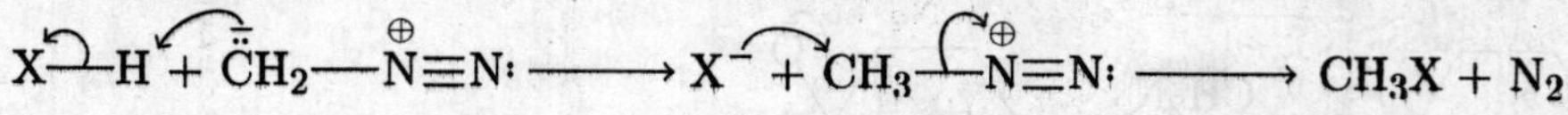

**4. Methylation of alcohols :** The H atom of the alcoholic hydroxyl group being not highly acidic, an alcohol cannot be methylated easily; however, in the presence of aluminium alkoxide, which acts as a Lewis acid, methylation can be done.

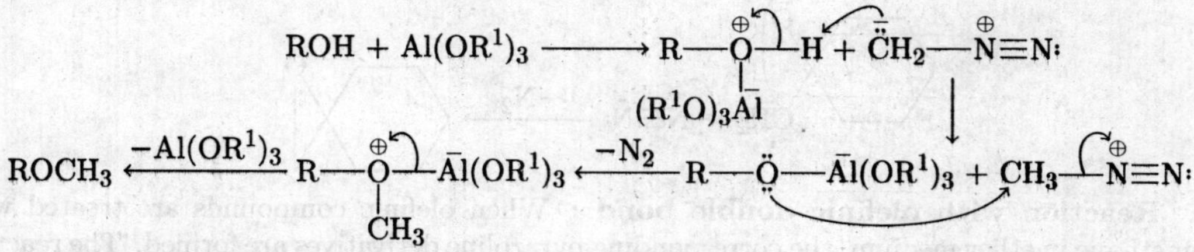

**5. The Wolff-rearrangement and the Arndt-Eistert synthesis :** This has already been discussed in detail in Sec. 14.10 (p. 730). Reactions 6, 7, 8, 9 of Chart 20.39 have also been discussed along with the Arndt-Eistert synthesis.

**6. Reaction with carbonyl compounds :** Diazomethane reacts with carbonyl compounds to form methyl and higher ketones. Epoxides are also formed as side products.

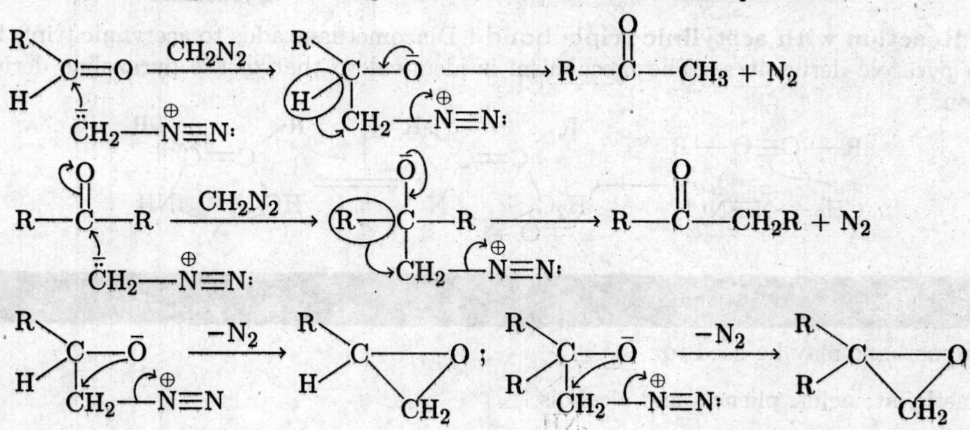

**7. Aromatic ring expansion reaction :** When an aromatic hydrocarbon is mixed with diazomethane in ether and irradiated with UV light, carbene is generated which adds to the aromatic hydrocarbon to form an alicyclic compound having a bigger ring than the parent aromatic hydrocarbon had.

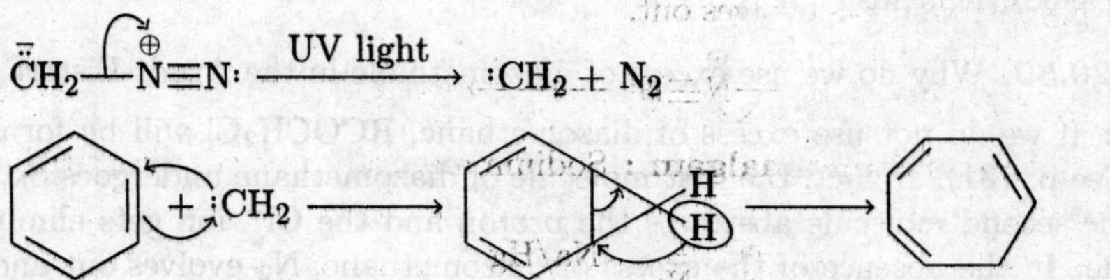

Benzene Cycloheptatriene

**8. Reaction with cyclic ketones :** A cyclic ketone reacts with diazomethane to form the higher cyclic ketone and an epoxide. When cyclopentanone is used, cycloheptanone is ultimately formed through cyclohexanone since the former is more stable than the latter.

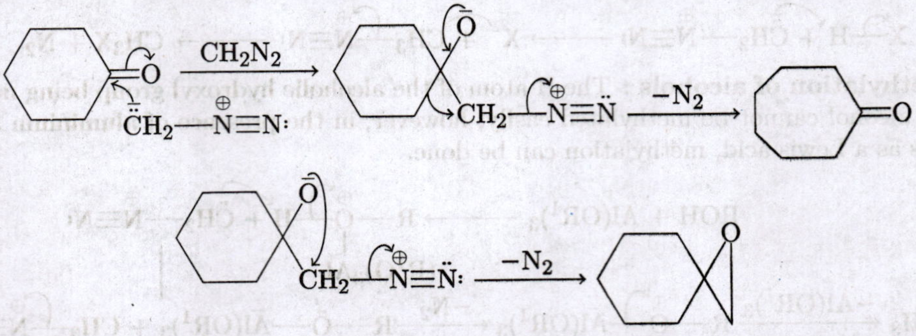

**9. Reaction with olefinic double bond :** When olefinic compounds are treated with diazomethane in ether medium, the corresponding pyrazoline derivatives are formed. The reaction is an example of 1, 3-dipolar addition and this may or may not occur in a concerted step.

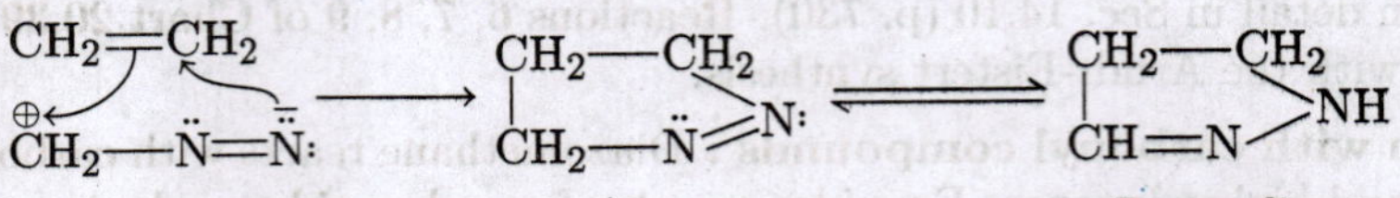

Pyrazoline

**10. Reaction with acetylinic triple bond :** Diazomethane adds to acetylinic triple bonds to form pyrazole derivatives. The mechanism is identical to that of the pyrazoline derivative formation.

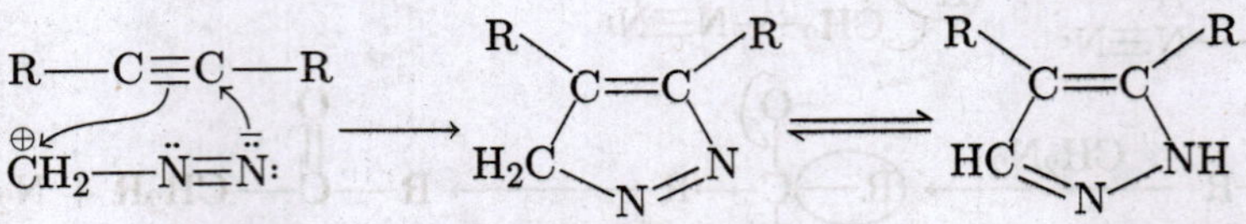

## 20.16.5 Uses

Diazomethane may be used to :

(i) methylate acids, phenols and alcohols
(ii) generate carbene
(iii) synthesise pyrazoles and pyrazolines
(iv) expand rings and
(v) ascend carboxylic acids

**Problem 20.30.** Why do we use excess of diazomethane in the Arndt-Eistert synthesis?

**Solution :** If we do not use excess of diazomethane, $RCOCH_2Cl$ will be formed instead of diazoketone (See p. 731). In fact, the first molecule of diazomethane undergoes $S_N$ reaction with RCOCl and the second molecule abstracts the proton and the $Cl^-$ ion gets eliminated to form the diazoketone. In the absence of the excess of diazomethane, $N_2$ evolves out and the Cl atom with bonded *es* migrates to the nitrogen bonded C atom to form $RCOCH_2Cl$.

## 20.17 Diazoacetic Ester

### 20.17.1 Introduction

Ethyl diazoacetate, $N_2CHCOOEt$, is popularly called diazoacetic ester. It is fairly stable and used as a good synthetic reagent.

### 20.17.2 Preparation of diazoacetic ester

When a cooled solution of hydrochloride of ethyl ester of glycine is treated with cold and dilute sodium nitrite solution in water, ethyl diazoacetate is formed.

$$CH_3COOH \xrightarrow[\text{sunlight}]{Cl_2} Cl{\cdot}CH_2COOH \xrightarrow[\text{conc. } H_2SO_4]{EtOH} Cl{\cdot}CH_2COOEt$$

$$Cl{\cdot}CH_2COOEt \xrightarrow{\text{conc. } NH_4OH} NH_2CH_2COOEt \xrightarrow{HCl} \overset{-}{Cl}\overset{\oplus}{N}H_3CH_2COOEt \xrightarrow[\text{aq. } NaNO_2]{\text{cold}} H_2O + N_2CHCOOEt + NaCl$$

### 20.17.3 Physical properties and structure

It is a yellow oil; b.p. 141 °C at 720 mm pressure. It is insoluble in water but soluble in alcohol and ether. It is a resonance hybrid of several resonating structures as shown below :

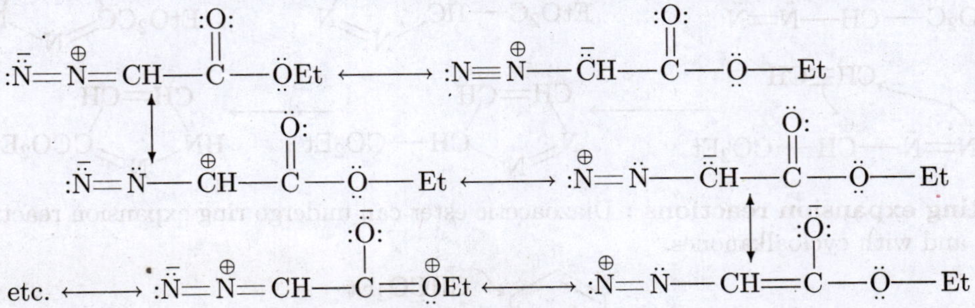

Since the resonance hybrid of diazoacetic ester involves a large number of resonating structures and since resonance occurs over a wide range, diazoacetic ester is fairly stable and it is more stable than diazomethane.

### 20.17.4 Reactions of diazoacetic ester

The following are the reactions given by diazoacetic ester.

**1. Reduction by zinc dust and acetic acid :** It is reduced by zinc dust and acetic acid to glycine ester and ammonia.

$$N_2CHCO_2Et \xrightarrow{Zn/HOAc} NH_3 + NH_2CH_2CO_2Et$$

**2. Hydrolysis by dilute halogen acids :** Dilute halogen acids hydrolyse ethyl diazoacetate to glycollic ester and nitrogen.

$$N_2CH\ CO_2Et + H_2O \xrightarrow{\text{dil. HX}} HOCH_2CO_2Et + N_2$$

**3. Action of concentrated halogen acids :** When ethyl diazoacetate is warmed with concentrated halogen acids, ethyl haloacetate and molecular nitrogen are formed.

$$N_2CHCO_2Et + HX\ (conc.) \xrightarrow{\text{warmed}} N_2 + XCH_2CO_2Et$$

**4. Action of diazoacetate on carboxylic acids, phenols and alcohols :** Diazoacetic ester replaces the active H of acids, phenols and alcohols and forms derivatives of glycollic ester.

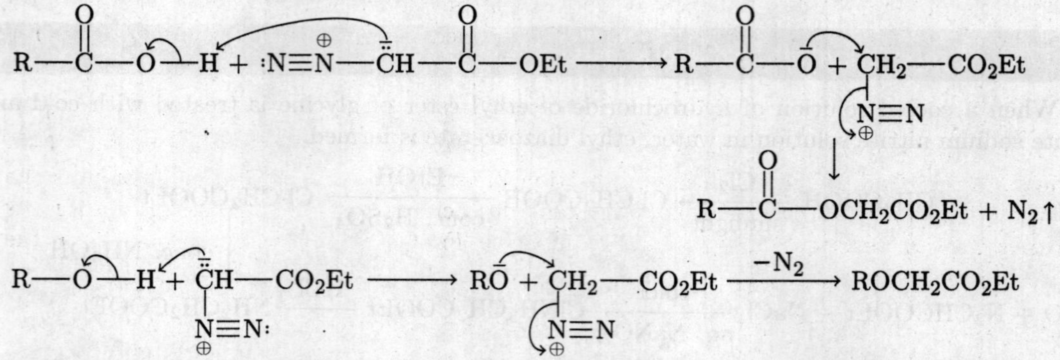

**5. Reaction with ethylenes and acetylenes :** Diazoacetic ester forms pyrazolines and pyrazoles with ethylenes and acetylenes respectively.

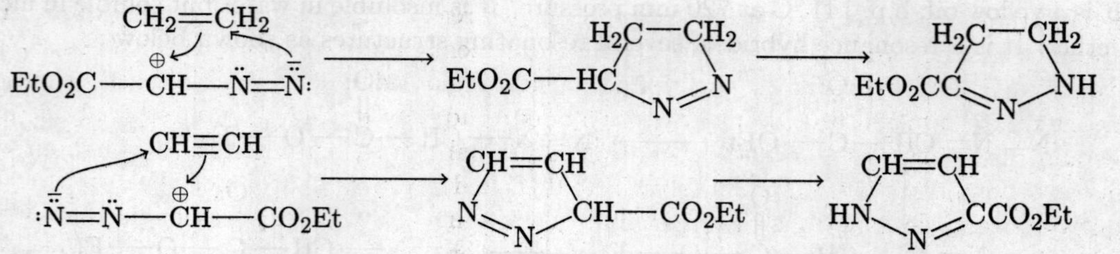

**6. Ring expansion reactions :** Diazoacetic ester can undergo ring expansion reaction with benzene and with cycloalkanones.

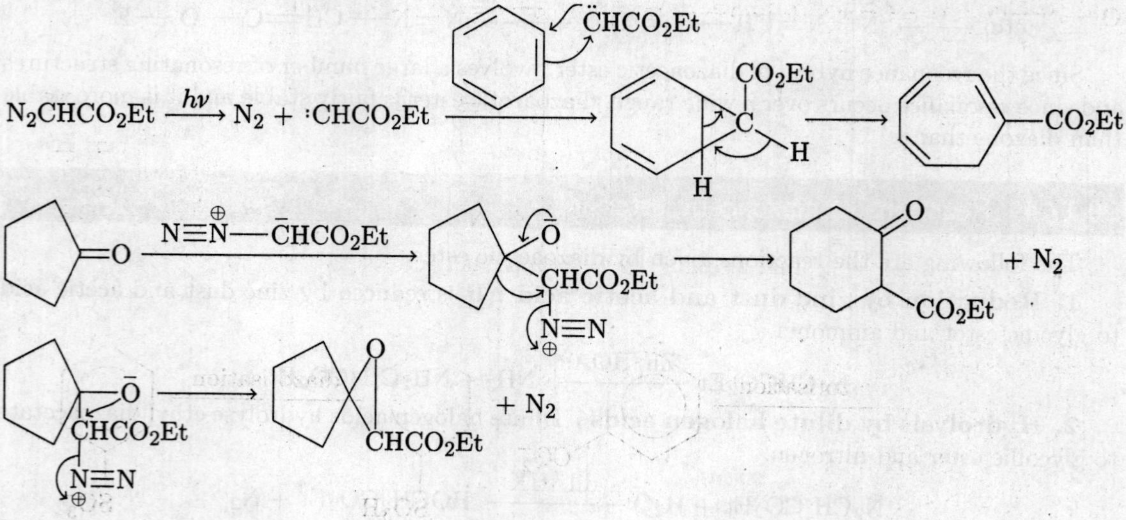

## 20.18 Aromatic Diazonium Salts

### 20.18.1 Introduction

Aromatic diazonium salts are usually represented by $Ar{-}N{=}N^{\oplus}X^{-}$ (where X=Cl, Br, $HSO_4$, etc.). In these compounds two N atoms (azo atoms) are present and one is directly bonded to the nuclear carbon and the other N atom is bonded to the former N atom by a multiple bond. Aromatic diazonium salts are good intermediate compounds for preparing a large number of aromatic compounds, dyes and drugs.

### 20.18.2 Preparation of aromatic diazonium salts

These are prepared by the diazotisation reaction. The formation of aromatic diazonium salts from aromatic primary amines by the action of nitrous acid at 0-5 °C in strong acid medium is called diazotisation reaction since a two-azo (nitrogen; French azote) system is bonded to the nuclear C by the process.

$$ArNH_2 \xrightarrow[\text{0-5 °C}]{HX} ArNH_2{\cdot}HX \xrightarrow[HX]{\text{aq. }NaNO_2} Ar{\cdot}\overset{\oplus}{N_2}X^-$$

$$PhNH_2 \xrightarrow[\text{0-5 °C}]{HX} PhNH_2{\cdot}HX \xrightarrow[HX]{\text{aq. }NaNO_2} Ph{\cdot}\overset{\oplus}{N_2}X^-; \quad NaNO_2 + HX \longrightarrow NaX + HNO_2$$

The equations show that 2 moles of mineral acid is required to diazotise 1 mole of an aromatic primary amine. In practice, however, one equivalent of the amine is dissolved in 2.5 to 3 equivalents of dilute hydrochloric or sulphuric acid and cooled to 0-5 °C. To the cold solution an aqueous dilute solution of sodium nitrite is added gradually during the course of about five minutes and the mixture is allowed to stand for additional five to ten minutes. Throughout the process the temperature is kept preferably above 0 °C and below 5 °C at which the speed of the reaction is appreciable as well as the diazonium ion is fairly stable and does not decompose to nitrogen. The diazonium salt so formed remains in solution and is used as such in situ. However, the mechanism of the reaction is as follows :

$$HO{-}N{=}O + \overset{\oplus}{H} \underset{}{\overset{\text{fast}}{\rightleftharpoons}} H{-}\overset{\oplus}{\underset{H}{\underset{|}{O}}}{-}N{=}O + H \overset{\text{fast}}{\rightleftharpoons} H_2O + \overset{\oplus}{N}O$$

$$Ar\ddot{N}H_2 + \overset{\oplus}{N}O \overset{\text{slow}}{\rightleftharpoons} Ar{-}\overset{\oplus}{\underset{H}{\underset{|}{N}H}}{-}NO \underset{\text{fast}}{\overset{H^{\oplus}}{\rightleftharpoons}} ArNH{-}N{=}O \underset{}{\overset{\text{fast}}{\underset{\text{tautomerisation}}{\rightleftharpoons}}} Ar\ddot{N}{=}\ddot{N}{-}OH$$

$$\Big\downarrow\!\!\uparrow \text{ fast } H^{\oplus} \text{ (fast)}$$

$$\text{etc.} \longleftrightarrow Ar{-}\ddot{N}{=}\overset{\oplus}{N}{:} \longleftrightarrow Ar{-}\overset{\oplus}{N}{\equiv}N{:} + H_2O \overset{\text{fast}}{\rightleftharpoons} Ar\ddot{N}{=}\ddot{N}{-}\overset{\oplus}{O}H_2$$

Some derivatives of 1° aromatic amines give anomalous products.

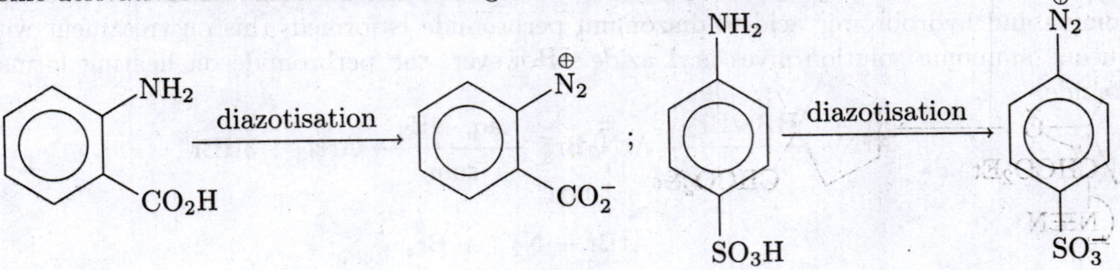

*o*-Phenylenediamine on diazotisation gives benztriazole, and *m*-phenylenediamine on similar reaction gives Bismark brown, a brown dye (See p. 1002, Problem 17.20 and, Problem 17.22, p. 1003).

Solid diazonium salts may be prepared by dissolving the amine salt in an acidic solution of alcohol and adding an alkyl nitrite, ethyl or isoamyl nitrite. The diazonium salt is then precipitated by the addition of ether and placing the reaction vessel in a freezing mixture of common salt and ice; the salt is separated by filtration.

$$ArNH_2 \cdot HCl + EtONO \longrightarrow Ar\overset{\oplus}{N}_2Cl^- + EtOH + H_2O$$

## 20.18.3 Physical properties

The solid salts are colourless and crystalline but darkens on exposure to air. Being ionic, they are soluble in water and insoluble in ether; they conduct electricity. When diazonium salts are heated or subjected to mechanical shock, they explode with evolution of nitrogen gas; this indicates their instability. However, the stability varies from salt to salt. *p*-Nitrobenzenediazonium salt is more stable than benzenediazonium salt because of the greater resonance stabilisation of the former than that of the latter. Double salts of zinc chloride or, 1, 5-naphthalenedisulphonic acid and aromatic diazonium salt are more stable than their mineral acid salts alone. Aromatic diazonium nitrates are comparatively less stable.

## 20.18.4 Reactions and applications of diazonium salts

Most of the reactions given by diazonium salts are found to be intermediate steps in synthesising several types of organic compounds, these and other reactions are listed in Chart 20.40.

**1. Formation of double salts and polyhalides :** Diazonium salts form double salts with zinc chloride, platinum chloride, etc.

$$[(Ar\ \overset{\oplus}{N}_2)_2\ ZnCl_4^{2-},\ (Ar\ \overset{\oplus}{N}_2)_2\ PtCl_6^{2-},\ (Ar\ \overset{\oplus}{N}_2)Cu_2Cl_3^-]$$

The double salts stabilise the diazonium ions. Diazonium salts also form polyhalides; e.g., with KI—I$_2$ they form $Ar\ \overset{\oplus}{N}_2I_3^-$ and $Ar\ \overset{\oplus}{N}_2ClBr_2^-$ with KCl—Br$_2$.

**2. Formation of diazohydroxide via diazonium hydroxide :** When diazonium salts are treated with silver oxide or made alkaline, diazonium hydroxides are formed; these being unstable, rapidly rearrange to diazohydroxides which remain in solution as salts known as diazoates.

$$[Ar-\overset{\oplus}{N}\equiv N:]\,Cl^- \xrightarrow{NaOH} \underset{\text{diazonium hydroxide}}{[Ar-\overset{\oplus}{N}\equiv N]\,\bar{O}H} \rightleftharpoons \underset{\text{diazohydroxide}}{Ar-\ddot{N}=\ddot{N}-OH} \longrightarrow \underset{\text{sodium diazoate}}{Ar-N=N-\bar{O}Na^{\oplus}}$$

**3. Formation of aryl azide :** When an aromatic diazonium salt solution is treated with bromine and hydrobromic acid, a diazonium perbromide is formed; this on treatment with an aqueous ammonia solution gives aryl azide. However, the perbromide on heating forms aryl bromide.

$$Ar\overset{\oplus}{N}_2 \xrightarrow{HBr/Br_2} Ar\overset{\oplus}{N}_2Br_3^- \xrightarrow[\text{soln.}]{\text{aq. }NH_3} ArN_3 + 3HBr$$

$$Ar\overset{\oplus}{N}_2Br_3^- \xrightarrow{\text{heat}} ArBr + N_2\uparrow + Br_2$$

**4. The diazonium coupling reaction :** See Sec. 17.16, p. 997. This reaction is a test of diazonium group.

$$\text{Ar·NH}_2\text{·HX} \xrightarrow[0\text{-}5\,^\circ\text{C}]{\text{HX/NaNO}_2} \text{Ar}\overset{\oplus}{\text{N}}_2\text{X}^- \qquad \text{X = Cl or HSO}_4$$

$$\xrightarrow{\text{HCl \& ZnCl}_2} (\text{Ar}\overset{\oplus}{\text{N}}_2)_2\ \text{ZnCl}_4^{2-}$$

$$\xrightarrow{\text{KI-I}_2} \text{Ar}\overset{\oplus}{\text{N}}_2\text{I}_3^-$$

$$\xrightarrow{\text{NaOH}} \text{Ar—N=N—}\bar{\text{O}}\text{Na}^{\oplus}$$

$$\xrightarrow{\text{1. Br}_2\text{ \& HBr 2. aq. NH}_3} \text{Ar·N}_3$$

$$\xrightarrow[\text{coupling reaction}]{\text{Ar}\bar{\text{O}}\text{ or ArNMe}_2} \text{Ar—N=N—Ar}\bar{\text{O}} \text{ or } \text{Ar—N=N—ArNMe}_2$$

$$\xrightarrow{\text{Zn/HCl}} \text{ArNHNH}_2 \xrightarrow{\text{Zn/HCl}} \text{ArNH}_2 + \text{NH}_3$$

$$\xrightarrow{\text{SnCl}_2\text{/HCl or SO}_2\text{ or NaHSO}_3} \text{Ar—NHNH}_2 \text{ as salt}$$

$$\xrightarrow[\text{or NaBH}_4\text{ or Na}_2\text{SnO}_2]{\text{HOPH}_2\text{O or EtOH/Cu}} \text{ArH} + \text{N}_2$$

$$\xrightarrow{\text{X}^-\text{ and heat}} \text{ArX} + \text{N}_2$$

$$\xrightarrow[\text{boiled}]{\text{dil. H}_2\text{SO}_4} \text{ArOH} + \text{N}_2$$

$$\xrightarrow{\text{KI}} \text{Ar—I} + \text{N}_2$$

$$\xrightarrow[\text{Sandmeyer reaction}]{\text{Cu}_2\text{X}_2\text{ (X = Cl, Br)}} \text{Ar—X} + \text{N}_2$$

$$\xrightarrow[\text{Gattermann reaction}]{\text{Cu powder and HX(X = Cl, Br)}} \text{Ar—X} + \text{N}_2$$

$$\xrightarrow{\text{HBF}_4\text{ \& heat}} \text{Ar—F} + \text{N}_2$$

$$\xrightarrow{\text{CuCN/KCN or K}_3\text{Cu(CN)}_4} \text{Ar—CN} + \text{N}_2$$

$$\xrightarrow{\text{NaNO}_2 + \text{NaHCO}_3\text{; Cu}^{++}} \text{Ar—NO}_2 + \text{N}_2$$

$$\xrightarrow{\text{Cu}^{++}\text{/EtOH}} \text{ArOEt} + \text{ArH} + \text{CH}_3\text{CHO} + \text{Ar—Ar}$$

$$\xrightarrow[\text{HCl}]{\text{Na}_3\text{AsO}_3\text{/CuSO}_4} \text{ArAsO}_3\text{H} + \text{N}_2$$

$$\xrightarrow{\text{C}_6\text{H}_6\text{/NaOH}} \text{Ar—Ph}$$

$$\xrightarrow[\text{2. heat}]{\text{1. HgCl}_2\text{ in Me}_2\text{CO or EtOH/Cu}} \text{ArHgCl} + \text{N}_2$$

$$\xrightarrow{\text{CuSCN; X = Cl}} \text{ArSCN} + \text{N}_2$$

***Contd.***

$$Ar\cdot NH_2\cdot HX \xrightarrow[0\text{-}5\,°C]{HX/NaNO_2} Ar\overset{\oplus}{N_2}X^-$$
$X = Cl$ or $HSO_4$

$$Ar\overset{\oplus}{N_2}X^- \xrightarrow{KNCO/Cu;\; X = Cl} ArNCO + N_2$$
$$Ar\overset{\oplus}{N_2}X^- \xrightarrow{(NH_4)_2S;\; X = Cl} Ar_2S$$
$$Ar\overset{\oplus}{N_2}X^- \xrightarrow{SO_2/Cu;\; X = HSO_4} ArSO_2H + N_2$$
$$Ar\overset{\oplus}{N_2}X^- \xrightarrow{NaSH} ArSH + N_2$$

**Chart 20.40** Reactions of aromatic diazonium ion.

**5. Reduction by zinc and hydrochloric acid :** When aromatic diazonium salts are reduced with strong reducing agents like zinc and hydrochloric acid, an aromatic primary amine is formed through arylhydrazine.

$$Ar\overset{\oplus}{N_2}Cl^- \xrightarrow{Zn/HCl} ArNHNH_2 \xrightarrow{Zn/HCl} ArNH_2\cdot HCl + NH_4Cl$$

**6. Reduction by stannous chloride and hydrochloric acid :** Stannous chloride and hydrochloric acid reduce aromatic diazonium salts to arylhydrazines. $SO_2$ and sodium bisulphite also reduce diazonium salts to the hydrazines.

$$[Ar\overset{\oplus}{N}{\equiv}N]\ Cl^- \xrightarrow{SnCl_2/HCl} ArNHNH_2\cdot HCl$$

$$[Ar\overset{\oplus}{N}{\equiv}N]\ HSO_4^- + 2SO_2 + 2H_2O \longrightarrow ArNH\cdot NH_2\cdot HSO_4^- + 2H_2SO_4$$

**7. Replacement of the diazonium group by H atom :** When a diazonium salt is treated with hypophosphorus acid (in the presence of cuprous, ferrous or manganous ions), an arene is obtained.

$$Ar\overset{\oplus}{N_2}Cl^- + H_3PO_2 + H_2O \longrightarrow ArH + N_2 + HCl + H_3PO_3$$

However, ethyl alcohol and a trace of Cu or alkaline sodium stannite or sodium borohydride can also reduce diazonium ions to arenes.

**8. Formation of aryl halide (chloride or bromide) by aromatic $S_N1$ reaction :** See Sec. 12.8, p. 593.

**9. Formation of phenol by aromatic $S_N1$ mechanism :** See Sec. 12.8, p. 593.

**10. Formation of aryl iodide :** See Sec. 20.1.2n, p. 1152.

**11. Formation of aryl chloride or bromide by the Sandmeyer reaction :** See Sec. 20.1.2n, p. 1151 and p. 1153.

**12. Formation of aryl chloride or bromide by the Gattermann reaction :** See Sec. 20.1.2n, p. 1151.

**13. Formation of aryl fluoride :** See Sec. 20.1.2n, p. 1152.

**14. Formation of aryl cyanide :** See Sec. 20.1.2n, p. 1151.

**15. Formation of nitroarene :** When aromatic diazonium salts are heated with sodium nitrite in the presence of sodium bicarbonate and $Cu^{++}$, nitroarene is obtained;

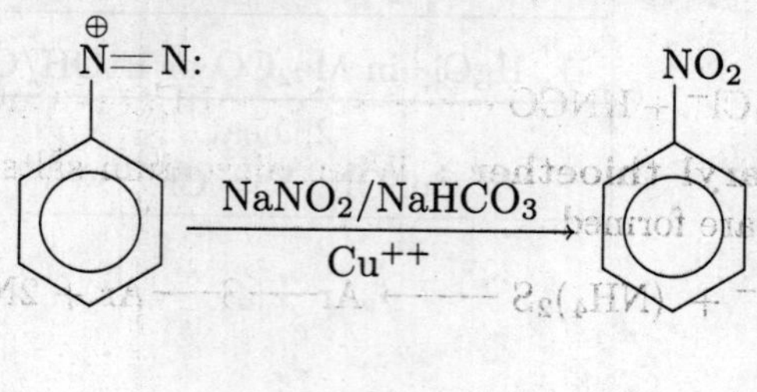

**16. Formation of biaryl by the extended Gattermann reaction :** When diazonium sulphates are treated with ethyl alcohol and copper powder, biaryls are formed; arenes are obtained as side products. This is a special case of the Gattermann reaction. Aryl ethyl ether is also formed by this reaction.

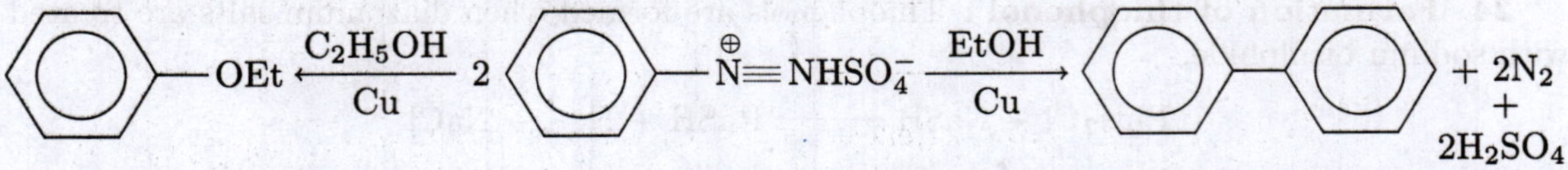

**17. Formation of biaryl by the Gomberg reaction :** When aromatic diazonium salts are treated with aromatic compounds in an alkaline medium, biaryls are formed and the reaction is called the Gomberg reaction. Whatever may be the nature of the substituent in the aromatic compound, *o*- or, *p*-substituted products are usually predominant.

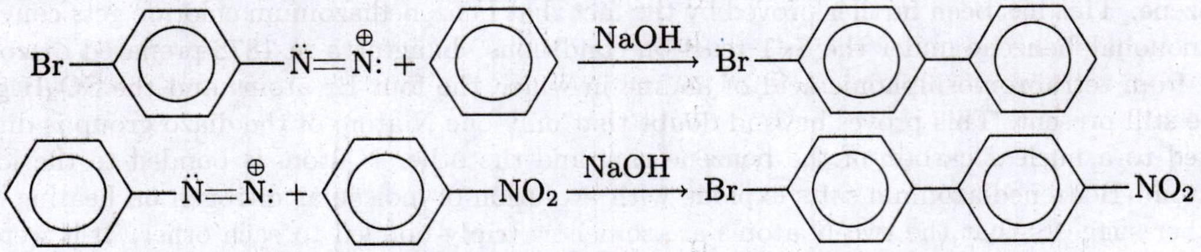

$$Ar\overset{\oplus}{N}_2Cl^- \xrightarrow{\bar{O}H} Ar-N{=}N-OH \longrightarrow Ar\cdot + N_2 + \cdot OH$$

$$\cdot OH + Ar\cdot + H-C_6H_4NO_2 \longrightarrow p\text{-}Ar-C_6H_4NO_2 + H_2O$$

**18. Formation of arylarsonic acid :** When aryl diazonium salts are decomposed with sodium arsenite in the presence of a copper salt and the product is treated with dilute hydrochloric acid, arylarsonic acid is obtained and the reaction is called the Bart reaction.

$$Ph\overset{\oplus}{N}_2Cl^- + Na_3AsO_3 \xrightarrow[-N_2]{CuSO_4} Ph-AsO_3Na \xrightarrow{HCl} Ph-AsO_3H$$

**19. Formation of arylmercuric chloride :** When mercuric chloride solution is added to a diazonium salt solution, a complex between the two salts form. When the complex is heated with copper powder in ethanolic or acetone medium, arylmercuric chloride is obtained.

$$Ar\overset{\oplus}{N}_2Cl^- + HgCl_2 \longrightarrow Ar\overset{\oplus}{N}_2HgCl_3^- \xrightarrow[heat]{Cu} ArHgCl + N_2 + 2CuCl$$

**20. Formation of aryl thiocyanate :** When aromatic diazonium salt solution is treated with cuprous thiocyanate, aryl thiocyanate is obtained.

$$Ph\overset{\oplus}{N}_2Cl^- + CuSCN \longrightarrow PhSCN + CuCl + N_2\uparrow$$

**21. Formation of aryl isocyanate :** This is a Gattermann type reaction involving potassium isocyanate.

$$Ph\overset{\oplus}{N}_2Cl^- + KNCO \xrightarrow{Cu} Ph-NCO + CuCl + N_2\uparrow$$

**22. Formation of aryl thioether :** When diazonium salts are treated with ammonium sulphide, aryl thioethers are formed.

$$2Ar\overset{+}{N}_2Cl^- + (NH_4)_2S \longrightarrow Ar-S-Ar + 2NH_4Cl + N_2\uparrow$$

**23. Formation of arylsulphinic acids :** Diazonium bisulphate solution gives arylsulphinic acid on treatment with sulphur dioxide in the presence of copper powder.

$$Ph\overset{\oplus}{N}_2HSO_4^- + SO_2 \xrightarrow{Cu} PhSO_2H + N_2\uparrow + CuSO_4$$

**24. Formation of thiophenol :** Thiophenols are formed when diazonium salts are treated with sodium bisulphide.

$$Ph\overset{\oplus}{N}_2Cl + NaSH \longrightarrow PhSH + N_2\uparrow + NaCl$$

## 20.18.5 Elucidation of the structure of benzenediazonium chloride

Analytical data suggests that the formula unit of benzene diazonium chloride is $C_6H_5N_2Cl$.

The salts form phenol on boiling with water. This tells us that it is a monosubstituted benzene. This has been further proved by the fact that benzenediazonium chloride gets converted to monohalobenzene under the $S_N1$ reaction conditions. Langfurth in 1878 prepared diazonium salt from tetrabromosulphonic acid of aniline in which the four Br atoms and the $SO_3H$ group were still present. This proves beyond doubt that only one N atom of the diazo group is directly linked to a nuclear carbon of the benzene ring and the other N atom is bonded to the former N atom. Benzenediazonium salts explode with evolution of molecular nitrogen on heating. This further suggests that the two N atoms are somehow triply-bonded to each other. If it were not so there would have been no evolution of nitrogen gas.

Benzenediazonium chloride is highly soluble in water but insoluble in ether. Its aqueous solution conducts electricity. Thus it is an ionic compound like ammonium chloride. So one N atom of the diazo group must bear a positive charge.

Benzenediazonium chloride couples with phenoxide to form the following :

$$C_6H_5—N{=}N—C_6H_4O^-$$

This indicates that benzenediazonium chloride contains $C_6H_5—N{=}N^{\oplus}$ unit.

Spectral analysis indicates that the C—N and N—N bond lengths in benzenediazonium chloride have some C—N double bond and N—N triple bond characters respectively.

In order to explain C=N and N≡N bond lengths the V.B. theory considers benzenediazonium ion as a resonance hybrid of several resonating structures.

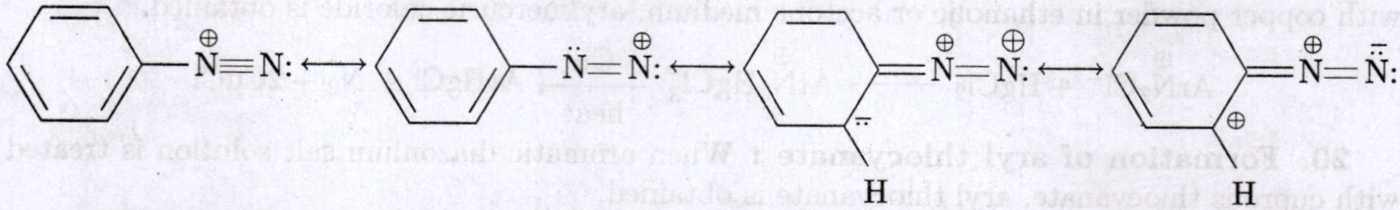

The M.O. theory tells us that the positive charge of the benzenediazonium ion gets delocalised through $\pi$ M.O.s involving all the six C atoms and two N atoms.

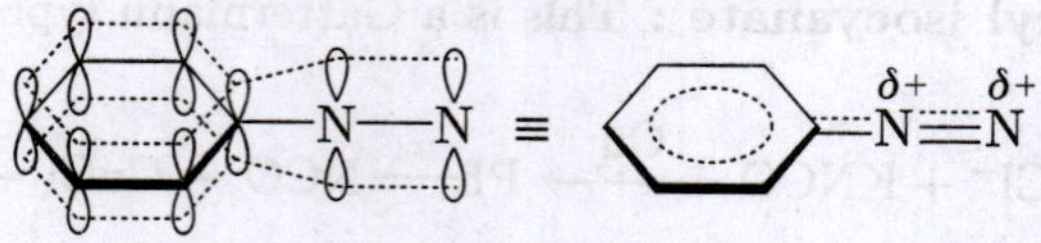

The V.B. and M.O. theories of bonding explain the greater stability of benzenediazonium chloride than aliphatic diazo compounds. These also explain the bond length of C—N and N—N bonds in the compound.

## 20.19 Urea

### 20.19.1 Introduction

Urea is a diamide of carbonic acid. The monoamide of urea is carbamic acid. The latter is very unstable. Urea is the chief ultimate product of nitrogen metabolism in mammals and it is

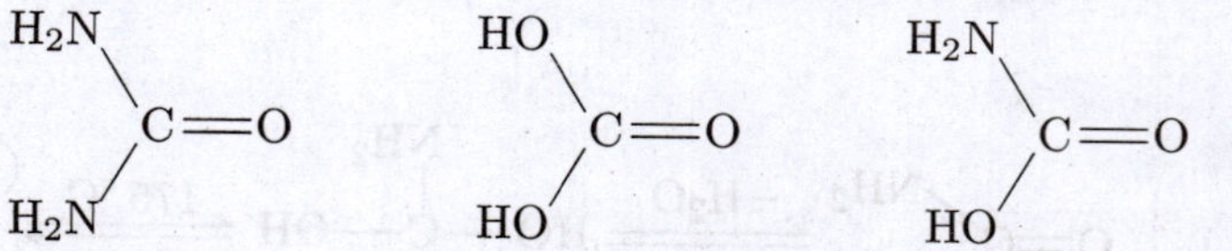

Urea stable | Carbonic acid | Carbamic acid unstable

eliminated in the urine. When a large volume of urine is evaporated to a small bulk and nitric acid is added, urea mononitrate precipitates. However, Wohler (1828) first synthesised urea in the laboratory by evaporating an aqueous solution of ammonium sulphate and potassium isocyanate.

$$(NH_4)_2SO_4 + KNCO = K_2SO_4 + NH_4NCO$$

$$NH_4NCO \rightleftharpoons NH_3 + H{-}N{=}C{=}O$$

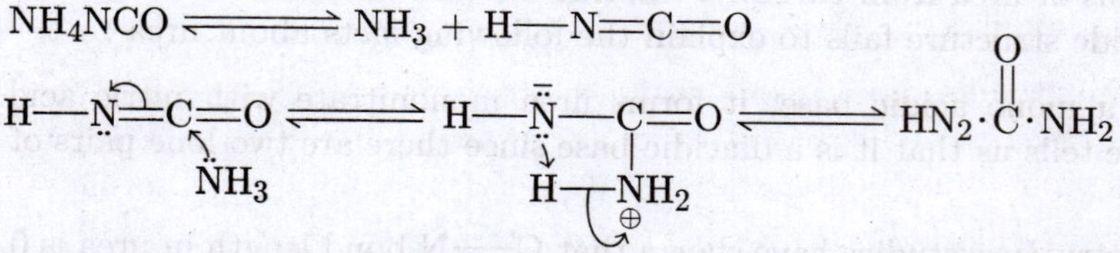

This synthesis bears a great importance in the history of the organic chemistry. Before this synthesis it was thought that organic compounds can only form in living bodies by the vital force and cannot be synthesised in the laboratory. Synthesis of urea thus ruled out the vital force theory and since then laks of organic compounds have been synthesised.

### 20.19.2 General methods of formation of urea

1. Urea may be prepared in the laboratory by the action of ammonia on carbonyl chloride, phosgen gas.

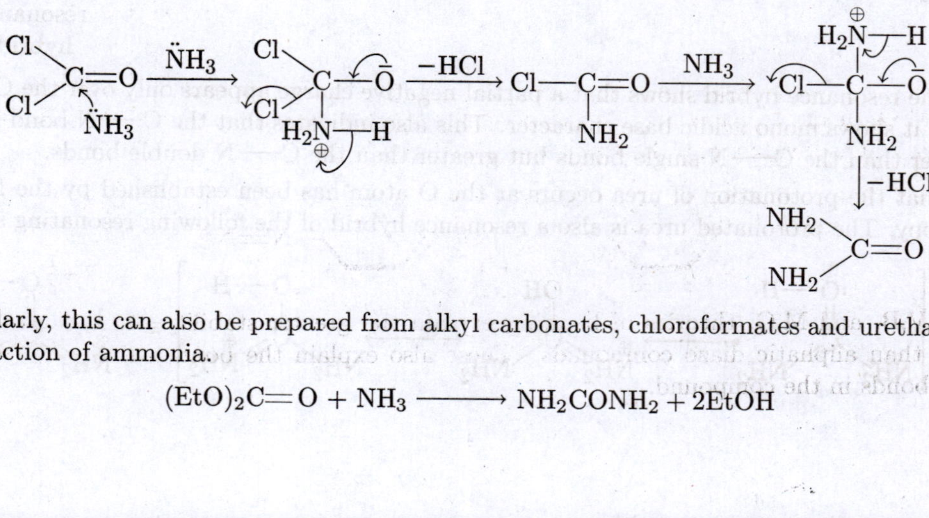

Similarly, this can also be prepared from alkyl carbonates, chloroformates and urethans by the action of ammonia.

$$(EtO)_2C{=}O + NH_3 \longrightarrow NH_2CONH_2 + 2EtOH$$

2. Urea is prepared commercially by allowing liquid carbon dioxide to react with liquid ammonia when ammonium carbonate forms; from this on heating at 175 °C urea is obtained.

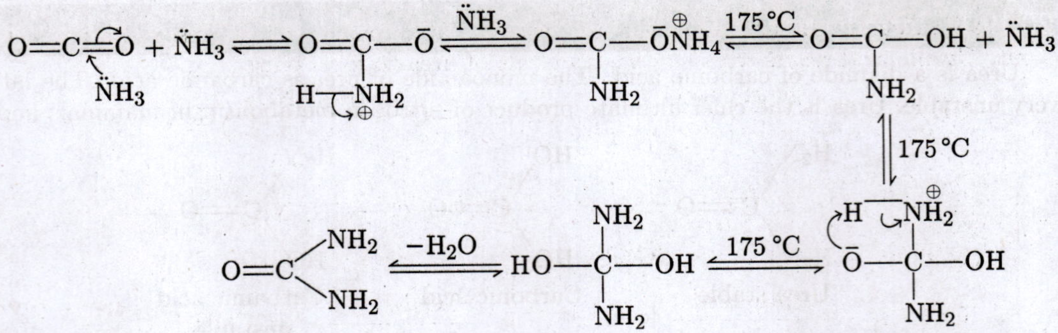

## 20.19.3 Physical properties and structure

Urea is a white crystalline solid and its m.p. is 132 °C. It is soluble in water as well as in alcohol but insoluble in ether.

The synthesis of urea from carbonyl chloride and ammonia indicates that urea is a diamide. However, diamide structure fails to explain the following facts about urea :

(i) Urea is a mono acidic base; it forms urea mononitrate with nitric acid. The diamide structure tells us that it is a diacidic base since there are two lone pairs of es, each on one N atom.

(ii) Crystal structure studies have shown that C—N bond length in urea is 0.137 nm instead of 0.147 nm, the latter is usually the C—N single bond length in aliphatic compounds. Thus the C—N bonds in urea have some C—N double bond character.

The V.B. theory, however, explains this by resonance. Urea is considered to be a resonance hybrid of the following three structures :

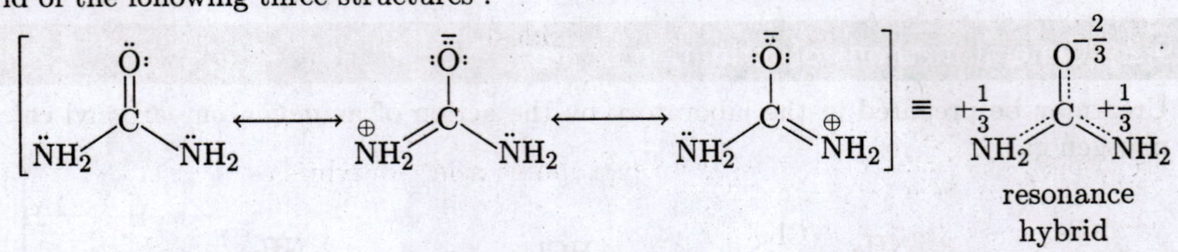

The resonance hybrid shows that a partial negative charge appears only over the O atom and hence it shows mono acidic base character. This also indicates that the C—N bond lengths are smaller than the C—N single bonds but greater than the C—N double bonds.

That the protonation of urea occurs at the O atom has been established by the NMR spectroscopy. The protonated urea is also a resonance hybrid of the following resonating structures :

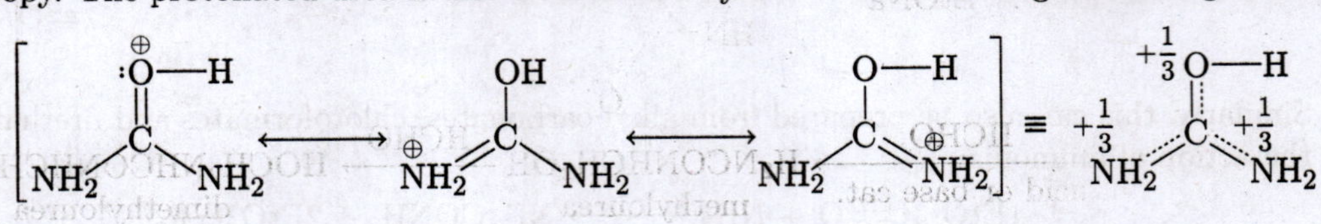

## 20.19.4 Chemical reactions of urea

These are listed in Chart 20.41.

**(a) Effect of heat :** When urea is gently heated, cyanic acid forms and ammonia evolves out. Cyanic acid reacts with another molecule of urea to form biuret.

$$NH_2CONH_2 \xrightarrow{\text{gentle heat}} NH_3 + HN{=}C{=}O \xrightarrow{NH_2CONH_2} \underset{\text{Biuret}}{H_2NCONH\cdot CONH_2}$$

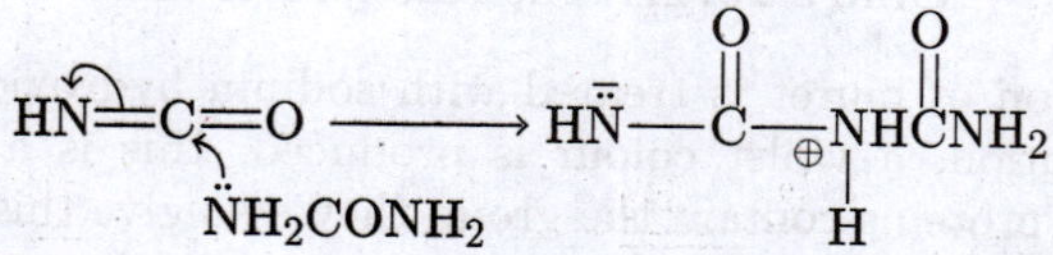

(a) $CO(NH_2)_2 \xrightarrow{\text{gentle heat}} NH_3 + HCNO \xrightarrow{CO(NH_2)_2} NH_2CONHCONH_2$

(b) $CO(NH_2)_2 \xrightarrow[\text{heating; } -NH_3]{\text{rapid}} HCNO \xrightarrow[\text{polymerisation}]{\text{rapid}} (HCNO)_3 \longrightarrow$ cyanuric acid

(c) $CO(NH_2)_2 \xrightarrow{H_3O^{\oplus}/\text{heat}} \overset{\oplus}{N}H_4 + CO_2$ (d) $CO(NH_2)_2 \xrightarrow[\text{heat}]{\bar{O}H/H_2O} CO_3^{2-} + NH_3$

(e) $CO(NH_2)_2 \xrightarrow{HNO_2} H_2O + CO_2 + N_2$

(f) $CO(NH_2)_2 \xrightarrow{\text{urease}/H_2O} NH_3 + CO_2$

(g) $CO(NH_2)_2 \xrightarrow[NaOBr/\bar{O}H]{\text{excess of}} N_2 + CO_3^{2-} + Br^-$

(h) $CO(NH_2)_2 \xrightarrow{NaOCl/\bar{O}H} NH_2{-}NH_2 + Na_2CO_3 + NaCl + H_2O$

(i) $CO(NH_2)_2 \xrightarrow{HNO_3} CO(NH_2)_2{\cdot}HNO_3 \xrightarrow[H_2SO_4]{\text{cold}} NH_2CONHNO_2 + H_2O$

(j) $CO(NH_2)_2 \xrightarrow{RCOCl} RCONHCONH_2$ (ureides)

(k) $CO(NH_2)_2$ 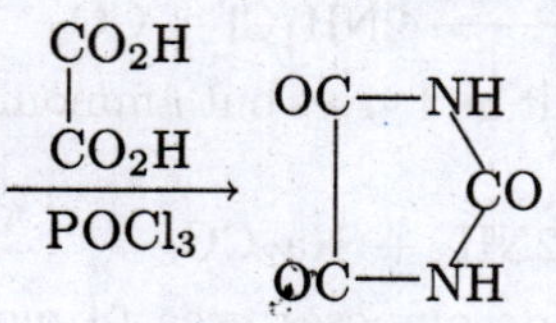 (parabonic acid) oxalylurea

(l) $CO(NH_2)_2$ 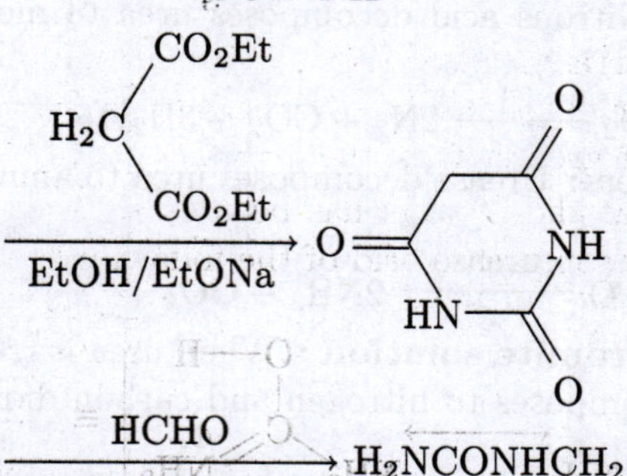 (malonylurea)

(m) $CO(NH_2)_2 \xrightarrow[\text{acid or base cat.}]{HCHO} \underset{\text{methylourea}}{H_2NCONHCH_2OH} \xrightarrow{HCHO} \underset{\text{dimethylourea}}{HOCH_2NHCONHCH_2OH}$

(n) $CO(NH_2)_2 \xrightarrow[H\bar{O}]{(CH_3)_2SO_4} CH_3-O-C(=NH)-NH_2$ (O-methyl urea)

(o) $CO(NH_2)_2 \xrightarrow{H_2S_2O_7} 2H_2NSO_3H + CO_2$

**Chart 20.41.** The reactions of urea.

When an aqueous solution of biuret is treated with sodium hydroxide solution and a drop of 1% copper sulphate solution, a violet colour is produced; this is a characteristic test for —CONH— group. Since proteins contain the group they also give this test.

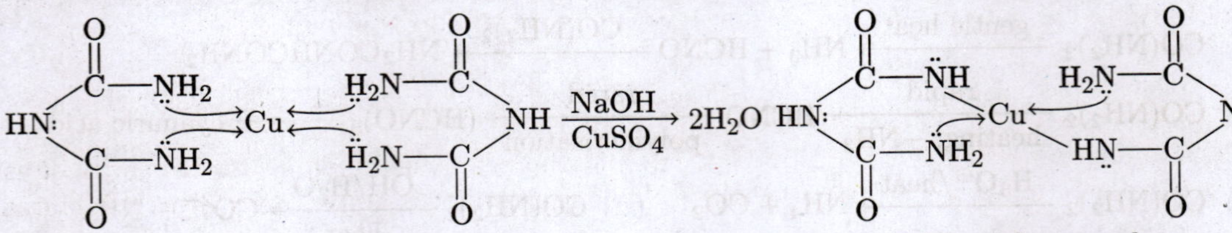

soluble violet complex

**(b) Effect of rapid heating :** On rapid heating the cyanic acid (so formed) polymerises to cyanuric acid.

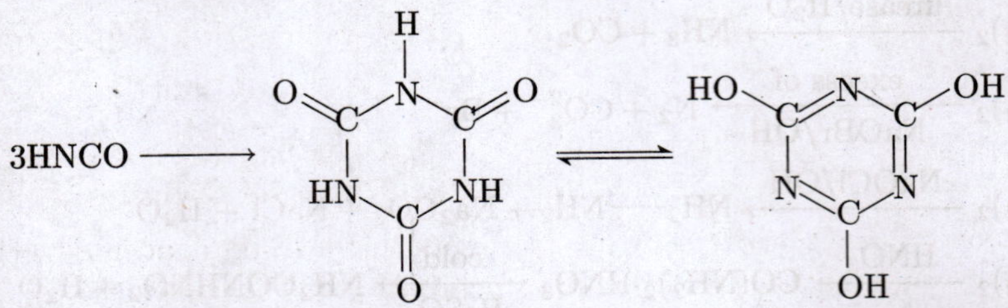

**(c) Acid hydrolysis :** On boiling with dilute acid urea gives up carbon dioxide and forms ammonium salt.

$$H_2O + NH_2CONH_2 + 2HCl\ (dil.) \longrightarrow 2NH_4Cl + CO_2$$

**(d) Alkaline hydrolysis :** Urea forms carbonate salt and gives out ammonia on boiling with dilute alkali.

$$NH_2CONH_2 + 2NaOH \longrightarrow 2NH_3 + Na_2CO_3$$

**(e) Reaction with nitrous acid :** Nitrous acid decomposes urea to molecular nitrogen, carbon dioxide and water.

$$NH_2CONH_2 + 2HNO_2 \longrightarrow 2N_2 + CO_2 + 3H_2O$$

**(f) Enzymetic decomposition of urea :** Urease decomposes urea to ammonia and carbon dioxide.

$$NH_2CONH_2 + H_2O \xrightarrow{urease} 2NH_3 + CO_2$$

**(g) Action of excess alkaline hypobromite solution :** When urea is treated with excess of alkaline hypobromite solution, urea decomposes to nitrogen and carbon dioxide which forms carbonate salt with alkali.

$$NH_2CONH_2 + 3NaOBr + 2NaOH \longrightarrow N_2 + Na_2CO_3 + 3NaBr + 3H_2O$$

**(h) Formation of hydrazine :** When equivalent amount of alkaline hypohalite solution is treated with urea, hydrazine forms by the Hofmann's haloamide reaction (See p. 724).

$$NH_2CONH_2 + NaOCl + 2NaOH \longrightarrow NH_2{\cdot}NH_2 + NaCl + Na_2CO_3 + H_2O$$

**(i) Formation of urea mononitrate and nitrourea :** When nitric acid is added to a concentrated solution of urea, urea nitrate is formed; this on treatment with cold concentrated sulphuric acid produces nitrourea. Nitrourea on electrolytic reduction forms semicarbazide.

$$NH_2CONH_2 + HNO_3 \longrightarrow \underset{\text{Urea nitrate}}{NH_2CONH_2{\cdot}HNO_3} \xrightarrow[\text{conc. } H_2SO_4]{\text{cold}} \underset{\text{Nitrourea}}{NH_2CONH{\cdot}NO_2}$$

**(j) Reaction with acyl halides :** Acyl halides react with urea to form ureides; acetic anhydride reacts to form diacetyl ureide.

$$CH_3CONHCONHCOCH_3 \xleftarrow{Ac_2O} NH_2CONH_2 \xrightarrow[-HCl]{CH_3COCl} CH_3CONHCONH_2$$

**(k) Reaction with dicarboxylic acids :** Dicarboxylic acids undergo reaction with urea in the presence of phosphoryl chloride to form cyclic ureides; for example, urea forms oxalylurea with oxalic acid.

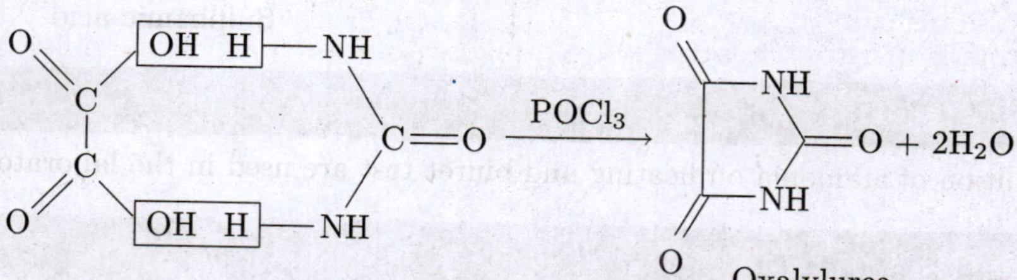


Oxalylurea

**(l) Formation of barbituric acid :** When malonic ester is condensed with urea in the ethanolic medium containing sodium ethoxide, barbituric acid is obtained.

Barbituric acid and 5, 5-disubstituted barbiturates are good hypnotics.

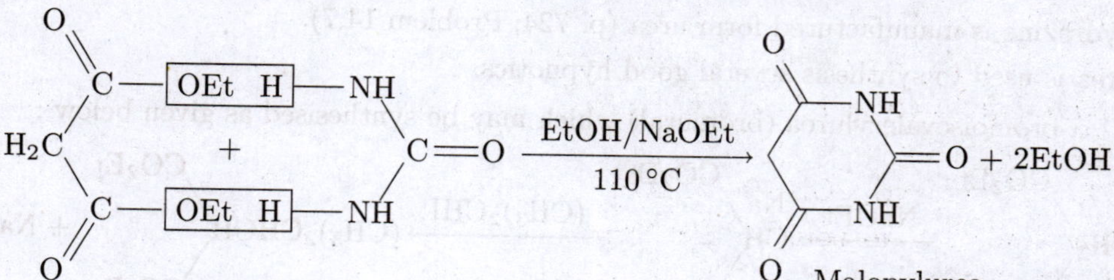


Malonylurea
Barbituric acid

**(m) Reaction with formaldehyde :** Urea reacts with formaldehyde to form the urea-formaldehyde resins which are highly important in moulded plastics.

Hydroxy derivatives lose water molecule to form aldimino derivatives and condense with formaldehyde and urea to give resins which have complex cyclic structures.

$$H_2C{=}O + NH_2CONH_2 \longrightarrow \overset{\ominus}{O}CH_2{-}\overset{\oplus}{N}HCONH_2 \longrightarrow \underset{\text{Methylourea}}{HOCH_2NHCONH_2}$$
$$\qquad\qquad\qquad\qquad\qquad\qquad\quad | \\ \qquad\qquad\qquad\qquad\qquad\qquad\quad H$$

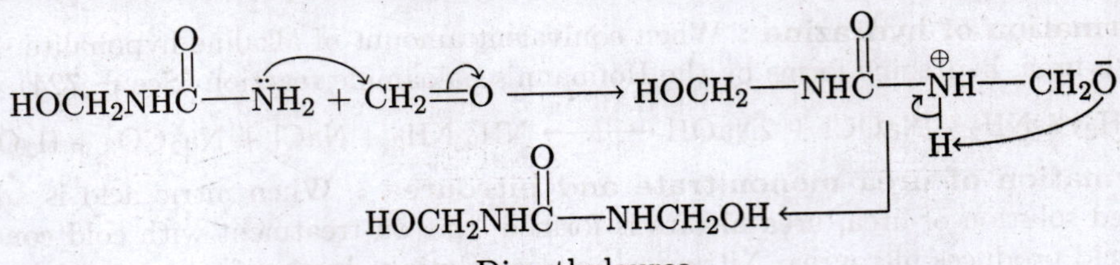

Dimethylourea

**(n) Reaction with dimethyl sulphate :** Dimethyl sulphate reacts with urea in the presence of alkali to form O-methyl urea.

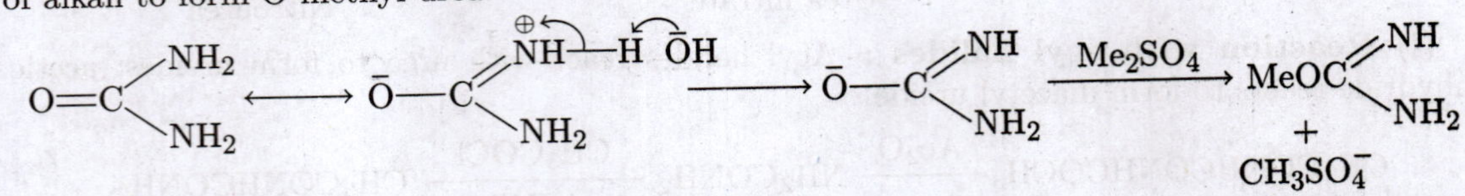

**(o) Reaction with fuming sulphuric acid :** Fuming sulphuric acid reacts with urea to form sulphamic acid.

$$H_2NCONH_2 + SO_3 + H_2SO_4 \longrightarrow \underset{\text{Sulphamic acid}}{2H_2N{\cdot}SO_3H} + CO_2$$

## 20.19.5 Test

Evolution of ammonia on heating and biuret test are used in the laboratory to identify urea.

## 20.19.6 Uses

(i) It is used to prepare urea-formaldehyde resins.

(ii) Excess of nitrous acid is removed from a reaction medium by decomposing it with urea.

(iii) It is used as a nitrogenous fertiliser.

(iv) Hydrazine is manufactured form urea (p. 724; Problem 14.7).

(v) Urea is used to synthesis several good hypnotics.

(a) $\alpha$-bromoisovalerylurea (bromural) which may be synthesised as given below :

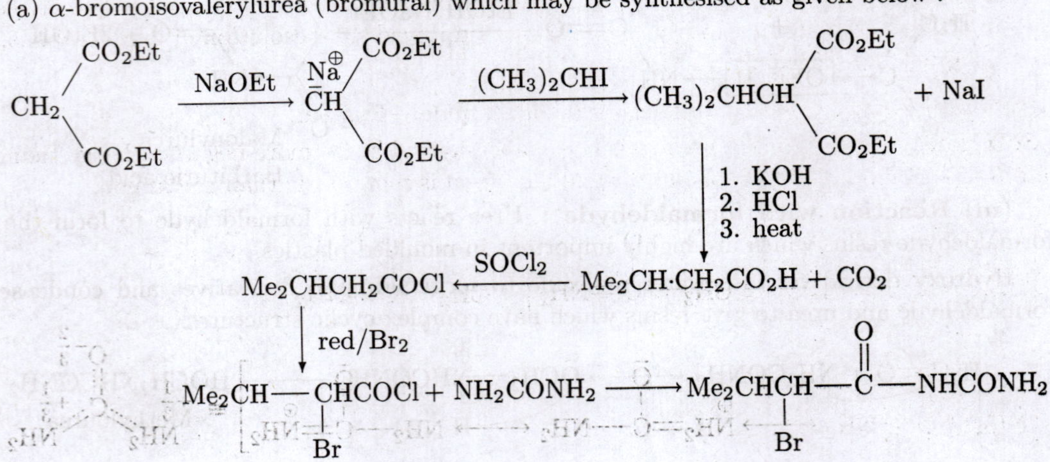

(b) Synthesis of 5, 5-diethylbarbituric acid (barbital) :

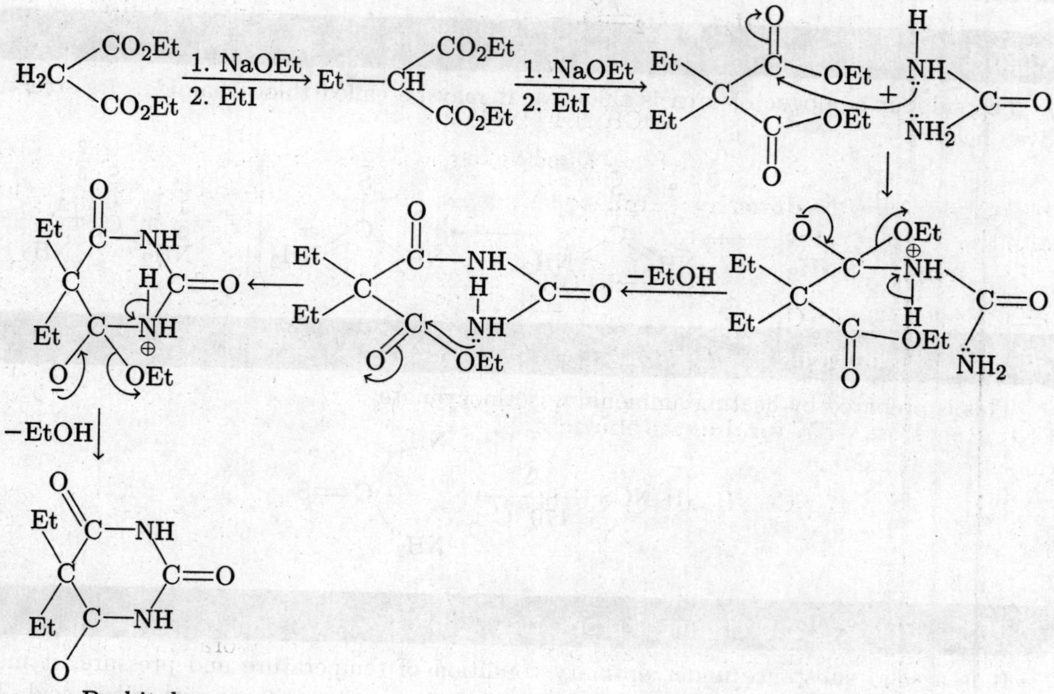


**(vi) Separation and resolution of hydrocarbons :** Urea has the important property of forming inclusion complexes. The urea cage has a diameter of 0.5 nm and the cage is dissymmetric. The cage can accommodate straight chain hydrocarbons but not the branched chain isomers. When a methanolic solution of hydrocarbons containing straight and branched isomers is mixed with saturated methanolic solution of urea, a crystalline precipitate of inclusion complex consisting of the straight chain hydrocarbon and urea is obtained. This on dissolving in water or extracting with ether gives the desired straight chain hydrocarbon. In this way the normal hydrocarbons may also be removed from petroleum and thus octane number of gasoline may be improved (for resolution of hydrocarbons, see Chapter 9, p. 362).

**Problem 20.31.** Urea is more basic than acetamide.—Explain.

**Solution :** The O atom of urea in its resonance hybrid bears more negative charge than the O atom of acetamide in its resonance hybrid. So, urea is more basic than acetamide.

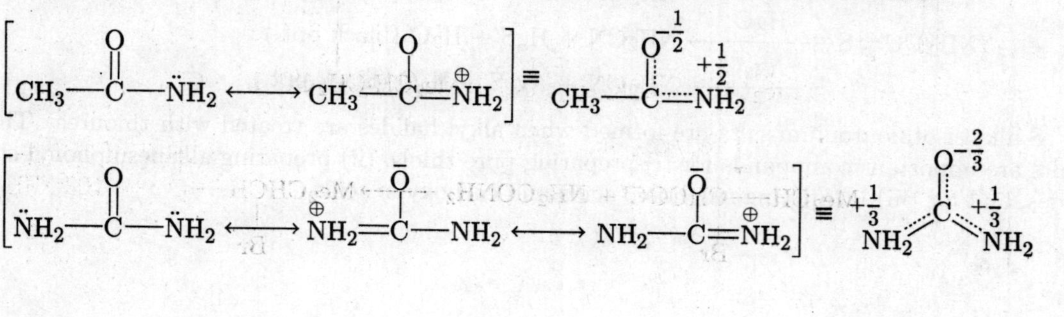

## 20.20 Thiourea

### 20.20.1 Introduction

The sulphur analogue of urea is thiourea, it may be called thiocarbamide. Its structure is given below :

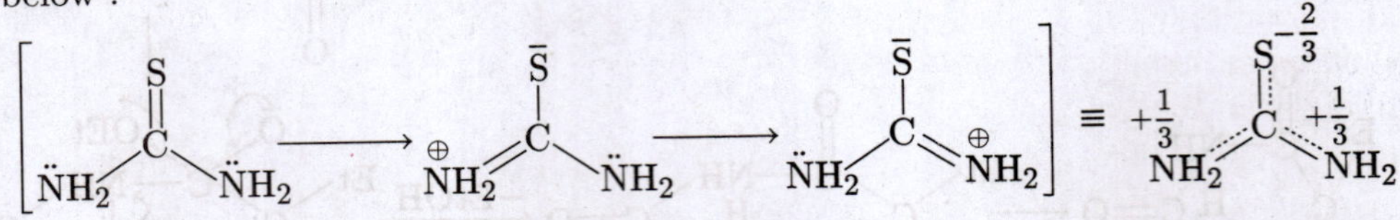

### 20.20.2 Formation

This is prepared by heating ammonium isothiocyanate.

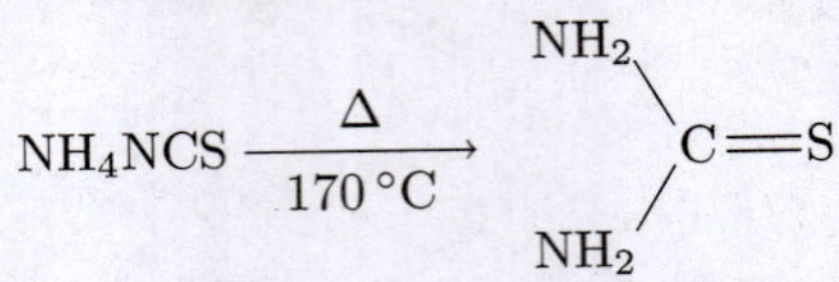

### 20.20.3 Physical and chemical properties

It is a solid substance under ordinary condition of temperature and pressure; it melts at 180 °C. Thiourea is a monoacidic base like urea. It forms mononitrate with nitric acid. It also forms inclusion complexes with branched chain alkanes and with cycloalkanes. The diameter of the channel being 0.7 nm (larger than that of urea), it cannot accommodate straight-chain alkanes.

Thiourea hydrolyses to carbonate, sulphide and ammonia when this is heated with dilute alkalis.

$$S{=}C(NH_2)_2 + 4\,NaOH + H_2O \longrightarrow Na_2S + Na_2CO_3 + 2\,NH_3 + 2H_2O$$

Thiourea gives urea on oxidation with alkaline potassium permanganate or alkaline hydrogen peroxide.

$$(NH_2)_2C{=}S + [O] \xrightarrow[\text{or alk. } H_2O_2]{\text{alk. } KMnO_4} (NH_2)_2C{=}O + S$$

Thiourea gives cyanamide on treatment with oxides of lead, silver or mercury.

$$(NH_2)_2C{=}S \xrightarrow{PbO} NH_2CN + PbS + H_2O \text{ (black ppt.)}$$

$$(NH_2)_2C{=}S \xrightarrow{HgO} NH_2CN + HgS + H_2O \text{ (black ppt.)}$$

$$(NH_2)_2C{=}S \xrightarrow{Ag_2O} NH_2CN + Ag_2S + H_2O \text{ (black ppt.)}$$

S-alkyl-isothiouronium salts are formed when alkyl halides are treated with thiourea. These salts are important compounds for (i) preparing pure thiols, (ii) preparing alkanesulphonyl chlorides, and for (iii) identifying sulphonic acids and carboxylic acids.

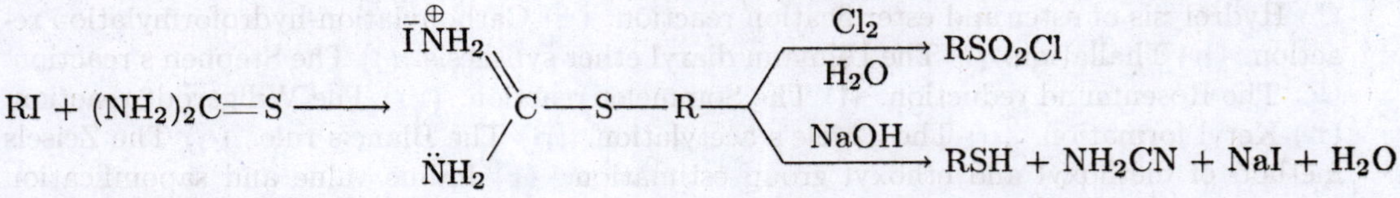

Thiourea is used to separate isomeric alkanes and to resolve (±)-alkanes.

In the laboratory S-benzylisothiouronium chloride is used to identify carboxylic acids. They form S-benzylisothiouronium salts which have sharp and specific melting points. Here chloride ion gets replaced by $RCO_2^-$ from $RCO_2Na$.

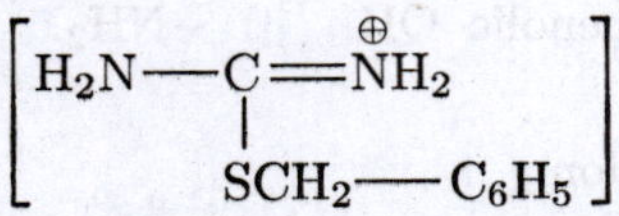

$Cl^-$ (S-benzylisothiouronium chloride)

**Problem 20.32.** Hydrogen chloride does not add to a carbonyl compound.—Explain.

**Solution :** The addition of hydrogen chloride to a carbonyl compound is a reversible reaction and the equilibrium lies far over to the left. So, as soon as the adduct chlorohydrin is formed it decomposes to hydrogen chloride and the substrate.

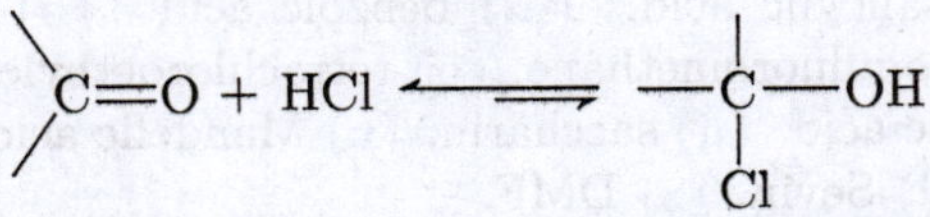

If we calculate the heat of formation of the adduct we shall find it $-25.12$ kJ $mol^{-1}$; but the reaction possesses a negative entropy of activation since two molecules, HCl and the carbonyl compound, are converted to a single molecule, the chlorohydrin. In this particular reaction perhaps $T\Delta S$ term is more negative than the $\Delta H_f$ term of the equation

$$\Delta G = \Delta H_f - T\Delta S.$$

Hence $\Delta G$ is always positive and the reaction does not take place. However, the value of $\Delta H_f$ of the reaction can be calculated from the bond dissociation energy data as given below :

| Bond energy of | | |
|---|---|---|
| | H—Cl | : 431.24 kJ $mol^{-1}$ |
| | >C=O | : 695.00 kJ $mot^{-1}$ |
| | >C—O | : 334.94 kJ $mol^{-1}$ |
| | C—Cl | : 351.69 kJ $mol^{-1}$ |
| | O—H | : 464.73 kJ $mol^{-1}$ |

$$\therefore \quad \Delta H_f = [431.24 + (695.00 - 334.94) - 351.69 - 464.73] \text{ kJ mol}^{-1}$$
$$= -25.12 \text{ kJ mol}^{-1}.$$

## Exercise

1. Write notes :

   (a) Williamson synthesis. (b) Gabriel phthalimide synthesis. (c) The Sandmeyer and the Gatterman reaction. (d) The Ullmann biaryl synthesis. (e) The Wurtz-Fittig reaction.

(f) Hydrolysis of ester and esterification reaction. (g) Carbonylation-hydroformylation reaction. (h) Thallation. (i) The Ullmann diaryl ether synthesis. (j) The Stephen's reaction. (k) The Rosenmund reduction. (l) The Sommelet reaction. (m) The Willgerodt reaction. (n) Ketyl formation. (o) The Thiele's acetylation. (p) The Blanc's rule. (q) The Zeisels method of methoxyl and ethoxyl group estimation. (r) Iodine value and saponification value. (s) The Thorpe nitrile condensation. (t) The carbylamine test. (u) The Victor Meyer's test for alcohols. (v) The Leuckart reaction. (w) The Ritter reaction. (x) The Etard reaction. (y) Diazotisation reaction. (z) Benzoylation of amines.

2. Write down the tests of the following groups :

(a) Chloro. (b) bromo. (e) enolic OH. (d) phenolic OH. (e) $-NH_2$. (f) $-NO_2$. (g) $-CHO$. (h) >CO. (i) $-CO_2H$. (j) $-COOEt$.

3. How can you estimate the following from their solution :

(a) formaldehyde. (b) acetone (c) acetic acid. (d) aniline. (e) phenol.

4. Write down the uses and synthesis of the following compounds :

(a) vinyl chloride. (b) chlorobenzene. (c) methyl alcohol. (d) ethyl alcohol. (e) benzyl alcohol. (f) formaldehyde. (g) glycol. (h) glycerol. (i) acetaldehyde. (j) acetone. (k) acetic acid. (l) salicylic acid. (m) benzoic acid. (n) ethylene oxide. (o) acetic anhydride. (p) dichlorodifluoromethane. (q) tetrachloroethylene. (r) phaltan. (s) terephthalic acid. (t) phthalic acid. (u) saccharin. (v) Mandelic acid. (w) N-bromosuccinimide. (x) sulphanilamide. (y) Sevin. (z) DMF.

5. Synthesise the following compounds :

(i) EtCH(Me)—COOH from ethyl acetoacetate

(ii) 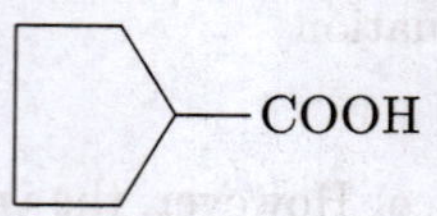 from ethyl malonate

(iii) 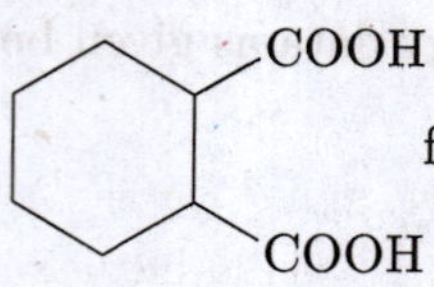 from butadiene

(iv) MeCH(Me)CH(Me)COOH from acetone and ethyl cyanoacetate

(v) anthranilic acid from naphthalene

(vi) *p*-benzoquinone from benzene

(vii) T.N.B. from toluene

(viii) *m*-nitrotoluene from toluene

(ix) 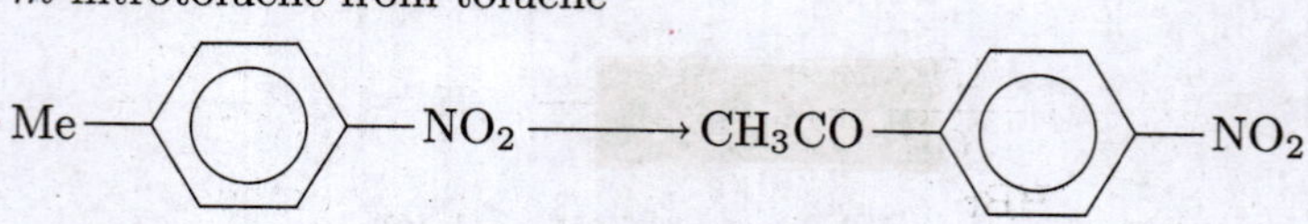

(x) 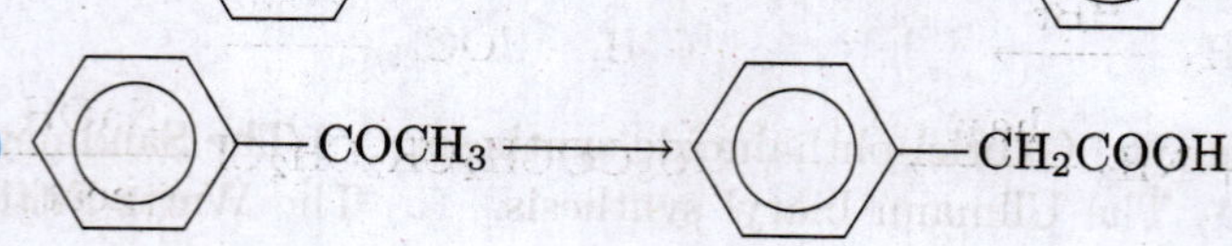

**(xi)** 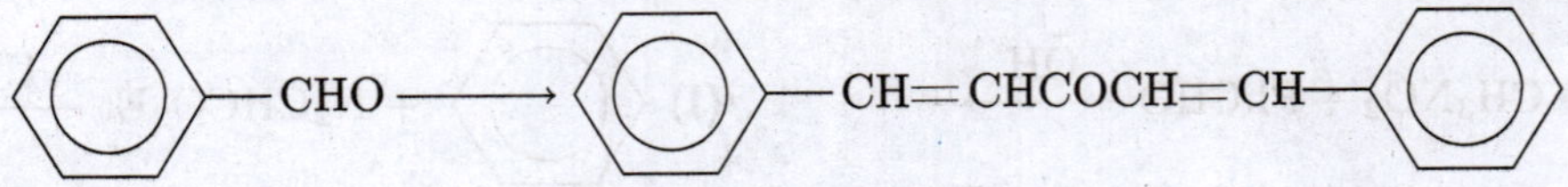

**(xii)** aniline $\longrightarrow$ *p*-dinitrobenzene

**(xiii)** *p*-nitrotoluene $\longrightarrow$ *m*-nitrotoluene

**(xiv)** *m*-nitrobenzene $\longrightarrow$ *m*-chloronitrobenzene

**(xv)** 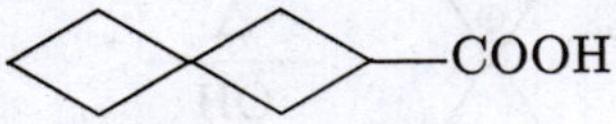 from diethyl malonate

**(xvi)** 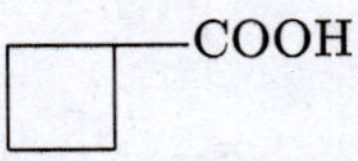 from diethyl malonate

**(xvii)** 4-oxopentanoic acid from acetoacetic ester

**(xviii)** *p*-methoxybenzoic acid from toluene

**(xix)** 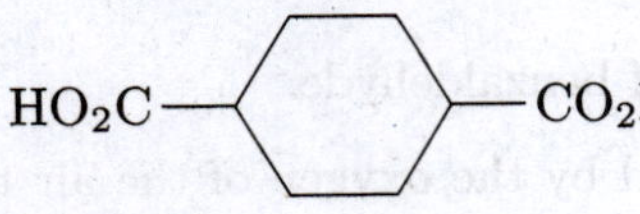 from diethyl malonate

**(xx)** 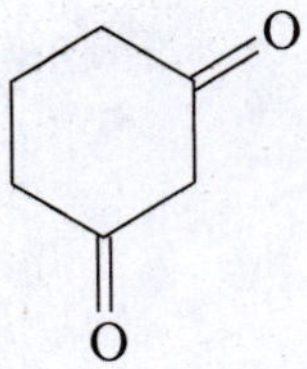 from ethyl acrylate

**(xxi)** 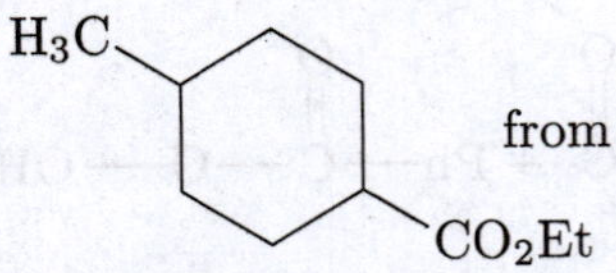 from 2-methyl-1, 3-butadiene

**(xxii)** PHCHO $\longrightarrow$ PhCH(Me)CHO

**6.** Predict the product/products in the following reactions with plausible mechanism :

**(a)** *m*-Chlorobenzaldehyde $\xrightarrow{\bar{C}N}$

**(b)** *m*-Chlorobenzaldehyde $\xrightarrow{\bar{O}H}$

**(c)** $CH_3CH_2CH(NH_2)-CH_3 \xrightarrow[H_2SO_4]{NaNO_2}$

**(d)** $CH_3CH_2-CH(\overset{\oplus}{N}Me_3)-CH_3 \xrightarrow[\bar{O}H]{150\,°C}$

**(e)** $PhCH_2CH_2\overset{\oplus}{N}(Me)_2-Et \xrightarrow[OH^-]{150\,°C}$

**(f)** $F_3C-CH=CH_2 \xrightarrow{HCl}$

**(g)** $CH_3CH_2CH=CHCH_3 \xrightarrow{HBr}$

**(h)** $EtOH + SOCl_2 \xrightarrow{ether}$

**(i)** phthalic anhydride + $NH_3 \xrightarrow{heat}$

**(j)** $CH_3COCH_2CH_2CH_2Cl \xrightarrow[heat]{NaOH}$

(k) $CH_3NO_2 + PhCHO \xrightarrow{\bar{O}H}$ (l) 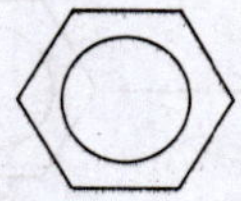 $+ N_2CHCO_2Et \xrightarrow{UV}$

(m) *m*-dinitrobenzene $\xrightarrow[\bar{O}H]{K_3Fe(CN)_6}$ (n) salicylaldehyde $\xrightarrow{H_2O_2/\bar{O}H}$

(o) 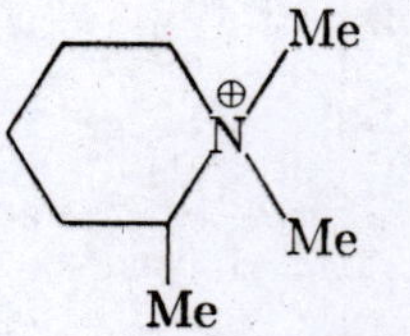 $\xrightarrow[\bar{O}H]{\Delta}$ (p) 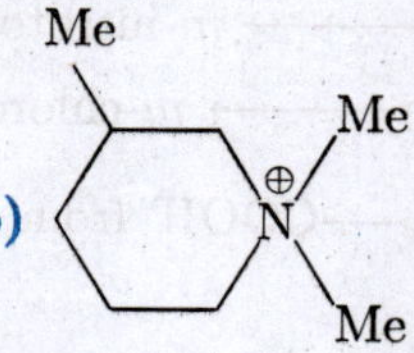 $\xrightarrow[\bar{O}H]{\Delta}$

(q) $RCOCl \xrightarrow{NH_2OH} A \xrightarrow{H^{\oplus}} B$ (r) $PhCHO + HCHO \xrightarrow{\bar{O}H}$

(s) Aniline + dil. HCl + aq. $NaNO_2 \xrightarrow{0\,°C} \xrightarrow{heat}$

7. Write down the mechanism of the auto-oxidation of benzaldehyde.

**Ans.** Benzaldehydes gets converted to benzoic acid by the oxygen of the air through free radical mechanism as shown below :

$$Ph—\overset{O}{\overset{||}{C}}—H \longrightarrow Ph—\overset{O}{\overset{||}{C}}\cdot + H\cdot$$

$$Ph—\overset{O}{\overset{||}{C}}\cdot + O_2 \longrightarrow Ph—\overset{O}{\overset{||}{C}}—O—O\cdot$$

$$Ph—\overset{O}{\overset{||}{C}}—H + Ph—\overset{O}{\overset{||}{C}}—O—O\cdot \longrightarrow Ph—\overset{O}{\overset{||}{C}}\cdot + Ph—\overset{O}{\overset{||}{C}}—O—OH$$

$$Ph—\overset{O}{\overset{||}{C}}—O—OH \longrightarrow Ph—\overset{O}{\overset{||}{C}}O\cdot + \cdot OH$$

$$Ph—\overset{O}{\overset{||}{C}}—O\cdot + Ph—\overset{O}{\overset{||}{C}}—H \longrightarrow Ph—\overset{O}{\overset{||}{C}}—OH + Ph\overset{O}{\overset{||}{C}}\cdot$$

$$Ph—\overset{O}{\overset{||}{C}}\cdot + \cdot OH \longrightarrow Ph—\overset{O}{\overset{||}{C}}—OH$$

8. Draw mechanistic steps for the following reactions :

(i) $ArCOOAg + Br_2 \longrightarrow Ar—Br + CO_2 + AgBr$

(ii) $ROH + PX_3 \longrightarrow RX + H_3PO_3$

(iii) $Me_2CHCH_2X \xrightarrow[\Delta]{AlCl_3} Me_3CX$

(iv) $Me_2SO + CH_3CH_2Cl \longrightarrow CH_3CHO + Me_2S$

**(v)** $(\pm)\cdot \text{Ph—}\underset{\text{Me}}{\overset{\text{OH}}{\text{CH}}} + \text{HCOOH} \longrightarrow (\pm)\ \text{PhCH(Me)OOCH}$

**(vi)** $RCH_2OH \xrightarrow[\text{conc. } H_2SO_4]{K_2Cr_2O_7} RCHO \xrightarrow[\text{conc. } H_2SO_4]{K_2Cr_2O_7} RCOOH$

**(vii)** $RCN \xrightarrow{H_3O^{\oplus}} RCOOH + \overset{\oplus}{N}H_4$

**(viii)** $RNC \xrightarrow{H_3O^{\oplus}} R\overset{\oplus}{N}H_3 + HCOOH$

**(ix)** $RCHOHCHOHR \xrightarrow{HIO_4} RCHO + RCHO + HIO_3$

**(x)** $PhOH + CH_2O \xrightarrow{H^{\oplus}} p\text{-HO}\cdot C_6H_4CH_2OH + o\text{-HOC}_6H_4CH_2OH$

**(xi)** $ROR + CO \xrightarrow[H_3O^{\oplus}]{BF_3} RCOOR$

**(xii)** $RCH_2CHO + SeO_2 \longrightarrow RCOCHO$

**(xiii)** 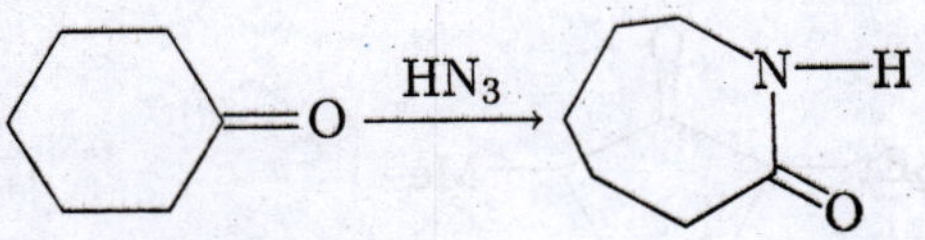

**(xiv)** 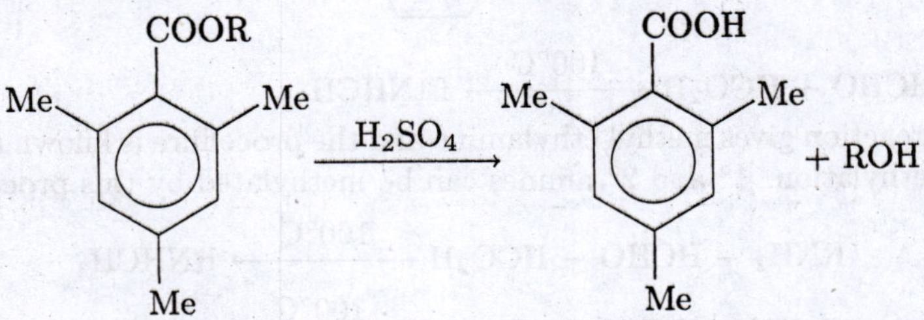

**(xv)** 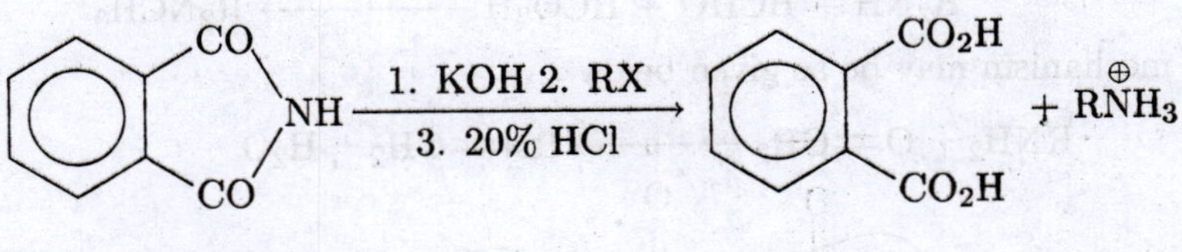

**(xvi)** $RNH_2 \xrightarrow{HNO_2} ROH + N_2 + H_2O$

**(xvii)** 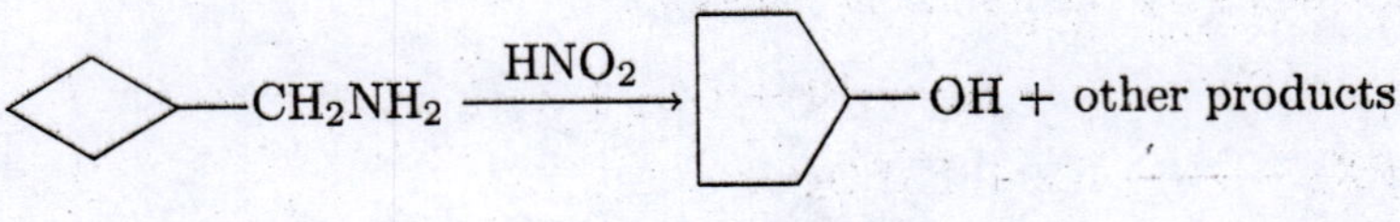

**(xviii)** 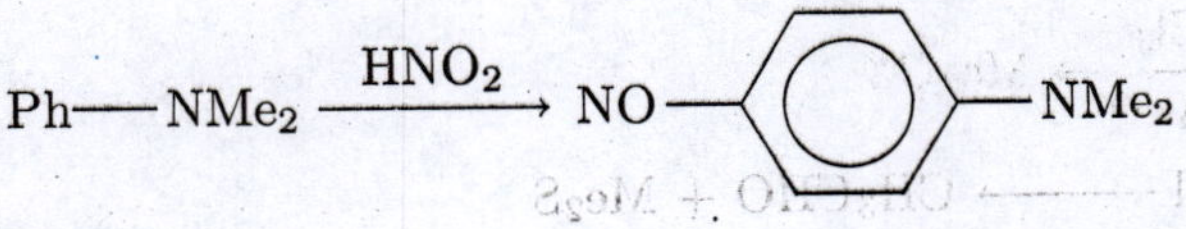

**(xix)** 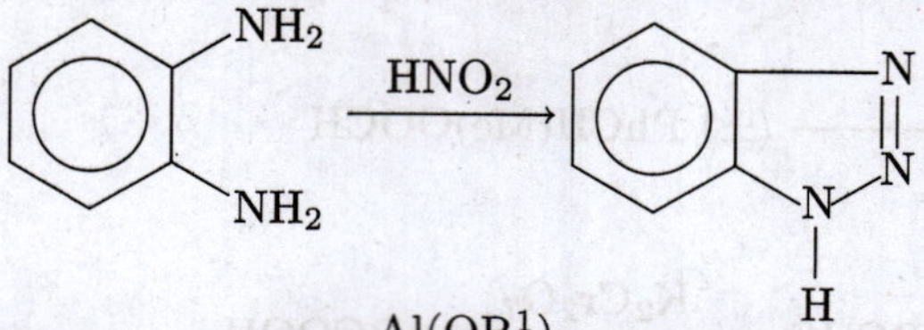

**(xx)** $ROH + CH_2N_2 \xrightarrow{Al(OR^1)_3} ROCH_3$

**(xxi)** 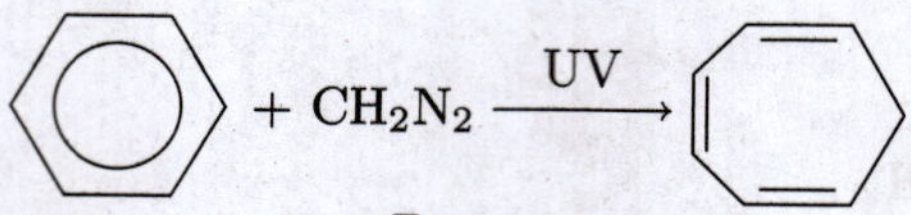

**(xxii)** $ClCH_2CHCl_2 \xrightarrow{\bar{O}H} CH_2{=}CCl_2$

**(xxiii)** $t\text{-}BuBr + KCN \xrightarrow[H_2O]{EtOH} Me_2C{=}CH_2$

**(xxiv)** $ClCH_2CO_2Et$ + 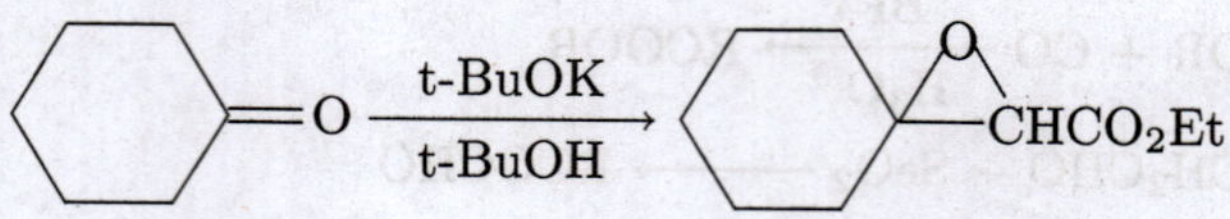

**(xxv)** 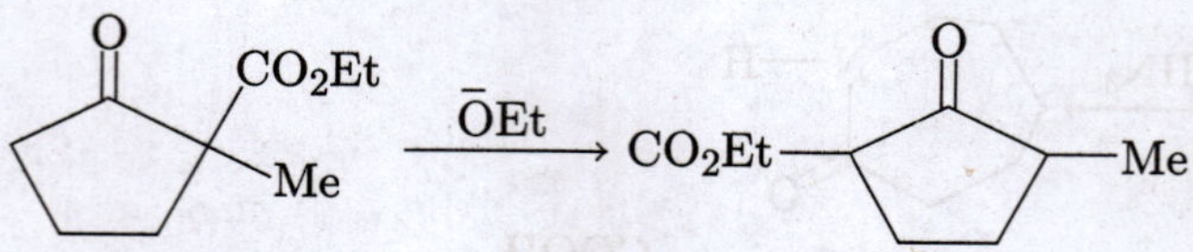

**(xxvi)** $EtNH_2 + HCHO + HCO_2H \xrightarrow{100°C} EtNHCH_3$

**Ans.** This reaction gives methyl ethylamine and the procedure is known as Eschwiler – Clarke methylation. 1° and 2° amines can be methylated by this process.

$$RNH_2 + HCHO + HCO_2H \xrightarrow{100°C} RNHCH_3$$

$$R_2NH + HCHO + HCO_2H \xrightarrow{100°C} R_2NCH_3$$

The mechanism may be as given below :

$$RNH_2 + O{=}CH_2 \longrightarrow RN{=}CH_2 + H_2O$$

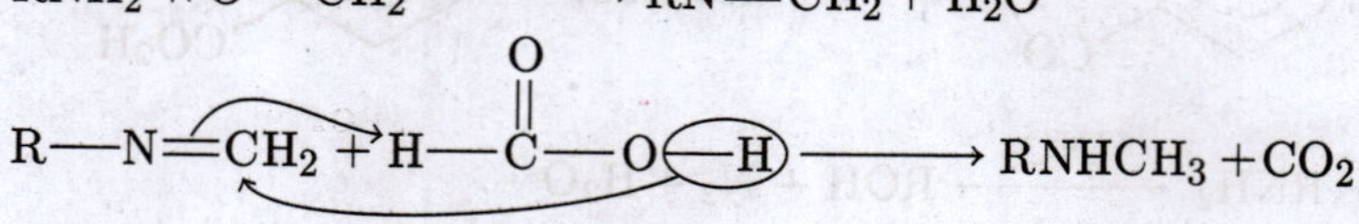

Chapter 21

# Planning a Synthesis

## Objectives

- To understand:
  - Disconnection approach [21.1] p. 1401
  - Synthons and synthetic equivalents [21.2] p. 1401
  - Where to disconnect [21.3] p. 1404
  - The synthetic routes of some common classes of compounds [21.3.1–21.3.7] p. 1406
  - Addition and removal of functional groups [21.4] p. 1421
- To know:
  - The reactions involving C—C bond formations [Table 21.3] p. 1417
  - Chemoselectivity [21.5.1] p. 1424–1435
  - Regioselectivity [21.5.2] p. 1437
  - Stereoselective reactions [21.5.3] p. 1438
  - Selective reducing and oxidising agents [Table 21.5 & 21.6] p. 1425 and 1431

- How to:
  - Protect the functional groups and remove the protecting groups [21.6] p. 1439
  - Activation of the functional groups [21.7] p. 1441
  - Reconnection of the synthons [21.8] p. 1444
- To know:
  - The use of specific reactions given by carbanions (Active methylenes) [21.9] p. 1450
  - The arrangement of events involved in aromatic syntheses [21.10] p. 1469
  - The disconnections for the synthesis of aromatic hetero cycles [21.11] p. 1476

## 21.1 Introduction

Synthetic organic compounds are of wide and varied types. Insofar as the carbon skeletons of the compounds are concerned we come across with small and large, simple and complex, straight and branched, open and cyclic, carbocyclic and heterocyclic, aromatic and non-aromatic carbon chains. The chains, in their turn, may bear one or more functional groups. So to synthesise all these compounds appropriate starting materials are to be chosen. The starting molecule thence is to be converted to the required product molecule called the Target Molecule (TM), through routes which will give the optimum yield of the target molecule with minimum of trouble and time. These routes involve the formation of several types of bonds, C—C, C—N, C—O, etc., ring-closure reactions and the introduction of functional groups and their interconversions. Most of the important reactions regarding this have already been discussed in the previous chapters. However, every synthetic route has its characteristic problems. In this chapter our intention is to find out the starting molecules and the routes through which the TMs can be synthesised. To do this we want to work backward and the process may be called synthetic analysis. Synthetic analysis is a process of stepwise breaking down of the TM to available starting materials by disconnection of bonds and groups and functional group interchange (FGI). This process gives us the back or reverse steps for the synthesis of the TM. However, every disconnected part is the idealised fragment and is called a synthon. We cannot get the Synthon straight-way from a bottle in the laboratory. Every synthon has its Synthetic Equivalent (SE). The process of converting one FG (Functional group) into another is called Functional Group Interchange (FGI). The process of adding a functional group to a moiety is called FGA (Functional Group Addition). To understand the process of synthetic analysis and the new terms such as Synthons, TM, FGI, FGA, etc. Let us consider the synthesis of phenylhydroxylamine (TM).

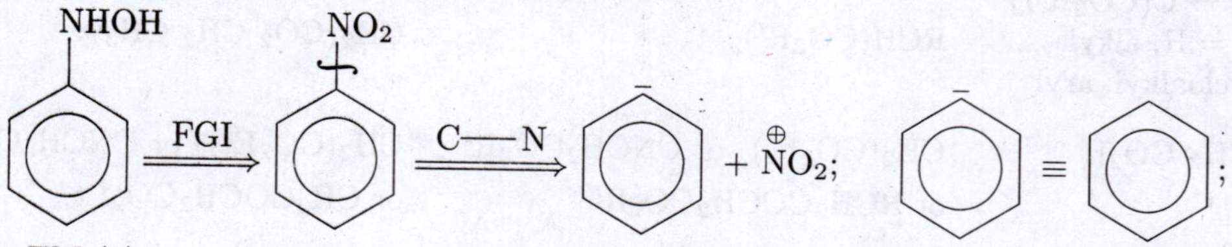


TM (1) Synthons SE

$\overset{\oplus}{N}O_2 \equiv HNO_3 + \text{conc. } H_2SO_4$ (SE)

⟹ Stands for a fragmentation (disconnection) process or FGI or FGA process.

## 21.2 Synthons and Synthetic Equivalents

Synthons may be cations, anions, radicals or actual reagents. Here phenyl anion and nitronium ion are two synthons. Since benzene undergoes electrophilic substitution reaction, a phenyl anion is equivalent to benzene, the synthetic equivalent of the former. Similarly, a mixture of conc. $HNO_3$ and conc. $H_2SO_4$ is a synthetic equivalent of $\overset{\oplus}{N}O_2$. Since —NHOH group has been converted to $-NO_2$ group in the synthetic analysis at the first step, it is a process of FGI. So the backward steps have been obtained for the synthesis of phenylhydroxylamine. The forward steps will be (i) nitration of benzene by mixed acid and (ii) reduction of nitrobenzene to the TM by Zn-$NH_4Cl$ in an aqueous or alcoholic suspension.

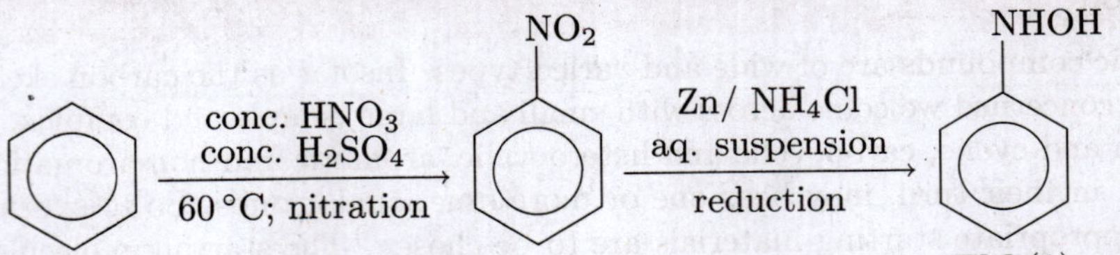

TM (1)

**Table 21.1** Some common nucleophilic and electrophilic carbon-synthons and their synthetic equivalents

| Synthons | Synthetic Equivalents | Examples |
|---|---|---|
| **(a) Nucleophilic** | | |
| 1. $R^-$ (alkyl or aryl or cycloalkyl) | $RLi$, $RMgX$, $R_2Cd$, $RCu$, $R_2CuLi$. | $CH_3Li$, $PhMgBr$, $(CH_3)_2Cd$, $CH_3Cu$, $(CH_3)_2CuLi$, etc. |
| $Ar^-$ | $ArH$ | $Ph{-}H$ |
| 2. $R{-}C{\equiv}C^-$ | $R{-}C{\equiv}C{-}Na$ (or Li), $R{-}C{\equiv}CMgX$, $R{-}C{\equiv}C{-}Cu$ | $CH_3{-}C{\equiv}C{-}Na(Li)$, $Ph{-}C{\equiv}C{-}MgBr$, $CH_3{-}C{\equiv}C{-}Cu$ |
| 3. $R{-}\bar{\ddot{C}}(CO_2R')_2$ R = H, alkyl, cycloalkyl, aryl | $RCH(CO_2R')_2$ | $CH_2(CO_2\,CH_2\,CH_3)_2$ |
| 4. $:\bar{C}H_2\,CO_2H$ | $CH_2(CO_2R')_2$ or $CNCH_2CO_2R'$ or $RCH_2COCH_2CO_2R'$ | $CH_2(CO_2Et)_2$ or $CNCH_2CO_2Et$ or $CH_3COCH_2CO_2Et$ |
| 5. $R\bar{C}HCHO$ | $RCH_2CH{=}NR'$ or $RCH_2CHO$ | $CH_3CH{=}NH$ $(R = R' = H)$ $PhCH_2CHO$ |
| 6. $\bar{C}H_2COR'$ | $CH_3COR'$ or $R'COCH_2CO_2R$ | $CH_3COCH_3$ or $CH_3COCH_2CO_2Et$ |
| 7. $R\bar{C}HCO_2R'$ | $RCH_2CO_2R'$ | $CH_3CO_2Et$ |
| 8. $RCH_2CO\bar{C}HCOR'$ | $RCH_2COCH_2COR'$ | $CH_3{-}\overset{O}{\overset{\Vert}{C}}{-}CH_2{-}\overset{O}{\overset{\Vert}{C}}{-}CH_3$ |
| 9. $R\bar{C}HCOCH_2COR'$ | $RCH_2COCH_2COR'$ | $CH_3COCH_2COCH_3$ |
| 10. $\bar{C}(R)(R')$ $(R = R' = H)$ | $Ph_3P{=}C(R)(R')$ | $Ph_3P{=}CHCH_3$ |

*Contd.*

| Synthons | Synthetic Equivalents | Examples |
|---|---|---|
| **(b) Electrophilic** | | |
| 1. $R^{\oplus}$ (alkyl or cycloalkyl) | RCl, RBr, RI, $ROSO_2R$ | $CH_3CH_2Cl$, $CH_3I$, $CH_3CH(Br)CH_3$ $CH_3OSO_2CH_3$ |
| | ROH, alkene | $CH_3CH_2OH$, $CH_2{=}CH_2$ |
| 2. $R^{\oplus}$ (aryl) | RBr, RI, $RN_2^{\oplus}X^-$ | PhBr, $PhN_2^{\oplus}$ |
| 3. $RCH{=}\overset{\oplus}{C}H$ | $RCH{=}CH{-}Br$ | $CH_3{-}CH{=}CH{-}Br$ |
| 4. $R{-}C{\equiv}\overset{\oplus}{C}$ | $RC{\equiv}C{-}Br$ | $CH_3{-}C{\equiv}C{-}Br$ |
| 5. $R{-}\overset{\oplus}{C}{=}O$ | RCOCl, $(RCO)_2O$, $RCO_2R'$ $RCONR'_2$ | PhCOCl, $CH_3CO{\cdot}O{\cdot}COCH_3$, $PhCO_2Et$, $CH_3CONEt_2$ |
| 6. $H{-}\overset{\oplus}{C}{=}O$ | $HCO_2R'$, $HCONR'_2$ $CH(OR)_3$ | $HCO_2Et$, $HCONMe_2$ |
| 7. $O{=}\overset{\oplus}{C}{-}OR'$ | $ClCO_2R'$, $R'OCOOR'$ | $ClCO_2Et$, $O{=}C(OEt)_2$ |
| 8. $O{=}\overset{\oplus}{C}{-}OH$ | $CO_2$ | $CO_2$ |
| 9. $\overset{\oplus}{C}H_2OH$ | HCHO | HCHO |
| 10. $R\overset{\oplus}{C}HOH$ (R= alkyl or cycloalkyl or aryl) | RCHO | $CH_3CHO$, PhCHO |
| 11. $(R)(R')\overset{\oplus}{C}{-}OH$ (R = R′; R ≠ R′) | RCOR′ | $CH_3COCH_3$, $PhCOCH_3$ |
| 12. $\overset{\oplus}{C}H_2{\cdot}CH_2OH$ |  |  |
| 13. $R_2\overset{\oplus}{C}{-}CH_2COR'$ R′ = R or R ≠ R′ | $R_2C{=}CH{-}COR'$ | $Me_2C{=}CHCOCH_3$ $Me_2C{=}CH{\cdot}CHO$ |
| 14. $R_2\overset{\oplus}{C}{-}CH_2CO_2H$ | $R_2C{=}CH{-}CO_2R'$ $R_2C{=}CH{-}CN$ | $PhCH{=}CH{-}CO_2Et$ $CH_2{=}CH{-}CN$ |
| 15. $RCOCH_2\overset{\oplus}{C}H_2$ | $RCOCH_2CH_2NR'_2$ | $CH_3COCH_2CH_2N(CH_3)_2$ |
| 16. :CHX | $CH_2X_2$ | $CH_2I_2$ |

**Table 21.2** Some of the common electrophilic and nucleophilic synthons other than carbon

| *For aromatic compounds* | | *For aliphatic compounds* | |
|---|---|---|---|
| **Synthons** | **Synthetic Equivalents** (i.e., reagents) | **Synthons** | **Synthetic Equivalents** (i.e., reagents) |
| $\overset{\oplus}{NO_2}$ | conc. $HNO_3$ + conc. $H_2SO_4$<br>dil. $HNO_3$ (for phenols).<br>$AcONO_2$, $N_2O_5$ | $Cl\cdot$<br>$Br\cdot$<br>$X^{\oplus}$(X=Cl, Br, I) | $Cl_2$<br>$Br_2$<br>$Cl_2$, $Br_2$, $I_2$, HOCl, HOBr. |
| $\overset{\oplus}{Cl}$ | $Cl_2/Fe$, $Cl_2/I_2$, $HOCl/\overset{\oplus}{H}$ | | |
| $\overset{\oplus}{Br}$ | $Br_2/Fe$, $HOBr/H^{\oplus}$ | $Cl^-$, $Br^-$, $I^-$ | NaCl, NaBr, NaI |
| $\overset{\oplus}{SO_2OH}$ | $H_2S_2O_7$, conc, $H_2SO_4$. | $\bar{C}N$ | KCN, AgCN, HCN. |
| $\overset{\oplus}{SO_2Cl}$ | $ClSO_2OH$ | $\bar{N}H_2$ | $NH_3$, $MNH_2/NH_3(l)$ |
| $\overset{\oplus}{NO}$ | $HNO_2 \Longrightarrow NaNO_2$ + acid | $H\bar{O}$ | $H_2O$, NaOH, KOH, AgOH |
| $Ar\overset{\oplus}{N_2}$ | $Ar\overset{\oplus}{N_2}$ | $R\bar{O}$ | RONa |
| $\bar{N}H_2$ | $KNH_2/NH_3$(l) | $NO_2^-$ | $NaNO_2$, $AgNO_2$ |
| $RO^-$ | $RO^-$, ROH | $NO_3^-$ | $NaNO_3$, $AgNO_3$ |
| $HO^-$ | $H_2O$ | $\overset{\oplus}{NO}$ | $HNO_2 \Longrightarrow NaNO_2 + H^{\oplus}$ |
| | | $\bar{S}H$ | NaSH |

## 21.3 Where to Disconnect C—C Bonds

Tables 21.1 and 21.2 supply most of the common synthons and the corresponding synthetic equivalents for the synthesis of simple TMs. However, the major problem in getting appropriate synthons is where to disconnect the TM. To get good disconnections the following guidelines may be followed.

(i) It is worth considering a disconnection adjacent to a functional group specially adjacent to —OH and >C=O.

$$\underset{\text{TM (2)}}{CH_3CHOH \,\nmid\, CH_2CHO} \Longrightarrow CH_3\overset{\oplus}{C}HOH + \bar{C}H_2CHO \longleftarrow \text{Synthons}$$

$$\qquad\qquad\qquad\qquad ||| \qquad\qquad ||| $$

$$\underset{\text{TM (2)}}{CH_3CHOHCH_2CHO} \overset{\bar{O}H/H_2O}{\rightleftharpoons} CH_3CHO + CH_3CHO \longleftarrow \text{Synthetic equivalents}$$

(ii) A C—C double bond may be converted to C(OH)—CH before disconnection if C=C is a part of a conjugated system involving a hetero atom.

$$CH_2{=}CH{-}\overset{O}{\overset{\|}{C}}{-}CH_3 \xrightarrow{\text{FGI}} CH_2OH \nmid CH_2{-}\overset{O}{\overset{\|}{C}}{-}CH_3 \Longrightarrow \overset{\oplus}{C}H_2OH + \bar{C}H_2{-}\overset{O}{\overset{\|}{C}}{-}CH_3$$

TM (3)

$$\text{TM (3)} \xleftarrow{\Delta/-H_2O} CH_2OH{-}CH_2{-}\overset{O}{\overset{\|}{C}}{-}CH_3 \xleftarrow{\bar{O}H/H_2O} H_2C{=}O + CH_3{-}\overset{O}{\overset{\|}{C}}{-}CH_3$$

(iii) In a conjugated or non-conjugated system the disconnection of double bonded Cs of C=C may also be done without adding $H_2O$ to the C=C part :

$$(CH_3)_2C \nmid CHCH_3 \Longrightarrow (CH_3)_2\overset{\oplus}{C}OH + \bar{C}H_2CH_3$$

TM (4)

$$\text{TM (4)} \longleftarrow (CH_3)_2C{=}O + Ph_3P{=}CH{\cdot}CH_3$$

(iv) The branch point is to be disconnected to get unbranched carbon-synthons which are more likely to be available.

$$\underset{\substack{|\\CH_3}}{CH_3CH_2CH} \nmid CH_2CH_2CH_3 \Longrightarrow \underset{\substack{|\\CH_3}}{CH_3CH_2\ddot{C}H} + \overset{\oplus}{C}H_2CH_2CH_3 \longleftarrow \text{Synthons}$$

TM (5)

$$\text{TM (5)} \xleftarrow[\text{synthesis}]{\text{Corey-House}} (\underset{\substack{|\\CH_3}}{CH_3CH_2CH}{-})_2CuLi + CH_3CH_2CH_2Br \longleftarrow \text{SE}$$

(v) It is better to disconnect towards the middle of the TM rather than cutting off one or two C atoms from the end.

$$CH_3CH_2CH_2 \nmid CH_2CH_2CH_3 \Longrightarrow CH_3CH_2\bar{C}H_2 + \overset{\oplus}{C}H_2CH_2CH_3 \longleftarrow \text{Synthons}$$

TM (6)

$$\text{TM (6)} \longleftarrow (CH_3CH_2CH_2)_2CuLi + BrCH_2CH_2CH_3 \longleftarrow \text{SE}$$

(vi) Branches must be disconnected from rings.

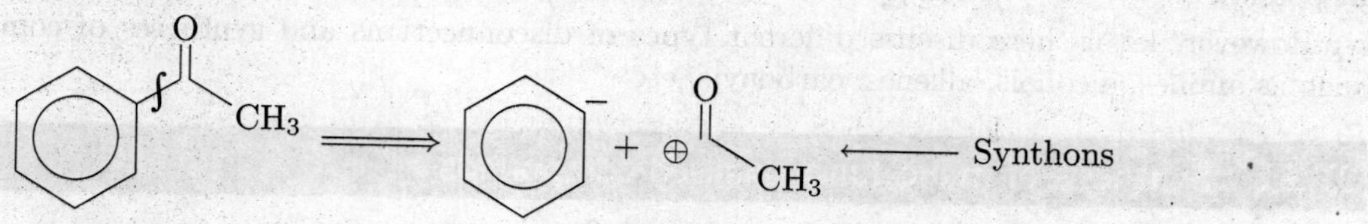


TM (7)

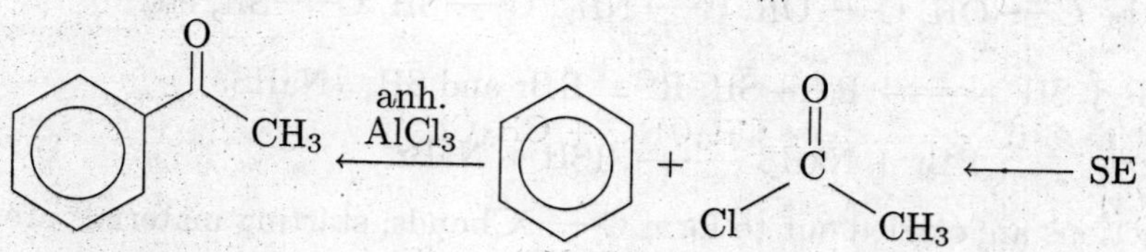


TM (7)

(vii) Symmetric TMs may be disconnected symmetrically.

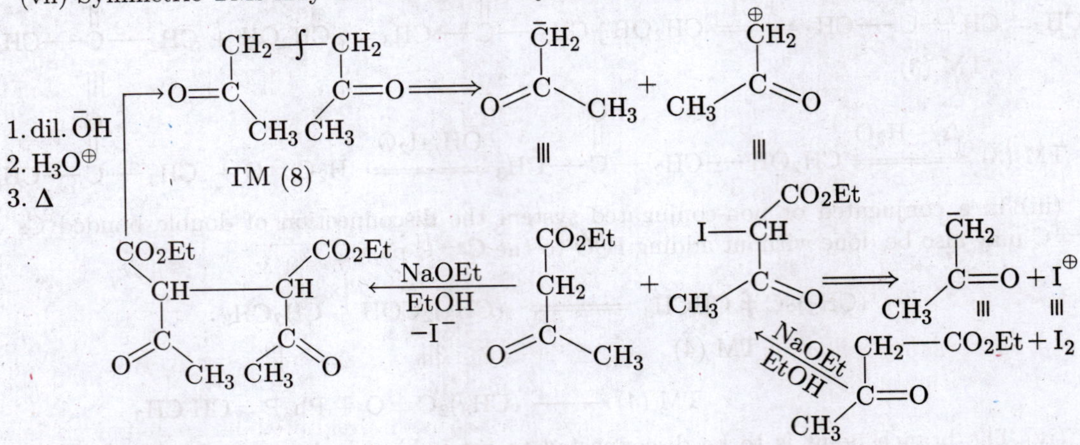

(viii) It is worth disconnecting a C—C bond between two functional groups separated by not more than 3 other carbons.

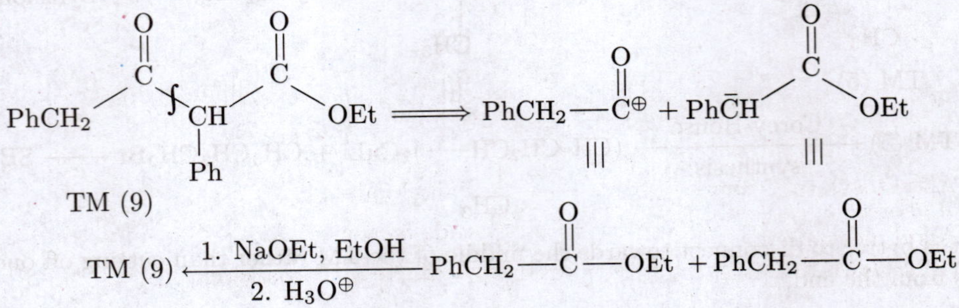

(ix) The disconnection should be in conformity with the reliable reactions. This is specially important when a C—C bond of a ring is disconnected.

(x) The C—C bond disconnection should be according to the FGs in the TM.

(xi) The disconnection should be done in such a way that the synthetic route will be as short as possible.

However, let us here discuss different types of disconnections and syntheses of compounds such as amines, alcohols, alkenes, carbonyls, etc.

## 21.3.1 Synthesis of mono-functionalised compounds

(a) *Through one group* C—X *disconnection (X is a hetero atom) :*

The C—X may be C—OH, C—OR, C—$NH_2$, C—SH, C—SR, etc.

$$R \,\text{⨍}\, SH \Longrightarrow R^{\oplus} + \bar{S}H;\; R^{\oplus} \equiv RBr \text{ and } \bar{S}H \equiv NaHS.$$

$$RBr + NaHS \longrightarrow RSH + NaBr.$$

In general $S_N$ reactions are carried out to form C—X bonds; starting materials are usually ROTs, R—X, $ROSO_2R'$, RCOCl, $(RCO)_2O$, RCOOR′ etc. and nucleophiles are ROH, $\bar{S}H$,

$NH_3$, $\bar{S}R$, $\bar{O}H$, $\bar{O}R$, etc. An example is given below :

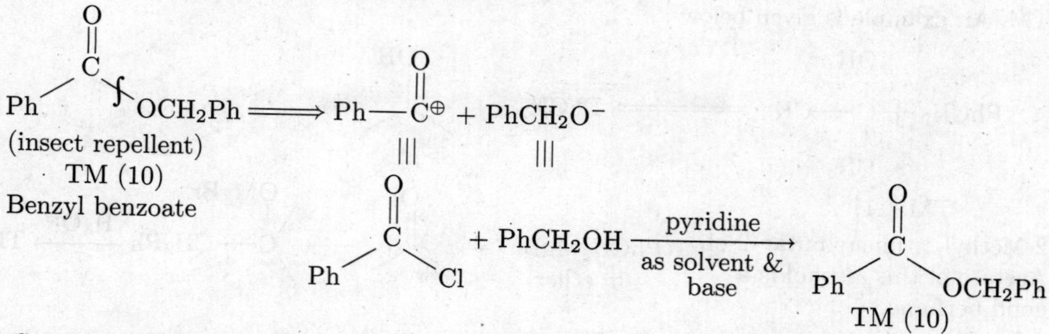


(b) *Synthons for mono-functionalised compounds can also be obtained through* C—C *disconnection :* See the synthesis of TM(11) given on the next page.

Alcohols are good starting materials for synthesising mono-functionalised compounds. Since they can be synthesised readily from alkenes and can be converted to several other compounds via alkyl halides, tosylates and mesylates as shown below :

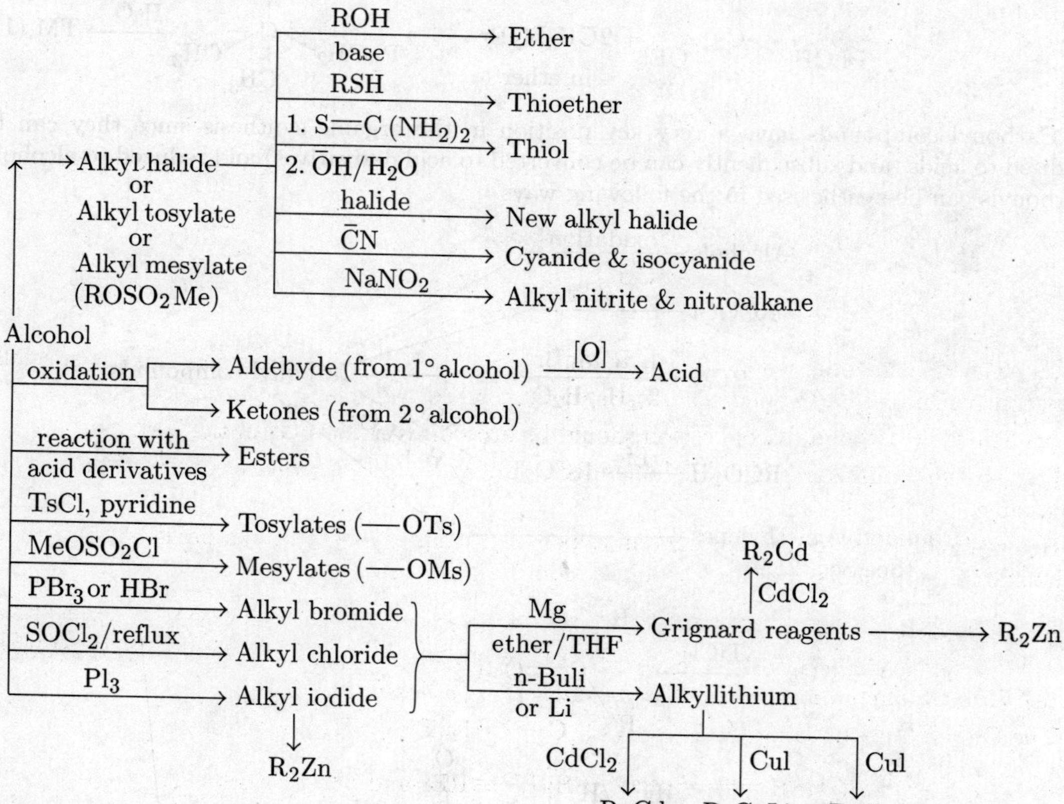


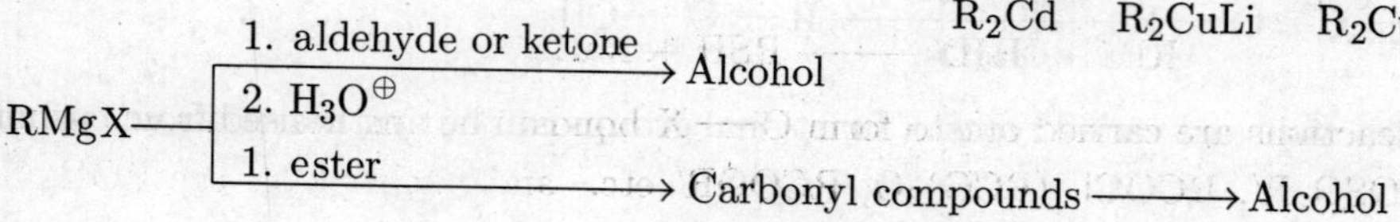

Since alcohols can readily be prepared, it is better to disconnect a C(OH)—C bond of a TM. An example is given below :

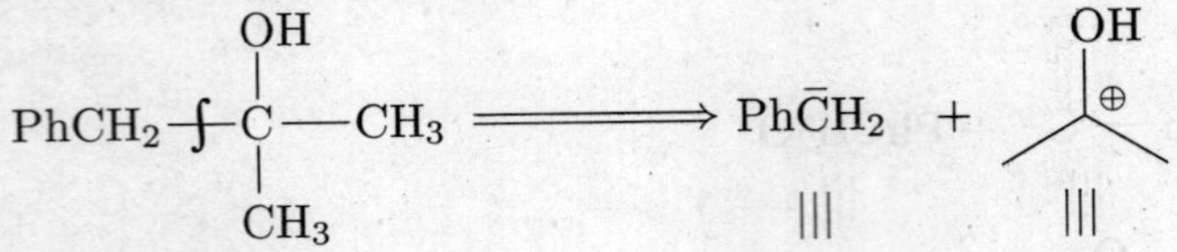

TM (11)

2-Methyl-1-phenylbutan-2-ol. Acetate of this alcohol is a good perfume.

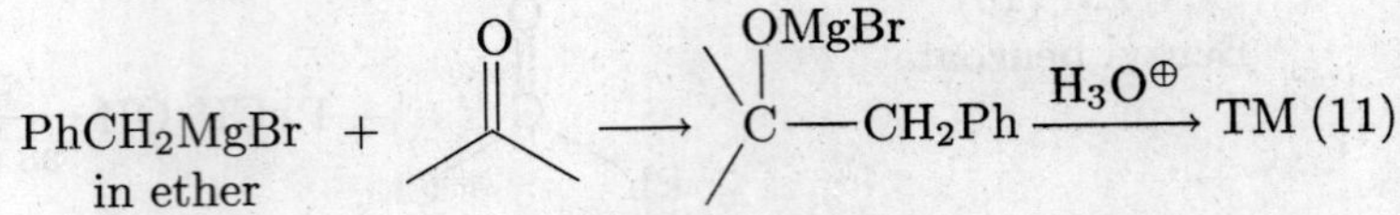

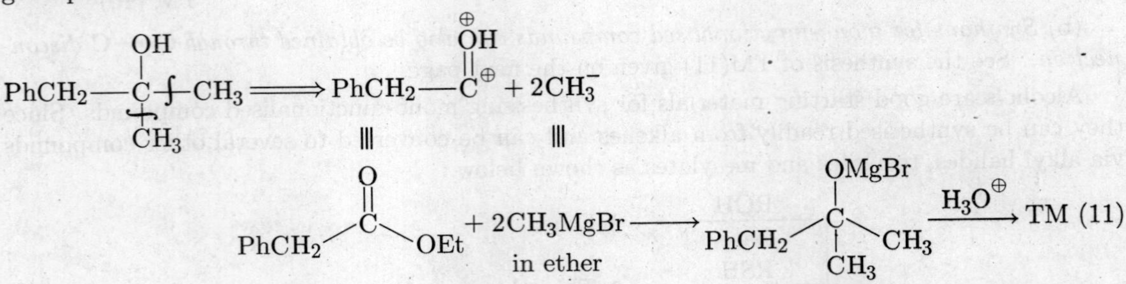

Carbonyl compounds have a very key position in the organic synthesis since they can be oxidised to acids (and subsequently can be converted to acid derivatives) and reduced to alcohols. Carbonyls can be synthesised in the following ways.

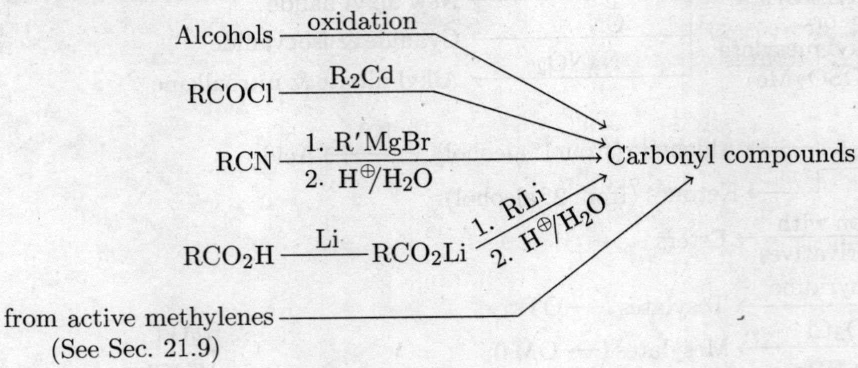


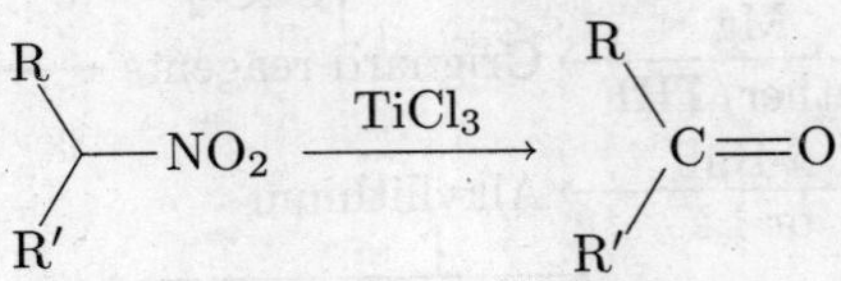

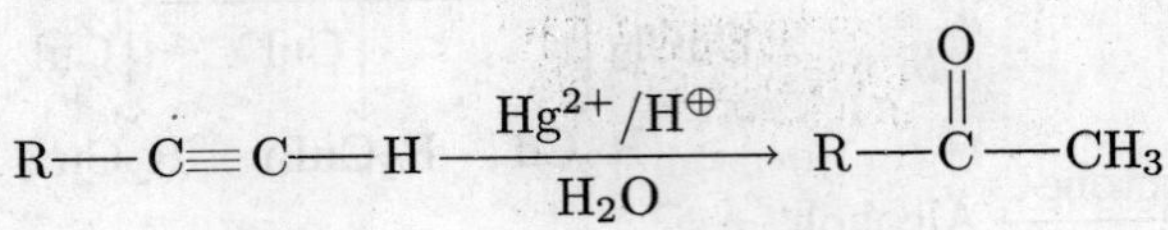

Compounds with wide and varied types of functional groups can be synthesised from carbonyl compounds.

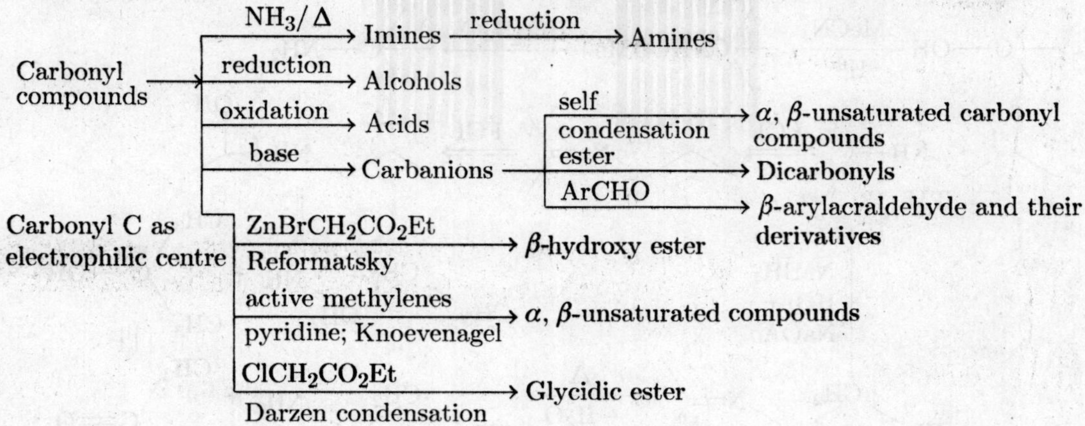


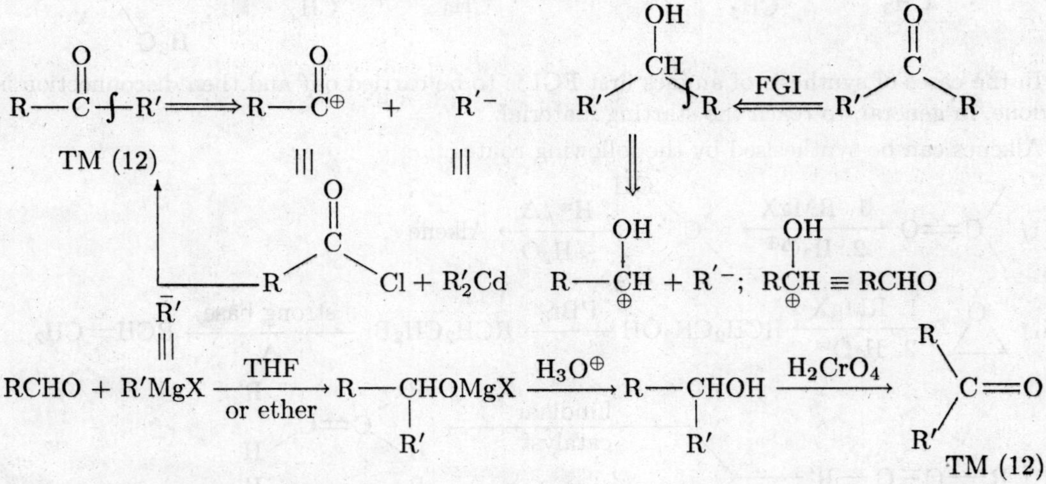


Amines can be prepared from alkyl halides by ammonolysis. But this is not a good method for 1° and 2° amines. Since amines are more reactive than $NH_3$, 3° amines and quaternary ammonium salts are usually formed along with 1° and 2° amines by the process. So it is better to synthesise 1° and 2° amines in such a way that the product will be less reactive than the starting materials. A few processes are shown below where alkyl halides have been avoided.

$$1°NH_2 : R{-}C{\equiv}N \xrightarrow[\text{or } H_2/PtO_2]{LiAlH_4/\text{ether}} RCH_2NH_2;\; RNO_2 \xrightarrow{Fe/H^{\oplus}} RNH_2;$$

$$ArCN \xrightarrow[H^{\oplus}]{H_2/Pd\text{-}C} Ar{\cdot}CH_2NH_2$$

$$2°NH_2 : \;>C{=}O + RNH_2 \xrightarrow{\Delta/-H_2O} \;>C{=}NR \xrightarrow[\text{ether}]{LiAlH_4} \;>CHNHR$$

$$3°C{-}NH_2 : \;-\!\!\!\!\!>C{-}MgCl \xrightarrow{ClNH_2} \;-\!\!\!\!\!>CNH_2 + MgCl_2$$

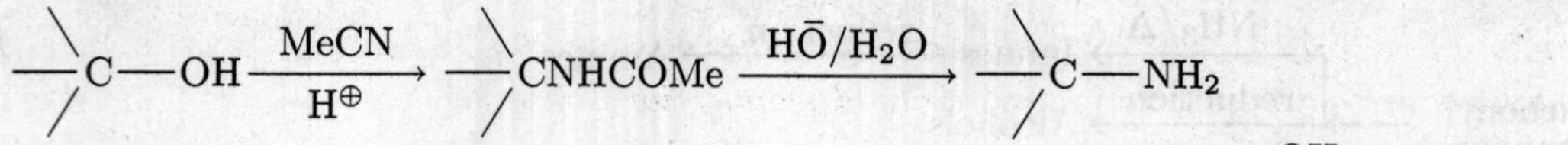

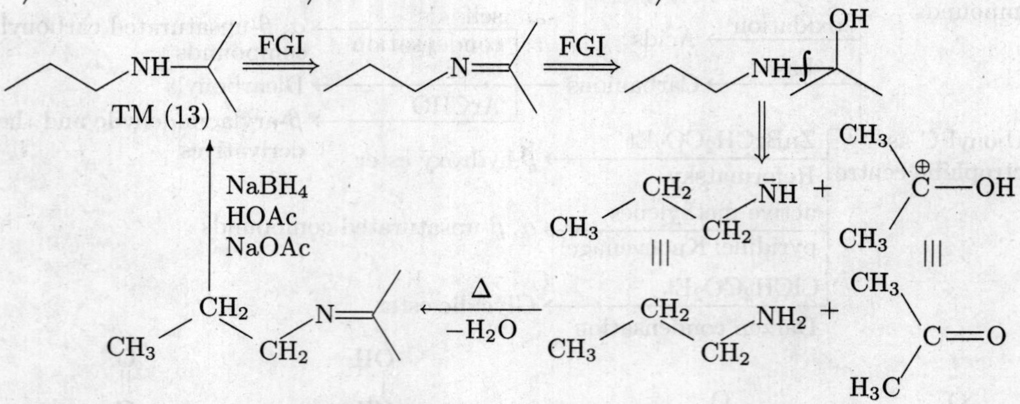

In the cases of synthesis of amines first FGI is to be carried out and then disconnection is to be done, in general, to reach the starting material.

Alkenes can be synthesised by the following routes :

(i) 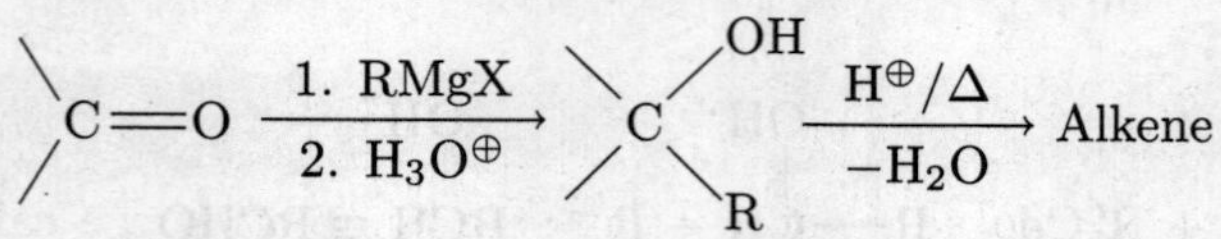

(ii) 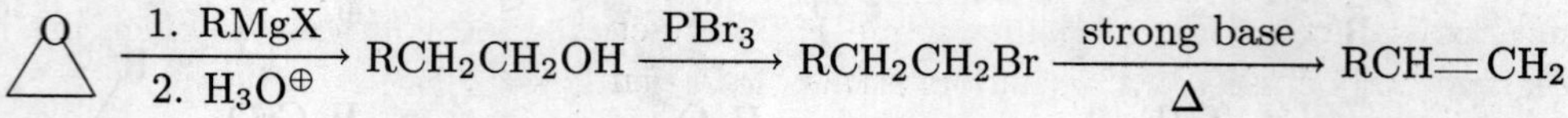

(iii) 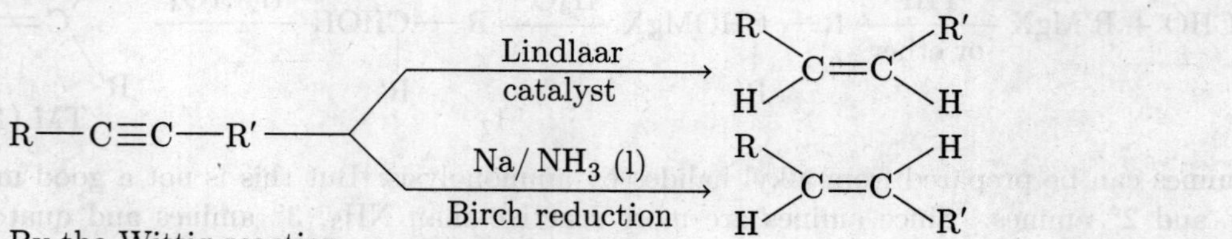

(iv) By the Wittig reaction.

This has already been discussed in section 16.8, p. 867.

Electron-withdrawing groups bonded to double bonded C in the ylids enhance the reaction rate (See Sec. 21.3.3).

The synthesis of $\alpha$-hydroxyketones :

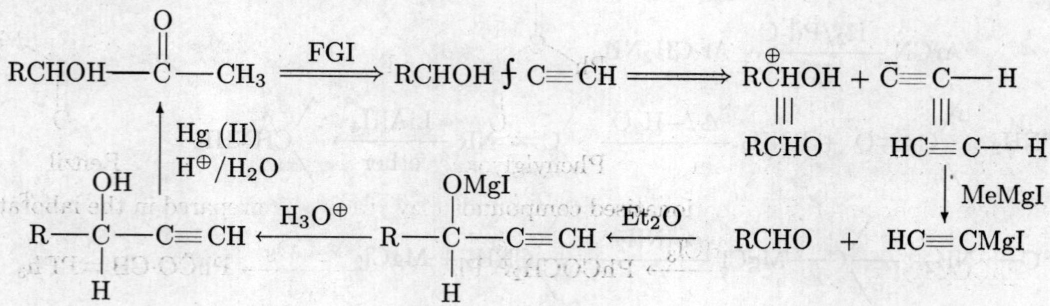

## 21.3.2 Synthesis of 1, 1-difunctionalised compounds

*Synthons are obtained through two-group disconnections on a single carbon* (1, 1-diX).

Acetals, cyanohydrins, $\alpha$-hydroxyacids, $\alpha$-aminoacids, halohydrins are usually prepared from aldehydes and ketones as starting materials. A TM of the above classes requires 1, 1-diX disconnection.

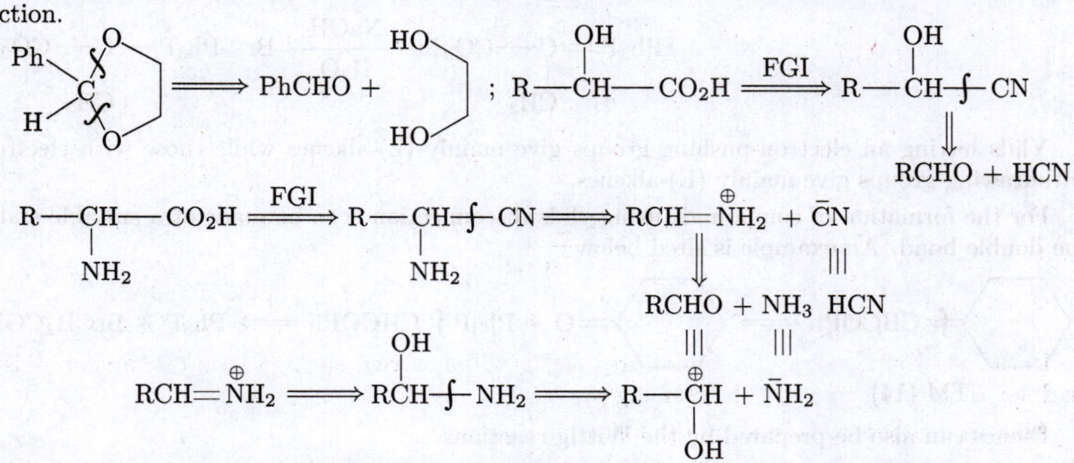

## 21.3.3 Synthesis of 1, 2-difunctionalised compounds

When two hetero atoms are present on two adjacent C atoms the compounds are called 1, 2-difunctionalised compounds. Some simple 1, 2-difunctionalised compounds which are readily available and can be used as starting materials, are given below :

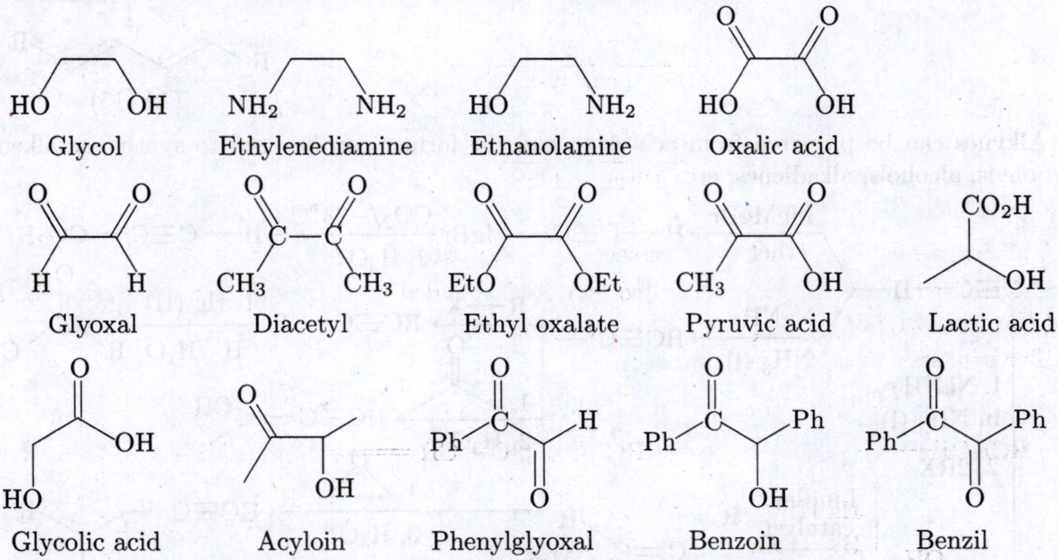


A few important 1, 2-difunctionatised compounds may readily be prepared in the laboratory :

$$PhCOCH_2Br + Ph_3P \xrightarrow{CHCl_3} PhCOCH_2\overset{\oplus}{P}Ph_3Br^- \xrightarrow[H_2O]{Na_2CO_3} PhCO{\cdot}CH{=}PPh_3$$

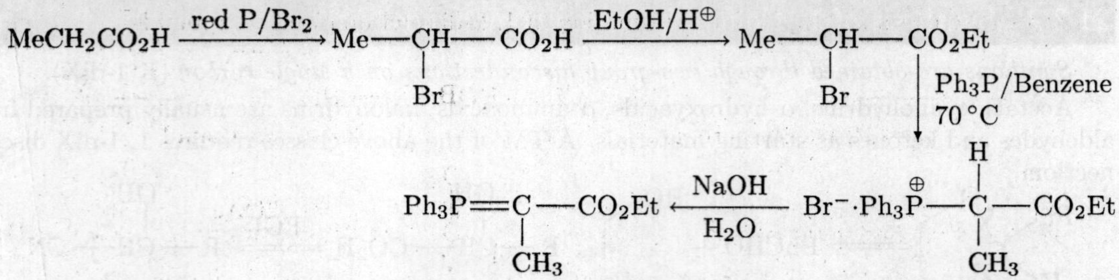

Ylids having an electron-pushing groups give mainly (Z)-alkenes while those with electron-withdrawing groups give mainly (E)-alkenes.

For the formation of compounds using ylids disconnection is to be made at a suitable end of the double bond. An example is sited below :

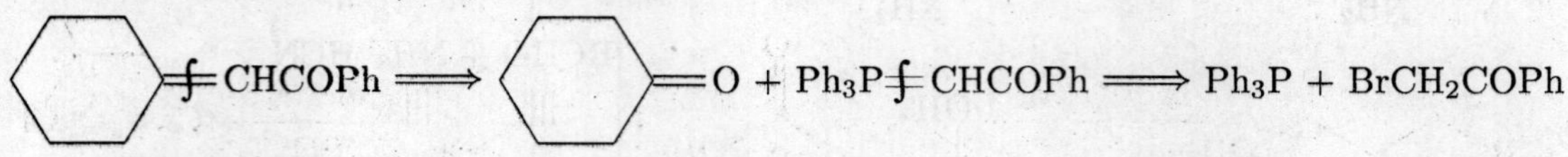

TM (14)

Dienes can also be prepared by the Wittig reactions.

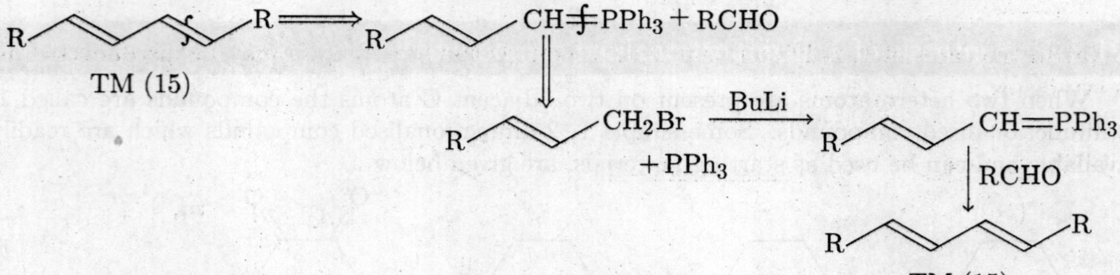

TM (15)

Alkynes can be prepared from acetylene and the former can be used to synthesise alkenes, carbonyls, alcohols, alkadienes, etc.

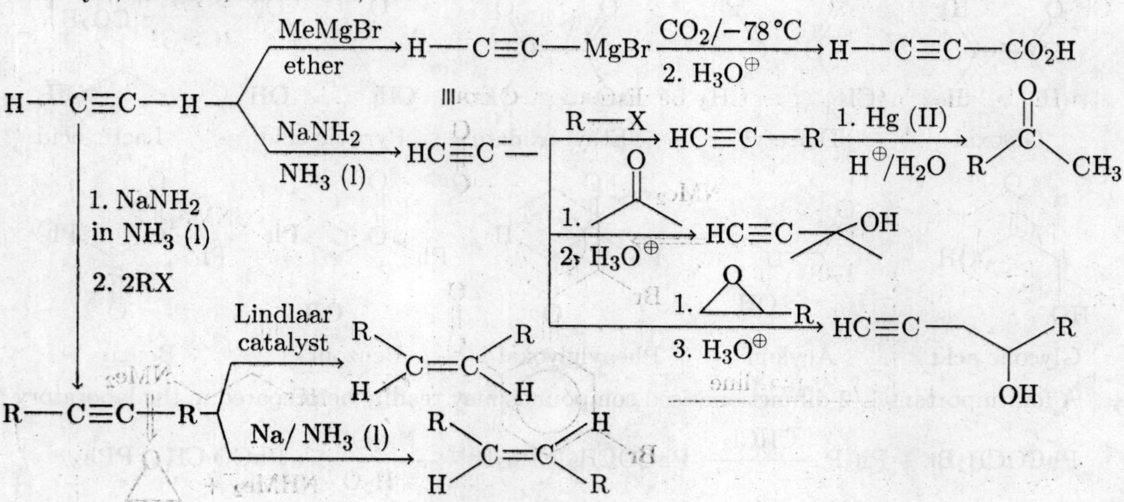

Let us take examples to show the formation of 1, 2-difunctionalised compounds.

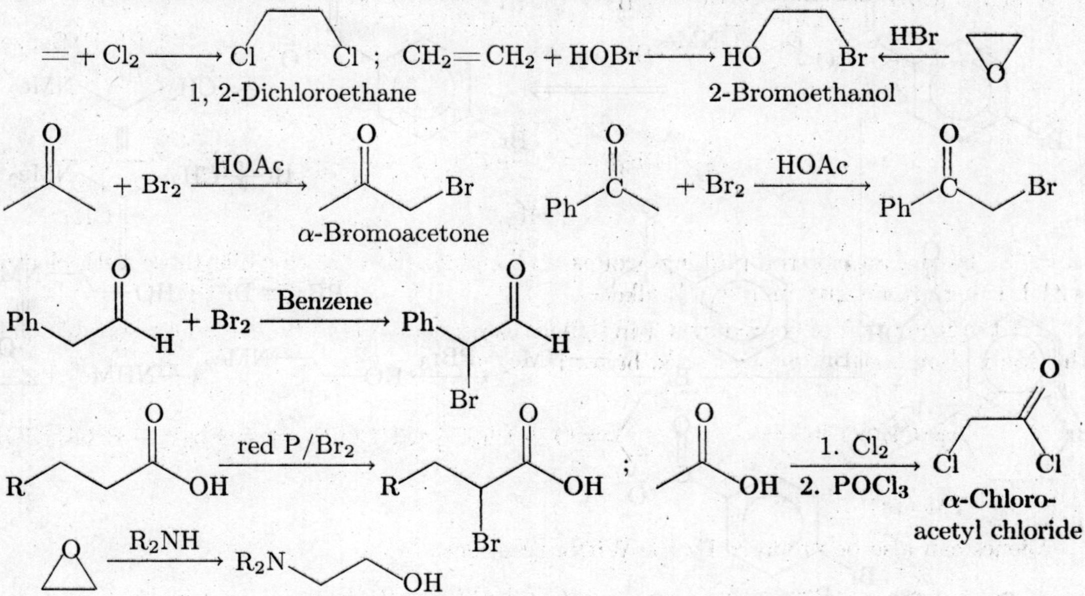


(a) In the cases of 1, 2-difunctionalised compounds one hetero atom may be disconnected as shown below :

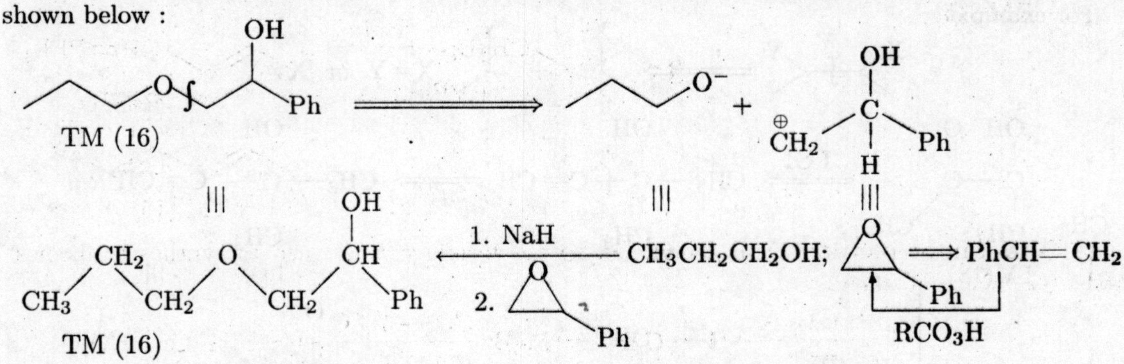


(b) 1, 2-difunctionalised part may be disconnected from the TM.

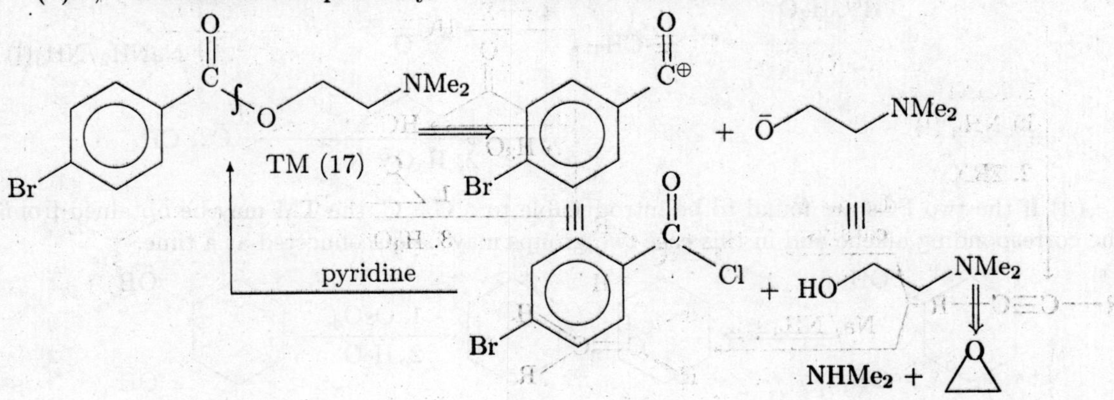

or,

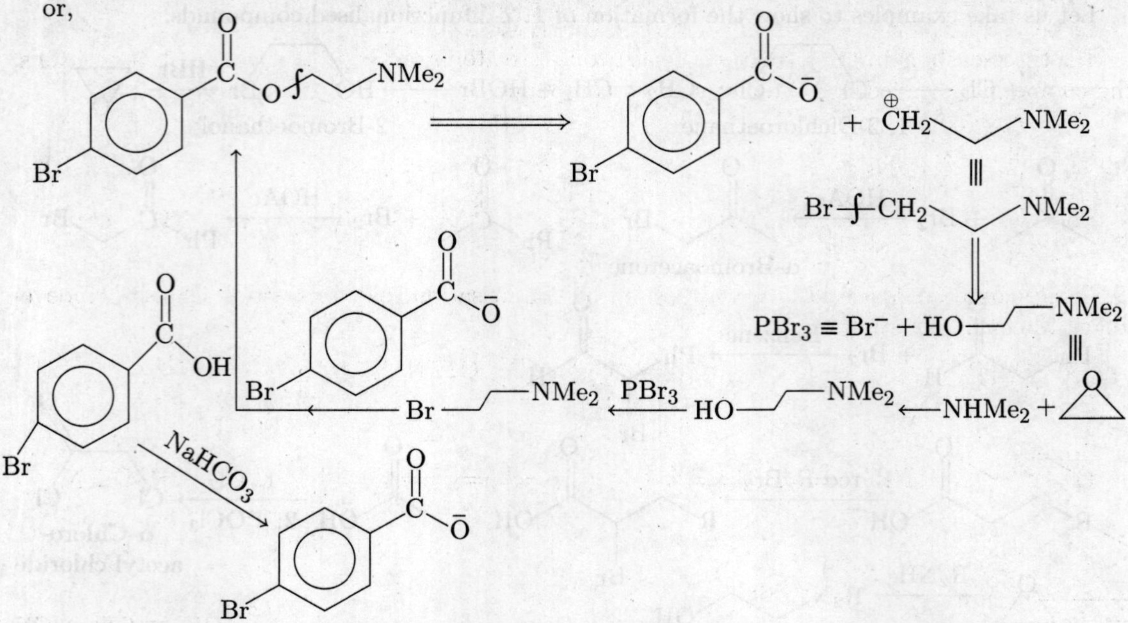

(c) One can disconnect a C—C bond which has one hetero atom on each C.

For example :

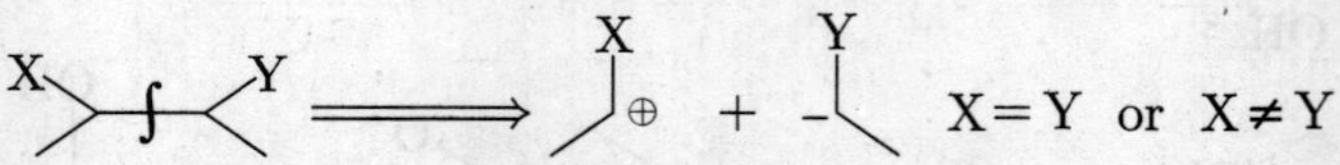

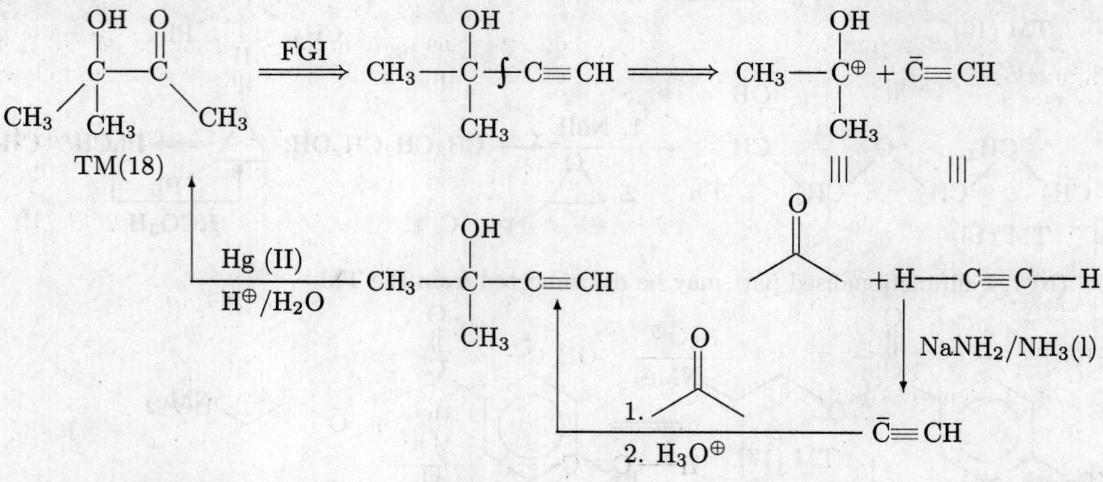

(d) If the two FGs are found to be introducable to a C=C, the TM may be obtained from the corresponding alkene and in this case two groups may be disconnected at a time.

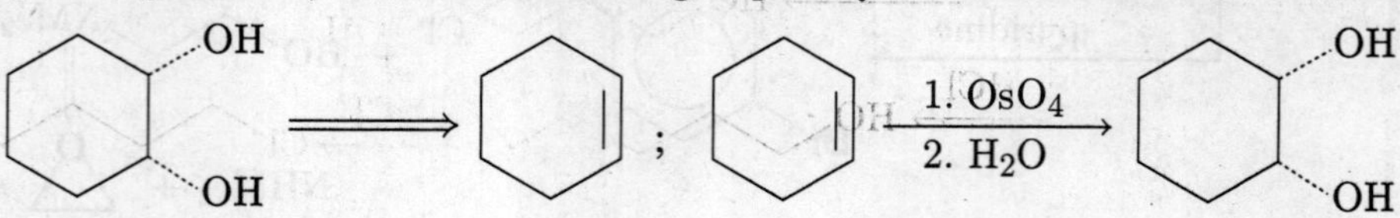

## 21.3.4 Synthesis of 1,3-difunctionalised compounds

If a three-carbon unit in a compound has two hetero atoms on the two carbons at 1, 3-positions, the compound is called 1, 3-difunctionalised compound.

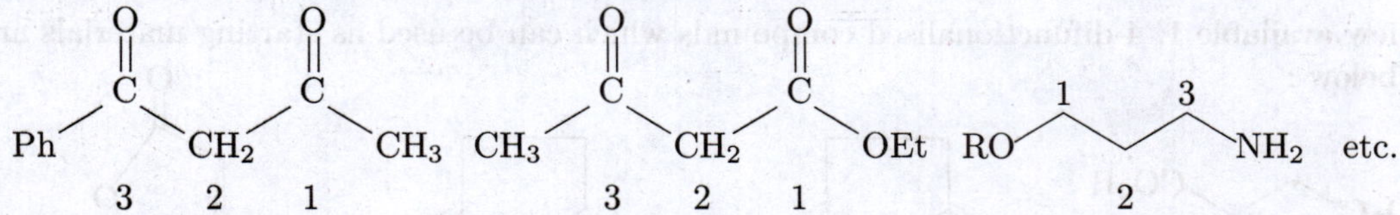

These compounds are usually synthesised by condensation reactions, aldol, Claisen, Knoevenagel, Michael, Reformatsky, etc.

For example,

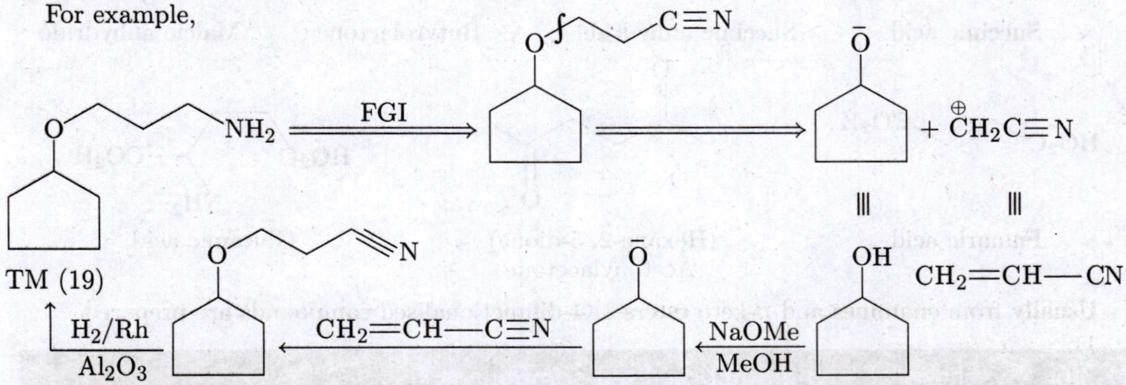

## 21.3.5 Synthesis of 1, 4-difunctionalised compounds

A four-carbon unit system holding two hetero atoms at 1 and 4 positions is called 1, 4-difunctionalised compounds. Some 1, 4-difunctionalised compounds may be synthesised as follows.

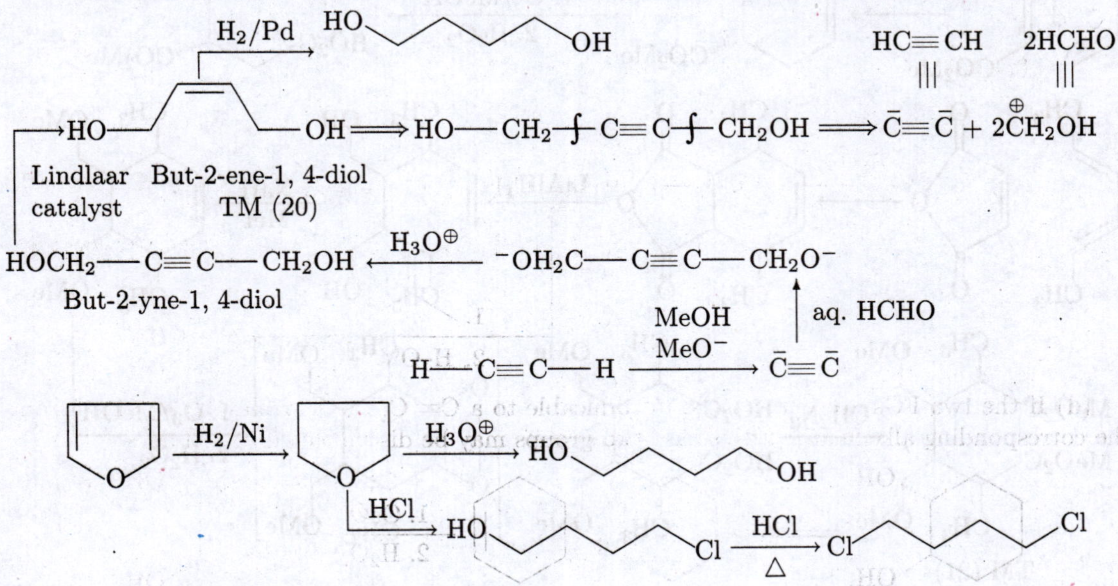

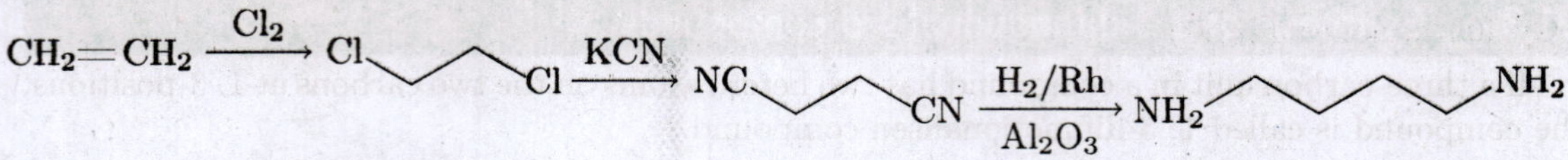

A few available 1, 4-difunctionalised compounds which can be used as starting materials are given below :

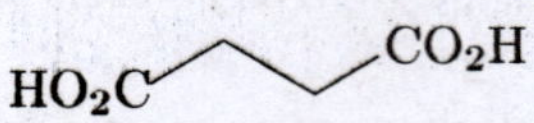

Succinic acid

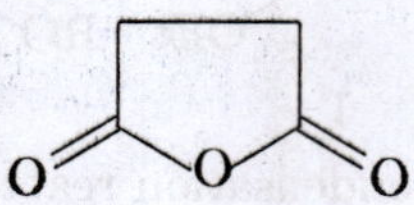

Succinic anhydride

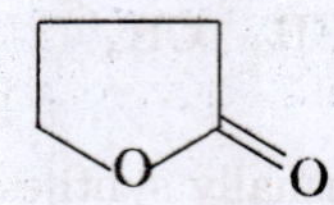

Butyrolactone

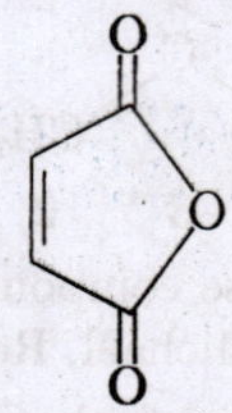

Maleic anhydride

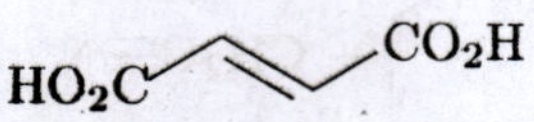

Fumaric acid

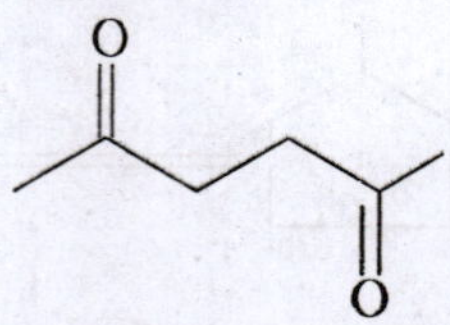

(Hexane-2, 5-dione)
Acetonylacetone

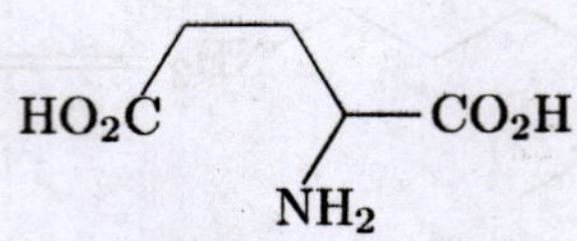

Glutamic acid

Usually from enamines and $\beta$-keto esters 1, 4-difunctionalised compounds are prepared.

## 21.3.6 Synthesis of 1, 6-difunctionalised compounds

These compounds are usually synthesised from Diels-Alder adducts.

For example,

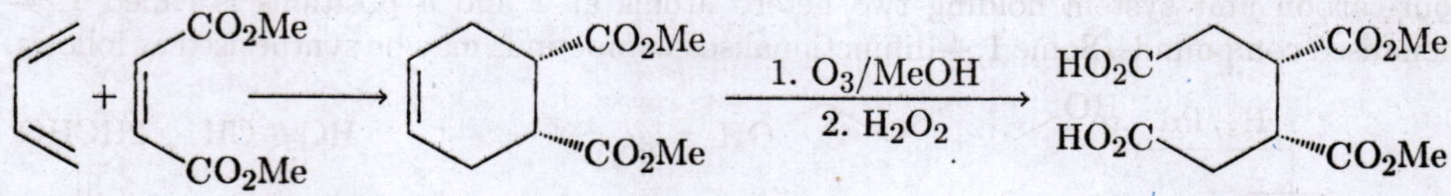

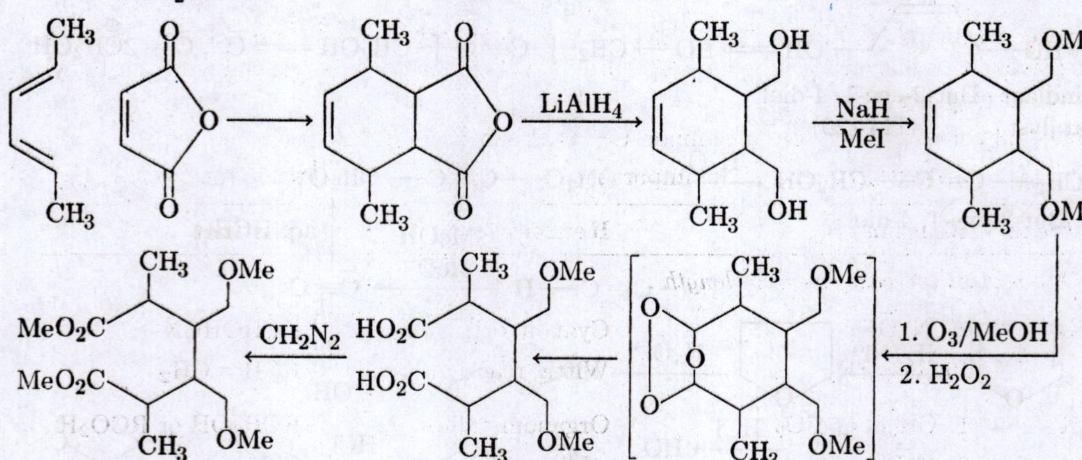

TM (21)

Cyclohexanone and their derivatives can also be converted to 1, 6-difunctionalised compounds :

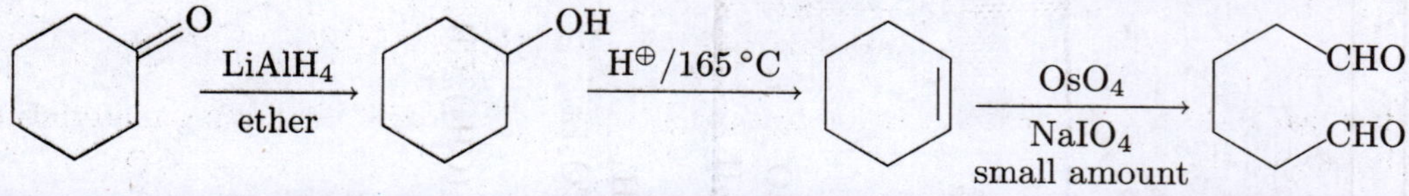


## 21.3.7 Synthesis of cyclic compounds

Table 21.3 contains the starting materials, reagents and reactions by which cyclic compounds can be synthesised. Here one example is shown below :

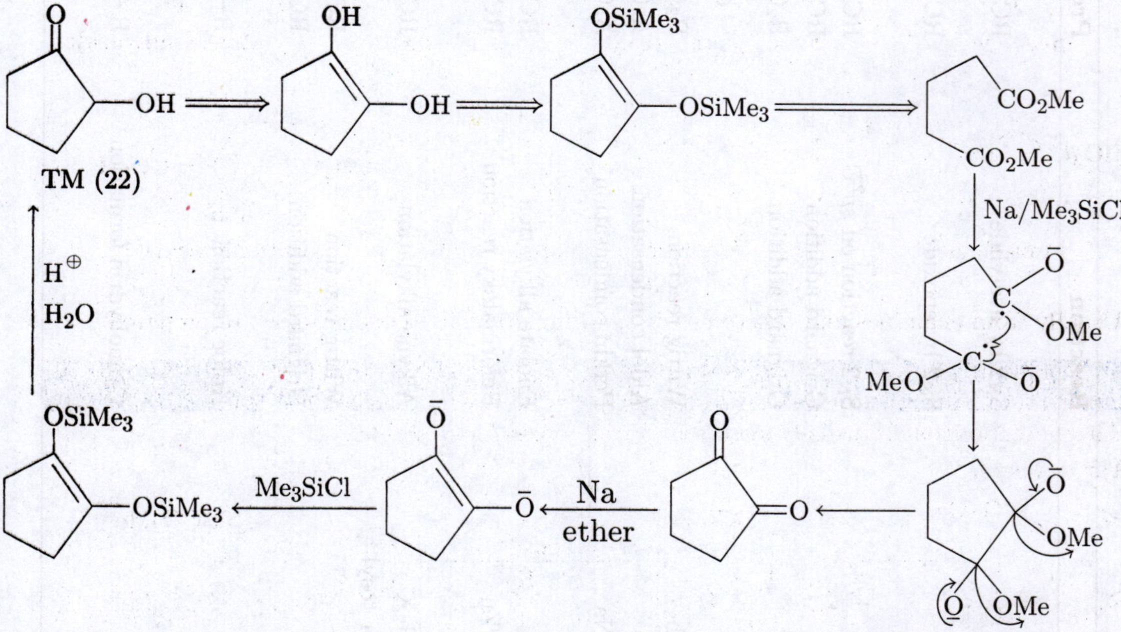


However, carbonyls (which contain weakly active methylene groups) and active methylenes are extensively used in synthesis. Section 21.9 p. 1450 will deal with those which involve carbonyls and active methylenes.

Table 21.3 contains some of the common C—C bond forming reactions.

**Table 21.3** Some of the important C—C bond forming reactions

| Substrate | Reagent | Reaction | Product |
|---|---|---|---|
| i. *One-carbon increase in chain length :* | | | |
| RCHO | $HCN/\bar{O}H$ | Cyanohydrin formation | RCHOHCN |
| | $Ph_3P{=}CH_2$ | Wittig reaction | $RCH{=}CH_2$ |
| RMgX | 1. $CH_2O$ or $CO_2$<br>2. $H_3O^{\oplus}$ | Organometallic addition | $RCH_2OH$ or $RCO_2H$ |

*Contd.*

| Substrate | Reagent | Reaction | Product |
|---|---|---|---|
| ii. *Two-carbon increase :* | | | |
| RX | $NaC\equiv CH$ | $S_N2$ by acetylide | $RC\equiv CH$ |
| RMgX | 1. <br>2. $H_3O^{\oplus}$ | $S_N2$ by epoxide | $RCH_2CH_2OH$ |
| RCOCl | $Et_2Cd$ | $S_N2$ reaction on $sp^2$C | $RCOCH_2CH_3$ |
| RMgX | $CH_3CHO$ | Grignard addition | $RCH(OH)CH_3$ |
| $R_2C=O$ | EtMgX | Grignard addition | $R_2C(OH)—CH_2CH_3$ |
| $R_2C=O$ | $CH_3{\cdot}CH=PPh_3$ | Wittig reaction | $R_2C=CH{\cdot}CH_3$ |
| ArCHO | $CH_3CHO$ | Aldol condensation | $ArCH=CH{\cdot}CHO$ |
| | 1. $Ac_2O/AcONa$<br>2 $H_3O^{\oplus}$ | Perkin condensation | $ArCH=CHCO_2H$ |
| $RCOCH_3/\bar{O}H$ | $C_2H_5X$ | Enolate alkylation | $RCOCH_2CH_2CH_3$ |
| RCHO | $BrCH_2CO_2Et/Zn$ | Reformatsky reaction | $RCH(OH)CH_2CO_2Et$. |
| iii. *Three or more carbon increase :* | | | |
| $RC\equiv C^-$ | $CH_3(CH_2)_nCH_2X$<br>X = halogen or tosyl gr. | Alkyne alkylation | $RC\equiv C—CH_2(CH_2)_nCH_3$ |
| RCHO | $Ph_3P=CH—$ | Wittig reaction | $RCH=CH—$ |
| | 1. R′MgX<br>2. $H_3O^{\oplus}$ | Grignard addition | RCH(OH)R′ |
| RCOOK | electrolysis | Kolbe reaction | R—R |
| iv. *Introduction of a branch in a C-chain :* | | | |
| Ketone | $HCN/\bar{O}H$ | Cyanohydrin formation | $R—C(OH)(R')—CN$ |

*Contd.*

| Substrate | Reagent | Reaction | Product |
|---|---|---|---|
| Aldehydes or ketones | 1. RMgX; 2. $H_3O^{\oplus}$ | Organometallic addition | |
| Esters, nitriles, carbonyls (aldehydes, ketones, $\beta$-diketones, etc.) | R—X | $S_N2$ involving carbanions | |
| $\alpha$, $\beta$-unsaturated, aldehydes ketones, esters, nitriles. | active methylenes and base | Michael or conjugate or 1, 4-addition | |
| $R—CH{=}CH—C(=O)—X$ | $Y—CH_2—Z$ | | $R—CH(—CH(Y)—Z)—CH_2—C(=O)—X$ |
| X = H, $CH_3$, OEt etc. | Y is Z = $CO_2Et$ or RCO | | |
| RCHO or $R_2CO$ | $\bar{O}H$ or $H_3O^{\oplus}$ | Aldol condensation | RCHOHCH(R)CHO |
| $RCH_2CO_2Et$ | 1. $\bar{O}Et$; 2. $H_3O^{\oplus}$ | Claisen condensation | $RCH_2—C(=O)—CH(R)—C(=O)—OEt$ |
| RCHO or $R_2CO$ | $Ph_3P{=}CH—$ | Wittig reaction | $RCH{=}CH—$<br>$R_2C{=}CH—$ |
| ArH | $RX/AlCl_3$ or HF<br>$ROH/H^{\oplus}$; alkene/$H^{\oplus}$ | Friedel-Crafts alkylation | Ar—R |
| ArH | $RCOCl/AlCl_3$ | Friedel-Crafts acylation | ArCOR |
| Ar—H | $CH_2O/HCl/ZnCl_2$ | Chloromethylation | $ArCH_2Cl$ |
| Ar—OH | 1. $CHCl_3/O\bar{H}$ 2. $H_3O^{\oplus}$ | Reimer-Tiemann | *o*-/*p*-formylphenol |

*Contd.*

| Substrate | Reagent | Reaction | Product |
|---|---|---|---|
| Aromatic compounds | $CO/HCl/AlCl_3$ | Gattermann Koch | Formylated derivative |
| | $Zn(CN)_2 + HCl$ | Gattermann | Formylated pdt. |
| | Me—NH—CHO; $POCl_3$ | Vilsmeier-Haak | Formylated pdt. |
| ArOM | 1. $CO_2$/pressure | Kolbe-Schmidt | *o* & *p* carboxylic acid |
| M = Na or K | 2. $H_3O^{\oplus}$ | | |
| ArOH | $CH_2OH/\bar{O}H$ or $H_3O^{\oplus}$ | Lederer-Manasse | *o* & *p* carbinol derivative |
| v. *3-membered carbon ring formation :* | | | |
| Haloketones, haloesters, Halonitriles | base | Internal alkylation | |
| Alkenes | carbenes, carbenoids $CH_2I_2 + Zn(Cu)$ | Addition of carbenc | |
| vi. *4-membered carbon-ring formation :* | | | |
| Haloketones, haloesters Halonitriles | base | Internal alkylation | |
| 1, 3-Dibromoalkane | base, active methylenes | Carbanion synthesis | exception : E.A.A |
| Alkenes | heat or light | Cycloaddition | |
| vii. *5-membered carbon ring formation :* | | | |
| Haloketones, haloesters, Halonitriles | base | Internal alkylation | |
| Diesters | Alkoxide base | Dieckmann | |
| | $Na/Me_3SiCl$ | Acyloin type | |
| Dinitriles | $R_2NLi$ | Thorpe | |
| | base | Internal aldol. | |
| Keto ester | base | Internal Claisen condensation. | |

*Contd.*

| Substrate | Reagent | Reaction | Product |
|---|---|---|---|
| viii. *6-membered carbon ring formation :* | | | |
| Haloketones, esters, nitriles | base | Internal alkylation. | |
| Diester | alkoxide | Dieckmann | |
| | Na/Trimethylsilyl chloride | Acyloin | |
| Dinitrile | $R_2NLi$ | Thorpe | |
| Conjugated diene and dienophile | heat | Diels-Alder | |
| Dicarbonyls | base | Internal aldol condensation. | |
| $\delta$-Keto-ester | alkoxide | Internal Claisen condensation | |
| Aromatic compounds | hydrogenating agents | Hydrogenation | |
| ix. *Long ring formation :* | | | |
| Dinitriles | $R_2NLi$ | Thorpe-Ziegler reaction | |
| Diesters | 1. Na/toluene; 2. $H_2O$<br>1. $NH_3$/Na; 2. $H_2O$ | Acyloin condensation. | |

Ring expansion and ring contraction reactions are also used for synthesising 4-, 5- and 6-membered rings.

## 21.4 FGA and its Removal

When we work back from a TM, we often require to add a functional group either to the TM or to any intermediate to get an appropriate synthon and its equivalent which can be utilised in a reliable reaction. The addition of functional group to a molecule or ion or radical is called FGA. Now the question is which FG is to be added. The following factors determine which FG is to be added.

(i) One which will give us the right disconnection to get the right synthon and (ii) one which will be removed readily. To understand the process of FGA two examples are given below.

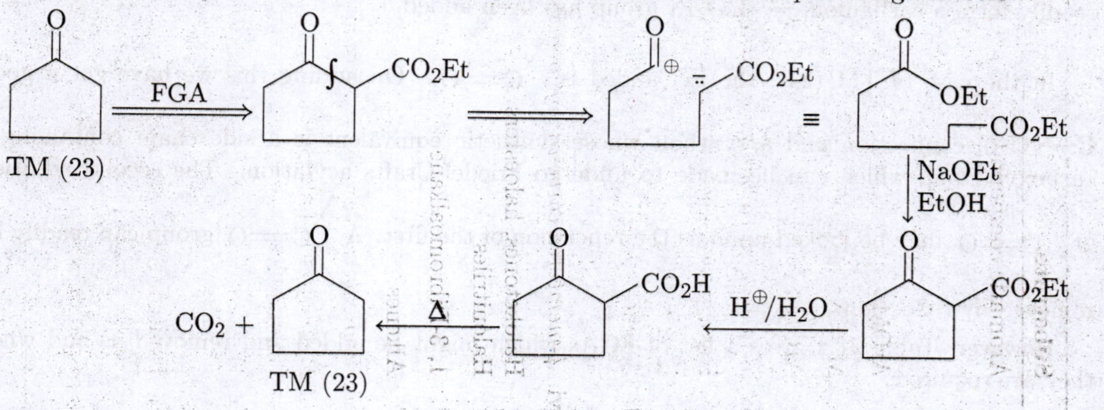

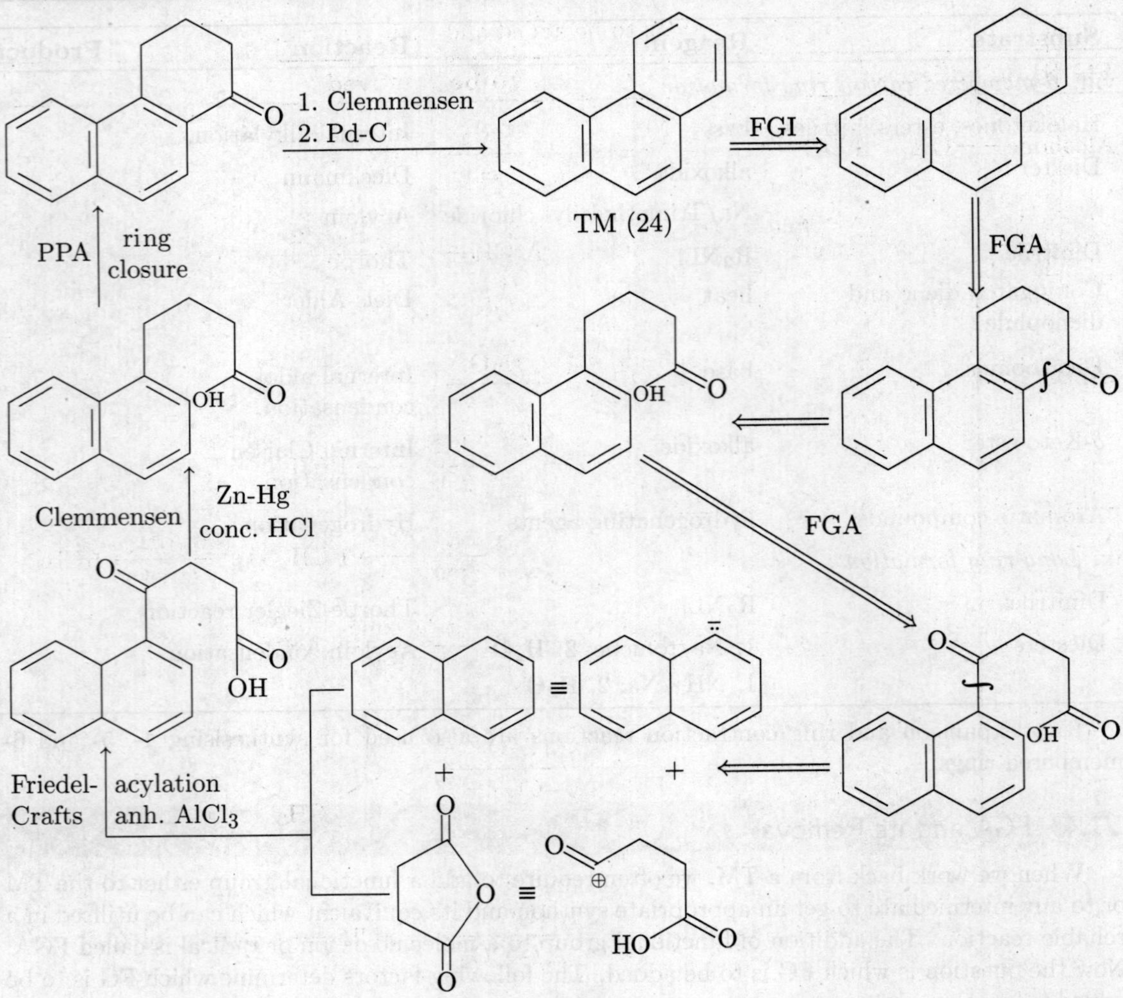


In the first case of TM (23), the FG added is —$CO_2Et$. This is removed by hydrolysis and decarboxylation. Since a $\beta$-keto acid decarboxylates readily and since the $\alpha$-C of an ester will readily form a carbanion, —$CO_2Et$ group has been added.

In the case of TM (24), the FG added is $>C{=}O$. On adding this we have got a good C—C disconnection and a synthon whose synthetic equivalent is a side chain containing a carboxyl group which can be made to undergo Friedel-Crafts acylation. The second addition of $>C{=}O$ may be looked upon as the repetition of the first. A $>C{=}O$ group can readily be removed by reduction.

However, Table 21.4 gives a list of FGAs which might be added and removed as and when they are required.

**Table 21.4** FGs to be added and their removal

| **FG to be added** | **To be removed** |
|---|---|
| Alcoholic —OH | $ROH \xrightarrow{H^{\oplus}/\Delta} \rangle C{=}C\langle \xrightarrow[\text{cat.}]{H_2} RH$ |
| Phenolic —OH & Benzylic —OH | $ArOH \xrightarrow[\text{dust dist.}]{\text{Zn}} ArH$; $ArCH(OH)\text{—} \xrightarrow{H_2/Pd\text{-}C} ArCH_2\text{—}$ |
| Cl, Br, I | $R\text{—}X \xrightarrow[\text{ether}]{\text{Mg}} RMgX \xrightarrow{H_2O} RH$ |
| —$CO_2R'$ | $RCO_2R' \xrightarrow[\text{ether}]{LiAlH_4} RCH_2OH$; $RCH_2OH \xrightarrow{1.\ H^{\oplus}/\Delta\ \ 2.\ H_2/\text{cat.}} RCH_3$; $RCH_2OH \xrightarrow[\text{pyridine}]{TsCl} RCH_2OTs \xrightarrow[\text{ether}]{LiAlH_4} RCH_3$ |
| | $ArCO_2R' \xrightarrow[\text{reflux}]{H^{\oplus}/H_2O} ArCO_2H \xrightarrow[\text{dist.}]{\text{soda-lime}} ArH$ |
| | 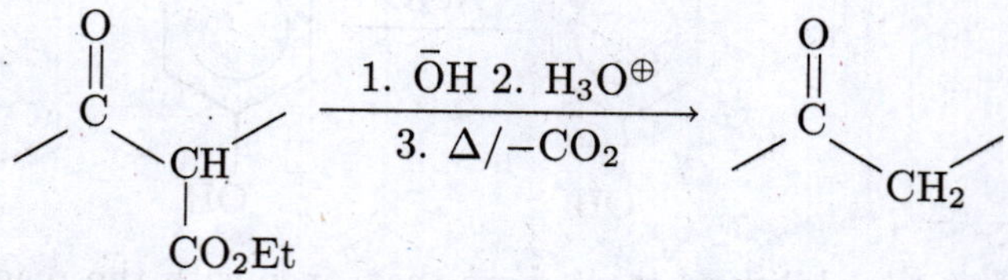 |
| | $\text{—}CH(CO_2Et)\text{—}CO_2Et \xrightarrow{1.\ H_3O^{\oplus}\ \ 2.\ \Delta/-CO_2} \text{—}CH_2\text{—}CO_2H$ |
| $\rangle C{=}O$ | 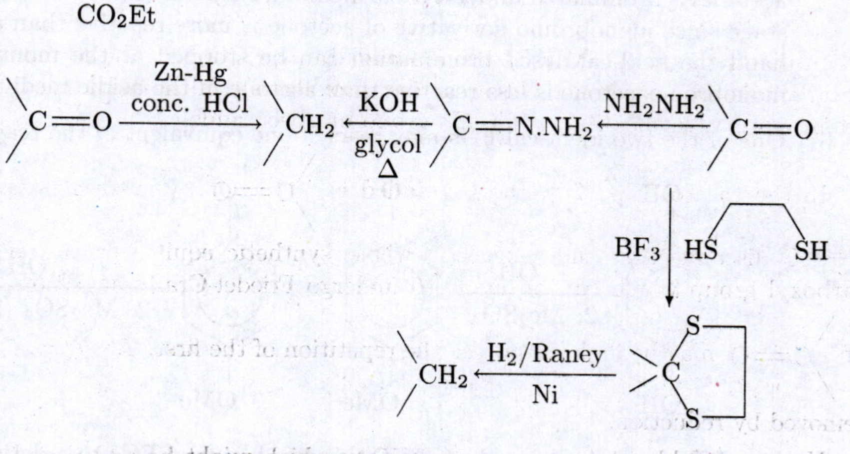 |
| —$NH_2$ | $ArNH_2 \xrightarrow{\text{diazotisation}} \xrightarrow{H_3PO_2/Cu(I)} ArH$ |

## 21.5 Selectivity

The functional group interchanging operations are no less important than the C—C bond forming operations. The FGI operations involve amongst others, the problem of finding selective reagents. In this regard there are three kinds of selectivities.

(i) Chemoselectivity, (ii) Regioselectivity and (iii) Stereoselectivity.

### 21.5.1 Chemoselectivity

Chemoselectivity of a reagent is its ability to react with one FG amongst two or more and the reagent is said to be a chemoselective reagent. Chemoselective reducing and oxidising agents are very important in organic synthesis. However, chemoselectivity is a means by which one FG is made to react keeping others intact. Selectivity in functional group regions may be utilised successfully in synthesis by understanding the following guidelines.

(i) The more reactive FG between two FGs of unequal reactivity can be made to react alone. For example, an amino group is more reactive than a hydroxyl group towards acetylation reaction. This selectivity in functional group reactions has been utilised successfully to prepare paracitamol, an analgesic, from *p*-aminophenol.

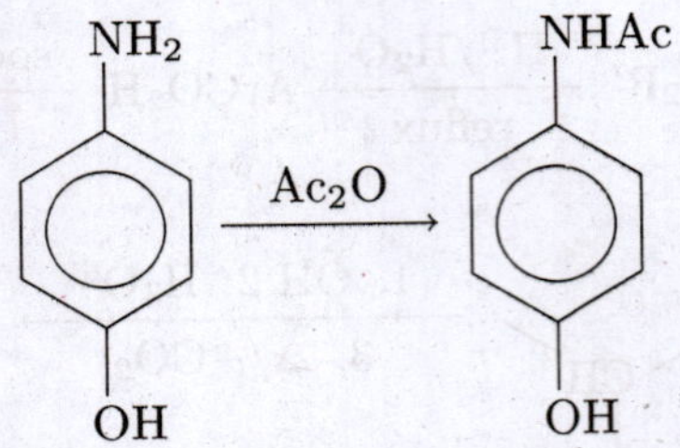


(ii) When an FG has the ability to react more than once with the reagents the first product and the starting molecule compete with each other. If the starting molecule is more reactive than the first product molecule, the reaction can be utilised successfully. When acetone, is brominated under a basic medium, the reaction does not stop at the monobromo stage since monobromo derivative of acetone is more reactive than acetone. On the other hand the acid-catalysed bromination can be stopped at the monobromo step since the monobromoacetone is less reactive than acetone in the acidic medium.

(iii) One of the two identical FGs may react if one equivalent of the reagent is used.

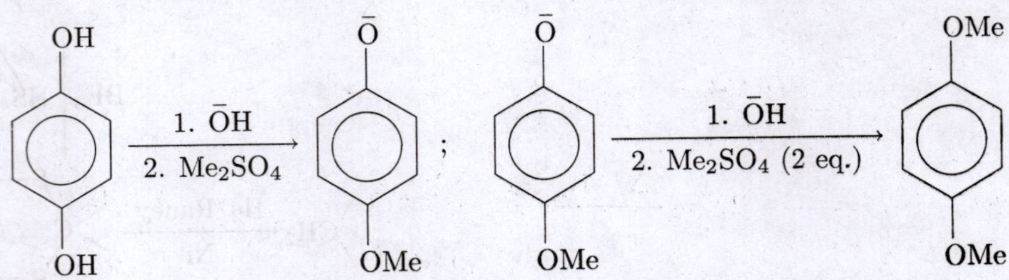


(iv) It is better to convert a compound with two identical FGs to a derivative which will react once with the reagent.

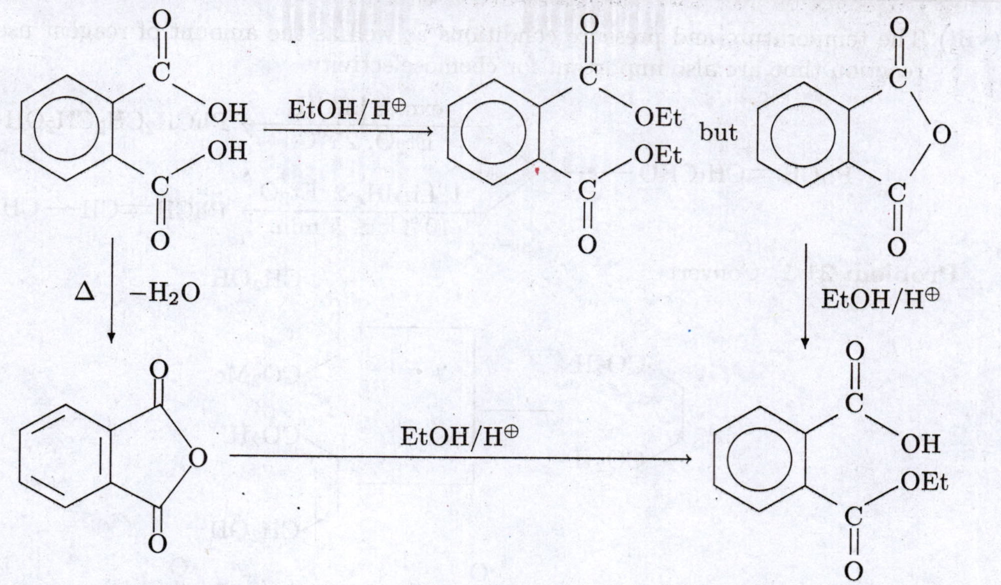

**(v)** If a compound contains two identical FGs having unequal reactivity, the more reactive FG will react preferentially.

$$Br{-}CH_2{-}CH_2{-}CH{=}CH{-}CH_2{-}Br \xrightarrow{Ac\bar{O}\overset{\oplus}{N}R_4} Br{-}CH_2CH_2{-}CH{=}CH{-}CH_2OAc$$

The allylic bromo group being more reactive than the other bromo group, the former has undergone the reaction.

**(vi)** Sometimes the steric effect make an FG to react but not others.

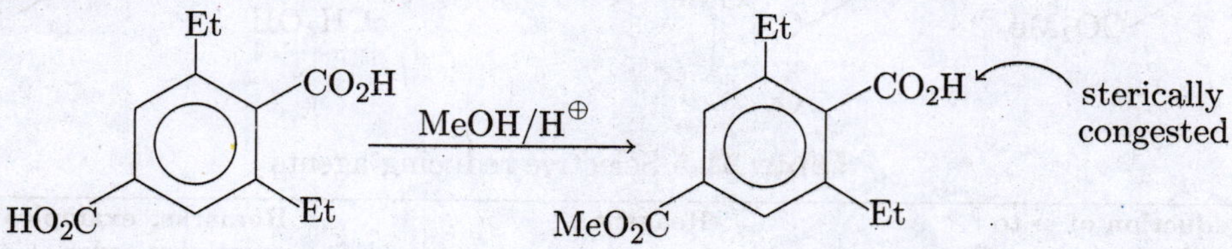

**(vii)** The choice of solvent plays a vital role to obtain the desired selectivity of a reaction. For example, sodium borohydride in diglyme reduces an aldehydic carbonyl group but it fails to reduce a ketonic carbonyl group while in the aqueous methanolic medium $NaBH_4$ reduces both the functional groups.

$$OHC{-}CH_2{-}CH_2{-}\overset{O}{\overset{\|}{C}}{-}CH_3 \xrightarrow[\text{diglyme}]{NaBH_4} CH_2OH{\cdot}CH_2CH_2{-}\overset{O}{\overset{\|}{C}}{-}CH_3$$

$$OHC{-}CH_2{\cdot}CH_2\overset{O}{\overset{\|}{C}}{-}CH_3 \xrightarrow[H_2O/MeOH]{NaBH_4} CH_2OH{-}CH_2CH_2{-}CHOH{\cdot}CH_3$$

**(viii)** The temperature and pressure conditions as well as the amount of reagent used and the reaction time are also important for chemoselectivity.

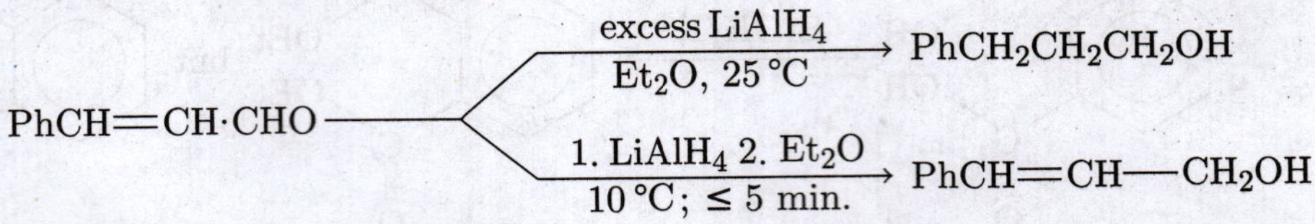

**Problem 21.1.** Convert

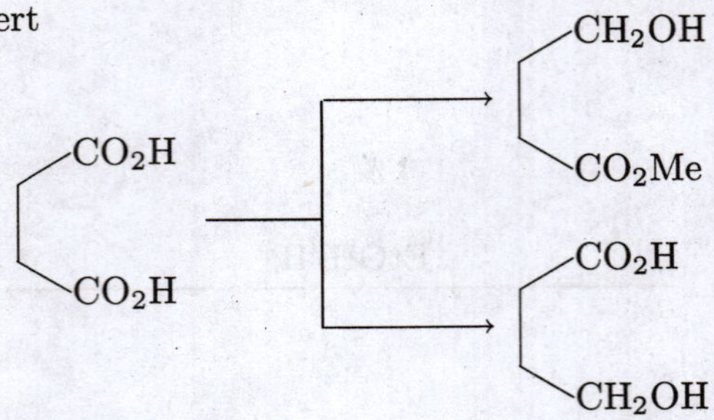

**Solution:**

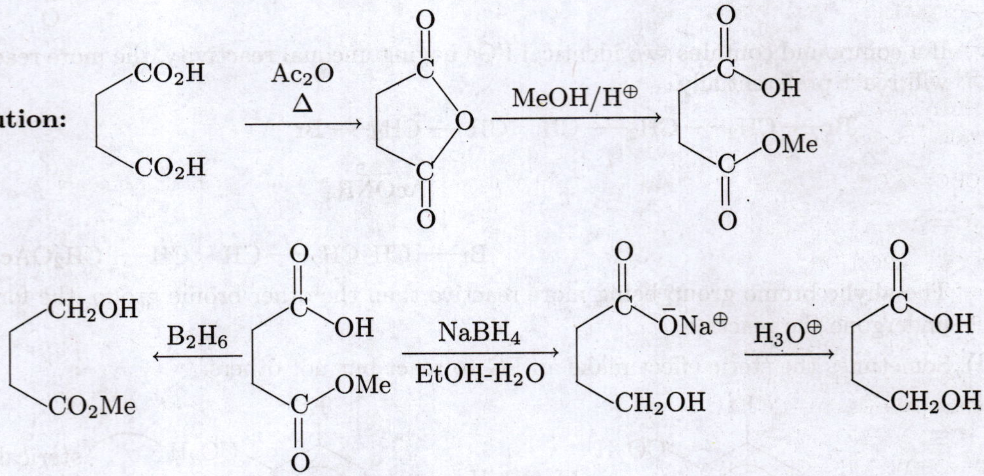

**Table 21.5** Selective reducing agents

| Reduction of → to (in the presence of) | Reagent | Remarks, examples and reactivity order |
|---|---|---|
| 1. $RCHO \rightarrow RCH_2OH$, $>C{=}O \rightarrow\ >CHOH$, $RCOCl \rightarrow RCH_2OH$, Lactone → $\alpha$, $\omega$-diol, Epoxide → 1, 2-diol, Ester/acid → alcohol, $RCONH_2 \rightarrow RCHO$, RCN and RCH=NOH $\rightarrow RCH_2NH_2$, $RNO_2 \rightarrow RNH_2$ | $LiAlH_4$ in ether or THF or diglyme. After the reaction $H_2O$ or $H_3\overset{\oplus}{O}$ is to be added. | —CHO > $>$C=O > —COCl > —$CO_2R$ > —$CO_2H$ > —$CONH_2$ > —CONHR > —C≡N > $>$CH=NOH —C(O)—$NO_2$ > —$CH_2Br$ > $>$CHBr. Does not reduce C=C, $ArNO_2$ and Ar—X. Exception : In excess reagent $PhCH{=}CHCHO \rightarrow Ph(CH_2)_2CH_2OH$ |

*Contd.*

| Reduction of → to (in the presence of) | Reagent | Remarks, examples and reactivity order |
|---|---|---|
| $RCO_2^- \rightarrow RCH_2OH$, $RX \rightarrow RH$<br>$ROSO_2R' \rightarrow RH$, $RC{\equiv}CR \rightarrow$<br>(E)-$RCH{=}CHR$, $CH_2{=}CHCH_2OTs$<br>$\rightarrow CH_2{=}CH{\cdot}CH_3$<br>(in the presence of C=C) | | Special Case :<br>$RCOOR' \xrightarrow[LiAlH_4]{BF_3\ etharate} RCH_2OR'$<br>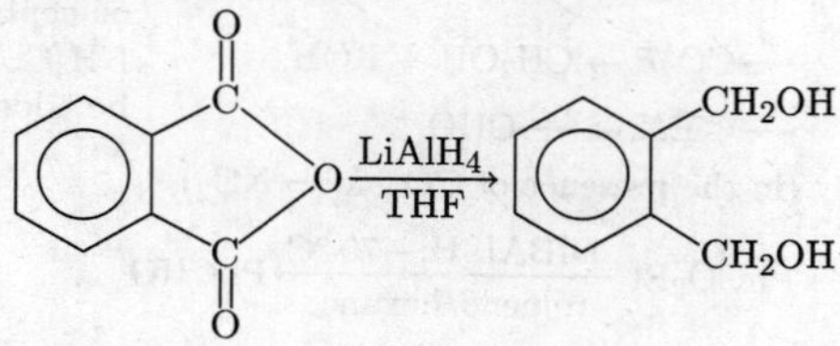 |
| 2. $RCHO \rightarrow RCH_2OH$,<br>>C=O, → >CHOH<br>—COCl, → —$CH_2OH$,<br>Epoxide → >CH—C(OH);<br>—$CO_2H$ and —$CO_2^- \rightarrow CH_2OH$;<br>—$CONR_2 \rightarrow$ —$CH_2NR_2$,<br>—C≡N → —$CH_2NH_2$,<br>—$CH_2X \rightarrow CH_3$<br>(in the presence of C=C) | $AlH_3$ in THF or $LiAlH_4$ and $AlCl_3$ in THF. After the reaction $H_2O$ or $H_3O^{\oplus}$ is to be added. | Reactivity order :<br>—CHO > >C=O > COCl > epoxide, > $CO_2H$ or $CO_2^-$ > —$CONR_2$ > —C≡N > —$CH_2X$ (slow) > —$NO_2$ (very slow). |
| 3. $RCHO \rightarrow RCH_2OH$,<br>>C=O → >CHOH<br>$RCOCl \rightarrow RCH_2OH$,<br>Lactone → α, ω-diol,<br>Epoxide → 1, 2-diol,<br>$RCO_2R \rightarrow RCH_2OH + ROH$,<br>$RCONH_2 \rightarrow RCHO$ (sometimes)<br>(In the presence of —$CONR_2$, —C≡N, —$NO_2$, C—X, C—$OSO_2R$, $CH_2{=}CH$—$CH_2OR$, —C≡C—, C=C)<br>ON | $LiBH_4$ in THF or diglyme. After the reaction $H_2O$ or $H_3O^{\oplus}$ is to be added. | Reduces an ester group more rapidly than $NaBH_4$. |
| 4. $RCHO \rightarrow RCH_2OH$, $R_2CO \rightarrow R_2CHOH$, $RCOCl \rightarrow RCHO$, $RCOOPh \rightarrow RCHO$,<br>(in the presence of C=C, C≡C, lactone, epoxide, $CO_2H$, —$CONR_2$, —C≡N, —$NO_2$, C—X) | $LiAlH[OBu(t)]_3$ in THF or diglyme. After the reduction $NH_4Cl/H_2O$ or $H_3O^{\oplus}$ is to added. | Reduces selectively >C=O in the presence of —$CO_2R$. |
| 5. —$CO_2H$, → —$CH_2OH$, —CHO → —$CH_2OH$, >C=O → >CHOH,<br>—$CO_2R \rightarrow CH_2OH + ROH$<br>(in the presence of —$CO_2^-$ and $NO_2$) | $B_2H_6$ in ether THF, Diglyme, DME (1, 2-dimethoxyethene) | Reactivity order —$CO_2H$ > —CH=CH— > CHO > >C=O > $CONR_2$ > C≡N, > C=NOH > —COCl, —COCl reduces very slowly —$CO_2H$ can be reduced in the presence of —$CO_2R$. |

*Contd.*

| Reduction of → to (in the presence of) | Reagent | Remarks, examples and reactivity order |
|---|---|---|
| 6. —$CO_2R$ → CHO, —CHO →<br>—$CH_2OH$, >C=O → >CHOH,<br>—$CO_2R$ → $CH_2OH$ + ROH;<br>—C≡N → —CHO.<br>(in the presence of $CO_2^-$ & —$NO_2$)<br>$PhCO_2Et \xrightarrow[\text{toluene/hexane}]{\text{DIBAL-H; } -70\,°C} PhCHO$<br>$\xrightarrow[\text{without boiling}]{\text{benzene, DIBAL-H}} PhCH_2OH$ | DIBAL—H<br>$AlH(CH_2CHMe_2)_2$ in toluene, diglyme, benzene or heptane. After reaction $NH_4Cl/H_2O$ or $H_3O^{\oplus}$ is to be added. | 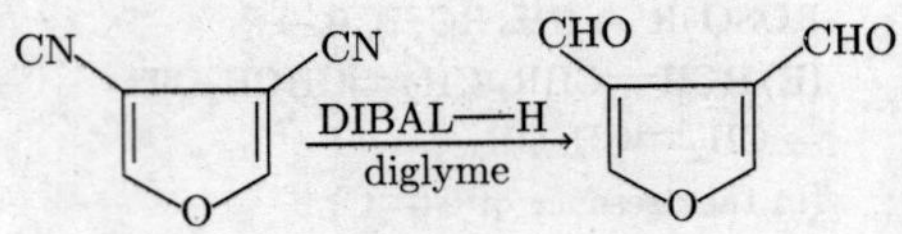 |
| 7. —$CO_2R$ → —$CH_2OH$<br>—$CO_2R$ → —$CH_3$<br>(in the presence of phenolic OH) | RED—AL<br>$NaAlH_2(OCH_2CH_2OCH_3)_2$<br>in benzene toluene, xylene | |
| 8. RCHO → $RCH_2OH$<br>>C=O → >CHOH;<br>RCOCl → $RCH_2OH$<br>(in the presence of C=C, C≡C, —$CO_2H$, —$CONR_2$, NO, —C≡N R—X, $ROSO_2R$, —COOCO—) | $NaBH_4$ in water, ethanol, diglyme. If the reaction is carried out in an organic medium, after reaction $H_2O$ is to be added. | It reduces the following functions very slowly —$CO_2R$, epoxide, lactone. |
| 9. RCHO → $RCH_2OH$,<br>$R_2C$=O → RCHOHR<br>(in the presence of C=C, C≡C) | $Al(OHCMe_2)_3$ in isopropyl alcohol | Reduces<br>Ar—C(H)=CH—CO—<br>↓<br>Ar—$CH_2CH_2$CHOH— |
| 10. Ar—X → Ar—H | $Ph_3SnH$ | 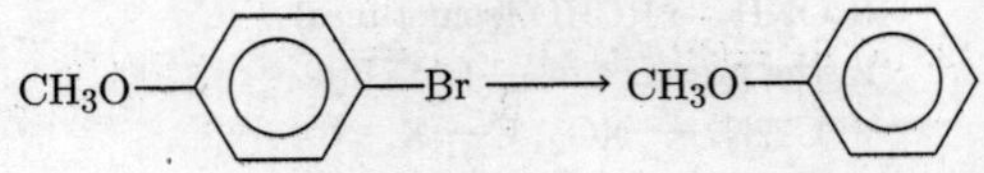 |
| 11. >C=C< → >CH—CH<<br>—C≡C → —$CH_2$—$CH_2$—<br>(in the presence of C=O, —CN, —$NO_2$) | $(Ph_3P)_3RhCl/H_2$<br>+ solvent(s)<br>↓<br>$[Ph_3P)_2Rh(s)Cl/H_2]$ | Mono and disubstituted Cs are more readily reduced than others.<br>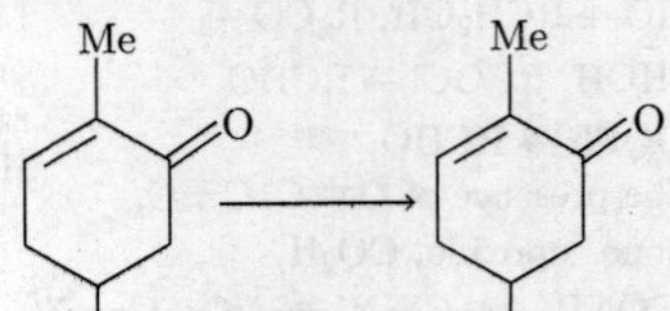 |
| 12. All the groups as stated in 1 as well as C=C, C≡C | Metal/$H_2$, $H_2$/Pd, $H_2$/Pt, $H_2$/Ni | The reactivity order —COCl > $NO_2$ > C≡C > —CHO > C=C > C=O > $ArCH_2OR$ > —C≡N > —COOR′ Ph—H → cyclohexane (difficult). |

*Contd.*

| Reduction of → to (in the presence of) | Reagent | Remarks, examples and reactivity order |
|---|---|---|
| 13. —COCl → CHO | $H_2$/Pd, $BaSO_4$ suspension, quinoline | Rosenmund reduction. |
| 14. —C≡C— → (Z) —CH=CH— | $H_2$/Pd, $CaCO_3$ or $BaSO_4$ poisoned with $Pb(OAc)_2$ or quinoline | Lindlaar Catalyst |
| 15. a. $PhCH_2OH \rightarrow PhCH_3$ | a. $H_2$/Pd; 25 °C, 3 atm | Does not reduce acetals or ketals. |
| b. $PhCH(OH)(CO_2H) \rightarrow PhCH_2CO_2H$ | b. $H_2$/Pd; 25 °C, 3.5 atm | |
| c. $(PhCH_2)_2NPh \rightarrow PhCH_3$ | c. $H_2$/Pd; 25 °C, 1 atm | |
| 16. —NH—$CH_2Ph$ → —$NH_2$ | $H_2$-Pd/C in ethanol | $HOCH_2—CH(NHCH_2Ph)—CO_2H \rightarrow HOCH_2—CH(NH_2)—CO_2H$ |
| α-epoxyester → α-hydroxyester | 75 °C – 80 °C/13.6 atm | |
| C=C (in the presence of CHO) | | 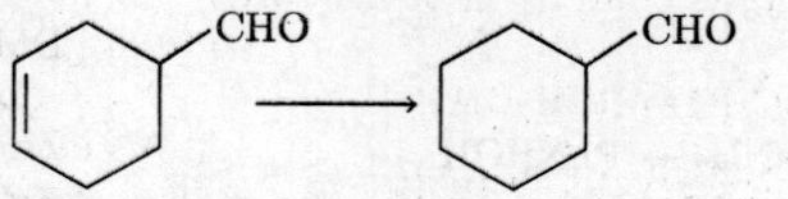 |
| 17. —CHO → —$CH_2OH$, >C=O → >CHOH, —$CO_2R$ → —$CH_2OH$ + ROH, —CH=NOH → —$CH_2$—$NH_2$ | Bouveanlt-Blanc reduction Na/alcohol (EtOH) | Does not reduce —$CO_2H$ but converts —$CO_2H$ → $CO_2Na$; should not be used on compounds containing —$CO_2H$. |
| 18. >C=O → >CHOH, —CHO → —$CH_2OH$ | Na-Hg/$H_2O$ | Direct reduction, should not be carried out in the presence of —$CO_2H$. |
| 19. (a) RC≡CR → (E) H(R)C=C(R)H | Li or Na in $NH_3$ (1) Birch reduction | The ratio of (Z) & (E) are temp. dependent |
| (b) Reduction of C=C bond | | 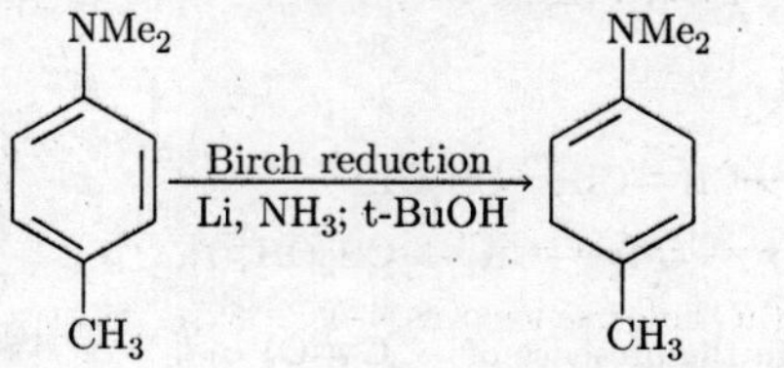 |
| 20. Anhydrides of dibasic acids to lactone | Zn/HOAc | 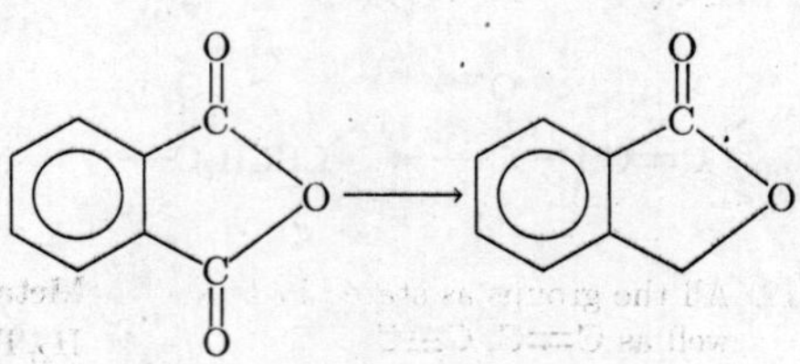 |
| 21. 2 >C=O → —CH(OH)—CH(OH)— | Mg-Hg/$H_2O$ | Bimolecular reduction |

*Contd.*

| Reduction of → to (in the presence of) | Reagent | Remarks, examples and reactivity order |
|---|---|---|
| 22. >C=O → —CH₂—;<br>—CHO → —CH₃<br>>C=CH—CO₂R → >CH—CH₂—CO₂R<br>>C=CH—CO₂H → >CH—CH₂—CO₂H<br>(In the presence of —CO₂R, —CO₂H) | Zn-Hg/conc. HCl<br>Clemmensen reduction | Does not reduce phenolic OH, —CO₂H, —CO₂R. 1, 3-dione undergo rearrangement; α-hydroxyketones give olefins. |
| 23. >C=O → >CH₂ | NH₂—NH₂/ŌEt/Δ or NH₂—NH₂/KOH in glycol 180 °C or NH₂—NH₂/KOBu(t) DMSO | Should not be used when —COCl, CO₂R or —CO₂H groups are present. |
| 24. PhNH₂ → PhNHOH | Zn/NH₄Cl in H₂O or EtOH | |
| 25. PhNO₂ → Ph—N=N—Ph<br>PhNO₂ → PhNH—NHPh | Zn-NaOH along with MeOH/H₂O as solvent | Zn : 2 moles<br>Zn : 3 moles |
| 26. RC≡N → RCHO, ArNO₂ → ArNH₂<br>R—X → RH | SnCl₂/conc. HCl | Reduces ArNO₂ selectively 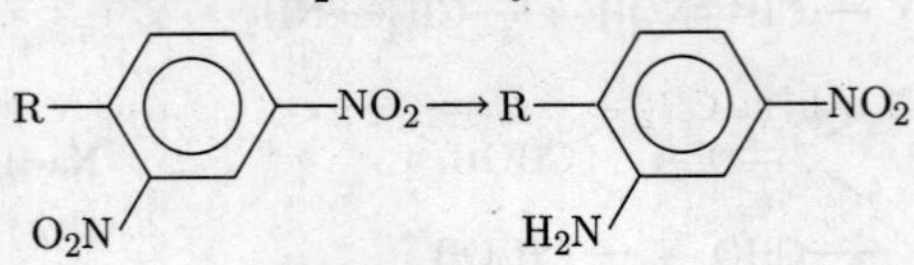 |
| 27. Selectively reduction of ArNO₂ → ArNH₂ | (NH₄)₂S | 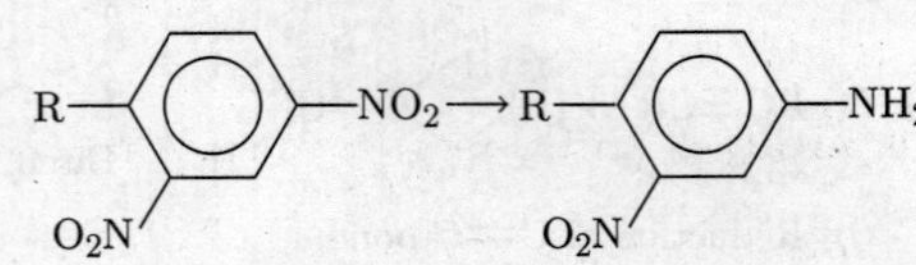 |
| 28. —CH=CH—CH₂OH → —CH₂—CH₂—CH₂OH<br>(In the presence of >C=O) | Azo-dicarboxylate AcOH/EtOH room temperature. | Moderately polar C=C reduces.<br>Conjugate double bonds of carbonyls do not reduce. |
| 29. >C=CH—C(=O)— → >CHCH₂C(=O)— | NH₂—NH₂ + H₂O₂<br>↓<br>H—N=N—H | Reduces C=C in the presence of >C=O 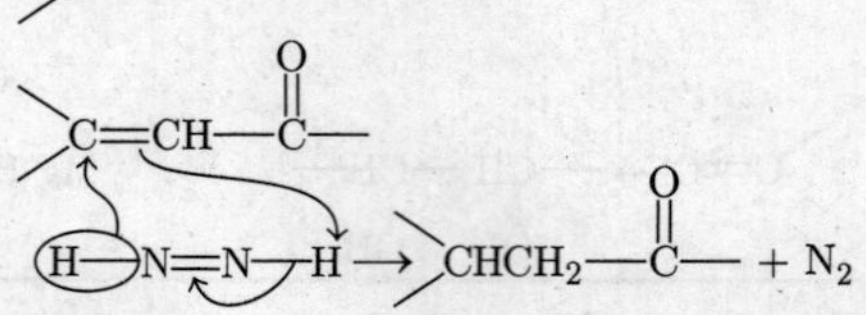 |

**Table 21.6** Selective oxidising agents

| Operation | Reagent | Remarks/Examples |
|---|---|---|
| 1. 3° R—H → R—OH | a. $\bar{O}H/KMnO_4$<br>b. $Na_2Cr_2O_7/AcOH/H_2SO_4$ | If the 3°C is chiral, it retains its configuration. unfavourable. |
| 2. allylic H → Allylic OH | a. i. $Pb(OAc)_4/Ac_2O$<br>ii. $H_3O^{\oplus}$<br>b. $SeO_2/EtOH$<br>c. i. NBS<br>ii. $\bar{O}H$ | 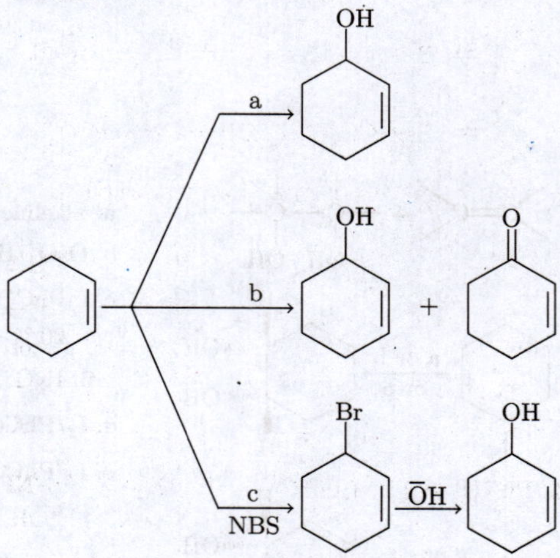 |
| 3. $Ph_3CH \rightarrow Ph_3C$—OH | $CrO_3/CH_3CO_2H$ | |
| 4. $PhCH_3 \rightarrow PhCH_2OH$ | i. $SO_2Cl_2$/peroxide<br>ii. $H_2O$ | |
| 5. Allylic $CH_2 \rightarrow$ >C=O | $CrO_3/AcOH—H_2O$ | 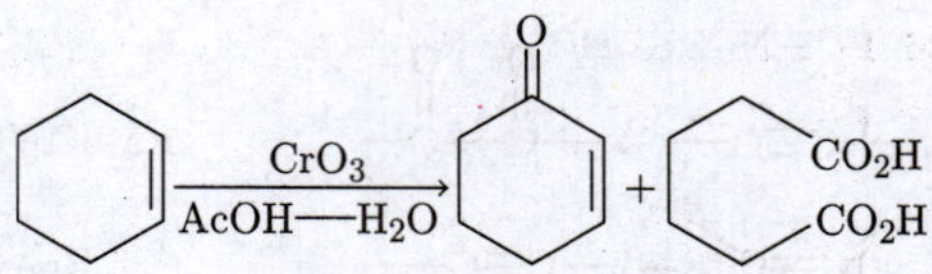 |
| 6. $ArCH_3 \rightarrow ArCHO$ | a. i. $2CrO_2Cl_2/CCl_4$ or $CS_2$<br>ii. $Na_2SO_3$<br>iii. $H_3O^{\oplus}$ | a. 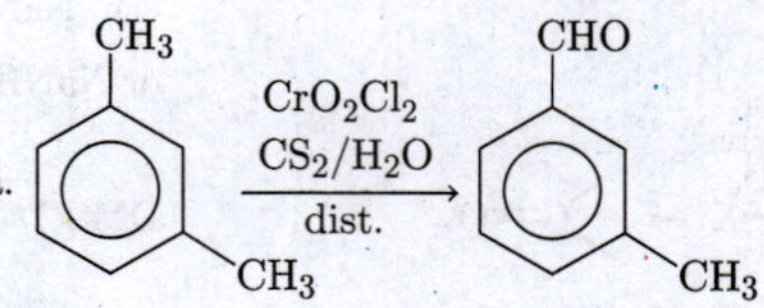 |
| | b. i. $CrO_3/Ac_2O$<br>ii. $H_3O^{\oplus}$ | b. 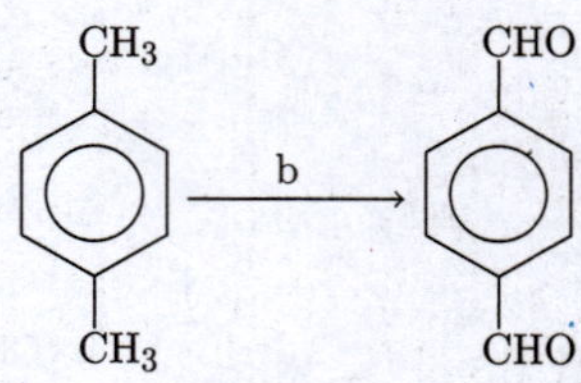 |
| | c. i. $2Br_2/CaCO_3—H_2O$<br>ii. $H_3O^{\oplus}$ | c. 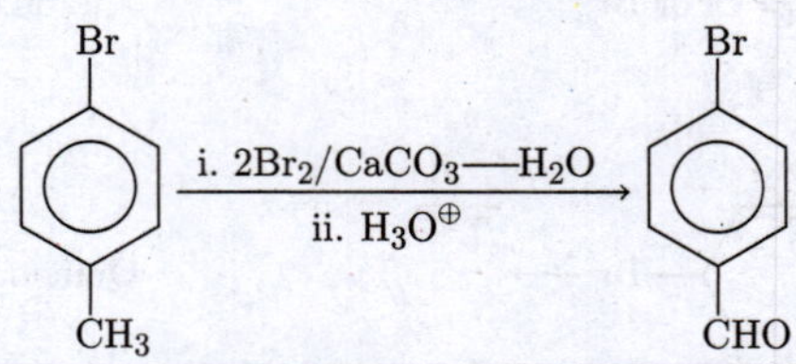 |

*Contd.*

| Operation | Reagent | Remarks/Examples |
|---|---|---|
| 7. $ArCH_3 \rightarrow ArCO_2H$ | a. $CrO_3$<br>b. alkaline $KMnO_4$<br>c. conc. $HNO_3$ | 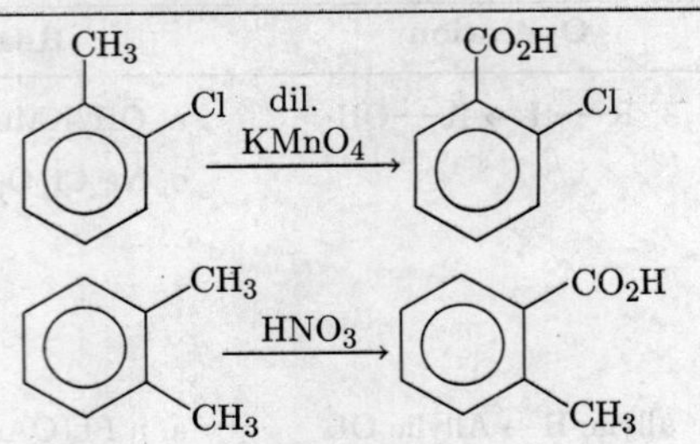 |
| 8. 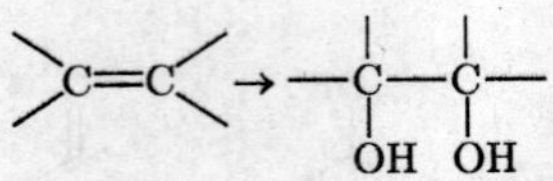<br>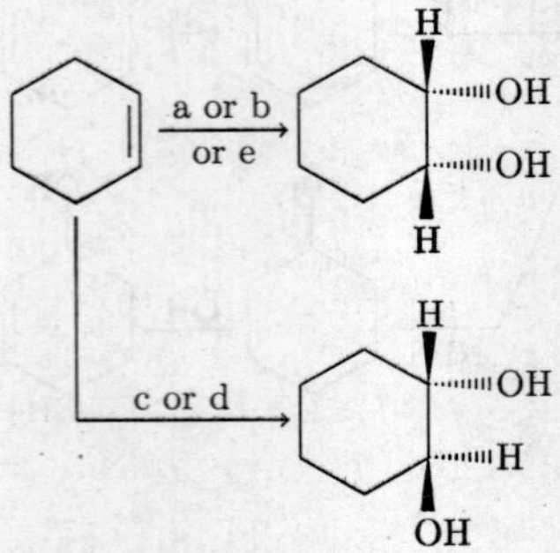 | a. alkaline $KMnO_4$<br>b. $OsO_4/H_2O$<br>c. i. $PhCO_2OH/CH_2Cl_2$ 20 °C − 30 °C<br>ii. $H_3O^{\oplus}$<br>d. $I_2/PhCO_2Ag$; dry benzene<br>e. $I_2/PhCO_2Ag$; $H_2O$ | a. & b. *cis*-addition<br>c. *trans*-addition<br>d. *trans*-addition<br>e. *cis*-addition |
| 9. 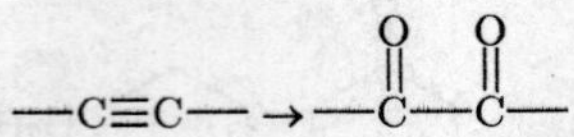 | $KMnO_4/KHCO_3$—$H_2O$ | Less used. |
| 10. 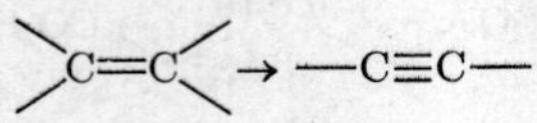 | i. $Br_2/CCl_4$<br>ii. $NaNH_2/NH_3(l)$<br>or $NaNH_2$/THF | 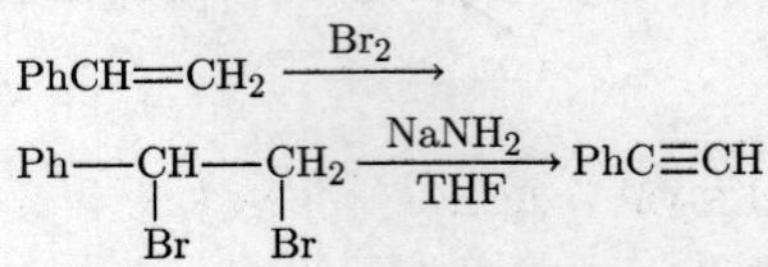 |
| 11. 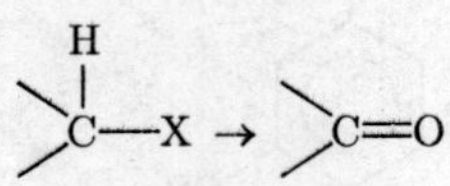<br>X = Cl, Br, I, OTs | DMSO/$Na_2CO_3$ | 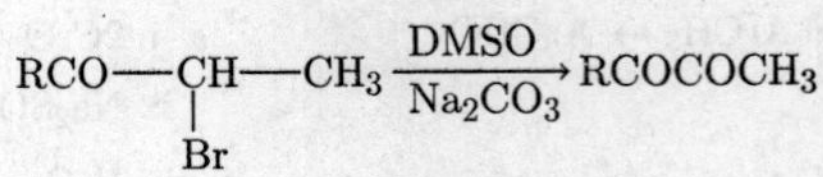<br>$CH_3(CH_2)_7CH_2I \xrightarrow[150\ °C/N_2\ atm]{DMSO} CH_3(CH_2)_7CHO$ |
| 12. $PhCH_2X \rightarrow PhCHO$<br>X = Cl or Br | i. $(CH_2)_6N_4/CHCl_3$<br>ii. 50% AcOH | c. 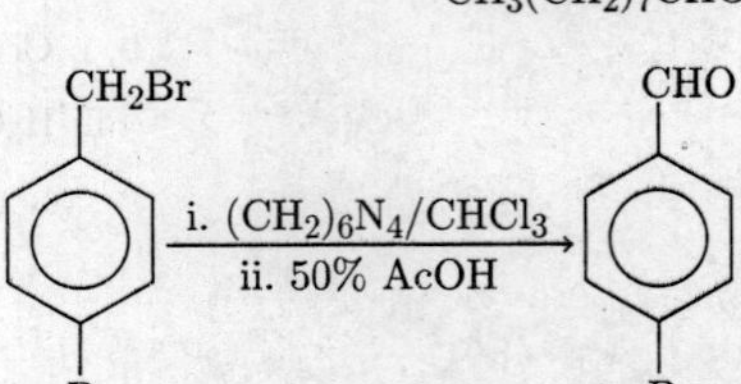 |
| 13. 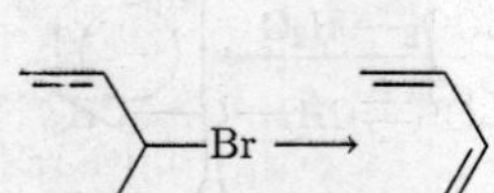 | Quinoline 160 °C | 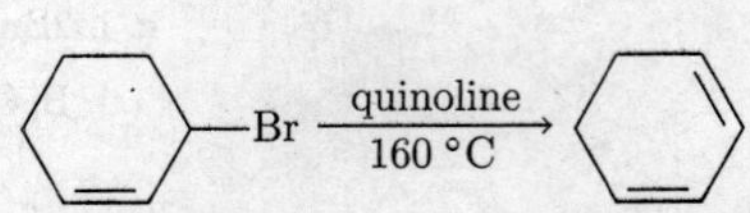 |

*Contd.*

| Operation | Reagent | Remarks/Examples |
|---|---|---|
| 14. $RCH_2—Br \rightarrow RCHO$ | 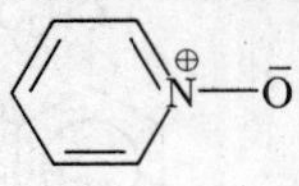 | 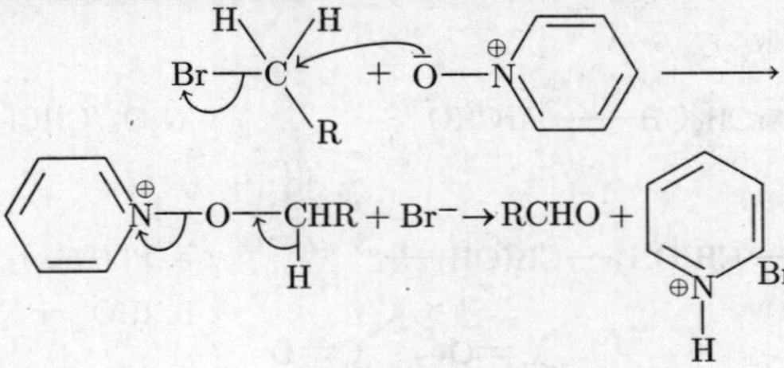 |
| 15. $R_2CHOH \rightarrow R_2CO$<br>$RCH_2OH \rightarrow RCHO$ | $K_2Cr_2O_7/H_2SO_4$ (conc.) | 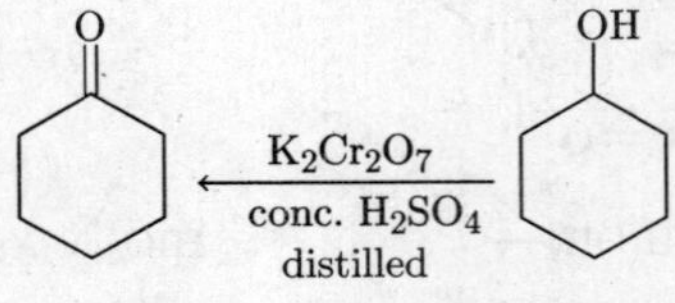 |
| 16. $R_2CHOH \rightarrow R_2C{=}O$ | $Al[O—Bu(t)]_3$/acetone<br>Oppenauer oxidation | $RCH_2CHOH{\cdot}CH_2R' \rightarrow RCH_2COCH_2R$<br>C=C does not get oxidised. |
| 17. $R_2CHOH \longrightarrow R_2CO$ | a. $K_2Cr_2O_7/H_2SO_4—H_2O$ | a. 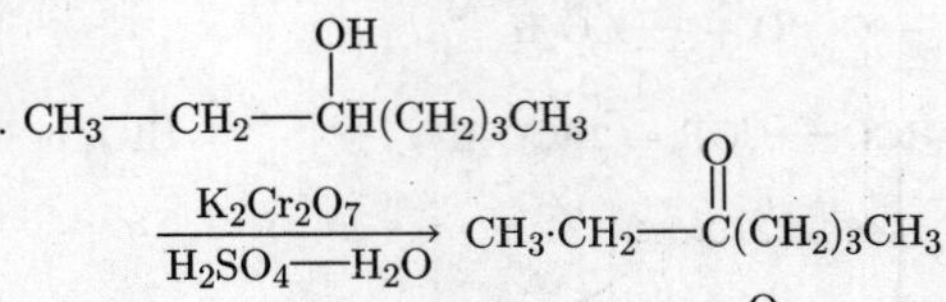 |
| $R_2CHOH \rightarrow R_2C{=}O$ | b. $CrO_3/AcOH—H_2O$ | b. 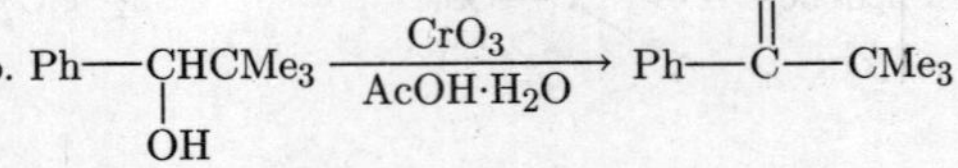 |
| | c. $CrO_3$/ pyridine $CH_2Cl_2$ | c. In this case, C=C and C≡C remain undisturbed. |
| 18. $RCH_2OH \rightarrow RCHO$<br>$R_2CHOH \rightarrow R_2C{=}O$ | a. Pyridinium chloro-chromate (PCC) | c. 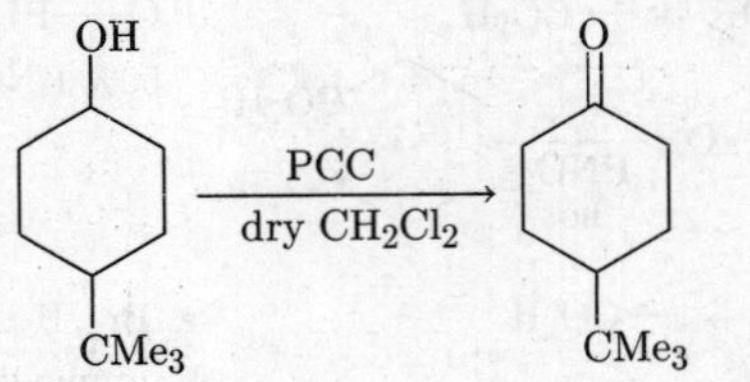 |
| | b. Pyridinium dichromate (PDC) | b. 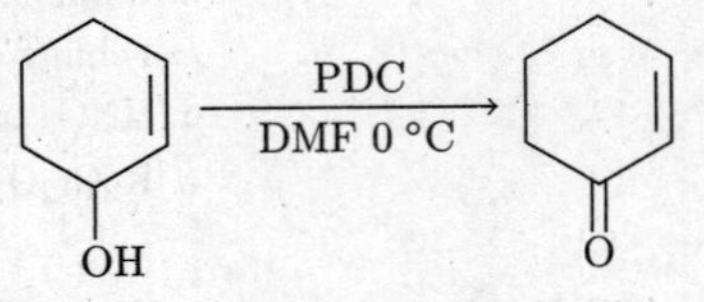 |
| 19. 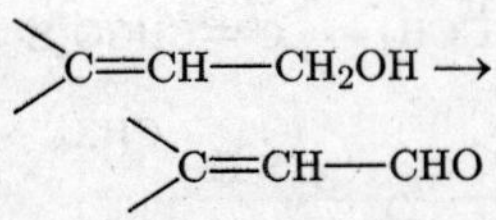 | a. Specially made $MnO_2$<br>b. Chloranil | Does not oxidise C=C and C≡C<br>$PhCH{=}CH—CH_2OH \xrightarrow[CH_2Cl_2,\ 15\ °C]{MnO_2} PhCH{=}CH{\cdot}CHO$ |
| 20. 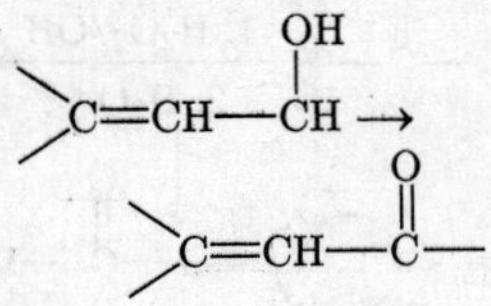 | t-Butyl chromate | 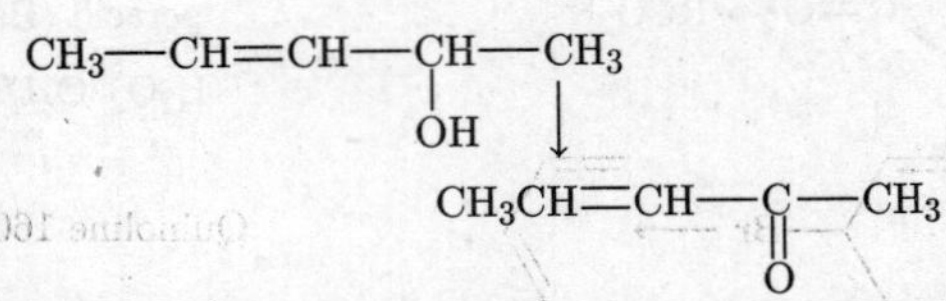 |

*Contd.*

| Operation | Reagent | Remarks/Examples |
|---|---|---|
| 21. $ArCH_2OH \longrightarrow ArCHO$ | $N_2O_4/CHCl_3/0\,°C$ | 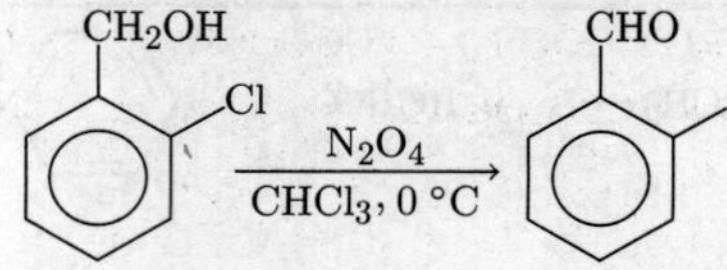 |
| 22. $—CH(OH)—CH(OH)— \downarrow >C{=}O + >C{=}O$ | a. $Pb(OAc)_4$/organic solvent<br>b. $HIO_4$ or $NaIO_4/H_2O$ | Very important in the carbohydrate chemistry |
| 23. $>C{=}O + >C{=}O \uparrow —C(OH)—CH(OH)—$<br>$—C(NH)—C(=O)— \downarrow —C{=}O + —CO_2H$ | $HIO_4$ or $NaIO_4/H_2O$ | |
| $RCO—COR \rightarrow 2RCO_2H$ | $HIO_4$ or $NaIO_4/H_2O$ | |
| 24. Phenolic OH $\rightarrow >C{=}O$ | $Na_2Cr_2O_7/H_2SO_4—H_2O$ | 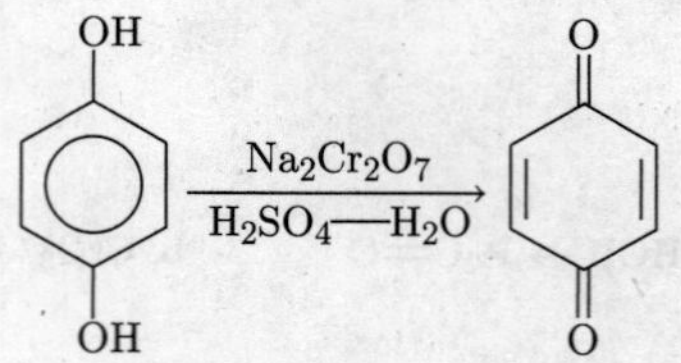 |
| 25. $—CH_2OH \rightarrow —CO_2H$<br>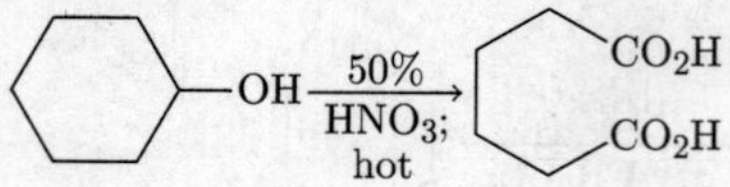 | $O_2—Pt/NaHCO_3, 35\,°C$<br>50% $HNO_3$; hot | $C(CH_2OH)_4 \xrightarrow[NaHCO_3,\ 35\,°C]{O_2—Pt} (CH_2OH)_3C—CO_2H$ |
| 26. $—CHO \rightarrow —CO_2H$ | a. $Br_2/H_2O$ or<br>b. ammoniacal $AgNO_3$<br>c. Fehling or<br>d. $HNO_3$ or<br>e. $K_2Cr_2O_7$/conc. $H_2SO_4$ (dist.) | $CH_2{=}CH{\cdot}CHO \xrightarrow{b\text{ or }c} CH_2{=}CH—CO_2^-$<br>b or c does not oxidise C=C or C≡C<br>b does not also oxidise $>C{=}O$, ArCHO |
| 27. $—COCH_3 \rightarrow CO_2H$ | i. $X_2$/NaOH; ii. $H_3O^{\oplus}$<br>X= Cl, Br, I | $>C{=}CH—C(=O)CH_3 \rightarrow >C{=}CHCO_2H + CHX_3$ |
| 28. $R_2C{=}O \rightarrow RCO_2R$ | peracid (Baeyer-Villigar)<br>$H_2O_2/\bar{O}H/H_2O$ (Dakin) | 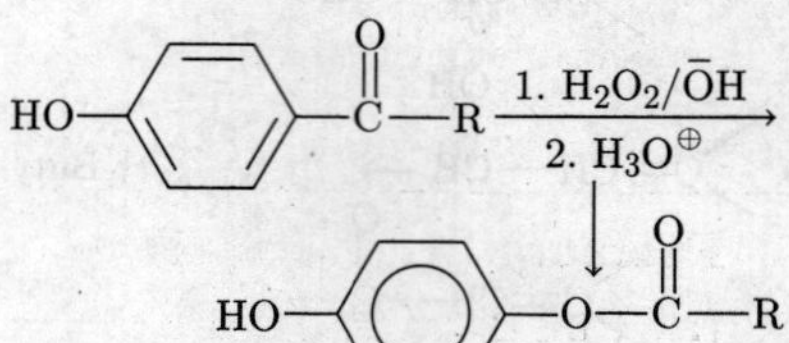 |

*Contd.*

| Operation | Reagent | Remarks/Examples |
|---|---|---|
| 29. $RCH_2COCH_3 \rightarrow PCOCOCH_3$ $\rightarrow RCH_2COCHO$ | $SeO_2/H_2O$ | $SeO_2$ preferentially oxidises —$CH_3$ to —CHO than $>CH_2$ to $>C{=}O$ <br> $RCH_2COCH_3 \xrightarrow[H_2O]{SeO_2} RCH_2COCHO$ (major) |
| | $RONO/H^{\oplus}$ 2. $H_2O$ via oxime | In this method $>CH_2$ group preferentially gets oxidised to $>C{=}O$, <br> $RCH_2COCH_3 \xrightarrow{RONO/H^{\oplus}} R—\underset{\Vert\atop NOH}{C}COCH_3 \xrightarrow{H_3O^{\oplus}} RCOCOCH_3$ (major) |
| 30. $RCO_2H \rightarrow RCO_2OH$ | $H_2O_2/H_2O$ | $PhCO_2H \xrightarrow{H_2O_2} PhCO_2OH$ |
| 31. $R_3N \rightarrow R_3\overset{\oplus}{N}—\bar{O}$ | $H_2O_2/H_2O$ | $Me_3N \xrightarrow{H_2O_2/H_2O} Me_3\overset{\oplus}{N}—\bar{O}$ |
| 32. $>NH \rightarrow >N—OH$ | $H_2O_2/HCO_2Me$ | 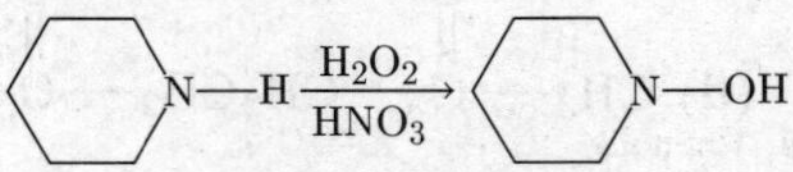 |
| 33. $—NH_2 \rightarrow —NO_2$ | $CF_3CO_2OH/CH_2Cl_2$ | 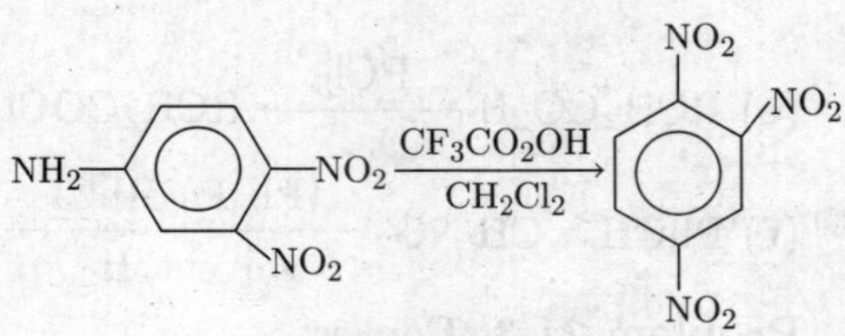 <br> $Et—\underset{\vert\atop Me}{CH}—NH_2 \xrightarrow[dry\ ClCH_2CH_2Cl]{CH_3CO_2OH} Et—\underset{\vert\atop Me}{CH}—NO_2$ |
| 34. $—NH_2 \rightarrow {=}N—OH$ | $H_2O_2$ or $H_2SO_5$ | $n\text{-}BuNH_2 \xrightarrow{H_2O_2} n\text{-}CH_3(CH_2)_2CH{=}N—OH$ |
| 35. $—NH_2 \rightarrow —NO$ | $(NH_4)_2S_2O_8—H_2SO_4$ | 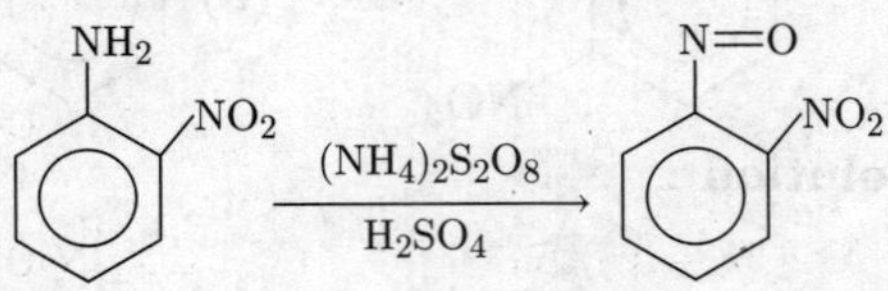 |
| 36. $—SH \rightarrow —S—S—$ | $CH_3CO_2OH$ | |
| 37. Cycloalkane → aromatic compound (aromatisation) | a. Pd–C <br> b. Se <br> c. S <br> d. Chloranil | 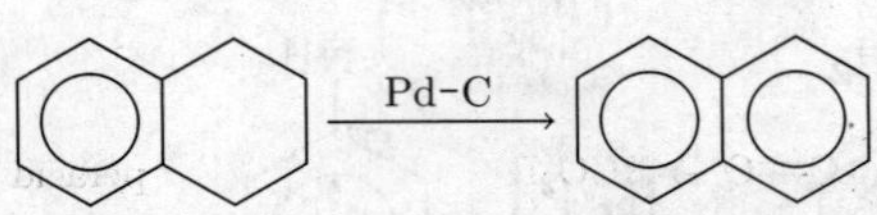 |

**Problem 21.2.** Convert

**(a)** $PhCH{=}CH{\cdot}CO_2Et \longrightarrow PhCH_2CH_2CO_2Et$

**(b)** $PhCH{=}CH{\cdot}CO_2Et \longrightarrow PhCH{=}CH{\cdot}CH_2OH$

(c) $CH_3CH{=}CH{-}\underset{\large OH}{\underset{|}{CH}}{-}CH_3 \longrightarrow CH_3CH{=}CH{-}\overset{\large O}{\overset{\|}{C}}{-}CH_3$

(d) $CH_3{-}\overset{O}{\overset{\|}{C}}{-}CH_2{-}CH_2{-}\underset{O}{\underset{\|}{C}}{-}OEt \longrightarrow CH_3\underset{OH}{\underset{|}{C}}HCH_2CH_2CO_2Et$

(e) $RCH_2CO_2H \longrightarrow RCH_2CHO$ f. $PhCH{=}CHNO_2 \longrightarrow Ph(CH_2)_2NO_2$.

**Solution :** The conversions can be done by using selective reducing and oxidising agents as given in Tables 21.5 and 21.6 respectively.

(a) $PhCH{=}CH{-}CO_2Et \xrightarrow[1\text{ atm}]{H_2/Pd} Ph{-}CH_2{-}CH_2CO_2Et$

(b) $PhCH{=}CHCO_2Et \xrightarrow[Et_2O]{LiAlH_4} PhCH{=}CHCH_2OH + EtOH$

(c) $CH_3CH{=}CH{-}\overset{OH}{\overset{|}{C}H}{-}CH_3 \xrightarrow{PDC/CH_2Cl_2} CH_3CH{=}CH{-}\overset{O}{\overset{\|}{C}}{-}CH_3$

(d) $CH_3{-}\overset{O}{\overset{\|}{C}}{-}CH_2CH_2{-}\overset{O}{\overset{\|}{O}}{-}OEt \xrightarrow[NaBH_4/H_2O]{1\text{ eq.}} CH_3{-}\underset{OH}{\underset{|}{C}}HCH_2CH_2{-}\overset{O}{\overset{\|}{C}}{-}OEt$

(e) $RCH_2CO_2H \xrightarrow{PCl_5} RCH_2COCl \xrightarrow[-78\,°C]{LiAl[O{-}Bu(t)]_3H} RCH_2CHO$

(f) $PhCH{=}CH{\cdot}NO_2 \xrightarrow[\text{solvent}/H_2]{(Ph_3P)_3RhCl} PhCH_2CH_2NO_2$

**Problem 21.3.** Convert

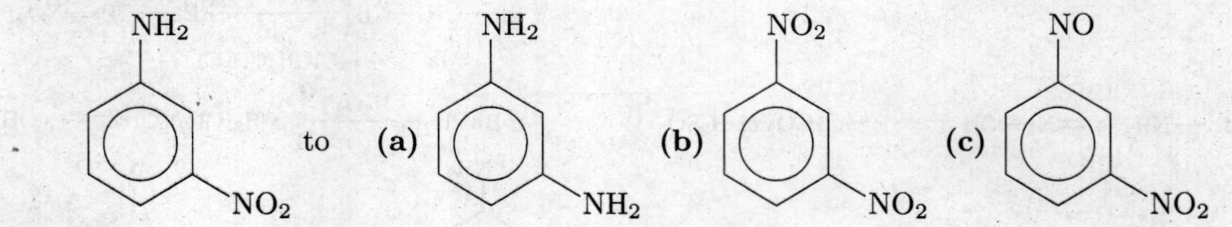


**Solution :**

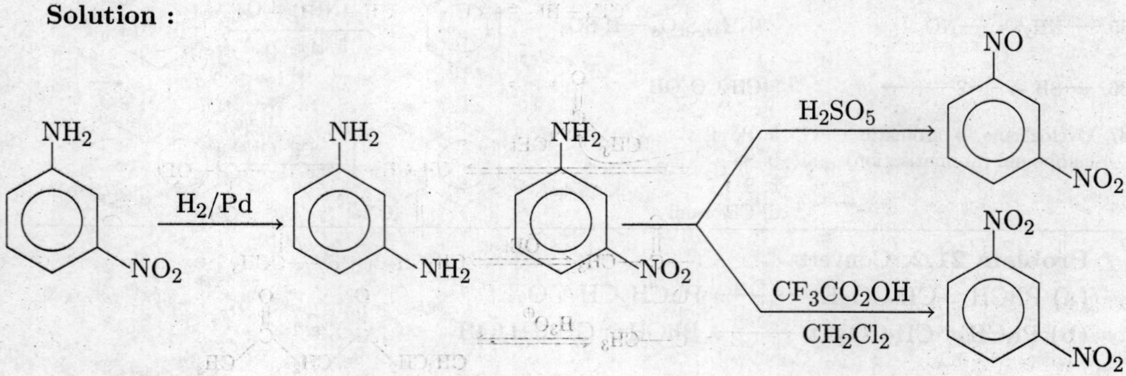

## 21.5.2 Regioselectivity

[Regio (L. regionem means direction)]

When in a reaction one constitutional isomer is formed predominantly amongst two or more constitutional isomers, the reaction is said to be regioselective. Markownikoff's addition, Saytzeff's and Hofmann's eliminations are regioselective reactions.

$$\mathrm{PhCH{=}CH_2 \xrightarrow{H_3O^{\oplus}} PhCHOH{\cdot}CH_3} \qquad \mathrm{PhCH{=}CH_2 \xrightarrow[\times]{H_3O^{\oplus}} PhCH_2CH_2OH}$$

The hydration of styrene to $\alpha$-phenylethyl alcohol exclusively (no $\beta$-phenylethyl alcohol forms) is not only a regioselective reaction, but it is also a regiospecific reaction which gives exclusively only one product between two or more possible constitutional isomers. Another good example of regioselective reaction is the Diels-Aider reaction. In Section 13.14.1 it has been shown that ortho and para types of products in the Diels-Alder reaction are major and meta type products are minor. The reaction can be made regiospecific by introducing a catalyst in the reaction medium as shown below:

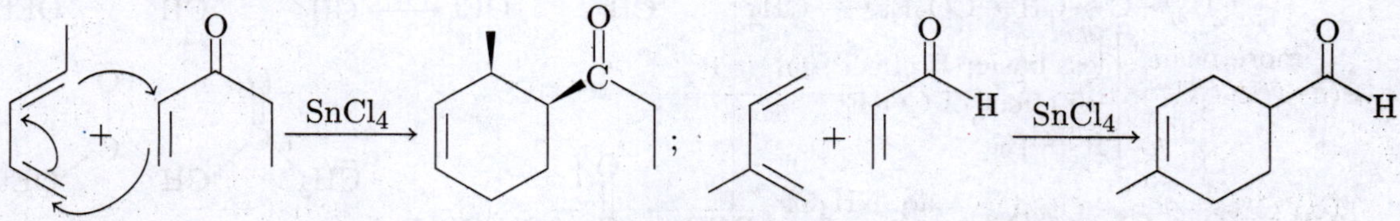

Acylation at two $\alpha$-Cs of a mixed ketone and alkylation of acetoacetic ester at two sites are the two important regioselective reactions from the standpoint of synthesis.

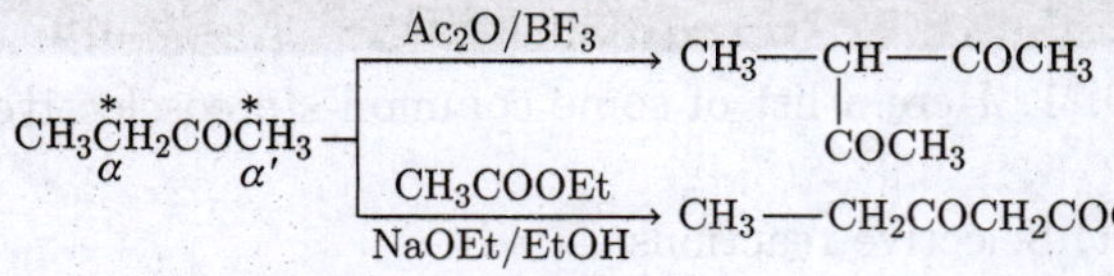

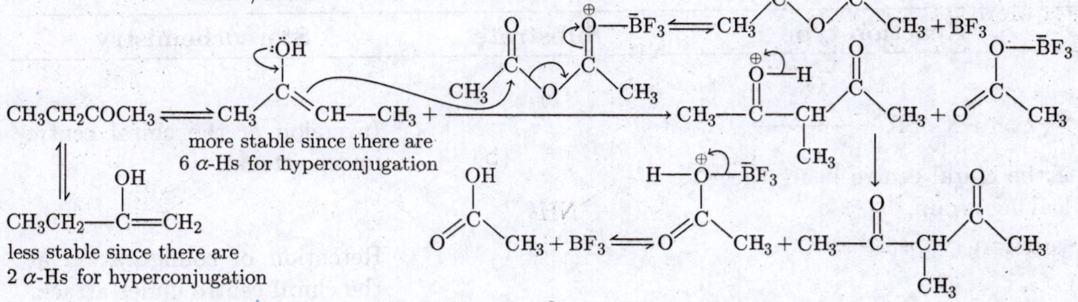

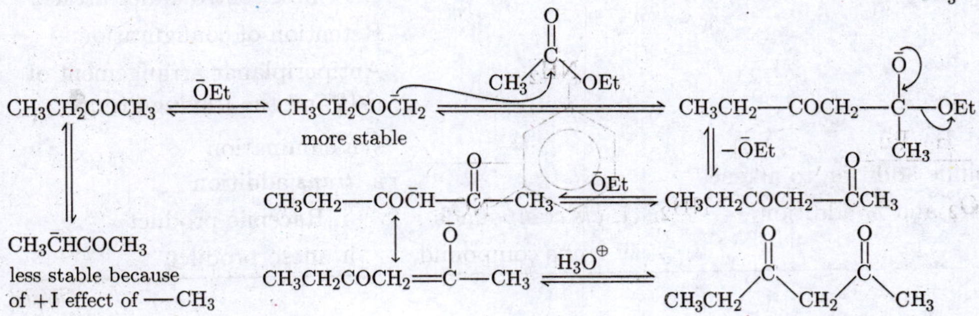

α, β-unsaturated carbonyls undergo regioselective 1, 2 and 1, 4 addition reactions with organometallies. These have already been discussed in Sec. 16.10. p. 876. Michael addition is another example of regioselectivity.

Regioselective alkylation of β-keto esters :

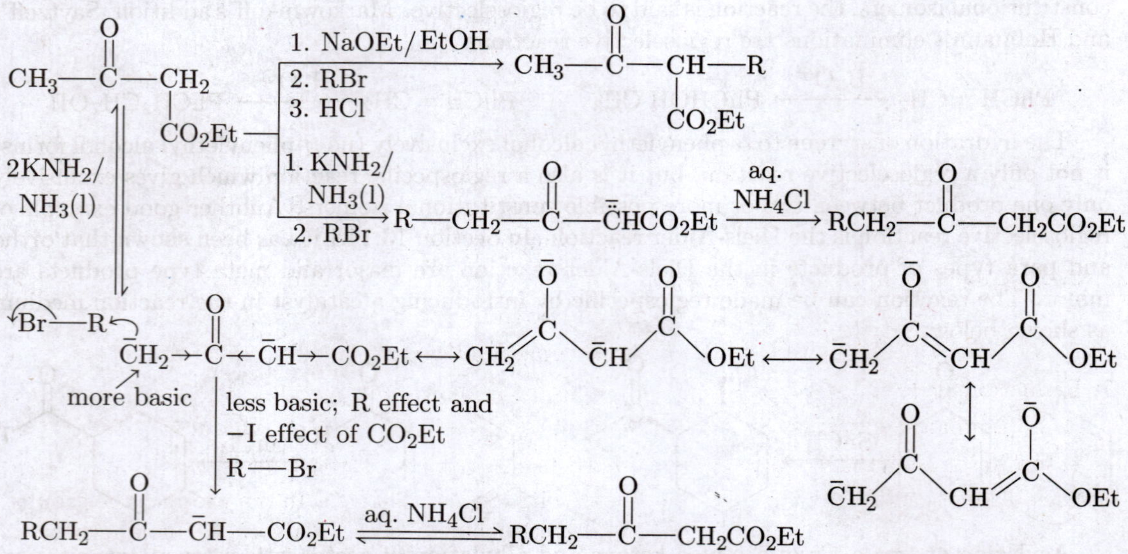

## 21.5.3 Stereoselectivity

This has already been discussed in Section 9.21. Here a list of some common stereoselective reactions is given in Table 21.7.

**Table 21.7** (Product) Selective reactions

| Reaction type | Substrate | Stereochemistry |
|---|---|---|
| 1. $S_N2$ on RX or R 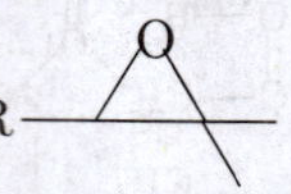 if the chiral centre bears the leaving group. | | Inversion of the chiral centre under attack. |
| 2. $S_N2$ with NGP | | Retention of configuration at the chiral centre under attack. |
| 3. $S_Ni$ | | Retention of configuration. |
| 4. E2 | | Antiperiplanar arrangement of β-H and the leaving group. |
| 5. Ei | | syn elimination |
| 6. Electrophilic addition to alkenes | | a. *trans* addition |
| a. $Cl_2$, $Br_2$ and $I_2$ addition | a. i. *cis* compounds | i. Racemic product |
| | ii. *trans* compound | ii. meso product |

*Contd.*

| Reaction type | Substrate | Stereochemistry |
|---|---|---|
| b. Epoxidation by $RCO_2OH$ | b. | b. *cis* addition |
| | i. *cis* compound | i. *cis* epoxide |
| | ii. *trans* compound | ii. *trans* epoxide |
| c. Hydroxylation by $KMnO_4/\bar{O}H$ or $OsO_4/H_2O$ | c. i. *cis* compound | c. *cis* addition<br>i. meso diol |
| | ii. *trans* compound | ii. Racemic diol |
| By Epoxidation and hydrolysis | | *trans* addition |
| | i. *cis* compound | i. Racemic diol |
| | ii. *trans* compound | ii. meso diol |
| d. Hydrogenation $H_2$/Pd cat. | | *cis* addition |
| | i. *cis* compound | i. meso product |
| | ii. *trans* compound | ii. Racemic product |
| 7. Reduction R—C≡C—R′ | | |
| i. by Lindlaar's catalyst | | i. *cis* product |
| ii. by Na or $Li/NH_3(l)$ | | ii. *trans* product |
| 8. Diels-Alder reaction | i. Diene with *trans-trans* 1, 4 substituents | i. In the product the groups are *cis* to one another |
| | ii. Diene with *cis-trans* 1, 4 substituents | ii. *trans* to one another |
| | iii. *cis* dienophiles | *cis* product |
| | iv. *trans* dienophiles | *trans* product |
| 9. Addition of singlet carbene to C═C | i. *cis* compounds<br>ii. *trans* compounds | i. *cis* cyclopropanes<br>ii. *trans* cyclopropanes |
| 10. Rearrangements : | i. Migrating group | i. Retention of configure-ation |
| Intramolecular 1, 2-shift carbanionic | ii. Migration terminus | ii. Inversion |
| 11. Reactions not involving the chiral centre | | Retention of configuration. |

## 21.6 Protection of Functional Groups

Chemoselective reagents constitute a very small group. So it is difficult to find a chemoselective reagent for an FGI in a polyfunctional compound. In practice an FG is protected from unwanted attack during reactions at other functions. The protection is given to the FG by converting it to a suitable derivative. The desired FGI is then carried out on the protected molecule and finally the protected FG is brought back to its original form. For example, if we want to carry out a nitration reaction on aniline to get *p*-nitroaniline, we cannot perform the reaction directly on aniline because undesirable products will be obtained owing to the activating effect of and coordination of a proton to —$NH_2$ group. The amino group is, therefore, protected before nitration. After the completion of nitration the —$NH_2$ group is deprotected. In this case, the amino group is protected by converting it to an amide, acetanilide, and deprotected by hydrolysing *p*-nitroacetanilide as shown below :

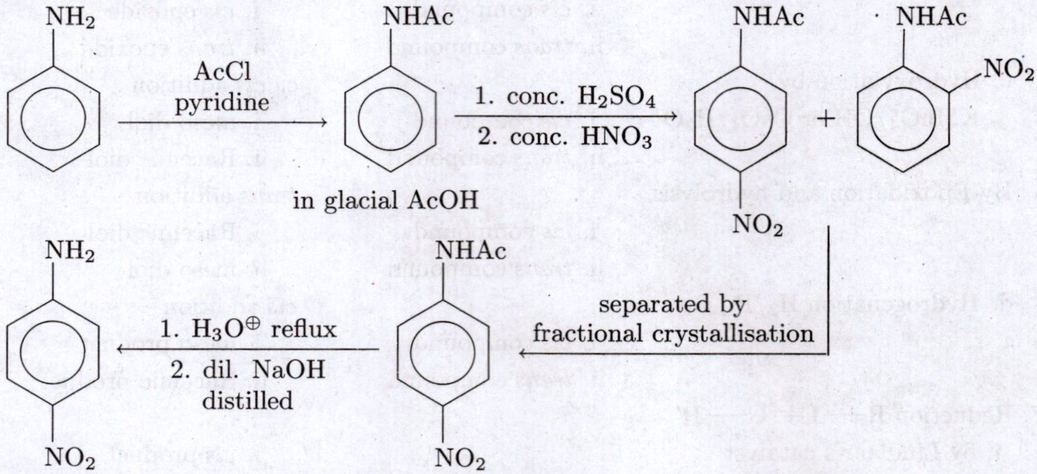


The following three characteristics are desirable for a functional group protecting operation.

(i) The protecting group(PG) must easily be attached to the substrate and yield of the product should be high.

(ii) The PG must be resistant to the conditions under which the FGI will be done.

(iii) After the introduction of desired FG, the PG must easily be removed, i.e., deprotected to the previous state. Table 21.8 gives us a list of PGs, their formation and removal, etc.

**Table 21.8** Protection of functional groups and their removal

| Group to be protected | Reagents for protection | Protecting group | Reagents for removal of PG | PG resists (under condition) |
|---|---|---|---|---|
| 1. Aldehyde and ketone | $ROH/H^{\oplus}$<br>$HOCH_2CH_2OH/H^{\oplus}$ | 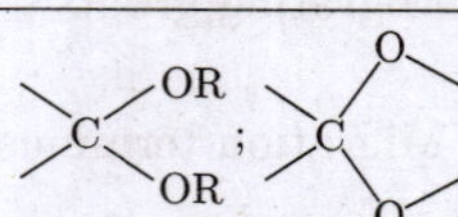<br>Acetal | $H^{\oplus}/H_2O$ | Nucleophiles, reducing agents (basic) |
| | $HS(CH_2)_2SH$<br>$BF_3$/ether | 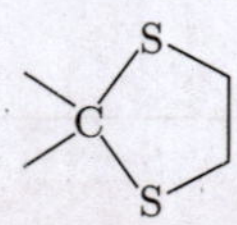<br>Thioacetal | $HgCl_2/CdCO_3/$ glycol : $H_2O$ or $CuO/CuCl_2{\cdot}H_2O/$ acetone | Nucleophiles (both basic and acidic medium) |
| | $RNH_2$ | $>C{=}NHR$<br>Imine | $H^{\oplus}/H_2O$ | Nucleophiles (basic) |
| 2. Acids | $MeOH/H^{\oplus}$ | $—CO_2Me$ | | Electrophiles (weakly basic) |
| | $EtOH/H^{\oplus}$ | $—CO_2Et$ | $H_3O^{\oplus}$ or $\bar{O}H$ strong | |
| | $t$-$BuOH/H^{\oplus}$ | $—CO_2{\cdot}Bu(t)$ | $H_3O^{\oplus}$ mild | |
| | $PhCH_2OH/H^{\oplus}$ | $—CO_2CH_2Ph$ | $H_2/Pd$ or HBr | |
| | $CCl_3CH_2OH/H^{\oplus}$ | $—CO_2CH_2CCl_3$ | Zn/MeOH | |

*Contd.*

| Group to be protected | Reagents for protection | Protecting group | Reagents for removal of PG | PG resists (under condition) |
|---|---|---|---|---|
| 3. Alcohol : monohydric | $PhCH_2Br$/Base | $ROCH_2Ph$ | $H_2$/cat. or HBr | Electrophiles, oxidising agents (basic medium) |
| 1, 2-diol | $CH_3COCH_3/H^{\oplus}$ | 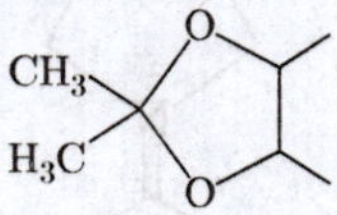 | $H_3O^{\oplus}$ | |
| 1, 3-diol | PhCHO | 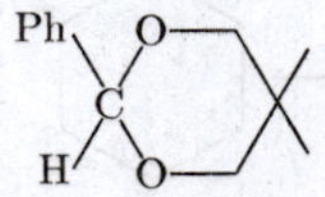 | $H_3O^{\oplus}$ | Nucleophiles, reducing agents (basic medium) |
| Phenols | $Me_2SO_4/K_2CO_3$ aq. | $PhOCH_3$ | HI or HBr or $BBr_3$ | Oxidation (basic medium) |
| | $MeOCH_2Cl$/Base | Ar—O—$CH_2$—OMe | HOAc/$H_2O$ | |
| $RNH_2$ | R′COCl/Pyridine | R′CONHR | $H\bar{O}/H_2O$ or $H_3O^{\oplus}$ | |
| | Phthalic anlydride | Phthalimide | $NH_2$—$NH_2$ | |
| | Trityl chloride | $Ph_3CCl$ | $Ph_3C$—NHR | HCl/aq. EtOH |
| 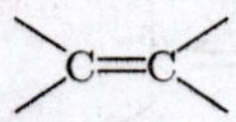 | $Br_2$ | 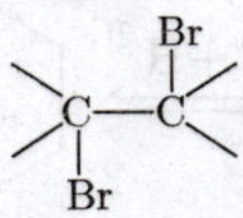 | Zn | |

## 21.7 Activation of FGs

Unreactive or weakly active FGs are required to be activated to perform reactions involving them. Those FGs are activated by converting them to other FGs which are more reactive than the former FGs and give the desired product. For example a —$CO_2H$ group being weakly reactive, it is converted to an acid halide for performing an aliphatic $S_N$ reaction bimolecular on the $sp^2$ C of the carbonyl group.

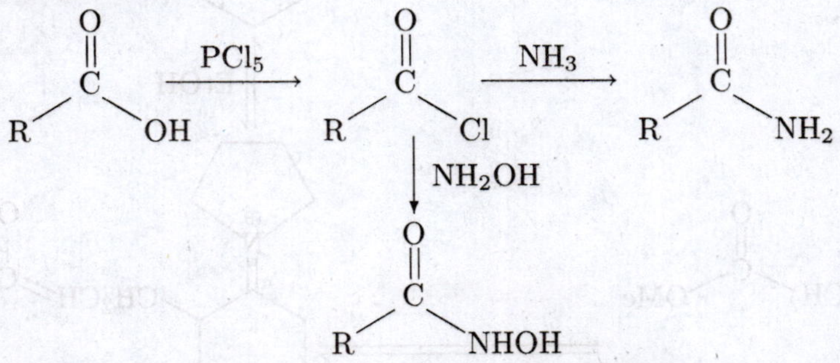

**Enamines** can be used for the purpose of activating $\alpha$-$CH_2$ group of a carbonyl compound. The activation process, in this case, involves the formation of the enamine of a carbonyl compound followed by the reaction with the reagent and hydrolysis of cyclic secondary amines such as pyrrolidine, piperidine, morpholine and their derivatives which are used for activating carbonyl functions.

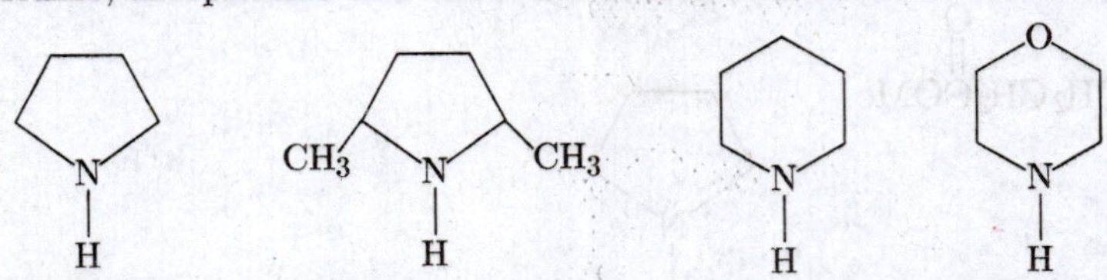

Pyrrolidine Pyrrolidine derivative Piperidine Morpholine

Though cyclohexanone is not an active methylene compound yet it can be made to undergo Michael addition with methyl acrylate via enamine formation as shown below :

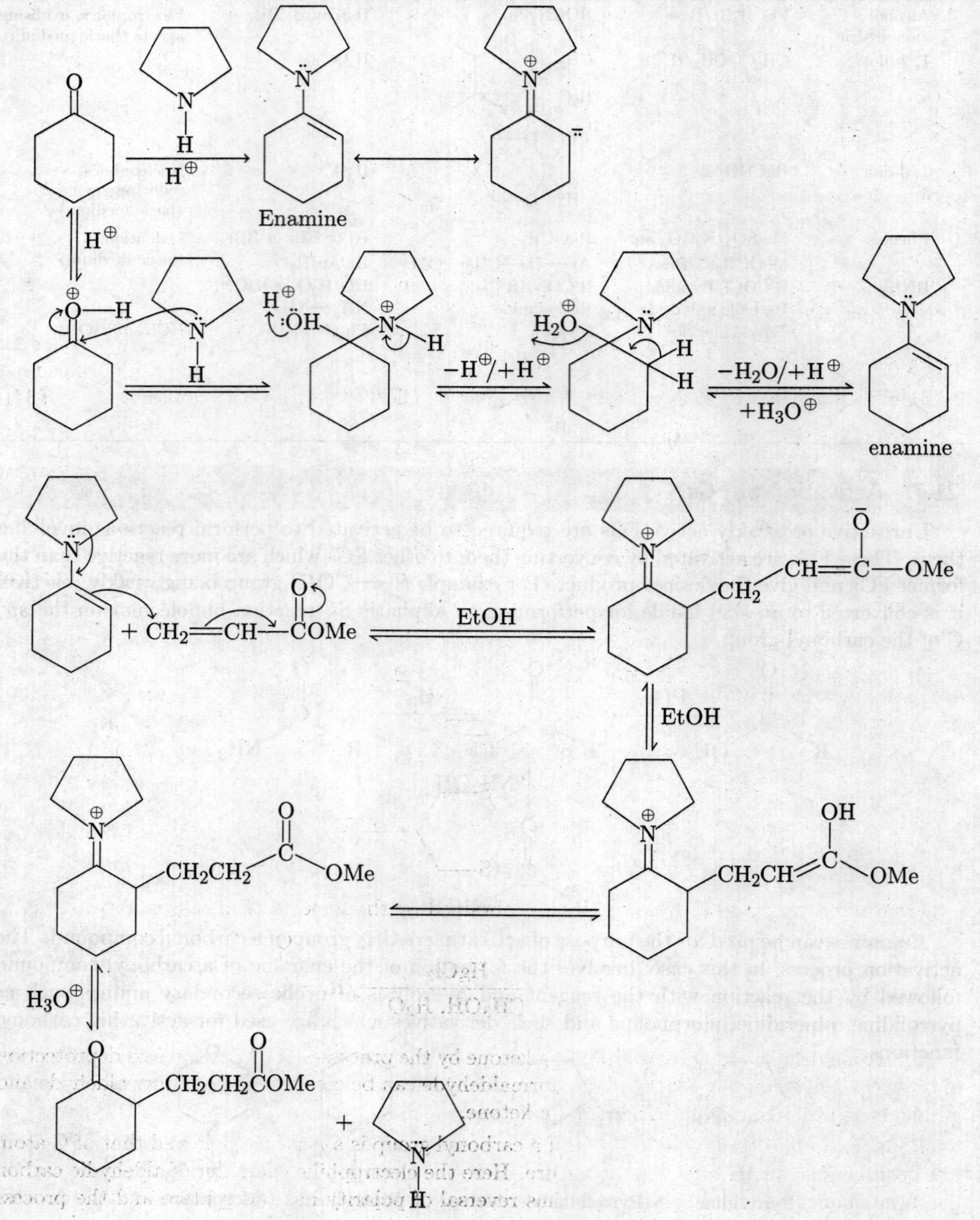

Similarly,

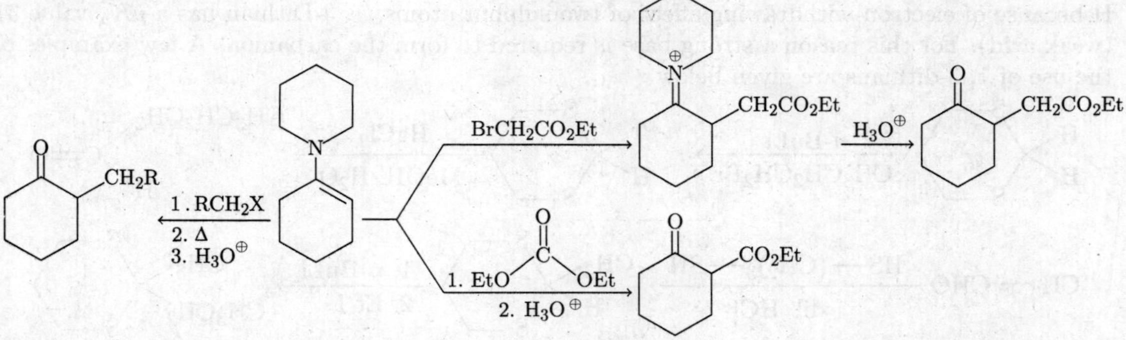

In this connection the activation of aldehydic H and conversion of aldehydes into ketones deserve special mention.

We have come to know that an aldehyde group could be protected by thiols to form thioacetal. The aldehyde group of formaldehyde may be protected by reacting the former with propane-1, 3-dithiol as given below. The protected compound is called 1, 3-dithian (for stability see p. 837).

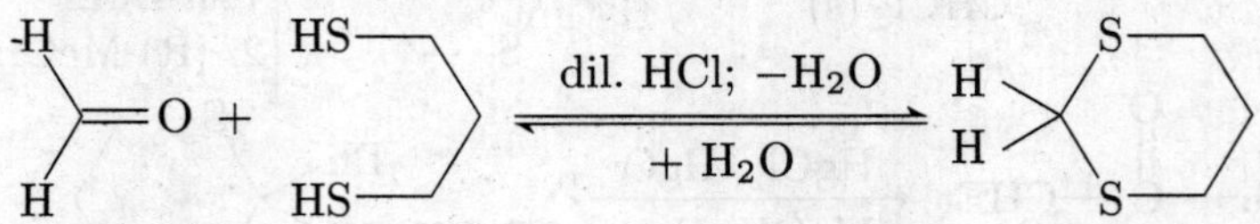

1, 3-Dithian

The aldehydic H atom can now be removed by a strong base to develop a carbanion and the latter may be subjected to alkylation ($S_N2$).

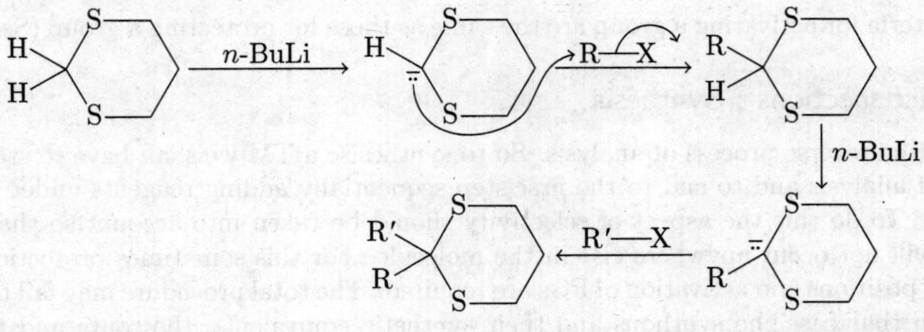

From the substituted dithian a ketone is obtained by the removal of dithiol moeity.

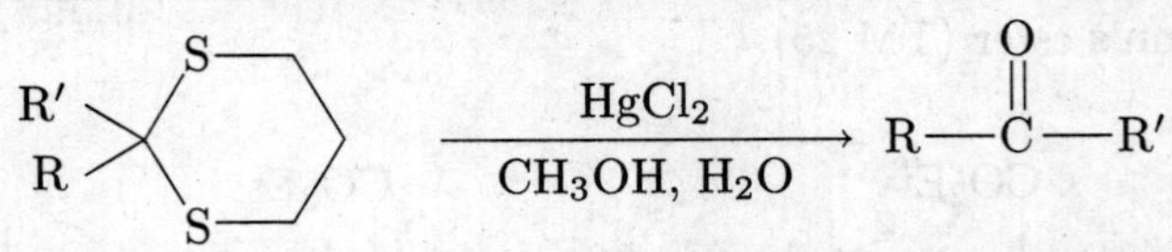

Thus an aldehyde can be converted to a ketone by the processes of protection and deprotection of the —CHO group. In the same way formaldehyde can be converted to a higher aldehyde and an aldehyde other than formaldehyde to a ketone.

However, we know that the O atom of a carbonyl group is a negative pole and that of C atom is a positive pole, i.e. electrophilic in nature. Here the electrophilic character of aldehydic carbon has been changed to nucleophilic one; thus reversal of polarity has taken place and the process is said to be Umpolung. Benzoin condensation also involves umploung. Thioacetals have acidic H because of electron-withdrawing effect of two sulphur atoms. 1, 3-Dithian has a $pK_a$ value 31 (weak acid). For this reason a strong base is required to form the carbanion. A few examples of the use of 1, 3-dithians are given below :

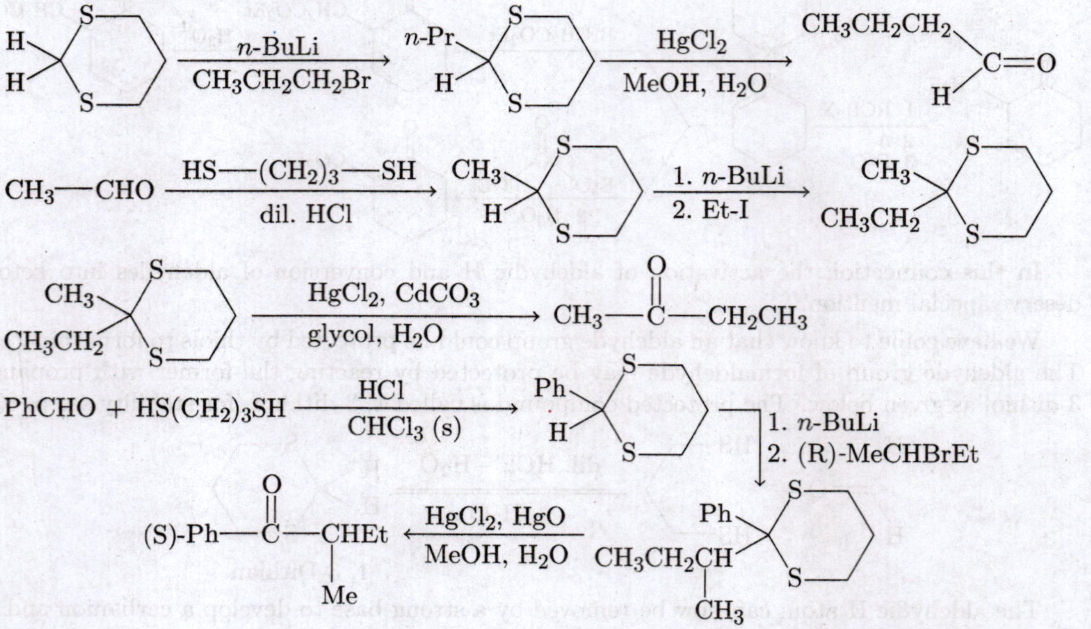

(S)-configuration of the chiral carbon

The criteria for activating a group are the same as those for protecting a group (See Sec. 21.6).

## 21.8 Reconnections : Synthesis

This is the reverse process of analysis. So to synthesise a TM we shall have to start from the last step of analysis and to end to the first step sequentially adding reagents under appropriate conditions. To do this the aspect of selectivity should be taken into account so that unwanted reactions will not occur anywhere else in the molecule. For this sometimes protection of group, blocking of positions and activation of FGs are required. The total procedure may fail to synthesise the TM. In that case the synthons and their synthetic equivalents, the route and the reaction conditions are to be checked and modified as they are required. For example, let us consider the synthesis of Hagemann's ester (TM 25).

**Analysis I.**

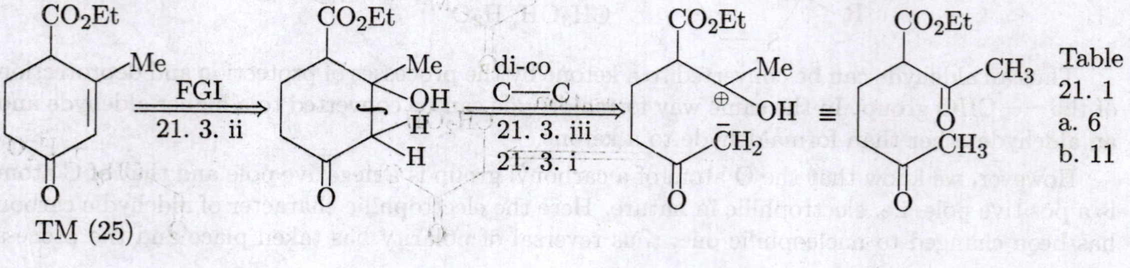

TM (25)

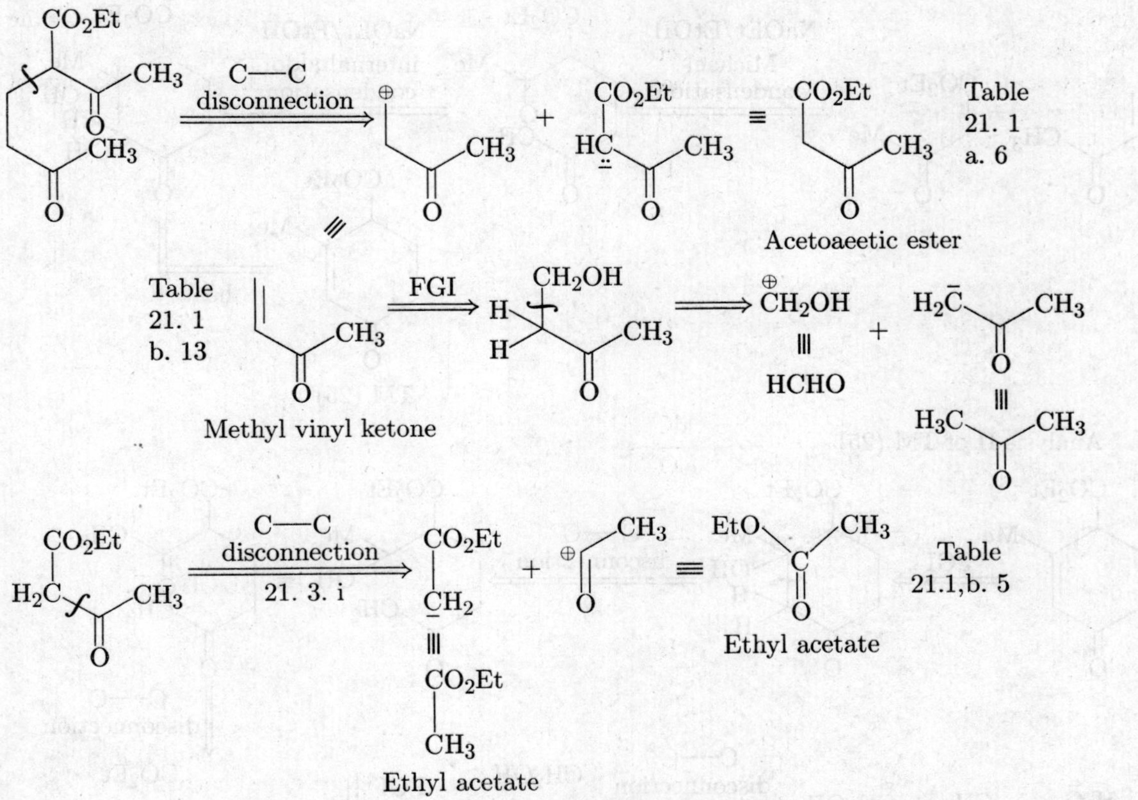

So, the starting materials for the synthesis of TM (25) might be formaldehyde (HCHO), acetone ($CH_3COCH_3$) and ethyl acetate ($CH_3CO_2Et$). Let us now find out the routes (i) for the synthesis of methyl vinyl ketone, (ii) ethyl acetoacetate and lastly TM (25). Ethyl acetoacetate could easily be synthesised by the Claisen ester condensation. Methyl vinyl ketone could also be synthesised from formaldehyde and acetone by aldol condensation. Methyl vinyl ketone and ethyl acetoacetate when subjected to Michael condensation followed by cyclisation give TM (25). However, both methyl vinyl ketone and ethyl acetoacetate are available in the market and we may not take the trouble of synthesising them in the laboratory. Anyway following are the steps for the synthesis of TM (25).

Synthesis of TM (25)

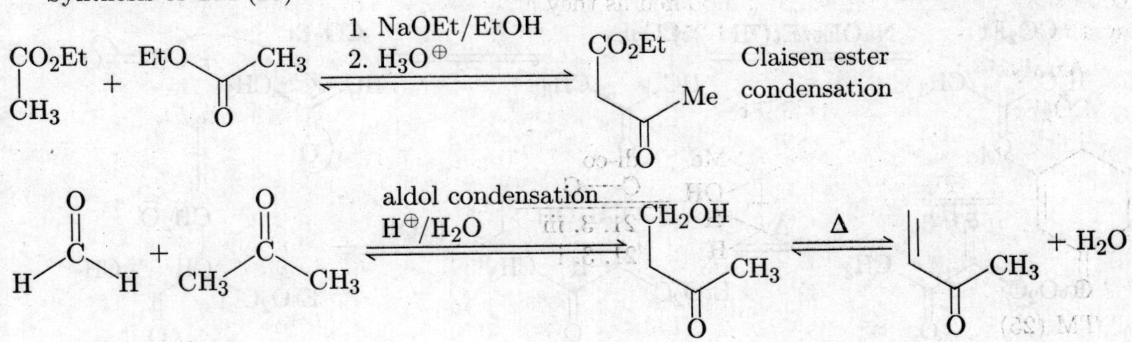

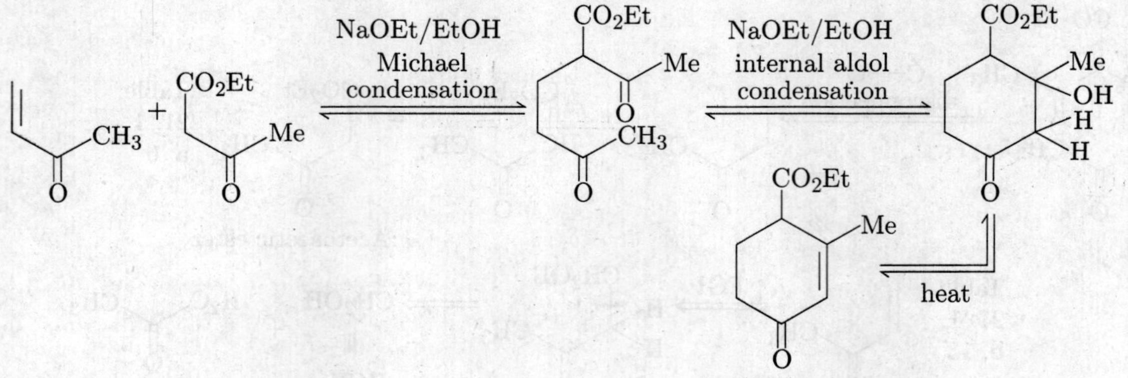

TM (25)

Analysis II of TM (25)

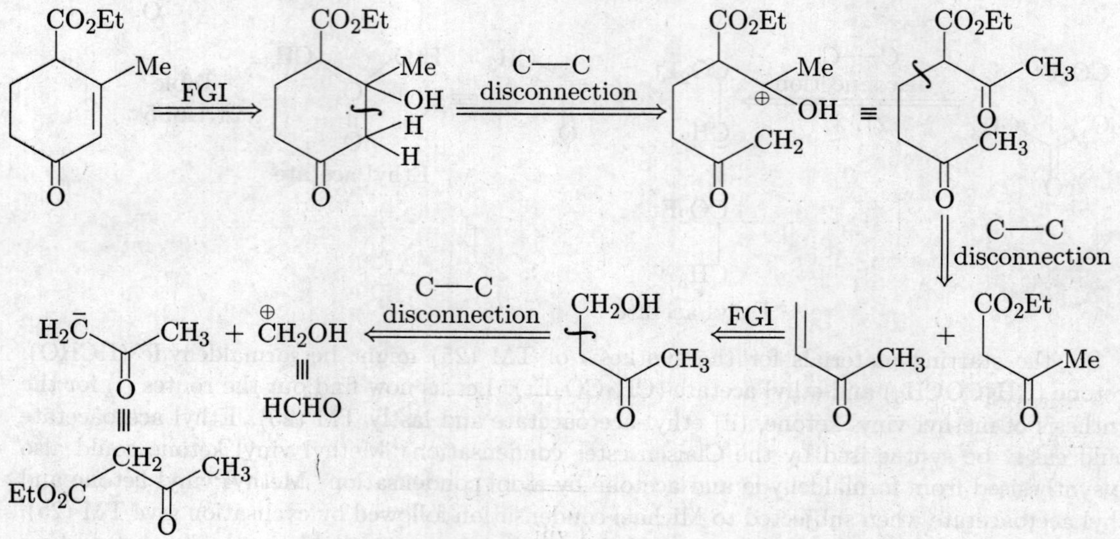

So, the two fragments are ethyl acetoacetic ester.

Here the SE of $\bar{C}H_2COCH_3$ is ethyl acetoacetate instead of acetone as in analysis I.

Synthesis II of TM (25)

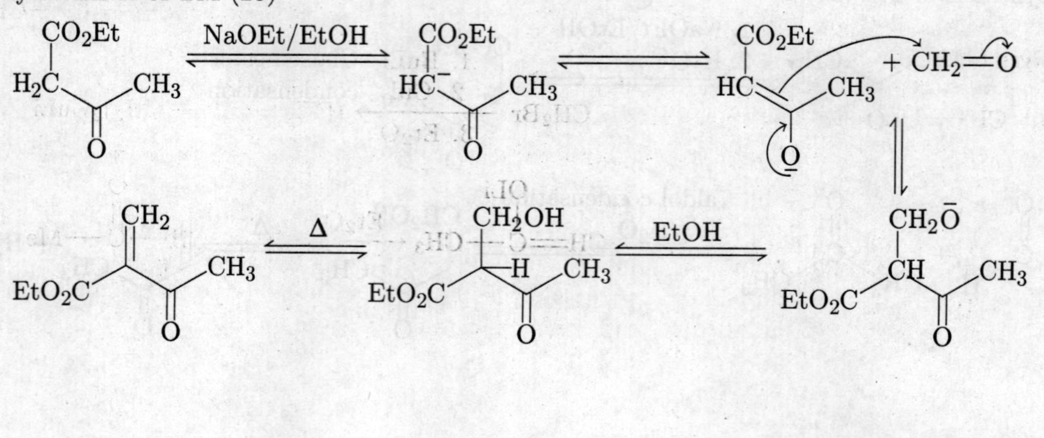

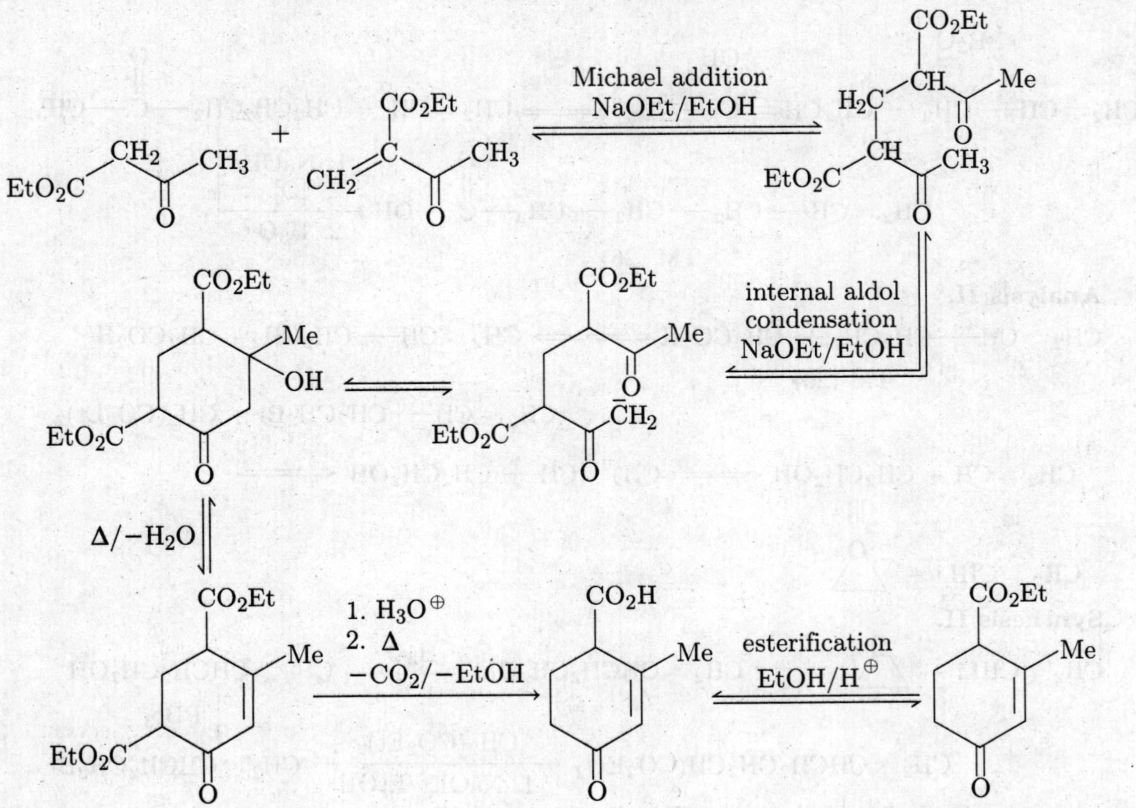

TM (25)

**Example 2:** Synthesis of $CH_2{=}CH{-}CH_2{-}CH_2{-}CH_2CO_2H$

**Analysis I.**

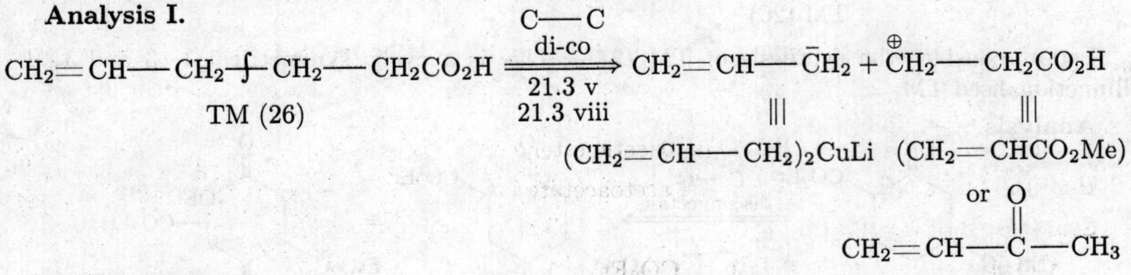

**Synthesis I.**

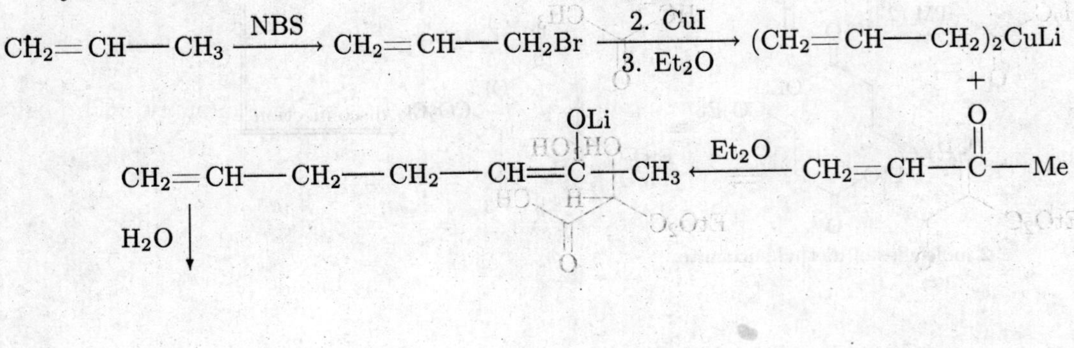

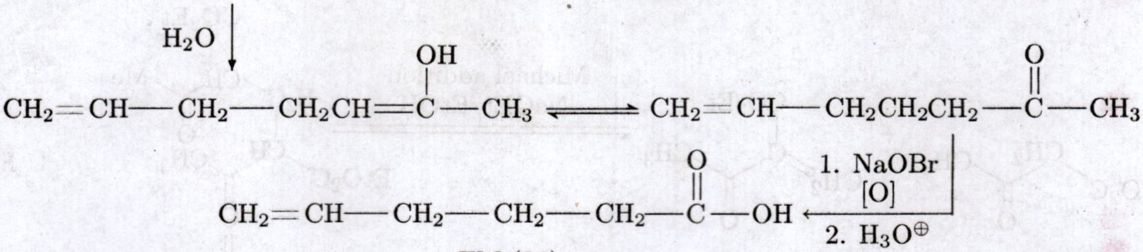

TM (26)

**Analysis II.**

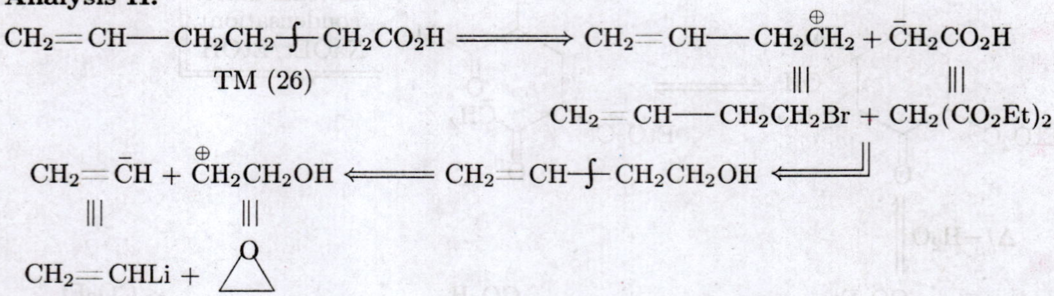

**Synthesis II.**

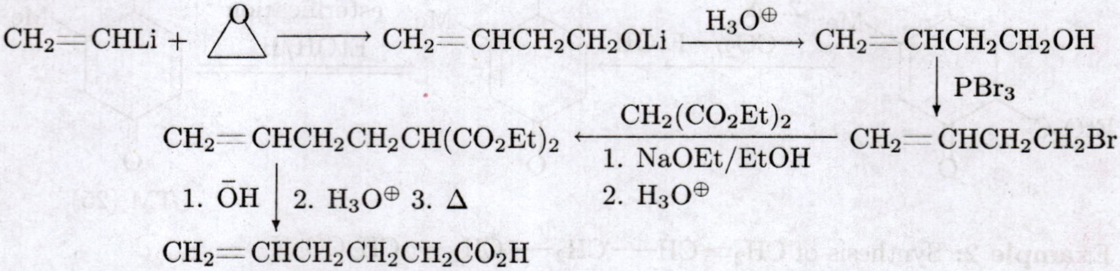

TM (26)

Let us consider the synthesis of TM (27). This is a cyclic symmetrical 1, 3 and 1, 4-difunctionalised TM.

**Analysis**

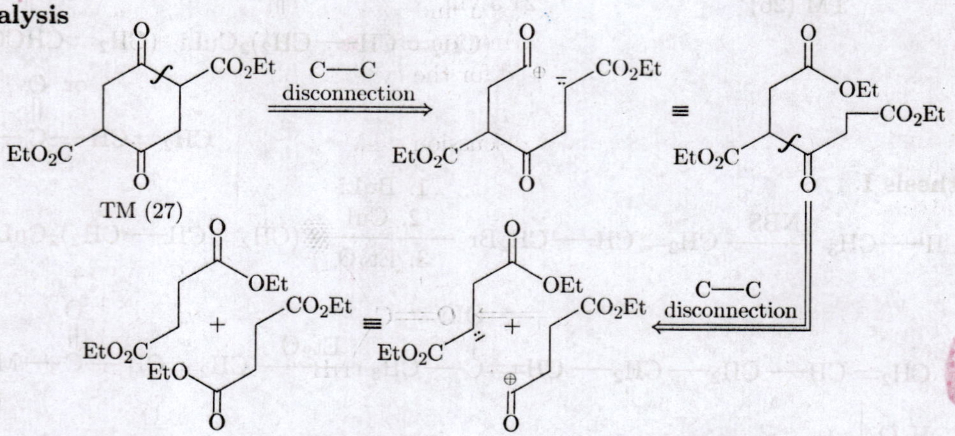

2 molecules of diethyl succinate

We can show the two C—C disconnections at a time.

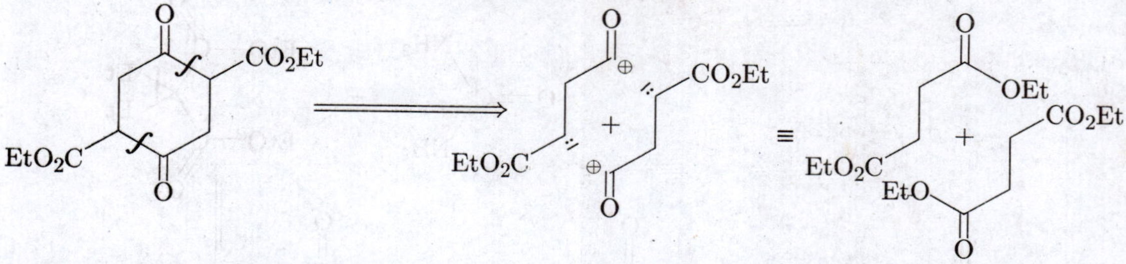

Since there are carbonyl groups, C—C disconnections (1, 3 di-co) may be carried out. Since the TM is a symmetrical one, the two identical disconnections (1, 4 di-co) may also be done.

**Synthesis**

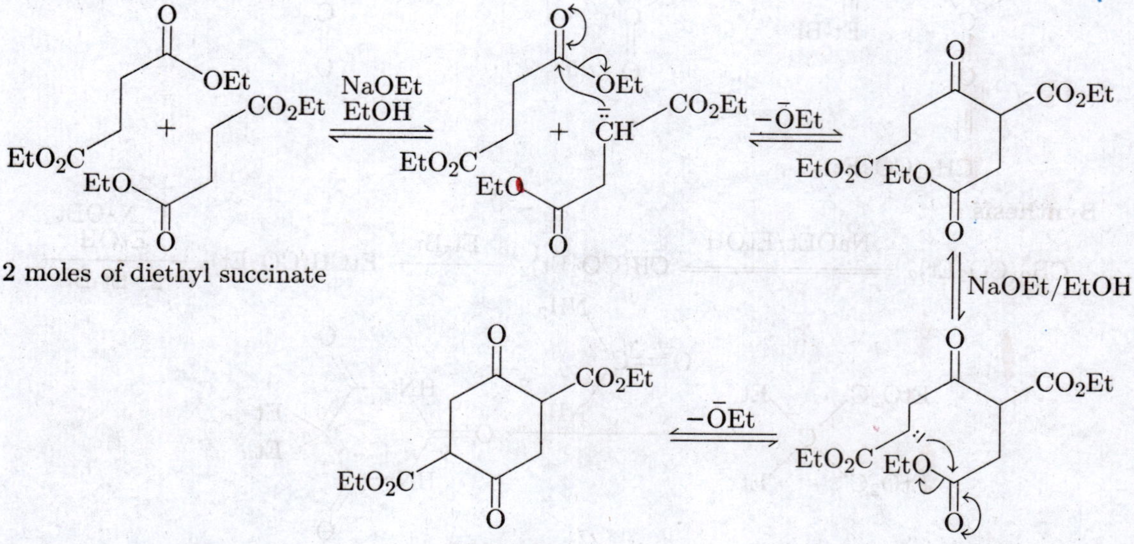

TM (27)

In this synthesis Claisen ester condensation and Dieckmann ester condensation have been used. Table 21.3 viii tells us that for the formation of six membered ring a diester is a starting material and Dieckmann reaction can be used for the cyclisation reaction. Diethyl succinate is a good 1, 4-difunctionalised starting material.

A second example may be taken under discussion—TM (28), a sedative called 5, 5 diethylemalonyl urea :

**Analysis**

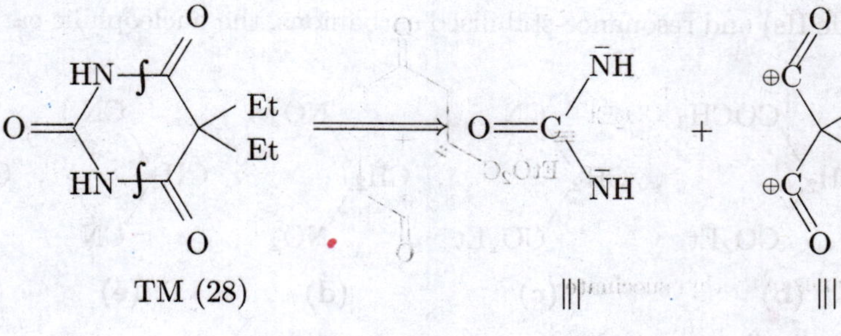

TM (28)

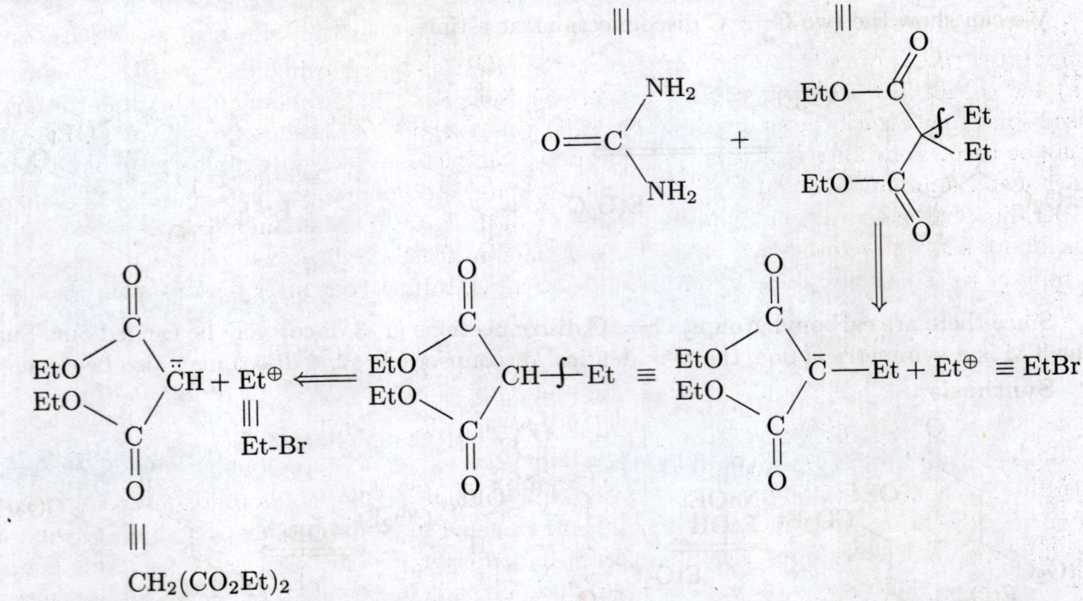

**Synthesis**

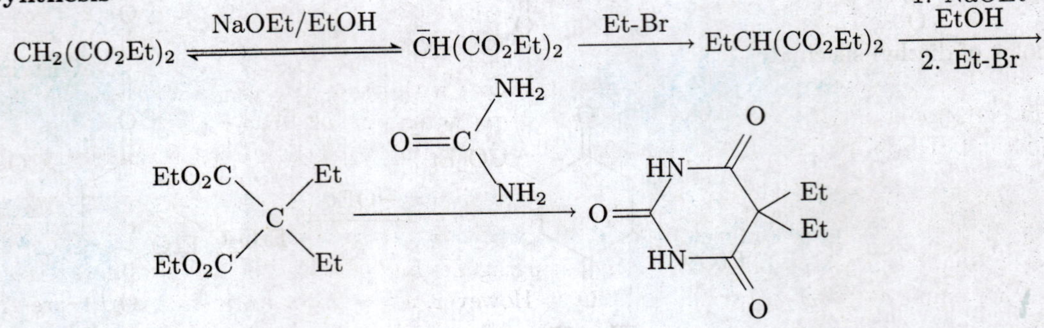

TM (28)

## 21.9 Active Methylene Compounds and C—C Bond Formation

An active methylene compound contains a —$CH_2$— or —CHR— group flanked by two strong electron-withdrawing (−I and −R) groups. Diethyl malonate, ethyl acetoacetate, ethyl cyanoacetate, dinitromethane, propane-dinitrile and 1, 3-diketones are the most important active methylene compounds. The methylenic H atoms of these compounds can be abstracted by bases (since these are acidic Hs) and resonance-stabilised carbanions, the nucleophilic carbon-synthons, can be formed.

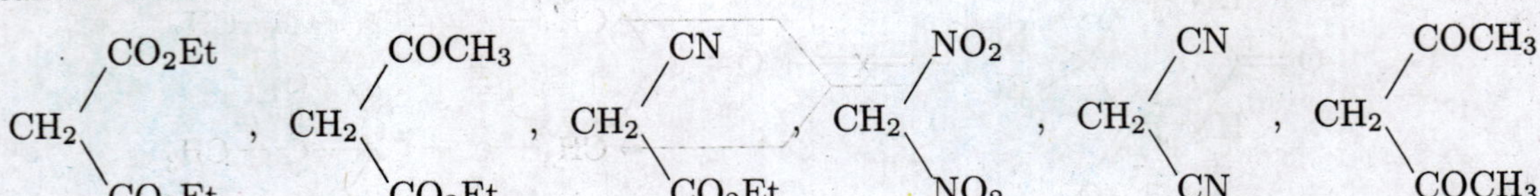

(a) (b) (c) (d) (e) (f)

However, the choice of base to generate the carbanions from the active methylenes is an important task. An alkali, $\bar{O}H$, will attack the electron-withdrawing groups also. Usually for (a) and (b) NaOEt in ethyl alcohol can safely be used. For cyano compound, alkoxides cannot be used since they attack the strongly electrophilic cyano-carbon; for them sodium hydride in DMSO may be used. For dinitromethane (p$K_a$ 3.6) and 1, 3-diketones (p$K_a$ 9) weak bases like potassium carbonate, ammonium acetate, piperidinium acetate, etc. may be used to generate carbanions; piperidine, secondary amines can also be used to get an appreciable amount of carbanions from the above said active methylenes. To develop dianions from 1,3-diketones and ethyl acetoacetate strong bases like sodamide, sodium hydride and alkyllithiums (one after another), etc. are used.

$$CH_3COCH_2COCH_3 \xrightarrow[NH_3(l)]{2NaNH_2} \bar{C}H_2CO\bar{C}HCOCH_3$$

$$CH_3COCH_2CO_2Et \xrightarrow[2.\ n\text{-BuLi}]{1.\ NaH} \bar{C}H_2CO\bar{C}HCOEt$$

Besides the above said bases, lithium diisopropylamide (LDA), potassium t-butoxide, magnesium diethoxide in dry ether, pyridine, sodiumtriphenyl methide, cyclohexylamine, etc, may be used according to necessity. The alkoxy part of the ester is to be used as the alkoxide base to generate a carbanion from the ester; if a different alkoxide base is used, a new ester will generate.

$$RCOCH_2\text{—}CO_2Et + Me\bar{O} \longrightarrow RCO\ddot{\bar{C}}HCO_2Et + \underset{\text{new ester}}{RCOCH_2\text{—}CO_2Me} + Et\bar{O} + MeOH$$

Aldehydes, ketones and esters ($\alpha$-haloesters also) with an $\alpha$-$CH_2$ group may also be looked upon as compounds with a methylene group activated by a single electron-withdrawing group. For example, $CH_3COCH_2CH_3$, $CH_3CHO$, $CH_3CH_2CO_2Et$, $PhCH_2CO_2Et$, $CH_3CN$, $CH_3NO_2$, etc.

## 21.9.1a Alkylation of active methylenes having two electron-withdrawing groups

Alkyl halides are generally used as alkylating agents and stoichiometric equivalents of strong bases are employed for generating carbanions. However, a few problems in this regard are given below.

(i) A resonance-stabilised carbanion from any one of the active methylene compounds is an ambident nucleophile. So, alkylation may occur at the hetero atom; for example, at the O atom of an enolate ion, or at the methylenic C atom.

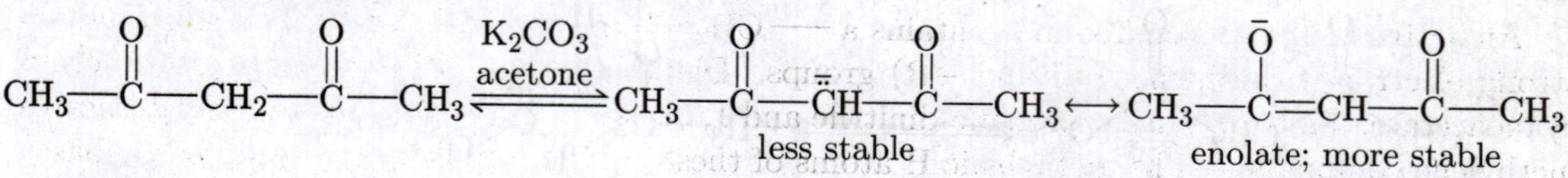


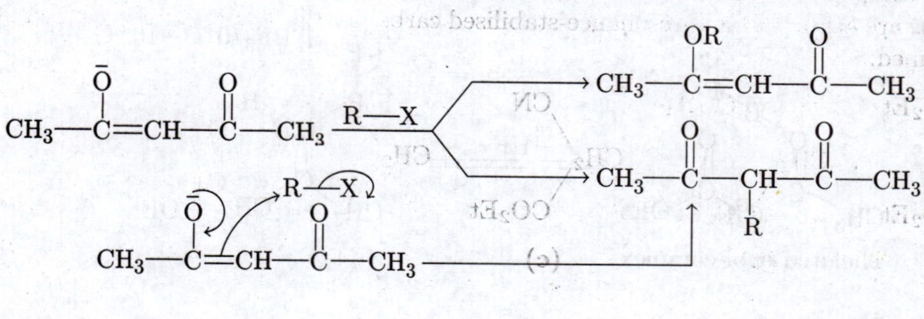

To get C-alkylated product, a sodium salt of a base, a protic solvent (e.g., alcohol; which can stabilise the anionic O of the enolate and thereby can deactivate the $O^-$) and alkyl halides as alkylating agents are usually employed. Amongst alkyl halides, alkyl iodides give the highest percentage of C-alkylated products. This is because this involves soft-soft interaction between the anionic C of the enolate and the cationic C of the halide, the decreasing order of hardness of the carbons amongst halides being C—F > C—Cl > C—Br > C—I.

(ii) If dialkylation is required, after the introduction of one alkyl group the methylene part becomes sterically congested and the alkyl group will exert electron-repelling effect. However—

(a) To introduce two same alkyl groups, we can use dimolar amount of the base at a time and dimolar amount of alkyl halide at a time for monomolar amount of the active methylene compound though the alkylation will occur one at a time.

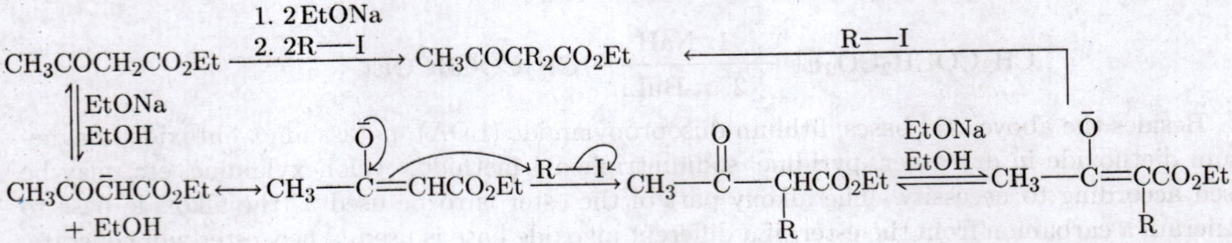

(b) To introduce two different alkyl groups, we shall have to add one mole of the base followed by one mole of the alkyl halide containing the alkyl group with lesser +I effect and then the other mole of the base followed by the alkyl halide having the alkyl group with greater +I effect.

$$CH_2(CO_2Et)_2 \xrightarrow[\text{2. } CH_3I]{\text{1. NaOEt/EtOH}} CH_3CH(CO_2Et)_2 \xrightarrow[\text{2. } C_2H_5Br]{\text{1. NaOEt/EtOH}} CH_3CH_2—\underset{CH_3}{\underset{|}{C}}(CO_2Et)_2.$$

For the introduction of the second alkyl group, it is better to use a base stronger than the base used for the introduction of the first alkyl group in order to overcome the electron-repelling effect of the first substituent.

(iii) To introduce a *t*-butyl group we cannot use a 3° alkyl halide since the *t*-alkyl halide undergoes the E2 reaction in the presence of a base to form an alkene. However, *t*-alkyl groups can be introduced to an active methylene group using *t*-alkyl halides or olefins in the presence of an acid catalyst as shown below :

(a)

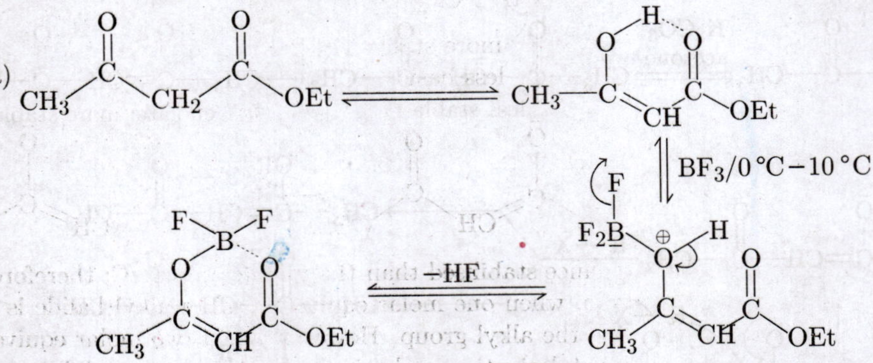

chelated enol-complex

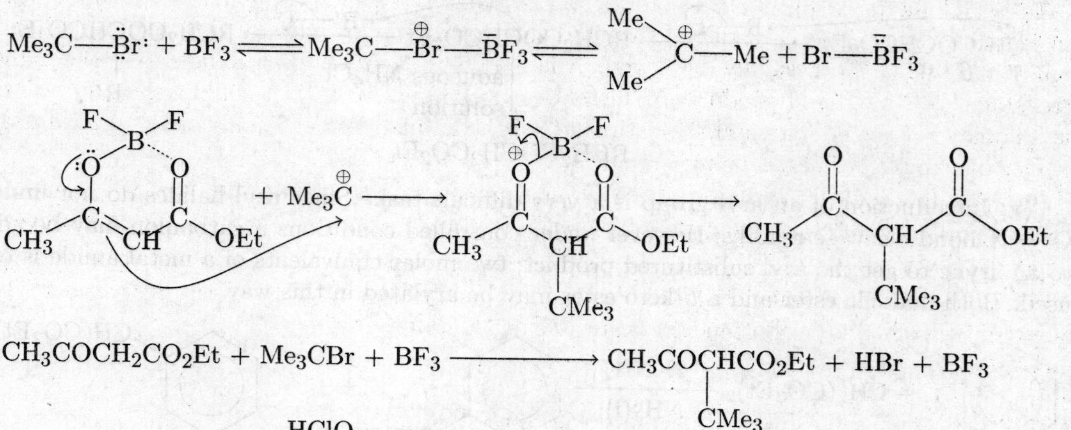

(b)

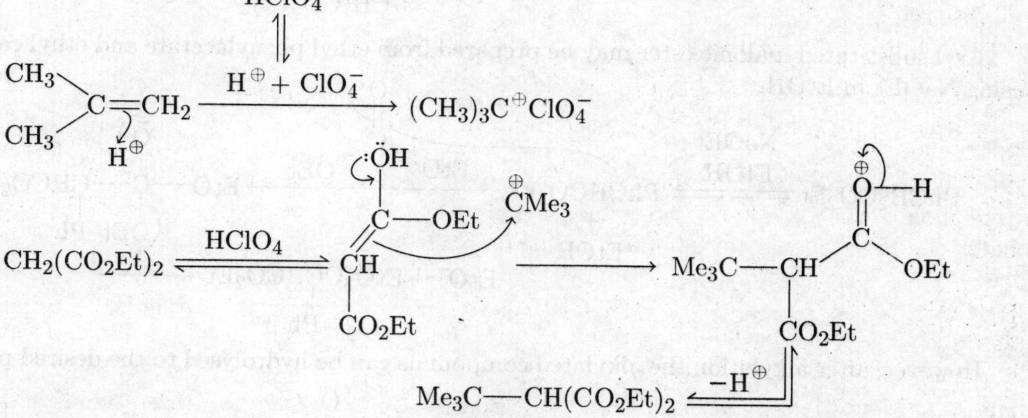

(iv) In the cases of $\beta$-keto esters and 1, 3-diketones, sometimes we require to alkylate a $\gamma$-methyl group or a less reactive methylene group. In such a case as stated two molar equivalents of a strong base are to be used to create a dianion followed by one molar equivalent of the alkyl halide.

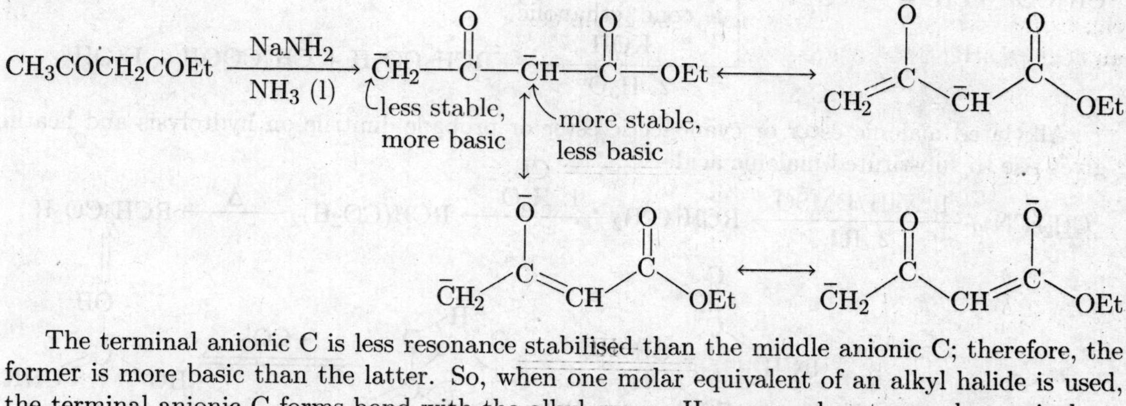

The terminal anionic C is less resonance stabilised than the middle anionic C; therefore, the former is more basic than the latter. So, when one molar equivalent of an alkyl halide is used, the terminal anionic C forms bond with the alkyl group. However, when two molar equivalents of an alkyl halide is used, an $\alpha$, $\gamma$-dialkyl derivative is obtained. In this way two different alkyl groups may be introduced.

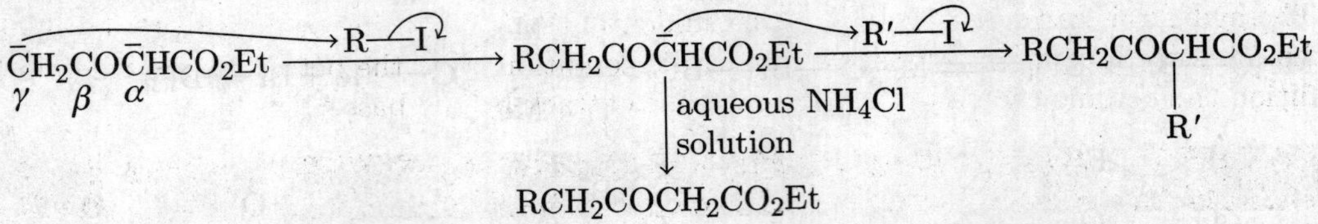

(v) Introduction of an aryl group is a very difficult task. Since aryl halides do not undergo C—X bond cleavage readily. However under controlled conditions a carbanion may be added to an aryne to get the aryl substituted product; two molar equivalents of a metal amide is to be used. Both malonic ester and a $\beta$-keto ester may be arylated in this way.

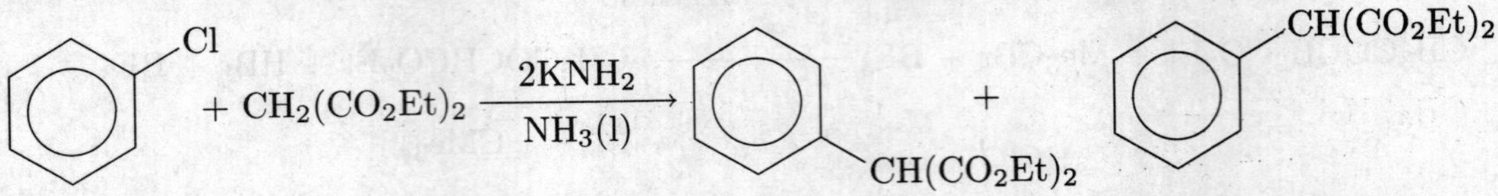

Aryl substituted malonic ester may be prepared from ethyl phenylacetate and ethyl carbonate using NaOEt in EtOH.

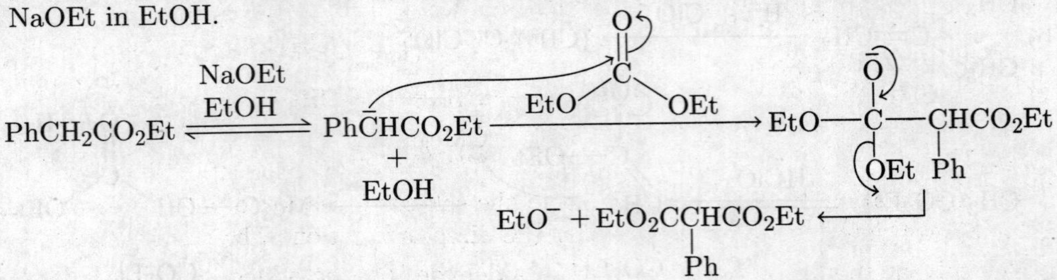

However, after alkylation the alkylated compounds can be hydrolysed to the desired products.

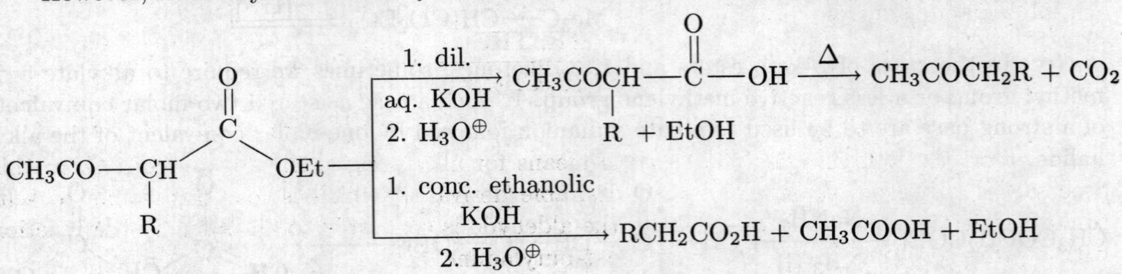

Alkylated malonic ester or cyanoacetic ester or propane-dinitrile on hydrolysis and heating gives rise to substituted malonic acid.

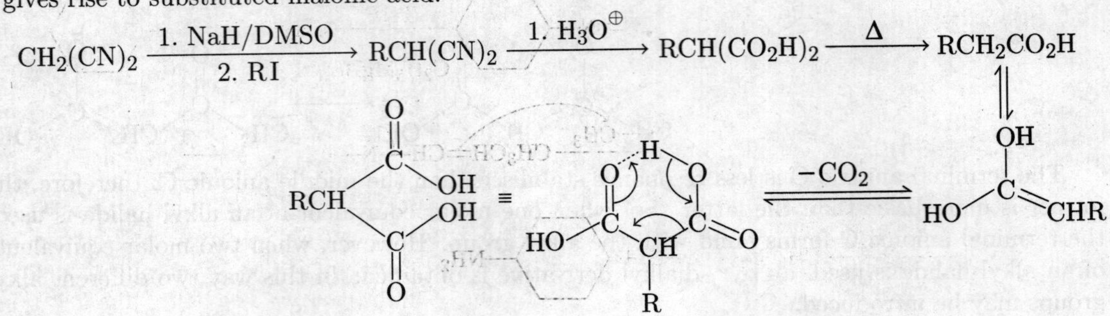

The hydrolysis can be carried out either under an acidic condition or under a basic condition. In the cases of dialkylated $\beta$-keto ester an acidic condition gives the better result. Under a basic condition the ketonic carbonyl C is susceptible to attack by the base.

## 21.9.1b $\alpha$-Alkylation of methylenes with one electron-withdrawing group

For alkylating alkanenitrile and esters no difficulty arises since the alkylation process is not regioselective and the self condensation can easily be stopped. On the other hand the $\alpha$-alkylation of aldehydes involves self condensation as a side reaction which is to be stopped, In the case of a mixed ketone, the alkylation usually involves regioselectivity which creates a problem. Nitroalkanes usually undergo O-alkylation instead of C-alkylation.

$$\mathrm{CH_3CH_2CN} \xrightarrow[\text{ii. } 2n\text{-Pr-Br, toluene, }\Delta]{\text{i. 2NaNH}_2} \mathrm{(CH_3CH_2CH_2)_2C(CH_3){-}C{\equiv}N}$$

$$\mathrm{CH_3CH_2CH_2COCH_2CH_3} \xrightarrow[\text{ii. CH}_3\text{CH}_2\text{Br}]{\text{i. LDA; THF}} \mathrm{CH_3CH_2CH(CH_2CH_3){-}COCH_2CH_3}$$

$$\mathrm{PhCOCH_2CH_3} \xrightarrow[\text{ii. CH}_3\text{CH}_2\text{I}]{\text{i. Ph}_3\text{CNa. ether}} \mathrm{PhCO{-}CH(CH_2CH_3){-}CH_2CH_3}$$

(i) The $\alpha$-alkylation of aldehydes : To avoid the self-condensation of aldehydes we shall have to employ appropriate reaction conditions for the direct alkylation. The conditions are—

(i) The solvent should be aprotic. (ii) The aldehyde is to be added slowly to a large excess of the base. (iii) The alkylating agent is to be added rapidly to the aldehyde-base mixture to achieve a kinetic control.

$$\mathrm{Me_2CHCHO} \xrightarrow[\text{ii. CH}_3\text{CH}_2\text{Br}]{\text{i. KH; THF}} \mathrm{Me_2C(CH_2CH_3)CHO}$$

However, the indirect process is a good means for alkylating aldehydes at the $\alpha$-C. In this process the aldehyde is converted to a desirable derivative which is less susceptible to self-condensation than the aldehyde. Usually the aldehyde is converted to an imine by the reaction with a primary amine like cyclohexylamine, $t$-butylamine, etc.

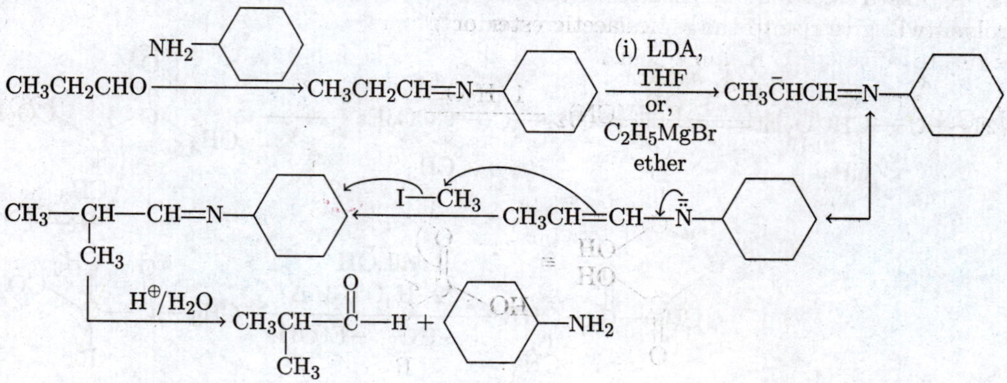

(ii) The α-alkylation of ketones : The direct α-alkylation of a ketone may be carried under conditions as stated above for aldehydes, However, by this process the regioselective α-alkylation on a mixed ketone is a trouble- some job. More than one product is usually obtained in this case. Through either kinetic control or thermodynamic control one product may be obtained in excess over the other. So the separation of the product mixture is also a difficult task. It is better to carry out the regioselective α-alkylation in indirect ways. A few examples are given below :

(i) The conversion of ketone to α, β-ketoaldehyde before alkylation and removal of formyl group after alkylation give a good indirect route for the regioselective α-alkylation.

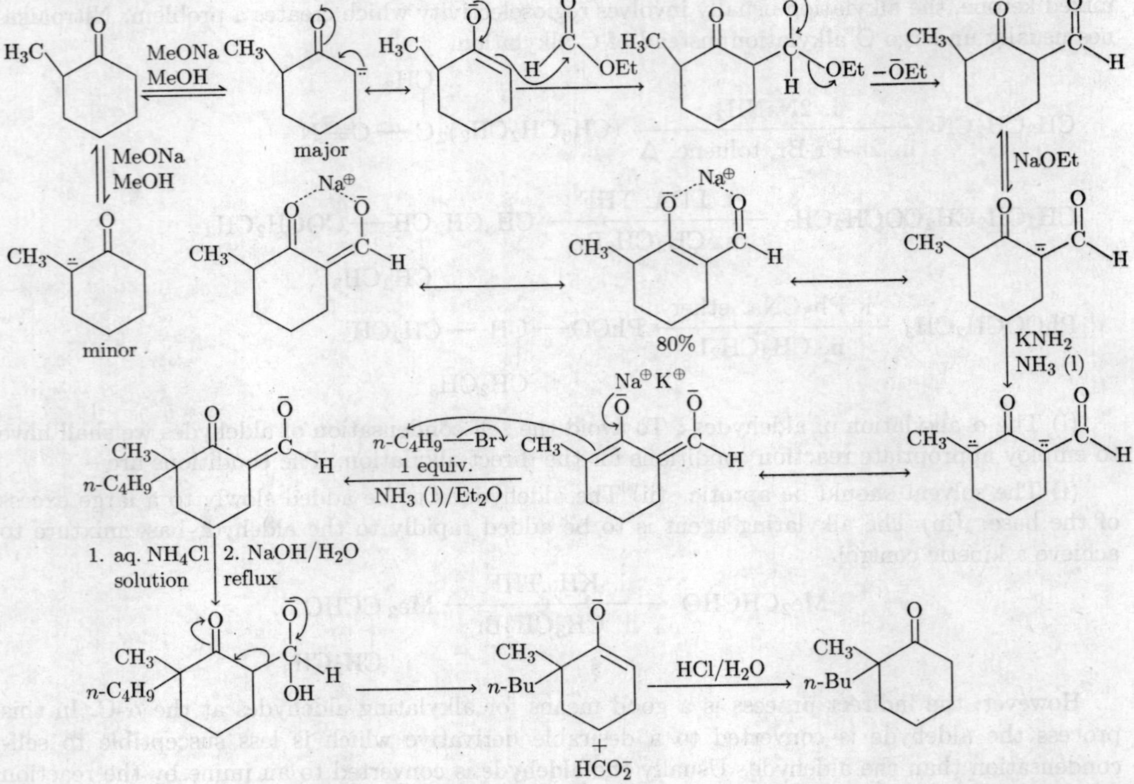

(ii) A β-keto ester may be used to alkylate either α- or the γ-C and subsequently the ketonic hydrolysis will give rise to the regioselective product.

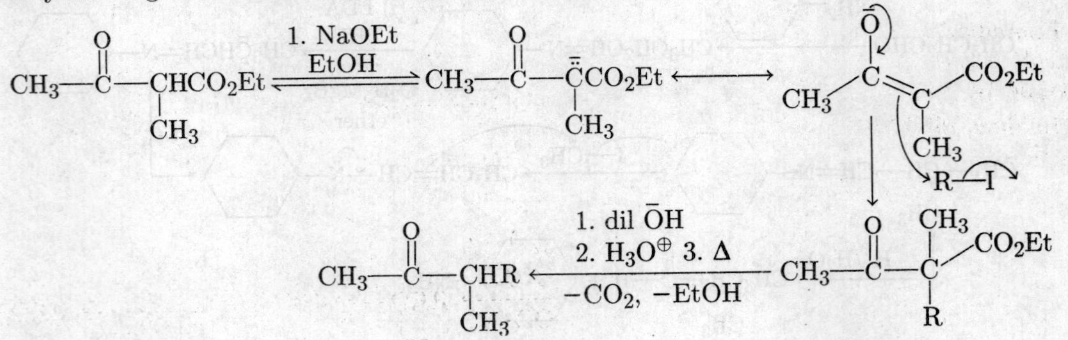

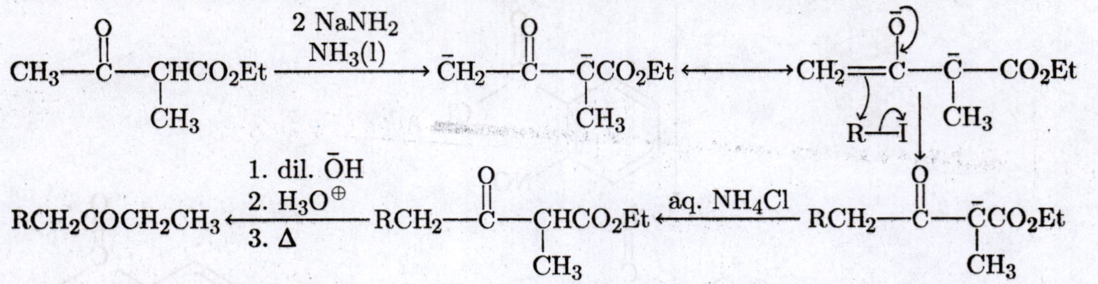

(iii) When an $\alpha$, $\beta$-unsaturated ketone is reduced with a dissolving metal-solvent system, a specific enolate is generated; this enolate undergoes the regioselective alkylation.

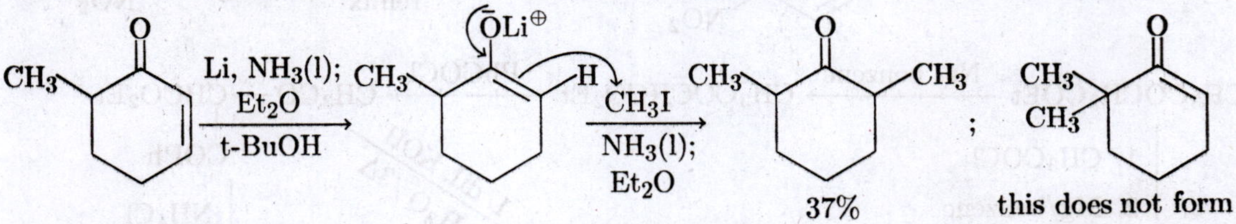

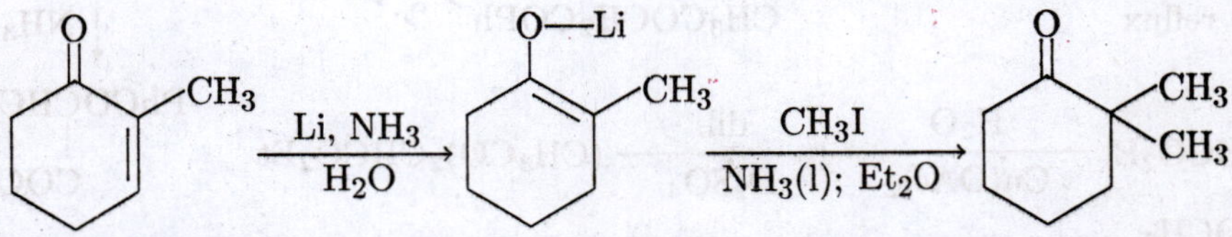

## 21.9.2 Acylation of the active methylene compounds

The acylation of active methylene compounds involves two problems :

(i) Protic solvents like alcohols cannot be used for the reaction since the acylating agent will acylate the solvent. This problem can be solved by replacing the protic solvent by an aprotic one. In this case sodium hydride in benzene (or toluene, etc.) may be used as a base.

(ii) The acylated derivative is more reactive towards a base than the substrate since the former has three electron-withdrawing groups but the latter has two. As a consequence side reactions will compete.

To avoid both the problems magnesium ethoxide in ether may be used as the base. After acylation, a ketone may be obtained by hydrolysis under an acidic condition. $\beta$-keto esters may also be acylated as shown below :

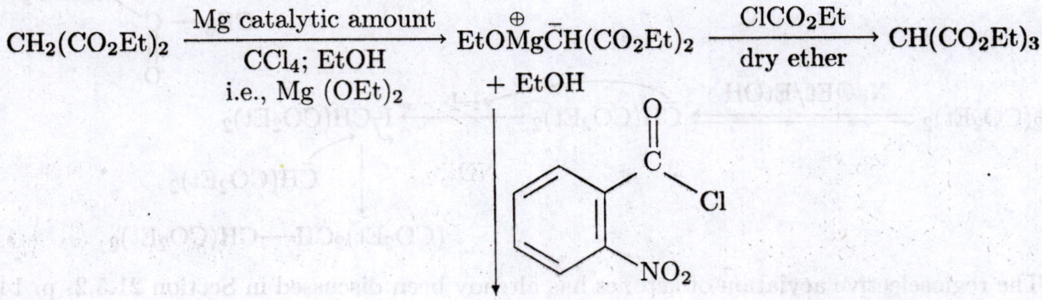

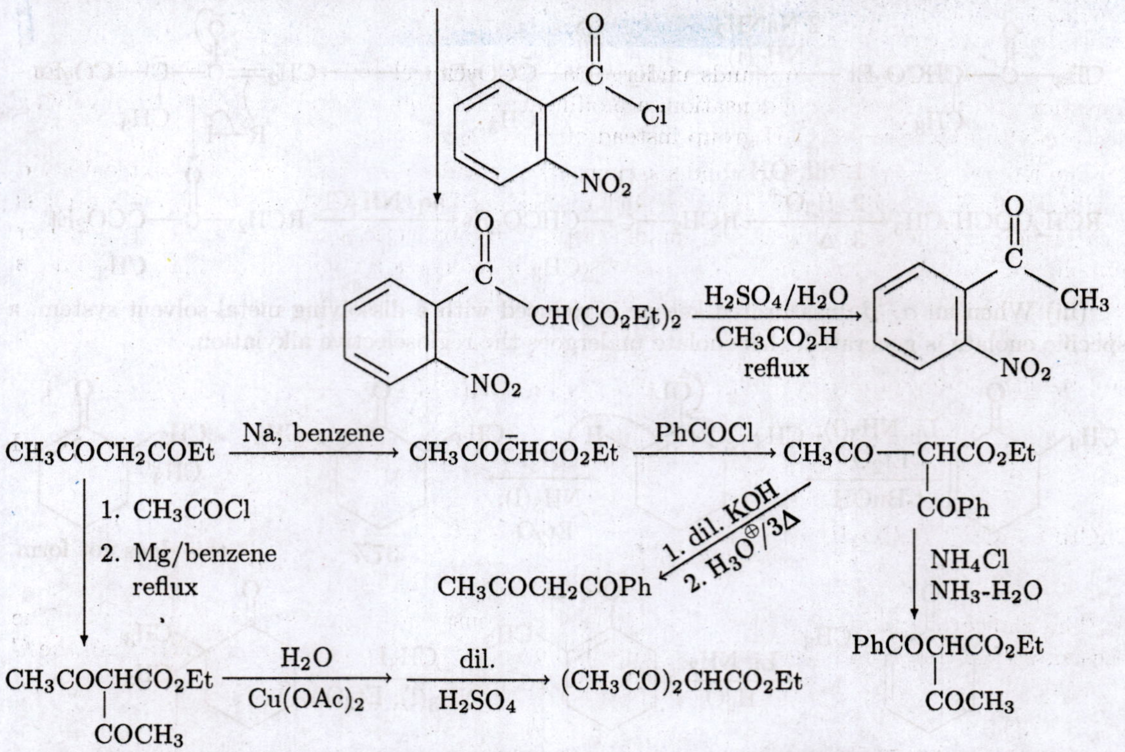


## 21.9.3 Two-unit Coupling

Two units of an active methylene compound can undergo coupling by the following procedure :

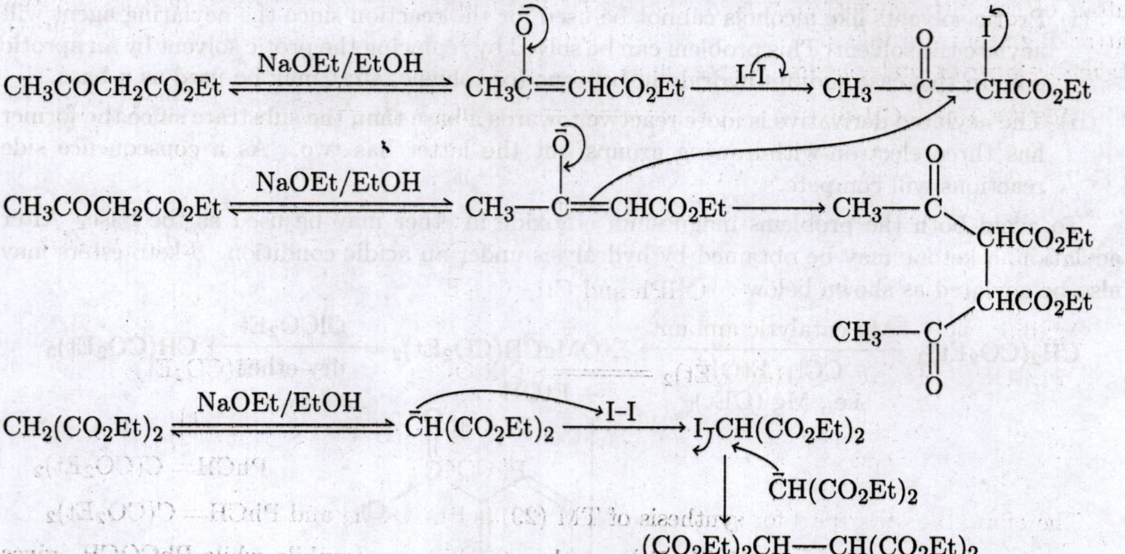


The regioselective acylation of ketones has already been discussed in Section 21.5.2. p. 1437.

## 21.9.4 Condensation reactions

The active methylene compounds undergo (a) Knoevenagel condensation, (b) Michael condensation and (c) Doebner condensation, a modification of Knoevenagel condensation, involving electron-withdrawing —$CO_2H$ group instead of —$CO_2R$ group.

The Knoevenagel and Michael condensations have already been discussed in the sixteenth and thirteenth chapters respectively. Here we shall discuss the Doebner condensation. In this process a carbanion is generated either from malonic acid or cyanoacetic acid using either pyridine or quinoline as a solvent which acts as a base also. In some cases a small amount of piperidine is also used as a strong base.

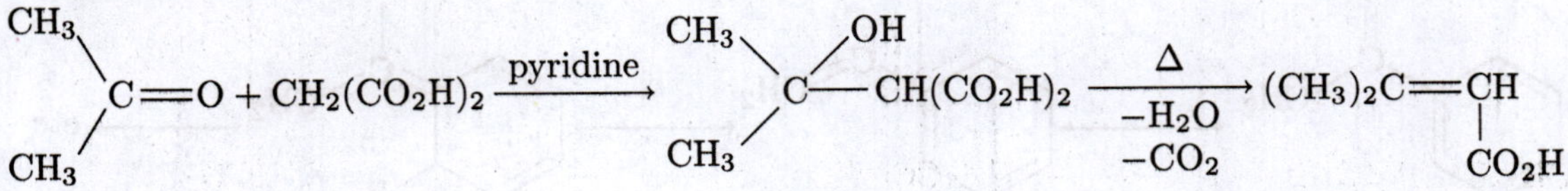

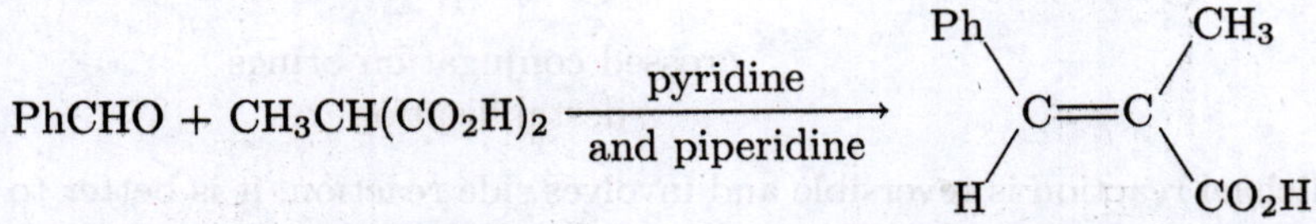

The choice of interacting pairs in a Michael condensation deserves special mention. The substrate of Michael condensation is a conjugated system. $\beta$-carbon is a softer centre than the C

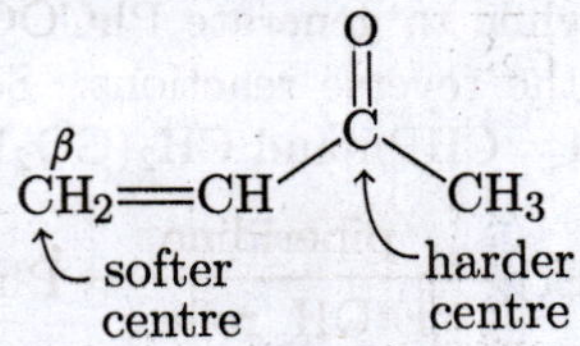

atom of the electron-withdrawing group as indicated above. A weaker base having a softer attacking atom will take part in the Michael addition. Let us see which of the two pairs will be taken as starting materials for the synthesis of TM(29).

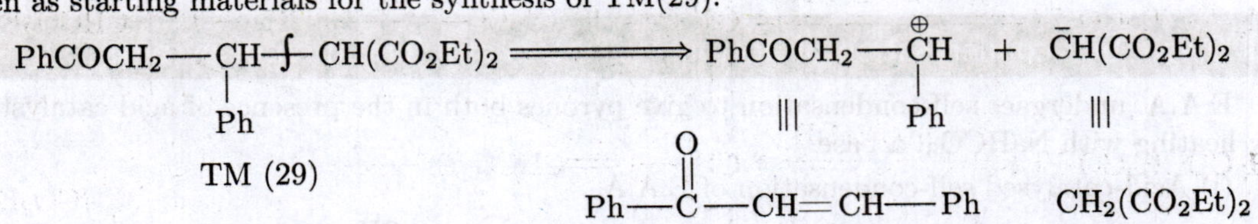

So, one pair will be PhCOCH=CHPh and $CH_2(CO_2Et)_2$

Again,

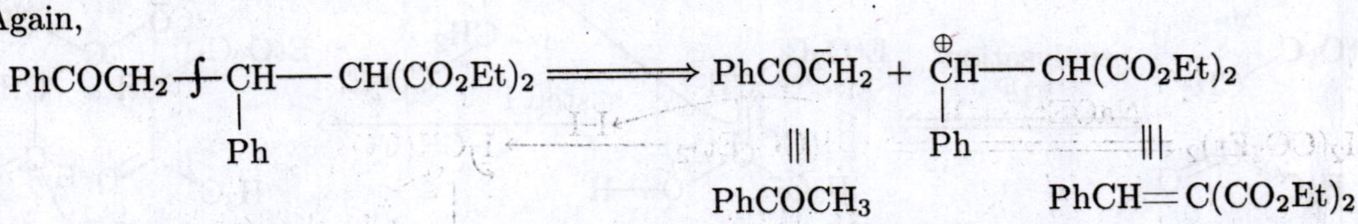

Therefore, the other pair for synthesis of TM (29) is $PhCOCH_3$ and $PhCH{=}C(CO_2Et)_2$

Now, $CH_2(CO_2Et)_2$ gives $\bar{C}H(CO_2Et)_2$ as the attacking nucleophile while $PhCOCH_3$ gives $PhCO\bar{C}H_2$. Between these two $\bar{C}H(CO_2Et)_2$ is a weaker nucleophile than $PhCO\bar{C}H_2$ because of the better resonance stabilisation of the negative charge of the former than that of the latter as shown below :

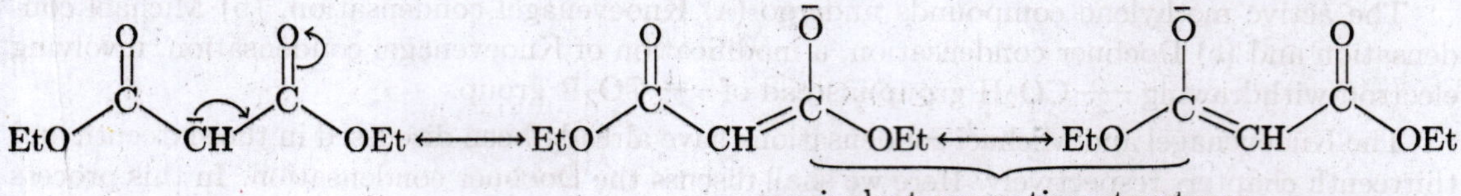

equivalent resonating structures bring extra stability

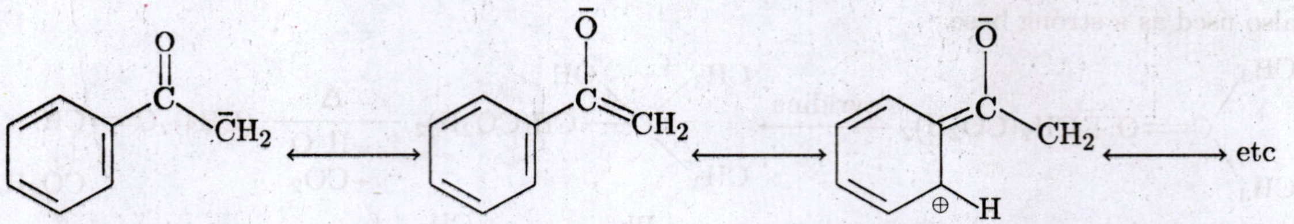

crossed conjugation brings destabilisation

Furthermore, since the Michael reaction is reversible and involves side reaction, it is better to avoid strong base to generate a weak nucleophile. The acidity of $CH_2(CO_2Et)_2$ (p*ka* 9) is much higher than that of $PhCOCH_3$ (p*Ka* 19). So to generate $\bar{C}H(CO_2Et)_2$, mild base like piperidine, pyridine, KOH, etc. may be used while to generate $PhCO\bar{C}H_2$ one needs a strong base which will help side reactions as well as the reverse reactions. So for the synthesis of TM (29) the appropriate choice will be $PhCOCH{=}CHPh$ and $CH_2(CO_2Et)_2$.

$$PhCOCH{=}CHPh + CH_2(CO_2Et)_2 \xrightarrow[\text{EtOH, reflux}]{\text{piperidine}} PhCOCH_2\underset{\substack{|\\ Ph}}{CH}{-}CH(CO_2Et)_2$$

TM (29)

## 21.9.4a Self-condernsation of acetoacetic esters and formation of pyrones

E.A.A. undergaes self-condensation to give pyrones both in the presence of acid catalyst and by heating with $NaHCO_3$, a base.

(i) Acid-catalysed self-condensation of E.A.A.

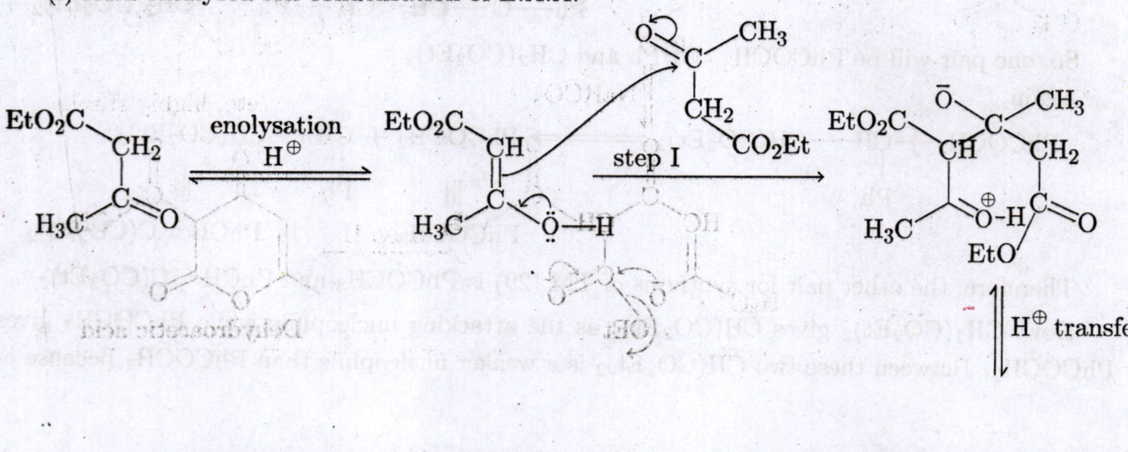

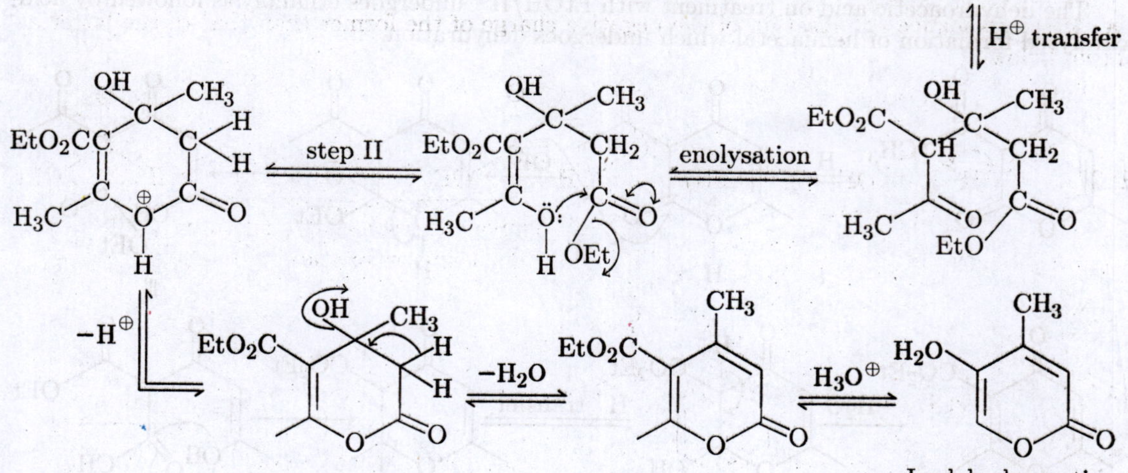

Isodehydroacetic acid

**Step I.** Soft-soft interaction to form the dimer. The carbonyl C is more reactive than ester C.

**Step II.** Internal lactone formation to give a stable ring structure by Claisen condensation.

(ii) Self condensation by heating with sodium bicarbonate at 200 °C

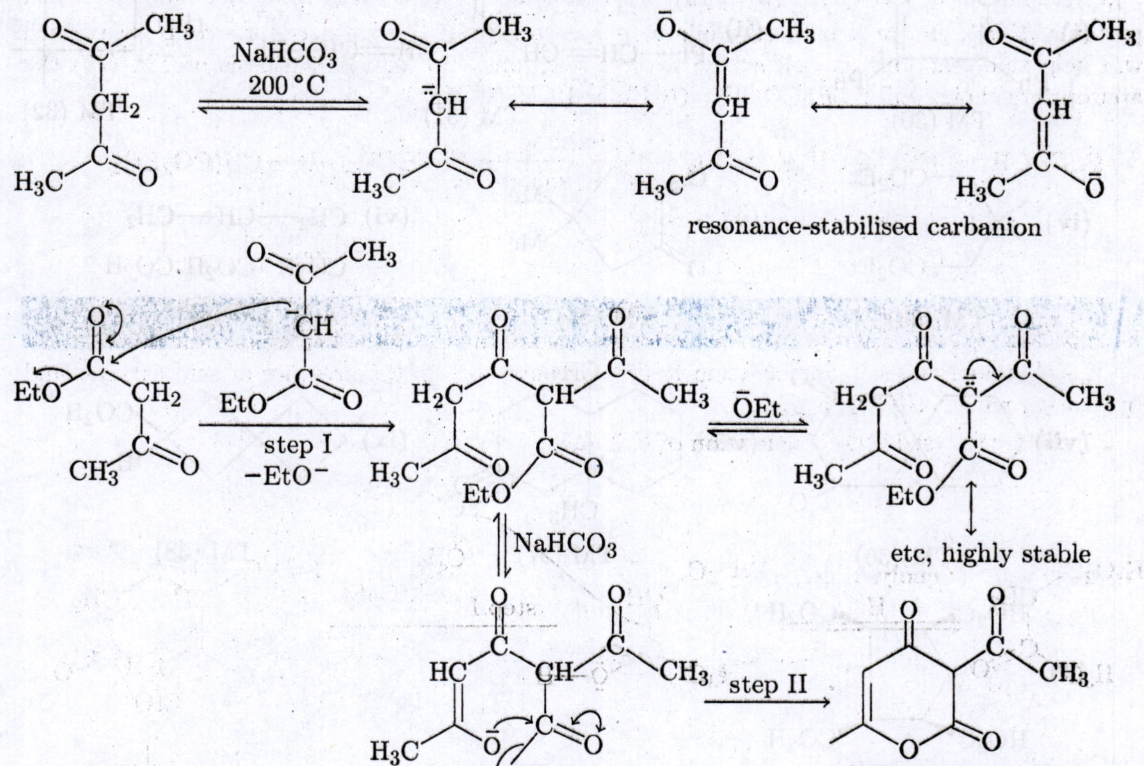

Dehydroacetic acid

The dehydroacetic acid on treatment with EtOH/H$^{\oplus}$ undergoes ethanolysis followed by acid-catalysed formation of hemiacetal which undergoes dehydration.

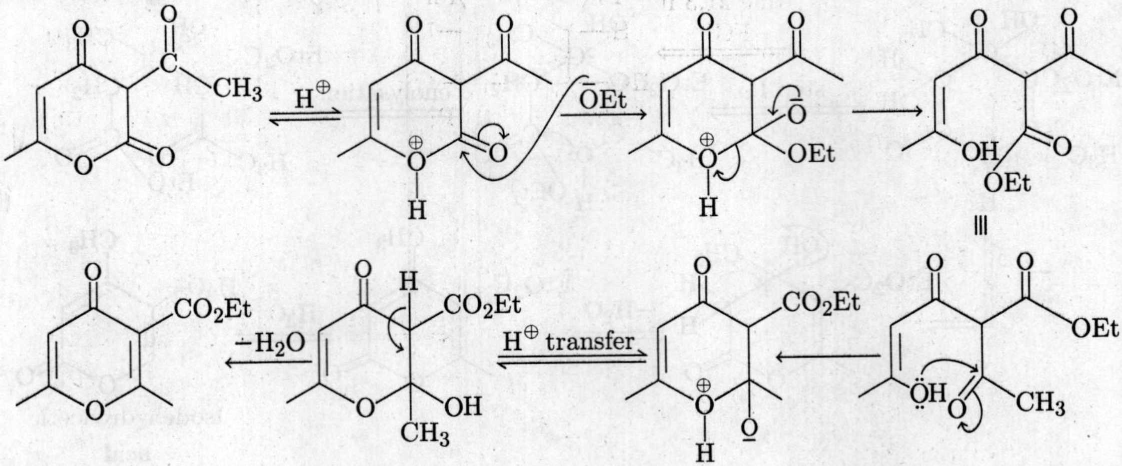

**Problem 21.4.** Working backward synthesise the following compounds :

(i) 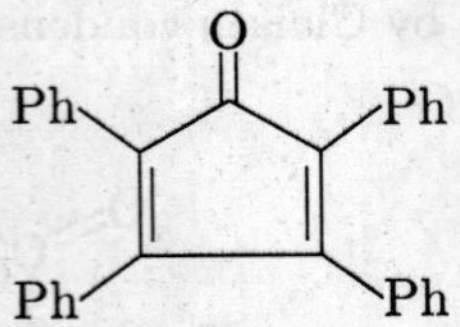

TM (30)

(ii) 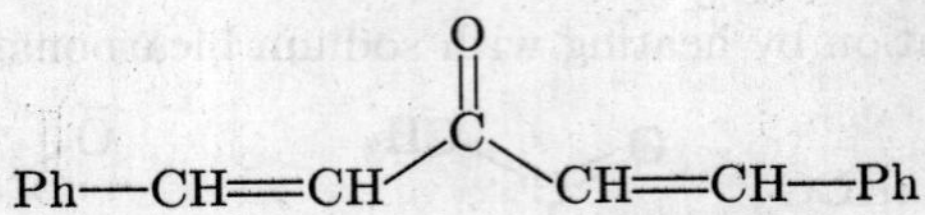

TM (31)

(iii) 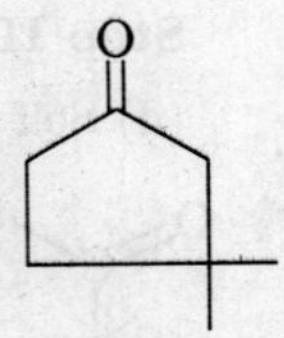

TM (32)

(iv) 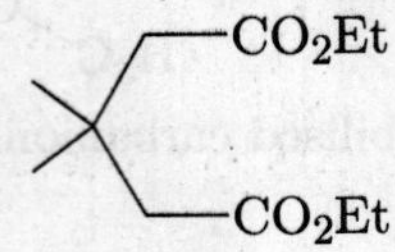

TM (33)

(v) 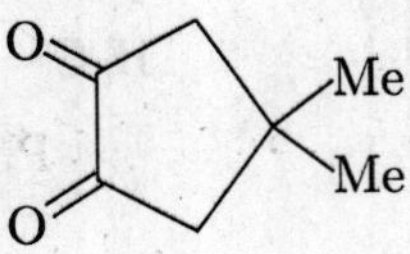

TM (34)

(vi) 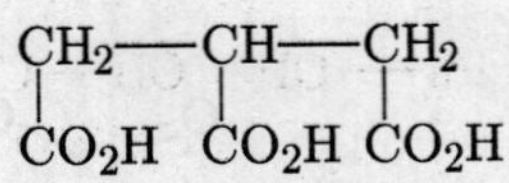

TM (35)

(vii) 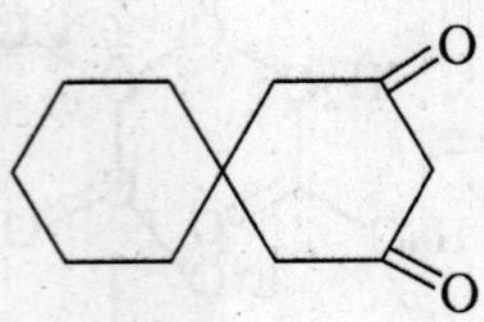

TM (36)

(viii) 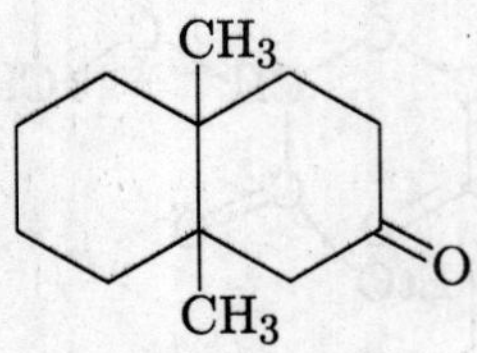

TM (37)

(ix) 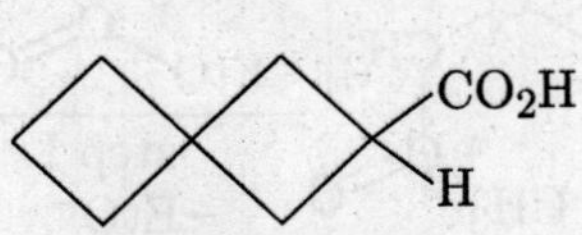

TM (38)

(x) 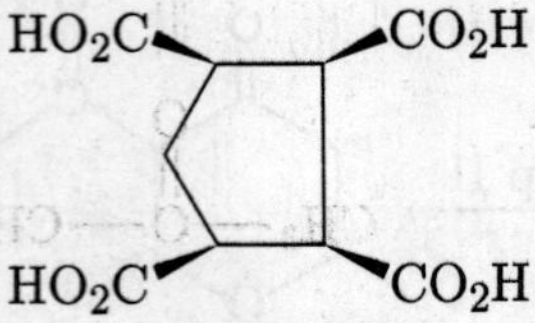

TM (39)

**Solution :**

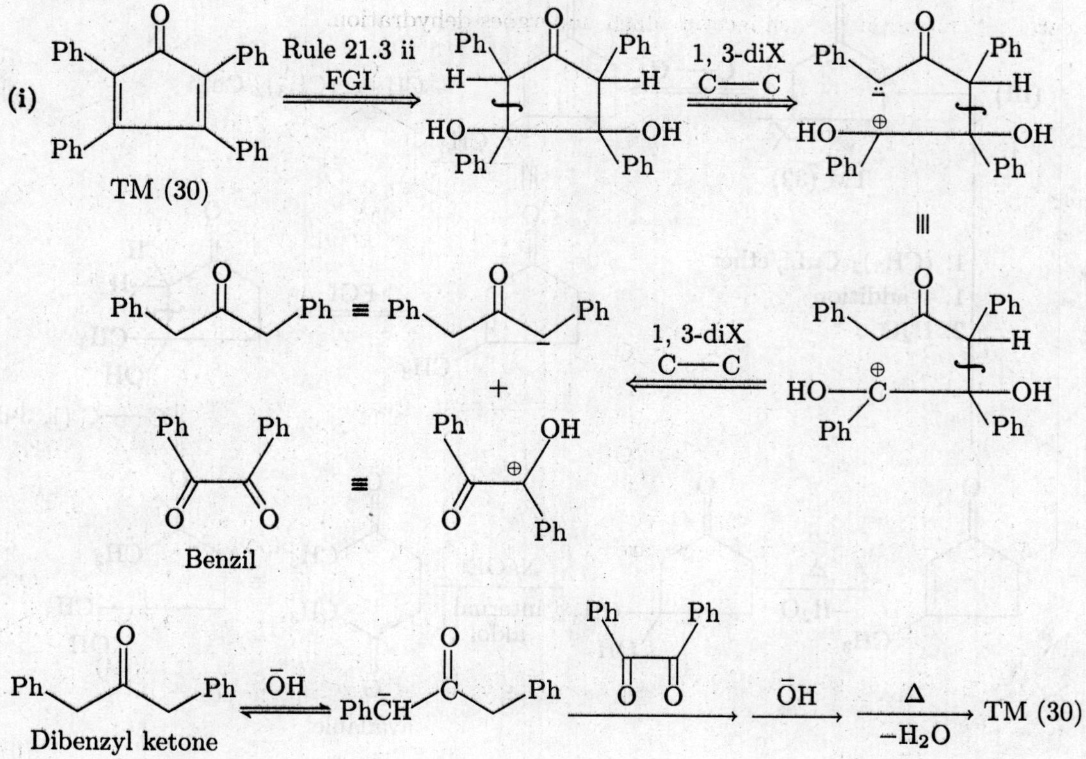

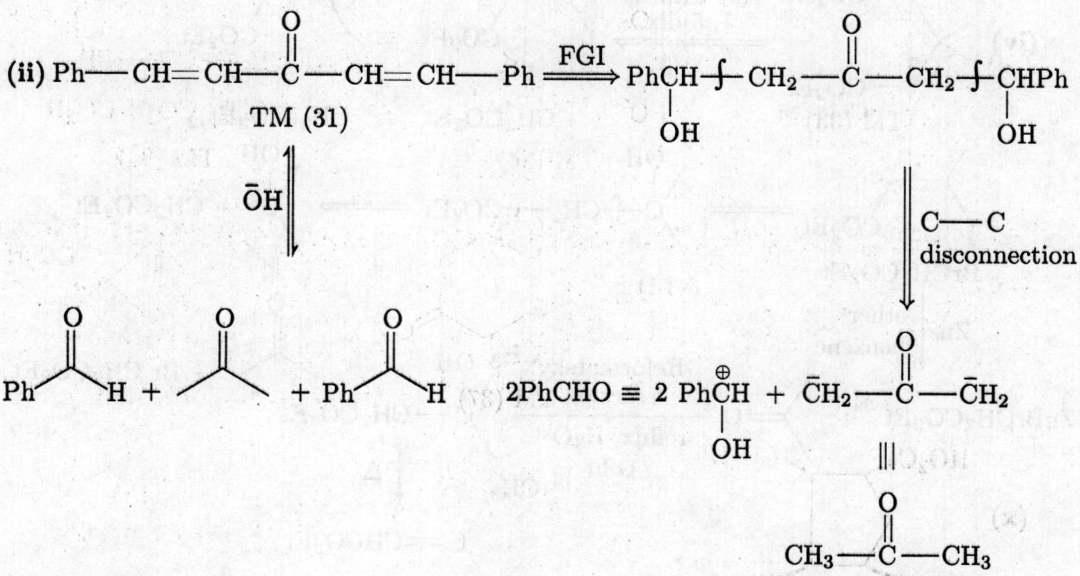

**(iii)**

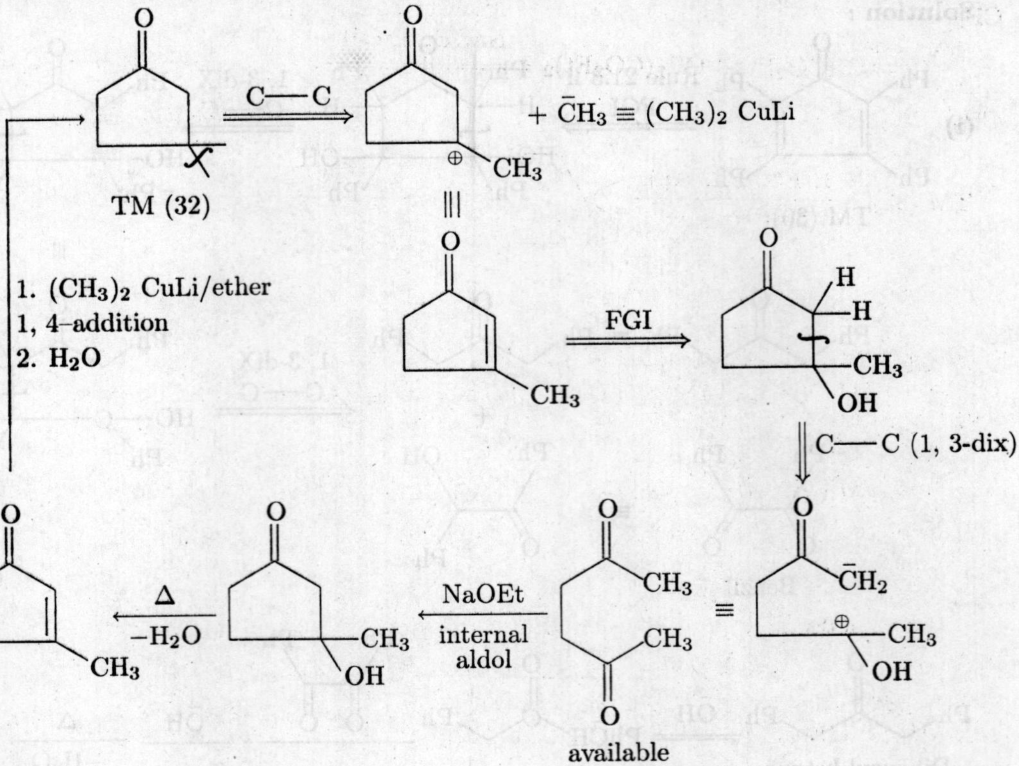
TM (32)
C—C
+ $\bar{C}H_3 \equiv (CH_3)_2$ CuLi
1. $(CH_3)_2$ CuLi/ether
1, 4–addition
2. $H_2O$
FGI
C—C (1, 3-dix)
Δ
$-H_2O$
NaOEt
internal
aldol
available

**(iv)**

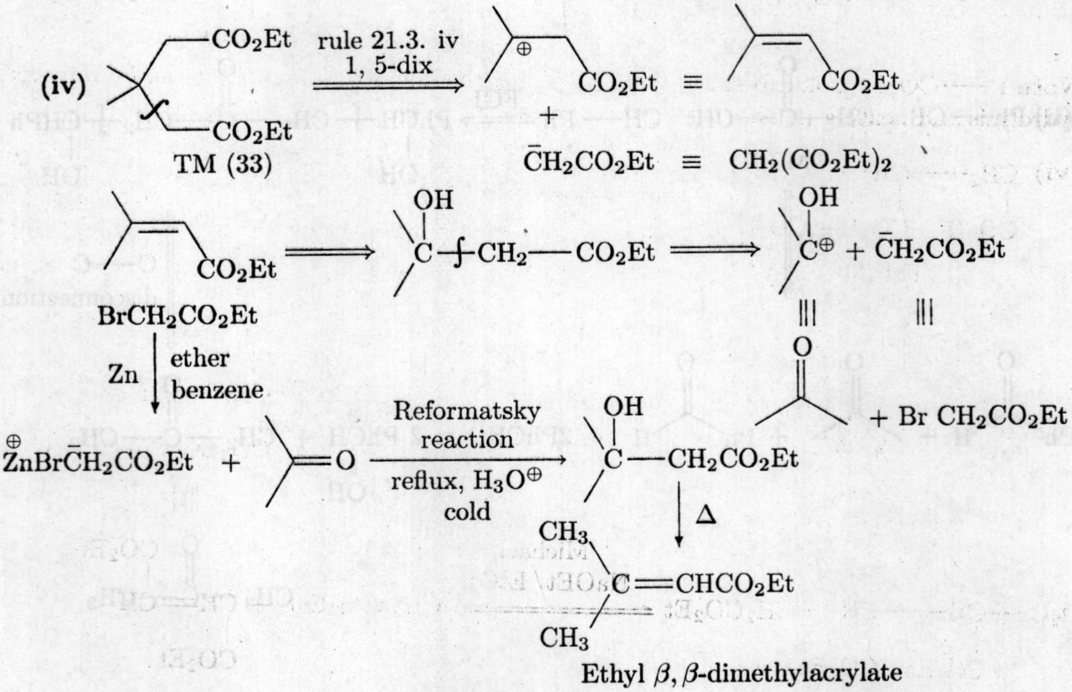
TM (33)
rule 21.3. iv
1, 5-dix
$CO_2Et$
+
$\bar{C}H_2CO_2Et \equiv CH_2(CO_2Et)_2$
$BrCH_2CO_2Et$
Zn
ether
benzene
$\overset{\oplus}{Zn}Br\bar{C}H_2CO_2Et$ +
Reformatsky
reaction
reflux, $H_3O^{\oplus}$
cold
$C—CH_2CO_2Et$
+ Br $CH_2CO_2Et$
Δ
$CH_3$
$C=CHCO_2Et$
$CH_3$
Ethyl $\beta,\beta$-dimethylacrylate

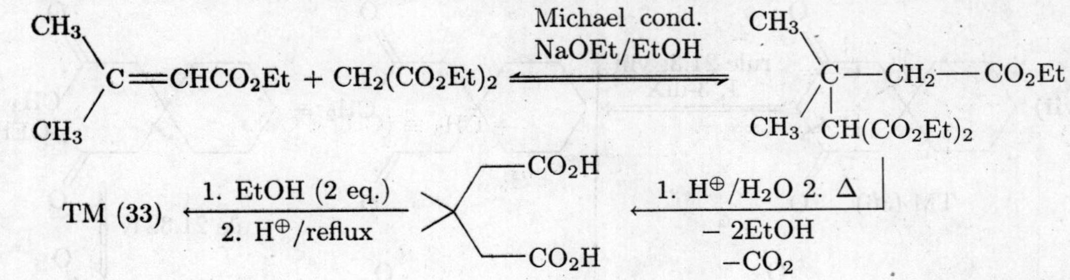


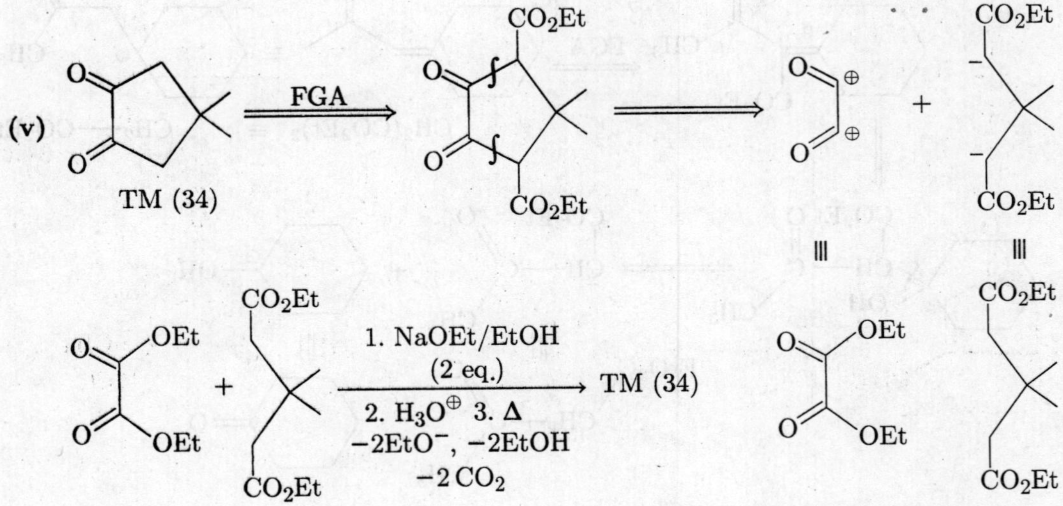


Ethyloxalate

**Note :** —$CO_2Et$ is added since it will help in carbanion formation and it can be removed by hydrolysis and heating (it is being a $\beta$-keto ester).

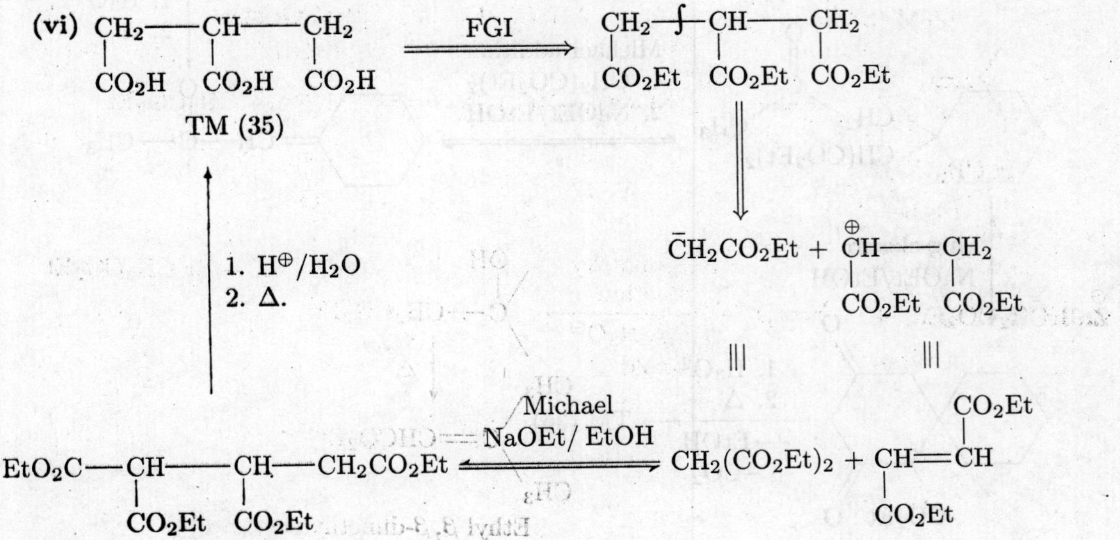

(vii)

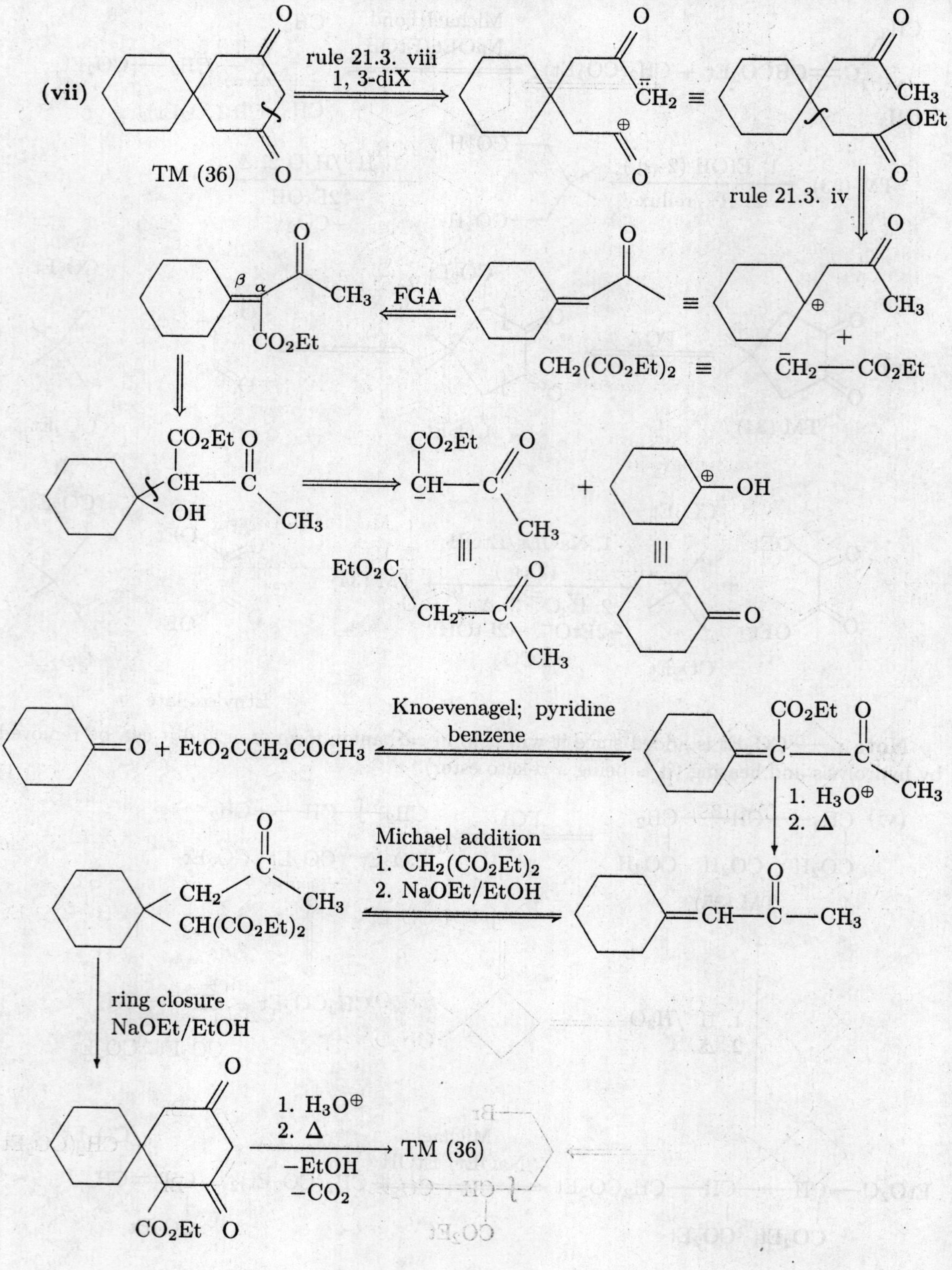
rule 21.3. viii
1, 3-diX
TM (36)
rule 21.3. iv
FGA
CH2(CO2Et)2 ≡
Knoevenagel; pyridine
benzene
O + EtO2CCH2COCH3
1. H3O⊕
2. Δ
Michael addition
1. CH2(CO2Et)2
2. NaOEt/EtOH
ring closure
NaOEt/EtOH
1. H3O⊕
2. Δ
−EtOH
−CO2
TM (36)

**(viii)**

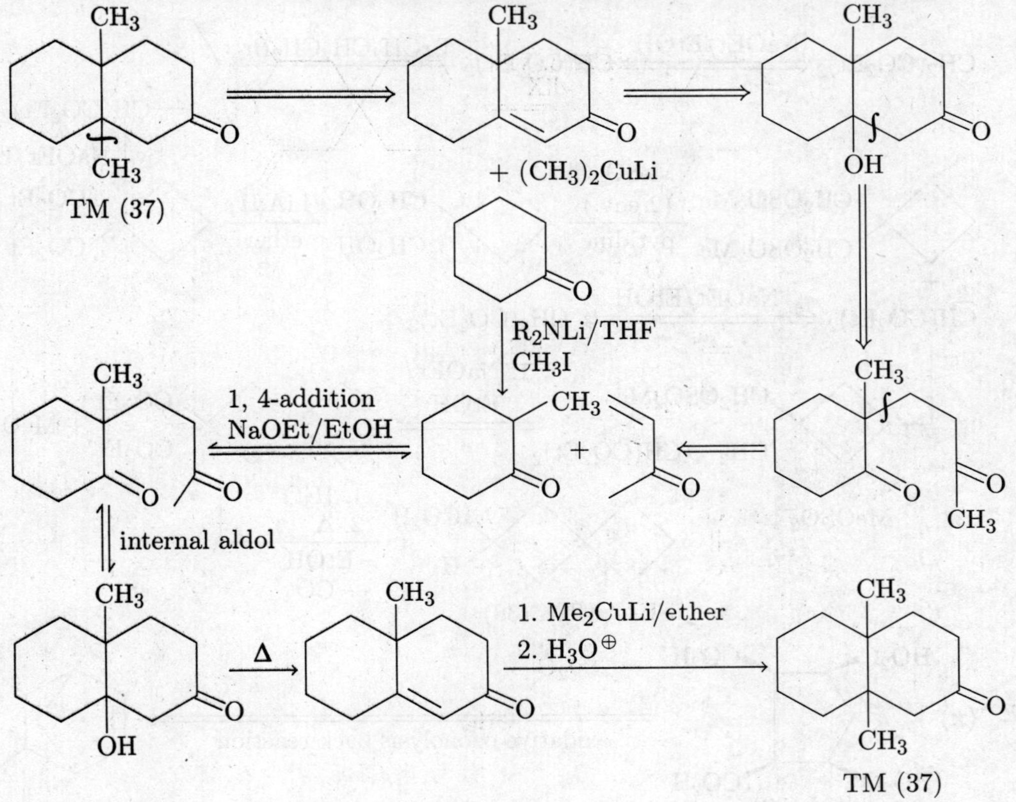
CH3
TM (37)
+ (CH3)2CuLi
OH
R2NLi/THF
CH3I
1, 4-addition
NaOEt/EtOH
internal aldol
Δ
1. Me2CuLi/ether
2. H3O⊕
TM (37)

**(ix)**

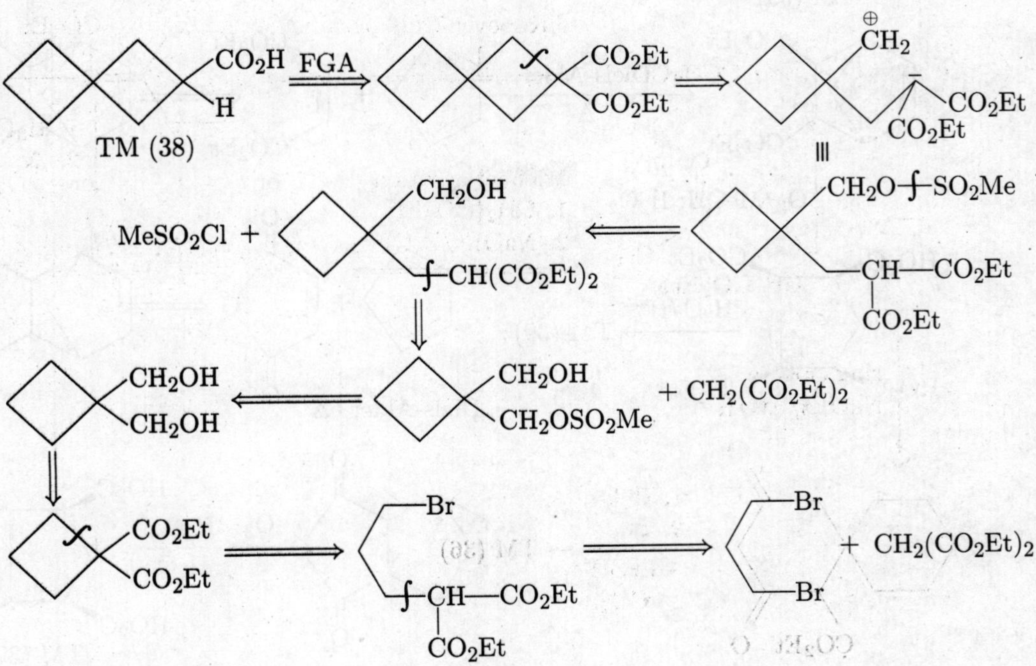
CO2H
H
TM (38)
FGA
CO2Et
CO2Et
⊕CH2
MeSO2Cl +
CH2OH
CH(CO2Et)2
CH2O—SO2Me
CH—CO2Et
CH2OSO2Me
+ CH2(CO2Et)2
Br
Br

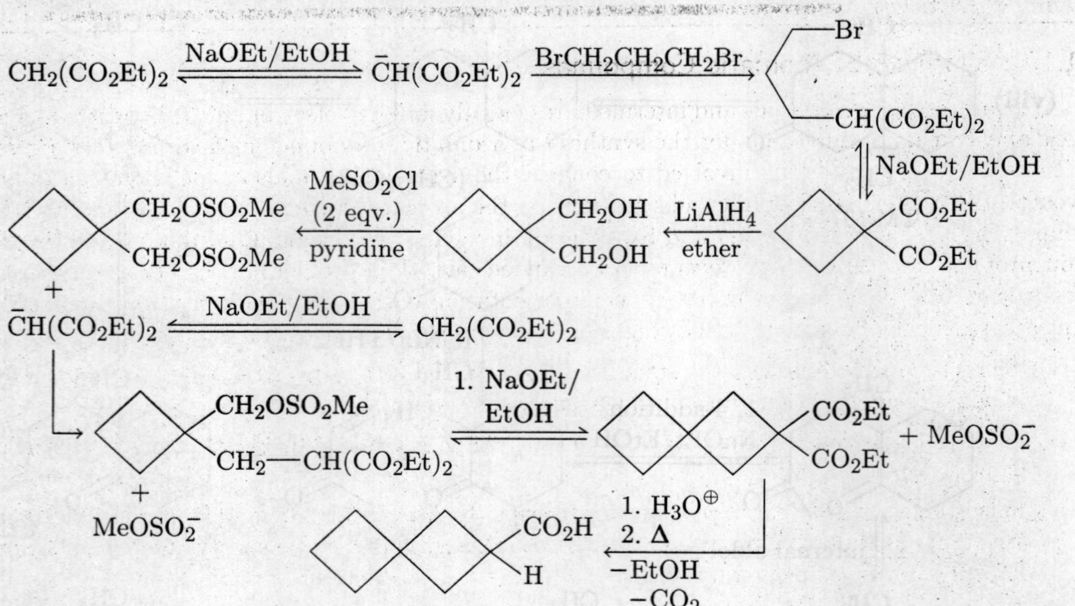

TM (38)

(x)

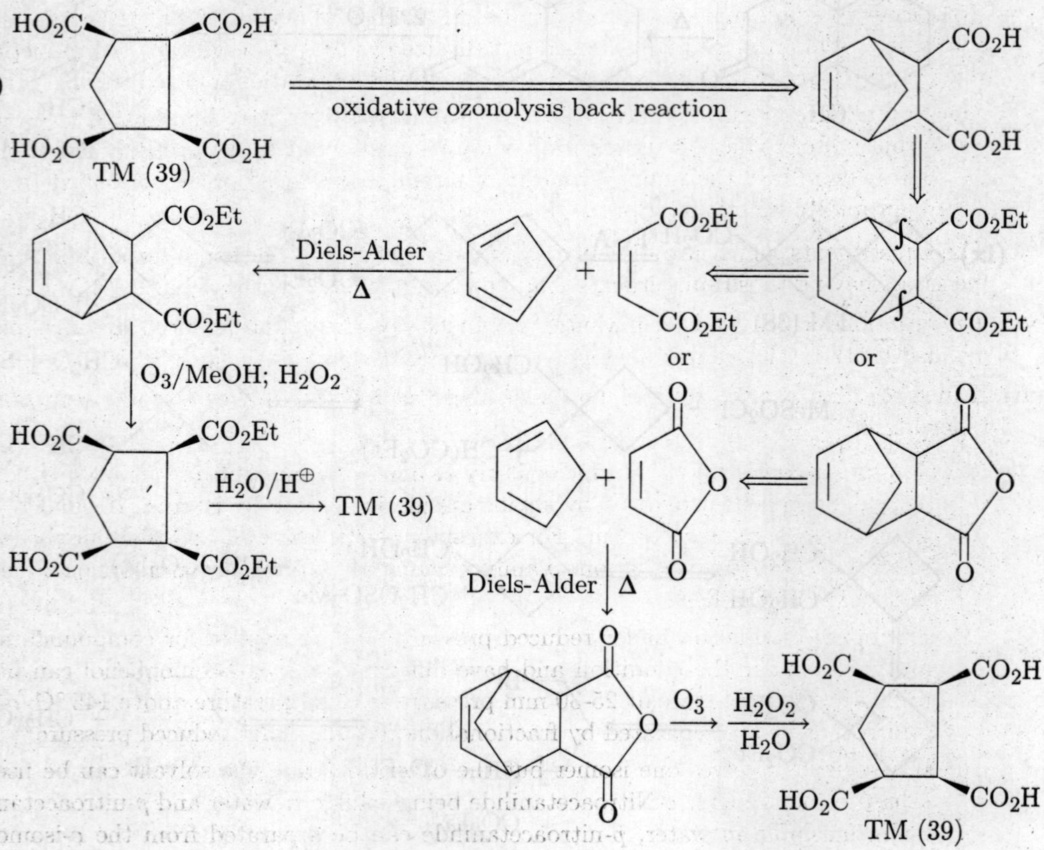

TM (39)

## 21.10 Synthesis of Aromatic Compounds

Aromatic starting materials and intermediates usually undergo electrophilic substitution. Synthons and synthetic equivalents for the synthesis of aromatic compounds have already been given in Table 21.2. The reactions involved to connect the synthons have also been discussed in the seventeenth chapter. Nucleophilic substitutions on the aromatic ring can be done by adding $NH_2$ group, carrying out diazotisation and $S_N$ reaction involving the diazonium group. However, the main problem is to arrange the events to be carried out. The problem arises because different substituents in an aromatic ring have specific orientation effects ($o/p$ and $m$). However for the right arrangement of events the following guidline may be followed with a flexible net work.

**(i)** The relationship amongst the substituents on the TM are to be ascertained to find the FG which directs the other to the right position.

**(ii)** The most electron-withdrawing substituent, which is to be added last during synthesis, is to be disconnected first.

**(iii)** Since there are *o*-/*p*- and *m*- directing groups, to get proper orientation of incoming substitutents FGI may be required. In this case new groups may be substituted either before FGI or after it, as per requirement.

**(iv)** To get substituents in right orientation :

**(a)** A nitro group may be introduced before the introduction of substituents and the nitro group may then be removed or converted to other functions by the reduction to —$NH_2$ group and latter may be removed by diazotisation and reduction.

**(b)** A nitro group may be introduced and the nitro group may be reduced to —$NH_2$ group; after reduction substitutions may be carried out to get properly oriented substituents. Next the amino group may either be removed or be converted to other appropriate functional group.

**(v)** The substituents which are difficult to introduce may not be disconnected and starting materials having those substituents may be chosen for the synthesis.

**(vi)** The reactions which lead to unwanted products along with the required product may be avoided if other reliable unambiguous reactions are available.

**(vii)** If in a reaction a mixture of *o*- and *p*- isomers are obtained, they may be separated by following processes.

**(a)** A mixture consisting of intermolecularly H-bonded compound and internal H-bonded compound can be separated by steam distillation where by internal H-bonded compound passes out with steam. For example, *o*-aminophenol, and *p*-aminophenol are separated by steam distillation. *o*-aminophenol distils out and *p*-aminophenol remains in the distillation flask.

**(b)** Fractional distillation under reduced pressure may be applied for compounds which are liquid under the condition and have different b.p.s. *p*-Bromophenol can be obtained in the pure state at 25-30 mm pressure at a temperature above 145 °C; *o*- and *p*-nitrotoluene are separated by fractional distillation under reduced pressure.

**(c)** If a solvent dissolves one isomer but the other does not, the solvent can be used to separate the isomers. *o*-Nitroacetanilide being soluble in water and *p*-nitroacetanilide being insoluble in water, *p*-nitroacetanilide can be separated from the *o*-isomer by dissolving the latter and filtering it down.

(d) Fractional crystallisation is another process for separating a mixture of *o*- and *p*-isomers when their solubilities are appreciably different at a particular temperature in a solvent.

(e) Nowadays chromatographic separation is used successfully. For example, syn and anit-azobenzenes can be separated by thin layer chromatography.

However, to understand the order of events in an aromatic synthesis a few illustrations are given below :

**Example 1.** Synthesis of 2, 6-dihydroxynitrobenzene [TM(40)].

In TM (40) nitro group is at ortho position with respect to two OH groups. Nitro group is a *m*-directing group. Hence the two OH groups cannot be introduced to nitrobenzene easily. On the other hand, OH groups are *o*/*p* directing groups. We know that $NO_2$ group can be introduced to resorcinol. However, the position of nitro group in the TM is a sterically congested one. The ortho position with respect to one OH group is also the para position with respect to the other OH group. They increase electron density at their *o*-/*p*- positions. So, these positions are more electron dense because of +R effect of the OH groups. Hence dilute nitric acid treatment of resorcinol will give 2, 4-dihydroxynitrobenzene in large excess over 2, 6-isomer. To avoid the formation of 2, 4-isomer, 2 and 4 positions of resorcinol are to be blocked, the nitration on the blocked molecule is to be carried out and then the molecule is to be deblocked. $SO_3H$ group can readily be introduced and removed. So, sulphonation reaction can be carried out on resorcinol to block 2 and 4 positions.

According to the guideline (ii) the most electron-withdrawing nitro group is to be disconnected. Before this, FGA of —$SO_3H$ is to be carried out as shown below :

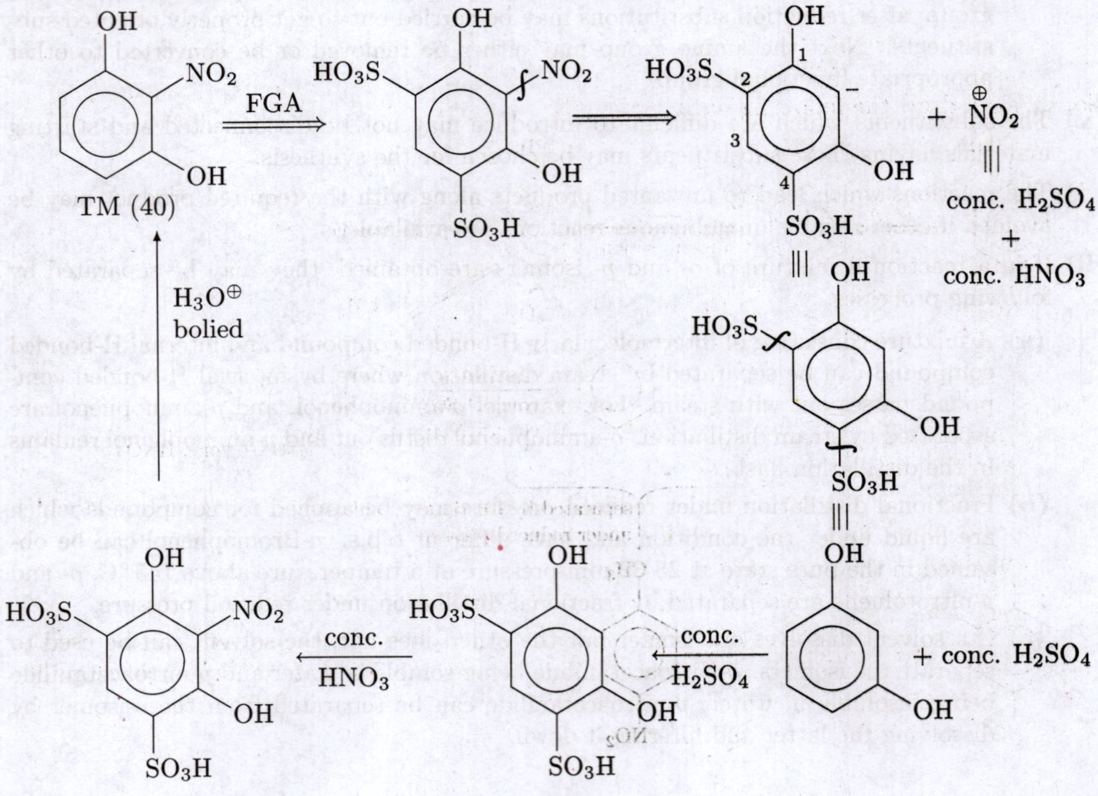

**Example 2.** Synthesis of 2-bromo-4-nitrobenzoic acid 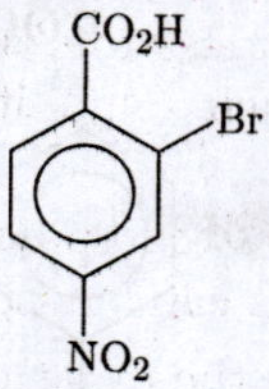 TM (41)

$CO_2H$ and $NO_2$ groups are *m*-directing but they are present in the TM at the *p*-position with respect to each other. Bromo group is a *o*-/*p*-director, but $NO_2$ group is meta to Br and $CO_2H$ is ortho to Br. So we cannot start either from benzoic acid or from bromobenzene or from nitrobenzene. If we convert $CO_2H$ to $CH_3$, the latter has the required directing ability. So, FGI is to be carried out first. Now which FG is to be disconnected first, Br or $NO_2$? Since FGI has to be carried out first, the last step of the synthesis will be the oxidation of $CH_3$ to $CO_2H$. Now if we disconnect Br first, it will be introduced at the last but one step. It will not be a problem since $CH_3$ is a *o*-/*p*-directing group and $NO_2$ is *m*-directing. The combined effect will give only one bromo derivative. But if $NO_2$ group is disconnected first, it will be introduced at the penultimate step. Bromo group is to be introduced before introducing the $NO_2$ group. In that case nitration of *o*-bromotoluene is to be carried out and we shall get a mixture of products since ($CH_3$ and Br) *o*-/*p*-directing ability. The separation of TM (41) from the mixture is a difficult task.

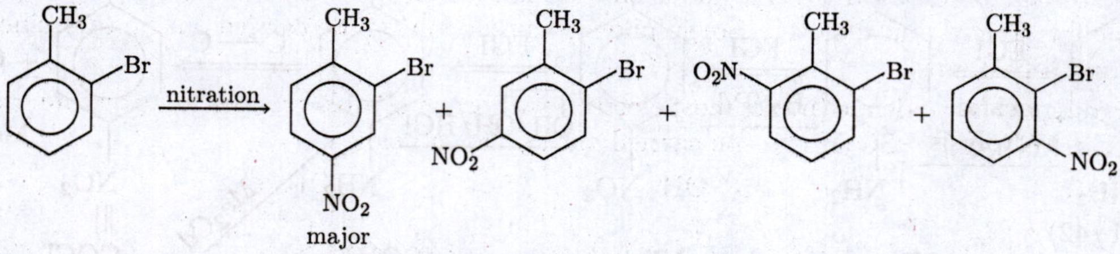


So, we shall draw the following backward steps.

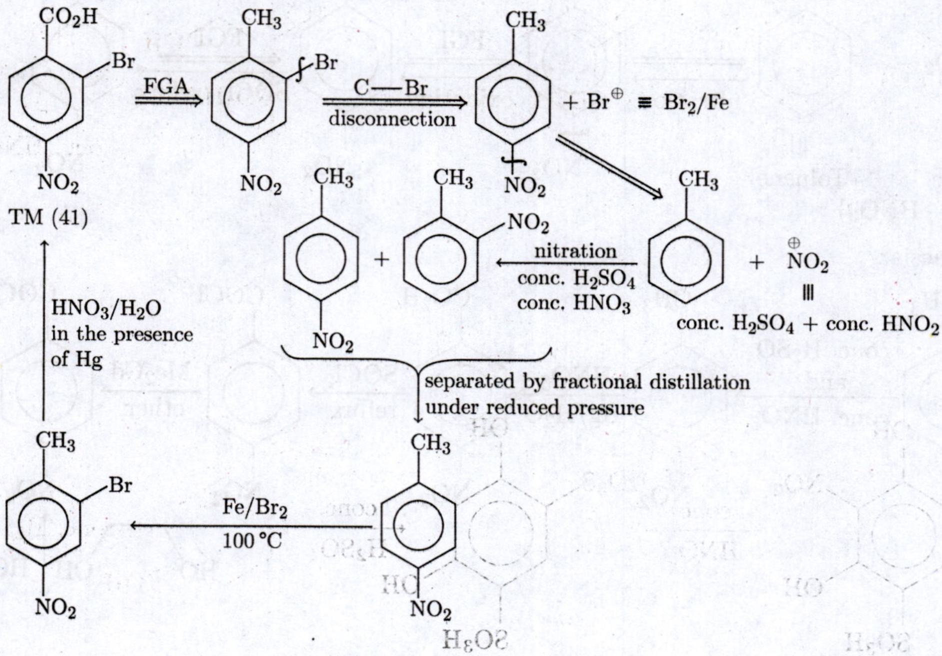

**Example 3.** Synthesis of 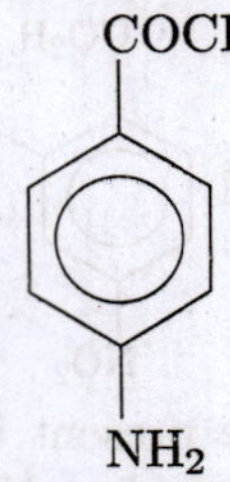 TM (42)

$COCH_3$ group is *m*-director but $NH_2$ group is *o-p* director. Introduction to benzene ring requires nitration followed by reduction. If these processes are carried out sequentially, nuclear acetylation will not be possible on aniline. acetylation not also be done on nitrobenzene. Nitration of acetophenone does not also give the right orientation of groups. Taking all these facts together the backward and forward steps are given below :

Analysis and synthesis together :

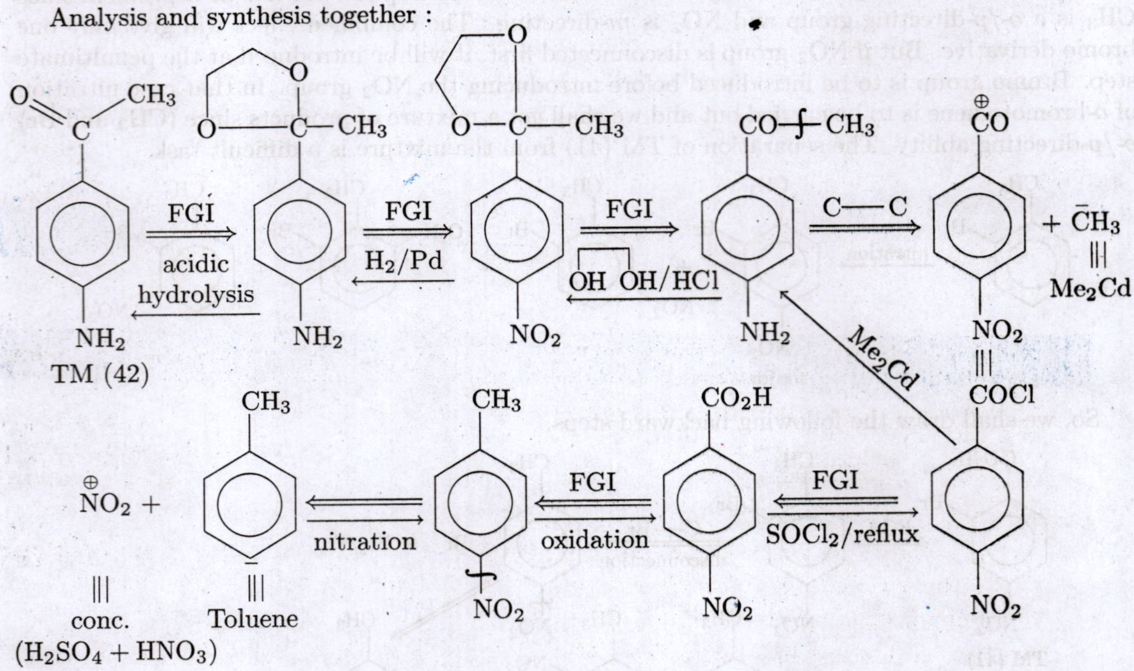


Synthesis :

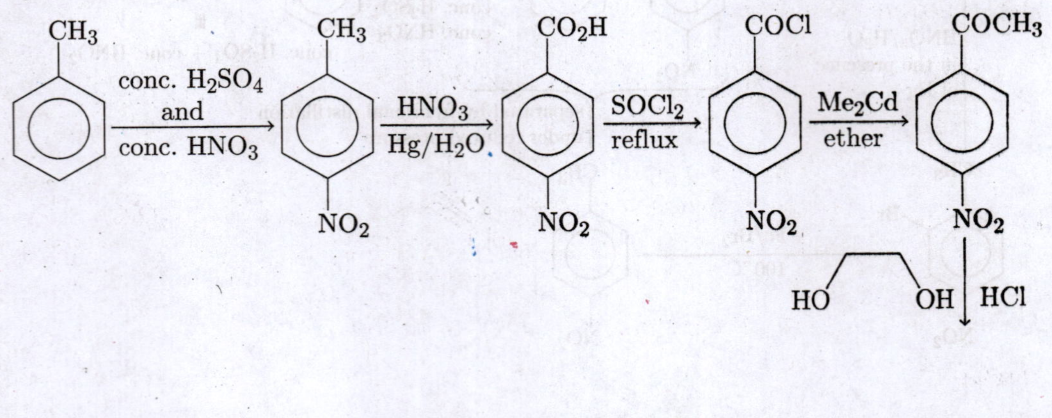

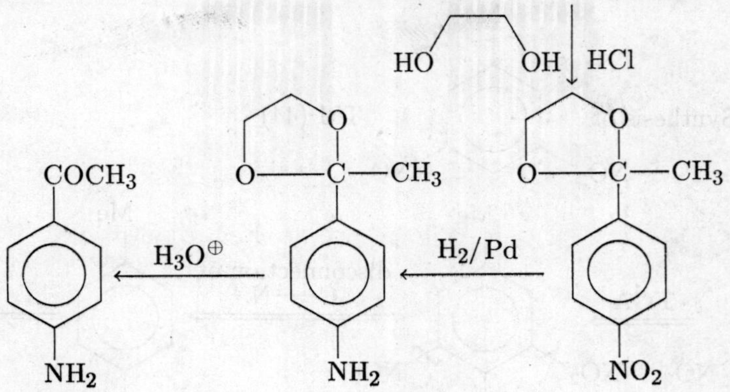

TM (42)

**Example 4.** Synthesis of 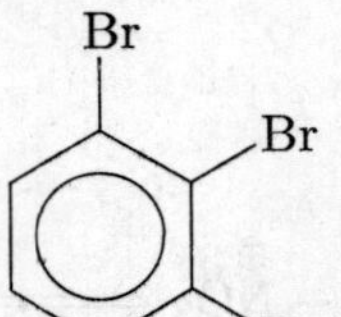 TM (43)

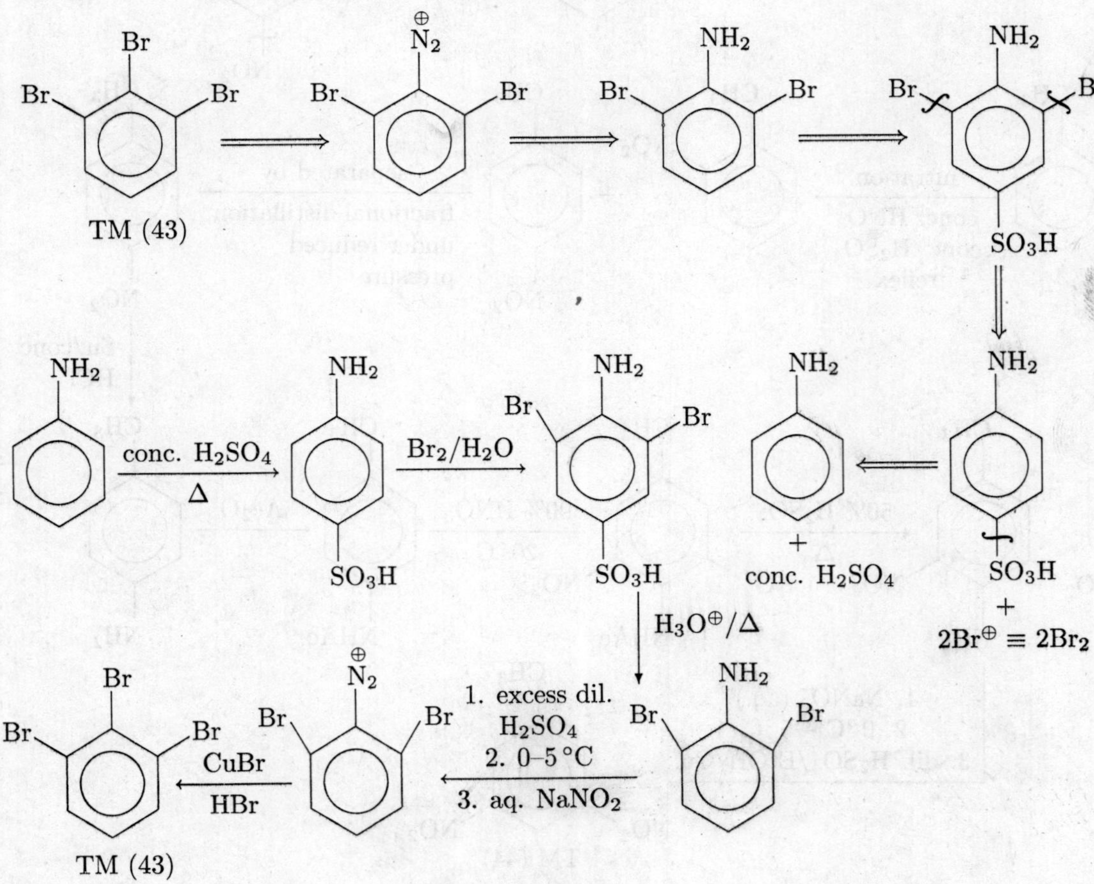

**Example 5.** Synthesis of 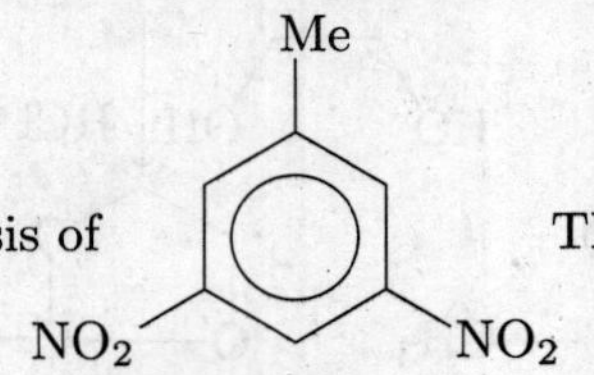 TM (44)

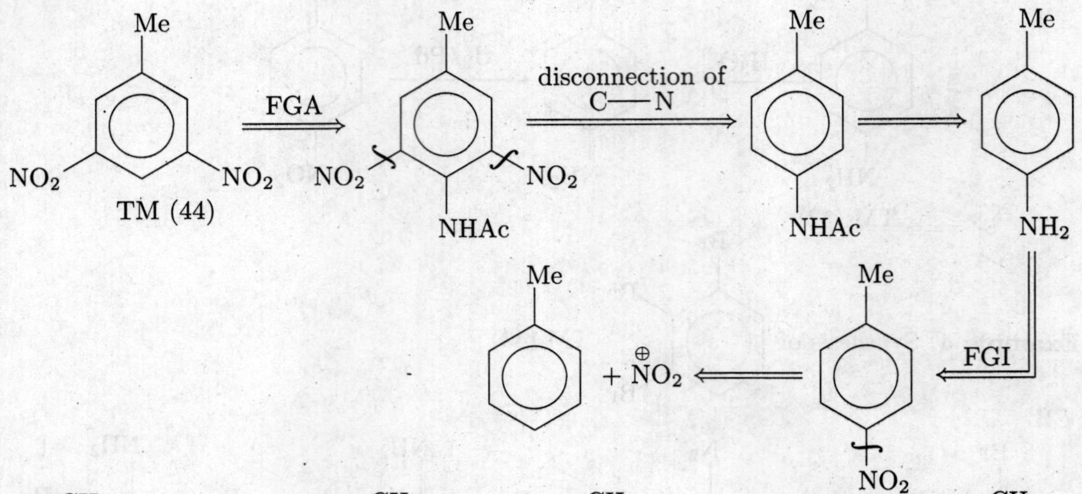


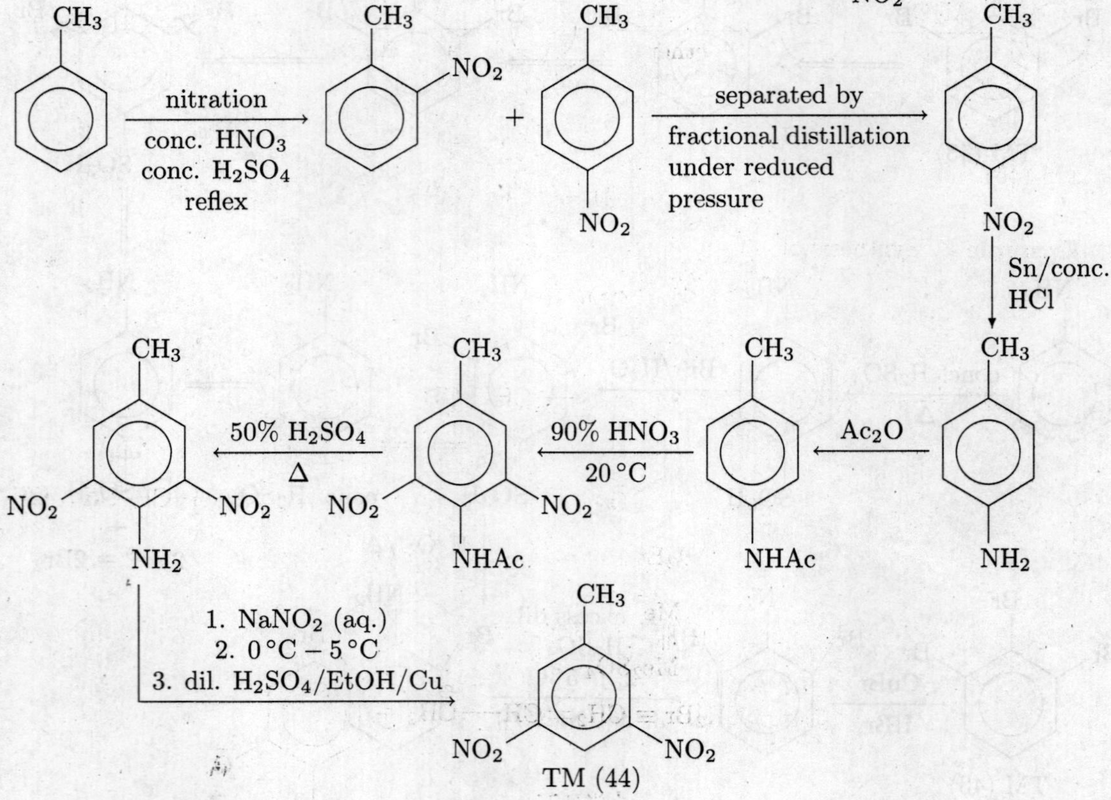

**Example 6.** Synthesis of 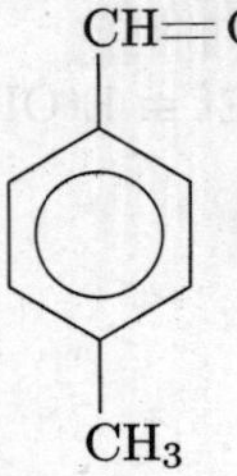 TM (45)

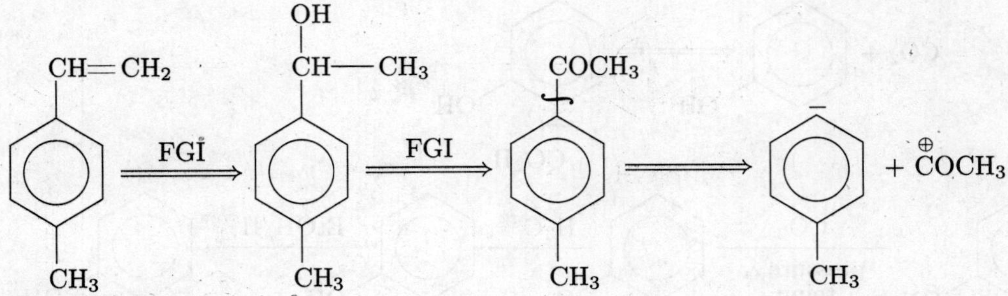

TM (45)

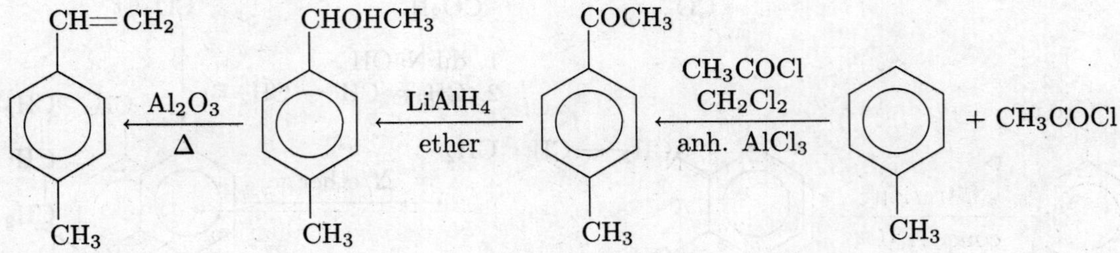

TM (45)

**Example 7.** Synthesis of 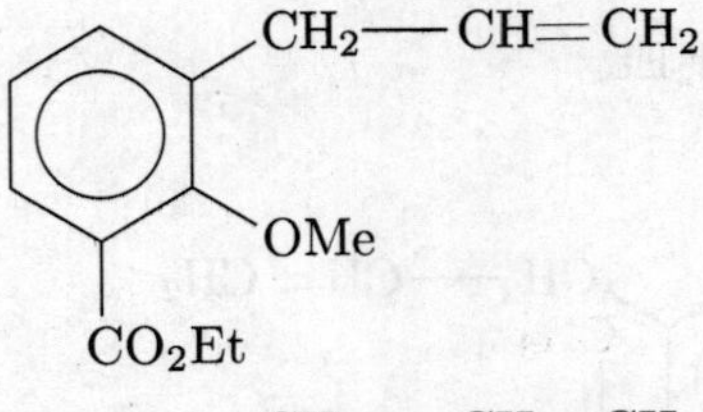 TM (46)

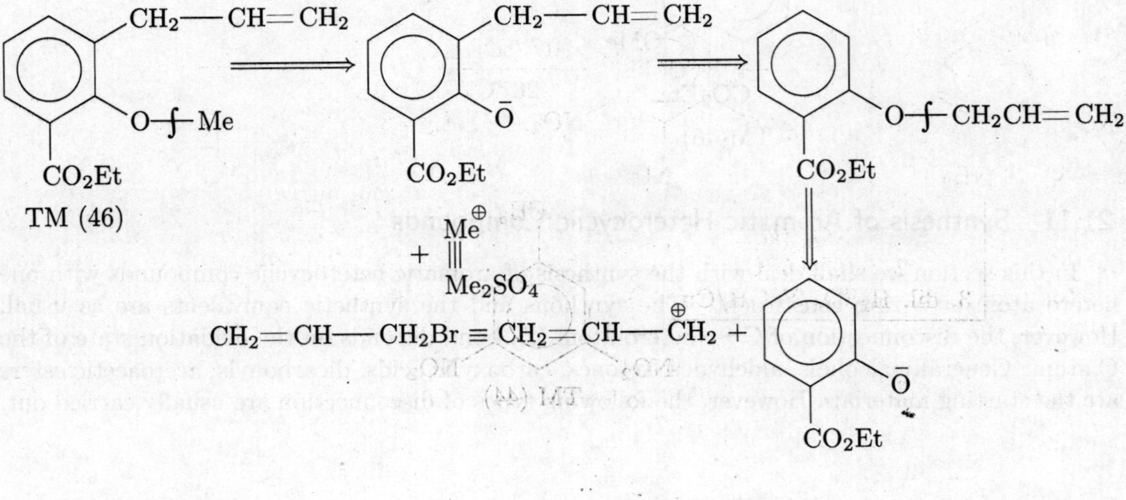

TM (46)

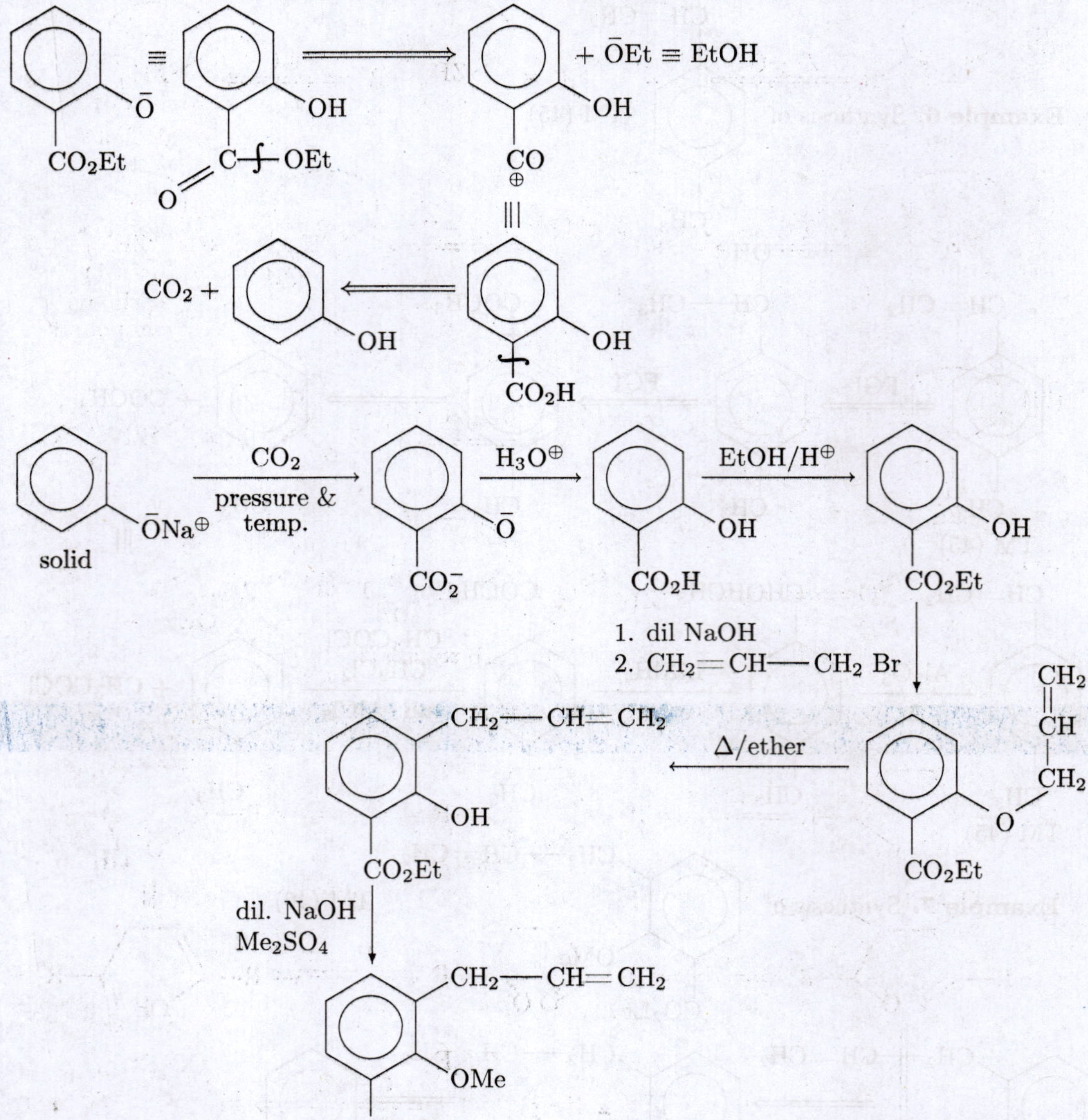


TM(46)

## 21.11 Synthesis of Aromatic Heterocyclic Compounds

In this section we shall deal with the synthesis of aromatic heterocyclic compounds with one hetero atom as a ring constituent. The synthons and the synthetic equivalents are as usual. However, the disconnection of C—hetero atom (Z) bond depends on the oxidation state of the C atom. Generally alcohols, aldehydes, ketones, carboxylic acids, dicarbonyls, acetoacetic esters are the starting materials, However, the following types of disconnection are usually carried out.

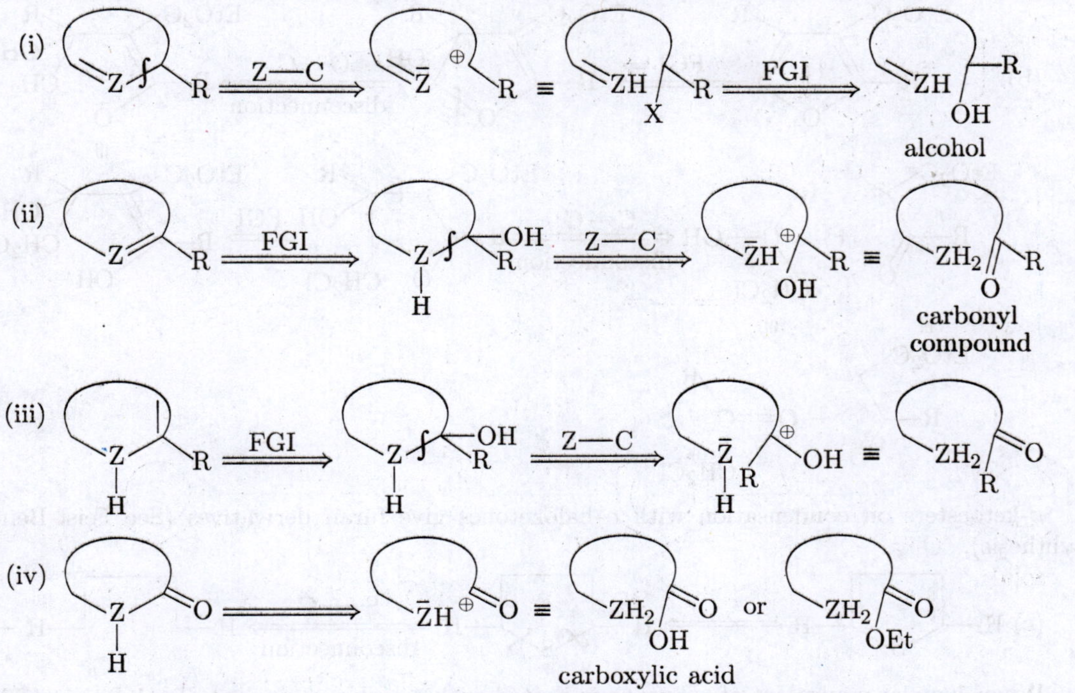

## 21.11.1 Synthesis of furan, pyrrole, thiophen, indole and their derivatives

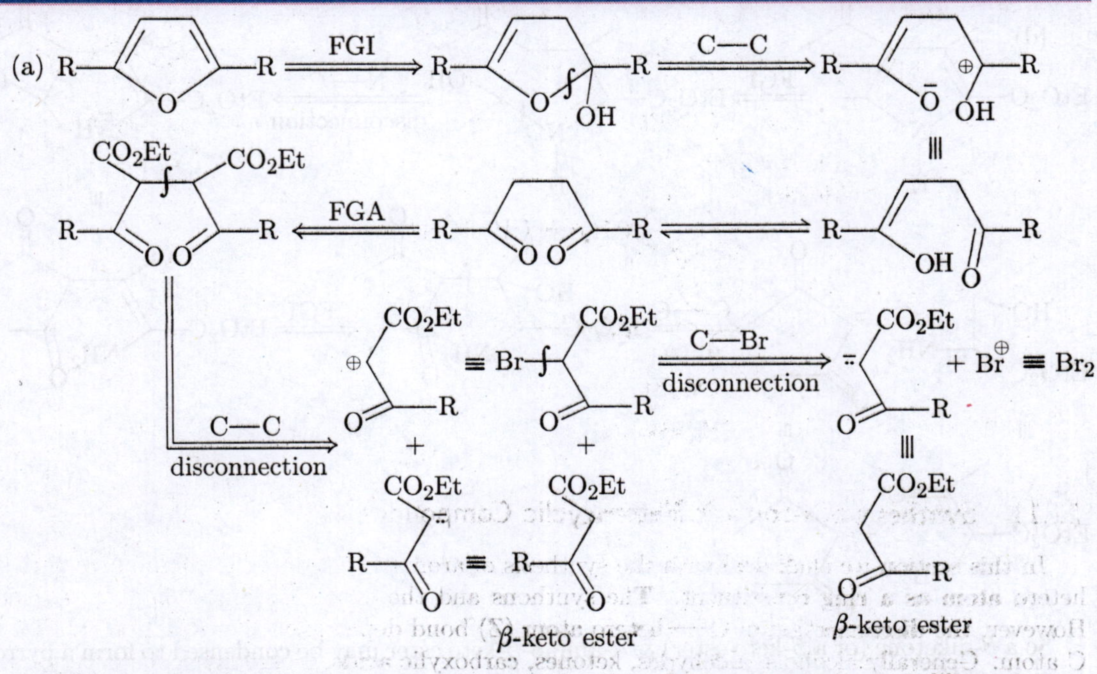

**So a furan derivative maybe synthesised from 2 moles of $\beta$-ketoesters via 1, 4-diketones.**

(b)

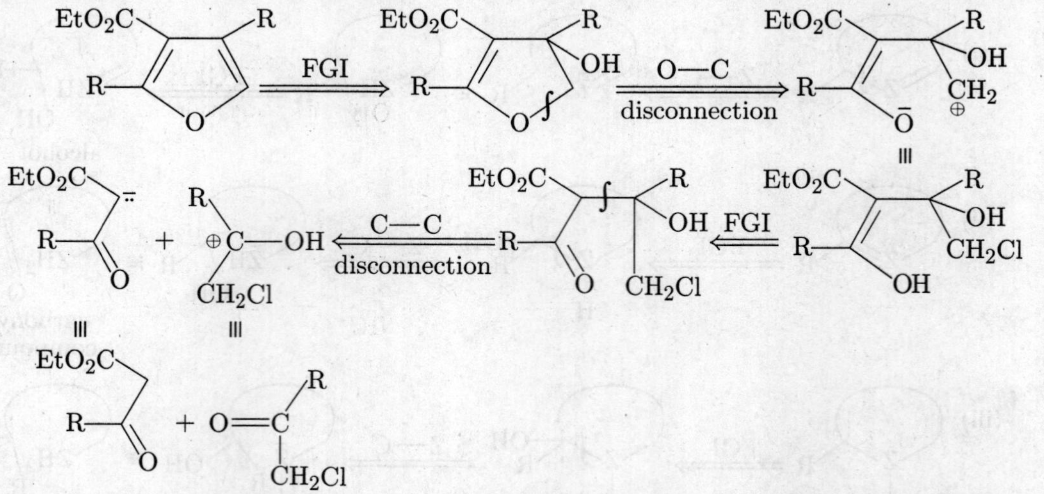

$\beta$-ketoesters on condensation with $\alpha$-haloketones give furan derivatives (See Feist-Benary synthesis).

(c)

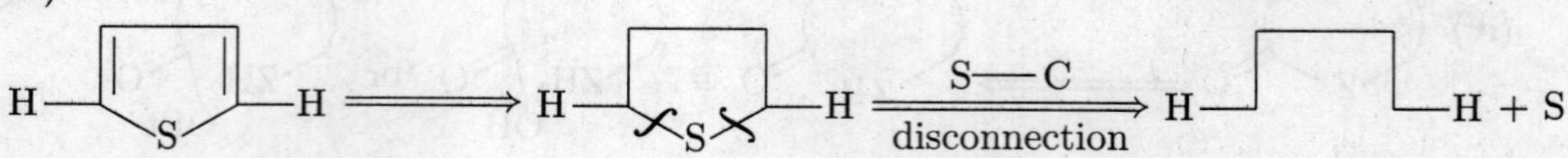

When 1 molar equivalent of $n$-butane is heated with 4 molar equivalent of sulphur at 600 °C thiophen is formed in the vapour state.

(d)

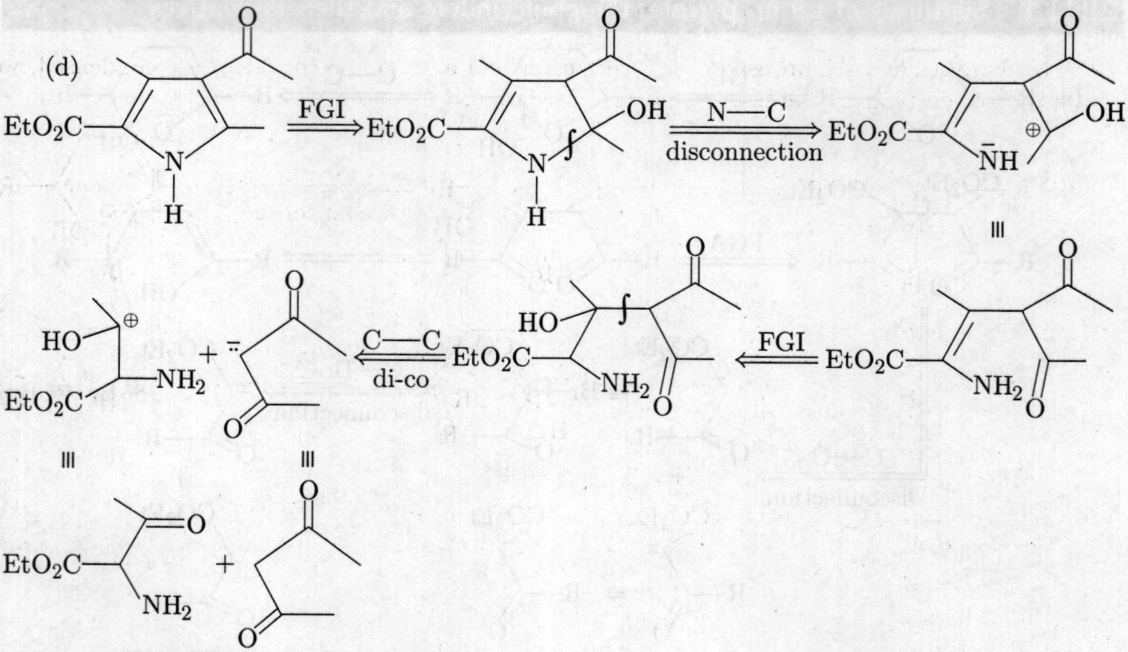

So a $\beta$-diketone (or a $\beta$-keto ester) & $\alpha$-amino-$\beta$-keto ester may be condensed to form a pyrrole derivative (Knorr synthesis of pyrrole).

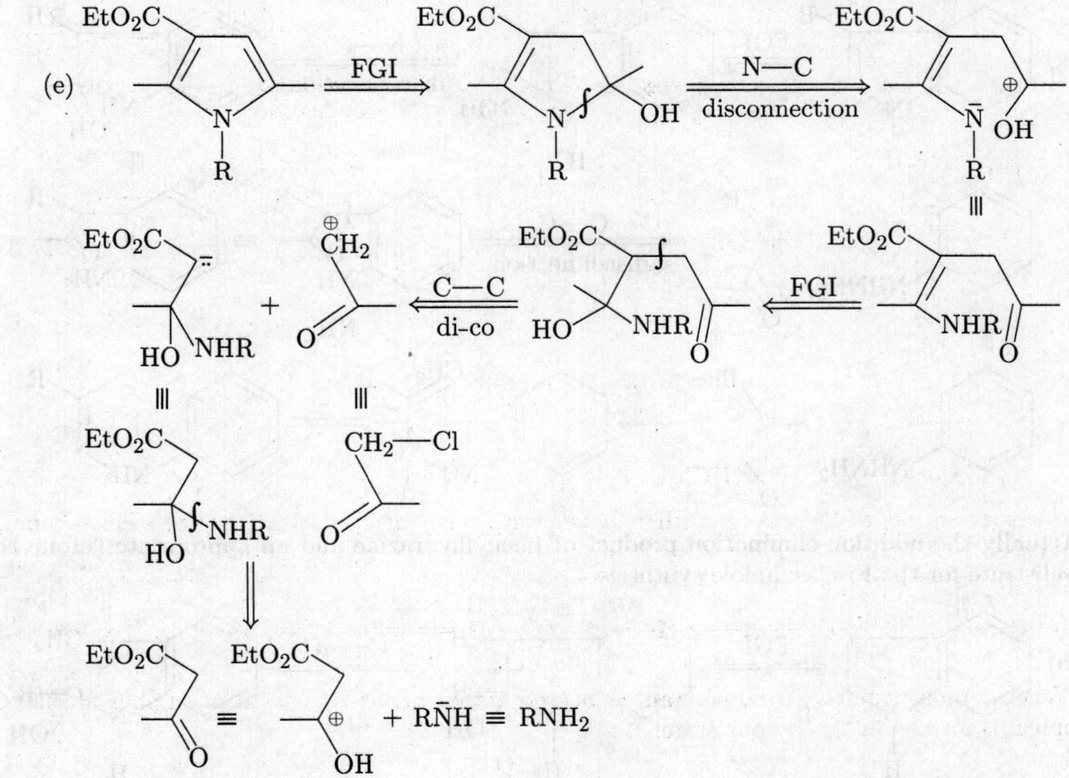


Therefore, when $\alpha$-chloroketone, a $\beta$-keto ester and a 1° amine (or $NH_3$) are condensed, we get a pyrrole derivative (Hantzsch synthesis of pyrrole).

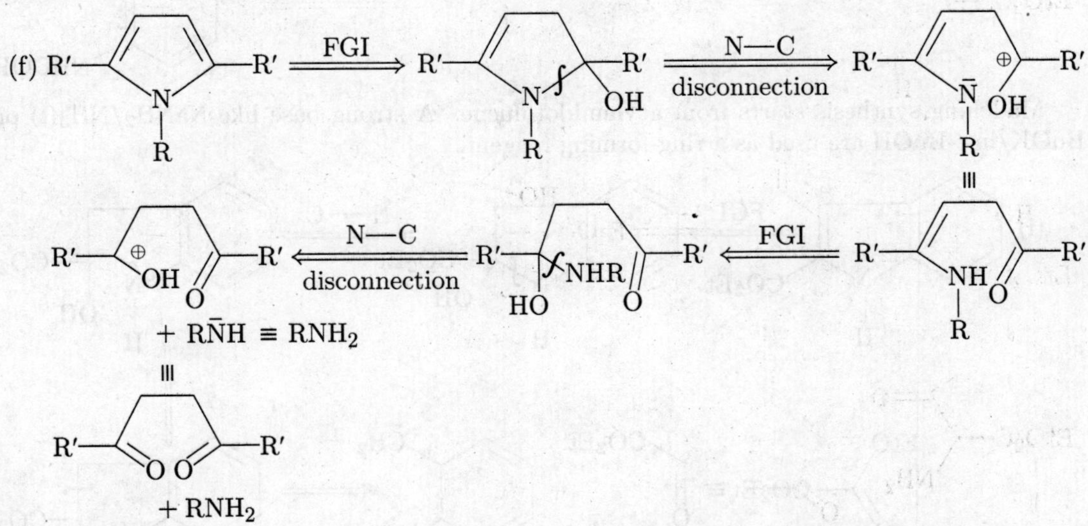


Paal-Knorr synthesis of pyrrole derivatives involves the treatment of 1, 4-diketones with ammonia or a primary amine.

(g)

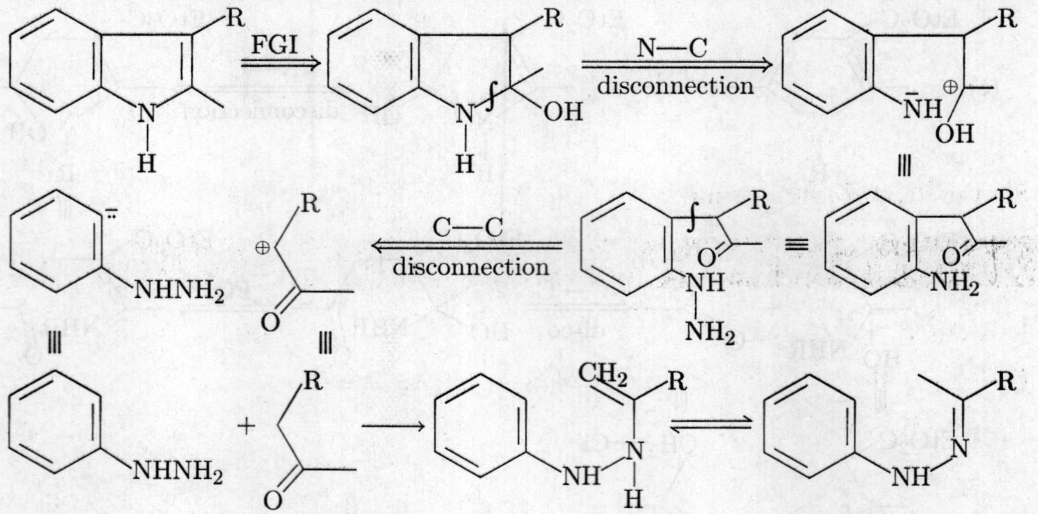

Actually the addition-elimination product of phenylhydrazine and an appropriate ketone is the substrate for the Fischer indole synthesis.

(h)

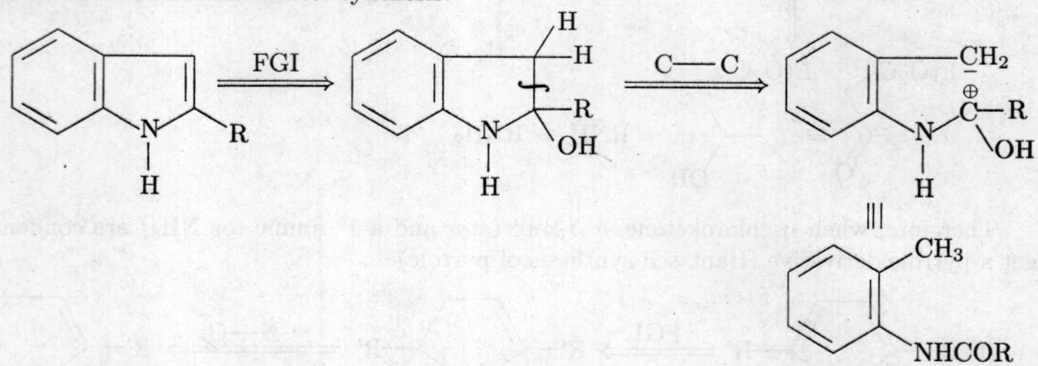

Madelung synthesis starts from acylamidotoluene. A strong base like $NaNH_2/NH_3(l)$ or *t*-BuOK/in-*t*-BuOH are used as a ring-forming reagent.

(i)

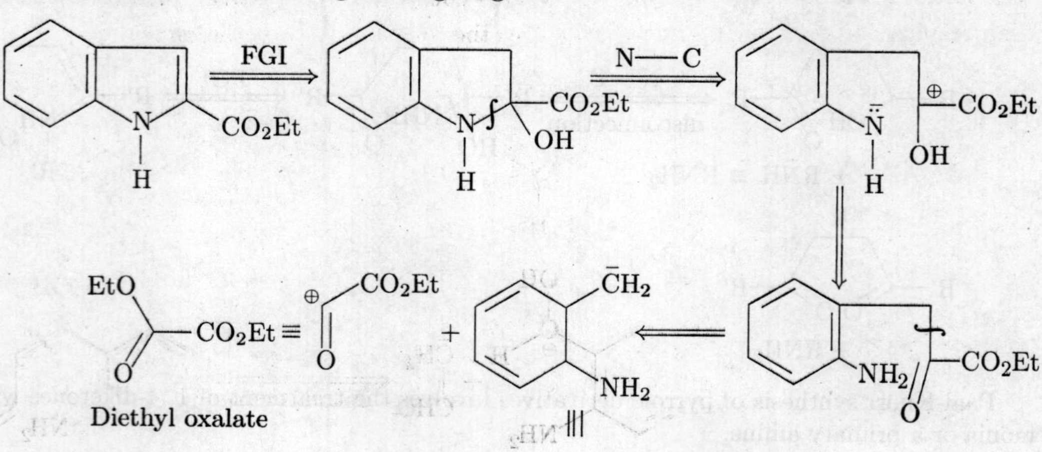

Diethyl oxalate

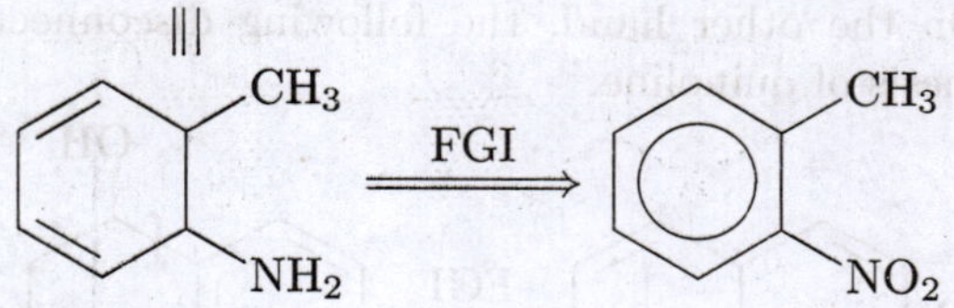

See the Reissert synthesis inside [Sec. 19.5.2(d)].

## 21.11.2 Synthesis of pyridine, quinoline and isoquinoline

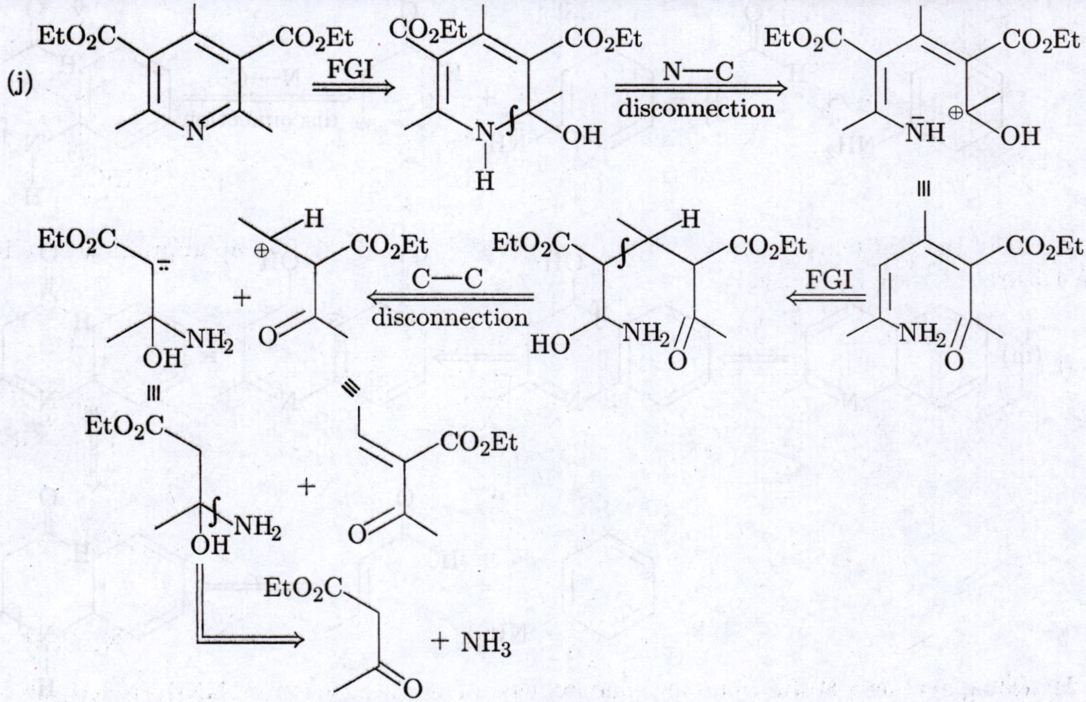

The forward reaction with appropriate reagents constitute the Hantzsch pyridine synthesis.
The Friedlander's quinoline synthesis involves the following type of disconnection :

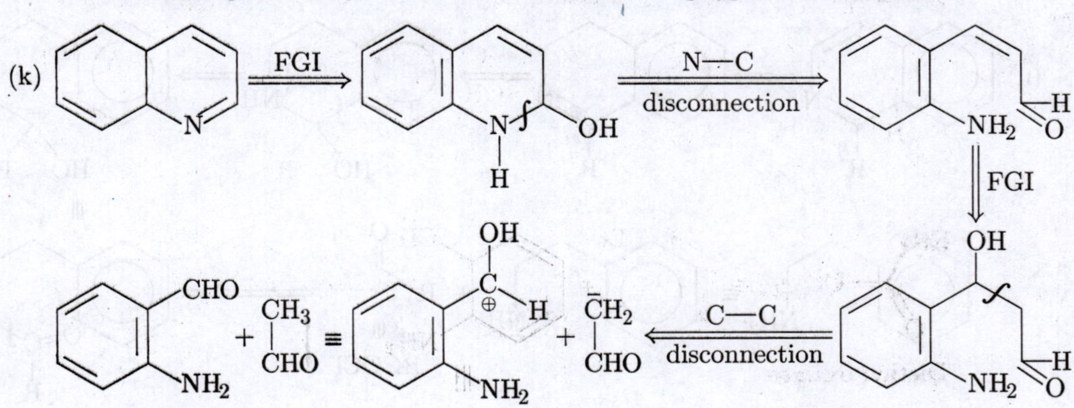

On the other hand, the following disconnection approach gives the route for the Skraup synthesis of quinoline.

(l)

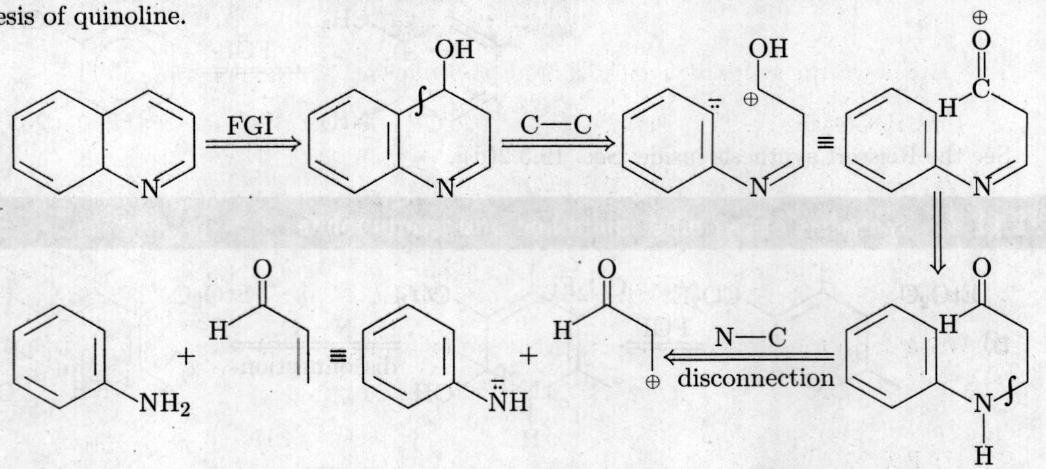


(m)

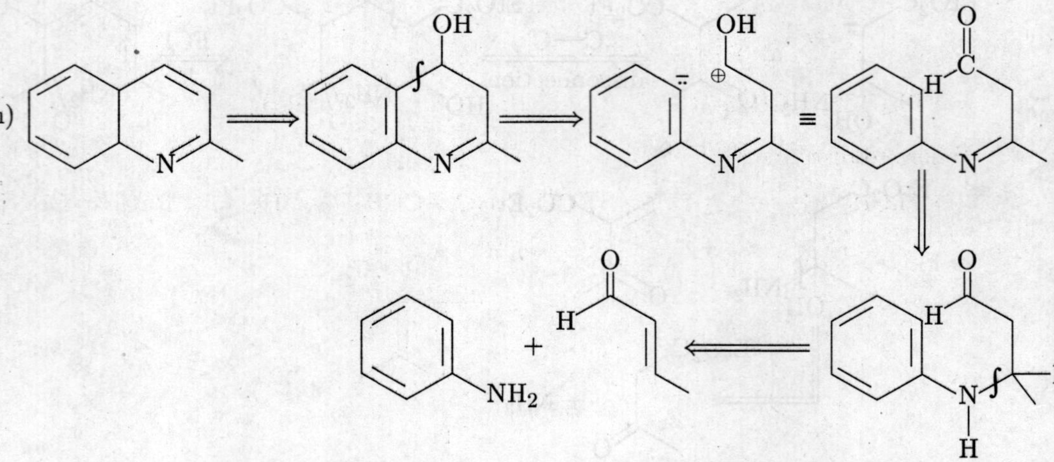


Isosquinoline is synthesised by Bischler-Napieralski reaction. The disconnection approach for the method is as follows :

(n)

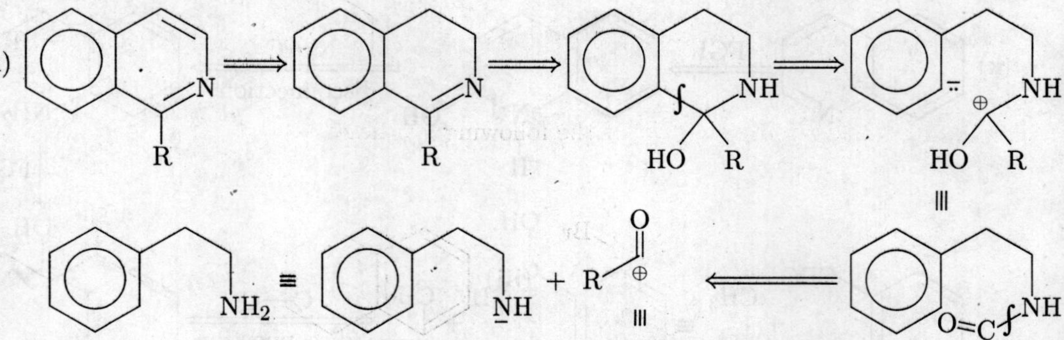

## Exercise

**1.** Write down the synthetic equivalents for the following synthons :

**(i)** $\overset{-}{C}H_2CO_2H$ **(ii)** $PhCH_2\overset{-}{C}H_2$ **(iii)** $CH_3—C{\equiv}C^{\oplus}$ **(iv)** $\overset{\oplus}{C}H_2CH_2OH$

**(v)** $—\underset{\underset{O}{\|}}{C}^{\oplus}$ **(vi)** $O{=}\overset{\oplus}{C}—OH$ **(vii)** $Ph^{\oplus}$ **(viii)** $\overset{-}{C}H_2COCH_2CH_3$

**(ix)** $\overset{\oplus}{N}O$ **(x)** $\overset{-}{N}HR$ **(xi)** $\overset{-}{S}R$ **(xii)** $Cl^{\oplus}$

**2.** Write down the probable synthons from the following systems :

**(i)** 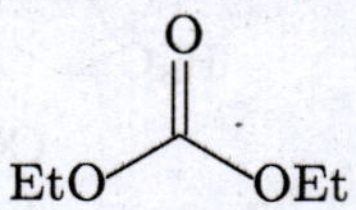 **(ii)** 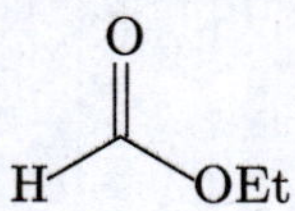 **(iii)** 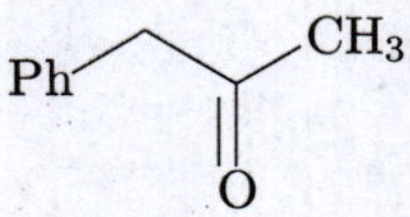 **(iv)** $CH_3—CH{=}NHR$

**(v)** $Ph_3PCH{=}CHCH_2CO_2Et$ **(vi)** $(CH_3)_2Cd$ **(vii)** PhMgBr **(viii)** $CH_3\underset{}{\overset{\overset{O}{\|}}{C}}CH_2Br$

**3.** Where to disconnect?

**(i)** $PhCH_2\underset{\underset{NH_2}{|}}{C}H—CO_2H$ **(ii)** $CH_3CH_2C{\equiv}CCH_2CH_2CH_3$ **(iii)** $PhCH{=}CHNO_2$

**(iv)** 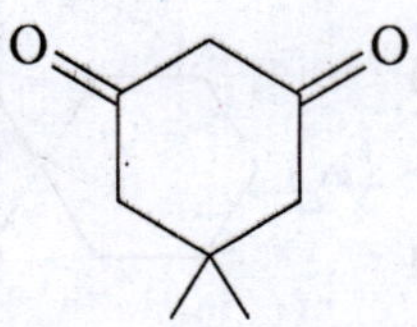 **(v)** 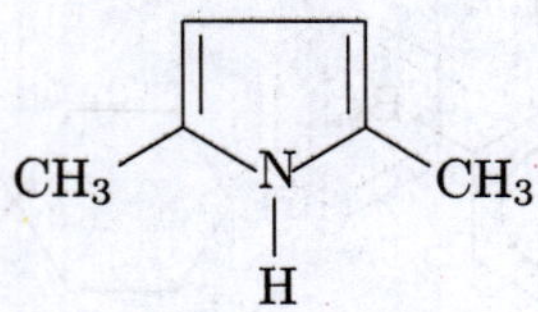 **(vi)** 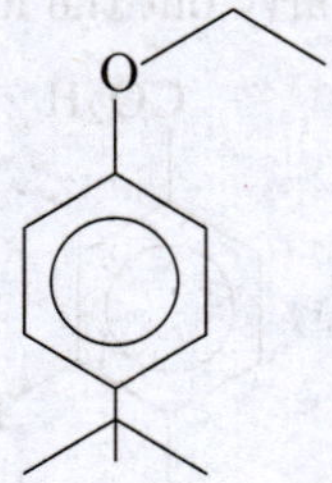

**(vii)** 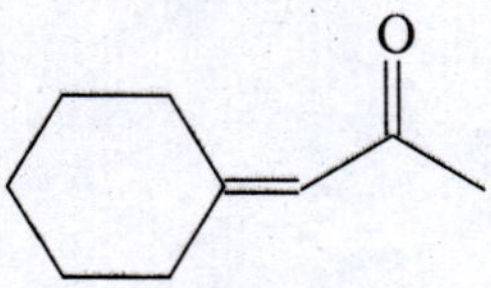 **(viii)** $Ph_3CH$ **(ix)** 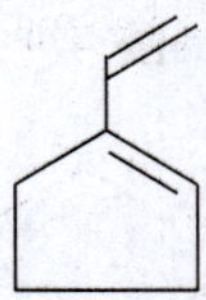 **(x)** $CH_3\overset{\overset{OH}{|}}{C}H—CH_2CH_3$

**4.** Add FG and remove the FG for the following syntheses :

**(i)** 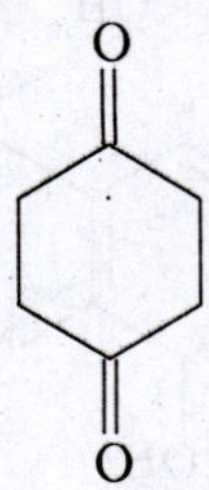 **(ii)** 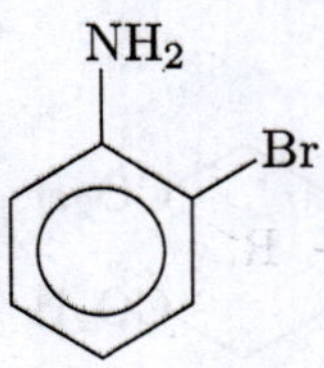 **(iii)** 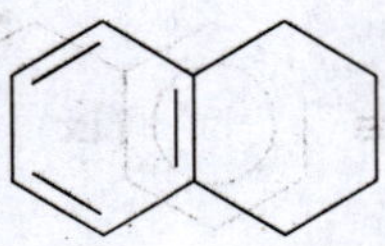 **(iv)** 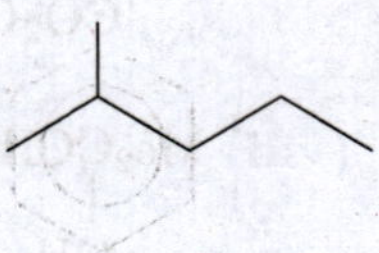

**5.** How would you carry out the following transformations?

(i) $PhCH{=}CH{-}CO_2H$ to $PhCH_2CH_2CO_2H$ (ii) $CH_3CH{=}CH_2$ to $CH_3CH_2CH_2OH$

(iii) 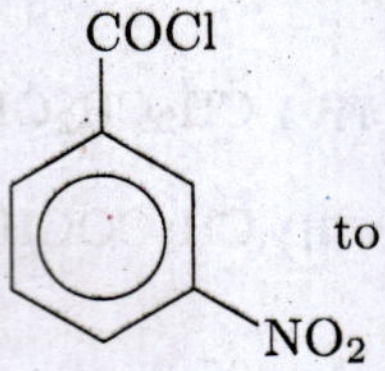 to 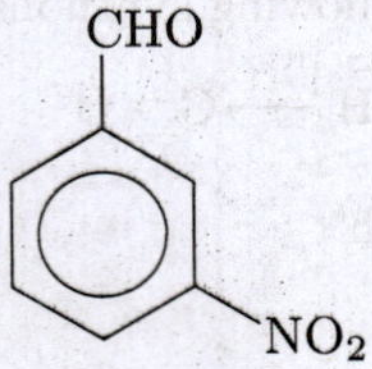

(iv) 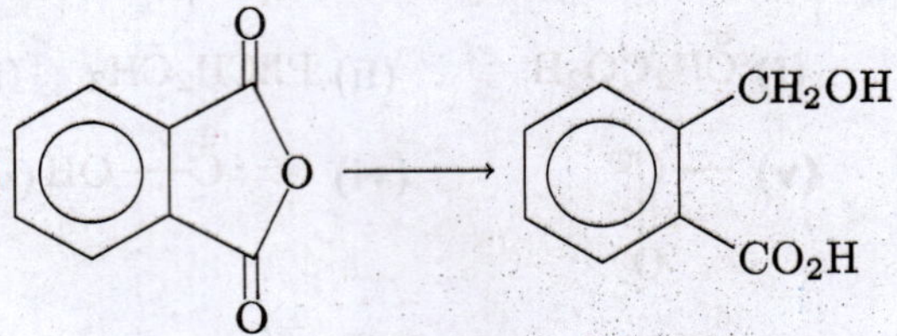

(v) $CH_3CH_2CH_2CHO \longrightarrow CH_3CH_2CH(CH_3)CHO$ (vi) $PhCHO \longrightarrow$ 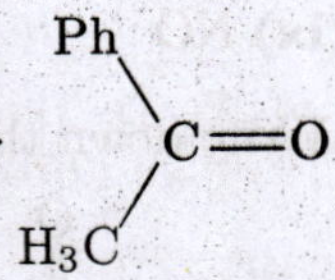

(vii) $CH_3CHOHCH_3 \longrightarrow CH_3CH_2CH_3$ (viii) 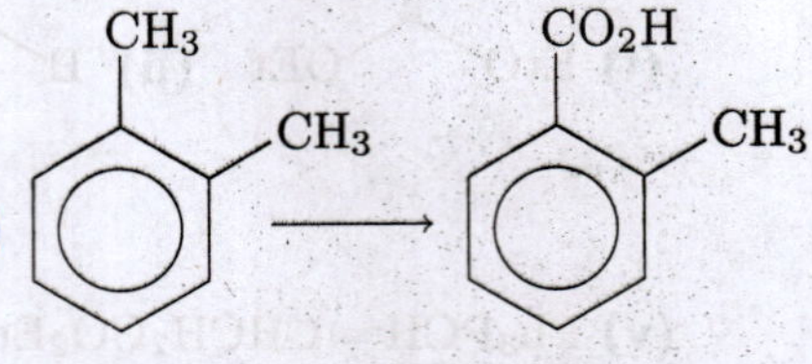

**6.** How to protect the following groups and how to remove the protecting groups?

(i) $-CO_2H$ (ii) $-NH_2$ (iii) $-OH$ (iv) $>C{=}O$ (v) $C{=}C$

**7.** Carry out the following syntheses working back :

(i) 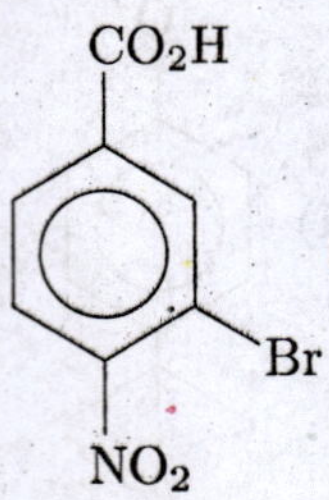 (ii) 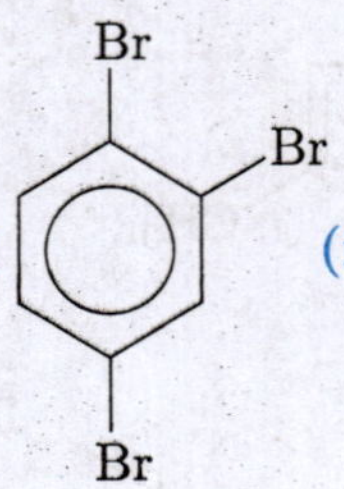 (iii) 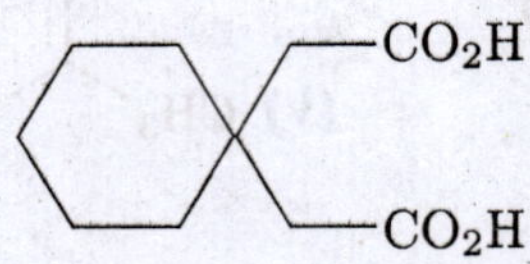 (iv) 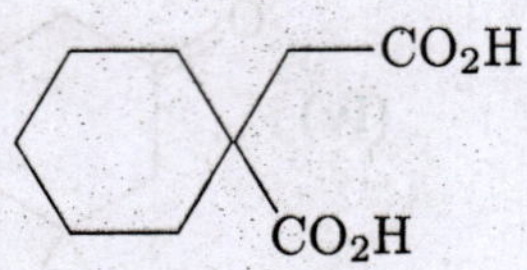

(v) 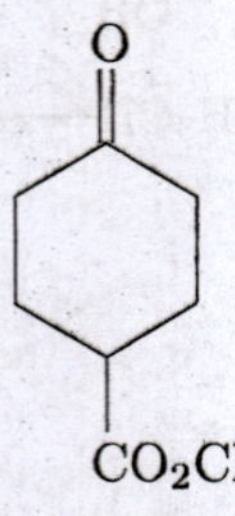 (vi) $(PhCH_2CH_2)_2CHNH_2$ (vii) $Me_3CC(OH)Me_2$

(viii) $Me_2C(OH)CH_2CH_2C(OH)Me_2$ (ix) 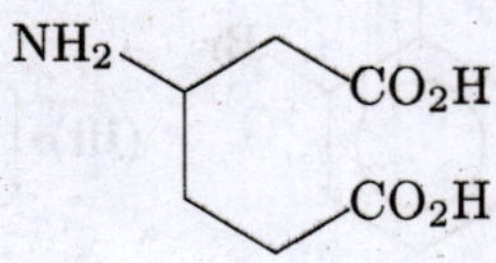 (x) 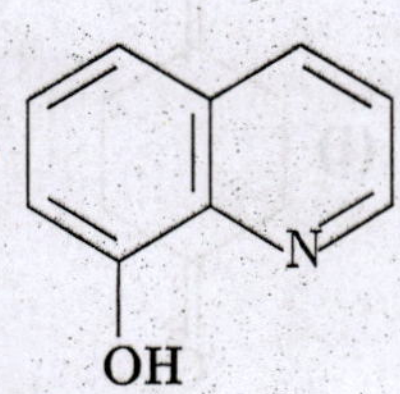

**8.** Convert

(i) 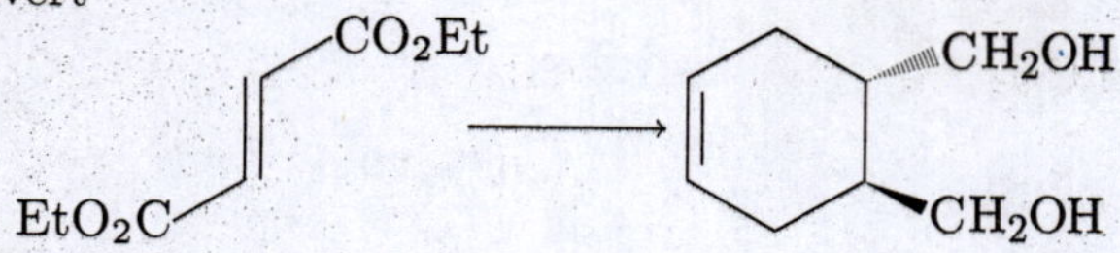


(ii) 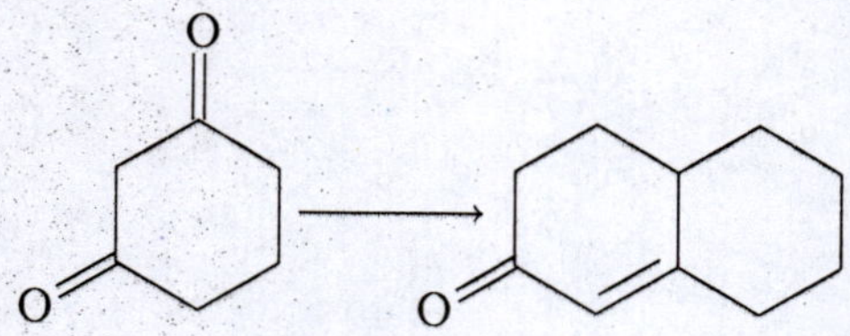


(iii) 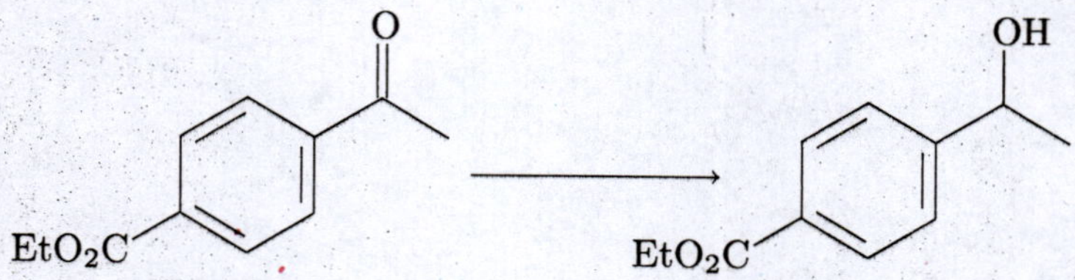


(iv) 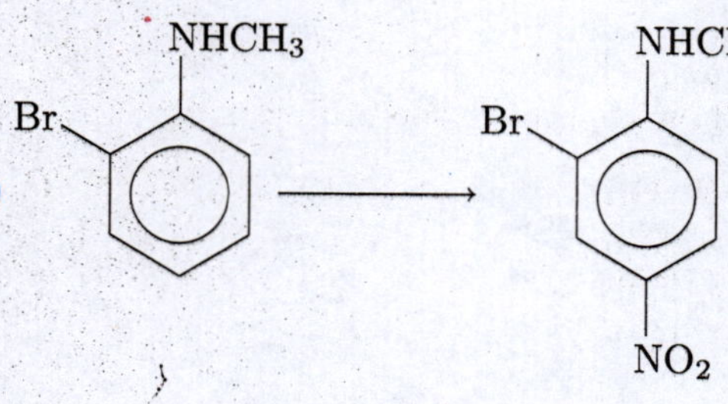

Chapter 22

# Chemistry of Amino Acids and Peptides

## Objectives

## 22.1 Introduction

The very name 'amino acid' indicates that a compound of the class contains at least one amino (—$NH_2$) function and one acidic functional group. The acidic function may either be carboxyl or sulphonic or arsonic or any other else. The amino function may have any position $\alpha, \beta, \gamma, \delta$, etc. with respect to the acidic function and accordingly amino acids may further be classified as $\alpha$-amino acids, $\beta$-amino acids, $\gamma$-amino acids and so on. Aminoethanoic acid (Glycine; $NH_2CH_2CO_2H$), 3-aminopropanoic acid ($NH_2CH_2CH_2CO_2H$) and 4-aminobutanoic acid ($NH_2(CH_2)_3CO_2H$) are examples of $\alpha$-, $\beta$- and $\gamma$-amino acids respectively. The $\alpha$-amino acids, which are obtained from naturally occurring proteins by the process of hydrolysis, are twenty two in number and they are the building units of proteins. Ten out of twenty two are considered to be essential for life. We shall mainly study the chemistry of these $\alpha$-amino acids. These $\alpha$-amino acids contain —$CO_2H$ group as acidic function and may further be classified, depending on the relative numbers of amino and acidic functions they contain, as follows :

**(a)** Neutral amino acids : A compound of this class contains the same number of amino and carboxyl groups; glycine (aminoacetic acid, $NH_2CH_2CO_2H$) is a neutral amino acid.

**(b)** Basic amino acids : The number of amino groups exceeds the number of carboxyl groups in a compound of the class. Lysine, $NH_2CH_2(CH_2)_3CH(NH_2)CO_2H$, is an example of basic amino acids.

**(c)** Acidic amino acids : Each compound of the class contains more carboxyl functions than amino functions; aspartic acid, $CO_2H{\cdot}CH_2CH(NH_2)CO_2H$, is a member of the class.

However, these $\alpha$-amino acids may be represented as $RCH(NH_2)CO_2H$, where R may be called a side chain other than H; when R is H, $RCH(NH_2)CO_2H$ becomes $CH_2(NH_2)CO_2H$ (glycine). Glycine is an achiral and, therefore, an optically inactive compound. Other $\alpha$-amino acids are chiral. Almost all naturally occurring $\alpha$-amino acids have one and the same configuration at the $\alpha$-C and that is L (relative to L-glyceraldehyde) or, S (absolute configuration) as indicated below :

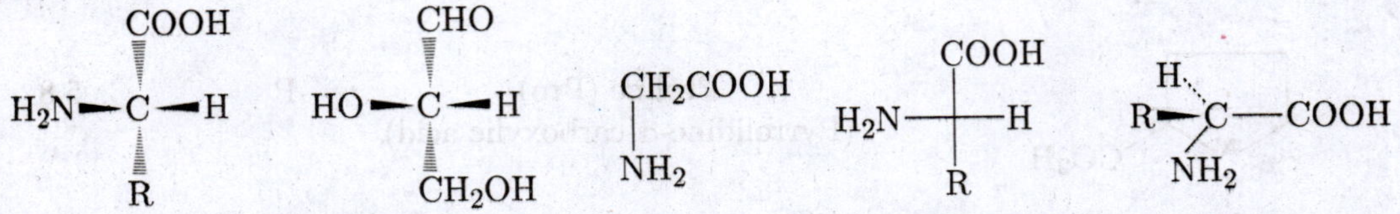

L-$\alpha$-amino acid L-Glyceraldehyde Glycine L-amino acids found in proteins

Depending on the nature of the side chain, R, of the general formula $RCH(NH_2)CO_2H$, the $\alpha$-amino acids may be classified under five heads. The names, abbreviated names and symbols along with isoelectric points are given in Table 22.1.

**Table 22.1** Names, abbreviated names, symbols and isoelectric points of $\alpha$-amino acids obtained from protein hydrolysis

| Constitution of R | Name (abbreviated) Substituted name | Symbol | Isoelectric point |
|---|---|---|---|
| (a) R is a neutral group | | | |
| H | Glycine (Gly) (amino acetic acid) | G | 6.0 |

*Contd.*

| Constitution of R | Name (abbreviated)<br>Substituted name | Symbol | Isoelectric point |
|---|---|---|---|
| $-CH_3$ | Alanine (Ala)<br>($\alpha$-amino propionic acid) | A | 6.1 |
| $-CH(CH_3)_2$ | Valine (Val)<br>($\alpha$-amino isovaleric acid) | V | 6.0 |
| $-CH_2CH(CH_3)_2$ | Leucine (Leu)<br>($\alpha$-amino isocaproic acid) | L | 6.0 |
| $-CHCH_2CH_3$<br>$\quad\vert$<br>$CH_3$ | Isoleucine (Ile)<br>($\alpha$-amino-$\beta$-methylvaleric acid) | I | 6.0 |
| 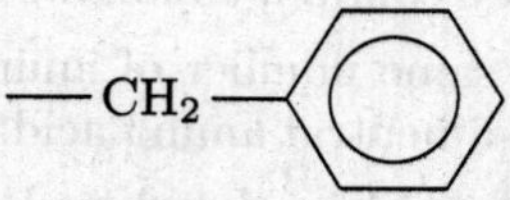 | Phenylalanine (Phe)<br>($\alpha$-amino-$\beta$-phenylpropionic acid) | F | 5.9 |
| $CH_2CONH_2$ | Asparagine (Asn)<br>($\alpha$-amino succinamic acid) | N | 5.4 |
| $-CH_2CH_2CONH_2$ | Glutamine (Gln)<br>($\alpha$-amino glutaramic acid) | Q | 5.7 |
| 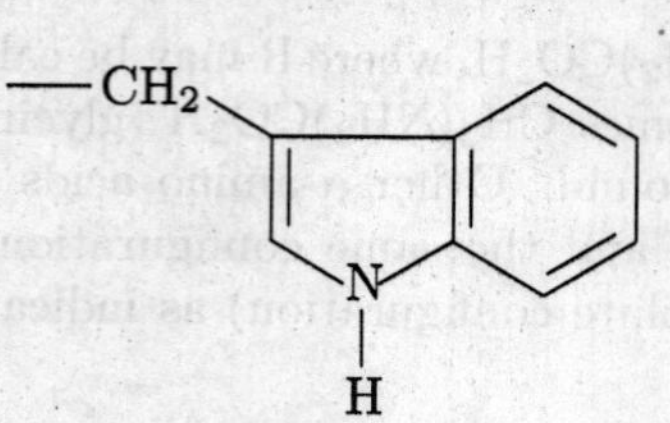 | Tryptophan (Try)<br>($\alpha$-amino-$\beta$-indolepropionic acid) | W | 5.9 |
| 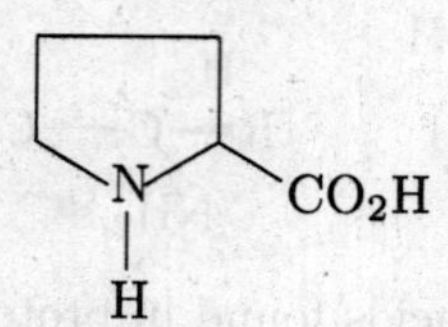 | Proline (Pro)<br>(Pyrrolidine-$\alpha$-carboxylic acid) | P | 5.8 |
| (b) R contains $-OH$ | | | |
| $-CH_2OH$ | Serine (Ser)<br>($\alpha$-amino-$\beta$-hydroxypropionic acid) | S | 5.7 |
| $-CHOH$<br>$\quad\vert$<br>$CH_3$ | Threonine (Thr)<br>($\alpha$-amino-$\beta$-hydroxylbutyric acid) | T | 5.7 |
| 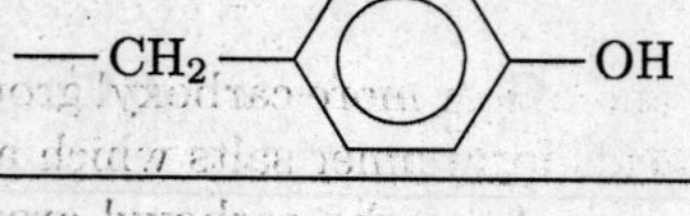 | Tyrosine (Tyr)<br>($\alpha$-amino-$\beta$-($p$-hydroxyphenyl)-propionic acid | Y | 5.4 |

*Contd.*

| Constitution of R | Name (abbreviated)<br>Substituted name | Symbol | Isoelectric point |
|---|---|---|---|
| 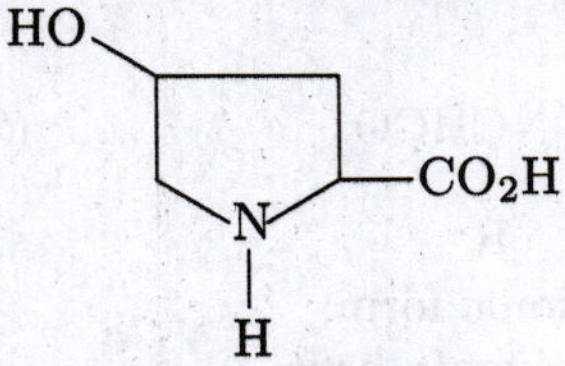 | Hydroxyproline (Hyp)<br>(γ-hydroxypyrrolidine-α-carboxylic acid) | | 5.8 |
| (c) R contains a basic amino group | | | |
| $—(CH_2)_4NH_2$ | Lysine (Lys)<br>(α, ε-diamino caproic acid) | K | 9.5 |
| $—(CH_2)_3NH—C(=NH)—NH_2$ | Argine (Arg)<br>(α-amino-δ-guanidinovaleric acid) | R | 10.8 |
| 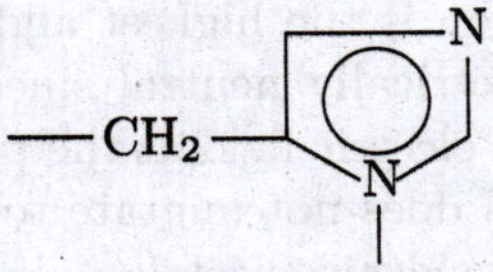 | Histidine (His)<br>(α-amino-β-imidazolylpropionic acid) | H | 7.6 |
| (d) R contains sulphur | | | |
| $—CH_2SH$ | Cysteine (Cys)<br>(α-amino-β-marcaptopropionic acid) | C | 5.1 |
| $—CH_2—S$<br>$—CH_2—S$ | Cystine (Cys—Cys)<br>(Bis-(α-aminopropionic acid)-β-disulphide) | | 5.0 |
| $—CH_2CH_2SCH_3$ | Methionine (Met)<br>(α-amino-γ-methylthio-n-butyric acid)<br>(β-methylthioethylglycine) | | 5.7 |
| (e) R contains a carboxyl group | | | |
| $—CH_2CO_2H$ | Aspartic acid (Asp)<br>α-aminosuccinic acid<br>(glycylacetic acid) | | 3.0 |
| $—CH_2CH_2CO_2H$ | Glutamic acid (Glu)<br>α-aminoglutaric acid<br>(glycylpropionic acid) | | 3.2 |

## 22.2 Dipolar Ions

Since amino acids contain one or more amino group ($—NH_2$) and one or more carboxyl group ($—CO_2H$), they are amphoteric. In the dry solid state, amino acids form inner salts which are called dipolar ions or zwitter ions or ampholytes. In the dipolar ion form the carboxyl group remains as a carboxylate ($CO_2^-$) group and the amino group exists as an ammonium ($-\overset{\oplus}{N}H_3$) group. However, in the aqueous medium, an equilibrium exists involving the dipolar ion, the anionic and the cationic forms of the amino acids

$$\underset{\substack{|\\ R}}{H_3\overset{\oplus}{N}CHCO_2H} \underset{H^{\oplus}}{\overset{-H^{\oplus}}{\rightleftharpoons}} \underset{\substack{|\\ R}}{H_3\overset{\oplus}{N}CHCO_2^-} \underset{H^{\oplus}}{\overset{\bar{O}H/-H^{\oplus}}{\rightleftharpoons}} \underset{\substack{|\\ R}}{H_2NCHCO_2^-} \tag{22.1}$$

| cationic form; conjugate acid of the dipolar ion (cA) predominant at pH < 2 | dipolar ion; predominant at the isoelectric point | anionic form; conjugate base of the dipolar ion (cB) predominant at pH > 11 |
|---|---|---|

The pH of the solution and the nature of the amino acid determine the predominant form of an amino acid in a solution. In a strongly acidic medium (pH < 2) all amino acids remain primarily as cations; while in a strongly basic solution, they remain as anions (at a pH > 11). At an optimum pH (2 < pH < 11), the concentration of the dipolar ion is the highest and the concentrations of anions and cations are equal. A dipolar ion is electrically neutral since its electric charge is zero; and consequently if an amino acid is placed in an electric field at the pH at which the concentration of its dipolar ion is maximum, the amino acid does not migrate to any electrode and the pH is called the isoelectric point. Some authors consider isoelectric point as the pH at which the dipolar ion of an amino acid has the maximal concentration in the presence of salts; while the pH at which the concentration of dipolar ion is maximum in the presence of hydrogen ion is called isoionic point. Every amino acid has its own isoelectric point (Table 22.1). However, the isoelectric point of a monoamino monocarboxylic acid, for example, of glycine, may be determined as follows.

At the isoelectric point the equilibrium (Eqn. 22.1) may be looked upon as composed of two separate equilibria.

$$\underset{\text{conjugate acid (cA)}}{\overset{\oplus}{N}H_3CH_2CO_2H} \rightleftharpoons \underset{\text{dipolar ion}}{H_3\overset{\oplus}{N}CH_2CO_2^-} + H^{\oplus} \tag{22.2}$$

$$\underset{\text{dipolar ion}}{\overset{\oplus}{N}H_3CH_2CO_2^-} \rightleftharpoons \underset{\text{conjugate base (cB)}}{NH_2CH_2CO_2^- + H^{\oplus}} \tag{22.3}$$

From Eqn. 22.2,

$$K_1 = \frac{[\text{Dipolar ion}]\ [H^{\oplus}]}{[cA]};\ \text{where } K_1 = \text{equilibrium constant.}$$

$$\therefore\ [cA] = \frac{[\text{Dipolar ion}]\ [H^{\oplus}]}{K_1} \tag{22.4}$$

From Eqn. 22.3,

$$K_2 = \frac{[cB]\ [H^{\oplus}]}{[\text{Dipolar ion}]};\ \text{where } K_2 = \text{equilibrium constant.}$$

$$\therefore\ [cB] = \frac{K_2[\text{Dipolar ion}]}{[H^{\oplus}]} \tag{22.5}$$

At the isoelectric point the concentration of the dipolar ion being maximum and the net charge of the ion being zero, [cA] = [cB]

or, $\frac{[\text{Dipolar ion}]\,[\text{H}^{\oplus}]}{\text{K}_1} = \frac{\text{K}_2[\text{Dipolar ion}]}{[\text{H}^{\oplus}]}$ (from Eqns. 22.4 and 22.5)

or, $[\text{H}^{\oplus}]^2 = \text{K}_1\text{K}_2$; the concentration terms of dipolar ion being cancelled.

or, $2 \times -\log_{10}[\text{H}^{\oplus}] = -\log_{10}\text{K}_1 - \log_{10}\text{K}_2$

or, $2\text{pH} = \text{pK}_1 + \text{pK}_2$

or, $\text{pH} = \frac{\text{pK}_1 + \text{pK}_2}{2} = p_i$, the isoelectric point. (22.6)

Now $\text{pK}_1$ and $\text{pK}_2$ of glycine have been found to be 2.4 and 9.6. Hence $p_i$ of glycine is the

$\text{pH} = \frac{(2.4 + 9.6)}{2}$; (from the Eqn. 22.6)

or $\text{pH} = p_i = 6.0$.

Here it is needless to mention that a neutral amino acid like glycine has two pK values, one as an acid and another as a base since it has two functional groups of acidic and basic characters.

The optical rotatory power differs from group to group because of their different polarisability. Therefore, the optical rotatory powers of a carboxyl ($-\text{CO}_2\text{H}$) group and the corresponding carboxylate ($-\text{CO}_2^-$) group are different. The same is the case with amino ($-\text{NH}_2$) group and ammonium ($-\overset{\oplus}{\text{N}}\text{H}_3$) group. As a consequence, the specific rotations of dipolar ion and those of the cation (cA) and the anion (cB) are different. So at different pHs an amino acid has different specific rotations. In an acidic medium, the amino acid primarily becomes $\text{RCH}(\overset{\oplus}{\text{N}}\text{H}_3)\text{CO}_2\text{H}$ and in the basic medium it becomes $\text{RCH}(\text{NH}_2)\text{CO}_2^-$, while at the isoelectric point the same remains primarily as a dipolar ion, $\text{RCH}(\overset{\oplus}{\text{N}}\text{H}_3)\text{CO}_2^-$, when the amino acid under consideration is $\text{RCH}(\text{NH}_2)\text{CO}_2\text{H}$. Therefore, the specific rotation of an amino acid is pH dependent and specific rotation of an amino acid is measured at its isoelectric point However, the presence of dipolar ion in the solid state has been proved by the fact that solid amino acids conduct electricity and they have high b.p., m.p. and solubility in polar solvents. It is interesting to note that the amino acids have normally the lowest solubility in water at their isoelectric points. Perhaps zwitter ions are less hydrated than the corresponding cations and anions. At $p_i$ the concentrations of cA an cB are equal but least, and that of the dipolar ion is highest ; the latter, however, does not get hydrated appreciably and thus an amino acid has the least solubility in water at $p_i$. At a pH at which the concentration of either cation (pH < 2) or anion (pH > 11) is maximum and thereby the amino acid in the cationic or in the anionic form gets maximum hydrated and becomes most soluble in water. This property of solubility of amino acids could be utilised to separate amino acids which have appreciably different $p_i$s. For example, a mixture of glycine ($p_i = 6.0$) and glutamic acid ($p_i = 3.20$) may be separated at a pH = 6, the $p_i$ of glycine, by fractional crystallisation, at which glycine is least soluble but glutamic acid is not. Therefore, the former could be filtered out. Similarly, a mixture of glutamic acid and lysine ($p_i = 9.5$) could be separated by fractional crystallisation at a pH = 9.5 whence lysine will remain mostly as insoluble solid and glutamic acid will remain soluble and, therefore, can be separated by the process of filtration. Electrophoresis may also be used for the purpose of separation.

**Problem 22.1.** **(a)** Explain why the isoelectric point ($p_i$) for glutamic acid is 3.2, that for alanine is 6.1 and that for lysine is 9.5. **(b)** Do all compounds containing —$NH_2$ and —$CO_2H$ groups remain as zwitter ions?

**Solution :** **(a)** At $p_i$ the concentration of electrically neutral dipolar ion is the highest and that of the corresponding cation and anion are equal and the least. These facts are to be considered to justify $p_i$s of amino acids of acidic, basic and neutral characters along with their equilibrium reactions in the aqueous medium at their $p_i$s as shown below :

Ion equlibria

(i) For alanine :

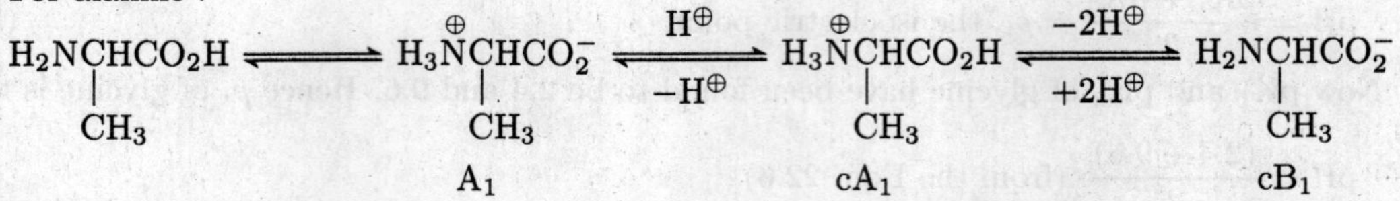


(ii) For glutamic acid :

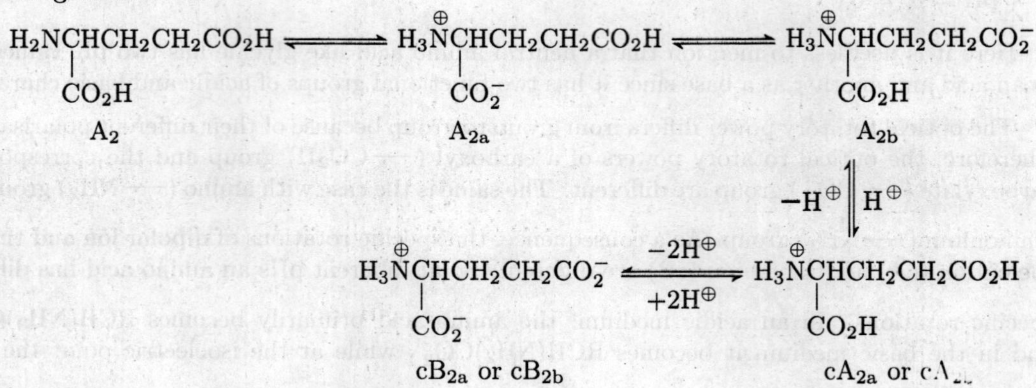


(iii) For lysine :

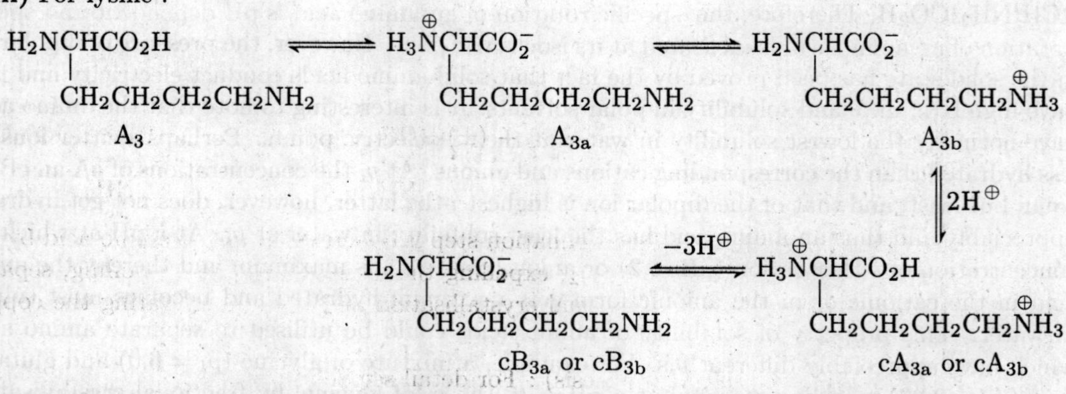


At $p_i$ of alanine the concentration of electrically neutral $A_1$ should be the highest and those of $cA_1$ and $cB_1$ should be equal but the least. Since carboxyl group is acidic to water at a pH = 7 (neutral pH) —$CO_2H$ group will convert to $CO_2^-$ but just to have —$NH_2$ group as —$\overset{\oplus}{N}H_3$, the solution must be slightly acidic in which $cA_1$ and $cB_1$ will have minimum but equal concentrations. So, the $p_i$ of alanine is just below 7.

In the case of glutamic acid, at a neutral pH, there will be a significant concentration of $cB_{2a}$ (or $cB_{2b}$) since carboxyl groups are acidic to water. To get the highest concentration of $A_{2a}$ or $A_{2b}$ the dissociation of one of the two carboxyl groups is to be suppressed to a point where the concentrations of $cB_{2a}$ and $cA_{2a}$ are equal. As a consequence the $p_i$ of glutamic acid will be at a much low pH (= 3.2).

However, for lysine the concentration of cation $cA_{3a}$ or $cA_{3b}$ is to be suppressed in order to get the maximum concentration of $A_{3a}$ and $A_{3b}$ and minimum but equal concentrations of $cA_{3a}$ and $cB_{3b}$; this is possible at a pH greater than 7. Naturally the $p_i$ of lysine is at a high pH (= 9.5).

**(b)** No. Compounds in which —$CO_2^-$ is less basic than —$NH_2$ will only develop zwitter ions. If —$CO_2^-$ is more basic than —$NH_2$, then —$\overset{\oplus}{N}H_3$ will be more acidic than —$CO_2H$ and the compound will not form a zwitter ion. In anthranilic acid and sulphanilic acid —$CO_2^-$ and —$SO_3^-$ groups are more basic than —$NH_2$. They do not form zwitter ion.

## 22.3 Formation of Amino Acids

The following are the general methods for the formation of amino acids in the racemic form (because the processes involve achiral conditions). However, each method can be used to synthesise a few members of the class with satisfaction.

### 22.3.1 Synthesis of (±) α-amino acids

(i) *From fatty acids* (Perkin et al.) : This method is good for the synthesis of glycine, alanine, serine, threonine, valine, leucine and isoleucine.

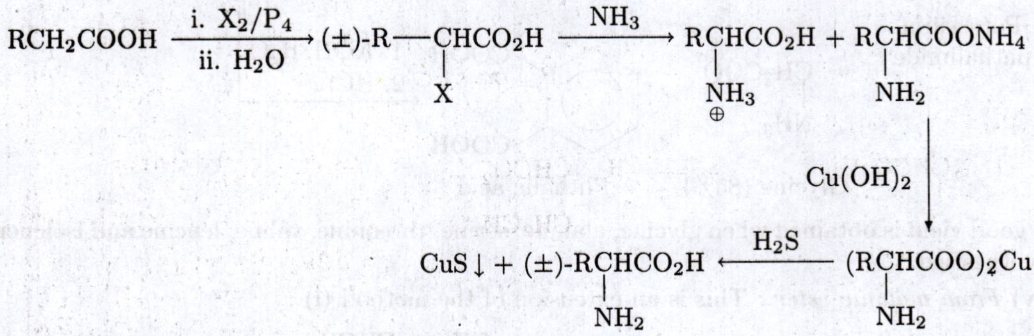

The ammonium salt formed in the amination step is converted to the α-amino acid by reacting the aminated product with $Cu(OH)_2$, expelling the excess ammonia by boiling, separating the cupric amino carboxylate by fractional crystallisation and then precipitating the copper as sulphide with hydrogen sulphide.

(ii) *From aldehyde* (Strecker's synthesis) : For detail see Sec. 16.13.8.

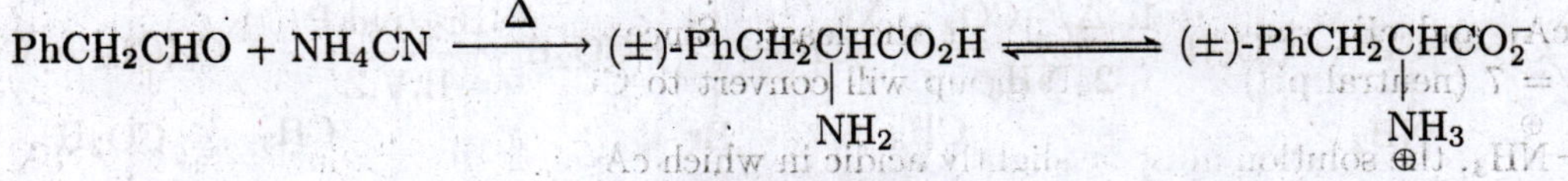

Phenylalanine

(iii) *From phthalimide* (Gabriel phthalimide synthesis) :

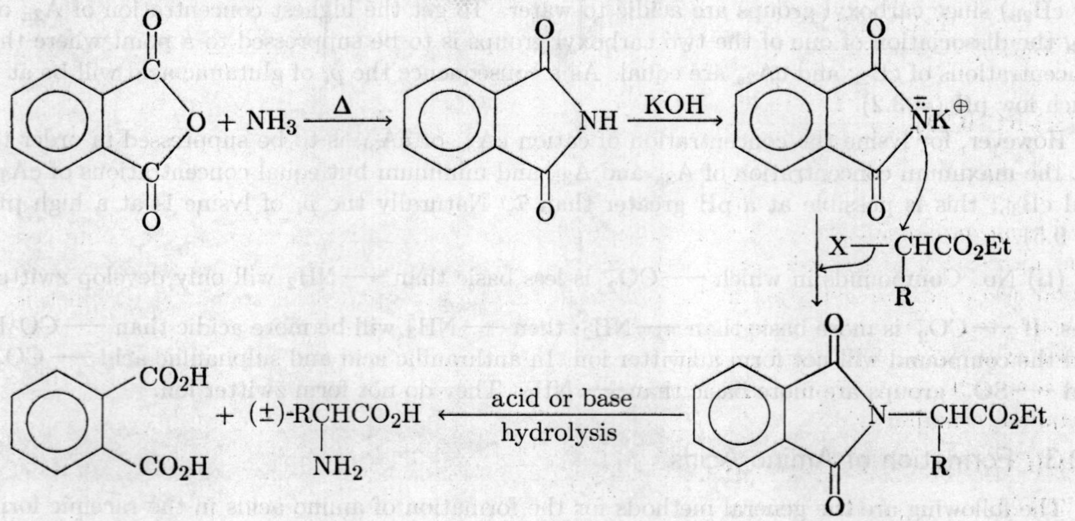

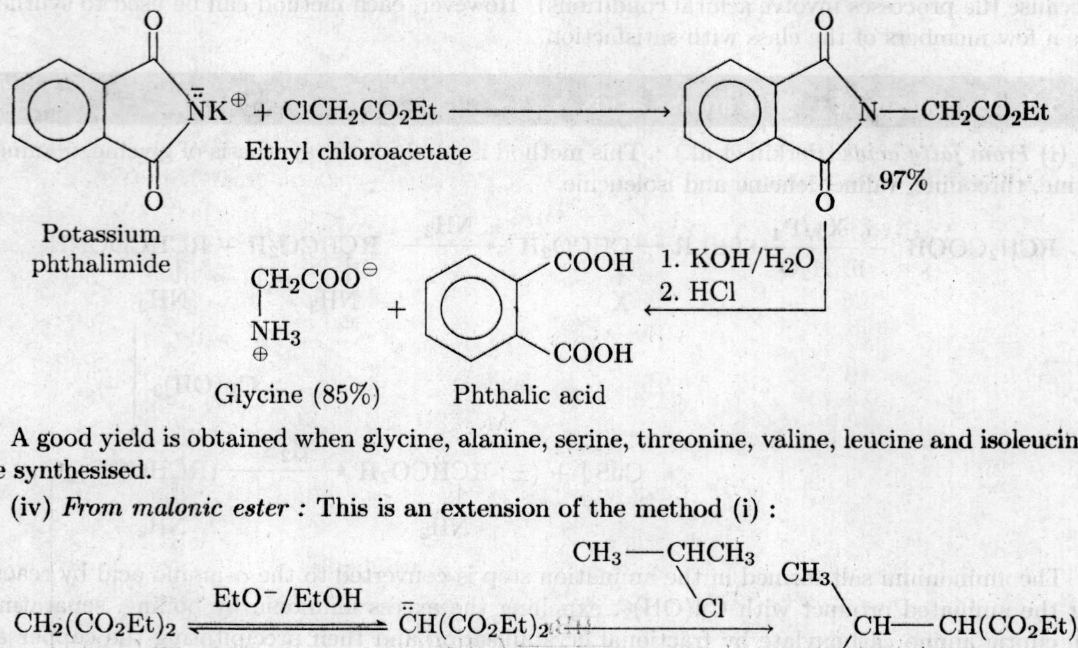

Potassium phthalimide

Ethyl chloroacetate

Glycine (85%)

Phthalic acid

**A good yield is obtained when glycine, alanine, serine, threonine, valine, leucine and isoleucine are synthesised.**

**(iv)** ***From malonic ester :*** **This is an extension of the method (i) :**

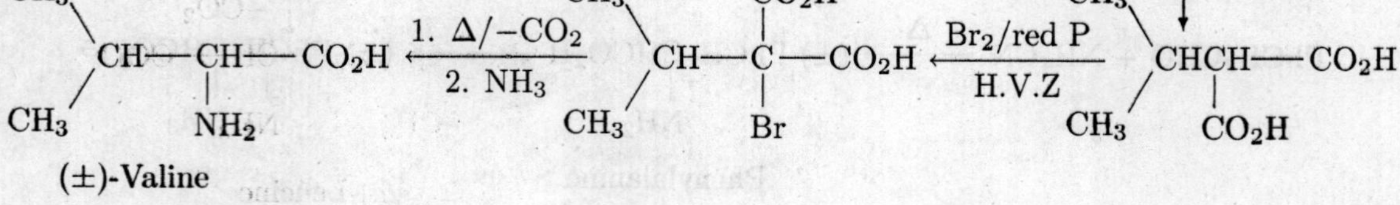

(±)-Valine

**In the general form :**

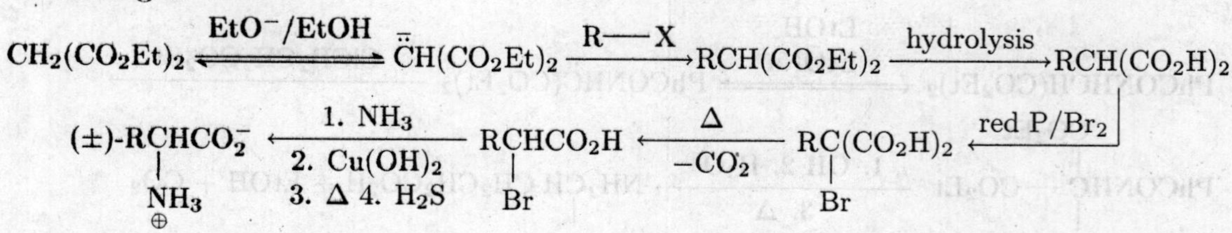

(v) ***From amidomalonic esters and imido malonic esters :*** Amidomalonic esters are obtained by the nitrosation of malonic ester with nitrous acid followed by its reduction with metallic zinc and acetic acid or hydrogen in the presence of nickel catalyst to the amino group which is then converted to an amido group by treating with acetic anhydride or benzoyl chloride in the presence of pyridine. The amidomalonic ester is then treated with a base and an alkyl halide to obtain a substituted amidomalonic ester which on hydrolysis followed by decarboxylation gives the required $\alpha$-amino acid.

$$\underset{\text{DEM}}{CH_2(CO_2Et)_2} + \underset{n\text{-Butyl nitrite}}{CH_3(CH_2)_3ONO} \xrightarrow[\text{ii. } H_2SO_4]{\text{i. } 0\,°C\text{-}20\,°C} \underset{\text{Diethyl oximinomalonate (87\%)}}{HON{=}C(CO_2Et)_2} + CH_3(CH_2)_3OH$$

$$\text{Diethyl oximinomalonate} \xrightarrow{H_2/Ni} \underset{\text{Diethyl aminomalonate (65\%)}}{H_2NCH(CO_2Et)_2}$$

$$\underset{\text{Diethyl aminomalonate}}{H_2NCH(CO_2Et)_2} + C_6H_5COCl \xrightleftharpoons{\text{Pyridine}} \underset{\text{Diethyl benzamidomalonate (90\%)}}{C_6H_5CONHCH(CO_2Et)_2}$$

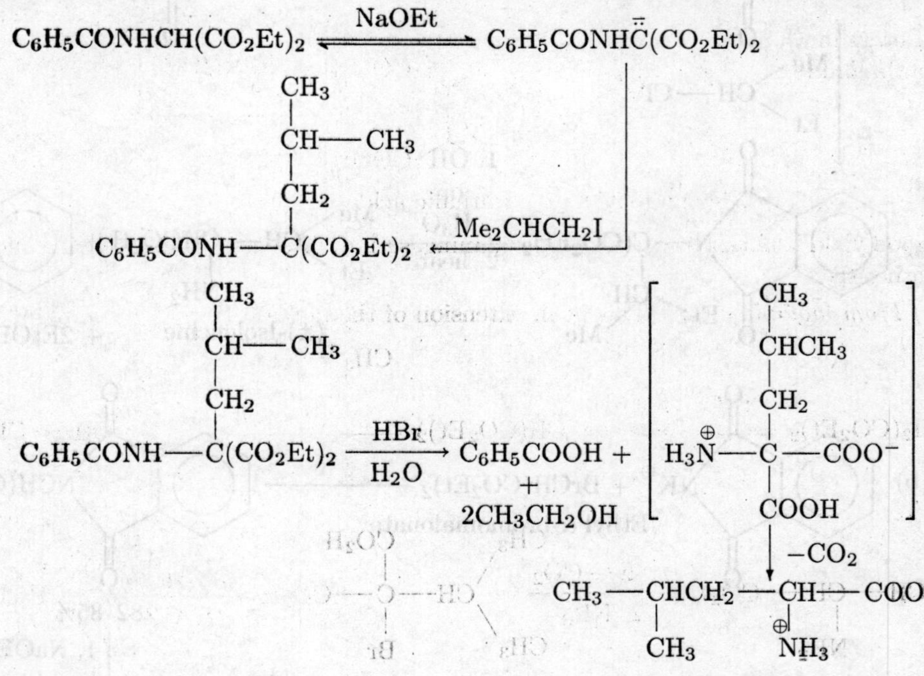

(±)-Leucine

An acidic α-amino acid may also be prepared from amidomalonic ester.

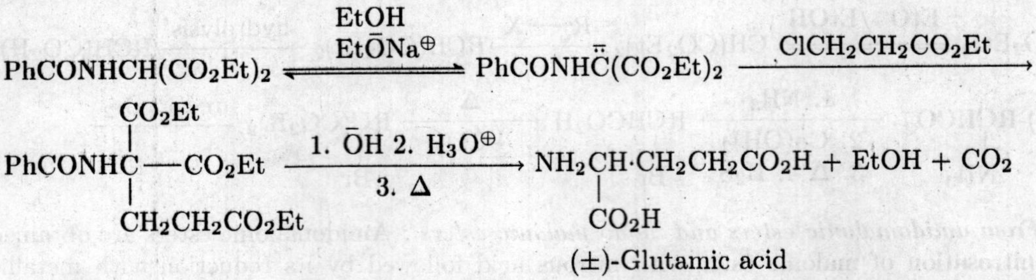


A variation of the amidoester procedure uses potassium phthalimide and bromomalonic ester to prepare an imidomalonic ester.

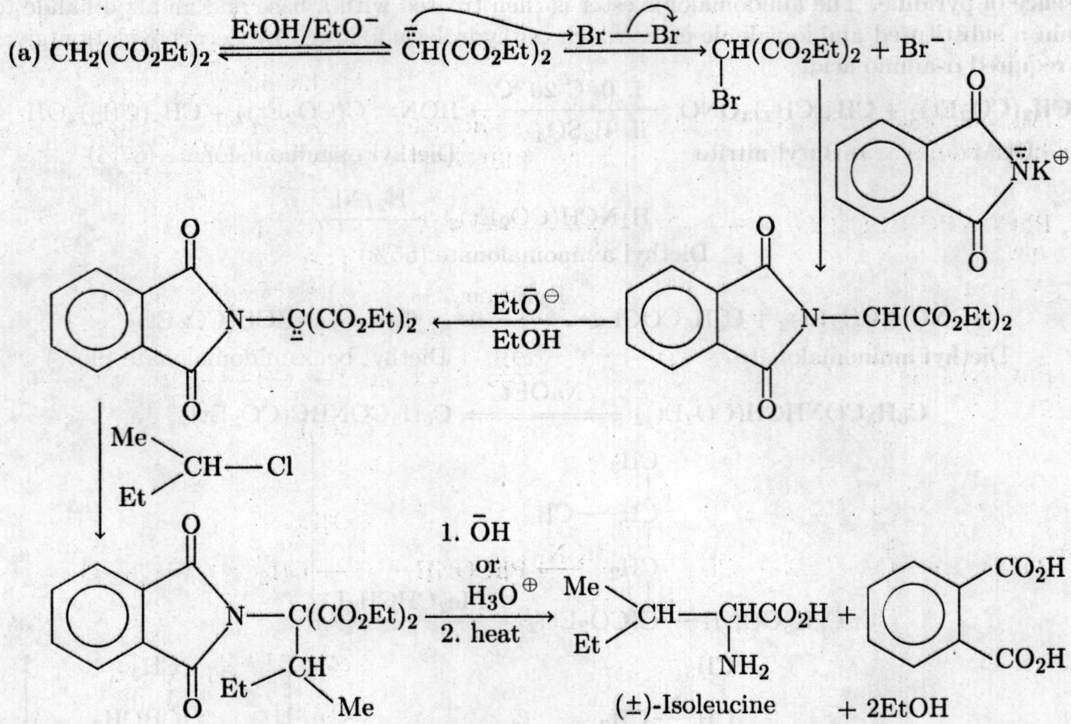


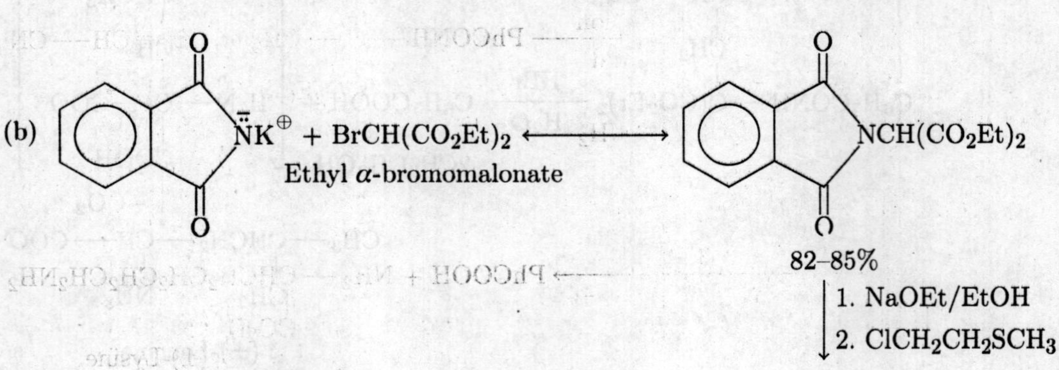

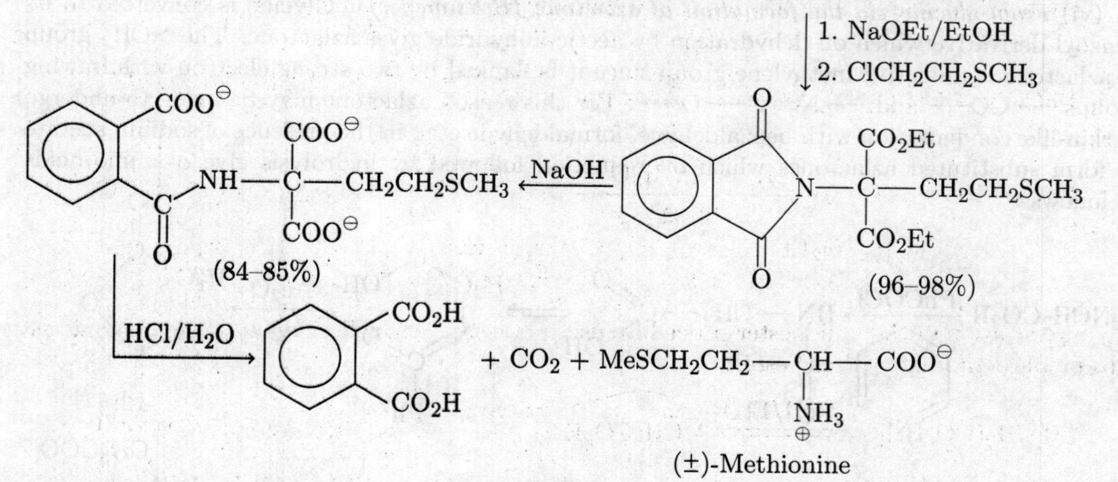

An amidomalonic ester may be subjected to Michael condensation to prepare an $\alpha$-amino acid; for example, a basic $\alpha$-amino acid, lysine is prepared as follows :

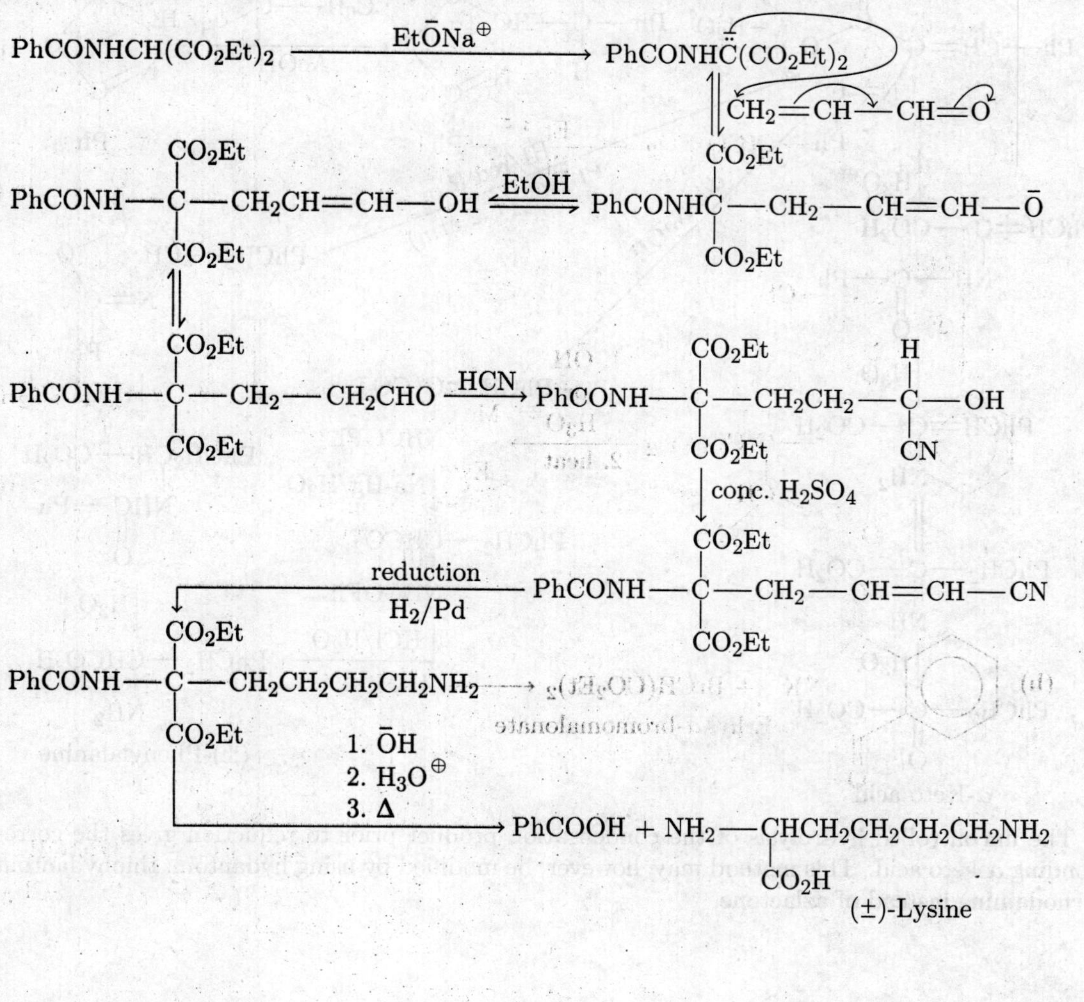

(vi) *From glycine via the formation of azlactone (Erlenmeyer)* : Glycine is converted to its benzoyl derivative which on dehydration by acetic anhydride gives azlactone. The >$CH_2$ group of azlactone is an active methylene group since it is flanked by two strong electron-withdrawing groups —CO— and —N=C—O—. For this reason azlactone may be made to undergo Perkin-like condensation with benzaldehyde, formaldehyde etc., in the presence of sodium acetate to form substituted azlactones which on reduction followed by hydrolysis give α-amino acids as follows :

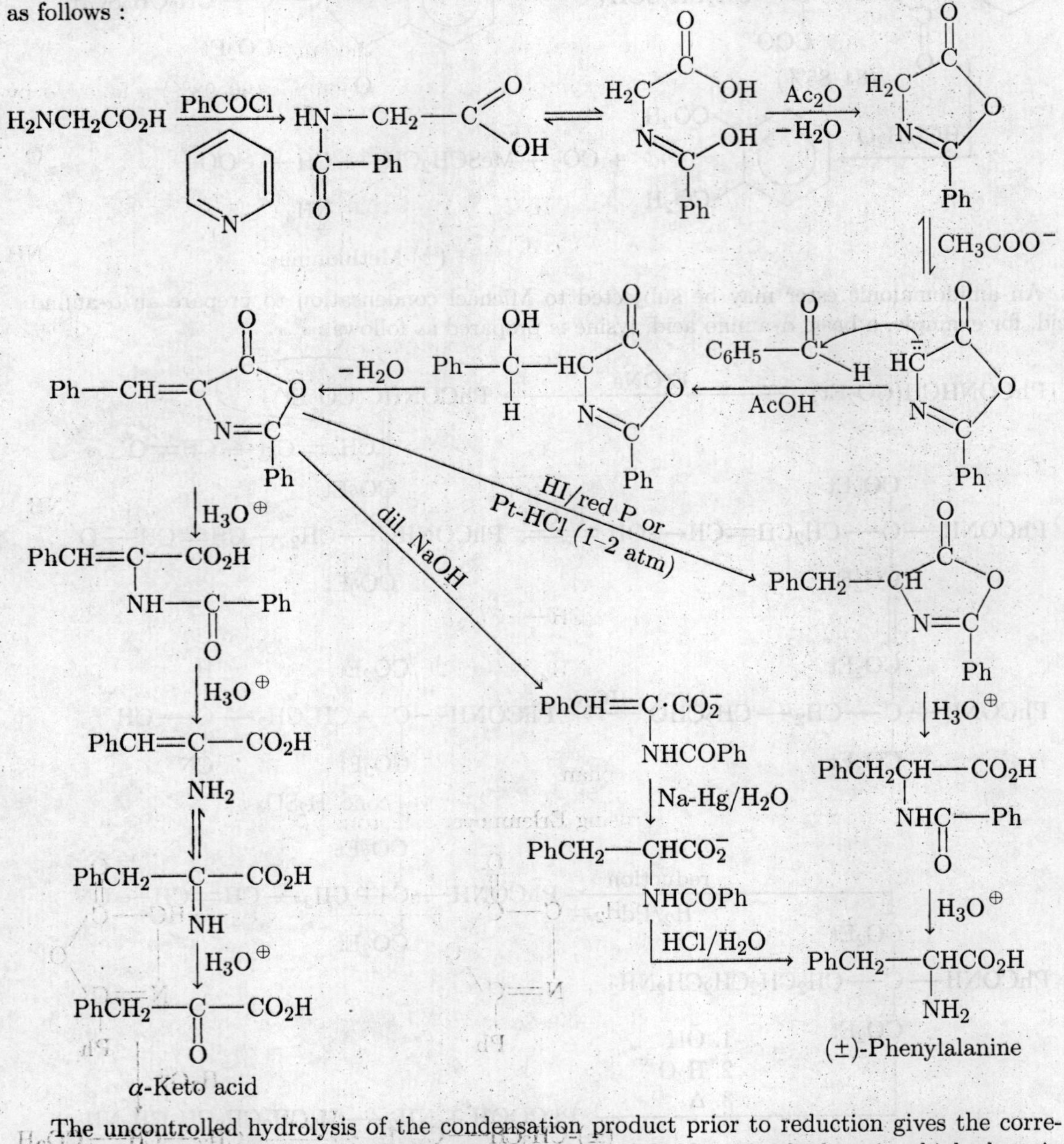

α-Keto acid

(±)-Phenylalanine

The uncontrolled hydrolysis of the condensation product prior to reduction gives the corresponding α-keto acid. This method may, however, be modified by using hydantoin, thiohydantoin or rhodamine instead of azlactone.

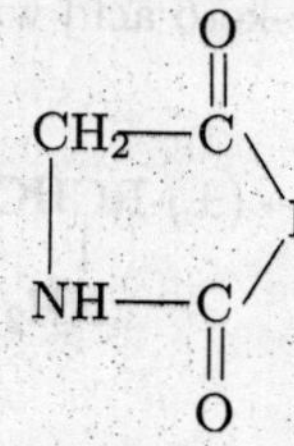

Hydantoin

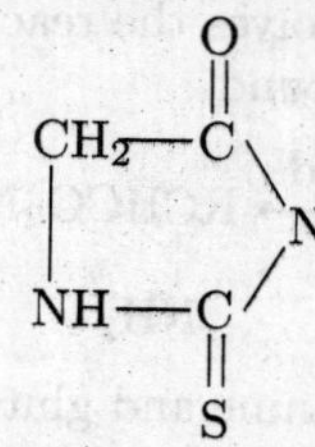

Thiohydantoin

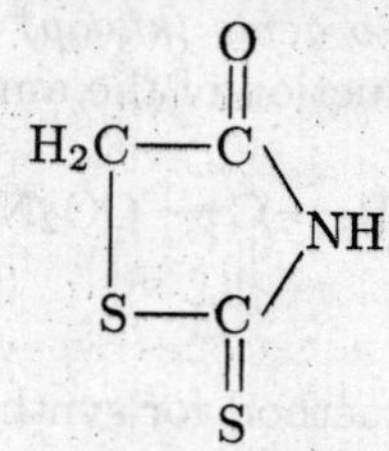

Rhodamine

3-Formylindole can be converted to tryptophan by the reaction with hydantoin followed by reduction and hydrolysis.

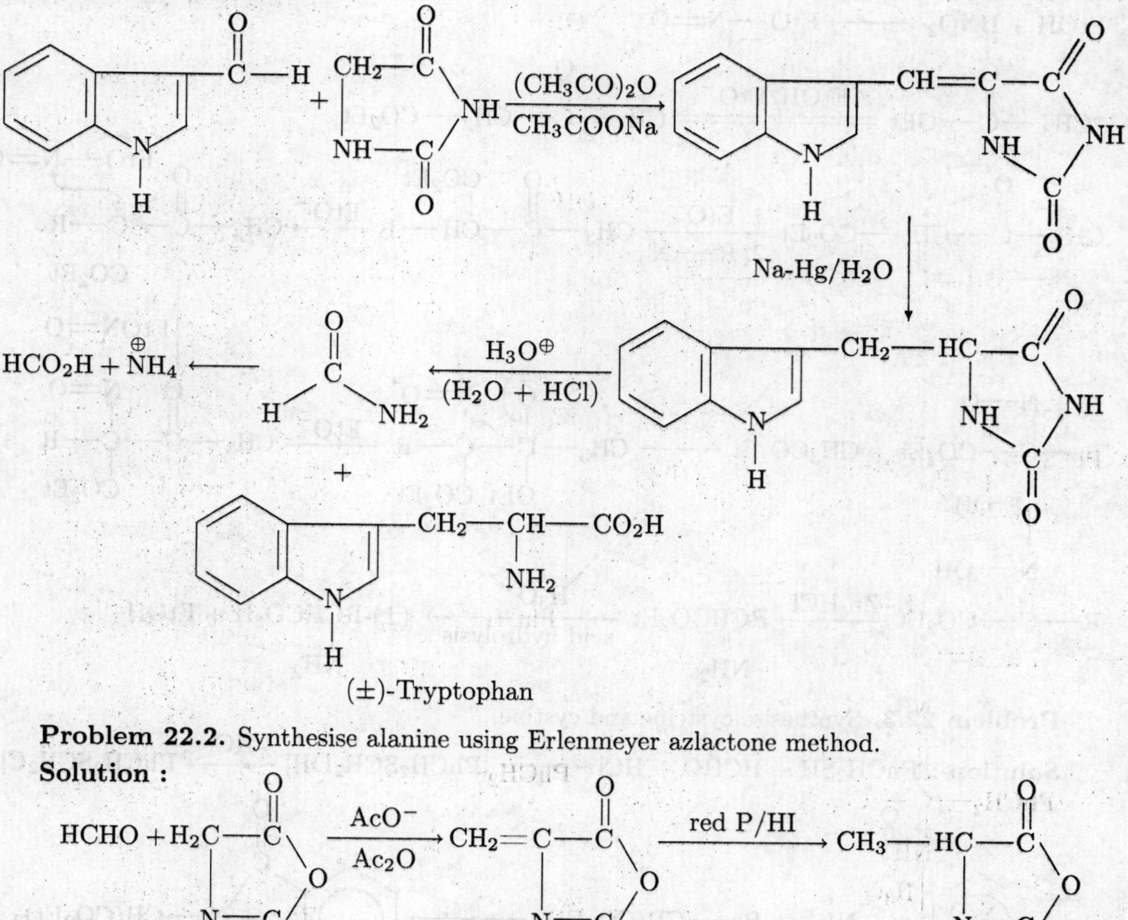

(±)-Tryptophan

**Problem 22.2.** Synthesise alanine using Erlenmeyer azlactone method.

**Solution :**

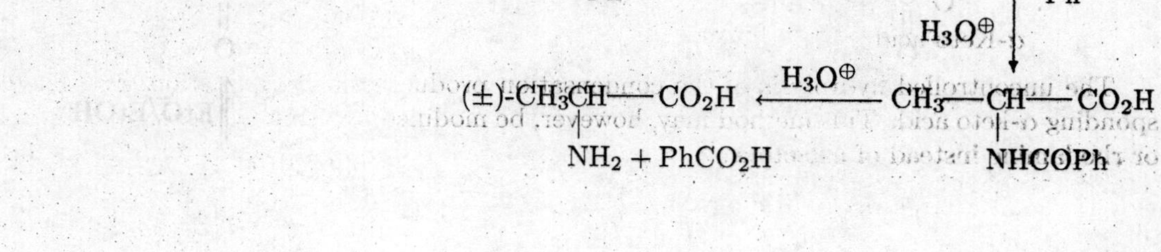

(vii) *From α-keto acids (knoop) :* This involves the reaction of an α-keto acid with ammonia and subsequent reduction of the amino compound.

$$RCOCO_2H \xrightarrow{NH_3} R-\underset{\underset{NH}{\parallel}}{C}-CO_2NH_4 \xrightarrow{H_2/Pd} R\underset{\underset{NH_2}{|}}{C}HCO_2NH_4 \xrightarrow{H_3O^{\oplus}} (\pm)\text{-}R\underset{\underset{NH_2}{|}}{C}HCO_2H + \overset{\oplus}{N}H_4$$

This is a good method for synthesising alanine and glutamic acid.

(viii) *From β-keto esters :* This involves the nitrosation of a derivative of acetoacetic ester by an alkyl nitrite; the nitroso derivative is then decomposed to an oxime which on reduction followed by hydrolysis gives an α-amino acid.

$$EtOH + HNO_2 \longrightarrow EtO-N=O$$

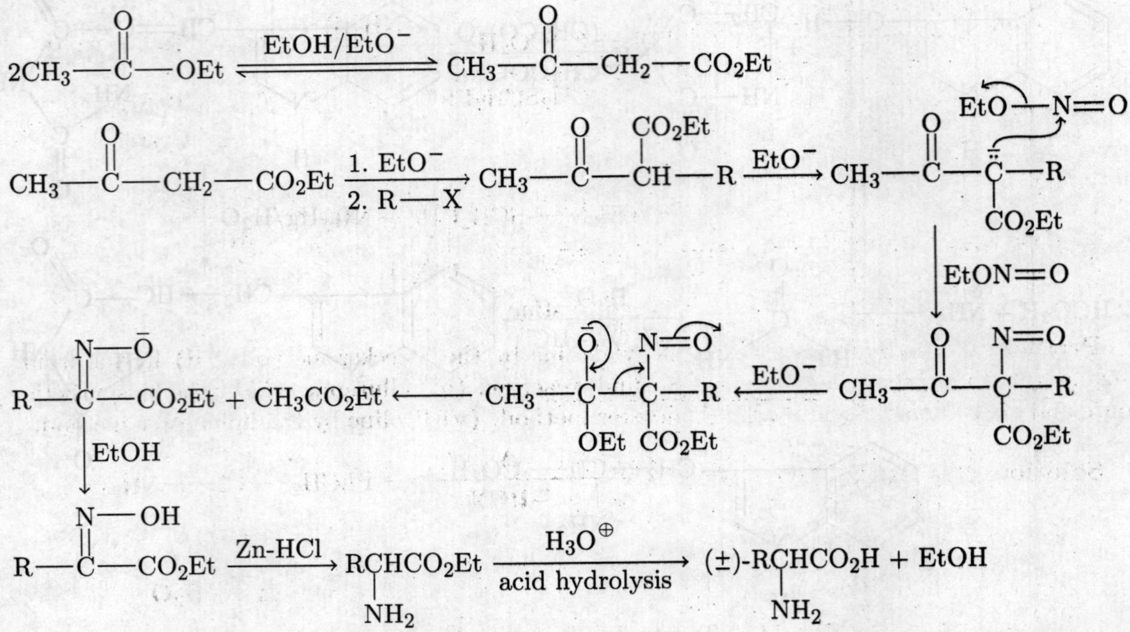

**Problem 22.3.** Synthesise cysteine and cystine.

**Solution :** $PhCH_2SH + HCHO + HCl \longrightarrow [PhCH_2SCH_2OH] \xrightarrow{HCl} PhCH_2SCH_2Cl$

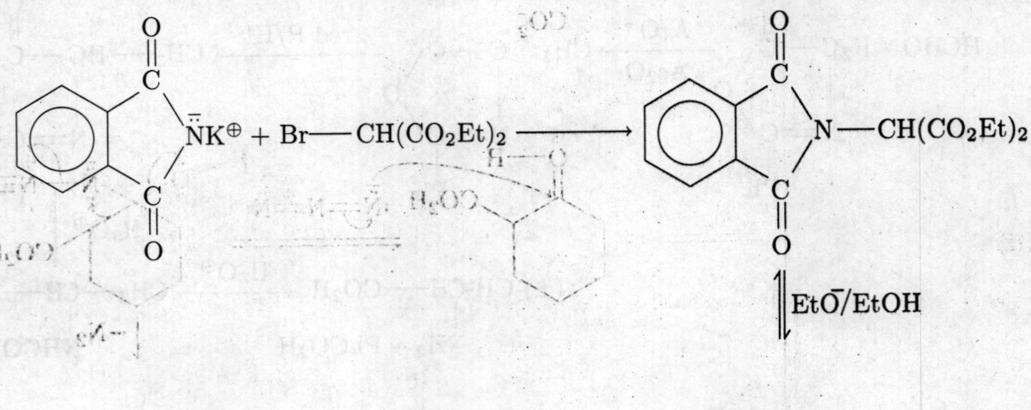

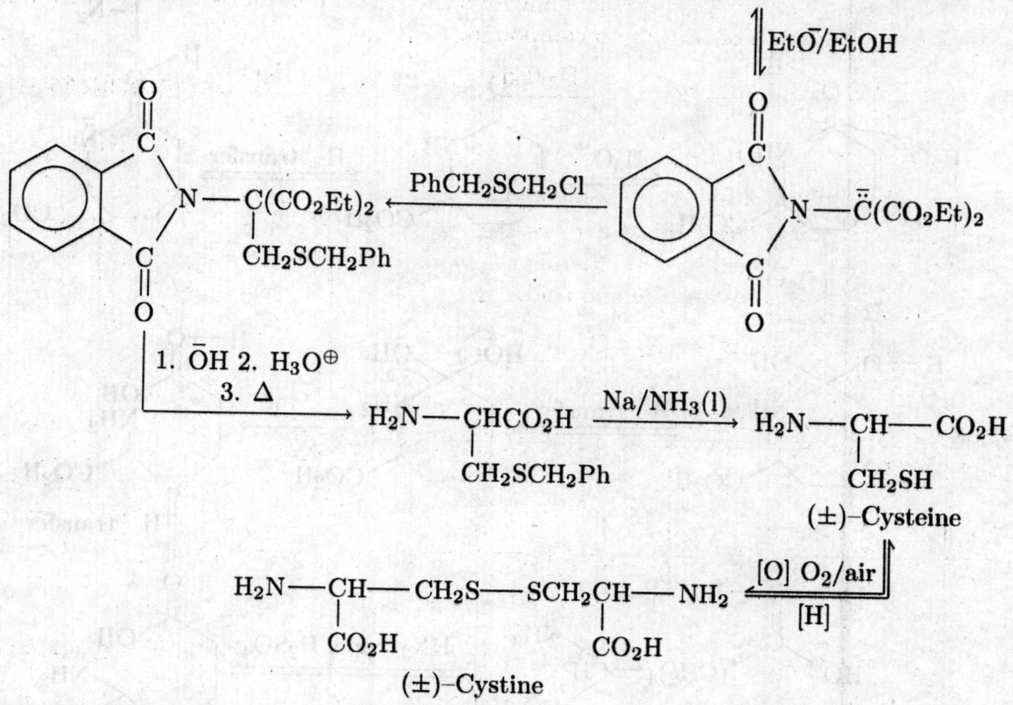

**Problem 22.4.** Synthesise : **(i)** Phenylalanine by the Strecker method. **(ii)** Lysine from 2-oxocyclohexanecarboxylic acid by the Schmidt reaction. **(iii)** Glutamic acid by the biosynthetic route and **(iv)** Aspartic acid by the imidoester method. **(v)** Proline by the imidoester method.

**Solution :** (i)

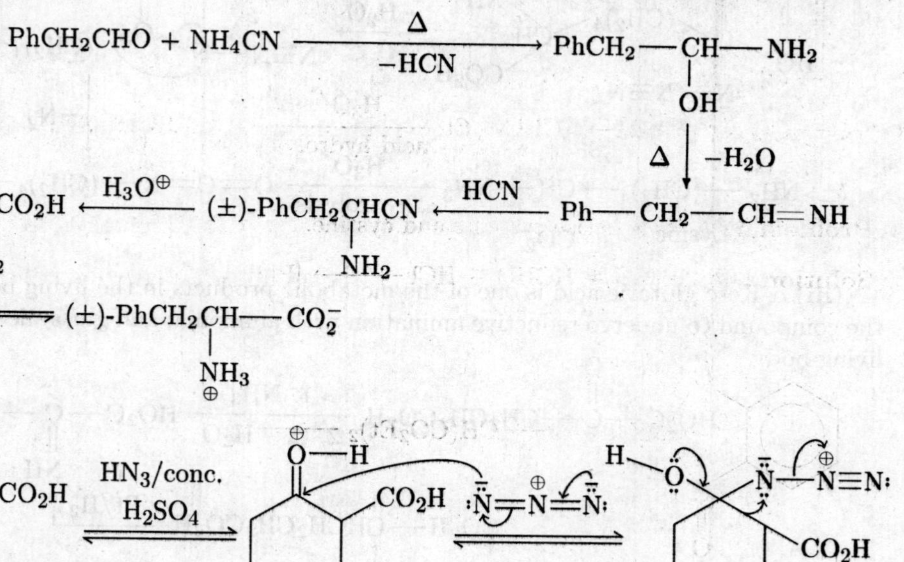

(ii)

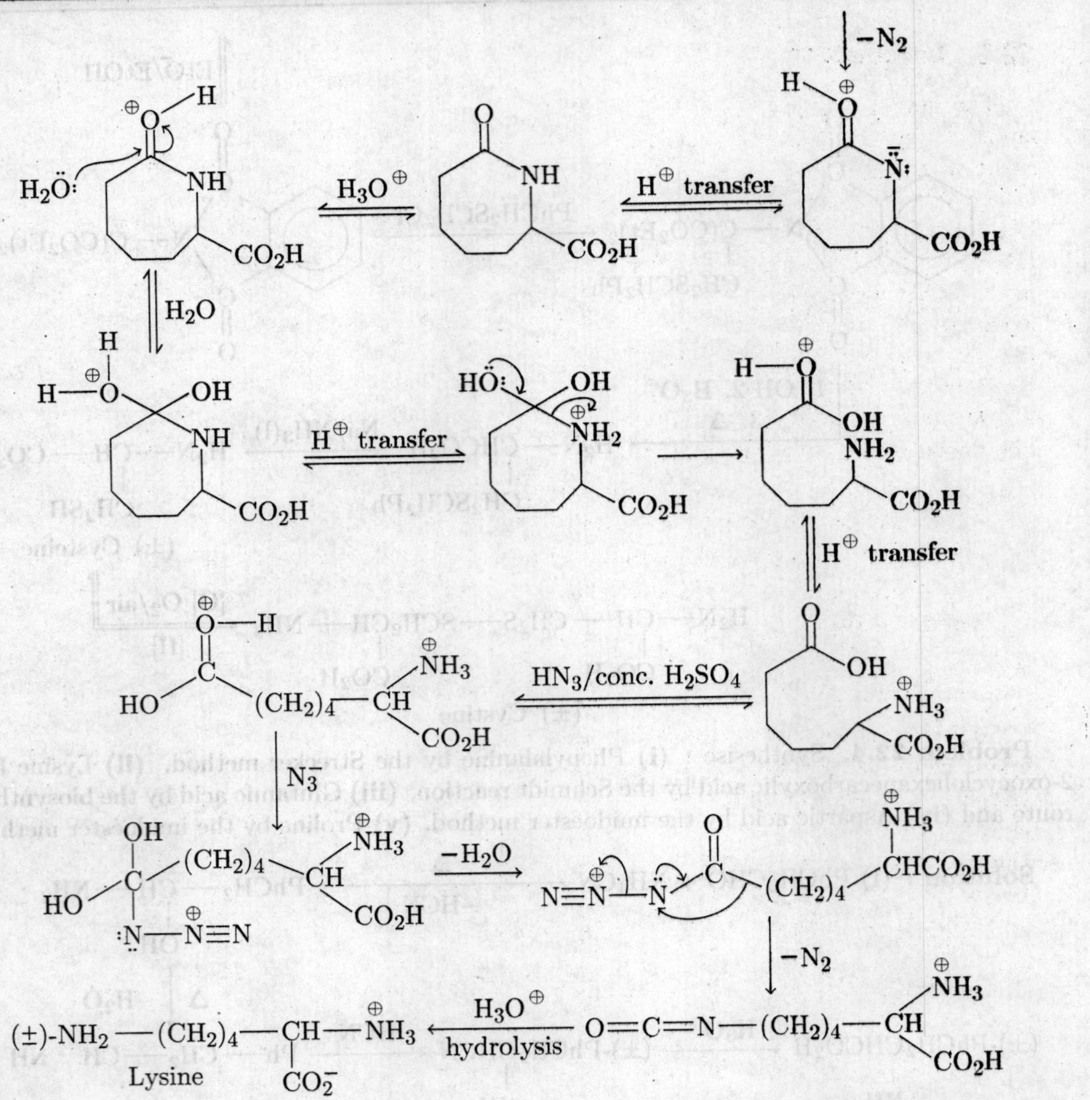

**(iii)** $\alpha$-Keto glutaric acid is one of the metabolic products in the living bodies. NADH makes the compound to undergo reductive amination with ammonia, and glutamic acid is formed in the living body.

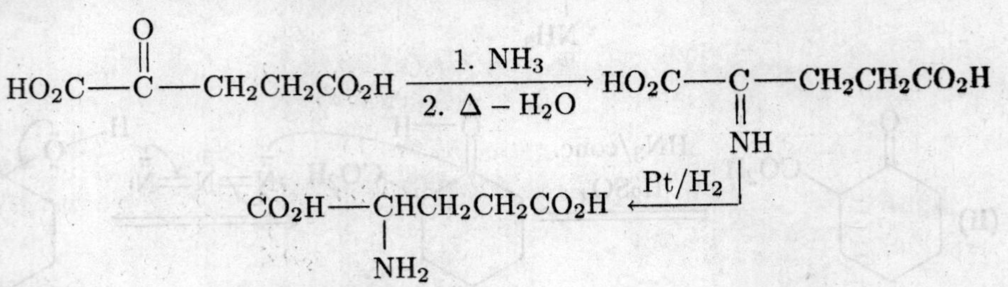

(±)-Glutamic acid

**(iv)** $CH_2(CO_2Et)_2 \underset{}{\overset{NaOEt/EtOH}{\rightleftharpoons}} :\bar{C}H(CO_2Et)_2 \xrightarrow{Br—Br} Br—CH(CO_2Et)_2 + Br^-$

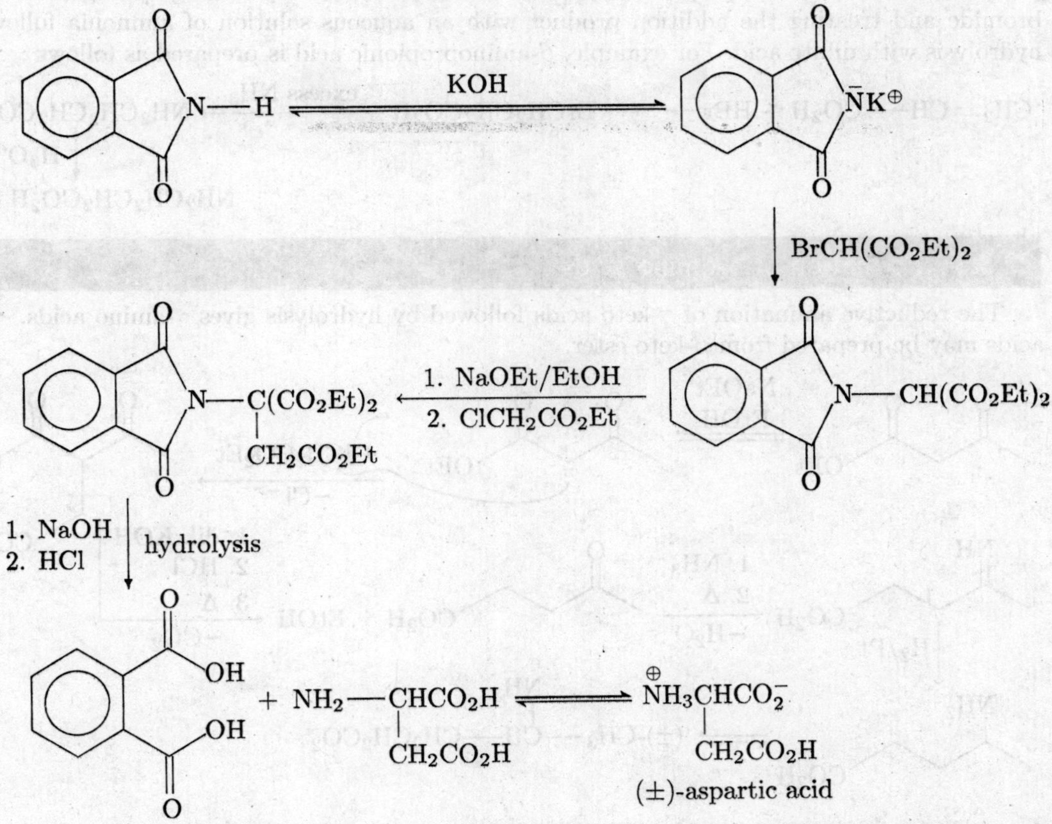

**(v)**

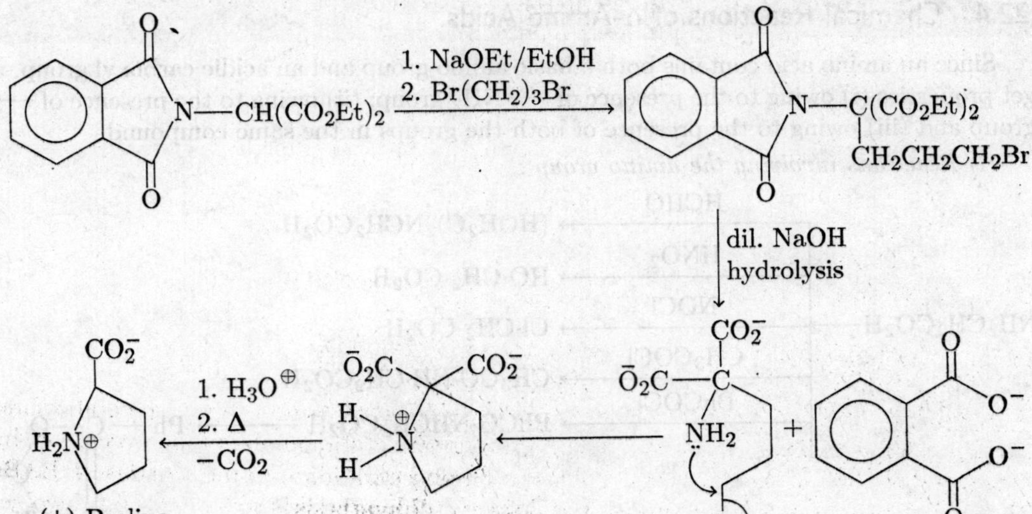

### 22.3.2 Formation of β-amino acids

β-Amino acids can be prepared from α, β-unsaturated acids by adding hydrogen chloride or bromide and treating the addition product with an aqueous solution of ammonia followed by hydrolysis with dilute acid. For example, β-aminopropionic acid is prepared as follows :

$$CH_2{=}CH{-}CO_2H + HBr \longrightarrow BrCH_2CH_2CO_2H \xrightarrow{\text{excess } NH_3} NH_2CH_2CH_2CO_2NH_4$$

$$NH_2CH_2CH_2CO_2NH_4 \xrightarrow{H_3O^{\oplus}} NH_2CH_2CH_2CO_2H + \overset{\oplus}{N}H_4$$

### 22.3.3 Formation of γ-amino acids

The reductive amination of γ-keto acids followed by hydrolysis gives γ-amino acids. γ-keto-acids may be prepared from β-keto ester.

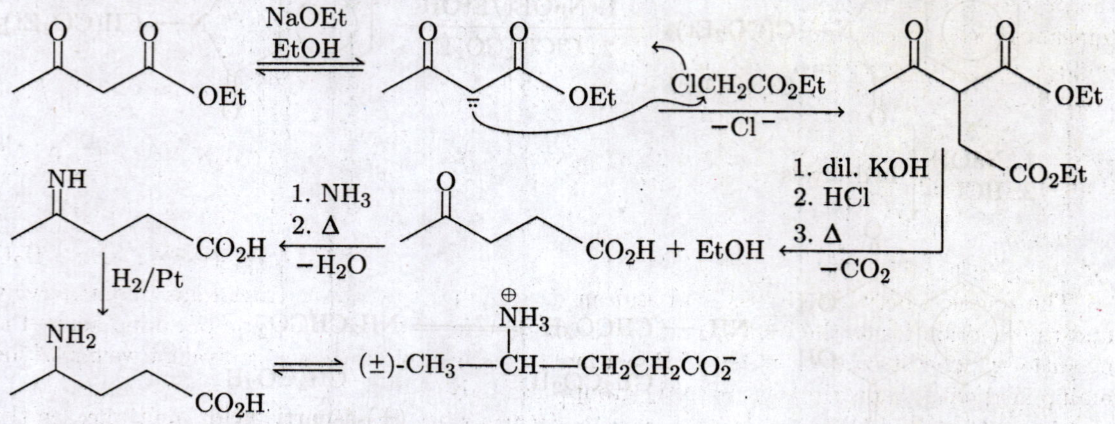


## 22.4 Chemical Reactions of α-Amino Acids

Since an amino acid contains both a basic amino group and an acidic carboxyl group, we shall get properties (i) owing to the presence of —$NH_2$ group; (ii) owing to the presence of —$CO_2H$ group and (iii) owing to the presence of both the groups in the same compound.

(i) *Reactions involving the amino group :*

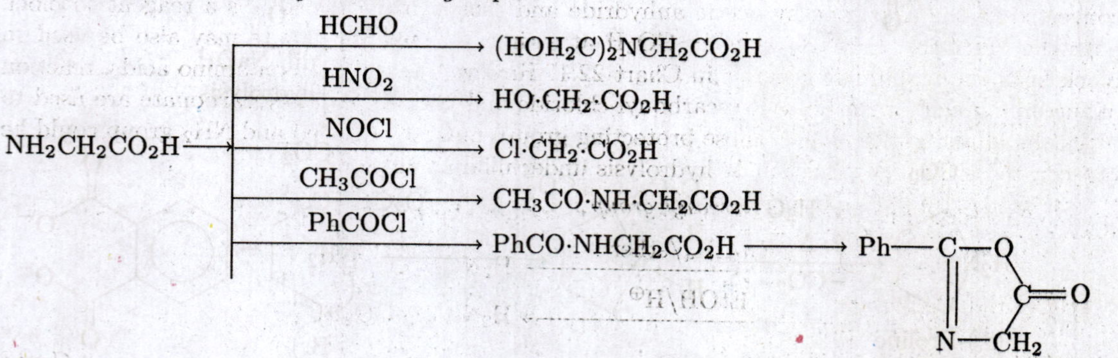


*Contd.*

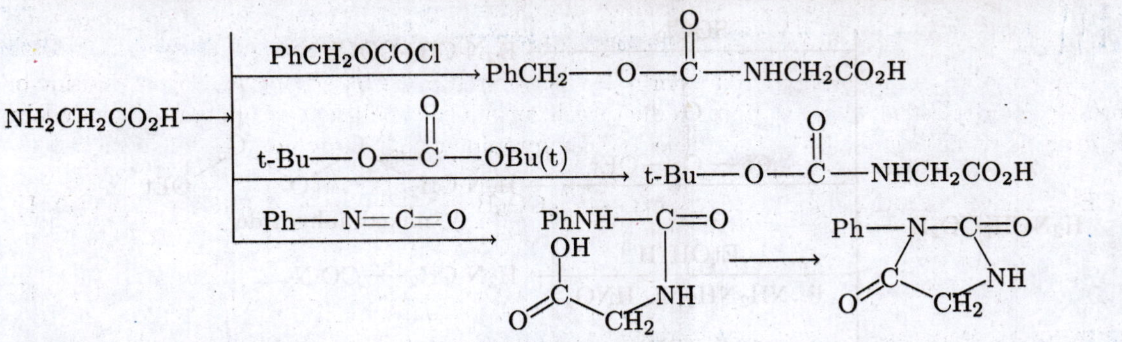

**Chart 22.1**

The reaction between an α-amino acid and formaldehyde is utilised in the estimation of α-amino acid. The reaction product is titrated against a standard sodium hydroxide solution and the process is known as Sörensen titration method. The blocked amino group (by reacting with formalin) makes the α-amino acid to act as a strong acid.

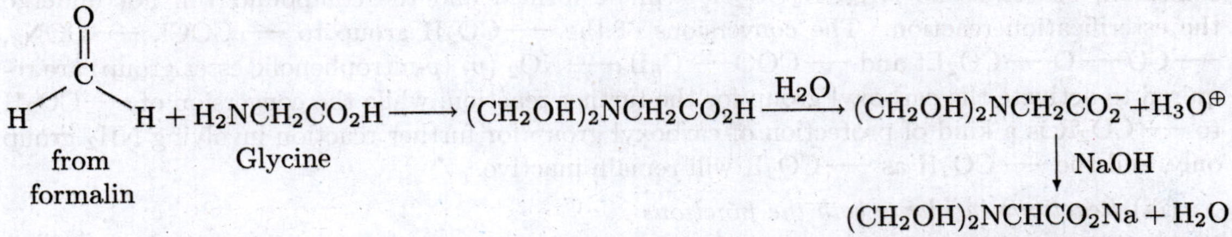

The reaction between nitrous acid and an α-amino acid evolves nitrogen gas quantitatively and the reaction forms the basis of van Slykes method for the determination of amino acid—the evolution of 22.4 litre nitrogen at N.T.P. per mole of amino acid gives the equivalent weight of the amino acid with respect to the —$NH_2$ group; naturally proline and hydroxyproline containing —NH— group do not undergo the reaction. However, the equivalent weight multiplied by the number of —$NH_2$ group present in one molecule of the amino acid gives molecular weight of the amino acid.

Acetylation and benzoylation reactions are carried out on the amino group by acetic anhydride and benzoyl chloride respectively to form the corresponding acetyl and benzoyl derivatives—the reaction medium, however, should be strongly alkaline; if it is not, there will be no free $NH_2$ group in the amino acid and the reaction will not take place. Benzoylated amino acid may be converted to the azlactone by acetic anhydride and thus benzoyl chloride is a reagent to block both the functions, —$NH_2$ and —$CO_2H$ at a time. Phenyl isocyanate may also be used to block both the functions as shown in Chart 22.1. However, acetylation on amino acids, reaction of the amino acids with benzyloxycarbonyl chloride and with di-*t*-butoxy carbonate are used to block the amino group alone. These protecting groups may be removed and $NH_2$ group could he regenerated either by reduction or hydrolysis under mild condition.

(ii) *Reactions involving the carboxyl group* :

$$H_2NCH_2CO_2H \xrightarrow{LiAlH_4/\text{ether}} H_2NCH_2CH_2OH$$

$$H_2NCH_2CO_2H \xrightarrow{EtOH/H^{\oplus}} H_2N{\cdot}CH_2{\cdot}CO_2Et$$

*Contd.*

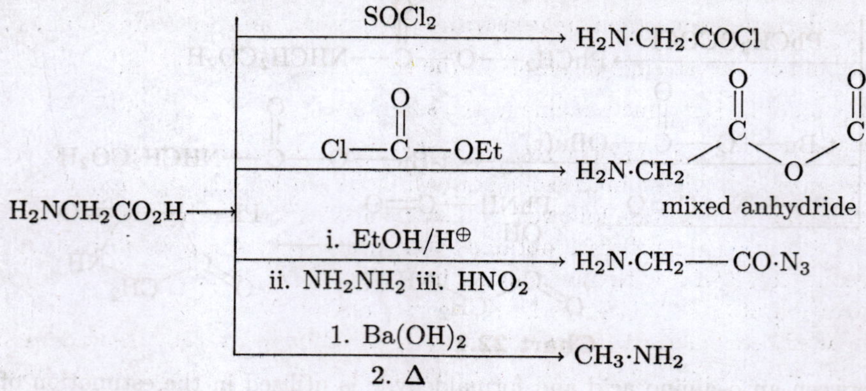

**Chart 22.2**

The esterification of amino acids is to be carried out under a moderately strong acidic condition, otherwise no $\overset{\oplus}{N}H_3CH_2C\overset{\oplus}{O}_2H_2$ will be formed and the compound will not undergo the esterification reaction. The conversions of the —$CO_2H$ group to —COCl, —$CON_3$, —CO—O—$CO_2Et$ and —COO—$C_6H_4$—$NO_2$ (*p*) (*p*-nitrophenolic ester group) are required to activate the carboxyl group for the further reaction; while the conversion of —$CO_2H$ to —$CO_2R$ is a kind of protection of carboxyl group for further reaction involving $NH_2$ group only, and the —$CO_2H$ as —$CO_2R$ will remain inactive.

(iii) *Reactions involving both the functions :*

$$H_3\overset{\oplus}{N}CH_2\cdot C\bar{O}_2 \xrightarrow{CH_3I} (CH_3)_3\overset{\oplus}{N}\cdot CH_2\cdot C\bar{O}_2$$

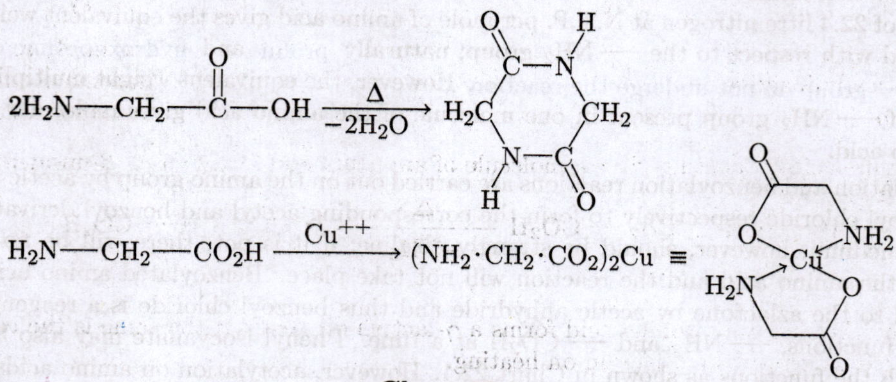

**Chart 22.3**

The formation of zwitter ion is responsible for the development of $p_i$ for an amino acid. Because of the formation of dipolar ion, amino acids conduct electricity and possess high b.p.s and m.p.s. For the same reason racemic forms of neutral amino acids are difficult to resolve into their enantiomers. The procedure for resolution of neutral $\alpha$-amino acids has already been discussed in the ninth chapter. However, the resolution may also be carried by protecting the amino group by acetylation or benzoylation.

Let us take the case of resolution of DL-alanine. The racemic form is subjected to benzoylation under basic condition. N-benzoyl derivatives of D- and L-alanines are formed. Thus the amino group is blocked. The derivatives are then treated with (−)-brucine. Two salts are obtained.

(i) N-benzoyl-D-alanine· (−)-brucine, insoluble in the medium and precipitates.

(ii) N-benzoyl-L-alanine· (−)-brucine remains in the solution.

The two salts are separated by filtration. The residue is of salt (i) : This on hydrolysis with dilute mineral acid gives N-benzoyl-D-alanine in the solid state while (−)-brucine.H$^{\oplus}$ salt remains in the dissolved state. Filtration of the solution followed by alkaline hydrolysis, heating and acidification gives optically pure D-alanine. The filtrate obtained from the separation of two salts (i) and (ii) contains the second salt along with first salt as impurity. This on sequential treatments as done for the separation of D-alanine gives L-alanine containing D-alanine as impurity. The separated L-alanine, after benzoylation, is treated with (−)-strychinine whereby insoluble salt of N-benzoyl-L-alanine· (−)-strychnine separated but N-benzoyl-D-alanine· (−)-strychnine remains in the solution in the dissolved state. The residue obtained by filtration is subjected to sequential procedures as done to get D-alanine whereby optically pure L-alanine is obtained.

When neutral $\alpha$-amino acids are heated, derivatives of diketopiperazines are obtained. Each ketopiperazine obtained remains as a mixture which consists of a racemic form and a meso form, both are optically inactive but the former one is resolvable. This means the products of heating a neutral $\alpha$-amino acid exhibit both enantiomerism and diastereomerism.

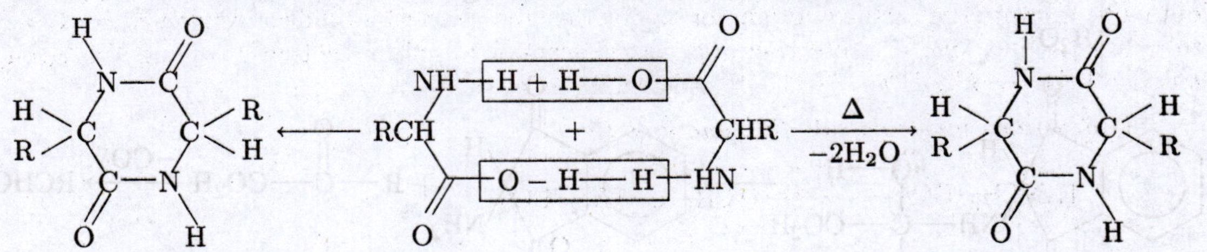

*trans*; it has *i* and $S_2$ axis. Optically inactive meso form

*cis*; no $S_n$ axis. Racemic form; inactive but resolvable

A $\beta$-amino acid on heating gives a molecule of ammonia and forms an $\alpha$, $\beta$-unsaturated acid.

$$CH_3—\underset{NH_2}{\underset{|}{CH}}—CH_2—CO_2H \xrightarrow[-NH_3]{\Delta} CH_3CH{=}CH—CO_2H$$

On the other hand a $\gamma$-amino acid forms a $\gamma$-lactum on heating; the same is the case with a $\delta$-amino acid which gives a $\delta$-lactum on heating.

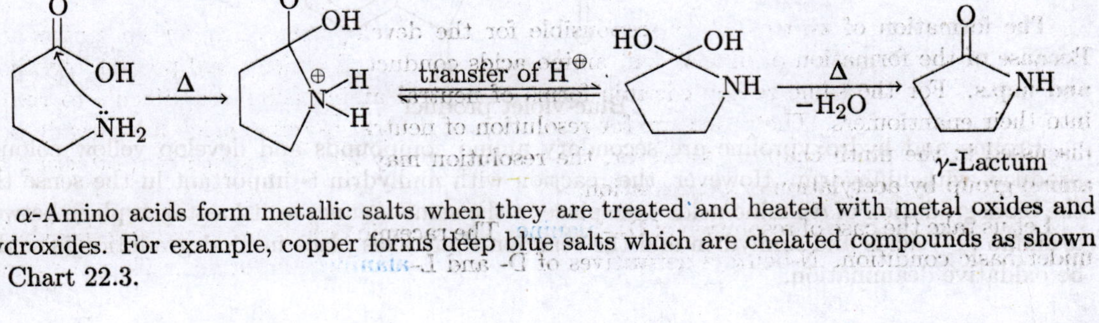

$\gamma$-Lactum

$\alpha$-Amino acids form metallic salts when they are treated and heated with metal oxides and hydroxides. For example, copper forms deep blue salts which are chelated compounds as shown in Chart 22.3.

## 22.5 Tests for Amino Acids

All $\alpha$-amino acids (proline and hydroxyproline are exceptions) develop a blue-violet coloured product on reaction with ninhydrin which is hydrated indane-1, 2, 3-trione.

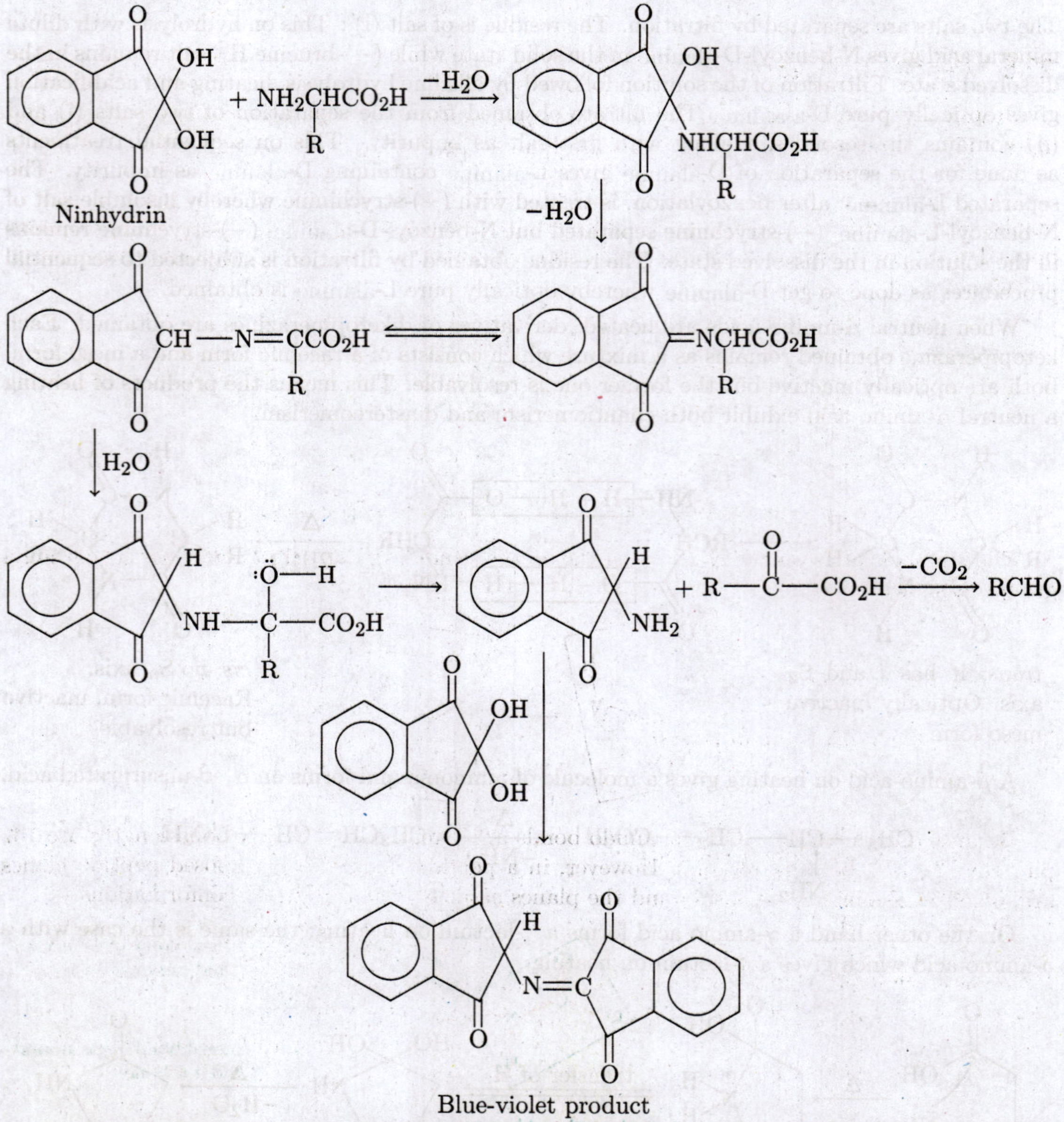


Blue-violet product

Proline and hydroxyproline are secondary amino compounds and develop yellow coloured products with ninhydrin. However, the reaction with ninhydrin is important in the sense that the biodegradation of $\alpha$-amino acids take place in the same manner as it has taken place between $\alpha$-amino (1°) acids and ninhydrin and the reaction that takes place in biodegradation is said to be oxidative deamination.

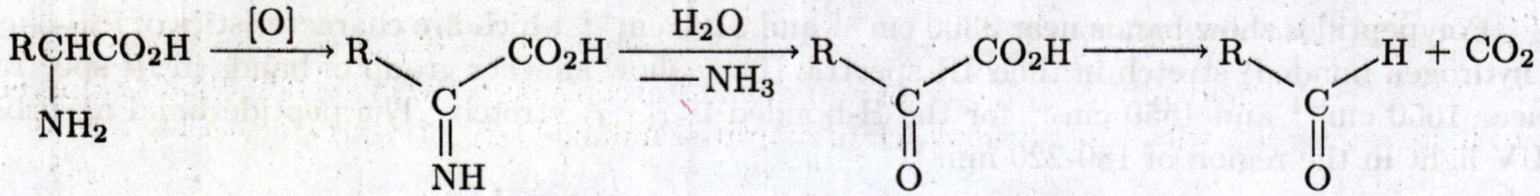

## 22.6 Peptide Linkage

Proteins (Gr. 'Proteios' meaning 'first') are compounds composed of $\alpha$-amino acids as building units. The amino acid units of proteins are bound together by a repeating sequence of amide bonds, $-\overset{O}{\overset{\|}{C}}-NH-$, called peptide linkages. In a petide linkage the O atom and the H atom bonded to N are trans to each other. The dotted portions of the following figure are peptide linkages.

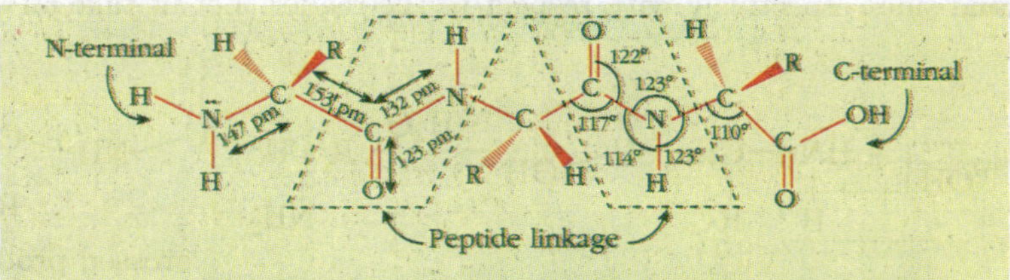


**Fig. 22.1. Description of bond angles and bond lengths in a peptide linkage**

The lone pair of *e*s on the N atoms of the amido bonds can delocalise involving the carbonyl groups as shown below :

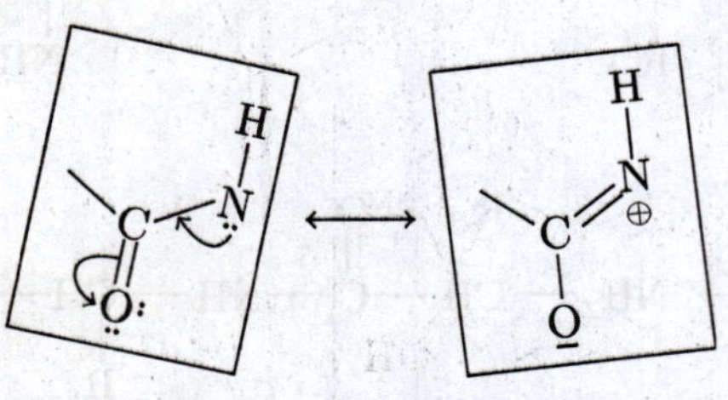

Because of this resonance effect, amido bonds are planar and OC—N bond lengths are 132 pm, HN—CHR bond is 147 pm. However, in a peptide chain two juxtaposed peptide planes are joined to a common $\alpha$-C atom and the planes assume wide and varied conformations.

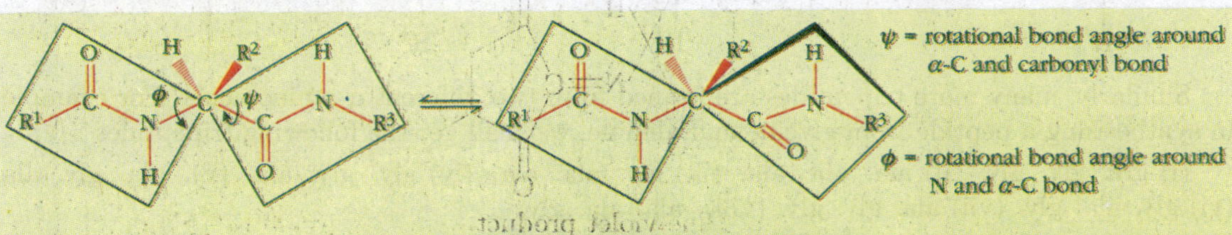


**Fig. 22.2. Conformation of juxtaposed peptide planes**

The presence of a peptide linkage may be identified by the biuret test as done in the case of urea. When an alkaline solution of a large peptide is treated with a drop of dilute copper sulphate solution, a blue-violet to pink colour develops. However, small peptides may not develop the colour.

Polypeptides show bands near 3300 cm$^{-1}$ and 3100 cm$^{-1}$ which are characteristics of $\bar{\nu}_{N—H}$ (hydrogen bonded) stretch in their IR spectra. They show another group of bands in IR spectra near 1650 cm$^{-1}$ and 1550 cm$^{-1}$ for the H-bonded $\bar{\nu}_{>C=O}$ stretch. The peptide bond absorbs UV light in the region of 180-220 nm.

## 22.7 Polypeptides and their Synthesis

A large number of α-amino acid molecules, different or same, join together to form very big molecules known as polypeptides. Usually a polypeptide consisting of hundred α-amino acid units is known as a protein unit. However, theoretically when two different α-amino acid units are joined together to form a dipeptide, actually a mixture of four dipeptides is obtained since an α-amino acid unit may undergo the reaction through both the ends, amino end (called N-terminal) and carboxyl end (called C-terminal). In this connection it is important to notė that the N-terminal and the C-terminal of an α-amino acid or of a peptide are written on the left-hand side and right-hand side respectively with respect to the reader. Let us take an example.

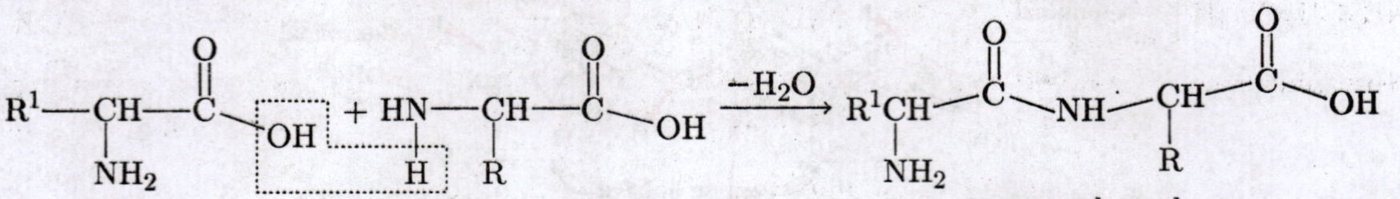

crossed product

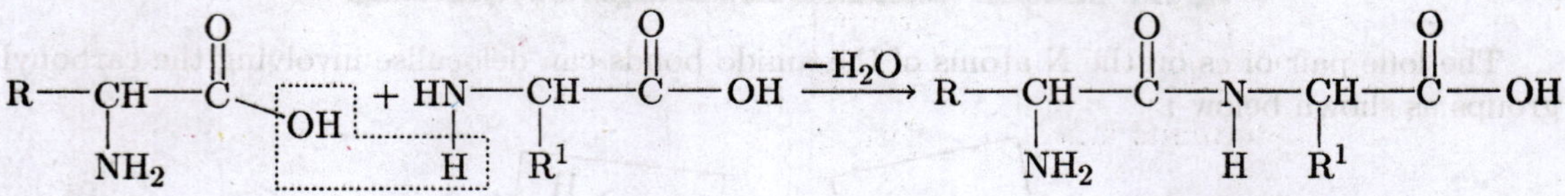

crossed product

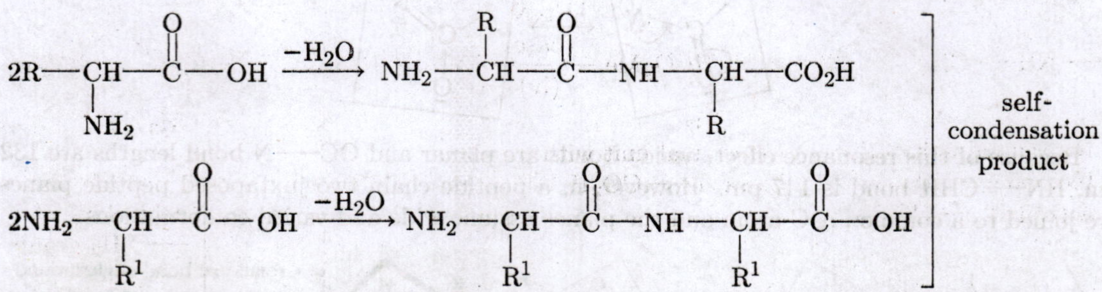


Similarly, many more tripeptides are formed from two different α-amino acids. For example, in synthesising a peptide from glycine and alanine, we shall get the following tripeptides :

(i) gly. gly. gly. (ii) ala. ala. ala. (iii) gly. ala. ala. (iv) gly. gly. ala. (v). ala. gly. ala. (vi) gly. ala. gly. (vii) ala. gly. gly. (viii). ala. ala. gly.

So to synthesise a particular peptide in good yield, we have to take care so that other peptides do not form from the amino acid units which are taken as starting materials. To reach this objective, we shall have to adopt the following sequential procedures

(i) The amino group of the amino acid unit which forms the N-terminal of the required peptide is to be blocked in order to make the amino group inactive.

(ii) The carboxyl group of the N-terminal amino acid unit is to be activated in order to make the carboxyl group more reactive than the carboxyl group of the next amino acid unit in the peptide chain.

(iii) The carboxyl group of the second amino acid unit of the peptide chain is to be deactivated by blocking the group.

(iv) The N-terminal amino acid unit with blocked amino group and activated carboxyl group is to be treated with the second amino acid unit with the blocked carboxyl group in a basic medium, and lastly :

(v) The amino group of the N-terminal unit and the carboxyl group of the C-terminal unit of the peptide chain are to be deblocked under mild conditions. For example, if we want to synthesise gly. ala, $NH_2CH_2CONHCH(CH_3)CO_2H$, we shall have to proceed a path as shown below :

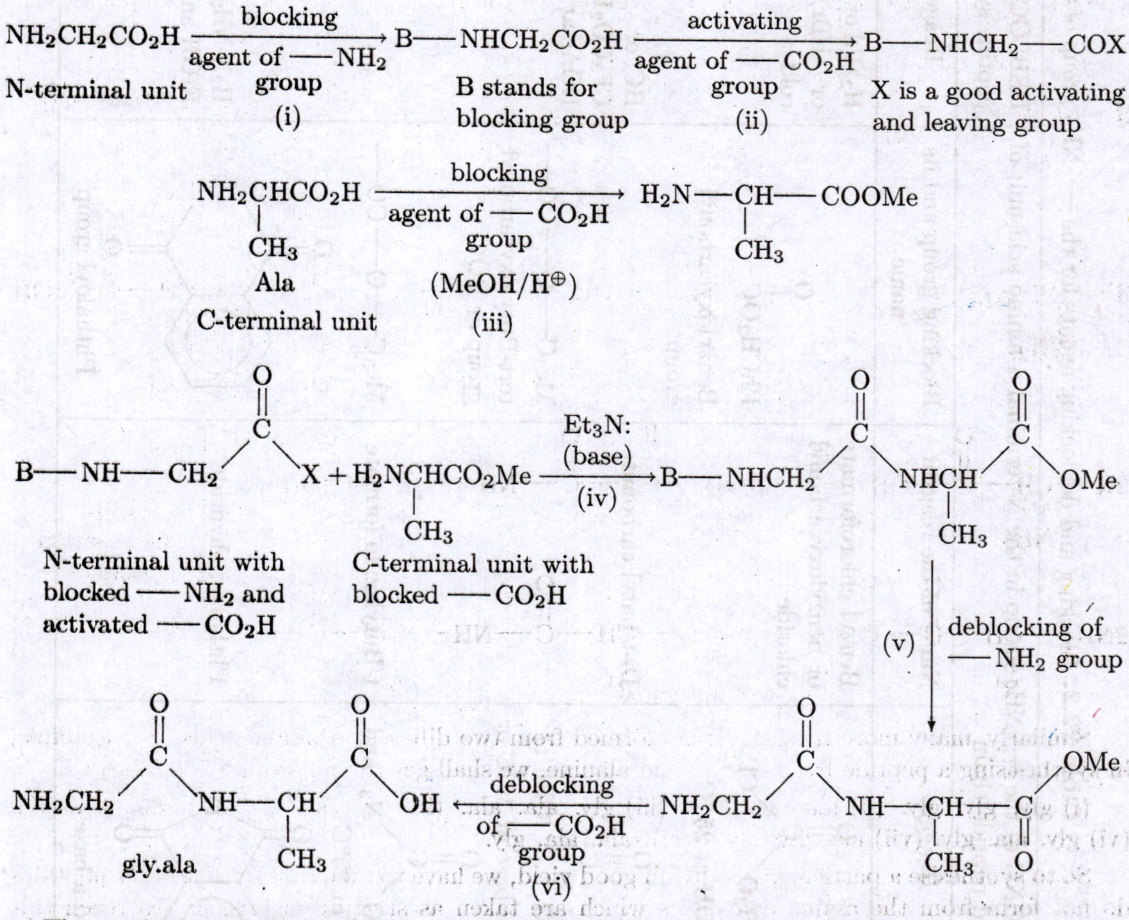

The methods and agents for blocking and deblocking of —$NH_2$ group are given in Table 22.2. Let us carry out the methods on an $\alpha$-amino acid as given in Table 22.2.

**Table 22.2.** Blocking and deblocking agents for the —$NH_2$ group of an $\alpha$-amino acid unit

| BLOCKING of the —$NH_2$ group in the N-terminal $\alpha$-amino acid unit of a desired peptide chain | | | DEBLOCKING of the —$NH_2$ group in the peptide synthesised | |
|---|---|---|---|---|
| Reagent | Name of the reagent | Blocking group and its name | Reagent | Products |
| 1. $PhCH_2O-C(=O)-Cl$ and $\bar{O}H$ at 25 °C | Benzyl chloroformate or benzyloxycarbonyl chloride | $PhCH_2OC(=O)-$ Benzyloxycarbonyl group | $H_2/Pd$ or $H_2/Pt$ or $HBr/HOAc$ cold | $PhCH_3 + CO_2 +$ peptide<br>$PhCH_2Br + CO_2 +$ peptide |
| 2. $t\text{-BuO}-C(=O)-OBu(t)$ and a base at 25°C or | Di-t-butyl carbonate | $Me_3C-O-CO-$ tert-Butyloxycarbonyl group or BOC | HCl or $CF_3CO_2H$ in HOAc at 25 °C | $(CH_3)_2C=CH_2 + CO_2 +$ peptide |
| $t\text{-BuO}-C(=O)-N_3$ and base at 25°C | t-Butylazido formate | $Me_3C-O-CO-$ | | |
| 3. 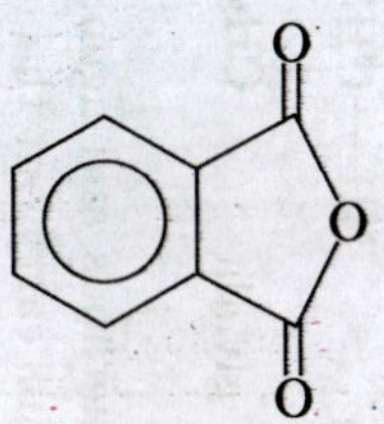 and a base | Phthalic anhydride | 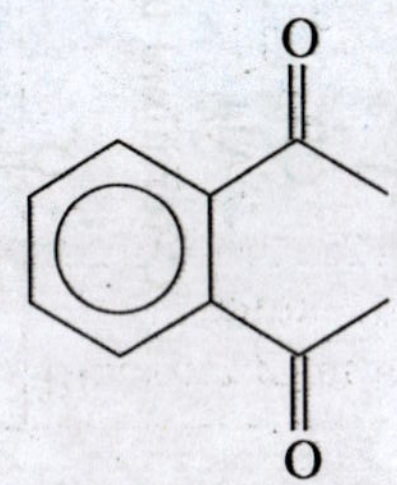 Phthaloyl group | $H_2N{\cdot}NH_2$ EtOH and HCl | 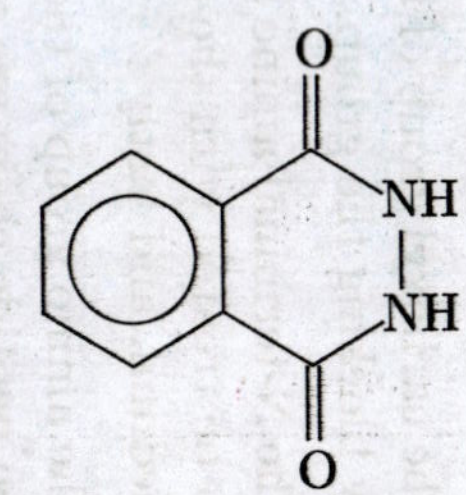 + peptide |

*Contd.*

| BLOCKING of the —$NH_2$ group in the N-terminal α-amino acid unit of a desired peptide chain | | | DEBLOCKING of the —$NH_2$ group in the peptide synthesised | |
|---|---|---|---|---|
| Reagent | Name of the reagent | Blocking group and its name | Reagent | Products |
| 4. 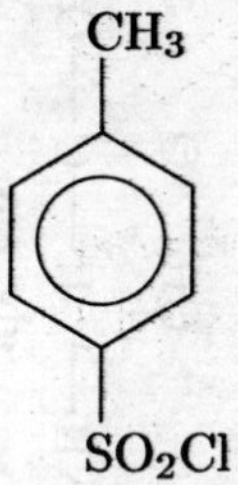 | p-Toluenesulphonyl chloride (Tosyl chloride) | 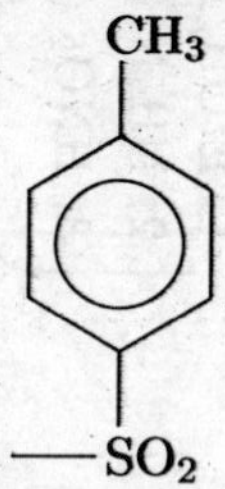 Tosyl group | $Na/NH_3(l)$ | 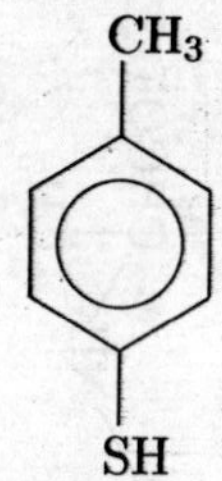 + peptide |
| 5. $Ph_3C$—Cl and a base | Trityl chloride or triphenylmethyl chloride | $Ph_3C$— Trityl or triphenylmethyl group | $H_2/Pd$ or aqueous acid | $Ph_3CH$ + peptide or $PH_3C$—X + peptide |
| ** 1. PhCOCl and a base | Benzoyl chloride protects both α-amino and —$CO_2H$ functions | 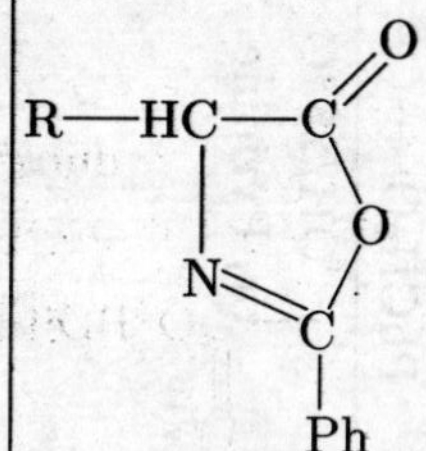 | $HCl/H_2O$ | $RCH(\overset{\oplus}{N}H_3)CO_2^-$ + $PhCO_2H$ |

** *For α-amino acids : See the synthesis of α-amino acids by azlactone method.*

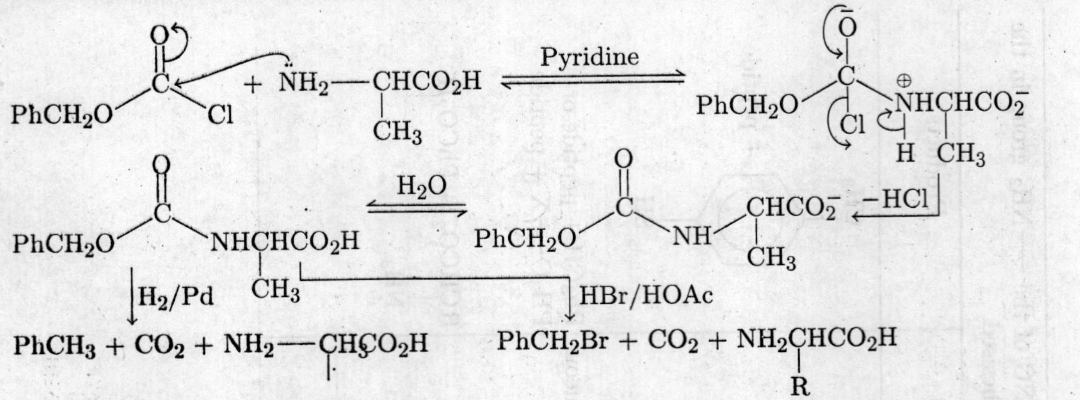

Activation of the carboxyl group of the amino-blocked amino acid unit may be done in several ways as given in Table 22.3. The blocking of the carboxyl group is usually done by esterification of the second unit of amino acid. The deblocking of the carboxyl group of the synthesised peptide is, therefore, done by the acidic hydrolysis of the peptide.

Let us synthesise gly. ala following the procedures stated above.

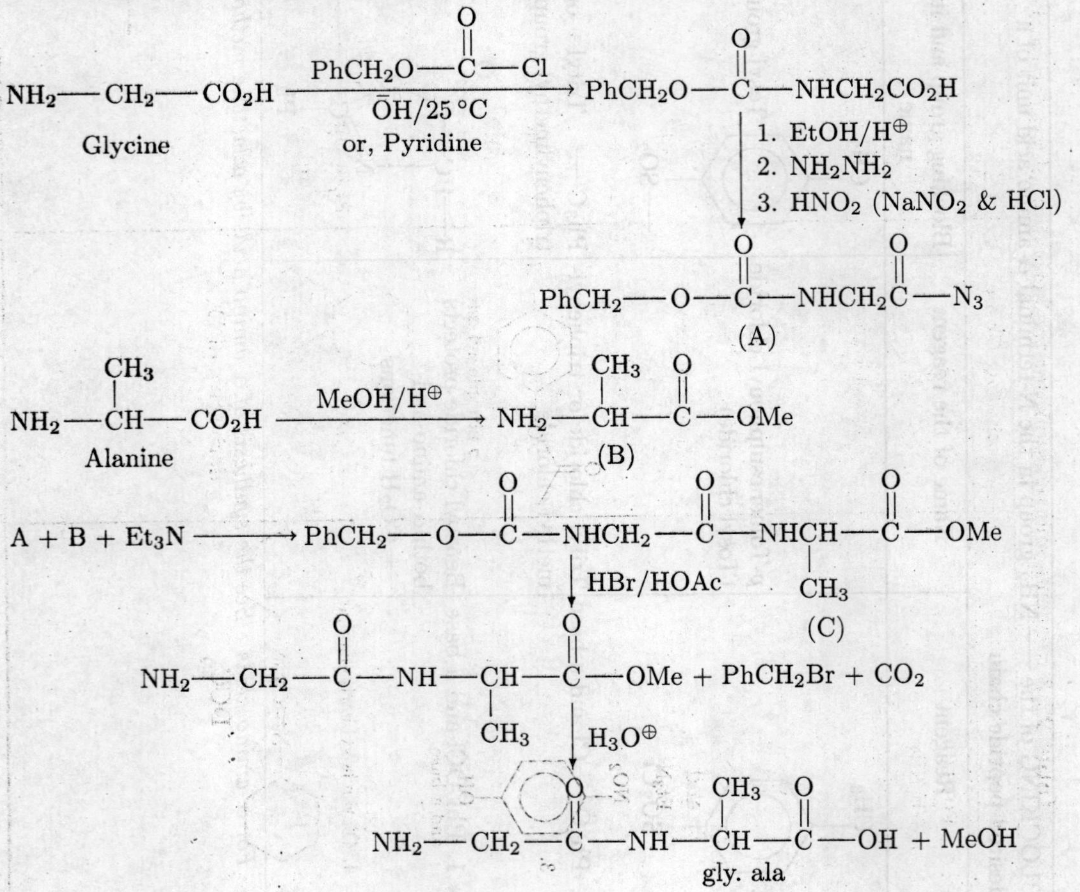

**Table 22.3** Reagents and reactions for the activation of the carboxyl group of the amino blocked amino acid unit

| Activating agents | Activating Group | Reactions |
|---|---|---|
| 1. i. $EtOH/H^{\oplus}$<br>ii. $NH_2NH_2$<br>iii. $NaNO_2/HCl$ | 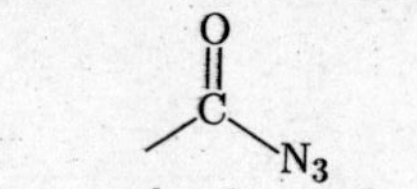<br>acylazido group | 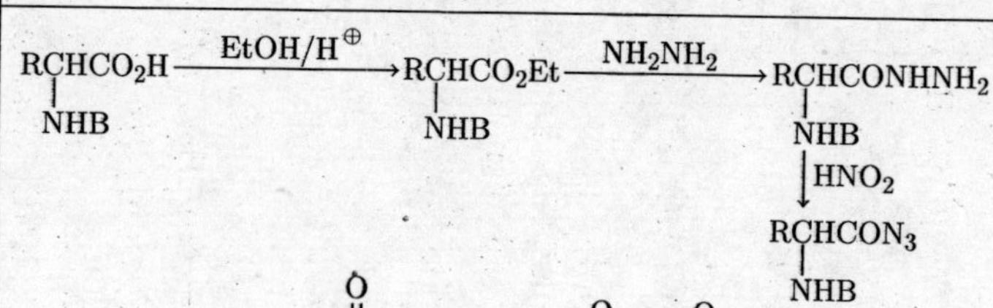 |
| 2. 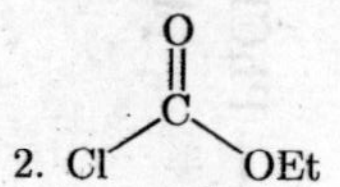<br>and<br>$Et_3N$ | 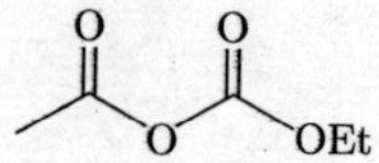<br>anhydride (mixed) | 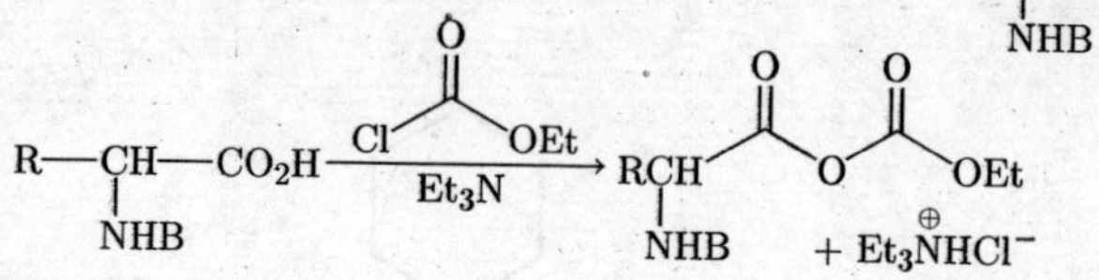 |
| 3. <br>and a base | 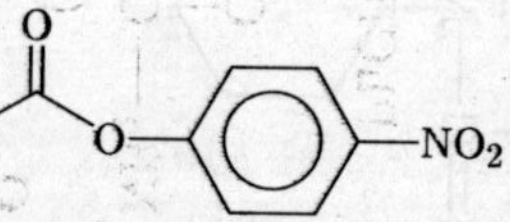<br>*p*-nitrophenyl ester | 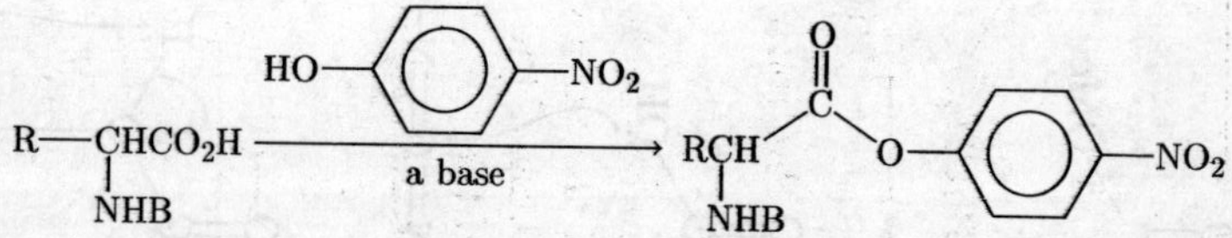 |
| 4. Dicyclohexyl carbodiimide<br>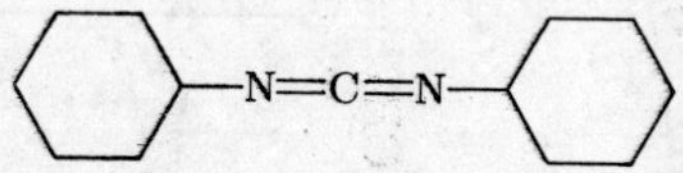<br>DCC | 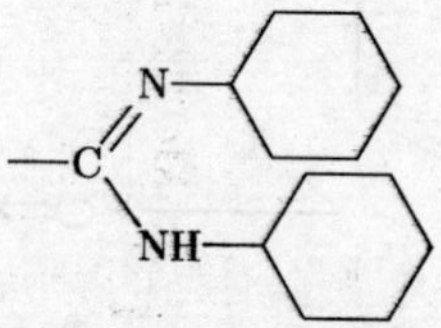<br>derivative of DCC | 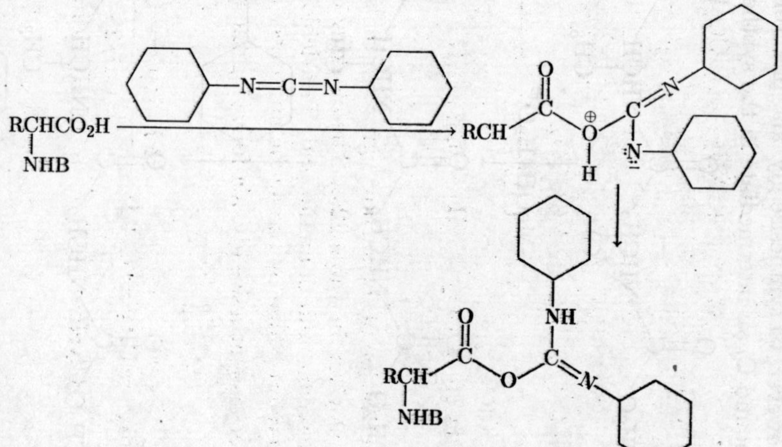 |

However, to synthesise gly. ala. phe from gly. ala., a different procedure may be used taking compound C, an intermediate in the synthesis of gly. ala., as the starting material :

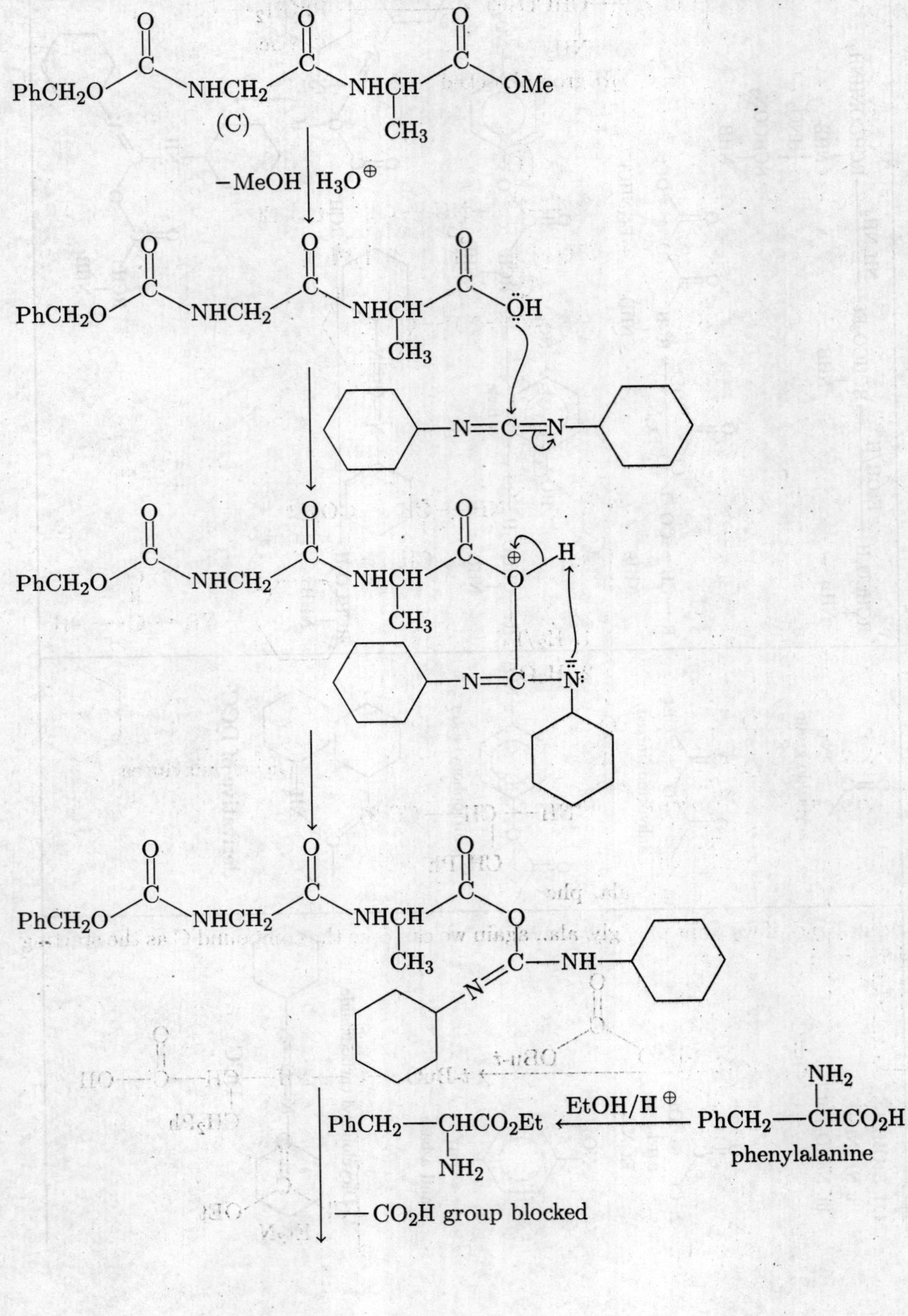

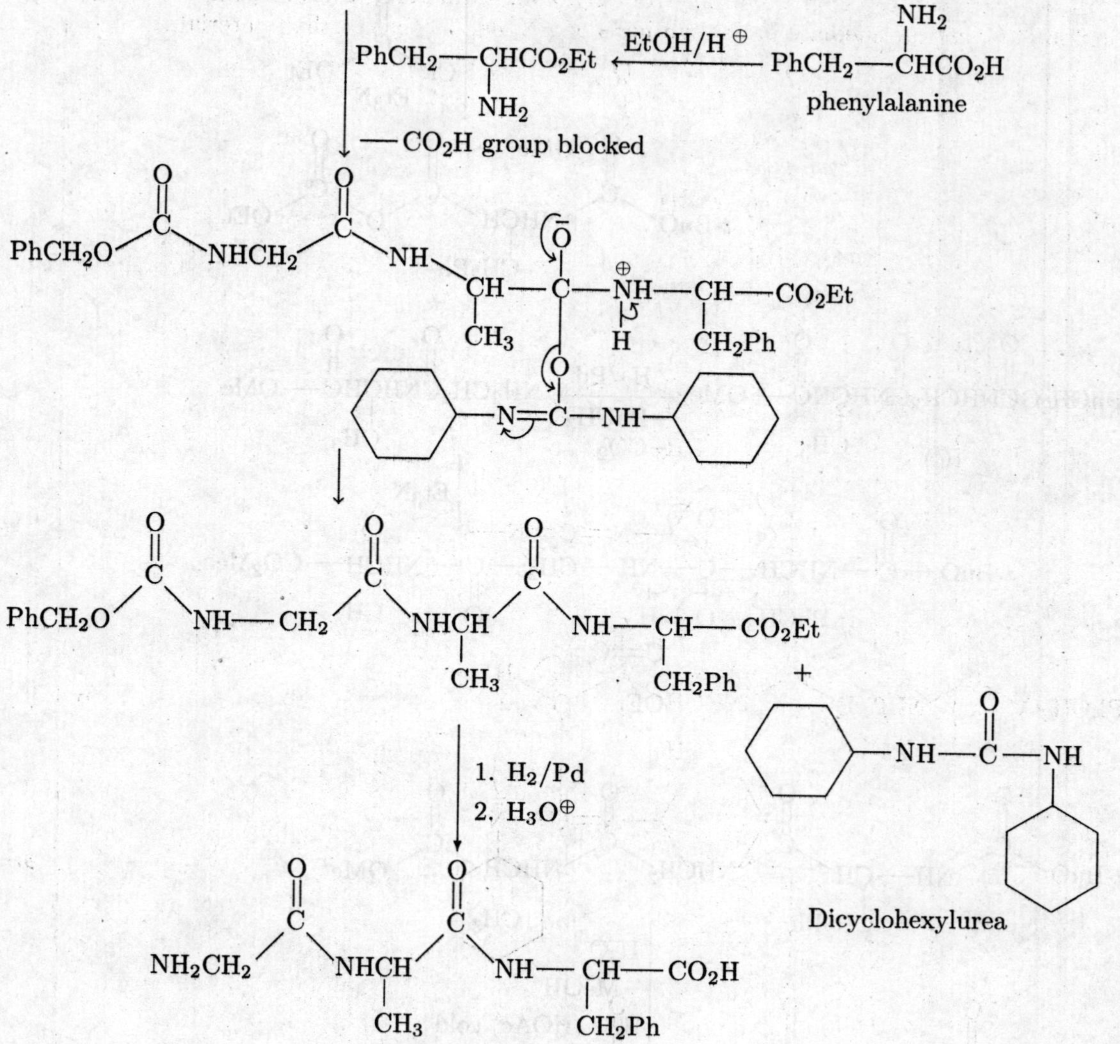


gly. ala. phe

On the other hand, if we want phe. gly. ala., again we can take the compound C as the starting material as follows :

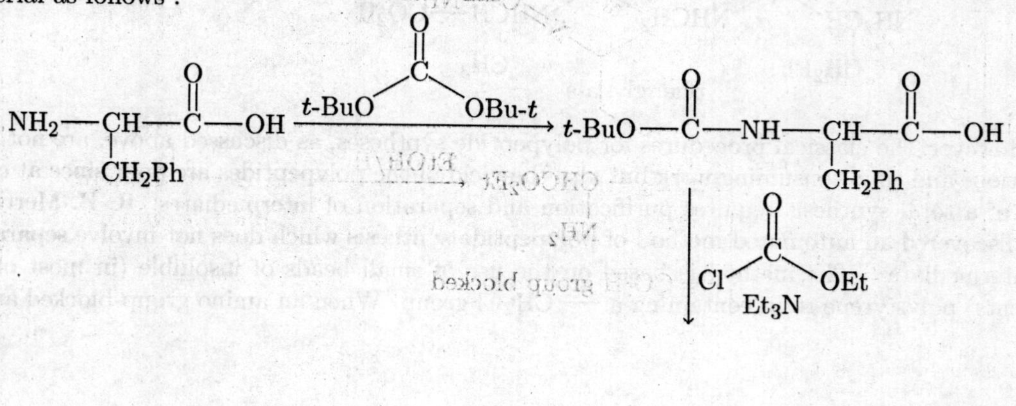

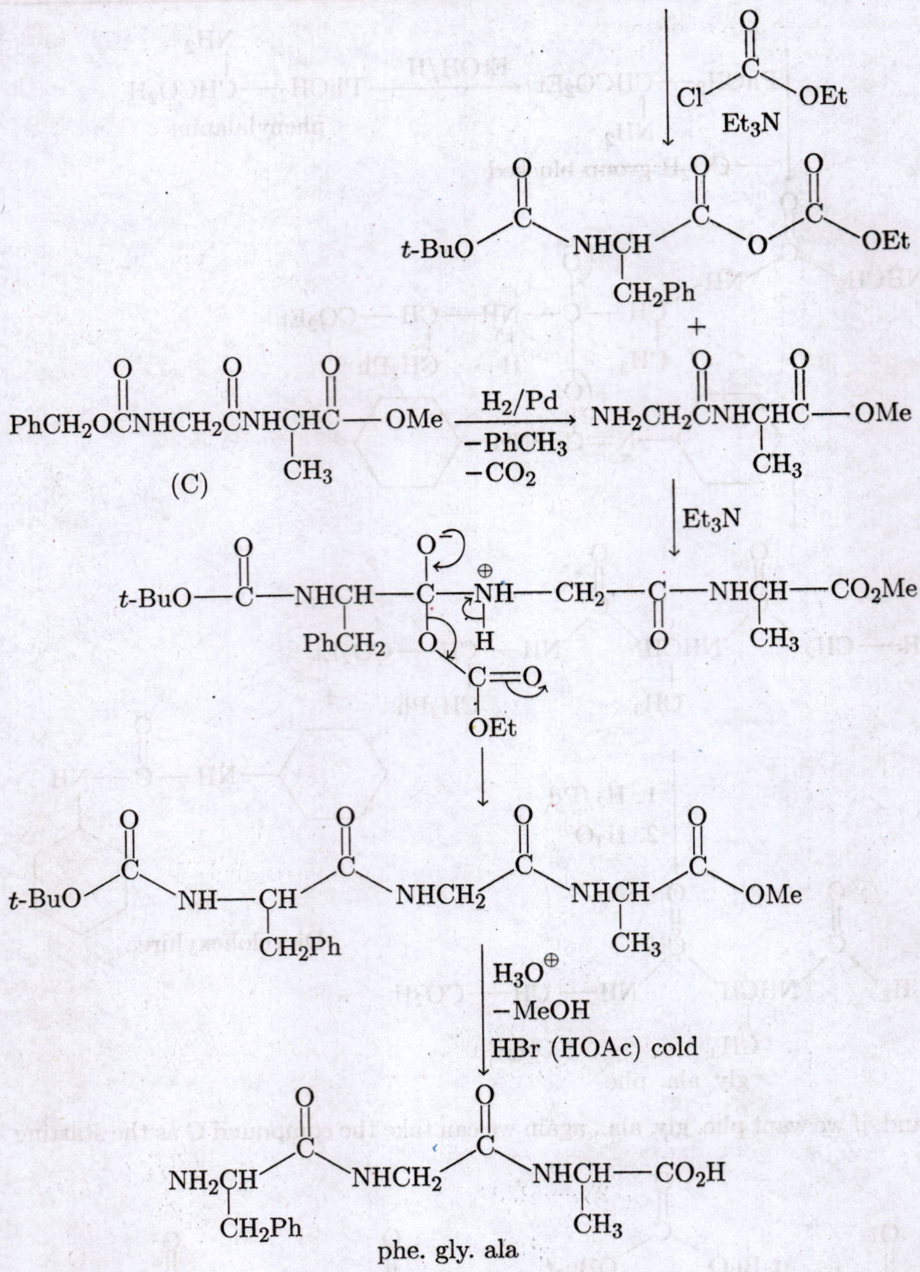

However, the classical procedures for polypeptide syntheses, as discussed above, are not only laborious and time consuming work but also the yield of the polypeptides are poor since at every step of a total synthesis requires purification and separation of intermediates. **R. B. Merrifield has discovered an automated method of polypeptide synthesis** which does not involve separation of intermediates. The method is based on the use of small beads of insoluble (in most of the solvents) polystyrene resin containing a —$CH_2Cl$ group. When an amino group-blocked amino acid is added to the resin under basic condition, the carboxyl group of the amino acid unit gets bonded to the resin by forming an ester linkage. Thus both the functions get blocked. The second step involves the purification of bonded resin by washing with an appropriate solvent and in the third step —$NH_2$ group is deblocked. The fourth step consists of addition of another amino acid unit, which has an activated C-terminal but blocked N-terminal.

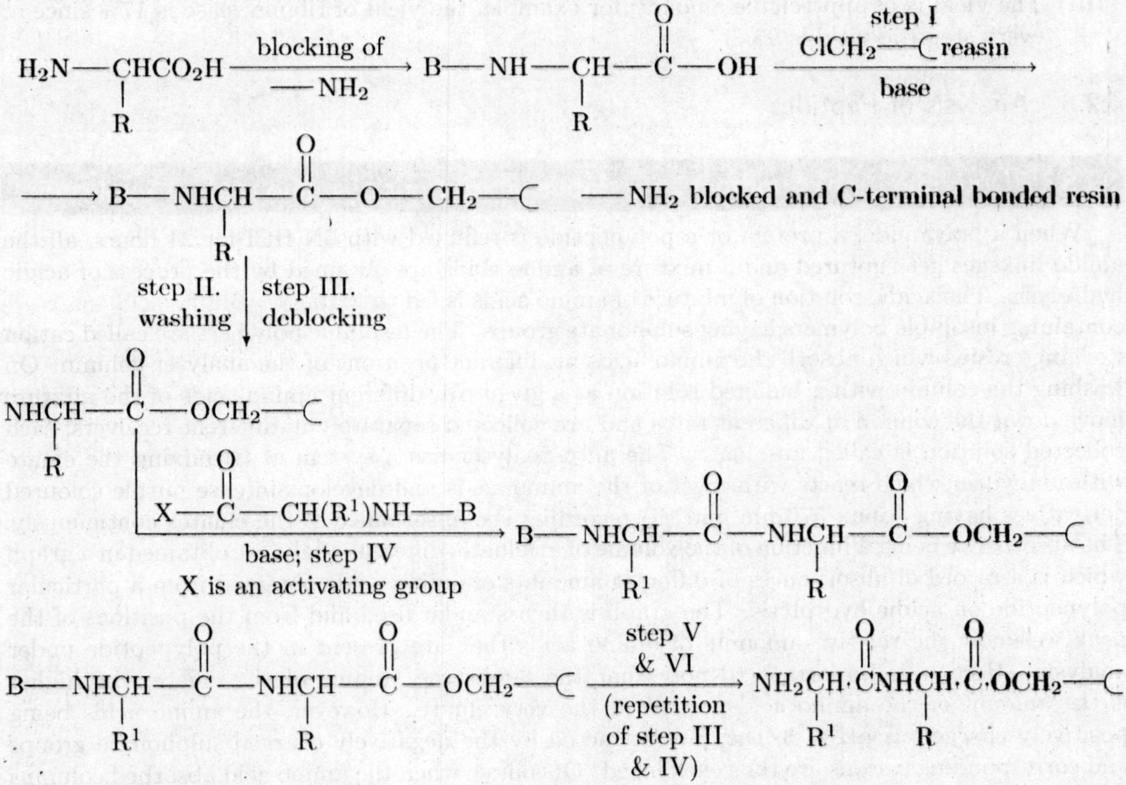

The blocking and deblocking of N-terminal as well as activation of C-terminal are carried out as usual. Through the repetition of steps II to IV a large polypeptide can be synthesised readily. Ribonuclease, a protein with 124 amino acid units, has been synthesised; this involves 359 chemical reactions with 11 931 steps. However, the last step of the Merrifield method is the delocking of the C-terminal of the polypeptide.

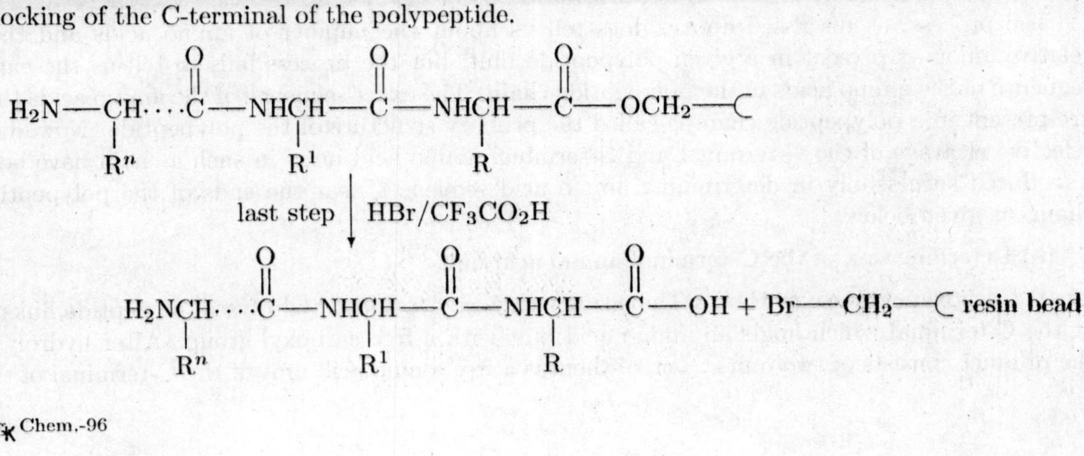

Merits of the Merrifield method are :

(i) The impurities are readily removed by washing with a solvent since they are not bonded to resin.

(ii) The separation of intermediates are thus not required.

(iii) The yield is of appreciable amount; for example, the yield of ribonuclease is 17% since at every step the yield is 99%.

## 22.8 Analysis of Peptides

### 22.8.1 Determination of number and relative proportions of amino acids in a polyamide

When a polyamide, a protein or a polypeptide is refluxed with 6N HCl for 24 hours, all the amido linkages get ruptured and a mixture of amino acids are obtained by the process of acidic hydrolysis. The acidic solution of mixture of amino acids is fed to automatic amino acid analysers containing insoluble polymers having sulphonate groups. The insoluble polymers are called cation exchange resins which absorb the amino acids at different positions of the analyser-column. On washing the column with a buffered solution at a given pH, different amino acids of the mixture move down the column at different rates and are collected separately in different receivers; each collected solution is called an eluate. The auto-analyser has a system of (i) mixing the eluate with ninhydrin which reacts with most of the amino acids and develops intense purple coloured derivatives having $\lambda$max 570 nm and (ii) recording the absorbance of the eluates continuously. The absorbance being a function of the volume of an eluate, different peaks are obtained in a graph which is a record of absorbances of different amounts of amino acids obtained from a particular polypeptide on acidic hydrolysis. The graph is then standardised and from the positions of the peak we know the relative amounts of amino acids that are present in the polypeptide under analysis. Here it is important to note that the larger the volume of the eluate, the higher is the amount of the amino acid present in the very eluate. However, the amino acids being positively charged at pH $\approx$ 3, they are attracted by the negatively charged sulphonate groups and corresponding cations are thus exchanged. Of course, when the amino acid absorbed columns are washed down with eluents, the amino acids get dissolved in the eluent and come down the column and are collected as eluates.

### 22.8.2 Determination of primary structure of a polyamide

The process, as discussed above, does tell us about the number of amino acids and their relative amounts present in a given polypeptide unit; but the process fails to tell us the exact sequence of the amino acids in the polypeptide chain. The exact sequence of the amino acids that are present in a polypeptide chain is called the primary structure of the polypeptide. Nowadays selective cleavage of the C-terminal and N-terminal amino acid units in such a chain have been introduced successfully in determining amino acid sequences near the ends of the polypeptide chains as given below.

**(a) Determination of the C-terminal amino acid unit :**

**(i) Carboxypeptidase method :** The enzyme carboxypeptidase hydrolyses the peptide linkage at the C-terminal which holds an amino acid unit with a free carboxyl group. After hydrolysis the product consists of two units : one of them is a free amino acid unit of the C-terminal of the polypeptide under analysis and the other unit is the rest of the polypeptide having one amino acid unit less than the parent peptide.

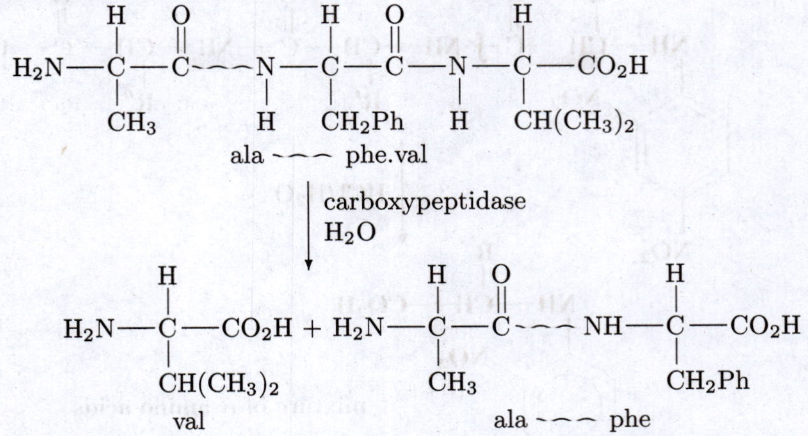

**(ii) Akabori method :** When a polypeptide is heated at 100 °C with anhydrous hydrazine, the C-terminal amino acid separates and other amino acids of the parent polypeptide chain form the corresponding $\alpha$-aminoacyl hydrazides.

$$\underset{\substack{|\\ R}}{NH_2CHCONHCHR'CONHCHR''CO_2H} \xrightarrow[100\,°C]{2N_2H_4} NH_2CHRCONHNH_2 + NH_2CHR'CONHNH_2 + NH_2CHR''CO_2H$$

**(b) Determination of the N-terminal amino acid unit :**

**(i) Sanger's method :** 2, 4-Dinitrofluorobenzene (DNFB) is the reagent for the method. The reaction between DNFB and the polyamide under consideration is carried out in a mildly basic solution of aqueous sodium bicarbonate. The free amino group (of the N-terminal) of the polyamide attacks the C atom of the benzene ring which holds the C—F bond, and a bimolecular aromatic nucleophilic substitution reaction takes place as shown below :

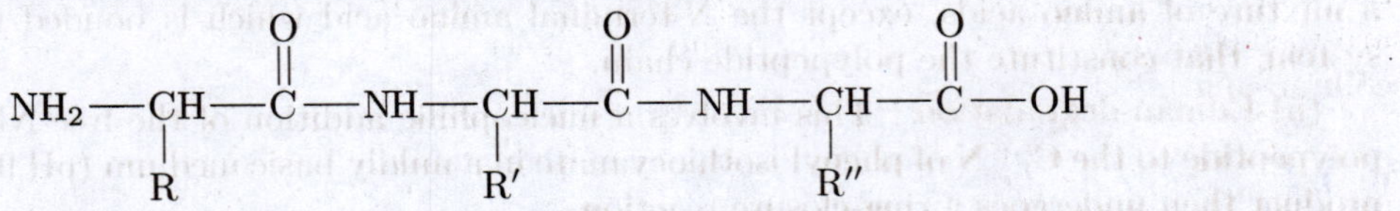

peptide

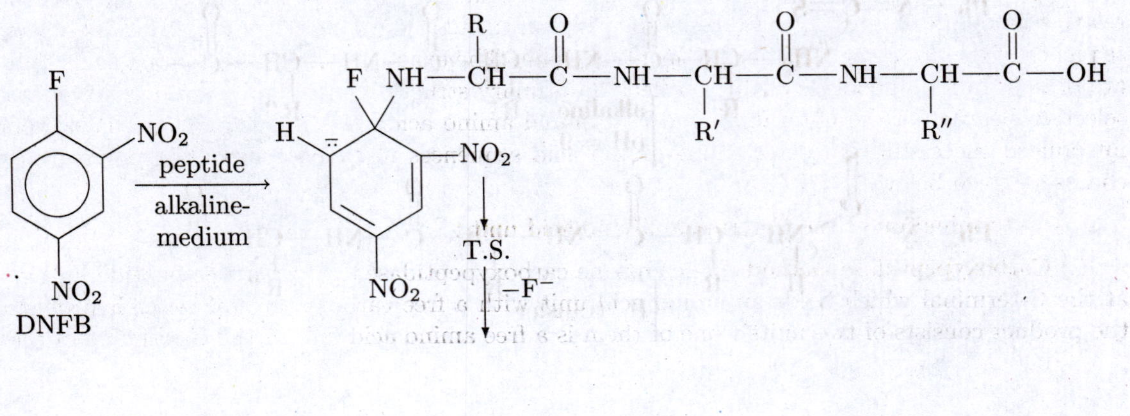

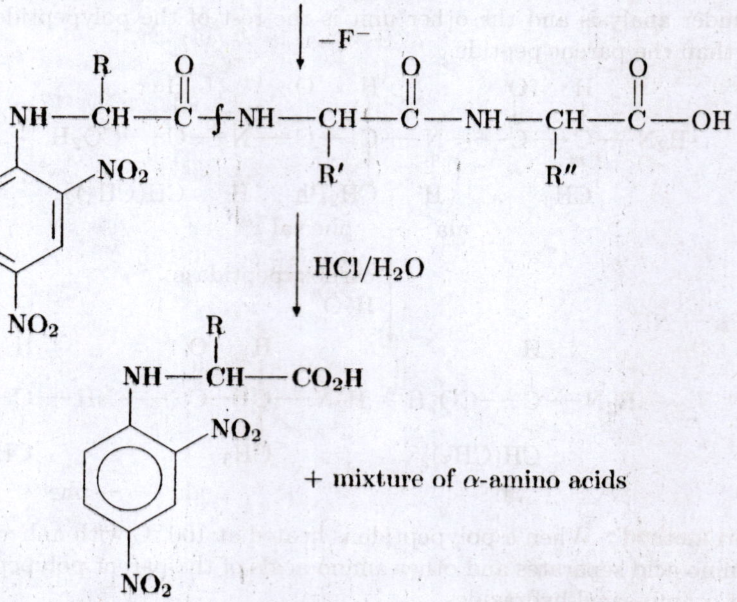

labelled N-terminal amino acid

The labelled N-terminal amino acid is then separated and identified. However, the strong $-I$ and $-R$ effects of two nitro groups at *o*- and *p*-positions with respect to the F group and also the strong $-I$ effect of the F group make the C atom bonded to F a site for nucleophilic attack and ultimately after aromatisation we get a labelled tripeptide which on hydrolysis gives the labelled N-terminal amino acid unit. The nuclear substituted —$\ddot{N}HCH(R)CO_2H$ unit has a partial double bond character because of the delocalisation of non-bonded electron-pair of the N atom involving the conjugated system of the ring and nitro groups (the resonance interaction) and the process of hydrolysis cannot break the ring C—NH bond but attacks the electron-deficient amido C. However, the hydrolysis reaction breaks all the amide links at a time and thus we get a mixture of amino acids, except the N-terminal amino acid which is bonded to the aromatic system, that constitute the polypeptide chain.

**(ii) Edman degradation :** This involves a nucleophilic addition of the free $NH_2$ group of the polypeptide to the C=N of phenyl isothiocyanate in a mildly basic medium (pH 9). The addition product then undergoes a ring-closure reaction.

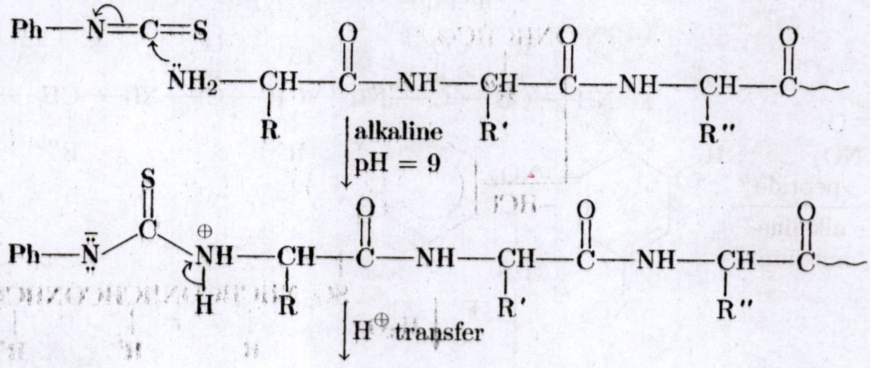

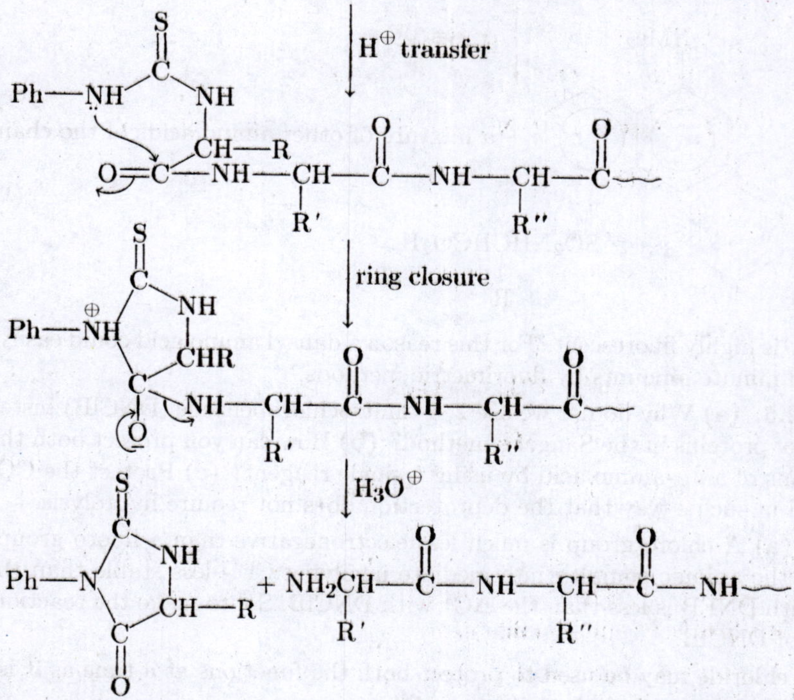

Thus in this reaction, the N-terminal amino acid unit forms a phenylthiohydantoin and detaches itself from the rest of the peptide which does not decompose but remains intact with all its sequences. The N-terminal unit can be identified by comparing the phenylthiohydantoin so formed with standard phenyl thiohydantoins. Since the residual polypeptide chain remains intact, the residual chain can further be subjected to the Edman degradation and all the amino acids of the starting polypeptide can be identified sequentially; here is the superiority of the Edman's method to the Sanger's method.

**(iii) Dansyl method :** This is a modification of DNFB method. In this method 5-(N, N-dimethyl)-aminonaphthalene-1-sulphonyl chloride (dansyl chloride : DNS-Cl) is used instead of DNFB.

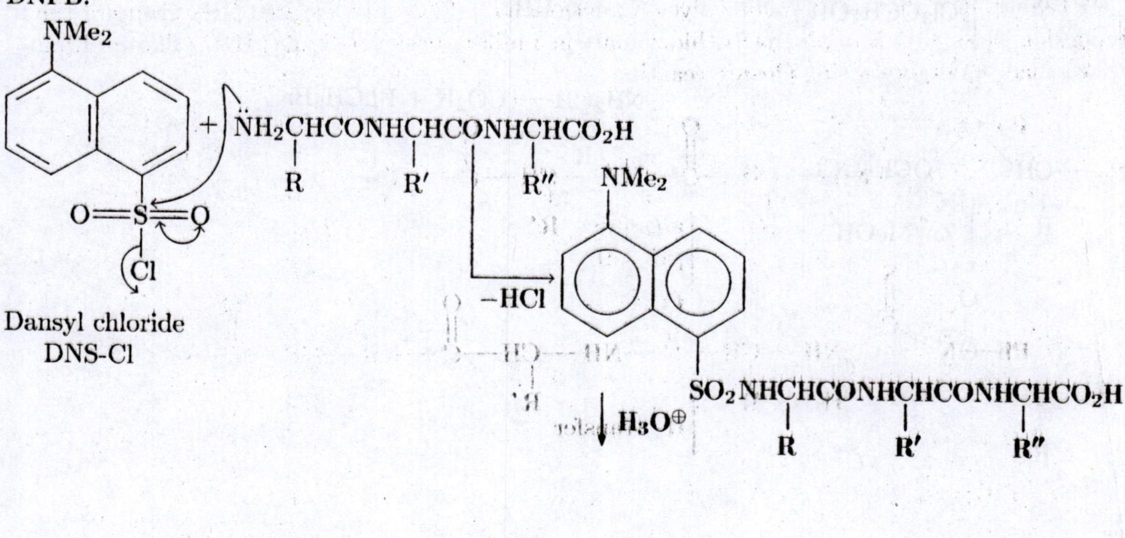

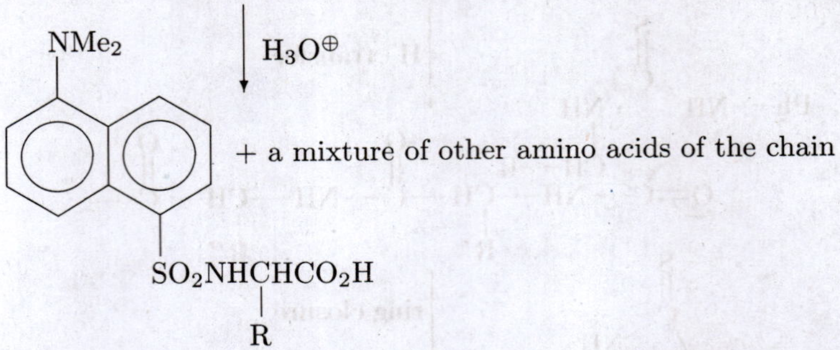


Dansyl group is highly fluorescent. For this reason a dansyl amino acid could easily be detected and identified in minute amounts by fluorimetric methods.

**Problem 22.5.** **(a)** Why do not we use 2, 4-dinitrochlorobenzene (DNClB) instead of DNFB for the analysis of proteins in the Sanger's method? **(b)** How can you protect both the amino and carboxyl functions of an $\alpha$-amino acid by using a single reagent? **(c)** Protect the $CO_2H$ group of an $\alpha$-amino acid in such a way that the deprotection does not require hydrolysis.

**Solution :** **(a)** A chloro group is much less electronegative than a fluoro group; because of this the T.S. of the anionic complex intermediate involving Cl is less stable than that involving F. Thus $\Delta G^{\ddagger}$ with DNFB is less than the $\Delta G^{\ddagger}$ with DNClB. So, to make the reaction fast DNFB is used instead of DNClB.

**(b)** Benzoyl chloride may be used to protect both the functions at a time as it is done in the case of Erlenmeyer azlactone synthesis (see p. 1506).

**(c)** The $CO_2H$ group of an amino acid may be protected by forming an ester using benzyl alcohol and acid or 2, 2, 2-trichloroethanol. The former ester can be converted to acid by the process of reduction using $H_2$/cat or HBr, while the latter ester can be converted to acid by Zn/MeOH.

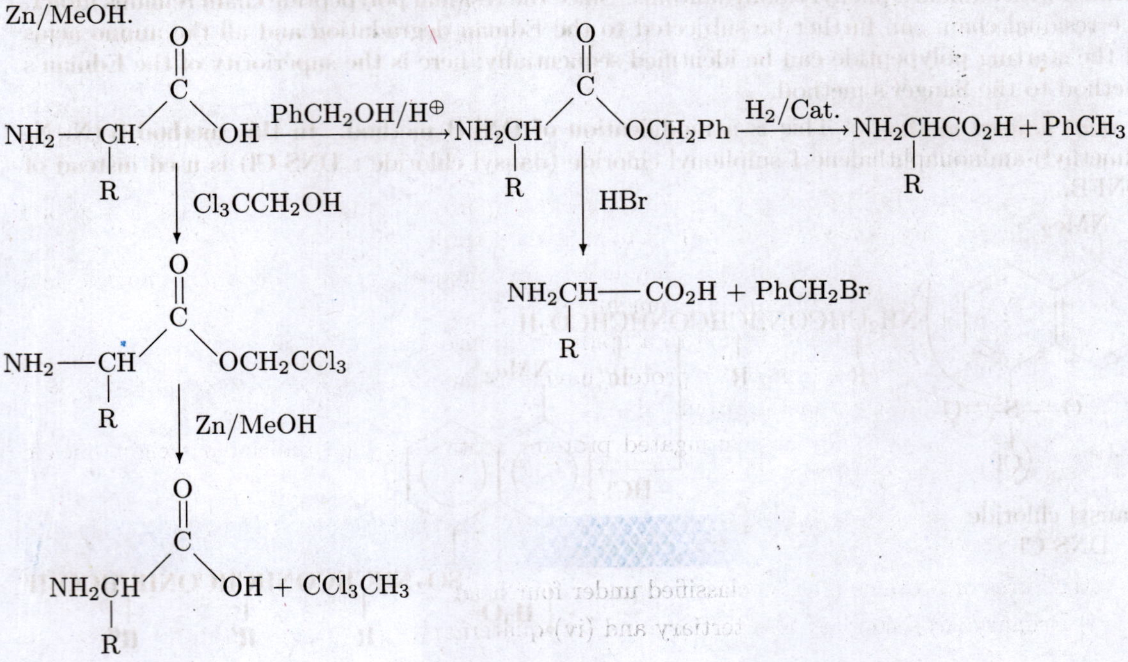

## 22.9 Proteins

### 22.9.1 Definition and classification

Proteins may be looked upon as polymers of amino acids, a polypeptide. Their molecular weight is very high (18 ,000 to 10 ,000 ,000). They contain C: 50-52%, H: 6.8-7.7%, N: 15-18%, O: 24%, S:0.5-2.0. They also contain small amounts of metal, halogen and phosphorus (conjugated proteins). Since they are polypetides, they have free carboxyl and/or amino groups. Naturally they are amphoteric and soluble in acids and alkalis. All proteins are optically active. Each protein has an isoelectric point. They give biuret and ninhydrin tests. Depending on physical properties and solubility.

(i) The simple proteins are classified as follows.

(a) Scleroproteins (abluminoids) : They are completely insoluble proteins. Hair, wool, horns, feathers, etc. contain **keratin**; skin contains **collagen** and silk contains **fibroin**. So, the examples of scleroptoriens are keratin, collagen and fibroin.

(b) Albumins : They are water-soluble proteins : Lactalbumin, ovalbumin and serum albumin are the examples. White of egg is ovalbumin. Lactalbumin is obtained from milk.

(c) Globulins : They are soluble in dilute aqueous salts, in alkalis and in acids. From yolk of egg we get **ovaglobulin** and **fibrinogen** from blood.

(d) Glutelins : They are insoluble in salt solutions but soluble in acids and alkalis and related to globulins. Example, **glutenin** (from wheat).

(e) Prolamins : They are soluble in 70% alcohol. **Gliadin**, obtained from wheat, and **zein**, obtained from maize, are the examples.

(f) Histones : The histone of the thymas gland is the example. Basic proteins which are soluble in water giving an alkaline reactions are histones.

(g) Protamins : Strongly basic and simpler than the above-said proteins but like histones they are soluble in water to give alkaline solutions. **Salmine** (from salmon) and **clupein** (from herring) are the examples of the class.

(ii) **Conjugated proteins :** A protein molecule bonded to a prosthetic group (a non-protein group) is a conjugated protein. The prosthetic group could often be separated from a conjugated protein by hydrolysis.

(a) Phosphoproteins : Proteins having phosphoric acid, in the from of ester, as prosthetic group is called a phosphoprotein. **Casein** of milk is an example.

(b) Mucoproteins or Glycoproteins : They have a carbohydrate group other than a nucleic acid as a prosthetic group. Example : **mucin**.

(c) Lipoproteins : A protein bonded to a lipid is a lipoprotein.

(d) Chromoprotein : A conjugated protein having a metal-containing prosthetic group is a chromoprotein , e.g., haemoglobin.

(e) Nucleoproteins : They are conjugated proteins containing high molecular weight nucleic acids (See Sec. 22.10).

### 22.9.2 Structures of proteins

Structures of proteins may be classified under four head :

(i) Primary (ii) secondary (iii) tertiary and (iv) quaternary.

**(i) Primary structure :** By the term 'primary structure' we mean the linear sequence of amino acid units in the polypeptide chain in which peptide linkage between two amino acids is the force which is taken into account and no other force or bond is considered. Primary structure of a protein is known from their analysis as we have done in the cases of polypeptides.

**(ii) Secondary structure :** By the secondary structure of a protein we mean regular folding patterns of contiguous parts of the polypeptide chain. There are three most abundant secondary structures of proteins : (a) $\alpha$-helix, (b) parallel $\beta$-pleated sheet and (c) antiparallel $\beta$-pleated sheet. The existence of these structures depends on (i) H-bonding between N—H and O=C< (—NH···O=C< ) of the polypetide chain and (ii) steric limitations on the rotations of bonds involving N—CH(R)—C=O moiety. Two rotational angles have been defined (a) $\psi$, for the bond between the $\alpha$-C and C of carbonyl group and the other (b) $\phi$, for the N of N—H to $\alpha$-C bond (Fig. 22.2). The question of rotation around N—C=O bond does not arise since the N—C bond has some double bond character owing to resonance involving lone pair electrons of the N atom and C—O double bond as shown on the page below Fig. 22.1. The O=C—N bond distances support the effect of resonance : N—CO bond length is 132 pm which is much less than N—C single bond length and the C=O length is 123 pm which is also greater them C=O bond length not involving resonance.

(a) The $\alpha$-helical structure (conformation) of proteins is given below :

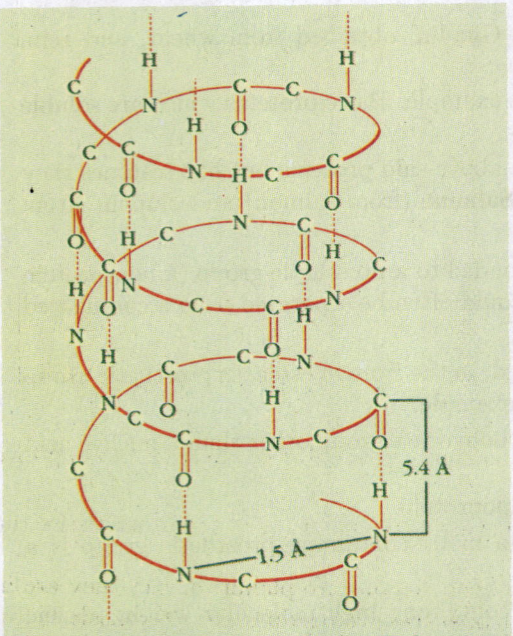


**Fig. 22.3. $\alpha$-Helical structure (conformation) of a protein. Thick lines indicate front surface.**

The helix of a protein is right handed, i.e., helical turns are clockwise. The description of the helix is as follows :

The length of one amino acid residue along the cylindrical surface in the $\alpha$-helix is 1.5 Å (no matter whether the amino acid residues contain aromatic or aliphatic or heterocyclic substituents). The diameter of the helix is about 10 Å. The distance between two successive turns, called the pitch of the helix, is 5.4 Å and each turn of the helix accommodate 3.6 amino acid residues. The intramolecular H-bonds in the $\alpha$-helix are formed between the H atom of NH group in one turn and the O atom of carbonyl group of the next turn, i.e., between every first and fourth imido (—C(=O)—NH) group since at these points N—H···O atoms become in linear orintation. Under these conditions the rotational angles (i) $\psi$ varies from 123° to 136° and (ii) $\phi$ varies from 113° to 132° and within these ranges of rotational angles the $\alpha$-helical structures are sterically faboured. However, both glycine and proline residues of protein chains are said to be '$\alpha$-helix breakers'. Since the H atoms in $\alpha$-C of glycine unit are too small to foster the helical structure having $\psi$ and $\phi$ values within the required range; while proline residue is big enough for not allowing the chain to assume $\alpha$-helix. The $\alpha$-helix-promoting residues are, however, phenylalanine, glutamic acid leucine, methionine, etc. Almost 80% myoglobin and haemoglobin possess α-helix structures while paramyosin and α-keratin are the examples of fibrous proteins that have α-helix structures.

(b) β-pleated sheets (conformations) are of two types : (i) parallel β-pleated sheets and (ii) antiparallel β-pleated sheets. A picture of simple pleated sheet is drawn in Fig. 22.4.

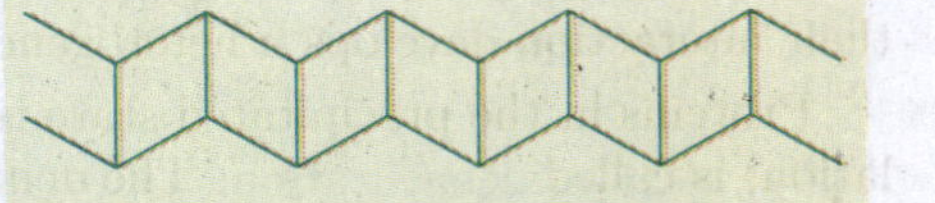

**Fig. 22.4. A simple pleated sheet.**

In these conformations the polypeptide chain is extended and chains are held together by intermolecule H-bonds.

(i) Parallel β-pleated sheets : In these all the chains extend in the same direction (Fig. 22.5a). Parallel chains are found in stretched keratin.

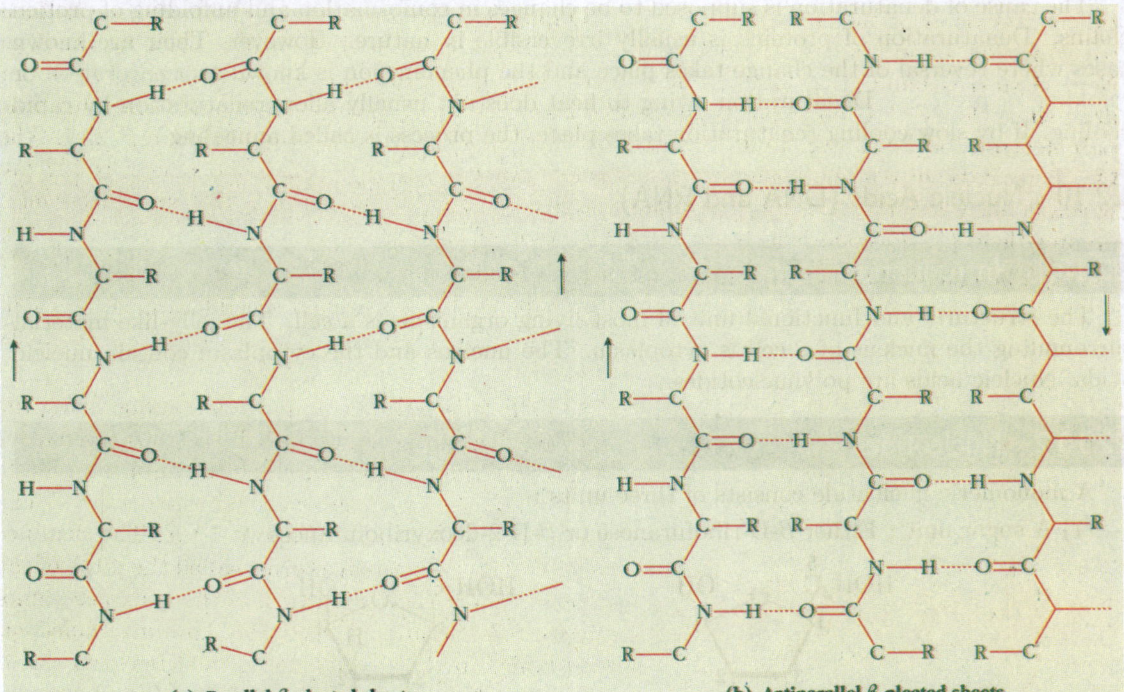

**(a) Parallel β-pleated sheets** **(b) Antiparallel β-pleated sheets**

**Fig. 22.5**

(ii) Antiparallel β-pleated sheets : In these all the chains extend alternately in the opposite directions (Fig. 22.5b). This type is found in silk.

For showing the H-bonds between N—H of one chain and O=C< of the other chain, the sheets are drawn in such a way that as if the atoms of the chains are planar; in fact they are in pleated sheets as shown in Fig. 22.4. The side chains of successive residues are alternately below and above the general plane of the structure.

The repeat distance, the distance between two —CHR groups on the same side of a chain is 7.0 Å in β-pleated sheets.

(iii) Tertiary structure : This tells about the shape of a molecule of a protein. Fibrous proteins have rod-like shape whereas globular proteins have spherical shape. The tertiary structures are thus folding of entire molecule. The driving force of the above shapes again are (i) steric effect, (ii) H-bonding, (iii) ionic, (iv) chenical and (v) hydrophobic bonds.

**(iv) Quaternary structure :** This results from interaction between separate polypeptide units of a protein having more than one subunit. When two such subunits are similar, the structure is called homogeneous quaternary structure. On the other hand, if the subunits are dissimilar, their interaction develops a heterogeneous quaternary structure.

Proteins in the precipitated state is called denatured and the process of precipitation (coagulation) is called denaturation. The denaturation can be carried out either by adding acids, alkalis, salts and organic solvents miscible with water or by heating.

The process occurs near the isoelectric point and optical activity gets changed. The loss of biological achivity also occurs by denaturation.

The cause of denaturation is supposed to be changes in conformation and unfolding of protien chains. Denaturation of proteins is usually irreversible in nature. However, Their are known cases where reversal of the change takes place and the phenomenon is known as renaturation or refolding of proteins. Denaturation owing to heat does not usually allow renaturation by rapid cooling. If by slow cooling renaturation takes place, the process is called **annealing**.

## 22.10 Nucleic Acids (DNA and RNA)

### 22.10.1 Introduction

The structural and functional unit of most living organisms is a **cell**. The jelly-like material surrounding the **nucleus** of a cell is cytoplasm. The nucleus and the cytoplasm contain nucleic acids. Nucleic acids are polynucleotides.

### 22.10.2 Nucleotides

A monomeric nucleatide consists of three units :

**(1) A sugar unit :** Either $\beta$-D-ribofuranose or $\beta$-D-2-deoxyribofuranose

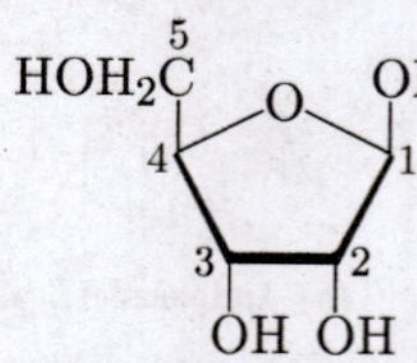


$\beta$-D-Ribofuranose

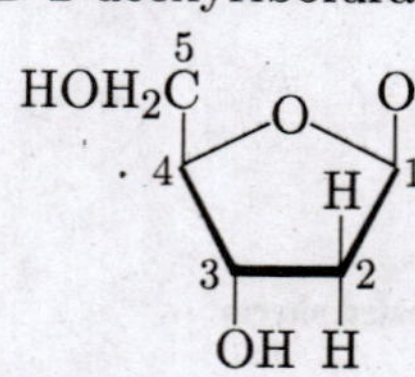


$\beta$-D-2-Deoxyribofuranose

**(2) A phosphoric acid unit :** Esterified with either $\beta$-D-ribofuranose or $\beta$-D-2-deoxyribofuranose.

**(3) A base unit :** Ether a pyrimidine or a purine base

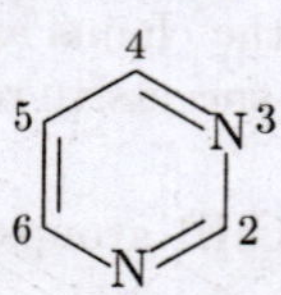


Pyrimidine

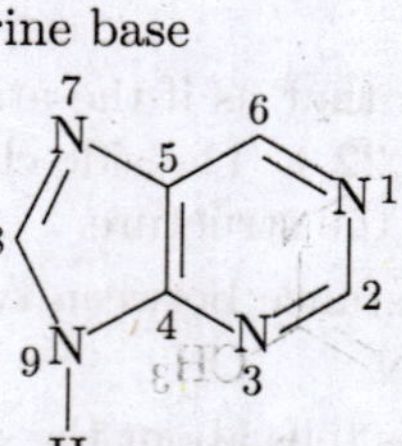


Purine

Three major pyrimidines which are found in nucleotides are :

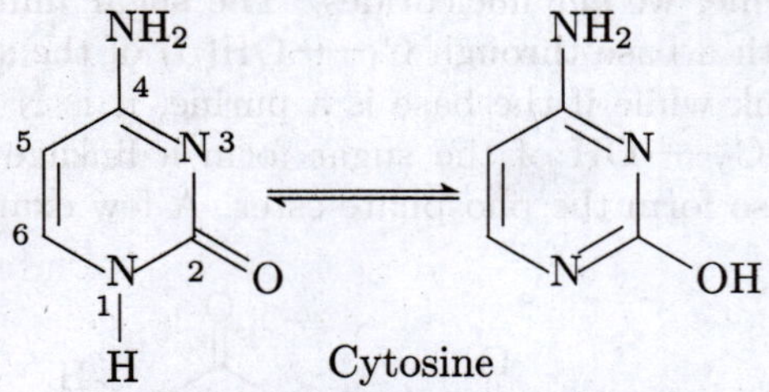

Cytosine
2-oxo-4-aminopyrimidine

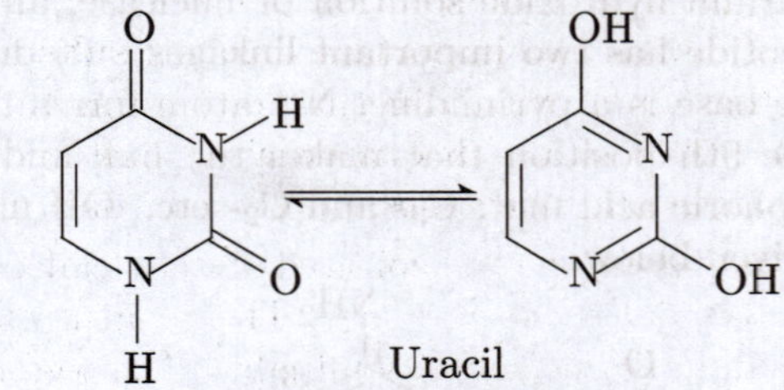

Uracil
2, 4-dioxopyrimidine

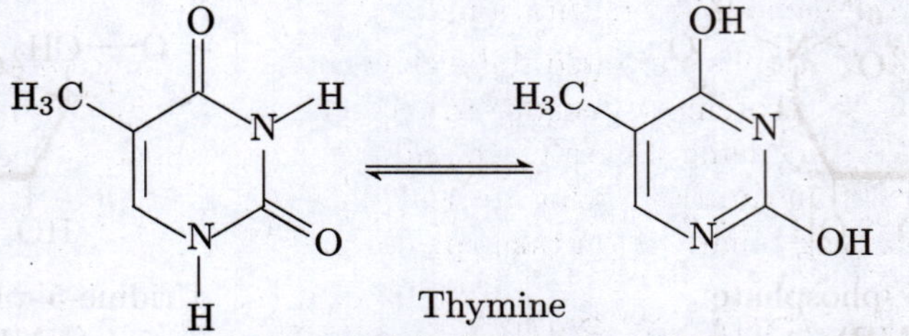

Thymine
5-methyl-2, 4-dioxopyrimidine

Two minor pyrimidines are :

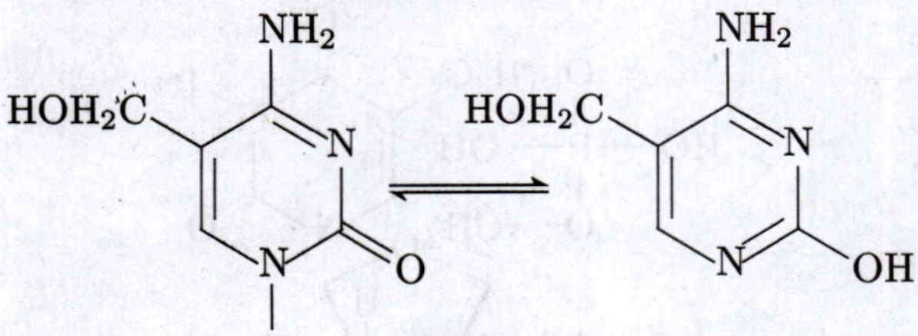

5-hydroxymethylcytosine

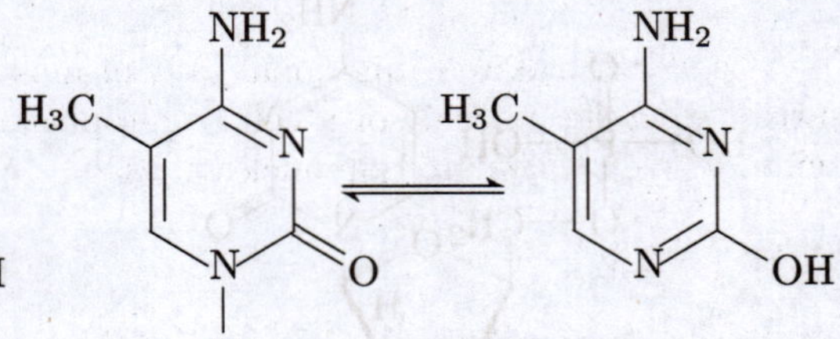

5-methylcytosine

Two major purines are :

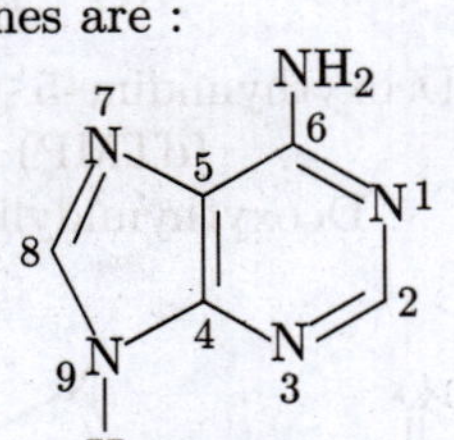

Adenine
6-aminopurine

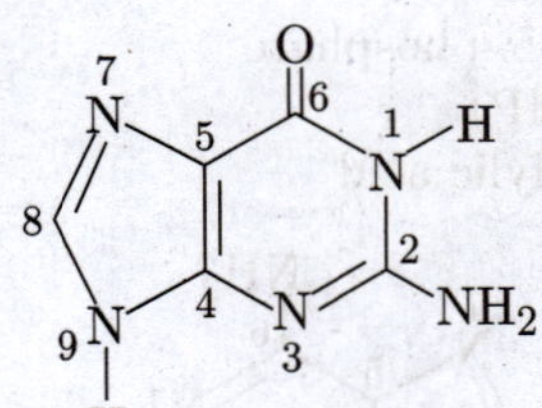

Guanine
2-amino-6-oxopurine

Two minor purines are :

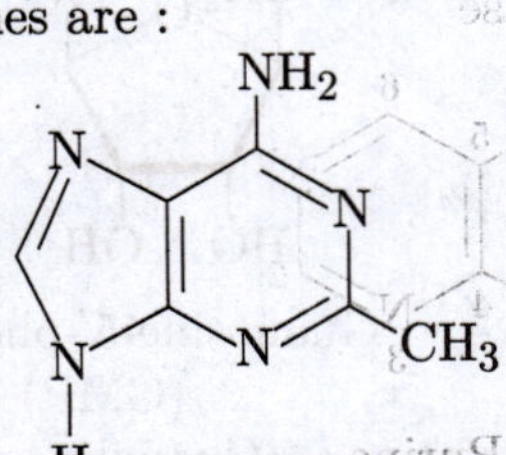

2-methyladenine

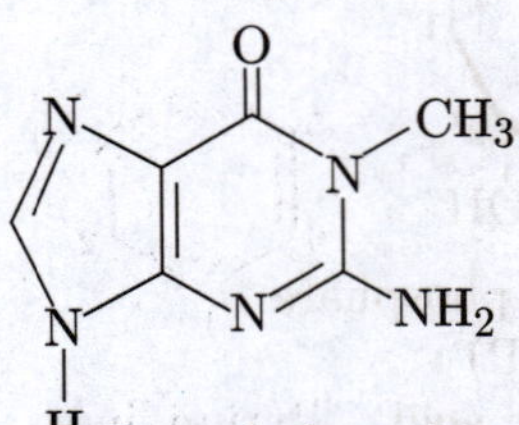

1-methylguanine

A nucleic acid being a polynucleotide, when it is treated with either aq. ammonia at 115 °C or barium hydroxide solution or nuclease, an enzyme, we get nucleotides. The sugar unit of a nucleotide has two important linkages : (i) one with a base through $C_1$—OH($\beta$) of the sugar; if the base is a pyrimidine, N-1 atom forms the link while if the base is a purine, it is N atom of the 9th position that makes the link and (ii) $C_5$—OH of the sugar form a linkage with phosphoric acid unit; $C_3$- and $C_2$- etc., OH may also form the phosphate ester. A few examples are given below.

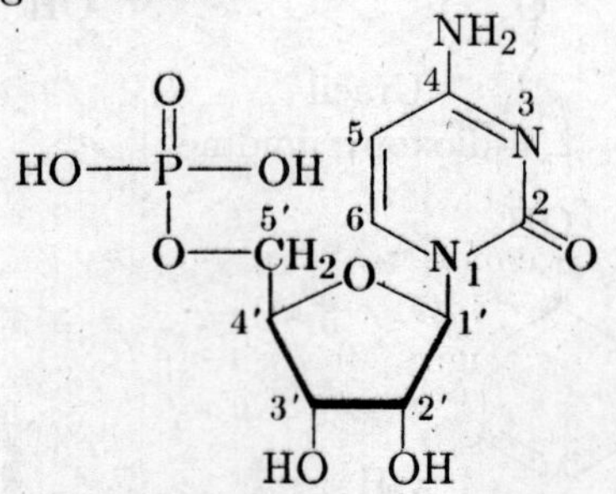

Cytidine-5′-phosphate
(CMP)
(Cytidine Monophosphate)

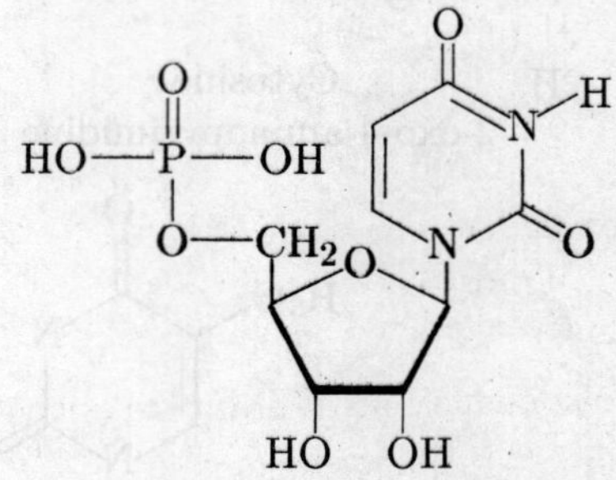

Uridine-5′-phosphate
(UMP)

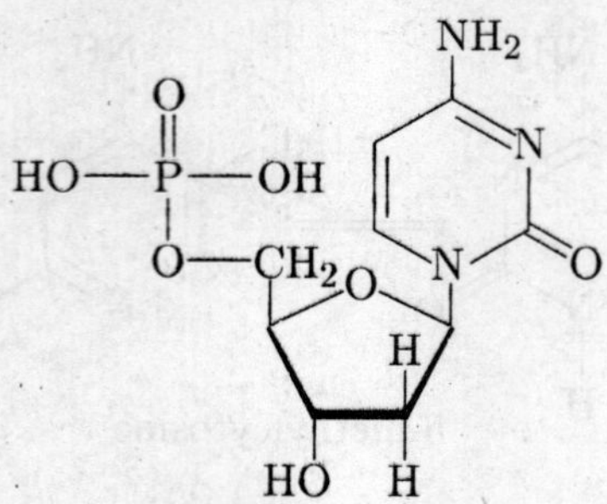

Deoxycytidine-5′-phosphate
(dCMP)
Deoxycytidylic acid

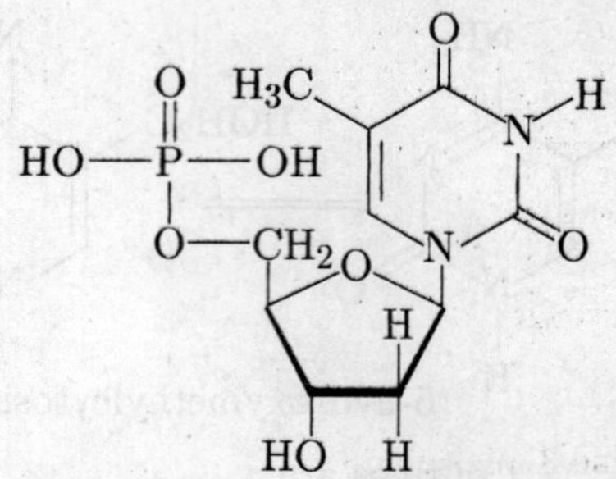

Deoxythymidine-5′-phosphate
(dTMP)
Deoxythymidylic acid

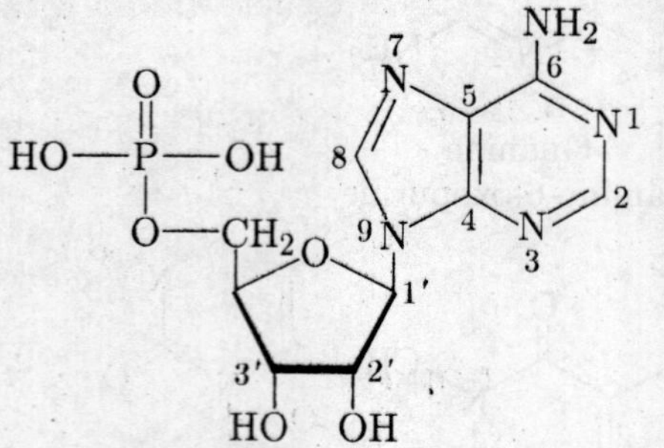

Adenosine-5′-phosphate
(AMP)
Adenylic acid

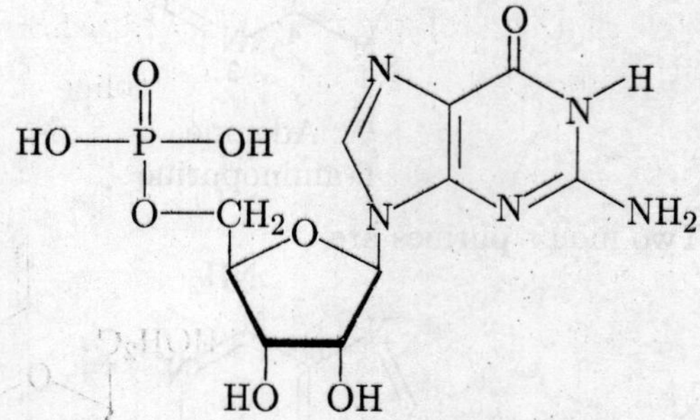

Guanosine-5′-phosphate
(GMP)
Guanylic acid

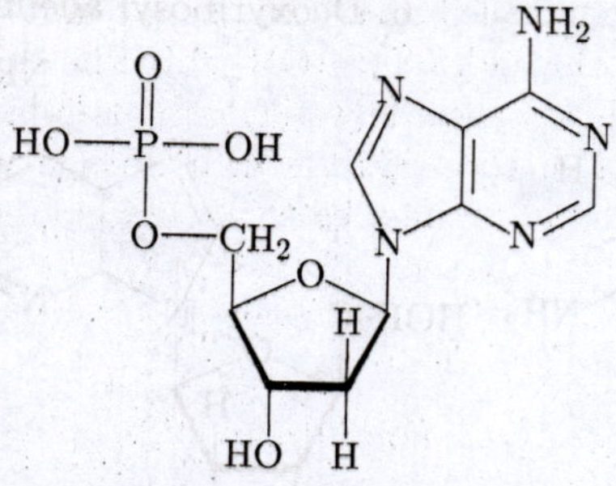

Deoxyadenosine-5′-phosphate (dAMP)
Deoxyadenylic acid

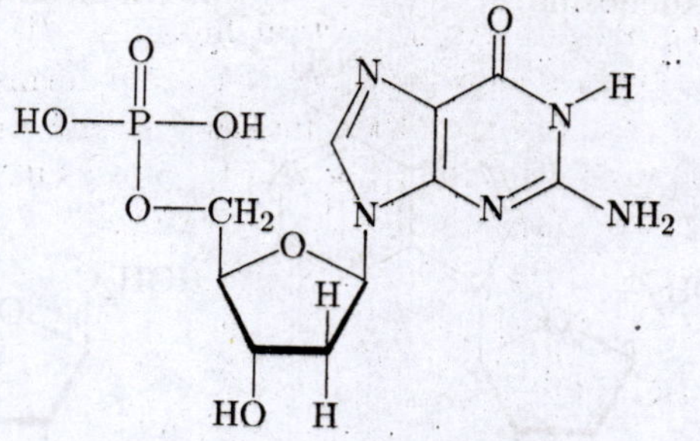

Deoxyguanosine-5′-phosphate (dGMP)
Deoxyguanylic acid

Nucleotides may have also phosphate linkages at 3′ and 2′, etc positions.

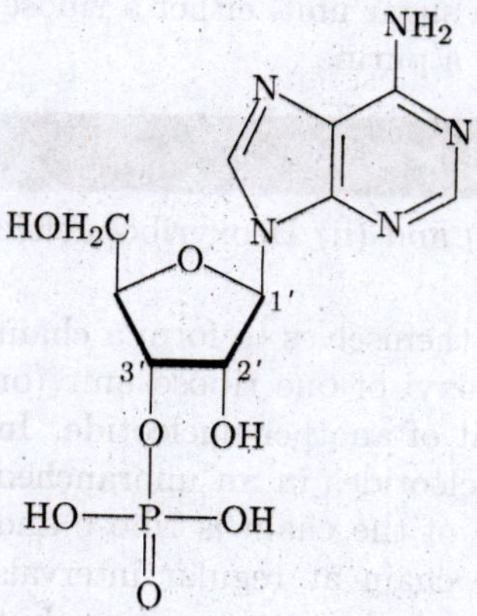

Adenosine-3′-phosphate

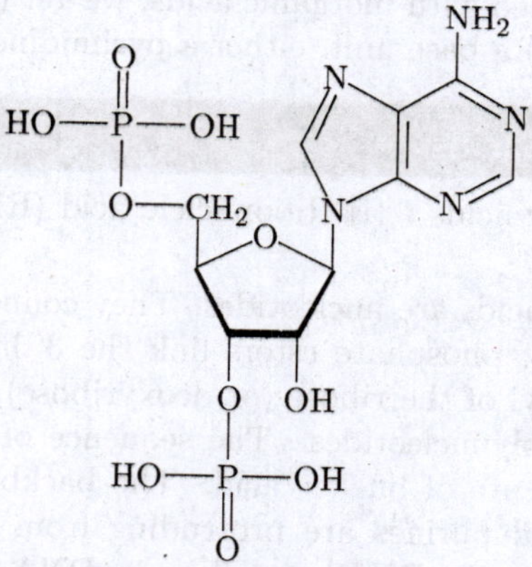

Adenosine-3′,5′-bisphosphate

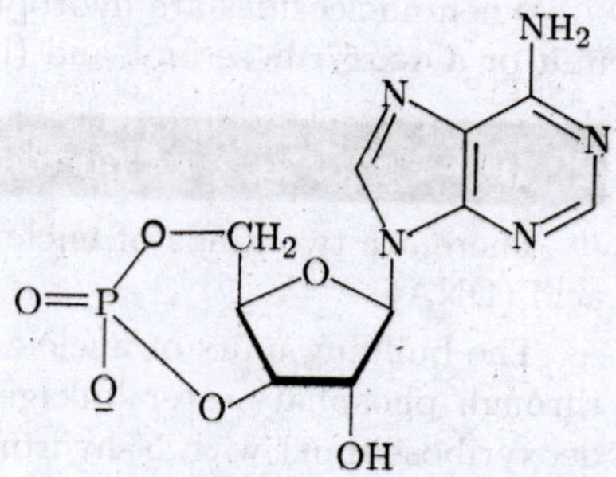

3′,5′-cyclicadenylic acid

## 22.10.3 Nucleosides

Dephosphated nucleotides are nucleosides. They are obtained by treating nucleotides with ammonia at 175 °C. Examples are given below.

1. Uridine

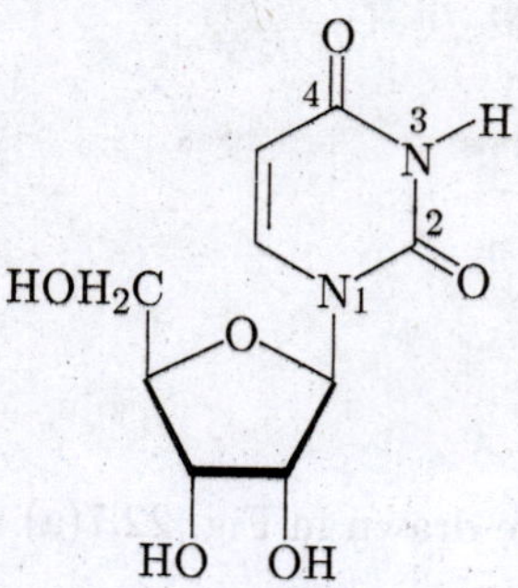

1-β-D-Ribofuranosyl uracil

2. Cytidine

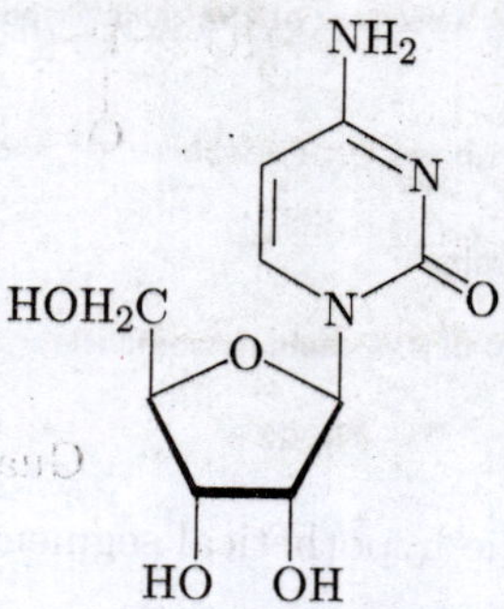

1-β-D-Ribofuranosyl cytosine

3. Deoxyribosylthymidine

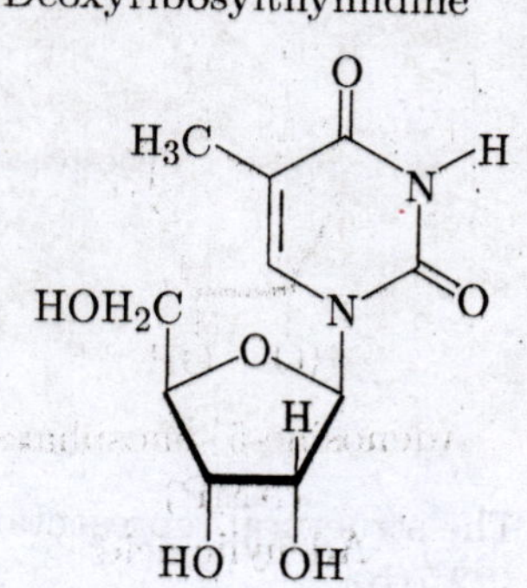

1-β-2-Deoxyribofuranosyl thymidine

4. Adenosine

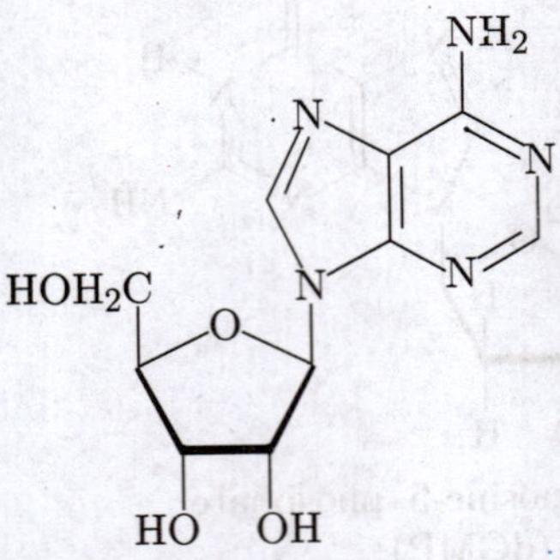

9-$\beta$-D-Ribofuranosyl adenine

5. Guanosine

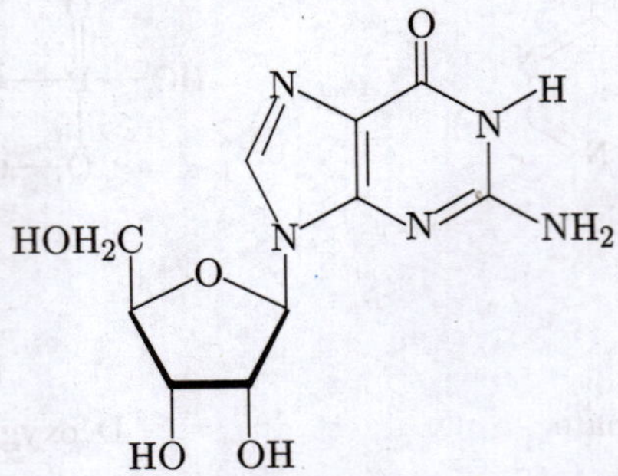

9-$\beta$-D-Ribofuranosyl guanine

6. Deoxyribosyl adenine

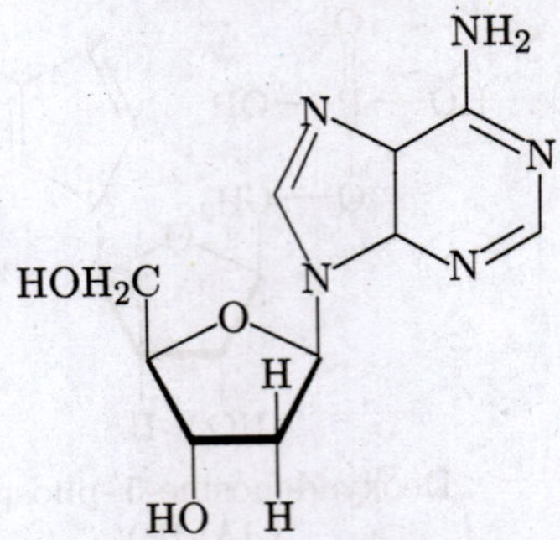

9-$\beta$-D-Deoxyribofuranosyl adenine

When nucleosides are hydrolysed with inorganic acids, we get (i) a sugar unit, either a ribose unit or a deoxyribose unit and (ii) a base unit, either a pyrimidine or a purine.

## 22.10.4 Structure of nucleic acids

There are two kinds of nucleic acids : (i) Ribonucleic acid (RNA) and (ii) Deoxyribonucleic acid (DNA).

The building units of nucleic acids are nucleotides. They connect themselves to form a chain through phosphate ester linkages; phosphate esters link the 3′-hydroxyl of one ribose unit (or deoxyribose unit) with 5′-hydroxyl of the ribose (or deoxyribose) unit of another nucleotide. In this way they form a chain of polynucleotides. The sequence of nucleotides in an unbranched chain is known as primary structure of nucleic acid. The backbone of the chain is ribose and phosphate units; pyrimidines and purines are protruding from the chain at regular intervals (Fig. 22.6). Common purine bases in RNA as well as in DNA are adenine and guanine; but common pyrimidine bases in DNA are cytosine and thymine while those in RNA are cytosine and uracil. Pictorial illustration of hypothetical segments of RNA and DNA chains are given in Fig. 22.6(a) and Fig. 22.6(b).

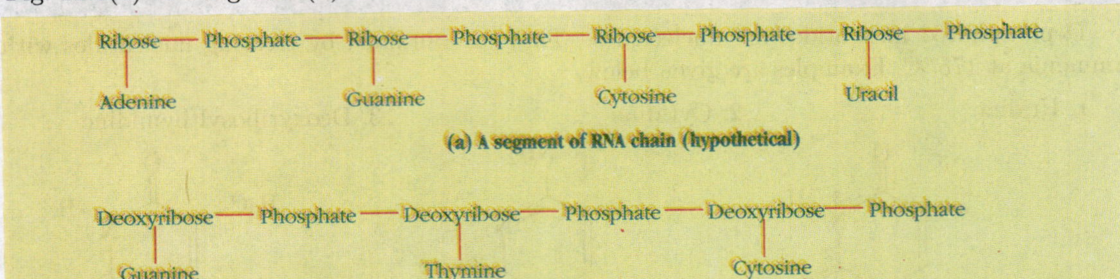


**(a) A segment of RNA chain (hypothetical)**

**(b) A segment of DNA chain (hypothetical)**

**Fig. 22.6**

The structural representations of the hypothetical segments are drawn in Fig. 22.7(a) and in Fig. 22.7(b).

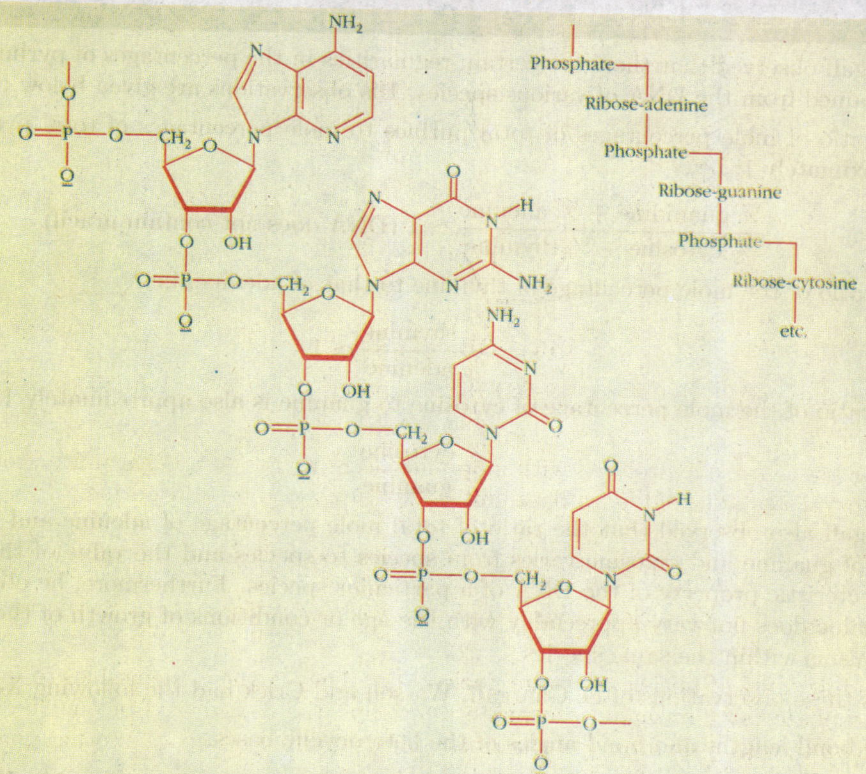


**Fig. 22.7(a) Structural representation of Fig. 22.6(a) (RNA)**

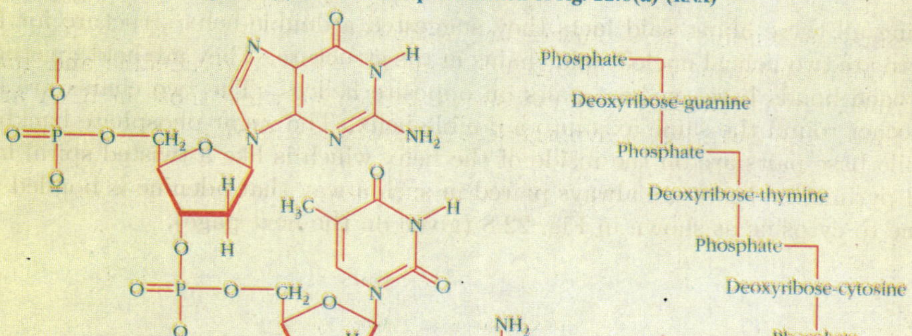


**Fig. 22.7(b) Structural representation of Fig. 22.6(b) (DNA)**

## 22.10.5 Secondary structure of DNA

E. Chargaff observed that there are certain regularities in the percentages of pyrimidines and purines obtained from the DNA of various species. His observations are given below :

1. The ratio of mole percentages of total purines to mole percentages of total pyrimidine is approximately 1; i.e.,

$$\frac{\%\text{ guanidine} + \%\text{ adenine}}{\%\text{ cytosine} + \%\text{ thymine}} \simeq 1 \text{ (DNA does not contain uracil)}$$

2. The ratio of the mole percentage of thymine to that of adenine, i.e.,

$$\frac{\%\text{ thymine}}{\%\text{ adenine}} \simeq 1$$

3. The ratio of the mole percentage of cytosine to guanine is also approximately 1, i.e.,

$$\frac{\%\text{ cytosine}}{\%\text{ guanine}} \simeq 1$$

4. Chargaff also observed that the ratio of total mole percentage of adenine and thymine to that of guanine and cytosine varies from species to species and the value of the ratio is a characteristic property of the DNA of a particular species. Furthermore, he observed that the value does not vary appreciably with the age or conditions of growth of the individual organisms within the same species.

Besides these observations of E. Chargaff, Watson and Crick had the following X-ray data.

(i) The bond lengths and bond angles of the heterocyclic bases.

(ii) Long repeat distance (34 Å) in natural DNA. They have this data from Ms. Wilkins.

Analysing all these above said facts they suggested a double-helix structure for DNA. This means, there are two helical nucleic acid chains in the structure. They are held together by inter-chain hydrogen bonds between base pairs on opposite helices. The two chains are interwound with each other round the same axis into a double-helix. The sugar-phosphate backbone on the outside while base pairs are on the inside of the helix which is like a twisted spiral ladder. The purine and pyrimidine bases are always paired in such a way that adenine is bonded to thymine and guanine to cytosine as shown in Fig. 22.8 (given on the next page).

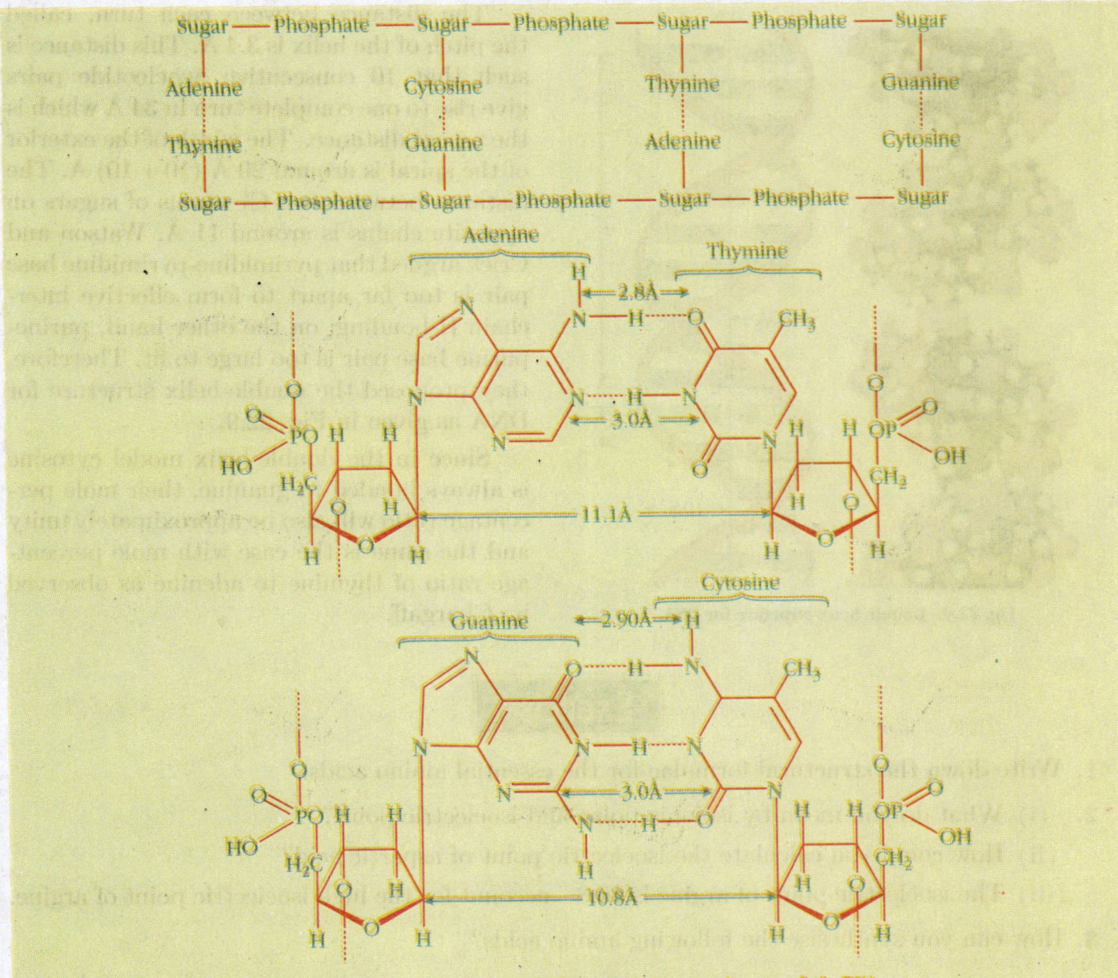


**Fig. 22.8. H-bonds between base pairs in opposite strands in DNA**

See next page further discussions.

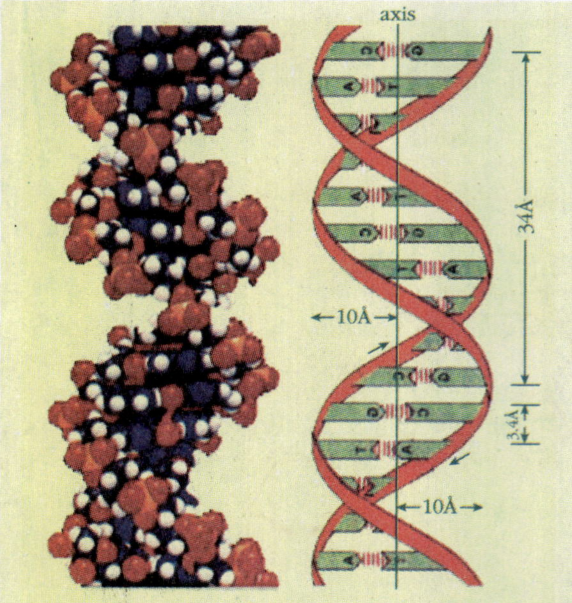


**Fig. 22.9. Double helix-structure for DNA**

The distance between each turn, called the pitch of the helix is 3.4 Å. This distance is such that 10 consecutive neocleotide pairs give rise to one complete turn in 34 Å which is the repeat distance. The width of the exterior of the spiral is around 20 Å (10 + 10) Å. The distance between two $C_1'$ atoms of sugars on opposite chains is around 11 Å. Watson and Crick argued that pyrimidine-pyrimidine base pair is too far apart to form effective inter-chain H-bonding; on the other hand, purine-purine base pair is too large to fit. Therefore, they proposed the double-helix structure for DNA as given in Fig. 22.9.

Since in the double-helix model cytosine is always bonded to guanine, their mole percentage ratio will also be approximately unity and the same is the case with mole percentage ratio of thymine to adenine as observed by Chargaff.

## Exercise

1. Write down the structural formulae for the essential amino acids.
2. (i) What do you mean by isoionic point and isoelectric point?
   (ii) How could you calculate the isoelectric point of aspartic acid?
   (iii) The isoelectric point of argine is 10.8—account for the high isoelectric point of argine.
3. How can you synthesise the following amino acids?
   (i) Isoleucine, (ii) Argine, (iii) Serine, (iv) Threonine.
4. Specific rotation of an $\alpha$-amino acid is pH dependent—Explain.
5. What happens when $\alpha$-, $\beta$- and $\gamma$-amino acids are heated separately? Comment on the optical activity of the products.
6. What are peptides? How peptides can be analysed to know their amino acid sequence?
7. Edman method of peptide analysis is superior to Sanger method—Comment.
8. Synthesise (i) phe. ala. gly, (ii) leu. gly. ala.
9. Write a short note on resolution of amino acids.
10. How can you identify the presence of a peptide linkage in a molecule?
11. Separate glycine, alanine and leucine from their mixture.

Chapter 23

# An Introduction to Carbohydrates

## Objectives

- To know :
  - The definition, nomenclature and classification and D- and L-series of carbohydrates [23.1–23.3] p. 1541
  - Elimentary facts about the structures, their projections and conformations of aldoses [23.4] p. 1542
  - Elimentary facts about the structures of ketoses and their projections [23.5] p. 1548
- To study the chemical properties of aldoses and ketoses [23.6] p. 1549
  - Mutarotation [23.6.a] p. 1549
  - Formation and hydrolysis of glycosides [23.6.b] p. 1551
  - Action of alkalis [23.6.c] p. 1553. (i) Reverse aldol reaction (p. 1553) and (ii) Lobry de Bruyn-van Ekenstein rearrangment (p. 1554)
  - Action of acids on monosaccharides [23.6.d] p. 1554
  - Blocking of alcoholic OH groups of saccharides [23.6.e] p. 1556

## 22.1 Introduction

Carbohydrates are polyhydroxy aldehydes or ketones and their derivatives including those which on hydrolysis give rise to polyhydroxy aldehydes or polyhydroxy ketones. They have the general formula $C_xH_{2y}O_y$. The name carbohydrates, i.e., hydrates of carbon were given to these compounds because the then known compounds contained hydrogen and oxygen in the same atomic ratio as in water. However, each compound of the class does not have the general formula $C_x(H_2O)_y$; for example, 6-deoxy-L-mannose, called L-rhamnose has the molecular formula $C_6H_{12}O_5$. A few of the carbohydrates are known to everybody, e.g., D-glucose, sucrose (cane sugar), starch, cellulose, etc. D-glucose is photosynthesised by green plants and it is the fuel molecule for most of the organisms. Cane sugar is our everyday table sugar. Starch is present in plants in high quantity where it is the chief form of fuel storage. The paper we use and the cotton cloths we wear are made up of cellulose, the predominant extracellular structural component of the woody and fibrous tissues of plants. The two simplest carbohydrates are D-glyceraldehyde (D-$CHOCHOHCH_2OH$) and dihydroxyacetone ($CH_2OHCOCH_2OH$). Cane sugar on hydrolysis gives D-glucose and D-fructose, of which the former contains an aldehydic carbonyl group and the latter a ketonic carbonyl group in their open chain forms.

## 22.2 Nomenclature and Classification

The name-ending of simpler carbohydrates is 'ose'. A single polyhydroxy aldehyde unit is called an aldose and the similar unit of ketone is said to be a ketose. The number of C atoms present in the main chain is indicated by prefixing tri, tetr, pent, hex, etc. before the name-ending. For example, D-glyceraldehyde contains 3 C atoms in the main chain and it is a dihydroxyaldehyde; it is, therefore, named as D-aldotriose. D-fructose contains a ketonic carbonyl group and five hydroxyl groups in a chain of six C atoms; so it is named as a ketohexose. Replacement of an OH group by one H atom is shown by the prefix 'deoxy'. When one OH group is replaced by an $NH_2$ group, 'the amino-deoxy' is prefixed. On the other hand, when the H atom of OH group is replaced by an alkyl group, we write 'O-alkyl'.

Carbohydrates are first classified as Sugars and Polysaccharides. Sugars have sweet tastes and they are soluble in water. D-Glucose, D-fructose and cane sugar belong to sugars. Polysaccharides are very big molecules, usually non-crystalline substances insoluble in water and do not taste sweet. Starch and cellulose are the examples of polysaccharides. However, sugars are divided into two classes :

(i) *Monosaccharides* and (ii) *Oligosaccharides.*

(i) Monosaccharides consist of a single polyhydroxyaldehyde or ketone unit which cannot be further hydrolysed to a still smaller carbohydrate unit. The most abundant monosaccharide is the D-glucose, a D-aldohexose. It is the basic building unit of the most abundant polysaccharides.

(ii) Oligosaccharides contain two to ten monosaccharide units joined together and on hydrolysis oligosaccharides give two to ten units of monosaccharides. Sucrose and lactose (milk sugar) are disaccharides which belong to oligosaccharides. The former on hydrolysis gives D-glucose and D-fructose while lactose on hydrolysis gives D-glucose and D-galactose.

Sugars are further classified under the heads

(a) *Reducing sugars* and (b) *Non-reducing sugars.*

(a) Reducing sugars are those which reduce ammoniacal silver nitrate solution, Fehling solution, etc. Monosaccharides are reducing sugars. D-Glucose and D-fructose are reducing sugars. An oligosaccharide may or may not be a reducing sugar. (+)-maltose and (+)-lactose are reducing disaccharides but (+)-sucrose is non-reducing.

(b) Non-reducing sugars cannot reduce ammoniacal silver nitrate solution, Fehling solution, etc. Cane sugar (sucrose) is a non-reducing sugar.

## 22.3 D- and L-Series

The simplest members of the series are glyceraldehydes, D and L-(1 and la respectively in Fig. 23.1). The simplest ketose is dihydroxyacetone (B).

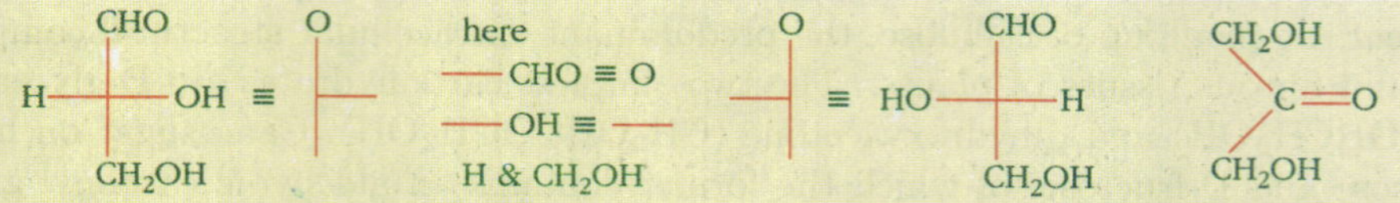


D-(+)-Glyceraldehyde **A (I)** L-(−)-Glyceraldehyde **1a** Dihydroxyacetone **B**

**Fig. 23.1. Glyceraldehydes and their short-hand representations**

Most abundant aldoses are D-ribose, D-glucose and D-mannose. The prefixes 'D' and 'L' are conventionally written before the trivial names of aldoses and ketoses. The prefixes 'D' and 'L' (relative to D and L-glyceraldehydes respectively) refer to the relative configuration of the chiral C atom farthest from the carbonyl carbon atom (or the highest numbered chiral C) in monosaccharides containing two or more chiral carbon atoms. D and L-glucoses, guloses are represented in Fig. 23.2 according to the convention discussed above.

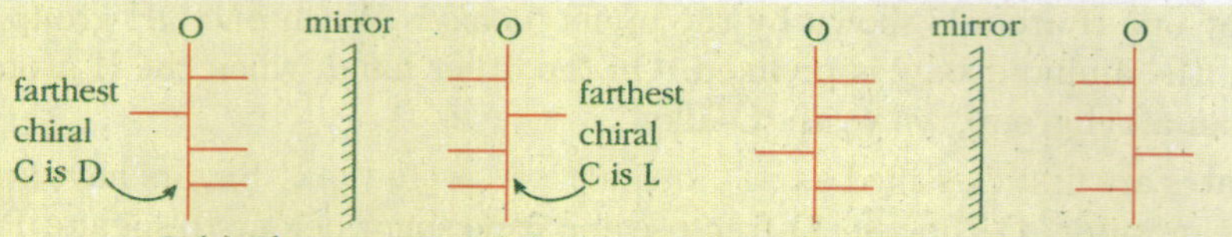


D-(+)-Glucose L-(−)-Glucose D-(−)-Gulose L-(+)-Gulose

**Fig. 23.2. D and L-monosaccharides**

## 22.4 Aldoses

The names of the D-aldoses up to six carbon atoms are given in the family tree (Fig. 23.3). The corresponding L-aldoses are the mirror images of D-aldoses. However, the names can be memorised by remembering the group of words (limerically) as given below :

<u>Go</u> <u>Er</u>nest <u>Th</u>omas! <u>Ri</u>ng <u>Ara</u>bian <u>Xy</u>lophones <u>L</u>oudly—<u>All</u> <u>Alt</u>ruist <u>Gl</u>adly <u>Ma</u>ke <u>Gu</u>ms <u>I</u>n <u>Ga</u>llon <u>Ta</u>nks. The letters underlined are the first one or two or more letters of the names given in the family tree (Fig. 23.3).

Two aldotetroses of the same series, D or L, constitute a pair of diastereomers : they are also called a pair of epimeric sugars. Epimers are configurational isomers that differ from each other in their configuration about one chiral carbon (stereocentre) containing more than one such carbon (centre). However, aldotrioses and aldotetroses usually occur in nature in the open chain forms as shown in the family tree. From aldopentoses onward all aldoses occur in nature in the

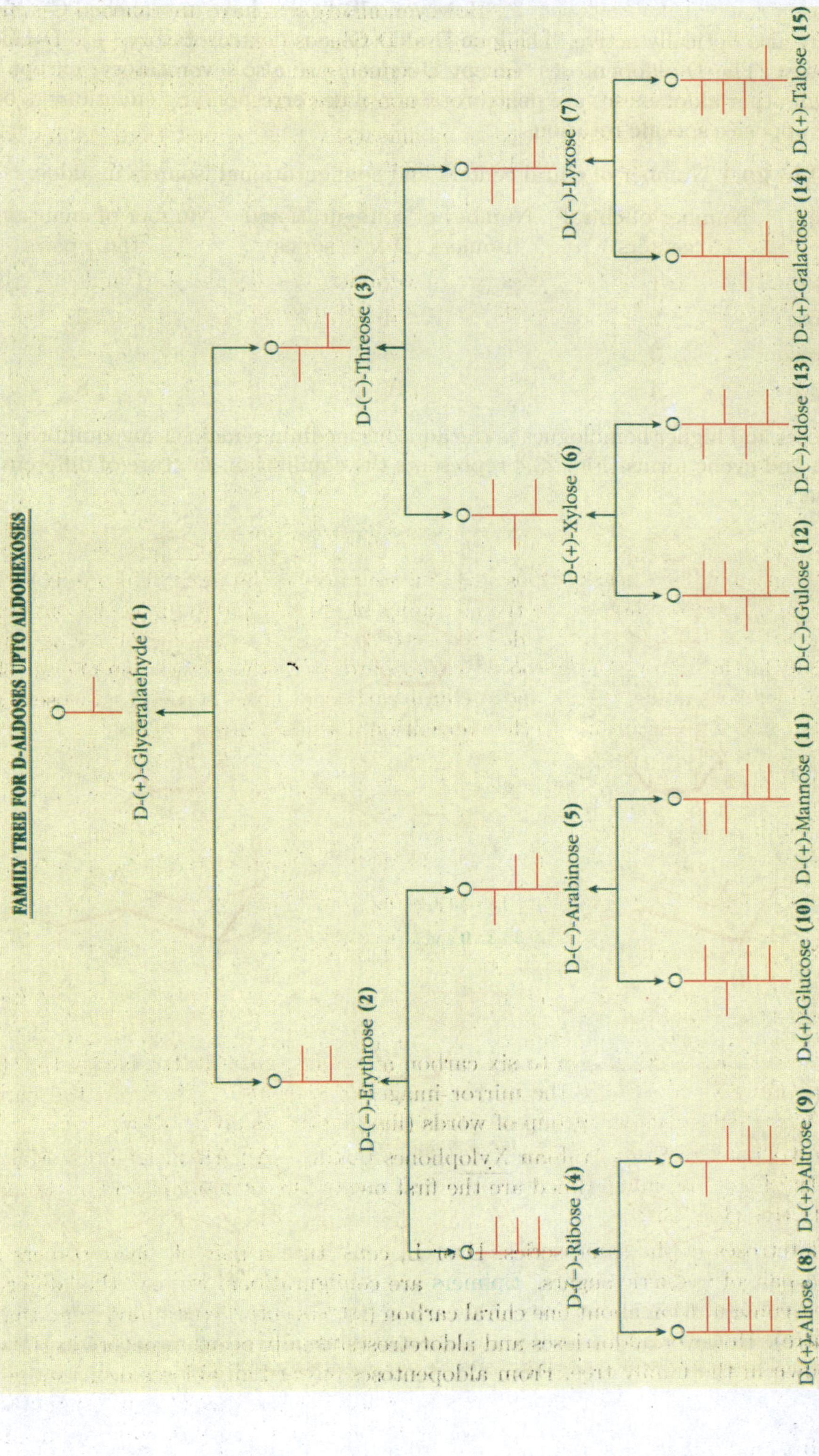
FAMILY TREE FOR D-ALDOSES UPTO ALDOHEXOSES
D-(+)-Glyceralaehyde (1)
D-(−)-Erythrose (2)
D-(−)-Threose (3)
D-(−)-Ribose (4)
D-(−)-Arabinose (5)
D-(+)-Xylose (6)
D-(−)-Lyxose (7)
D-(+)-Allose (8)
D-(+)-Altrose (9)
D-(+)-Glucose (10)
D-(+)-Mannose (11)
D-(−)-Gulose (12)
D-(−)-Idose (13)
D-(+)-Galactose (14)
D-(+)-Talose (15)

**Fig. 23.3**

cyclic forms as intramolecular hemiacetals. However, all of them have unbranched C-skeleton. All the aldoses are also optically active. The lone D-aldotriose is dextrorotatory (+). D-aldotetroses are levorotatory. The D-aldopentoses, except D-xylose, are also levorotatory; except D-gulose and D-idose all other aldohexoses are dextrorotatory. The corresponding enantiomers of L-series have just the opposite specific rotations.

**Table 23.1** Number of chiral centers and configurational isomers in aldoses

| Aldose | Number of chiral centers | Number of configurational isomers (D & L series) | Number of enantiomeric (DL) pairs |
|---|---|---|---|
| Aldotriose | 1 | 2 | 1 |
| Aldotetrose | 2 | 4 | 2 |
| Aldopentose | 3 | 8 | 4 |
| Aldohexose | 4 | 16 | 8 |

Aldopentoses and higher homologues in the aqueous medium remain as an equilibrium mixture of open chain and cyclic forms. Fig. 23.4 represents the equilibrium mixture of different forms of D-(+)-glucose.

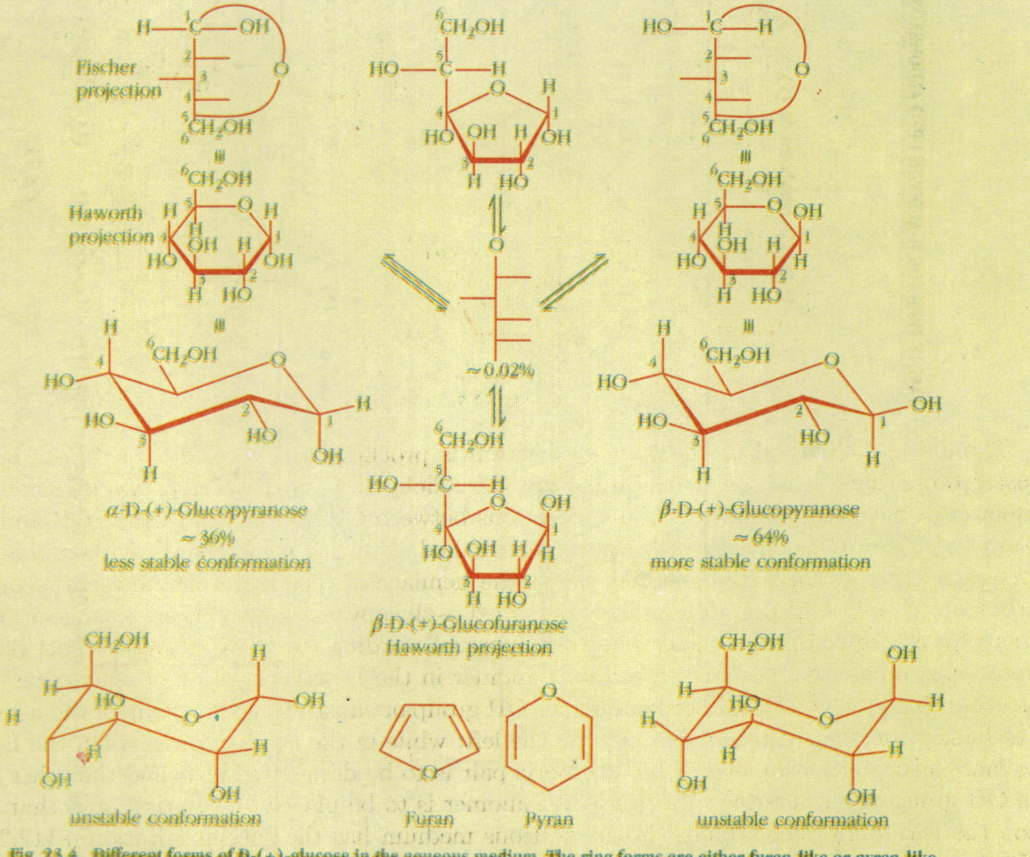


**Fig. 23.4.** Different forms of D-(+)-glucose in the aqueous medium. The ring forms are either furan-like or pyran-like. The energy difference between the two conformations of the same anomer is around 25 kJ $mol^{-1}$

Haworth projections are commonly used to indicate the configurations of chiral carbons in cyclic forms; the bold edge indicates that this is nearest to the reader. Actually the rings are not planar as indicated by the Haworth projections but they are puckered as shown by the most stable chair conformation. The formation of ring structures from the open chain forms is a hemiacetal formation reaction involving the aldehydic carbonyl group and either the $C_5$—OH or the $C_4$—OH. When the $C_5$—OH is involved, we get a pyran-like ring structure, called glucopyranose; but the intramolecular hemiacetal formation reaction involving $C_4$—OH gives rise to a furan-like structure known as glucofuranose.

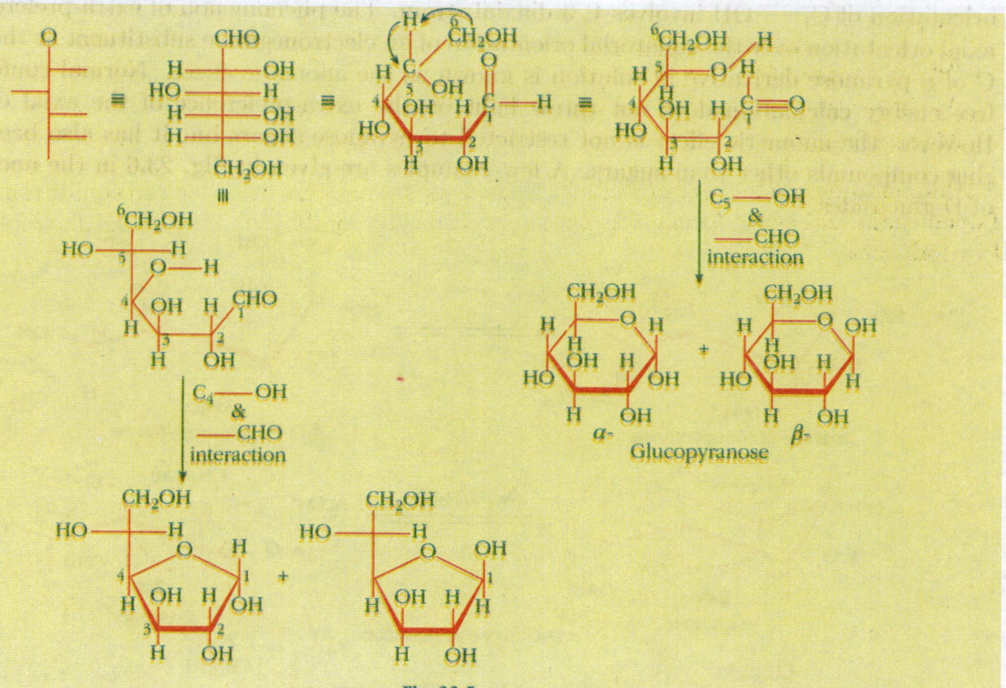


**Fig. 23.5.**

The aldehydic carbon in the open chain form is prochiral and —HC=O moiety has two diastereotopic faces. In the hemiacetal form the aldehydic C is chiral and two diastereomeric hemiacetals develop from each of the interactions between CHO group and $C_5$—OH and CHO group and $C_4$—OH. The two diasteriomers are called anomers. Anomers are epimers that differ from each other in their configuration about the hemiacetal (hemiketal in the cases of ketoses) carbon atom only. One pair of anomers are named $\alpha$-glucopyranose and $\beta$-glucopyranose and the other pair as $\alpha$-glucofuranose and $\beta$-glucofuranose. According to C.S. Hudson rule, between the two anomeric pair the more dextrorotatory anomer in the D-series is to be designated as $\alpha$ and the other $\beta$, and in the Fischer projection the OH group or substituted OH group of the $\alpha$-anomer is to be kept on the right and that of $\beta$ on the left; while in the cases of anomers in the L-series the more levorotatory anomer of the anomeric pair is to be designated as $\alpha$ and the other $\beta$ and the OH group or substituted OH group of $\alpha$-anomer is to be placed on the right and that of the $\beta$ on the left. $\alpha$-D-glucopyranose in the aqueous medium has the specific rotation +112.2° and $\beta$-D-glucopyranose +18.7°. However, in converting an open chain form into a ring, the left-hand OH of any chiral carbon is to be placed above the plane of the ring and the right-hand OH below the plane (L-lift and D-down). In the chair form of glucose the anomeric OH of the $\alpha$-anomer occupies the axial bond while the equatorial bond is occupied by the $C_1$—OH in the $\beta$-anomer.

The percentage composition of $\alpha$-D- and $\beta$-D-glucoses in the aqueous medium shows that the conformations have different stabilities; the $\beta$-anomer is more stable than the $\alpha$-anomer in the aqueous medium. But this is not the fact for all cases. There are a large number of examples in the carbohydrate chemistry where pyranose sugars with a strong electronegative (electron-withdrawing) substituent, such as halogen, alkoxy, etc., at $C_1$ are more often more stable when the substituent has an axial orientation rather than equatorial one though the axial orientation of $C_1$—OH involves 1, 3-diaxial strain. The phenomenon of extra-preference for an axial orientation over the equatorial orientation of an electronegative substituent at the anomeric C of a pyranose derivative in solution is known as the anomeric effect. Normal conformational free energy calculation does not throw light on the extra-preference of the axial orientation. However, the anomeric effect is not restricted to pyranose sugars but it has also been found in ring compounds other than sugars. A few examples are given in Fig. 23.6 in the anomerisation of D-glucosides.

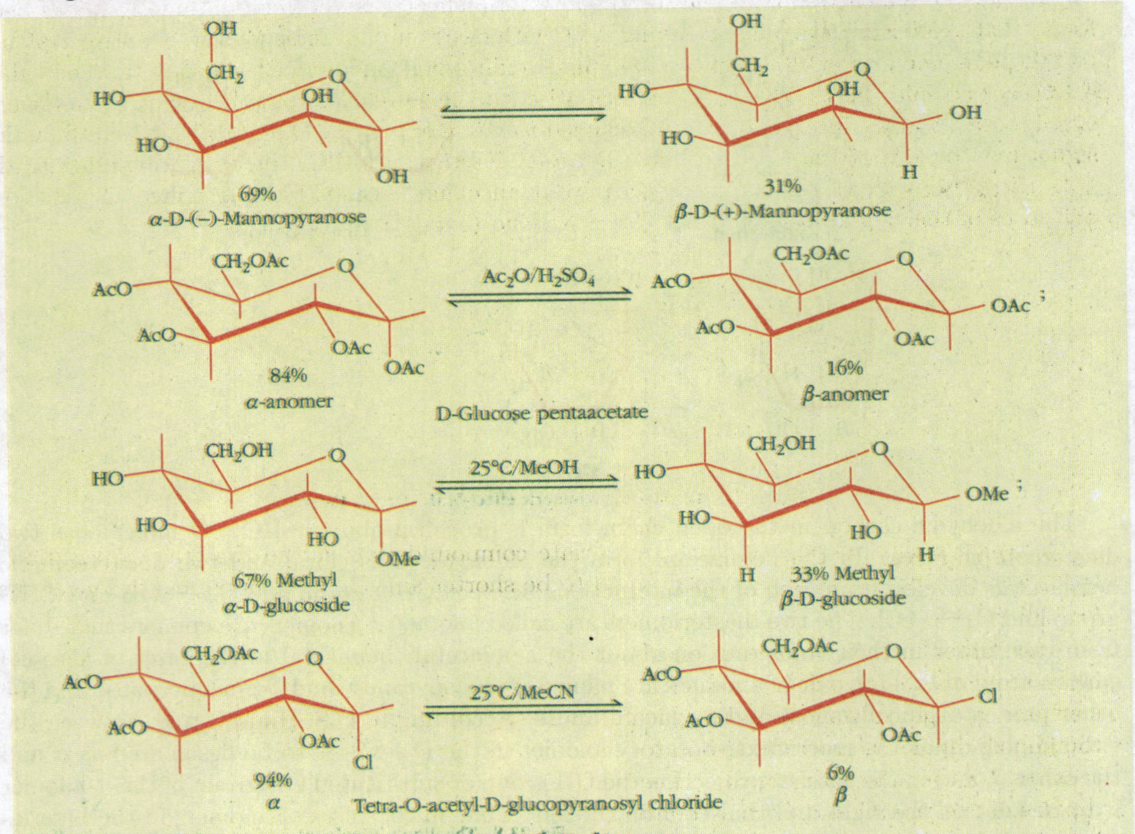


**Fig. 23.6. % of Anomers**

The anomeric effect depends very much on the nature of the electronegative substituents. The decreasing order of anomeric effect amongst the following is

$$\text{Halogens} > PhCO_2^- > AcO > RO > RS > HO > NH_2 > CO_2Me > \overset{\oplus}{N}$$

When the anomeric substituent is N-pyridinium group, the anomeric effect is negative, i.e., the pyridinium group preferentially occupies the equatorial bond rather than the axial bond — the equatorial isomer at any time is more populated than expected from its space occupying bulk. This is known as reverse anomeric effect and is characteristic of substituents attached to the anomeric carbon through a positively charged nitrogen atom. In the case of an alkoxy group, an increase in bulk of alkyl group (Me → Et → i-Pr → t-Bu) has a negligible effect on the magnitude of the anomeric effect.

The anomeric effect has been found to be less in the solvents of high dielectric constant, whereas the effect is more important when the solvent of high dielectric constant is replaced by the solvent of low dielectric constant. For example, 2-methoxy-6-methyltetrahydropyran.

$$\beta \rightleftharpoons \alpha$$

$$K_{eq} = 3.4 \text{ in } CCl_4\ (\varepsilon = 2.2)$$
$$K_{eq} = 1.8 \text{ in MeCN } (\varepsilon = 37.5)$$

No fully convincing explanation for the anomeric effect has yet been obtained. It has been found that there is always a non-bonding A.O. with electron-pair and the A.O. is anti-periplanar to the polar anomeric C-X bond in compounds exhibiting anomeric effect; so it may be called a stereoelectronic effect. The major contribution to the stereoelectronic factor of the anomeric effect comes from the interaction of the occupied $p$ orbital of the ring O atom with the antibonding sigma ($\sigma^*$) orbital of the C—X bond (i.e, $n$-$\sigma^*$ interaction; M.O. theory). According to the V.B. theory, it is a case of coupling effect of resonance and no-bond resonance involving the lone pair of $e$s of the ring O atom and the $C_1$—X bond (axial) as shown below.

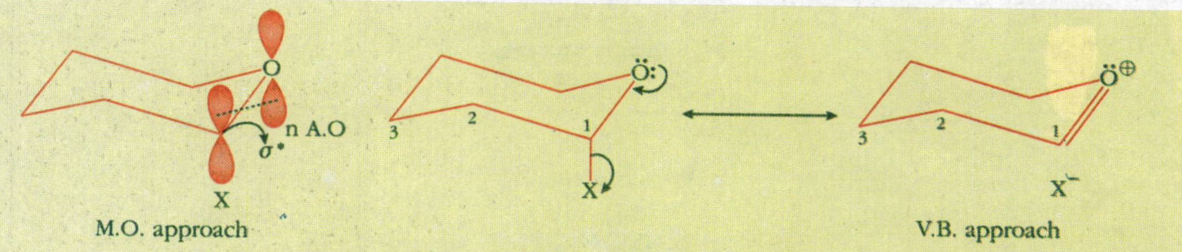


**Fig. 23.7. The cause for anomeric effect-M.O. and V.B. theories**

X-ray crystallographic study on appropriate compounds has shown that the explanation is correct since O—$C_1$ bond has been found to be shorter and $C_1$—X bond length longer than normal values.

The fact that the anomeric effect increases with the decreasing dielectric constant of the solvent tells us that the dipole-dipole interaction also contributes to the anomeric effect. The dipoles of the ring O and C—X bond in the equatorial form being parallel, the interaction is very strong and the equatorial form loses stability while in the axial isomer, the two dipoles being divergent, the dipole-dipole interaction is weak and the axial form gets stability. This is further coroborated by the fact that equatorial pyridinium-type substituents are more stable than our expectation (reverse anomeric effect).

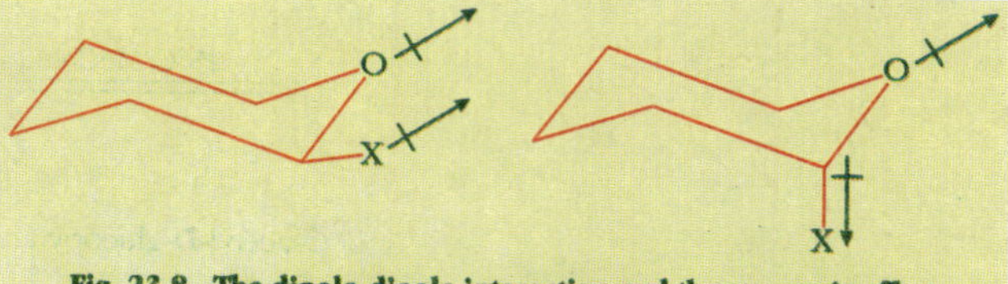


**Fig. 23.8. The dipole-dipole interaction and the anomeric effect**

**Table 23.2** Physical properties of some aldoses

| Aldose | Source (Chief) | m.p. (in °C) | Change of specific rotation of $\alpha$-anomer in water (°) |
|---|---|---|---|
| D-(+)-Xylose | Xylan | 145 | $+94 \longrightarrow +19$ |
| L-(+)-Arabinose | Gums | 160 | $+190 \longrightarrow +105$ |
| D-(+)-Glucose | Starch | 146 | $+113 \longrightarrow +52.5$ |
| D-(+)-Mannose | Mannan | 133 | $+29 \longrightarrow +14$ |
| D-(+)-Galactose | Gums | 167 | $+150 \longrightarrow +80$ |

## 22.5 Ketoses

The family tree for D-ketoses is given in Fig. 23.9. The enantiomeric L-ketoses are mirror images of D-ketoses in configuration.

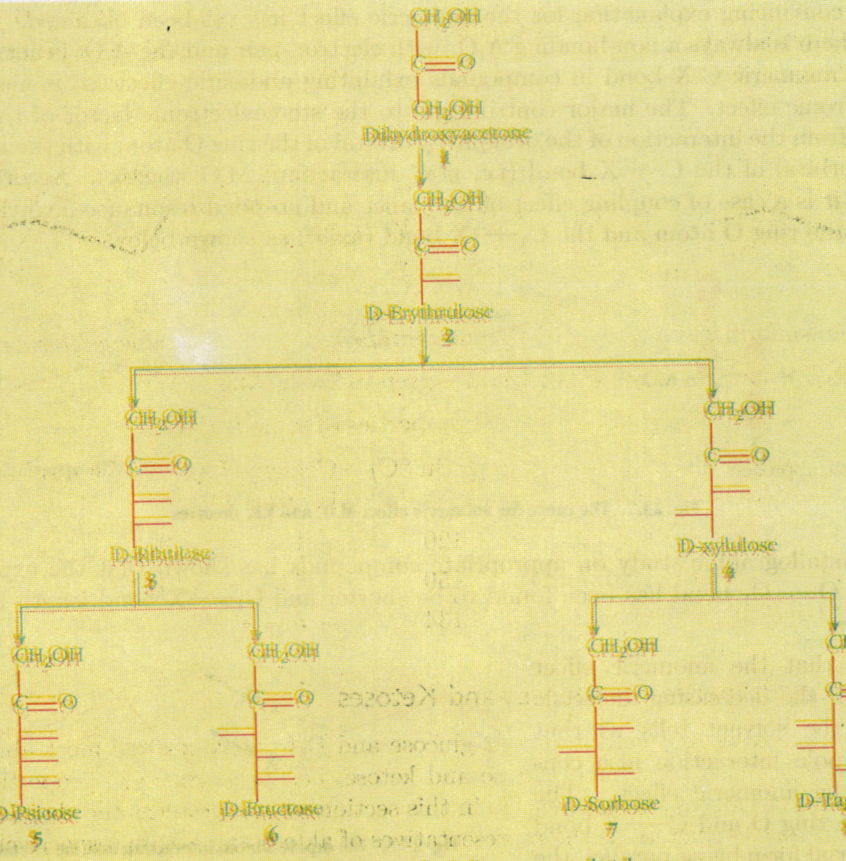


Fig. 23.9. The family tree for D-ketoses up to ketohexoses

Dihydroxyacetone is optically inactive but other ketoses are optically active. D-(−)-Fructose is the most abundant ketose. It remains in nature in the hemiketal form. The aqueous solution of D-fructose is an equilibrium mixture of its different forms.

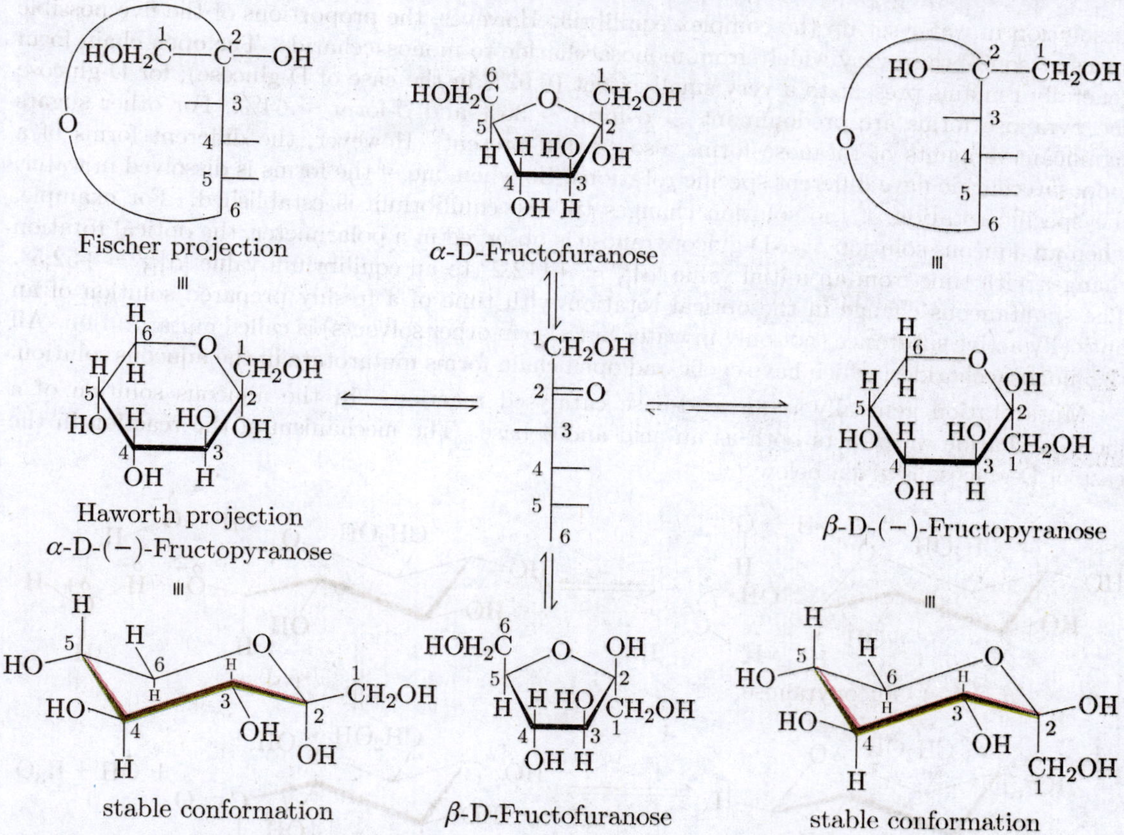


The physical properties of a few ketoses are given in Table 23.3.

**Table 23.3** Physical properties of some ketoses

| Ketose | Source (Chief) | m.p. (in °C) | Change of specific rotation in water (°) |
|---|---|---|---|
| D-(−)-Fructose | Insulin | 120 | $-132 \longrightarrow -94$ |
| L-(−)-Sorbose | Rowans | 159 | $-44 \longrightarrow -43$ |
| D-(−)-Tagatose | Gums | 134 | $-3 \longrightarrow -5$ |

## 22.6 Chemical Properties of Aldoses and Ketoses

Amongst aldohexoses and ketohexoses D-glucose and D-fructose are the most abundant respectively. They are also the typical aldose and ketose, i.e., their chemical properties are the general properties of aldoses and ketoses. So in this section we shall discuss the chemical properties of D-glucose and D-fructose as the representatives of aldose and ketose respectively.

### 23.6.a Action of water on aldoses and ketoses : Mutarotation

It has already been represented in Fig. 23.4 that when α- or β-D-glucopyranose is dissolved in water, a complex equilibrium is set up. Not only glucose but also other aldoses and ketoses on dissolution in water set up the complex equilibria. However, the proportions of the five possible forms at equilibrium vary widely from monosaccharide to monosaccharide. The open chain form generally remains present to a very small extent (0.02% in the case of D-glucose); for D-glucose the pyranose forms are predominant — $\alpha$-form ~ 36% and $\beta$-form ~ 64%. For other sugars significant amounts of furanose forms also remain present. However, the different forms of a monosarccharide have different specific rotations. So when one of the forms is dissolved in water, the specific rotation of the solution changes till the equilibrium is established. For example, when an aqueous solution of $\alpha$-D-glucopyranose is observed in a polarimeter, the optical rotation changes with time from an initial value $[\alpha]_D^{20} = +112.2°$ to an equilibrium value $[\alpha]_D^{20} = +52.5°$. The spontaneous change in the optical rotation with time of a freshly prepared solution of an optically active substance (not only in water but also in other solvents) is called mutarotation. All the monosaccharides, which have cyclic and open chain forms mutarotate in the aqueous solution.

Mutarotation generally is an acid-base catalysed reaction. In the aqueous solution of a monosaccharide water acts both as an acid and a base. The mechanism of the reaction in the case of D-glucose is given below :

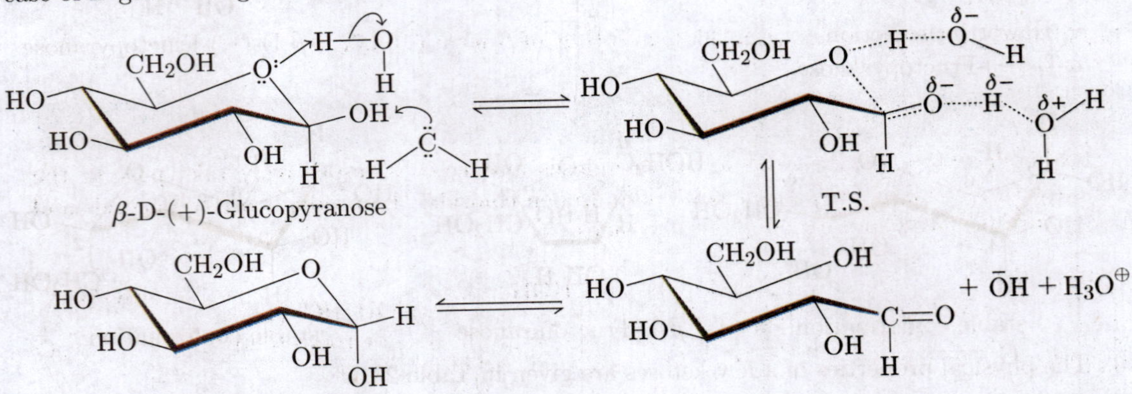


The acid-base catalysed mechanism of the reaction has been proved by the fact that the mutarotation of D-glucose in an aprotic solvent does not occur in the presence of pyridine alone or phenol alone; when both phenol and pyridine are present together, the mutarotation of D-glucose takes place.

The mutarotation occurs owing to spontaneous configurational or spontaneous structural change of a compound. Thus the mutarotation of D-glucoses mainly occur owing to configurational change (i.e., anomeric change).

The mutarotation of $\delta$-gluconolactone ($[\alpha]_D^{20} = +63.5°$) and that of $\gamma$-gluconolactone ($[\alpha]_D^{20} = +67.5°$) in the aqueous medium is $[\alpha]_D^{20} = +12°$; this involves spontaneous structural change via gluconic acid ($[\alpha]_D^{20} = -6°$) as shown below :

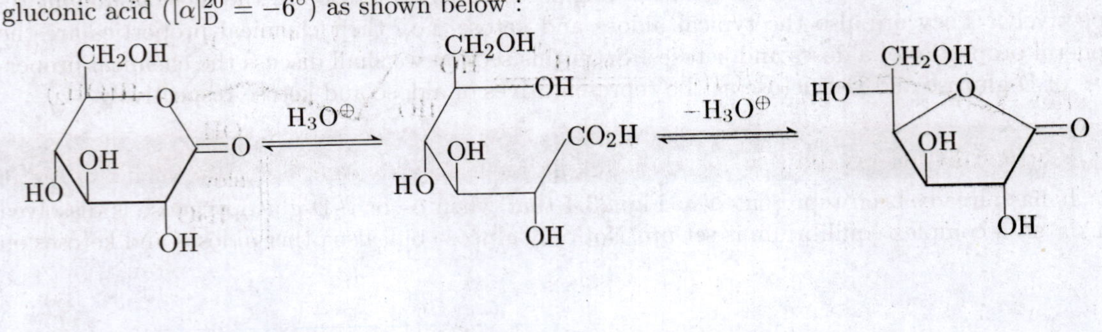

Generally, the rate of mutarotation depends on temperature, solvent and catalyst. When the mutarotation involves only two components, the reaction may follow the first order kinetics. In the complex cases the mutarotation does not follow the first order kinetics.

**Problem 23.1.** The specific rotation of $\alpha$-D-(+)-glucopyranose is +112° in the aqueous medium and that of $\beta$-D-(+)-glucopyranose is +18.7°. The specific rotation of the equilibrium mixture is +52.5°. Find out the % composition of $\alpha$- and $\beta$-D-glucopyranoses at equilibrium.

**Solution :** Let the % of $\alpha$-D-(+)-glucopyranose is $x$ at equilibrium. The percentage of $\beta$-D-(+)-glucopyranose is, therefore, $100 - x$, the % of the open chain form and furanose forms being negligible.

$$\therefore \quad x \times 112^\circ + (100 - x)18.7^\circ = 100 \times 52.5^\circ$$

$$\text{or,} \quad 112^\circ x + 1870^\circ - 18.7^\circ x = 5250^\circ$$

$$\text{or,} \quad 93.3^\circ x = 5250^\circ - 1870^\circ$$

$$\text{or,} \quad x = \frac{3380^\circ}{93.3^\circ x} = 36.23.$$

$\therefore$ at equilibrium the % of $\alpha$-form is 36.23 and that of $\beta$-form is 63.77.

## 23.6.b Reaction of alcohols (in the presence of dry hydrogen chloride) with aldoses a

The hemiacetal and hemiketal forms of aldoses and ketoses respectively take part in the reaction with alcohol in the presence of dry hydrogen chloride; the products are full acetals and full ketals respectively and called alkyl glycosides.

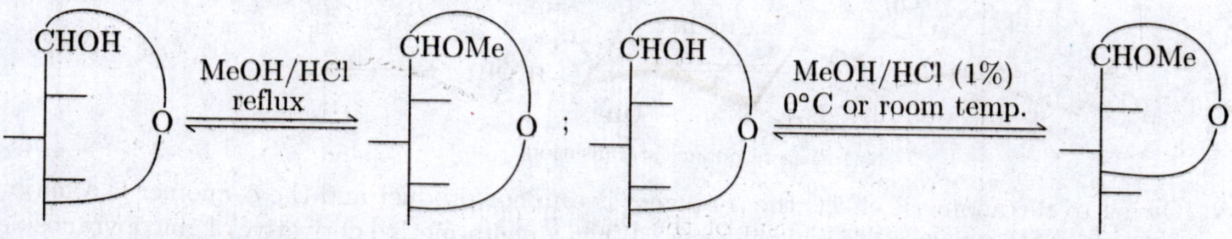


$\alpha$- or $\beta$-D-glucopyranose — Methyl $\alpha$- & $\beta$-D-glucopyranosides ; $\alpha$- or $\beta$ — Methyl $\alpha$- & $\beta$-D-glucofuranosides

Now, the question is ; why does the anomeric —OH group undergo methylation but others do not? The answer is : when the anomeric OH is removed as $H_2O$, a resonance-stabilised carbocation is formed. If any one of the other hydroxyl groups is removed, a highly unstable carbocation is formed. So the reaction occurs on the anomeric carbon preferentially because its carbocation is the most stable amongst others. The mechanism of the reaction is given below :

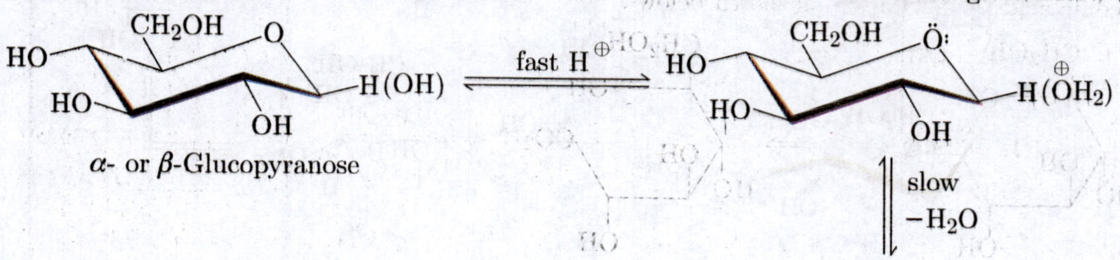

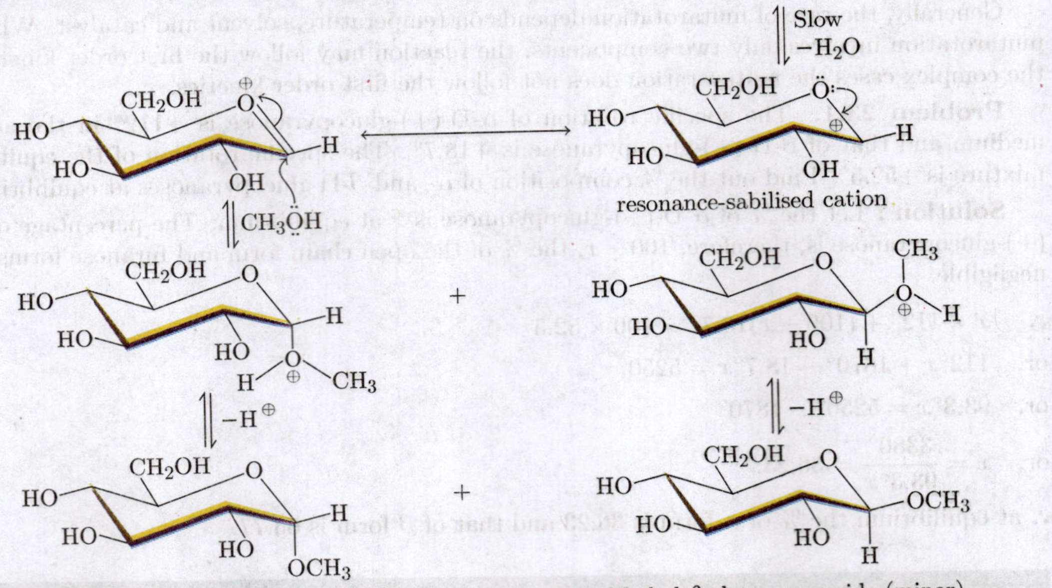


Methyl $\alpha$-glucopyranoside (major) Methyl $\beta$-glucopyranoside (minor)

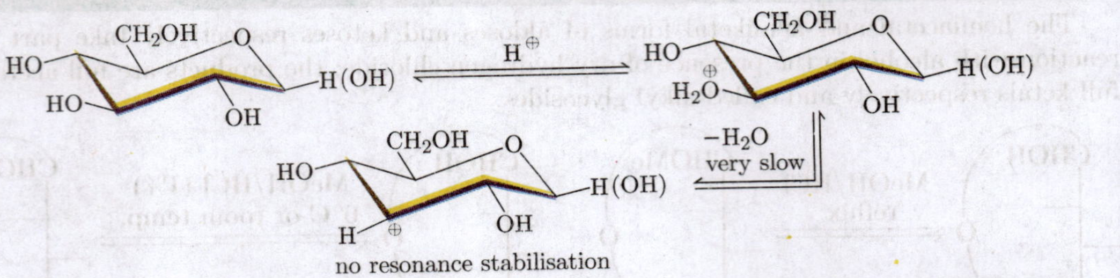


Owing to the anomeric effect, the $\beta$-anomer is a minor product and the $\alpha$-anomer is a major one. However, the reaction is a reversible one. When a penta-methyl derivative of glucopyranoside is hydrolysed, it is the anomeric OMe group that gets hydrolysed; this is because the reaction takes place with the formation of a carbocation and if the anomeric C bears the positive charge, the carbocation gets stability by the resonance effect. The hydrolysis of furanosides, however, takes place in a different way.

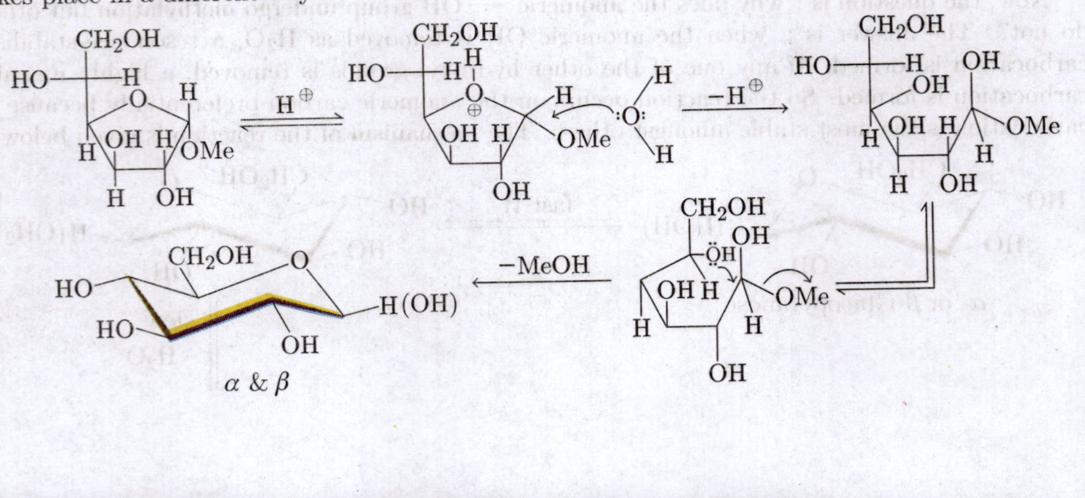

## 23.6.c Action of alkalis on monosaccharides

Monosaccharides, aldoses and ketoses (hemiacetals and hemiketals), are not stable towards alkalis. When a monosaccharide is treated with a dilute alkali, two different types of reactions take place on the open chain forms.

(i) Reverse aldol reaction : Usually the compound breaks down to simple hydroxyaldehydes and hydroxyketones by the reverse aldol reaction which follows the ElcB pathway. D-glucose on treatment with dilute alkali breaks down to glycolaldehyde.

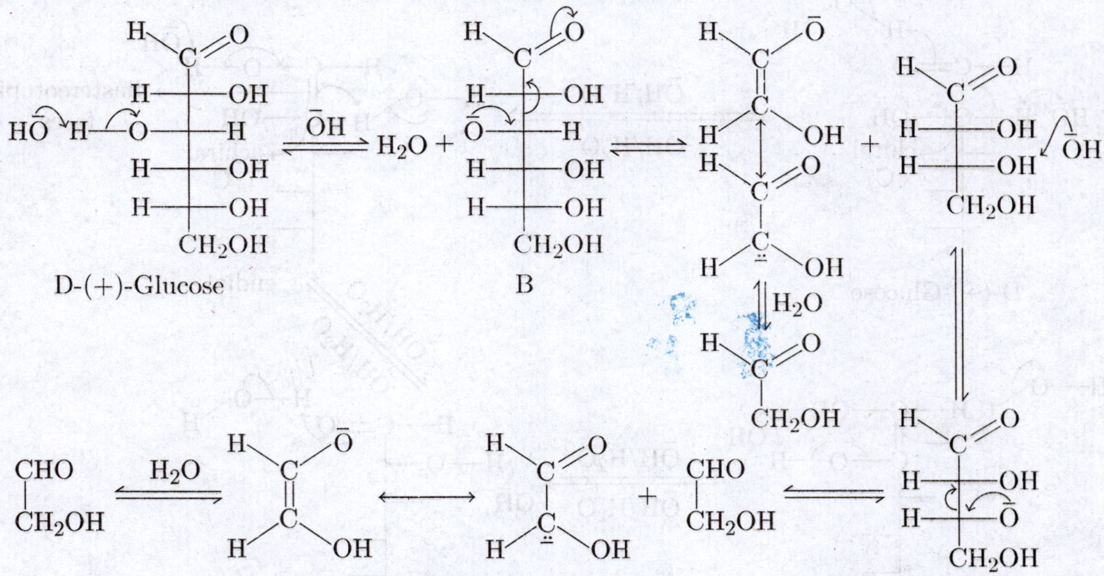


In fact, when 3 moles of glycolaldehyde are subjected to aldol condensation, glucose is obtained. However, by the reverse aldol reaction D-(−)-fructose gives dihydroxyacetone, glycolaldehyde and formaldehyde.

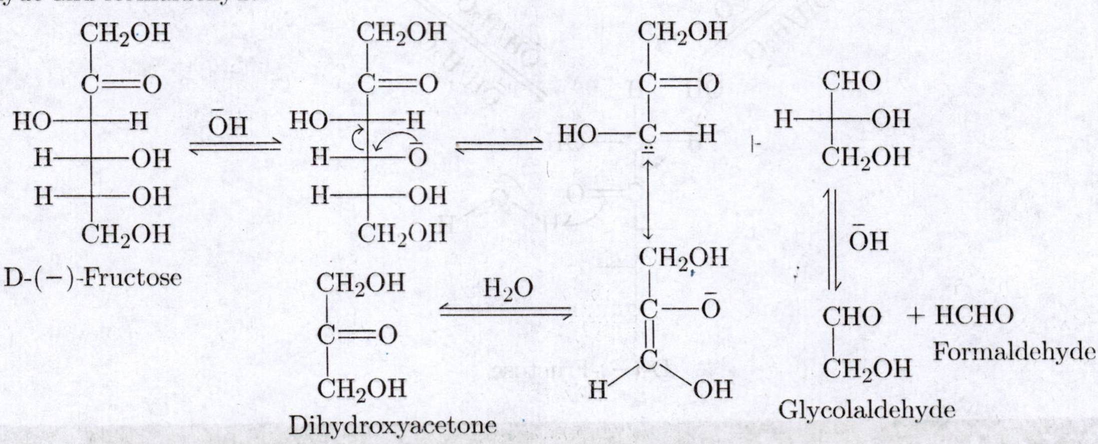


The hydroxide ion abstract the H atom of that hydroxyl group which gives a resonance-stabilised carbanion on $C_\alpha$—$C_\beta$ bond cleavage (with respect to the carbonyl group) in the next step.

(ii) Lobry de Bruyn-van Ekenstein rearrangement : In an alkaline medium monosaccharides undergo a rearrangement reaction which involves epimerisation and aldo-keto isomerisation simultaneously via an endiol having diastereotopic faces and the rearrangement is called Lobry de Bruyn-van Ekenstein rearrangement. D-glucose undergoes the rearrangement and an equilibrium mixture of D-glucose, D-mannose and D-fructose is obtained. D-fructose and D-mannose also undergo the rearrangement in an alkaline medium to give the same equilibrium mixture as D-glucose gives.

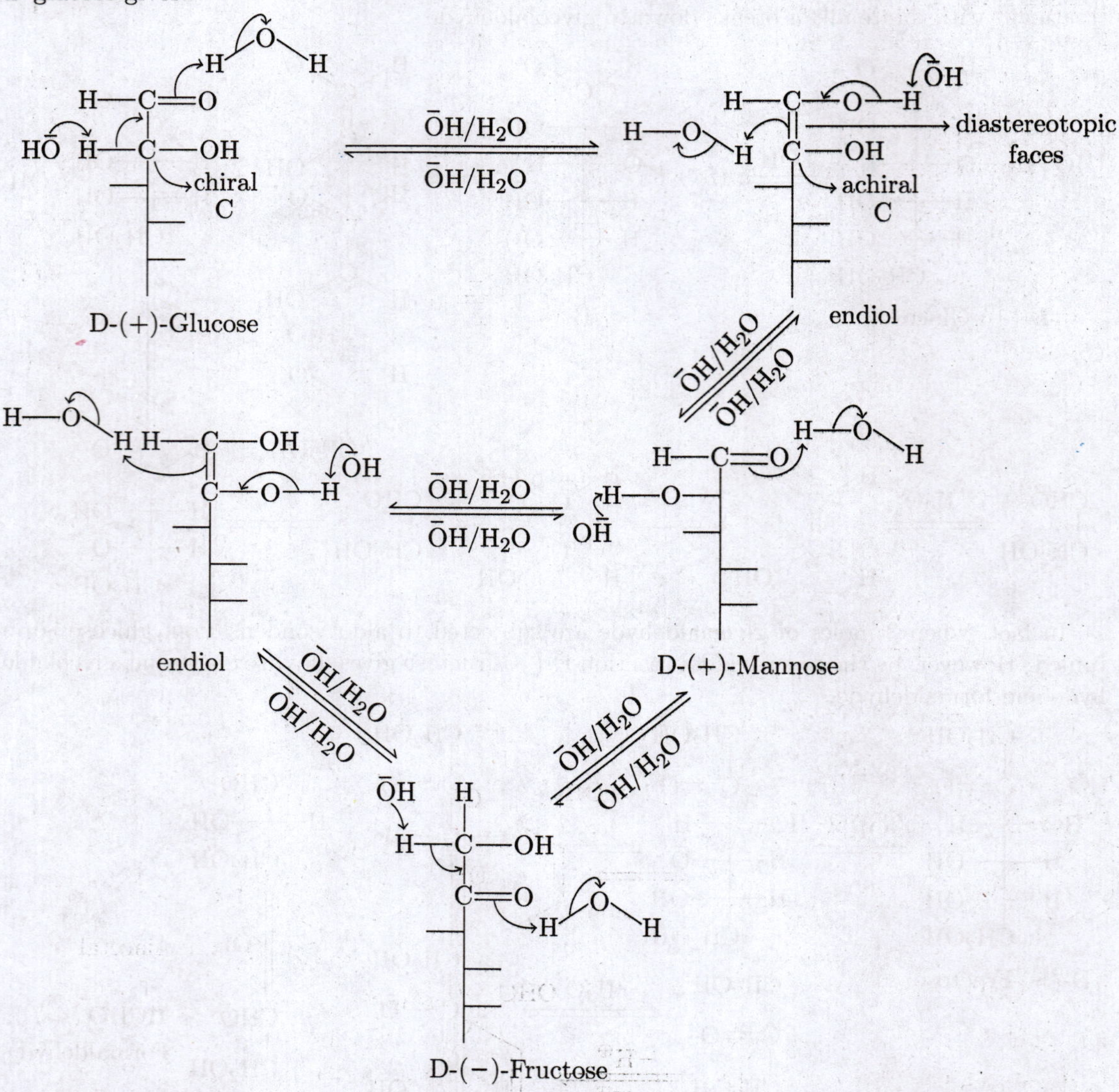


## 23.6.d Action of mineral acids

Dilute mineral acids at room temperature convert β- or α-anomer into a mixture of α- and β-anomers only and no other effect on monosaccharides is observed. However, hot mineral acids produce complex changes on aldopentoses and hexoses involving dehydration. All pentoses on distillation with 12% hydrochloric acid give approximately theoretical yields of furfural, the ultimate source of furans. The dehydration is supposed to take place by a series of $\beta$-elimination reactions followed by a cyclic hemiacetal formation with dehydration.

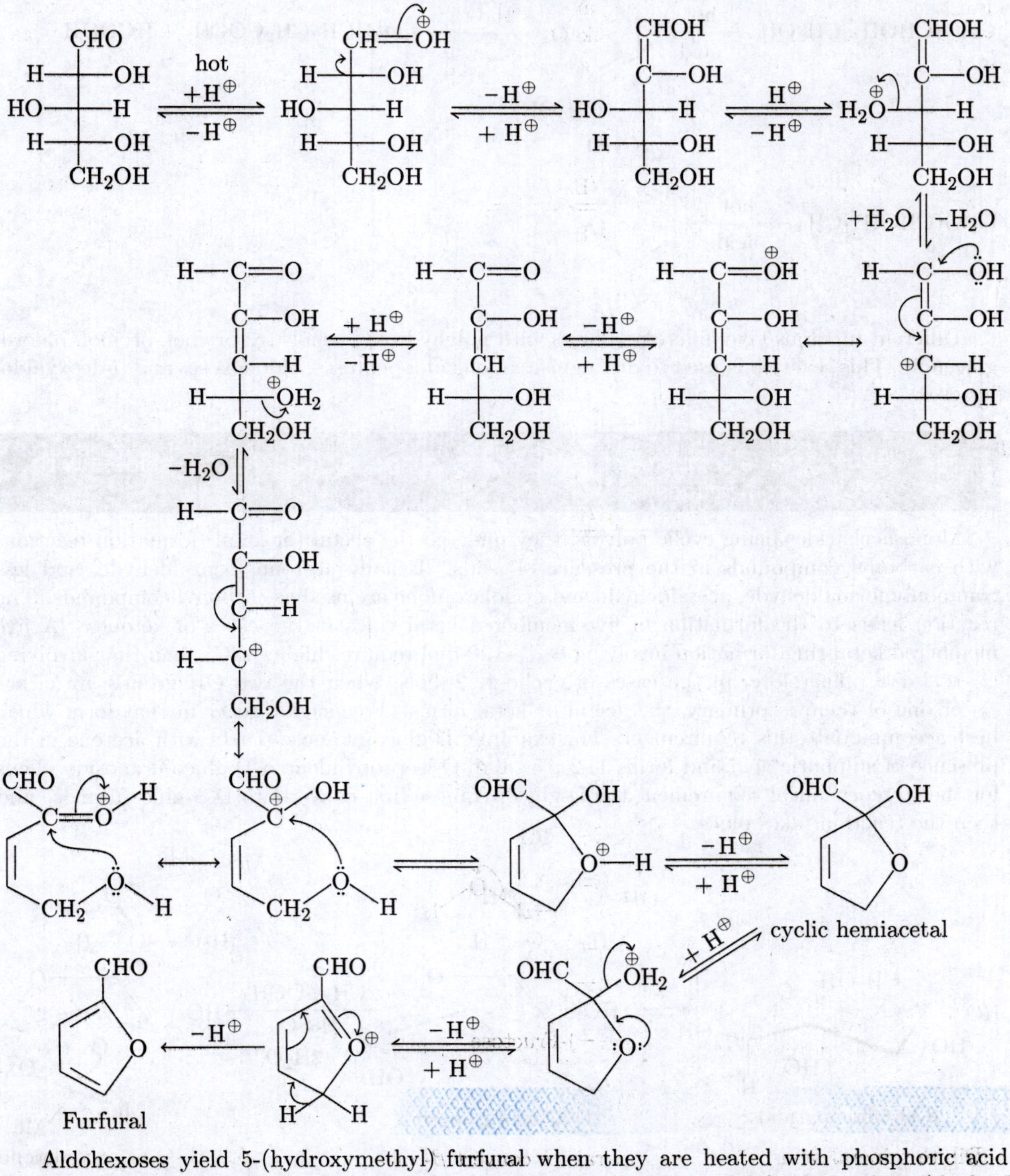

Aldohexoses yield 5-(hydroxymethyl) furfural when they are heated with phosphoric acid buffered with ammonium phosphate (pH4). Hot mineral acids also yield 5-(hydroxymethyl)furfural from aldohexoses but the product being not readily separable reacts with hot acid and gets converted to levullinic acid and some insoluble condensation products.

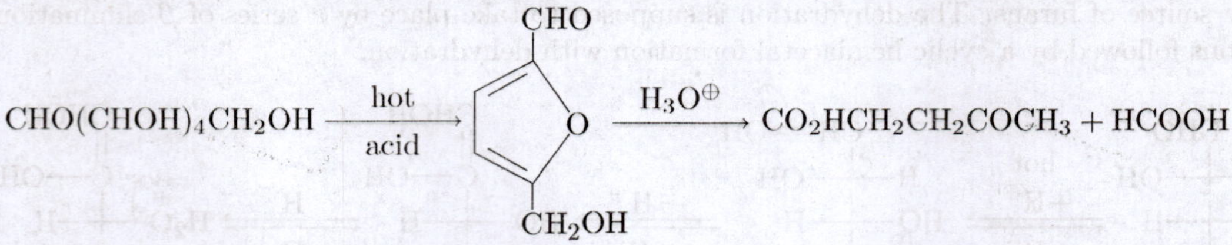


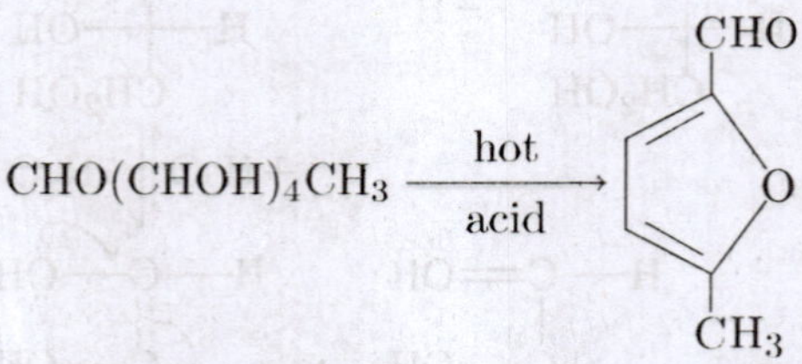


Different furfurals give different colours with polyhydroxyphenols (resorcinol, orcinol, phloroglucenol). This reaction is used to distinguish among aldopentoses, aldohexoses and $\beta$-deoxyaldohexoses.

## 23.6.e Blocking of alcoholic OH groups of saccharides : Reaction of aldoses with carbonyl compounds

Monosaccharides being cyclic polyols, they undergo the acetal (or ketal) formation reactions with carbonyl compounds in the presence of acids. Usually acetone, benzaldehyde, and less commonly formaldehyde, acetaldehyde and cyclohexanone are used as carbonyl compounds. The reaction leads to the formation of five-membered ketal rings in the cases of ketones. A five membered ketal ring formation involving trans-1, 2-diol requires higher $\Delta G^{\ddagger}$ than that involving cis-1, 2-diol. Therefore, in the cases of cyclic 1, 2-diols, when the two OH groups are either cis or one of them is primary, the acetal or ketal forms. Free sugars react in that form which best accommodate this requirement. For example, D-glucopyranose reacts with acetone in the presence of sulphuric acid and forms 1, 2 : 5 , 6-di-O-isopropylidene-$\alpha$-D-glucofuranoside. Thus for the stereochemical requirement the D-glucopyranose first converts to D-$\alpha$-glucofuranose and then the reaction takes place.

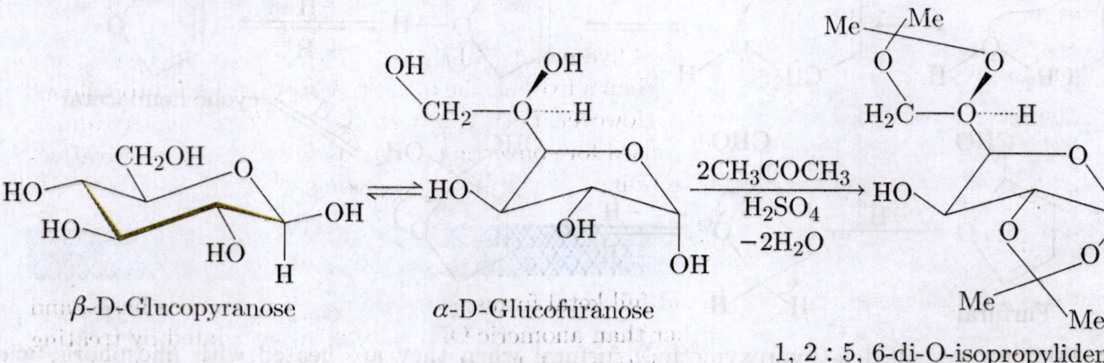


$\beta$-D-Glucopyranose $\alpha$-D-Glucofuranose 1, 2 : 5, 6-di-O-isopropylidene -$\alpha$-D-glucofuranoside α-D-glucopyranose can form mono-O-isopropylidene-α-D-glucopyranoside since $C_1OH$ and $C_2OH$ are cis, but it cannot form di-O-isopropylidene derivative. However, D-galactose has two pairs of cis hydroxyl groups and it forms 1, 2: 3, 4-di-O-isopropylidene-α-D-galatopyranoside.

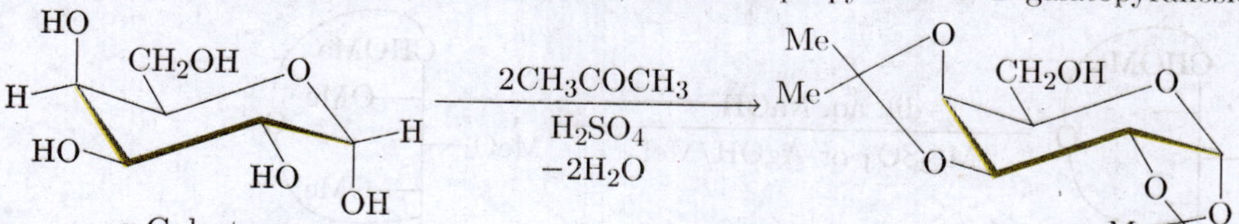


α-Galactose

1, 2 : 3, 4-di-O-isopropylidene -α-D-galactopyranoside

This reaction is used to block hydroxyl groups. To block C-4 and C-6 hydroxyl groups of aldohexoses benzaldehyde is used. When methyl α-D-glucopyranoside is treated with benzaldehyde in the presence of anhydrous zinc chloride, methyl 4, 6-O-benzylidene-α-D-glucopyranoside is obtained.

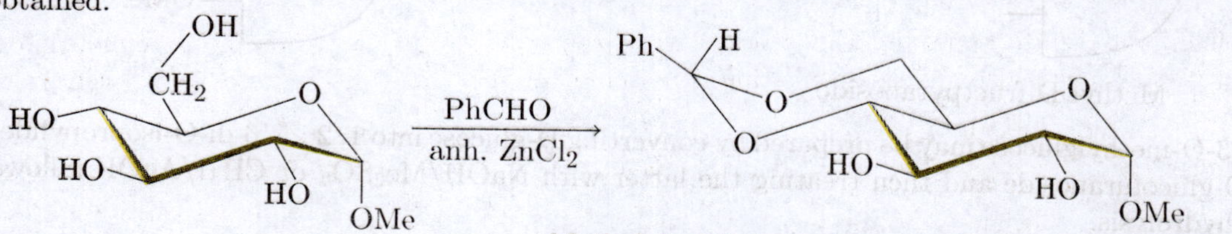


Methyl 4, 6-O-henzylidene-α-D-glucopyranoside

However, 1, 2: 5, 6-di-O-isopropylidene-α-D-glucofuranoside under very mild condition hydrolyses to yield 1, 2-O-isopropylidene-α-D-glucofuranoside.

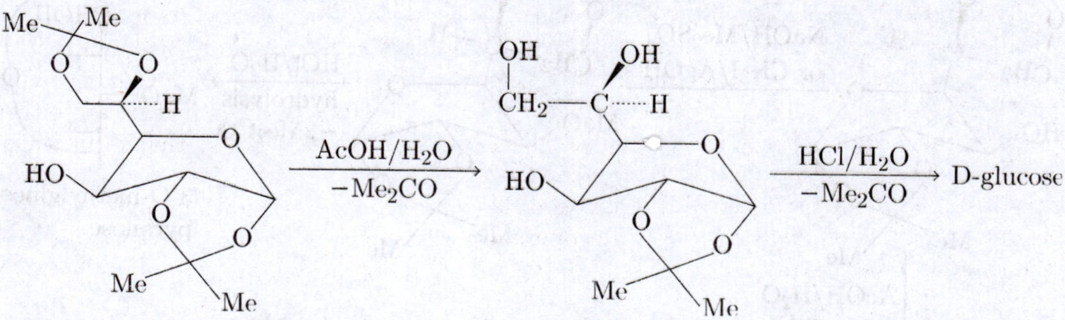


In this case a system having two cis-fused five-membered rings (1, 2-O-isopropylidene ring and α-D-glucofuranose ring) being more stable than a five-membered ketal ring (5, 6-O-isopropylidene ring), the latter hydrolyses preferentially. However, both 1, 2-O-isopropylidene derivative and 1, 2 : 5, 6-di-O-isopropylidene derivative are used for converting D-glucose into its other derivatives which will be discussed later on in the section.

## 23.6.f Methylation of hydroxyl groups other than anomeric OH

Since glycosides exist in full acetal and full ketal forms, they are resistant towards bases; and as a consequence any free OH group, other than anomeric OH, is usually methylated by treating glycosides with dilute sodium hydroxide solution and dimethyl sulphate and then hydrolysing the product with dilute acids; methylation may also be done by methyl iodide and moist silver oxide. Any specific OH group may, however, be methylated by blocking other OH groups. The reactions are given below :

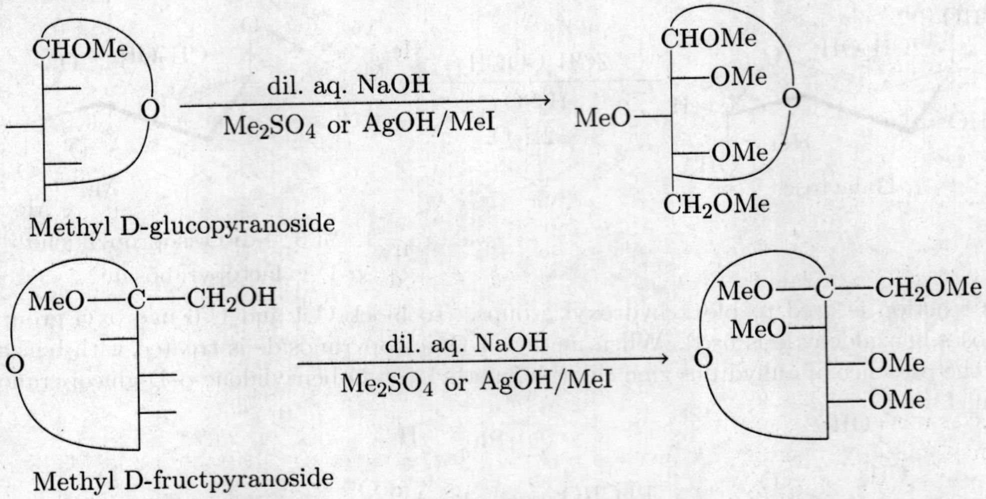

3-O-methylglucose may be prepared by converting D-glucose into 1, 2 : 5, 6-di-O-isopropylidene-$\alpha$-D-glucofuranoside and then treating the latter with $NaOH/Me_2SO_4$ or $CH_3I/AgOH$ followed by hydrolysis.

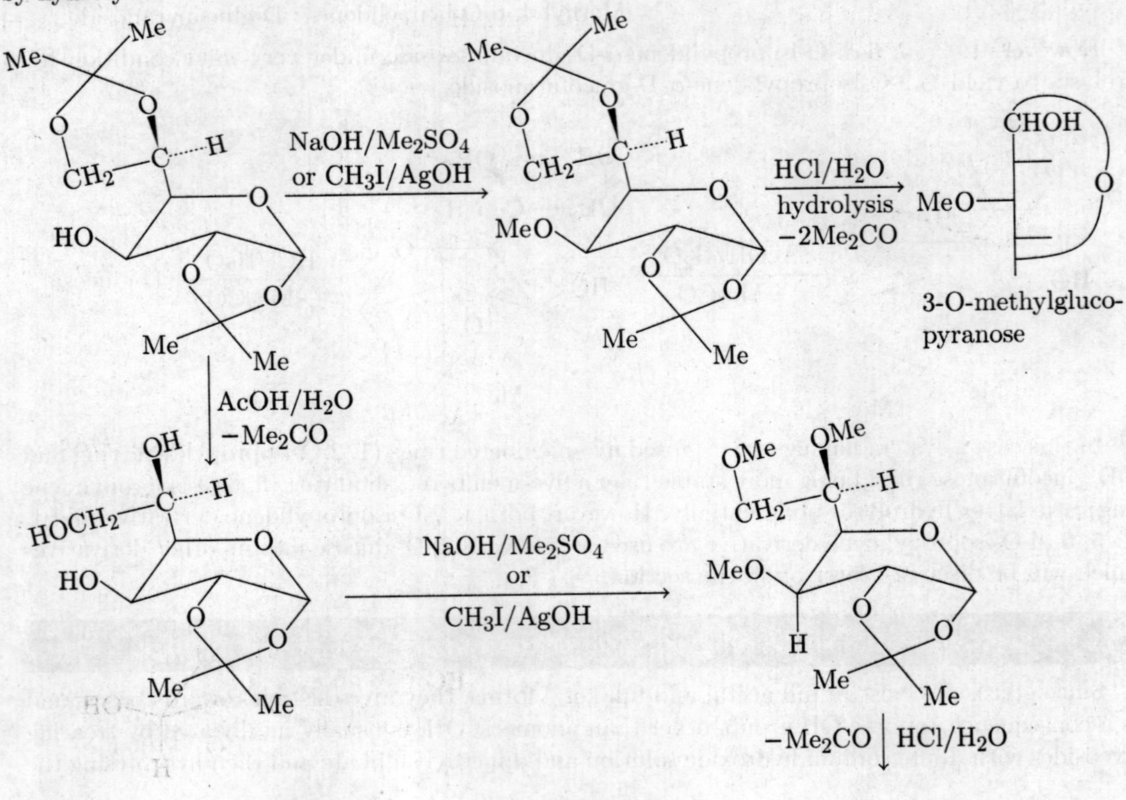

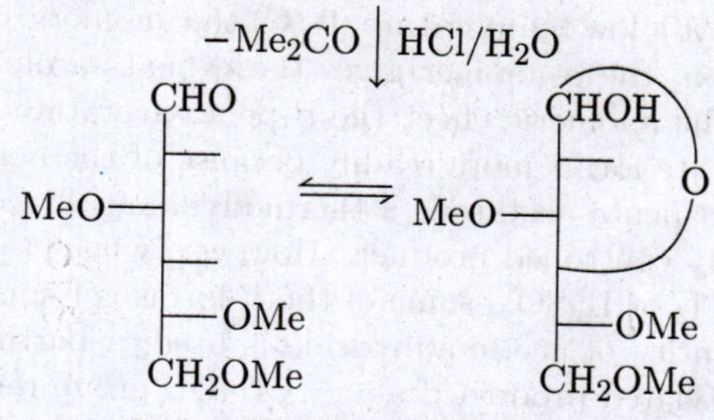


Similarly, 3, 5, 6-tri-O-methylglucose may be prepared treating 1, 2-O-isopropylidene-$\alpha$-D-glucofuranoside with aq. sodium hydroxide solution and dimethyl sulphate (or $AgOH/CH_3I$) and then hydrolysing the product with dilute HCl under very mild conditions.

## 23.6.g Esterification of monosaccharides

Since monosaccharides are polyhydroxylic compounds, they can be readily esterified by the usual methods of esterifying alcohols. However, acetate, benzoate, tosylates and nitrobenzoates are important esters of monosaccharides which are used for identification of monosaccharides and their conversion to other derivatives.

(i) Acetate : Acetic anhydride is used to prepare them in the presence of either pyridine or sodium acetate or anhydrous zinc chloride or sulphuric acid or perchloric acid; the most commonly used reagent is, however, pyridine. When the anomeric OH group is free, either the $\alpha$- or $\beta$-acetate may be formed depending on the nature of the catalyst and reaction conditions.

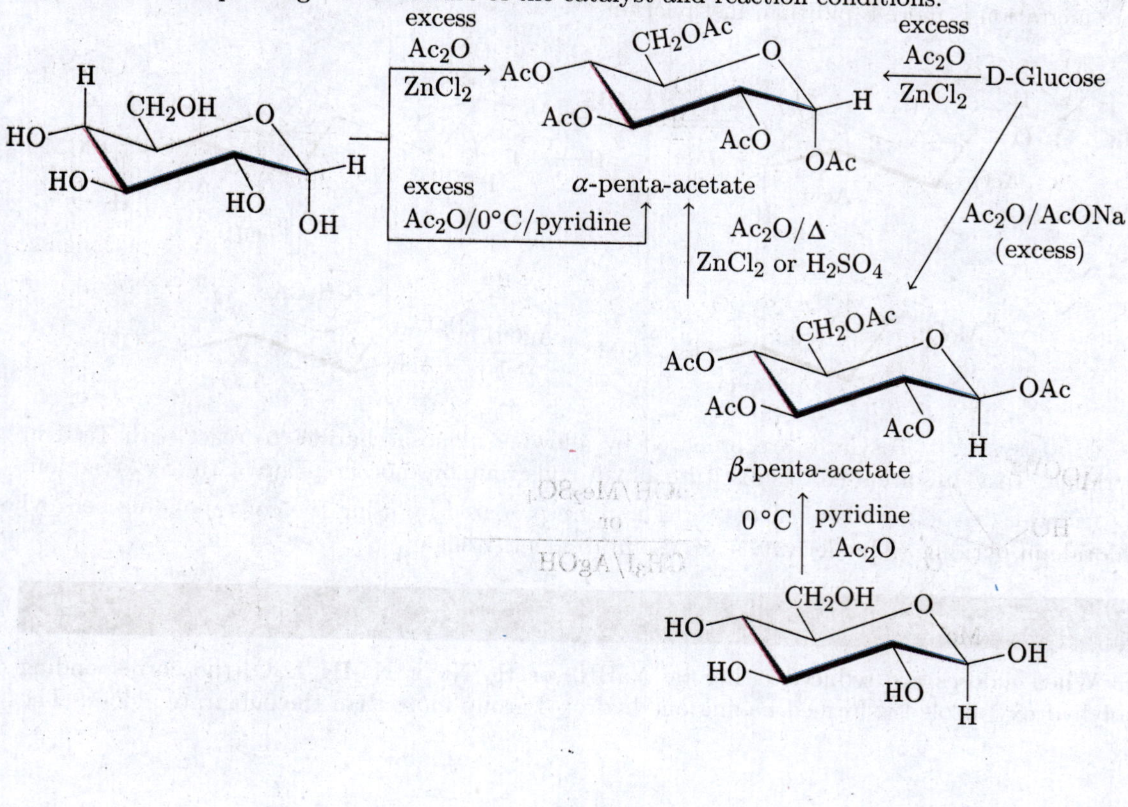

At a low temperature (0 °C) the anomeric change being much slower than the process of acetylation, the α-anomer gives the α-penta-acetate and the β-anomer the β-penta-acetate. Owing to the anomeric effect the α-penta-acetatate is more stable than β-penta-acetate; but β-penta-acetate forms more readily because of the better availability of space at the equatorial position. So α-penta-acetate is a thermodynamically controlled product while β-penta-acetate is a kinetically controlled product. However, when β-penta-acetate is heated in $Ac_2O$ in the presence of $ZnCl_2$ or $H_2SO_4$, some of the β-forms get converted to α-form ($K_{eq} = 6.7$). Again, when limited quantity of acetic anhydride is used, a partially acetylated product is obtained. In a partially acetylated product if a —$CH_2OH$ group remains free, an acetyl group migrates from a nearby

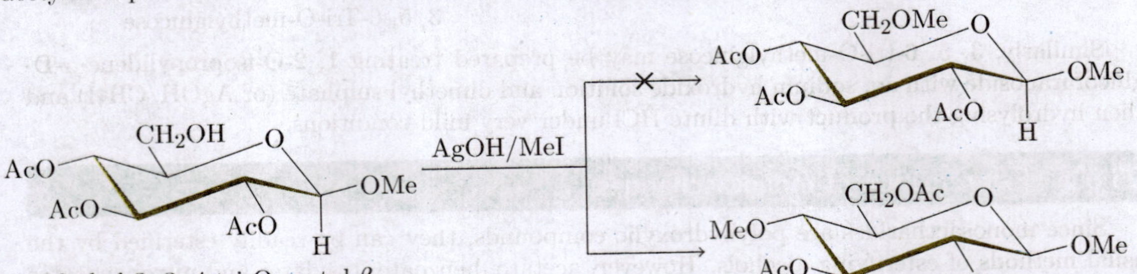


Methyl 2, 3, 4-tri-O-acetyl-β-D-glucopyranoside

Methyl 2, 3, 6-tri-O-acetyl-4-O-methyl-β-D-glucopyranoside

carbon atom; this occurs in a very mild alkaline condition. It has been said that prior to methylation the 4-O-Ac group migrates to $C_6$-OH and after the migration methylation of $C_4$-OH occurs; here migration is more rapid than methylation.

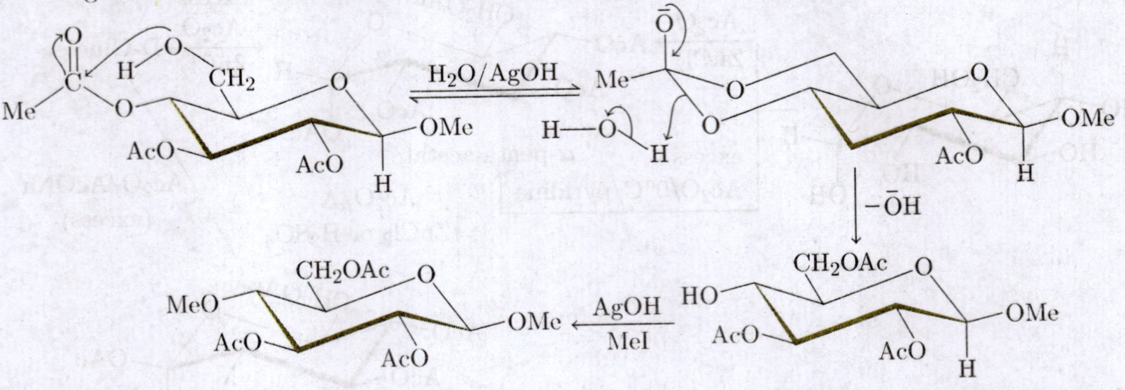


(ii) Tosylates : Tosylates are prepared by allowing monosaccharides to react with TsCl in pyridine. Tosylates are used to substitute a hydroxyl group by another group by the $S_N2$ reaction.

(iii) Benzoates and nitrobenzoates : These are prepared by using the corresponding benzoyl chloride in pyridine. The derivatics are useful to identify sugars.

## 23.6.h Reduction

When aldoses are reduced either by $NaBH_4$ or $H_2/Ni$ or $Na\text{-}Hg/H_2O$, the corresponding polyhydroxy alcohol is formed having one hydroxyl group more than the substrate aldose. The aldehyde group gets reduced by the process. The polyhydroxy alcohol formed is called alditol. The suffix 'ose' is to be replaced by 'itol' to name the alditol of an aldose. Thus the alditol of D-glucose is glucitol and that of D-mannose is mannitol. The reaction probably occurs on the trace amount of the open chain form which is present in the equilibrium mixture. As the small amount of the open chain form is removed in the reaction, the equilibrium shifts continually until whole of the aldose is consumed. D-glucose and L-gulose give one and the same alditol, glucitol, on reduction. Since hydrogen is added to the aldehydic group by the process of reduction and since the aldehydic C holds one H atom we get an alcohol on reduction of an aldose.

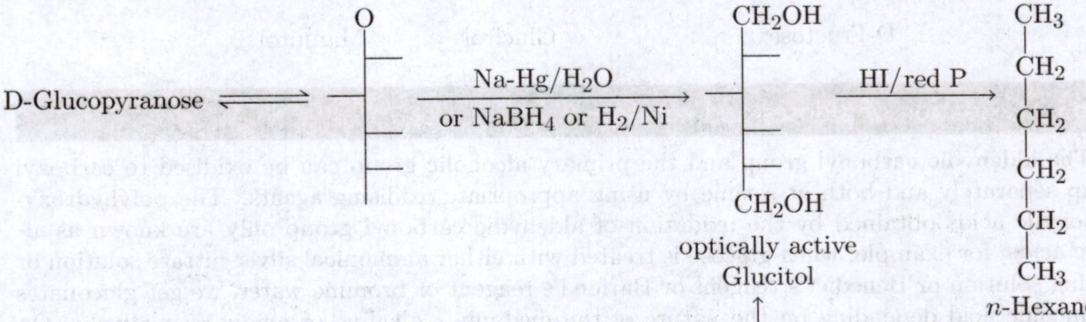


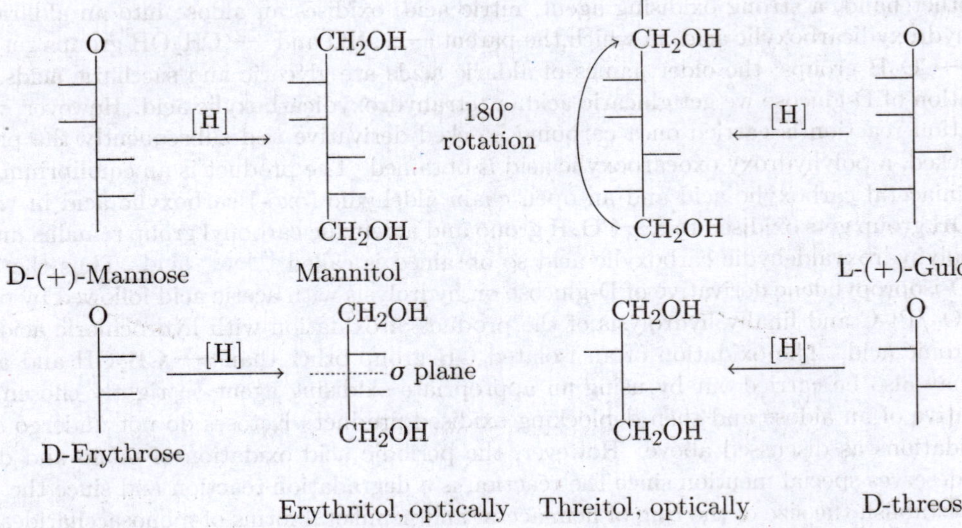


When an alditol is further reduced by red phosphorous and hydroiodic acid, the corresponding *n*-alkane is obtained. These two sequential reactions on an aldose show that the carbon skeletons of the aldose is unbranched. However, two isomeric aldoses, depending on their configurations, may give identical alditols (as in the cases of D-glucose and L-gulose) or they may give two different alditols as in the cases of D-erythrose and D-threose or D-mannose and D-glucose; the reduction may thus be used to identify the isomeric aldoses with the help of some other experiments. However, D-altrose and D-talose also give one and the same alditol.

Except dihydroxyacetone all other ketoses give a mixture of diastereomeric ketitols (alditols) on reduction; because (i) the ketonic carbonyl carbon is prochiral, (ii) the carbonyl group has two diastereotopic faces and (iii) the carbonyl carbon does not hold any H atom. However, one alditol predominates over the other.

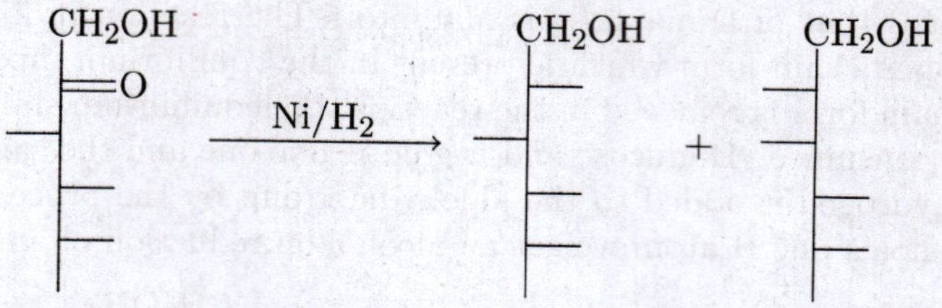


D-Fructose Glucitol Mannitol

## 23.6.i Oxidation of aldoses and ketoses

The aldehydic carbonyl group and the primary alcoholic group can be oxidised to carboxyl group separately and both at a time by using appropriate oxidising agents. The polyhydroxy-carboxylic acids obtained by the oxidation of aldehydic carbonyl group only are known as aldonic acids; for example, when glucose is treated with either ammonical silver nitrate solution or Fehling solution or Benedict's reagent or Barfoed's reagent or bromine water, we get gluconates or gluconic acid depending on the nature of the medium— alkaline or acidic respectively. On the other hand, a strong oxidising agent, nitric acid, oxidises an aldose into an aldaric acid, a polyhydroxydicarboxylic acid, in which the parent —CHO and —$CH_2OH$ groups get oxidised to —$CO_2H$ groups; the older names of aldaric acids are glycaric and saccharic acids. By the oxidation of D-glucose we get glucaric acid, a tetrahydroxydicarboxylic acid. However, when the oxidation reaction is carried on a carbonyl blocked derivative and subsequently the product is deblocked, a polyhydroxy oxocarboxylic acid is obtained. The product is an equilibrium mixture of hemiacetal carboxylic acid and an open chain aldehydic (oxo-) carboxylic acid in which the $CH_2OH$ group gets oxidised to —$CO_2H$ group and aldehydic carbonyl group remains unreacted; the polyhydroxyaldehydic carboxylic acid so obtained is called Uronic acid. Thus the 1, 2 : 5, 6-di-O-isopropylidene derivative of D-glucose on hydrolysis with acetic acid followed by oxidation with $O_2$/Pt-C and finally hydrolysis of the product of oxidation with hydrochloric acid, we get glucuronic acid. The oxidation of an isolated OH group other than —$CH_2OH$ and anomeric OH may also be carried out by using an appropriate oxidising agent on rightly chosen blocked derivative of an aldose and then deblocking oxidised product. Ketoses do not undergo all kinds of oxidations as discussed above. However, the periodic acid oxidation of mono and disaccharides deserves special mention since the reaction is a degradation reaction and since the reaction helps establish the size of the ring of hemiacetal and hemiketal forms of monosaccharides. In this section our aim is to study all the oxidation processes stated above in detail as given below :

(i) Oxidation by copper (II) complexes : Three alkaline and one acidic copper (II) complexes deserve mention. All these are mild oxidising agents and oxidise the aldehydic carbonyl group in the open chain forms and convert it to carboxylate group or carboxyl group as the nature of the reagent is alkaline or acidic respectively. Ketoses themselves do not react directly but they first get converted to aldoses in the alkaline medium and then react to form aldonates. The reactions are practically irreversible.

As the open chain form reacts, the equilibrium gets disturbed and ring forms gradually change to the open chain form, and whole of the aldose reacts.

a. Benedict's reagent : An alkaline solution of cupric citrate complex
b. Fehling solution : An alkaline solution of cupric tartrate complex
c. Tollen's reagent : An ammoniacal silver nitrate complex
d. Barfoed's reagent : Cupric acetate in 1% acetic acid

However, copper(II) complexes get reduced to red cuprous oxide while silver mirror is formed in the case of Tollen's reagent.

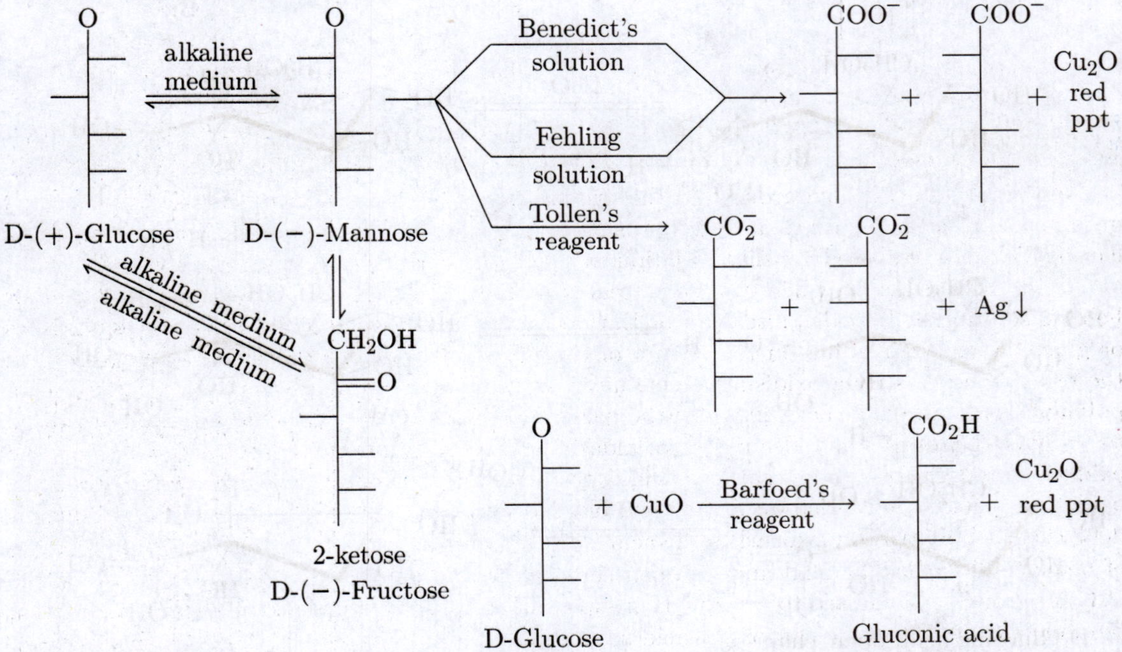


Since 2-ketoses isomerise to two epimeric aldoses in the alkaline medium by Lobry de Bruyn-van Ekenstein rearrangement, they also give positive Benediet's, Fehling's and Tollen's tests. On the other hand Barfoed's reagent oxidises aldoses to aldonic acids. Ketoses do not respond to this test.

**(ii) Oxidation by bromine water :** Bromine water being a mild oxidising agent, it oxidises the carbonyl C atom of the aldoses in the hemiacetal form to $\delta$-lactone which via aldonic acid gets changed to $\gamma$-lactone. The free —$CH_2OH$ group does not get oxidised by the reagent. The reagent also fails to oxidise ketoses. The $\beta$-anomers of D-aldoses react much faster than the corresponding *alpha*-anomers. However, the reaction mechanism is given below taking $\beta$-D-glucopyranose as substrate.

$$Br_2 + H_2O \longrightarrow HOBr + HBr \longrightarrow H_2\overset{\oplus}{O}\text{—}Br + Br^-$$

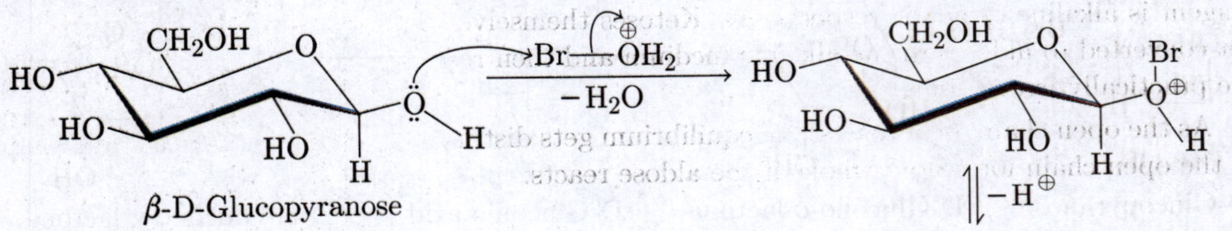

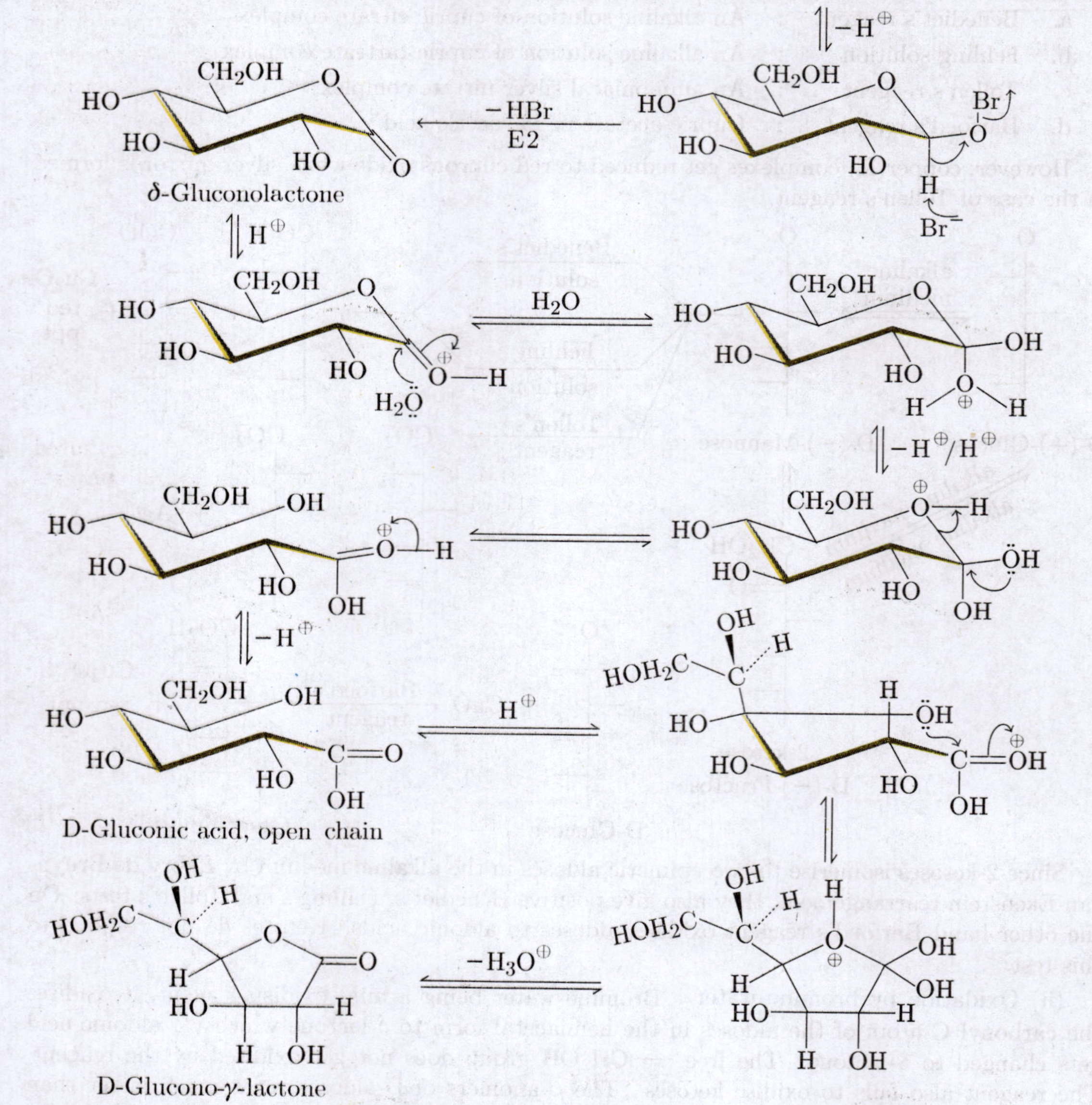


Overall reaction :

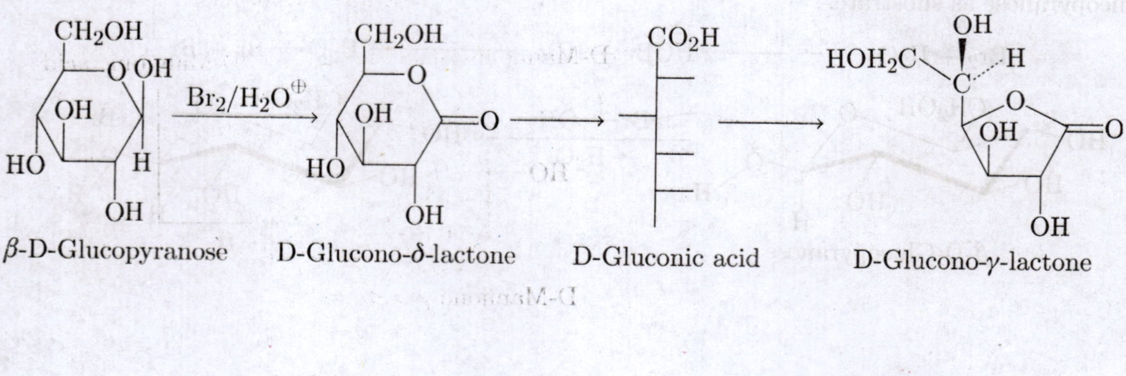

β-D-Glucopyranose D-Glucono-δ-lactone D-Gluconic acid D-Glucono-γ-lactone

The stereoelectronic requirement for an E2 reaction is the anti-periplanar arrangement of the two leaving groups. In the case of $\beta$-D-glucose the stereoelectronic condition for the E2 reaction involving $C_1$—H and O—Br is fulfilled. H of $C_1$, and Br of O—Br are the two leaving groups. In the case of $\alpha$-D-glucose this stereoelectronic condition does not exist because of the 1, 3-diaxial interaction involving $C_1$—O—Br/$C_3$—H($a$) and $C_1$—O—Br/$C_5$—H($a$) as shown below :

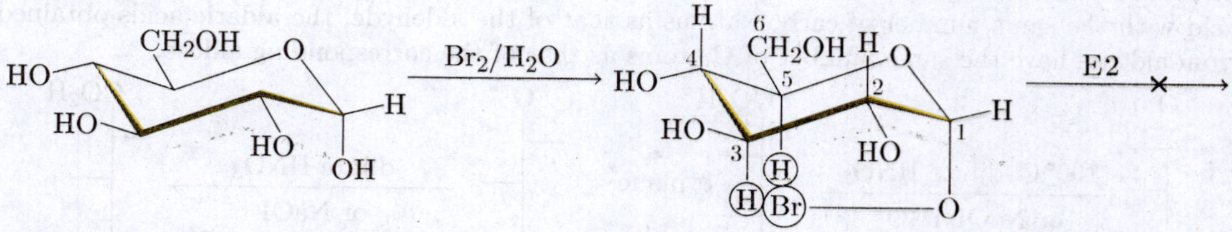


It has been reported that $\beta$-D-glucopyranose undergoes bromine-water oxidation two hundred and fifty times faster than $\alpha$-D-glucopyranose. It is stated that $\alpha$-D-glucopyranose first converts to $\beta$-D-glucopyranose via the open chain form and then the oxidation reaction occurs.

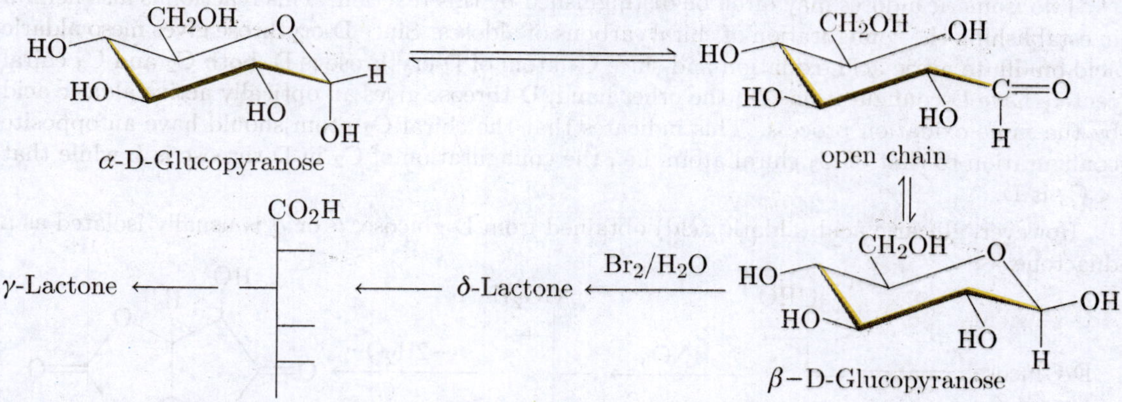


Like $\beta$-D-glucopyranose $\beta$-D-mannopyranose gets oxidised to D-mannono-$\gamma$-lactone via D-mannono-$\delta$-lactone and open chain D-mannonic acid by bromine-water.

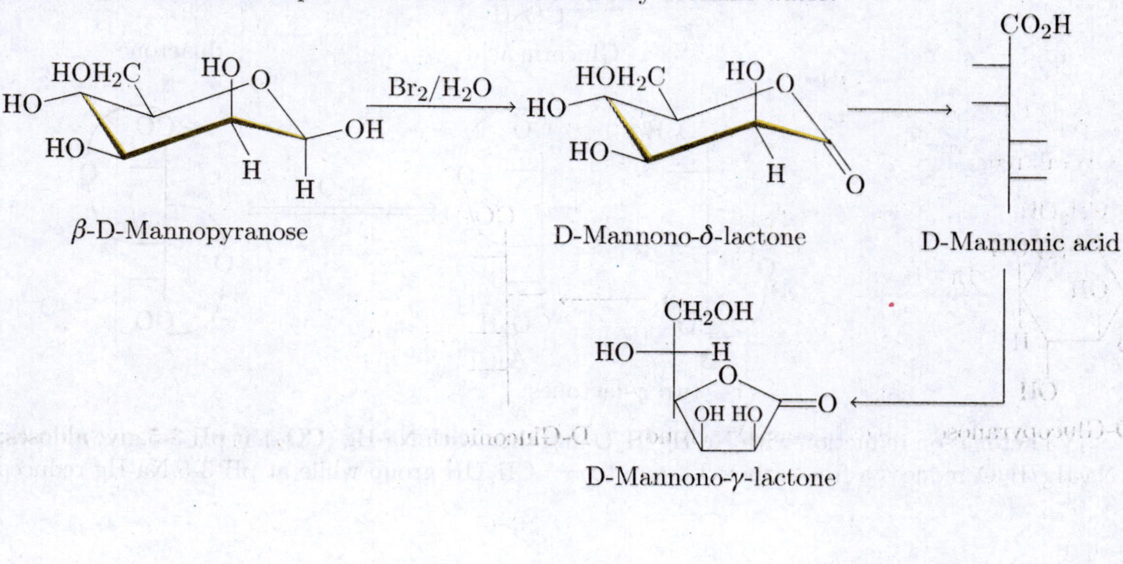

Ketoses do not get oxidised by bromine-water. This is because the anomeric carbon does not have an H atom (instead there is a $CH_2OH$ group) and hence the question of dehydrobromination does not arise.

(iii) Oxidation by hot dilute nitric acid (or hypohalite) : These reagents oxidise aldoses to polyhydroxydicarboxylic acids, called aldaric acids (older names are glycaric acids or saccharic acids). The reactions take place on open chain isomers. Since an aldehyde on oxidation forms an acid with the same number of carbon atoms as that of the aldehyde, the aldaric acids obtained from aldoses have the same number of C atoms as that of the corresponding aldose.

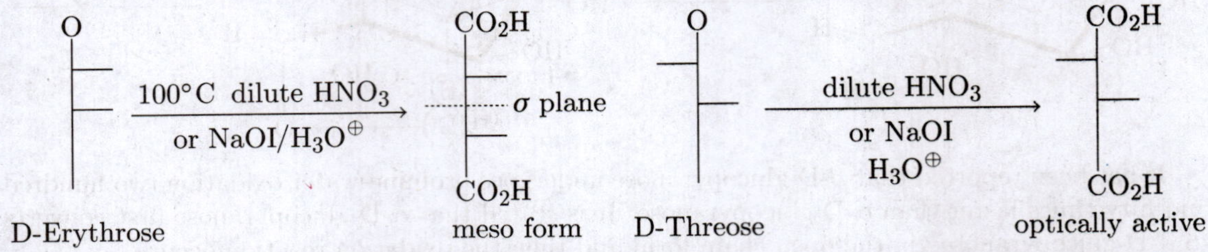


Two isomeric aldoses may often be distinguished by this reaction. This reaction is also helpful in establishing the configuration of chiral carbons of aldoses. Since D-erythrose gives meso aldaric acid on dilute nitric acid oxidation and since $C_3$ atom of D-erythrose is D, both $C_2$ and $C_3$ chiral centers have D-configuration. On the other hand, D-threose gives an optically active aldaric acid by the same oxidation process. This indicates that the chiral $C_2$ atom should have an opposite configuration to that of $C_3$ chiral atom, i.e., the configuration of $C_2$ in D-threose is L while that of $C_3$ is D.

However, glucaric acid (aldaric acid) obtained from D-glucose, α or β is usually isolated as a dilactone.

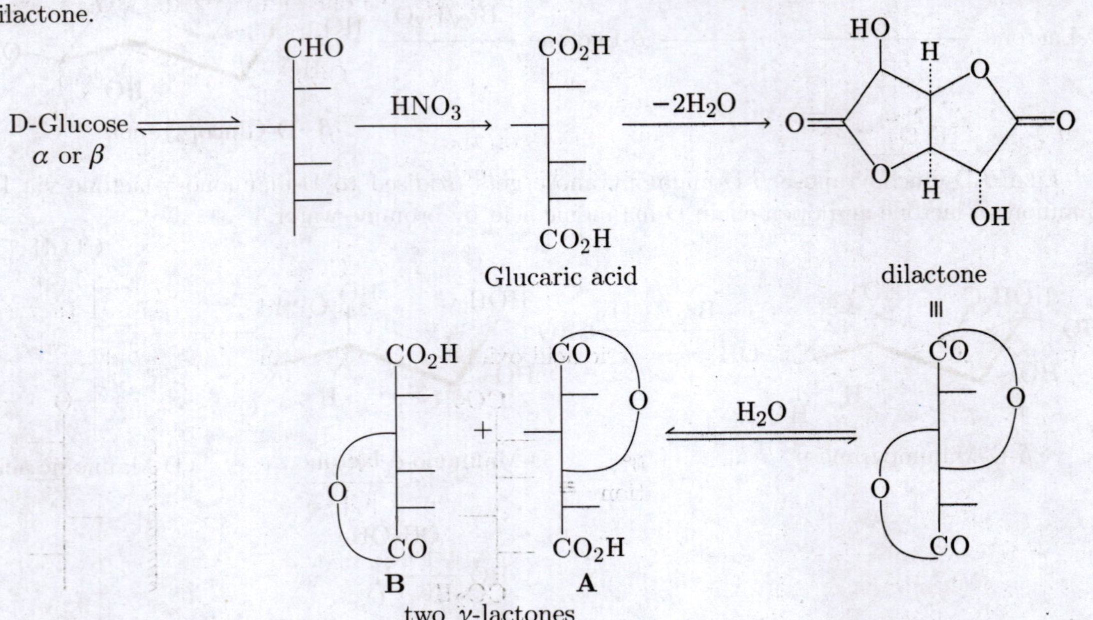


two γ-lactones

γ-Lactones on reduction with $Na\text{-}Hg/H_2O$ and then with Na-Hg ($CO_2$) at pH 3-5,give aldoses; $Na\text{-}Hg/H_2O$ reduces a free carboxyl group to —$CH_2OH$ group while at pH 3-5 Na-Hg reduces a lactone group into an aldehydic carbonyl group. These are shown below by taking A and B as substrates.

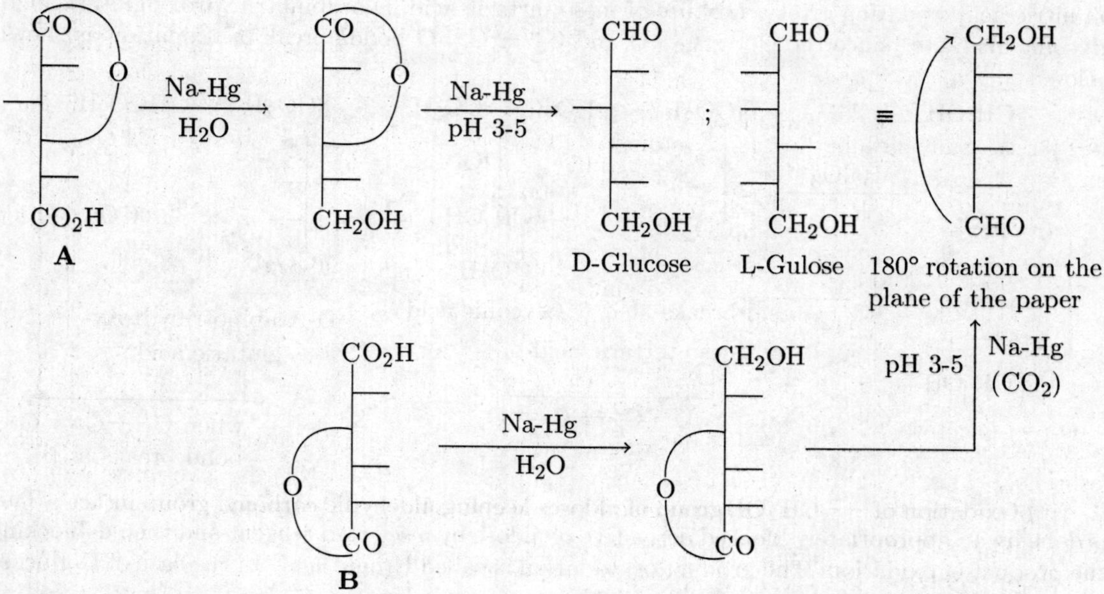


Thus by the nitric acid oxidation followed by Na-Hg/$H_2O$ and Na-Hg/pH 3-5 reductions D-glucose can be converted to L-gulose.

However, amongst D-aldohexoses D-altrose and D-talose give one and the same aldaric acid (as also they give the same alditol on reduction).

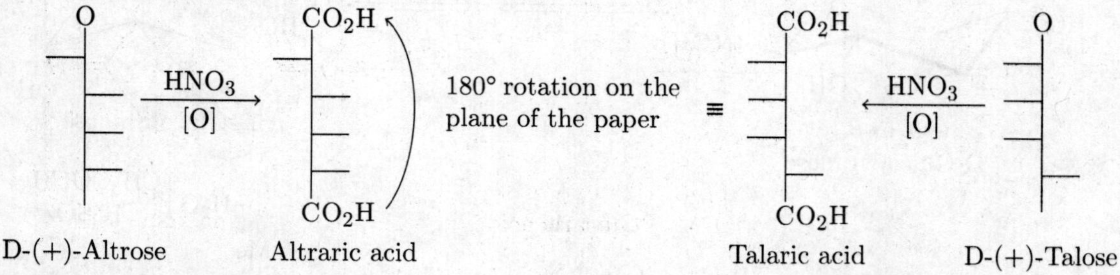


Similarly, D-glucose and L-gulose on nitric acid oxidation give the same aldaric acid.

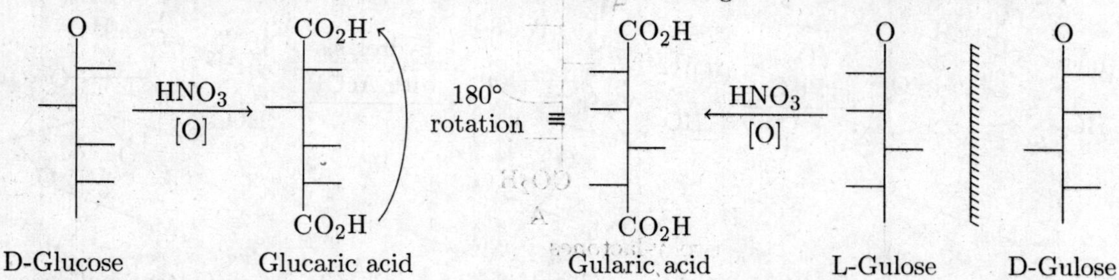


So, nitric acid oxidation is a good means to establish configurational relationship amongst aldoses.

Dilute nitric acid oxidises a ketose into a mixture of dicarboxylic acids, each of which contains a fewer number of carbon atoms than that of the parent ketose. For example, D-(−)-fructose on nitric acid oxidation gives a mixture of meso tartaric acid, D-arabinotrihydroxy glutaric acid, glycollic acid, etc. since the $C_1$—C=O and $C_3$—C=O bonds break on oxidation as shown below :

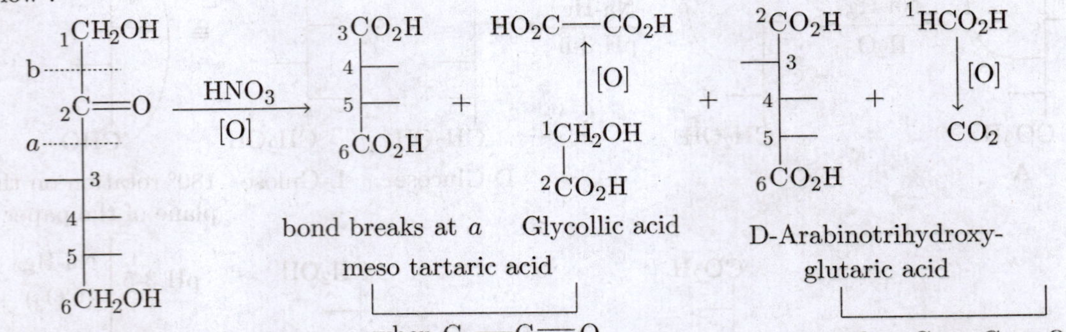


D-(−)-Fructose

(iv) **Oxidation of —$CH_2OH$ group of aldoses keeping aldehydic carbonyl group intact :** This is done on an appropriately blocked derivative of aldose by a selected reagent and then deblocking the product of oxidation. The product of oxidation is called Uronic acid. In the case of D-glucose and some other aldoses the sequential procedure is given below :

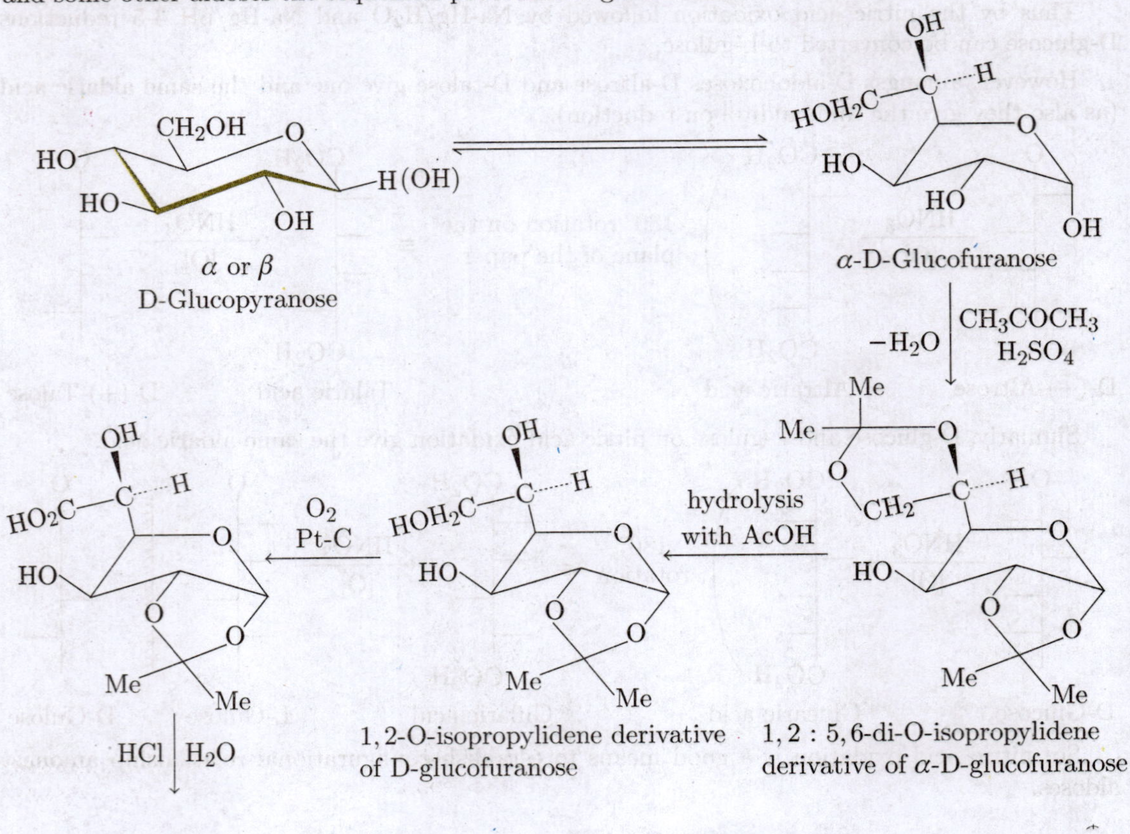

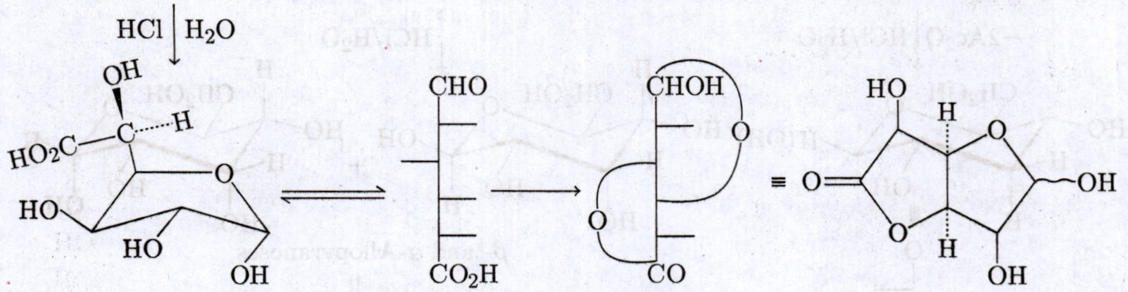

Glucuronic acid $\quad$ γ-lactone

The glucuronic acid is isolated as γ-lactone in which there is a cis-fused five-membered furanose ring arrangement.

(v) Oxidation of an —OH group other than —$CH_2OH$ of an aldose : This can be carried out by blocking all other functional groups (except the OH group to be oxidised) and then carrying out the process of oxidation by a mild oxidising agent like DMSO/DCC (dicyclohexyl carbodimide) or DMSO/$Ac_2O$ or DMSO/$P_4O_{10}$ or ruthenium tetroxide prepared in situ from $RuO_2/KIO_4$.

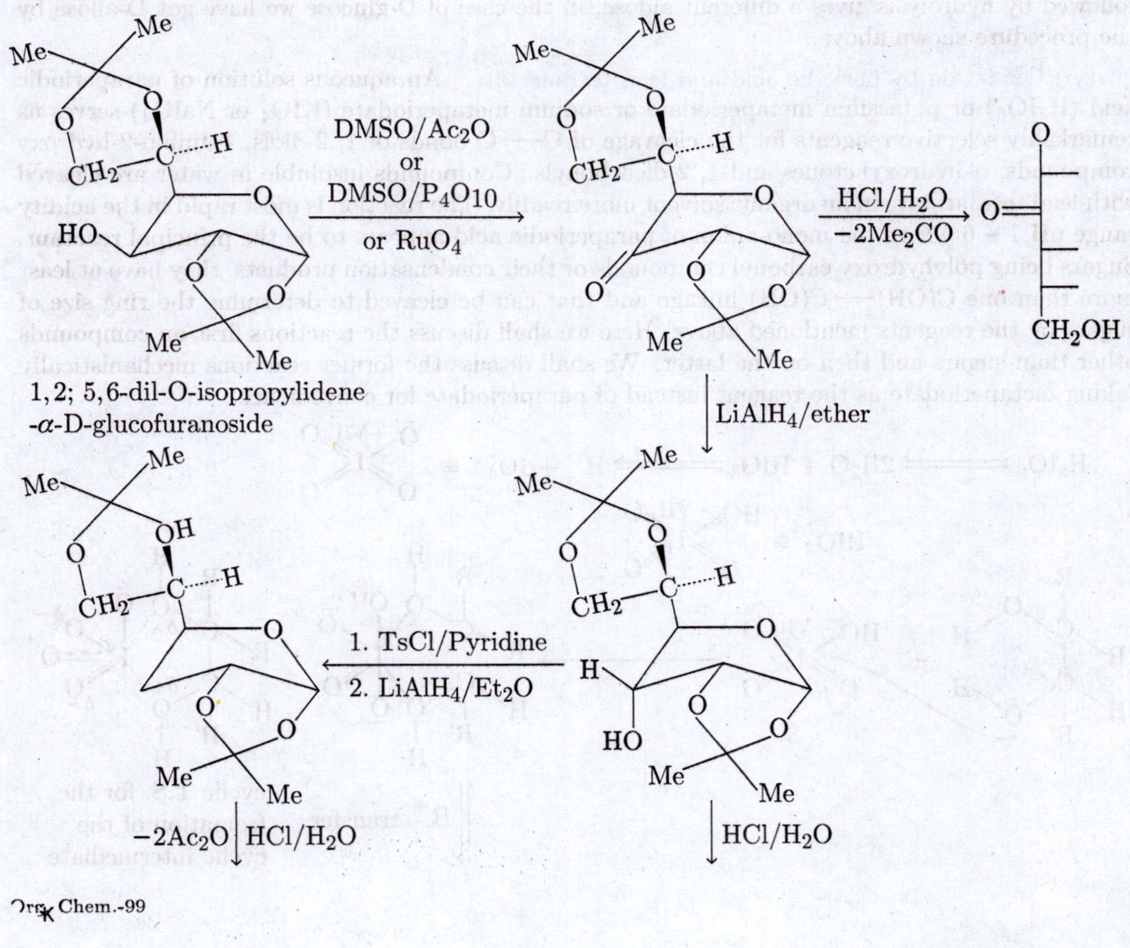

1, 2; 5, 6-dil-O-isopropylidene -α-D-glucofuranoside

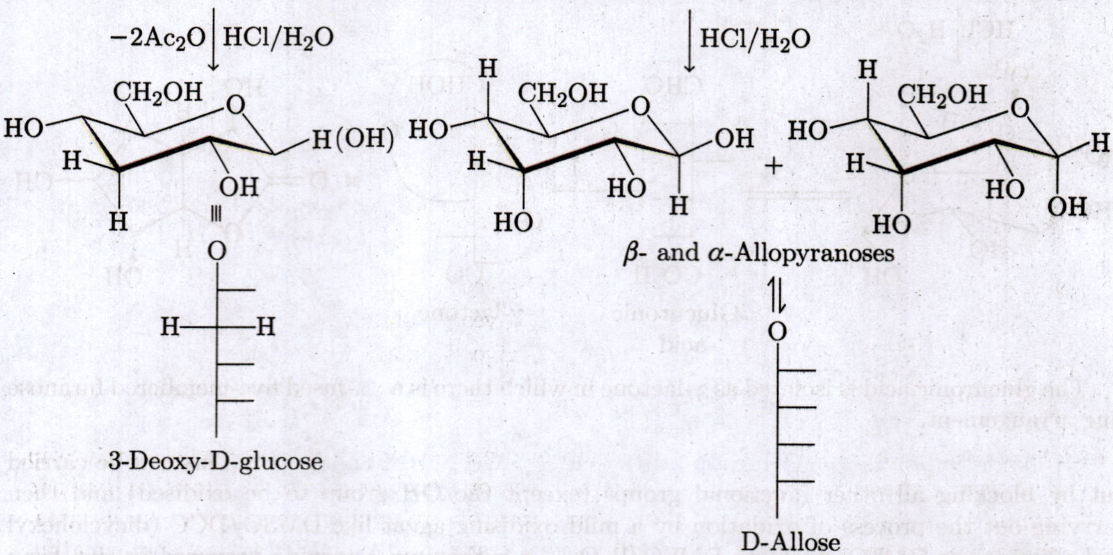

However, the oxidised and blocked derivative of the aldose on reduction with $LiAlH_4$/ether followed by hydrolysis gives a different aldose; in the case of D-glucose we have got D-allose by the procedure shown above.

(vi) Oxidation by periodic acid and lead tetraacetate : An aqueous solution of paraperiodic acid ($H_5IO_6$) or potassium metaperiodate or sodium metaperiodate ($KIO_4$ or $NaIO_4$) serves as remarkably selective reagents for the cleavage of C—C bonds of 1, 2-diols, 1-amino-2-hydroxy compounds, α-hydroxyketones and 1, 2-dicarbonyls. Compounds insoluble in water are cleaved with lead tetraacetate in an organic solvent more readily. The reaction is most rapid in the acidity range pH 1 – 6 where the mono-anion of paraperiodic acid appears to be the principal reactant. Sugars being polyhydroxy carbonyl compounds or their condensation products, they have at least more than one C(OH)—C(OH) linkage and that can be cleaved to determine the ring size of sugars by the reagents mentioned above. Here we shall discuss the reactions first on compounds other than sugars and then on the latter. We shall discuss the former reactions mechanistically taking metaperiodate as the reagent instead of paraperiodate for convenience.

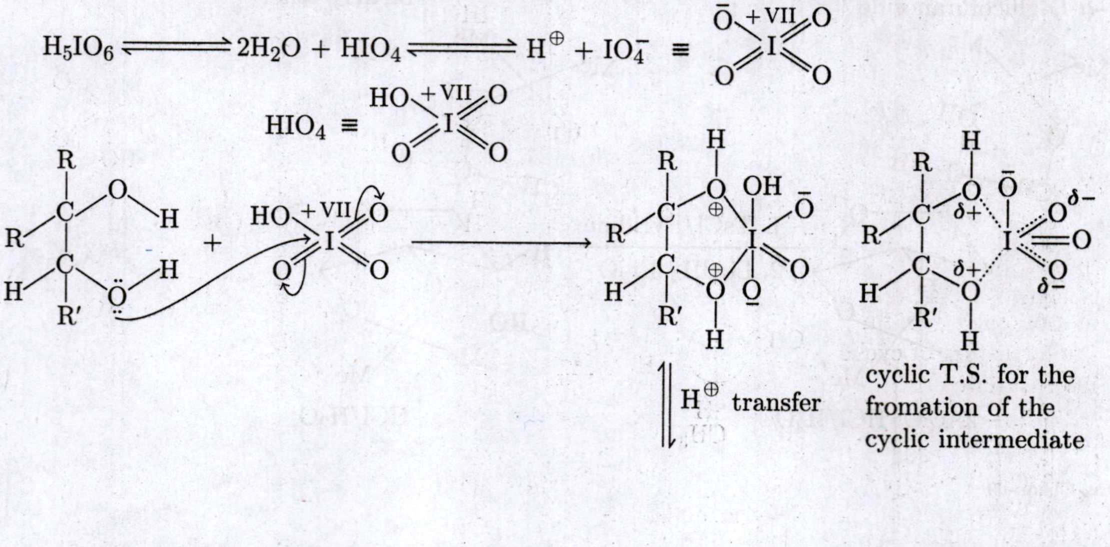

cyclic T.S. for the fromation of the cyclic intermediate

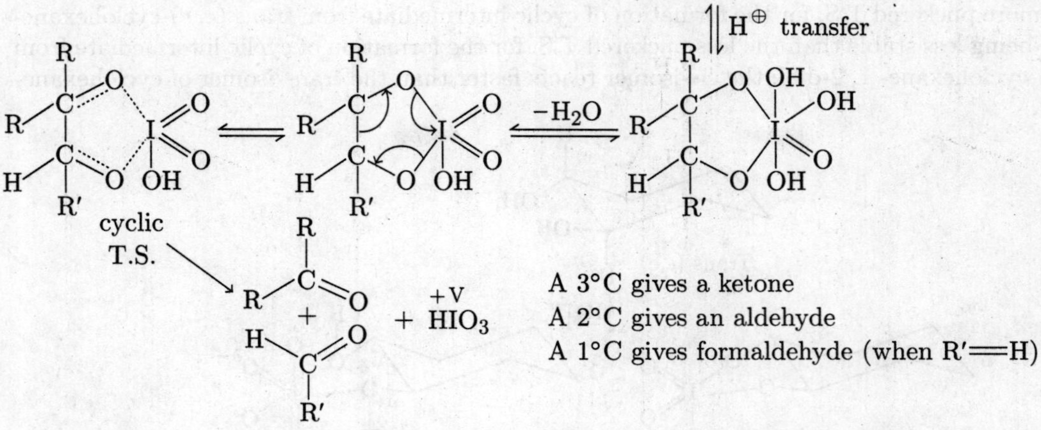


In this reaction two —OH groups interact from the same side of the C—C bond to form the cyclic intermediate and the T.S. by a concerted step or not. The rate of the reaction depends on the stability of the cyclic T.S. since the formation of the cyclic intermediate is the rate-determining step. So, in the case of a cyclic diols, the reaction takes place on that conformation which has two OH groups eclipsed. The more stable conformation having the OH groups eclipsed will have higher population at any instant and will react faster than the less stable conformation. For example, erythro form of butane-2, 3-diol reacts slower than the threo forms because in the former case the two eclipsed methyl groups increase the strain and the energy of the cyclic T.S.; while in the case of a threo form this strain factor being absent, the cyclic T.S. is more stable than that of the erythro form. Therefore, $\Delta G^{\ddagger}$ threo < $\Delta G^{\ddagger}$ erythro and naturally the threo form reacts faster than the erythro form. The reactions are shown below :

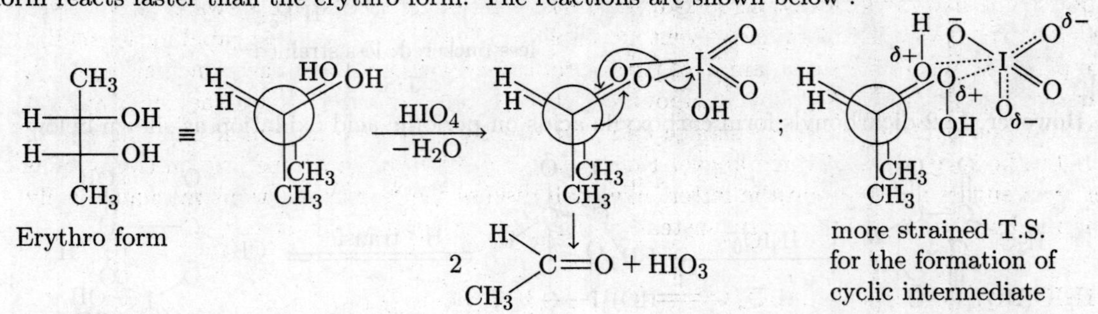


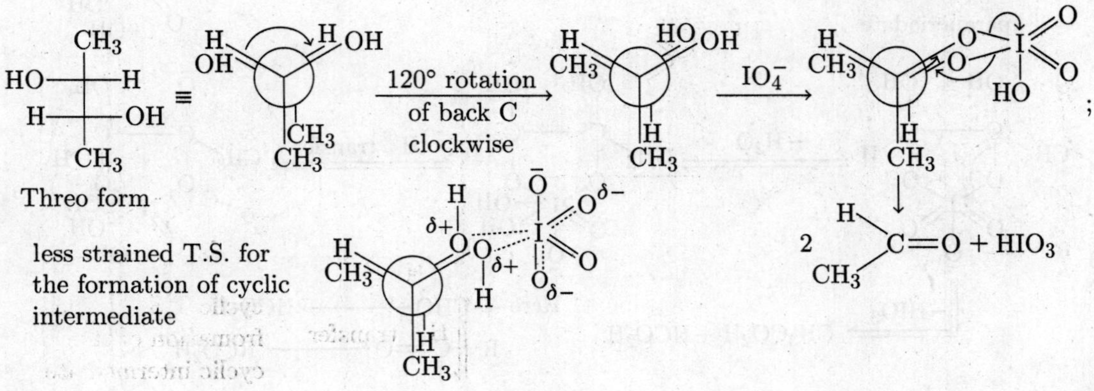

The more puckered T.S. for the formation of cyclic intermediate from trans $(e, e)$-cyclohexane-1, 2-diol being less stable than the less puckered T.S. for the formation of cyclic intermediate from cis $(a, e)$-cyclohexane- 1, 2-diol, the cis-isomer reacts faster than the trans-isomer of cyclohexane-1, 2-diol.

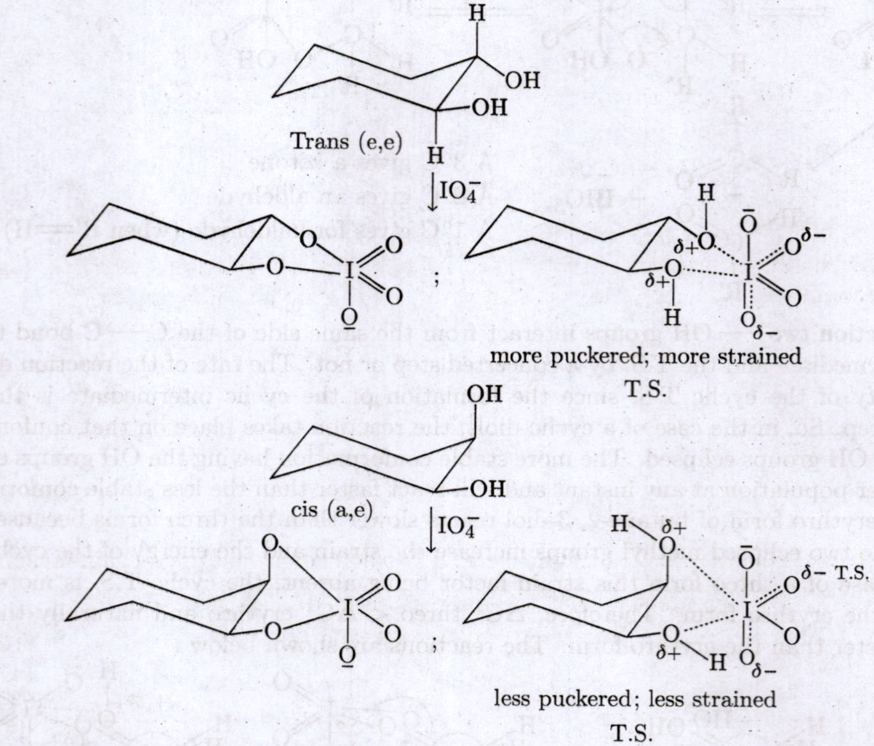


However, 1. 2-dicarbonyls form carboxylic acids on periodic acid oxidation as shown below :

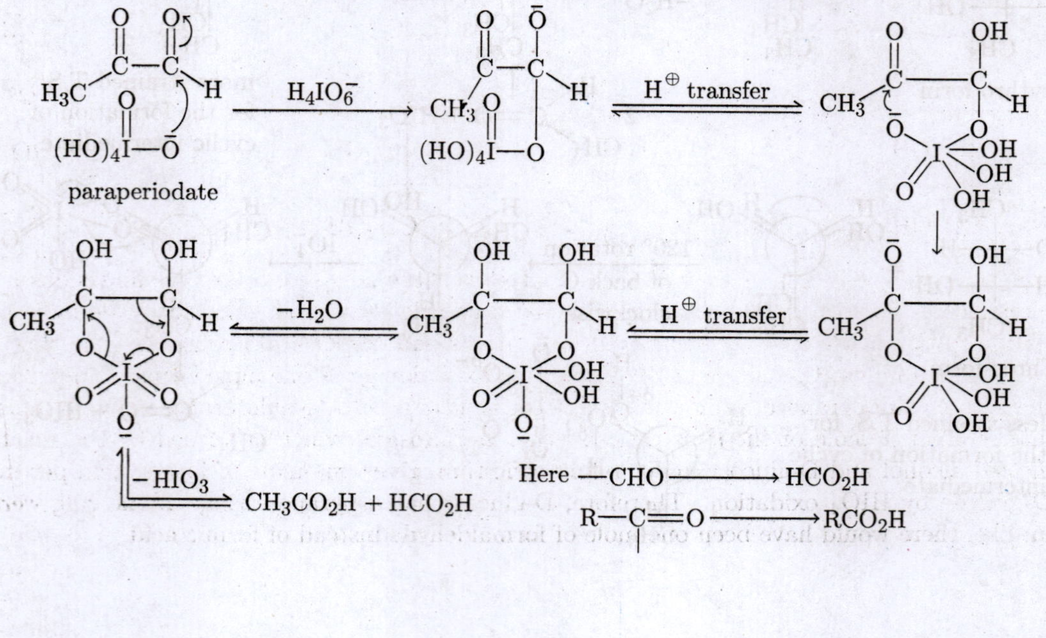

When there are three juxtaposed C—OH groups, the middle C—OH group is oxidised to formic acid on periodic acid oxidation. If there are $n$(—CHOH) groups in the substrate molecule, $(n-1)$ moles of periodates react and $(n-2)$ moles of formic acid form per mole of the substrate.

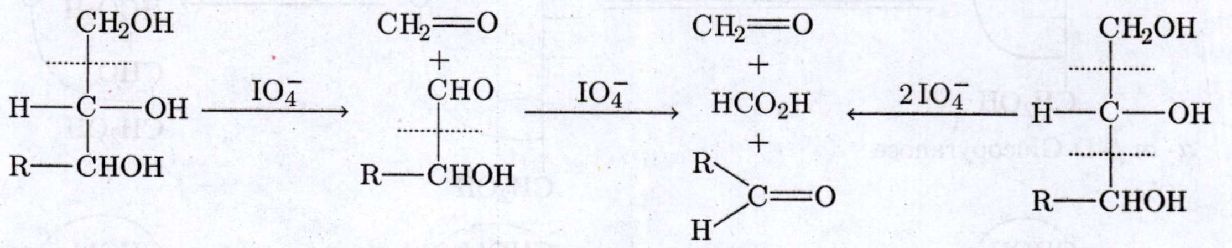

On the other hand, a carbonyl group flanked by two —CHOH groups changes to carbonic acid, which decomposes to carbon dioxide and water, by the oxidation process.

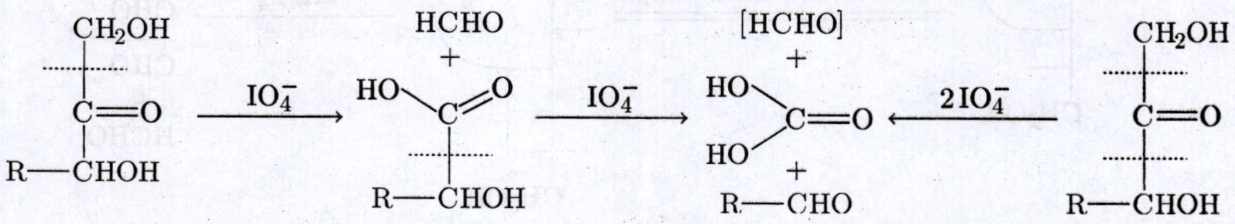

The above discussions tell us that D-glucose and D-fructose in their open chain forms give the following products :

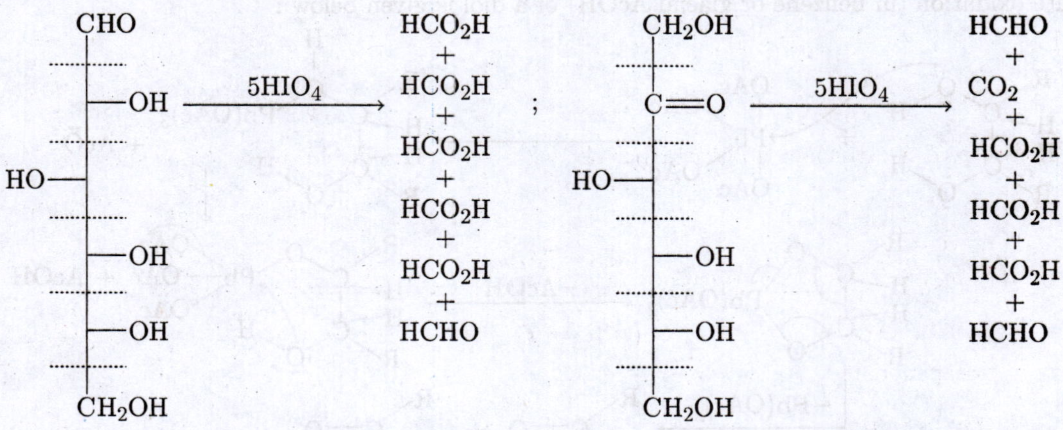

Ring-size of sugars and $HIO_4$ oxidation : However, the ring-size of ketoses and aldoses (in the hemiacetal form) can be determined by the glycoside formation followed by periodic acid oxidation. A pyranose ring(six membered; 5 C atoms and one O atom constitute the ring) on conversion to a full acetal followed by periodic acid oxidation gives one mole formic acid per mole of the keto or aldopyranose, while the same treatment of a keto or aldofuranose gives a mole of formaldehyde per mole of the substrate. Methyl D-glucoside, which is formed by the reaction of methyl alcohol and D-glucose under reflux condition, gives one mole of formic acid per mole of D-glucose by $HIO_4$ oxidation. Therefore, D-glucose has a pyranose ring. If the ring were a furan-like, there would have been one mole of formaldehyde instead of formic acid.

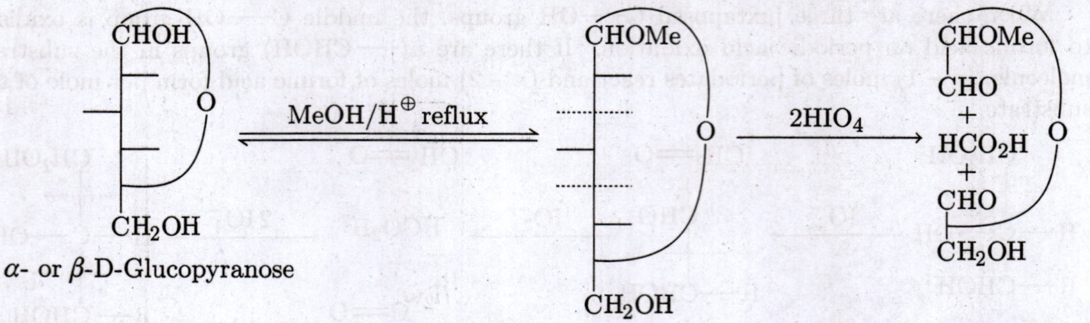

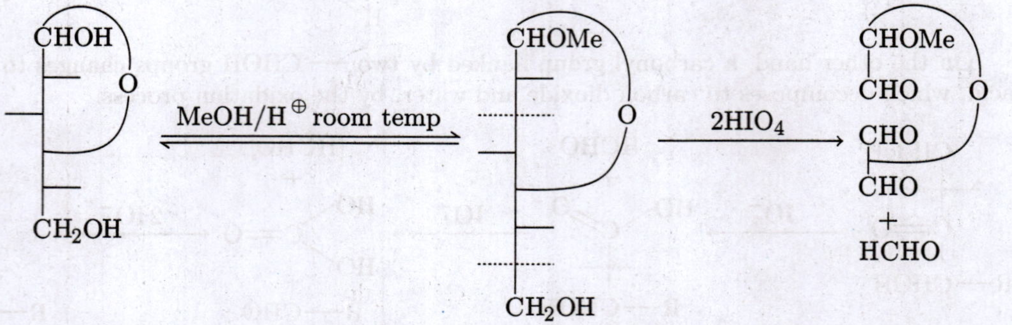

In the similar way the ring-size of a ketose can also be determined.

Mechanism for the lead tetraaceteate oxidation : The mechanistic steps for the lead tetraacetate oxidation (in benzene or glacial AcOH) of a diol is given below :

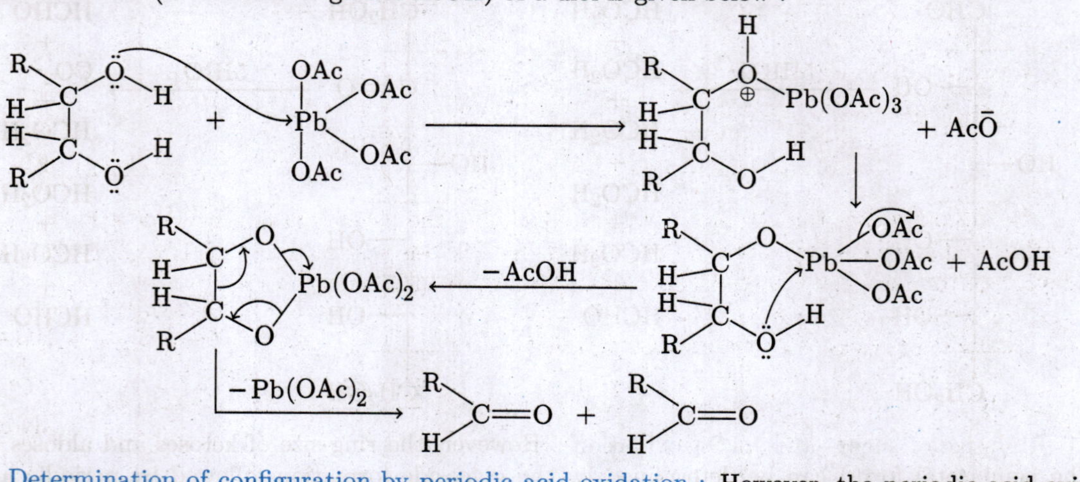

Determination of configuration by periodic acid oxidation : However, the periodic acid oxidation of monosarccharides is also useful in determining the configuration of anomeric OH and that of the farthest removed chiral C. When the periodic acid oxidation product of methyl $\alpha$-D-glucopyranoside is further oxidised by bromine-water and the solution is neutralised with strontium carbonate and recrystallised, a strontium salt is obtained. This salt is optically active and has a definite specific rotation. All the $\alpha$-D-aldohexoses on similar sequential treatments give the same salt with the same specific rotation. This process tells us, beyond doubt, that all the $\alpha$-anomers of D-aldohexoses have the same configuration about the anomeric C since all other chiral carbons except $C_5$ (which is D by convention for all D-series aldohexoses) and $C_1$ through the reactions as shown below :

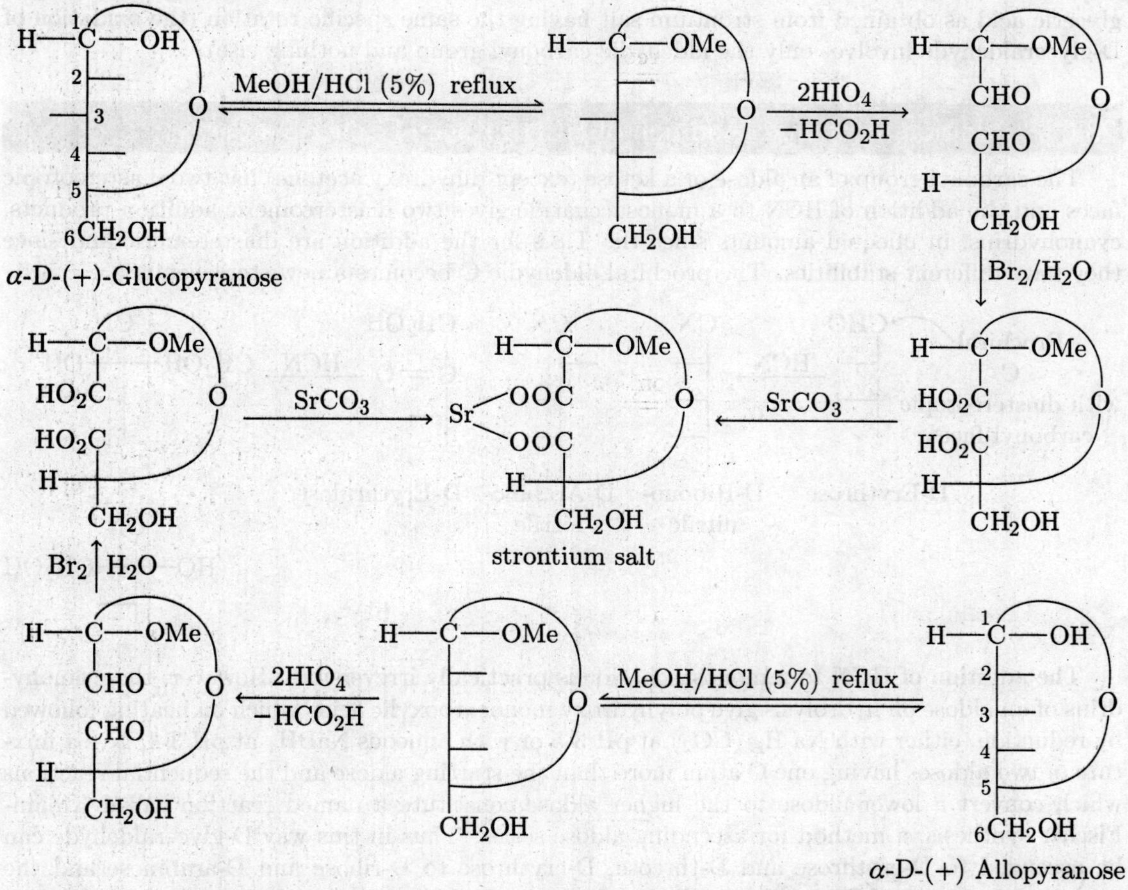

By decomposing the strontium salt and cleaving the ether linkages by hydrolysing the salt with dil HCl, we get two products and one of them is D-glyceric acid. This is a good evidence in favour of the fact that $C_5$ atom has a D-configuration in D-aldohexoses.

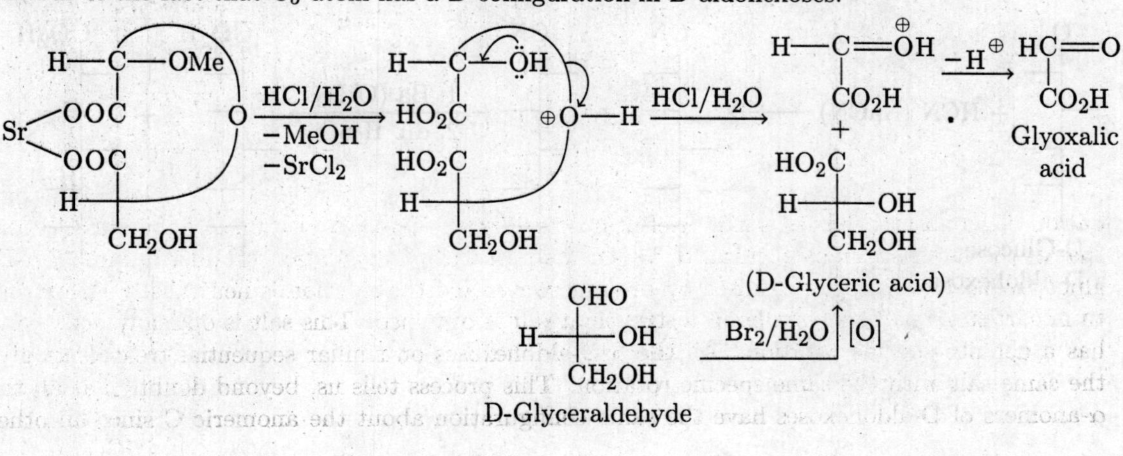

The glyceric acid liberated by the hydrolysis of the strontium salt is of D configuration that has been confirmed by the fact that D-glyceraldehyde on bromine-water oxidation gives the same glyceric acid as obtained from strontium salt having the same specific rotation (the oxidation of D-glyceraldehyde involves only the aldehydic carbonyl group and nothing else).

## 23.6.j Addition reactions : Addition of HCN : Kiliani-Fischer Synthesis

The carbonyl group of an aldose or a ketose (except dihydroxy acetone) has two diastereotopic faces and the addition of HCN to a monosaccharide gives two diastereomeric addition products, cyanohydrins, in unequal amounts since the T.S.s for the addition are diasteremeric and since they have different stabilities. The prochiral aldehydic C becomes a new stereocentre.

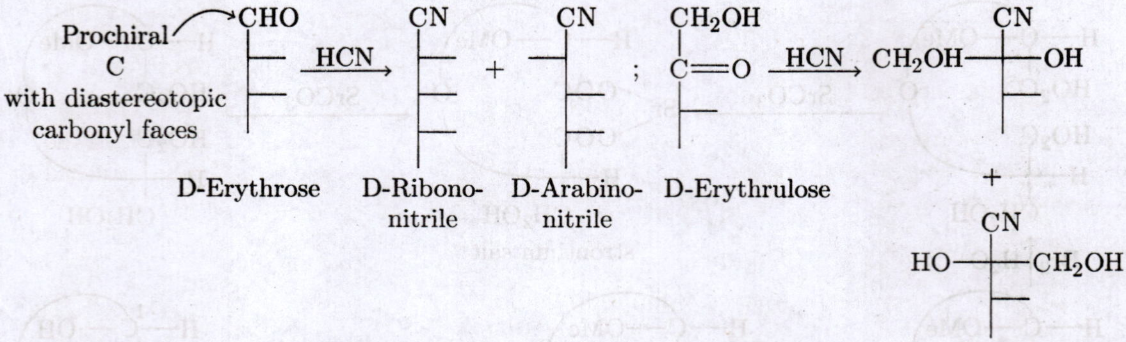


The addition of HCN to a monosaccharide is practically irreversible. However, the cyanohydrins of an aldose on hydrolysis give polyhydroxy monocarboxylic acids which on heating followed by reduction, either with Na-Hg ($CO_2$) at pH 3-5 or with aqueous $NaBH_4$ at pH 3-4, give a mixture of two aldoses having one C atom more than the starting aldose and the sequential reactions which convert a lower aldose to the higher aldose constitute a named reaction called Kiliani-Fischer synthesis, a method for ascending aldose series. Thus in this way D-glyceraldehyde can be ascended to D-erythrose and D-threose, D-erythrose to D-ribose and D-arabinose and the latter to a mixture of D-glucose and D-mannose. An L-aldose may similarly be ascended to the next higher C-2 epemeric L-aldoses.

**Fischer-Kiliani Synthesis :**

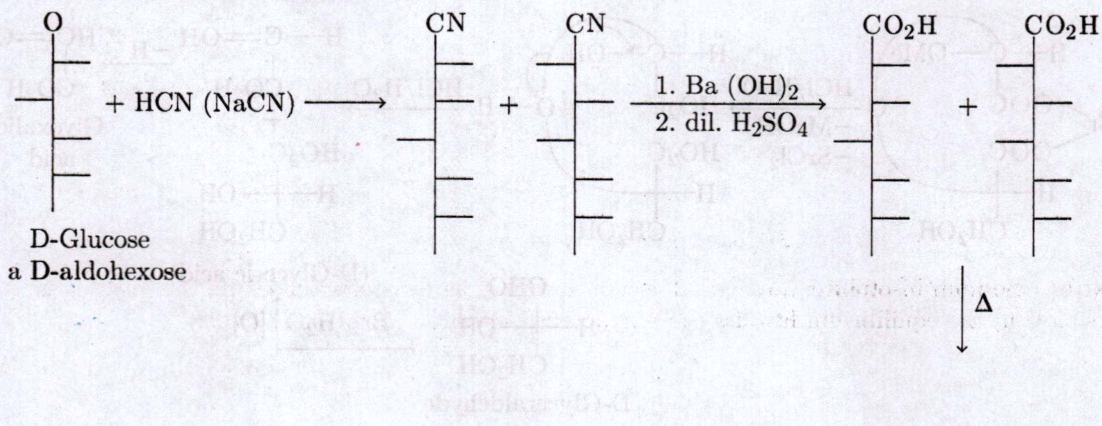

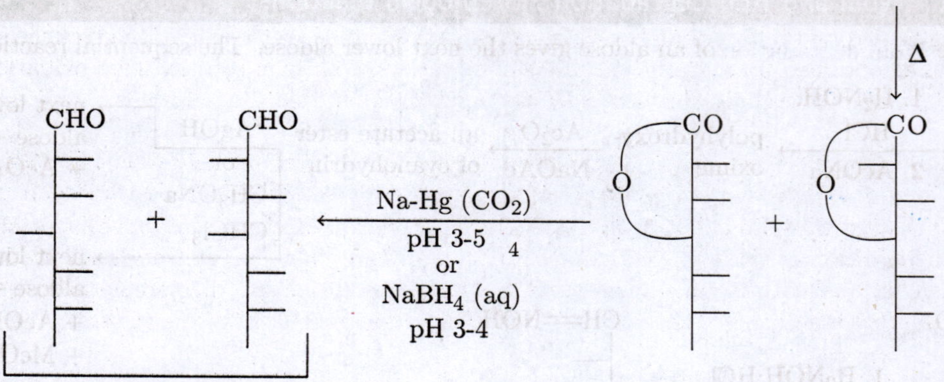

A pair of epimeric
D-aldoheptoses

However, the carboxylic acids on reduction with HI/Red P give a monocarboxylic acid. In the case of an aldose the acid is unbranched one, while that in the case of a ketose is a $C_2$—branched acid as shown below :

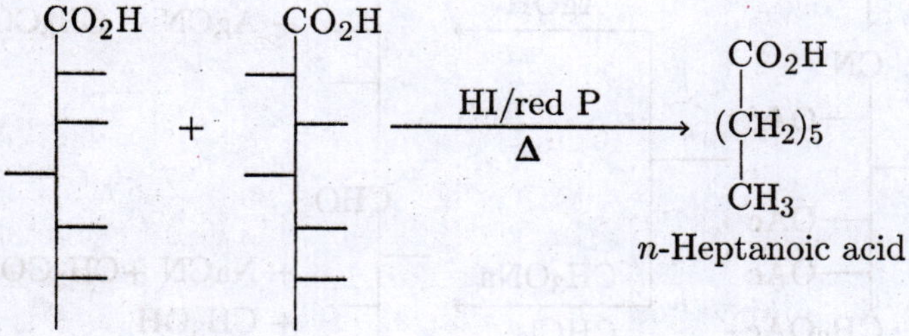

*n*-Heptanoic acid

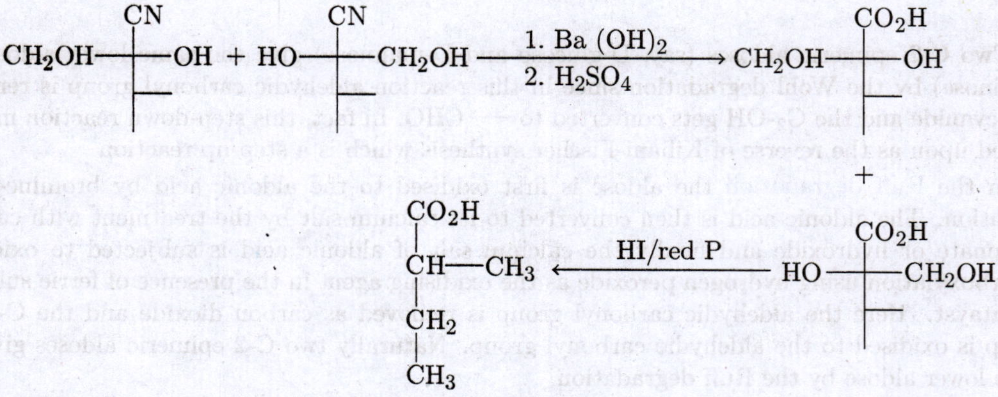

2-Methylbutanoic acid

**Note : Sodium bisolphite**, however, does not add to monosaccharides because the reaction is reversible and the equilibrium favours the unreacted hemiacetal (hemiketal) side.

## 23.6.k Wohl and Ruff degradation : Step-down reactions

The Wohl degradation of an aldose gives the next lower aldose. The sequential reactions are :

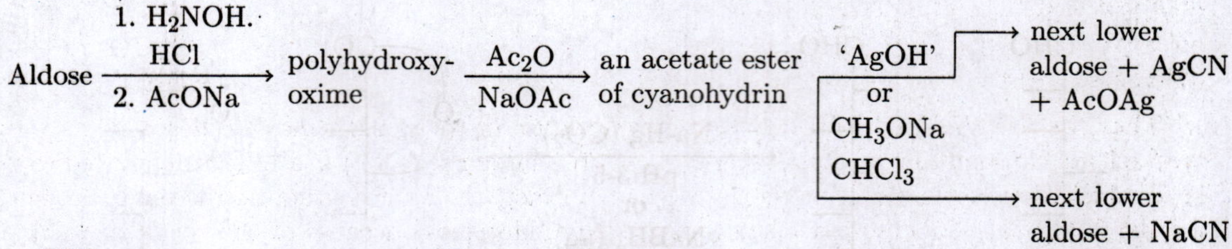

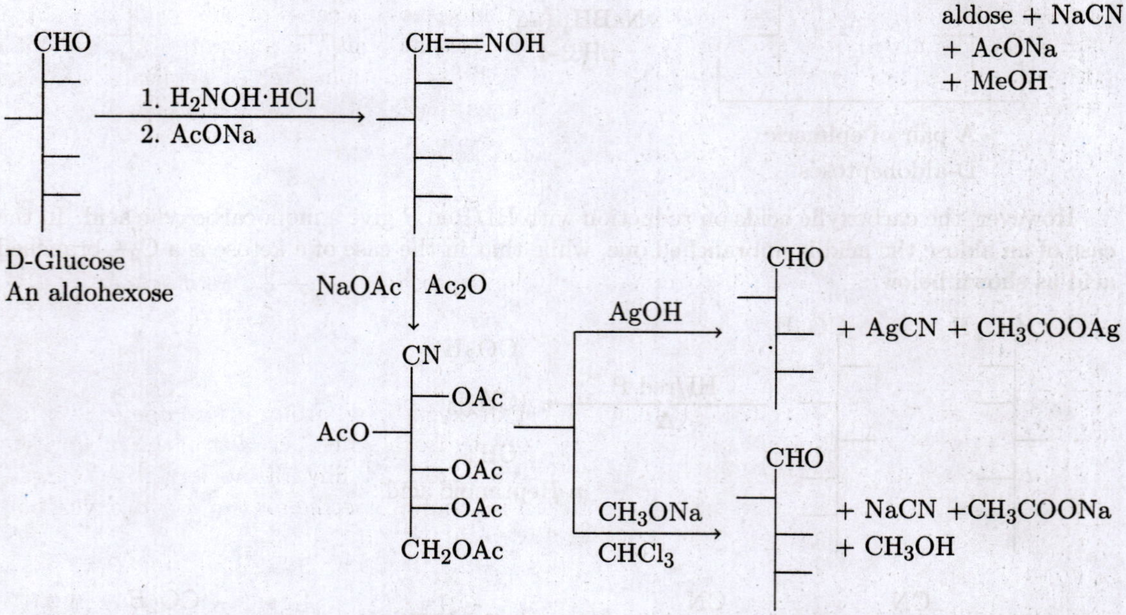

Two C-2 epimeric aldoses (say, D-glucose and D-mannose) give the same lower aldose (D-arabinose) by the Wohl degradation since in this reaction aldehydic carbonyl group is removed as a cyanide and the $C_2$-OH gets converted to —CHO. In fact, this step-down reaction may be looked upon as the reverse of Kiliani-Fischer synthesis which is a step-up reaction.

In the Ruff degradation the aldose is first oxidised to the aldonic acid by bromine-water oxidation. The aldonic acid is then converted to its calcium salt by the treatment with calcium carbonate or hydroxide and finally the calcium salt of aldonic acid is subjected to oxidative decarboxylation using hydrogen peroxide as the oxidising agent in the presence of ferric sulphate as catayst. Here the aldehydic carbonyl group is removed as carbon dioxide and the C-2 OH group is oxidised to the aldehydic carbonyl group. Naturally two C-2 epimeric aldoses give the same lower aldose by the Ruff degradation.

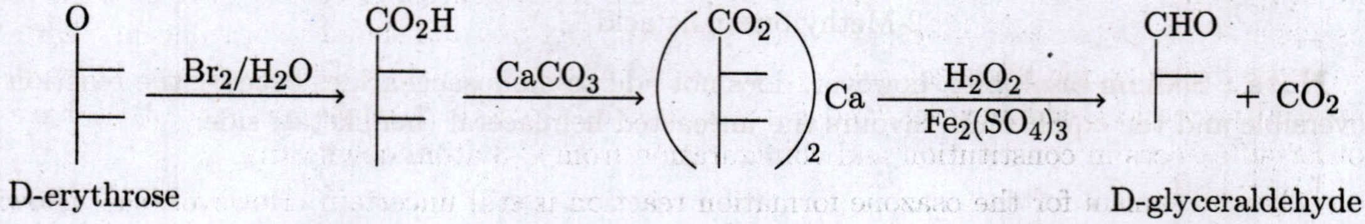

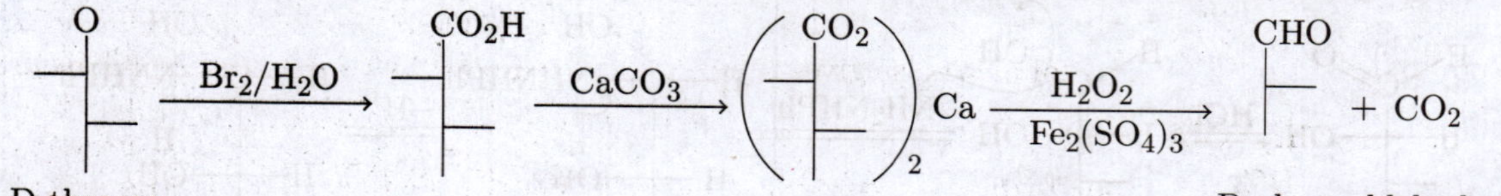

D-threose D-glyceraldehyde

Thus, by the Wohl or Ruff degradation an aldose of any length may be descended to glyceraldehyde through several steps. However, when a large number of C atoms are removed to get glyceraldehyde the yield will be negligibly low. In fact, the Ruff degradation is not even useful for degrading an aldopentose to the corresponding aldotetrose because of the very low yield. In the Wohl degradation the yield of an aldotetrose is also low but the amount is not negligible. However, the Kiliani synthesis and the Wohl and Ruff degradations are not applicable to ketose series. A ketose is usually stepped up or stepped down through their conversions to aldoses.

A ketohexose ⟶ the corresponding aldohexose

↓ step-down (Wohl or Ruff) ↓ ; step-up (Kiliani) ↓

ketopentose ⟵ aldopentose     2 aldoheptoses ⟶ 2 ketoheptoses

### 23.6.1 Reaction with phenylhydrazine : Osazones

Glucose and other reducing sugars having $\alpha$-hydroxycarbonyl group in the open-chain form react with three molar equivalents of phenylhydrazine hydrochloride (per mole of the substrate) in sodium acetate/water or AcOH mixture at about 100 °C to form dihydrazino derivatives of reducing sugars called osazones. The osazone formation reaction is specific for the $\alpha$-hydroxycarbonyl compounds.

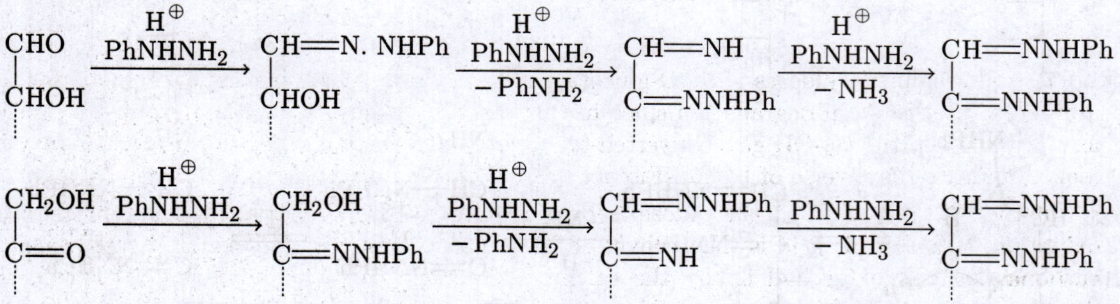

In this reaction only the carbonyl group and an $\alpha$-hydroxy group take part. The other groups and atoms do not get affected through this reaction. Therefore, C-2 epimeric reducing aldoses and C-2 isomeric ketoses having differences in carbonyl group and in configuration of the $\alpha$-C and no other differences in constitution or in configuration of other chiral carbons will give the same osazone. Thus D-glucose, D-mannose and D-fructose give one and the same osazone on treatment with phenylhydrazine hydrochloride under the usual condition since D-glucose and D-mannose are C-2 epimeric D-aldohexoses, and D-fructose is an isomeric 2-ketose and since they have no other differences in constitution and configuration from C-3 atom downward.

The mechanism for the osazone formation reaction is still uncertain. However, the following pathway has received importance :

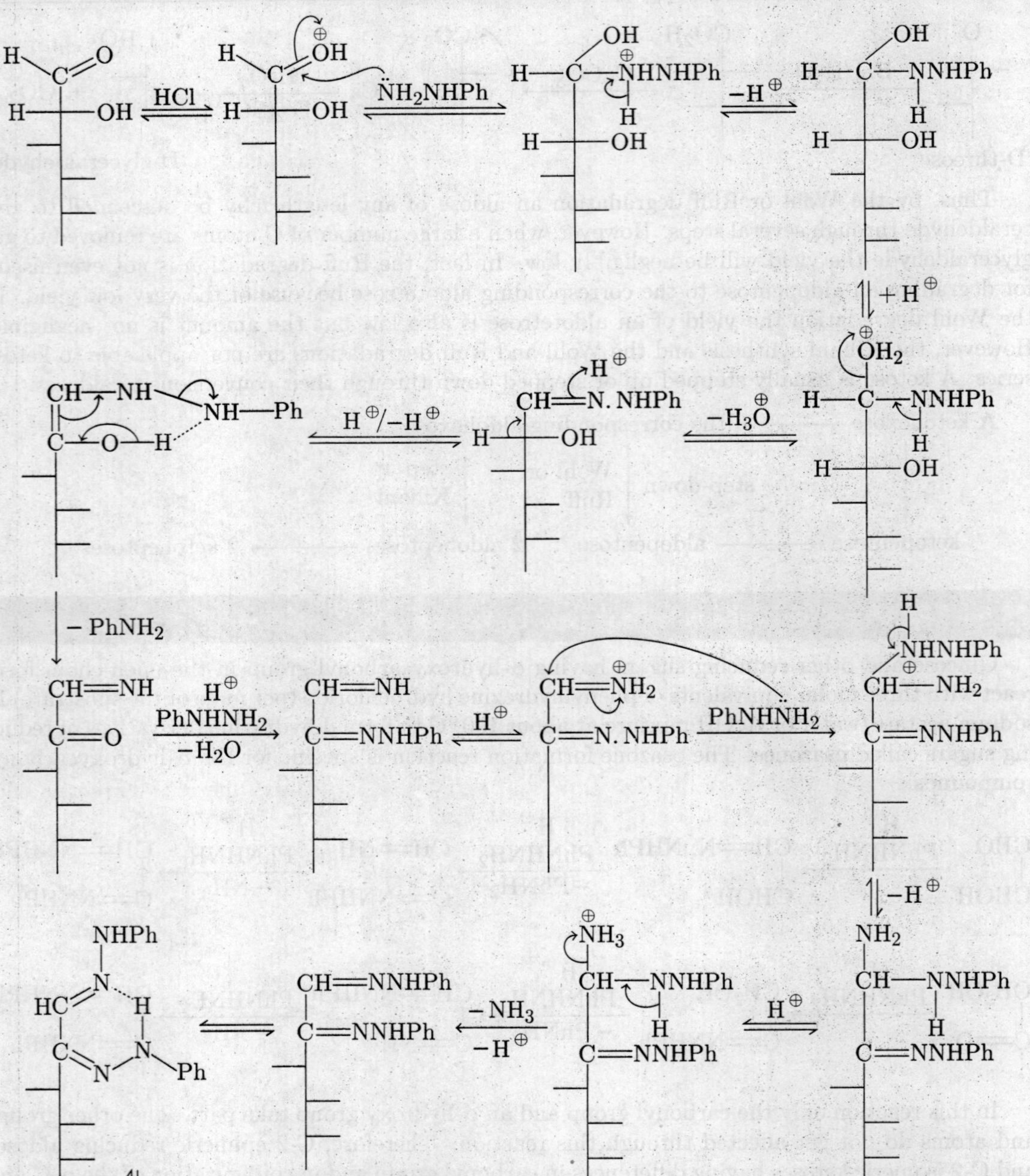

D-Glucosazone

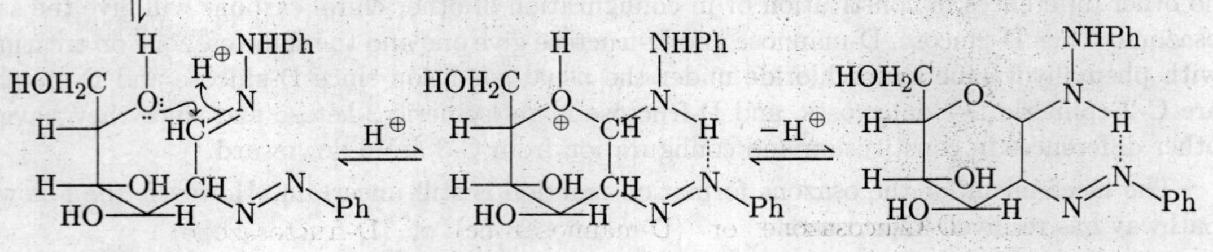

This bicyclochelated compound remains in the solid state and in the solution in equilibrium with the open chain compound. However, one pertinent question arises: why do not other OH groups ($\beta$, $\gamma$, etc.) along with $\alpha$-OH group convert to C=NNHPh moiety? It has been found that when D-glucose is reacted with $\alpha$-N-methylphenylhydrazine PhN(Me)$NH_2$, the hexahydrazino derivative of glucose forms. This indicate that the internal H-bonding involving two N atoms before the formation of chelated bicyclo derivative being enough stable, it does not interact further.

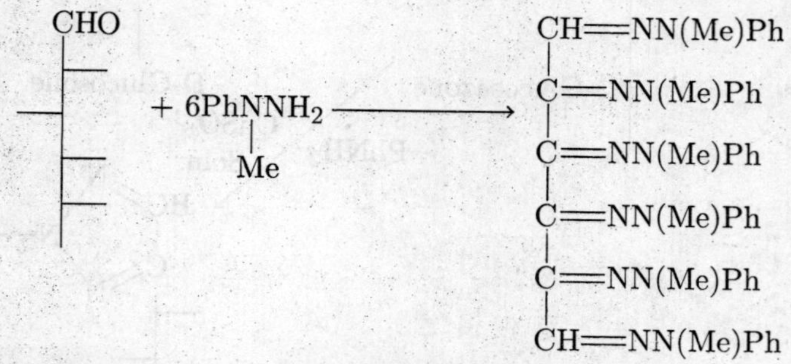

However, osazones have wide and varied applications :

(i) The sharp m.p.s, characteristic colours and crystalline shapes of osazones help to identify reducing sugars. Before the discovery of osazones by Fischer, the identification was done by m.p.s only.

(ii) The configurational relationship amongst two C-2 epimeric aldoses and isomeric ketose can be established since they give one and the same osazone. Since D-glucose D-mannose and D-fructose (a 2-ketohexose) give the same osazone, the configurations of $C_3$, $C_4$ and $C_5$ atoms of the three sugars are the same and from $C_3$ downward they do not also differ in constitution; the three differ from each other because of their differences either in the carbonyl function or in the configuration about $C_2$ atom.

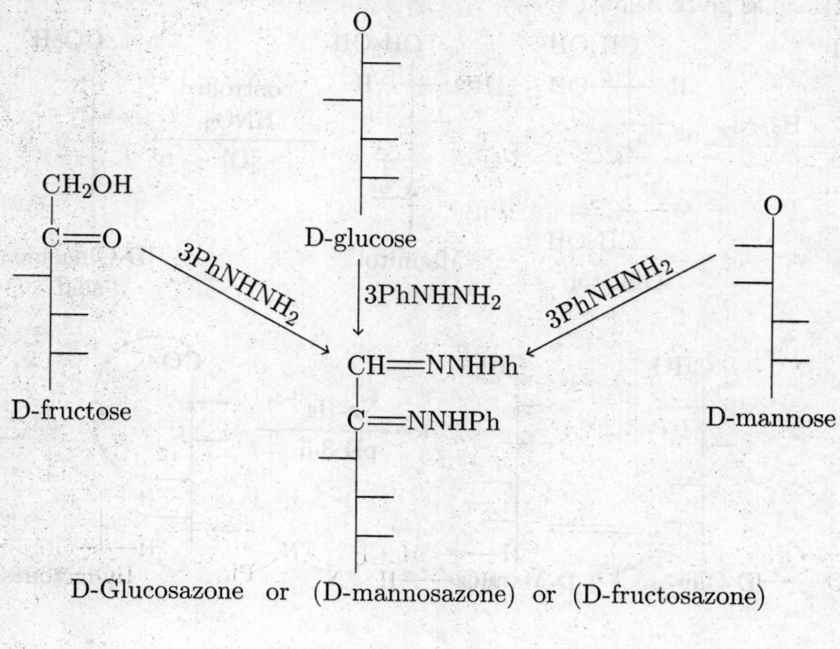

D-Glucosazone or (D-mannosazone) or (D-fructosazone)

**(iii)** Osazones on hydrolysis give osones, polyhydroxy-1, 2-dicarbonyls :

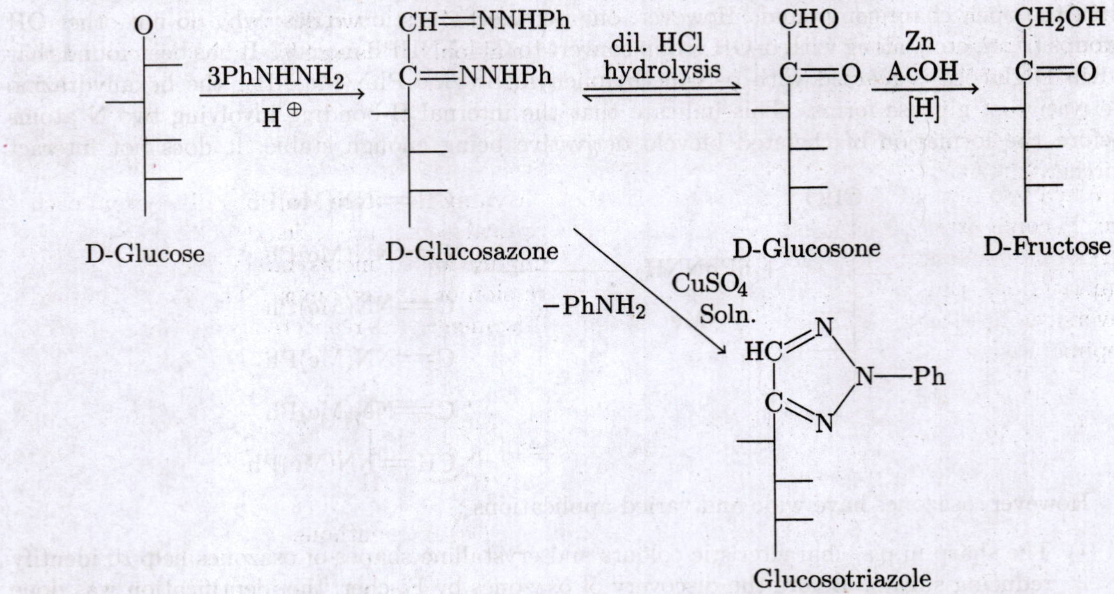


Osones also have sharp m.p.s, characteristic colours and crystalline shapes and, therefore, they are also employed to identify reducing sugars. Osones on reduction by Zn/AcOH convert to aldoses; —CHO group preferentially reduces to —$CH_2OH$. In this way D-glucose is converted to D-fructose as shown above.

## 23.6.m Conversion of a ketose to the corresponding aldoses

This is done as given below :

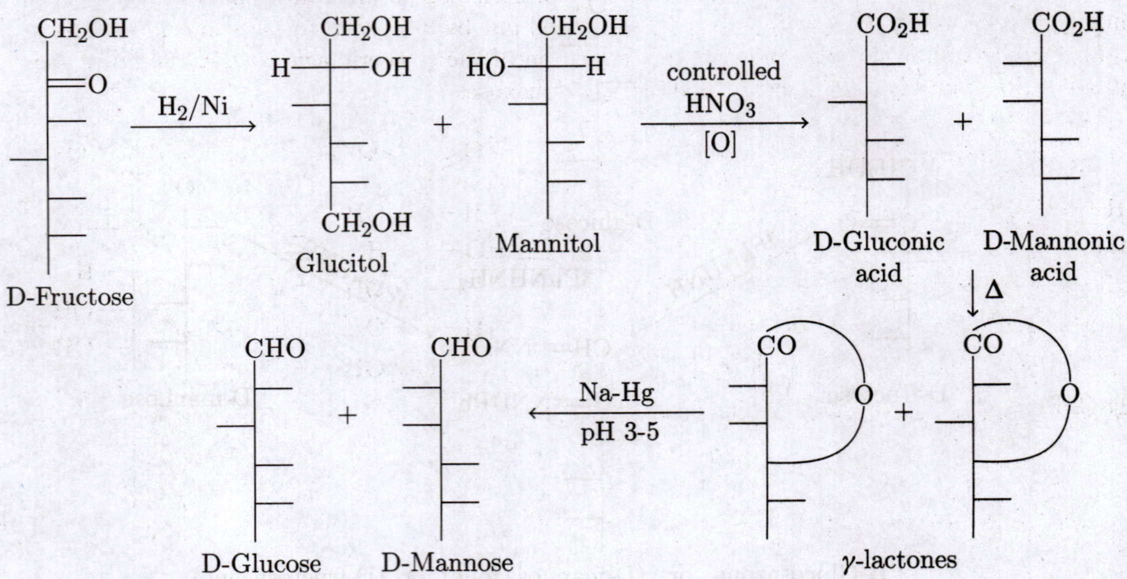

The carbonyl C in ketoses is prochiral and the carbonyl group has two diastereotopic faces. On reduction, a new stereocentre generates ($C_2$; the carbonyl C) and two diastereomers are formed in unequal amounts. Their conversion to $\gamma$-lactones via aldonic acids and finally Na-Hg reduction (at pH 3-5) of the $\gamma$-lactones give the epimeric aldoses.

## 23.6.n Epimerisation

When two optical isomers having more than one asymmetric atom (carbon) differ from each other in configuration about one such atom, the two optical isomer are called epimers; the conversion of on epimer into the other by changing the configuration of one asymmetric atom is called epimerisation—thus epimerisation involves interconversion of diastereomers. The base-catalysed conversion of (+)- and (−) menthones to (−)- and (+)-isomenthones respectively are the examples of epimerisation.

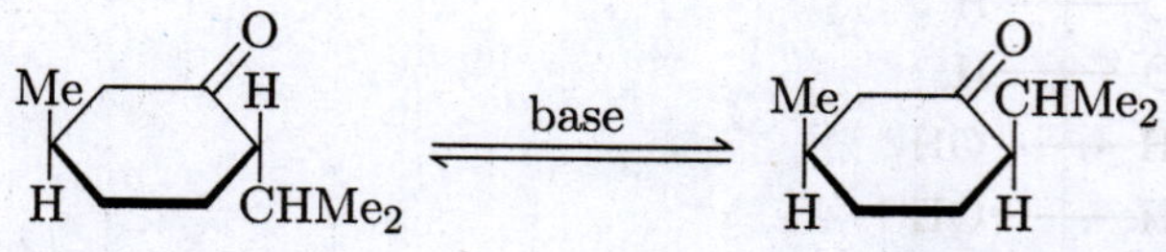

(−)-Menthone (+)-Isomenthone

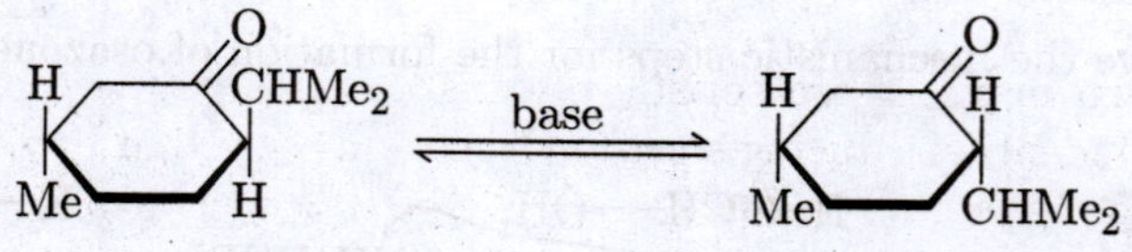

(+)-Menthone (−)-Isomenthone

The terms 'epimers' and 'epimerisation' were originally used in the sugar chemistry. Aldoses which produce the same osazones, are called epimers since such aldoses differ only in configuration about $\alpha$-C atom; and the interconversion of epimeric sugars are called epimerisation. The epimerisation of aldohexoses is, however, best carried out by oxidation to aldonic acid, epimerisation of the aldonic acid with pyridine, lactonisation of the epimeric acid and reduction of the lactone of the epimeric aldonic acid to epimeric aldohexose.

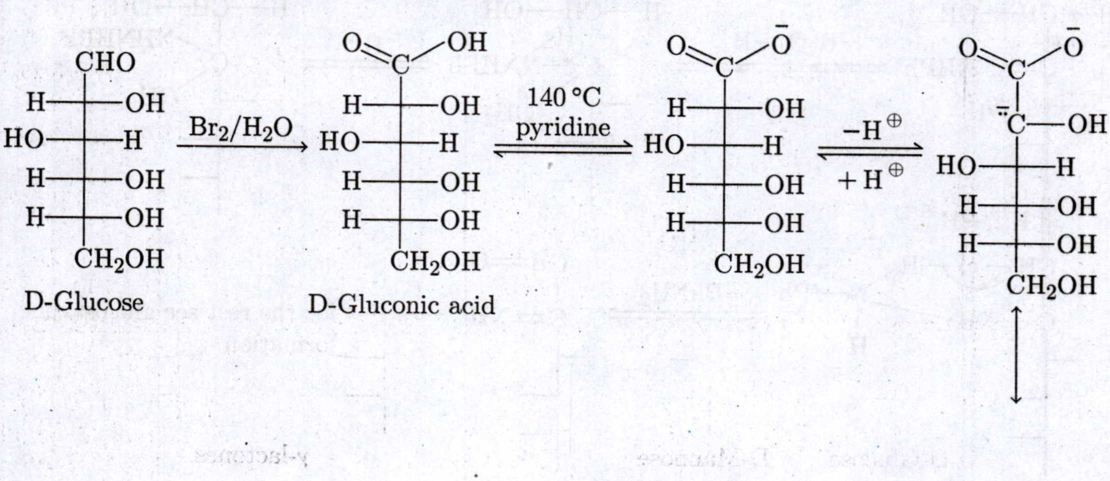

D-Glucose D-Gluconic acid

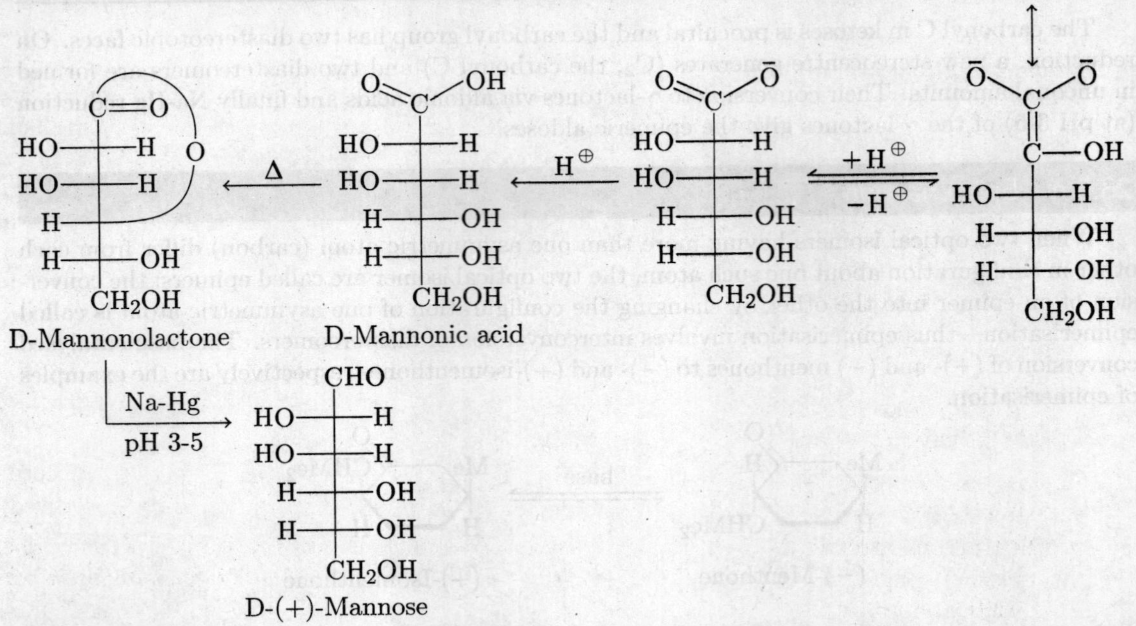

**Problem 23.2.** Give the mechanistic steps for the formation of osazone form D-fructose.

**Solution :**

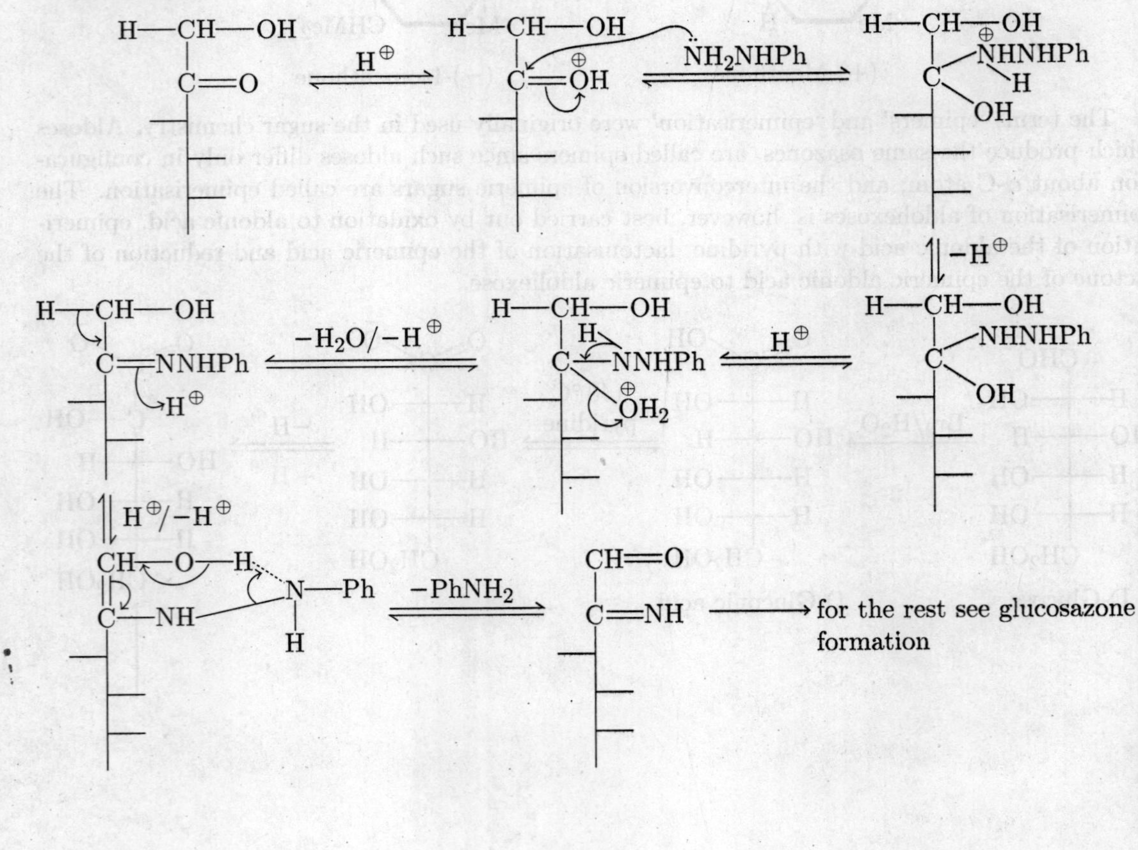

## 23.7 Determination of Configurational Relationship Amongst Monosaccharides

The reactions discussed in Section 23.6 are helpful to establish the configurational relation amongst monosaccharides. However, we shall first present the brilliant method by which Fischer established the configurations of D-arabinose, D-mannose and D-glucose although very little of the chemistry of carbohydrates was known to the then scientists. Emil Fischer, however, arbitrarily assigned D-configuration to the C-5 chiral C of D-glucose and later on it has been found to be correct. If the configuration of C-5 of D-glucose were considered by Fischer as L, he would have not been successful. However, besides the arbitrary assignment of configuration to C-5 as D, the rest of the procedure followed by Fischer to establish the configuration of other three chiral carbons (C-2, C-3 and C-4) of D-glucose is really superb, logical and deserve masterly honour. His procedure is noted below in brief as far as practicable.

### FISCHER'S PROOF OF THE CONFIGURATIONS OF CHIRAL CARBONS OF D-GLUCOSE, D-MANNOSE AND D-ARABINOSE

Though Fischer arbitrarily assigned D-configuration to the C-5 atom of D-(+)-glucose yet depending upon the following handful of facts about the carbohydrate chemistry he proved the configuration of D-(+)-glucose using brilliant logic and performing intelligent experiments :

*The facts were :*

(i) D(+)-Glucose and D (+)-mannose are aldohexoses.

(ii) D (−)-Arabinose is an aldopentose.

(iii) Aldohexoses have four chiral carbon atoms.

(iv) Aldopentoses have three chiral carbon atoms.

(v) Aldoses on oxidation by nitric acid give polyhydroxydicarboxylic acids (aldaric acids).

(vi) Whether a compound is optically active or not that can be determined by experiments.

(vii) An aldose can be stepped-up by Kiliani-Fischer synthesis to two C-2 epimeric higher aldoses.

(viii) (−)-Arabinose on stepping up give (+)-glucose and (+)-mannose. The latter two on stepping down gives (−)-arabinose.

*The arguments on facts :*

(1) As Fischer arbitrarily assigned D-configuration to (+)-glucose, the configuration of 3 other chiral carbons ($C_2$, $C_3$ and $C_4$) are to be assigned. Naturally there will be $2^3$ or 8 configurational isomers for 3 chiral carbons and one of them will be (+)-glucose. The eight configurational isomers having D-configuration at C-5 are A, B, C, D, E, F, G and H. These can be stepped up by the Kiliani-Fischer synthesis from four aldopentoses, I, J, K and L; one of which is (−)-Arabinose.

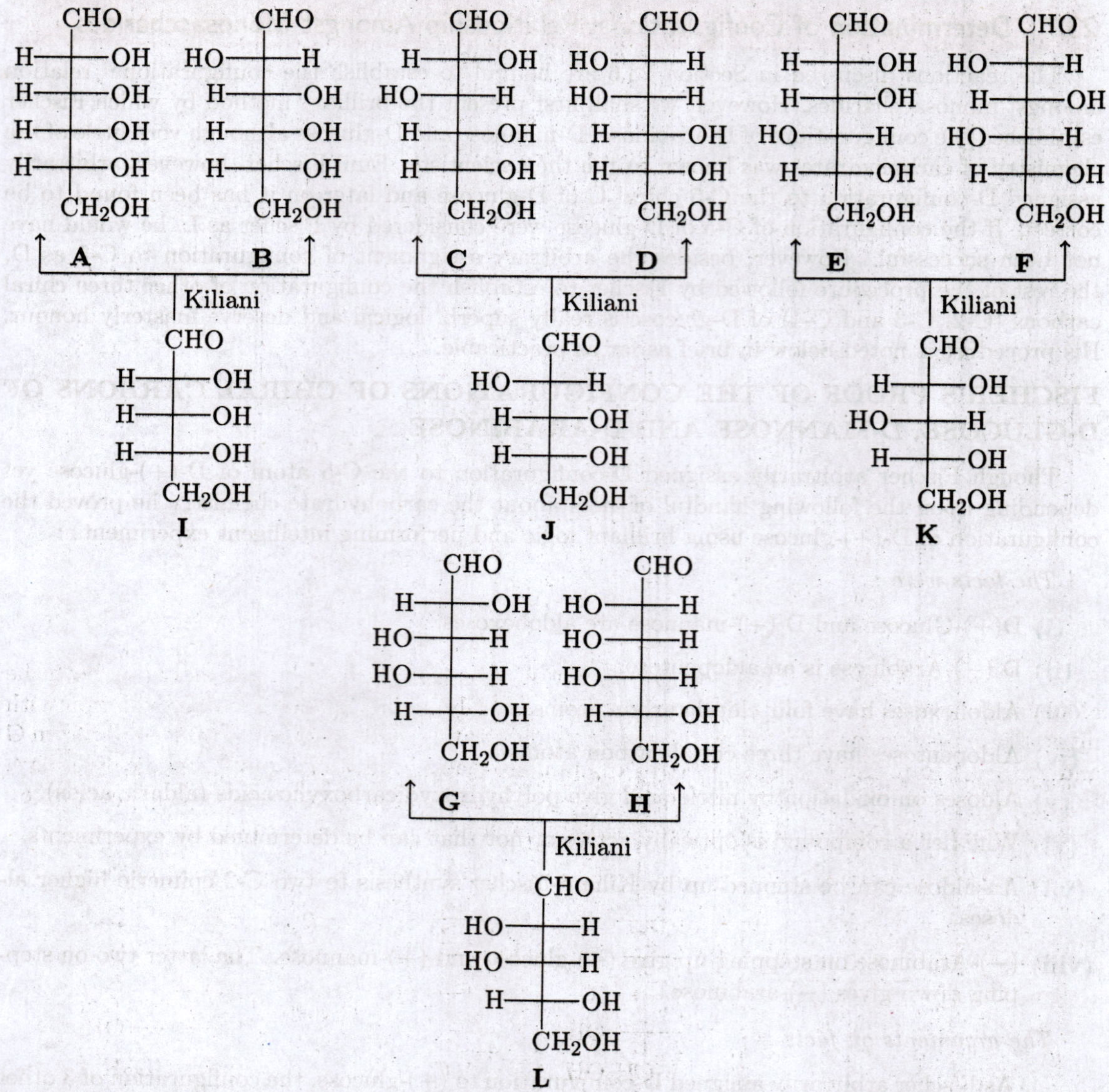

(2) The nitric acid oxidation of (+)-glucose gives an optically active aldaric acid. So, A or G cannot be (+)-glucose since they on nitric acid oxidation give optically inactive meso acids.

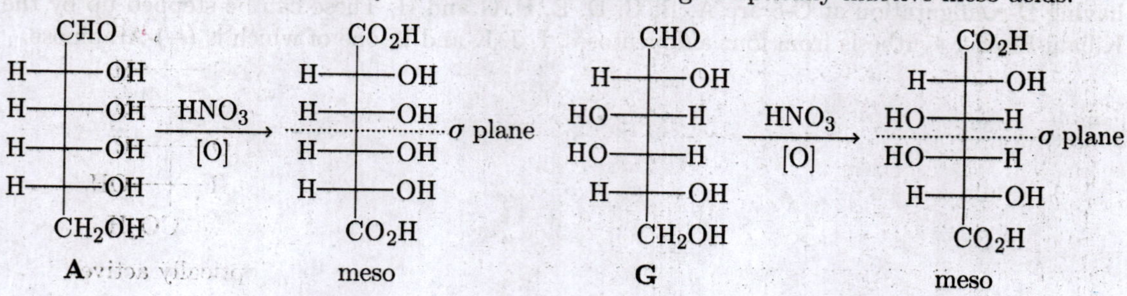

**(3)** The stepping down of (+)-glucoses gives (−)-arabinose which on nitric acid oxidation gives an optically active aldaric acid. I and K on nitric acid oxidation give optically inactive meso aldaric acids. So none of I and K represents (−)-arabinose. (−)-Arabinose is either J or L. This also tells us that none of B, E and F could be (+)-glucose since these are obtained from I and K which cannot represent (−)-arabinose. So, (+)-glucose is one of C, D and H.

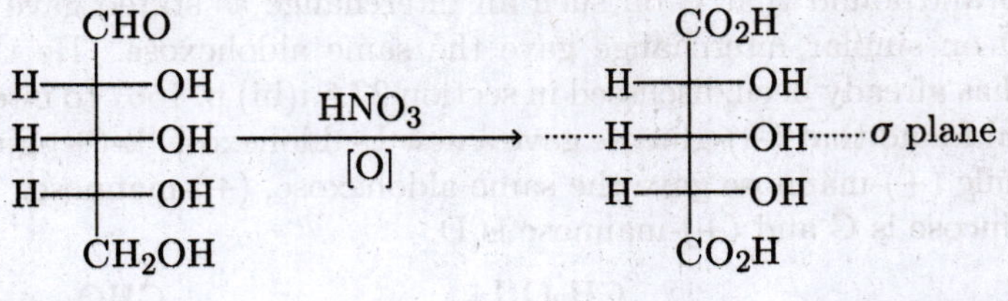

I meso

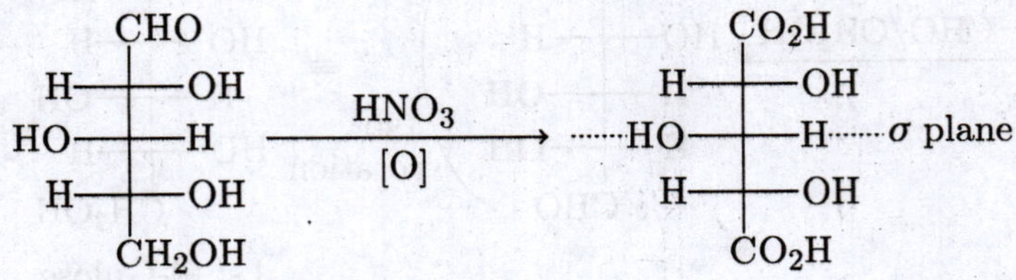

K meso

**(4)** A Kiliani-Fischer synthesis on (−)-arabinose gives (+)-glucose and (+)-mannose; both the aldohexoses, (+)-glucose and (+)-mannose, give optically active aldaric acids on oxidation with nitric acid. L on stepping up followed by nitric acid oxidation gives two aldaric acids —one from G and the other from H, and the acid from G is optically inactive but that from H is optically active. Hence L does not represent (−)-arabinose and consequently H does not represent (+)-glucose. So (−)-Arabinose is represented by J and, therefore, either C or D represents (+)-glucose.

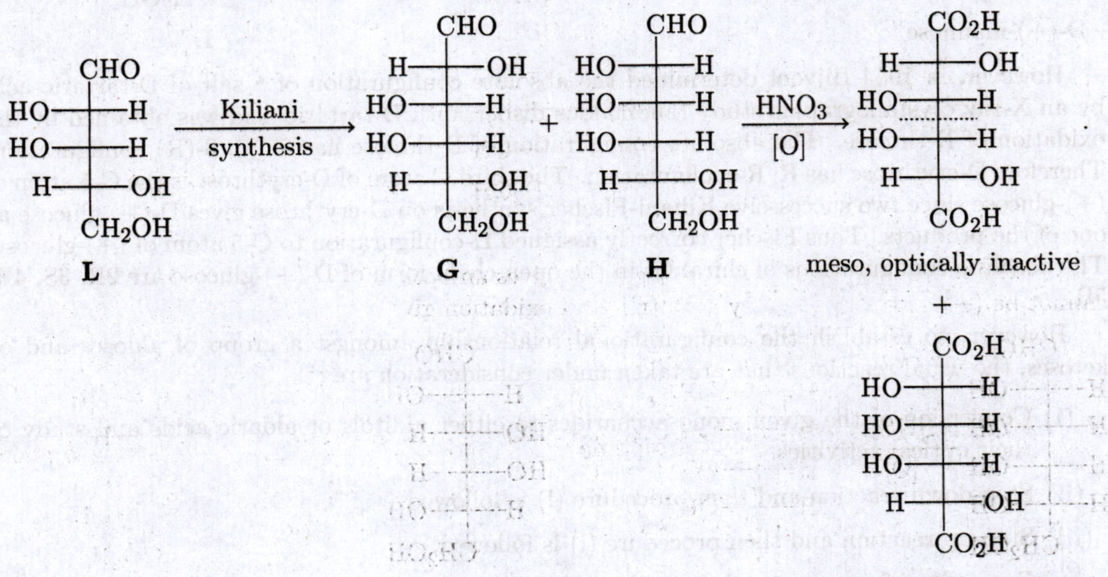

L G H meso, optically inactive + optically active

(5) Since (−)-Arabinose (J) on stepping up gives a mixture of (+)-glucose and (+)-mannose and since J on stepping up gives C and D, one of C and D is (+)-glucose and the other is (+)-mannose.

(6) Now to determine which one is (+)-glucose and which one is (+)-mannose of C and D Fischer did the following mental exercise. He interchanged the —CHO and —$CH_2OH$ groups of each of C and D and found that C on such an interchange as stated gave a new aldohexose of L-series while D on similar interchange gave the same aldohexose. He then undertook an experiment, which has already been discussed in section 23.6.i(iii) p. 1567 to interchange —CHO and —$CH_2OH$ and found that (+)-glucose gave a new L-aldohexose [L-(+)-gulose, mirror image of D-(−)-gulose] while (+)-mannose gave the same aldohexose, (+)-mannose. Therefore, Fischer inferred that (+)-glucose is C and (+)-mannose is D.

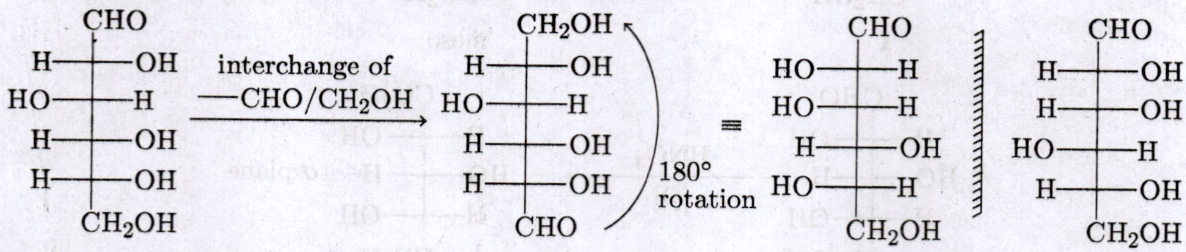


C [(+)-glucose] L-(+)-Gulose D-(−)-Gulose

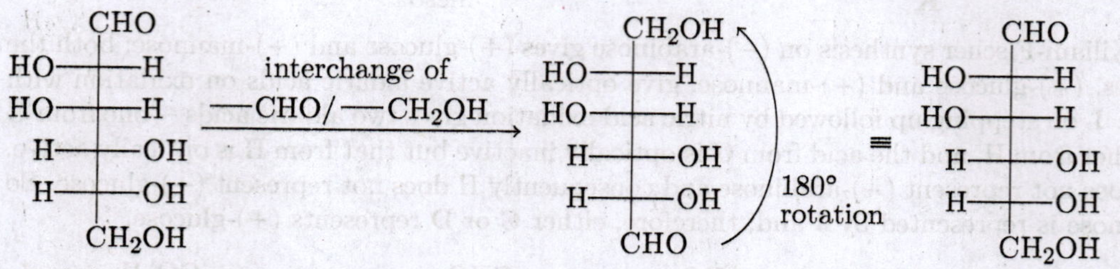


D-(+)-mannose D

However, in 1954 Bijvoet determined the absolute configuration of a salt of D-tartaric acid by an X-ray crystallographic study (anomalous dispersion); D-tartaric acid was obtained by the oxidation of D-threose. The absolute configuration of D-threose has 2-(S), 3-(R) configuration. Therefore D-erythrose has R, R configuration. The third C atom of D-erythrose is the C-5 atom of (+)-glucose since two successeive Kiliani-Fischer synthesis on D-erythrose gives D-(+) glucose as one of the products. Thus Fischer correctly assigned D-configuration to C-5 atom of (+)-glucose. The absolute configurations of chiral Cs in the open chain form of D-(+)-glucose are 2R, 3S, 4R, 5R.

However, to establish the configurational relationship amongst a group of aldoses and or ketoses, the usual reactions that are taken under consideration are :

(i) Conversion of the given monosaccharides to either alditols or aldaric acids and study of their optical activities.

(ii) Step-down reaction and then procedure (i) is followed.

(iii) Step-up reaction and then procedure (i) is followed.

(iv) Formation of osazones and their identification.

(v) Interchange of —CHO and —$CH_2OH$ groups.

(vi) Periodic acid oxidation and subsequent reactions as discussed in the section 23.6.i(vi) p. 1570.

We shall discuss the general procedure framed above by solving a few problems.

**Problem 23.3.** An aldopentose (1) is oxidised by nitric acid to an optically active aldaric acid (2). On Ruff degradation of (1) an aldotetrose (3) is obtained which is oxidised to optically inactive alderic acid (4). Assuming the aldepentose (1) to be a D-sugar, write down the configurations of 1 to 4.

**Solution :** Since (1) is of D-series, it may have any one of the following configurations, **a, b, c** or **d.**

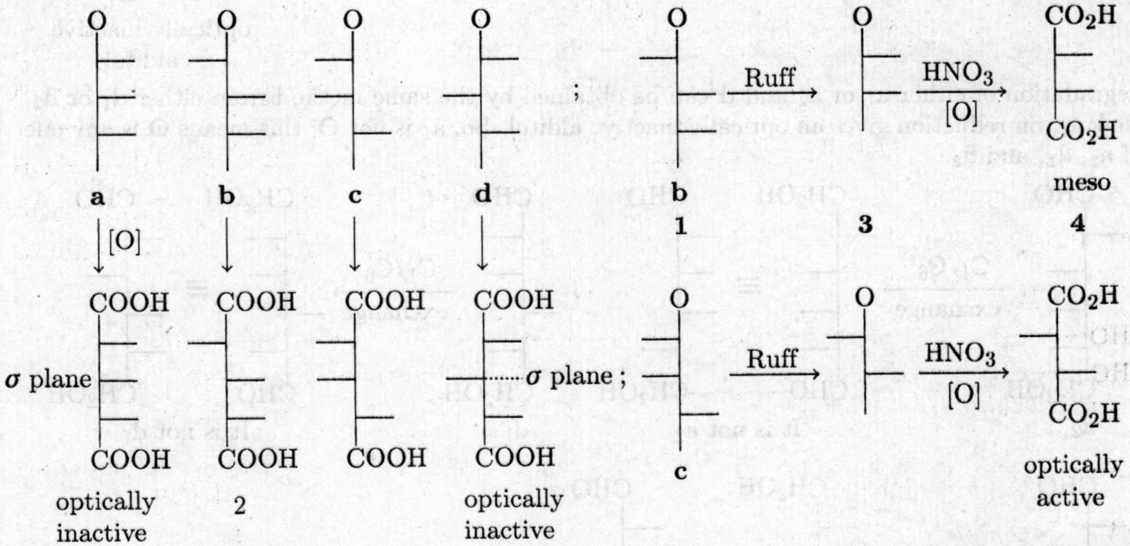


Thus **a** and **d** give meso aldaric acids; so (1) is either **b** or **c.** Again **b** on Ruff degradation followed by nitric acid oxidation gives a meso acid; but **c** on the same treatment gives an optically active acid. According to the problem the Ruff degradation product of (1) on oxidation gives optically inactive acid. Hence (1) is **b**, D-arabinose (2) is arabinaric acid, (3) is D-erythrose and (4) is meso tartaric acid.

**Problem 23.4.** Assuming **O, P, Q, R** as D-sugars write down their configurations from the following facts;

1. Aldohexose (**O**) $\xrightarrow[\text{pH 3-4}]{NaBH_4}$ Alditol (**P**); (optically active)

2. **O** $\xrightarrow[\text{degradation}]{\text{Wohl}}$ an aldopentose (**Q**)

3. **Q** $\xrightarrow[\text{[O]}]{HNO_3}$ Saccharic acid (**R**); optically inactive

4. **O** $\xrightarrow[\text{exchange}]{\text{C-1/C-6}}$ **O**

**Solution :** There are four D-aldopentoses, **a, b, c** and **d** as in solution 23.3; **a** and **d** give optically inactive saccharic (aldaric) acids. So **Q** is either **a** or **d**. Now **a** can be obtained by Wohl

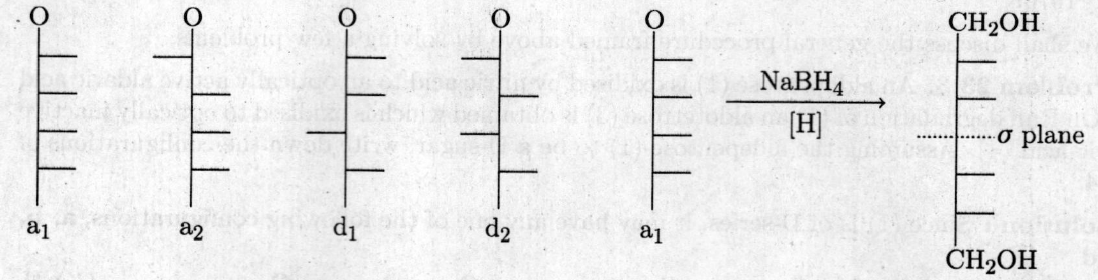

$a_1$ $a_2$ $d_1$ $d_2$ $a_1$ optically inactive alditol

degradation of either $a_1$ or $a_2$ and **d** can be obtained by the same method from either $d_1$ or $d_2$. Now $a_1$, on reduction gives an optically inactive alditol. So, $a_1$ is not **O**; this means **O** is any one of $a_2$, $d_1$, and $d_2$.

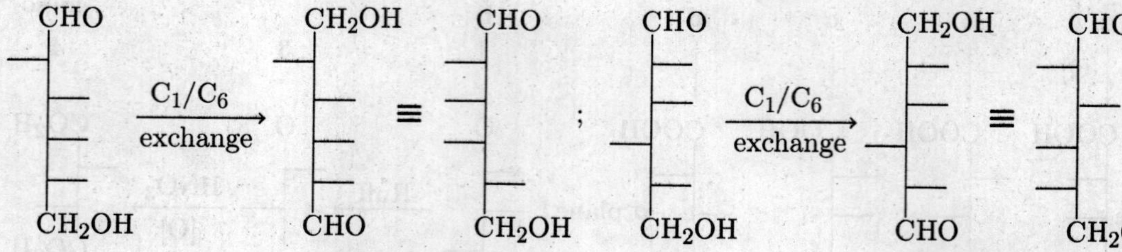

$a_2$ It is not $a_2$ ; $d_1$ It is not $d_1$

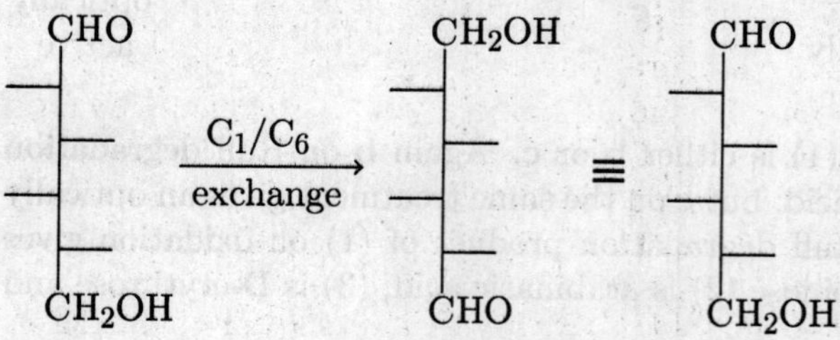

$d_2$ It is $d_2$

So, **O** is $d_2$. This is further coroborated by the fact O on reduction gives an optically active alditol as shown below :

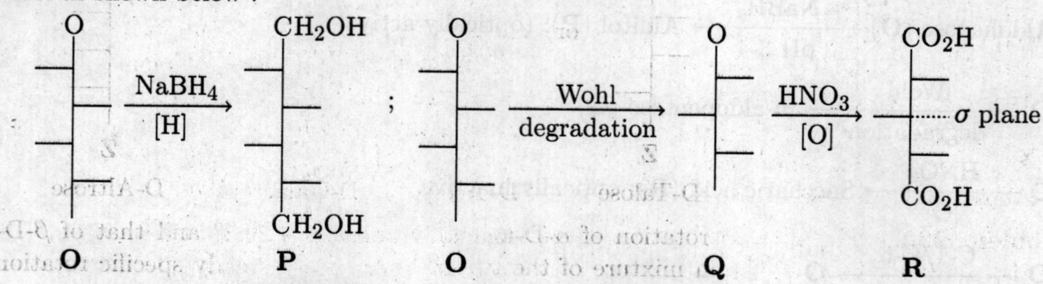

**O** **P** optically active ; **O** **Q** **R**

Wohl degradation of O followed by oxidation to optically inactive saccharic acid also supports the configuration of O is as shown above.

**Problem 23.5. X, Y** and **Z** are three D-aldohexoses. **X** and **Y** yield the same osazone but different aldaric acids on nitric acid oxidation; while **Y** and **Z** give the same aldaric acid but different osazones. Give structures and names of **X, Y** and **Z.**

**Solution.** There are eight D-aldohexoses (**1-8**) as given below.

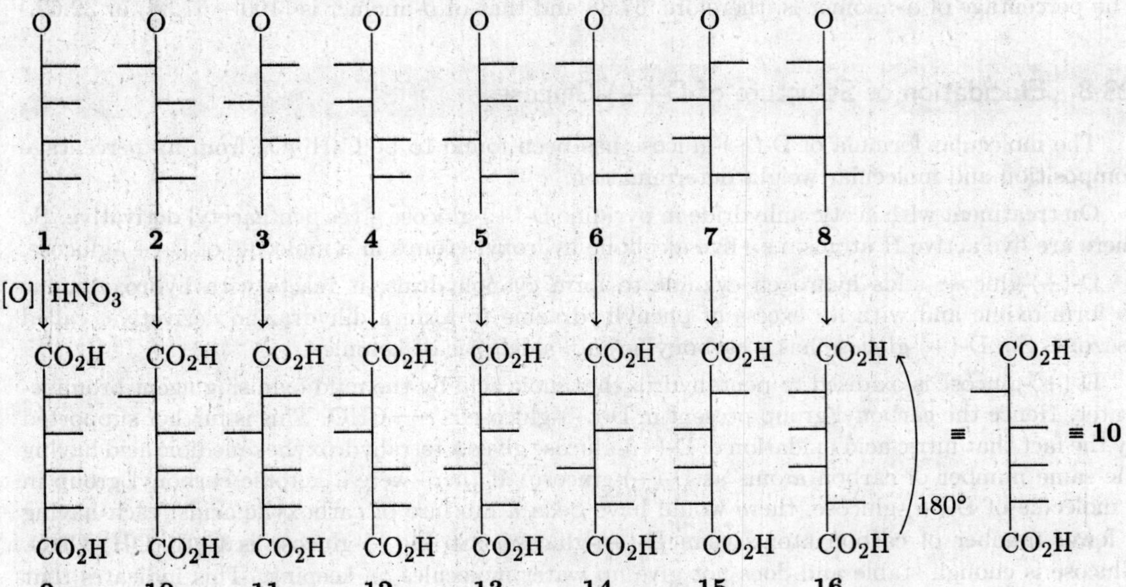

**1** **2** **3** **4** **5** **6** **7** **8** ; [O] $HNO_3$ ; **9** **10** **11** **12** **13** **14** **15** **16** ; 180° ; ≡ 10

They give eight aldaric acids (**9-16**). However, all are not different. **10, 12, 14** and **16** on rotating 180° on the plane of the paper will give configurations having C-5 D because their C-2 atoms are of L-configuration. Amongst these four 10 and 16 are identical. Therefore, 2 and 8 give the same aldaric acid. 2 and 8 are not C-2 epimeric. C-2 epimeric aldohexoses give the same osazone. So, if 2 is **Y**, **8** is **Z**. But 2 is C-2 epimeric with **1**. **X** and **Y** yield the same osazone. **1** and **2** yield the same osazone but different aldaric acids. Hence **2** is **Y** and **1** is **X**. On the other hand if **8** is **Y, 2** is **Z** and **7** is **X.**

So,

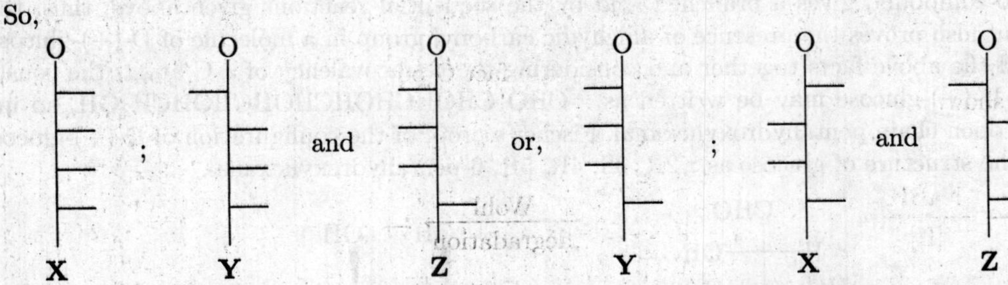

**X** D-Allose ; **Y** D-Altrose and **Z** D-Talose or, **Y** D-Talose ; **X** D-Galactose and **Z** D-Altrose

**Problem 23.6.** The specific rotation of $\alpha$-D-mannopyranose is +29.3° and that of $\beta$-D-anomer is −17.0°. The equilibrium mixture of the two in water has a steady specific rotation +14.2°. Calculate the percentage composition of the equilibrium mixture.

**Solution :** Let $x$ be the percentage of $\alpha$-anomer. Then

$$x \times (+29.3°) + (100 - x)(-17.0°) = 100 \times (+14.2°)$$

or, $29.3°x - 1700° + 17°x = +1420°$.

or, $46.3°x = 3120°$, $\therefore$ $x = \dfrac{3120°}{46.3°} = 67.38$.

The percentage of $\alpha$-anomer is, therefore, 67.38 and that of $\beta$-anomer is (100 − 67.38) or 32.62.

## 23.8 Elucidation of Structure of D-(+)-Glucose

The molecular formula of D-(+)-glucose has been found to be $C_6H_{12}O_6$ from its percentage composition and molecular weight determination.

On treatment with acetic anhydride in pyridine D-(+)-glucose gives pentaacetyl derivative. So there are five active H atoms, i.e., five alcoholic hydroxyl groups in a molecule of D-(+)-glucose.

D-(+)-glucose adds hydrogen cyanide to form cyanohydrins, it reacts with hydroxylamine to form oxime and with an excess of phenylhydrazine to form a dihydrazino derivative, called osazone. So, D-(+)-glucose has a carbonyl group, aldehydic or ketonic.

D-(+)-glucose is oxidised to pentahydroxyhexanoic acid by the mild oxidising agent bromine-water. Hence the carbonyl group present in D-(+)-glucose is —CHO. This is further supported by the fact that nitric acid oxidation of D-(+)-glucose gives a tetrahydroxyhexanedioic acid having the same number of carbon atoms as D-(+)-glucose. If there were a ketonic carbonyl group in a molecule of D-(+)-glucose, there would have been a mixture of carboxylic acids, each having a fewer number of carbon atoms than D-(+)-glucose. So D-(+)-glucose is $C_5H_6(OH)_5CHO$. Glucose is enough stable and does not give up water molecules on keeping. This indicates that not a single carbon atom holds more than one OH group; so each C atom, excepting the C atom of the —CHO group, bears one OH group.

When glucose is successively reduced with Na-Hg/$H_2O$ and HI/red P, $n$-hexane (a straight chain compound) is obtained. Therefore, D-(+)-glucose is a straight chain compound. This is also supported by the following fact :

$$\text{Glucose} \xrightarrow{\text{HCN}} \text{cyanohydrin} \xrightarrow{H_3N^{\oplus}} \xrightarrow{\text{HI/red P}} CH_3(CH_2)_5CO_2H \ (n\text{-heptanoic acid}).$$

A keto-compound gives a branched acid by the sequential reactions given above; thus, this experiment also proves the presence of aldehydic carbonyl group in a molecule of D-(+)-glucose. Taking all the above facts together and considering the tetracovalency of a C atom, the constitution of D-(+)-glucose may be written as : CHO CHOHCHOHCHOHCHOHCH$_2$OH, an unbranched open chain pentahydroxyhexanal, Fischer's proof of the configuration of D-(+)-glucose gives us the structure of glucose as : 2R, 3S, 4R, 5R, 6-pentahydroxyhexanal.

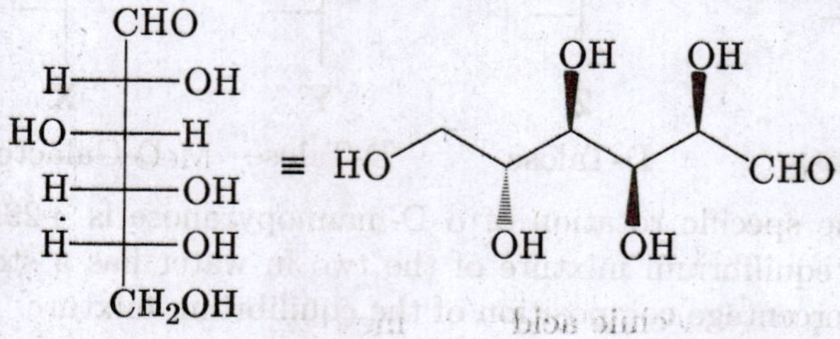

The structure of a compound should be capable of explaining all the properties of the compound. The unbranched open chain structure of D-(+)-glucose fails to explain many of its properties. For example,

(i) There are two kinds of naturally occurring D-(+)-glucose, $\alpha$ and $\beta$.

(ii) There are two kinds of glucosides in nature, $\alpha$- and $\beta$-D-(+)-glucoside.

(iii) It adds hydrogen cyanide but it does not add sodium bisulphite.

(iv) It reduces Fehling solution and Tollen's reagent but does not restore the pink colour of the decolourised fucshine salt solution (Schiff's reagent).

(v) In the aqueous solution two kinds of D-(+)-glucose mutarotate, one downward and the other upward directions.

(vi) **Only one molecule of an alcohol reacts with one molecule of D-(+)-glucose to give alkyl glucoside, a full acetal.**

If D-(+)-glucose has a free aldehyde group, it should not have two isomers, $\alpha$ and $\beta$, and it should add sodium bisulphite and two molecules of alcohol to form a full acetal. It should also give the Schiff's test.

Since glucose adds only one molecule of alcohol to form a full acetal, it may exist in nature as a hemiacetal. This is not impossible because one of its 5 OH groups may undergo internal hemiacetal formation with its aldehydic carbonyl group. In that case it must be a cyclic compound instead of an open chain one; the aldehydic C being prochiral, the interaction will develop a new stereocentre and the CHO group with two diastereotopic faces will produce two diastereomers, the two hemiacetals, $\alpha$- and $\beta$-D-(+)-glucoses. Each of the two diastereomeric $\alpha$- and $\beta$-D-(+)-glucoses, $\alpha$- and $\beta$-anomers, in their turn, can react with one molecule of an alcohol to form $\alpha$- and $\beta$-glucosides, respectively. The hemiacetal ring by the ring-chain tautomerism could remain as an equilibrium mixture of ring and open chain glucoses in the solution.

However, if D-(+)-glucose has a cyclic structure, then what will be the size of the ring? Because of the participation of an OH group to form the hemiacetal internally, an O atom must be the constituent of the ring. But how many carbons are the ring constituents? Fischer arbitrarily assumed that the ring of a D-(+)- glucose unit is five-membered furan-like.

Haworth and Hirst by the classical oxidative degradation reaction were able to prove that naturally occurring D-(+)-glucose possesses a six-membered pyran-like ring structure. The flowsheet for the sequential reactions (which were carried out by Haworth and Hirst) is given below :

$$\text{D-(+)-glucose} \xrightarrow[\text{reflux}]{\text{MeOH/HCl}} \text{Methyl D-glucoside} \xrightarrow{\text{NaOH/Me}_2\text{SO}_4} \text{Methyl tetra-O-methyl-}$$

$$\text{D-glucoside} \xrightarrow{\text{H}_3\text{O}^{\oplus}} \text{Tetra-O-methyl-D-glucose} \xrightarrow[\text{[O]}]{\text{HNO}_3} \text{optically active dimethoxysuccinic acid}$$

and meso or xylotrimethoxyglutaric acid, the structures of which are noted below.

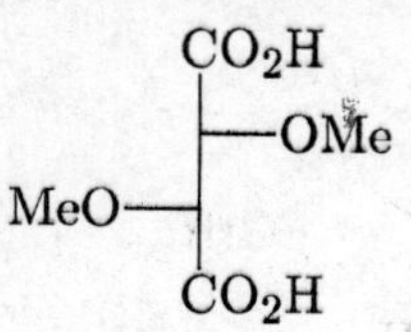

optically active Dimethoxysuccinic acid

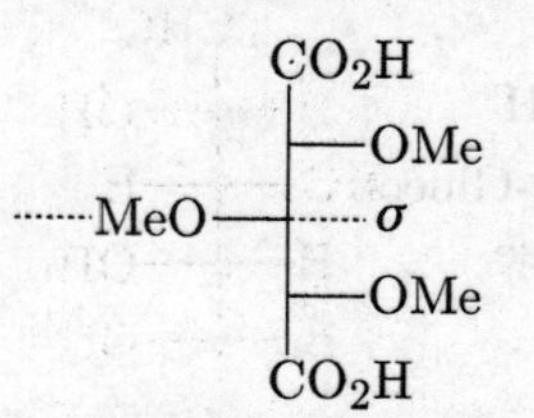

meso or Xylotrimethoxyglutaric acid

Haworth and Hirst correctly argued that a pyranose structure for D-(+)-glucose could only give rise to the products mentioned above by the sequential reactions involving oxidative degradation. A furanose ring could not give rise to meso or xylotrimethoxyglutaric acid.

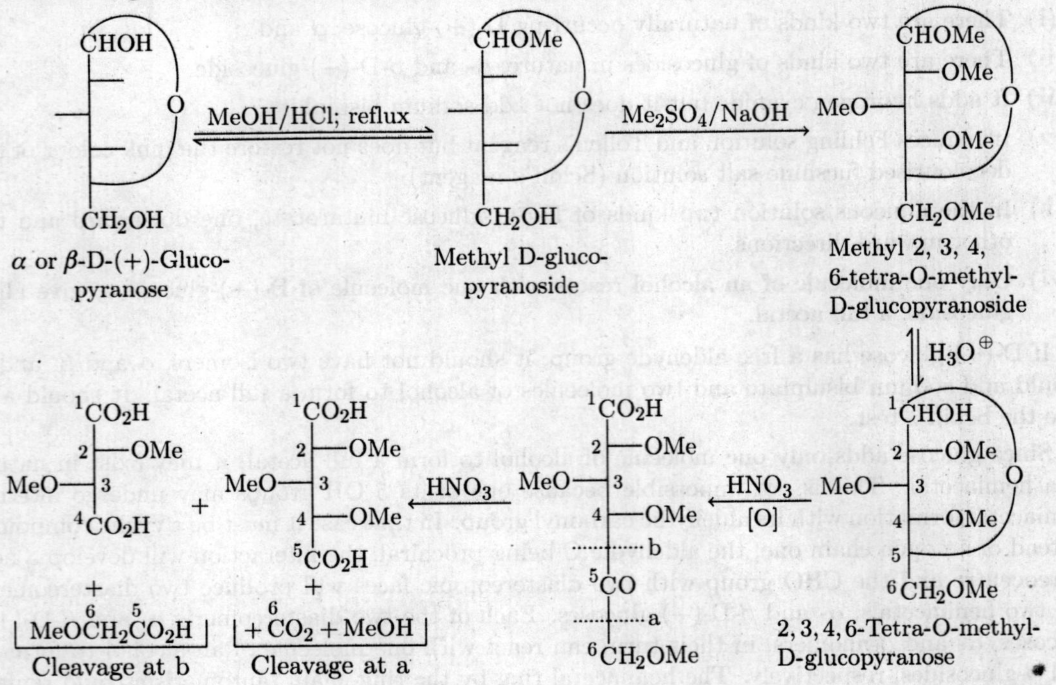


Since $C_1—O—C_5$ linkage gets converted to $C_1O_2H$ and $C_5{=}O$ and since on oxidation $C_4—C{=}O$ or $O{=}C_5—C_6$ linkage could cleave, the possible products are respectively optically active dimethoxysuccinic acid and meso or xylotrimethoxyglutaric acid from two types of cleavage; $C_4$ becomes $—CO_2H$ in the former cleavage (at b), while $C_5$ becomes $CO_2H$ in the latter type (at a).

Now the question is : what would have been the products if D-(+)-glucose had been a furanose ring structure? Let us see the possible products by drawing the similar pathway as drawn for the pyranose ring structure.

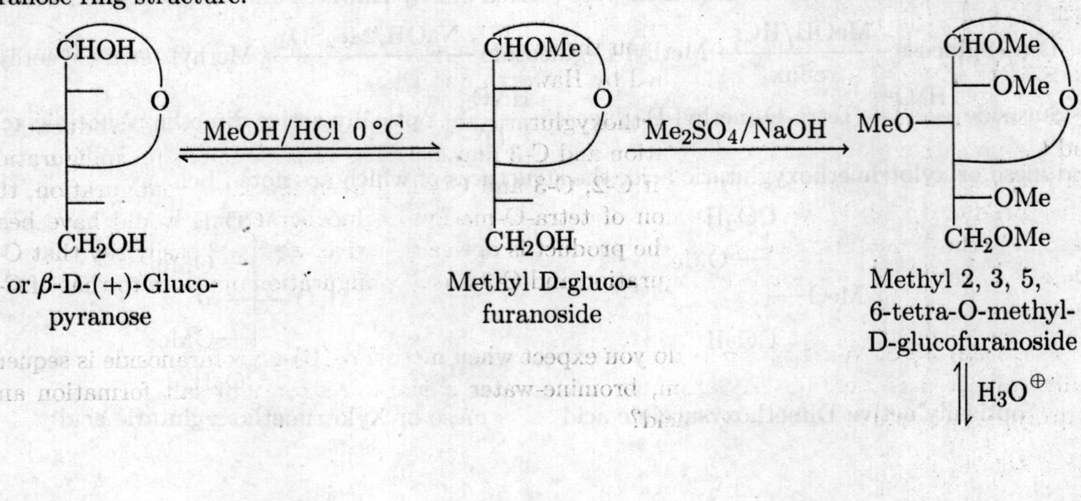

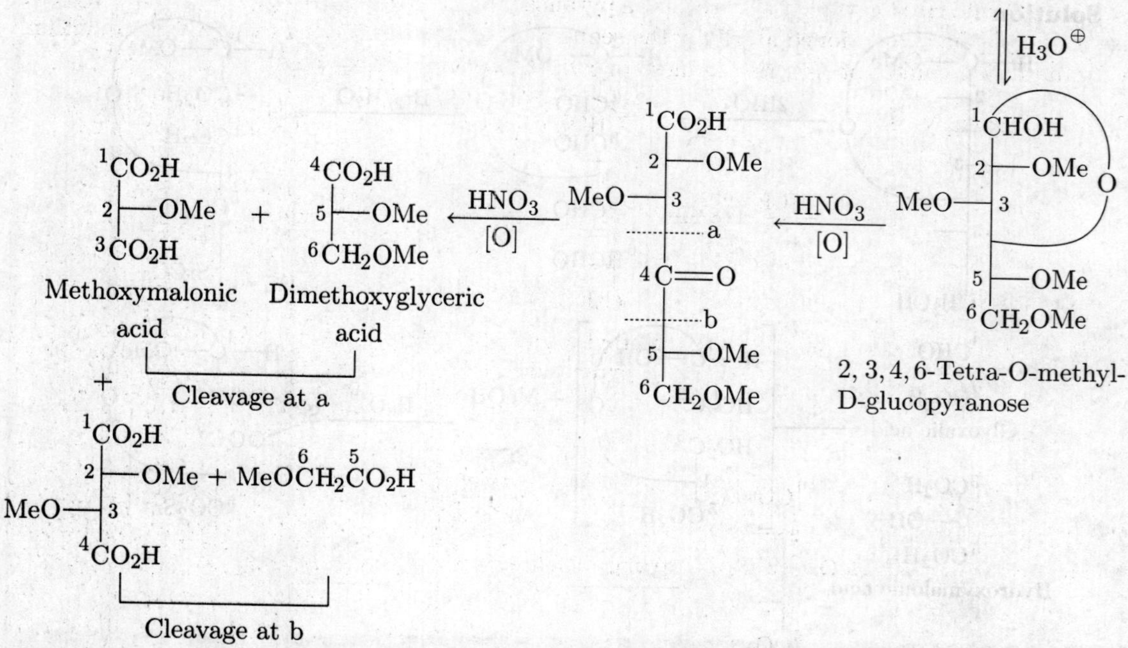


Thus, we see that $C_3$—$C_4$=O bond cleavage gives rise to methoxymalonic acid and O=$C_4$—$C_5$ bond cleavage gives rise to optically active dimethoxysuccinic acid and xylo or meso trimethoxyglutaric acid could not form from the furanose ring.

However, nowadays the size of the ring is determined by the periodic acid oxidation of glycosides as shown in Section 23.6.i(vi).

A methyl aldofuranoside (methyl glucofuranoside) on periodic acid oxidation gives no formic acid since the system does not contain C(OH)—C(OH)—C(OH) grouping (instead the system contains C(OH)—C(OH) group); on the other hand an alkyl aldopyranoside gives a molecule of formic acid per molecule of the substrate because of the presence of C(OH)—C(OH)—C(OH) moiety in the system. Thus when an alkyl glycoside on $HIO_4$ oxidation gives formic acid, the ring is a six-membered pyranose ring; and if no formic acid is found as one of the products, the ring is a five-membered furanose ring.

**Problem 23.7.** What light could you throw on the configuration of D-(+)-glucose from the oxidative degradation products obtained by Haworth and Hirst?

**Solution :** Since meso or xylotrimethoxyglutaric acid is obtained as one of the products C-2 and C-4 must have the same configuration and C-3 may have the same or opposite configuration with respect to C-2 and C-4. Again, if C-2, C-3 and C-4 had been the same configuration, the other product of oxidative degradation of tetra-O-methyl-D-glucopyranoside would have been meso dimethoxy succinic acid. Since the product is optically active, we could easily say that C-2 and C-4 have one and the same configuration and C-3 has a configuration opposite to that of C-2 and C-4.

**Problem 23.8.** What products do you expect when methyl $\alpha$-(D)-glucofuranoside is sequentially subjected to periodic oxidation, bromine-water oxidation, strontium salt formation and hydrolysis with dilute hydrochloric acid?

**Solution :**

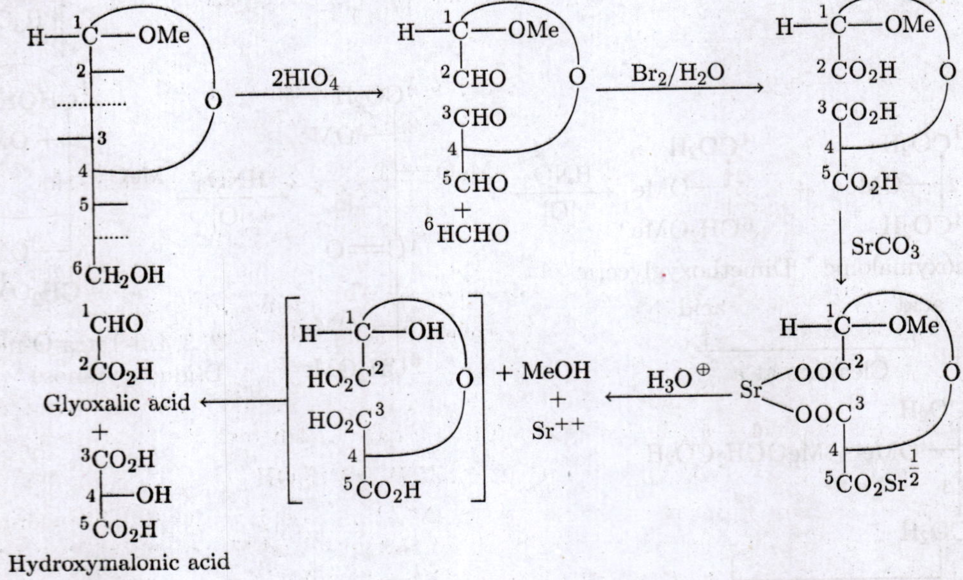

## 23.9 Elucidation of Structure of D-(−)-Fructose

From analysis and molecular weight determination the molecular formula for D-(−)-Fructose has been found to be $C_6H_{12}O_6$.

Fructose on successive reduction with Na-Hg and HI/red P forms *n*-hexane. Therefore, it contains an unbranched six-carbon chain.

Fructose reacts with acetic anhydride in pyridine and gives a pentaacetate which indicates the presence of five OH groups in a molecule of D-(−)-fructose. Fructose is enough stable and does not give up even a molecule of water on keeping. This tells us that no single carbon atom holds more than one OH group, i.e., each of the five carbon atoms bears one OH group.

D-(−)-fructose reacts with hydroxylamine to form an oxime; it forms dihydrazino derivative (D-glucosazone) on reaction with phenylhydrazine. So, there is a carbonyl group in the molecule. However, D-(−)- Fructose is resistant towards bromine-water oxidation. This indicates that the carbonyl group is not aldehydic but ketonic one. This is further proved by the fact that D-(−)-fructose on nitric acid oxidation gives a mixture of acids, each of which contains a fewer number of carbon atoms than the parent fructose; the nitric acid oxidation products are meso tartaric acid, glycolic acid and D-arabinotrihydroxyglutaric acid [See Sec. 23.6.i (iii) p. 1567].

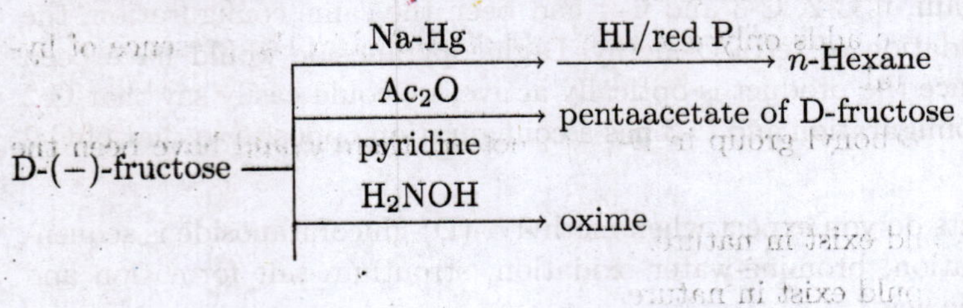

*Contd.*

D-(−)-fructose —
- $\xrightarrow{PhNHNH_2}$ dihydrazino derivative (D-glucosazone)
- $\xrightarrow{Br_2/H_2O}$ no oxidation
- $\xrightarrow[{[O]}]{HNO_3}$ Glycollic acid + meso tartaric acid + D-arabinotrihydroxyglutaric acid
- $\xrightarrow{HCN}$ $\xrightarrow{H_3O^{\oplus}}$ $\xrightarrow{HI/red\ P}$ 2-Methylhexanoic acid
- $\xrightarrow[reflux]{MeOH/HCl}$ Methyl fructoside $\xrightarrow{2HIO_4}$ Formic acid (one mol.)

Furthermore, fructose adds one molecule of HCN to give cyanohydrin which on hydrolysis followed by HI/red P reduction gives 2-methylhexanoic acid. The formation of a C-2 branched carboxylic acid indicates that the carbonyl carbon is C-2.

Taking all the above facts under consideration and keeping in mind the tetracovalency of a carbon atom, the unbrached open chain constitution for D-(−)-fructose can be written as

$$CH_2OHCOCHOHCHOHCHOHCH_2OH$$

Since D-(−)-Fructose forms the same osazone as D-(+)-glucose, it must be C-2 isomeric (with D-(+)- glucose) ketose having the configurations of C-3, C-4 and C-5 identical with those of D-(+)-glucose. Hence the open chain structure for D-(−)-fructose is

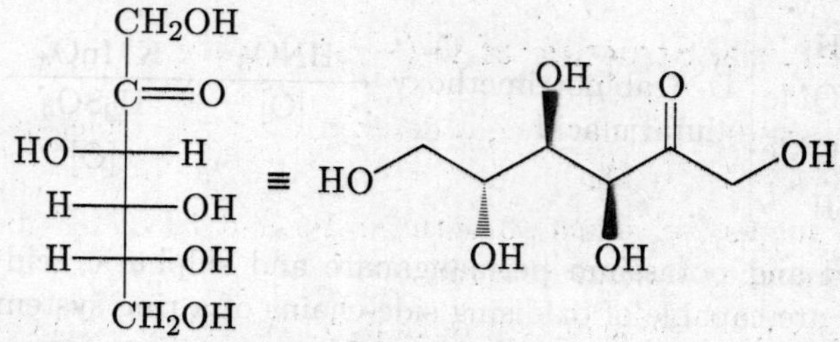

1, 3S, 4R, 5R, 6-pentahydroxyhexan-2-one

However, the open chain structure as written here is unable to explain a good many properties of D-(−)- fructose and thus it cannot be accepted as the exact structure for D-(−)-fructose. For example,

(i) There are two kinds of naturally occuring D-(−)-fructose, $\alpha$ and $\beta$.

(ii) There are two kinds of alkyl fructosides, $\alpha$ and $\beta$ forms.

(iii) It mutarotates.

(iv) It does add HCN but does not add sodium bisulphite.

(v) It reduces Fehling solution and Tollen's reagent.

(vi) One molecule of D-(−)-fructose adds only one molecule of alcohol in the presence of hydrochloric acid to form full ketal.

Had there been a free ketonic carbonyl group in D-(−)-fructose, there would have been the following information.

(a) Only one kind of fructose would exist in nature.

(b) Only one kind of fructoside would exist in nature.

**(c)** It would not mutarotate.

**(d)** It should add $NaHSO_3$.

**(e)** It should add two molecules of an alcohol to form a full ketal via hemiketal.

So, D-(−)-fructose probably does not contain a free ketonic carbonyl group and on the other hand, it may be a hemiketal since it adds only one molecule of an alcohol to form a full ketal. This is not impossible because the carbonyl group may undergo internal hemiketal formation with any one of its five hydroxyl groups. In such an eventuality the hemiketal must be a cyclic one having one O atom as the ring constituent. Haworth and Hirst's classical oxidative degradation and periodic acid oxidation of methyl D-fructosides have proved beyond doubt that there are six-membered pyran-like rings in D-fructoses. However, fructose obtained by the hydrolysis of sucrose has a five-membered furan-like ring. So, normally D-(−)-fructose is $\alpha$ or $\beta$-D-(−)-fructopyranose; but that in sucrose is $\beta$-D-(−)-fructofuranose. The classical oxidative degradation reactions and periodic acid oxidations on pyranose and furanose structures of D-(−)-fructose are shown below.

A. (1) The classical oxidative degradation reaction on a fructopyranose follows the following sequential path and ultimately gives the product as noted below :

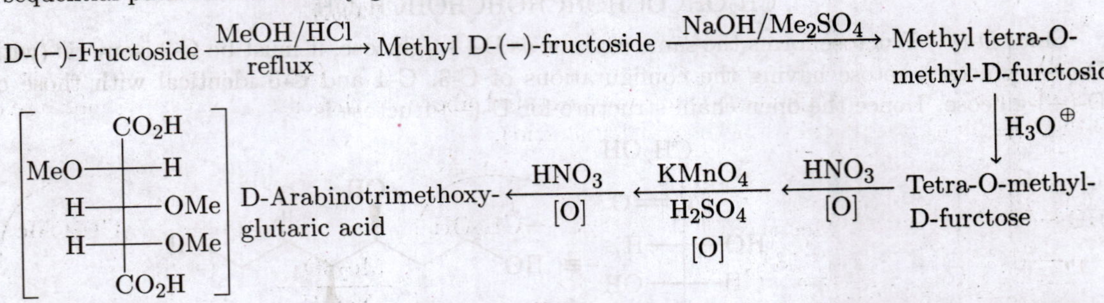


Nitric acid and potassium permanganate and sulphuric acid together are strong oxidising agents. They are capable of oxidising side-chains of a ring system to carboxyl groups which get further oxidised to carbon dioxide by the acidic potassium permanganate solution. Since a dicarboxylic acid having five carbon atoms in a chain is obtained from tetra-O-methyl-D-fructose, it may be concluded that the D-fructose possesses a six-membered pyranose ring structure. Furthermore the oxidative degradation product, D-arbinotrimethoxyglutaric acid, can only be reconciled if we consider D-fructose as a D-fructopyranose. The process is discussed with $\beta$-D-(−)-fructose.

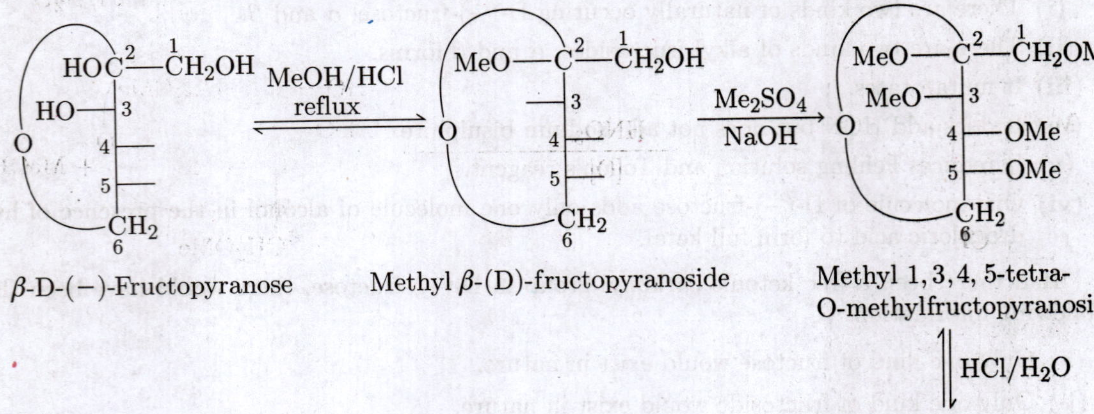

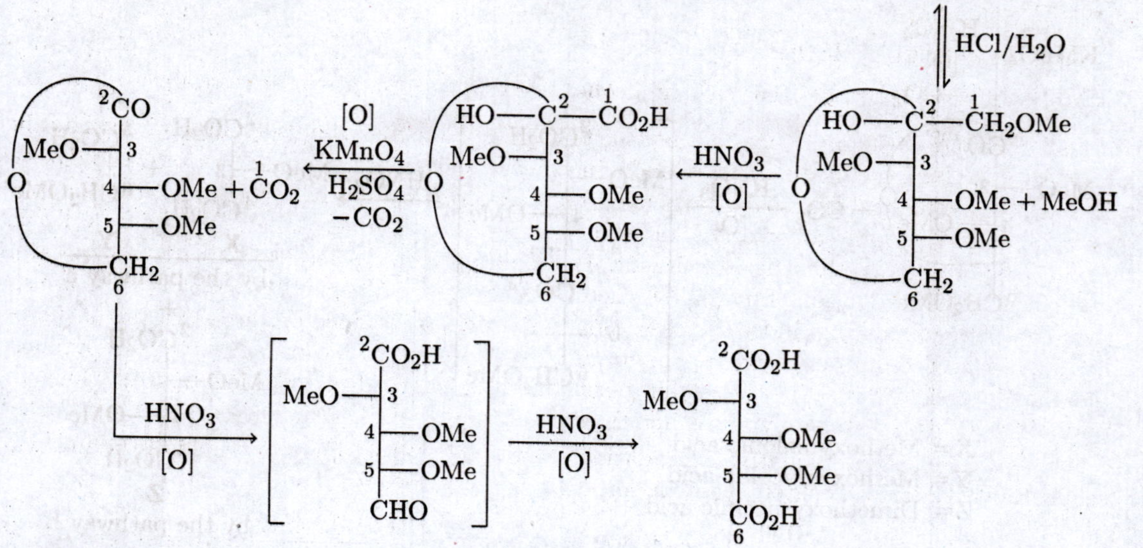

(2) The classical oxidative degradation on a fructofuranose ring (say, obtained by the hydrolysis of sucrose) : The sequential reactions as stated above in the case of fructopyranose give D-di-O-methyltartaric acid (dimethoxysuccinic acid).

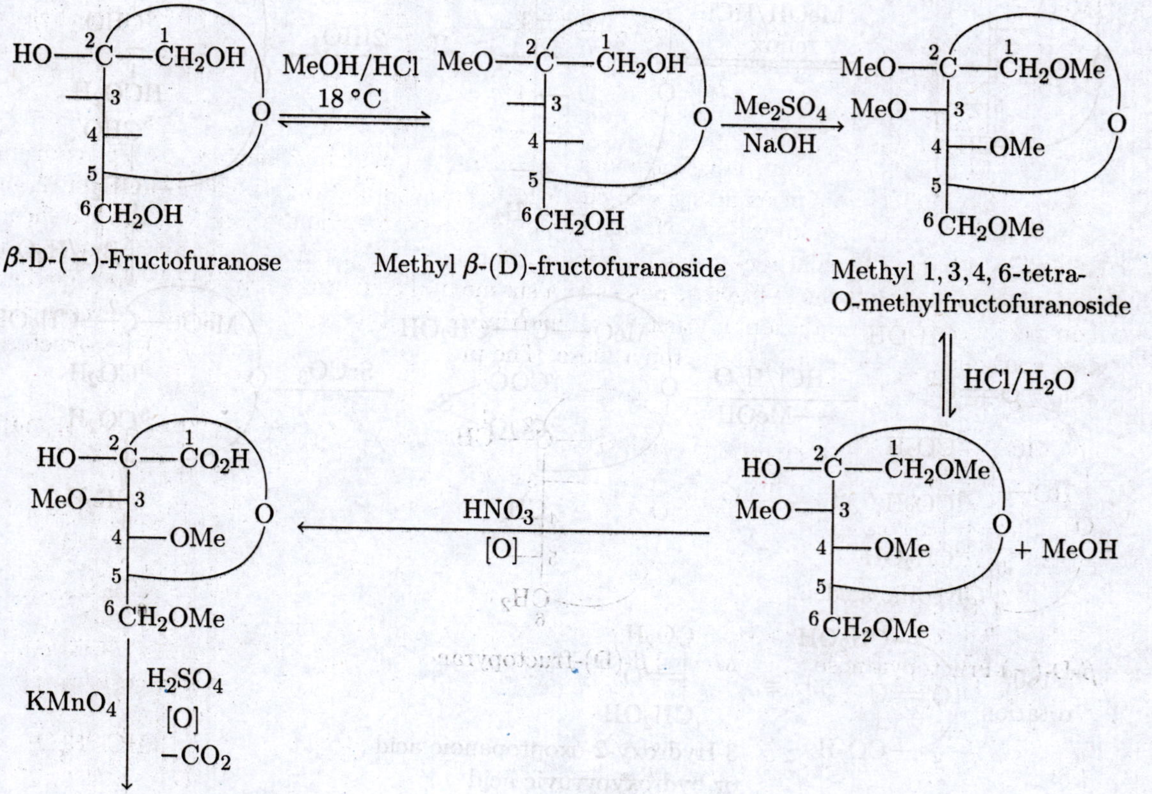

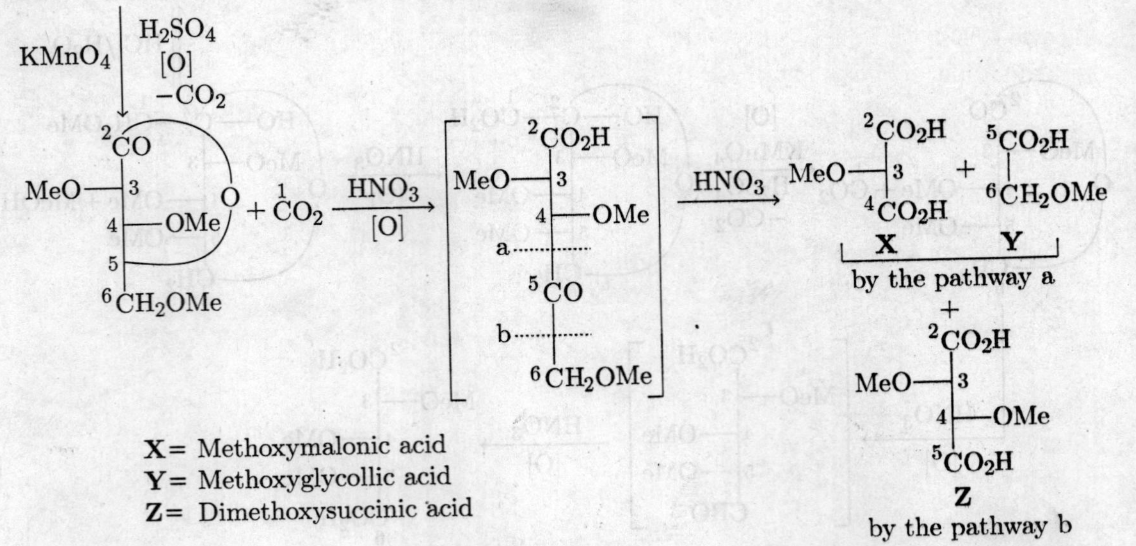

**X** = Methoxymalonic acid
**Y** = Methoxyglycollic acid
**Z** = Dimethoxysuccinic acid

B. (1) Periodic acid oxidation of D-(−)-fructopyranose :

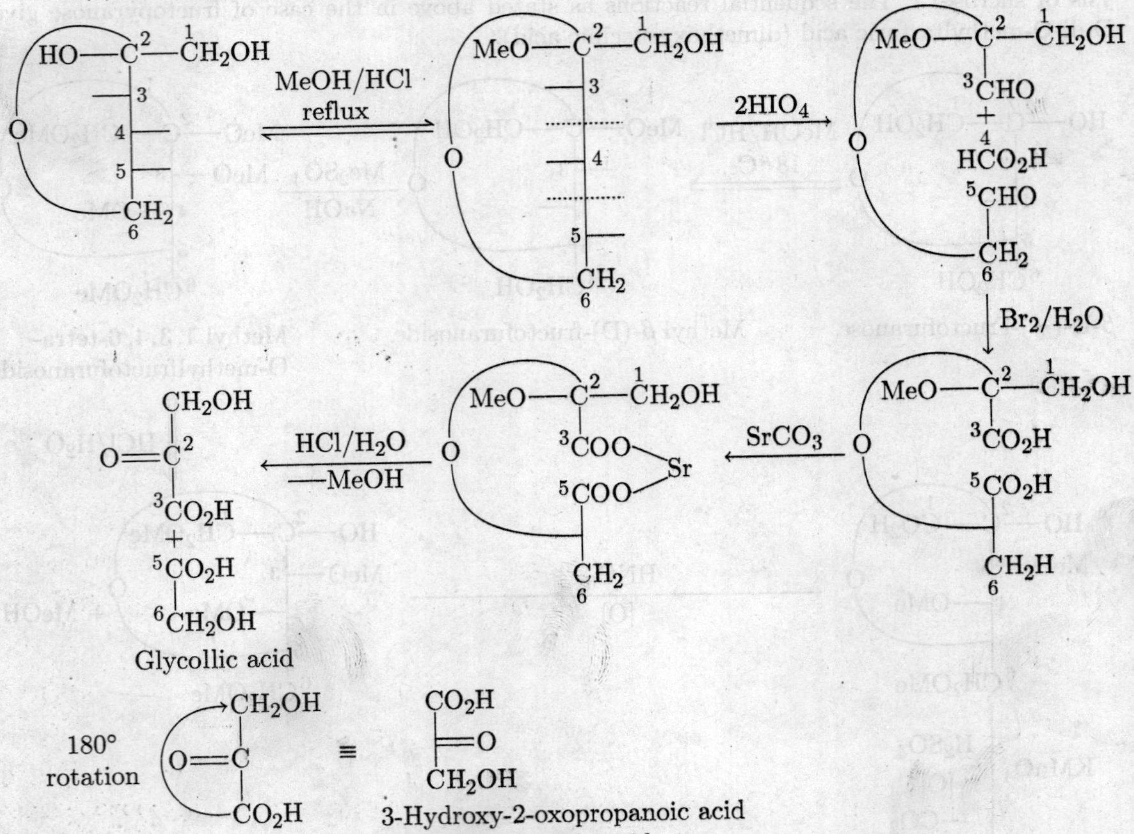

Glycollic acid

3-Hydroxy-2-oxopropanoic acid
or hydroxypyruvic acid

Here it is needless to mention that the formation of one molecule of formic acid from one molecule of D-fructose indicates the pyranose structure of fructose.

(2) Periodic acid oxidation of D-fructofuranose :

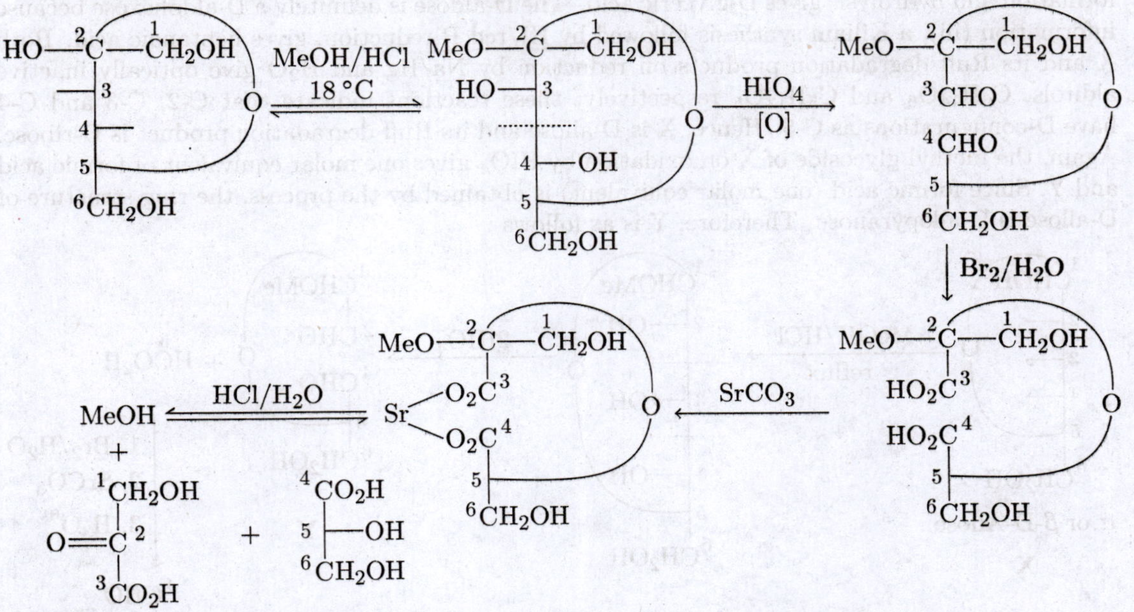

3-Hydroxy-2-oxo-propanoic acid or hydroxypyruvic acid (after rotating the molecule 180° on the plane of the paper) D-Glyceric acid

Since no formic acid is formed by the periodic acid oxidation of methyl fructoside, the starting fructose has a furanose structure.

**Problem 23.9.** A compound (X) of the molecular formula $C_6H_{12}O_6$ has the following properties

(i) X is optically active

(ii) $X \xrightarrow[\text{solution}]{\text{Fehling}} Cu_2O$ (red ppt.)

(iii) $X \xrightarrow{HCN} \xrightarrow{H_3O^{\oplus}} \xrightarrow[\Delta]{HI/\text{red P}}$ Heptanoic acid

(iv) $X \xrightarrow[H_2O]{Na/Hg}$ optically inactive $C_6H_{14}O_6$

(v) $X \xrightarrow[\text{degradation}]{\text{Ruff}} C_5H_{10}O_5 \xrightarrow[H_2O]{Na/Hg} C_5H_{12}O_5$, optically inactive

(vi) $X \xrightarrow[\text{reflux}]{MeOH/HCl} \xrightarrow{HIO_4}$ one molar equivalent formic acid + Y

(vii) $Y \xrightarrow{Br_2/H_2O} \xrightarrow{SrCO_3} \xrightarrow{HCl/H_2O}$ D-Glyceric acid

Suggest structures for X and Y.

**Solution :** The whole set of properties tells us that compound X is a carbohydrate and more particularly a monosaccharide. Since information (iii) tells us that the product is an unbranched acid, the monosaccharide is an aldose. The aldose is of D-series since Y on oxidation, salt formation and hydrolysis gives D-glyceric acid. The D-aldose is definitely a D-aldohexose because information (iii), a Kiliani synthesis followed by HI/red P reduction, gives heptanoic acid. Both X and its Ruff degradation products on reduction by Na/Hg and $H_2O$ give optically inactive alditols, $C_6H_{14}O_6$ and $C_5H_{12}O_5$ respectively; these reactions indicate that C-2, C-3 and C-4 have D-configurations as C-5. Hence, X is D-allose and its Ruff degradation product is D-ribose. Again, the methyl glycoside of X on oxidation by $HIO_4$ gives one molar equivalent of formic acid and Y. Since formic acid (one molar equivalent) is obtained by the process, the ring structure of D-allose is D-allopyranose. Therefore, Y is as follows :

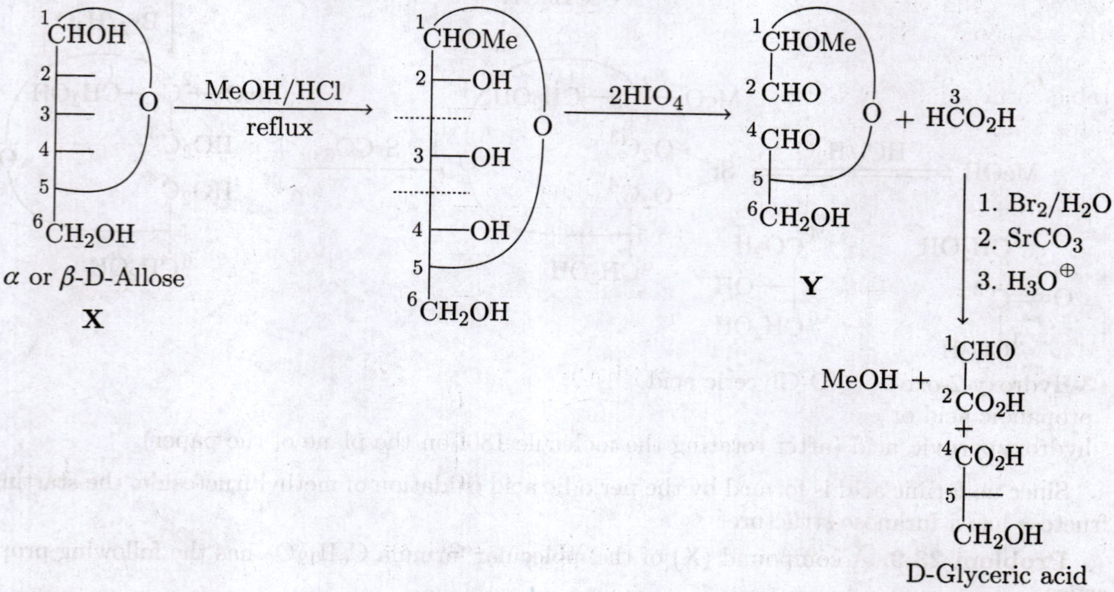


## 23.10 Disaccharides

Disaccharides usually have either one of the OH groups other than anomeric OH of one monosaccharide bonded to the anomeric carbon of another monosaccharide or the two anomeric carbons of the two monosaccharides are bonded through an oxygen atom. In other wards, disaccharides are glycosides. For example, (+)-sucrose is a disaccharide of $\alpha$-D-glucopyranose and $\beta$-D-fructofuranose and it is named either as $\alpha$-D-glucopyranosyl $\beta$-D-fructofuranoside or as $\beta$-D-fructofuranosyl $\alpha$-D-glucopyranoside since the glucose and the fructose units are bonded to each other by the anomeric carbons through an oxygen atom. (+)-Maltose has two D-glucose units, one involves the $\alpha$-OH of anomeric carbon and the other C-4 OH and accordingly (+)-maltose is named 4-O-($\alpha$-D-glucopyranosyl)-$\beta$-D-glucopyranose. Similarly, (+)-lactose is 4-O-($\beta$-D-galactopyranosyl)-$\beta$-D-glucopyranose since it is formed by the condensation of $\beta$-OH of anomeric carbon of D-galactose and C-4 OH of $\beta$-D-glucose. However, we shall discuss here the three disaccharides; (+)-sucrose (cane sugar or beat sugar), (+)-maltose (malt sugar) and (+)-lactose (milk sugar).

## 23.10.1 (+)-Sucrose (α-D-glucopyranosyl β-D-fructofuranoside or β-D-fructofuranosyl α-D-glucopyranoside)

The following chart represents a few of the properties of sucrose :

Sucrose $C_{12}H_{22}O_{11}$
$[\alpha]_D^{25} = +66.5°$ ($H_2O$)
m.p. 184-185 °C
stable form soluble in water

- hydrolysis by dilute acid or by the enzyme, invertase → D-(+)-glucose +D-(−)-fructose; Invert sugar $[\alpha]_D^{15} = -19.9°$
- Fehling's or Tollen's reagent → no reduction of $Cu^{++}$ or $Ag^+$
- Does not mutarotate
- $PhNHNH_2$ → no osazone
- $Br_2/H_2O$ [O] → no oxidation product
- $Me_2SO_4/NaOH$ → octa-O-methylsucrose
- $3HIO_4$ → one molar equivalent of $HCO_2H$
- $Ac_2O$/Pyridine → octaacetyl derivative
- $H_2NOH{\cdot}HCl$ → does not form oxime
- HCN → does not give addition product

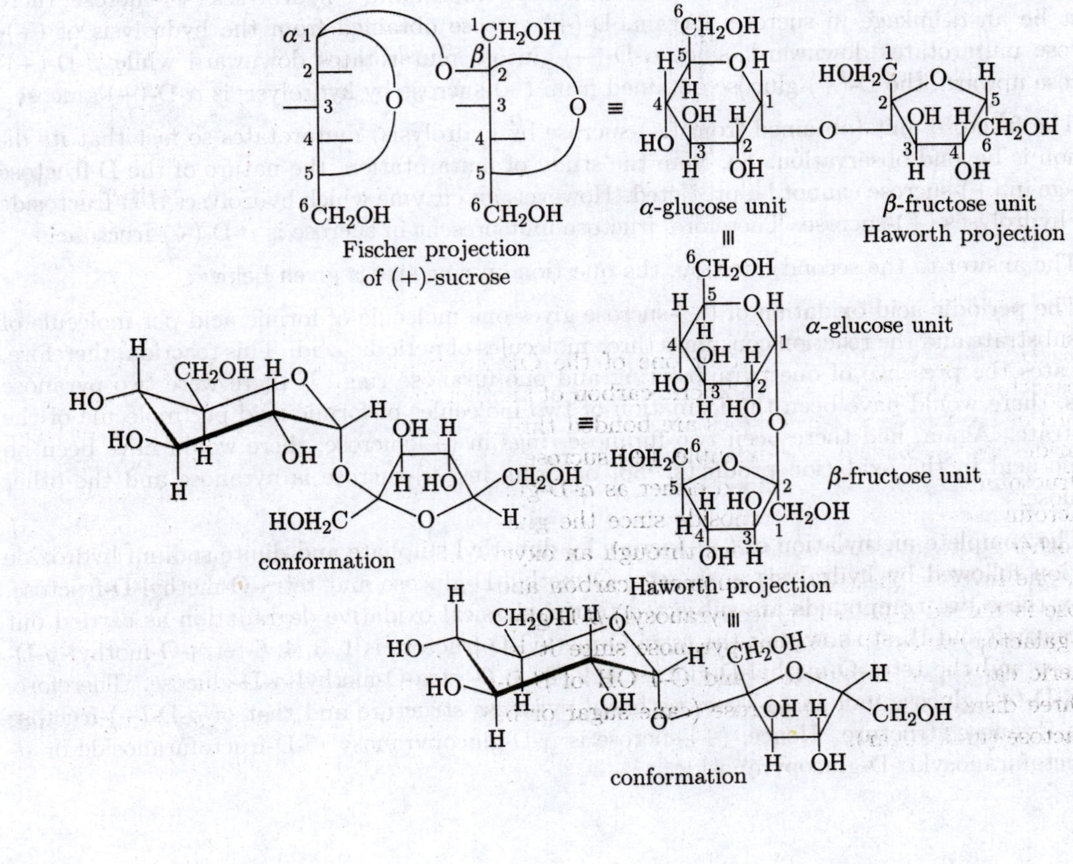

**Elucidation of the structure of (+)-sucrose :** From analysis and from molecular weight determination the molecular formula of (+)-sucrose has been found to be $C_{12}H_{22}O_{11}$.

On dilute acid hydrolysis or enzymatic hydrolysis of one molecule of (+)-sucrose, one molecule of D-(+)-glucose and one molecule of D-(−)-fructose are obtained. So, (+)-sucrose is a disaccharide formed by the condensation of D-(+)-glucose and D-(−)-fructose.

(+)-Sucrose does neither form oxime nor cyanohydrin nor even osazone by the reaction with hydroxylamine or hydrogen cyanide or phenylhydrazine respectively. Hence in the aqueous medium it has no free aldehydic and/or ketonic carbonyl groups. It does not also mutarotate in any medium. Therefore, it is neither a hemiacetal nor a hemiketal. If it were so, there would have been mutarotation, formation of oxime, cyanohydrin and osazone and the exhibition of reducing properties involving Fehling's reagent, Tollen's reagent, etc. Thus we conclude that the anomeric OH of D-glucose and that of the D-fructose take part in the reaction to develop the full acetal (ketal).

At this stage two questions regarding the structure of (+)-sucrose remain unsolved : (1) What is the nature of the linkage? Is that $\alpha$, $\alpha$(if $\alpha$-anomer of glucose combines with $\alpha$-anomer of fructose), or $\alpha, \beta$ or $\beta, \alpha$ or $\beta, \beta$?

(2) What are the sizes of the D-glucose ring and the D-fructose ring?

The answer to the first question is as follows :

Maltase, an enzyme, which hydrolyses an $\alpha$-linkage also hydrolyses (+)-sucrose. Since enzymes are optically active highly stereospecific reagents and since maltase hydrolyses (+)-sucrose, there must be an $\alpha$-linkage in sucrose. Again, D-(+)-glucose obtained from the hydrolysis of (+)-sucrose mutarotates downward; since $\alpha$-D-(+)-glucose mutarotates downward while $\beta$-D-(+)-glucose upward, the D-(+)-glucose obtained from (+)-sucrose by hydrolysis is $\alpha$-D-(+)-glucose.

The fructose unit (obtained from (+)-sucrose by hydrolysis) mutarotates so fast that its direction is beyond observation. So, from the study of mutarotation, the nature of the D-fructose linkage in (+)-sucrose cannot be predicted. However, an enzyme which hydrolyses $\beta$-D-fructoside also hydrolyses (+)-sucrose. Therefore, fructose unit present in sucrose is $\beta$-D-(−)-fructose.

The answer to the second question, the question on ring size is given below :

The periodic acid oxidation of (+)-sucrose gives one molecule of formic acid per molecule of the substrate and the reaction consumes three molecules of periodic acid. This reaction, therefore, indicates the presence of one pyranose ring and one furanose ring. If there were two pyranose rings, there would have been the formation of two molecules of formic acid per molecule of the substrate. Again, had there been two furanose rings in (+)-sucrose, there would have been no formic acid in the oxidation products. So, one ring in (+)-sucrose is pyranose and the other furanose.

The complete methylation of (+)-sucrose by dimethyl sulphate and dilute sodium hydroxide solution followed by hydrolysis gives tetra-O-methyl-D-glucose and tetra-O-methyl-D-fructose. When these two compounds are subjected to the classical oxidative degradation as carried out by Haworth and Hirst show that the tetra-O-methyl-D-fructose is 1, 3, 4, 6-tetra-O-mothyl-$\beta$-D-fructose and the tetra-O-methyl-D-glucose is 2, 3, 4, 6-tetra-O-methyl-$\alpha$-D-glucose. Therefore, the $\alpha$-D-(+)-glucose unit in (+)-sucrose has a pyranose structure and that of $\beta$-D-(−)-fructose has a furanose structure. Hence, (+)-sucrose is $\alpha$-D-glucopyranosyl $\beta$-D-fructofuranoside or $\beta$-D-fructofuranosyl $\alpha$-D-glucopyranoside.

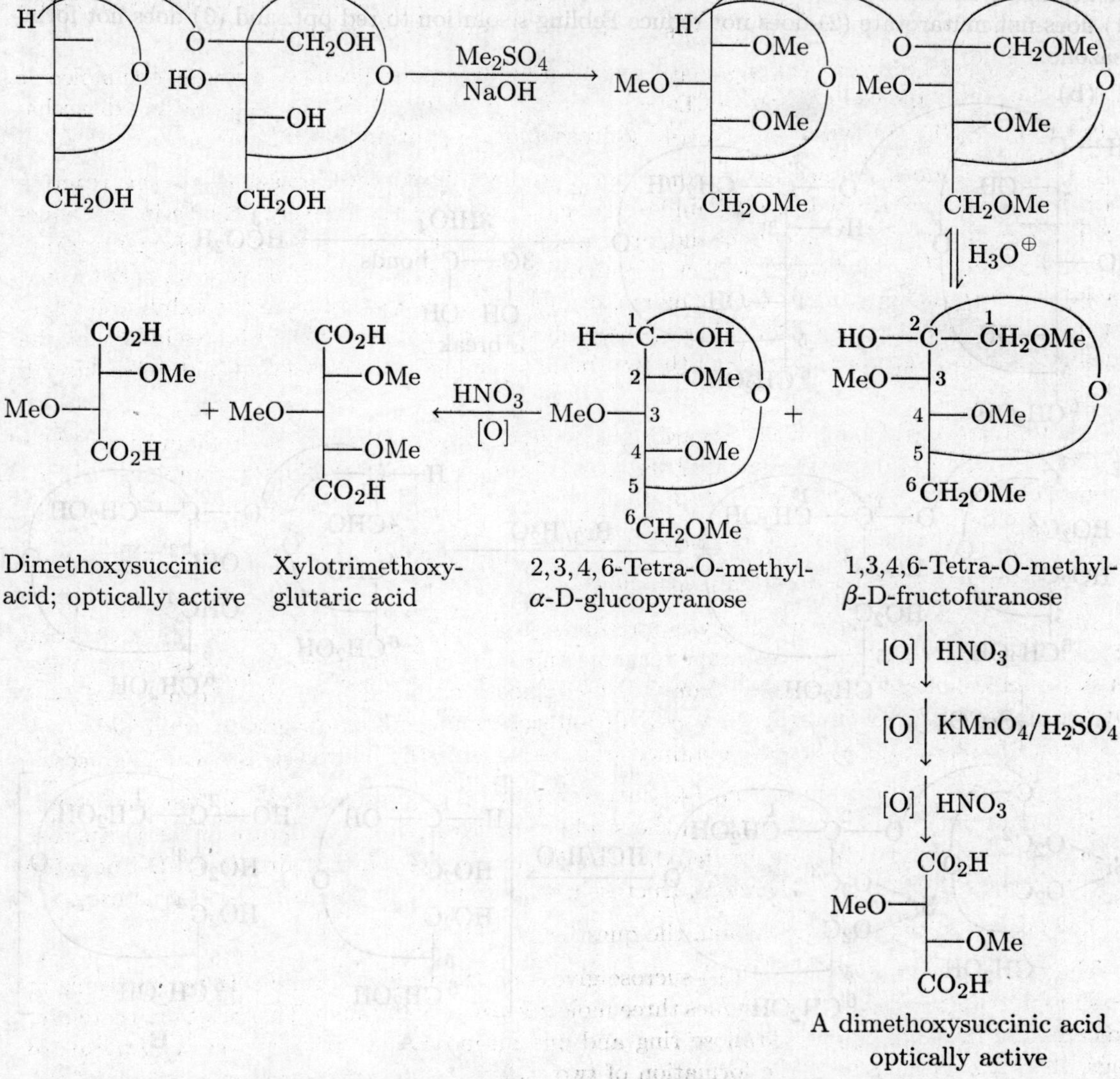


The isolation of meso or xylotrimethoxyglutaric acid from the D-glucose unit and dimethoxysuccinic form the D-fructose unit tell us that the glucose and fructose units are D-glucopyranose and D-fructofuranose respectively.

**Problem 23.10.** **(a)** Distinguish by a few chemical and physical tests between $\alpha$-D-(+)-glucose and methyl $\alpha$-D-glucoside. (b) What products are obtained when (+)-sucrose is subjected to periodic acid oxidation followed by $Br_2/H_2O$ oxidation, strontium salt formation and complete hydrolysis of the salt? **(c)** 2-Hydroxypyridine may be used as a catalyst for mutarotation of aldoses—give an explanation with mechanism.

**Solution :** **(a)** In the aqueous medium $\alpha$-D-(+)-glucose mutarotates, reduces Fehling's reagent and forms osazone with $PhNHNH_2$. Methyl $\alpha$-D-glucoside being a cyclic full acetal, it does not undergo ring-chain tautomerism in the aqueous or alkaline medium and as a consequence no open chain form having a free aldehyde group is obtained. So methyl α-D-glucoside (1) does not mutarotate (2) does not reduce Febling's solution to red ppt. and (3) does not form osazone.

**(b)**

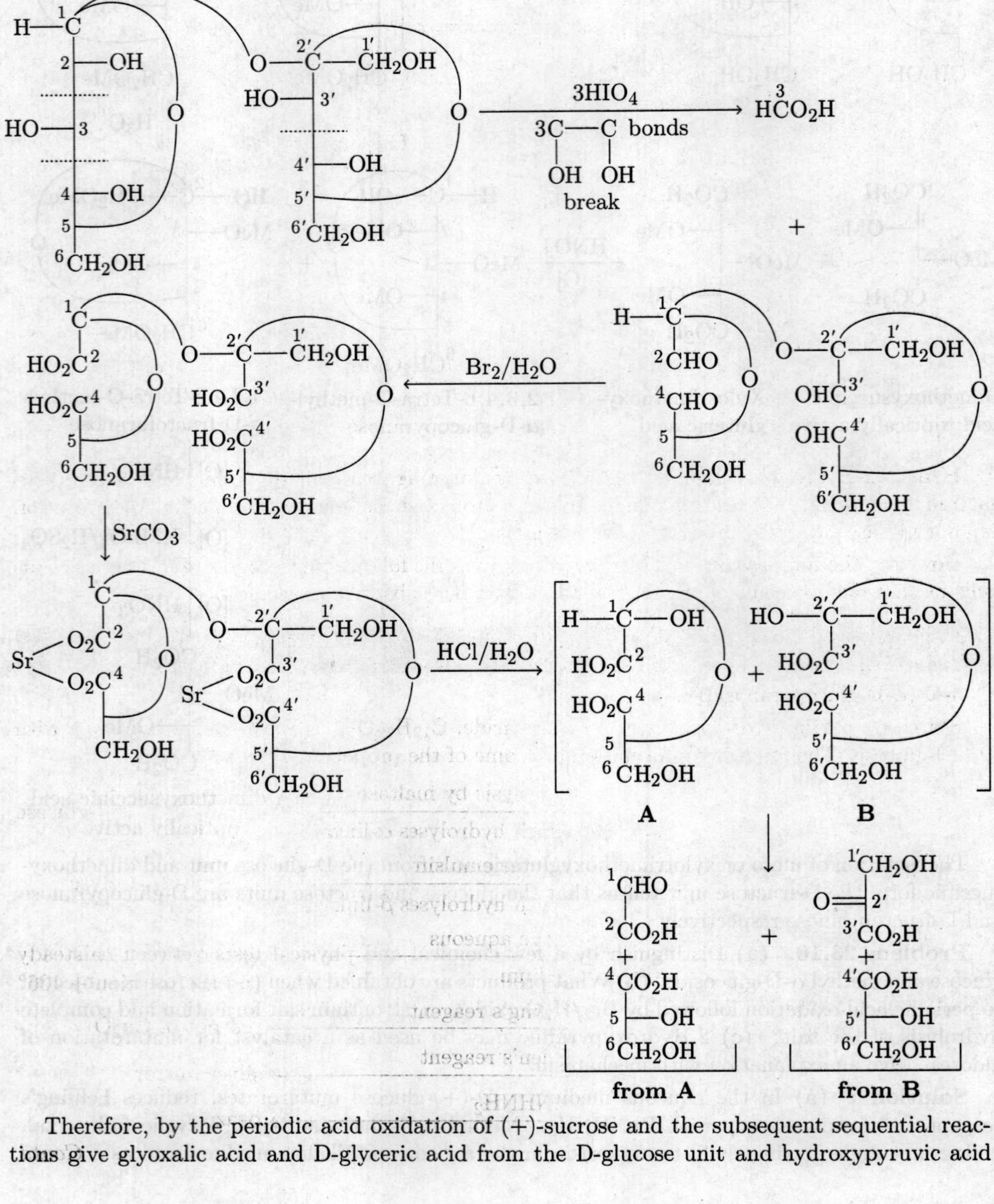


Therefore, by the periodic acid oxidation of (+)-sucrose and the subsequent sequential reactions give glyoxalic acid and D-glyceric acid from the D-glucose unit and hydroxypyruvic acid and D-glyceric acid from the fructose unit. Here it is important to mention that (+)-sucrose being a glycoside, a glucoside as well as a fructoside, (+)-sucrose does not require methylation by MeOH/HCl (it has no such scope also) prior to periodic acid oxidation.

**(c)**

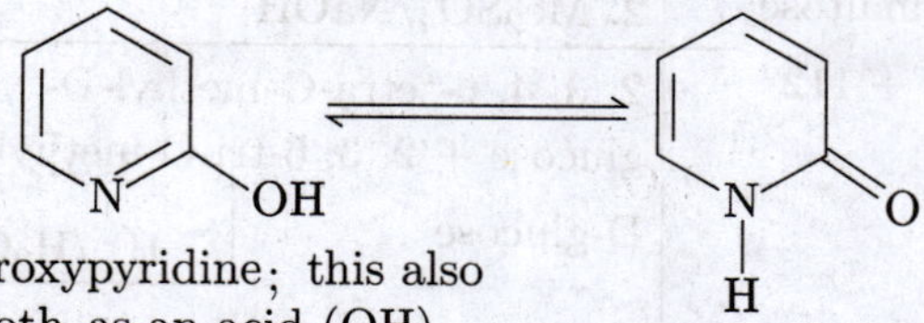

2-Hydroxypyridine; this also acts both as an acid (OH) and a base (N)

Lactum

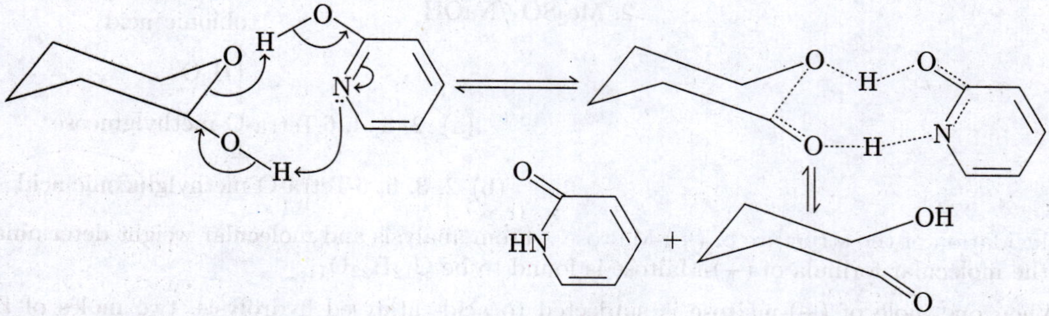

**Hint :** 2-Hydroxypyridine acts as a base by donating lone-pair of *es* on the N atom; as a lactum it can donate the H atom bonded to the N atom and thereby acts as an acid. Mutarotation is an acid-base catalysed process (draw mechanism).

However, the mutaration involving 2-hydroxypyridine follows the second order kinetics. This tells us that one molecule of the reagent takes part per substrate molecule.

## 23.10.2 (+)-Maltose

4-O-($\alpha$-D-glucopyranosyl)-D-glucopyranose.

Starch on partial hydrolysis gives a disaccharide, $C_{12}H_{22}O_{11}$, called (+)-maltose along with D-(+)-glucose. The following information are some of the properties of (+)-maltose.

(+)-Maltose
m.p. 102–103 °C
$\alpha$-(+)-maltose $\rightleftharpoons$ $\beta$-(+)-maltose
$[\alpha]_D = +168°$  $[\alpha]_D^{20} = +112°$
Equilibrium mixture has
$[\alpha]_D = +136°$

- hydrolysis by maltase (which hydrolyses $\alpha$-linkage) → 2 units of D-(+)-glucose
- enzyme emulsin (which hydrolyses $\beta$-linkage) → does not hydrolyse
- in the aqueous medium → mutarotates to a steady specific rotation + 136°
- Fehling's reagent → red ppt of $Cu_2O$
- Tollen's reagent → silver mirror
- $PhNHNH_2$ → osazone

*Contd.*

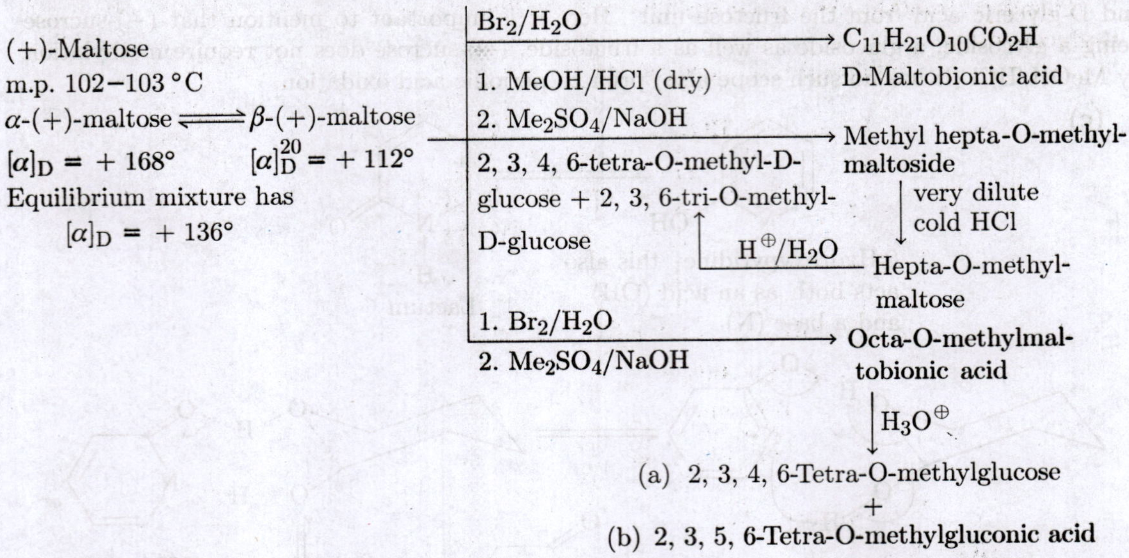


**Elucidation of the structure of (+)-Maltose :** From analysis and molecular weight determination the molecular formula of (+)-Maltose is found to be $C_{12}H_{22}O_{11}$.

When one mole of (+)-maltose is subjected to acid-catalysed hydrolysis, two moles of D-glucose are obtained. So, it is a condensation product of two units of D-glucose. (+)-Maltose reduces Fehling solution and Tollen's reagent; it also mutarotates. Therefore, there is at least one hemiacetal linkage.

(+)-Maltose gets oxidised to a monocarboxylic acid, D-maltobionic acid. This indicates that there is only one hemiacetal linkage.

When (+)-maltose is treated with methyl alcohol and hydrogen chloride, methyl maltoside is formed. Methyl maltoside on treatment with dimethyl sulphate and sodium hydroxide solution, methyl hepta-O-methylmaltoside is formed; the latter on hydrolysis by very dilute acid gives hepta-O-methylmaltose which on hydrolysis under stronger condition gives 2, 3, 4, 6-tetra-O-methyl-D-glucose and 2, 3, 6-tri-O-methyl-D-glucose (known by the classical oxidative degradation reaction). This has also been supported by the fact that D-maltobionic acid, which is obtained by the bromine-water oxidation of (+)-maltose, on complete methylation with dimethyl sulphate and dilute sodium hydroxide solution followed by hydrolysis with acid catalyst gives 2, 3, 4, 6-tetra-O-methyl-D-glucose (a) and 2, 3, 5, 6-tetra-O-methyl-D-gluconic acid (b). This experiment indicates that the first product (a) is obtained from a pyranose ring which has a full acetal linkage because C-5 OH of the compound (a) is free and it does not get oxidised to gluconic acid. This experiment also tells us that the second compound (b) bears a free C-4 OH group, and a carboxyl group. This is possible if this is obtained from a glucopyranose unit in the hemiacetal form and linked to the first glucose unit through its C-4 OH group. Thus the glycosidic linkage is formed by the reaction between anomeric OH of one D-glucopyranose and C-4 OH of another D-glucopyranose, the latter unit is in the hemiacetal form while the former is a full acetal.

Lastly (+)-maltose is hydrolysed by maltase, an enzyme which hydrolyses α-glycosidic linkage and (+)-maltose does not hydrolyse by emulsin, which hydrolyses a β-glycosidic linkage. Therefore, (+)-maltose contains an α-glycosidic linkage and no β-glycosidic linkage.

**Therefore, from the above facts we can conclude that (+)-maltose is 4-O-($\alpha$-D-glucopyranosyl)-D-glucopyranose.**

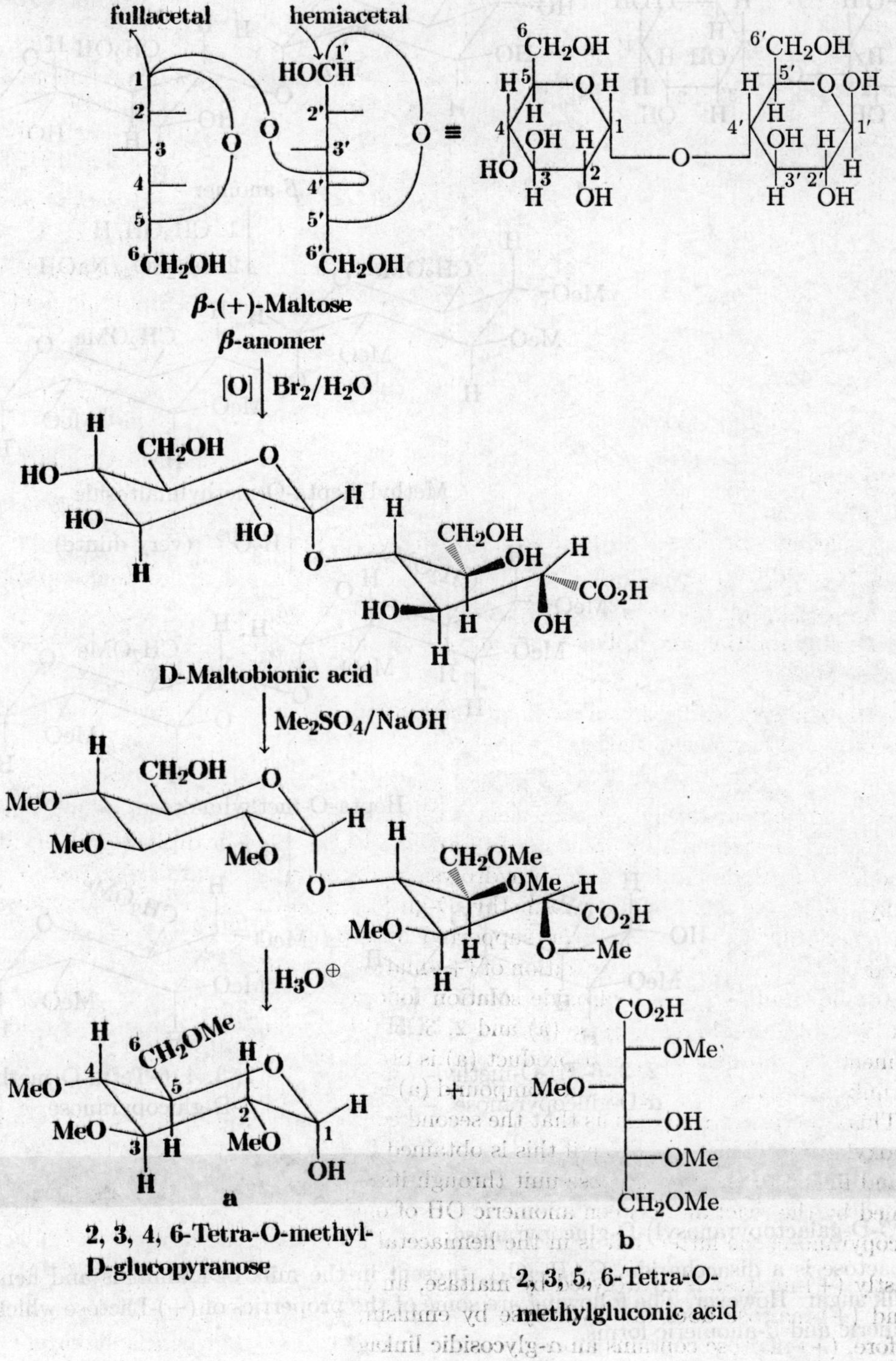

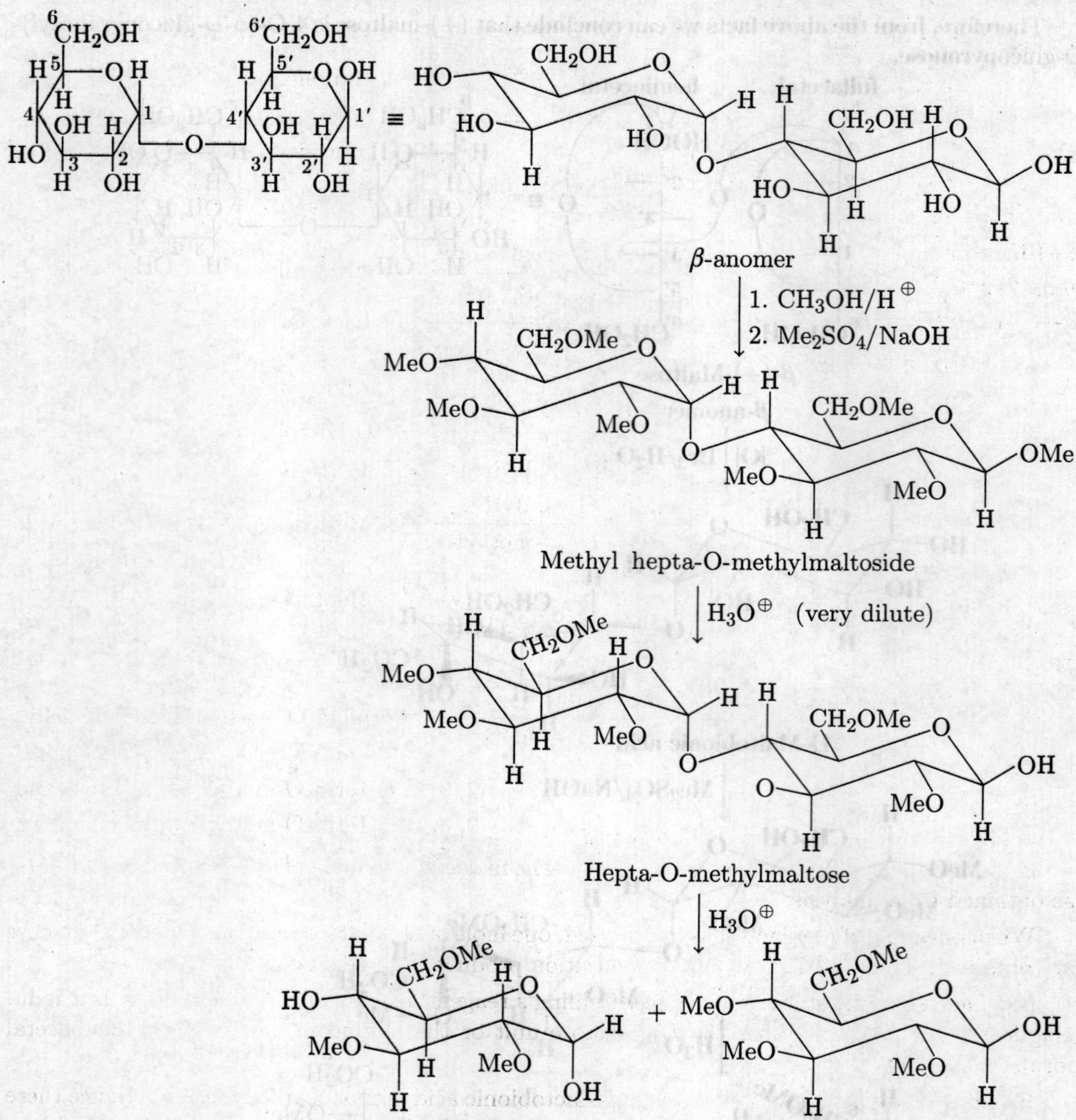

β-anomer

1. $CH_3OH/H^{\oplus}$
2. $Me_2SO_4/NaOH$

Methyl hepta-O-methylmaltoside

$H_3O^{\oplus}$ (very dilute)

Hepta-O-methylmaltose

$H_3O^{\oplus}$

2, 3, 6-Tri-O-methyl-α-D-glucopyranose + 2, 3, 4, 6-Tetra-O-methyl-β-D-glucopyranose

## 23.10.3 (+)-Lactose

4-O-(β-D-galactopyranosyl)-D-glucopyranose.

(+)-Lactose is a disaccharide, $C_{12}H_{22}O_{11}$, present in the milk of mammals and hence it is called milk sugar. However, The following are some of the properties of (+)-Lactose which exists in α-anomeric and β-anomeric forms.

(+)-Lactose
m.p. 252 °C
$\alpha$-(+)-Lactose $\rightleftharpoons$ $\beta$-(+)-Lactose
$[\alpha]_D^{20} = +52.5°$

- hydrolysis by emulsin which hydrolyses $\beta$-linkage → one unit of $\beta$-(D)-galactose and one unit of D-glucose
- the aqueous medium → mutarotates to a steady specific rotation : + 52.5°
- Fehling's reagent → red ppt of $Cu_2O$
- Tollen's reagent → silver mirror
- $PhNHNH_2$ → Lactosazone $\xrightarrow{H_3O^{\oplus}}$ D-Glucosazone + $\beta$-(D)-Galactose
- $Br_2/H_2O$ → D-Lactobionic acid ($C_{11}H_{21}O_{10}CO_2H$)
- 1. MeOH/HCl (dry) 2. $Me_2SO_4$/NaOH → Methyl hepta-O-methyllactoside $\xrightarrow{H_3O^{\oplus}\ (\text{dilute})}$ Hepta-O-methyllactose + MeOH $\xrightarrow{H_3O^{\oplus}}$ 2, 3, 6-Tri-O-methyl-D-glucose + 2, 3, 4, 6-Tetra-O-methyl-D-galactose
- 1. $Br_2/H_2O$ 2. $Me_2SO_4$/NaOH → Octa-O-methyllactobionic acid $\xrightarrow{H_3O^{\oplus}}$ 2, 3, 4, 6-Tetra-O-methyl-D-galactose and 2, 3, 5, 6-Tetra-O-methyl-D-gluconic acid

**Elucidation of the structure of (+)-lactose :** The molecular formula of (+)-lactose is $C_{12}H_{22}O_{11}$ as obtained from analysis.

When one mole of (+)-lactose is hydrolysed, one mole of D-galactose and one mole of D-glucose are obtained. Thus (+)-lactose is a condensation product of D-glucose and D-galactose.

(+)-Lactose mutarotates and reduces Fehling's reagent and Tollen's reagent. So, it is a reducing sugar and both or either the D-galactose unit or the D-glucose unit is in the hemiacetal form.

Bromine-water oxidises (+)-lactose to D-lactobionic acid, a monocarboxylic acid. Hence there is one hemiacetal unit, either the D-galactose unit or the D-glucose unit. Lactobionic acid on complete methylation with dimethyl sulphate and sodium hydroxide solution followed by the hydrolysis with an acidic catalyst gives two compounds, 2, 3, 4, 6-tetra-O-methyl-D-galactose (x) and 2, 3, 5, 6-tetra-O-methyl-D-gluconic acid (y). Therefore, the D-glucose unit is in the hemiacetal form and the D-galactose unit is in the full acetal form.

This is also proved by the fact that lactosazone on hydrolysis produces D-galactose and D-glucosazone. However, the hydrolysis of octa-O-methyllactobionic acid and identification of the compounds $x$ and $y$ as above indicate that (1) D-galactose has a pyranose ring in the full acetal form because C-5 OH is free and (2) D-glucose unit is bonded to D-galactose unit by C-4 OH since this OH is free.

(+)-Lactose on complete methylation gives methyl hepta-O-methyllactoside which on stepwise acid-catalysed hydrolysis gives 2, 3, 6-tri-O-methyl-D-glucose and 2, 3, 4, 6-tetra-O-methyl-D-galactose indicating that both the rings are pyranose rings since in both the cases C-5 OH groups are free in addition to C-4 OH of the D-glucose unit.

(+)-Lactose gets hydrolysed by the enzyme lactase (which has been said to be equivalent to emulsin) and emulsin which hydrolyses a $\beta$-glycosidic linkage. Hence, D-galactose is $\beta$-D-galactose because the D-galactose unit is in the full acetal form as it has been concluded from the hydrolysis of the completely methylated lactobionic acid.

Taking all the above facts under consideration we can say that (+)-lactose has the following structure :

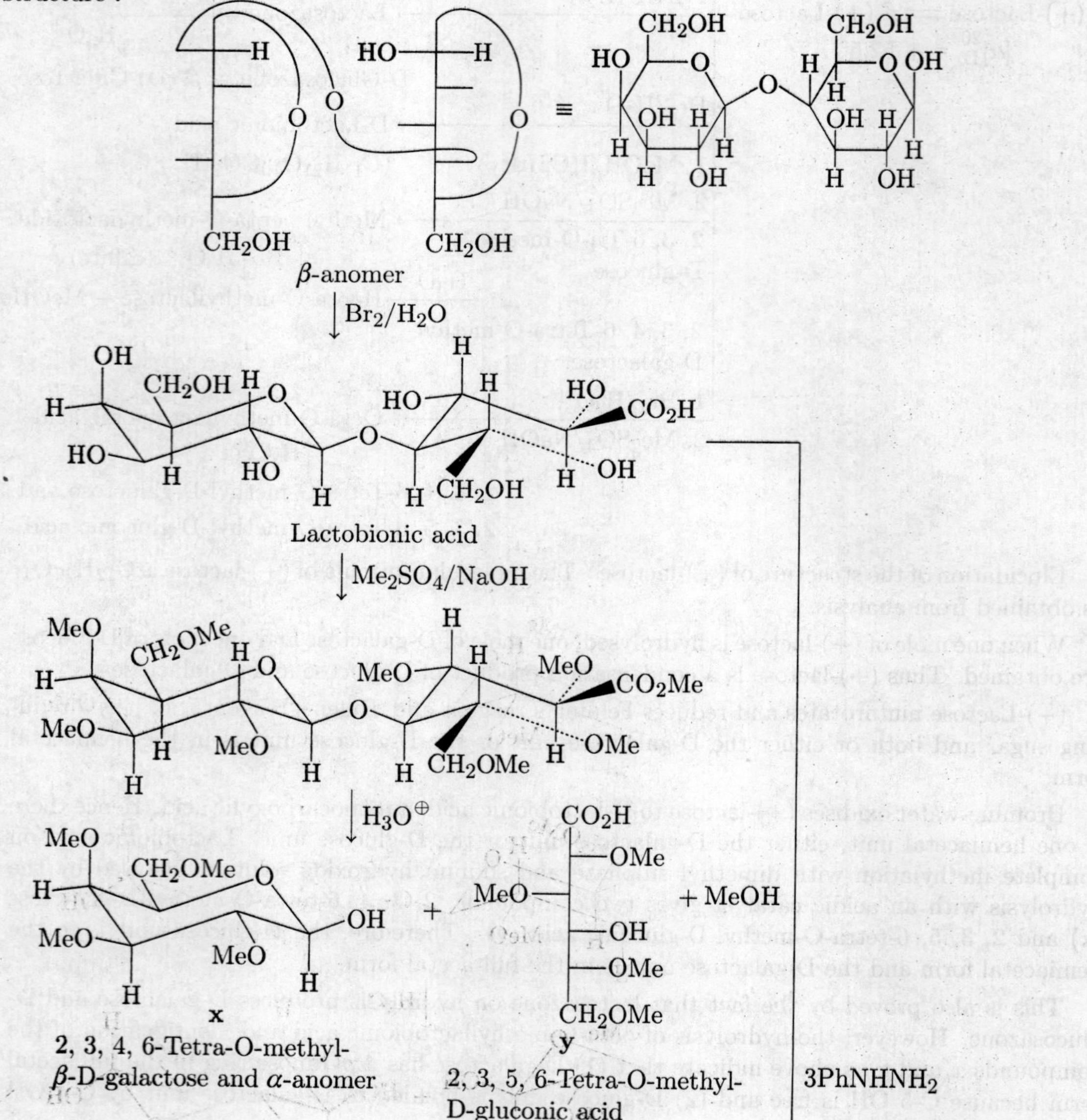


2, 3, 4, 6-Tetra-O-methyl-$\beta$-D-galactose and $\alpha$-anomer

2, 3, 5, 6-Tetra-O-methyl-D-gluconic acid

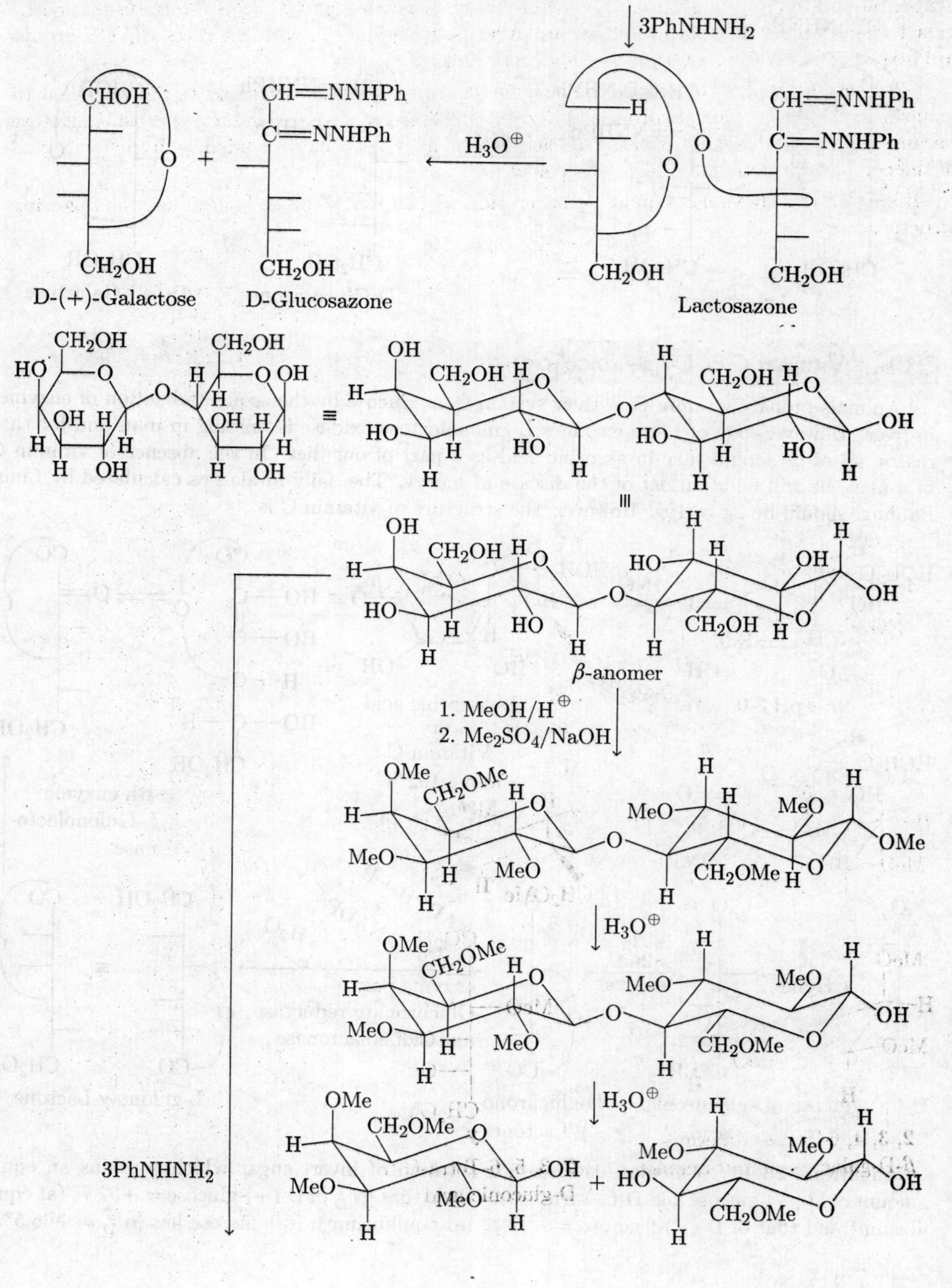
3PhNHNH2
H3O⊕
D-(+)-Galactose
D-Glucosazone
Lactosazone
β-anomer
1. MeOH/H⊕
2. Me2SO4/NaOH
H3O⊕
H3O⊕
3PhNHNH2

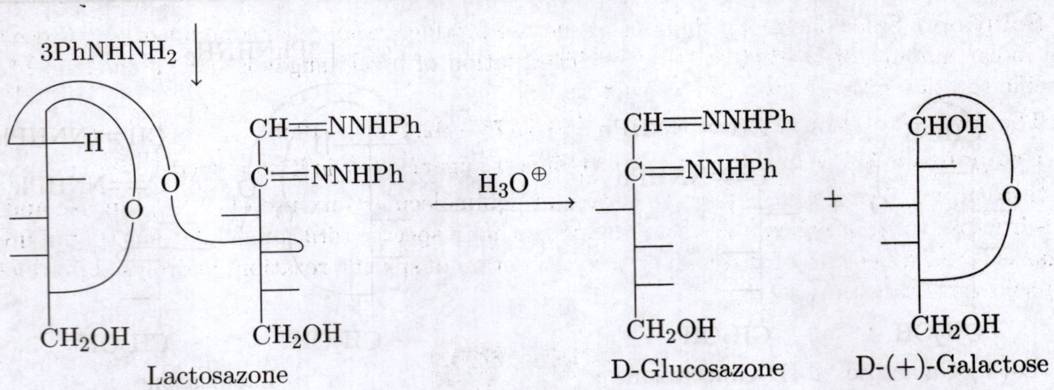


## 23.11 Vitamin C or L-Ascorbic Acid

Animals prepare vitamin C in their system from glucose by the sequential action of enzymes in liver. However, one of these enzymes L-gulonolactose oxidase is missing in man and for that reason all of us should take in ascorbic acid as a part of our diet. In the absence of vitamin C in a man, he will be a patient of the disease of scurvy. The daily intakes as calculated by Linus Pauling, should be 2g to 10g. However, the structure of vitamin C is :

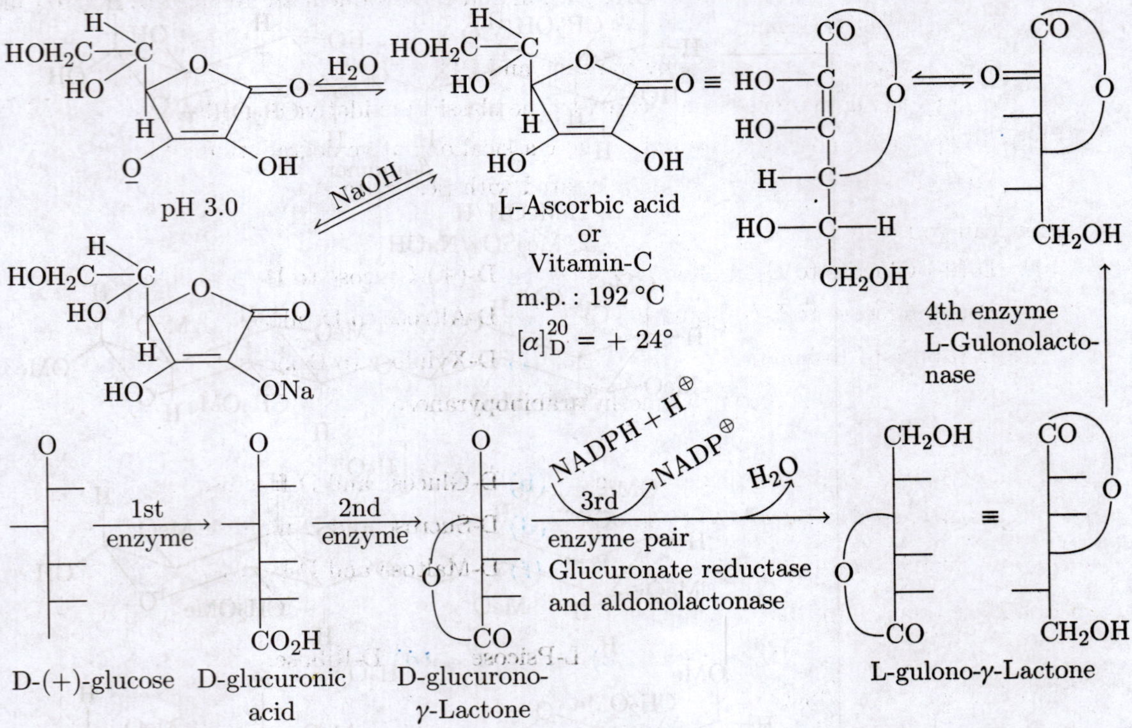


**Problem 23.11.** Calculate the specific rotation of invert sugar which contains an equal amount of D-(+)-glucose and D-(−)-fructose. Given that $[\alpha]_D^T$ of D-(+)-glucose = +52.7° (at equilibrium) and that of D-(−)-fructose = −92.4° (at equilibrium); (+)-sucrose has $[\alpha]_D^T$ = +66.5°.

**Solution :** Since one molar amount of sucrose produces one molar amount of D-glucose and one molar amount of D-fructose, the specific rotation of invert sugar is half of the sum of the specific rotation of D-glucose and D-fructose.

The specific rotation of invert sugar $= \frac{1}{2}(+52.7° - 92.4°) = -19.9°$.

**Problem 23.12.** What do you mean by invert sugar? Why is it so named?

**Solution :** Sucrose on hydrolysis forms an equimolecular mixture of D-(+)-glucose and D-(−)- fructose which is called invert sugar. Sucrose has a specific rotation +66.5° but that of invert sugar is −19.7°; because of the change in the direction of specific rotation, hydrolysed product of (+)-sucrose is called invert sugar.

## Exercise

1. Draw the family tree for L-aldoses up to six C atoms and name them.
2. What happens when the following reactions are carried out?
   (a) L-(+) Arabinose is treated with an alkali solution (give mechanism).
   (b) L-(+)-Arabinose is subjected to Kiliani synthesis and the products are oxidised by dilute and hot nitric acid.
   (c) L-Talose is subjected to Ruff degradation and the products are reduced by $NaBH_4$ in the aqueous medium.
   (d) D-(−)-Gulose is epimerised by pyridine and HCl.
   (e) α-D-(−)-Gulopyranose is subjected to the classical oxidative degradation.
   (f) β-D-psicofuranose is subjected to the classical oxidative degradation.
   (g) Methyl α-D-arabinopyranoside is treated with periodate.
3. How can you convert?
   (a) D-(+)-Glucose to D-(+)-mannose (b) D-(+)-Glucose to D-(+)-allose
   (c) D-(+)-Glucose to L-(+)-gulose (d) D-Altrose to D-talose
   (e) D-Ribose to D-ribulose (f) D-Xylulose to D-xylose
   (g) D-Arabinose to 2, 3, 4-tri-O-methylarabinopyranose.
4. Distinguish between
   (a) D-Glucose and methyl D-glucoside (b) D-Glucose and D-fructose
   (c) D-Glucose and sucrose (d) D-Sucrose and D-maltose
   (e) D-Sucrose and D-lactose (f) D-Maltose and D-lactose.
5. Elucidate the structure of
   (a) D-Arabinose (b) D-Altrose (c) L-Psicose (d) D-Ribose.
6. Write down the products
   (a) D-Maltose $\xrightarrow{3PhNHNH_2}$ Maltosazone $\xrightarrow{H_3O^{\oplus}}$
   (b) α-D-Glucofuranosyl β-D-fructopyranoside $\xrightarrow{3HIO_4}$




   



# Appendix A

# Isolation and Purification of Organic Compounds

The aim of this chapter is to discuss some of the more common methods of securing a pure organic compound from natural source or from reaction product mixture. They never contain a single chemical substance; instead a large number of different compounds, more often families of similar compounds are found. Therefore, the processes are highly technical and need theoretical background.

Isolation of a pure compound from its source consists of three steps : (1) isolation, (2) purification, and (3) test of purity.

Isolation of an organic compound from its source (plant, animal, coal tar, petroleum, etc.) means its separation or extraction in a crude form containing many other substances. Dialysis, diffusion, ultracentrifugation, electrophoresis, zone melting are the modern processes, and this book has no scope to discuss them. On the other hand, the more common Soxhlet extraction and subsequent liquid-liquid extraction or the more modern chromatographic separation will be discussed in this book.

The substances which contaminate a desired compound are called **impurities** and the process of removal of impurities from the concerned compound is called **purification**. Without purification no compound could be subjected to structure elucidation. The more common methods of purification are: (1) Liquid-liquid extraction, (2) Chromatographic separation, (3) Recrystallisation, (4) Sublimation, (5) Distillation at atmospheric pressure, (6) Distillation under reduced pressure, (7) Steam distillation, etc. The last four processes are also used in isolation step.

Therefore, our sequence of discussing the processes are : (I) Solvent extraction using Soxhlet extraction apparatus, (II) Liquid-liquid extraction, (III) Chromatography, (IV) Recrystallisation, (V) Sublimation, (VI) Distillation at atmospheric pressure, (VII) Distillation under reduced pressure, (VIII) Steam distillation, and (IX) Chemical method.

These processes necessitate an extensive knowledge on the relation between the structure and properties of a compound, intermolecular forces in organic compounds, and their solubilities, melting points, boiling points, etc.

So, the students are required to go through the fourth and seventh chapters of this book before studying in detail about the isolation and purification processes.

## Solvent Extraction

### Soxhlet extraction

By this process nonvolatile organic solids and liquids are separated from the dead plant and animal bodies. Organic compounds being soluble in organic solvents, a hot volatile organic solvent is allowed to percolate through the dried sample to dissolve out the soluble organic compounds leaving other insoluble compounds behind. Choice of the solvent depends upon the nature of the compound to be isolated. The commonly used solvents for this purpose are given :

**Some common organic solvents** : 1. Carbon tetrachloride, 2. *n*-Hexane, 3. Petroleum ether, 4. Benzene, 5. Ethyl ether, 6. Chloroform, 7. Acetone, 8. Ethyl acetate, 9. Ethyl alcohol, 10. Toluene, etc.

Generally a very large amount of a raw material contains a fraction of the desired compound. Naturally the isolation of it in a reasonable quantity requires a fantastic volume of the expensive solvent. Soxhlet extraction apparatus (Fig. 1) needs only a small volume of the solvent because continuous extraction of the sample is done by refluxing the solvent. **Reflux** means flow back. The solvent boils and its vapour gets condensed and flows back to the liquid above the raw material.

**This hot liquid then percolates through the raw material** extracting the desired compound mixed **with other soluble substances** and accumulates in the boiling flask as a solution. The solvent **being continuously volatilised,** recondensed and repercolated through the sample, a concentrated solution of the compound and other substances is obtained. The extract is, therefore, a crude mixture.

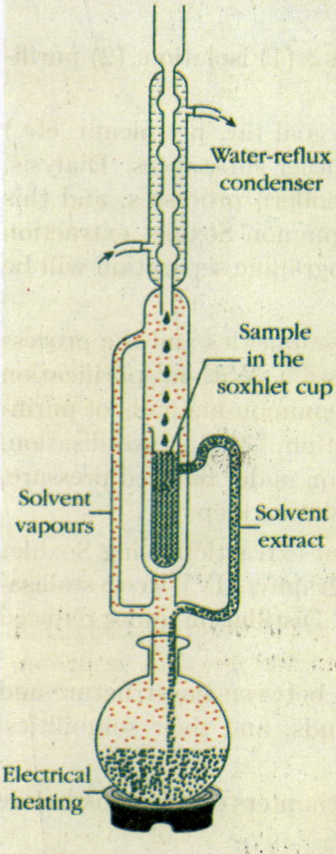


**Fig. 1. Soxhlet extraction**

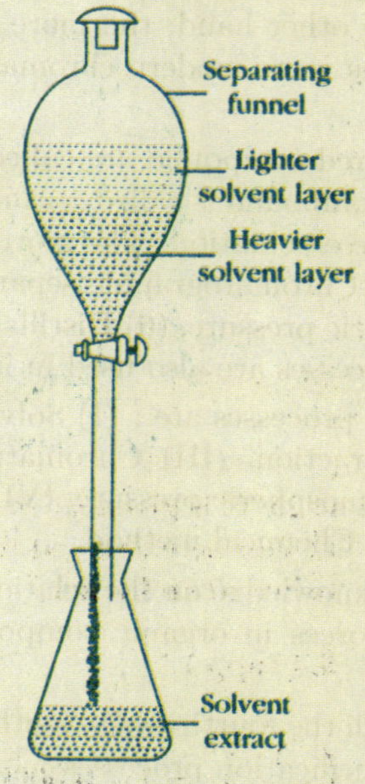


**Fig. 2. Small scale solvent extraction**

A crude mixture, once obtained, from a natural source or from a synthetic process must then be subjected to further separations.

### Liquid-liquid extraction

This is used for the further separation of the crude mixture and involves the shaking of the crude mixture in an organic solvent with an aqueous solution or water to remove water soluble impurities.

The process utilises the partitioning of substances between two immiscible liquids. When a compound is shaken with a mixture of two immiscible liquids, the ratio of the concentration of the compound in one layer to that in the other layer is a constant quantity and the value of the constant is called the partition or distribution coefficient ($K_D$) for the two liquids and is characteristic of the compound. It is actually the equilibrium constant of the equilibrium,

$$\text{Compound (aq. layer)} \rightleftharpoons \text{Compound (org. layer)}$$

$$K_D = \frac{C_{\text{compound}}(org.)}{C_{\text{compound}}(aq.)} \qquad (1)$$

**Thus nonionic compounds, being more soluble in an organic solvent than in water, will concentrate in the organic layer. On the other hand, ionic compounds will concentrate in the aqueous layer.**

**If two or more compounds of different** $K_D$**s** are present in the crude mixture, the composition **of the mixtures will be different in different layers** from that of the original mixture and so partial **separation is possible.** Now on separation of the two layers by separating funnel (Fig. 2) and **shaking up each layer** with the solvent of the other layer repeatedly an appreciable amount of **separation can be done.** This is known as counter-current distribution. The fraction extracted **from aqueous layer after** $n$**th extraction using equal volumes of organic solvent and aqueous solution is given by the equation :**

$$f_{org} = 1 - \left(\frac{1}{K_D + 1}\right)^n \qquad (2)$$

Ordinarily liquid-liquid extraction is done in a separating funnel (Fig. 2). For a large scale separation continuous solvent extraction arrangement is employed (Fig. 3).

**Problem 1.** How can you separate a mixture of benzyl alcohol (neutral) and phenol (acidic) in ether using liquid-liquid extraction?

**Solution :** Both being highly soluble in ether and slightly soluble in water, they cannot be separated form their etherial solution and water. Instead, if a dilute solution of caustic soda is shaken with the etherial solution of the mixture in a separating funnel, the phenol, being acidic, will form a salt with NaOH and will be much more soluble in water. Thus the upper ether-layer will then be rich in benzyl alcohol and the lower aqueous layer in phenolate ion. The aqueous layer, on treatment with the excess of dilute acid and distillation, gives the phenol and the etherial layer on evaporation gives benzyl alcohol.

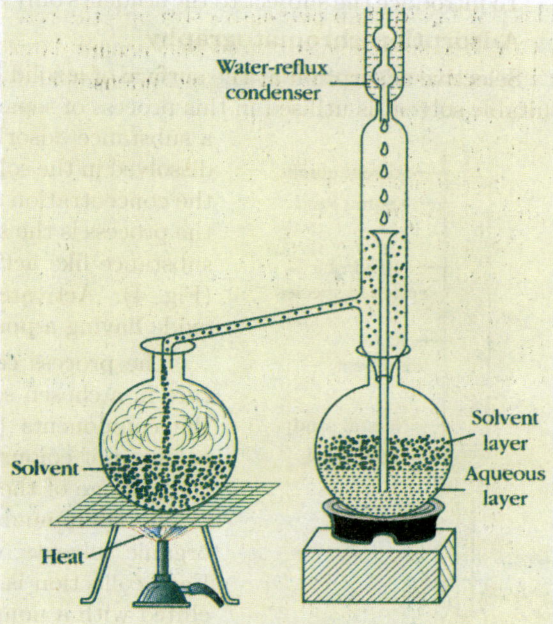

**Fig. 3. Large scale solvent extraction**

**Problem 2. The solubility of a compound at 25 °C is given as 10.4 g/l in water and 98 g/l in ether. Find out the weight of the compound that can be extracted from 100 ml of its saturated solution in water by two successive extractions using 100 ml ether.**

**Solution :** From Eqn. 1, $K_D = \dfrac{98 \text{ g/l}}{10.4 \text{ g/l}} = \mathbf{9.4 \text{ (app.)}}$

From Eqn. 2, $f_{org} = 1 - \left(\dfrac{1}{9.4+1}\right)^2 = 1 - \left(\dfrac{1}{10.4}\right)^2 = 1 - 0.0099 = \mathbf{0.9901}$

$\therefore$ weight of the extracted compound $= 0.9901 \times 10.4 = \mathbf{10.297 \text{ g}}$

## Chromatography

This is a valuable modern technique not only for the further separation of the crude mixture but also for the purification and identification of organic compounds. The process has also two advantages. One is that it can be applied for both macro and micro separation and the other is its rapidness. It takes a minimum of time in comparison to other processes of separation.

The distribution of the crude mixture between two phases, one moving and the other fixed, is the basis of chromatography. The moving phase may be a liquid or a gas and the fixed phase may be a column of solid or a paper strip. There are several types of chromatography, e.g.,

(i) Adsorption chromatography.

(ii) Thin layer chromatography.

(iii) Partition chromatography.

(iv) Ion-exchange chromatography.

To introduce the subject to the students only adsorption chromatography will be discussed here.

**Adsorption chromatography**

Selective adsorption at the surface of a solid of the components of a crude mixture taken in a suitable solvent is utilised in this process of separation. It depends upon the equilibrium between a substance adsorbed on the surface of the solid and the material dissolved in the solvent in contact with the solid. Adsorption means the concentration of a substance on a surface. The moving phase of the process is the solution and the fixed phase is the column of solid substance like activated alumina, silica gel, etc., called adsorbent (Fig. 4). Activated alumina is a specially dehydrated aluminium oxide having a porous structure of high surface area.

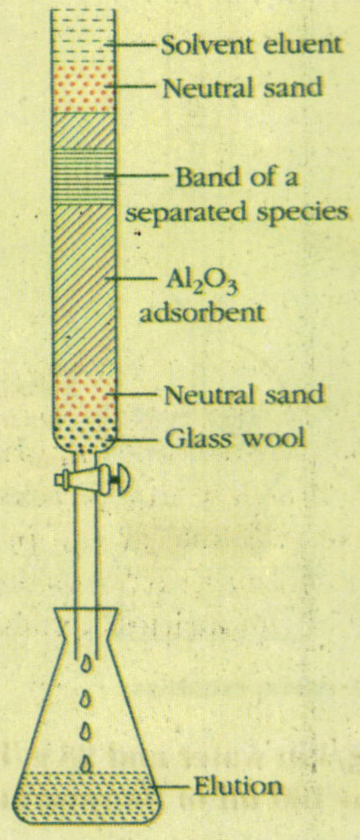


**Fig. 4. Adsorption chromatography**

The process consists in allowing a solution of the crude mixture in a chosen solvent to pass slowly down an adsorbent. When the components (adsorbates) are selectively adsorbed, separate coloured or colourless bands or zones (Chrometograms) result and the passage of the solution is then stopped. The individual bands of the compounds are then eluted (washed) several times with organic solvents, called eluents and collected in separate vessels. Each collection is technically called elutions. When the column is eluted with a nonpolar eluent, the more soluble and least strongly bound adsorbate is washed down first. By eluting the column with pure solvents or their mixture of increasing polarity, sequential elutions of the components of the crude mixture may be collected. The list of commonly used solvents with increasing polarity is given in Table 1.

**Table 1.** Commonly used solvents for adsorption chromatography

| | |
|---|---|
| 1. Hexane | 5. Chloroform |
| 2. Hexane/Benzene | 6. Chloroform/Methanol |
| 3. Benzene | 7. Methanol/Glacial Acetic acid |
| 4. Benzene/Chloroform | |

If the bands are colourless, a quartz column is to be used instead of a glass column; this, when placed in the light of mercury lamp, helps reveal the bands by their fluorescence. Colourless bands may also be detected by colouring them with a suitable colouring agent, Adsorption chromatography may be used for gas mixtures too. Gas-solid chromatography (GSC) and gas-liquid chromatography (GLC) are for them.

**Problem 3.** A benzene solution of phenol and naphthalene is chromatographed on alumina using hexane as an eluent. Which one will come out in the first elution and why?

**Solution :** Naphthalene will come down in the first elution. Phenol is strongly adsorbed on $Al_2O_3$ by hydrogen bonding and less soluble in hexane, a nonpolar solvent. On the other hand, naphthalene is less adsorbed and more soluble in hexane than phenol; so is the answer.

## Recrystallisation

The process is very useful for the solid compounds with low percentage of impurities. The basis of this procedure is that the solubility of organic compounds usually increases greatly with increasing temperature. Therefore, when saturated solution of a solid in a suitable solvent at its boiling point is cooled, crystals of the solid come down. As the fitting of molecules in a crystal lattice involves highly selective forces, usually very pure crystals are obtained by crystallisation. The solvents commonly used are given in Table 2. Mixed solvents are also used for the purpose.

**Table 2.** Commonly used solvents for recrystallisation

| | | | |
|---|---|---|---|
| 1. | Acetic acid | 7. | Chloroform |
| 2. | Water (distilled) | 8. | Ethyl ether |
| 3. | Rectified spirit | 9. | Petroleum ether |
| 4. | Absolute alcohol | 10. | Benzene |
| 5. | Methyl alcohol | 11. | Carbon tetrachloride |
| 6. | Acetone | | |

Briefly, it consists in dissolving the crude material taken in a conical or round bottomed flask with a reflux water-condenser in a minimum of the boiling solvent, single or mixed (Fig. 5). The solution is then filtered hot by means of a hot water-jacketed funnel to remove insoluble impurities (Fig. 6). Sometimes the filtrate requires boiling with a small amount of a very dry powder of pure animal charcoal for a very short period and filtering hot in a buchner funnel at the pump (Fig. 7) to eliminate coloured impurities. The filtrate is then cooled to room temperature or below that. The desired compound crystallises; more soluble and less abundant impurities remain in the solution.

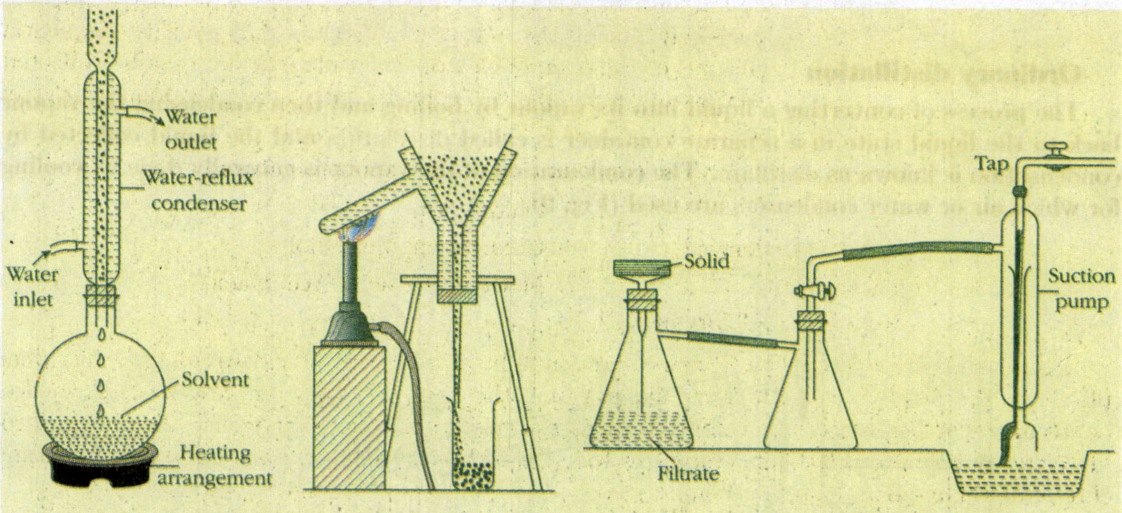


**Fig. 5. Reflux apparatus** **Fig. 6. Hot filtration** **Fig. 7. Vacuum filtration**

The crystals are then separated by filtration using buchner funnel, and repeated crystallisation ultimately gives pure crystals.

A solvent that dissolves an appreciable amount of solid when hot but a small amount when cold is the suitable solvent. Before a large scale crystallisation a proper solvent is to be selected, and a solvent/solute ratio is to be determined by trial and error approach.

Sometimes a solvent in which the desired compound is insoluble is added to the saturated solution of the compound for crystallisation. Scratching of the inside wall of the glass ware containing the solution creates nuclei for the process and helps crystallisation.

**Problem 4. Why do less soluble impurities remain in the solution in the process of recrystallisation?**

**Solution : Both the solvent and the desired compound being present in far larger quantity than the less soluble impurity in the mixture, the solution is saturated with respect to the compound and not with respect to the impurity. Hence the impurity remains in the solution on cooling and the compound recrystallises down.**

## Sublimation

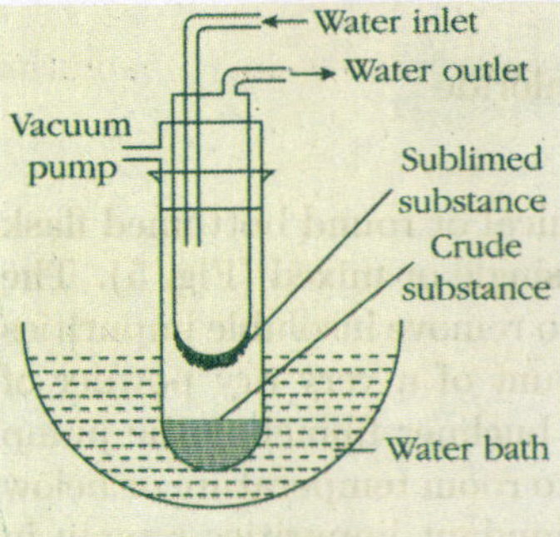


**Fig. 8. Sublimation**

**The process is used for the separation of the less volatile impurities from the more volatile compounds having a reasonably high vapour pressure in the solid state and comparatively weak intermolecular forces.**

**Such a compound when heated is vaporised directly or via liquid state but recondenses to the solid state directly.**

**Typically, the crude material is heated below its melting point at atmospheric pressure (or under reduced pressure for compounds with low vapour pressure) until all the desired compound is vaporised and the vapour recondensed to solid on a cooled surface (Fig. 8). Naphthalene is commercially purified by sublimation.**

## Distillation

### Ordinary distillation

**The process of converting a liquid into its vapour by boiling and then condensing the vapour back to the liquid state in a separate container is called distillation and the liquid collected by condensation is known as distillate. The condensation of the vapour is generally done by cooling for which air or water condensers are used (Fig. 9).**

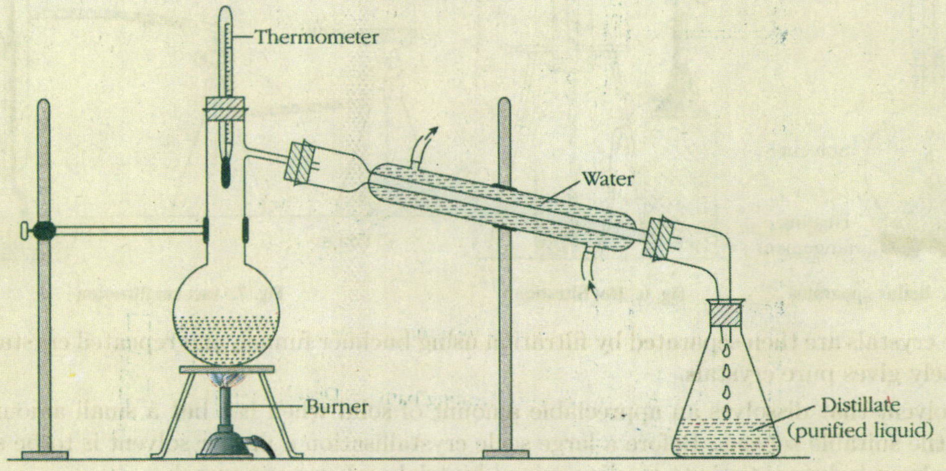


**Fig. 9. Distillation**

**The intermolecular attractive forces being weak, the most of the organic liquids can be converted into their vapours at their boiling points without decomposition. Boiling point, a characteristic**

**property of a liquid, may be defined as a temperature at which its vapour pressure is equal to the external pressure. The boiling points of different liquids are different as they have different intermolecular forces. When a liquid mixture boils, the composition of the vapour is not the same as that of the mixture except in the case of azeotropic mixture where the composition of both is one and the same. Owing to the difference in boiling points, the vapour is richer in lower boiling, i.e., more volatile component than others in the mixture. Hence the lower boiling component distils out first leaving behind the higher boiling one. Nevertheless the distillate is not free form the other. It always contains certain precentage of the higher boiling component and the composition of the distillate can be calculated from the Rault's law. On repeated distillation the purity of the distillate can be increased.**

**The method is applied for the separation of (i) a liquid from a non-volatile solid in solution or (ii) two liquids whose boiling points differ by 30 °C or more.**

### Fractional distillation

**If two liquids in a mixture differ in boiling points by not less then 3 °C the separation may effectively be done by using a special cooling device, called fractionating column (Fig. 10). It is a long vertical column containing rings of plates or bubble caps which are attached to still (a glass apparatus where vapours are condensed to liquid drops) for reflux of the vapours within the column.**

**As a result of the internal reflux gradual separation of two or more liquids is possible using fractionating column and the process is called fractional distillation.**

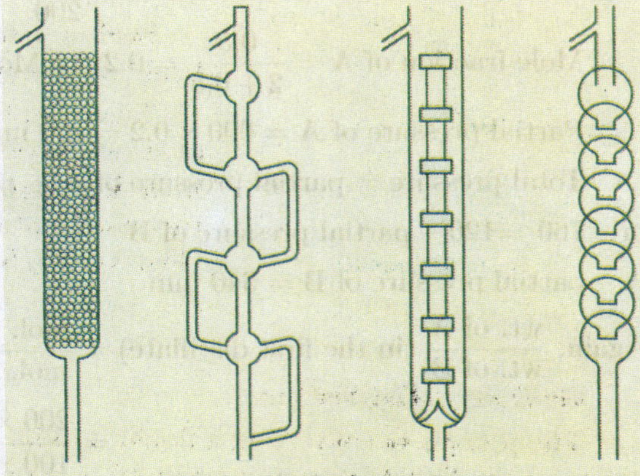

**Fig. 10. Fractionating columns**

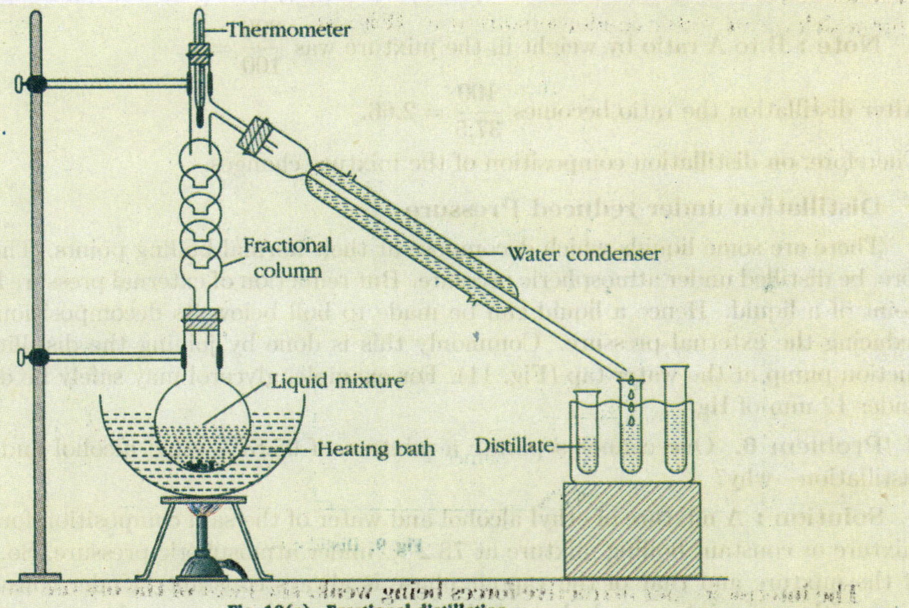


**Fig. 10(a). Fractional distillation**

When vapour rises through the fractionating column, a certain amount of it gets condensed and descends as a reflux. The hot vapour which meets the colder descending reflux, gets slightly cooled, whereby the higher boiling component of the vapour flows back to the boiling flask by condensation and the lower boiling component of the reflux, taking the heat of condensation of the higher boiling vapour, vaporises and distils out. The boiling mixture in the flask is thus gradually enriched with the higher boiling liquid and on repetition of the process for several times the liquid mixtures can be separated completely.

**Problem 5.** A liquid mixture composed of two compounds A and B boils at 80 °C at 760 mm of Hg. The vapour pressure of pure A at 80 °C is 600 mm of Hg. If the mixture contains 100 g of A and 200 g of B, what amount of B will be obtained in the first distillate weighing 137.5 g if the mol. wts. of A and B are 200 and 100 respectively?

**Solution :** Number of moles of A $= \frac{100}{200} = 0.5$ Number of moles of B $= \frac{200}{100} = 2$

Mole fraction of A $= \frac{0.5}{2 + 0.5} = 0.2$ Mole fraction of B $= \frac{2}{2.5} = 0.8$

Partial pressure of A $= 600 \times 0.2 = 120$ mm.

Total pressure = partial pressure of A + partial pressure of B

or, 760 = 120 + partial pressure of B

∴ partial pressure of B = 640 mm

Again, $\frac{\text{wt. of A}}{\text{wt. of B}}$ (in the first distillate) $= \frac{\text{mol. wt. of A} \times \text{partial pressure of A}}{\text{mol. wt. of B} \times \text{partial pressure of B}}$

$$= \frac{200 \times 120}{100 \times 640} = \frac{0.375}{1}$$

∴ 1.375 g of the distillate contains 1 g of B; so, 137.5 g of the distillate contains 100 g of B.

**Note :** B to A ratio by weight in the mixture was $\frac{200}{100} = 2$.

After distillation the ratio becomes $\frac{100}{37.5} = 2.66$.

Therefore, on distillation composition of the mixture changes.

### Distillation under reduced Pressure

There are some liquids which decompose at their normal boiling points. They cannot, therefore, be distilled under atmospheric pressure. But reduction of external pressure lowers the boiling point of a liquid. Hence a liquid can be made to boil below its decomposition temperature by reducing the external pressure. Commonly this is done by joining the distilling flask with the suction pump at the water-tap (Fig. 11). For example, glycerol may safely be distilled at 180 °C under 12 mm of Hg.

**Problem 6.** One cannot separate a mixture of 95.87% ethyl alcohol and 4.13% water by distillation—why?

**Solution :** A mixture of ethyl alcohol and water of the said composition forms an azeotropic mixture or constant boiling mixture at 78.2 °C under atmospheric pressure. So, the composition of the mixture and that of the vapour phase is always constant. Under constant pressure the composition of mixture and that of the vapour remain the same for any constant boiling or azeotropic mixture. But the composition of the both at different pressures is different. Therefore, by distillation a mixture of 95.87% EtOH and 4.13% $H_2O$ cannot be separated.

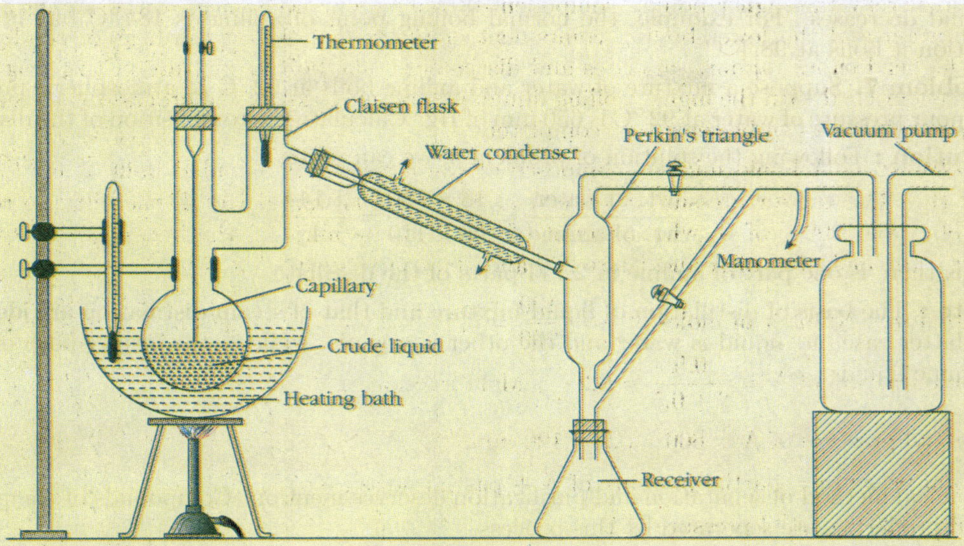


**Fig. 11. Distillation under reduced pressure**

## Steam distillation

Liquids contaminated with non-volatile and water-insoluble impurities may be purified by steam distillation. It consists in allowing the hot steam to enter the distilling flask where it takes up the vapour of the organic liquids, and co-distillation occurs (Fig. 12). The desired liquid may then be separated by a separating funnel (if immiscible with water) or by solvent extraction procedure (if miscible).

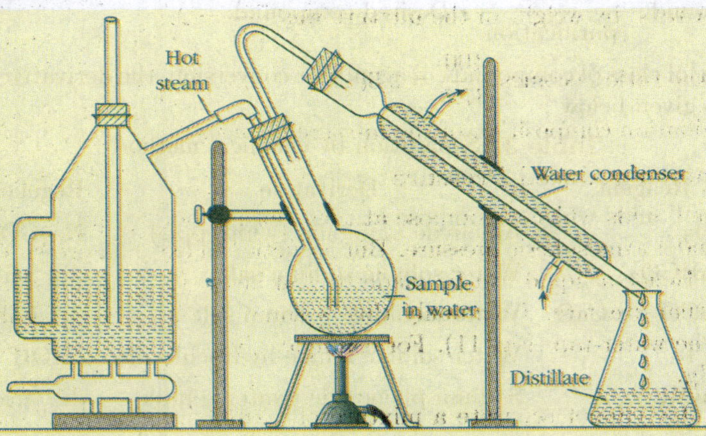


**Fig. 12. Steam distillation**

As in the vapour-phase both steam and the liquid vapour co-exist at the boiling point, the total pressure of the vapour-phase is equal to the atmospheric pressure.

But total pressure = partial pressure of $H_2O$ + partial pressure of B (org. comp.)

∴ atmospheric pressure = partial pressure of $H_2O$ + partial pressure of B

Hence, lowering of vapour pressure occurs during the steam distillation. So the boiling point of the liquid decreases. For example, the normal boiling point of aniline is 184 °C but in steam distillation it boils at 98.4 °C.

**Problem 7.** Suppose a mixture of water and aniline boils at 92 °C at atmospheric pressure. The vapour pressure of water at 92 °C is 650 mm of Hg. Calculate the composition of the distillate.

**Solution :** Following the solution of problem 5, we can write,

$$\frac{\text{wt. of water}}{\text{wt. of aniline}} = \frac{18 \times 650}{93 \times 110} = \frac{1.144}{1}$$

Thus there is one part of aniline in 2.144 parts of the distillate.

**Note :** The basis of distillation of liquid mixture and that of steam distillation are identical. In the latter case one liquid is water and the other is organic, while in the former both of them are organic liquids.

## Chemical Method

Chemical method of separation and purification deserves mention. Compounds of comparable physical properties are separated by this process.

The principle involved in the process is the conversion of the desired compound to a separable derivative by reagent, separation of it and regeneration of the original compound from the derivative. Sometimes impurities are decomposed by chemical reaction. However, the compound obtained by chemical separation contains other impurities which are removed by physical methods like solvent extraction, distillation, recrystallisation, etc.

$$\text{Crude mixture} \xrightarrow{\text{reagent}} \text{Derivative} \xrightarrow{\text{separation}} \text{Separated derivative} \xrightarrow{\text{reagent}}$$

$$\text{Original compound} \xrightarrow[\text{purification}]{\text{further}} \text{Desired compound.}$$

A list consisting of class of compounds, reagents for conversion, the derivatives and the reagents for regeneration is given below :

**Table 3.** Separation by chemical method

| Class | Reagent | Derivative | Regenerating reagent |
|---|---|---|---|
| Aldehyde | $NH_3$ | Aldehyde ammonia (white ppt.) | $Na_2CO_3$ soln. |
| Acid | $NaHCO_3$ | Water soluble sodium salt | dil. acid |
| Acid | $CaCO_3$ or CaO | Water insoluble calcium salt | dil. $H_2SO_4$ |
| Amine | HCl | Water soluble amine hydrochloride | NaOH |
| Phenol | NaOH | Sodium phenoxide (water soluble) | dil. acid |

## Drying of Organic Compounds

Organic compounds, solid or liquid, obtained by the above methods are found to be contaminated with water or organic solvents in tracer amount. Before testing the purity of the substance they require drying.

Solid substances are generally dried by pressing them between folds of filter paper or spreading over a porous plate. They may also be dried by heating them in an air oven at a temperature below their melting point. But these methods have disadvantages too. The compound may be contaminated with filter-paper fibre, complete removal of moisture may not be possible and on heating, the compound may decompose. So it is better to use an atmospheric desiccator (Fig. 13(a)) or a vacuum desiccator (Fig. 13(b)) for drying.

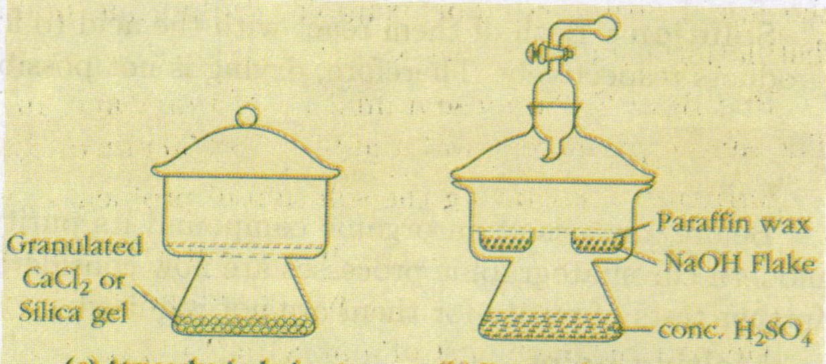

(a) Atmospheric desiccator (b) Vacuum desiccator

Fig. 13

It consists of a glass vessel with a close fitting ground lid and a perforated glass plate or wire gauze on which pots containing the sample to be dried are kept. At the bottom of the vessel a hygroscopic substance, called drying agent, remains. Common drying agents of the atmospheric desiccator are either fused calcium chloride, uniform fragments of silica gel coloured blue with a cobalt salt. When the colour of the gel turns red the silica gel loses its dehydrating power. Paraffin wax is also kept in the desiccator for absorbing organic solvents. The spent silica gel is regenerated by strong heating.

Most efficient drying is possible with a vaccum desiccator containing conc. $H_2SO_4$ at the bottom and caustic soda flake with paraffin was in small pots below the wire gauze as shown in the figure. Explosive solids like triazenes are dried in the vacuum desiccator by silica gel instead of conc. $H_2SO_4$.

Liquids are mostly dried either by keeping or refluxing them in contact with some powdered dehydrating agents with which they do not react. In either case the liquid should at once be distilled after filtering off the dehydrating agent. Table 4 contains a few commonly used drying agents with their uses.

**Table 4.** Common drying agents for liquids

| Drying agents | Uses | Drying agents | Uses |
|---|---|---|---|
| 1. Granulated $CaCl_2$ | May be used for hydrocarbons, ethers, carbonyl compounds. Should not be used for alcohols, phenols or amines as they combine with $CaCl_2$. Acidic liquids may react with $Ca(OH)_2$ formed by the hydrolysis of $CaCl_2$. | 4. Anhydrous $Na_2SO_4$ powder | Used for almost all types of compounds. It dehydrates very slowly. |
| 2. Quicklime (CaO) | Used for alcohols; should not be used for acidic compounds, ester, acyl halides, amides, cyanides, anhydrides, etc. | 5. Metallic Na | Used for ethers; should not be used to dry acids, alcohols, phenols, etc. with which sodium react. |
| 3. Solid KOH | Used for amines; should not be used for acids, phenols, esters, acyl halides, amides, cyanides, anhydrides, etc. | 6. $CaSO_4$ | Used for all liquids. |

**Problem 8.** Why cannot liquid acids be dried by metallic sodium or quicklime?

**Solution :** Both of them react with the acid to form salt. Hydrogen and water are the other products respectively. Therefore, drying is not possible.

## Test of Purity

Before analysis of an organic compound its purity must be tested. Compounds obtained by modern chromatographic processes are now considered to the highest grade of pure samples and further tests of purity for them are not required.

### Melting point

Ordinarily, melting point, being a characteristic property of a solid, is determined as a criterion of purity. The constant temperature at which the solid and liquid phase of a compound are in equilibrium at a given pressure is called its **melting point**. It is normally quoted for 760 mm pressure. A pure compound has usually a sharp melting point. It melts entirely within a range of 1 °C. Hence if a compound melts sharply it is pure. On the other hand, if it melts indefinitely without decomposition, the compound is impure. Melting point may be determined in a metal bath or in a conc. sulphuric acid bath (Fig. 14).

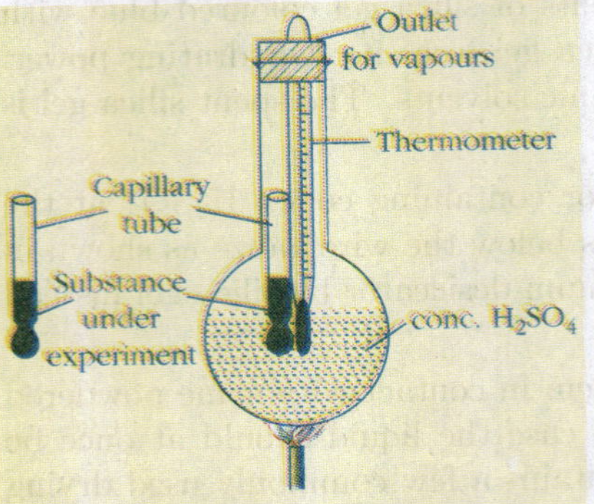

Fig. 14. Determination of melting point

In both the cases the powdered sample is taken in a small capillary tube sealed at one end.

If the concerned compound is a known one, its purity may be ascertained by **mixed melting point** determination. A pure variety of the compound is to be mixed with the isolated product and the mixture is to be subjected to melting point determination. A sharp melting point of the mixture indicates the purity of the isolated compound.

Nowadays **zone melting** is a good test of purity.

**Problem 9.** Normal melting point of a compound is 132 °C. But a student found it to be 140 °C sharp. Why is it so?

**Solution :** Presence of impurity lowers the melting point of a compound and fluctuation of the atmospheric pressure practically has no effect on melting point. The reason for elevation of melting point, therefore, is the wrong calibration of the thermometer used.

### Boiling point

A pure liquid has a sharp boiling point which remains constant at a constant pressure until the whole of the liquid boils off. Of course **constant boiling mixtures** possess sharp boiling point. But when such a solution is distilled at different external pressures, different boiling points and different compositions of the distillates result. On the other hand, under different pressures the compositions of the distillates of a pure compound remain the same. The boiling point is commonly determined by the apparatus given in the Fig. 15.

### Paper chromatography

A pure compound, when subjected to **paper chromatography**, moves as a single spot. Hence purity of a compound may be ascertained by paper chromatography. The rate of movement of the spot to the rate of movement of the solvent is a constant ratio and this a characteristic property of a compound. An impure compound produces more than one spot and the rate ratio is not a constant one.

The method consists in allowing a Whatman chromatographic direction paper number four impregnated with a liquid to soak the solution of the compound in a chosen solvent. Cellulose of the paper impregnated with a liquid is the fixed phase or adsorbent and solution is the moving phase. Elution occurs by capillary action or by capillary and gravity flow.

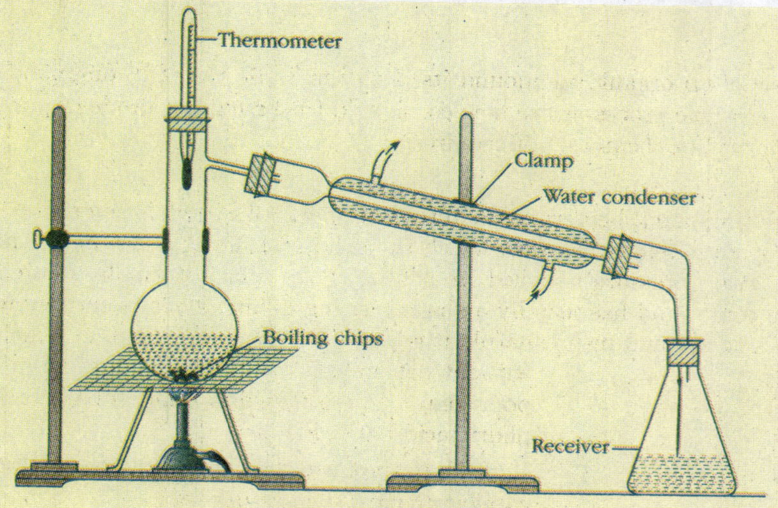


**Fig. 15. Determination of boiling point**

Appendix B

# Important Reagents Used in Organic Chemistry

**1. Acetic anhydride :** $CH_3CO{\cdot}O{\cdot}COCH_3$ ($Ac_2O$)

**(i)** Manufacturing processes :

**(a)** When acetaldehyde is oxidised by air at 50 °C in the presence of cobalt acetate and copper acetate catalyst, acetic anhydride is formed.

$$2CH_3CHO + O_2 \longrightarrow (CH_3CO)_2O + H_2O$$

The water is removed by ethyl acetate.

**(b)** When acetic acid is decomposed at 700-720 °C in the presence of a small amount of ethyl phosphate, ketene is formed; ketene may also be prepared non-catalytically by decomposing acetone at 700 °C. The ketene is then absorbed in acetic acid in a scrubbing tower.

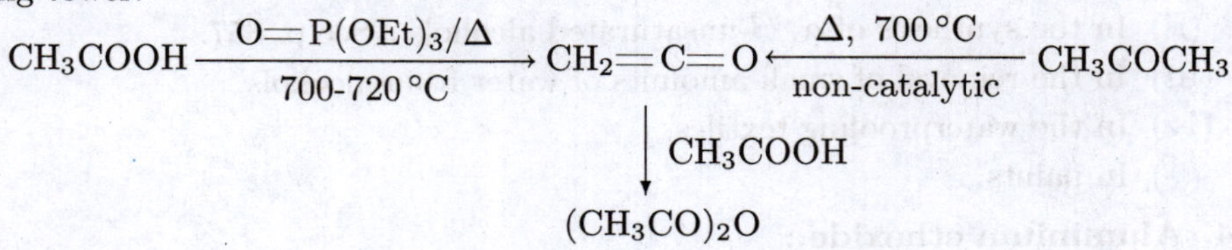

**(ii)** Uses :

**(a)** To prepare acetate ester from an alcohol where other methods fail. It is used in the manufacture of cellulose acetate.

**(b)** To manufacture aspirin : See p. 1213.

**(c)** To prepare acetyl derivative of phenols and amines in the laboratory for their identification.

**(d)** To manufacture vinyl acetate : See p. 1037.

**(e)** As a dehydrating agent : To convert aldoxime to cyanides; this is an intermediate step in the Wohl method of descending the sugar series.

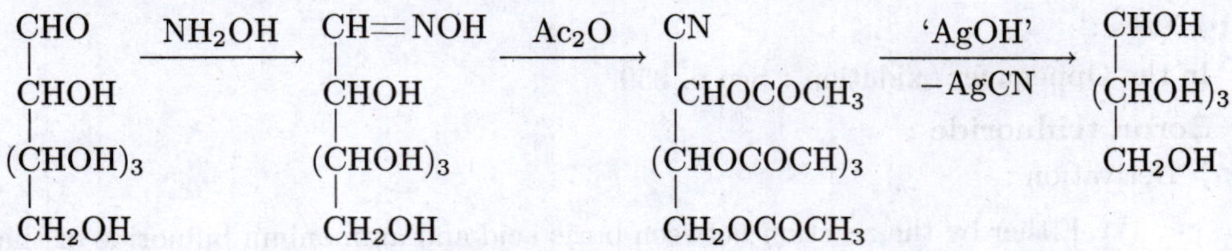

**(f)** In the Friedel-Crafts acylation : See p. 942.

**(g)** In the synthesis of cinnamic acid by the Perkin reaction : See p. 903.

**(h)** In the Erlenmeyer azlactone synthesis : See p. 1500.

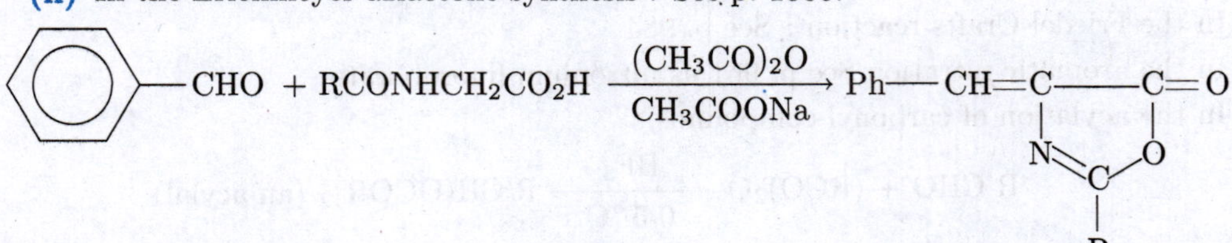

An azlactone

**2. Aluminium chloride (anhydrous) :**

**(a)** Derivation : By passing purified chlorine gas through molten aluminium.

(b) Uses :

(i) In the Friedel-Crafts reaction : See p. 975-976.
(ii) In the Fries rearrangement : See p. 768.
(iii) In the isomerisation of alkanes : See p. 1025.
(iv) In the Gattemann—Koch aldehyde synthesis : See p. 988.

**3. Aluminium isopropoxide :**

(a) Derivation : By the action of isopropyl alcohol on aluminium amalgam.

(b) Uses :

(i) In the Meerwein-Pondorf-Varley reduction : See p. 857.
(ii) In the synthesis of $\alpha$, $\beta$-unsaturated alcohols : See p. 857.
(iii) In the removal of small amounts of water from alcohols.
(iv) In the waterproofing textiles.
(v) In paints.

**4. Aluminium ethoxide :**

(a) Derivation : By the action of ethanol on amalgamated aluminium or by the reaction of ethanol and aluminium catalysed by $HgCl_2$ and iodine.

(b) Uses :

(i) In the Tischenko reaction : See p. 862.
(ii) As a catalyst for polymerisation.
(iii) In methylation of alcohols by diazomethane : See p. 1184.

**5. Aluminium tert-butoxide :**

(a) Derivation : By the action of tert-butyl alcohol on amalgamated aluminium.

(b) Uses :

In the Oppenauer oxidation : See p. 859.

**6. Boron trifluoride :**

(a) Derivation :

(i) Either by the reaction between boric acid and ammonium bifluoride or that between boran and hydrofluoric acid and then treating the reaction product with cold fuming sulphuric acid.

(b) Uses :

(i) In the Friedel-Crafts reaction : See p. 984.
(ii) In the aromatic nitration See p. 961 as nitronium fluoroborate.
(iii) In the acylation of carbonyl compounds.

$$R'CHO + (RCO)_2O \xrightarrow[0\text{-}5\,°C]{BF_3} R'CH(OCOR)_2 \text{ (an acylal)}$$

(iv) In the addition to multiple bond : carbamates add to alkenes and conjugated dienes in the presence of boron trifluoride-etharate and form N-substituted urethans.

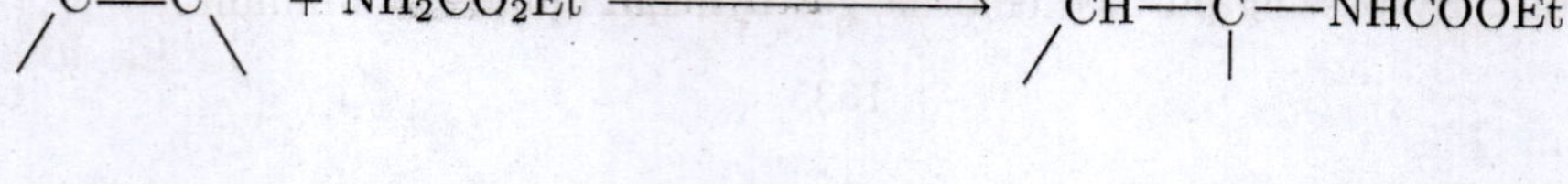

**(v)** In the formylation of aromatic hydrocarbons and chloro- and fluorobenzenes with formyl fluoride,

$$PhH \xrightarrow[BF_3]{FCHO} PhCHO$$

**(vi)** In the conversion of ethers to esters :

$$R—O—R + CO \xrightarrow[125\text{-}180\,°C]{BF_3/5\text{ atm}} RCOOR$$

**7. Carbon monoxide :**

**(a)** Derivation : When a mixture of oxygen and carbon dioxide is kept in contact with incandescent graphite or coke, carbon monoxide is formed. It may also be obtained by the dehydration of formic acid with conc. sulphuric acid.

**(b)** Uses:

**(i)** In the synthesis of methanol : See p. 1193.
**(ii)** In the manufacture of formic acid : See p. 1301.
**(iii)** In the conversion of ether to ester : See above.
**(iv)** In the conversion of alkene to amide : See p. 669. Prob. 13.11(ii).
**(v)** In the Gattermann—Koch aldehyde synthesis : See p. 988.
**(vi)** In the manufacture of phosgene :

$$CO + Cl_2 \xrightarrow[200\,°C]{\text{activated C}} COCl_2$$

**(vii)** In the preparation of tertiary alcohol by the process of hydroboronation-carbonylation and oxidation : See p. 651.

**(viii)** In the manufacture of acrylic acid.

$$C_2H_2(g) + CO(g) + H_2O(g) \xrightarrow[\text{heat}]{Ni} CH_2{=}CH—COOH$$

**8. Carbonyl chloride (Phosgene) :**

**(a)** Derivation : It is prepared by passing a mixture of carbon monoxide and chlorine gas over activated carbon.

**(b)** Uses :

**(i)** To synthesise carbobenzoxy chloride, a reagent for protecting the amino group in the synthesis of peptide linkage.

$$PhCH_2OH + COCl_2 \xrightarrow{-HCl} PhCH_2OCOCl + NH_2—CH_2—COOH$$

$$\downarrow$$

$$PhCH_2OCONH—CH_2—COOH$$

**(ii)** In the synthesis of isocyanates which are used to make polyurethan rubber and plastics.

$$\underset{(Ar)}{R—NH_2} + COCl_2 \xrightarrow{-HCl} \underset{(At)}{R—NHCOCl} \xrightarrow[\Delta]{-HCl} \underset{(Ar)}{R—N{=}C{=}O}$$

**(iii)** In the synthesis of alkyl carbonates (mainly ethyl carbonate, a useful solvent) and alkyl chloroformates :

$$COCl_2 + EtOH \longrightarrow \underset{\text{Ethyl chloroformate}}{ClCO_2Et} + HCl$$

$$\mathrm{ClCO_2Et + EtOH \longrightarrow \underset{\text{Ethyl carbonate}}{O{=}C(OEt)_2} + HCl}$$

**(iv)** In the manufacture of certain dye-intermediates such as Michler ketone in the manufacture of crystal violet.

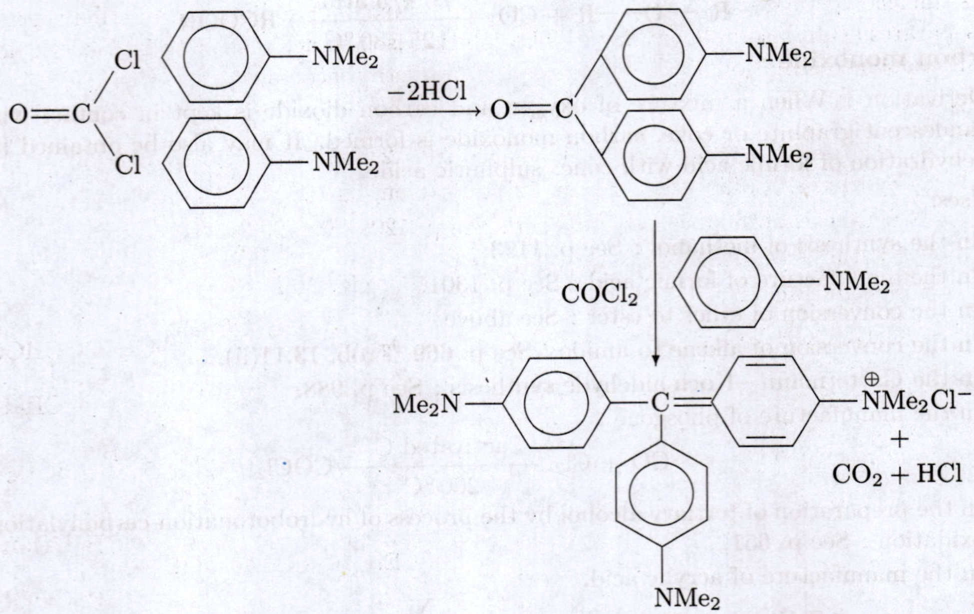

Crystal violet

**(v)** In the synthesis of urea : See p. 1385.

**(vi)** In the synthesis of benzoyl chloride by the Friedel-Crafts reaction :

$$\mathrm{PhH + COCl_2 \xrightarrow[-HCl]{\text{anh. } AlCl_3} PhCOCl}$$

**9. Dimethyl sulphate :**

**(a)** Derivation : By adding fuming sulphuric acid to methyl alcohol and then distilling in vacuo.

**(b)** Uses : It is used as methylating agent for phenols, sugars, thiophenol amines, etc.

**(i)** Methylation of phenols.

$$\mathrm{PhOH + NaOH \xrightarrow{-H_2O} PhONa \xrightarrow{Me_2SO_4} PhOMe}$$

**(ii)** Methylation of thiophenols.

$$\mathrm{PhSH \xrightarrow[NaOH]{Me_2SO_4} PhSMe + NaMeSO_4}$$

**(iii)** Methylation of alcoholic OH groups in carbohydrates : See p. 1557.

**13. Ethyl acetoacetate :**

(a) Preparation : See p. 894. Sec. 16.13.2, The Claisen condensation. In practice 2 mole of anhydrous ethyl acetate and 1 mole of sodium wire and a little ethyl alcohol are taken together in a refluxing flask. At the beginning, reaction is slow but after a while vigorous reaction takes place with evolution of hydrogen gas and the mixture starts boiling. The flask is then slowly heated in a water bath and then refluxed. A sodium salt of ethyl acetoacetate separates as an insoluble oily layer. The solution is then acidified with 1 : 1 acetic acid and subsequent distillation in vacuum gives the pure ethyl acetoacetate.

(b) Uses :

(i) Synthesis of ketone : See p. 1248, Sec. 20.7.2. (xv) Prob. 20.16. p. 1250.

(ii) Synthesis of diketones : See p. 1250, Problem 20.16.

(iii) Synthesis of monocarboxylic acid : See p. 1292, Sec. 20.9.2 (o).

(iv) Synthesis of dicarboxylic acid :

Succinic acid :

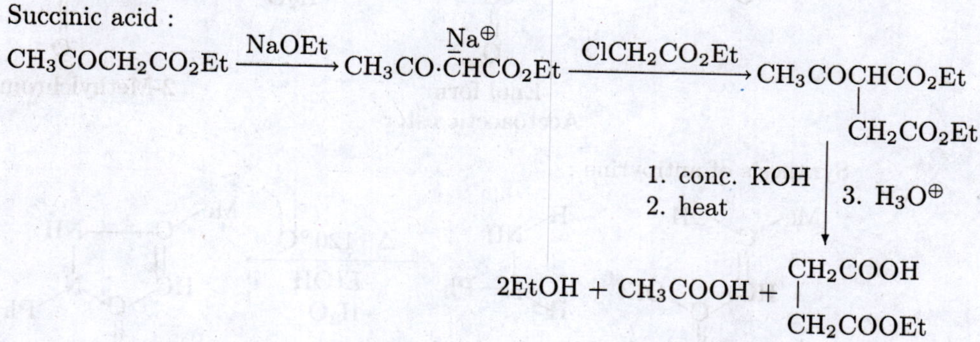


Adipic acid :

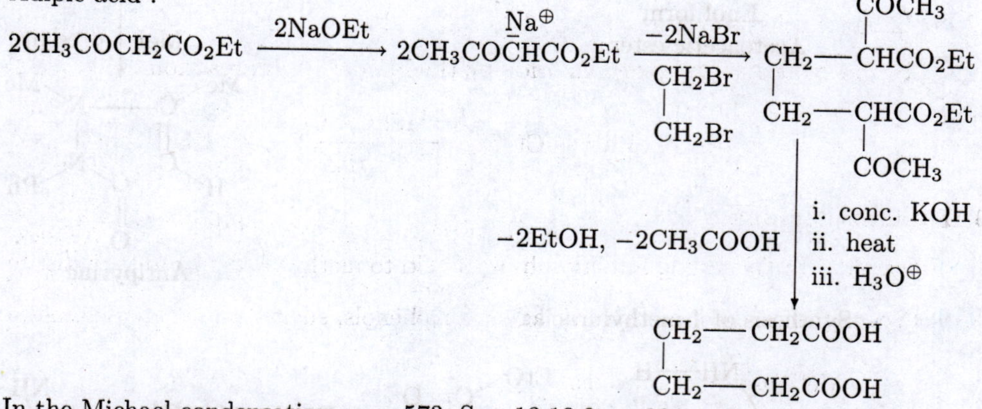


(v) In the Michael condensation ee p. 573, Sec. 13.13.2. p. 664.

(vi) Synthesis of heterocycles :

Synthesis of quinoline : The Knorr quinoline synthesis : See p. 1132, Sec. 19.7.2(c).

Synthesis of Furan : The Feist Benary synthesis : See p. 1097, Sec. 19.2.2(e) and 19.2.2(d).

Synthesis of pyridine derivative, Hantzsch synthesis : See p. 1121, Sec. 19.6.2(f) and

Synthesis of pyrrole : See p. 1106, Sec. 19.4.2(e), the Knorr synthesis.

Synthesis of coumarins and chromones :

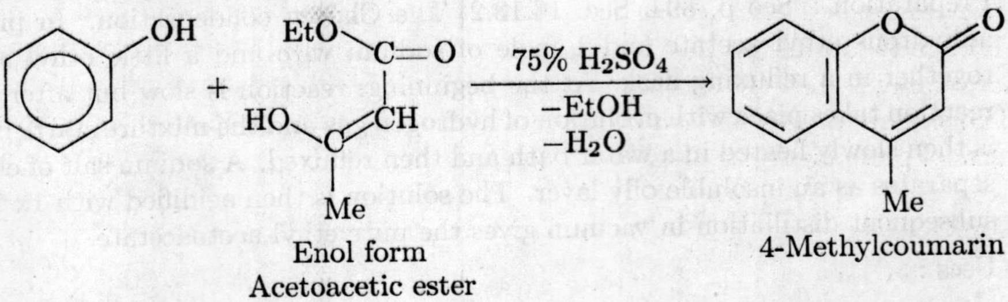


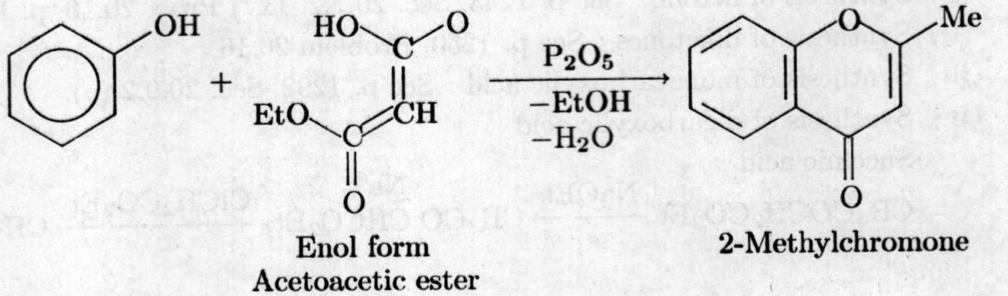


Synthesis of antipyrine :

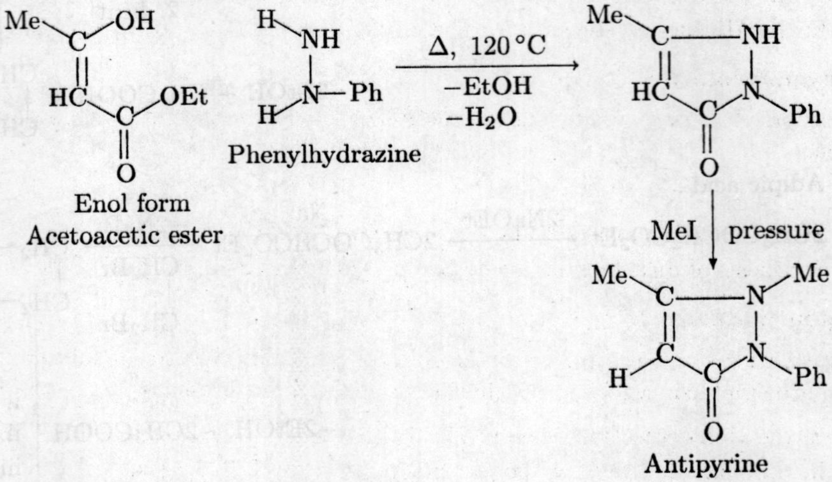


Synthesis of 4-methyluracil :

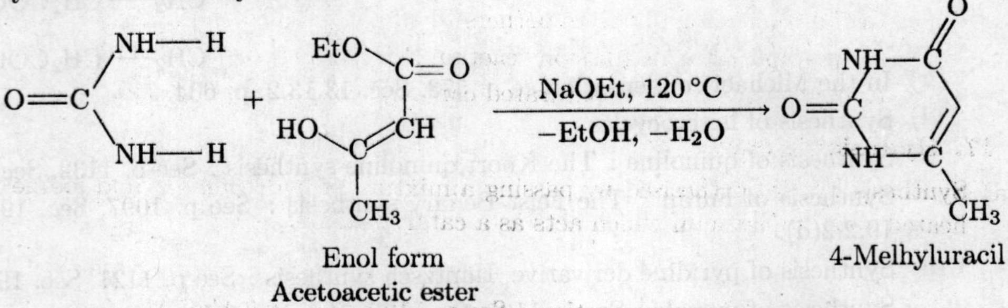

**14. Malonic ester :**

(a) Preparation : acetic acid is first converted to monochloroacetic acid by passing chlorine through the former at 100 °C. The solution is neutralised with aqueous potassium carbonate solution and then treated with potassium cyanide when potassium cyanoacetate is formed. This on hydrolysis with dilute hydrochloric acid and subsequent esterification with ethyl alcohol gives diethyl malonate (malonic ester).

$$CH_3COOH \xrightarrow[-HCl]{Cl_2,\ 100\,°C} ClCH_2COOH \xrightarrow{K_2CO_3} ClCH_2COOK \xrightarrow{KCN} CNCH_2COOCN$$

$$CNCH_2COOCN \xrightarrow[-NH_4Cl]{dil.\ HCl} HOOC—CH_2—COOH$$

$$EtOOCCH_2COOEt \xleftarrow[conc.\ H_2SO_4]{2EtOH} HOOC—CH_2—COOH$$

(b) Uses :

(i) Synthesis of monocarboxylic acid : See p. 1293. Sec. 20.9.2(p).
(ii) Synthesis of dicarboxylic acid : See p. 1293, Sec. 20.9.2(p)(i).
(iii) Synthesis of alicyclic compounds : See p. 1049, Sec. 18.5.3(d).
(iv) Synthesis of barbituric acid : See p. 1389, Sec. 20.19.4(l).
(v) Synthesis of barbital and bromural (good hypnotics) : See p. 1390-1391. Sec. 20.19.6.(v).
(vi) In the Knovenagel condensation : See p. 777, Sec. 16.13.7.
(vii) In the Michael condensation : See p. 664, Sec. 13.13.2.

**15. Ethyl cyanoacetate :**

(a) Preparation : It is prepared by the interaction of ethyl chloroacetate with potassium cyanide solution.

(b) Uses :

(i) Synthesis of monocarboxylic acid : See p. 1294, Sec. 20.9.2.(r).
(ii) Synthesis of dicarboxylic acid : See p. 1294, Sec. 20.9.2.(q (iii)).

**16. Hydrogen peroxide :**

(a) Preparation : By the electrolysis of ice cold 50% sulphuric acid and subsequent hydrolysis with dilute sulphuric acid and distillation under reduced pressure.

(i) In the Dakin reaction : See p. 775,
(ii) In the manufacture of glycerol : See p. 1193, Sec. 20.2.9.(ix).
(iii) In the synthesis of benzoyl peroxide and peracids : See p. 1302, Sec. 20.9.7.(iv).
(iv) In the conversion of the thioethers to sulphones : See p. 1234, Sec. 20.5.10(c).
(v) In the piperidine ring fission reaction : See 19.6.4(vii)(a). p. 1128.
(vi) Epoxydation of $\alpha, \beta$-unsaturated carbonyl compounds : p. 640.

**17. Hydroiodic acid :**

(a) Synthesis : It is synthesised by passing a mixture of hydrogen gas and iodine vapour over heated spongy platinum which acts as a catalytic agent.

(b) Uses :

(i) In the ring cleavage reaction : See p. 1129, Sec. 19.6.4(viic).

(ii) In the cleavage of ether linkage and estimation of alkoxy group. See p. 1227, Sec. 20.5.4(c).

(iii) As a reducing agent

$$CH_2OH\ (CHOH)_4\ CH_2OH \xrightarrow{HI/red\ P} CH_3\ (CH_2)_4\ CH_3$$

**18. Iodine :**

Uses :

(i) As a halogen carrier in the chlorination and bromination of aromatic compounds : See p. 968, Sec. 17.9.1.

(ii) In the detection of ketomethyl group and in the preparation of iodoform : See p. 885, Sec. 16.11.

(iii) In the synthesis of naphthoic acid : See p. 1072, Sec. 18.7.5.

**19. Lithiumaluminium hydride :** See p. 854, Sec. 16.6.1.

**20. Lead tetraacetate :**

(a) Derivation : When red lead ($Pb_3O_4$) is treated with glacial acetic acid in the presence of acetic anhydride, lead tetraacetate is formed.

(b) Uses :

(i) Oxidative decarboxylation :

$$RCOOH \xrightarrow{Pb(OAc)_4} ROAc \xrightarrow{H_3O^{\oplus}} ROH$$

(ii) Cleavage of $\alpha$-glycolic linkage : See p. 1191, Sec. 20.2.6.(c).

**21. Potassium cyanide :**

(a) Preparation : It is prepared by absorbing hydrogen cyanide gas in potassium hydroxide; HCN is synthesisted catalytically from a mixture of air, ammonia and methane or natural gas.

(b) Uses :

(i) In the conversion of diazonium ion to aryl cyanides : See p. 1151, Sec. 20.1.2.(n).

(ii) in the preparation of alkyl cyanides : See p. 1322, chart 20.31.

(iii) In the benzoin condensation : See p. 900, Sec. 16.13.5.

(iv) In the Streker synthesis : See p. 915, Sec. 16.13.8.

**22. Metallic copper :**

Uses :

(i) In the dehydrogenation of alcohols : See p. 1179, Sec. 20.2.5(i).

(ii) in the Gattermann reaction : See p. 1151, Sec. 20.1.2(n).

(iii) In the Ullman biaryl synthesis : p. 1163.

**23. Metallic Zinc :**

Uses :

(i) In the conversion of phenols to arenes : See p. 1056, Sec. 18.6.3.(f)

(ii) In the conversion of nitro compunds to phenyl hydroxyl amine : See p. 1334, Sec. 20.14.4.(c).

(iii) In the Reformatsky reaction : See p. 847, Sec. 16.4.6.

(iv) In the synthesis of pyrrole : See p. 1105, Sec. 19.4.2.(a).

(v) In the dehalogenation of dibromoalkanes : See p. 812, Sec. 15.5.

(vi) In the dehydroxylation of phenols : See p. 1056 and see p. 1067, Sec. 18.7.3(a).

**24. Periodic acid :**

(a) Derivation : By the interaction of perchloric acid and iodine.

(b) Uses :

(i) In the cleavage of $\alpha$-glycolic linkage : See p. 1191, Sec. 20.2.6.(c).

(ii) In the determination of the size of sugar rings : See p. 1191, Sec. 20.2.6.(c).

**25. Phosphorus pentachloride :**

(a) Preparation : By the action of excess chlorine on phosphorus.

(b) Uses :

(i) In the Beckmann rearrangement : See. p. 715, Sec. 14,5.

(ii) In the conversion of *p*-benzoquinone to *p*-dichlorobenzene : See. p. 1277, Sec. 20.8.4.

(iii) In the conversion of alcohols to alkyl chlorides : See p. 1148, Sec. 20.1.2(c)(i).

(iv) In the conversion of acids to acid chlorides : See p. 1297, Sec. 20.9.5(h).

**26. Palladium :**

Uses :

(i) In the catalytic hydrogenation : See p. 645.

(ii) In the aromatisation of alicyclic compounds : See p. 1075, Sec. 18.7.3(b).

(iii) In the Rosenmund reduction : See p. 1244, Sec. 20.7.2(i).

**27. Raney nickel :**

(a) Preparation : An alloy of Ni and Al (1 : 1) on digestion with 25% sodium hydroxide solution at 100 °C gives Raney nickel.

(b) Uses : It is a good reducing agent.

(i) It reduces alkenes to alkanes.

(ii) Nitriles to amines.

(iii) Aromatic and heterocyclic compounds to alicyclic compounds.

(iv) It may also be used to desulphurise thioacetals, thioketals, tetrahydrothiophene, etc.

**28. Sodium borohydride :**

Preparation : It is prepared by the reaction of trimethyl borate and sodium hydride at 250 °C.

Uses : See p. 856, Sec. 16.6.1. (note).

**29. Selenium dioxide :**

(a) Preparation : It is prepared by heating selenium metal in an atmosphere of oxygen in the presence of nitrogen dioxide which acts as a catalyst.

(b) Uses :

(i) In the conversion of $—CH_2CO$ compounds to dicarbonyls : See p. 1254, Sec. 20.7.5(vi).

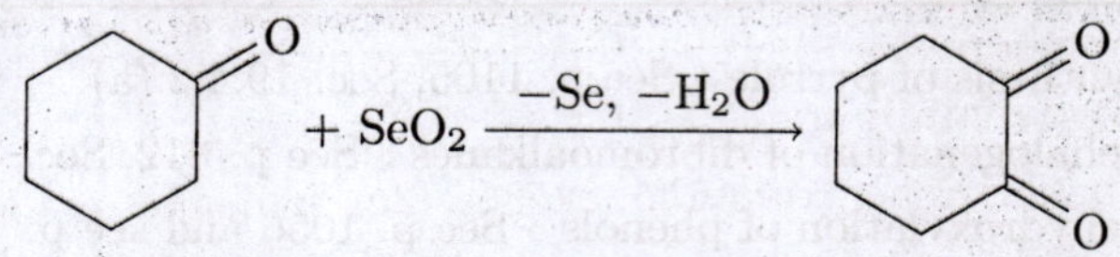

**(ii)** It is also used in the conversion of thioethers to sulphones.

$$Et_2S \xrightarrow{SeO_2} Et_2S{=}O \xrightarrow{SeO_2} Et_2S(=O)_2$$

**(iii)** Conversion of toluene to benzaldehyde.

**30. Sodamide :**

**(a)** Preparation : By passing dry ammonia gas on sodium at 350 °C.

**(b)** Uses :

- **(i)** In the synthesis of isoprene form acetone and acetylene : See P. 831, Sec. 16.4.2.
- **(ii)** In the synthesis of alkyl acetylenes from acetylene : See p. 1036, Sec. 18.3.2(f).
- **(iii)** In the Chichibabin reaction : See P. 596, Sec. 12.9.1.
- **(iv)** In the alkylation of alkyl cyanides : See p. 1326, Sec. 20.12.4.(j).
- **(v)** In the Darzen glycidic ester condensation : see p. 899, Sec. 16.13.4.

**31. Sodium ethoxide :**

**(a)** Preparation : By adding small amount of metallic sodium to absolute alcohol at 10 °C and then heating at 37.7 °C and again adding sodium and cooling to 10 °C and at last adding absolute alcohol.

**(b)** Uses :

- **(i)** In the Claisen condensation : See p. 894, Sec. 16.13.2.
- **(ii)** In the Dieckmann condensation : See p. 897, Sec. 16.13.2.
- **(iii)** In the Michael condensation : See p. 666, Sec. 13.13.2.
- **(iv)** In the Williamson synthesis of ethyl alkyl or ethyl aryl ethers : See p. 997, Sec. 20.1.4(xvii)
- **(v)** In the synthesis of indole : See p. 1114, Sec. 19.5.2.(b).

**32. Sodium nitrite :**

Uses :

- **(i)** In the diazotisation of 1° aromatic amines. ; See p. 1379, Sec. 20.18.2.
- **(ii)** In the Demyanov rearrangement : See p. 711, Sec. 14.4.
- **(iii)** In the conversion of diazonium group to nitro group : See p. 1382, Sec. 20.18.4.15
- **(iv)** In the preparation of nitroalkanes from $\alpha$-alkyl halides : See p. 998, Sec. 20.1.4(xx).
- **(v)** In the synthesis of 1, 3-indanedione from 1, 4-naphthaquinone. See p. 1283.

**33. Ozone :**

Uses : In the ozonolysis of alkenes and alkynes, and in determination of the position of C—C multiple bonds : See p. 1341, Sec. 13.8.

**34. Potassium permanganate :**

**(a)** Preparation : By the oxidation of potassium manganate in an alkaline electrolytic cell.

**(b)** Uses :

**(i)** In the detection of the presence of C—C double bond : See p. 635 Sec. 13.7.2.

**(ii)** In the process of cis-glycolisation of alkenes : See p. 635, Sec. 13.7.2.

**(iii)** In the degradation of alkenes by conc. $H_2SO_4$ and potassium permanganate : See p. 1033, Sec. 18.2.4.(t)

**(iv)** In the conversion of quinoline to pyridine : See p. 1134, Sec. 19.7.4.(b).

**(v)** In the conversion of thioethers to sulphones : See p. 1234, Sec. 20.5.10.(c).

**(vi)** In the oxidation of side chains of aromatic compounds : See p. 1661, benzaldalyde.

**35. Osmium tetroxide :**

**(a)** Preparation : It is prepared by heating powdered osmium in air or by heating it with aquaregia or nitric acid.

**(b)** Uses : It is used for cis-glycolisation of alkenes : See p. 633, Sec. 13.7.1.

**36. Zinc chloride (anhydrous) :**

**(a)** Preparation : By the action of hydrochloric acid on metallic zinc or zinc oxide and then heating to dryness :

**(b)** Uses :

**(i)** In the synthesis of diphenylamine : $PhOH + H_2NPh \xrightarrow[260\,°C]{ZnCl_2} Ph_2NH + H_2O$.

**(ii)** In the Gattermann aldehyde synthesis : See p. 989, Sec. 17.12.

**(iii)** In the chloromethylation of arenes : See p. 986, Sec. 17.10.

**(iv)** In the Houben-Hoesch synthesis : See p. 986, Sec. 17.9.2.

**(v)** In the synthesis of triphenyl methane.

Appendix C

# Chemistry of A Few Classified Compounds

## HYDROCARBONS

### Liquid Compounds

**1. *n*-Hexane :** $C_6H_{14}$; b.p. 69 °C; density (at 20 °C) 0.659 g/cc.

Colourless mobile liquid, highly inflammable and petrol-like odour.

When it is heated at 480–550 °C under pressure in the presence of chromic oxide on alumina support gives benzene.

Manufacturing process : By fractional distillation of petroleum.

Used as : Alcohol denaturant; a medium for polymerisation reaction, solvent for vegetable oils, etc.

**2. Cyclohexane :** b.p. 80 °C; density (at 20 °C) 0.778 g/cc.

Colourless liquid and odour is petrol-like : (i) No reaction with cold fuming $HNO_3$ but gives adipic acid (m.p. 151 °C) on boiling. (ii) It solidifies in freezing mixture. (iii) Fuming sulphuric acid converts it to benzenesulphonic acid.

Manufacturing process : catalytic hydrogenation of benzene.

Uses : Fungiside, solvent; solid fuel; and in nylon manufacturing.

**3. Cyclohexene :** b.p. 83 °C; density (at 20 °C) 0.810 g/cc.

Colourless liquid. Decolourises bromine in $CCl_4$ and gives adipic acid with conc. $HNO_3$.

Manufacturing process : By dehydrating cyclohexanol with conc. $H_2SO_4$.

Uses : Adipic and maleic acid manufacturing, oil extraction.

**4. *cis*-Decalin :** $C_{10}H_{18}$; b.p. 194 °C.

Colourless liquid.

Manufacturing process : By the hydrogenation of naphthalene.

Uses : Solvent for resin, rubber, oils, fats, waxes, etc; substitute for turpentine; cleaning machinery; stain-remover; shoe-creams; floor waxes, etc; cleaning fluids; lubricants.

**5. *trans*-Decalin :** $C_{10}H_{18}$; b.p. 185 °C.

Same as cis-Decaline given above.

**6. Benzene :** b.p. 80 °C; density (at 20 °C) 0.879 g/cc.

Colourless mobile liquid having characteristic aromatic odour. Insoluble in water but soluble in organic solvents.

Derivatives : (i) *m*-Dinitrobenzene. m.p. 90 °C. (ii) picrate, m.p. 84 °C.

Manufacturing process : (a) Hydrodealkylation of toluene, (b) catalytic reforming of petroleum, (c) fractional distillation of coal tar.

Used as : Ethylbenzene (for styrene monomer); cyclohexane (for nylon); dodecyl-benzene (for detergents); phenol; nitrobenzene (for aniline); maleic anhydride; chlorobenzene; diphenyl; benzenesulfonic acid; solvent.

**7. Toluene :** $PhCH_3$; b.p. 111 °C; density (at 20 °C); 0.866 g/cc.

Colourless mobile liquid having benzene-like odour. Insoluble in water but soluble in organic solvents.

Derivatives—(a) 2, 4-Dinitrotoluene, m.p. 71 °C. (b) 2, 4, 6-Trinitrotoluene, m.p. 82 °C.

On oxidation with alkaline $KMnO_4$ yields benzoic acid, m.p. 121 °C. When it is heated with pinch of $MnO_2$ and 1 ml conc. $H_2SO_4$ in the presence of $CuSO_4$, benzaldehyde is formed.

Manufacturing process : (a) By catalytic reforming of petroleum (b) By fractional distillation of light oil from coal tar.

Used as : Aviation gasoline and high octane blending stock; solvent for paints and coatings, gums, resins; rubber, vinyl organosols; explosives (TNT); toluene sulfonate (detergents).

**8. Ethylbenzene :** PhEt; b.p. 135 °C; density (at 20 °C) 0.867 g/cc.

Colourless mobilie liquid with a smell of benzene. Miscible with organic solvents but immiscible with water. Oxidation with alkaline $KMnO_4$ yields benzoic acid, m.p. 121 °C.

Derivative—2, 4, 6-Trinitroethylbenzene, m.p. 37 °C.

Manufacturing process : By beating benzene and ethylene in presence of aluminium chloride, at 95 °C with subsequent distillation.

Used as : Intermediate in production of styrene; solvent.

**9. *p*-Xylene :** *p*-$MeC_6H_4Me$; b.p. 138 °C; density (at 20 °C) 0.861 g/cc.

Colourless liquid having characteristic aromatic smell. Immiscible with water but miscible with alcohol and ether. On oxidation with alkaline $KMnO_4$ yields terephthalic acid, m.p. 300 °C and addition of bromine at 130 °C yields bromo-*p*-xylene.

Derivatives—(a) 2, 3, 5-Trinitro-*p*-xylene, m.p. 139 °C. (b) Bromo-*p*-xylene, m.p. 144 °C.

Manufacturing process : From coal tar by fractional distillation.

Used in the synthesis of terephthalic acid for polyester resins and fibres; vitamin and pharmaceutical syntheses, insecticides.

**10. *m*-Xylene :** *m*-$MeC_6H_4Me$; b.p. 139 °C; density (at 2 0°C) 8.864 g/cc.

Colourless liquid with benzene-like odour. Immiscible with water but soluble in organic solvents. On oxidation with alkaline $KMmO_4$ gives isophthalic acid. m.p. 348 °C.

Derivatives—(a) 2, 4, 6-Trinitro-*m*-xytene, m.p. 182 °C. (b) Monobromo, m.p. 76 °C.

Manufacturing process : Same as *p*-xylene.

Used as : Solvent, intermediate for dyes and organic synthesis, especially isophthalic acid; insecticides; aviation fuel.

**11. *o*-Xylene :** *o*-$MeC_6H_4Me$; b.p. 144 °C; density (at 20 °C) 0.880 g/cc.

Colourless mobile liquid having benzene-like smell. Insoluble in water but soluble in organic solvent. On Oxidation with alkaline $KMnO_4$ gives phthalic acid, m.p. 195 °C.

Derivatives—(a) 4, 5-Dinitro-*o*-xylene, m.p. 71 °C. (b) Monobromo-*o*-xylene, m.p. 76 °C. (c) Readily sulphonated by cold conc. $H_2SO_4$.

Manufacturing process : From coal tar by fractional distillation.

Used : in the manufacture of phthalic anhydride; vitamin and pharmaceutical syntheses, dyes; insecticides.

**12. Styrene :** Ph·CH=$CH_2$, b.p. 146 °C; density (at 20 °C) 0.907 g/cc.

Colourises liquid, benzene-like odour, sparingly soluble in water but completely soluble in organic solvent.

(i) Decolourises bromine water to give dibromide. (ii) Polymerises to a glassy mass with conc. $H_2SO_4$. (iii) Benzaldehyde is formed on heating with acidic $KMnO_4$. (iv) Reaction with alkaline $KMnO_4$ gives benzoic acid.

Derivative—Dibromide, m.p. 73 °C.

Manufacturing process : Catalytic dehydrogenation of ethyl benzene at 630 °C.

Used in the manufacture of polystyrene, protective coatings; styrenated polyesters; rubber-modified polystyrene;

**13. Cumene :** (isopropylbenzene), $PhCH(Me)_2$; b.p. 151 °C; density (at 20 °C) 0.862 g/cc.

Colourless liquid miscible with organic solvents. Oxidation by alkaline $KMnO_4$ gives benzoic acid (m.p. 121 °C).

Derivative—2, 4, 6-Trinitrocumene, m.p. 109 °C.

Manufacturing process : Alkylation of benzene with propylene (phosphoric acid catalyst).

Uses : Production of phenol, acetone, and alpha-methylstyrene, solvent.

**14. Mesitylene :** (1, 3, 5-trimethylbenzene); b.p. 164 °C, density (at 20 °C) 0.864 g/cc.

Colourless liquid miscible with organic solvents. (i) Oxidation by alkaline $KMnO_4$ gives trimeric acid, m.p. 300 °C. (ii) Very readily sulphonated by conc. $H_2SO_4$ to give sulphonic acid derivative which on boiling with HCl regenerates mesitylene.

Derivative—Trinitromesitylene, m.p. 230 °C.

Manufacturing process : acid catalysed condensation and dehydration of acetone.

Used as : Intermediate, including anthraquinone vat-dyes; ultraviolet oxidation stabilisers for plastics.

**15. *p*-Cymene :** $p$-$MeC_6C_6H_4CHMe_2$; b.p. 177 °C; density (at 20 °C) 0.857 g/cc.

Oily liquid having characteristic smell of thymol and eucalyptus oil (i) Oxidation by alkaline $KMnO_4$ gives terephthalic acid. (ii) Sulphonates slowly by warming with conc. $H_2SO_4$.

Manufacturing process : Dehydrogenation of monocyclic terpenes.

Used as : Solvents; in synthetic resin manufacture; metalpolishes; in the preparation of pure para-cresol and carvacrol.

**16. Tetralin (Tetrahydronapthalene) :** b.p. 207 °C.

Tetralin is used as a solvent for varnishes, lacquer, etc. (i) Oxidation by alkaline $KMnO_4$ gives phthatic acid. (ii) Bromination at 100 °C yields dibromo derivative, m.p. 70 °C, substitution occurs on the alicyclic ring. (iii) On catalytic (Pt-C) dehydrogenation gives naphthalene.

Derivative—Dibromo, m.p. 70 °C.

Manufacturing process : Hydrogenation of naphthalene in the presence of catalyst at 150 °C.

Used as : Solvent for greases, fats, oils. waxes; substitute for turpentine.

**17. α-Methylnaphthalene :** b.p. 240 °C.

Colourless liquid, miscible with organic solvent. (i) When it is treated with bromine water, 1-bromo-4-methyl naphthalene is formed. (ii) When it is treated with fuming $HNO_3$ and conc. $H_2SO_4$, 1-methyl-4-nitronaphthalene is formed.

Derivative—Picrate, m.p. 141 °C.

Manufacturing process : From coal tar.

Used as : Starting material in organic synthesis.

**18. Diphenylmethane :** $PhCH_2Ph$; b.p. 262°; m.p. 26 °C.

Colourless liquid, faint organic smell.

(i) Bromination of diphenylmethane gives diphenylmethylbromide. (ii) When it is oxidised with chromic acid, it form benzophenone. (iii) When its vapour is passed through red hot tube, it forms fluorene.

Derivative—Tetranitro derivative, m.p. 172 °C.

Manufacturing process : Condensation of benzyl chloride or methylene chloride and benzene in the presence of aluminium chloride.

Uses in : Organic synthesis; dyes; perfumery.

## Solid compounds

**19. β-Methylnaphthalene :** m.p. 34 °C, b.p. 240 °C.

White solid insoluble in water.

(i) On bromination in dark gives α-bromo-β-methylnaphthalene. (ii) Oxidation by conc. $HNO_3$ gives β-naphthoic acid, m.p. 182 °C. (iii) When it is sulphonated above 100 °C, it gives 6-sulfonic acid derivative.

Derivative—Picrate, m.p. 115 °C.

Manufacturing process : From coal tar.

Used : In organic synthesis of insecticides.

**20. Dibenzyl :** $PhCH_2{\cdot}CH_2Ph$; m.p. 52 °C; b.p. 294 °C.

White crystalline solid, soluble in alcohol and ether. (i) On treatement with cold conc. $HNO_3$ it forms dinitro derivative. (ii) When boiled with $KClO_3$ in HCl yield stilbene, m.p. 124 °C. (iii) Benzoic acid is formed when it is oxidised with alkaline $KMnO_4$.

Derivative—Dinitro, m.p. 180 °C.

Manufacturing process : (a) By treating benzyl chloride with metallic sodium. (b) Action of benzyl chloride on benzylmagnesium chloride.

Used in : Organic synthesis.

**21. Diphenyl :** Ph—Ph; m.p. 70 °C, b.p. 255 °C.

White crystalline solid having characteristic aromatic smell. Immiscible with water but miscible with alcohol and ether. When oxidised with chromic acid in glacial acetic acid gives benzoic acid and also ozonolysis of diphenyl at −20 °C gives 86% of benzoic acid, m.p. 121 °C.

Derivatives—(i) *p*, *p*′-Dinitro, m.p. 234 °C. (ii) *p*, *p*—Dibromo, m.p. 164 °C.

Manufacturing process : (a) By slowly passing benzene through a red hot iron tube. (b) By heating bromobenzene and sodium, with subsequent distillation.

Uses : Organic synthesis; heat-transfer agent; fungistat in packaging of citrus fruit; plant disease control; mfg. of benzidine; dyeing assistant for polyesters.

**22. Naphthalene :** m.p. 80 °C; b.p. 218 °C.

White shining crystalline solid. It has a characteristic odour and it is very volatile and used as an insecticide, insoluble in water but soluble in organic solvents; (i) oxidation by conc. $H_2SO_4$ in $HgSO_4$ gives phthalic acid. (ii) A green colour develops when dry $AlCl_3$ is added to a small amount of naphthalene in $CHCl_3$.

Derivative—Picrate, m.p. 150 °C.

Manufacturing process : (a) From coal-tar oils boiling between 200 to 250 °C followed by crystallisation and distillation. (b) From petroleum fractions.

Uses : Intetmediate; moth repellent; fungicide; smokeless powder; cutting fluid; lubricant; synthetic resins; textile chemicals; emulsion breakers; antiseptic.

**23. Triphenylmethene :** $(C_6H_5)_3CH$, m.p. 93 °C, b.p. 360 °C.

Colourless crystalline solid, insoluble in water but soluble in organic solvents and acidic in nature. (i) When oxidised with $CrO_3$ gives triphenyl carbinol, m.p. 165 °C. (ii) When brominated, it forms triphenylmethyl bromide.

Derivatives—(i) Trinitro, m.p. 206 °C. (ii) Monobromo, m.p. 152 °C.

Manufacturing process : From benzene and chloroform using $AlCl_3$ cat.

Used in : The formation of sodium triphenylmethide which is a very strong base.

**24. Acenaphthene :** m.p. 95 °C.

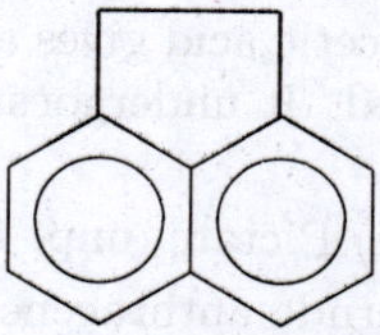

Colourless needle shaped crystal, faint aromatic smell. Sparingly soluble in alcohol but completely miscible with benzene and toluene. When it is oxidised by acid dicromate, naphthalic acid is formed.

Derivatives—(i) Mononitro, m.p. 101 °C. (ii) Picrate, m.p. 162 °C.

Manufacturing process : From coal tar.

Uses : Dye intermediate; pharmaceuticals; insecticide; fungicide; plastics.

**25. Phenanthrene :** m.p. 100 °C, b.p. 340 °C.

White shining crystal having faint aromatic smell. Immiscible with water but miscible with organic solvents. (i) Ozonolysis followed by oxidation or oxidation with peracetic acid gives diphenic acid. (ii) Oxidation with $Na_2Cr_2O_7$ or $CrO_3$ in glacial acetic acid gives phenanthraquinone. (iii) In benzene, it shows blue fluorescence.

Derivatives—(i) Monobromo, m.p. 65 °C. (ii) Yellow picrate. m.p. 143 °C.

Manufacturing process : Fractional distillation of high-boiling coal tar oils, with subsequent recrystallisation from alcohol.

Used : In manufacturing dyestuffs; explosives; synthesis of drugs; biochemical research; phenanthraquinone.

**26. Fluorene :** m.p. 114 °C.

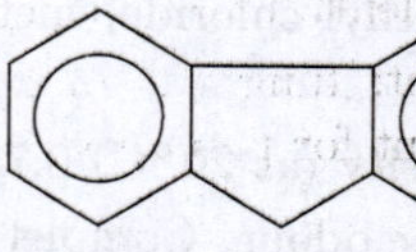

Colourless crystalline solid having faint aromatic smell. Insoluble in water but miscible with organic solvents. (i) Oxidation by cromic acid in glacial acetic acid, gives fluorenone, m.p. 84 °C. (ii) In warm state with conc. $H_2SO_4$, produces a blue colour. (iii) Oxidation by alkaline $KMnO_4$ gives phthalic acid, m.p. 195 °C.

Derivatives—(i) Picrate, m.p. 79 °C (unstable). (ii) Fluorenone, m.p. 84 °C.

Manufacturing process : By reduction of fluorenone with zinc; from coal tar.

Uses : Resinous products; dyestuffs.

**27. *trans*-Stilbene :** PhCH=CHPh; m.p. 124 °C. b.p. 307 °C.

Colourless crystalline solid. Insoluble in water but soluble in organic solvent. (i) On oxidation with acid permanganate gives benzaldehyde. (ii) When it is reduced by $Na/C_2H_5OH$, it yields dibenzyl.

Derivative—Dibromide, m.p. 237 °C.

Manufacturing process : By passing toluene over hot lead oxide.

Uses : In manufacturing of dyes and optical bleaches; crystals are used as phosphors and scintillators.

**28. Anthracene :** m.p. 216 °C; b.p. 354 °C.

Colourless solid having faint aromatic smell. Insoluble in water, but sparingly soluble in organic solvents. Oxidation by cromic acid in acetic acid gives anthraquinone, m.p. 286 °C which when reduced with Zn/HCl, bianthryl is formed. It undergoes Diels-Alder reaction with maleic anhydride.

Derivatives—(i) Dibromide, m.p. 221 °C. (ii) Picrate, m.p. 138 °C (red).

Manufacturing process : (a) By distilling crude anthracene oil with alkali carbonate in iron retorts the distillate containing only anthracene and phenanthrene. (a) The latter is removed by $CS_2$. (b) By salting out from crude anthracene oil, and draining.

Uses : Dyes; alizarin; carbazole; calico printing; as an ingredient of smoke screens; as scintillation counter crystals; organic semiconductor research.

## ALCOHOLS

### Liquid compounds

**1. Methyl alcohol :** $CH_3OH$; b.p. 64.5 °C; density (at 20 °C) 0.793 g/cc.

Colourless liquid with a characteristic smell. Completely soluble with water and organic solvents but almost insoluble in petroleum ether. It is highly volatile.

(i) When a red hot copper wire is dipped into 2-3 ml of methyl alcohol, formic acid is formed.

Derivatives—(i) *p*-Nitrobenzoyl, m.p. 96 °C. (ii) 3.5 Dinitrobenzoyl, m.p. 96 °C. (iii) Phenylurethan, m.p. 48 °C. (iv) Dimethyl phthalate, m.p. 82°C.

Manufacturing process : (a) By high-pressure catalytic synthesis from carbon monoxide and hydrogen, (b) Partial oxidation of natural gas, hydrocarbons, (c) Destructive distillation of wood.

Uses : Manufacture of formaldehyde and dimethyl terephthalate; antifreeze; solvent for nitrocellulose, ethylcellulose, manila resin, dyes; dehydrating agent for natural gas, denaturant for ethyl alcohol; chemical synthesis (methyl amines, methyl chloride, methyl methacrylate, automotive fuels).

**2. Ethyl alcohol :** $C_2H_5OH$; b.p. 78.3 °C, density (at 20 °C) 0.789 g/cc.

Colourless liquid with characteristic liquor like odour. Completly soluble in water and organic solvents. it is also volatile. (i) With $CH_3CO_2H$ and conc. $H_2SO_4$, It forms ethyl acetate. (ii) With $I_2/NaOH$ yields iodoform, m.p. 119 °C. (iii) Oxidation by $K_2Cr_2O_7/H_2SO_4$ gives acetaldehyde.

Derivatives—(i) 3, 5-Dinitrobenzoate, m.p. 94 °C. (ii) *p*-Nitrobenzoate, m.p. 56 °C. (iii) Phenylurethan, m.p. 54 °C.

Manufacturing process : (a) From ethylene by direct catalytic hydration or with ethyl sulfate as intermediate, (b) Fermentation of sugar-rich organic wastes (black-strap molasses, whey, degraded corn, etc.).

Uses : Solvent for resins, fats, fatty acids, oils, hydrocarbons; extraction medium, manufacture of acetaldehyde, acetic acid, ethylene, butadiene; detergent, cleaning preparations, surface coatings.

**3. iso-Propyl alcohol :** $(CH_3)_2CHOH$, b.p. 8.25 °C; density (at 20 °C) 0.789 g/cc.

Colourless liquid with alcoholic smell. Soluble in water and organic solvents; (i) gives the iodofom test. (ii) Oxidation with acid dichromate gives acetone.

Derivatives—(i) *p*-Nitrobenzoate, m.p. 111 °C; (ii) 3, 5-Dinitrobenzoate, m.p. 122 °C; (iii) Phenylurethan, m.p. 86 °C.

Manufacturing process : By treatment of propylene with sulphuric acid and hydrolysing.

Uses : In the manufacture of acetone and its derivatives; in the manufacture of glycerol and isopropyl acetate; solvent for essential and other oils, alkaloids, gums, resins, etc; latent solvent for cellulose manufacturing processes; coatings solvent; dehydrating agent; preservative; lotions; denaturant.

**4. tert-Butyl alcohol :** $(CH_3)_3C(OH)$; b.p. 83 °C; density (at 20 °C) 0.789 g/cc.

Colourless liquid, soluble in water. It does not give iodoform test.

Derivatives—(i) Carbanilate, m.p. 136 °C. (ii) 3, 5-Dinitrobenzoate, m.p. 142 °C. (iii) *p*-Nitrobenzoate, m.p. 26 °C. (iv) $\alpha$-Naphthyl carbamate, m.p. 100 °C.

Manufacturing process : Absorption of isobutylne, from cracking petroleum or natural gas, in sulphuric acid with subsequent hydrolysis by steam.

Uses : Alcohol denaturant, solvent for pharmaceuticals; dehydrating agent; perfumery; chemical intermediate; paint removers; manufacture of methyl methacrylate.

**5. *n*-Propyl alcohol :** $C_3H_7OH$; b.p. 97 °C; Density (at 20 °C) 0.804 g/cc.

Colourless liquid with a smell of alcohol. Completely miscible with water, ethyl alcohol and ether. Oxidation by $K_2Cr_2O_7/H_2SO_4$ gives propionaldehyde.

Derivatives—(i) Carbanilate, m.p. 58 °C. (ii) *p*-Ntrobenzoate, m.p. 35 °C. (iii) 3, 5-Dinitrobenzoate, m.p. 75 °C. (iv) $\alpha$-Naphthyl carbamate, m.p. 105 °C.

Manufacturing process : From oxidation of natural gas hydrocarbons; also form fusel oil.

Uses : Organic synthesis and chemical intermediate; solvent for waxes, vegetable oils, natural and synthetic resins and ethers; polishing compositions, brake fluids; solvent degreasing; antiseptic.

**6. Allyl alcohol :** $H_2C{=}CHCH_2OH$; b.p. 97 °C; density (at 20 °C) 0.855 g/cc.

Colourless liquid having pungent odour with lachrymatory properties. Soluble in water. (i) Decolourises bromine-water. (ii) On oxidation with $K_2Cr_2O_7/H_2SO_4$ gives acrolein and on oxidation with dilute alkaline $KMnO_4$ yields glycerol.

Derivatives—(i) Phenylurethan, m.p. 70 °C. (ii) 3, 5-Dinitrobenzoate, m.p. 50 °C.

Manufacturing process : (a) hydrolysis of allyl chloride (from propylene) with dilute caustic soda; (b) Isomerisation of propylene oxide over lithium phosphate cat. at 230–270 °C; (c) dehydration of propylene glycol.

Uses : Esters for use in resins and plasticisers; intermediate for pharmaceuticals; and other organic chemicals; manufacture of glycerol and acrolein; poison gas; herbicide.

**7. sec-Butyl alcohol :** $CH_3CHOHC_2H_5$; b.p. 99.5 °C; density (at 20 °C) 0.806 g/cc.

Colourless liquid, miscible with water. (i) Gives iodoform test. (ii) Oxidation by acid dichromate gives methyl ethyl ketone.

Derivatives—(i) 3, 5-Dinitrobenzoate, m.p. 75 °C. (ii) Phenylurethan, m.p. 64 °C.

Manufacturing process : Absorption of butene, from cracking petroleum or natural gas, in sulphuric acid with subsequent hydrolysis by steam.

Uses : Preparation of methyl ethyl ketone; lacquer solvent; organic synthesis; paint remover; industrial cleaners.

**8. tert-Amyl alcohol :** $(CH_3)_2C(OH)C_2H_5$; b.p. 102 °C; density (at 20 °C) 0.809 g/cc.

Colourless liquid, miscible with water.

Derivatives—(i) *p*-Nitrobenzoate, m.p. 85 °C. (ii) 3, 5-Dinitrobenzoate, m.p. 118 °C. (iii) Phenylurethan, m.p. 42 °C.

Manufacturing process : Fractional distillation of the mixed alcohols resulting from the chlorination of pentanes and alkaline hydrolysis of tert-amyl chloride.

Uses : Solvent; floatation agent; organic synthesis; medicine (sedative).

**9. iso-Butyl alcohol :** $(CH_3)_2CHCH_2OH$; b.p. 108 °C; density (at 20 °C) 0.802 g/cc.

Colourless liquid, miscible with water, alcohol, ether.

Derivatives—(i) Carbanilate, m.p. 80 °C. (ii) *p*-Nitrobenzoate, m.p. 68 °C. (iii) 3, 5-Dinitrobenzoate, m.p. 88 °C. (iv) $\alpha$-Naphthyl carbamate, m.p. 104 °C.

Manufacturing process : Bye product of synthetic methanol production, purified by rectification.

Uses : Organic synthesis; latent solvent in paints and lacquer; intermediate for amino coating resins; substitute for *n*-butyl alcohol. Paint removers; in fluorometeric determinations, liquid chromatography. In fruit flavour concentration.

**10. *n*-Butyl acohol :** $C_4H_9OH$; b.p. 118 °C; density (at 20 °C) 0.810 g/cc.

Colourless liquid miscible with about 10 volumes of water.

Derivatives—(i) $\alpha$-Naphthyl carbamate, m.p. 70 °C. (ii) *p*-Nitrobenzoate, m.p. 36 °C. (iii) 3, 5-Dinitrobenzoate, m.p. 64 °C. (iv) Phenylurethan, m.p. 61 °C.

Manufacturing process : (a) Hydrogenation of butyraldehyde, obtained in the Oxo process, (b) Condensation of acetaldehyde to form crotonaldehyde, which is then hydrogenated (aldol condensation).

Uses : Preparation of esters; solvent for resins and coatings, plasticisers; dyeing assistant; hydraulic fluids; detergent formulations; dehydrating agent.

**11. Methyl cellosolve :** $CH_2(OH)CH_2OCH_3$; b.p. 124 °C.

Colourless liquid soluble in water.

Derivative—*p*-Nitrobenzoate, m.p. 50 °C.

Manufacturing process : From ethylene oxide.

Uses : Solvent for nitrocellulose, natural and synthetic resins, lacquers; solvent mixtures; enamels; varnishes; leather; perfume; fixative; wood stain; sealing moisture-proof cellophane, jet fuel; de-icing; additive.

**12. iso-Amyl alcohol :** $(CH_3)_2CHCH_2CH_2OH$; b.p. 131 °C; density (at 20 °C) 0.813 g/cc.

Colourless liquid, sparingly soluble in water.

Derivatives—(i) Phenylurethan, m.p. 56 °C. (ii) 3, 5-Dinitrobenzoatc, m.p. 63 °C.

Manufacturing process : Distillation of fusel oil or the mixed alcohols resulting from the chlorination followed by hydrolysis of pentane.

Uses : Photographic chemicals; organic synthesis, pharmaceutical product; solvent; determination of fat in milk; microscopy.

**13. Ethyl cellosolve :** $CH_2(OH)CH_3OC_2H_5$; b.p. 135 °C.

Colourless liquid, insoluble in water.

Derivative—3, 5-Dinitrobenzoate, m.p. 76 °C.

Manufacturing process : From alkali cellulose and ethyl chloride or sulfate; from cellulose and ethyl alcohol in presence of dehydrating agents.

Uses : Hot-melt adhesives and coatings for cables, paper, textiles, etc; extrusion wire insulation; protective coatings; pigment-grinding bases; toughening agent for plastics; printing inks; moulding powders; proximity fuses; vitamin preparations; food and feed additive.

**14. *n*-Amyl alcohol :** $C_5H_{11}OH$; b.p. 138 °C; density (at 20 °C) 0.817 g/cc.

Colourless liquid, insoluble in water.

Derivatives—(i) Phenylurethan, m.p. 46 °C. (ii) 3, 5-Dinitrobenzoate, m.p. 46 °C.

Manufacturing process : Fractional distillation of the mixed alcohol resulting from the chlorination of pentane and subsequent alkaline hydrolysis of *n*-amyl chloride.

Uses : Raw material for (i) pharmaceutical preparations; in organic synthesis; solvent.

**15. Cyclohexanol :** b.p. 161.5 °C; density (at 20 °C) 0.962 g/cc.

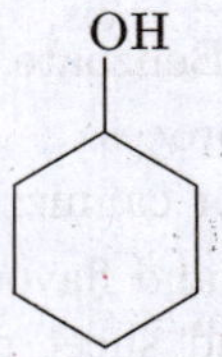


Colourless liquid, sparingly soluble in water, but soluble in organic solvents. Oxidation by hot conc. $HNO_3$ yields adipic acid, m.p. 151 °C.

Derivatives—(i) 3, 5-Dinitrobenzoate, m.p. 112 °C. (ii) m.p. 82 °C. (iii) carbamate, m.p. 12 °C. (iv) *p*-Nitrobenzoate, m.p. 50 °C.

Manufacturing process : Phenol is reduced with hydrogen over active nickel at 71 ° to 76.6 °C.

Uses : Soap making, source of adipic acid for nylon; textile finishing; solvent for alkyd and phenolic resins, cellulosic substance; blending agent; lacquers; paints and varnishes; finish removers.

**16. *n*-Heptyl alcohol :** $C_7H_{15}OH$; b.p. 176 °C; density (at 20 °C) 0.822 g/cc.

Colourless liquid, immiscible with water.

Derivatives—(i) Carbanilate, m.p. 65 °C. (ii) 3,5-Dinitrobenzoate, m.p. 48 °C.

Manufacturing process : From heptaldehyde by reduction.

Uses : Organic intermediate in synthesis, solvent, cosmetic formulation.

**17. Propylene glycol :** $CH_3CH(OH)CH_2OH$; b.p. 187 °C; density (at 20 °C) 1.040 g/cc.

Colourless liquid, soluble in water.

Derivative—(i) *p*-Nitrobenzoate, m.p. 128 °C. (ii) Carbnilate. m.p. 153 °C.

Manufacturing process : By hydration of propylene oxide.

Uses : Organic synthesis, especially polypropylene, glycol and polyester resins, cellophane; antifreeze solution; solvent for fats, oils, waxes, resins, flavouring extracts, perfumes, coloured soft drink syrups, hygroscopic agent; antioxidants; coolant in refrigeration systems; plasticisers; hydraulic fluids; bactericide; textile conditioners; emulsifier.

**18. Ethylene glycol :** $CH_2(OH)CH_2OH$. b.p. 197 °C. Density (at 20 °C) 1.113 g/cc.

Colourless liquid; soluble in water, alcohol and acetone but insoluble in benzene and ether. (i) On treatment with $HIO_4$ or $Pb(OAc)_4$ gives HCHO. (iii) When it is heated with $KHSO_4$, acetaldehyde is formed.

Derivatives—(i) Dibenzoate, m.p. 73 °C. (ii) Dicarbanilate, m.p. 157 °C. (iii) *p*-Nitrobenzoate, m.p. 119 °C. (iv) 3, 5-Dinitrobenzoate, m.p. 178 °C. (v) α-Naphthyl carbamate, m.p. 176 °C.

Manufacturing process : (i) Air oxidation of ethylene followed by hydration of the ethylene oxide formed; (ii) Acetoxylation, (iii) From carbon monoxide and hydrogen from coal gasification; (iv) Oxirane process.

Uses : Coolant and antifreeze; asphaltemulsion paint; heat-transfer agent; low pressure laminates; brake fluids; glycol diacetate; polyester fibres and films; low freezing dynamite; solvent; textile processing; tobacco.

**19. Benzyl alcohol :** $C_6H_5CH_2OH$; b.p. 205 °C; density (at 20 °C) 1.046 g/cc.

Colourless liquid, soluble in excess water. On oxidation with $KMnO_4$ yields benzoic acid, m.p. 121 °C.

Derivatives—(i) Benzoate, m.p. 21 °C. (ii) *p*-Nitrobenzoate, m.p. 310 °C.

Manufacturing process : (a) By hydrolysis of benzyl chloride; (b) From benzaldehyde by catalytic reduction or cannizzaro reaction.

Uses : Perfumes and flavours; photographic developer for coloured movie films; dyeing nylon filament, textiles and sheet plastics; solvent for dyestuffs, cellulose esters, casein, waxes, etc; heat-sealing polyethylene films; intermediate for benzyl esters and ethers; cosmetics, ointments, emulsions; bacteriostat; bull point pen ink; stencil inks.

**20. Glycerol :** b.p. 290 °C; density (at 20 °C) 1.261 g/cc.

Colourless liquid, soluble in water and alcohol but immiscible with benzene, $CHCl_3$ and ether. It is highly viscous. (i) It reduces Tollen's reagent. (ii) When it is heated with $KHSO_4$, acrolein is obtained (pungent smell).

Derivatives—(i) Tribenzoate, m.p. 76 °C. (ii) *p*-Nitrobenzoate, m.p. 188 °C. (iii) Tricarbanilate, m.p. 180 °C. (iv) α-Naphthyl carbamate, m.p. 191 °C.

Manufacturing process : (i) By-product of soap manufacture; (ii) From propylene and chlorine to form allyl chloride, which is converted to the dichlorohydrin with hypochlorous acid; this is then saponified to glycerol with caustic solution; (iii) Hydrogenation of carbohydrates with nickel catalyst.

Uses : Alkyd resins; dynamite; ester gums; pharmaceuticals; perfumery; cosmetics; food-stuffs; plasticiser for regenerated cellulose; liquors; solvent; printer's ink rolls; etc.

## Solid compounds

**21. Pinacol :** $(CH_3)_2C(OH){\cdot}C(OH)(CH_3)_2$; m.p. 35 °C.

White solid. Sparingly soluble in cold water but completely soluble in hot water, alcohol and ether.

Derivatives—(i) When it is distilled with conc. $H_2SO_4$, pinacolone is formed. (ii) Gives iodoform test. (iii) On cooling its hot aqueous solution crystals of pinacol monphydrate, m.p. 46 °C, separate out.

Manufacturing process : From acetone with amalgamated Mg or Al or Na in moist ether by reduction.

Uses : To form pinacolone.

**22. *l*-Menthol :** $C_{10}H_{10}OH$; m.p. 43 °C.

Colourless solid having a smell of peppermint. Immiscible with water but miscible with alcohol and ether. On heating with $K_2Cr_2O_7/H_2SO_4$ gives menthone, b.p. 207 °C.

Derivatives—(i) Phenylurethan, m.p. 112 °C. (ii) *p*-Nitrobenzoate, m.p. 62 °C. (iii) 3, 5-Dinitrobenzoate, m.p. 153 °C. (iv) Benzoate, m.p. 54 °C.

Manufacturing process : By freezing from peppermint oil; by hydrogenation of thymol.

Uses : Perfumery; cigaretts; liquors; flavouring agent; chewing gum; chest rubs; cough drops.

**23. Cinnamyl alcohol :** $PhCH{=}CHCH_2OH$; m.p. 33 °C, b.p. 257.5 °C.

Whitish solid, insoluble in water. (i) Oxidation by alkaline $KMnO_4$ gives benzoic acid. (ii) Decolourises bromine-water.

Derivatives—(i) Phenylurethan, m.p. 91 °C. (ii) 3, 5-Dinitrobenzoate, m.p. 121 °C. (iii) *p*-Nitrobenzoate, m.p. 78 °C.

Manufacturing process : (a) From oil of cassia or oil of cinnamon. Occurs as an ester. (b) Reduction of cinnamic aldehyde.

Uses : Perfumery, particularly for lilac and other floral scents, flavouring agent; cosmetics; soaps.

**24. *d*-Mannitol :** $HOCH_2(CHOH)_4CH_2OH$; m.p. 166 °C.

White solid. Soluble in water but insoluble in alcohol and ether. (i) A blue ppt. appears when a very small amount of *d*-mannitol is added to a clear blue solution of ammoniacal $CuSO_4$. (ii) On heating with $PCl_5$ gives hexachlorohexane, m.p. 137 °C.

Derivatives—(i) Hexabenzoate, m.p. 124 °C. (ii) Hexaacetate, m.p. 110 °C. (iii) Phenylurethan, m.p. 303 °C.

Manufacturing process : (i) A natural plant exudate. (ii) By hydrogenation of corn sugar or glucose.

Uses : Base for dietetic foods; diluent; determination of boron; pharmaceutical products; medicine; thickener and stabiliser in food products.

**25. Triphenylcarbinol :** $(C_6H_5)_3C(OH)$; m.p. 163 °C.

Colourless solid, insoluble in water. (i) Forms intense yellow solution in conc. $H_2SO_4$. (ii) Forms a Yellow solution in conc. HCl, from which triphenylchloramethane, m.p. 108 °C separates out.

Derivative—Acetate, m.p. 99 °C.

Manufacturing process : By the action of phenylmagnesium bromide on ethyl benzoate.

Uses : Dyestuffs

**26. Pentaerythritol :** $C(CH_2OH)_4$; m.p. 253 °C.

White solid, sparingly soluble in cold water.

Derivatives—(i) Tetrabenzoate, m.p. 99 °C. (ii) Tetraacetate, m.p. 84 °C.

Manufacturing process : Reaction of acetaldehyde with an excess of formaldehyde in an alkaline medium.

Uses : Alkyd resins, rosin and tall oil esters; special varnishes; pharmaceuticals, plasticisers; insecticides-, synthetic lubricants; explosives; paint swelling agents.

## ALDEHYDES

### Liquid compounds

**1. Formalin :** (40% aq. solution of HCHO) b.p. 98 °C.

Colourless liquid having a pungent odour. (i) It shows no colouration with sodium nitroprusside. (ii) A white ppt. of paraformaldehyde is obtained, when it is evaporated. (iii) Its very dilute solution gives reddish-brown ppt. with Nessler's reagent which turns grey on standing. (iv) Reduces Tollen's reagent and Fehling's solution.

Derivatives—(i) Semicarbazone, m.p. 169 °C. (ii) 2. 4-Dinitrophenylhydrazone. m.p. 155 °C.

Manufacturing process : Oxidation of synthetic methanol or low-boiling petroleum gases such as propane and butane, silver, copper, or an iron-molybdenum oxide are the most common catalysts.

Uses : Urea and melamine resins; preservative.

**2. Acetaldehyde :** $CH_3CHO$; b.p. 21°.

Colourless liquid having disagreeable and suffocating smell. It is a gas at room temperature. Its dilute aqueous solution has a smell of apple. Miscible with water and organic solvents. (i) It shows wine red colouration with sodium nitroprusside in NaOH solution. (ii) A yellow colour is obtained when aq. solution of acetaldehyde is added to a dilute solution of benzidine hydrochloride. (iii) It reduces Tollen's reagent and Fehling's solution.

Derivatives—(i) 2, 4-Dinitrophenylhydrazone, m.p. 162 °C. (ii) Semicarbazone. m.p. 163 °C.

Manufacturing process : (a) Oxidation of ethylene; (b) vapour-phase oxidation of ethanol; (c) vapour-phase oxidation of propane and butane; (d) catalytic reaction of acetylene and water.

Uses : Manufacture of acetic acid and acetic anhydride, *n*-butanol, 2-ethylhexanol, peracetic acid, aldol, pentaerythritol, pyridines, chloral, 1, 3-butylene glycol; synthetic flavours.

**3. Propionaldehyde :** $C_2H_5CHO$; b.p. 49 °C.

Colourless liquid. Soluble in water. It has a disagreeable odour. It reduces Tollen's reagent and Fehling's solution.

Derivatives—(i) Semicarbazone m.p. 89 °C. (ii) 2, 4-Dinitrophenylhydrazone, m.p. 156 °C.

Manufacturing process : (a) Oxidation of propyl alcohol with dichromate. (b) By passing propyl alcohol over copper at elevated temperatures.

Uses : In the manufacture of propionic acid, polyvinyl and other plastics; synthesis of rubber chemicals; disinfectant; preservative.

**4. *n*-Butyraldehyde :** $CH_3CH_2CH_2CHO$; b.p. 76 °C.

Colourless liquid, miscible with water. Reduces Fehling's solution and Tollen's reagent.

Derivatives—(i) Semicarbazone, m.p. 123 °C. (ii) 2, 4-Dinitrophenylhydrazone, m.p. 108 °C.

Manufacturing process : (a) Oxo process. (b) Dehydrogenating butanol vapours over a catalyst, the butyraldehyde being separated by distillation. (c) Selective reduction of crotonaldehyde.

Uses : Plasticisers, rubber accelerators, solvents, high polymers.

**5. Acrolein :** $H_2C{=}CHCHO$; b.p. 52 °C.

Colourless liquid having pungent irritating smell. Miscible with water. (i) Reduces Tollen's reagent and Fehlings's solution. (ii) Discharges the colour of bromine water and alkaline $KMnO_4$. (iii) With bromine water and subsequent oxidation by conc. $HNO_3$ gives $\alpha$, $\beta$-dibromopropionic acid. m.p. 64 °C.

Derivatives—(i) 2, 4-Dinitrophenylhydrazone m.p. 165 °C. (ii) Semicarbazone, m.p. 127 °C.

Manufacturing process : (a) Oxidation of allyl alcohol or propylene. (b) By beating glycerol with magnesium sulphate. (c) From propylene with bismuth-phosphorus-molybdenum catalyst.

Uses : Intermediate for synthetic glycerol, polyurethane, and polyester resins, methionine, pharmaceuticals; herbicide; warning agent in gases.

**6. Furfural :** b.p. 161 °C. 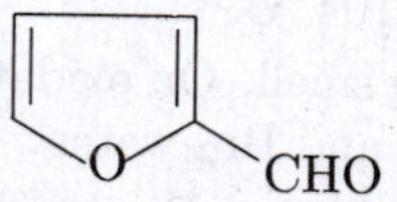

Colourless liquid turns brown in the presence of air and light. It has bitter almond-like odour. Soluble in water and organic solvents and heavier than water. It also reduces Fehling's solution and Tollen's reagent.

Derivatives—(i) Semicarbazone, m.p. 201 °C. (ii) 2, 4-Dinitrophenylhydrazone, m.p. 207 °C.

Manufacturing process : From oat hulls, rice hulls, corn cobs, bagasse and other cellulosic waste materials by steam-acid digestion.

Uses : Solvent for refining of lubricating oils, butadiene, rosin and other organic materials; solvent for nitro-cellulose, cellulose acetate, shoe dye; intermediate for tetrahydrofuran and furfuryl alcohol; road construction; production of lysine, refining of rare earths and metals; flavouring and analytical reagent.

**7. Benzaldehyde :** b.p. 179 °C.

Colourless oil-like liquid having a smell of bitter almonds. Insoluble in and heavier than water, but miscible with organic solvents. (i) Oxidation by alkaline $KMnO_4$ gives benzoic acid. (ii) On oxidation with conc. $HNO_3$ or chromic acid gives benzoic acid. (iii) It does not reduce Fehling's solution but it reduces Tollen's reagent very slowly giving a grey ppt.

Derivatives—(i) Phenylhydrazone, m.p. 74-77 °C. (ii) 2, 4-Dinitrophenylhydrazone, m.p. 237 °C. (iii) Oxime m.p. 87 °C. (iv) Semicarbazone, m.p. 167 °C.

Manufacturing process : (a) Air oxidation of toluene with uranium or molybdenum oxides as catalysts, (b) reaction of benzyl dichloride with lime; (c) extraction from oil of bitter almonds.

Uses : Chemical intermediate for dyes, perfumes, flavouring materials, and aromatic alcohols; solvent for oils, resins, some cellulose ethers, cellulose acetate and nitrate; synthetic perfumes; in the manufacture of cinnamic acid, benzoic acid; pharmaceuticals; photographic chemicals.

**8. Phenylacetaldehyde :** $PhCH_2CHO$; b.p. 194 °C.

Colourless liquid, heavier than water. On oxidation with alkaline $KMnO_4$ yields benzoic acid. Reduces Fehling's solution and Tollen's reagent.

Derivatives—(i) Semicarbazone, m.p. 231 °C. (ii) 2, 4-Dinitrophenylhydrazone, m.p. 261 °C.

Manufacturing process : From phenyl-alpha-chloroacetic acid, by action of alkalies; oxidation of $\beta$-phenylethyl alcohol.

Uses : Perfumes; flavouring and laboratory reagent.

**9. Cinnamaldehyde :** PHCH═CHCHO; b.p. 221 °C.

Colourless oil-like liquid having a smell of cinnamon. immiscible with water. (i) In the presence of air and lights, it is oxidised to cinnamic acid, m.p. 133 °C. (ii) When it is warmed with conc. KOH, it gives cinnamic acid. (iii) On oxidation with alkaline $KMnO_4$, yields benzoic acid. (iv) Decolourises bromine-water. (v) Reduces Tollen's reagent.

Derivatives—(i) Phenylhydrazone, m.p. 168 °C. (ii) 2, 4-Dinitrophenylhydrazone, m.p. 227 °C. (iii) Semicarbazone, m.p. 215 °C. (iv) Oxime, m.p. 137 °C.

Manufacturing process : (a) From Ceylon and Chinese cinnamon oils. (b) By condensation of benzaldehyde and acetaldehyde.

Uses : Flavours; spice perfumes.

**10. Crotonaldehyde :** $CH_3CH$═CHCHO; b.p. 104 °C.

Colourless liquid, miscible with water, irritating smell. On oxidation with $Ag_2O$ gives crotonic acid, m.p. 72 °C and decolourises alkaline $KMnO_4$ and $Br_2$-water.

Derivatives—(i) Phenylhydrazone, m.p. 56 °C. (ii) 2, 4-Dinitrophenylhydrazone, m.p. 145 °C. (iv) Oxime, m.p. 119 °C.

Manufacturing process : Aldol condensation of two molecules of acetaldehyde.

Uses : Intermediate for n-butyl alcohol and 2-ethylhexyl alcohol synthesis; solvent; preparation of rubber accelerators; purification of lubricating oils, insecticides; tear gas; fuel-gas warning agent; organic synthesis; leather tanning; alcohol denaturant.

## Solid compounds

**11. Piperonal :** m.p. 37 °C; b.p. 263 °C.

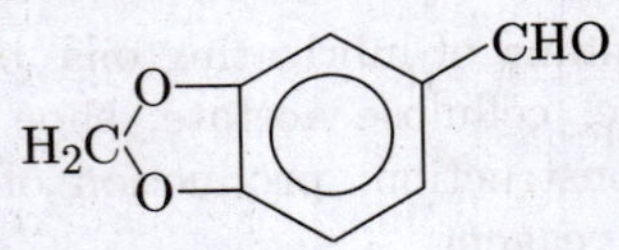

Colourless crystals having a smell of heliotrope. Insoluble in water but soluble in organic solvents. (i) With aqueous $KMnO_4$ gives piperonylic acid, m.p. 228 °C. (ii) It does not restore the colour of Schiff's reagent

Derivatives—(i) 2, 4-Dinitrophenylhydrazone, m.p. 265 °C. (ii) Semicarbazone, m.p. 234 °C. (iii) Oxime, m.p. 110 °C.

Manufacturing process : By oxidation of isosafrole.

Uses : Perfumery; mosquito repellent; laboratory reagent; flavouring agent.

## HYDROXY (PHENOLIC) ALDEHYDES

### Liquid compounds

**1. Salicylaldehyde :** $o$-$HOC_6H_4CHO$; b.p. 197 °C; m.p. 2 °C.

Colourless, oil-like liquid, miscible slightly with water. (i) Shows violet colouration with alcoholic $FeCl_3$ solution. (ii) On oxidation with alkaline $KMnO_4$ yields salicylic acid, m.p. 158 °C.

Derivatives—(i) Triacetyl derivative, m.p. 101 °C. (ii) Phenylhydrazone, m.p. 142 °C. (iii) 2, 4-Dinitro- phenylhydrazone, m.p. 151 °C. (iv) Semicarbazone, m.p. 231 °C. (v) Oxime, m.p. 57 °C.

Manufacturing process : Interaction of phenol and chloroform in the presence of aqueous alkali.

Uses : Analytical chemistry; perfumery (violet); synthesis of coumarin; auxiliary fumigant; flavouring agent.

### Solid compounds

**2. *m*-Hydroxybenzaldehyde :** m.p. 10 °C.

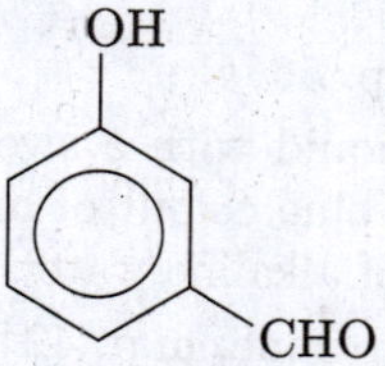


Colourless crystals, slightly soluble in water but soluble is organic solvents. With alcoholic solution of $FeCl_3$, it gives yellowish violet colouration.

Derivatives—(i) Semicarbazone, m.p. 198 °C. (ii) 2, 4-Dinitrophenylhydrazone, m.p. 259 °C.

Manufacturing process : May be prepared from *m*-nitrobenzaldehyle through Sn/HCl reduction followed by diazotisation and boiling with acidic water.

Uses : Intermediate for dyes, plastics, pharmaceuticals and bactericides; colour reagent for Schiff's reagent; sensitising agent in photographic emulsions.

**3. *p*-Hydroxybenzaldehyde :** m.p. 116 °C.

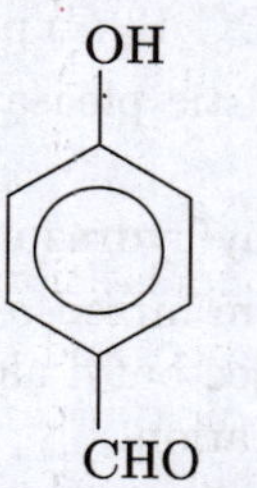


Very light pink coloured crystalline solid, soluble in hot water and organic solvents. Gives light violet colouration with aqueous $FeCl_3$ solution.

Derivatives—(i) 2, 4-Dinitrophenylhydrazone, m.p. 157 °C. (ii) Semicarbazone, m.p. 225 °C.

Manufacturing process : By Reimeir Tiemann reaction on phenol and separation of the *o*-/*p*-isomers by steam distillation.

Uses : In pharmaceuticals.

**4. Vanillin :** m.p. 80 °; b.p. 285 °C.

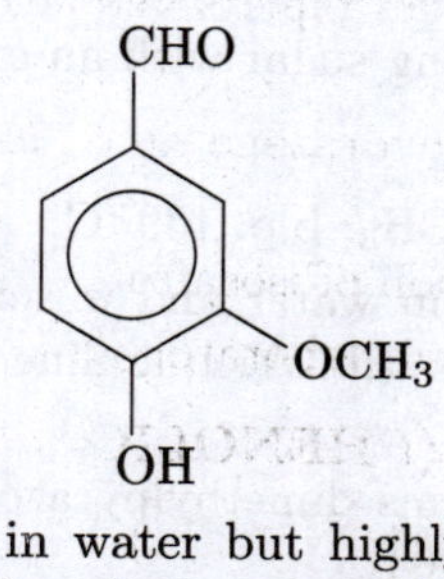


White crystals, sparingly soluble in water but highly soluble in organic solvents. It has a smell of vanila. (i) With aqueous $FeCl_3$ solution gives blue colour. (ii) It shows pink colour with Schiff's reagent. (iii) On oxidation by alkaline $KMnO_4$ gives vanillic acid, m.p. 207 °C.

Derivatives—(i) Benzoate, m.p. 77 °C, (ii) Oxime, m.p. 118 °C. (iii) Bromo derivative, m.p. 160 °C. (iv) Triacetyl, m.p. 87 °C. (v) Semicarbazone, m.p. 228 °C.

Manufacturing process : (a) By extraction of the vanilla bean; (b) from lignin contained in sulphite waste pulp liquor.

Uses : Perfumes; flavouring agent; pharmaceuticals; laboratory reagent; source of L-dopa.

## KETONES

### Liquid compounds

**1. Acetone :** $CH_3COCH_3$; b.p. 56 °C.

Colourless highly volatile liquid with a sweet smell. Soluble in water and organic solvents. (i) Gives iodoform test. (ii) A blue colour of indigo is formed when acetone reacts with *o*-nitrobenzaldehyde in the presence of alkaline water.

Derivatives—(i) Phenylhydrazone, m.p. 42 °C. (ii) 2, 4-Dinitrophenylhydrazone, m.p. 128 °C (iii) Semicarbazone, m.p. 187 °C. (iv) Oxime, m.p. 59 °C.

Manufacturing process : (a) Oxidation of cumene (b) dehydrogenation or oxidation of isopropyl alcohol with metallic catalyst; (c) vapour-phase oxidation of butane ; (d) byproduct of synthetic glycerol production.

Uses : Chemicals (methyl isobutyl ketone, methyl isobutyl carbinol; methyl methacrylate); paint varnish and lacquer solvent; solvent for potassium iodide and permanganate; delusterant for cellulose acetate fibers.

**2. Methyl ethyl ketone :** $CH_3COC_2H_5$; b.p. 80 °C.

Colourless liquid with characteristic pleasant odour. Soluble in water and organic solvents. Gives the iodoform test.

Derivatives—(i) 2, 4-Dinitrophenylhydrazone, m.p. 114 °C. (ii) Semicarbazone, m.p. 140 °C.

Manufacturing process : (a) From mixed *n*-butylenes and sulphuric acid to cause hydration, followed by distillation to separate sec-butyl alcohol, which is dehydrogenated. (b) By controlled oxidation of butane. (c) By fermentation.

Uses : Solvent in nitrocellulose coatings and vinyl films; "glyptal" resins; paint removers; adhesives; organic synthesis; cleaning fluids; printing; in the manufacture of smokeless powder; catalyst carrier.

**3. Diethyl ketone :** $C_2H_5COC_2C_5$; b.p. 102 °C.

Colourless liquid, miscible with water and organic solvents. Does not give iodoform test.

Derivatives—(i) 2, 4-Dinitrophenylhydrazone, m.p. 154 °C. (ii) Semicarbazone, m.p. 139 °C.

Manufacturing process : By distilling sugar with an excess of lime.

Uses : Medicine; organic synthesis.

**4. Acetylacetone :** $CH_3COCH_2COCH_3$; b.p. 139 °C.

Colourless liquid, slightly soluble in water and miscible with organic solvents. (i) It gives orange red colour with $FeCl_3$ solution. (ii) With aniline it gives "anil". m.p. 52 °C. (iii) Forms metal chetate complexes.

Derivatives—(i) Semicarbazide forms dimethylpyrazolecarboxylic imide, m.p. 109 °C. (ii) 2, 4-Dinitro-phenylhydrazone, m.p. 209 °C.

Manufacturing process : By condensing ethyl acetate with acetone.

Uses : Solvent for cellulose acetate; intermediate; chelating agent for metals; paint driers; lubricant additives; pesticides.

**5. Acetonylacetone :** $CH_3COCH_2CH_2COCH_3$; b.p. 194 °C.

Derivatives—(i) Semicarbazone, m.p. 221 °C. (ii) 2, 4-Dinitrophenylhydrazone, m.p. 257 °C.

Manufacturing process : Byproduct in the production of acetaldehyde from acetylene.

Uses : Solvent for cellulose-acetate, roll-coating inks, lacquers, stains; intermediate for pharmaceuticals and photographic chemicals; electroplating.

**6. Pinacolone :** $CH_3COC(CH_3)_3$; b.p. 106 °C.

Colourless liquid, slightly miscible with water. It has a smell of peppermint. (i) It does not form bisulphite adduct. (ii) On oxidation with chromic acid or iodine in NaOH solution yields trimethyl acetic acid, m.p. 35 °C.

Derivative—(i) Semicarbazone, m.p. 157 °C. (ii) 2, 4-Dinitrophenylhydrazone, m.p. 125 °C. (iii) Oxime, m.p. 74 °C.

Manufacturing process : From pinacol through pinacol-pinacolone rearrangement.

**7. Mesityl oxide :** $CH_3COCH{=}C(CH_3)_2$; b.p. 130 °C.

Colourless liquid having a smell of peppermint. Slightly soluble in water. (i) Decolourises bromine-water. (ii) When it is boiled with dil. $H_2SO_4$, acetone is formed. (iii) Gives "iodoform test".

Derivatives—(i) Semicarbazone, m.p. 156 °C. (ii) 2. 4-Dinitrophenylhydrazone, m.p. 203 °C. (iii) Oxime, m.p. 49 °C.

Manufacturing process : Dehydration of acetone or diacetone alcohol.

Uses : Solvent for cellulose esters and ethers, vinyl resins, lacquers; roll-coating inks; stains; ore floatation; paint and varnish-removers; insect repellent; etc.

**8. Cyclohexanone :** 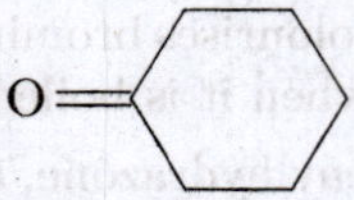

, b.p. 156 °C.

Colourless liquid, miscible with water. (i) When it is heated with conc. $HNO_3$, it yeilds adipic acid. (ii) When it is boiled with dil. NaOH, it gives acetone.

Derivatives—(i) Semicarbazone, m.p. 166 °C. (ii) 2, 4-Dinitropyenylhydrazone, m.p. 161 °C. (iii) Oxime, m.p. 57 °C.

Manufacturing process : By passing cyclohexanol over copper with air at 280 °F; also by oxidation of cyclohexanol with chromic acid or oxide.

Uses : Organic synthesis, particularly of adipic acid and caprolactam; polyvinyl chloride and its copolymers, and methacrylate ester polymers; wood stains; degreasing of metals; paint and varnish removers; spot removers; polishes; levelling agent in dyeing and delustering silk.

**9. *l*-Menthone :** 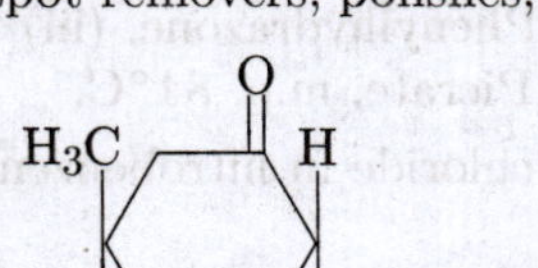

, b.p. 209 °C.

Colourless liquid having a smell of peppermint. Immiscible with water.

Derivatives—(i) Semicarbazone, m.p. 189 °C. (ii) 2, 4-dinitrophenylhydrazone, m.p. 147 °C.

Manufacturing process : The ketone is extracted from oil of peppermint.

Uses : Flavouring agent.

**10. Acetophenone :** b.p. 202 °C; m.p. 21 °C.

Colourless liquid having a sweet smell. Slightly miscible with water but completely miscible with organic solvents. (i) It gives orange colour in conc. $H_2SO_4$. (ii) Gives iodoform test. (iii) When reacts with alkaline $KMnO_4$, it gives benzoic acid.

Derivatives—(i) Semicarbazone, m.p. 161 °C. (ii) Phenylhydrazone, m.p. 105 °C. (iii) 2, 4-Dinitrophenylhydrazone, m.p. 250 °C. (iv) Oxime, m.p. 59 °C.

Manufacturing process : (a) Friedel-Crafts process with benzene and acetic anhydride or acetyl chloride; (b) byproduct from the oxidation of cumene (c) oxidation of ethylbenzene.

Uses : Perfumery; solvent; intermediate for pharmaceuticals, resins, etc; flavouring agent; polymerisation catalyst; organic synthesis.

**11. *p*-Methylacetophenone :** $H_3C$—  —$COCH_3$, b.p. 224 °C.

Colourless liquid, miscible with water and organic solvents. Responds to "iodoform test".

Derivatives—(i) Semicarbazone, m.p. 204 °C. (ii) 2, 4-Dinitrophenylhydrazone, m.p. 258 °C.

Manufacturing process : Action of acetic anhydride on toluene.

Uses : Perfumery; flavouring.

### Solid compounds

**12. Phorone :** $(CH_3)_2C{=}CHCOCH{=}C(CH_3)_2$; m.p. 28 °C; b.p. 197 °C.

Light yellow coloured solid, insoluble in water. (i) Decolourises bromine water to yield a yellow ppt. of tetrabromide, b.p. 88 °C. (ii) Acetone is formed when it is boiled with very dil. $H_2SO_4$.

Derivatives—(i) Semicarbazone, m.p. 234 °C. (ii) Phenylhydrazone, m.p. 74 °C.

Manufacturing process : Three molecules of acetone are condensed in the presence of dry HCl and $ZnCl_2$ or $AlCl_3$ to produce phorone.

Uses : Solvent for nitrocellulose; coating composition; stains; intermediate (organic synthesis).

**13. Methyl β-Naphthyl ketone :** 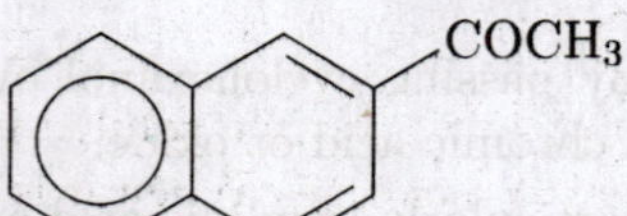

, m.p. 52 °C.

Colourless crystals, insoluble in water.

Derivatives—(i) Semicarbazone, m.p. 234 °C. (ii) Phenylhydrazone, (iii) 2. 4-Dinitrophenylhydrazone, m.p. 261 °C. (iv) Oxime, m.p. 141 °C. (v) Picrate, m.p. 84 °C.

Manufacturing process : (a) Reaction with acetyl chloride in nitrobenzene solution on naphthalene, gives methyl β-naphthyl ketone.

**14. *d*-Camphor :** $C_{10}H_{16}O$; m.p. 178 °C.

White solid having a characteristic smell. Insoluble in water. On oxidation with conc. $HNO_3$ gives camphoric acid, m.p. 187 °C.

Derivatives—(i) Semicarbazone, m.p. 237 °C. (ii) 2, 4-Dinitrophenylhydrazone, m.p. 177 °C.

Manufacturing process : Steam distillation of the camphor tree wood and crystallisation. This product is called natural camphor and is dextrarotatory. Synthetic camphor, most of which is optically inactive, may be made from pinene, which is converted into camphene, which by treatment with acetic acid and nitrobenzene becomes camphor, turpentine oil is also used.

**Uses :** Medicine (internal and external); plasticiser for cellulose nitrate, other explosives and lacquers; insecticides, moth and mildew preventives; tooth powder; flavouring; embalming; pyrotechines; intermediate.

**15. Benzalacetone :** PhCH═CHCOCH$_3$, m.p. 42 °C.

Light yellow crystals, insoluble in water. (i) It decolourises bromine-water. (ii) An orangish red colour is produced when it is dissolved in conc. $H_2SO_4$. (iii) Benzaldehyde is formed when it is reacted with acidic $KMnO_4$.

Derivatives—(i) Semicarbazone, m.p. 185 °C. (ii) 2, 4-Dinitrophenylhydrazone, m.p. 227 °C.

Manufacturing process : condensation of benzaldehyde and acetone.

Uses : Organic synthesis; perfumery; fixative, flavouring.

**16. Benzophenone :** PhCOPh. 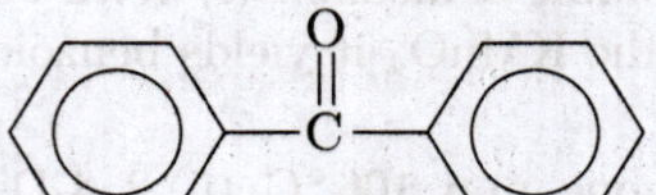 ; m.p. 49 °C.

Colourless crystals, insoluble in water but soluble in organic solvents. (i) When it is fused with metallic sodium, a deep blue colour is produced. (ii) In conc. $H_2SO_4$ it develops yellow solution. (iii) Benzoic acid is formed when it is fused with KOH followed by acidification.

Derivatives—(i) Phenylhydrazone, m.p. 136 °C. (ii) 2, 4-Dinitrophenylhydrazone, m.p. 238 °C. (iii) Semicarbazone, m.p. 164 °C. (iv) Oxime, m.p. 141 °C.

Manufacturing process : Friedel-Crafts reaction from benzene and benzoyl chloride.

Uses : Organic synthesis; odour fixative, derivatives are used as ultraviolet absorbers; flavouring; soap fragrance; pharmaceuticals; polymerisation inhibitor for styrene,

**17. Benzil :** PhCOCOPh 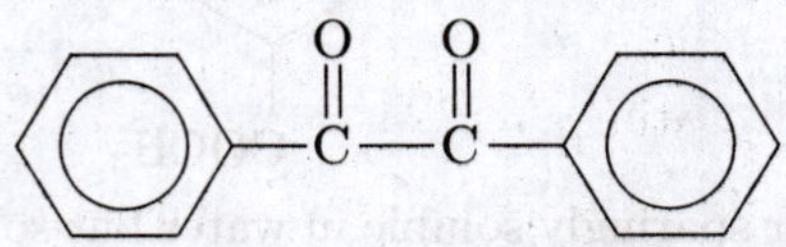 ; m.p. 95 °C.

Pale yellow crystals, insoluble in water but soluble in alcohol, ether, etc. (i) On oxidation with alkaline $KMnO_4$ yields benzoic acid. (ii) When it is boiled with alcoholic KOH and treated with excess HCl, it yields benzilic acid, m.p. 151 °C.

Derivatives—(i) 2, 4–Dinitrophenylhydrazone, m.p. 189 °C. (ii) Diphenylhydrazone, m.p. 224 °C. (iii) $\alpha$-dioxime, m.p. 237 °C.

Manufacturing process : From benzoin by oxidation with nitric acid.

Uses : Organic synthesis; insecticide.

## HYDROXY AND PHENOLIC KETONES

### Liquid compounds

**1. Diacetone alcohol :** $(CH_3)_2C(OH)\ CH_2COCH_3$; b.p. 166 °C.

Colourless liquid miscible with water and organic solvents. When it is boiled with dilute NaOH, it gives acetone.

Derivatives—2, 4-dinitrophenylhydrazone, m.p. 202 °C.

Manufacturing process : Condensation of acetone.

Uses : Solvent for nitrocellulose, cellulose acetate, various oils, resins, waxes, fats, dyes, tar; lacquers, dopes, coating compositions; wood preservatives; stains; imitation gold leaf, rayon and artificial leather; dyeing mixture; antifreeze mixtures; extraction of resins and waxes; preservative for animal tissue; metal-cleaning compounds; hydraulic compression fluids; laboratory reagent; stripping agent (textiles).

**Solid compounds**

**2. Benzoin :** PhCOCHOHPh 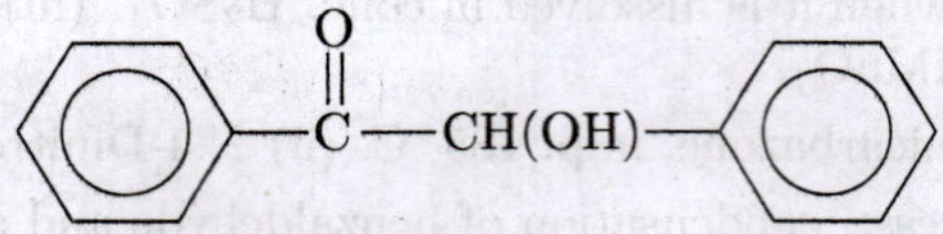 ; m.p. 130 °C.

Colourless crystals but on standing it acquires yellow colour. Insoluble in water, slightly soluble in ether but completely soluble in alcohol. (i) With Fehling solution, it gives a red ppt. of $Cu_2O$. (ii) On oxidation with acidic $KMnO_4$ it yields benzoic acid. (iii) When oxidised by conc. $HNO_3$ gives benzil, m.p. 95 °C.

Derivatives—(i) Phenylhydrazone, m.p. 106 °C. (ii) 2, 4-Dinitrophenylhydrazone, m.p. 245 °C. (iii) Benzoyl derivative, m.p. 125 °C. (iv) Semicarbazone, m.p. 205 °C. (v) Acetyl derivative, m.p. 82 °C. (vi) Oxime, m.p. 157 °C.

Manufacturing process : Condensation of benzaldehyde in an alkaline cyanide solution.

Uses : Organic synthesis; intermediate photopolymerisation catalyst.

**3. *p*-Hydroxyacetophenone :** *p*-$HOC_6H_4COMe$ 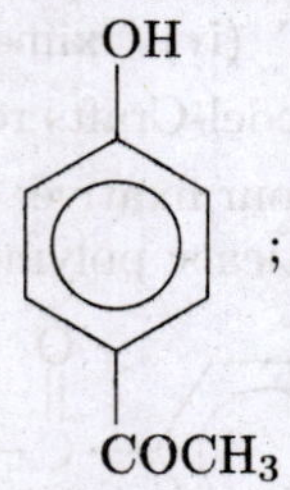 ; m.p. 109 °C.

Colourless cyrstals. Insoluble or sparingly soluble in water but soluble in organic solvents. On oxidation with acidic $KMnO_4$ yields *p*-hydroxybenzoic acid, m.p. 213 °C.

Derivatives—2, 4-Dinitrophenylhydrazone, m.p. 261 °C.

Manufacturing process : By the Fries rearangement of phenyl acetate in the carbon disulphide medium in the presence of anhydrous aluminium chloride and then extracting with toluene.

## QUINONES

**Solid compounds**

**1. *p*-Benzoquinone :** 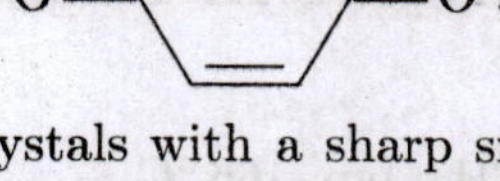, m.p. 117 °C.

Yellow prism-like crystals with a sharp smell. It sublimes when it is heated. It is slightly soluble in cold water but soluble in organic solvents. It is steam volatile. It turns brown on exposure to light. (i) Reduces Tollen's reagent. (ii) It adds on hydrogen chloride to form mainly chloroquinol (chlorohydroquinone), m.p. 106 °C. (iii) When it reacts with acetic anhydride in the presence of $H_2SO_4$, it yields hydroxyquinol triacetate.

Derivatives—(i) Oxime, m.p. 241 °C. (ii) Semicarbazone, m.p. 243 °C.

Manufacturing process : By oxidation of aniline with chromic acid, extraction with ether and distillation.

Uses : In the manufacture of dyes and hydroquinone; fungicides; analytical reagent; photography; oxidizing agent.

**2. α-Naphthaquinone** 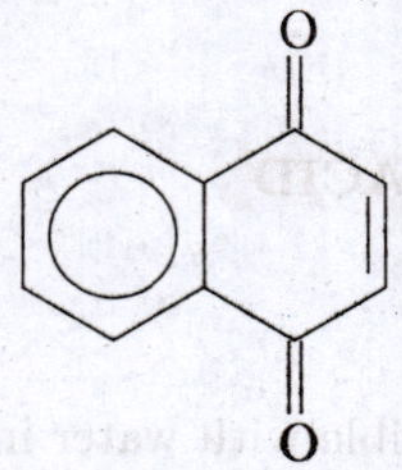 , m.p. 125 °C.

Yellow crystals with a pronounced odour. Sparingly soluble in water but soluble in organic solvent. (i) On oxidation with conc. $HNO_3$ yields phthalic acid. (ii) Shows reddish brown colour with dil. NaOH. (iii) It undergoes Deils-Alder reaction.

Derivatives—(i) Semicarbazone, m.p. 246 °C. (ii) Oxime, m.p. 198 °C.

Manufacturing process : By the oxidation of 4-amino-1-naphthol.

Uses : Polymerisation regulator for rubber and polyester resins; synthesis of dyes and pharmaceuticals; fungicide; algicide.

**3. Anthraquinone :** 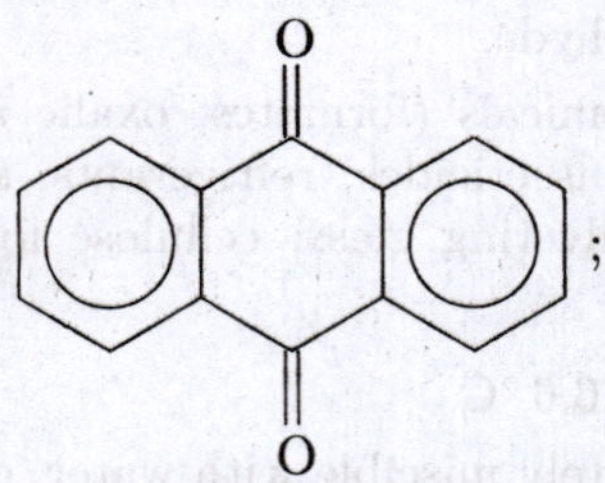 ; m.p. 286 °C.

Pale yellow crystal which sublimes to needles. Insoluble in water, sparingly soluble in benzene but soluble in alcohol. It has no smell. Not very volatile. (i) When distilled with Zn dust anthracene is formed. (ii) Remains unchanged in conc. $H_2SO_4$. (iii) Anthrone is formed when reduced with Sn/HCl in acetic acid.

Derivatives—Oxime, m.p. 225 °C.

Manufacturing process : (a) By heating phthalic anhydride and benzene in the presence of $AlCl_3$ and dehydrating the product; (b) by condensation of 1, 4-naphthaquinone with butadiene.

Uses : Intermediate for dyes; organic inhibitor; bird repellent for seeds.

**4. Phenanthraquinone :** 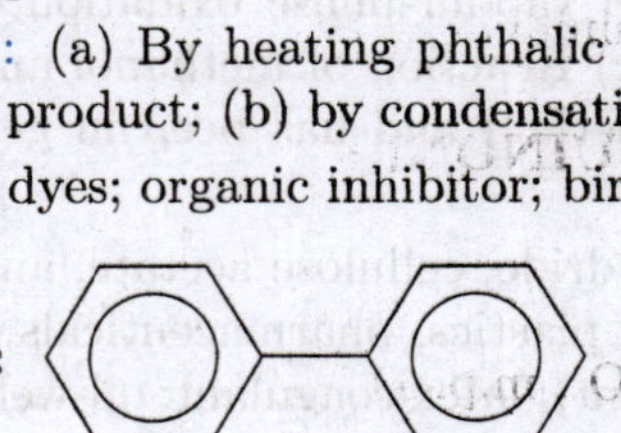 ; m.p. 208 °C.

Odourless orange crystals, slightly soluble in water. (i) When it is distilled with Zn dust phenanthrene is formed. (ii) It undergoes benzilic acid rearrngement (iii) With acid $KMnO_4$ gives diphenic acid, m.p. 228 °C.

Derivatives—(i) Semicarbazone, m.p. 221 °C. (ii) Oxime, m.p. 157 °C.

Manufacturing process : By oxidation of a boiling solution of phenanthrene in glacial acetic acid with chromic acid solution in sodium bisulphite, precipitation by means of hydrochloric acid and recrystallisation.

Uses : Organic synthesis; dyes.

## CARBOXYLIC ACID

### Liquid compounds

**1. Formic acid :** $HCO_2H$; b.p. 100.5 °C; m.p. 8 °C.

Colourless volatile liquid with pungent smell. Miscible with water in all proportions. (i) Decolourises acidic $KMnO_4$. (ii) Reduces $HgCl_2$, Tollen's reagent and Fehling's solution. (iii) When the neutral soln. of formic acid is heated with $AgNO_3$, a white ppt. is obtained which turns grey. (iv) When it is heated with conc. $H_2SO_4$, the issuing gas (CO) burns with a blue flame on ignition.

Derivatives—(i) Anilide, m.p. 47 °C. (ii) Ammonium salt, m.p. 114 °C. (iii) *p*-Toluidide, m.p. 53 °C.

Manufacturing process : (a) Treatment of sodium formate with sulphuric acid at a low temperature, and distilling in vacuo. (b) By acid hydrolysis of methyl formate. (c) As a byproduct in the manufacture of acetaldehyde and formaldehyde.

Uses : Dyeing and finishing of textiles; chemicals (formates, oxalic acid, organic esters); Leather treatment; manufacture of fumigants, insecticides, refrigerants, solvent for perfumes, lacquers; electroplating; brewing (antiseptic); silvering glass; cellulose formate; natural latex coagulant; vinyl resin plasticisers, ore floatation.

**2. Acetic acid :** $CH_3CO_2H$; b.p. 118 °C; m.p. 16.6 °C.

Colourless liquid with pungent smell. Completely miscible with water, ethanol and ether.

A disagreeable smell of cacodyl oxide is obtained when a neutral soln. of acetic acid evaporates to dryness and the residue is heated with $As_2O_3$.

Derivatives—(i) Anilide, m.p. 103 °C. (ii) Amide, m.p. 79 °C. (iii) Acid ammonium salt, m.p. 54 °C. (iv) *p*-Toluidide, m.p. 147 °C.

Manufacturing process : (a) Liquid and vapour-phase oxidation of petroleum gases (with catalyst). (b) Oxidation of acetaldehyde. (c) Reaction of methanol and carbon monoxide (with catalyst). This is the most cost-efficient method and has been in general use for some years; (d) fermentative oxidation of ethanol.

Uses : In the manufacture of acetic anhydride, cellulose acetate, and vinyl acetate monomer; acetic ester; chloroacetic acid; production of plastics, pharmaceuticals, dyes, insecticides, photographic chemicals, etc; food additive (acidulant); latex coagulant; oil-well acidiser; textile printing.

**3. Propionic acid :** $CH_3CH_2CO_2H$; b.p. 141 °C.

Colourless liquid miscible with water. It has acrid odour.

Derivatives—(i) Anilide, m.p. 103 °C. (ii) Amide, m.p. 79 °C. (iii) Acid ammonium salt, m.p. 54 °C. (iv) *p*-Toluidide, m.p. 124 °C.

Manufacturing process : By reaction of ethyl alcohol with carbon monoxide, using boron trifluoride catalyst; also by the reaction of carbon monoxide with hydrogen and olefins or alcohols.

Uses : Propionates, some of which are used as mould inhibitors in bread and fungicides in general; herbicides; preservative for grains and wood chips; solutions for electroplating nickel; perfume esters; emulsifying agents; artificial fruit flavours; pharmaceuticals; cellulose propionate plastics.

**4. Acrylic acid :** $H_2C{=}CHCO_2H$, b.p. 141 °C.

Colourless liquid, miscible with water. (i) Decolourises bromine-water to form dibromo derivative, m.p. 64 °C. (ii) Acrylic acid on standing slowly polymerises to a solid.

Derivatives—(i) Dibromo, m.p. 64 °C. (ii) Anilide, m.p. 104 °C. (iii) *p*-toluidide, m.p. 141 °C.

Manufacturing process : (a) Condensation of ethylene oxide with hydrocyanic acid followed by reaction with sulphuric acid at 320 °F. (b) Acetylene, carbon monoxide and water, with nickel catalyst. (c) Propylene vapour is oxidised to acrolein, which is oxidised to acrylic acid at 300 °C with molybdenum-vanadium catalyst. (d) Hydrolysis of acrylonitrile.

Uses : Monomer for polyacrylic and polymethacrylic acids and other acrylic polymers.

**5. *n*-Butyric acid :** $CH_3(CH_2)_2CO_2H$; b.p. 163 °C.

It is a viscous unpleasant smelling liquid, completely miscible with water, alcohol and ether.

Derivatives—(i) Anilide, m.p. 90 °C. (ii) Amide, m.p. 115 °C.

Manufacturing process : Occurs as glyceride in animal milk fats. Produced as a byproduct in hydrocarbon synthesis, by oxidation of butyraldehyde, and by butyric fermentation of molasses or starch.

Uses : Synthesis of butyrate ester perfume and flavour ingredients; pharmaceutical; deliming agent; disinfectants; emulsifying agents; sweetening gasolines.

**6. iso-Butyric acid :** $(CH_3)_2CHCO_2H$; b.p. 154 °C.

Colourless liquid, miscible with water.

Derivatives—(i) Anilide, m.p. 105 °C. (ii) Amide, m.p. 129 °C.

Manufacturing process : Dehydrogenation of isobutyl alcohol followed by oxidation.

Uses : In the manufacture of esters for solvents, tanning agents, flavours and perfume bases.

## Solid compounds

**7. Palmitic acid :** $CH_3(CH_2)_{14}CO_2H$; m.p. 62 °C; b.p. 269 °C.

It is a very weak acid. Waxy solid, immiscible with water but miscible with organic solvents.

Derivatives—(i) Anilide, m.p. 93 °C. (ii) Amide m.p. 106 °C. (iii) Methyl ester, m.p. 30 °C. (iv) Ethyl ester, m.p. 24 °C. (v) *p*-Toluidide, m.p. 98°C.

Manufacturing process : From palm oil; hydrolysis of natural fats.

Uses : In the manufacture of metallic palmitates; soaps; lube-oils; waterproofing; food grade additives.

**8. Stearic acid :** $CH_3(CH_2)_{16}CO_2H$; m.p. 69 °C; b.p. 287 °C.

Very weak acid. Waxy solid, insoluble in water, less soluble in alcohol.

Derivatives—(i) Anilide, m.p. 93 °C. (ii) Amide, m.p. 108 °C. (iii) Methyl ester, m.p. 38°C. (iv) Ethyl ester, m.p. 33°C. (v) *p*-Toluidide, m.p. 102°C.

Manufacturing process : (a) From high grade tallows and yellow grease stearin by washing, hydrolysis with the twitchell or similar reagent, boiling, distilling, cooling and pressing. (b) From oleic acid by hydrogenation.

Uses : Chemicals, especially stearates and stearate driers; lubricants; soaps; pharmaceuticals and cosmetics, dispersing agent and softener in rubber compounds; shoe and metal polishes; coatings; food packaging, suppositories and ointments.

**9. Glutaric acid :** $HOOC(CH_2)_3COOH$; m.p. 97 °C.

Odourless white solid, soluble in water, alcohol and ether, slightly soluble in benzene. (i) Gives the yellowish-green fluorescence test. (ii) On heating with aniline yields "anil", m.p. 144 °C.

Derivatives—(i) Anilide, m.p. 224 °C. (ii) Amide, m.p. 174 °C. (iii) *p*-Toluidide, m.p. 218 °C.

Manufacturing process : From cyclopentanone.

Uses : Organic synthesis; biochemical research.

**10. Phenylacetic acid :** $C_6H_5CH_2CO_2H$; m.p. 76 °C; b.p. 266 °C.

White crystalline solid with a perfume-like pleasant smell. Soluble in hot water. On oxidation with acid permanganate, benzoic acid is formed.

Derivatives—(i) Anilide, m.p. 117 °C. (ii) Amide, m.p. 154 °C. (iii) *p*-Toluidide, m.p. 136 °C.

Manufacturing process : From benzyl cyanide refluxed with dilute HCl.

Uses : Perfume; precursor in manufacturing of penicillin G, fungicide; flavouring; laboratory reagent.

**11. Citric acid :** $CH_2CO_2H$
|
$C(OH)CO_2H.\ H_2O$; m.p. 100 °C:
|
$CH_2CO_2H$

Crystalline solid, soluble in water and alcohol, less soluble in ether. (i) To the aqueous solution of the acid when Denige's reagent is added, boiled and treated with $KMnO_4$ solution drop by drop, the purple colour disappears and white ppt. is obtained. (ii) With $\beta$-naphthol in conc. $H_2SO_4$, blue colour appears, which does not change to green on warming. (iii) To the aqueous solution of the acid, a few drops of $AgNO_3$ solution are added, a white ppt. of silver citrate is obtained.

Derivatives—(i) Anilide, m.p. 199 °C. (ii) Amide, m.p. 215 °C. (iii) Methyl ester, m.p. 79°C. (iv) *p*-Toluidide, m.p. 189 °C.

Manufacturing process : By mould fermentation of carbohydrates, including deep fermentation; from lemon, lime, pineapple juice, molasses.

Uses : Preparation of citrates, confections, flavouring extracts, soft drinks, effervescent, salts; acidifier; dispersing agent; medicines; acidulants and antioxidant in foods; water-conditioning agent and detergent builder; cleaning and polishing of stainless steel and other metals.

**12. Oxalic acid :** $CO_2H$ , $2H_2O$; m.p. 101 °C.
|
$CO_2H$

Crystalline colourless solid, soluble in water, moderately soluble in alcohol and sparingly soluble in ether. It looses water of crystallisation on fusion. (i) When it is heated with conc. $H_2SO_4$ and the issuing CO gas is burnt, a blue flame is seen. (ii) The purple colour of $KMnO_4$ gets decolourised when the acid soln. is treated with dil. $H_2SO_4$ and $KMnO_4$ drop by drop. (iii) A white ppt. appears when the neutral solution of the acid is treated with $CaCl_2$ solution.

Derivatives—(i) Anilide, m.p. 245 °C. (ii) *p*-Toluidide, m.p. 267 °C.

Manufacturing process : Occurs naturally in many plants (wood sorrel, rhubarb, spinach) and can be made by alkali extraction of sawdust. Now manufactured by reaction of sodium formate with sodium hydroxide followed by distillation of the resulting dehydrated crystal or by reaction of CO with concentrated NaOH.

Uses : Automobile radiator cleanser; general metal and equipment cleaning; purifying agent; leather tanning; catalyst; laboratory reagent; stripping agent for permanent press resins; bleaching of textiles; rare-earth processing.

**13. Malonic acid :** $H_2C(CO_2H)_2$; m.p. 133 °C.

Colourless crystalline solid, completely soluble in water, alcohol and ether. (i) When warmed with $Ac_2O$ gives orange coloured solution with green fluorescene. (ii) On heating above its m.p. it gives acetic acid and $CO_2$. (iii) Reaction with bromine-water gives tribromo acetic acid, m.p. 135 °C, and $CO_2$.

Derivatives—Amide, m.p. 170 °C.

Manufacturing process : From monochloroarcetic acid by reaction with potassium cyanide, followed by hydrolysis.

Uses : Intermediate for barbiturates and pharmaceuticals.

**14. Ethylmalonic acid :** $C_2H_5CH(CO_2H)_2$; m.p. 111 °C.

Colourless crystalline solid, soluble in water. On heating above its m.p. it gives *n*-butyric acid, b.p. 163 °C, and $CO_2$.

Derivatives —*p*-nitrobenzyl ether, m.p. 75 °C.

Manufacturing process : From ethyl malonate by the base catalysed ethylation followed by hydrolysis.

Uses : Organic synthesis.

**15. Methylmalonic acid :** $CH_3CH(CO_2H)_2$; m.p. 135 °C.

Colourless solid, soluble in water. On heating above its m.p. it gives propionic acid, b.p. 140 °C.

Manufacturing process : From ethyl malonate by base catalysed methyl methylation followed by hydrolysis.

Uses : Organic synthesis.

**16. Succinic acid :** $CH_2CO_2H$ ; m.p. 185 °C.

$$\begin{array}{l} CH_2CO_2H \\ | \\ CH_2CO_2H \end{array}$$

Colourless crystalline solid, miscible with water, alcohol and acetone but sparingly soluble in ether. (i) It sublimes when it is fused in a test tube. (ii) It gives fluorescence test. (iii) When small amount of acid is heated with Zn dust and $NH_4OH$ and then a match stick soacked with conc. HCl is placed in the vapour coming out from the heated mixture, the stick turns red.

Derivatives—(i) Anilide, m.p. 226 °C. (ii) Amide, m.p. 242 °C. (iii) *p*-Toluidide, m.p. 220 °C.

Manufacturing process : Fermentation of ammonium tartrate.

Uses : Organic synthesis; manufacture of lacquers, dyes, esters for perfumes, photography; in foods as a sequestrant buffer, neutralising agent.

**17. Adipic acid :** $(CH_2)_4(CO_2H)_2$; m.p. 151 °C.

Colourless crystalline solid, sparingly soluble in cold water and ether but fairly in alcohol. (i) It gives a violet-red colour in fluorescence test (ii) $FeCl_3$ solution gives violet colouration with it.

Derivatives—(i) Anilide, m.p. 235 °C. (ii) Amide, m.p. 220 °C.

Manufacturing process : Oxidation of cyclohexane, cyclohexanol, or cyclohexanone with air or nitric acid.

Uses : Manufacture of nylon and of polyurethane foams; preparation of esters for use as plasticisers and lubricants; food additive (acidulant); baking powders; adhesives.

**18. Camphoric acid :** $C_8H_{14}(CO_2H)_2$; m.p. 187 °C.

Colourless crystalline solid, optically active, soluble in water. When it is heated with conc. $H_2SO_4$, CO is evolved, which burns with a blue flame on ignition.

Derivatives—(i) Anilide, m.p. 226 °C. (ii) Amide, m.p. 192 °C, (iii) Anil, m.p. 117 °C.

Manufacturing process : By oxidising camphor with nitric acid.

Uses : Pharmaceuticals; medicine.

**19. β-phenylpropionic acid :** $PhCH_2CH_2CO_2H$, m.p. 148 °C.

Colourless crystalline solid, insoluble in water. On heating with acid permanganate, benzoic acid is formed.

Derivatives—(i) Anilide, m.p. 98 °C. (ii) Amide, m.p. 105 °C. (iii) *p*-Toluidide, m.p. 135 °C.

Manufacturing process : Reduction of cinnamic acid with sodium amalgam.

Uses : Fixative for perfumes; flavouring.

**20. Cinnamic acid :** $PhCH{=}CHCO_2H$; m.p. 133 °C.

Shining cream coloured crystals, sparingly soluble in water and $CHCl_3$ but soluble in alcohol and ether. (i) Decolourises bromine-water on warming with the acid, forming a dibromide, m.p. 195 °C. (ii) Decolourises the purple colour of $KMnO_4$ and a brown ppt. is obtained in warming with it and a odour of bitter almonds of benzaldehyde is perceived. (iii) On oxidation with alkaline $KMnO_4$ followed by acidification, it yilds benzoic acid, m.p. 121 °C.

Derivatives—(i) Anilide, m.p. 153 °C. (ii) Amide, m.p. 141 °C. (iii) *p*-Toluidide, m.p. 168 °C.

Manufacturing process : By heating benzaldehyde with sodium acetate in presence of a dehydrating agent (acetic anhydride) or by heating benzyl chloride with sodium acetate in an autoclave. Occurs naturally in tobacco and some balsams.

Uses : Medicine(anthelmintic; destroy intestinal worms); perfumes; intermediate.

**21. Maleic acid :** $HO_2C{\cdot}CH{=}CHCO_2H$(cis); m.p. 130 °C.

White crystalline solid, soluble in water. (i) When it is oxidised by dilute alkaline $KMnO_4$, it yields meso tartaric acid. (ii) Decolourises bromine-water on heating and yields iso-dibromosuccinic acid, m.p. 160 °C. (iii) It gives an orange red flouorescence test.

Derivatives—(i) Anilide, m.p. 187 °C (ii) Amide, m.p. 181 °C. (iii) *p*-Toluidide, m.p. 142 °C.

Manufacturing process : Same as for maleic anhydride, with special recovery conditions.

[Maleic anhydride—(a) By passing a mixture of benzene vapour and or over a vanadium oxide catalyst at about 450 °C. (b) Byproduct from manufacture of phthalic anhydride from naphthalene. (c) Catalytic oxidation of butylenes]

Uses : Organic synthesis (malic, succinic, aspartic, tartaric, propionic, lactic, malonic and acrylic acids); and finishing of cotton, wool and silk; perservative for oils and fats.

**22. Fumaric acid :** $HO_2CCH{=}CHCO_2H$ (trans); m.p. 287 °C.

It is a colourless crystalline solid, sublimes at 200 °C. Slightly soluble in water, ether and acetone but soluble in benzene. Isomeric to maleic acid but does not give fluorescence test.

Derivatives—(i) Anilide, m.p. 314 °C. (ii) Amide, m.p. 266 °C. (iii) *p*-Nitrobenzyl ester, m.p. 150 °C.

Manufacturing process : (a) isomerisation of maleic acid; (b) catalytic oxidation of benzene.

Uses : Modifier for polyester, alkyd, and phenollic resin coatings, in foods, as a substitute for tartaric acid; as acidulant arid flavoring agent; mordant, printing inks.

**23. *d*-Tartaric acid :** $HO_2C(CHOH)_2CO_2H$; m.p. 169 °C.

White crystalline solid, highly soluble in water and alcohol but less soluble in ether. (i) Charring occurs when it is heated in a dry tube. (ii) A smell of burnt sugar is obtained when it is heated with conc. $H_2SO_4$. (iii) Responds to Fenton's test. (iv) A green colour is obtained when it is heated with *β*-naphthol in conc. $H_2SO_4$, which remains unchanged on warming. (v) It decolourises the purple colour of $KMnO_4$ when aqueous solution is heated with Denige's reagent and then $KMnO_4$ is added dropwise. (vi) With acetyl chloride yields diacetyltartaric anhydride, m.p. 126 °C.

Derivatives—(i) Anilide, m.p. 264 °C. (ii) Amide. m.p. 196 °C. (iii) Methyl ester, m.p. 48 °C.

Manufacturing process : Occurs naturally in wine lees; made synthetically from maleic anhydride and hydrogen peroxide and by an enzymatic reaction with a succinic acid derivative.

Uses : Chemicals (cream of tartar, tartar emetic, acetaldehyde); sequestrant; tanning, effervescent beverages; baking powder; fruit esters; ceramics; galvanoplastics; photography; textile industry.

**24. Benzoic acid :** $PhCO_2H$; m.p. 121 °C.

White shining crystals, slightly soluble in cold water but highly soluble in hot water and organic solvents. (i) Buff coloured ppt. is obtained with $FeCl_3$ solution. (ii) On nitration gives *m*-nitrobenzoic acid, m.p. 140 °C. (iii) Decarboxylation with soda-lime gives the vapour of benzene.

Derivatives—(i) Anilide, m.p. 161 °C. (ii) Amide, m.p. 128 °C. (iii) *p*-Nitrobenzyl ester, m.p. 89 °C. (iv) *p*-Toluidide. m.p. 158 °C.

Manufacturing process : (a) Decarboxylation of phthalic anhydride in the presence of catalysts. (b) Chlorination of toluene to yield benzotrichloride, which is hydrolysed to yield bezoic acid. (c) From benzoin resin. (d) Oxidation of toluene.

Uses : Sodium and butyl benzoates plasticisers, benzoyl chloride, alkyd resins; food preservative; seasoning tobacco; flavours; perfumes; standard in analytical chemistry; antifungal agent.

**25. *α*-Naphthoic acid :**

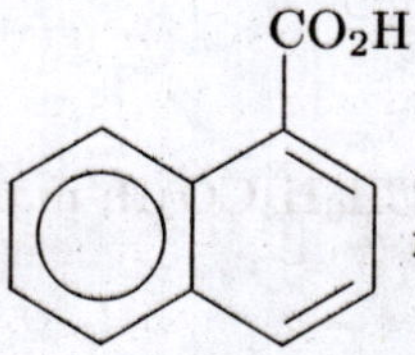


m.p. 161 °C.

Colourless crystalline solid, insoluble in cold water but soluble in hot water. Decarboxylation with soda-lime gives naphthalene, m.p. 80 °C.

Derivatives—(i) Anilide, m.p. 163 °C. (ii) Amide, m.p. 202 °C.

Manufacturing process : (i) By hydrolysis of $\alpha$-nitrile. (ii) By oxidation of $\alpha$-methylnaphthalene with dilute nitric acid. (iii) From $\alpha$-bromonaphthalene through the Grignard reagent.

**26. $\beta$-Naphthoic acid :** 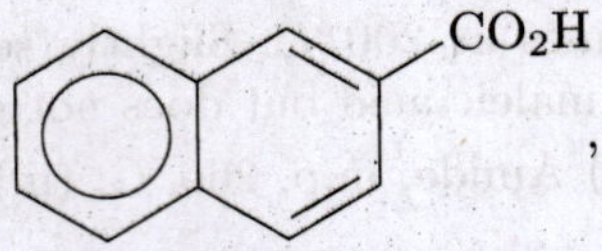 , m.p. 185 °C.

Colourless crystalline solid, insoluble in water but soluble in alcohol. Decarboxylation with soda-lime yields naphthalene.

Derivatives—(i) Anilide, m.p. 170 °C. (ii) Amide, m.p. 192 °C. (iii) *p*-Toluidide, m.p. 191 °C.

*Manufacturing process :* **(i) By the hydrolysis of $\beta$-nitrile. (ii) From sulphonic acid by fusion with sodium cyanide followed by hydrolysis. (iii) From amine through diazo reaction.**

**27. Phthalic acid :** *o*-$CO_2HC_6H_4CO_2H$; m.p. 195 °C to 213 °C.

White crystals, soluble in hot water and $CHCl_3$ but slightly soluble in ether. (i) Forms anhydride by loosing water on heating above its m.p. (ii) It gives fluorescence test.

Derivatives—(i) Phthalimide, m.p. 238 °C. (ii) *p*-Nitrobenzyl ester, m.p. 153 °C.

Manufacturing process : Catalytic oxidation of *o*-toluic acid and oxidation of *o*-xylene.

Uses : Dyes; phenolphthalein; phthalimide; anthranilic acid; synthetic perfumes; laboratory reagent.

**28. Terephthalic acid :** 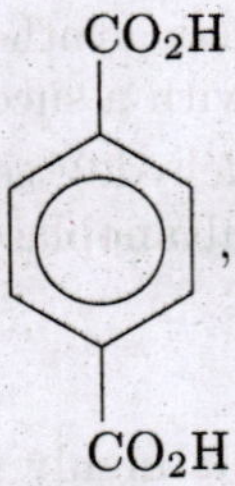 , *p*-$CO_2HC_6H_4CO_2H$; m.p. above 300 °C sublimes.

White crystalline solid, insoluble in water, ether and $CHCl_3$ but soluble in hot alcohol. Barium terephthalate is obtained when the acid is treated with $BaCl_2$ solution.

Derivatives—(i) Anilide, m.p. 337 °C. (ii) Amide, m.p. > 251 °C. (iii) *p*-Nitrobenzyl ester, m.p. 264 °C.

Manufacturing process : (i) Oxidation of *p*-xylene or of mixed xylenes and other alkyl aromatics. (ii) Reaction of benzene and potassium carbonate over a cadmium catalyst.

Uses : Production of linear, crystalline polyester resines, fibres and films by combination with glycols; reagent for alkali in wool; additive to poultry feeds.

**29. Isophthalic acid :** 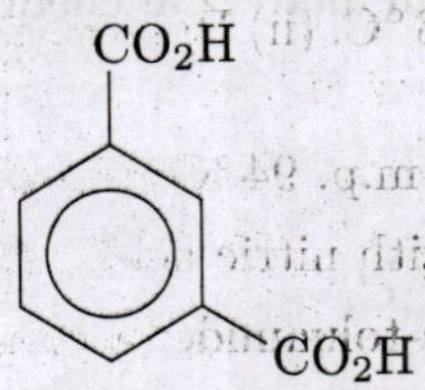 , *m*-$CO_2CH_6H_4CO_2H$; m.p. 348 °C.

White needle-type crystals slightly soluble in water.

Derivatives—(i) Anilide, m.p. 250 °C. (ii) Amide, m.p. 280 °C.

**Manufacturing process :** (a) Oxidation of *m*-xylene. (b) Liquid-phase oxidation of mixed xylenes. (c) Direct oxidation of mixed alkyl aromatics with heavy metal salts and bromine as catalysts.

**Uses :** Polyester, alkyd, polyurethane, and other high polymers; plasticisers.

**30. Furoic acid :** 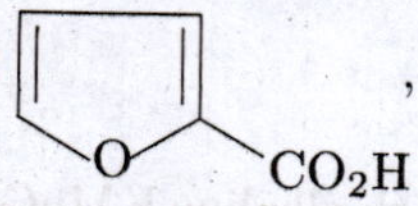 , m.p. 133 °C.

Colourless crystalline solid, soluble in water, alcohol and ether. (i) Furoic acid is heated with $NH_4OH$ and a match stick soaked with conc. HCl is kept in the vapour coming out from the mixture, the stick turns red. (ii) On decarboxylation with soda-lime, furan is obtained.

**Derivatives**—(i) Anilide, m.p. 123 °C. (ii) Amide, m.p. 141 °C. (iii) *p*-Nitrobenzyl ester, m.p. 133 °C.

**Manufacturing process :** By Cannizzaro reaction on furfural; oxidation of furfural.

**Uses :** Preservative; bactericide; furoates for perfume and flavoring; fumigant; textile processing; chemical intermediate.

**31. *o*-Toluic acid :** 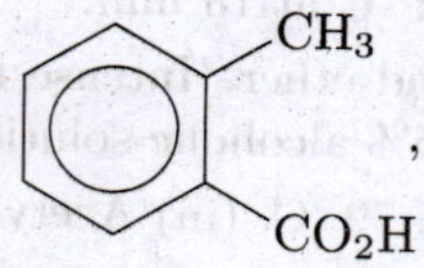 , *o*-$MeC_6H_4CO_2H$; m.p. 105 °C; b.p. 259 °C.

Prism like crystals, sparingly soluble in water but miscible with hot water and alcohol. (i) On oxidation with alkaline $KMnO_4$ yields phthalic acid, m.p. 195 °C. (ii) Decarboxylation with soda-lime, yields toluene, the vapours of which, on ignition, burns with a smoky flame.

**Derivatives**—(i) Anilide, m.p. 125 °C. (ii) Amide, m.p. 143 °C. (iii) *p*-Toluidide, m.p. 145 °C.

**Manufacturing process :** By oxidation of *o*-xylene.

**Uses :** Conversion to phthalic acid.

**32. *m*-Toluic acid :** 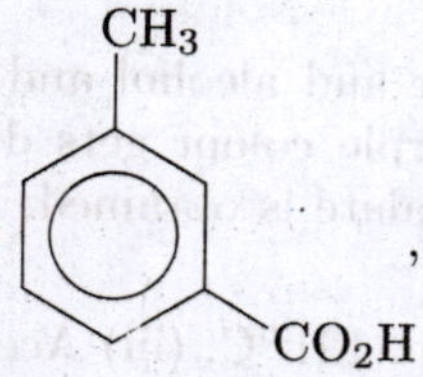 , *m*-$MeC_6H_4CO_2H$; m.p. 11°C; b.p. 263 °C.

Prism like crystals, slightly soluble in water but soluble in alcohol and ether, (i) Oxidation with alkaline $KMnO_4$ yields iso-phthalic acid, m.p. 348 °C. (ii) By decarboxylation with soda-lime, yields toluene.

**Derivatives**—(i) Anilide, m.p. 126 °C. (ii) Amide, m.p. 94 °C. (iii) *p*-Toluidide, m.p. 118 °C.

**Manufacturing process :** Oxidation of *m*-xylene with nitric acid.

**Uses :** Organic synthesis; to form N, N-diethyl–*m*-toluamide, a broad-spectrum insect repellent.

**33. *p*-Toluic acid :** 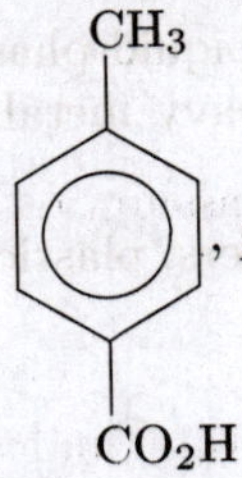 , *p*-$MeC_6H_4CO_2H$; m.p. 178°C; b.p. 275 °C.

Solid crystals miscible with hot water. (i) By reaction with alkaline $KMnO_4$ yields terephthalic acid, m.p. > 300 °C. (ii) Decarboxylation with soda-lime yields toluene, m.p. 110 °C.

Derivatives—(i) Anilide, m.p. 140 °C. (ii) Amide, m.p. 159 °C. (iii) *p*-Toluidide, m.p. 161 °C.

Manufacturing process : By treating cymene or turpentine with nitric acid.

Uses : Agricultural chemicals; animal feed supplement.

## HYDROXY ACIDS

### Liquid compounds

**1. Lactic acid :** $CH_3CHOHCO_2H$; m.p. 18 °C; b.p. 121 °C at 15 mm.

Colourless viscous liquid, soluble in water, alcohol and ether. Intense red colour is obtained when it is warmed with conc. $H_2SO_4$ in the preseme of 5% alcoholic solution of guaiacol.

Derivatives—(i) Anilide, m.p. 58 °C. (ii) Amide, m.p. 79 °C. (iii) Acetyl m.p. 57 °C.

Manufacturing process : (a) By fermenting starch, milk whey, molasses, potatoes, etc. and neutralising the acid as soon as formed with calcium or zinc carbonate. The solution of lactates is concentrated and decomposed with sulphuric acid. (b) Synthetically by hydrolysis of lactonitrile.

Uses : Cultured diary products; as acidulant; chemicals (salts, plasticisers, adhesives, pharmaceuticals); mordant in dyeing wool; general-purpose food additive; manufacture of lactates.

### Solid compounds

**2. Glycollic acid :** $HOCH_2CO_2H$; m.p. 79 °C.

Colourless crystalline solid, highly soluble in water and alcohol and moderately soluble in ether. (i) On oxidation with alkaline $KMnO_4$ the purple colour gets discharged and then on addition of $CaCl_2$ solution, a white ppt. of calcium oxalate is obtained. (ii) See the test under lactic acid.

Derivatives—(i) Anilide, m.p. 98 °C. (ii) Amide, m.p. 121 °C. (iii) Acetyl, m.p. 66 °C.

Manufacturing process : From chloroacetic acid by reaction with NaOH, or by reduction of oxalic acid. Occurs naturally in suger cane syrup.

Uses : pH control; leather dyeing and tanning; textile dyeing, cleaning, polishing and soldering compounds; copper picking, adhesives; electroplating; breaking of petroleum emulsions; chelating agent for iron, chemical milling.

**3. Citric acid :** See p. 1672.

Manufacturing process : From cell lemon or by the fermentation of molasses or starch with Aspergillus niger at pH 3-5

Uses : For cleaning metals and an acidulent in foods and soft drinks.

**4. Tartaric acid :** See p. 1675.

Manufacturing process : From grape juice as cream of tartar.

Uses : As the acid component of baking powder, sodium potassium tartrate (Rochelle salt; a constituent of Felhing solution) used as purgative.

**5. Malic acid :** $HO_2C\ CH_2CHOH\ CO_2H$; m.p. 101 °C.

Colourless crystalline solid, hygroscopic in nature. Soluble in water and alcohol but less soluble in ether. (i) When equal amounts of the acid and $\beta$-naphthol are heated together in the presence of conc. $H_2SO_4$ and kept in boiling water, a green-yellow colour with fluorescence is obtained. (ii) It gives blue fluorescrence with resorcinol and conc. $H_2SO_4$; the solution becomes red on making alkaline with NaOH in cold water. (iii) A blue colour is obtained when aqueous solution of the acid is treated with cobalt nitrate solution in the presence of excess NaOH solution.

Derivatives—(i) Anilide, m.p. 197 °C. (ii) Amide, m.p. 158 °C. (iii) *p*-Nitrobenzyl ester, m.p. 125 °C. (iv) *p*-Toluidide, m.p. 207 °C.

Manufacturing process : Occurs naturally in unripe apples, and other fruits, made synthetically by catalytic oxidation of benzene to maleic anhydride, which is converted to malic acid by heating with steam under pressure.

Uses : Manufacture of various esters and salts; wine manufacture; chelating agent; food acidulant; flavoring.

**6. *dl*-Mandelic acid :** $PhCHOHCO_2H$; m.p. 118 °C.

White crystals soluble in water, alcohol and ether. (i) On oxidation with acid permanganate, benzaldehyde is formed (confirmed by its bitter almonds smell). (ii) Gives the test no. (i) under malic acid when a wine red colour is developed.

Derivatives—(i) Anilide, m.p. 151 °C. (ii) Amide, m.p. 134 °C. (iii) Acetyl, m.p. 72 °C.

Manufacturing process : Hydrolysis of the cyanohydrin formed from benzaldehyde, sodium bisulphite, and sodium cyanide.

Uses : Organic synthesis; medicine (urinary antiseptic).

**7. Benzilic acid :** $Ph_2C(OH)\ CO_2H$; m.p. 150 °C.

Light yellow solid, soluble in hot water. (i) On chromic acid oxidation it yields benzophenone, m.p. 48 °C. (ii) In conc. $H_2SO_4$ it gives a red colour. (iii) It forms diphenylacetic acid when it is refluxed in acetic acid solution with HI and red phosphorus.

Derivatives—(i) Anilide, m.p. 175 °C, (ii) Amide, m.p. 154 °C. (iii) Acetyl, m.p. 97 °C.

Manufacturing process : By the benzilic acid rearrangement of benzil with an aqueous solution of caustic alkali.

Uses : Chemical intermediate.

**8. Mucic acid :** $HO_2C(CHOH)_4CO_2H$; m.p. 214 °C.

White solid, slightly soluble in water but insoluble in alcohol. When it is heated with $NH_4OH$ and a match stick shoaked with conc. HCl is kept in contact with vapours coming out from the mixture, the stick turns red.

Derivatives—(i) *p*-Nitrobenzyl ester, m.p. > 310 °C. (ii) Tetraacetyl, m.p. 267 °C.

Manufacturing process : Oxidation of lactose or similar carbohydrate with nitric acid.

Uses : Substitute for tartaric acid; sequesterant for metal ion (calcium, iron); retards hardening of concrete; intermediate for synthesis of heterocyclic compounds (pyrroles).

## AROMATIC HYDROXY (PHENOLIC) ACIDS

**9. Salicylic acid :** 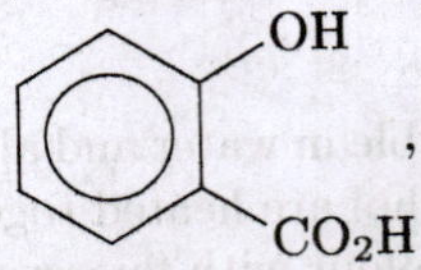

, m.p. 158 °C.

Colourless needle type crystals, slightly soluble in cold water but completely soluble in hot water, alcohol, benzene. (i) With $FeCl_3$ solution shows violet colouration. (ii) With methyl alcohol and conc. $H_2SO_4$, a characteristic smell of "oil of winter green" (methyl salicylate) is developed. (iii) When it is heated with soda-lime, a smell of carbolic acid is obtained due to the formation of phenol.

Derivatives—(i) Anilide, m.p. 134 °C. (ii) Amide, m.p. 139 °C. (iii) Acetyl, m.p. 135°C.

Manufacturing process : Reacting a hot solution of sodium phenolate with $CO_2$, and acidifying the sodium salt thus formed.

Uses : Manufacture of aspirin and salicylates; resins; dyestuff intermediate prevulcanisation inhibitor;

**10. *o*-Coumaric acid :** 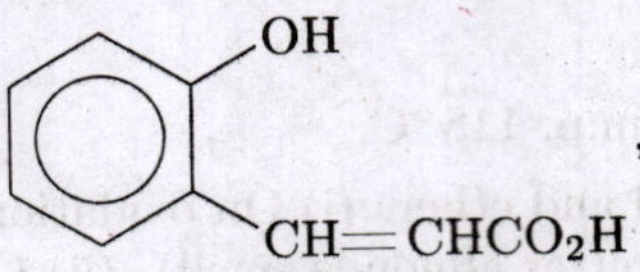

, *o*-$HOC_6H_4CH{=}CHCO_2H$; m.p. 208 °C.

*trans*

Solid crystals, slightly soluble in water. (i) On fusion with KOH yields salicylic acid after acidification, m.p. 158 °C. (ii) With $FeCl_3$ gives a ppt. (iii) It gives the reactions of unsaturated acids.

Derivative—Acetyl, m.p. 146 °C.

Manufacturing process : From salicylaldehyde by the Perkin condensation in the presence of triethylamine and then separating the product from coumarin followed by hydrolysis.

Its *cis*-isomer is coumarinic acid. It in unstable and readily forms δ-lactone (m.p. 69 °C).

**11. *m*-Hydroxybenzoic acid :** 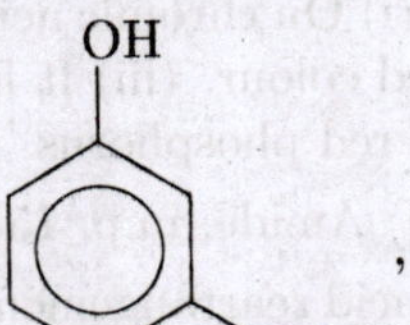

, m.p. 201 °C.

Solid crystals, sparingly soluble in cold water but soluble in hot water, alcohol, ether and also in acetone. It gives no colour with $FeCl_3$ solution.

Derivatives—(i) Acetyl, m.p. 127 °C. (ii) Anilide, m.p. 157 °C. (iii) Amide, m.p. 168 °C.

Manufacturing process : Nitration of ethyl benzoate to *m*-nitroderivative, hydrolysis and reduction of the latter to *m*-aminobenzoic acid which after diazotisation and boiling with acidulated water gives the compound.

Used as : Intermediate for plasticisers; resins; light stabilisers; petroleum additives; pharmaceuticals.

**12. *p*-Hydroxybenzoic acid :** 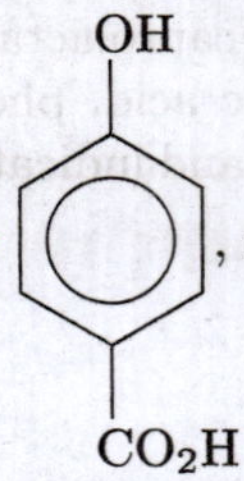, m.p. 213 °C.

Colourless needle-shaped crystals, sparingly soluble in water and benzene but miscible with alcohol acetone and ether. (i) With alcoholic $FeCl_3$ solution shows reddish violet colour. (ii) On methylation, gives anisic acid, m.p. 184 °C. (iii) On decarboxylation with soda-lime gives phenol.

Derivatives—(i) Anilide, m.p. 197 °C. (ii) Amide, m.p. 161 °C. (iii) Acetyl, m.p. 185 °C.

Manufacturing process : Interaction of *p*-aminobenzoic acid and nitrous acid.

Uses : Intermediate; synthetic drugs; food preservative.

**13. Resorcylic acid :** 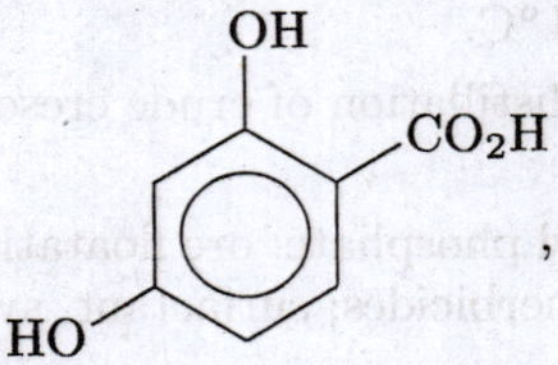, 2, 4-dihydroxybenzoic acid; m.p. 213 °C.

Solid crystals, low solubility in water. On heating with soda-lime gives resorcinol, m.p. 110 °C.

Derivatives—(i) Anilide, m.p. 127 °C. (ii) Amide, m.p. 220 °C.

Manufacturing process : By Kolbe-Schmidt synthesis on sodium salt of resorcinol.

Uses : Dyestuff and pharmaceutical intermediate; chemical intermediate in synthesis of fine organic chemicals; reagent for iron.

**14. 3-Hydroxy-2-naphthoic acid :** m.p. 223 °C.

Dirty white crystals, sparingly soluble in cold water. With $FeCl_3$ solution gives blue colour.

Derivatives—(i) Anilide, m.p. 246 °C. (ii) Amide, m.p. 218 °C. (iii) Acetyl, m.p. 185 °C.

Manufacturing process : By treating sodium 2-naphtholate with $CO_2$ under pressure.

Uses : Dyes and pigments.

## PHENOLS

### Liquid compounds

**1. Phenol :** PhOH, b.p. 181 °C; m.p. 14 °C.

Colourless liquid but turns pinkish in the presence of air and light. It has a characteristic carbolic odour. Slightly soluble in water but soluble in organic solvents. (i) With $FeCl_3$ solution, it gives violet colouration. (ii) Responds to "Phthalein test". (iii) It gives Libermann's reaction. (iv) On nitration with $HNO_3$ (fuming) and conc. $H_2SO_4$, yields picric acid, m.p. 122 °C.

Derivatives—(i) Tribromophenol, m.p. 93 °C. (ii) *p*-Nitrobenzoate, m.p. 127 °C. (iii) Benzoate, m.p. 68 °C. (iv) Acetate, m.p. 197 °C. (v) Picrate, m.p. 83 °C. (vi) *p*-Toluenesulphonate, m.p. 96 °C.

Manufacturing process : Oxidation of cumene, yielding acetone as a byproduct.

Uses : Phenolic resins; epoxy resins; nylon-6 (caprolactam); selective solvent for refining lubricating oils; manufacture of adipic acid, salicylic acid, phenolphthalien, pentachlorophenol, acetophenetidine, picric acid; germicidal paints; dyes and indicators; laboratory reagent; slimicide; general disinfectant.

**2. *o*-Cresol** 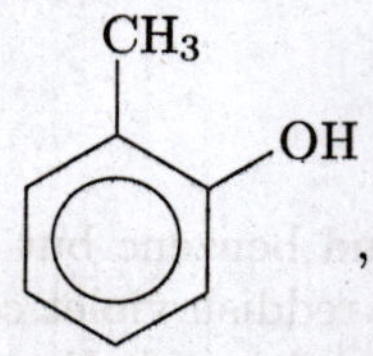 , *o*-$HOC_6H_4Me$; m.p. 31 °C; b.p. 191 °C.

Colourless liquid but turns pinkish in the presence of light and air. Slightly soluble in water, but soluble in organic solvents. (i) It can be distinguished from phenol by its immiscibility with $NH_4OH$. (ii) Gives Libermann's reaction.

Derivatives—(i) Dibromo derivative, m.p. 56 °C. (ii) Picrate, m.p. 87 °C. (iii) *p*-Toluenesulphonate, m.p. 55 °C. (iv) *p*-Nitrobenzoate, m.p. 94 °C.

Manufacturing process : (a) By fractional distillation of crude cresol from coal tar. (b) Interaction of methanol and phenol.

Uses : Disinfectant; phenolic resins; tricresyl phosphate; ore floatation; textile scouring agent; organic intermediate, coumarin synthesis and herbicides; surfactant, synthetic food flavours.

**3. *m*-Cresol :** *m*-HO—$C_6H_4Me$ 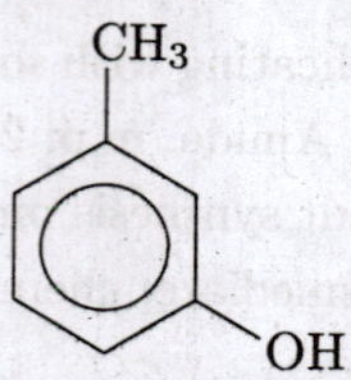 ; m.p. 11 °C; b.p. 201 °C.

Colourless liquid, turns pinkish in the presence of air and light. It is heavier than water. Slightly soluble in water but soluble in organic solvents. With fuming $HNO_3$ and conc. $H_2SO_4$ yields trinitrocresol, m.p. 105 °C.

Derivatives—(i) Benzoate, m.p. 55 °C. (ii) Bromo derivative, m.p. 84 °C. (iii) *p*-Nitrobenzoate, m.p. 91 °C. (iv) Pierate, m.p. 88 °C. (v) *p*-Toluenesulphonate, m.p. 56 °C.

Manufacturing process : (a) By fractional distillation of crude cresol (from coal tar); also synthetically.

Uses : Same as *o*-cresol.

**4. *p*-Cresol :** *p*-HO—$C_6H_4$—Me 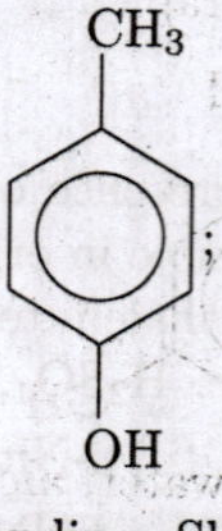 ; m.p. 35 °C; b.p. 202 °C.

Colourless liquid, turns dark on standing. Slightly soluble in water but completely soluble in organic solvents. (i) With aqueous $FeCl_3$ solution, it shows blue colouration. (ii) Responds to Libermann reaction.

Derivatives—(i) Tribromo derivative, m.p. 108 °C. (ii) Benzoate, m.p. 71 °C. (iii) *p*-Nitrobenzoate, m.p. 98 °C. (iv) Acetate, m.p. 214 °C. (v) *p*-Toluenesulphonate, m.p. 71 °C.

Manufacturing process : (a) By fractional distillation of crude cresol. (b) From benzene by the cumene process.

Uses : same as *o*-cresol.

**5. 1, 3, 4-Xylenol :** (2, 4-dimethylphenol) 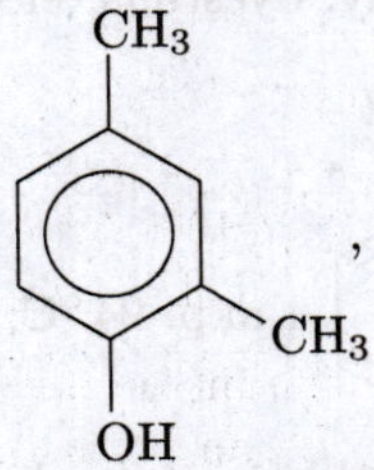

, b.p. 211 °C; m.p. 26 °C.

Colourless liquid, sparingly soluble in water. (i) It gives transient green blue colour with $FeCl_3$ in alcoholic solution. (ii) With $FeCl_3$ in aqueous solution shows transient blue-violet colouration.

Derivatives—(i) Tribromo derivative, m.p. 179 °C. (ii) benzoate, m.p. 38 °C.

Manufacturing process : Cresylic acid or tar acid fraction of coal tar.

Uses : Disinfectants; solvents, pharmaceuticals, insecticides and fungicides; plasticisers; rubber chemicals; additives to lubricants and gasolines; manufacture of polyphenylene oxide.

### Solid compounds

**6. Catechol :** 1, 2-dihydroxybenzene 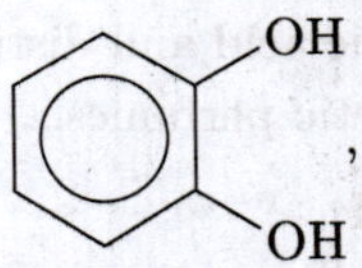

, m.p. 105 °C.

Colourless solid but darkens on exposure to air due to oxidation. It is a powerful reducing agent with faint phenolic odour. Highly soluble in water and organic solvents. (i) It reduces cold $AgNO_3$ solution and warm Fehling's solution. (ii) Shows green colour with $FeCl_3$ solution and turns red on addition of $Na_2CO_3$ solution. (iii) Solution of catechol shows brown colouration in the presence of NaOH solution on standing. (iv) A white ppt. is obtained when aqueous solution of catechol is treated with $Pb(CH_3CO_2)_2$ solution.

Derivatives—(i) Tribromo derivatice, m.p. 192 °C. (ii) Benzoate, m.p. 84 °C. (iii) *p*-Nitrobenzoate, m.p. 169 °C. (iv) Diacetate, m.p. 63 °C. (v) Picrate, m.p. 121 °C.

Manufacturing process : (a) By fusion of ortho-phenolsulphonic acid with caustic potash at 350 °C. (b) By heating guaiacol with hydroiodic acid.

Uses : Antiseptic; organic synthesis; photography; dyestuffs; electroplating; special inks; antioxidants.

**7. Resorcinol :** 1, 3-dihydroxybenzene 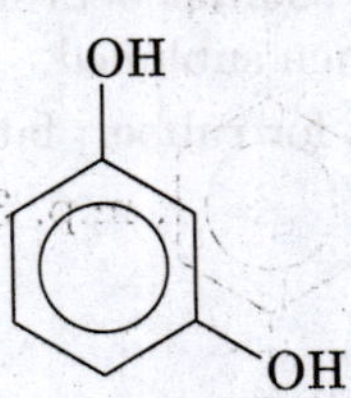

, m.p. 110 °C.

Purple-white coloured crystals. Soluble in water, alcohol, ether, $CHCl_3$ but slightly soluble in benzene. (i) It shows bluish-violet colouration with $FeCl_3$ solution. (ii) It gives fluorescence test. (iii) It reduces $AgNO_3$ solution and Fehling's solution on heating. (iv) With nitrous acid, it forms dinitrosoresorcinol.

Derivatives—(i) Dibenzoate, m.p. 117 °C. (ii) *p*-Nitrobenzoate, m.p. 182 °C. (iii) Diacetate, m.p. 278°C. (iv) *p*-Toluenesulphonate, m.p. 201°C.

Manufacturing process : (a) By fusion of meta-phenolsulphonic acid with caustic potash at 350 °C. (b) By heating guaiacol with hydroiodic acid.

Uses : Antiseptic; photography; dyestuffs; electroplating; specially in inks, antioxidants; organic synthesis.

**8. α-Naphthol :** 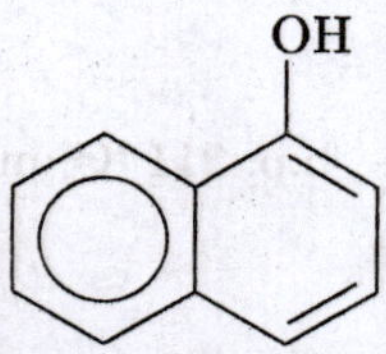



, m.p. 94 °C; b.p. 280 °C.

Light coloured crystals. Insoluble in water but highly soluble in organic solvents. (i) With $FeCl_3$ solution, it gives no colouration. (ii) It shows blue colour when it is warmed in dil. NaOH with $CCl_4$ and Cu powder. (iii) When it is warmed with titanic acid in conc. $H_2SO_4$, a green colour is obtained. (iv) It reduces ammoniacal $AgNO_3$ solution and is oxidised by alkaline $KMnO_4$ to phthalonic acid and by chromic acid to 1, 2- or β-napthaquinone.

Derivatives—(i) Benzoate, m.p. 56 °C. (ii) Picrate, m.p. 189 °C. (iii) 2, 4-Dibromo, m.p. 105 °C. (iv) Aetate, m.p. 47 °C. (v) *p*-Nitrobenzoate, m.p. 143 °C. (vi) *p*-Toluenesulphonate, m.p. 87 °C.

Manufacturing process : By fusing sodium alpha-naphthalene sulphonate and caustic soda. The melt is decomposed with hydrochloric acid and distilled.

Uses : Dyes; organic synthesis; synthetic perfumes.

**9. β-Naphthol :** 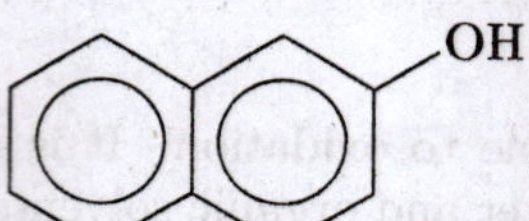



, m.p. 123 °C; b.p. 286 °C.

Colourless crystalline solid, crude crystals are brownish pink. Insoluble in water but soluble in organic solvents. It has a faint phenolic odour. (i) With $FeCl_3$ solution, a green ppt. of β-binaphthol is formed. (ii) It shows a blue colour when it is heated to 50 °C with $CHCl_3$ in KOH. (iii) It gives no colouration when it is warmed in KOH with $CCl_4$ and Cu-powder. (iv) With nitrous acid, it forms 1-oxime of β-naphthaquinone.

Derivatives—(i) 1-bromo-2-naphthol, m.p. 84 °C. (ii) Benzoate, m.p. 107 °C. (iii) *p*-Nitrobenzoate, m.p. 169 °C. (iv) Picrate, m.p. 156 °C. (v) Acetate, m.p. 71 °C. (vi) *p*-Toluenesulphonate, m.p. 125 °C.

Manufacturing process : By fusing sodium beta-naphthalene sulphonate with caustic soda. The product is distilled in vacuo and then sublimed.

Uses : Dyes; pigments, antioxidants for rubber; fats, oil; insecticides; synthesis of fungicides, pharmaceuticals, perfumes; antiseptic.

**10. Quinol or hydroquinone :** 1,4-dihydroxybenzene 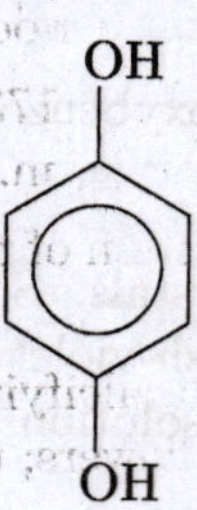



; m.p. 169 °C; b.p. 286 °C.

Colourless needle-shaped crystals, soluble in water, ether, alcohol and $CHCl_3$. (i) It gives transient blue colour with $FeCl_3$ solution which on boiling yields yellowish green crystals of quinhydrone after cooling. (ii) It is a powerful reducing agent. (iii) When warmed with excess of conc. $FeCl_3$ solution, yields benzoquinone, m.p. 116 °C.

**Derivatives**—(i) Dibenzoate, m.p. 199 °C. (ii) Diacetate, m.p. 123 °C. (iii) *p*-Toluenesulphonate, m.p. 159 °C. (iv) *p*-Nitrobenzoate, m.p. 251 °C.

**Manufacturing process :** Aniline is oxidised to quinone by manganese dioxide and is then reduced to hydroquinone.

**Uses :** Photographic developer (except coloured film); dye intermediate; stabiliser in paints and varnishes; motor fuels and oils; antioxidants for fats and oils; inhibitor of polymerisation.

**11. Pyrogallol :** 1, 2, 3-trihydroxybenzene 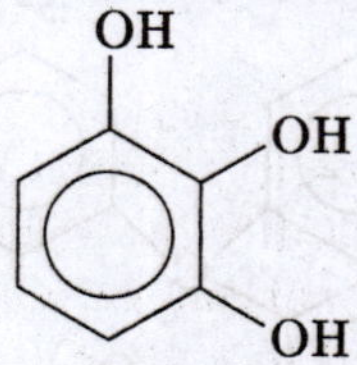 ; m.p. 133 °C.



Colourless, odourless solid, turns black on exposure to light and air. Highly soluble in water, ether, alcohol, etc. (i) It shows red colouration with $FeCl_3$ solution. (ii) With freshly prepared solution of $FeSO_4$, it shows blue-violet colouration. (iii) When it reacts with $Pb(CH_3CO_2)_2$ solution, a yellow emulsion is formed which changes to white ppt. (iv) Reduces Tollen's reagent and Fehling solution.

**Derivatives**—(i) Tribenzoate, m.p. 89 °C. (ii) Triacetate, m.p. 161 °C. (iii) *p*-Toluenesulphonate, m.p. 177 °C. (iv) benzoate, m.p. 152 °C.

**Manufacturing process :** By heating gallic acid with three times its weight of water in an autoclave.

**Uses :** Protective colloid in preparation of metallic colloidal solutions; photography; dyes; synthetic drugs; medicine(antibacterial); process engraving; laboratory reagent; reducing agent; antioxidant in lubricating oils.

**12. Phloroglucinol :** 1, 3, 5-trihydroxybenzene 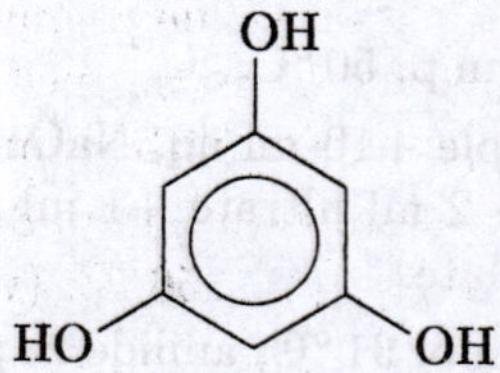 ; m.p. 218 °C.



Colourless, odourless crystalline solid, soluble in water, alcohol, ether and pyridine. (i) Gives blue-violet colouration with $FeCl_3$ solution. (ii) An alkaline solution of phloroglucinol rapidly darkens on exposure to air due to oxidation. (iii) It reduces Tollen's reagent and Fehling solution. (iv) It shows deep red colouration on wood or jute just moistened with conc. HCl.

**Derivatives**—(i) Tribenzoate, m.p. 173 °C. (ii) *p*-Nitrobenzoate, m.p. 284 °C. (iii) Triacetate, m.p. 105 °C. (iv) Tribromo derivative, m.p. 151 °C. (v) *p*-Toluenesulphonate, m.p. 159 °C.

**Manufacturing process :** By fusion of resorcinol with caustic soda; by reduction of trinitrobenzene.

**Uses :** Analytical chemistry;decalcifying agent for bones; preparation of pharmaceuticals and dyes, resins; preservative for cut flowers; textile dyeing and printing.

**13. Thymol :** 5-methyl-2-*i*-propylphenol, m.p. 51 °C.

Colourless crystals with characteristic smell of thyme. Slightly soluble in hot water but soluble in organic solvents. (i) Alcoholic solution of thymol gives transient green colour with $FeCl_3$ solution. (ii) A violet-red colour is produced when it is shaken with 50% acetic acid in the presence of conc. $H_2SO_4$.

Derivatives—(i) *p*-Toluenesulphonate, m.p. 77 °C. (ii) *p*-Nitrobenzoate, m.p. 70 °C.

Manufacturing process : From thyme oil or other oils, synthetically from meta cresol and isopropyl chloride by the Friedel-Crafts method at −10 °C.

Uses : Perfumery; mould and mildew preventive microscopy; antioxidant; preservative; flavouring and; laboratory reagent; synthetic menthol.

**14. Phenolphthalein :** 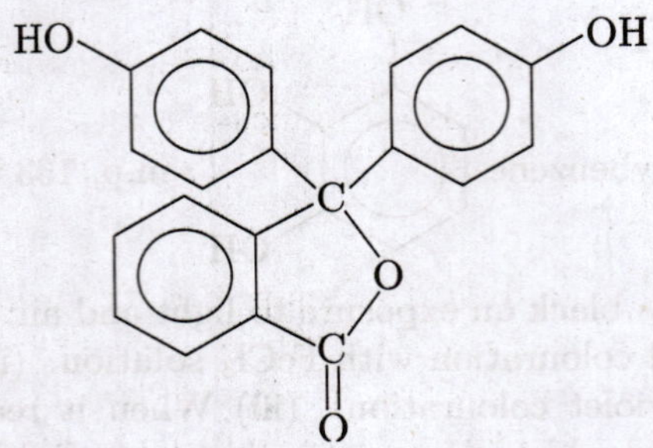 , m.p. 250 °C.

White powder, insoluble in water but soluble in alcohol. Used as an indicator. (i) It shows no colouration with $FeCl_3$ solution. (ii) Its alcoholic solution shows pink colour in NaOH solution but the colour discharges in excess conc. NaOH.

Derivatives—(i) Dibenzoyl derivative, m.p. 169 °C. (ii) Diacetyl derivative, m.p. 143 °C

Manufacturing process : Interaction of phenol and phthalic anhydride in sulphuric acid.

Uses : Acid-base indicator; laboratory reagent; laxative.

## HALO COMPOUNDS SOLID

**1. Bromoacetic acid :** m.p. 50 °C.

Soluble in water, sample +10 ml aq. NaOH ⟶ boiled (1 min) +0.5 g $KMnO_4$ and shaken vigorously and filtered → 2 ml filtrate +1 ml glacial acetic acid +a few drops of aq. $CaCl_2$ → white ppt. of calcium oxalate.

Derivatives—Amide, m.p. 91 °C; anilide m.p. 130 °C.

Manufacturing process : By heating acetic acid and bromine.

Uses : Organic synthesis; abscission of citrus fruit in harvesting.

**2. Trichloroacetic acid :** m.p. 55 °C.

Soluble in water; 1 g resorcinol +2 ml aq. NaOH and a pinch of the sample ⟶ heated to boiling ⟶ red soln., appears violet-red on shaking.

Derivatives—Amide, m.p. 141 °C; anilide m.p. 94 °C.

Manufacturing process : (a) Treating chloral hydrate with fuming nitric acid; (b) from glacial acetic acid by the action of chlorine in presence of sunlight, ultraviolet radiation or catalysts.

Uses : Organic synthesis; reagent for detection of albumin; medicine; pharmacy; herbicides.

**3. Chloroacetic acid :** m.p. 63 °C.

Soluble in water. Test: Same as in the case of bromoacectic acid.

Derivatives—Amide, m.p. 119 °C; Anilide, m.p. 134 °C.

Manufacturing process : Action of chlorine on acetic acid in the presence of acetic anhydride, phosphorus, or sulphur.

Uses : Herbicides; preservative; bacteriostat; intermediate in production of carboxymethyl cellulose, ethyl chloroacetate, glycine, synthetic caffeine, sarcosine, EDTA.

**4. *o*-Chlorobenzoic acid :** m.p. 137 °C.

Soluble in hot water. Yields benzoic acid with Na/Hg.

Derivatives—Amide, m.p. 142 °C; Anilide m.p. 114 °C.

Manufacturing process : By the oxidation of *o*-chlorotoluene by $KMnO_4$.

Uses : Intermediate for the preparation of dyes, fungicide, pharmaceuticals and other organic chemicals; preservative for adhesives and paints.

**5. *o*-Bromobenzoic acid :** m.p. 150 °C.

Soluble in hot water. It yields dinitro derivative (m.p. 213 °C) on heating with conc. $HNO_3$ and conc. $H_2SO_4$ at 140 °C.

Derivatives—Amide, m.p. 155 °C; Anilide m.p. 141 °C.

Manufacturing process : By the $KMnO_4$ oxidation of bromotoluene.

Uses : Organic synthesis.

**6. *m*-bromobenzoic acid :** m.p. 155 °C.

Sparingly soluble in water.

Derivatives—Amide, m.p. 155 °C; Anilide, m.p. 137 °C.

Manufacturing process : From *m*-aminobenzoic acid by diazotisation followed by Sandmeyer reaction or bromination of benzoic acid.

Uses : Organic synthesis.

**7. *m*-chlorobenzoic acid :** m.p. 158 °C.

Sparingly soluble in cold water.

Derivative—Amide, m.p. 134 °C.

Manufacturing process : Chlorination of benzoic acid.

Uses : Same as *o*-chlorobenzoic acid.

**8. *p*-chlorobenzoic acid :** m.p. 236 °C.

Sparingly soluble in hot water.

Derivatives—Amide, m.p. 179 °C; Anilide, m.p. 194 °C.

Manufacturing process : By the oxidation of *p*-chlorotoluene.

Uses : Same as *o*-chlorobenzoic acid.

**9. *p*-bromobenzoic acid :** m.p. 251 °C.

Sparingly soluble in hot water.

Derivatives—Amide, m.p. 189 °C; Anilide, m.p. 197 °C.

Manufacturing process : From *p*-bromotoluene by oxidation.

Uses : Organic synthesis; detection of strontium.

**10. *p*-Bromophenol :** m.p. 63 °C.

Insoluble in water; with bromine-water yields 2, 4, 6-tribromophenol; m.p. 95 °C, $FeCl_3$ +aq. sample ⟶ violet.

Derivatives—Benzoyl, m.p. 102 °C.

Manufacturing process : Controlled halogenation in anhydrous solvent like carbon disulphide on phenol.

Uses : Disinfectant.

**11. *s*-Trichlorophenol :** m.p. 67 °C.

Insoluble in water, $FeCl_3$ gives no colour with it.

Derivative—Benzoyl, m.p. 70 °C, methyl ether.

Manufacturing process : Action of chlorine on phenol.

Uses : Fungicide, herbicide.

**12. *s*-Tribromophenol :** Insoluble in water; $FeCl_3$ gives no colour.

Derivatives—Acetyl, m.p. 82 °C; Benzoyl, m.p. 81 °C.

Manufacturing process : Action of bromine on phenol.

**13. *p*-Dichlorobenzene :** m.p. 53 °C.

Manufacturing process : Chlorination of monochlorobenzene.

Uses : Moth repellent; general insecticide; germicide; space odorant; in the manufacture of 2, 5-dichloroaniline; dyes; intermediates; pharmacy; agriculture (fumigating soil).

**14. *p*-Bromotoluene :** m.p. 28 °C. Oxidation by alkaline $KMnO_4$ and subsequent acidification gives benzoic acid; m.p. 251 °C.

Derivative—*p*-bromobenzoic acid, m.p. 251 °C.

Manufacturing process : By bromination of toluene.

Uses : Intermediate.

**15. *p*-Dibromobenzene :** m.p. 89 °C.

Derivatives—2, 5-Dibromonitrobenzene, m.p. 84 °C.

Manufacturing process : Nitration of bromobenzene followed by reduction, diazotisation and Sandmeyer reaction.

Uses : Organic synthesis of dyestuffs and drugs; intermediates; fumigant.

## LIQUIDS

**16. Methyl iodide :** b.p. 43 °C.

Derivative—β-naphthyl ether, m.p. 72 °C.

Manufacturing process : Interaction of methyl alcohol, sodium iodide and sulphuric acid, with subsequent distillation.

Uses : Organic synthesis; microscopy; testing for pyridine.

**17. Ethyl iodide :** b.p. 72 °C.

Derivative—β-naphthyl ether; m.p. 37 °C.

Manufacturing process : By digesting red phosphorus with absolute ethyl alcohol, after which iodine is added, and the mixture is distilled.

Uses : Medicine; organic synthesis.

**18. *i*-Propyl iodide :** b.p. 89 °C.

Derivative—Picrate of $\beta$-naphthyl ether; m.p. 92 °C.

Manufacturing process : Addition of hydrogen iodide to propene.

Uses : Organic synthesis; pharmaceuticals.

**19. *n*-Propyl iodide :** b.p. 102 °C.

Derivative—$\beta$-naphthyl ether, m.p. 39 °C.

Manufacturing process : By the reaction of iodide on *n*-propyl bromide.

**20. Chloroform :** b.p. 61 °C.

Immiscible with water. Gives red or violet-red colouration with resorcinol and dil. NaOH on boiling.

Manufacturing process : (a) Reaction of chlorinated lime with acetone, acetaldehyde, or ethyl alcohol; (b) byproduct from the chlorination of methane,

Uses : Fluorocarbon refrigerants; fluorocarbon plastics; solvent; analytical chemistry; fumigant; insecticides.

**21. Carbon tetrachloride :** b.p. 77 °C.

Immiscible with water. Gives carbylamine test with 2 ml of alcohol, 1 drop of aniline, and solid bead of NaOH and heat (to boiling).

Manufacturing process : (a) Interaction of $CS_2$ and chlorine in presence of iron. (b) chlorination of methane or higher hydrocarbons at 250-400 °C.

Uses : Refrigerants; metal degreasing; agricultural fumigant; chlorination of organic compounds; production of semiconductors; solvent (fats, oil, rubber, etc.).

**22. Benzyl chloride :** b.p. 179 °C.

Very irritating odour.

Derivative—$\beta$-naphthyl ether, m.p. 99 °C.

Manufacturing process : By passing chlorine over boiling toluene until it is increased 39% in weight. The product is washed with water and separated by fractional distillation.

Uses : Dyes; intermediates; benzyl compounds; synthetic tannins; perfumery.

**23. Ethyl bromide :** b.p. 35 °C.

Derivative—$\beta$-naphthyl ether, m.p. 37 °C.

Manufacturing process : From ethyl alcohol or ethylene and hydrogen bromide. One process uses gamma radiation to initiate the combination.

Uses : Organic synthesis; medicine (anaesthetic); refrigerant; solvent; grain and fruit fumigant.

**24. *i*-Propyl bromide :** b.p. 60 °C.

Derivative—$\beta$-naphthyl ether, m.p. 92 °C.

Manufacturing process : By the addition of HBr to propylene

Uses : Synthesis of pharmaceuticals, dyes, other organics.

**25. *n*-Propyl bromide :** b.p. 71 °C.

Derivative—$\beta$-naphthyl ether, m.p. 39 °C.

**26. Ethylene dibromide :** b.p. 131 °C.

Manufacturing process : Action of bromine on ethylene.

Uses : Scavenger for lead in gasoline; grain fumigant; general solvent; waterproofing preparation; organic synthesis; fumigant for tree crops.

**27. Bromoform :** b.p. 151 °C. Gives carbylamine test.

Manufacturing process : By heating acetone or ethyl alcohol with bromine and alkali hydroxide, and recovery by distillation.

Uses : Intermediate inorganic synthesis; geological assaying; solvent for waxes, greases and oils; medicine.

**28. Benzyl bromide :** b.p. 198 °C. Lachrymatory.

Derivative—$\beta$-naphthyl ether, m.p. 99 °C.

Manufacturing process : (a) Bromination of toluene, (b) interaction of benzyl alcohol and hydrobromic acid.

Uses : In making foaming and frothing agents; organic synthesis.

## AMIDES AND ANILIDES

**1. Acetamide :** $CH_3CONH_2$; m.p. 82 °C.

Soluble in water. On heating with aniline yields acetanilide, m.p. 114 °C. On distillation with conc. $H_2SO_4$ forms acetic acid, b.p. 115 °C.

Derivatives—Acetanilide, m.p. 114 °C. Picrate, m.p. 107 °C.

Manufacturing process : Interaction of ethyl acetate and ammonium hydroxide.

Uses : Organic synthesis (reactant, solvent, peroxide stabiliser); general solvent; lacquers; explosives; soldering flux; hygroscopic agent; wetting agent; penetrating agent.

**2. Semicarbazide :** ($NH_2CONHNH_2$); m.p. 96 °C.

Soluble in water, reduces Fehling soln. and Tollen's reagent.

Derivative—Hydrochloride, m.p. 173 °C; isopropenyl derivative (with acetone), m.p. 187 °C.

Manufacturing process : From hydrazine sulphate, potassium or sodium cyanate and sodium carbonate or electrolytically by the reduction of nitrourea.

Uses : Reagent for aldehydes and ketones; isolation of hormones and isolation of certain fractions from essential oils.

**3. Acetanilide :** ($CH_3CONHPh$); m.p. 114 °C.

Soluble in hot water. On boiling with dilute HCl forms acetic acid and aniline.

Derivatives—Bromo derivative, m.p. 167 °C, *p*-nitro derivative, m.p. 210 °C.

Manufacturing process : Acetylation of aniline with glacial acetic acid.

Uses : Rubber accelerator; inhibitor in $H_2O_2$; stabiliser for cellulose ester coatings; manufacture of intermediates (*p*-nitroaniline, *p*-nitroacetanilide; *p*-phenylenediamine); synthetic camphor; pharmaceutical chemicals; dyestuffs; precursor in penicillin manufacture; medicine(antiseptic).

**4. Chloroacetamide :** ($ClCH_2CONH_2$), m.p. 119 °C.

Soluble in water. On boiling with HCl yileds chloroacetic acid.

Derivative—Chloroacetic acid, m.p. 63 °C.

Manufacturing process : By the reaction between chloroacetyl chloride and $NH_4OH$.

Uses : Intermediate.

**5. Succinimide :** m.p. 125 °C.

Soluble in water. On heating with aniline forms succinanil and ammonia. On hydrolysis with HCl under boiling condition yields succinic acid.

Derivatives—Succinic acid, m.p. 185 °C; succinanil, m.p. 156 °C.

Manufacturing process : Conversion of succinic acid to succinamide, followed by heating; $NH_3$ splits off to give succinimide.

Uses : Growth stimulants for plants; organic synthesis.

**6. Benzamide :** ($PhCONH_2$); m.p. 128 °C.

Sparingly soluble in water. On hydrolysis with conc. NaOH or $H_2SO_4$ yields sodium benzoate and benzoic acid respectively. On heating with aniline gives ammonia and benzanilide.

Derivatives—Benzoic acid, m.p. 121 °C; Anilide, m.p. 160 °C.

Manufacturing process : From benzoyl chloride and ammonia or ammonium carbonate.

Uses : Organic synthesis.

**7. Urea :** ($NH_2CONH_2$); m.p. 132 °C.

Soluble in cold water. On heating at 150 °C gives biurate and ammonia.

Derivatives—Nitrate, m.p. 163 °C; Oxalate m.p. 171 °C.

Manufacturing process : Liquid ammonia and liquid carbon dioxide at 1750-3000 psi and 160-200 °C react to form ammonium carbamate, $NH_4CO_2NH_2$, which decomposes at lower pressure (about 80 psi) to urea and and water.

Uses : Fertiliser; animal feed; plastics; chemical intermediate; stabiliser in explosives; medicine (diuretic); adhesives, separation of hydrocarbons (as urea adducts); pharmaceuticals; cosmetics; sulphamic acid production; flameproofing agents; viscosity modifier for starch or casein-based paper coatings; preparation of biuret.

**8. Salicylamide :** ($o$-$HOC_6H_4CONH_2$); m.p. 139 °C.

Sparingly soluble in water. $FeCl_3$ solution : violet colour; boiling with HCl gives ammonium chloride and salicylic acid.

Derivative—Salicylic acid, m.p. 155 °C; methyl ether, m.p. 128 °C (prepared by adding $Me_2SO_4$ to the solution of the sample in sq. NaOH, boiling for 1 minute cooling and shaking and recrystallising from aq. alcohol).

Manufacturing process : Treatment of methyl salicylate with dry ammonia gas.

Uses : Medicine (analgesic).

**9. Benzanilide :** (PhNHCOPh); m.p. 161 °C.

Insoluble in water. On boiling with HCl, it hydrolyses to benzoic acid and anilinium chloride.

Derivative—Benzoic acid, m.p. 121 °C; $o$-nitro: m.p. 294 °C; $p$-nitro derivative 199 °C.

Manufacturing process : From benzoic anhydride and aniline with caustic soda.

Uses : Intermediate in the synthesis of dyes, drugs and perfumes.

**10. *p*-Aminoacetanilide :** (*p*-$NHC_6H_4NHCOCH_3$); m.p. 162 °C.

Soluble in hot water. On boiling with HCl form *p*-phenylenediamine and acetic acid.

Derivative—Diacetyl; m.p. 304 °C.

Manufacturing process : Acetylation of *p*-phenylenediamine.

Uses : Intermediate for azo dyes; and pharmaceuticals.

## NITRO HYDROCARBONS AND ETHERS

### Liquid compounds

**1. Nitromethane :** $CH_3NO_2$; b.p. 101 °C.

Colourless, pleasant smelling liquid. Sparingly soluble in water but soluble in organic solvents. In aqueous solution it is acidic in nature. (i) An intense red colour is obtained when nitromethane in NaOH is treated with excess $NaNO_2$ solution in the presence of dilute $H_2SO_4$. (ii) When it is treated with chloral hydrate and $Na_2CO_3$, it yields trichloronitro-iso-propyl alcohol. m.p. 41 °C.

Manufacturing process : By reaction of methane or propane with nitric acid under pressure.

Uses : Solvent for cellulosic compounds; polymers. waxes, fats, etc; chemical synthesis; rocket fuel; gasoline additive.

**2. Nitroethane :** $CH_3CH_2NO_2$; b.p. 114 °C.

Colourless, pleasant smelling liquid. Sparingly soluble in water but completely miscible in organic solvents. It is acidic in nature. (i) Responds to test no. (i) given under nitromethane, (ii) On standing with benzaldehyde and a trace of ethylamine gives $\beta$-nitro-$\alpha$-phenylpropylene, m.p. 64 °C. (iii) Benzeneazonitroethane, m.p. 141°C, is obtained when nitroethane in NaOH (dil.) is treated with diazobenzene.

Manufacturing process : By reaction of propane with nitric acid under pressure.

Uses : Solvent for nitrocellulose, cellulose acetobutyrate, vinyl, alkyd, and many other resins, waxes, fats and dyestuffs; Friedel-Crafts synthesis; propellant research; fuel additive.

**3. Tetranitromethane :** $C(NO_2)_4$; b.p. 126 °C.

Colourless liquid. Immiscible with water but miscible with organic solvents. (i) With unsaturated hydrocarbons, it shows faint yellow colour. (ii) When it is treated with $C_2H_5ONa$ in alcohol, it yields trinitromethane, m.p. 16 °C. (iii) It shows an orangish red colour with pyridine in alcohol.

Manufacturing process : By the action of fuming nitric acid on benzene, acetic anhydride or acetylene.

Uses : Rocket fuel, as an oxidant or monopropellant; qualitative test for unsaturated compounds; diesel fuel booster, organic reagent.

**4. Nitrobenzene :** $PhNO_2$; b.p. 211 °C.

Pale-yellow liquid having a smell of bitter almonds. immiscible with water but miscible with organic solvents. It is heavier than water. (i) When it is reduced with Sn/HCl, aniline is produced. (ii) When it is boiled with sodium arsenite, azoxybenzene is formed, m.p. 35 °C.

Derivatives—(i) *m*-Dinitrobenzene, m.p. 90 °C. (ii) *m*-Bromo derivative, m.p. 54 °C.

Manufacturing process : From benzene by nitrating with nitric acid-sulphuric acid mixture.

Uses : In the manufacture of aniline; solvent for cellulose ethers; modifying esterification of cellulose acetate; ingredient of metal polishes and shoe polishes; manufacture of benzidine, quinoline, azobenzene, etc.

**5. *o*-Nitrotoluene :** *o*-$(CH_3)C_6H_4(NO_2)$ 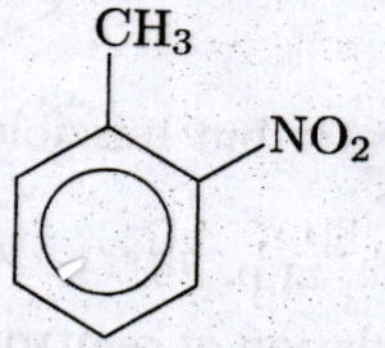 , b.p. 222 °C.

Light pale-yellow liquid having a smell of bitter almonds. It is heavier than water; insoluble in water but soluble in organic solvents. (i) When it is reduced with iron and HCl, *o*-toluidine is produced. (ii) When it is oxidised with dilute $KMnO_4$, it yields *o*-nitrobenzoic acid, m.p. 147 °C. (iii) It shows deep red colour with NaOH solution.

Derivatives—(i) 2, 4-Dinitrotoluene, m.p. 71 °C. (ii) 2, 4, 6-Trinitrotoluene, m.p. 82°C.

Manufacturing process : From toluene by nitration and separation by fractional distillation.

Uses : For production of toluidine, fuchsin and various synthetic dyes.

**6. *m*-Nitrotoluene :** *m*-$MeC_6H_4NO_2$ 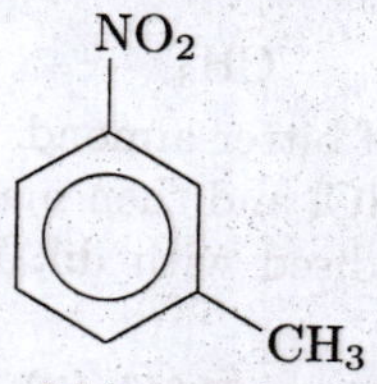 ; b.p. 230 °C. m.p. 16 °C.

Dark pale-yellow liquid having a smell of bitter almond. Immiscible with water but miscible with organic solvents. It is heavier than water. (i) When it is reduced with Sn/HCl, it yields *m*-toluidine. (ii) On oxidation with alkaline $KMnO_4$ or $K_2Cr_2O_7/H_2SO_4$ yields *m*-nitrobenzoic acid, m.p. 140 °C.

Manufacturing process : Nitration of benzaldehyde with fuming $HNO_3$ and conc. $H_2SO_4$ followed by reduction by Wolff-Kishner method.

Uses : Organic synthesis.

**7. Phenylnitromethane :** $PhCH_2NO_2$; b.p. 227 °C. Yellow liquid, immiscible with water but miscible with organic solvents. (i) When it is boiled with alkaline $KMnO_4$ it yields benzoic acid on acidification. (ii) Responds the test no. (i) given under nitromethane. (iii) When it is treated with benzaldehyde and a trace of methylamine in alcohol, it gives α-nitrostilbene, m.p. 75 °C.

Manufacturing process : By the action of aqueous alcoholic silver nitrate on benzyl chloride.

**8. *o*-Nitroanisole :** *o*-$(OCH_3)\ C_6H_4(NO_2)$ 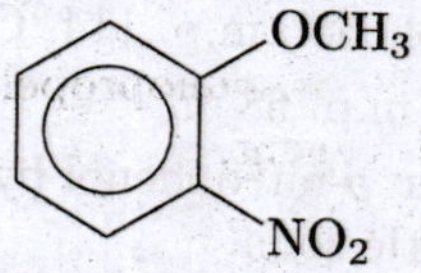 ; m.p. 10 °C; b.p. 265 °C.

Yellow coloured liquid, immiscible with water but miscible with organic solvents. When it is reduced with Sn/HCl, it yields *o*-anisidine.

Derivative—2, 4-Dinitroanisole, m.p. 94 °C.

Manufacturing process : From *o*-nitrophenol by methylation or from *o*-nitrochlorobenzene by action of methanol and caustic soda.

Uses : Organic synthesis; intermediate for dyes and pharmaceuticals.

**9. *o*-Nitrophenetole :** *o*-$C_6H_4(NO_2)\ (OC_2H_5)$ 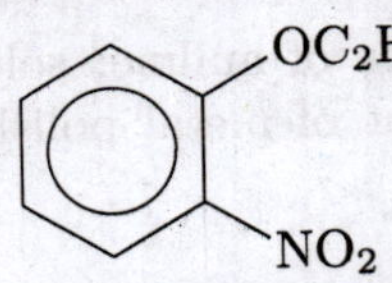 ; m.p. 2 °C; b.p. 267 °C.

Yellow liquid, immiscible with water but miscible with organic solvents. When it is reduced with Sn/HCl, it yields *o*-phenetidine.

Derivative—2, 4-Dinitrophenetole, m.p. 86 °C.

Manufacturing process : By ethylation of *o*-nitrophenol with ethyl chloride.

Uses : Dyes and other intermediates.

**Solid compounds**

**10. *p*-Nitrotoluene :** *p*-$MeC_6H_4NO_2$ 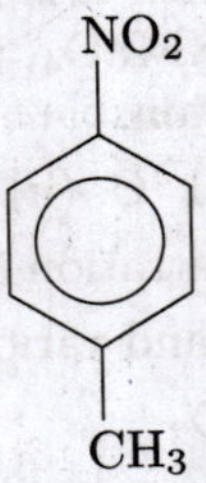 ; m.p. 54 °C.

Pale-yellow crystals, it has a smell of bitter almond. Insoluble in water and soluble in organic solvents. When it is reduced with Sn/HCl and then made alkaline with NaOH solution, it yields *p*-toluidine, m.p. 45 °C. (ii) When oxidised with dil. $KMnO_4$ yields *p*-nitrobenzoic acid, m.p. 241 °C.

Derivative—(i) 2, 4-Dinitrotoluene, m.p. 71 °C. (ii) 2, 4, 6-Trinitratoluene, m.p. 82 °C.

Manufacturing process : From toluene by nitration and separation by fractional distillation.

Uses : For production of toluidine, fuchsin, and various synthetic dyes.

**11. *p*-Nitroanisole :** *p*-$CH_3OC_6H_4NO_2$ 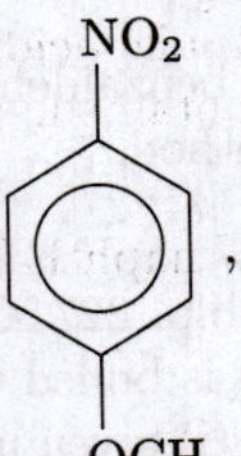 , m.p. 54 °C.

Light yellow crystals, steam volatile. Insoluble in water but soluble with organic solvents. (i) When it is reduced with tin and HCl, it yields *p*-anisidine, m.p. 57°C. (ii) When it is boiled with 48% HBr, it yields *p*-nitrophenol, m.p. 114 °C.

Derivative—Tirinitroanisole, m.p. 68 °C.

Manufacturing process : From *p*-nitrophenol by methylation or from *p*-nitrochlorobenzene by the action of methanol and caustic soda.

**12. *p*-Nitrophenetole :** *p*-$C_2H_5OC_6H_4NO_2$, 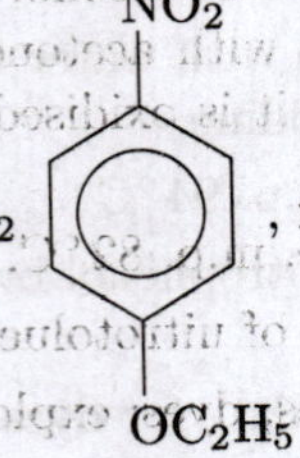 , m.p. 59 °C.

Light yellow crystals. Insoluble in water and alcohol but soluble in ether. When it is reduced with Sn/HCl it yields *p*-phenetidine.

Derivative—Trinitrophenetole, m.p. 78 °C.

Manufacturing process : By ethylation of *p*-nitrophenol with ethyl chloride.

Uses : Dyes and other intermediates.

**13. α-Nitronaphthalene :** α-$C_{10}H_7NO_2$ 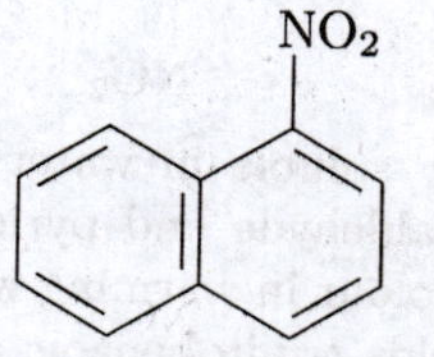 ; m.p. 60 °C.

Yellow solid crystals, insoluble in water but soluble in organic solvents. (i) When it is reduced with Sn/HCl, it yields α-naphthylamine, m.p. 50 °C. (ii) It shows dark red colour when it is dissolved in conc. $H_2SO_4$. (iii,) When it is treated with $CrO_3$ in acetic acid, it yields 3-nittophthalic acid, m.p. 219 °C.

Derivative—1, 5-Dinitronaphthalene, m.p. 214 °C.

Manufacturing process : Action of a mixture of nitric and sulphuric acids on finely ground naphthalene.

Uses : Dyes; naphthylamine; added to mineral oils to mask fluorescence.

**14. β-Nitronaphthalene :** β-$C_{10}H_7NO_2$ 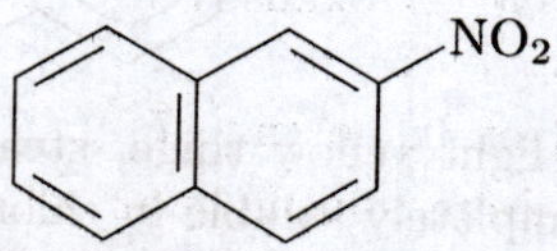 ; m.p. 79 °C.

Bright yellow crystals insoluble in water but soluble in organic solvents. (i) When it is oxidised by $CrO_3$ in acetic acid, it yields 4-nitrophthalic acid, m.p. 164 °C. (ii) When it is reduced with Sn/HCl, β-naphthylamine, m.p. 113 °C is obtained. (iii) When it is treated with $PCl_3$, it yields β-chloronaphthalene.

Manufacturing process : By nitration of naphthalene with a mixture of nitric and sulphuric acid, yields 5% β-product.

**15. 2, 4-Dinitrotoluene :** 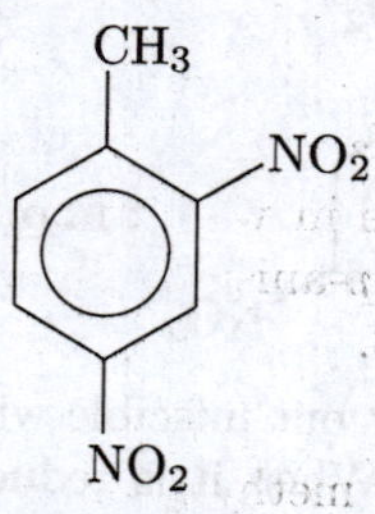 ; m.p. 71 °C.

Yellow crystals, slightly soluble in cold alcohol and ether, but soluble in benzene. (i) A blue colour is produced when it is reacted with acetone in KOH. This blue colour changes to violet on addition of acetic acid. (ii) When it is oxidised with $KMnO_4$, 2, 4-dinitrobenzoic acid, m.p. 182 °C is produced.

Derivative—2, 4, 6-Trinitrotoluene, m.p. 82 °C.

Manufacturing process : Nitration of nitrotoluene with mixed acid.

Uses : Organic synthesis, toluidines; dyes; explosives.

**16. 2, 4, 6-Trinitrotoluene :** 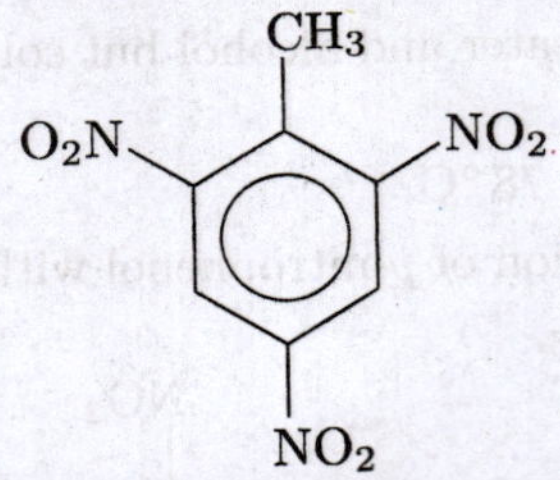 , m.p. 82 °C.

Pale yellow crystals, sparingly soluble in water but completely soluble in organic solvent. (i) When it is treated with benzaldehyde and pyridine, it yields trinitrostilbene, m.p. 158 °C. (ii) It shows deep reddish brown colour in warming with NaOH solution. (iii) When it is oxidised with $CrO_3$ in conc. $H_2SO_4$, it yields trinitobenzoic acid, m.p. 220 °C. (iv) With naphthalene in alcohol, it forms an addition compound, m.p. 96°C.

Manufacturing process : Nitration of toluene with mixed acid. Small amounts of the 2, 3, 4- and 2, 4, 5-isomers are produced which may be removed by washing with aqueous sodium sulphite solution.

Uses : Explosive; intermediate in dyestuffs and photographic chemicals.

**17. *o*-Dinitrobenzene :** *o*-$C_6H_4(NO_2)_2$ 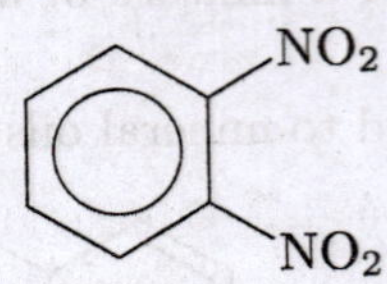 ; m.p. 119 °C.

Colourless crystals with light yellow tinge, steam volatile, sparingly soluble in hot water, alcohol and benzene, but completely soluble in chloroform. (i) When it is reduced with $NH_4SH$ in alcohol, it yields *o*-nitroaniline, m.p. 71 °C. (ii) When it is treated with aniline at 100 °C, it yields *o*-nitroaniline, m.p. 74 °C. (iii) When it is boiled with dil. NaOH, it yields *o*-nitrophenol, m.p. 44 °C.

Manufacturing process : Nitration of nitrobenzene with hot mixed acid.

Uses : Organic synthesis; dyes; camphor substitute in cellulose nitrate.

**18. *m*-Dinitrobenzene :** *m*-$C_6H_4(NO_2)_2$ 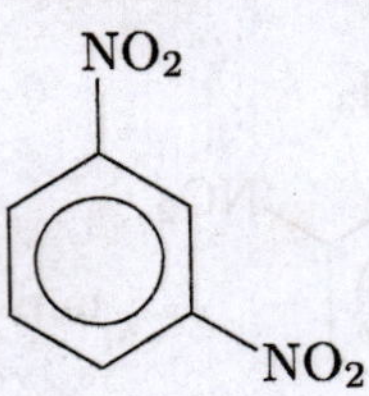 ; m.p. 90 °C.

Pale-orangish yellow crystals, insoluble in water but miscible with organic solvents. (i) Gives red colour when boiled with NaOH solution. (ii) When it is reduced with alcoholic solution of $NH_4SH$, it yields *m*-nitroaniline, m.p. 114 °C. (iii) Shows violet colour when treated with $SnCl_2$ solution or when boiled with glucose solution.

Derivatives—1, 3, 5-Trinitrobenzene, m.p. 122 °C.

Manufacturing process : Same as *o*-Dinitrobenzene.

Uses : As an intermediate for organic syntheses.

**19. sym-Trinitrobenzene :** 1, 3, 5-$C_6H_3(NO_2)_3$, 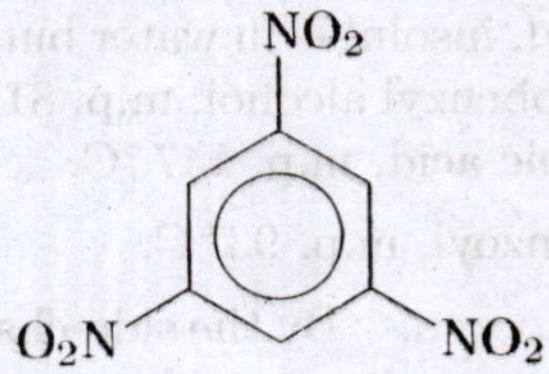 ; m.p. 122 °C.

Pale yellow solid crystals, sparingly soluble in hot water, alcohol and ether but completely soluble in benzene and $CHCl_3$. (i) When it is warmed with $Na_2CO_3$ and $K_3Fe(CN)_6$, picric acid, m.p. 122 °C, is formed. (ii) Gives red colour in NaOH solution. (iii) With naphthalene in alcohol, it forms addition compound, m.p. 152 °C.

Manufacturing process : From trinitrotoluene by removal of methyl group.

Uses : Explosive compositions.

**20. *p*-Dinitrobenzene :** *p*-$C_6H_4(NO_2)_2$ 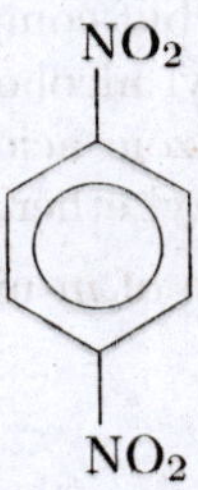 ; m.p. 172 °C.

Yellow crystals, steam volatile, slightly soluble in hot water, alcohol and benzene. (i) When it is reduced with $NH_4SH$/alcohol, it yields *p*-nitroaniline, m.p. 147 °C. (ii) When it is boiled with 5% NaOH solution, *p*-nitrophenol, m.p. 114 °C, is produced. (iii) It forms addition compound with naphthalene, m.p. 118 °C.

Manufacturing process : Same as *o*-dinitrobenzene.

Uses : as an intermediate for organic synthesis.

**21. 1, 8-Dinitronapthalene :** 1, 8-$C_{10}H_6(NO_2)_2$ 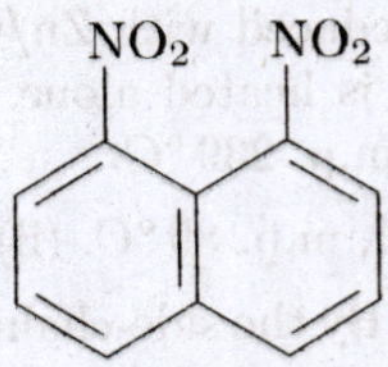 ; m.p. 173 °C.

Yellow crystals, sparingly soluble in benzene, hot water and $CHCl_3$. When it is reduced with Sn/HCl it yields 1, 8-diaminonaphthalene, m.p. 66 °C.

Derivative —1, 3, 8-Trinitronaphthalene, m.p. 217 °C.

Manufacturing process : By dissolving alpha-nitronaphthalene in sulphuric acid and adding nitric acid.

Uses : Dyes.

## NITRO ALCOHOLS

### Solid compounds

**1. *o*-Nitrobenzyl alcohol :** *o*-$(NO_2)C_6H_4(CH_2OH)$ 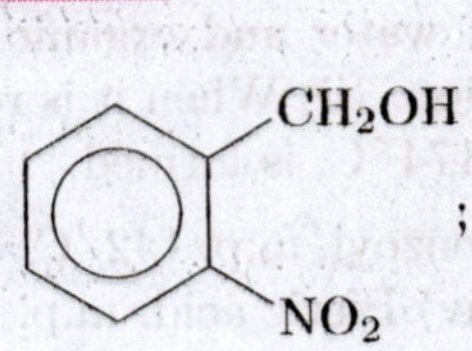 ; m.p. 74 °C.

Crystalline solid, insoluble in water but soluble in organic solvents. (i) When it is reduced with $Zn/CaCl_2$, *o*-aminobenzyl alcohol, m.p. 81 °C, is formed. (ii) On oxidation with $K_2Cr_2O_7/H_2SO_4$ yields *o*-nitrobenzoic acid, m.p. 147 °C.

Derivative—benzoyl, m.p. 93 °C.

Manufacturing process : By the side-chain chlorination of *o*-nitrotoluene followed by hydrolysis.

**2. *m*-Nitrobenzyl alcohol :** *m*-$(NO_2)C_6H_4(CH_2OH)$ 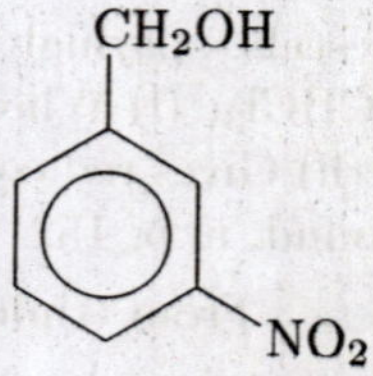 ; m.p. 27 °C.

Crystalline solid, slightly soluble in water but completely soluble in organic solvents. (i) When it is reduced with $Zn/CaCl_2$, *m*-aminobenzyl alcohol, m.p. 97 °C is obtained. (ii) When it is oxidised with $K_2Cr_2O_7/H_2SO_4$, *m*-nitrobenzoic acid, m.p. 141 °C is formed. (iii) When it is heated alone at 250 °C, it yields *m*-nitrobenzyl ether, m.p. 114 °C.

Manufacturing process : By the reduction of *m*-nitrobenzaldehyde with $NaBH_4/H_2O\text{-}Et_2O$.

**3. *p*-Nitrobenzyl alcohol :** *p*-$(NO_2)C_6H_4(CH_2OH)$ 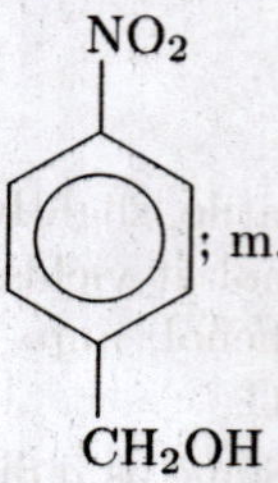 ; m.p. 193 °C.

Crystalline solid, sparingly soluble in cold water but readily soluble in hot water, alcohol, ether, etc. (i) When it is reduced with $Zn/CaCl_2$, *p*-aminobenzyl alcohol, m.p. 64 °C is formed. (ii) It decomposes when it is heated alone at 250 °C. (iii) On oxidation with $K_2Cr_2O_7/H_2SO_4$ yields *p*-nitrobenzoic acid, m.p. 239 °C.

Derivatives—(i) Benzoyl, m.p. 89 °C. (ii) Acetyl, m.p. 77 °C.

Manufacturing process : By the side-chain chlorination of *p*-nitrotoluene followed by hydrolysis.

## NITRO PHENOLS

### Solid compounds

**1. *o*-Nitrophenol :** *o*-$NO_2C_6H_4OH$ 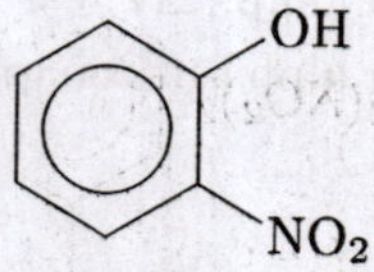 ; m.p. 45°C; b.p. 216 °C.

Bright yellow prism-like solid having a tarry smell. Sparingly soluble in cold water but completely soluble in hot water and organic solvents. It is steam volatile. (i) It shows dark red colour in NaOH solution. (ii) When it is reduced with $Sn/CaCl_2$ or $NaHSO_3$ in alkaline solution, *o*-aminophenol, m.p. 174 °C, is formed.

Derivatives—(i) Benzoyl, m.p. 142 °C. (ii) Acetyl derivative, m.p. 40 °C. (iii) 4, 6-dibromoderivative, m.p. 117 °C. (iv) Picric acid, m.p. 122 °C (v) *p*-Toluenesulphonate, m.p. 83 °C.

Manufacturing process : Action of dilute nitric acid on phenol at low temperature; para-nitrophenol formed at the same time. They are separated by steam distillation.

Uses : Intermediate in organic synthesis; indicator; reagent for glucose.

**2: *m*-Nitrophenol :** *m*-$NO_2C_6H_4OH$ 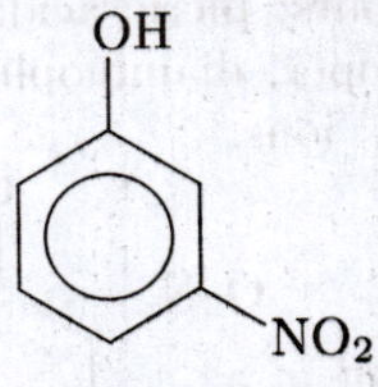 ; m.p. 97 °C.

Very light yellow crystals. It has no odour. Completely soluble in hot water and organic solvent. It is not steam volatile. (i) Shows orange yellow colour when it is dissolved in NaOH solution. (ii) When it is reduced with Zn/$CaCl_2$, *m*-aminophenol, m.p. 122 °C, is produced. (iii) With a few drops of $FeCl_3$ solution, a violet-red colour is produced.

Derivatives—(i) Benzoate, m.p. 95 °C. (ii) Acetyl derivative, m.p. 56 °C. (iii) Bromo derivative, m.p. 91 °C. (iv) *p*-Toluenesulphonate, m.p. 113 °C.

Manufacturing process : Diazotised meta-nitroaniline is boiled with water and sulphuric acid.

Uses : Indicator.

**3. *p*-Nitrophenol :** *p*-$NO_2C_6H_4OH$  ; m.p. 114 °C.

Colourless crystalline solid becomes greyish yellow when it is impure. It has no odour. Soluble in water and organic solvents. (i) Develops bright yellow colour in NaOH solution. (ii) When it is reduced with Zn/$CaCl_2$ solution, *p*-aminophenol, m.p. 184 °C is obtained.

Derivatives—(i) Benzoate, m.p. 142 °C. (ii) Acetate, m.p. 81 °C. (iii) Bromo derivative (2,6), m.p. 142 °C. (iv) Picric acid, m.p. 122 °C. (v) *p*-Toluenesulphonate, m.p. 96 °C.

Manufacturing process : (a) From *p*-chloronitrobenzene; (b) See *o*-nitrophenol.

Uses : Intermediate in organic synthesis; production of parathion; fungicide for leather; indicator.

**4. 2, 4-Dinitrophenol :** 1, 2. 4-$HOC_6H_3(NO_2)_2$ 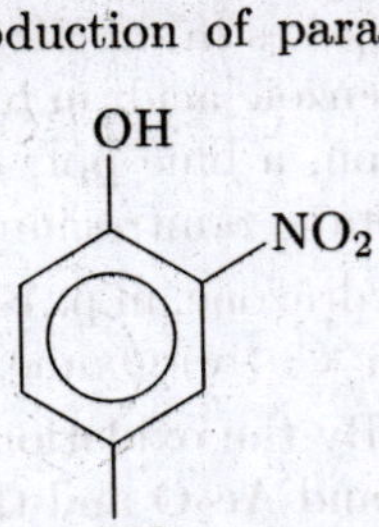 ; m.p. 114 °C.

Almost colourless crystals, soluble in hot water and organic solvents. (i) Develops bright yellow colour in NaOH solution. (ii) On reduction with Zn/$CaCl_2$ yields 2, 4-diaminophenol, m.p. 79 °C.

Derivatives—(i) Benzoate, m.p. 131 °C. (ii) Acetate, m.p. 72 °C. (iii) Bromo derivative (6-bromo), m.p. 147 °C. (iv) Picric acid, m.p. 122 °C. (v) *p*-Toluenesulphonate, m.p. 120 °C.

Manufacturing process : (a) By heating phenol with dilute sulphuric acid, cooling the product and then nitrating, keeping the temperature below 50 °C. (b) By nitration with mixed acid with careful temperature control.

Uses : Dyes, especially sulphur colours; picric acid; picramic acid; preservation of lumber; manufacture of the photographic developer, diaminophenol hydrochloride; explosives manufacture; indicator; reagent for $K^{\oplus}$ and $NH_4^{\oplus}$ ions.

**5. Picric acid :**, 2, 4, 6-$HOC_6H_2(NO_2)_3$ 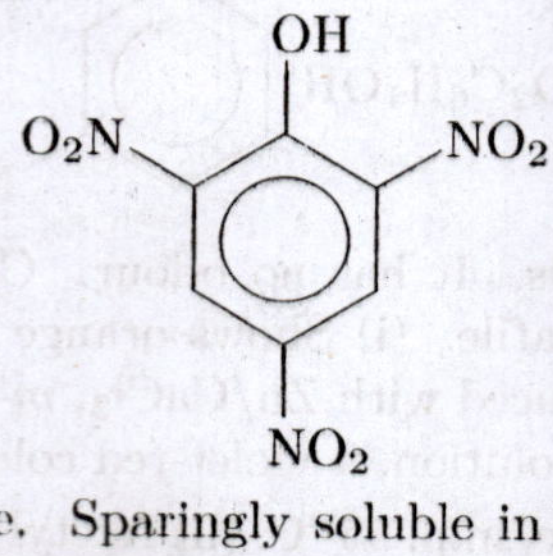 ; m.p. 122 °C.



Light yellow crystals with a bitter taste. Sparingly soluble in cold water but dissolves on boiling, sparingly soluble in ether but soluble in other organic solvents. (i) Aqueous or alcoholic solution of picric acid is yellow but the colour disappears on addition of dil. HCl. (ii) When it is reduced with Sn dust in $NH_4OH$, it yields picramic acid, m.p. 168 °C. (iii) With $PCl_5$ yields picryl chloride, m.p. 83 °C. (iv) With naphthalene, it forms an addition compound, m.p. 149 °C.

Derivative—Acetyl, m.p. 76 °C.

Manufacturing process : Nitration of phenolsulphonic acid, obtained by heating phenol with concentrated sulphuric acid.

Uses : Explosives; matches; electric batteries; etching copper; mordant in textile dyeing; reagent, picrates.

## NITRO ALDEHYDES AND KETONES

### Solid compounds

**1. *o*-Nitrobenzaldehyde :** *o*-$NO_2C_6H_4CHO$ 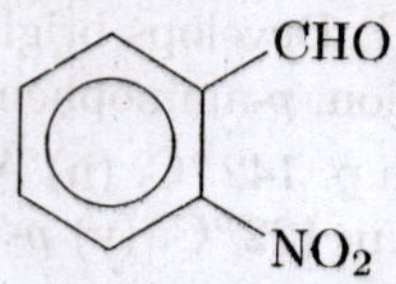 ; m.p. 44 °C.



Yellow crystals, steam volatile. Sparingly soluble in hot water but completely soluble in organic solvent. (i) It decomposes in dil. HCl to produce orangish red colour. (ii) On oxidation with $KMnO_4$ yields *o*-nitrobenzoic acid, m.p. 144 °C. (iii) When *o*-nitrobenzaldehyde in acetone is treated with NaOH solution, a blue ppt. of indigotin is formed. (iv) When it is boiled with $(CH_3CO)_2O$ and sodium acetate, *o*-nitrocinnamic acid, m.p. 240 °C, is formed.

Derivatives—(i) Phenylhydrazone, m.p. 156 °C. (ii) 2, 4-Dinitrophenylhydrazone, m.p. 192 °C. (iii) Semicarbazone, m.p. 256 °C. (iv) Oxime, m.p. 103 °C.

Manufacturing process : By the oxidation of *o*-nitrotoluene with chromium trioxide and sulphuric acid in glacial AcOH and $Ac_2O$ and then hydrolysis of the diacetate.

Uses : Pschorr synthesis of phenanthrene.

**2. *m*-Nitrobenzaldehyde :** *m*-$NO_2C_6H_4CHO$ 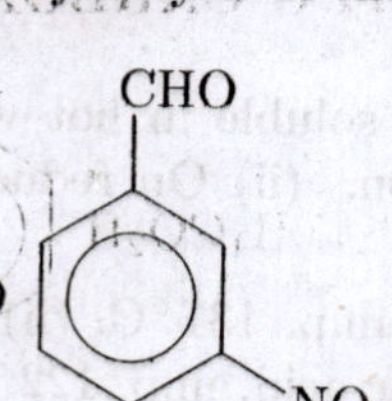 ; m.p. 58 °C.

Pale yellow crystals, slightly soluble in water but completely soluble in organic solvents. (i) Produces yellow colour in NaOH solution by decomposition. (ii) When it is oxidised with $KMnO_4$ solution, *m*-nitrobenzoic acid, m.p. 140 °C, is produced. (iii) *m*-Nitrocinnamic acid, m.p. 196 °C, is formed when it is heated with $(CH_3CO)_2O$ and $CH_3CO_2Na$.

Derivatives—(i) 2, 4-dinitrophenylhydrazone, m.p. 268 °C. (ii) Phenylhydrazone, m.p. 121 °C. (iii) Semicarbazone, m.p. 246 °C. (iv) Oxime, 92 °C.

Manufacturing process : By nitrating benzaldehyde with fuming nitric acid and conc. sulphuric acid.

Uses : Synthesis of dyes, pharmaceuticals, surface active agents; vapour-phase corrosion inhibitor; antioxidant for chlorophyll; mosquito repellent.

**3. *p*-Nitrobenzaldehyde :** *p*-$NO_2C_6H_4CHO$ 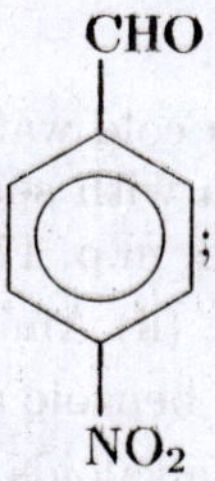 ; m.p. 106 °C.



Very pale-yellow crystals, slightly soluble in water but completely soluble in organic solvents. (i) Decomposes in dil. NaOH solution. (ii) On oxidation with $K_2Cr_2O_7/H_2SO_4$ yields *p*-nitrobenzoic acid, m.p. 241 °C. (iii) *p*-Nitrocinnamic acid; m.p. 285 °C is produced when it is heated with $(CH_3CO)_2O$ and $CH_3CO_2Na$.

Derivatives—(i) Benzoate, m.p. 141 °C. (ii) Acetate, m.p. 81 °C. (iii) Phenylhydrazone, m.p. 159 °C. (iv) 2, 4-Dinitrophenylhydrazone, m.p. 319 °C. (v) Bromo derivative, m.p. 142 °C. (vi) Oxime, m.p. 132 °C.

Manufacturing process : By the oxidation of *p*-nitrotoluene with $CrO_3/H_2SO_4$ in glacial $AcOH/Ac_2O$ followed by hydrolysis of diacetate.

**4. *m*-Nitroacetophenone :** *m*-$NO_2C_6H_4COCH_3$ 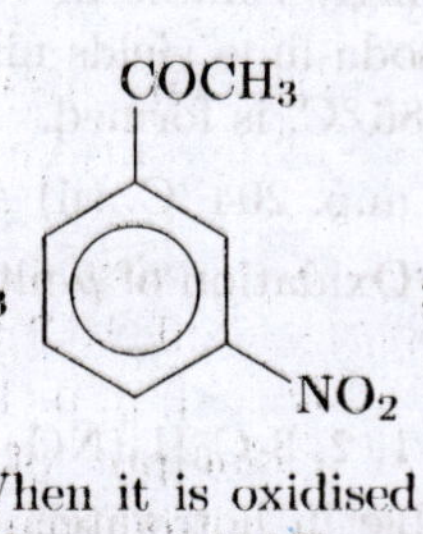 ; m.p. 81 °C.



Yellow crystals, soluble in organic solvents. When it is oxidised with alkaline $KMnO_4$ and then on acidification, *m*-nitrobenzoic acid is formed, m.p. 140 °C.

Derivatives—(i) Oxime, m.p. 131 °C. (ii) Phenylhydrazone, m.p. 135 °C. (iii) 2, 4-Dinitrophenylhydrazone, m.p. 227 °C.

## NITRO CARBOXYLIC ACIDS

### Solid compounds

**1. *o*-Nitrobenzoic acid :** *o*-$NO_2C_6H_4CO_2H$ 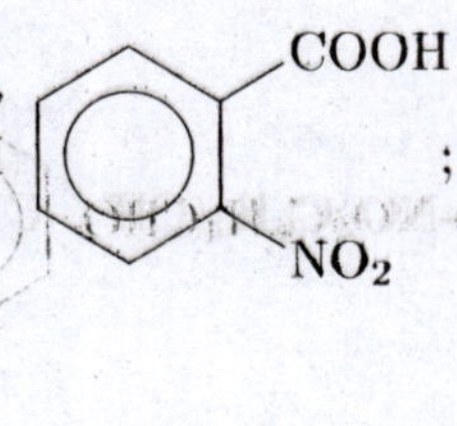 ; m.p. 144 °C.

Colourless crystals, soluble in boiling water and organic solvents. (i) On decarboxylation with soda-lime yields nitrobenzene. (ii) When it is reduced with Sn/HCl, it yields anthranilic acid, m.p. 144 °C.

Derivatives—(i) Anilide, m.p. 155 °C. (ii) Amide, m.p. 174 °C.

Manufacturing process : Oxidation of *o*-nitrotoluene with $MnO_2$ and $H_2SO_4$.

Uses : Organic synthesis.

**2. *m*-Nitrobenzoic acid :** *m*-$NO_2C_6H_4CO_2H$ 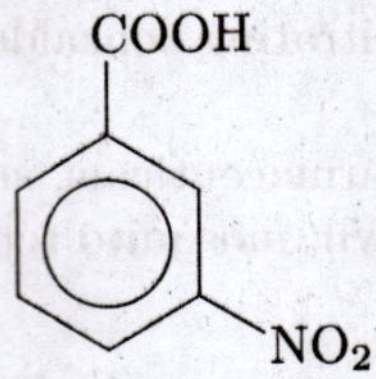 ; M.p. 41 °C.

Colourless crystals, slightly soluble in cold water but completely soluble in boiling water and organic solvents. (i) On decarboxylation with soda-lime yields nitrobenzene. (ii) On reduction with Sn/HCl yields *m*-aminobenzoic acid, m.p. 174 °C.

Derivatives—(i) Anilide, m.p. 153 °C. (ii) Amide, m.p. 142 °C. (iii) *p*-Toluidide, m.p. 161 °C.

Manufacturing process : Nitration of benzoic acid.

Uses : Dye intermediate; reagent for alkaloids.

**3. *p*-Nitrobenzoic acid :** *p*-$NO_2C_6H_4CO_2H$ 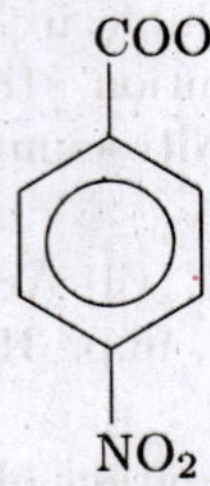 ; m.p. 241 °C.

Colourless crystals, sparingly soluble in cold water and benzene but completely soluble in ether. (i) On heating with soda-lime yields nitrobenzene. (ii) When it is reduced with Sn/HCl, *p*-aminobenzoic acid, m.p. 186 °C, is formed.

Derivatives—(i) Anilide, m.p. 204 °C. (ii) Amide, m.p. 200 °C. (iii) *p*-Toluidide, m.p. 203 °C.

Manufacturing process : Oxidation of *p*-nitrotoluene by hot chromic acid mixture.

Uses : Organic synthesis.

**4. 3-nitrophthalic acid :** 1, 2, 3-$C_6H_3(NO_2)(CO_2H)_2$; m.p. 217 °C.

Colourless crystals, soluble in hot water, ether and alcohol but insoluble in $CHCl_3$. (i) It forms ethyl ester, m.p. 45 °C. (ii) When it is heated above its m.p., 3-nitrophthalic anhydride, m.p. 163 °C, is formed.

Derivatives—(i) Anilide, m.p. 234 °C. (ii) Amide, m.p. 200 °C. (iii) *p*-Toluidide, m.p. 222 °C.

Manufacturing process : By oxidation with $KMnO_4$ of 1-Nitronaphthalene.

**5. *o*-Nitrocinnamic acid :** *o*-$NO_2C_6H_4(CH{=}CHCO_2H)$ 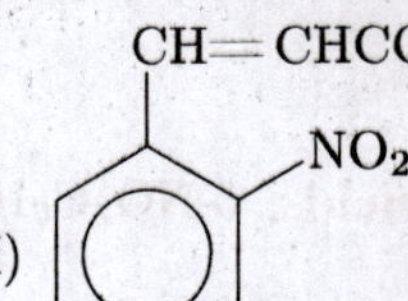 ; m.p. 237 °C.

Colourless solid, insoluble in water but soluble in alcohol. (i) When it is treated with $KMnO_4$, *o*-nitrobenzaldehyde, m.p. 44 °C, is produced with disappearance of pink colour. (ii) When it is oxidised with $K_2Cr_2O_7/H_2SO_4$, *o*-nitrobenzoic acid, m.p. 144 °C is obtained. (ii) It decolourises bromine-water to form a dibromide, m.p. 180 °C.

Derivative—Amide, m.p. 184 °C.

Manufacturing Process : By the nitration of cinnamic acid with concentrated nitric acid and then seperating the mixture of *o*- and *p*-nitrocinnamic acid.

**6. *m*-Nitrocinnamic acid :** *m*-$NO_2C_6H_4(CH{=}CHCO_2H)$ 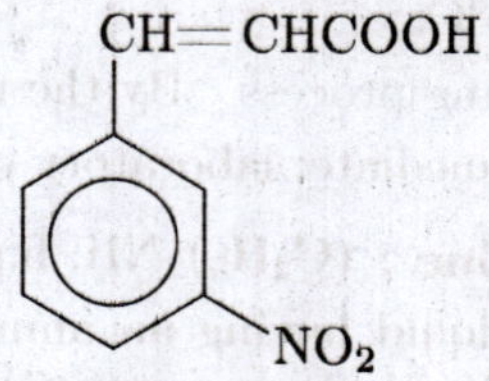 ; m.p. 205 °C.

Colourless solid, soluble in alcohol and benzene. (i) When it is oxidised with dilute $KMnO_4$, *m*-nitrobenzoic acid, m.p. 141 °C is produced. (ii) It decolourises bromine water to form a dibromide.

Derivatives—Amide, m.p. 195 °C.

Manufacturing process : By the Perkin reaction on *m*-nitrobenzaldehyde.

**7. *p*-Nitrocinnamic acid :** *p*-$NO_2C_6H_4(CH{=}CHCO_2H)$ 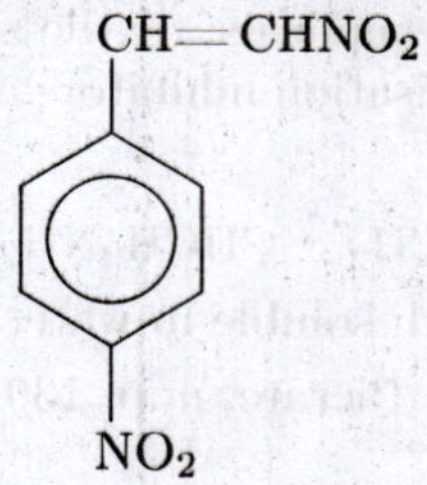 ; m.p. 287 °C.

Colourless solid, insoluble in water and organic solvents like alcohol, ether. (i) On oxidation with alkaline $KMnO_4$ followed by acidification yields *p*-nitrobenzoic acid, m.p. 241 °C. (ii) Slowly decolourises bromine-water to form a dibromide, m.p. 216 °C.

Derivative—Amide, m.p. 217 °C.

Manufacturing process : Same as *o*-nitrocinnamic acid.

**8. 5-Nitrosalicylic acid :** 1, 2, 5-$(CO_2H)(OH)C_6H_3(NO)_2$ 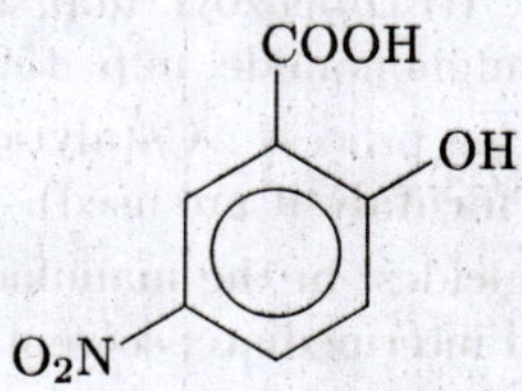 ; m.p. 230 °C.

Colourless solid, insoluble in cold water but soluble in hot water and alcohol. (i) Forms picric acid, m.p. 122 °C, when it is boiled with conc. $HNO_3$. (ii) On reduction with Sn/HCl, 5-aminosalicylic acid is formed. (iii) Shows red colour with $FeCl_3$ solution.

Derivatives—(i) Anilide, m.p. 223 °C. (ii) Amide, m.p. 225 °C.

Manufacturing process : Nitration of salicylic acid with dilute nitric acid.

Uses : Intermediate; azo dyes.

## AMINES

### Liquid compounds

**1. *n*-Propylamine :** $C_2H_5CH_2NH_2$; b.p. 49 °C.

Colourless liquid with characteristic smell, soluble in water. A brown ppt. is obtained with Nessler's reagent.

Derivatives—(i) Picrate, m.p. 135 °C. (ii) *p*-Toluenesulphonamide, m.p. 51 °C. (iii) N-phthalimide, m.p. 65 °C.

Manufacturing process : By the reduction of ethyl cyanide.

Uses : Intermediate; laboratory reagent.

**2. Diethylamine :** $(C_2H_5)_2NH$; b.p. 55 °C.

Colourless liquid having an ammoniacal fishy smell. Soluble in water. (i) A white ppt. is obtained with Nessler's reagent. (ii) A yellow ppt. is obtained when it is treated with $HgCl_2$ in KI solution.

Derivatives—(i) Hydrochloride, m.p. 215 °C. (ii) Picrate, m.p. 155 °C. (ii) Acetyl, m.p. 103°C. (iv) 2, 4-Dinitrophenyl, m.p. 80 °C. (v) *p*-Toluenesulphonamide, m.p. 59 °C.

Manufacturing process : From ethyl chloride and ammonia under heat and pressure.

Uses : Rubber chemicals; textile special-ties; selective solvent; dyes; floatation agents; resins; pesticides; polymerisation inhibitors; pharmaceuticals; petroleum chemicals; electoplating; corrosion inhibitors.

**3. Allylamine :** $CH_2{=}CHCH_2NH_2$; b.p. 55 °C.

Colourless liquid, soluble in water. It decolourises bromine-water to form an oily liquid.

Derivatives—(i) Picrate, m.p. 139 °C. (ii) *p*-Toluenesulphonamide, m.p. 64 °C. (iii) N-Phthalimide, m.p. 69 °C.

Manufacturing process : By the amination of allyl bromide formed from propylene by NBS.

Uses : Pharmaceutical intermediate; organic synthesis.

**4. Ethylene diamine :** $H_2NCH_2CH_2NH_2$; b.p. 116 °C. Colourless liquid, soluble in water but immiscible with ether and benzene.

Derivatives—(i) Dibenzoyl, m.p. 249 °C. (ii) Diacetyl, m.p. 171 °C. (iii) Picrate, m.p. 233 °C. (iv) *p*-Toluenesulphonamide, m.p. 159 °C.

Manufacturing process : Catalytic reaction of ethylene glycol or ethylene dichloride and ammonia (Ni or Cu catalysit are used).

Uses : Fungicides; in the manufacture of chelating agents (EDTA); dimethylolethylene-urea resins; chemical intermediate; solvent; emulsifying agent; textile lubricants; antifreeze inhibitor.

**5. Aniline :** $C_6H_5NH_2$; b.p. 183 °C.

Oily liquid having a faint characteristic smell. It has a light straw colour when it is freshly distilled but turns reddish brown in the presence of air and light. Sparingly soluble in water but soluble in acids and organic solvents. (i) A violet colouration is obtained when it is treated with water suspension of bleaching powder. (ii) A blue or black colour is obtained when treated with $K_2Cr_2O_7$ in conc. $H_2SO_4$.

Derivatives—(i) Benzoyl, m.p. 161 °C. (ii) Acetyl, m.p. 113 °C. (iii) sym-Tribromo derivative, m.p. 118 °C. (iv) *p*-Toluenesulphonamide, m.p. 103 °C.

Manufacturing process : (a) By catalytic vapour-phase reduction of nitrobenzene with hydrogen. (b) Reduction of nitrobenzene with iron filings (HCl acid as catalyst). (c) Catalytic reaction of chlorobenzene and aqueous ammonia. (d) ammonolysis of phenol.

Uses : Rubber accelerators and antioxidants, dyes and intermediates; photographic chemicals (hydroquinone); isocyanates for urethan foams; pharmaceuticals; explosives; petroleum refining; diphenylamine; herbicides; fungicides.

**6. *o*-Toluidine :** *o*-$CH_3C_6H_4NH_4$ 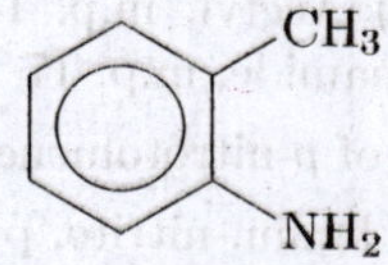 ; b.p. 197 °C.

Colourless oily liquid turns reddish brown in the presence of air and light. Sparingly soluble in water but highly soluble in alcohol and ether. (i) A yellowish brown colour is obtained when aqueous solution of *o*-toluidine is treated with bleaching powder. (ii) A blue colour is obtained when it is treated with $K_2Cr_2O_7$ in conc. $H_2SO_4$. This blue colour changes to purple on dilution with water. (iii) Responds to "Carbylamine test".

Derivatives—(i) Benzoyl, m.p. 125 °C. (ii) Acetyl, m.p. 65 °C, (ii) Picrate, m.p. 213 °C. (iv) *p*-Toluenesulphonamide, m.p. 109 °C.

Manufacturing process : By the reduction of *o*-nitrotoluene or obtained admixed with *p*-toluidine by the reduction of crude nitrotoluene.

Uses : Textile printing dyes; vulcanisation acceleraor; organic synthesis.

**7. *m*-Toluidine :** *m*-$CH_3C_6H_4NH_2$ 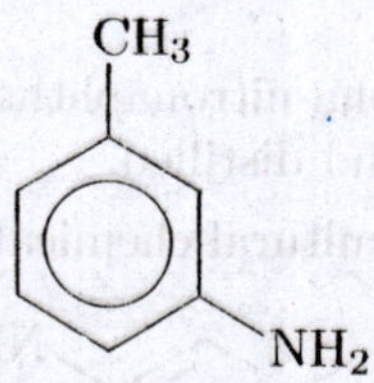 , b.p. 199 °C.

Colourless oily liquid turns pink-brown on standing. Slightly soluble in water but soluble in alcohol and ether. (i) When ethereal solution of *m*-toluidine is treated with bleaching powder suspension, water layer becomes brown while ethereal layer becomes red. (ii) A red colouration is obtained when it is treated with (1 : 1) $H_2SO_4$ and conc. $HNO_3$.

Derivatives—(i) Benzoyl, m.p. 125 °C. (ii) Acetyl, M.p. 65 °C. (ii) Picrate, m.p. 213 °C. (iv) *p*-Toluenesulphonamide, m.p. 109 °C.

Manufacturing process : Reduction of meta-nitrotoluene.

Uses : Dyes; manufacture of organic chemicals.

## Solid compounds

**8. *p*-toluidine :** *p*-$CH_3C_6H_4NH_2$ 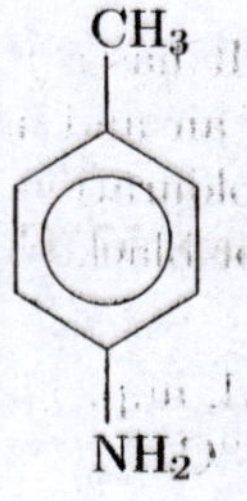 ; m.p. 45 °C.

Light coloured solid having characteristic odour, it becomes grey on standing. Sparingly soluble in water but completely soluble in alcohol and ether. (i) When *p*-toluidine is dissolved in (1 : 1) $H_2SO_4$ and conc. $HNO_3$, a blue colour is obtained which gradually changes to violet, then red and finally brown. (ii) A yellow colour is obtained when it is dissolved in 50% $H_2SO_4$ and then $K_2Cr_2O_7$ is added. (iii) It develops no colour with bleaching powder. It develops a yellow colour which changes to red when it is treated with $FeCl_3$ in HCl.

Derivatives—(i) Benzoyl, m.p. 158 °C. (ii) Acetyl, m.p. 148 °C. (iii) Dibromo, m.p. 73 °C. (iv) Picrate, m.p. 181 °C. (v) *p*-toluenesulphonamide, m.p. 17 °C.

Manufacturing process : By the reduction of *p*-nitrotoluene with iron and hydrochloric acid.

Uses : Dyes; organic synthesis; reagent for lignin, nitrite, pholoroglucinol.

**9. α-Naphthylamine :** 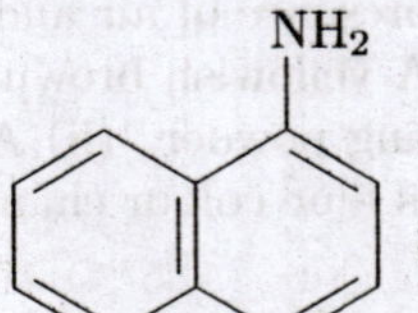

; m.p. 50 °C.

Colourless crystals, develops dark-brown-black colours on standing. Insoluble in water but soluble in organic solvents. (i) A blue ppt. is obtained when the substance in dil. HCl is treated with $FeCl_3$ solution. (ii) Reduces Tollen's reagent.

Derivatives—(i) Benzoyl, m.p. 260 °C. (ii) Acetyl, m.p. 159 °C. (iii) Picrate, 161 °C. (iv) *p*-Toluene sulphonamide, m.p. 157°C.

Manufacturing process : Reduction of alpha-nitronaphthalene with iron and hydrochloric acid. The mass is then mixed with milk of lime and distilled.

Uses : Dyes and dye intermediates; agricultural chemicals.

**10. β-Naphthylamine :** $C_{10}H_7NH_2$ 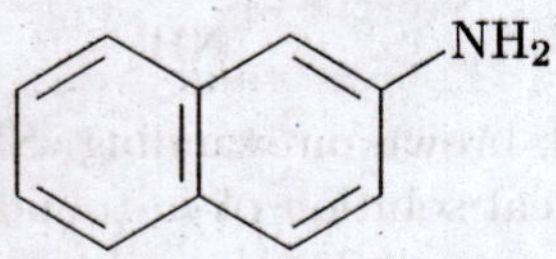

; m.p. 111 °C.

Pink coloured crystals, odourless, sparyingly soluble in hot water. (i) Reduces Tollen's reagent. (ii) Gives no colouration with $FeCl_3$ solution. (iii) With oxidising agent gives no coloured product.

Derivatives—(i) benzoyl, m.p. 162 °C. (ii) Acetyl, m.p. 132 °C. (iii) Picrate, m.p. 195 °C. (iv) *p*-Toluene sulphonamide, m.p. 132 °C.

Manufacturing process : From beta-naphthol by heating in an autoclave with ammonium sulphite and ammonia (Bucherer reaction).

Uses : Dyes; organic synthesis.

**11. *p*-Anisidine :** *p*-$(CH_3O)C_6H_4NH_2$ 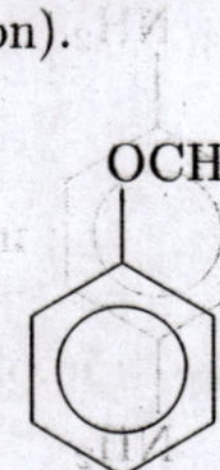

; m.p. 57 °C.

Colourless crystals develops dark colour on keeping. Sparingly soluble in hot water but soluble in organic solvents. (i) When it is boiled with 50% HBr, it yields *p*-aminophenol, m.p. 184 °C. (ii) Gives violet colour with $FeCl_3$ solution.

Derivatives—(i) Benzoyl, m.p. 154 °C. (ii) Acetyl, m.p. 127 °C. (iii) *p*-Toluenesulphonamide, m.p. 254 °C.

Manufacturing process : (a) Reduction of para-nitroanisole with iron filings and hydrochloric acid. (b) Methylation of *p*-aminophenol.

Uses : Azo dyestuffs; intermediate.

**12. *o*-Phenylenediamine :** *o*-$C_6H_4(NH_2)_2$ 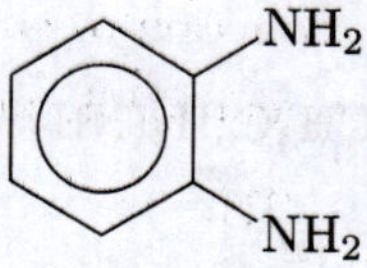 ; m.p. 102 °C.

Colourless crystal turns dark on standing. Soluble in hot water and alcohol. A red ppt. is obtained with $FeCl_3$ in HCl.

Derivatives—(i) Dibenzoyl, m.p. 301 °. (ii) Diacetyl, m.p. 185 °C. (iii) Picrate, m.p. 195 °C.

Manufacturing process : Reduction of *o*-dinitrobenzene or nitroaniline with iron and hydrochloric acid. Purification by crystallisation.

Uses : In the manufacture of dyes; photographic developing agent; organic synthesis; laboratory reagent.

**13. *m*-Phenylenediamine :** *m*-$C_6H_4(NH_2)_2$ 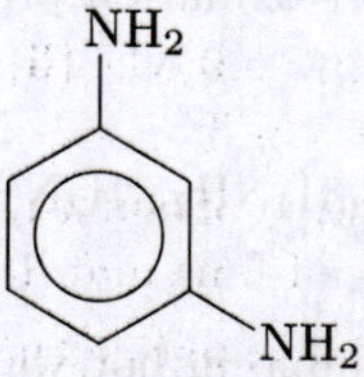 ; m.p. 63 °C.

Colourless crystals rapidly darken on keeping. Highly soluble in hot water. (i) On warming the substance with dil. HCl and adding 1% acetaldehyde solution in 50% alcohol, it gives a yellow colour with green fluorescence. (ii) Gives a brown ppt. with $NaNO_2$ in HCl.

Derivatives—(i) Dibenzoyl, m.p. 240 °C. (ii) Diacetyl, m.p. 191 °C. (iii) Picrate, m.p. 183°C. (iv) Tribromo, m.p. 158°C. (v) *p*-Tolenesulphonamide, m.p. 173 °C.

Manufacturing process : Reduction of *m*-dinitrobenzene or nitroaniline with iron and hydrochloric acid. Purification by crystallisation.

Uses : In dyestuff manufacturing; detection of nitrite, as textile developing agent; laboratory reagent; vulcanising agent; ion-exchange resins; block polymers and in photography.

**14. *p*-Phenylenediamine :** *p*-$C_6H_4(NH_2)_2$ 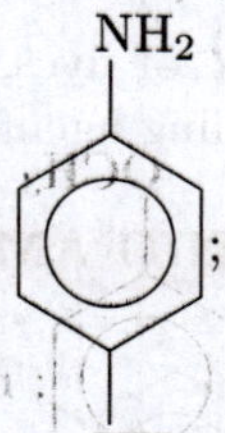 ; m.p. 140 °C.

Colourless crystals turns dark in the presence of air and light. Soluble in hot water and organic solvents. On oxidation with $K_2Cr_2O_7/H_2SO_4$, yields *p*-benzoquinone.

Derivatives—(i) Dibenzoyl, m.p. > 300 °C. (ii) Diacetyl, m.p. 304 °C. (iii) *p*-Toluenesulphonamide, m.p. 265 °C.

Manufacturing process : Reduction of *p*-dinitrobenzene or nitroaniline with iron and hydrochloric acid. Purified by crystallisation.

Uses : Azo dye intermediate; photographic developing agent; photochemical measurements; intermediate in manufacture of antioxidants and accelerators for rubber; laboratory reagent; dyeing hair and fur.

**15. *m*-Toluenediamine :** 1, 2, 4-$CH_3C_6H_2(NH_2)_2$ 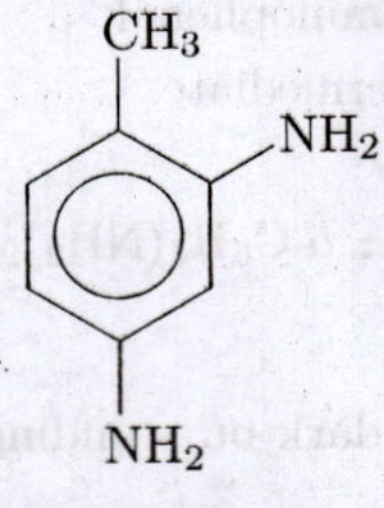 ; m.p. 99 °C.

Colourless solid turns dark in the presence of air. Soluble in hot water.

Derivatives—(i) Dibenzoyl, m.p. 224 °C. (ii) Diacetyl. m.p. 222 °C. (iii) *p*-Toluenesulphonamide, m.p. 191 °C.

Manufacturing process : Reduction of *m*-dinitrotoluene with iron and hydrochloric acid.

Uses : Dye intermediate; direct oxidation product blacks fur and hair; source of toluene-2, 4-diisocyanate.

**16. Benzidine :** $H_2NC_6H_4—C_6H_4NH_2$, 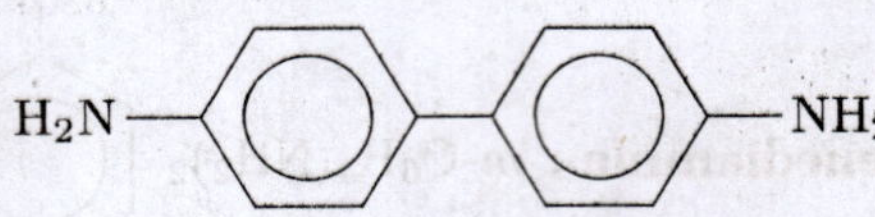 ; m.p. 126 °C.

Colourless solid. Sparingly soluble in hot water but completely soluble in organic solvents. (i) On oxidation with dil. $H_2SO_4$ and $MnO_2$ yields *p*-benzoquinone. (ii) When bromine water is added gradually to the substance in $CS_2$ the aqueous layer first turns blue, then green and finally colourless and $CS_2$, layer becomes red. (iii) When it is diazotised with dil. $H_2SO_4$ and then boiled, *p*-diphenol, m.p. 271 °C, is obtained.

Derivatives—(i) Dibenzoyl, m.p. 352 °C. (ii) Diacetyl, m.p. 317°C. (iii) *p*-Toluenesulphonamide, m.p. 242 °C.

Manufacturing process : (a) By reducing nitrobenzene with zinc dust in alkaline solution followed by distillation. (b) By electrolysis of nitrobenzene, followed by distillation. (c) Nitration of diphenyl followed by reduction of the product with zinc dust in alkaline solution, with subsequent distillation.

Uses : Organic synthesis; manufacture of dyes, especially of congo red; detection of blood stains; stain in microscopy; reagent; stiffening agent in rubber compounding.

## NITRO-AMINES

### Solid compounds

**1. *o*-Nitroaniline :** *o*-$NH_2C_6H_4NO_2$, 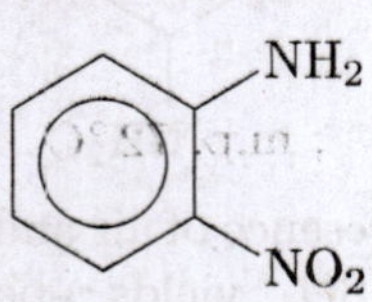 , m.p. 71 °C.

Orange-yellow crystals volatile in steam. Soluble in hot water and organic solvents. On heating with Sn/NaOH in alcohol, forms *o*-phenylenediamine, m.p. 102 °C.

Derivative—(i) Benzoyl derivative, m.p. 94 °C. (ii) Acetyl derivative, m.p. 92 °C. (iii) Picrate m.p. 72 °C. (iv) *p*-Toluenesulphonamide, m.p. 110 °C.

Manufacturing process : From aniline by nitration after acetylation, with subsequent removal of the acetyl group by hydrolysis; from *o*-dinitrosobenzene.

Uses : Dye intermediate; synthesis of photographic antifogging agent, ortho-phenylenediamine; coccidiostats.

**2. *m*-Nitroaniline :** *m*-$NH_2C_6H_4NO_2$ 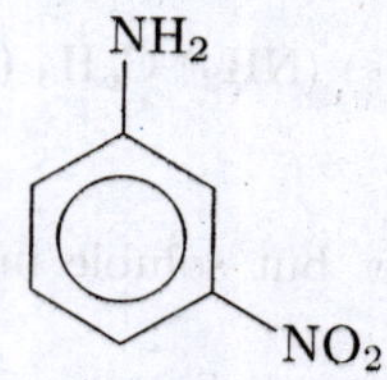 ; m.p. 114 °.

Yellow crystals, volatile in steam, soluble in hot water and organic solvents. When it is reduced with Sn/HCl, *m*-phenylenediamine, m.p. 63 °C, is produced.

Derivatives—(i) Benzoyl, m.p. 157 °C. (ii) Acetyl m.p. 155 °C. (iii) *p*-Toluenesulphonamide, m.p. 138 °C.

Manufacturing process : From aniline by nitration after acetylation, with subsequent removal of the acetyl group by hydrolysis; from *m*-nitrobenzoic acid.

Uses : Dye intermediate.

**3. *p*-Nitroaniline :** *p*-$NH_2C_6NO_2$ 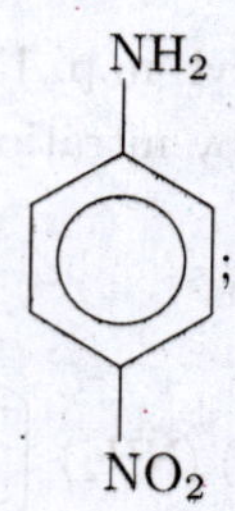 ; m.p. 147 °C.

Greenish yellow crystals not volatile in steam, soluble in hot water and organic solvents. (i) When it is reduced with tin and HCl, *p*-phenylenediamine, m.p. 140 °C, is obtained. (ii) When it is boiled with conc. NaOH for a long time, ammonia gas is evolved.

Derivatives—(i) Benzoyl, m.p. 199 °C (ii) Acetyl, m.p. 215 °C. (iii) *p*-Toluenesulphonamide, m.p. 191 °C.

Manufacturing process : (a) From *p*-chloronitrobenzene. (b) From aniline by nitration after acetylation. (c) From acetanilide.

Uses : Dye intermediate, especially *p*-nitroaniline red; intermediate for antioxidants, gasoline gum inhibitors;

**4. *o*-Nitro-*p*-toluidine :** 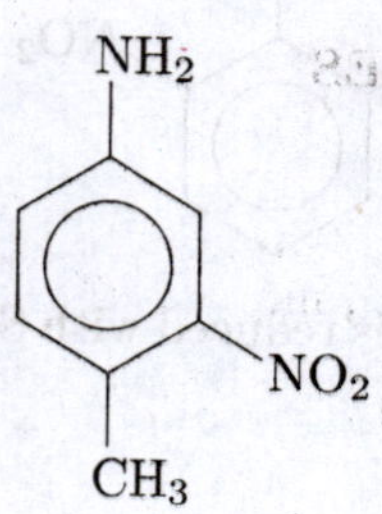 ; m.p. 72 °C.

Yellow crystals, soluble in hot water, alcohol and ether, etc. When it is reduced with Sn/HCl, 2, 4-diaminotoluene, m.p. 99 °C, is formed.

Derivatives—(i) Benzoyl derivative, m.p. 171 °C. (ii) Acetyl derivative, m.p. 93 °C.

Manufacturing Process : Selective reduction of 2, 4-dinitrotoluene by ammonium bisulphide.

Uses : Intermediates for dyes and pigments.

**5. *m*-Nitro-*o*-toluidine :** 1, 2, 3-$(CH_3)$ $(NH_2)$ $C_6H_3$ $(NO_2)$ 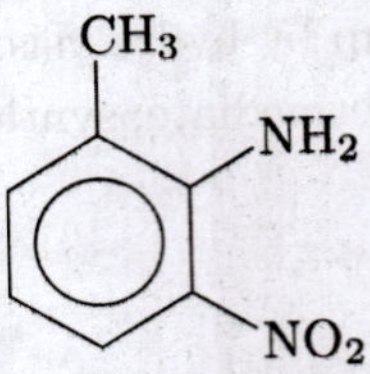 ; m.p. 97 °C.

Orange solid, insoluble in hot water but soluble in organic solvents. When reduced with Sn/HCl, 2, 3-diaminotoluene is formed.

Derivative—Acetyl, m.p. 157 °C.

Manufacturing process : By the selective reduction of 2, 3-dinitrotoluene by stannous chloride and hydrochloric acid.

Uses : Same as in *o*-nitro-*p*-toludine.

**6. *p*-Nitro-*o*-toluidine :** $(CH_3)$ $(NH_2)C_6H_3$ $(NO_2)$, m.p. 107 °C.

Yellow solid, soluble in organic solvents. When it is reduced with tin and HCl, 2, 4-diaminotoluene, m.p. 99 °C, is obtained.

Derivatives—(i) Benzoyl, m.p. 185 °C. (ii) Acetyl, m.p. 151 °C.

Manufacturing process : From ortho-toluidine by nitration.

Uses : Intermediate for dyes and pigments.

**7. 1-Nitro-2-naphthylamine :** 1, 2-$C_{10}H_6(NO_2)$ $(NH_2)$ 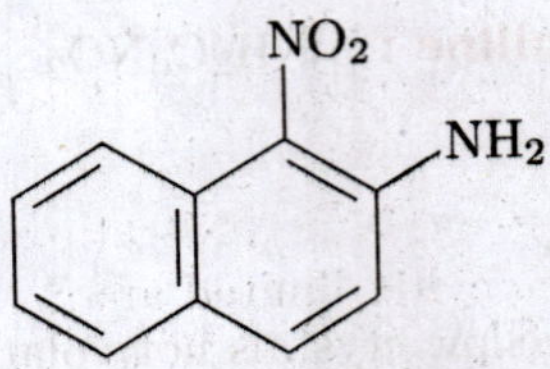 ; m.p. 126 °C.

Yellow crystals, soluble in organic solvents. On reduction with Sn/HCl yields 1, 2-diaminonaphthalene.

Derivatives—(i) Benzoyl, m.p. 167 °C. (ii) Acetyl, m.p. 123 °C. (iii) *p*-Toluenesulphonamide, m.p. 159 °C.

Manufacturing process : By nitration of acetylated product of $\beta$-naphthylamine followed by removal of acetyl group.

**8. 2, 6-Dinitroaniline :** $NH_2C_6H_3(NO_2)_2$ 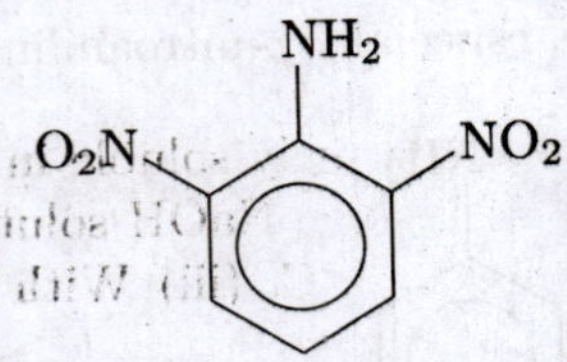 ; m.p. 138 °C.

Yellow solid, soluble in organic solvents. When it is reduced with Sn/HCl, 1, 2, 6-triaminobenzene, m.p. 130 °C, is obtained.

Derivatives—Acetyl derivative, m.p. 196 °C.

**9. 2, 4-Dinitroaniline :** 1, 2, 4-$(NH_2C_6H_2)C_6H_3(NO_2)_2$ 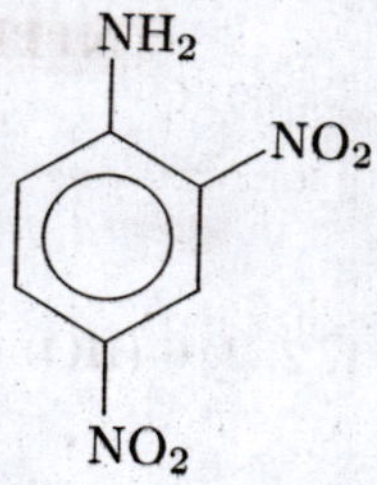 ; m.p. 176 °C.

Yellow solid with blue reflex, sparingly soluble in hot water but completely soluble in organic solvents. (i) On reduction with tin and HCl yields 1, 2, 4-triaminobenzene. (ii) On boiling with conc. NaOH, ammonia is evolved and 2, 4-dinitrophenol, m.p. 114 °C, is formed.

Derivatives—(i) Benzoyl, m.p. 220 °C. (ii) Acetyl, m.p. 120 °C. (iii) *p*-Toluenesulphonamide, m.p. 218 °C.

Manufacturing process : 2, 4-dinitrochlorobenzene is ammonolysed by boiling with ammonia and gives 2, 4-dinitro-aniline.

Uses : Ovicide, effective against moths, furniture and carpet beetles, cockroaches, and body lice.

**10. 2, 4, 6-Trinitroaniline :** (Picramide), 1, 2, 4, 6-$(NH_2)C_6H_2(NO_2)_3$ 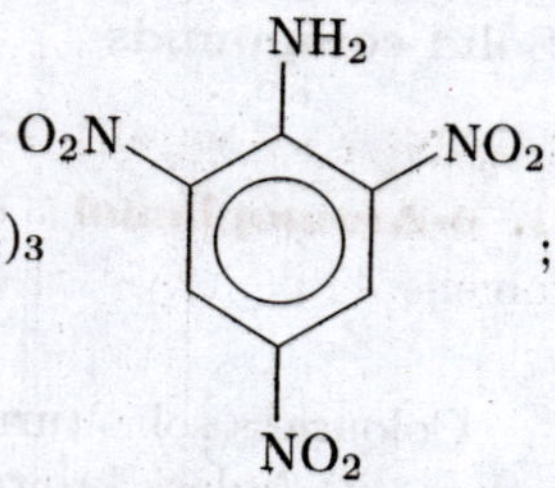 ; m.p. 188 °C.

Yellow crystals, soluble in organic solvents. (i) It can be diazotised only in the presence of conc. $H_2SO_4$ solution. (ii) When it is boiled with dilute NaOH, ammonia gas is evolved and picric acid, m.p. 122 °C, is formed. (iii) With naphthalene, It forms an addition compound, m.p. 167 °C.

Manufacturing process : Nitrating aniline in glacial acetic acid solution or by the use of mixed nitric-sulphuric acid in limited amounts.

Uses : Intermediate for azo pigment, toner pigment inks.

**11. 2, 4-dinitrophenylhydrazine :** *p*-$NO_2C_6H_4NHNH_2$ 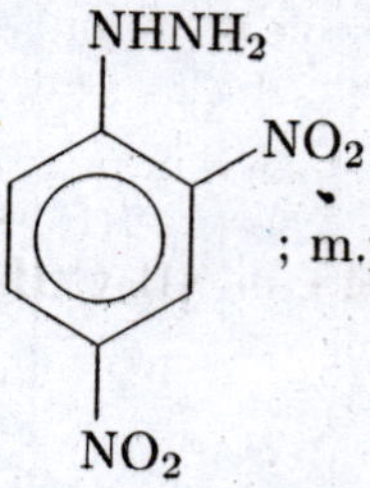 ; m.p. 200 °C.

Shining red crystalline solid, sparingly soluble in water. (i) 2, 4-Dinitrophenol, m.p. 114 °C, is obtained when it is boiled with conc. NaOH solution. (ii) On reduction with Sn/HCl yields, 1, 2, 4-triaminobenzene, m.p. < 100 °C. (iii) With acetone it forms isopropylidine derivative, m.p. 128 °C.

Manufacturing process : When 2, 4-dinitrochlorobenzene is hydrolysed by boiling with hydrazine, 2, 4-dinitrophenylhydrazine is obtained.

Uses : Explosive; reagent for aldehydes and ketones.

## NITRO-AMINOPHENOLS

**Solid compounds**

**1. Picramic acid :** 1, 2, 4, 6-(HO) $(NH_2)C_6H_2(NO_2)_2$ 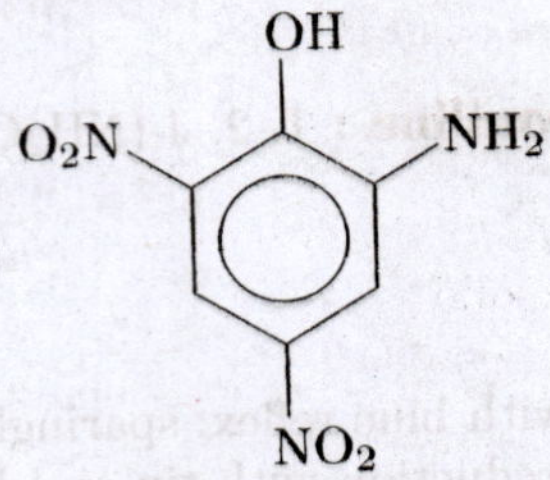 ; m.p. 168 °C.

Derivatives—(i) Benzoyl, m.p. 229 °C. (ii) Acetyl, m.p. 200 °C. (iii) *p*-Toluenesulphonate, m.p. 190 °C.

Manufacturing process : By partial reduction of picric acid.

Uses : Azo dyes; reagent for albumin.

## AMINOPHENOLS

**Solid compounds**

**1. *o*-Aminophenol :** *o*-$(OH)C_6H_4NH_2$ 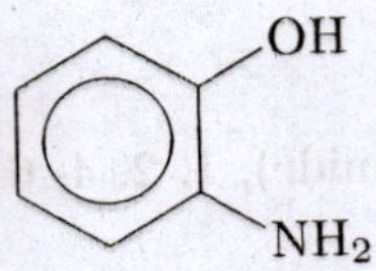 ; m.p. 174 °C.

Colourless solid turns dark on exposure to air. Slightly soluble in cold water but soluble in ether. (ii) A dark brown ppt. is obtained with $FeCl_3$ solution. (ii) With $AgNO_3$ solution gives a yellowish brown coloured solution on warming.

Derivatives—(i) Benzoyl, m.p. 18 °C. (ii) Monoacetyl, m.p. 201 °C. (iii) Diacetyl, m.p. 124 °C.

Manufacturing process : By reduction of *o*-nitrophenol mixed with aqueous ammonia by means of a stream of $H_2S$.

Uses : Dyeing fur and hair; dye intermediate for azo sulphur dyes; pharmaceuticals.

**2. *m*-Aminophenol :** *m*-$NH_2C_6H_4(OH)$ 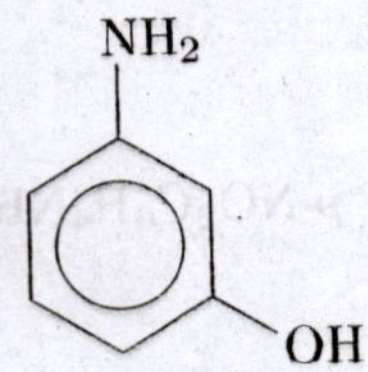 ; m.p. 122 °C.

Light coloured crystals turn dark on keeping. Slightly soluble in cold water and benzene. With $FeCl_3$, it gives no colour but the solution darkens on warming.

Derivatives—(i) Benzoyl, m.p. 153 °C. (ii) Diacetyl, m.p. 101 °C.

Manufacturing process : Fusion of *m*-sulphanilic acid with caustic soda and subsequent extraction of the melt with ether.

Uses : Dye intermediate; intermediate for *p*-aminosalicylic acid.

**3. *p*-Aminophenol :** *p*-$NH_2C_6H_4OH$ 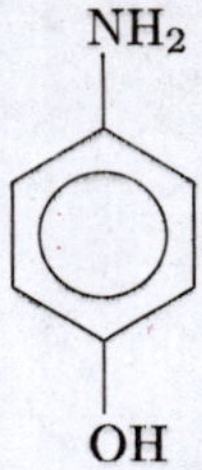 ; m.p. 184 °C.

Colourless solid; turns deep black on exposure to air and light. Sparingly soluble in water and ether. Gives purple colour with $FeCl_3$ in cold but on warming *p*-benzoquinone is formed.

Derivatives—(i) Tribenzoyl, m.p. 234 °C. (ii) Monoacetyl, m.p. 167 °C. (iii) Diacetyl, m.p. 150 °C.

Manufacturing process : (a) By reduction of *p*-nitrophenol with iron filings and hydrchloric acid. (b) By electrolytic reduction of nitrobenzene in concentrated sulphuric acid and treatment with an alkali to free the base.

Uses : Dyeing textiles, hair, fur, feather; photographic devoloper; pharmaceuticals; antioxidants, oil additive.

**4. 2, 5-Diaminophenol :** 1, 2, 5-(OH) $(NH_2)C_6H_3(NH_2)$ 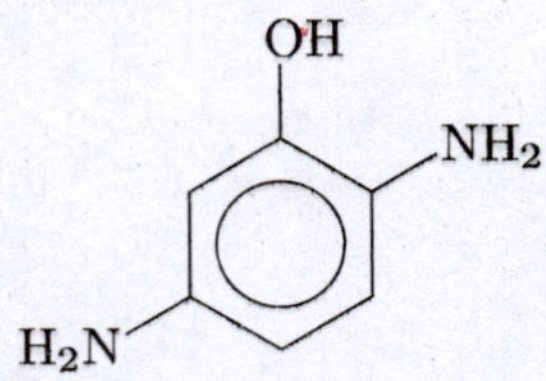 ; m.p. 68°C.

Colourless crystals, rapidly darkens in the presence of air and light. Soluble in water and alcohol; gives intense red colour with aq. $FeCl_3$ or $AgNO_3$ solution.

Derivatives—(i) Tribenzoyl, m.p. 231 °C. (ii) Triacetyl, m.p. 179 °C.

Manufacturing process : By reduction of 2, 5-dinitrophenol.

Uses : Organic synthesis.

# Appendix D

# Green Chemistry

It goes without saying that chemicals are essential for human life and society. Nowadays the methods followed to manufacture chemicals are more or less harmful to human health and environment. Even the laboratory experiments, chemical test and small scale synthesis, are hazardous. The purpose of Green Chemistry is to minimise or exclude the uses of chemicals and procedures which are harmful to human society and environment. To achieve the goal of green chemistry laboratories and industries require to invent beffiting designs, developments and applications of procedures and chemicals which are less harmful or harmless to human health and environment whenever practicable. To understand this the present laboratory procedure for the detection of element N, S and halogens of organic compounds may be taken as an example. Chemical laboratories usually follow Lassaigne's method of sodium fusion instead of following Middleton's sodium carbonate and Zinc dust method (published in the year 1935) : The sodium fusion method sometimes cause accidents but Middleton's method does not. So, the latter method is greener than the former method.

Principles to be followed for Green Chemistry.

1. The use of solvents is to be avoided as far as possible. solid state reactions are to be preferred.
2. Waste to be prevented as far as practicable.
3. Synthetic methods are to be designed in such a way that all of the reactants get converted to products, i.e., atom to atom conversion are to be followed.
4. Toxic Chemicals as reactants and as reaction products should be avoided whenever possible. Solvents being toxic, their use are to be restricted.
5. Energy to be used economically without (or less) causing disturbance to the environment. Microwave and ultrasound wave induced reactions are superior to conventional methods. It is also better to carry out the reactions under ambident temperature and pressure.
6. Chemical products prepared by green procedure should be effective enough but they should have minimal toxicity or no toxicity.
7. Catalytic reagents should have priority over stoichiometric reagents.
8. Synthetic products should be designed in such a way that after their use they should break down into innocuous components and should not persist in the environment.
9. Reactants of chemical reactions are to be chosen in a way so that release of toxic gases, accidents and break out of fire could be minimised or eleminated.
10. Simple and straight way procedures with rapid reaction rate are better than methods involving unnecessary blocking and deblocking procedures.
11. Utillisation of solar energy instead of gas burner should be followed whenever possible in designing a synthesis.

However, a few examples which can be practiced in the laboratory are given below :

1. Preparation of acetanilide from aniline

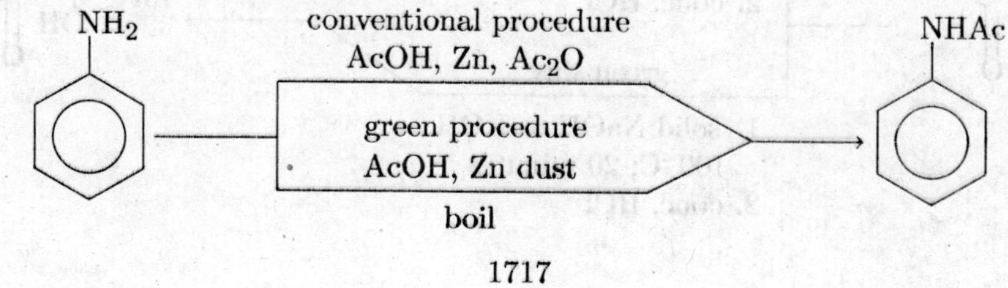

**Note :** $Ac_2O$ is costly and not always available.

2. Preparation of *p*-bromoacetanilide.

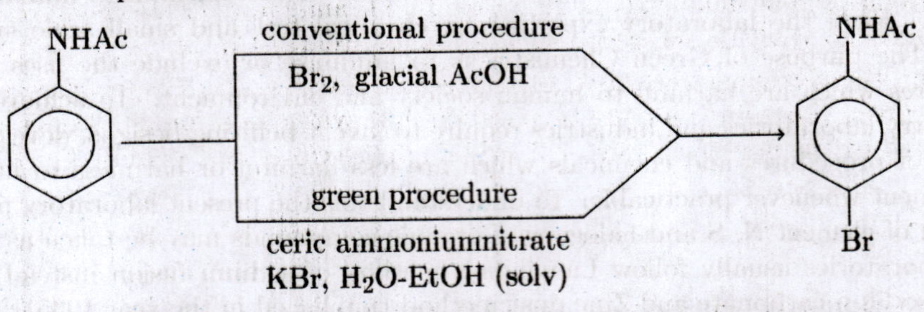


**Note :** (a) Corrosive liquid bromine is replaced by a new brominating agent.

(b) The reaction is carried out in the aqueous medium instead of glacial acetic acid.

(c) the green way takes less time than the conventional way.

3. synthesis of benzion.

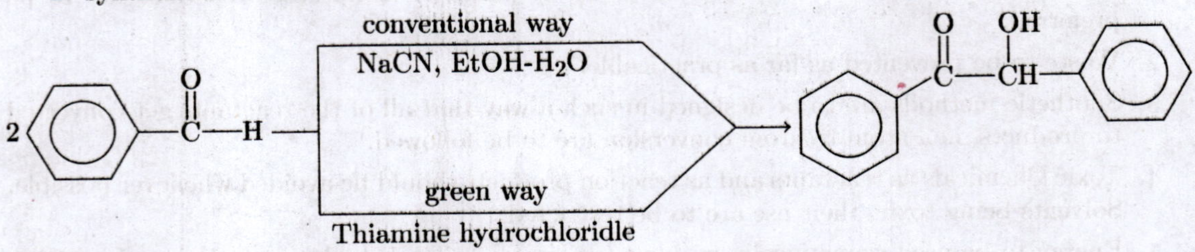


**Note :** Poisonous and hazardous sodium cyanide is avoided

4. Diels-Alder reaction.

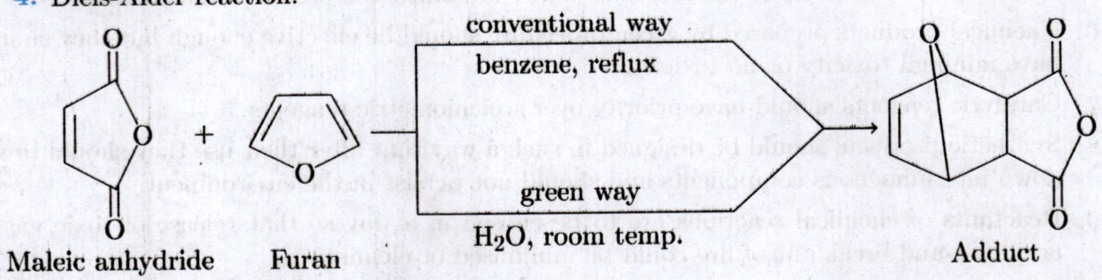


**Note :** (a) Being a toxic solvent, benzene is not used.

(b) The reaction is carried out at room temperature.

5. Preparation of benzilic acid using solvent.

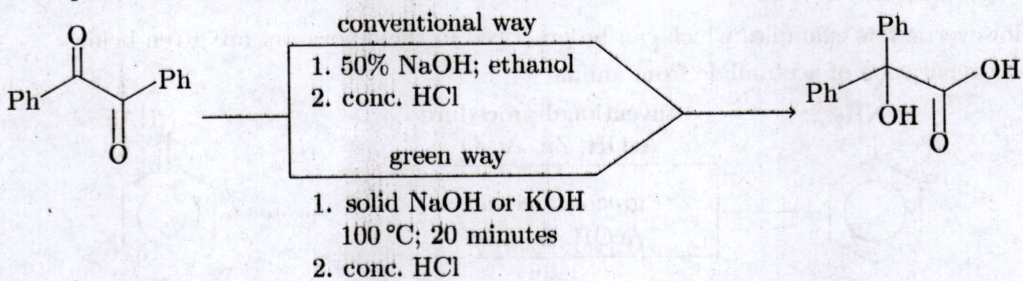

Polypeptide synthesis by Merrifield's protocol and analysis of polypeptides using polyner supported beads are green methods in the sense that the reactions are rapid, yields are good and macro molecules can be handled (Ch. 22). Use of PTC and crown ethers are also good procedures since they give good yields and reactions are less hazardous (Ch. 7).

However, these examples are cited to set the minds of the future researchers so that they try to innovet green procedures for the preparation and synthesis of organic compounds which are now prepared by the non-green way. This will minimise cost, time, pollution and health hazards.

[**Ref :** UGC sponsored National Seminar-Cum-Workship
on
Green chemistry : Its Implications in UG curriculam and in Industry
Held on : March 27 and 28, 2008
Organised by : The Department of Chemistry
New Alipore College
Block–L, New Alipore
Kolkata – 700053.]

Appendix E

# An Elementary Idea of Mass Spectra (MS)

## E.1 Introduction

From the mass spectra of an organic compound we get the following information :

(a) The exact molecular weight and from this, exact molecular formula.

(b) The bonds at which the molecule prefers to undergo fragmentation (in which a species cleaves into different other species).

## E.2 Procedure and Theory

In a simple spectroscopy a neutral organic molecule is bombarded with high energy electrons (energy higher than the ionisation energy for most of the molecules) of around 8 to 13 eV (775 to 1250 kJ mol$^{-1}$). Usually a beam of electrons accelerated to 70 eV is used in mass spectroscopy. This is in large excess of the general bond dissociation energy required in bond cleavage. Thus, the bond fragmentations are extensive. At the same time the molecules get sufficiently energised to eject out an electron owing to the impact of one electron from the electron beam; this process leads to the formation of radical cation as represented below :

$$\underset{\text{Neutral molecule}}{\text{R:G} + e^-} \longrightarrow 2e^- + \underset{\text{Mother radical cation or molecular ion}}{\text{R}\cdot\text{G}^{\oplus} \text{ or } \text{RG}^{+\cdot}} \longrightarrow \underset{\text{daughter ion}}{\text{R}\cdot + \text{G}^{\oplus} \text{ or } \text{R}^{\oplus} + \text{G}\cdot}$$

Daughter cations may further undergo fragmentation to develop other cations, neutral species and radicals. The molecular ions, the fragment ions and the fragmented radicals are separated by deflection in a variable magnetic field according to the ratio of their mass ($m$) and charge ($e$) and develop an ion current at the collector in proportion to their relative abundance[1]. A mass spectrum is a plot of relative abundance along ordinate against the ratio of mass to charge ($m/e$) along abscissa. Neutral molecules or radicals (e.g., G· and R·) cannot be detected directly in the mass spectrometer.

It is needless to mention that the molecular ion is formed by the ejection of electron from the lowest ionisation energy (i.e., from the highest energy MO). However, the energy requirement to remove an electron varies in the order :

Lone pair < conjugated diene < non-conjugated diene < $\sigma$ and C—C $\sigma$ < C—H $\sigma$ bond.

An example for the molecular ion formation and subsequent fragmentation are given here :

$$\text{CH}_3\ddot{\text{O}}\text{H} \longrightarrow \underset{\substack{\text{molecular ion} \\ (m/e = 32)}}{[\text{CH}_3\text{OH}]^{+\cdot} \text{ or } \text{CH}_3\text{OH}]^{+\cdot}} \xrightarrow{-\text{H}\cdot} \underset{(m/e = 31)}{[\text{CH}_2{=}\text{OH}]^+ \text{ or } \text{CH}_2{=}\text{OH}]^+} \xrightarrow{-\text{H}_2} \underset{(m/e = 29)}{[\text{CH}{\equiv}\text{O}]^+ \text{ or } \text{CH}{\equiv}\text{O}]^+}$$

$$[\text{CH}_3\text{OH}]^{+\cdot} \xrightarrow{-\dot{\text{O}}\text{H}} \underset{(m/e = 15)}{[\text{CH}_3]^+ \text{ or } \text{CH}_3]^+}$$

The fragments are usually represented within [ ] or ].

[1]The ions have to travel a considerable distance through magnetic field from ion source to collector; for this reason very low pressures (about $10^{-6}$ to $10^{-7}$ mm Hg) must be maintained using diffusion pump. The life-time of molecular ion and daughter species is around $10^{-5}$ s. Some decompose during the passage.

## E.3 Mass Spectrum and Base Peak

It has already been stated that to construct the graphic representation of a mass spectrum of a compound we take relative abundance (i.e., intensity) in ordinate and $m/e$ (mass to charge ratio) in abscissa. The most abundant peak is called the base peak; its intensity is taken as 100. The intensities of other peaks are given their proportionate values relative to the intensity of the base peak (taken as 100). The usual practice is to represent the MS graph in the form of a bar; for each $m/e$ value we get a bar (may be called signal). Fig. E.1 is a representative of the system. MS may also be presented in tabular forms.

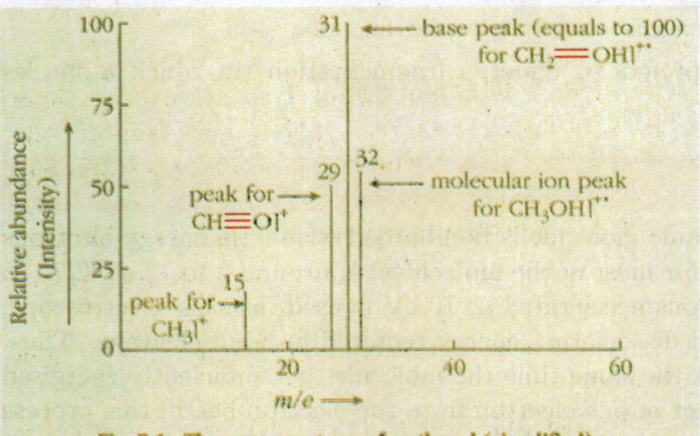

**Fig. E.1. The mass spectrum of methanol (simplified)**

Here it is important to note that the molecular ion peak is not always the base peak. Since it has the highest $m/e$ value. it is the molecular weight of the compound (the weight of the ejected electron being negligible). This. however. overlooks the presence of naturally occurring isotopes in the mother compound. since the chances of finding a $^{13}C$ atom in an organic molecule are 1.11%. Owing to the presence of $^{13}C$ containing molecule. a small peak at $m_{RG}+1$ appears. For $^{18}O$ a very small peak will appear at $m_{RG}+2$. The presence of 18-oxygen atom in a molecule is 0.2% and that of $^2H$ is negligible (0.015%).

In general. a peak for $m/e_{RG}+1$ may be due to the presence of any one of $^{13}C$, $^2H$, $^{15}N$ in RG and $m/e_{RG}+2$ may appear owing to the presence of O. Cl. Br. S in RG. Table E.1 is to be consulted for natural isotopic mass of elements as given here.

**Table E.1** Relative isotope abundances of common elements

| Isotope | Relative abundance (%) | Isotope | Relative abundance (%) | Note |
|---|---|---|---|---|
| $^{13}C$ | 1.11 | $^{18}O$ | 0.20 | $^{12}C$. $^1H$, $^{14}N$. $^{16}O$. $^{19}F$ |
| $^2H$ | 0.015 | $^{30}Si$ | 3.35 | $^{28}Si$, $^{31}P$, $^{32}S$ |
| $^{15}N$ | 0.38 | $^{34}S$ | 4.40 | $^{35}Cl$, $^{79}Br$, $^{127}I$, $^{16}O$. $^{19}P$ have relative abundance 100 each |
| $^{17}O$ | 0.04 | $^{37}Cl$ | 32.5 | |
| $^{29}Si$ | 5.10 | $^{81}Br$ | 98.0 | |
| $^{33}S$ | 0.78 | | | |

So far we have calculated $m/e$ values for molecular ions and fragmented species considering mass number of the elements. This will give neither 'exact molecular weight' of the mother molecule nor of the fragmented species. High resolution mass spectrometer (HRMS) can give their 'exact weight'. This may also be calculated from the atomic weights of the involved elements in atomic mass unit (amu). For example, a $^1H$ atom has actual mass in amu 1.007825 instead of 1; such is the case with $^{16}O$ which has actual mass 15.994914. Thus, 'exact mass' can be compared with the masses either from tables available for the same or by calculation of the different probable molecular formulae to obtain the molecular formula of the compound under study.

## E.4 Isotope Abundance and Mass Spectrum

If an ion contains $n$ carbon atoms, there is a possibility that about $1.1n$ per cent of these atoms will be $^{13}C$ (since natural abundance of $^{13}C$ is 1.1%). As an example, we can consider the case of $n$-butane; its molecular ion is associated with $m/e+1$ ion whose intensity is approximately 4.4 per cent that of the $[C_4H_{10}]^{+\cdot}$; while the molecular ion contains only $^{12}C$ atom the $m/e+1$ ion contains $^{13}C$ atom. However, if $n$-butane shows $m/e+2$ peak, we may assume three possibilities :

(a) the presence of two $^{13}C$ atoms or (b) the presence of two $^{2}H$ atoms or (c) the presence of one $^{13}C$ and one $^{2}H$ atoms in the compound.

These possibilities can be calculated which may be helpful in deciding the molecular formula of an ion in the absence of 'exact mass' measurement.

Bromine has two isotopes ($^{79}Br$ and $^{81}Br$) almost of equal abundance. The presence of bromine atom in an ion creates a dramatic effect in the mass spectrum of the ion. The same is the case with the presence of Cl ($^{35}Cl$ and $^{37}Cl$) atom in an ion. Their natural abundance is 3 : 1.

The binomial expression $(x+y)^n$ helps us calculate relative contributions of $m+1, m+2, m+4$, etc. peaks, where $x$ and $y$ are the relative abundances of the isotopes and $n$ is the number of these atoms present in the ion under consideration. Thus for the presence of three Cl atoms in an ion, the expression will be $(x+y)^3 = x^3+3x^2y+3xy^2+y^3$. Here let us consider that $x$ is $^{35}Cl$ and $y$ is $^{37}Cl$; $x^3$ contains 3 $^{35}Cl$, its intensity is 27 (since the coefficient of $x^3$ is 1, and the value of $x$ is 3) $3x^2y$ contains 2 $^{35}Cl$ and one $^{37}Cl$ and its coefficient is 3. So. $3x^2y$ becomes $3\times 3^2\times 1 = 27$. Similarly, $3xy^2 = 3\times 3\times 1^2$ or 9 and $y^3 = 1$.

So, there will be four peaks of intensities 27, 27, 9 and 1.

If we use the same formula for an ion containing Br atom, the expression becomes $(x+y)^2 = x^2+2xy+y^2$ and we shall see three peaks of intensities 1 : 2 : 1.

## E.5 Structure of Molecular Ion

It has already been stated in Sec. E.2. It is not worthful as well as not possible to write the exact structure. However, some examples are given here :

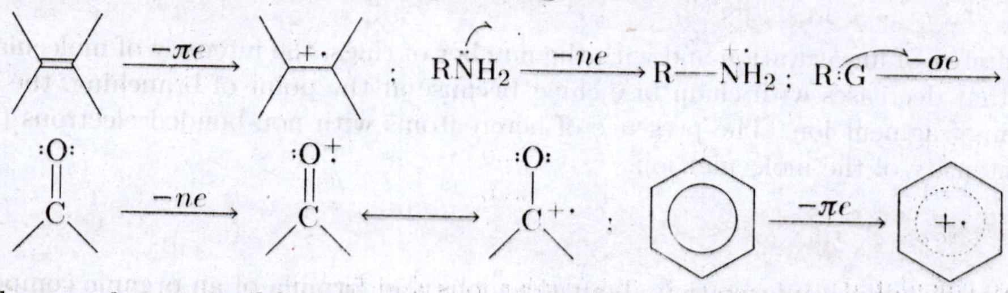

The use of crochets ([ ]) and ] as written in Sec. E.2 are better representations.

## E.6 Recognition of Molecular Ions

The correct identification of $m$, $m+1$, $m+2$. $m+4$, etc. is necessary as far as possible in order to ascertain the molecular formula of a compound. If, however, molecular ion peak is very weak, it is difficult to identify $m$, $m+1$, $m+2$, etc. within the molecular cluster of molecular ion ($m, m+1, m+2$, etc.). However, there are a number of tests which will tell us whether an ion is molecular ion or not.

1. Since a molecular ion is formed by ejection of one electron, it might be an odd electron species (other fragmented parts and rearranged parts also may have odd electron).

   The index of hydrogen deficiency (IHD) is another test. IHD is the number of pairs of H atoms which must be removed from the saturated open-chain formula to give the observed formula. This is possible if the elemental composition of the ion can be determined. However, for a molecule $a_y\, b_n\, c_z\, d_x$ : the index of hydrogen deficiency $= x - y/2 + z/2 + 1$,

   where $a$ = any monovalent atom
   $b$ = O, S or any divalent atom
   $c$ = N, P or any trivalent atom
   $d$ = C, Si or any tetravalent atom.

   and $y, n, z, x$ are the corresponding number of atoms.

   For example, furan $C_4H_4O$, 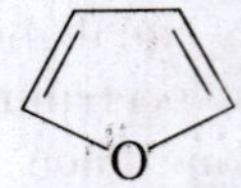 has an index of hydrogen deficiency = 3

   Here, $a$ = H, $y = 4$; $b$ = O, $n = 1$; $c = 0$, $z = 0$ and $d$ = C, $x = 4$.

   So, $\quad \text{IHD} = 4 - \frac{4}{2} + \frac{0}{2} + 1 = 3 \left(\text{according to } a_y\ b_n\ c_z\ d_x \text{ and } x - \frac{y}{2} + \frac{z}{2} + 1\right)$

   The IHD must be a whole number for an odd-electron ion. For an even-electron ion the value will be non-integral. Actually IHD is equivalent to D.B.E (for D.B.E see p. 1057)

2. An examination of the fragment ion peaks in vicinity of ion under consideration may give us clue whether the ion is a molecular ion. Mass losses between 3 and 15 and between 20 and 26 are unlikely, and there observation would suggest that the supposed molecular ion is in fact a fragment ion.

3. The change of instrumental condition may also provide evidence to confirm the recognition of the molecular ion. Very weak molecular ion peak will be shown if maximum sensitivity of the instrument is used. Again, with the decrease of the energy of the electron beam, the intensity of fragment ion will decrease relative to molecular ion.

## E.7 Intensity of the Molecular Ion

With the number of unsaturation and with the number of rings, the intensity of molecular ion increases, but that decreases with chain branching because at the point of branching, the bond breaks and forms fragment ion. The presence of heteroatoms with non-bonded electrons (*n* es) increases the intensity of the molecular ion.

## E.8 Fragmentation

Fragment ions give much information about the molecular formula of an organic compound; they form by the decomposition of molecular ion. Some discussion on fragment ions are given below :

1. Odd-electron ions are more significant than even-electron ions having similar mass or intensity, since they are usually formed via a rearrangement reaction which may be characteristic of a particular class of compounds.

2. Ions of higher mass are likely to give more useful information than those of lower mass because the former ions are more likely to have been formed through a simple rational fragmentation.

3. Metastable ions (ions of very low stability) give significant information on the nature of fragmentation process.

Two factors which determine the relative abundance of fragment ion in MS are : (1) the stability of the ion and (2) the energy relationships of bond cleaved and formed in the reaction leading to the ion. However, the following are the common fragmentations which give rise to stable carbenium ions.

(a) Formation of a tertiary carbenium ion. It is needless to mention that the stability of saturated carbenium ions decreases in the following order ; 3° > 2° > 1°

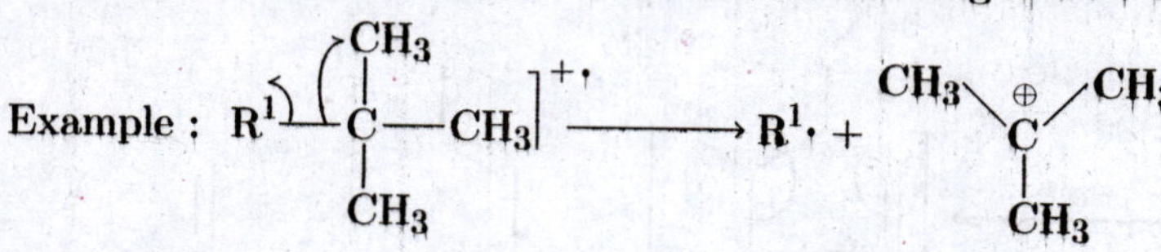

(b) Formation of a resonance-stabilised allylic cation.

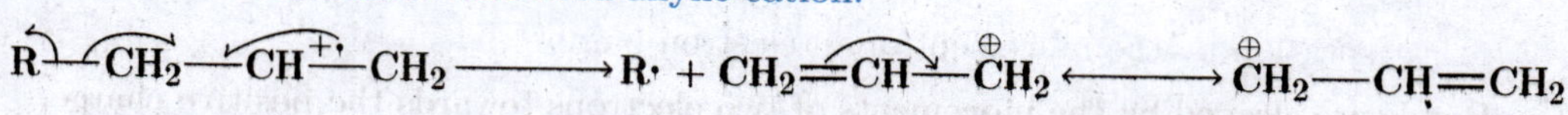

(c) Formation of seven-membered tropylium ion (aromatic stability)

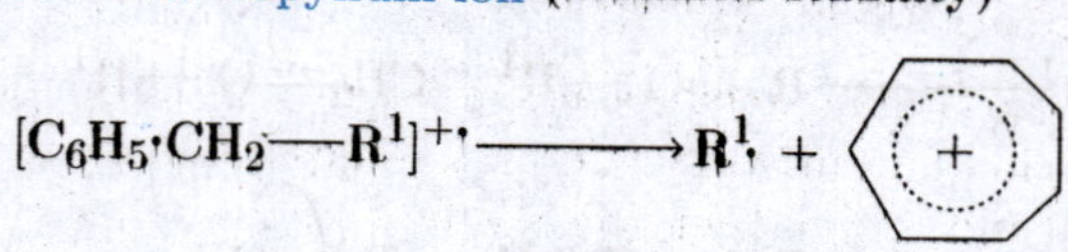

(d) Formation of a stable carbenium ion owing to $p\pi - p\pi$ bonding involving a hetero atom with *n* es.

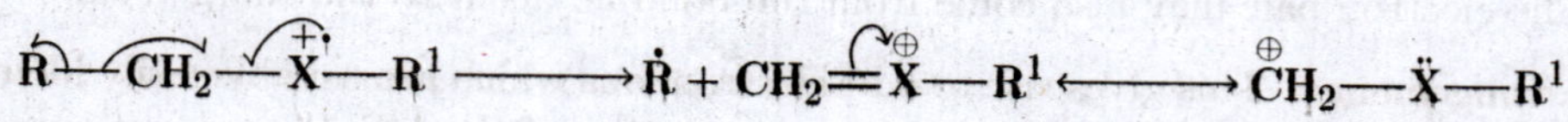

(e) By $\sigma$ bond cleavage.

(f) $\alpha$-bond cleavage.

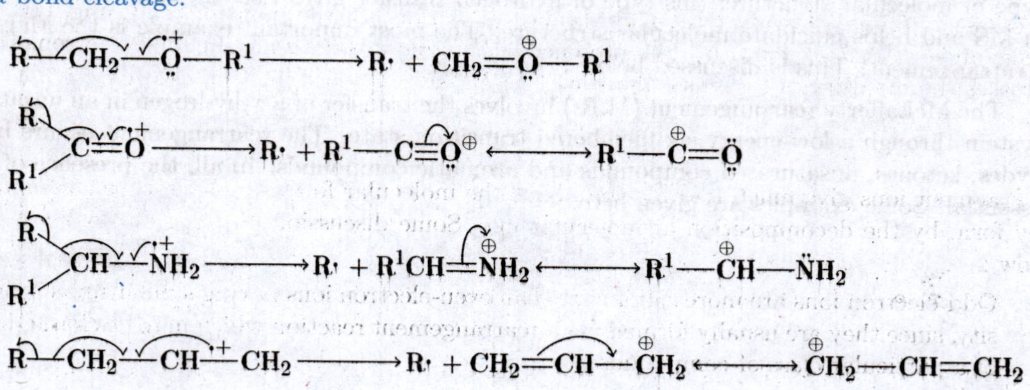

**In all these fragmentations an uncharged radical is generated enabling the residual electron to couple with that from which it formed to form even-electron ionic species.**

(g) By multiple $\sigma$ bond rupture and generation of neutral molecule.

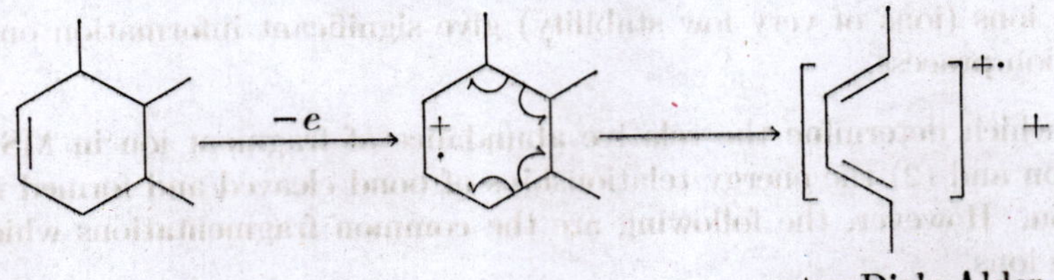

retro-Diels-Alder

**This involves one-electron mechanism.**

Two-electron mechanism

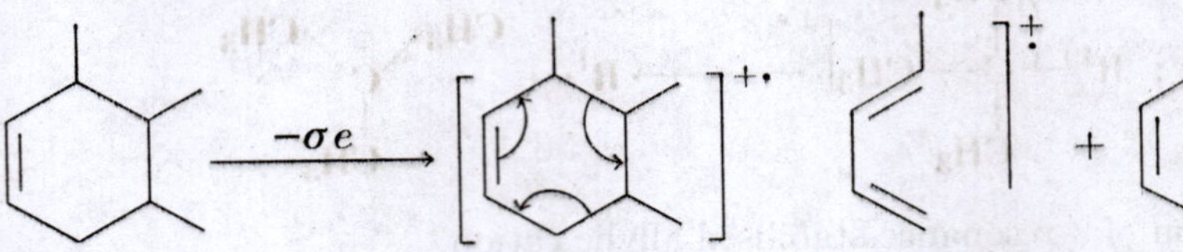

(h) Fragmentation by the movement of an electron pair.

**Bonds are cleaved by the movements of two electrons towards the positive charge.**

$$R-\overset{+\cdot}{\ddot{O}}-R^1 \longrightarrow \overset{\oplus}{R} + \cdot\ddot{O}-R^1 \; ; \quad CH_2=\overset{\oplus}{\ddot{O}}-R^1 \longrightarrow CH_2=O + \overset{\oplus}{R^1}$$

$$R^1-C\equiv\overset{\oplus}{O}: \longrightarrow \overset{\oplus}{R^1} + :C=\ddot{O}: \; ; \quad R-\overset{\oplus}{O}H_2 \longrightarrow \overset{\oplus}{R} + H_2O$$

**The electron pair may also come from the bond adjacent to the charged side.**

**Rearrangements : This gives rise to odd-electron ions which are usually easily identified in the spectrum. So, they are useful aids in the interpretation of the MS. Since large excess of energy normally available in the ion source, molecular rearrangements are very much common. The rearrangements may be ramdom in nature. As a consequence redistribution of certain atoms in the molecule takes place; frequently transfer of hydrogen atoms, which are characteristic of certain type of molecular structure; this type of hydrogen transfer gives rise to easily recognisable ions in MS and helps elucidate molecular structure. The most important example is the McLafferty Rearrangement. This is discussed below :**

**The McLafferty rearrangement (M.R.) involves the transfer of a $\gamma$-hydrogen in an unsaturated system through a low-energy six-membered transition state. The rearrangement occurs in aldehydes, ketones, unsaturated compounds and aromatic compounds. In all, the presence of $\gamma$-H is essential. Some examples are given here.**

General :

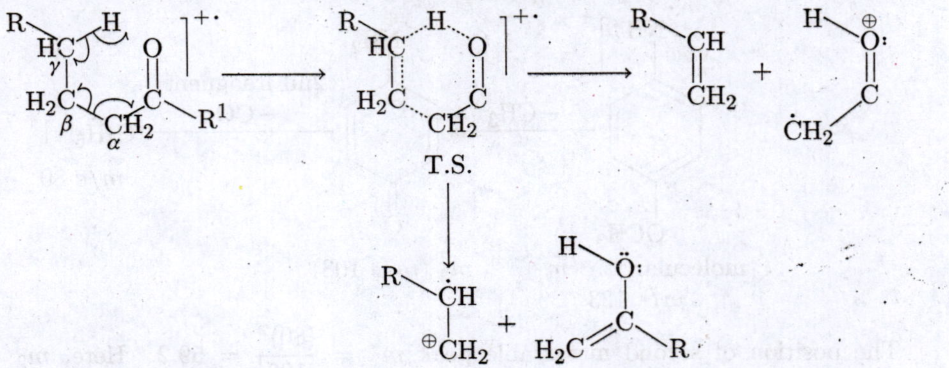

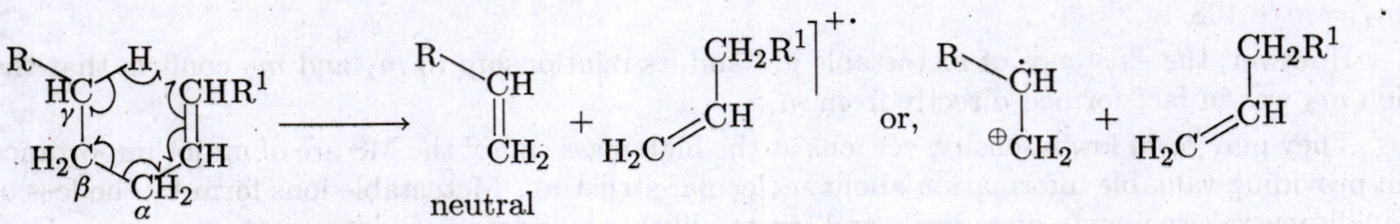

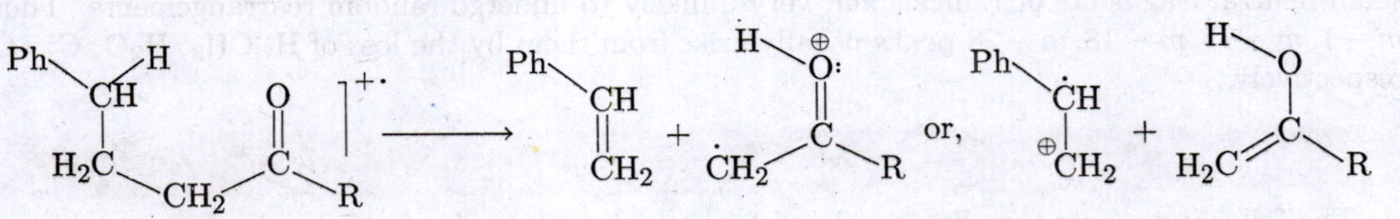

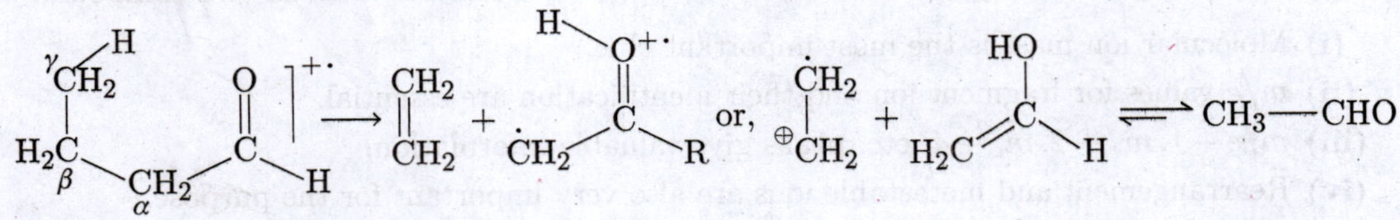

### Formation and fragmentation of metastable ion.

Metastable technically means unstable but so long-lived as to be used for practical purposes. Metastable ions arise from fragmentation of ions which have already been accelerated out of the ion source but have not reached the magnetic field. They show broad peaks of weak intensity and they have non-integral masses. Their positions in the spectrum are displayed according to their true masses. The position of the metastable peak ($m^*$) is related to the masses of the ion from which it is formed ($m_1$) and the mass of the ion formed by its fragmentation ($m_2$) by the equation,

$$m^* = \frac{(m_2)^2}{m_1}. \qquad \text{(E.1)}$$

An example is given here. $m/e$ *p*-aminoanisole is 123 ($m_1$) on losing methyl radical (loss of 15 mass units) it produces a daughter ion of mass 108 ($m_2$). The position of the metastable peak will be

$$m^* = \frac{(108)^2}{123} = 94.8.$$

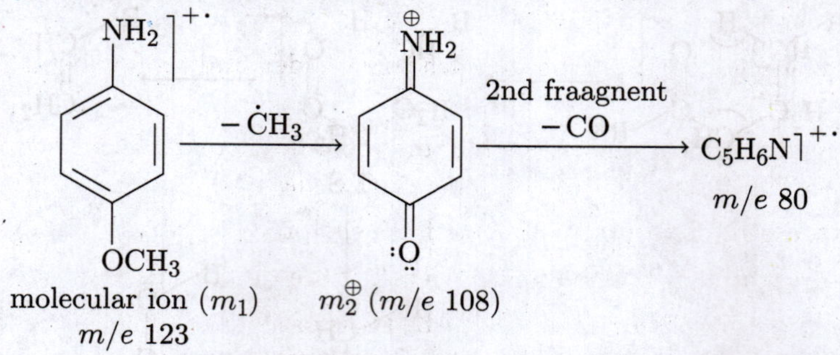


The position of second metastable peak $m^* = \dfrac{(80)^2}{108} = 59.2$. Here, $m_2 = m/e$ 80 and $m_1 = m/e$ 108.

However, the existence of metastable ion and its relationship to $m_1$ and $m_2$ confirm that the ion $m_2$ was in fact formed directly from $m_1$.

They may be of low intensity, yet ions at the high mass end of the MS are of major importance in providing valuable information about molecular structure. Metastable ions form by the loss of small neutral fragments or radicals and very unlikely to undergo random rearrangements. Thus $m-1, m-15, m-18, m-28$ peaks usually arise from them by the loss of H, $CH_3$, $H_2O$, C=O respectively.

## E.8 Assignment of Molecular Formula

The following points may be considered for assigning molecular formula of a compound :

**(i)** Molecular ion mass is the most important clue.

**(ii)** $m/e$ values for fragment ion and their identification are essential.

**(iii)** $m/e+1, m_e+2, m_e+4$, etc. peaks give valuable information.

**(iv)** Rearrangement and metastable ions are also very important for the purpose.

**(v)** Hydrocarbons and their molecular ions must have even number of hydrogen atoms since hydrocarbons have general formulae $C_nH_{2n+2}$, $C_nH_{2n}$, $C_nH_{2n-2}$ and atomic mass of carbon atom is even (12). This excludes the presence of atoms having relative natural abundance. This rule holds good for compound containing C, H, O, S also, since mass numbers of O and S are even.

**(vi)** The presence of Br and Cl atoms gives a good clue(See Sec. E.4)

**(vii)** Nitrogen rule : It states that a molecule of an even membered molecular ion mass must contain either no nitrogen atom or even number of nitrogen atoms. An odd numbered molecular ion mass tells the presence of odd number of nitrogen atoms. The rule holds good for any compound containing C, H, N, O, S and halogen atoms. This nitrogen rule is often helpful to rule out many peaks from the MS as possible molecular ions because reasonable structural requirements do not fulfil.

## E.9 Interpretation of the Mass Spectrum

The following points are suggested for the interpretation of MS. It is needless to mention that MS of each compound has its own characteristic.

**(i)** Molecular ions to be identified.

**(ii)** Elemental composition and hydrogen deficiency index and double bond equivalent are to be determined.

**(iii)** From the general appearance of the MS and identification of fragment ion series and from characteristic ions, possible deductions are to be made.

**(iv)** Possible structures of natural fragments from the presence of high mass are to be noted.

**(v)** From odd-electron ions possible rearrangements are to be considered.

**(vi)** A possible structure of the molecule on the basis of MS and any other evidences are to be suggested. An MS is to be predicted for the suggested structure and is to be compared with the spectrum of the compound under study. And modification for the purpose, if required, is to be made. The mass spectral behaviour of compounds of similar structures (by consulting reference collections) is to be checked.

**(vii)** Other spectral data (UV, IR, NMR) along with MS give us the right direction for elucidation of structure of a given compound.

## E.10 Substitution of an isotope and determination of reaction mechanism by MS

The mechanism of a number of reactions could relatively easily be followed by mass spectroscopy. Less abundant isotopes are incorporated either to the substrate or to the reagent or to the solvent. The isotopes that might be used include $^2$H, $^{13}$C, $^{18}$O, $^{15}$N, $^{37}$Cl, etc. However, because of their high cost the cheapest $^2$H has been used maximum. $^{13}$C and $^{18}$O are also used.

Two examples will be cited here.

**1.** Hydrolysis of esters (See Sec. 11.8.2, p. 548). Here water is labelled with $^{18}$O.

**2.** Reversibility of a reactions (See Sec. 11.8.2 p. 548). Here one $^2$H is incorporated in methylene group of ethyl acetoacetate. It is a good evidence that methylinic H atoms in the $\beta$-ketoester is enolisable.

It is needless to mention that kinetic isotope effects are studied by substituting heavier isotope in place of lighter isotope (See Sec. 11.8.3, p. 549 and Sec. 20.2.5(viii), p. 1183). However, we shall have to discuss the role of MS in determining the mechanism. In the case of ester hydrolysis, the products, the acid and alcohol are to be separated and each is to be subjected to mass spectral study. The alcohol or acid which one gives a low intensity peak at the right side of the molecular ion peak with $m/e+2$ value will show the presence of $^{18}$O. In the case of enolisation the deuterated product will show a peak of low intensity at the right side of the molecular ion peak with $m/e+1$ value.

## E.11 Disadvantages of MS

So far we have discussed all merits of MS. It is a good tool to determine the molecular formula and structure of organic compounds. Every system has its disadvantage though negligible they may be. However, they are discussed below :

**(a)** The lowest amount of the organic sample used in MS is $10^{-12}$ g. Usually less than $10^{-3}$ g is used. However small amount it may be, it cannot be recovered after the experiment since the sample gets destroyed. In other spectral study, the sample employed could be recovered.

**(b)** It is difficult to introduce the sample so small enough into the high vacuum system needed to handle the ionic species involved.

**(c)** The cost of mass spectroscope is very high.

**Problem E.1.** What structural formula containing only C and H atoms be assigned to a cation with $m/e$ value 43?

**Solution :** If we divide the $m/e$ value 43 by 12 (the mass number of carbon), we get the presence of 3 C atoms in the cation. After division the remainder is 7. Thus 7 H atoms are there. So, the structural formula is $C_3H_7^{\oplus}$.

**Problem E.2.** In a mass spectral analysis less than $10^{-3}$ g sample is used. Give an explanation for such a small amount of use of the sample.

**Solution :** A very small amount of sample contains small number of molecule. Thus this prevents collision and reactions amongst fragments. Combination of fragments might lead to ions with larger mass making it impossible to ascertain the molecular mass, and the fragment patterns also would become confusing.

**Problem E.3.** Show the various fragment patterns of *s*-butyl alcohol.

**Solution :**

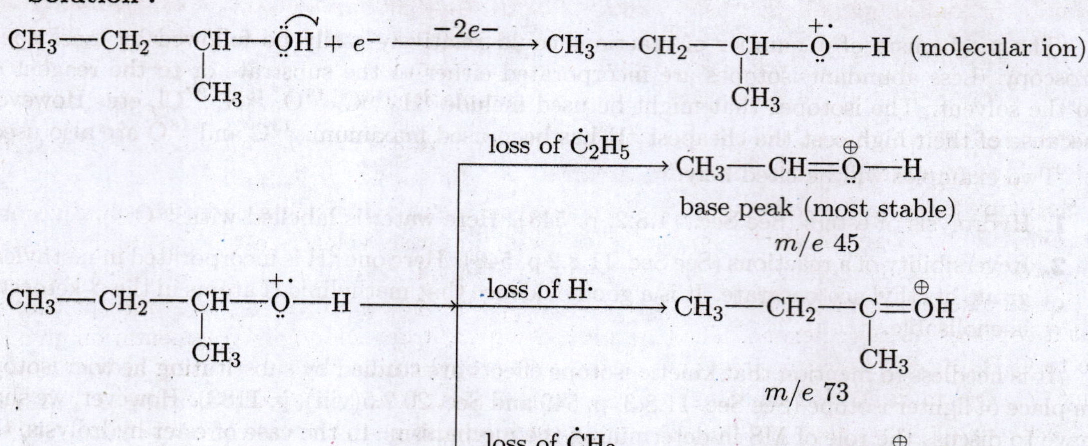


**Problem E.4.** Show the fragmentations of *n*-butylbenzene.

**Solution :** *n*-butylbenzene is

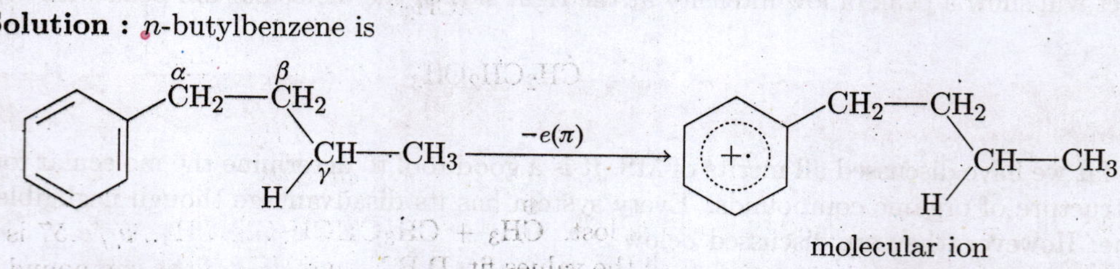


molecular ion

If contains a γ-H with respect to benzene ring. So, it will first undergo McLafferty rearrangement involving hydrogen transfer from γ-position to the aromatic benzene ring as follows :

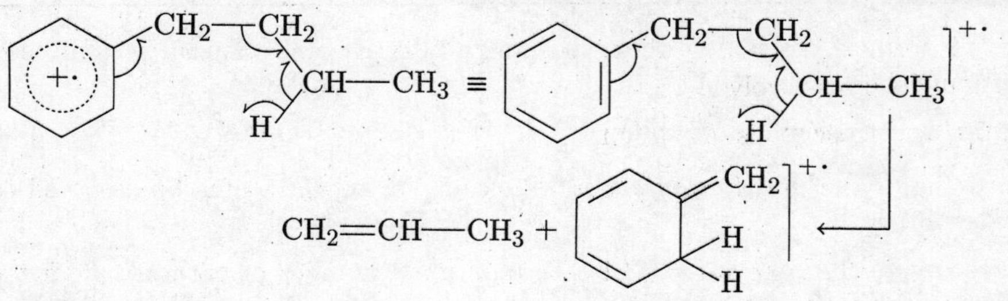

$m/e$ 92

McLafferty rearrangement

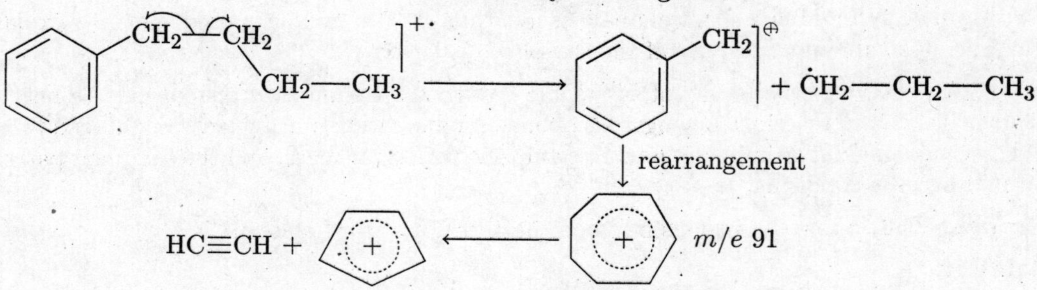

$m/e$ 65 Cyclopento dienyl cation

Cycloheptatrienyl cation Tropilium ion (aromatic stability)

**Problem E.5.** A hydrocarbon has the following $m/e$ values : 86, 71, 57, 43 (base peak), 29 and 15. Assign a structure for the hydrocarbon.

**Solution :** Highest $m/e$ being 86, it is the molecular peak. It being a hydrocarbon, division of the $m/e$ 86 by 12 gives 7 as the number of C atoms. To satisfy valences of carbon atom we have only two as remainder. Number of H atoms in that case will be either 16 (for saturated one) or 14 (for an alkene or cycloheptane) or 12 (for an alkyne or cycloheptene). So, it is most probable that it contains isomeric hexane. The base peak being 43, it has undergone fragmentation giving rise to $C_3H_7]^+$. This may be isopropyl cation or $n$-propyl cation. Fragmentation at branching point develops more stable carbenium ion. So the base peak is due to isopropyl cation (2°) which gets stability by hyper conjugation. This fragmentation also gives rise to $n$-propyl cation. Thus the structure :

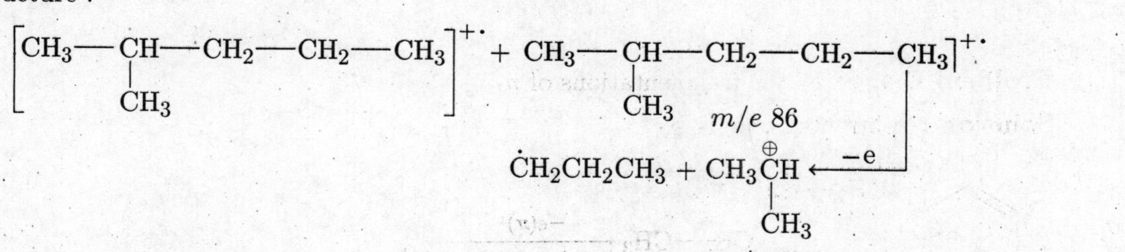

$m/e$ 43

$m/e$ 86 – 15 = 71. So, a methyl moity is lost. $\overset{\oplus}{\dot{C}}H_3 + CH_3\dot{C}HCH_2CH_2CH_3$. $m/e$ 57 is for $C_4H_9]^+$ and $m/e$ 29 is for $C_2H_5]^+$. Since all the values fit, D.B.E. and IHD of the compound are also zero. The compound is 2-methylpentane. Students are requested to calculate D.B.E. and IHD for the compound.

# Books Consulted


| | |
|---|---|
| Albert L. Lehninger | *Biochemistry* |
| Alexander | *Principles of Ionic Organic Reactions* |
| Allinger and Allinger | *Structure of Organic Molecules* |
| B.A. Marples | *Elementary Organic Stereochemistry and Conformational Analysis* |
| B.S. Furniss and others (revised by) | *Vogel's Text Book of Practical Organic Chemistry* |
| Bansal, R.K. | *Organic Reaction Mechanism* |
| Bhal, B.S. and Bhal Arun | *Advanced Organic Chemistry* |
| Brown, G.I. | *New Guide to Modern Valency Theory* |
| C.D. Gutsche | *The Chemistry of Carbonyl Compounds* |
| C.H. Depuy and O.L. Chapman | *Molecular Reactions and Photochemistry* |
| Cartmell, E.D. and Fowler, G.W.A | *Valency and Molecular Structure* |
| Cotton, F.A. and Wilkinson, G. | *Advanced Inorganic Chemistry* |
| D. Nasipuri | *Stereochemistry of Organic Compounds* |
| Dyer | *Applications of Absorption Spectroscopy of Organic Compound* |
| Eliel, E.L. | *Stereochemistry of Carbon Compounds* |
| Eric E. CONN and Stumff | *Outlines of Biochemistry* |
| F.A. Cotton | *Chemical Applications of Group Theory* |
| Ferguson, L.N | *The Modern Structural Theory of Organic Chemistry* |
| Finar, I.L. | *Organic Chemistry (Vol. 1)* |
| Gabril, E. Wing | *Instrumental Methods of Chemical Analysis* |
| Gessner G. Hawley (Revised by) | *The Condensed Chemical Dictionary (Tenth Edition)* |
| Gould, E.S. | *Mechanism and Structure in Organic Chemistry* |
| Guthric and Honeyman | *An Introduction To Chemistry of Carbohydrate* |
| Hendrickson Cram and Hammond | *Organic Chemistry* |
| Hine Jack | *Physical Organic Chemistry* |
| House | *Modern Synthetic Reactions* |
| Huheey, J.E. | *Inorganic Chemistry* |
| I.S. D'mitrev | *Symmetry in the Worlds of Molecules* |
| K. Mislow | *Introduction to Stereochemistry* |
| Kemp, William | *Organic Spectroscopy* |
| King, Edward L. | *How Chemical Reactions Occur* |
| Mann and Saunders | *Practical Organic Chemistry* |

| | |
|---|---|
| March, Jerry | *Advanced Organic Chemistry (Reactions, Mechanisms and Structure)* |
| Morrison, R.T. and Boyd, R.N. | *Organic Chemistry* |
| Mukherjee and Singh | *Reaction Mechanism in Organic Chemistry* |
| Nesmeyanov, A.N., Nesmeyanov, N.A. | *Fundamentals of Organic Chemistry (Vols. 3 and 4)* |
| Niels Isaac | *Physical Organic Chemistry* |
| Noller, C.R. | *Text Book of Organic Chemistry* |
| Norman | *Organic Synthesis* |
| O'Connor Rod | *Fundamentals of Chemistry* |
| P.S. Kalsi | *Spectroscopy of Organic Compounds* |
| Palit, S.R. | *Physical Chemistry* |
| Paul K. Stumff and others | |
| Potapov | *Stereochemistry of Organic Compounds* |
| Rakshit, P.C. | *Physical Chemistry* |
| Ray Brewster and William E. McEwen | *Organic Chemistry* |
| Richardson and Lowry | *Mechanism and Theory in Organic Chemistry* |
| Silverstein, Bassler and Morrill | *Spectrometric Identification of Organic Compounds* |
| Simpson, J.E. Hanch, C. and Helmkamp G.K. | *An Outline of Organic Chemistry* |
| Smith and Mackie | *Guide Book To Organic Synthesis* |
| Streitwieser and Heathcock | *An Introduction to Organic Chemistry* |
| Stuart Warren | *Organic Synthesis* |
| Subrata Sengupta | *Basic Stereochemistry of Organic Molecules* |
| Sykes, Peter | *A guidebook to mechanism in organic chemistry* |
| T.W Graham Soloman | *Organic Chemistry* |
| V.K. Ahluwalia and Renu Aggarwal | *Organic Synthesis: Special Techniques* |
| Vishnoi, N.K. | *Advanced Practical Organic Chemistry* |
| W. Carruthers | *Modern Method of Organic Synthesis (Third Edition)* |
| W. Reusch | *An Introduction to Organic Chemistry* |
| Whitfield | *Organic Reaction Mechanism* |

# Subject Index

## A

**C**

## D

## G

## I

## K

## L

## N

## P

## Q

## S

## T

## U

## V



## W

**X**



**Y**



**Z**

# Named Topic Index

# Compound Index

## B

## C

## D

**E**

**F**

## G

**H**



**I**

J



K



L



M

## O



## P

## Q



## R

## T

## U



## V



## W



## X

## Z